Library of
Davidson College

Papers in
Biochemical Genetics
Second Edition

Papers in Biochemical Genetics
Second Edition

Edited by
GEOFFREY L. ZUBAY
Columbia University
and
JULIUS MARMUR
Yeshiva University

Holt, Rinehart and Winston, Inc.
New York • Chicago • San Francisco • Atlanta • Dallas • Montreal • Toronto • London

575.2
Z93p

Copyright © 1973 by Holt, Rinehart and Winston, Inc.
All rights reserved

Library of Congress Cataloging in Publication Data

Zubay, Geoffrey L. comp.
   Papers in biochemical genetics.

   Includes bibliographical references.
   1. Bacterial genetics. 2. Biochemical genetics.
I. Marmur, Julius, 1926-   joint comp. II. Title.
[DNLM: 1. Genetics, Biochemical—Collected works.
2. Genetics, Microbial—Collected works. QH431 Z92p
1973]
QH434.Z8 1973   575.2'1   73-12317
ISBN 0-03-085197-1

Printed in the United States of America
456789 006 98765432

CONTENTS

I. DNA Synthesis — 1

1. J. Marmur, C. L. Schildkraut and P. Doty (1962). Biological and Physical Chemical Aspects of Reversible Denaturation of Deoxyribonucleic Acids. *In* "The Molecular Basis of Neoplasia". pp. 9-43. University of Texas Press, Austin. — 14

2. R. Okazaki, T. Okazaki, K. Sakabe, K. Sugimoto and A. Sugino (1968). Mechanism of DNA Chain Growth, I. Possible Discontinuity and Unusual Secondary Structure of Newly Synthesized Chains. Proc. Natl. Acad. Sci. U.S. **59** 598-605. — 49

3. A. Kornberg (1969). Active Center of DNA Polymerase. Science, **163**, 1410-1418. — 57

4. M. Goulian, A. Kornberg and R. L. Sinsheimer (1967). Enzymatic Synthesis of DNA, XXIV. Synthesis of Infectious Phage ϕX174 DNA. Proc. Natl. Acad. Sci. U.S. **58**, 2321-2328. — 66

5. D. Brutlag and A. Kornberg (1972). Enzymatic Synthesis of Deoxyribonucleic Acid. XXXVI. A Proofreading Function for the $3' \longrightarrow 5'$ Exonuclease Activity in Deoxyribonucleic Acid Polymerases. J. Biol Chem. **247**, 241-248. — 74

6. D. Brutlag, R. Schekman and A. Kornberg (1971). A Possible Role for RNA Polymerase in the Initiation of M13 DNA Synthesis. Proc. Natl. Acad. Sci. U.S. **68**, 2826-2829. — 82

7. P. DeLucia and J. Cairns (1969). Isolation of an *E. coli* Strain with a Mutation affecting DNA Polymerase. Nature, **224**, 1164-1166. — 86

8. H. Schaller, B. Otto, V. Nüsslein, J. Huf, R. Herrman and F. Bonhoeffer (1972). Deoxyribonucleic Acid Replication *in vitro*. J. Mol. Biol. **63**, 183-200. — 89

9. M. L. Gefter, Y. Hirota, T. Kornberg, J. A. Wechsler and C. Barnoux (1971). Analysis of DNA Polymerases II and III in Mutants of *Escherichia coli* Thermosensitive for DNA Synthesis. Proc. Natl. Acad. Sci. U.S. **68**, 3150-3153. — 107

10. R. E. Moses and C. C. Richardson (1970). Replication and Repair of DNA in Cells of *Escherichia coli* Treated with Toluene. Proc. Natl. Acad. Sci. U.S. **67**, 674-681. — 111

11. M. Gellert, J. W. Little, C. K. Oshinsky and S. B. Zimmerman (1968). Joining of DNA Strands by DNA Ligase of *E. coli*. Cold Spring Harbor Symp. Quant. Biol. **33**, 21-26.   119

12. B. M. Alberts and L. Frey (1970). T4 Bacteriophage Gene 32: A Structural Protein in the Replication and Recombination of DNA. Nature, **227**, 1313-1318.   125

13. R. B. Inman and M. Schnös (1971). Structure of Branch Points in Replicating DNA: Presence of Single-stranded Connections in λ DNA Branch Points. J. Mol. Biol. **56**, 319-325.   131

14. D. Dressler (1970). The Rolling Circle for φXDNA Replication, II. Synthesis of Single-Stranded Circles. Proc. Natl. Acad. Sci. U.S. **67**, 1934-1942.   139

15. M. Meselson and R. Yuan (1968). DNA Restriction Enzyme from *E. coli*. Nature, **217**, 1110-1114.   148

16. H. Echols (1970). Integrative and Excisive Recombination by Bacteriophage λ: Evidence for an Excision-specific Recombination Protein. J. Mol. Biol. **47**, 575-583.   153

17. E. Cassuto and C. M. Radding (1971). Mechanism for the Action of λ Exonuclease in Genetic Recombination. Nature New Biology, **229**, 13-16.   162

18. M. S. Fox (1966). On the Mechanism of Integration of Transforming Deoxyribonucleate. J. Gen. Physiol. **49**, 183-196.   166

19. P. Howard-Flanders, W. D. Rupp, B. M. Wilkins and R. S. Cole (1968). DNA Replication and Recombination after UV Irradiation. Cold Spring Harbor Symp. Quant. Biol. **33**, 195-208.   180

20. S. R. Kushner, J. C. Kaplan, H. Ono and L. Grossman (1971). Enzymatic Repair of Deoxyribonucleic Acid. IV. Mechanism of Photoproduct Excision. Biochem. **10**, 3325-3334.   193

21. M. Osborn, S. Person, S. Phillips and F. Funk (1967). A Determination of Mutagen Specificity in Bacteria using Nonsense Mutants of Bacteriophage T4. J. Mol. Biol. **26**, 437-447.   203

22. F. H. C. Crick, L. Barnett, S. Brenner and R. J. Watts-Tobin (1961). General Nature of the Genetic Code for Proteins. Nature, **192**, 1227-1232.   215

23. B. C. Westmoreland, W. Szybalski and H. Ris (1969). Mapping of Deletions and Substitutions in Heteroduplex DNA Molecules of Bacteriophage Lambda by Electron Microscopy. Science, **163**, 1343-1348.   229

II. RNA Synthesis   235

24. R. R. Burgess, A. A. Travers, J. J. Dunn and E. K. F. Bautz (1969). Factor Stimulating Transcription by RNA Polymerase. Nature, **221**, 43-46.   241

25. F. R. Blattner and J. E. Dahlberg (1972). RNA Synthesis Startpoints in Bacteriophage λ DNA. Nature New Biology, 237, 227-232. — 245

26. F. R. Blattner, J. E. Dahlberg, J. K. Boettinger, M. Fiandt and W. Szybalski (1972). Distance from a Promoter to an RNA Synthesis Startpoint in Bacteriophage λ DNA. Nature New Biology, 237, 232-236. — 250

27. U. Maitra and J. Hurwitz (1965). The Role of DNA in RNA Synthesis, IX. Nucleoside Triphosphate Termini in RNA Polymerase Products. Proc. Natl. Acad. Sci. U.S. 54, 815-822. — 255

28. J. W. Roberts (1969). Termination Factor for RNA Synthesis. Nature, 224, 1168-1174. — 263

29. N. Morikawa and F. Imamoto (1969). Degradation of Tryptophan Messenger. Nature, 223, 37-40. — 270

30. J. E. Darnell, L. Philipson, R. Wall and M. Adesnik (1971). Polyadenylic Acid Sequences: Role in Conversion of Nuclear RNA into Messenger RNA. Science, 174, 507-510. — 274

31. W. Colli, I. Smith and M. Oishi (1971). Physical Linkage Between 5S, 16S and 23S Ribosomal RNA Genes in *Bacillus subtilis*. J. Mol. Biol. 56, 117-127. — 278

32. B. A. Hamkalo and O. L. Miller, Jr. (1971). Visualization of Genetic Transcription. Submitted to Proc. Int. Symp. Protein Synthesis and Nucleic Acids (XI Latin American Symposium, La Plata). — 289

33. S. Altman (1971). Isolation of Tyrosine †RNA Precursor Molecules. Nature New Biology, 229, 19-21. — 310

34. G. Zubay, L. Cheong and M. Gefter (1971). DNA-Directed Cell-Free Synthesis of Biologically Active Transfer RNA: su^+_{III} Tyrosyl-†RNA. Proc. Natl. Acad. Sci. U. S. 68, 2195-2197. — 315

35. M. Chamberlain, J. McGrath and L. Waskell (1970). New RNA Polymerase from *Escherichia coli* infected with Bacteriophage T7. Nature, 228, 227-231. — 318

36. N. R. Pace and S. Spiegelman (1966). In Vitro Synthesis of an Infectious Mutant RNA with a Normal RNA Replicase. Science, 153, 64-67. — 323

37. G. Feix, R. Pollet and C. Weissmann (1968). Replication of Viral RNA. XVI. Enzymatic Synthesis of Infectious Viral RNA with Noninfectious Qβ Minus Strands As Template. Proc. Natl. Acad. Sci. U.S. 59, 145-152. — 327

38. M. Kondo, R. Gallerani and C. Weissmann (1970). Subunit Structure of Qβ Replicase. Nature, 228, 525-527. — 335

39. I. M. Verma, N. L. Meuth, E. Bromfeld, K. F. Manly and D. Baltimore (1971). Covalently Linked RNA-DNA Molecule as Initial Product of RNA Tumor Virus DNA Polymerase. Nature New Biology, 233, 131-134. — 338

III. Protein Synthesis   343

40. M. B. Hoagland, M. L. Stephenson, J. F. Scott, L. I. Hecht and P. C. Zamecnik (1958). A Soluble Ribonucleic Acid Intermediate in Protein Synthesis. J. Biol. Chem. 231, 241-257.   353

41. H. M. Dintzis (1961). Assembly of the Peptide Chains of Hemoglobin. Proc. Natl. Acad. Sci. U.S. 47, 247-261.   370

42. M. R. Capecchi (1966). Initiation of *E. coli* Proteins. Proc. Natl. Acad. Sci. U.S. 55, 1517-1524.   386

43. M. Noll and H. Noll (1972). Mechanism and Control of Initiation in the Translation of R17 RNA. Nature New Biology, 238, 225-228.   394

44. J. Waterson, G. Beaud and P. Lengyel (1970). The S_1 Factor in Peptide Chain Elongation. Nature, 227, 34-38.   398

45. E. Scolnick, R. Tompkins, T. Caskey and M. Nirenberg (1968). Release Factors Differing in Specificity for Terminator Codons. Proc. Natl. Acad. Sci. U.S., 61, 768-774.   402

46. M. W. Nirenberg and J. H. Matthaei (1961). The Dependence of Cell-Free Protein Synthesis in *E. coli* Upon Naturally Occurring or Synthetic Polyribonucleotides. Proc. Natl. Acad. Sci. U.S. 47, 1588-1602.   409

47. M. Nirenberg and P. Leder (1964). RNA Codewords and Protein Synthesis. Science, 145, 1399-1407.   424

48. M. Nirenberg, T. Caskey, R. Marshall, R. Brimacombe, D. Kellogg, B. Doctor, D. Hatfield, J. Levin, F. Rottman, S. Pestka, M. Wilcox and F. Anderson (1966). The RNA Code and Protein Synthesis. Cold Spring Harbor Symp. Quant. Biol. 31, 11-24.   433

49. F. H. C. Crick (1966). Codon-Anticodon Pairing: The Wobble Hypothesis. J. Mol. Biol. 19, 548-555.   447

50. M. Fuller and A. Hodgson (1967). Conformation of the Anticodon Loop in †RNA. Nature, 215, 817-821.   455

51. D. Nathans, G. Notani, J. H. Schwartz and N. D. Zinder (1962). Biosynthesis of the Coat Protein of Coliphage f2 By *E. coli* Extracts. Proc. Natl. Acad. Sci. U.S. 48, 1424-1431.   460

52. P. G. N. Jeppesen, J. A. Steitz, R. F. Gesteland and P. F. Spahr (1970). Gene Order in the Bacteriophage R17 RNA: 5'-A Protein-Synthetase-3'. Nature, 226, 230-237.   468

53. G. Streisinger, J. Emrich, Y. Okada, A. Tsugita and M. Inouye (1968). Direction of Translation of the Lysozyme Gene of Bacteriophage T4 Relative to the Linkage Map. J. Mol. Biol. 31, 607-612.   476

54. A. Garen (1968). Sense and Nonsense in the Genetic Code. Science, 160, 149-159.   482

55. M. Nomura, S. Mizushima, M. Ozaki, P. Traub, and C. V. Lowry (1969). Structure and Function of Ribosomes and Their Molecular Components. Cold Spring Harbor Symp. Quant. Biol. 34, 49-61. 493

IV. Regulation of Nucleic Acid and Protein Synthesis 507

56. S. Cooper (1972). Relationship of Flac Replication and Chromosome Replication. Proc. Natl. Acad. Sci. U.S. **69**, 2706-2710. 517

57. W. Gilbert and B. Müller-Hill (1966). Isolation of the *Lac* Repressor. Proc. Natl. Acad. Sci. U.S. **56**, 1891-1898. 522

58. W. Gilbert and B. Müller-Hill (1967). The *Lac* Operator in DNA. Proc. Natl. Acad. Sci. U.S. **58**, 2415-2421. 530

59. K. Ippen, J. H. Miller, J. Scaife and J. Beckwith (1968). New Controlling Element in the *Lac* Operon of *E. coli*. Nature **217**, 825-827. 537

60. J. Beckwith, T. Grodzicker and R. Arditti (1972). Evidence for Two Sites in the *Lac* Promoter Region. J. Mol. Biol. **69**, 155-159. 540

61. G. Zubay, D. Schwartz and J. Beckwith (1970). Mechanism of Activation of Catabolite-Sensitive Genes: A Positive Control System. Proc. Natl. Acad. Sci. U.S. **66**, 104-110. 546

62. L. Eron and R. Block (1971). Mechanism of Initiation and Repression of *In Vitro* Transcription of the *Lac* Operon of *Escherichia coli*. Proc. Nat. Acad. Sci. U.S. **68**, 1828-1832. 553

63. A. D. Riggs, G. Reiness and G. Zubay (1971). Purification and DNA-Binding Properties of the Catabolite Gene Activator Protein. Proc. Natl. Acad. Sci. U.S. **68**, 1222-1225. 558

64. E. Englesberg, C. Squires and F. Meronk (1969). The L-Arabinose Operon in *Escherichia coli* B/r: A Genetic Demonstration of Two Functional States of the Product of A Regulator Gene. Proc. Natl. Acad. Sci. U.S. **62**, 1100-1107. 562

65. W. A. Haseltine, R. Block, W. Gilbert and K. Weber (1972). MSI and MSII made on Ribosome in Idling Step of Protein Synthesis. Nature, **238**, 381-384. 570

66. H. F. Lodish (1968). Bacteriophage f2 RNA: Control of Translation and Gene Order. Nature, **220**, 345-350. 574

67. M. Schachner, W. Seifert and W. Zillig (1971). A Correlation of Changes in Host and T4 Bacteriophage Specific RNA Synthesis with Changes of DNA-Dependent RNA Polymerase in *Escherichia coli* Infected with Bacteriophage T4. Eur. J. Biochem. **22**, 520-528. 580

68. F. W. Studier (1972). Bacteriophage T7. Science, **176**, 367-375. 589

69. R. Losick, A. L. Sonenshein, R. G. Shorenstein and C. Hussey (1970). Role of RNA Polymerase in Sporulation. Cold Spring Harbor Symp. Quant. Biol. 35, 443-450. 599

70. M. Ptashne (1971). Repressor and Its Action. *In* "The Bacteriophage Lambda". Ed. by A. D. Hershey, Chapter 11, pp. 221-237. Cold Spring Harbor Laboratory. 606

PREFACE TO THE 2ND EDITION

This second edition has been so extensively revised that it should be regarded as another book with the emphasis on recent publications. The number of readings has been expanded from 48 to 70, only of which 12 appeared in the first edition. Fifty-one of the readings postdate 1968. The organization has been changed so that there are four main headings: DNA Replication, Recombination, Repair and Mutagenesis; RNA Synthesis; Protein Synthesis; and Regulations of Nucleic Acid and Protein Synthesis. By limiting the scope of this book, each of the above areas can be dealt with in greater depth. Because of space limitations and a desire to emphasize papers dealing with fundamental advances, this volume has been restricted (with a few exceptions) to papers dealing with bacteria and bacteriophages. Clearly, many of the most exciting biological discoveries in the near future will involve applying what has been learned from these microorganisms to the unique and fascinating problems of higher forms.

The main criterion for selecting papers is that they represent crucial experiments whose execution and significance have been carefully explained. In this extremely broad and rapidly moving field it has been impossible in many cases to represent all the important publications. The authors have cited the most recent experimental papers or reviews in the introduction to each section, so that the reader could find therein earlier references to classical and related publications. We consider this book a worthwhile compilation for all those interested in molecular biology and biochemical genetics and particularly suitable for use in advanced undergraduate and graduate courses covering these subjects.

Papers in
Biochemical Genetics
Second Edition

SECTION I

DNA Synthesis

Role of Nucleic Acid as Genetic Material

The emergence of the contemporary approach to biochemical genetics came with the discovery by Avery and his colleagues (Avery et al., *J. Exptl. Med.* 1944, **79**, 137) that nucleic acid, not protein carries the hereditary information. Avery and his coworkers purified DNA from a genetically marked donor strain of *Diplococcus pneumoniae* (pneumococcus), exposed a second (recipient) strain to the DNA, and thereby transformed a small fraction of recipient pneumococcal cells with respect to the trait (capsule formation) carried by the donor strain. Although this type of DNA-mediated transformation experiment has been successfully executed with only a small number of microorganisms, the significance of this discovery, which showed that DNA is the ultimate carrier of genetic information, cannot be overemphasized. Related experiments demonstrated that the genetic information of DNA-containing bacterial viruses (bacteriophages, or simply phages) was carried exclusively by the DNA component of the virus. Thus, Hershey and Chase (*J. Gen. Physiol.* 1952, **36**, 39) showed by suitable labeling of the parental virus component with ^{35}S-SO_4 and ^{32}P-PO_4 that only the DNA of the virus was transferred from one generation to the next; the sulfur labels the protein component exclusively via the sulfur containing amino acids and the phosphorus label enters the DNA component almost exclusively. About half the phosphorus of a ^{32}P-labeled virus is transferred from parent to progeny virus, while less than one percent of the sulfur of an ^{35}S-labeled virus is transferred from parent to progeny.

In later experiments Guthrie and Sinsheimer (*J. Mol. Biol.* 1960, **2**, 297) were able to show that purified bacteriophage DNA produces whole viruses when the DNA is added to bacterial cells that lack cell walls (spheroplasts) so as to allow the entry of high polymer DNA. Earlier still, the pioneering efforts of Gierer and Schramm (*Nature*, 1956, **177**, 702) as well as of Fraenkel-Conrat and Singer (*Biochim. Biophys. Acta*, 1957, **24**, 540) demonstrated that for an RNA-containing virus, tobacco mosaic virus (TMV), the RNA was the genetic component. The viral progeny from cells infected with intact, purified viral (RNA or DNA) genomes reflects the same genetic constitution of the infectious nucleic acid. The conclusion from such experiments is that either DNA or (in the case of some animal, plant or bacterial viruses), RNA can carry genetic information. DNA is the repository for all hereditary information in all cells but in different viruses either DNA or RNA is used to carry genetic information.

Physical and Chemical Characterization

Watson and Crick (*Cold Spring Harbor Symp. Quant. Biol.* 1953, **18**, 123) have given a detailed description of DNA double helix structure and an account of the important physical and chemical evidence that led to the construction of their molecular model. The two most significant of these evidences were (1) the base analyses by Chargaff and his colleagues which demonstrated that DNA contained four bases, adenine (A), guanine (G), cytosine (C), and thymine (T), and showed that the molar ratio of A to T and of G to C are equal to unity; and (2) the x-ray diffraction studies of the Wilkins and Franklin groups which showed that all diffraction patterns of DNA could be interpreted in terms of a double helix structure (Wilkins, *Science*, 1963, **140**, 941).

Some DNAs, however, do not occur in the double helix form. The first of these to be discovered was the single-stranded DNA of the *Escherichia coli* spherical bacteriophage called ϕX174. When isolated from ϕX174 phage preparations, the DNA is circular with a molecular weight of 1.6×10^6 daltons. The physical characterization of the ϕX174 DNA helped to explain earlier observations that the incorporation of high levels of ^{32}P in the DNA of the phage leads to a higher rate of lethality (suicide) than when ^{32}P is incorporated into phages with double-stranded DNA.

For a long time it was generally believed that the molecular weights of DNA isolated from bacterial and animal sources centered around the values of $5-20 \times 10^6$ daltons. It was eventually realized that in addition to degradation which might result from nucleases, conventional pipetting, blending or stirring procedures used in isolation could generate physical forces capable of shearing the double helix. Once this was appreciated it was found that all DNA-containing viruses thus far examined contained a single molecule (or genome) of DNA (Thomas and MacHattie, *Ann Rev. Biochem.* 1967, 36, 485). The characterization of the physical properties of phage genomes lead, as in the case of ϕX174 DNA, to a better understanding of the biological properties of different phages. In some cases double-stranded viral DNA has been found to have short, single-stranded, complementary termini. In the case of *E. coli* phage λ DNA where this is found, it enables the DNA to assume a circular structure, which, under some circumstances, can become secured by covalent linkages. In infected cells the circular configuration of λ DNA is important in replication and genetic interaction with other circular λ genomes or with the host chromosome. Circular DNA structures (Helinski and Clewell, *Ann. Rev. Biochem.* 1971, 40, 899) have also been found for most bacterial episomes and plasmids (Clowes, *Bact. Revs.* 1972, 36, 361). Mitochondrial DNA from various sources (Borst, *Ann. Rev. Biochem.* 1972, 41, 333) as well as from some animal viruses and from one bacterial virus (PM2, which grows on *Pseudomonas*), have also been shown to consist of covalently closed, double-stranded, circular structures.

Another interesting feature of some linear, double-stranded phage DNAs is that a short double-stranded sequence, usually several percent of the total genome, is terminally redundant. In the case of the T-even *E. coli* phages, in addition to the redundant regions, the phage genomes are circularly permutted. These unusual features of phage DNA account very nicely for the presence of some types of phage heterozygotes in genetic crosses (due to terminal redundancy) and for the fact that even though the T-even phage genome is linear, crosses between genetically marked parents generate a circular map.

Just as most DNAs occur in the double helix form, most RNAs occur in the single polynucleotide chain form. RNAs serve more functions than DNAs and it is more difficult to generalize about their properties. RNAs contain mainly four bases, similar to the bases contained in DNA except that uracil replaces thymine. There are three types of cellular RNA, all of which are DNA transcription products that are synthesized in bacteria (such as *E. coli*) by a single RNA polymerase. The RNA species include ribosomal RNA (3 sedimenting species—5S, 16S and 23S—all found in association with ribosomal proteins), messenger RNA and transfer RNA. The transfer RNA molecules, the smallest of the RNA species are also unique in the numerous minor bases that they contain (Dirheimer *et al.*, *Biochimie*, 1972, 54, 127). The fourth species of RNA are those that carry the genetic information found in viruses. In addition to their genetic role, the viral RNAs often serve as templates for protein synthesis. Some animal viral RNAs possess the double helix structure like DNA (Gomatos and Tamm, *Proc. Nat. Acad. Sci. U.S.* 1963, 50, 878). A *Pseudomonas* phage, ϕ6, has now been isolated that has double-stranded RNA (Koski and Etten, *J. Virol.* 1973, 11, 799). (In Section III the structure of the cellular RNA species will be discussed.)

A good deal of information about nucleic acid structure and the function of nucleic acids as messengers in directing polypeptide synthesis has come from the extensive work done on synthetic polyribonucleotides. The discovery of the enzyme polynucleotide phosphorylase (Grunberg-Manago and Ochoa, *J. Am. Chem. Soc.* 1955, 77, 3165) made possible the synthesis of simple homopolymers such as polyriboadenylic acid and polyribouridylic acid. These can form a 1:1 double helix structure similar to DNA or a 1:2 triple helix structure, depending upon the conditions under which they are mixed. Although a variety of other double and triple helix structures have been observed using other synthetic polyribonucleotides (Michelson *et al.*, *Prog. Nuc. Acid Res. Mol. Biol.* 1967, 6, 83), as yet no biologically significant triple helix structure has been isolated. The synthesis of a variety of polydeoxyribonucleotides using the Kornberg *E. coli* polymerase (Burd and Wells, *J. Mol. Biol.* 1970, 53, 435) has made available deoxyribonucleotide polymers that have proven to be extremely useful in physical studies as well as acting as templates to generate RNA transcripts of specified sequence.

The stability of the double helix structure is due largely to hydrogen-bonding between the bases, although the stacking free energy between the bases is also believed to play a significant role. The higher the GC content of the DNA the higher the midpoint of its thermal denaturation profile or melting temperature (T_m) (see *paper 1*). This is partly due to the fact that the GC base pair in DNA makes three hydrogen bonds instead of two as the AT base pair does.

Semi-conservative Synthesis *in vivo* *

Clearly, the cellular genome must be duplicated once for every cell division. Consideration of the complementary duplex nature of the DNA structure and the necessity for doubling of DNA prior to cell division led to the suggestion that DNA synthesis *in vivo* occurs by the unwinding of the double helix, and the absorption of complementary mononucleotides to each parental polynucleotide strand which is then followed by polymerization to generate progeny strands. In this mode of semi-conservative replication, newly duplicated DNA should consist of one old strand and one newly synthesized strand. Evidence that this is the case for bacterial DNA was provided by Meselson and Stahl (*Proc. Nat. Acad. Sci. U. S.* 1958, 44, 671). They labeled DNA of one density by growing cells for several generations on a medium containing $^{15}NH_4Cl$ as the sole nitrogen source, and then transferred it to a medium of the normal $^{14}NH_4Cl$. They then allowed the DNA to duplicate for one or more generations. DNA of different densities can be separated and distinguished in the ultracentrifuge by a technique called density gradient centrifugation. A small amount of DNA in a concentrated solution of cesium chloride is centrifuged until equilibrium is closely approached. The opposing processes of sedimentation and diffusion produce a stable concentration gradient of the cesium chloride. The concentration and pressure gradient cause a continuous increase of density along the direction of centrifugal force. The macromolecules of DNA present in this density gradient are driven by the centrifugal field into the region where the solution density is equal to their own buoyant density. This concentrating tendency is opposed by diffusion, with the result that at equilibrium a single species of DNA is distributed over a band whose width is inversely related to the molecular weight of that species as well as to the base compositional heterogeneity of the DNA sample under analysis. Bands of varying intensity appear at points corresponding to pure ^{15}N DNA, pure ^{14}N DNA and hybrid $^{15}N-^{14}N$ DNA. After precisely one generation time in ^{14}N, only the band corresponding to $^{15}N-^{15}N$ hybrid DNA is visible. These data argue strongly in favor of the semi-conservative mode of DNA replication. Similar evidence was soon provided for the semi-conservative replication of human DNA using 5-bromodeoxyuridine as the density label (Djordjevic and Szybalski, *J. Exp. Med.* 1960, 112, 509). In addition, the fact that only hybrid DNA was found after one cell doubling in unsynchronized bacterial cells implied that continuous replication proceeded from a small number of replication forks (now thought to be two) since the DNA being analyzed for its density was extensively fragmented. Thus, in bacteria, the replicating unit, or replicon, is the entire bacterial chromosome, which in *E. coli* is about 1.2 mm in length. The rate of replication of the *E. coli* chromosome is about 30μ per minute.

Complexities of Explaining DNA Chain Growth*

It is customary to imagine that the double helix molecule gradually unwinds *in vivo* to serve as a template for the newly growing DNA chains. This requirement for unwinding introduces two fundamental complications that are not yet completely resolved. (1) Due to the antiparallel nature of the double helix and the unique directional sense observed for DNA synthesis *in vitro* (see below), progressive synthesis during unwinding could occur only on the one template strand with the 3'-OH terminus. The copying of the other DNA strand would have to wait until the double helix was unfolded. This problem would become more and more serious as the DNA molecule to be replicated increases

---

*Smith, *Prog. Biophys. Mol. Biol.* 1973, 26, 321.

*Goulian, *Prog. Nuc. Acid Res. Mol. Biol.* 1972, 12, 29.

in size. (2) Unwinding of a DNA double helix requires that the ends undergo rotation around each other. The larger the DNA molecule the more cumbersome this becomes. With circular chromosomes an entire chromosome must rotate once every time ten bases are replicated; the cumbersome nature of such a scheme compels one to examine other possibilities.

Recent observations may help to resolve these difficulties. Thus, studies by Okazaki (*paper 2*) on very briefly labeled DNA show that the newly synthesized strands of DNA exist as small discontinuous pieces rather than as a continuous part of a long DNA chain. With time these small pieces eventually become connected into one long DNA molecule. Several mechanisms for the details of this process have been proposed. The most important point to be gained from the observation is that DNA appears to be synthesized initially in relatively short segments of about 1000 to 2000 bases in length. Since the synthesis of such pieces only takes about a second it becomes possible to think in terms of synthesis in the 5′ to 3′ direction on both parental, complementary chains; first, synthesis occurs on one chain as the parent double helix unfolds and, after a section has unfolded, in the opposite direction on the antiparallel, complementary chain. The interrupted strands eventually become connected together by polymerases (if there are gaps), followed by enzymes called ligases (see below) which could serve in the final sealing step. Ligases have been found in uninfected cells and new ones appear in phage-infected bacteria; this will be discussed in more depth later in this section.

These considerations still do not solve the problem of unwinding the parental DNA. Transiently formed single-strand breaks in the parent double helix, judiciously spaced along the DNA at some distance from the point of synthesis, would enable the parental DNA to unwind by rotating around the single bonds in the remaining intact polynucleotide strand. This would provide us with an escape from the second dilemma but no evidence for such breaks in *E. coli* has yet been found (recent evidence for breaks ahead of the replicating fork in mammalian chromosomes has been reported by Taylor, *Proc. Nat. Acad. Sci. U. S.* 1973, **70**, 1083). Nor does the initial hypothesis of the synthesis of "Okazaki pieces" satisfy the requirement that all the known DNA polymerases of *E. coli* require not only a template but also a primer (Wickner, Ginsberg and Hurwitz, *J. Biol. Chem.* 1972, **247**, 498; Kornberg and Gefter, *J. Biol. Chem.* 1972, **247**, 5369). RNA polymerase, which requires a template but can initiate RNA synthesis without a primer, could then serve to provide the necessary polynucleotide primer for DNA synthesis (*paper 6*). Current observations show that transcription is indeed involved in DNA replication, and Okazaki and his coworkers (*Proc. Nat. Acad. Sci. U. S.* 1972, **69**, 1863; 1973, **70**, 88) have found that the small, nascent DNA fragments are covalently bound to RNA. The RNA primers would have to be removed prior to the generation of intact parental DNA strands, possibly by the action of an enzyme such as ribonuclease H (Henry *et al.*, *Biochem. Biophys. Res. Comm.* 1973, **50**, 603).

The First of Several DNA Polymerases

The Watson-Crick proposal of a complementary duplex structure for DNA stimulated a search for a DNA polymerase with certain implied properties. The enzyme should require an intact DNA chain to serve as a template for the absorption of complementary bases, and the *de novo* synthesized DNA should be a complement of one of the template DNA chains. Kornberg and his colleagues found a DNA synthesizing enzyme that satisfied these requirements and named it DNA polymerase (*paper 3*). It is now known as DNA polymerase I, or the Kornberg enzyme. The enzyme, isolated from *E. coli*, has been extensively purified; similar enzymes coded by viral genomes have been isolated from virus infected bacterial cells (for example see Aposhian and Kornberg, *J. Biol. Chem.* 1962, **237**, 519). In the pure state, the Kornberg enzyme requires a DNA template, the four commonly occurring deoxynucleotide triphosphates, and magnesium ion for synthesizing DNA. The enzyme catalyzes the addition of mononucleotides to a 3′-end of a growing chain. The reaction probably occurs as a nucleophilic attack by the 3′-hydroxyl group of the terminal mononucleotide residue at the growing end of the chain on the β-phosphorus atom of the entering nucleoside 5′-triphosphate, thus causing displacement of its pyrophosphate group and formation of the internucleotide linkage. If no primer with a free 3′-end is available (as when closed circular single-stranded DNA template is used) the enzyme must initiate synthesis from mononucleotides. In this case, either there is no DNA synthessis or else a long lag period will occur in the initiation of synthesis, generating polymers with defined sequences (Baldwin, in *The Bacteria*, ed. by I. C. Gunsalus and R. Y. Stanier, Academic Press, New York 1964, vol. V, 327). The bases added to the growing chain when a primer is present are determined by the sequence of bases in the DNA template. Complementary bases are added so that the single-stranded DNA template gradually becomes converted to a double helix. Ordinarily cell-free DNA synthesis proceeds beyond the double helix stage at a considerably reduced rate. Physiochemical studies have shown that the enzyme has a complex surface with specific attachment sites for template chain, growing primer chain, and monomer triphosphate. The enzyme is highly selective since it only adds a base that is properly paired to the template strand. As the primer chain becomes lengthened by synthesis the enzyme must move along the template one base at a time. The newly synthesized DNA has been shown to be a faithful complementary replica of the template strand in a number of ways: by gross base composition analysis; by melting curve profile of the hybrid formed between the template strand and the newly synthesized primer strand; and finally, through a series of manipulations, by the *de novo* synthesis of genetically active DNA. Using infectious circular φX174 DNA as a template, DNA was newly synthesized which proved infectious in a spheroplast assay (*paper 4*).

DNA Polymerase I Is Not Essential for *in vivo* Chromosomal DNA Synthesis

Some ten years after Kornberg had discovered DNA polymerase I, DeLucia and Cairns (*paper 7*) after laboriously scanning several thousand colonies of a heavily mutagenized *E. coli* culture were able to isolate a mutant that contained almost no Kornberg DNA polymerase. Cells with this mutation grow normally under optimal conditions in rich media but are more sensitive to ultraviolet radiation than the parent strain. These properties strongly suggest that the Kornberg enzyme is not indispensable but may be involved in repairing chromosome damage resulting, for example, from ultraviolet radiation. The discovery of this polymerase mutant has had a profound effect upon the thinking and on experimental design in studies on DNA biosynthesis (*papers 8, 9, 10*). An important general principle of biochemical research is underscored by this unexpected finding: One cannot necessarily assign a function to an enzyme merely on the basis of its *in vitro* properties or its abundance. A genetic approach using mutants make meaningful *in vivo* correlates possible. Such a genetic approach to the study of the enzymes involved in DNA in *E. coli* has now been exploited, and has taken the direction of isolating mutants that cannot synthesize DNA at non-permissive, elevated temperatures (Hirota *et al., Fed. Proc.* 1972, 31, 1422; Gross, *Curr. Topics Microbiol. Immunol.* 1972, 57, 29; Wechsler *et al., J. Bact.* 1973, 113, 1381). Several such different mutant types, defective in 6 or 7 different genes, have now been isolated and have been labeled *dnaA* through *dnaG*. In a parallel biochemical approach using cell-free extracts from an *E. coli* strain which contains no DNA polymerase I, two additional DNA polymerizing enzymes have been detected—DNA polymerase II and III (Moses *et al., Fed. Proc.* 1972, 31, 1415). Preliminary genetic evidence suggests that DNA polymerase III is the DNA synthesizing enzyme (*papers 8, 9*). This evidence is derived from the finding that the mutant *dnaE* contains an altered polymerase III enzyme, while *E. coli* mutants lacking polymerases I and II can still divide and replicate host DNA, most episomes, and phage DNA (Campbell, Soll and Richardson, *Proc. Nat. Acad. Sci. U. S.* 1972, 69, 2090; Hirota, Gefter and Mindich, *Proc. Nat. Acad. Sci. U. S.* 1972, 69, 3238). It has been found, however, that *E. coli* mutants lacking polymerase I cannot replicate some plasmids (Kingsbury and Helinski, *J. Bacteriol.* 1973, 114, 1116). Another temperature-sensitive gene product affecting DNA synthesis has been identified. Mutants in *dnaF* have an altered ribonucleotide diphosphate reductase (Fuchs *et al., Nature New Biology*, 1972, 238, 69), the first enzyme unique to DNA synthesis since its activity is necessary for the synthesis of DNA precursors. The *dnaG* gene product has recently been purified; it has not yet been assigned any enzymatic role or function in DNA synthesis (Wickner, Wright and Hurwitz, *Proc. Nat. Acad. Sci. U. S.* 1973, 70, 1613).

The Joining Enzyme—Polynucleotide Ligase

As mentioned earlier, discontinuous DNA synthesis necessitates the existence of an enzyme for joining newly synthesized segments which have nicks between them. Such an enzyme has been found in a variety of cell types and is called polynucleotide ligase (*paper*

11). Polynucleotide ligase requires DNA-adenylate with a pyrophosphate linkage at the 5′-phosphoryl terminus, and a receptive 3′OH terminal chain properly juxtaposed for reaction. If there is no phosphate on the 5′-end, an enzyme called polynucleotide kinase can transfer one from ATP. The following overall mechanism has been proposed for the sealing of a nick (or gap) in double stranded DNA: (1) E-AMP is formed by a reaction of enzyme with ATP (in the case of the T-even phage induced ligase), or NAD (in the case of ligase normally present in *E. coli*); (2) AMP is transferred to the 5′-phosphoryl terminus of a DNA chain to form a new pyrophosphate bond; (3) the activated 5′-phosphate group is attached to the 3′-hydroxyl group of the adjacent DNA chain, displacing the AMP and producing a new phosphodiester bond linking the two chains (Richardson, *Ann. Rev. Biochem.* 1969, 38, 795).

DNA-Binding Protein

A new type of protein, essential for DNA replication and recombination, has been isolated from T4-infected cells of *E. coli* (*paper 12*); as it is coded by gene 32 of this phage it has been called the gene 32 protein. This protein binds cooperatively to single-stranded DNA and catalyzes the denaturation and renaturation of DNA *in vitro*. Although the exact function of this protein in DNA synthesis is not known, it is known that *in vitro* it stimulates DNA synthesis when using the T4 DNA polymerase (Huberman, Kornberg and Alberts, *J. Mol. Biol.* 1971, 62, 39). A similar binding protein has been isolated from both *E. coli* infected with the filamentous single-stranded DNA phages (Oly and Knippers, *J. Mol. Biol.* 1972, 68, 125; Alberts, Frey and Delius, *J. Mol. Biol.* 1972, 68, 139) and from uninfected *E. coli*. Again, although these DNA binding proteins are involved in DNA replication, their role is not yet understood nor do they all necessarily act in the same way.

Chromosome Duplication*

What are the models that have been presented to explain the replication of bacterial and of phage chromosomes? The DNA of bacterial chromosomes, of phages, and of episomes and plasmids are either linear or circular duplexes and can control their own replication (that is, they are *replicons*). Genetic, biochemical, and cytological studies are consistent with the idea that duplication of the circular *E. coli* chromosome starts at a single origin and proceeds in *both* clockwise and counterclockwise directions to a common terminus. The most convincing evidence of this was obtained by the genetic technique of measuring gene frequency during replication (Masters and Broda, *Nature New Biology*, 1971, 232, 137; Hohlfield and Vielmetter, *Nature New Biology*, 1973, 242, 130). Because the position of many genes on the circular chromosome is precisely known (Taylor, in *Handbook of Biochemistry*, ed. H. A. Sober, The Chemical Rubber Co., Cleveland, Ohio, 1970, pp. 121—31) the origin, terminus and rates of chromosome duplication can be determined with a high degree of certainty. Recent radioautographic evidence has confirmed that DNA replication in *E. coli* is bidirectional (Rodriguez *et al.*, *J. Mol. Biol.* 1973, 74, 599). In addition, it has been recently shown that the *Bacillus subtilis* chromosome (Gyurasits and Wake, *J. Mol. Biol.* 1972, 73, 55) replicates bidirectionally.

The elegant electron microscopic observations of Inman and Schnös (*paper 13*) indicate that after infection the double helical bacterial virus λ replicates in a bidirectional manner like the *E. coli* chromosome. When λ phage infects *E. coli* it injects its single chromosome as a linear piece of DNA. The original linear chromosome then forms a covalent circle through the annealing of short single-stranded complementary 5′ ends and the subsequent joining action of polynucleotide ligase. Electron micrographs of the λ DNA isolated early from infected bacterial cells show a high percentage of the molecules in an intermediate replicative form with the configuration of the Greek letter Θ (Cairns saw a similar form, *J. Mol. Biol.* 1963, 6, 208, during the replication of the *E. coli* genome). This configuration most likely represents the replicating circular genome with two branch points at the growth points. In the case of λ DNA closer examination of the branch points shows that frequently the DNA is still single-stranded on one side of the branch point. This observation is consistent with the mechanism of staggered growth suggested by Okazaki. Single- and double-stranded DNA are distinguished in the electron microscope by a greater contrast shown by double-stranded

---
*Klein and Bonhoeffer, *Ann. Rev. Biochem.* 1972, 41, 301.

DNA (see *paper 23*). Later on in λ phage infection, long linear phage DNA molecule called concatamers, which are longer than unit genome length, predominate.

Although in bacteria DNA synthesis normally occurs throughout the cell division cycle, DNA synthesis in animal and plant cells takes place only during a defined portion of the cell cycle. Autoradiographic studies in animal and plant chromosomes show that semi-conservative replication of the DNA also takes place here (Callan, *Proc. Roy. Soc. London*, 1972, 181, 19). These measurements were made by incorporating labeled tritiated thymidine for a brief period into the DNA *in vivo*. Radioautographic observation of the chromosomes was then made at different times after labeling. This was done by flattening the chromosomes against a piece of film and then allowing enough time for the radioactive disintegration of the incorporated radioactive isotope to produce a picture on the film. In these much larger eukaryotic chromosomes, growth starts at many points in tandem along the chromosome and proceeds along the chromosome in both directions. Many of the tandem replicating units, which are about 50μ in length, replicate simultaneously at a rate of about 1μ/minute. Hogness has pointed out that the pattern of replication of eukaryotic DNA is similar to that which would be produced if a variety of λ-like rings (which may vary in length) were cut and then attached, one to another, to produce a linear array of bidirectional replicating units.

The replication of a single-stranded DNA molecule, such as if found in the bacteriophage ϕX174, presents special problems (*paper 14*). The parental DNA molecule is a covalent circular single strand and is referred to as the plus strand. Soon after infection this single strand is converted to a circular Watson-Crick double helix by host cell enzymes that synthesize the complementary minus strand. The double helix form that is generated is known as the replicative form (RF); in conjunction with virus encoded enzymes the RF serves as the template on which progeny plus strands are synthesized. Before this occurs, however, a number of copies of the original RF double helices are produced. It is believed that the progeny strands are produced on the minus strand template by a process of continuous extension in the usual 5′ to 3′ direction with displacement of pre-existing plus strand. This results in a gradual lengthening of the plus strand which eventually is split and circularized to produce progeny plus circular strands. This is known as the *rolling circle model* of DNA replication. This model also accounts nicely for the concomitant transfer of a single strand of DNA from donor to recipient and DNA synthesis in both cells during *E. coli* conjugal transfer of episomes, as well as during transfer of chromosomal DNA (Vapnek and Rupp, *J. Mol. Biol.* 1971, 60, 413). Recent evidence indicates that the rolling circle model of DNA replication can account for the amplification of ribosomal RNA cistrons in frog oocytes.

In addition to the model we have used to describe the early stage of replication of λ DNA and *E. coli* DNA, and the model illustrating the replication of ϕX174, there is a third model of DNA replication which is the most classical model of all. This is the Y-fork model for linear genomes such as the DNA of phage T7 (Wolfson and Dressler, *Proc. Nat. Acad. Sci. U. S.* 1972, 69, 2682). Here again, as in the case of λ DNA, replication actually proceeds bidirectionally. However, the replicating structure is not circular. Initiation of replication begins at an internal site about 17% from the left end of the linear DNA molecule, and proceeds in opposite directions. As soon as replication is complete in the lefthand direction, a Y-shaped replicating structure is generated until replication is complete in the righthand direction.

Of all the genomes examined thus far, only those in animal cell mitochondria have been shown to replicate unidirectionally (Vinograd and Kasamatsu, *Nature New Biology*, 1973, 241, 103). In this case, replication is initiated at a unique site on the covalently closed circular DNA using the light strand (defined by its buoyant density) as template and displacing a short segment of heavy strand. A newly synthesized light strand occurs on the displaced heavy strand after 0.6 or more genome units of displacement replication have taken place.

Role of Cell Membrane in DNA Replication

The cell membrane appears to contain attachment sites for bacterial as well as viral chromosomes that replicate within the cell. The DNA of bacteriophages λ and ϕX174 have mutually exclusive attachment sites (Salivar and Sinsheimer, *J. Mol. Biol.* 1969, 41, 39) and there seem to be 60 to 80 of these sites for each of these viruses. While only a few of the attachment sites are functional in replication of φX174 parental RF, most of them can replicate λ DNA. The φX174 RFs continue to synthesize progeny RFs until they have produced more progeny RFs than there are available membrane attachment sites. The RFs which are membrane bound are never released, and in the case of φX174 (though not in the filamentous, single-stranded phages) they do not function in progeny plus strand synthesis. The replication of the filamentous single-stranded DNA phage of E. coli, such as fd and M13 have certain features in common with φX174 phage DNA replication as well as some differences (Pratt, *Ann. Rev. Genet.* 1969, 3, 343; Marvin and Hohn, *Bact. Revs.* 1969, 33, 172).

The role of membrane in DNA replication in E. coli is obscure (Hirota et al., *Fed. Proc.* 1972, 31, 1422). Evidence that attachment sites also exist for the bacterial chromosome is cytological; it is also based on observations in pulse labeling experiments that newly synthesized cellular DNA is attached to a cell structure having the sedimentation properties of membrane fragments (Osborn in *Structure and Function of Biological Membranes*, ed. L. I. Rothfield, New York: Academic Press, 1971, 343). In B. subtilis (Sullivan and Sueoka, *J. Mol. Biol.* 1972, 69, 237) it has been found that genetic markers located at the origin and terminus of replication are associated with membranous material. In addition, the DNA of a number of bacteriophages such as T4, Salmonella phage P22, and φ29 of B. amyloliquefaciens has been shown to be associated with a rapidly sedimenting complex containing host cell membrane components. Recently, Kornberg and his associates (*Proc. Nat. Acad. Sci. U. S.* 1973, 70, 205) have shown that the coat protein of the single-stranded E. coli phage M13 participates in the membrane-oriented synthesis of phage DNA. The phage protein is believed to provide a link of the phage DNA to the replicative machinery located in or near the host membrane.

DNA Initiation and Post-replicative Modification

In the course of examining chromosome replication in bacteria we have discussed the role of four enzymes: DNA polymerases I, II, III, and polynucleotide ligase. DNA III, most likely in association with other proteins, probably forms the basic skeleton of the DNA. In addition to the deoxyribonucleotide triphosphate precursors, ATP is also required for DNA synthesis, as well as DPN to serve as a cofactor for ligase. In addition to the DNA polymerase special proteins (such as an RNA polymerase and a gene 32-type product) are most likely required for initiation and propagation steps in chromosome synthesis and perhaps for termination. This has been implied from two types of experiments. In the first type of experiment while in the presence of a protein inhibitor, such as chloramphenicol, bacterial DNA can complete its round of replication but reinitiation will not occur. In the second type, temperature-sensitive mutants in at least two separate loci, presumably coding for different proteins, have been isolated in E. coli which behave in the same manner when grown at the elevated, non-permissive temperature: that is, rounds of replication are completed but not reinitiated under non-permissive conditions.

Enzymes are also involved in the post-replicative modification of individual bases with methyl and glycosyl (in the case of the T-even E. coli phages) groups. When any of the T-even bacteriophages infect E. coli, host DNA synthesis is inhibited in favor of phage DNA synthesis. Concomitant with this change, the pathway for the synthesis of dCTP is interfered with and instead, its hydroxymethyl derivative is made, resulting in a replacement of cytosine by hydroxymethylcytosine in the bacteriophage DNA. The CTP is first hydrolyzed to CMP, followed by addition of the $-CH_2OH$ moiety and pyrophosphorylation. Some of the hydroxymethylcytosine groups are further modified by glucosylation after the DNA is polymerized. The glucosyl donor is UDPG and the acceptor is the 5-hydroxylmethyl group. Failure to glucosylate the T-even phage DNA exposes the nucleic acid to specific restrictions by nucleases present in certain hosts (Revel and Luria, *Ann. Rev. Genet.* 1970, 4, 177). Some of the deoxyribonucleases coded for by the T-even phage genome can indeed degrade host DNA of infected cells, but do not degrade glucosylated phage DNA.

DNA methylases have been found in bacteria which methylate adenine and cytosine in native double helical DNA giving 6-methylaminopurine (6-MAP) and 5-methylcytosine (5-MC). The methyl group donor, as in the methylation of RNA, is S-adenosylmethionine (SAM).

Host Restriction and Modification Enzymes*

Most strains of E. coli carry at least one set of specific DNA restriction and modification activities (paper 15). Phages such as PI may carry their own set of restriction and modification activities. The modification activity is exerted by a specific DNA methylase that methylates a very limited number of adenine and cytosine residues at specific sites in the double helix (Bouche and Dubert, Europ. J. Biochem. 1972, 27, 53). The restriction activity is caused by an endonuclease, present in E. coli or coded for by some of its phages or plasmids, that produces two phosphodiester cleavages in opposing strands leading to double strand breaks. The restriction enzyme acts within a specific DNA sequence except when the sequence has been specifically modified by the methylase from the same strain. The highly specific restriction enzyme makes a very limited number of breaks in susceptible heterologous DNA, but DNA thus cleaved is subsequently susceptible to other nucleases in the cell and is rapidly broken in vivo to low molecular weight compounds. The restrictive endonucleases of E. coli as well as that from Hemophilus all require Mg^{++}; in addition the E. coli restriction reactions require ATP and SAM. The splitting of the ATP that accompanies restriction of unmodified DNA is thought to provide activation energy for the configurational change necessary for the specific recognition of unmodified DNA (Yuan, et al., Nature New Biology, 1972, 240, 42).

A genetic analysis of the E. coli strains B and K restriction and modification systems show that they are determined by three closely linked genes. It is believed that one gene specifies the restriction peptide; another codes for the peptide involved in modification; and the third codes for a peptide used by the modification and restriction peptides, determining the specificity of the DNA nucleotide sequence that is recognized.

The restriction enzymes from Hemophilus species have now been studied in great detail with respect to the nucleotide sequences of the sites on the DNA where they act. The H. influenzae endonuclease (endonuclease R) breaks the DNA duplex in the middle of a specific symmetrical sequence of six nucleotide pairs (Smith and Wilcox, J. Mol. Biol., 1970, 51, 393). The very demanding specificities of the restricting enzymes as well as their wide distribution has currently made them extremely useful tools in the characterization of viral, bacterial, and animal DNA (Dana and Nathans, Proc. Nat. Acad. Sci. U. S. 1971, 68, 2913). The restriction endonuclease (endonuclease RI) from E. coli carrying the drug resistant factor RTF-1 has been used as a tool in constructing hybrid DNA molecules (Marx, Science, 1973, 180, 482). Berg and his associates (Proc. Nat. Acad. Sci. U. S. 1973, 69, 2904, 3365) have used the RI endonuclease to convert SV40 viral DNA and a λ derivative to linear forms before the addition of complimentary homopolymeric extensions which are subsequently annealed and ligased to form covalently closed hybrid DNA molecules. Since it has now been shown that the RI enzymes as well as other restriction enzymes make staggered cuts on the restriction sites to generate single stranded complementary ends, the fragments originating from homologous or heterologous sources can be reassociated in a variety of ways using appropriate conditions.

Deoxyribonucleases and Ribonucleases

A number of phosphodiesterases exist which are specific for DNA (Koerner, Ann. Rev. Biochem. 1971, 39, 291). These are classified according to whether they attack DNA chains exclusively at the ends (exonucleases) or internally (endonucleases). All DNases which have been isolated from E. coli produce $5'$-phosphoryl and $3'$-OH termini. Exonucleases with interesting and biochemically useful specificities exist; for instance, exonuclease I only attacks single-stranded DNA, while exonuclease III attacks only double-stranded DNA, degrading each strand about half-way. Both of these enzymes attack the chains from the $3'$-ends. Nucleases specific for single stranded DNA are very useful in renaturation or hybridization experiments to digest and eliminate unhybridized or mismatched sequences.

The ribonuclease that has received the greatest current attention is ribonuclease H originally found in animal cells. Ribonuclease H, since it is able to digest RNA in DNA-RNA hybrids, might be the enzyme that removes the priming RNA which is found covalently linked to DNA (Keller, Proc. Nat. Acad. Sci. U. S. 1972, 69, 1560).

---

*Meselson, Yuan and Heywood, Ann. Rev. Biochem. 1972, 41, 401.

Enzyme Mechanisms Involved in Genetic Recombination

Recombination, the exchange of genetic material between two homologous chromosomes, is one of the most important means of hereditary change. The process of recombination usually involves chromosome breakage, DNA synthesis, and rejoining between chromosomes. It is carried out in such a way that homologous regions are exchanged. Since this must occur in several steps, different types of enzymes would be expected to be involved. The discrete steps which have been postulated take into account many experimental facts obtained mostly from bacteriophage recombination studies. Such studies have shown that recombination can occur without gross DNA synthesis taking place, especially when phage and host DNA replication is restricted by use of appropriate mutants, and that recombinant genomes contain macromolecular components from each parent (Meselson, *J. Mol. Biol.* 1964 9, 734). The first step in recombination is postulated to invole (*a*) single, or (*b*) double strand breaks on the two chromosomes made by endonucleases. These molecules are either unwound further, or attacked by exonucleases to expose (*c*) single-stranded recognition lengths. The single stranded DNA regions of each parent anneal to form (*d*) a hydrogen-bonded, heteroduplex recombinant molecule; the heteroduplex overlap region may vary in size, and if, as in the case of T-even phage crosses, the heteroduplex structure is packaged *before* segregation, these phages can be recognized genetically as partial heterozygotes. The annealing step explains why recombination usually takes place between closely related chromosomal regions or sites. Only such regions could be expected to form stable annealed regions between the appropriate strands. (Extensive base sequence homology between recombining DNA might be unnecessary if a site specific recombination enzyme is involved, such as the one involved in the integration of λ DNA at a special site in the *E. coli* chromosome.) Residual gaps following heteroduplex formation are eliminated by repair enzymes, and overlapping strands are eliminated by (*e*) exonucleases. Finally, the segments are covalently joined by DNA ligase. Most models of recombination incorporate the above features, but the sequence of events may be somewhat altered (Whitehouse, *Biol. Revs.* 1970, 45, 265; Clark, *Ann. Rev. Microbiol.* 1971, 25, 437).

Although it is reasonably certain that at least some of the above steps must be involved in general recombination, the enzymes involved in the processes have not been unambiguously identified. In *E. coli*, at least three cistrons have been identified as essential for general recombination (*paper 19*); they probably code for a portion of the enzymes necessary for the steps described above. The characterization of the physical structures and the enzymological steps involved in recombination has been studied most intensively and productively with bacteriophages such as λ (*paper 17*) where recombination defective host and phage mutants can be readily obtained, gene products identified, and intact phage genomes in various stages of recombination can be isolated and handled with relative ease. What is even more important is that the intact complementary λ DNA strands can be separated and fractionated and then used to construct heteroduplexes *in vitro* by annealing strands from different genetically marked parents (see *paper 23*). Such heteroduplexes, which may serve as models of intermediates in λ recombination *in vivo*, can be used to infect spheroplasts to ascertain the generation of recombinants.

Lysogeny and Phage Induction

Upon infection some phages, such as λ, can either give rise to a lytic infection and consequent cell death or else become integrated (now called prophages) into the host and reside in a dormant state without harming the cell until special conditions lead to phage multiplication (Echols, *Ann. Rev. Genet.* 1972, 6, 157). Lysogenization, in the case of bacteriophages λ and P22, involve the integration of the phage DNA within specific sites on the host chromosome. In the case of the generalized transducing *E. coli* phage P1, the prophage is extrachromosomally located, existing as a plasmid. A number of steps involved in the integration of λ DNA into the *E. coli* chromosome are already fairly well understood; they are similar in many respects to the steps postulated for general recombination. One major difference is that only specific sites on the phage and bacterial chromosomes—called the integration or attachment (*att*) loci—are involved (*paper 16*). The bacterial *att* sites do not appear to show any extensive base sequence homology with the phage *att* sites (Hradecna and Szybalski, *Virology*, 1969, 38, 473). Initially, in the replication and/or integration sequence, the complementary overlapping 5′

ends of the linear λ chromosome can anneal (step 1). The resulting nicked circle is subsequently transformed into a covalent circle, presumably by a DNA ligase (step 2). The *att* locus on the phage and that on the bacterial chromosome form a covalent union by a process analogous to the one suggested for general recombination (step 3). The integrated prophage then forms a linear part of the bacterial chromosome. A phage coded repressor (a product of the λ gene *c*I) maintains the λ DNA in its integrated state and also makes the lysogenized cells immune to subsequent infections by λ or related (homoimmune) phages. When the prophage is induced, it is excised from the host chromosome as a whole circle and replicates. Although two λ phage genes, *int* and *xis*, have to be expressed for phage excision, only the *int* gene product is required for integration leading to lysogenization. As in general recombination, the actual enzymes associated with these processes have not yet been identified.

An interesting parallel to lysogeny is found in animal cells when they interact with a number of DNA cancer causing viruses. The simple DNA virus SV40, for example, interacts in quite different ways depending on the cell type that it infects (Green, *Ann. Rev. Biochem.* 1970, 39, 701). In mouse cells SV40 reproduces many times leading to cell death. This reaction is comparable to the lytic infection of *E. coli* by bacteriophage. In hamster cells (which are non-permissive) the SV40 DNA, which is already in the form of a covalently closed circular molecule in the mature viral particle, becomes incorporated into the cellular chromosome(s) resulting in (viral) transformed, malignant cells. The reaction of SV40 viral DNA with the hamster cell genome would appear to be similar in some respects to the lysogenization by bacteriophage λ or, even more likely, to phage Mu-1 which can be inserted at almost any site on the *E. coli* chromosome (Bukhari and Zipser, *Nature New Biology*, 1972, 236, 240). Induction of the integrated viral genome, which is brought about by fusing a transformed cell with a permissive cell, is not as efficient as in the case of λ nor do animal cells that carry the integrated viral genomes have λ type immunity.

Genetic Exchange in Bacteria

Genetic exchange in bacteria has been shown to take place by three mechanisms: transformation, transduction, or conjugation. These are mediated, respectively, by way of free DNA, phage, or cell-to-cell contact. All three means of exchange are characterized by the transfer of only a small segment of the donor DNA to the recipient cell. *E. coli* is the only microorganism known where all three means of genetic exchange can be demonstrated. The genetic interaction between the segment of donor DNA and recipient genome has best been studied most intensively in the case of transformation where the donor DNA can be labeled genetically as well as with radioactive and heavy isotopes, and what is most important, the recombinant DNA can be extracted and characterized not only physically but for its genetic configuration as well. The latter can be performed by reintroducing the DNA into a suitably marked recipient cell (*paper 18*). The recent important discovery that Ca^{++} treated cells of *E. coli* that lack the ATP-dependent DNase are capable of being transformed (Cosloy and Oishi, *Proc. Nat. Acad. Sci. U. S.* 1973, 70, 84) should provide an extremely important tool in further studies on the molecular biology of this organism. The recombination between host DNA introduced by generalized transducing phage and recipient cell has been investigated by Ebel-Tsipis *et al.*, (*J. Mol. Biol.* 1972, 71, 433, 449). They have found that in the case of phage P22 mediated generalized transduction, the major portion of the integrated DNA is double stranded, in contrast to that found in DNA mediated transformation. Using density and radioactively labelled donor and recipient cells, it has been found that the single-stranded DNA transferred to the recipient by means of conjugation in *E. coli* crosses is inserted exclusively and covalently linked to the newly formed strand of the recipient DNA (Siddiqi and Fox, *J. Mol. Biol.* 1973, 77, 101).

Repair of Chromosome Damage

Whereas breakage and mending of DNA are involved in recombination and lysogenization, it is now apparent that there are enzyme systems that can repair a much wider variety of chromosome damage (*papers 19, 20*). Chemical abnormalities in the DNA structure that would otherwise lead to chromosome damage and cell death are detected, excised, and replaced by normal DNA (Hanawalt, *Endeavour*, 1972, 31, 83). Attention has been centered around the repair of DNA damaged by exposure to ultraviolet irradiation (Lomant and Fresco,

Prog. Nuc. Acid Res. Mol. Biol. 1972, 12, 2). Such DNA contains a particularly frequent occurrence of pyrimidine dimers (the most common being thymine dimers) with covalent C–C bonds formed with very high efficiency between adjacent pyrimidines on the same DNA strand. Since the C–C bond length is about 1.5 A° and since the connected regions in DNA would normally be 6 to 7 A° distant, such pyrimidine dimer formation produces an obvious localized distortion of the double helix structure. Even one such dimer per chromosome can produce cell death in mutants that lack a repair system. One repair process involves the recognition and excision of the region containing the imperfection, followed by repair. For small amounts of ultraviolet damage genetic analysis suggests that the bacterial recombination (*paper 19*) system is adequate to the task. For extensive damage a special repair system involving 2 or 3 genes is called into play. This is probably related in some way to the fact that extensive ultraviolet damage severely inhibits DNA synthesis.

In the bacterium *Micrococcus lysodeikticus* a pair of enzymes have been isolated which appear to be involved in chromosome repair (*paper 20*). The first of these enzymes introduces single-stranded breaks into irradiated DNA but does not act on unirradiated native or single-stranded DNA. Photoproduct excision is absolutely dependent on a second enzyme which removes about 5 nucleotides including the ultraviolet damaged region.

Recently Kornberg and his coworkers (*Nature*, 1969, 244, 495) found that DNA polymerase I can act *in vitro* on damaged DNA after the initial scission has been made by the ultraviolet-specific endonuclease. Polymerase I mends the DNA as it excises the damaged region. A polynucleotide ligase is required for the final step in the mending operation. The ability of polymerase I to act in this way implicates this enzyme in repair functions (but not exclusively so, see Masker and Hanawalt *Proc. Nat. Acad. Sci. U.S.* 1973, 70, 129). The exonuclease ($3' \rightarrow 5'$) activity associated with DNA polymerase I can also remove mispaired bases (*paper 5*), acting as an "editor" that could potentially remove errors in repair or replication, thus preventing mutations which would result if they remained. The efficiency of the editing function of the phage T4 DNA polymerase has been shown to be related to the mutator and antimutator properties of mutants in the phage DNA polymerase gene (Muzyczka et al., *J. Biol. Chem.* 1972, 247, 7116).

A second repair mechanism monomerizes pyrimidine dimers (but only of the thymine-thymine type) without removing them from the DNA backbone. The enzyme that carries this out is called the photoreactivating enzyme. It is coded for by a single gene in *E. coli* and carries out its function only in the presence of light.

Biochemical Aspects of Mutations and Mutagenesis*

Mutations can be classified as falling into two general types. The first class are point mutations resulting from single base pair changes (AT $\rightleftharpoons$ GC). This can generate codons, called *missense* mutations, which alter the amino acid in a particular position in the protein; or else the generated codon does not code for any amino acid, in which case the mutations are of the *nonsense* variety. A second class of mutations called phase shift mutations involve a *deletion* of one (or sometimes the addition of one) or many bases resulting in drastic alterations of the genetic information within genes or of whole sets of genes. Watson and Crick suggested a mechanism for *spontaneous* mutation based upon their model for the DNA structure. Their hypothesis was that a tautomeric shift in any of the four DNA bases would permit the alternative purine-pyrimidine pairing to occur between template and adsorbed mononucleotides during synthesis. Since the normally occurring tautomeric forms are strongly favored, the resulting mutation rate should be low. It is still not known if the tautomeric shift in base structure is a frequent cause of so-called spontaneous mutation, but the Watson and Crick suggestion was important because it introduced one to thinking about *induced* mutation in terms of nucleic acid structure. Subsequently, Freese (*Brookhaven Symp. Biol.*, 1959, 12, 63) as well as Brenner and his coworkers (*J. Mol. Biol.* 1961, 3, 121) provided an explanation for the mutagenic character of two important classes of chemicals, the so-called base analogues class typified by 2-aminopurine (AP), 5-bromodeoxyuridine (BU), and the acridines. BU as well as AP are incorporated into DNA after their conversion to the triphosphate; acridines are neither modified further nor covalently incorporated.

Members of the base analogue class are believed to function by being incorporated into

---

*Drake, *The Molecular Basis of Mutation*, San Francisco: Holden-Day, 1970.

the DNA structure and producing a higher than spontaneous frequency of base mispairing in the double helix structure (*paper 21*). Members of the acridine class are believed to become sandwiched or intercalated between existing bases or base pairs producing either deletions or additions (possibly during DNA replication or recombination) of one or more bases of the genome (*paper 22*). This explanation accounts for the observation that the phase shift mutants produced by a mutagen belonging to one of this group are induced to revert by a mutagen of the same group but not by one of the other group. It is of interest that the potent carcinogen 3:4-benzpyrine is believed to interact with DNA by intercalations.

A third type of mutagen which can generate mutations results from covalent modification to a base already in the DNA structure or of the DNA structure itself. Physical agents such as radiation, or chemical mutagens such as nitrous acid, nitrosoguanidine (NG), ethylmethane sulfonate (EMS), or hydroxylamine belong to this class. Whereas nitrous acid mutagenizes by deamination, NG acts primarily at the DNA replication point, leading to a clustering of mutations.

Finally, "biological" mutagens exist that can greatly increase the mutation rate of bacteria anywhere along the genome. Mutator genes have been shown in several bacterial genera to greatly augment the mutation rate. In the case of *E. coli* one of these mutator genes (*mut T*, isolated by Treffers) has been shown to cause (Cox and Yanofsky, *Proc. Nat. Acad. Sci. U. S.* 1967, **58**, 1895) AT → CG *transversions*, the substitution of a pyrimidine for a purine, or vice-versa; *transitions* involve the substitution of a purine for a purine or pyrimidine for a pyrimidine. As well as causing chromosomal mutations, these can also cause mutations in infecting phages such as λ as well. Another biological mutagen that has been found to be extremely useful in analyzing the genetic organization of the *E. coli* chromosome is the temperate phage Mu-1. Unlike the DNA of λ, the DNA of this phage, can insert itself randomly into the *E. coli* chromosome, apparently requiring very little, if any, sequence homology between the phage and the host DNA (Bukhari and Zipser, *Nature New Biology* 1972, **236**, 240). By being able to insert themselves within any gene, thereby destroying its integrity, mutations can be generated practically anywhere on the host chromosome. The elegant use of mutants generated by phage Mu-1 has been used by Nomura and his colleagues to show that the synthesis of *E. coli* ribosomal proteins are coordinately controlled (Nomura and Engback, *Proc. Nat. Acad. Sci. U. S.* 1972, **69**, 1526; Davies and Nomura, *Ann. Rev. Genet.* 1972, **6**, 203). Small units of DNA (800—1400 nucleotide pairs), designated Is or "intertosomes" can also cause polar mutations analagous to those caused by phage Mu-1 (Malamy *et al.*, *Mol. Gen. Genet.* 1972, **119**, 207).

The location and extent of large deletions and substitutions of chromosomal DNA can be detected by studying *in vitro* generated heteroduplexes using electron microscopy (*paper 23*). This technique is particularly applicable to phage DNA whose genomes (and moreover their complementary strands) can be isolated intact. By annealing *in vitro* the DNA strand isolated from the genome of one phage strain with the complementary strand of another, the mismatched regions loop out and can be readily detected. The location of the mismatch can be calculated and has been found to correlate with the position estimated from genetic crosses. This has proven to be the most direct method of demonstrating the colinearity of the genetic and physical map of a genome.

Biological and Physical Chemical Aspects of Reversible Denaturation of Deoxyribonucleic Acids

J. Marmur, C. L. Schildkraut,* and P. Doty

Graduate Department of Biochemistry, Brandeis University, Waltham, Massachusetts; and Department of Chemistry, Harvard University, Cambridge, Massachusetts

The Watson-Crick model for the structure of deoxyribonucleic acid (DNA) has provided a basis for many studies of its genetic role and its physical chemical properties. The implication that the complementary strands separate and that each serves as a template for the synthesis of new complementary strands has been supported by the experiments of Levinthal and Thomas (1957) and Meselson and Stahl (1958). The latter workers did not define the basic replicating unit of DNA. However, subsequent work by Marmur and Lane (1960), Doty, Marmur, Eigner, and Schildkraut (1960), and Schildkraut, Marmur, and Doty (1961) on the reversible denaturation of DNA has presented further data consistent with the notion that the replicating units are indeed the complementary DNA strands themselves.

Reversible denaturation of DNA has provided a unique tool for studying some of its physical chemical properties and a new vehicle for investigation of the molecular basis of some aspects of genetics.

Experiments and Results

DNA strand separation and subsequent renaturation by exposure to elevated temperatures below the melting temperature (T_m) can

---

* Present address: Department of Biochemistry, Stanford University, Palo Alto, California.

be studied by biological means as well as by physical chemical techniques. The thermally induced denaturation of DNA, which can be compared to the melting of a crystal, occurs within a narrow temperature range. The helix-to-coil transition of the DNA can be followed in a number of ways: the loss of transforming activity, the increase in immunological reactivity (Levine, Murakami, Van Vunakis, and Grossman, 1960), an increase in relative absorbance at 260 mμ, an increase in the buoyant density in a cesium chloride gradient, the separation of the strands of hybrid N^{14}–N^{15} labeled DNA in density gradient ultracentrifugation, the fall in molecular weight by a factor of two (measured by sedimentation and viscosity) and changes in shape observed by the electron microscope (Doty, Marmur, Eigner, and Schildkraut, 1960). The renaturation also can be followed by each or all of the above techniques.

The denaturation and renaturation of DNA is readily followed by assaying the transforming activity of *Diplococcus pneumoniae* DNA after the various thermal exposures, since the helical form is necessary for activity. The transforming activity is plotted in Figure 1 for quickly cooled aliquots taken from a solution held at 100°C, and then at various times during the slow cooling (shown at the right of the vertical dashed line in Figure 1) as the temperature of

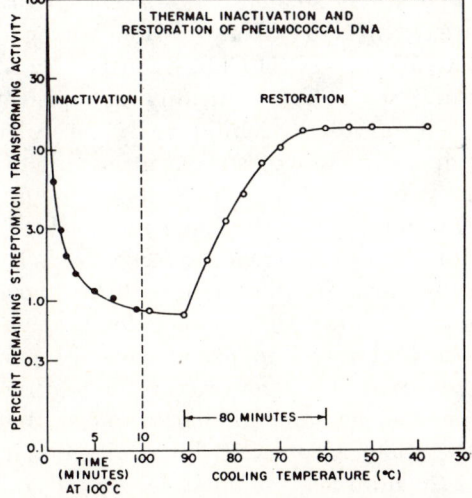

Fig. 1. Thermal inactivation and restoration of the transforming activity of *D. pneumoniae* DNA. Pneumococcal DNA at 20 µg/ml. in 0.15 M NaCl plus 0.015 M Na citrate was preheated for one and one-half minutes at 85.5°C (no loss in biological activity), then transferred to a boiling water bath at 0 minutes. At the times shown, samples were removed to an equal volume of 1.5 M NaCl plus 0.15 M Na citrate in an ice-water bath. After 10 minutes' exposure at 100°C, an equal volume of hot (100°C) 1.5 M NaCl plus 0.15 M Na citrate was mixed with the DNA solution, the mixture transferred to a large water bath and then cooled slowly. During the gradual descent of the temperature, samples were removed (shown at the right of the dashed vertical line) to prechilled tubes in an ice-water bath and then assayed for the ability to transform with respect to the streptomycin resistance marker.

the bath decreased. Thus, it is seen that restoration of the biological activity begins when the temperature of the slow cooling is about 90°C and continues until about 65°C. While in this experiment the recovery of the biological activity is approximately 15 per cent, values as high as 50 to 60 per cent have been attained using improved conditions (Marmur and Doty, 1961).

The remainder of this presentation will be devoted to (1) some factors which influence strand separation, (2) those influencing renaturation, (3) the thermal stability of renatured DNA, (4) studies on the formation of molecular hybrids between homologous and heterologous DNA, and (5) the genetic and taxonomic aspects of renatured hybrid formation.

Some Factors Influencing Strand Separation of DNA

The most useful methods in following strand separation have been the observation of the hyperchromic effect at 260 mμ, the buoyant density in ultracentrifugation and the loss of transforming activity.

Effect of Temperature and Time of Exposure

When native DNA is exposed to temperatures several degrees above the T_m, strand separation ensues. Using hybrid N^{14}–N^{15} DNA preparations isolated from *Escherichia coli* (grown for many generations in $N^{15}H_4Cl$ as the sole nitrogen source, followed by one generation in $N^{14}H_4Cl$ as described by Meselson and Stahl (1958), strand separation is readily demonstrated (Figure 2) using the density gradient technique. By carefully controlling the time and temperature of exposure, it is possible to denature part of the population of DNA molecules. These denatured, single strands, each containing N^{14} or N^{15}, have an increased buoyant density in cesium chloride and are readily recognized. By varying the time of exposure to a constant temperature above the T_m, increasing proportions of the DNA molecules become denatured. These experiments not only demonstrate strand separation, but also confirm the contention that denaturation by heating and fast cooling is an all-or-none phenomenon. Furthermore, they also clearly demonstrate DNA heterogeneity. By partially denaturing the DNA at 93.8°C for 20 minutes, the adenine plus thymine rich molecules are the first to melt out, leaving behind relatively richer guanine plus cytosine containing DNA molecules that can be recognized by their higher buoyant density.

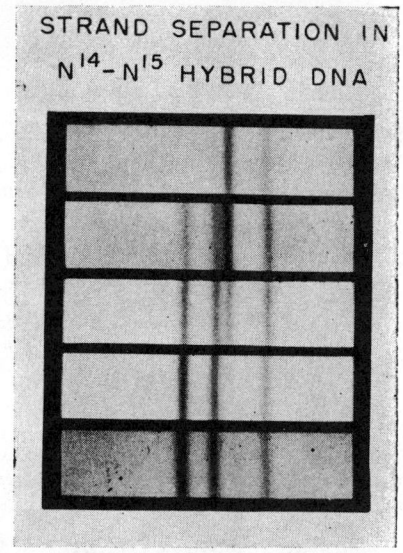

FIG. 2. Thermally induced strand separation of hybrid N^{14}–N^{15} *E. coli* B DNA. The hybrid DNA was prepared from *E. coli* grown for many generations in a synthetic medium with $N^{15}H_4Cl$ as the sole nitrogen source followed by one generation in $N^{14}H_4Cl$ containing medium, according to the method described by Meselson and Stahl, 1958. The ultraviolet absorption photographs, taken after centrifugation to equilibrium in CsCl at 44,770 rpm, show different stages of the thermally induced strand separation. In each strip, the band at the far right, which was introduced into the centrifuge cell as a standard, is DNA from *D. pneumoniae*. In the top photograph, the other band is native, biologically formed hybrid N^{14}–N^{15} *E. coli* DNA. The second photograph shows the stability of the hybrid to a 20 minute exposure in 0.15 M NaCl plus 0.015 M Na citrate at 93.8°C. At 100°C the number of molecules separating increases with time of exposure, as shown in the next three photographs. The samples were heated in the same solvent at 20 μg./ml. for 30 seconds, one, and 10 minutes, respectively.

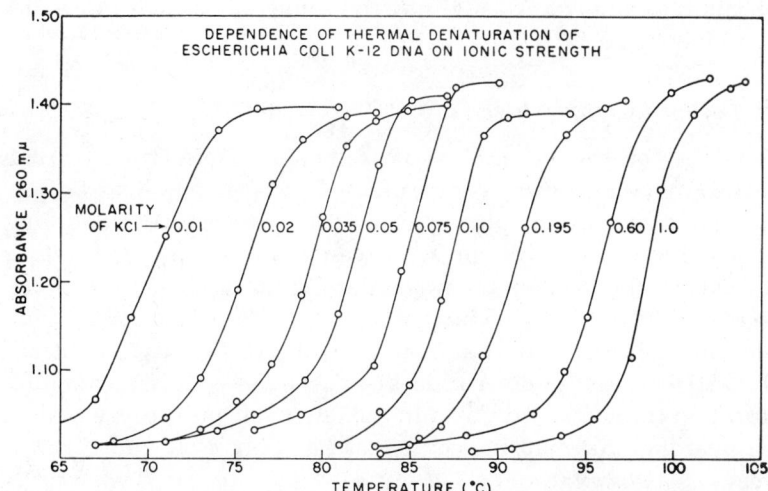

FIG. 3. Dependence of thermal denaturation of *E. coli* K-12 DNA on ionic strength. *E. coli* DNA, suspended in various concentrations of KCl in glass stoppered quartz cuvettes, was heated in the Beckman spectrophotometer chamber and the relative absorbance (corrected for thermal expansion) was measured at the elevated temperatures.

Effect of Ionic Strength

It would be expected that raising the salt concentration of the suspending medium would produce a shielding of the repulsive forces of the charged phosphate groups of the DNA molecule which require higher temperatures for denaturation. This is shown to be the case for *E. coli* K-12 DNA heated in various concentrations of KCl (Figure 3). The increase in relative absorbance (measured at the elevated temperatures) of DNA at various ionic strengths was followed as a function of the temperature to which it was exposed in the chamber of the Beckman spectrophotometer. The T_m (temperature after 50 per cent of the total increase in relative absorbance) for DNA dissolved in 0.01 M KCl is approximately 70°C and increases to approximately 98°C in 1.0 M KCl. The T_m is also dependent on the guanine plus cytosine content of the DNA (see *Effect of Base Composition*, below), on the pH of the solvent, and has recently been shown to be increased by the presence of certain polyamines (Mahler, Mehrotra, and Sharp, 1961). Small amounts of protein associated with purified DNA preparations or the presence of small quantities of divalent ions (when the denaturation is carried out in the presence of saline-citrate) have little or no detectable effect on the T_m. DNA degraded by sonic vibration to molecular weights of 300,000 has the same T_m as high molecular weight DNA (10–20×10^6).

Effect of Base Composition

The T_m for the thermal denaturation of DNA from various sources, varying in base composition, has been found to be proportional to its guanine plus cytosine (or hydroxymethylcytosine) content (Figure 4). This can be explained in part by the presence of an extra hydrogen bond between the guanine and cytosine (Pauling and Corey, 1956). The curve obtained at the lower ionic strength is parallel to that in 0.15 M NaCl plus 0.015 M Na citrate, and has been used to determine the T_m values of DNA whose denaturation cannot be estimated in the latter solvent because of their high guanine plus cytosine contents (e.g., poly d-GC and *Streptomyces viridochromogenes*).

The base compositions of the DNA samples shown in Figure 4 were obtained from published reports (Belozersky and Spirin, 1960; Chargaff, 1955). However, with the relation well established, it can be used to estimate the base compositions of the new DNA

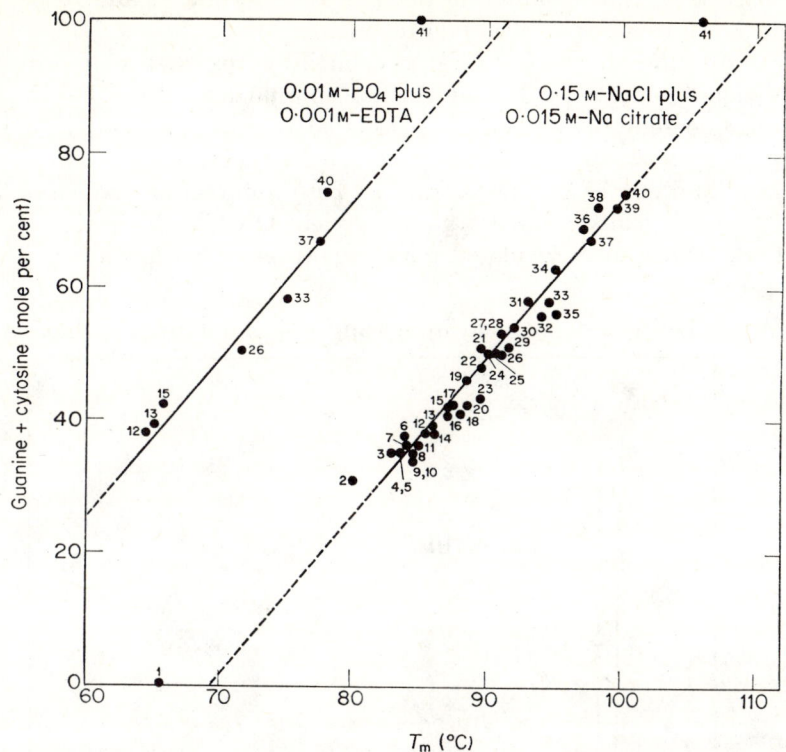

Fig. 4. Dependence of the denaturation temperature, T_m, on the guanine plus cytosine content of various samples of DNA. Native DNA samples (isolated by the method of Marmur, 1961) were dissolved in either of the two solvents shown in the figure. The T_m represents the midpoint of the thermal transition curve (carried out and plotted as shown in Figure 3) and has been plotted as a function of the guanine plus cytosine content (determined chemically and obtained from published reports). The numbers next to each T_m value in the figure refer to the DNA extracted from the following organisms: 1, poly d-GC; 2, *Streptomyces viridochromogenes*; 3, *Micrococcus lysodeikticus*; 4, *Pseudomonas aeruginosa*; 5, *Mycobacterium phlei*; 6, *Aerobacter aerogenes*; 6a, *Azotobacter vinelandii*; 7, *Serratia marcescens*; 8, *Brucella abortus*; 9, *Salmonella typhimurium*; 10, *Shigella dysenteriae*; 11, *Escherichia coli*; 11a, phage T_3; 12, *Bacillus licheniformis*; 13, phage T_7; 13a, *Vibrio cholerae*; 14, *Bacillus subtilis*; 15, salmon sperm; 16, calf thymus; 17, *Hemophilus influenzae*; 18, *Diplococcus pneumoniae*; 19, *Bacillus megaterium*; 20, bakers' yeast; 21, phage T_2r+; 22, phage T_4r+; 23, phage T_6r+; 24, *Bacillus thuringiensis*; 25, *Micrococcus pyogenes* var. *aureus*; 26, *Bacillus cereus*; 27, *Clostridium perfringens*; 28, poly d-AT. The first and last samples, the enzymatically prepared polymers, were obtained through the generosity of A. Kornberg, J. Josse, and J. Adler. The curve on the right was fitted to the points by the method of least squares.

samples simply from measuring their T_m (Marmur and Doty, in preparation). Extrapolation of the T_m curve beyond 25 and 75 per cent G + C remains uncertain at the moment.

Substitution of cytosine by glucosylated hydroxymethylcytosine (in bacteriophages T_2, T_4 and T_6) has no influence on the T_m. Denaturation temperature of DNA isolated from a thermophile, *B. stearothermophilus*, grown at an elevated temperature, is not abnormal and can be predicted from its base composition (Marmur, 1960). Genetically related organisms yield DNA with similar T_m values; taxonomically related microorganisms do not always.

Effect of pH

When DNA is exposed to increasingly acid conditions, the T_m

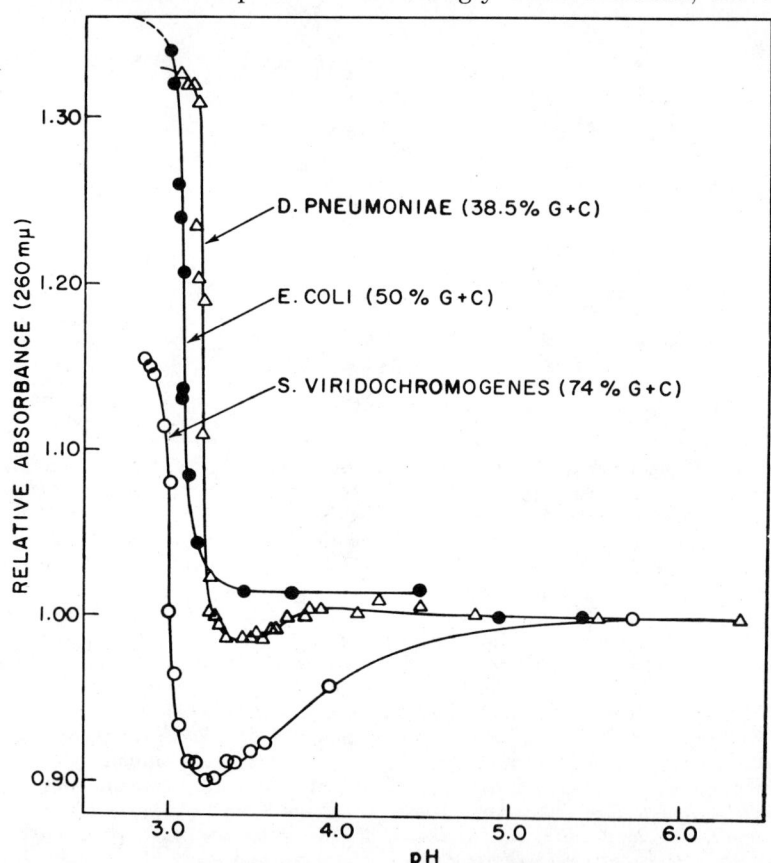

FIG. 5. Effects of acid on the denaturation of various DNA samples in 0.05 M NaCl at 25°C. Samples of DNA dissolved in 0.05 M NaCl were titrated with HCl at 25°C and the pH and relative absorbance measured after each addition.

is lowered until it passes below room temperature. Denaturation can be detected by changes in optical density (Figure 5) or the separation of the strands of hybrid N^{14}–N^{15} DNA (Figure 6). Figure 5 clearly shows that if the DNA has a higher composition of guanine plus cytosine, lower pH values must be reached in order to cause denaturation. The most acid (and temperature) resistant naturally occurring DNA thus far encountered has been isolated from the actinomycetes *Streptomyces viridochromogenes* and *Streptomyces albus*, each possessing approximately 74 per cent guanine plus cytosine (Belozersky and Spirin, 1960).

When hybrid N^{14}–N^{15} *E. coli* DNA was exposed to pH 2.8 (in 0.05 M NaCl at room temperature), neutralized and then centrifuged in a CsCl gradient, two bands were obtained with buoyant densities corresponding to that expected for N^{14} and N^{15} single-stranded *E. coli* DNA (Figure 6). The same situation was found true for native hybrid *E. coli* DNA exposed to high pH values, e.g., 11.5, in experiments carried out in collaboration with R. Cox.

Effect of Formamide

High concentrations of formamide have been shown to denature DNA (Helmkamp and Ts'o, 1961; Marmur and Ts'o, 1961). By increasing the concentration of formamide to which *D. pneumoniae* DNA is exposed, a concentration range is reached (55 to 65 per cent, in 0.02 M NaCl plus 0.002 M Na citrate at 37°C) where the transforming activity abruptly decreases. These conditions provide a mild method for the denaturation of DNA and may be found useful where little or no degradation or depurination (induced, in part, by heat treatments) is desired.

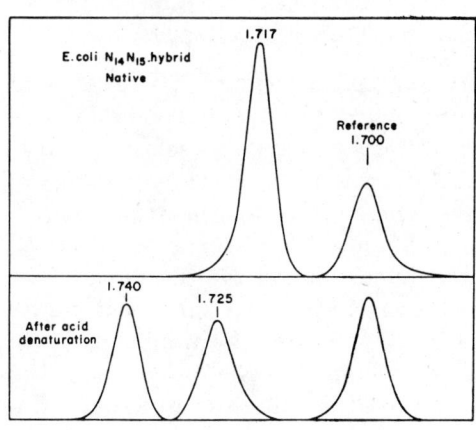

Fig. 6. Acid-induced strand separation of *E. coli* DNA. Hybrid N^{14}—N^{15} *E. coli* B DNA was exposed to pH 2.8 in 0.05 M NaCl at 25°C, re-neutralized and centrifuged for 24 hours at 44,770 rpm in a CsCl gradient in the Spinco ultracentrifuge. The microdensitometer tracings represent the DNA concentration (ordinate) as a function of distance from the axis of rotation. The native reference DNA is *D. pneumoniae* which has a density of 1.700 g./cm.^{3}

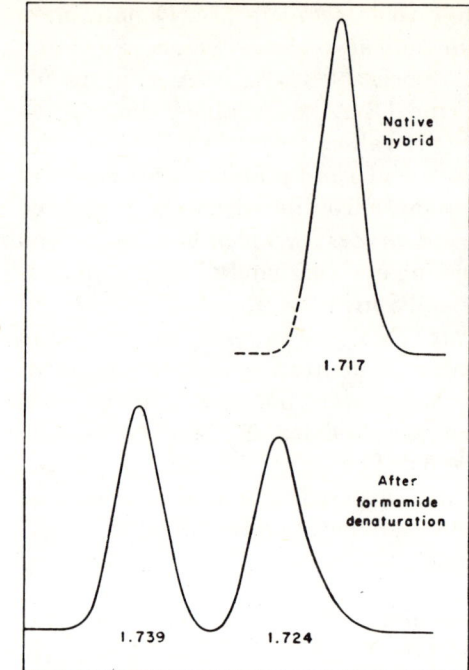

Fig. 7. Formamide-induced strand separation of *E. coli* DNA. Hybrid N^{14}—N^{15} *E. coli* B DNA was exposed to 95 per cent formamide in 0.02 M NaCl plus 0.002 M Na citrate at 37°C for 10 minutes, dialyzed free of formamide, and centrifuged for 24 hours at 44,770 rpm in a CsCl gradient in the Spinco ultracentrifuge. The microdensitometer tracings represent the DNA concentration (ordinate) as a function of the distance from the axis of rotation.

Denaturation is also detectable by treating hybrid N^{14}–N^{15} *E. coli* DNA with formamide, dialyzing out the denaturing agent and centrifuging the treated DNA in a CsCl gradient. Strand separation is again detected by the presence of two bands (Figure 7) whose buoyant densities correspond to denatured N^{14} and N^{15} *E. coli* DNA (Marmur and Ts'o, 1961).

Maximal Thermal Inactivation of Transforming Activity

If the transforming activity of *D. pneumoniae* DNA necessitates its presence in the double-strand helical form, then it should be possible to destroy completely the biological activity by selecting proper conditions of DNA concentration, ionic strength, temperature, and quick cooling. Thus, *D. pneumoniae* DNA was heated in 0.01 M phosphate plus 0.001 M ethylenediaminetetra aceate (EDTA), pH 6.8, at 100°C and, at varying times, aliquots were removed with prechilled pipettes and diluted into solvent of the same composition kept at ice-bath temperature. The loss of transforming activity for the streptomycin marker is seen in Figure 8. The initial sharp drop in biological activity within the first few minutes can be related to

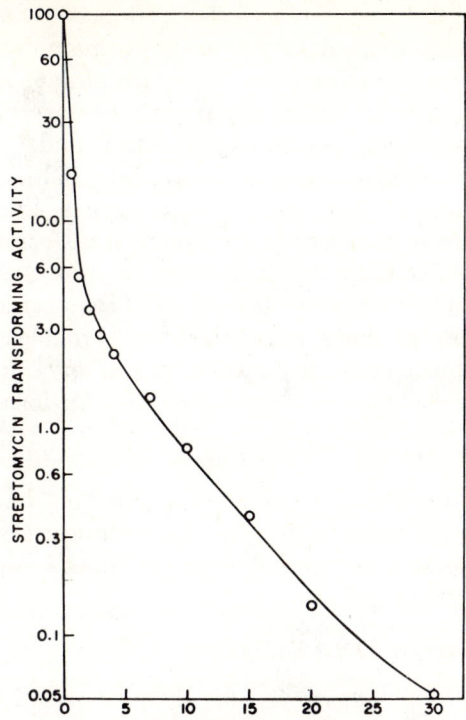

Fig. 8. Maximal thermal inactivation of *D. pneumoniae* transforming DNA. Conditions described in text.

strand separation; the slower component is in all likelihood due to inactivation by depurination and/or strand scission which have been shown to take place at elevated temperatures (Greer and Zamenhof, 1959; Doty, Marmur, Eigner, and Schildkraut, 1960). Since the T_m values of *D. pneumoniae* DNA and poly d-GC in this solvent are 65°C and 84°C, respectively (Figure 4), then all base pairs should have separated by exposing the transforming DNA to 100°C. The residual activities exhibited beyond the stage of base pair separation are unlikely to be due to renaturation, since little or no effect is noted if the same experiment is repeated at various DNA concentrations. They are also unlikely to be due to any extraordinarily resistant DNA, since DNA isolated from cells which have been transformed by this residual activity exhibits the same sensitivity to heat as the original DNA.

The most likely reason for the small residual activity is the probability that a small portion of the denatured DNA possesses transforming activity. Residual activities have been noted when strand separation has been induced by acid or alkali (Marmur and Lane, unpublished data) or formamide (Marmur and Ts'o, 1961). Since denaturation by formamide is performed under mild conditions, the residual activities are higher than by thermal denaturation, as no depurination and/or strand scission would be expected to take place. If the residual activity is indeed due to single-strand DNA, it is not clear at this time whether this may be due to the presence of some transforming molecules which have folded back on themselves to mimic the double-strand structure enough to satisfy the minimal requirements for cell entry. It is also possible that some of the transformable cells are in a certain physiological state (e.g., spheroplasts or L-forms, Madoff and Dienes, 1958) that can adsorb and be transformed by single-strand DNA.

Some Factors Influencing Renaturation

In studying the renaturation of denatured DNA, it has been found useful to employ the techniques of transformation of genetic markers and to follow the changes in optical density after various thermal exposures.

Effect of DNA Source

Although the T_m for strand separation of DNA samples having similar base compositions is independent of their source (except for the breadth of the transition), the extent to which renaturation proceeds is found to be greatly influenced by the type of DNA being studied. This is explicable by the observation that renaturation is dependent upon the concentration of specific complementary strands during the renaturation (Marmur and Lane, 1960). Thus, since there are about 5,000 times more molecules of DNA per cell in mammalian cells than in bacteria (Vendreley, 1958; Crampton, Lipshitz, and Chargaff, 1954; Brown and Brown, 1958), it is to be expected that at the same DNA concentration, renaturation will hardly take place in mammalian DNA, while it proceeds far toward completion in the case of bacterial DNA. Under similar conditions, it would be expected to proceed faster and farther in the case of even simpler sources, e.g., bacteriophage.

In order to explore this matter, renaturation was studied by following the optical density at 260 mμ. This was done by denaturing and quickly cooling the appropriate samples. In this denatured form, about 60 per cent of the hydrogen bonds have reformed in a nonspecific manner, determined from the optical density change.

When the temperature is raised again, to about 65°C, these nonspecific bonds dissociate and the strands are free to renature. If renaturation occurs, the optical density at 260 mµ will fall. Results of experiments of this type are shown in Figure 9. It can be seen that after the melting of the weak, intramolecular hydrogen bonds of the denatured, fast-cooled DNA, renaturation of some of the DNA sample occurs. The rate and extent of renaturation of DNA is seen to be dependent upon its source. Calf thymus DNA (as well as DNA from salmon sperm and tobacco leaves) does not renature; *D. pneumoniae* DNA (as DNA from most bacterial sources) readily renatures; whereas T_6r+ (as well as T_2 and T_4) and *Mycoplasma gallisepticum* (avian PPLO) whose DNA is more homogeneous

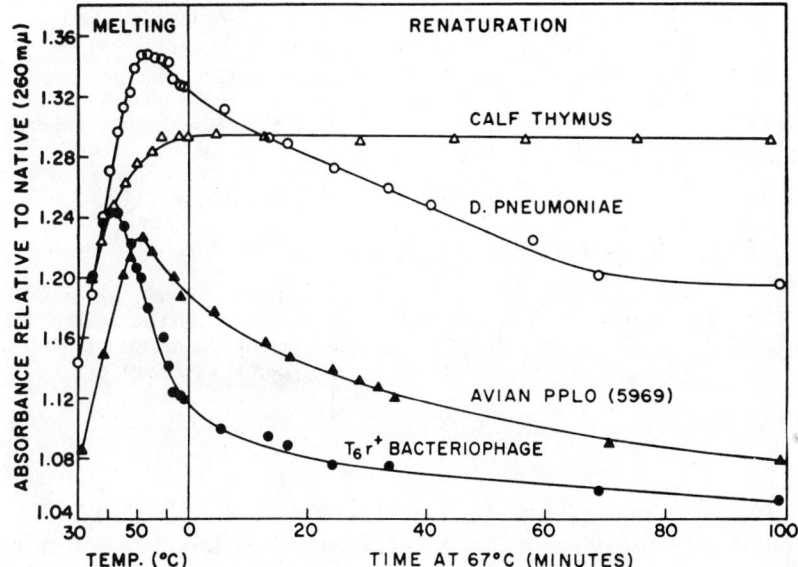

Fig. 9. Effect of source on the renaturation of thermally denatured DNA. DNA at 20 µg./ml. was heated at 100°C for 10 minutes in 0.3 M NaCl plus 0.03 M Na citrate and quickly cooled in an ice bath. The samples of DNA (in ground glass stoppered cuvettes) at the same concentration and in the same solvent, were then placed in the Beckman spectrophotometer which was prewarmed at 67°C. The initial part of the graph represents the *increase* in optical density at 260 mµ (melting) as the denatured DNA samples reach temperature equilibrium. The curves to the right of the vertical line (renaturation) represent the *decrease* in optical density of the temperature equilibrated, renaturing samples as a function of time of exposure at 67°C. The DNA samples employed were isolated from the organisms shown in the figure.

(Burgi and Hershey, 1960; Guild, Morowitz, and Castro, 1960; Guild, 1961), renature the most readily.

Effect of Temperature

Renaturation of DNA by slowly cooling a sample in a large water bath has led to somewhat varying results due to the difficulty of controlling accurately the rate of cooling. In order to find the optimal renaturation temperature of *D. pneumoniae*, denatured samples were exposed at different constant temperatures for varying periods of time. The rate and extent of renaturation was followed by either the decrease in optical density or the recovery of transforming activity.

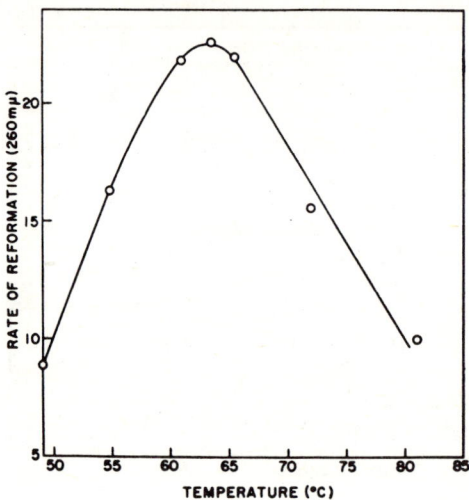

Fig. 10. Effect of temperature on the renaturation rate of denatured *D. pneumoniae* DNA in 0.3 M NaCl plus 0.03 M Na citrate, by optical density study. Conditions are similar to those described under Figure 9. The initial rates of *decrease* in optical density after temperature equilibrium is reached are plotted against the renaturation temperature.

It has been shown above that the renaturation of DNA can be followed by the decrease in optical density after the intramolecular hydrogen bonds of denatured, fast-cooled DNA are melted out. When the initial rate of decrease in the optical density is plotted against the renaturation temperature, the curve shown in Figure 10 is obtained. The optimum renaturation temperature for *D. pneumoniae* DNA is approximately 65°C and is dependent on the base composition of the DNA, increasing gradually for DNA samples with higher guanine plus cytosine contents (Marmur and Doty, 1961).

When the course of renaturation of *D. pneumoniae* DNA is followed by the restoration of the ability to transform with respect to the streptomycin resistance marker, the results are found to depend very much on the temperature. As shown in Figure 11, the recovery of biological activity displays an initial rapid rate, followed by a slower one. At the optimum temperature, the slow increase continues for at least five hours. When a plateau is reached in the

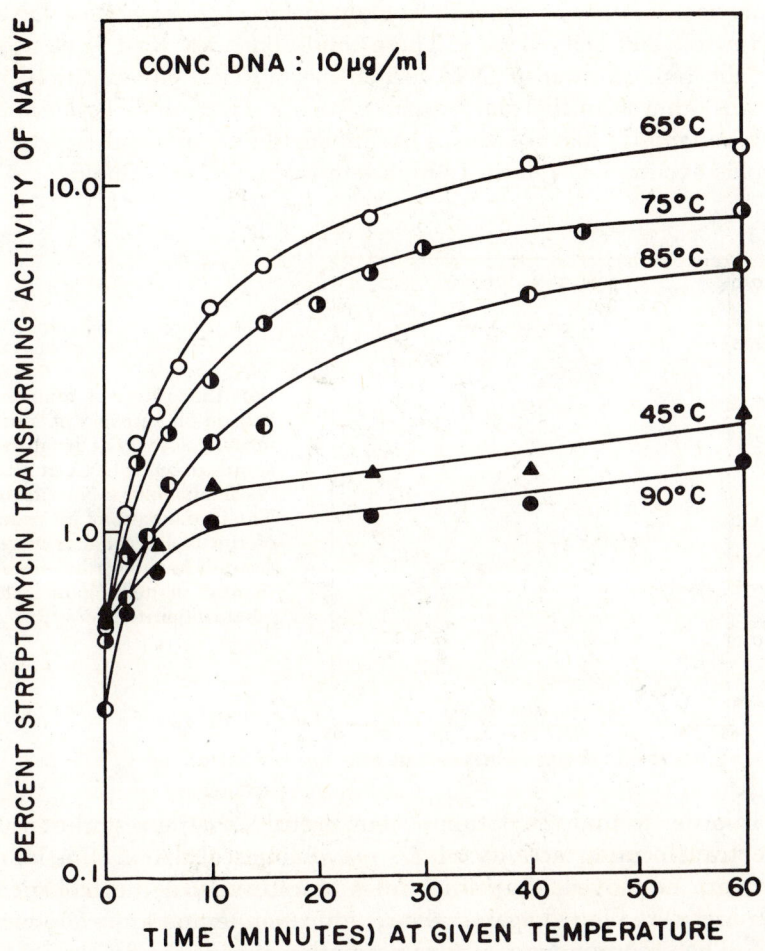

FIG. 11. Effect of temperature on the recovery of streptomycin-transforming activity of thermally denatured *D. pneumoniae* DNA. DNA from streptomycin-resistant cells of *D. pneumoniae* was denatured at 10 μg per ml. by heating for 10 minutes at 100°C in 0.3 M NaCl plus 0.03 M Na citrate. Aliquots, at the same DNA concentration and in the same solvent, were then distributed and exposed to various temperatures. Samples were removed and quickly cooled at different times and assayed for streptomycin-transforming activity.

restoration of the biological activity, subsequent cooling to room temperature results in an additional increment in transforming activity (10 to 50 per cent), depending on the previous temperature. By exposing denatured DNA to the optimum renaturation temperature for two to three hours, as well as a subsequent slow cooling to room temperature, restoration of 40 to 50 per cent of the original biological activity is generally obtained for *D. pneumoniae* DNA (Marmur and Doty, 1961). These conditions may readily be used for the renaturation of DNA which has suffered denaturation by means other than thermal exposure, such as formamide (see *Effect of Formamide*, above); thus, a mild method for renaturation which avoids exposure to relatively high temperatures is available.

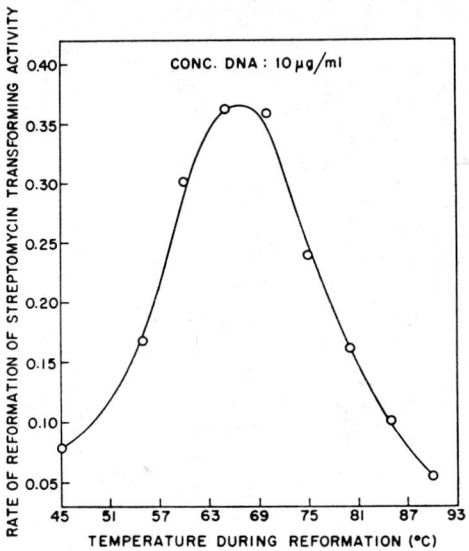

FIG. 12. Effect of temperature on the recovery of transforming activity of denatured *D. pneumoniae* DNA in 0.3M NaCl plus 0.03M Na citrate. The initial rates of recovery of the biological activity of the DNA shown in Figure 11 are plotted against the renaturation temperature.

In order to find the optimum temperature for the restoration of the transforming activity of *D. pneumoniae* DNA at the ionic strength employed, the initial rates of recovery of the biological activity were plotted against the renaturation temperature (Figure 12). The optimum temperature lies within the range of 65 to 70°C. The slightly higher value obtained by this method than that obtained by the optical density method may be a reflection of the observation that the streptomycin resistance marker (used in the biological assay for the renaturation) is probably associated with a DNA molecule which has a higher guanine plus cytosine content than the average composition of the total DNA population (Marmur and Lane, 1960).

Effect of Molecular Weight

In order to compare the rate and extent of the renaturation of denatured DNA as a function of the molecular weight, *D. pneumoniae* DNA was sonically degraded, in the presence of 2 aminoethylisothiuronium bromide hydrogen bromide (AET) to prevent free radical damage (Litt, Marmur, Ephrussi-Taylor, and Doty, 1958), from a molecular weight of 10×10^6 to 0.55×10^6. The degraded material retained 0.24 per cent of its transforming activity with respect to the streptomycin resistance marker. The renaturation of thermally denatured samples of both the undegraded and sonically treated DNA samples was studied by the restoration of the biological activity under optimum conditions of temperature and ionic strength.

In Figure 13 is shown the difference in the rate and extent of renaturation of the two samples. The transforming DNA was first denatured by heating the samples in 0.3 M NaCl plus 0.03 M Na citrate at 100°C for 10 minutes and then incubated at 65°C at 15 μg/ml. in the same solvent. Aliquots, removed at various times and quickly cooled, were assayed for the ability to transform sensitive cells of *D. pneumoniae* to streptomycin resistance. It can readily be seen that both the rate and extent of the recovery of biological activity are dependent on the molecular weight. Similar results are obtained when renaturation is followed by the optical density method (Marmur and Doty, 1961).

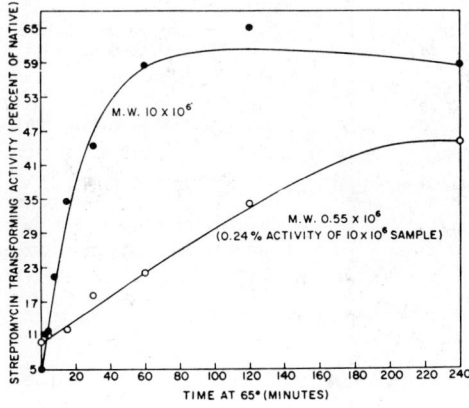

Fig. 13. Effect of molecular weight on the recovery of streptomycin-transforming activity of denatured *D. pneumoniae* DNA in 0.3 M NaCl plus 0.03 M citrate. Conditions are described in the text and are similar to those employed for Figure 11.

Thermal Stability of Renatured DNA

If renaturation is indeed the restoration of hydrogen bonding between complementary strands whose bases are in register, then the thermal stability of renatured DNA should mimic that of native DNA. This was tested by exposing both renatured and native DNA to a temperature several degrees above the T_m and following their transforming activity as a function of time. This has been shown previously (Doty, Marmur, and Sueoka, 1959) to result in an easily measurable, time-dependent loss of transforming activity due to denaturation of the molecules carrying the genetic marker.

Renatured DNA was prepared as follows: *D. pneumoniae* was heated at 100°C for 10 minutes in 0.3 M NaCl plus 0.03 M Na citrate and then exposed to 65°C at 20μg/ml. in the same solvent for varying periods of time. Samples were removed at several times, quickly cooled, and dialyzed against 0.15 M NaCl plus 0.015 M Na citrate.

The thermal inactivation curves at 89.2°C in 0.15 M NaCl plus 0.015 M Na citrate of the native and renatured, dialyzed samples are shown in Figure 14. DNA renatured for three hours (41 per cent restoration of biological activity) as well as those renatured for 25 minutes (33 per cent restoration) and 60 minutes (36 per cent restoration) showed the same thermal stability as native DNA. Data for the latter two renatured samples are not shown in the figure. DNA renatured for four minutes (7 per cent restoration of biological activity) shows an abnormally higher thermal resistance

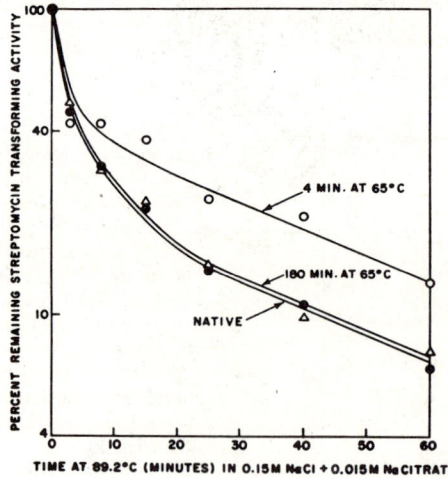

Fig. 14. Stability of renatured and native *D. pneumoniae* DNA to thermal inactivation of streptomycin transforming activity. Samples were exposed to 0.15 M NaCl plus 0.015 M Na citrate for various times at 89.2°C and aliquots removed and quickly cooled for biological activity at various times. Conditions are described in text.

which may be a reflection of the presence of a resistant component selected during the initial (100°C) thermal exposure.

It is suggested that this (or the optical density-temperature profile) criterion of the thermal stability of the material should be used as a test of its nativelike structure to rule out the possibility of nonspecific aggregation.

The optical density-temperature profile can also be employed as a criterion to check the structure of renatured DNA. The profiles of native and denatured DNA differ significantly (Doty et al., 1959). Whereas native DNA melts very sharply, resulting in a steep rise in optical density in the T_m region, denatured DNA gives rise to a broad profile.

Because of the high degree of homogeneity of the DNA from phage and PPLO, an attempt was made to see to what extent DNA from these sources could be renatured and to examine their thermal stability. T_6r+ DNA and avian PPLO DNA were denatured and then renatured under optimum conditions in 0.3 M NaCl plus 0.03 M Na citrate. Each was then compared to its corresponding native preparations for thermal stability in the same solvent. From Figures 15a and 15b it is seen that a large portion of each pair of curves may be superimposed in the T_m region and can be readily accounted for if one assumes that 70 to 80 per cent of the denatured molecules have renatured when exposed to the optimum renaturation conditions. We have recently found (Cordes, Epstein, and Marmur, 1961) that denatured DNA from phage α can be renatured to the extent of approximately 90 per cent by using the criterion

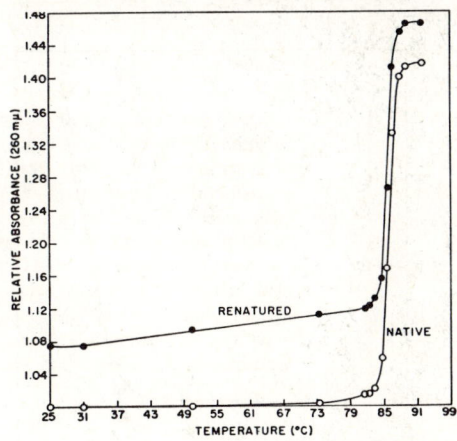

Fig. 15a. Thermal stability of renatured and native T_6r+ DNA in 0.3 M NaCl plus 0.03 M Na citrate. Conditions are described in text.

of the extent of the superimposing of the optical density-temperature curves in the T_m region.

Studies on Molecular Hybrid Formation

The preceding work has demonstrated two reliable routine methods for studying renaturation. It has shown that as the degree of renaturation of a given sample increases there is a continuous decrease in absorbance as well as a parallel increase in transforming activity. A third, very clear way to follow strand separation and recombination is by the buoyant density of the DNA in the CsCl gradient.

Renaturation of Denatured DNA

When a solution of bacterial DNA in 0.30 M NaCl and 0.030 M Na citrate is heated and quickly cooled, and then adjusted to 5.7 M in CsCl and centrifuged, it is found that the buoyant density has increased about 0.015 g./cm.³ (Sueoka, Marmur, and Doty, 1959). If, before centrifugation, an aliquot of the cooled solution is placed at 68°C for two hours and then slowly cooled to room temperature at about 5°C every 15 minutes, the renatured DNA band is only 0.004 g./cm.³ heavier than the native sample. In addition, some material whose density corresponds to that of the fast-cooled sample still remains. It is believed that the slight difference in density between the native and renatured molecules is due to the latter having unpaired ends. This should be expected as a direct consequence of the thermal depolymerization that unavoidably occurs during

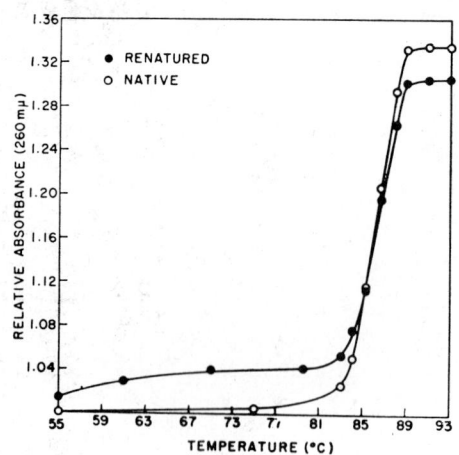

FIG. 15b. Thermal stability of renatured and native *Mycoplasma gallisepticum* (avian PPLO 5969) DNA in 0.3 M NaCl plus 0.03 M Na citrate. Conditions are described in text.

strand separation and renaturation and/or the random breakage of the double-strand DNA in its isolation from the cell. Single strands broken at somewhat different places would be produced during thermal exposure and these would not match completely during

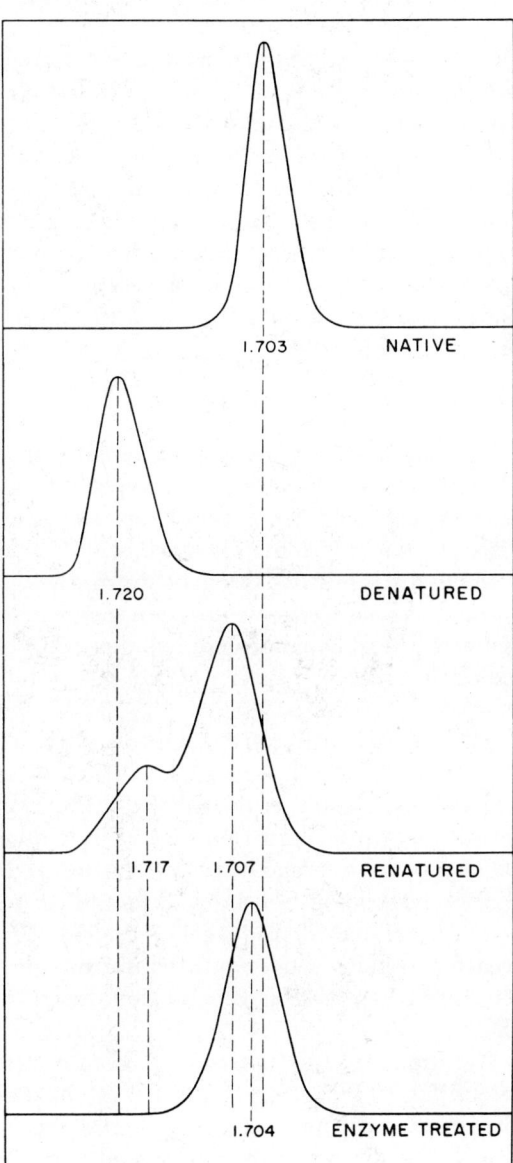

FIG. 16. Renaturation of *B. subtilis* DNA. When a solution of native DNA (band profile shown in top tracing) is heated for 10 minutes at 10 μg per ml. at 100°C in 0.285 M NaCl plus 0.0258 M Na citrate and quickly cooled, the density increases 0.017 g./cm.³ (second tracing). An aliquot of this solution was slowly cooled and, as is evident from the third tracing from the top, about 80 per cent of the DNA renatures. Treatment with the *E. coli* phosphodiesterase (Lehman, 1960) causes the complete disappearance of the denatured band and a decrease in the buoyant density of the renatured band (bottom tracing).

renaturation. These overlapping regions at the ends are thought to be responsible for the slightly higher densities. Results for *Bacillus subtilis* DNA are shown in Figure 16.

Recently, an enzyme that appears to have single-strand DNA as its optimum substrate has been isolated from *E. coli* (Lehman, 1960). Further evidence that this enzyme attacks only single-strand DNA has been obtained by incubation with the slowly cooled *B. subtilis* mixture whose band profile is shown in Figure 16. The resultant band profile strongly indicates that there remains only one component which is very similar to native *B. subtilis* DNA. This is also shown in Figure 16 where an additional change is evident: the density of the renatured material has decreased. This confirms the hypothesis that single-strand ends were present in the renatured DNA. The only change observed as a result of the action of the phosphodiesterase on the renatured DNA is the removal of these unpaired ends.

Molecular Hybrid Formation

It is possible, by means of density gradient centrifugation, to show that renaturation does take place by the union of two complementary single strands that were not previously paired together. This is done by using heavy-labeled N^{15}-deuterated DNA which can be distinguished clearly from the corresponding unlabeled material. These two native samples are separated by a sufficient distance to allow unambiguous detection of hybrid double helical molecules, each consisting of one heavy-labeled and one unlabeled strand.

The heating and slow cooling (after two hours at 68°C) of a mixture of 5 μg./ml. each of heavy-labeled and unlabeled *B. subtilis* DNA should result in the labeled and unlabeled forms each giving rise to denatured and renatured DNA as well as a hybrid whose density is the average of the two renatured samples. Thus, centrifugation of this solution in CsCl should produce five bands. A typical tracing is shown in Figure 17 (upper). The bands, in order of increasing density are: renatured normal, denatured normal, hybrid, renatured heavy-labeled, and denatured heavy-labeled *B. subtilis* DNA.

An aliquot of this same slow-cooled mixture was dialyzed against the buffer (0.067 M glycine, pH 9.2) in which the *E. coli* phosphodiesterase treatment is carried out. Incubation with this single-strand attacking enzyme eliminated two bands, as expected, leaving

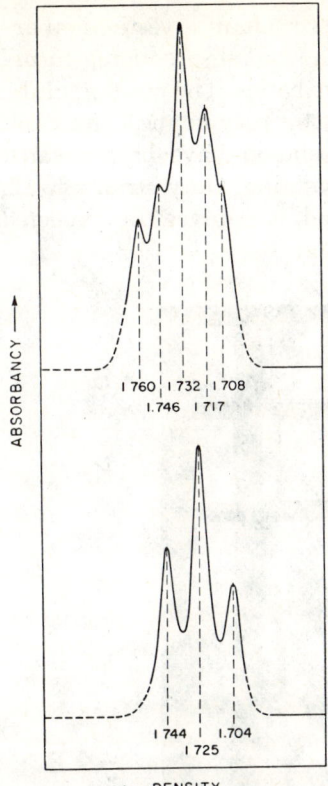

Fig. 17. Effect of *E. coli* phosphodiesterase on a slowly cooled mixture of heavy-labeled and normal *B. subtilis* DNA. Treatment with the phosphodiesterase eliminates the two denatured bands, decreases the density of the renatured bands so as to approach the buoyant densities of native material and greatly increases the resolution of the CsCl density gradient technique. Since the area of each band is proportional to the amount of DNA it contains, it is clear that there is approximately twice as much hybrid as either of the uniformly labeled components.

the three types of double helical molecules: both strands unlabeled, density 1.704 g./cm.^{3}; both strands heavy-labeled, density 1.744 g./cm.^{3}; and one strand unlabeled, one heavy-labeled, density 1.725 g./cm.^{3}. These results are shown in the lower tracing of Figure 17. Moreover, the density of these renatured and "cleaned up" molecules matches that of the native material instead of being about 0.004 units heavier as heretofore. The same studies were repeated using labeled and normal DNA isolated from *E. coli* B and these results were similar to those just described for *B. subtilis*.

Test of Homologies

It is to be expected from their close taxonomic, physiological and, in some cases, genetic (*E. coli* B recombines with strain K-12) relationships that all strains identified as *E. coli* should have DNA whose base sequences are for the most part similar. Thus, any *E. coli* strain should yield DNA which will form a five-band pattern in CsCl after heating and slow cooling with an equal amount of N^{15}

deuterated E. coli B DNA. The results of such an investigation are shown in Figure 18. All strains except the "alkali-producing-form" mutant (II–IV–4) of E. coli I formed five bands. The position of the hybrid band varies somewhat; however, this may not be significant. Ideally, each five-component mixture should have been treated with E. coli phosphodiesterase before banding. Only strains K-12, B and W have been checked so far, and each gives the expected three-band pattern after enzyme digestion.

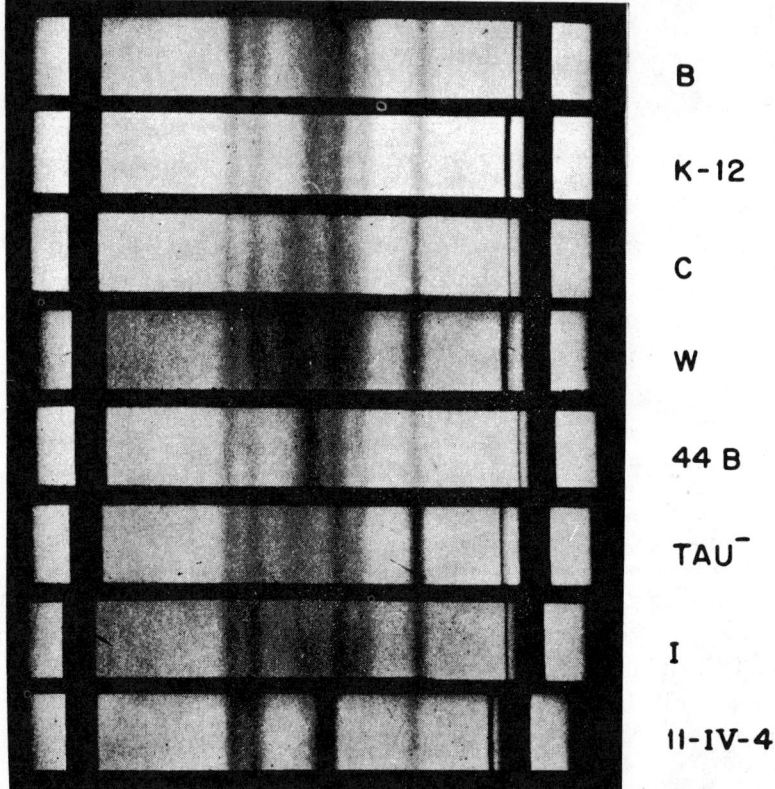

FIG. 18. Hybrid formation between E. coli B and other E. coli strains. E. coli DNA labeled with N^{15} and deuterium was mixed with DNA from each of the strains listed above and heated and slowly cooled in separate experiments. The concentrations were 5 µg./ml. each and the denaturation and renaturation conditions were the same as those described in Figure 20. Each of the eight different ultracentrifuge runs is represented in the above figure by a typical ultraviolet absorption photograph. Six DNA bands appear in all but the last experiment. The photographs have been lined up according to the position of the standard at the far right which is DNA from Cl. perfringens.

Nucleic Acids

Detection of Nonspecific Aggregates

Many investigators have reported effects with DNA that can best be explained by the hypothesis that the molecules under observation are entangled or connected by some type of nonchemical bond such as hydrogen or hydrophobic bonds. A vivid example is gel formation observed after heating and slowly cooling salmon sperm DNA at concentrations as low as 260 µg./ml. (Eigner, 1960). Since aggregation is a reality, it should be asked why it does not occur during hybrid formation. How is it possible to distinguish a native-like, hybrid, double helical structure from an aggregate of renatured or denatured normal and heavy-labeled DNA?

It seemed likely that any nonhelical aggregate would be degraded by the *E. coli* phosphodiesterase. With this in mind, equal amounts of DNA from organisms having significantly different GC contents were heated together at high concentration and slowly cooled. When examined in the CsCl gradient, bands of intermediate density and high diffusion coefficients were observed. When aliquots of these mixtures were treated with the phosphodiesterase and centrifuged again, the intermediate band disappeared. Examples of some of these aggregates, formed at high concentrations and banding at intermediate densities, and their elimination are seen in Figure 19.

It was also possible to form aggregates between *Bacillus subtilis* DNA and *Bacillus brevis* or *Bacillus macerans* DNA using concentrations of 40 and 50 µg./ml., respectively. The GC base composition of *B. subtilis* DNA is similar to that of *B. brevis*, while that of *B. macerans* is somewhat higher. The density of the aggregates was considerably higher than the average of the two renatured "parent" samples, as expected if renaturation was not complete. Again, the intermediate bands were eliminated by the *E. coli* phosphodiesterase treatment. When *B. subtilis* and calf thymus DNA were heated together and slowly cooled at a total concentration of 100 µg./ml., the heavy-labeled *B. subtilis* DNA reformed completely, while the calf thymus DNA maintained at the denatured density. No intermediate bands were visible. Thus, the conditions under which false intermediate bands can form have been clarified. The use of low concentrations and checks with *E. coli* phosphodiesterase will avoid artifacts.

Interspecies Hybridization

The minimal requirement for molecular hybrid formation appears to be close similarity in base composition. When DNA samples of quite different base composition are heated and slowly cooled, even at double the usual concentration, no hybrid is formed. In addition, the results of slow cooling DNA from two genetically unrelated species of *Bacillus* having very similar base compositions show

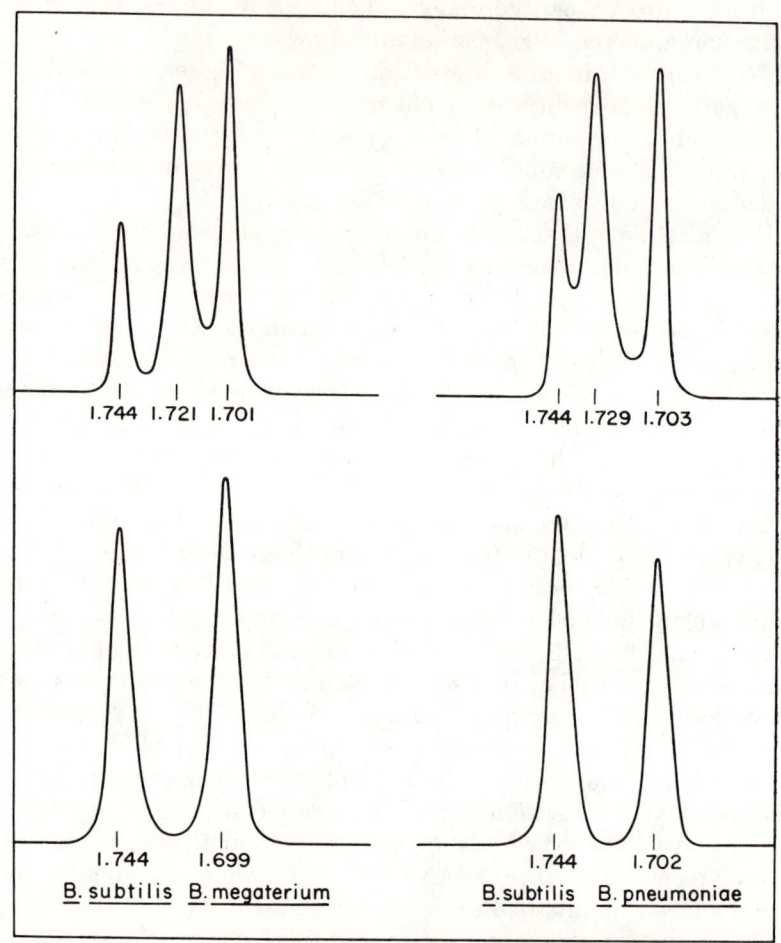

Fig. 19. Nonspecific aggregation in the renaturation of high concentrations of denatured DNA. N^{15}-deuterated DNA from *B. subtilis* was heated with either DNA from *B. megaterium* or *D. pneumoniae* at a concentration of 50 μg./ml. each in 0.285 M NaCl plus 0.025 M Na citrate. The mixtures were then slowly cooled under the usual conditions. Aliquots were centrifuged in CsCl before (upper tracings) and after (bottom tracings) treatment with the *E. coli* phosphodiesterase. Microdensitometer tracings were made on the ultraviolet photos of the bands when equilibrium had been reached in the density gradient centrifugations using the Spinco ultracentrifuge at 44,770 rpm.

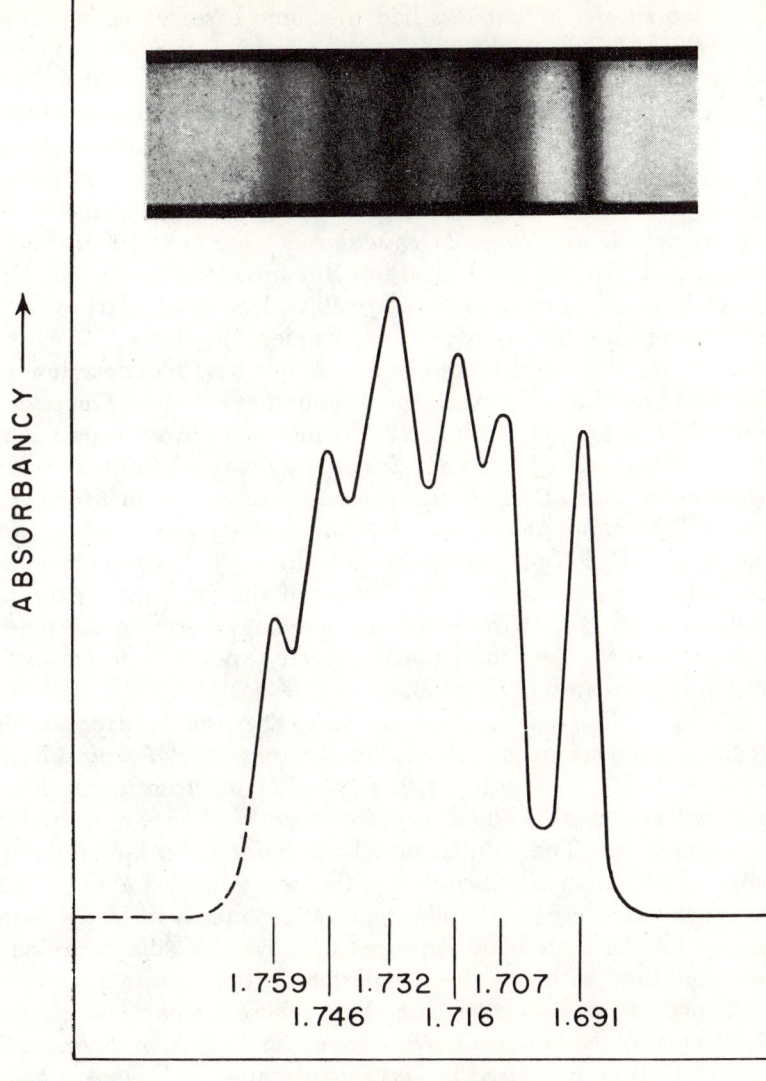

Fig. 20. Hybrid formation between the DNA of *B. subtilis* and *B. natto*. A mixture of 5 μg./ml. of each DNA was denatured (10 minutes at 100°C) and renatured (two hours at 68°C, then cooled to 25°C over a period of two hours) in 0.285 M NaCl plus 0.285 M Na citrate. The material was then centrifuged at 44,770 rpm in a CsCl gradient. The ordinate of the microdensitometer tracings represents the concentration of DNA in the centrifuge cell. Buoyant densities

no hybrid formation. Hybrid formation was possible, however, when two species of *Bacillus* had the same base composition and were related by the ability of one to transform the other. This is shown for the DNA of *B. subtilis* and *B. natto* in Figure 20.

Three members of the family *Enterobacteriaceae* having similar base compositions (Lee, Wahl, and Barbu, 1956) and which are genetically related have also been examined. *E. coli* K-12 shows a high degree of genetic exchange by conjugation and transduction with *E. coli* B and *Shigella dysenteriae* (Lennox, 1955; Luria, Adams, and Ting, 1960; Luria and Burrous, 1957), whereas *Salmonella typhimurium* mates fairly well with K-12 but is transduced only to a very limited extent, if at all (Zinder, 1960).

In an effort to see if there are any major base sequence homologies between the DNA of *E. coli* B and that of either *Shigella* or *Salmonella*, attempts to form hybrid molecules were begun. Unlabeled DNA isolated from *Sh. dysenteriae* was substituted for the unlabeled *E. coli* DNA in the procedure discussed in *Molecular Hybrid Formation*, above, and after the slowly cooled mixture was treated with *E. coli* phosphodiesterase, three bands were observed. The hybrid band was less than either of the renatured parental bands, indicating that the extent of homology between the heterologous strains is less than that between two DNA preparations from the same strain of *E. coli* B.

Similar studies were carried out with DNA isolated from different species and strains of *Salmonella*. Heating heavy *E. coli* B DNA at 100°C and slow cooling with DNA isolated from each of the *Salmonella* species, *typhosa*, *typhimurium*, *ballerup*, and *arizona* was carried out. The results for all the *Salmonella* samples were similar. The *Sal. typhimurium* and the heavy-labeled *E. coli* DNA concentrations were at 10 and also at 20 µg./ml. each. In the latter case, an intermediate band appeared; however, this did not remain after digestion with the phosphodiesterase. No intermediate band at all formed in the former case. Thus, there seems to be little, if any, sequence complementary between the DNA of *Salmonella* and *E. coli* B as measured by hybrid formation. The experiments were also repeated using the *Sal. typhimurium* DNA and the

---

(expressed in g./cm.[3]) correspond to the following DNA molecular species: 1.691–*Cl. perfringens* (used as a standard); 1.707 and 1.716–renatured and denatured *B. natto* DNA, respectively; 1.732–hybrid *B. natto–B. subtilis* DNA; 1.746 and 1.759–renatured and denatured N[15]–deuterated *B. subtilis* DNA, respectively.

heavy-labeled DNA from *E. coli* K-12. The latter behaved exactly like *E. coli* B DNA.

It should be noted that these results do not rule out the possibility of close homology only between several molecules of the heterologous total DNA populations, or between small sequences within their molecules. This might be enough to satisfy the minimal requirement for genetic exchange by conjugation. Very low concentrations of hybrids would not be observable in the analytical ultracentrifuge, but could be isolated by using larger amounts of interacting DNA and working with the preparative swinging-bucket rotor.

Genetic and Taxonomic Aspects of Molecular Hybrid Formation

It has recently been pointed out that there is a relationship between the taxonomic and genetic relationships of microorganisms and the base compositions of their DNA (Lee, Wahl, and Barbu, 1956; Belozersky and Spirin, 1960; Lanni, 1960). Those organisms which are genetically related most likely possess homologous base sequences; a minimum requirement is that they have similar base compositions. The question can now be asked, how molecular hybrid formation is related to the taxonomy and genetics of microorganisms as well as to their DNA base compositions. Which is a better test of homology, the extent of molecular hybrid formation or of genetic exchange? At a more practical level, we can ask if molecular hybrid formation can be used as a tool in studying the taxonomic and genetic relationship of microorganisms.

It has already been pointed out in the previous section that molecular hybrid formation takes place only between organisms possessing similar base compositions. Table 1 summarizes the results to date where such genetic exchanges have been observed; in each case where molecular hybrid formation takes place, the organisms are related taxonomically and genetically. Wherever the base compositions of the DNA differ (e.g., *E. coli* and *Serratia marcescens* or *Aerobacter aerogenes*), even though the organisms belong to the same taxonomic group, or if they belong to the same taxonomic group but are genetically unrelated (*B. subtilis* and *B. brevis*), no molecular hybrid formation can be detected.

In order to compare molecular hybrid formation and genetic compatibility as a measure of homology, it is first necessary to mention briefly the various methods of genetic exchange in bacteria.

The four main routes of genetic transfer are: F-duction, conjugation, transduction, and transformation (Ravin, 1958). F-duction (Jacob, Schaeffer, and Wollman, 1960) is mediated by an episomal element, and since this method of genetic exchange can take place between bacteria with dissimilar base composition (Falkow et al., 1961; Nakaya, Nakahura, and Murata, 1961) it would appear to be the poorest test of DNA homologies. This is borne out by the fact that the transferred material may or may not be integrated even between closely related organisms (Jacob, Schaeffer, and Wollman, 1960) and, in the case of F-duction between *E. coli* and *S. marcescens*, the incoming material can be detected as a separate band in CsCl density gradient centrifugation of the DNA isolated from the hybrid and thus presumed not to be integrated into the host chromosome (Marmur et al., 1961).

The second method of genetic exchange, conjugation, would appear to be less demanding than molecular hybrid formation (or transduction) for sequence homologies. Thus, Zinder (1960) found that whereas genetic exchange between *Sal. typhimurium* and *E. coli* K-12 can take place by conjugation, little or no exchange of certain markers is achieved by transduction. It should be possible however to transduce the markers in the regions where the homologies exist that satisfy the criteria for genetic exchange by conjugation. As seen in Table 1, no molecular hybrid formation is noted between the DNA of *E. coli* B and of various *Salmonella* species indicating that very little homology exists.

It would appear that transformation, transduction, and molecular hybrid formation are the best criteria for homology, and it remains to be shown which is the most demanding. This question might be answered by studying the extent of interspecific molecular hybrid formation among the DNA's isolated from various *Hemophilus* species (Table 1). Since they transform one another with varying degrees of efficiency (Schaeffer, 1958), it would be of great interest to see whether the extent of hybrid formation can be predicted by the transformation efficiency. This has not yet been accomplished because of the difficulty of obtaining heavy isotope labeled *Hemophilus* DNA, but can probably be carried out with *B. subtilis* (Spizizen, 1959). Preliminary experiments (Marmur, Seaman, and Levine, in preparation) with two *Bacillus* species may be of interest in this connection. *Bacillus polymyxa* and *Bacillus niger* DNA transforms *B. subtilis* at very low efficiencies with respect to the indole marker, yet the extent of molecular hybrid formation

TABLE 1

Genetic, Taxonomic, and DNA Composition Relationships of Several Groups of Microorganisms

Taxonomic group	Per cent G+C† of DNA of related organisms		Representative organism	Genetically related to representative organism	DNA molecular hybrid formation with representative organism
Lactobacillaceae	D. pneumoniae	39	D. pneumoniae	Streptococcus	Not examined
	Str. salivarius	39			
	L. acidophilus	39			
Brucellaceae	H. influenzae	38	H. influenzae	H. parainfluenzae	Not examined
	H. parainfluenzae	38		H. aegyptius	
	H. suis	38		H. suis	
	H. aegyptius	39			
Enterobacteriaceae	E. coli B, C, W, K-12, TAU⁻, 44-B, I	50	E. coli B	E. coli K-12	E. coli B, C, W, K-12, TAU⁻, 44-B, I
	E. freundii (17)	50		Sh. dysenteriae	
	Sh. dysenteriae	50		Sal. typhosa	E. freundii (17)
	Sal. arizona	50		Sal. typhimurium	Sh. dysenteriae
	Sal. ballerup	51.5			
	Sal. typhosa	50			
	Sal. typhimurium	51.5			
	Erwinia carotovora	51.5			
Bacillaceae	B. subtilis	43	B. subtilis	B. natto	B. subtilis
	B. natto	43		B. subtilis var. atterimus	B. natto
	B. subtilis var. atterimus	43			B. subtilis var. atterimus
	B. macerans	50			
	B. licheniformis	46			
	B. polymyxa	44			
	B. stearothermophilus	44			
	B. brevis	43			
	B. firmus	41			
	B. laterosporus	40			
	B. pumilus	40			
	B. sphaericus	37			

(Continued on next page.)

Table 1. Continued from preceding page.

Taxonomic group	Per cent G+C† of DNA of related organisms		Representative organism	Genetically related to representative organism	DNA molecular hybrid formation with representative organism
Bacillaceae	B. megaterium	37		B. polymyxa**	B. polymyxa**
	B. circulans	35			
	B. megaterium-cereus	34			
	B. thuringiensis	34			
	B. cereus	33			
	B. cereus var. mycoides	33			
	B. alvei	33			
E. coli viruses	T_2	34.8*	T_4 bacteriophage	T_2, T_6	T_2, T_6
	T_4	34.8*			
	T_6	34.8*			
	T_7	48			
	T_3	49.6*			
	T_5	39*			

† The guanine plus cytosine content of the DNA was estimated from the T_m (melting temperature) since it was found to give more self-consistent values for the strains of bacteria used in this study (Figure 4).
* Value obtained from the literature (Sinsheimer, 1960).
** Interacts at reduced levels.

between their DNA and that of *B. subtilis*, while not equal to the amount formed between homologous strains, is greater than would be expected from the transformation efficiencies. From this preliminary result one might conclude that transformation is a better test of homology, within certain limited regions of the DNA molecule, than is hybrid formation. Molecular hybrid formation is of course a measure of the entire population of DNA molecules. When the interspecific transformation efficiency of *B. subtilis* by DNA from heterologous species (e.g., *B. natto* and *B. subtilis* var. *atterimus*) is high, the extent of molecular hybrid formation (Table 1) is also high.

Since there is a close relationship between *in vitro* molecule hybrid formation, taxonomy, and genetic compatibility of microorganisms, it is proposed that organisms yielding DNA that form hybrids should belong to the same taxonomic group. Thus, if one wishes to utilize this technique to examine taxonomic (and possibly the genetic) relationship among a group of microrganisms, its

DNA should first be examined for the identity of base composition; and if this minimal criterion is met, heterologous pairs of DNA samples should be examined for their ability to form molecular hybrids.

It should be pointed out that since unfractionated samples of DNA's of higher plants and animals do not renature (see *Effect of DNA Source*) under conditions in which microbial nucleic acids will renature, the principles discussed above are applicable only to microorganisms. It is possible, however, that homogeneous fractions of the DNA of one higher species might hybridize with the corresponding fraction of the DNA of a closely related species, employing the techniques described for DNA from microorganisms.

Summary

The complementary strands of the DNA molecule which are held together by lateral hydrogen bonds between the base pairs adenine-thymine and guanine-cytosine can be separated by heat, acid, base, and formamide. The factors influencing the thermally induced transition from the native double-stranded form to the denatured single-stranded state as well as the renaturation of complementary strands have been studied by a number of techniques, such as the measurement of the biological activity of transforming factor DNA, the variation of ultraviolet absorption and cesium chloride density gradient centrifugation.

The separation of the DNA strands and their specific renaturation on subsequent cooling at temperatures below the melting temperature have made it possible to form "hybrid" molecules between the DNA of closely related strains of microorganisms. The molecular hybrids are recognized by their buoyant density in CsCl which is intermediate between the two "reactants": heavy (labeled with N^{15} and deuterium) and light, normal DNA. Minimum requirements for hybrid formation to occur are that the base compositions of the DNA samples be similar and that they have relatively little composition heterogeneity. Thus far it has been found that only microorganisms which are genetically related possess DNA which yields molecular hybrids *in vitro* on thermal denaturation and subsequent slow cooling. It is proposed that genetic compatibility and *in vitro* hybrid formation are dependent on homologous base sequences and that hybrid formation is indicative of and should predict close genetic and taxonomic relationships.

ACKNOWLEDGMENTS

The authors are grateful for the valuable discussions and suggestions offered by Drs. N. Sueoka, S. Falkow, E. Seaman and D. Green, as well as Mr. R. Rownd. The technical assistance of Miss D. Lane, Mr. W. Torrey and Miss M. Cahoon is gratefully acknowledged. The experiments on hybrid formation with the *E. coli* bacteriophages were carried out in collaboration with Dr. K. L. Wierzchowski and Dr. D. Green. The authors are also grateful to Dr. L. Grossman for his generous gift of the *E. coli* phosphodiesterase.

This work was supported by Grant C-2170 from the U.S. Public Health Service.

REFERENCES

Belorzersky, A. N., and A. S. Spirin. 1960. "Chemistry of the Nucleic Acids of Microorganisms," *The Nucleic Acids*, E. Chargaff and J. N. Davidson, Eds., Vol. 3, pp. 147–185. New York, New York: Academic Press, Inc.

Brown, G. L., and A. V. Brown. 1958. Fractionation of Deoxyribonucleic Acids and Reproduction of T2 Bacteriophage. *Symposia of the Society for Experimental Biology*, 12:6–30.

Chargaff, E. 1955. "Base Composition of Deoxypentose and Pentose Nucleic Acids in Various Species," *The Nucleic Acids*, E. Chargaff and J. N. Davidson, Eds., Vol. 1, pp. 521–531. New York, New York: Academic Press, Inc.

Cordes, S. A., H. Epstein, and J. Marmur. 1961. Some Properties of the Deoxyribonucleic Acid of Phage Alpha. *Nature, London*, 191: 1097–1098.

Crampton, C. W., R. Lipshitz, and E. Chargaff. 1954. Studies on Nucleoproteins. II. Fractionation of Deoxyribonucleic Acids Through Fractional Dissociation of their Complexes with Basic Proteins. *Journal of Biological Chemistry*, 211:125–142.

Doty, P., H. Boedtker, J. R. Fresco, R. Haselkorn, and M. Litt. 1959. Secondary Structure in Ribonucleic Acids. *Proceedings of the National Academy of Sciences of the U.S.A.*, 45:482–499.

Doty, P., J. Marmur, J. Eigner, and C. Schildkraut. 1960. Strand Separation and Specific Recombination in Deoxyribonucleic Acids: Physical Chemical Studies. *Proceedings of the National Academy of Sciences of the U.S.A.*, 46:461–476.

Eigner, J. 1960. *The Native, Denatured and Renatured States of Deoxyribonucleic Acid*. Ph.D. Dissertation. Harvard University, Cambridge, Massachusetts.

Falkow, S., J. Marmur, W. F. Carey, W. M. Spilman, and L. S. Baron. 1961. Episomic Transfer Between *Salmonella typhosa* and *Serratia marcescens*. *Genetics*, 46:703–706.

Greer, S., and S. Zamenhof. 1959. Loss of Purines from DNA Heated in Mutagenic Conditions at Physiological pH. (Abstract) *Federation Proceedings*, 18:939.

Guild, W. R. 1961. Fractionation of Microbial DNA by Density. (Abstract) *Fifth Annual Meeting of the Biophysical Society*, FB9.

Guild, W. R., H. J. Morowitz, and E. Castro. 1960. Some Properties of DNA from PPLO. (Abstract) *Fourth Annual Meeting of the Biophysical Society*, p. 19.

Helmkamp, G. K., and P. O. P. Ts'o. 1961. The Secondary Structures of Nucleic Acids in Organic Solvents. *Journal of the American Chemical Society*, 83:138–142.

Hershey, A. D., and E. Burgi. 1960. Molecular Homogeneity of the Deoxyribonucleic Acid of Phage T2. *Journal of Molecular Biology*, 2:143–152.

Jacob, F., P. Schaeffer, and E. L. Wollman. 1960. Episomic Elements in Bacteria. *Microbial Genetics* (Tenth Symposium of the Society for General Microbiology at the Royal Institution, London) pp. 67–91. London, England: Cambridge University Press.

Lanni, F. 1960. Genetic Significance of Microbial DNA Composition. *Perspectives in Biology and Medicine*, 3:418–432.

Lee, K. Y., R. Wahl, and E. Barbu. 1956. Contenu en bases puriques et pyrimidiques des acides desoxyribonucleiques des bacteries. *Annales de l'Institut Pasteur*, 91:212–224.

Lehman, I. R. 1960. The Deoxyribonucleases of *Escherichia coli*. I. Purification and Properties of a Phosphodiesterase. *Journal of Biological Chemistry*, 235:1479–1487.

Lennox, E. S. 1955. Transduction of Linked Genetic Characters of the Host by Bacteriophage P1. *Virology*, 1:190–206.

Levine, L., W. T. Murakami, H. Van Vunakis, and L. Grossman. 1960. Specific Antibodies to Thermally Denatured Deoxyribonucleic Acid of Phage T4. *Proceedings of the National Academy of Sciences of the U.S.A.*, 46:1038–1043.

Levinthal, C., and C. A. Thomas, Jr. 1957. Molecular Autoradiography: The β-Ray Counting from Single Virus Particles and DNA Molecules in Nuclear Emulsions. *Biochimica et biophysica acta*, 23:453–465.

Litt, M., J. Marmur, H. Ephrussi-Taylor, and P. Doty. 1958. The Dependence of Pneumococcal Transformation on the Molecular Weight of Deoxyribose Nucleic Acid. *Proceedings of the National Academy of Sciences of the U.S.A.*, 44:144–152.

Luria, S. E., J. N. Adams, and R. C. Ting. 1960. Transduction of Lactose-Utilizing Ability Among Strains of *E. coli* and *S. dysenteriae* and the Properties of the Transducting Phage Particles. *Virology*, 12:348–390.

Luria, S. E., and J. N. Burrous. 1957. Hybridization Between *Escherichia coli* and *Shigella*. *Journal of Bacteriology*, 74:461–476.

Madoff, S., and L. Dienes. 1958. L Forms from Pneumococci. *Journal of Bacteriology*, 76:245–250.

Mahler, H. R., B. D. Mehrotra, and C. W. Sharp. 1961. Effects of Diamines on the Thermal Transition of DNA. *Biochemical and Biophysical Research Communications*, 4:79–82.

Marmur, J. 1960. Thermal Denaturation of Deoxyribonucleic Acid Isolated from a Thermophile. *Biochimica et biophysica acta*, 38:342–343.

———. 1961. A Procedure for the Isolation of Deoxyribonucleic Acid from Micro-organisms. *Journal of Molecular Biology*. 3:208–218.

Marmur, J., and P. Doty. 1959. Heterogeneity in Deoxyribonucleic Acids. I.

Dependence on Composition of the Configurational Stability of Deoxyribonucleic Acids. *Nature, London*, 183:1427–1429.

———. 1961. Thermal Renaturation of Deoxyribonucleic Acids. *Journal of Molecular Biology*. 3:585–594.

———. Determination of the Base Composition of Deoxyribonucleic Acid from its Thermal Denaturation Temperature. (In preparation).

Marmur, J. R. Rownd, S. Falkow, L. S. Baron, C. Schildkraut, and P. Doty. 1961. The Nature of Intergenus Episomal Infection. *Proceedings of the National Academy of Science of the U.S.A.*, 47:972–979.

Marmur, J., and D. Lane. 1960. Strand Separation and Specific Recombination in Deoxyribonucleic Acids: Biological Studies. *Proceedings of the National Academy of Sciences of the U.S.A.*, 46:453–461.

———. Unpublished data.

Marmur, J., E. Seaman, and J. Levine. Interspecific Transformation in *Bacillus*. (In preparation).

Marmur, J., and P. O. P. Ts'o. 1961. Denaturation of Deoxyribonucleic Acid by Formamide. *Biochimica et biophysica acta*, 51:32–36.

Meselson, M., and F. Stahl. 1958. The Replication of DNA in *Escherichia coli*. *Proceedings of the National Academy of Sciences of the U.S.A.*, 44:671–682.

Nakaya, R., A. Nakahura, and Y. Murata. 1960. Resistance Transfer Agents in *Shigella*. *Biochemical and Biophysical Research Communications*, 3:654–659.

Pauling, L., and R. B. Corey. 1956. Specific Hydrogen-Bond Formation Between Pyrimidines and Purines in Deoxyribonucleic Acids. *Archives of Biochemistry and Biophysics*, 65:164–181.

Ravin, A. W. 1958. Bacterial Genetics. *Annual Review of Microbiology*, 12:309–364.

Schaeffer, P. 1958. Interspecific Reactions in Bacterial Transformation. *Symposia of the Society for Experimental Biology*, 12:60–74.

Schildkraut, C. L., J. Marmur, and P. Doty. 1961. The Formation of Hybrid DNA Molecules and Their Use in Studies on DNA Homologies. *Journal of Molecular Biology*, 3:595–617.

Sinsheimer, R. L. 1960. "The Nucleic Acids of Bacterial Viruses," *The Nucleic Acids*. E. Chargaff and J. N. Davidson, Eds., Vol. 3, pp. 187–244. New York, New York: Academic Press, Inc.

Spizizen, J. 1959. Genetic Activity of Deoxyribonucleic Acid in the Reconstitution of Biosynthetic Pathways. *Federation Proceedings*, 18:957–965.

Sueoka, N., J. Marmur, and P. Doty. 1959. Heterogeneity in Deoxyribonucleic Acids. II. Dependence of the Density of Deoxyribonucleic Acids on Guanine-Cytosine. *Nature, London*, 183:1429–1431.

Vendreley, R. 1958. La Notion d'espece à travers quelques données biochimiques recentes et le cycle L. *Annales de l'Institut Pasteur*, 94:142–166.

Zinder, N. D. 1960. Hybrids of *Escherichia* and *Salmonella*. *Science*, 131:813–815.

MECHANISM OF DNA CHAIN GROWTH, I. POSSIBLE DISCONTINUITY AND UNUSUAL SECONDARY STRUCTURE OF NEWLY SYNTHESIZED CHAINS

BY REIJI OKAZAKI, TUNEKO OKAZAKI, KIWAKO SAKABE, KAZUNORI SUGIMOTO, AND AKIO SUGINO

INSTITUTE OF MOLECULAR BIOLOGY AND DEPARTMENT OF CHEMISTRY, FACULTY OF SCIENCE, NAGOYA UNIVERSITY, NAGOYA, JAPAN

Communicated by Rollin D. Hotchkiss, December 14, 1967

In vivo studies[1-7] of chromosome replication have led to the inference that both daughter strands of chromosomal DNA grow continuously, the direction of synthesis being 3' to 5' on one strand and 5' to 3' on the other (Fig. 1A). No enzymatic mechanism for the biosynthesis of deoxypolynucleotide in the 3' to 5' direction has been demonstrated, although 5' to 3' *in vitro* synthesis of DNA is accomplished by DNA polymerase.[8] If discontinuous synthesis of DNA could occur *in vivo* (Figs. 1B–D), short stretches could be synthesized by a reaction in the 5' to 3' direction and subsequently connected to the growing polynucleotide chain by formation of phosphodiester linkages.

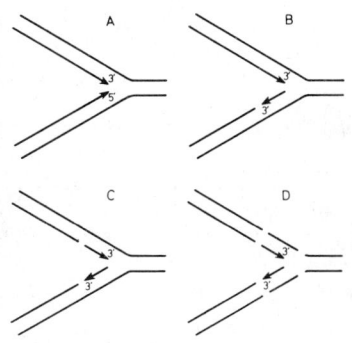

FIG. 1.—Models for the possible structure and reaction in the replicating region of DNA.

It is possible to distinguish between continuous and discontinuous chain growth by elucidating the structure of the most recently replicated portion of the chromosome; that is, that portion selectively labeled by an extremely short radioactive pulse. If the chromosome replicates discontinuously by one of the mechanisms shown in Figures 1B, C, or D, a large portion of the radioactive label would be found in unconnected short chains which can be isolated, after denaturation, from the large DNA chains derived from the other portion of the chromosome. No such difference in the molecular size between the pulse-labeled and bulk DNA would be expected from a mechanism of continuous synthesis (Fig. 1A).

Our results to be described here, together with those reported previously,[9] indicate that in a variety of bacterial systems and in one bacteriophage system most of the recently synthesized portion of the chromosome can be obtained after denaturation as small DNA molecules with a sedimentation coefficient of about 10S. This supports the prediction of those mechanisms by which two daughter strands are synthesized in a discontinuous fashion (Fig. 1C or D). It is also shown that the secondary structure of the chromosomal region containing these newly synthesized chains may differ from that of ordinary double-stranded DNA.

Materials and Methods.—Organisms used were as follows: *Escherichia coli* strains B, 15T⁻, W3110, and 1100 (endonuclease I-deficient strain provided by Dr. H. Hoffman-

Berling), *Bacillus subtilis* strain SB 19, and bacteriophages T4 (wild-type) and δA (provided by Dr. I. Watanabe).[10]

Reagents: The following commercial products were used: H^3-thymidine and C^{14}-thymidine (New England Nuclear); crystalline pancreatic DNase and RNase and egg-white lysozyme (Worthington); Pronase P (Kaken Chemical). *E. coli* exonuclease I was a gift of Dr. I. R. Lehman. *B. subtilis* nuclease was fraction I-A described previously.[11] Bacterial α-amylase was provided by Dr. F. Fukumoto. Hydroxylapatite was prepared according to Miyazawa and Thomas.[12] C^{14}-*E. coli* DNA used as standard substrate for DNase was prepared as described previously.[11] DNA from phage δA was obtained by phenol extraction.[10]

Culture media: Medium *A:* glucose salt medium containing 0.1 *M* potassium phosphate buffer, pH 7.3, 1 mM $MgSO_4$, 0.02 *M* $(NH_4)_2SO_4$, 0.002 mM $Fe(NH_4)(SO_4)_2$ and 1% glucose; medium *B:* medium *A* supplemented with 0.5% casamino acids, 0.01% cysteine and DL-tryptophan, and 1.2×10^{-5} *M* thymidine; medium *C:* M9 synthetic medium supplemented with 0.5% casamino acids. Media *A*, *B*, and *C* were used for experiments with *E. coli* B, *E. coli* 15T⁻, and T4 phage-infected *E. coli* B, respectively. Medium *B* containing no thymidine was used for *E. coli* W3110 and 1100.

Pulse labeling: To pulse label the bacteria with no thymine requirement or T4 phage-infected cells, H^3-thymidine (14 mc/μmole) was added to the stirred culture (5×10^8 cells/ml) at 20° (at 30° with *B. subtilis*) to a concentration of 10^{-7} *M*. After allowing the cells to incorporate H^3-thymidine for a desired time, the culture was poured onto crushed ice and KCN (to 0.02 *M*), and the cells were collected by centrifugation at 0°. To pulse label *E. coli* 15T⁻, cells grown in medium containing thymidine were precipitated and resuspended in a small volume of fresh medium at 0° containing no thymidine. The cell suspension was poured into a larger volume of stirred medium at 20°. H^3-thymidine (10^{-7} *M*) was added and the reaction was stopped by KCN and ice.

Extraction of DNA: (a) *Extraction of native DNA by the Thomas procedure*[13] *(Figs. 3, 4, 7, and 8; Table 1):* This was carried out as described previously[9] except that in some experiments sodium dodecyl sulfate (SDS) treatment was at 37° and the DNA solution was concentrated by filtration through a collodion membrane. DNA from 1 ml of culture was finally obtained in a volume of 0.5–1 ml. In *E. coli* B, recovery of DNA labeled by various lengths of pulse was greater than 90%. With *E. coli* 15T⁻, recovery varied from 30 to 60% but no systematic difference was found between the pulse- and uniformly labeled DNA's in parallel experiments.

(b) *Extraction of denatured DNA by NaOH-EDTA (Figs. 2, 5, and 6):* The cells were suspended in ice-cold 0.1 *N* NaOH containing 0.01 *M* ethylenediaminetetraacetic acid (EDTA) at a concentration of 5×10^9 cells/ml. The suspension was incubated at 37° for 20 min with occasional gentle stirring with a glass rod, and the insoluble material was removed by low-speed centrifugation. More than 80% of *E. coli* DNA and 50–80% of pulse-labeled DNA from T4 phage-infected cells were recovered by this procedure.

Denaturation of DNA: DNA extracted in the native state was denatured by incubation in 0.1 *N* NaOH containing 1 mM EDTA at room temperature for 20 min.

Zone sedimentation in sucrose gradients: (a) *Alkaline sucrose gradient:* Either the SW25.1 or SW25.3 rotor of a Spinco L or L2 centrifuge was used. With the SW25.1 rotor, 1 ml of DNA sample in 0.1 *N* NaOH containing 0.01 *M* EDTA was layered on a 29-ml 5–20% linear sucrose gradient containing 0.1 *N* NaOH, 0.9 *M* NaCl, and 1 m*M* EDTA. With the SW25.3 rotor, the volumes of the sample and gradient were 0.3 and 16 ml, respectively. Five mμmoles of DNA from bacteriophage δA was added to each sample as internal reference. After centrifugation, fractions were collected from the bottom of the tube. Radioactive DNA in each fraction was counted in a Tri-Carb liquid scintillation spectrometer after repeated precipitation with cold 5% trichloroacetic acid (TCA) and solubilization by 5% TCA at 90°. Distribution of δA DNA among fractions was determined by assaying aliquots for infectivity in *E. coli* protoplasts.[10] Distance of sedimentation was shown relative to the distance from the meniscus to the band of δA

DNA. Sedimentation coefficients were calculated from the value of 19S for this marker DNA, obtained by boundary sedimentation in 0.1 N NaOH–0.9 M NaCl.

(b) *Neutral sucrose gradient:* Centrifugation was carried out in the SW25.1 rotor, layering 1 ml of DNA sample over a 29-ml 5–20% sucrose gradient, pH 7.0, containing 0.15 M NaCl, 0.015 M sodium citrate, and 1 mM EDTA.

Recovery of DNA from alkaline and neutral sucrose gradients was more than 90%.

Other methods: Chromatography of DNA on hydroxylapatite was carried out according to Bernardi.[14] Recovery of DNA from the column was 60–65%. Formation of acid-soluble product by enzymatic degradation of labeled DNA was measured as described by Lehman.[15]

Results.—*Nature of the replicating region as revealed by alkaline sucrose gradient sedimentation:* To facilitate labeling of a small portion near the growing end of the daughter strands, all the pulse-labeling experiments with *E. coli* (normal or T4 phage-infected) were carried out at 20°. The rate of macromolecular synthesis at 20° is estimated to be about one sixth of the rate at 37°, since at 20° the generation time (and doubling time of DNA) of *E. coli* is about 3 hours and the lysis by T4 phage occurs about 140 minutes after infection.

In the experiment presented in Figure 2, growing cells of *E. coli* B were exposed

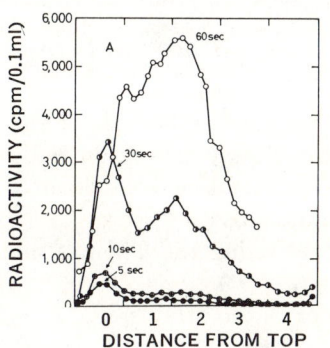 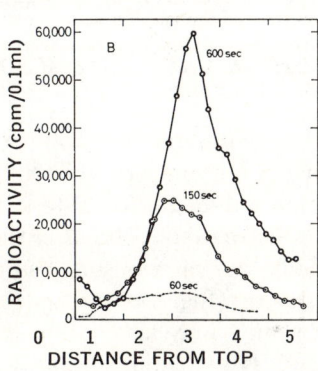

Fig. 2.—Alkaline sucrose gradient sedimentation of pulse-labeled DNA from *E. coli* B. Cells were grown at 37° to a titer of 3 × 10⁸ cells/ml and then at 20° to 5 × 10⁸ cells/ml. and the 10-ml culture was pulse-labeled with 10^{-7} M H³-thymidine at 20° for the indicated time. DNA was extracted by NaOH–EDTA treatment and sedimented in the SW25.3 rotor for 10 hr at 22,500 rpm and 4°. Distance from top is relative to that of infective DNA from phage δA (19S, reference).

to H³-thymidine for various times. DNA was extracted in the denatured state by the NaOH–EDTA treatment and sedimented in alkaline sucrose gradients. Infectious DNA from phage δA used as internal reference had a sedimentation coefficient of 19S in 0.1 N NaOH–0.9 M NaCl. Most of the radioactivity incorporated into DNA during the five-second pulse was recovered in a distinct component with an average sedimentation rate of 11S. Some radioactivity was found in material sedimenting at faster rates. Increasing the pulse time to 10 or 30 seconds increased the radioactivity in the "11S component" as well as the radioactivity in the fast-sedimenting DNA. Further increasing the pulse time resulted in large increases of the radioactivity in the fast-sedimenting DNA with little or no increase in the radioactive "11S component." The presence of the

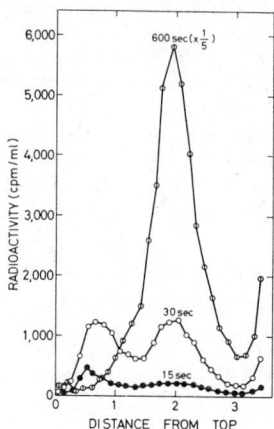
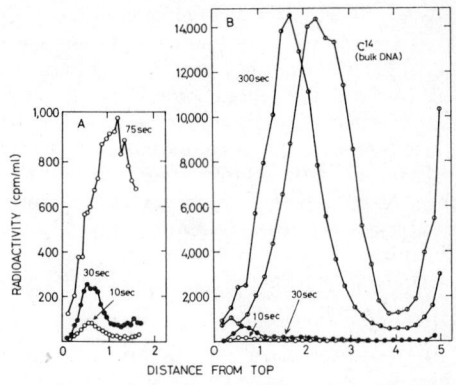

FIG. 3.—Alkaline sucrose gradient sedimentation of pulse labeled DNA from *E. coli* B. A 25-ml culture was pulse-labeled as in Fig. 2. DNA was extracted by the Thomas method. An aliquot was denatured in alkali and sedimented in the SW25.1 rotor for 15 hr at 20,500 rpm and 8°.

FIG. 4.—Alkaline sucrose gradient sedimentation of pulse-labeled DNA from *E. coli* 15T⁻. A 5-ml culture was pulse-labeled at 20° for the indicated period. To the 300-sec sample, a small amount of culture uniformly labeled by C^{14}-thymidine was added before DNA extraction by the Thomas method. Sedimentation was carried out in the SW25.1 rotor at 8° and 20,500 rpm for (*A*) 25 hr or (*B*) 10 hr.

latter was obscure after the 150- or 600-second labeling because of the possible trailing of the high molecular DNA containing a large amount of radioactivity. The average sedimentation rate of the fast-sedimenting component increased gradually and was about 50S after the ten-minute pulse.

Essentially the same result was obtained by using the Thomas method for DNA extraction (Fig. 3).

Similar results were also obtained with other *E. coli* strains, i.e., *E. coli* 15T⁻, W3110, and 1100 (endonuclease I-deficient) (Figs. 4 and 5), and *B. subtilis* strain SB 19. The initial label of H^3-thymidine always appeared in the DNA component with an average sedimentation rate of 10–11S.

That the "11S component" is really DNA was substantiated by several facts. It is degraded by the action of pancreatic DNase or by *E. coli* exonuclease I at the same rate as the denatured *E. coli* DNA routinely used as standard DNase substrate.[11] It is also completely degraded by *B. subtilis* nuclease[11] but not by alkali, pancreatic RNase, or bacterial α-amylase.

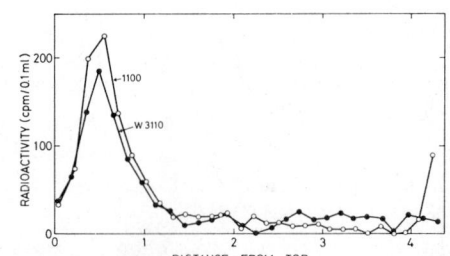

FIG. 5.—Alkaline sucrose gradient sedimentation of a 10-sec pulse DNA of *E. coli* W3110 and 1100. Experiments were carried out as in Fig. 2.

Figure 6 shows a result obtained with T4 phage-infected *E. coli* B. Cells were pulse-labeled after 70 minutes of infection at 20°, when phage DNA is being synthesized actively. After a two-second pulse the radioactive label incorporated

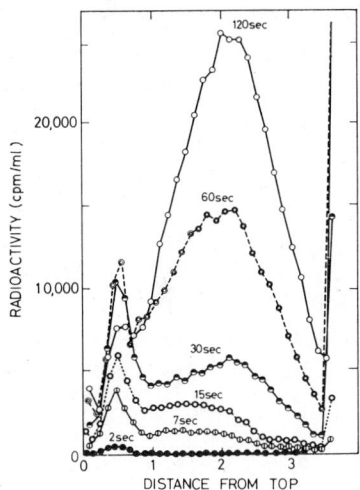

FIG. 6.—Alkaline sucrose gradient sedimentation of pulse-labeled DNA from T4 phage-infected *E. coli* B. Cells grown at 37° to 5×10^8 cells/ml were suspended in M9 medium containing no glucose at 10^9 cells/ml and incubated for 15 min at 37°. Following addition of DL-tryptophan (40 μg/ml), the cells were infected with T4 phage (MOI = 10). After 5 min at 37°, the culture was cooled to 20° and an equal volume of M9 medium containing twice as much glucose and casamino acids as medium *C* was added. After incubation with stirring at 20° for 70 min, the 10-ml culture was pulse-labeled with H^3-thymidine for the indicated time. DNA was extracted by NaOH–EDTA treatment and sedimented in the SW25.1 rotor for 15 hr at 20,500 rpm and 8°.

was recovered almost exclusively in DNA component with a sedimentation coefficient of 9*S*. After a longer period of labeling, the radioactivity was found also in faster-sedimenting material. The radioactivity in the "9*S* component" increased quickly and reached a maximum in about 30 seconds, whereas the radioactivity in the fast-sedimenting component increased almost linearly and in two minutes attained a level ten times higher than the radioactivity in the "9*S* component." The sedimentation rate of the fast component increased gradually as in growing bacterial cells. The average rate was about 40*S* after the two-minute pulse. In other experiments average rates of 45 and 50*S* were obtained for five- and ten-minute pulse DNA, respectively.

In these experiments the pulse labeling was stopped by KCN and ice, cells were precipitated, and denatured DNA was obtained by either (*a*) extraction by the Thomas method followed by alkali denaturation, or (*b*) extraction with NaOH–EDTA. The following changes in these procedures did not alter the essential feature of the results: (1) omission of the phenol step from (*a*), (2) addition of a pretreatment with lysozyme to (*b*), (3) directly adding NaOH–EDTA to the culture with or without prior addition of KCN and ice in (*b*), (4) denaturation with formamide in (*a*), and (5) extraction by the method of Nomura et al.[16] followed by alkali denaturation.

Secondary structure of the replicating region: Pulse-labeled DNA, isolated by the Thomas procedure but not subjected to denaturation, was analyzed by sedimentation in neutral sucrose gradients. A result obtained with *E. coli* B is shown in Figure 7. While most of the DNA isolated from the cells labeled with H^3-thymidine for ten minutes sedimented at a rate

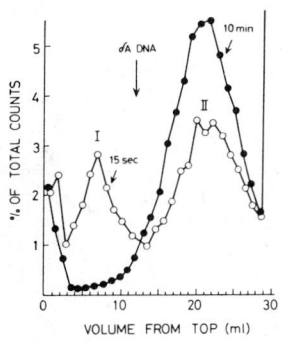

FIG. 7.—Neutral sucrose gradient sedimentation of pulse-labeled DNA from *E. coli* B. Native DNA samples of Fig. 3 were sedimented for 15 hr at 20,500 rpm and 8°.

faster than δA DNA, having a sedimentation coefficient of 29S in 0.5 M NaCl, pH 7.0, a considerable fraction of 15-second pulse DNA was recovered in a band sedimenting at a much slower rate. It was shown in other experiments that the fraction of the radioactivity found in the slowly sedimenting band decreases with increasing pulse time.

On the other hand, a large fraction of the DNA labeled by a short pulse was found to be susceptible to degradation by *E. coli* exonuclease I, which specifically hydrolyzes single-stranded DNA[17] (Table 1). Approximately the same fraction of the labeled DNA was eluted from hydroxylapatite at the relatively low phosphate concentration expected for single-stranded DNA and was found to be completely susceptible to the action of exonuclease I (Fig. 8 and Table 1). The susceptibility of unfractionated pulse DNA to exonuclease I and the fraction eluted from hydroxylapatite at low phosphate concentrations decrease with

TABLE 1. *Susceptibility of pulse-labeled DNA to E. coli exonuclease I prior to denaturation treatment.*

Pulse time	Unfractionated	Extent of Degradation by Exonuclease I (%)		Neutral Sucrose Gradient Fraction	
		Hydroxylapatite Fraction			
		I	II	I	II
5 Sec	45				
10 Sec	32*			77*	24*
15 Sec	30	96	24		
30 Sec	24, 24*	96, 88*	18, 15*		
10 Min	4, 0*	78	2		

E. coli B was pulse-labeled as in Fig. 2. Extraction and fractionation of labeled DNA were carried out as in Figs. 3, 7, and 8. SDS treatment for DNA extraction was 37°* or at 60°. Hydroxylapatite fractions I and II are shown in Fig. 8, and neutral sucrose gradient fractions I and II in Fig. 7.

For susceptibility to exonuclease, the 150-μliter reaction mixture, containing 10 μmoles glycine–KOH buffer, pH 9.2, 1 μmole MgCl₂, 0.15 μmole 2-mercaptoethanol, 60-μliter DNA sample (300–12,000 cpm), and 3 units of *E. coli* exonuclease I (DEAE-cellulose fraction), was incubated at 37°. After 60 min, 3 units of enzyme were added to the mixture and the incubation was continued for another 60 min. Acid-soluble and insoluble counts were determined at 0, 60, and 120 min. More than 85% of the radioactive DNA degraded during the 120-min period was already acid soluble at 60 min.

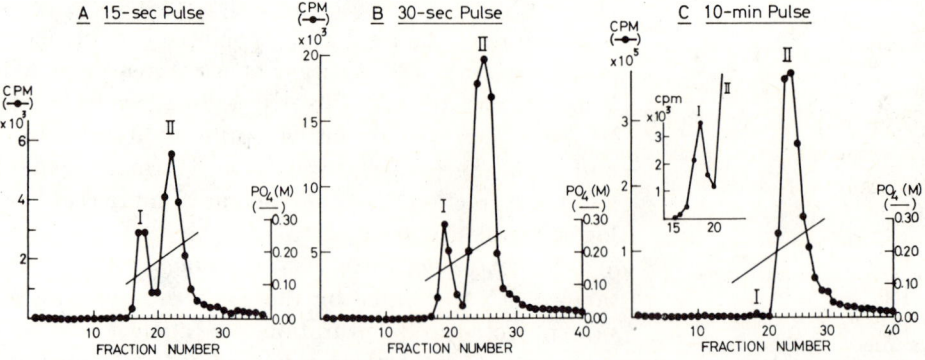

FIG. 8.—Hydroxylapatite chromatography of pulse-labeled DNA of *E. coli* B. The native DNA samples of Fig. 3 were dialyzed against 0.01 M potassium phosphate buffer, pH 6.8. Elution was achieved with a linear 0.01–0.7 M gradient of the same buffer (total vol 140 ml). Fractions of 2.5 ml were collected at 30-min intervals.

FIG. 9.—Schematic illustration of a possible structure of the daughter strand in the vicinity of the growing end.

increasing pulse time (Table 1 and Fig. 8). Furthermore, the slowly sedimenting component of pulse DNA recovered from the neutral sucrose gradient was shown to be highly susceptible to exonuclease I, while the fast-sedimenting component had a low susceptibility to the enzyme (Table 1).

Thus an appreciable fraction of the newly synthesized material as isolated appears to be single-stranded, and this fraction is sedimented slowly in the neutral sucrose gradient.

Discussion.—Average chain growth rate of *E. coli* chromosome is estimated to be about 400 nucleotides per second at 20°. Therefore, a 5 second pulse would label the stretches of about 2000 nucleotides or a 0.05 per cent portion of the whole chromosome. Our experiments show that the portion of the chromosome, labeled by such a short radioactive pulse is separable in alkali from the bulk of chromosomal DNA as small molecules. Observations described in this and a previous paper[9] indicate that this represents an intermediary state in the formation of chromosomal DNA. This result conforms to the prediction from the replication mechanisms by which two daughter strands are synthesized discontinuously (Figs. 1C and D). The replication mechanism by which only one of the two daughter strands is synthesized discontinuously (Fig. 1B) is less likely, because virtually all the label is recovered in the slowly sedimenting component after the very short pulse. The sedimentation coefficient of the initially labeled material is 10–11S in various bacterial systems and 8–9S in the T4 phage system, suggesting that the length of the "unit" may be 1,000–2,000 nucleotides. This corresponds to the dimension of cistron.

Figure 9 illustrates a possible structure of the daughter strands in the vicinity of the growing end. "Units" synthesized at the growing point would be joined together by phosphodiester bonds to form longer strands located in the nonterminal position. The number of "units" and of chains with intermediate lengths would be determined by the relative rates of synthesis and of joining.

An alternative interpretation of our results is that artificial breaks may be introduced selectively in the newly replicated region during DNA extraction. This possibility, which in any case suggests selective weakness in the newly replicated region, is diminished by the fact that similar results are obtained using different methods in a number of different systems (including an endonuclease I-deficient *E. coli* strain).

Our results on native DNA do not distinguish clearly between the two mechanisms for discontinuous chain growth shown in Figures 1C and D. Although a fraction of the pulse-labeled DNA sediments at a much slower rate than the bulk of DNA in the neutral sucrose gradient, this material proved to be single-stranded. The remaining portion, which is in a duplex form, is not separated from the bulk DNA by sedimentation. The fact that an appreciable fraction of the pulse DNA is isolated in the single-stranded form would imply either that most of the newly formed "units" exist as single strands in the cell or that the secondary structure of the replicating region containing these "units" is abnormally unstable. It may represent a unique state during replication or might indicate functioning of the newly synthesized "units" or the complementary portions of the parental strands as templates for RNA synthesis.

Our hypothesis of discontinuous DNA chain growth is encouraged by the discovery of polynucleotide-joining enzyme (ligase) in normal and T4 phage-infected *E. coli*.[18-21] The enzyme is encoded in one of the T4 genes previously implicated as a structural gene controlling DNA synthesis.[22] It has been used in *in vitro* synthesis of biologically active circular DNA in conjunction with DNA polymerase.[23-24] The synthesis and joining of the "units" assumed in our hypothesis may be carried out by DNA polymerase and polynucleotide ligase, respectively. A similar idea has recently been suggested by Kornberg and co-workers.[24,25] Further support for such hypotheses will await proof of the following: (1) the "units" are synthesized in the cell only by a reaction in the 5' to 3' direction; (2) the "units" are joined in the cell by the ligase reaction.

Note added in proof: Recent studies indicate that cells infected with temperature-sensitive mutants of T4 phage defective in ligase accumulate a large amount of the newly synthesized short DNA chains at 42°.

We are grateful to Dr. K. Gordon Lark and Dr. Rollin D. Hotchkiss for helpful discussions during preparation of the manuscript.

* Supported by the Research Fund of the Ministry of Education of Japan and by a grant from the Jane Coffin Childs Memorial Fund for Medical Research. This work was presented at the Symposium on Nucleic Acid Synthesis, Tokyo, March 1967 (Okazaki, R., T. Okazaki, K. Sakabe, and K. Sugimoto, *Jap. J. Med. Sci. Biol.*, **20**, 255 (1967)), and at the Seventh International Congress of Biochemistry, Tokyo, August 1967 (Okazaki, R., and K. Sakabe, *Abstract B*-10 (International Union of Biochemistry, 1967)).

[1] Cairns, J., *J. Mol. Biol.*, **6**, 208 (1963).
[2] Cairns, J., in *Cold Spring Harbor Symposia on Quantitative Biology*, vol. 28 (1963), p. 43.
[3] Nagata, T., these PROCEEDINGS, **49**, 551 (1963).
[4] Yoshikawa, H., and N. Sueoka, these PROCEEDINGS, **49**, 559 (1963).
[5] *Ibid.*, **49**, 806 (1963).
[6] Bonhoffer, F. B., and A. Gierer, *J. Mol. Biol.*, **7**, 534 (1963).
[7] Lark, K. G., T. Repko, and E. J. Hoffman, *Biochim. Biophys. Acta*, **76**, 9 (1963).
[8] Mitra, S., and A. Kornberg, *J. Gen. Physiol.*, **49**, 59 (1966).
[9] Sakabe, K., and R. Okazaki, *Biochim. Biophys. Acta*, **129**, 651 (1966).
[10] A filamentous bacteriophage specific to male strains of *E. coli*. The DNA extracted from this phage with phenol has a single-stranded circular structure and is infective to *E. coli* protoplasts (Okazaki, R., M. Morimyo, and K. Sugimoto, in preparation).
[11] Okazaki, R., T. Okazaki, and K. Sakabe, *Biochim. Biophys. Res. Commun.*, **22**, 611 (1966).
[12] Miyazawa, Y., and C. A. Thomas, Jr., *J. Mol. Biol.*, **11**, 223 (1965).
[13] Thomas, C. A., Jr., K. I. Berns, and T. J. Kelley, Jr., in *Procedures in Nucleic Acid Research* (New York: Harper and Row, 1966), p. 535.
[14] Bernardi, *Nature*, **22**, 779 (1965).
[15] Lehman, I. R., *J. Biol. Chem.*, **235**, 1479 (1960).
[16] Nomura, M., K. Matsubara, K. Okamoto, and R. Fujimura, *J. Mol. Biol.* **5**, 535 (1962).
[17] Lehman, I. R., and A. L. Nussbaum, *J. Biol. Chem.*, **239**, 2628 (1964).
[18] Gellert, M., these PROCEEDINGS, **57**, 148 (1967).
[19] Olivera, B. M., and I. R. Lehman, these PROCEEDINGS, **57**, 1426 (1967).
[20] Gefter, M. L., A. Becker and J. Hurwitz, these PROCEEDINGS, **58**, 241 (1967).
[21] Weiss, B., and C. C. Richardson, these PROCEEDINGS, **57**, 1021 (1967).
[22] Fareed, G. C., and C. C. Richardson, these PROCEEDINGS, **58**, 665 (1967).
[23] Goulian, M., and A. Kornberg, these PROCEEDINGS, **58**, 1723 (1967).
[24] Goulian, M., A. Kornberg, and R. L. Sinsheimer, these PROCEEDINGS, **58**, 2321 (1967).
[25] Mitra, S., P. Richard, R. B. Inman, L. L. Bertsch, and A. Kornberg, *J. Mol. Biol.*, **24**, 429 (1967).

Active Center of DNA Polymerase

The operations are localized and arranged in multiple sites within a single area of the molecule.

Arthur Kornberg

DNA polymerases have now been isolated from a variety of bacterial and animal cells. These enzymes, including those produced specifically in response to virus infection, catalyze the addition of mononucleotide units to the 3′-hydroxyl terminus of a primer DNA chain. Synthesis therefore proceeds in the direction of 5′ to 3′ (Fig. 1) (*1*). There is an absolute requirement for a DNA template, and errors in copying the template are very infrequent. The synthesis of DNA proceeds rapidly, at rates near 1000 nucleotides per minute per molecule of enzyme, with the production of chains several million in molecular weight.

The polymerases are remarkable enzymes. A polymerase takes instructions as it goes along to build a chain according to specifications by a template. Bacterial DNA polymerase will make animal DNA and animal polymerase will make bacterial DNA.

The author is professor and chairman of the department of biochemistry, Stanford University School of Medicine, Stanford, California 94305.

The DNA polymerase from *Escherichia coli* has additional catalytic properties. It may degrade DNA progressively from either end (5′ or 3′) of the chain by hydrolysis to produce deoxyribonucleoside monophosphates. Or it may degrade a chain by pyrophosphorolysis with inorganic pyrophosphate to produce deoxyribonucleoside triphosphates.

Until recently we understood little about how this enzyme works because we did not know enough about its physicochemical properties. We knew very little because our supplies of homogeneous enzyme were meager. Now with a simpler method for purification (*2*) and the invaluable use of the large-scale facilities of the New England Enzyme Center, we have had available 600 milligrams of homogeneous DNA polymerase obtained from 200 pounds (90 kilograms) of *E. coli*. The purpose of this article is twofold: (i) to assemble in a brief form the new physicochemical and functional observations concerning the pure enzyme; and (ii) to attempt an interpretation of these data in a model for the active center of the enzyme. This model is of course speculative, but it has helped us reconcile many hitherto confusing details and continues to suggest useful experiments.

Physicochemical Properties

The molecular weight of the homogeneous polymerase, determined by sedimentation equilibrium, is 109,000 (*2*). This large molecular weight and the presence of both polymerase and multiple nuclease activities suggest a subunit structure. However, the molecular weight measured by sedimentation equilibrium under denaturing and reducing conditions ($6M$ guanidine hydrochloride and $0.3M$ mercaptoethanol) was found to be the same as that of the native protein. Optical rotatory dispersion and velocity sedimentation studies showed that polymerase loses ordered structure in $6M$ guanidine hydrochloride, and would therefore be expected to be fully dissociated in this solvent.

More than 95 percent of the protein migrated as a single band on polyacrylamide-gel electrophoresis at pH 3.5, pH 8, and pH 11 (with or without $7M$ urea at pH 3.5 and pH 8). This result is most consistent with a structure composed of either a single polypeptide chain or of two or more identical subunits. However, the possibility of multiple, identical subunits is ruled out by the fact that polymerase contains, per 109,000 molecular weight, a single sulfhydryl group and a single disulfide group. The sulfhydryl group is probably not part of the active site because it can be modified either with iodoacetate or mercuric ion to give derivatives with full polymerase and exonuclease activity (3). The reaction with mercuric ion will give either a polymerase monomer, with one mercury atom per protein molecule, or, in the presence of a molar excess of enzyme, a dimer, with two protein molecules linked through a mercury atom. The dimer also has full activity. (The reaction of polymerase with a single atom of ^{203}Hg provides a convenient way of incorporating a radioactive label of about 10,000 counts per minute per microgram of enzyme, without affecting enzymatic activity. This label has served as a marker in DNA binding studies which will be described below.)

Amino acids (approximately 1000 per enzyme molecule) account for the dry weight. There is no evidence for any prosthetic group. The amino terminal residue, as determined by both the cyanate and the fluorodinitrobenzene procedures, is methionine.

Although these experiments indicate that DNA polymerase is a single polypeptide chain, the possibility still exists that there are subunits joined by nonpeptide linkages that resist disruption by guanidine hydrochloride–mercaptoethanol, or by urea, or by extremes of pH. The presence of blocked amino termini has also not been ruled out. There have been two recent reports of E. coli polymerase preparations with molecular weights in the range of 30,000 and 50,000, and these have been designated as possible subunits by Cavalieri (4) and by Lezius (5). However, their preparations are of relatively low specific activity and have not yet been fully characterized. If DNA polymerase is assumed to be roughly spherical, its diameter is calculated to be near 65 angstroms. The diameter of a DNA helix is about 20 angstroms.

Table 1. Influence of DNA structure on binding of DNA to DNA polymerase.

Conformation	Per DNA molecule	
	Nicks or ends	Polymerase molecules bound
d(AT)$_{12}$ oligomer		
Hairpin	1	1
ØX174 DNA		
Circular, single strand	0	20
Closed circular, duplex	0	< 0.1
Plasmid DNA		
Irreversibly denatured	0	21
3'-Hydroxyl nick	1	1
3'-Phosphate nick	5	6
T7 DNA		
Linear duplex	2.5	2.6
Single strand	2	240

DNA Binding Site

DNA binding to DNA polymerase (6) was studied with a variety of DNA structures (Fig. 2). The alternating copolymer of deoxyadenylate and deoxythymidylate (dAT) was partially digested by deoxyribonuclease. Oligomers were isolated from these digests either by gel filtration or by polyacrylamide-gel electrophoresis. Preparations were obtained with chain lengths of approximately 24 and 40 nucleotide residues [d(AT)$_{12}$ and d(AT)$_{20}$]; they were induced to assume "hairpin" conformations by melting and quick cooling at low ionic strength.

Binding was measured by sucrose density-gradient centrifugation of mixtures containing DNA labeled with ^{3}H or ^{32}P and polymerase labeled with ^{203}Hg. The mixtures were layered on top of the gradients, and the enzyme-DNA complexes were identified after sedimentation.

With excess enzyme, d(AT)$_{12}$ sedimented almost quantitatively with the enzyme, an indication of a very high binding affinity. The polymerase-dAT complex sedimented at 7.7S, compared to 6.1S for the free enzyme. With the dAT oligomer present in excess, all the enzyme sedimented as a complex

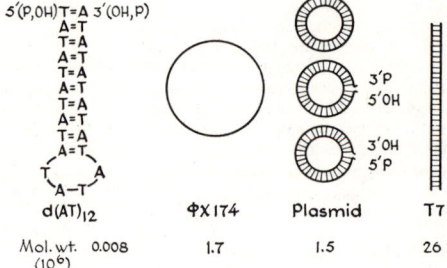

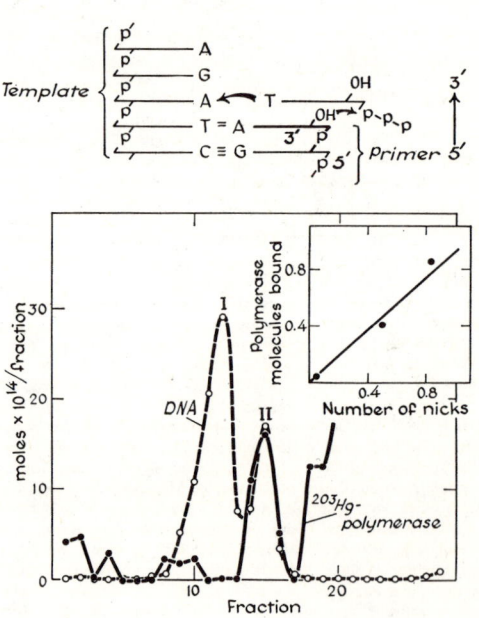

Fig. 1 (top left). Direction of chain growth catalyzed by DNA polymerase. Abbreviations are: A, adenine; C, cytosine; G, guanine; and T, thymine.

Fig. 2 (above). Various DNA structures used in studies of enzyme binding.

Fig. 3 (left). Binding of nicked plasmid DNA to the enzyme. Peak I contains intact, duplex, circular DNA in the supercoiled form; peak II contains the nicked form which is bound by polymerase. The inserted graph gives the number of polymerase molecules bound as a function of the number of nicks introduced by pancreatic deoxyribonuclease.

at 7.7S with an equimolar amount of dAT, an indication that DNA polymerase contains a single binding site for the dAT oligomer. Results identical to these were obtained with $d(AT)_{20}$. No difference was detected in the binding of oligomers terminated with a 3'-hydroxyl as compared with a 3'-phosphate group.

Binding of single-stranded circular ϕX174 DNA by polymerase resulted in about 20 enzyme molecules per molecule of DNA in the complex. A double-stranded, closed, circular plasmid DNA (7) was not bound at all by polymerase. However, when these duplex forms were denatured to make them single-stranded, they were bound by polymerase in proportion to their length and to the same extent as the single-stranded viral ϕX174 DNA.

We have introduced nicks into the plasmid DNA or the ϕX174 replicative form with pancreatic deoxyribonuclease. Such nicks have 3'-OH and 5'-P termini and are active points for replication (8). Nicks introduced with micrococcal nuclease produce 3'-P and 5'-OH termini and are not active points for replication; they inhibit replication (9). Regardless of the kind of nick, polymerase molecules formed complexes in numbers exactly equivalent to the number of nicks (Fig. 3, insert). The sedimentation pattern (Fig. 3) indicates that the polymerase molecules bound the nicked forms (II) and not the intact double circular forms of the plasmid (I).

The binding of DNA structures to polymerase is summarized in Table 1. There is no binding at all at helical regions. There is binding along single-stranded chains and to nicks and ends. In the case of the linear duplex DNA of bacteriophage T7, our preparation contained, on the average, one nick per two molecules; this explains why 2.6 molecules of the enzyme are bound per T7 DNA molecule.

Is binding of DNA at ends or nicks simply a consequence of fraying and single-strandedness in these regions, or is there more specific binding directed to the nucleotide termini at these points? We will assess this question in the formulation of a model for the active center of the enzyme.

Deoxyribonucleoside Triphosphate Binding Site

Is one, or more than one, molecule of deoxyribonucleoside triphosphate bound to a polymerase molecule? This and related questions are crucial to understanding the nature of the active center and the mechanism of polymerase action.

Triphosphate binding was studied by equilibrium dialysis (10). Scatchard plots for triphosphate binding showed that there is one binding site for each triphosphate and that the dissociation constants for the enzyme-triphosphate complexes were 12, 33, 81, and 147 micromoles per liter for deoxyguanosine, deoxyadenosine, deoxythymidine, and deoxycytidine triphosphates, respectively. Although the four triphosphates differ in the affinity of their binding, the interpretation of these particular values is necessarily limited. DNA template and primer were not present, and their influence on binding, which is likely to be profound, has yet to be assessed.

Is there a separate site for each of the four triphosphates, or does the enzyme have a single site for which all four triphosphates compete? Equilibrium dialyses were run with each of the six possible combinations of two triphosphates, and each of the pair was labeled distinctively. These competition experiments established that there is a single binding site on the enzyme for which all four triphosphates compete. Further explorations (10) of the specificity of binding in this site emphasize the primary importance of the triphosphate moiety and the only secondary importance of the sugar and base components.

The effects of DNA template and primer on the binding of triphosphates are difficult to study experimentally. An active template and primer invariably promote polymerization or are degraded, and binding measurements are thus complicated. Among the questions to be answered are whether a template confers specificity on triphosphate binding and what influence the primer terminus exerts on the entry and orientation of the triphosphate in the site.

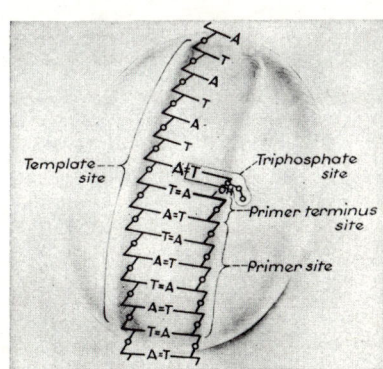
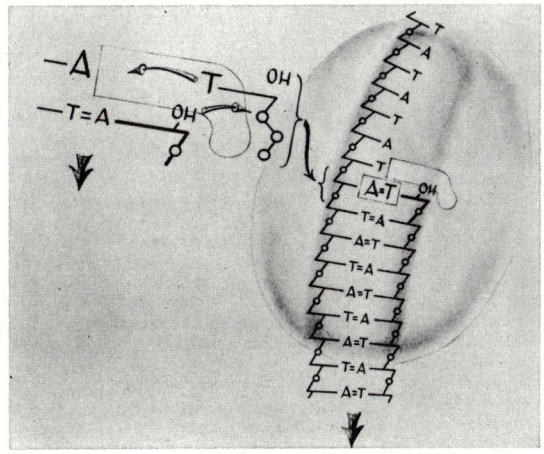

Fig. 4 (above). Sites in the active center of DNA polymerase. Fig. 5 (right). Mechanism of the polymerization step.

Multiple Functions of the Enzyme

At this point I would like to relate the results of the binding experiments to what we have learned about the catalytic properties of the enzyme in order to begin constructing a model of the many operations that may take place in the active center of the enzyme. These operations include: (i) the $5' \rightarrow 3'$ growth of a DNA chain by the polymerization of nucleotides; (ii) hydrolysis of a DNA chain from the 3'-OH end ($3' \rightarrow 5'$ direction); (iii) hydrolysis of a DNA chain from the 5' end ($5' \rightarrow 3'$ direction); (iv) pyrophosphorolysis of a DNA chain from the 3' end; and (v) exchange of inorganic pyrophosphate (PP_i) with the terminal pyrophosphate group of a deoxyribonucleoside triphosphate.

We picture the active center of the enzyme as some specially adapted polypeptide surface that recognizes and accommodates several nucleotide structures (Fig. 4). We will present evidence that, within the active center, there are at least five major sites. (i) There is a site for a portion of the template chain. This area binds the chain where a base pair is formed and for a distance of several nucleotides on either side of it. The chain is oriented with a particular polarity. It seems likely that this is the site where circular, single-stranded DNA is bound, but we are uncertain whether this site recognizes an extended or a tightly stacked, helical conformation. (ii) There is a site for the growing chain, the primer. The primer is oriented with a polarity opposite that of the template. (iii) There is a site with special recognition for the 3'-OH group of the terminal nucleotide of the primer, the primer terminus. We shall discuss this region later as a site for the hydrolytic and pyrophosphorolytic cleavage of the 3'-OH–terminated primer chain ($3' \rightarrow 5'$ direction). (iv) There is a site for a triphosphate, and (v) there is an additional site, to be considered later, which provides for hydrolytic cleavage of the 5'-P–terminated chain ($5' \rightarrow 3'$ direction).

The Polymerization Step

When a linear duplex is partially degraded from each 3'-OH end, as for example by the action of certain exonucleases, these denuded portions are repaired with great facility by all DNA polymerases (*11*). How is this accomplished?

The triphosphate is bound adjacent to the 3'-OH group on the terminal nucleotide of the primer and oriented so that it can be brought into direct contact and form a base pair with the template (Fig. 5). When the correct base pair is formed, a nucleophilic attack by the 3'-OH of the primer terminus on the innermost phosphate of the triphosphate takes place. A plausible model, for reasons to be mentioned below, assumes that movement or translation of the chain relative to the enzyme is concurrent with diester bond formation. As the primer terminus loses its 3'-OH group during transformation into a diester bond, it is no longer held in the primer terminus site. Through movement of the entire chain, the old primer terminus is replaced by the newly added nucleotide, which has a terminal 3'-OH group and is therefore held in the primer terminus site. (The new primer terminus is now ready to attack another triphosphate and add the next nucleotide.) Inorganic pyrophosphate is displaced only as formation of the diester bond is being completed, and the chain movement is translating the newly added nucleotide into the primer terminus site.

The possibility has been raised that the interaction between template and triphosphate is not direct, but rather allosteric in nature. However, since there is only one triphosphate binding site, it is difficult to imagine this site assuming four conformations, each absolutely specific for one of the triphosphates.

The basis for specificity of DNA polymerase very likely is not in the recognition by the enzyme of an incoming triphosphate, but rather in its demand for one of the four base pairs. All of the Watson-Crick base pairs contain regions of identical dimensions and geometry and are symmetrical. When the correct base pair is within the active site, the enzyme may respond, possibly by a change in conformation, so that the subsequent catalytic steps can then proceed. If an

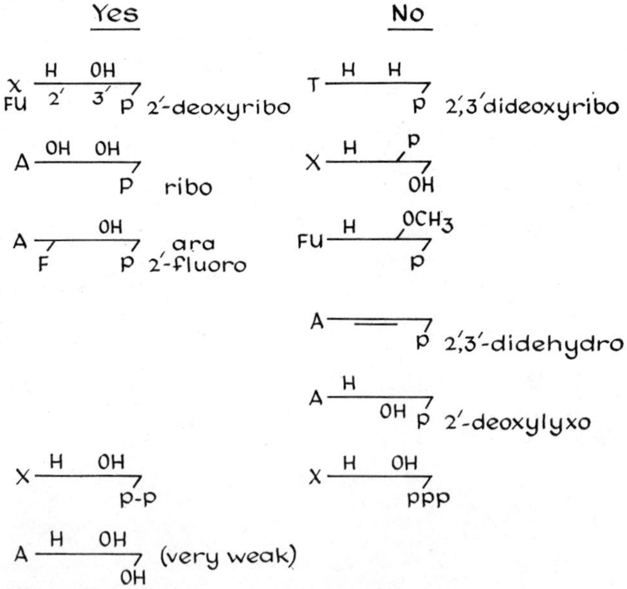

Fig. 6. Binding of monophosphates to DNA polymerase. Abbreviations are: X, the bases A, T, G, and C, as in the case of the 2'-deoxyribonucleoside 5'-monophosphate ($X \xrightarrow{H \ OH} P$); FU, fluorouracil; F, fluorine; ara, arabinosyl; and lyxo, lyxosyl.

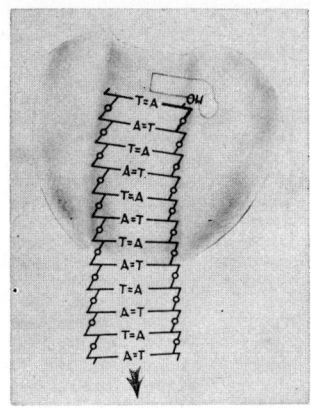
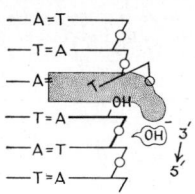
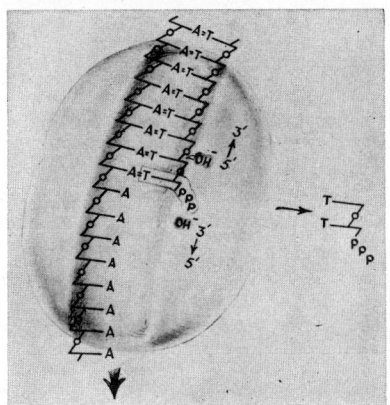

Fig. 7 (left). Binding of the end of a linear helix to the enzyme.

Fig. 8 (above). Binding of a nicked region of DNA to the enzyme.

Fig. 9 (right). Scheme for 5′ → 3′ exonucleolytic degradation by the enzyme.

incorrect triphosphate were to bind to the enzyme, the correct base pair could not be formed, there would be no conformational change, and the triphosphate would be rejected.

Primer Terminus Site

Evidence for the specificity of the primer terminus site comes from studies of the binding and functional behavior of several nucleotide analogs. Among the analogs studied are those that lack a 3′-OH group. Such an analog, if added to a chain, would prevent further chain growth (12, 13). Dideoxythymidine triphosphate is one example. It binds to the triphosphate site. One nucleotide is added per chain. The chain terminated with such a dideoxynucleotide is inert to further elongation; it is relatively inert to exonuclease action at the primer terminus end of the chain. In such a chain, as will be mentioned presently, attack by PP_i is inhibited. However, the chain can be degraded from the 5′ end (5′ → 3′ direction).

We interpret these results to mean that, without a 3′-OH group, the chain cannot bind properly in the primer site and is therefore not an effective substrate for exonuclease action. In keeping with this interpretation are the results of the following studies of the binding of monophosphate in what may prove to be the primer terminus site (14).

Each of the four common deoxyribonucleoside monophosphates binds to and competes for a single site on the enzyme. This site is, however, entirely distinct from the triphosphate site. Replacement of the 3′-OH group—by hydrogen as in dideoxythymidine monophosphate, by esterification with phosphate, or by O-methylation—prevents binding to this site. However, many other alterations in the nucleotide can be tolerated, provided that the 3′-OH group is in the "ribo" configuration (Fig. 6). For example, a certain arabinosyl nucleotide binds to the monophosphate site, whereas lyxosyl nucleotide does not.

In support of the interpretation that the 3′-OH monophosphates bind at the primer terminus site is the finding that these monophosphates inhibit hydrolysis of polynucleotides from the 3′ end of the chain. Other lines of evidence (15) have made it clear that there is only one primer terminus site serving for either polymerization or 3′ → 5′ hydrolysis.

Completion of a Linear or Circular Duplex

Picture a duplex in which template copying has been completed. What potential does such an intact, linear, double-stranded helix have for further replication? Consider a polymerase molecule which binds the end of such a linear duplex (Fig. 7). The primer strand is in its site with the 3′-OH group oriented in the primer terminus site; the strand is hydrogen-bonded to its complementary strand which is in the template site. But the template strand extends only as far as the primer terminus (Fig. 7). When a triphosphate enters the triphosphate site, there is no purine or pyrimidine base to serve as a template and thus no replication can take place. An intact linear duplex must therefore be inert.

Thus it appears that the replication of a linear duplex from its terminus, as pictured in the original Watson and Crick model, should not apply to DNA polymerase action in vitro, even for one of the strands. In recent studies (16) the DNA of phage T7 was prepared with care to avoid any internal breaks in this linear duplex and this DNA was essentially inert in supporting replication.

A special case of template copying is the replication of single-stranded circular DNA, such as the viral DNA of ϕX174 (17, 18). The circle provides no primer terminus and initiation of new strands by the enzyme does not take place readily (18). Therefore addition of an oligonucleotide which can anneal to the circle promotes replication by furnishing the necessary primer terminus (19). Copying then proceeds rapidly around the circle. The product is an incomplete circle. However, if a joining enzyme, called ligase, is present, the diester bond between the 3′-OH and 5′-P termini is made, and a fully covalent, double-stranded circle is produced (20). This synthetic molecule, as well as the double, circular molecules isolated from nature (Replicative Form I) do not bind to polymerase and are inert for replication.

Nicked Helix as the Functional Template Primer

Circular duplexes serve in vivo as chromosomes in bacteria (*E. coli* and *Bacillus subtilis*) and in viruses (polyoma), as replicative intermediates for other viruses (ϕX174, M13, and λ), and as bacterial episomes (*21*). Studies of ϕX174 replication indicate that a nicked form is the active replicative intermediate in vivo (*22*). Whereas the closed circular form is not replicated by DNA polymerase in vitro, the introduction by pancreatic deoxyribonuclease of one single-strand nick enables this DNA to bind to a polymerase molecule and converts the DNA to a favorable template and primer for replication (*16*). After replication has been initiated by introduction of a nick, the product, early or later in replication, is associated with the nicked form and is covalently linked to it (*16*).

As indicated earlier, an intact linear duplex, such as the DNA of phage T7, does not support replication. Upon the introduction of nicks by pancreatic deoxyribonuclease, the binding of these nicked duplexes to polymerase molecules and the appearance of productive sites for replication increase in direct proportion to the number of nicks introduced. In this instance, too, at least 90 percent of the DNA product is covalently attached to the primer (*16*).

How can the binding of a nicked region be visualized in the active center of the enzyme? The template and primer sites are filled. But the triphosphate site may not be vacant, and growth of a chain from the primer terminus is obstructed by the presence of the 5′-P–terminated strand, hydrogen-bonded to the template strand (Fig. 8). This dilemma might be resolved temporarily by the 3′ → 5′ exonuclease activity of the polymerase. Hydrolytic removal of the primer terminus nucleotide, accompanied by movement of the chain upward one nucleotide would open the triphosphate site for insertion and addition of a triphosphate. However, were this succession of events to take place nothing more would be achieved than the restoration of the original nicked region.

For progressive replication of the template to take place, the 5′-P–terminated strand must be displaced. Our evidence indicates that during the first phase of replication there is in fact a burst of hydrolysis of the template-primer, roughly matching the extent of replication (*23*). This hydrolysis is predominantly from the 5′ end of the DNA and entails degrading the chain from 5′ → 3′. The locus of this hydrolytic function appears to be distinct from that responsible for 3′ → 5′ hydrolysis.

Distinctive Sites for 3′ → 5′ and 5′ → 3′ Hydrolysis

Klett, Cerami, and Reich (*24*) were the first to recognize that polymerase preparations contained an exonuclease activity which degraded from the 5′ end of a chain. Their conclusions were based on studies with a synthetic block polymer resistant to hydrolysis from the 3′ end of the chain. A similar discovery was made independently when we tried to explain how DNA with 3′-P termini (introduced by micrococcal nuclease) and presumably insensitive to 3′ → 5′ exonuclease action was nevertheless extensively degraded (*25*). It became clear that such a 3′-P–terminated chain is degraded exclusively from the 5′ end. The principal products of extensive hydrolysis proved to be mononucleotides and an oligonucleotide which bore the 3′-P terminus.

With the recognition of this new property of DNA polymerase, an interesting possibility was raised. If a DNA chain were initiated *de novo* by DNA polymerase, its starting 5′ terminal should, as in the case of RNA polymerase, be marked by the initiating triphosphate. Yet attempts to identify such a triphosphate initiation point have not succeeded. It seemed possible that the 5′ → 3′ exonuclease activity of polymerase might act also on a chain terminated in a 5′-triphosphate and would therefore have erased a terminal triphosphate group even if it had been present initially.

In order to test this possibility, a polydeoxythymidylate of about 300 residues was synthesized containing a ^{32}P-labeled triphosphate group at the 5′ terminus and ^{3}H in the thymidine residues (*26*). This polynucleotide was degraded by polymerase from the 3′ end but *not* from the 5′ end. However, when annealed to form a helix wtih a polydeoxyadenylate chain, it was degraded rapidly from both ends. After an incubation period limited to only 10 seconds, 90 percent of the ^{32}P was liberated from the polymer, whereas less than 5 percent of the ^{3}H was released. Most remarkably, the principal ^{32}P product proved to be not deoxythymidine triphosphate as expected, but instead a dinucleoside tetraphosphate (Fig. 9).

We interpret these results as follows. There is a hydrolytic site for progressive 5′ → 3′ release of mononucleotides from the 5′ end of a strand in a location just above the triphosphate site. When the chain is terminated in a triphosphate, the close resemblance of this terminus to a deoxyribonucleoside triphosphate directs its binding in the triphosphate site. As a consequence, the initial product is not a mononucleotide but rather a dinucleotide, the dinucleoside tetraphosphate. Subsequently as the chain moves downward one nucleotide at a time the products are principally mononucleotide residues.

The fact that the 5′ → 3′ degradation

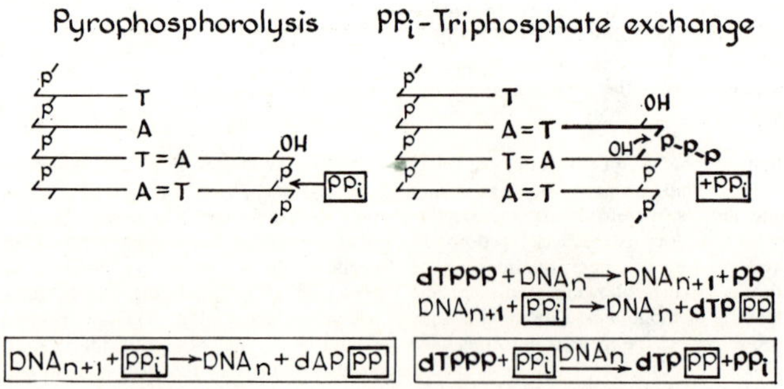

Fig. 10. Formulation of the pyrophosphorolysis and PP$_i$-triphosphate exchange reactions.

requires a helical structure indicates that this site cannot properly orient a single-stranded oligonucleotide chain. Because the site is occupied when such a chain is annealed to a complementary strand, we infer that the enzyme accommodates this complementary strand in the upper region of the template site.

The failure of the enzyme to degrade a 5′-terminated chain (from 5′ → 3′) when it is single stranded may help explain the stage in replication of helical DNA when synthesis proceeds without concomitant 5′ → 3′ hydrolysis. As the 3′-OH chain advances by growth along the template, the 5′ chain may be displaced from the active site for a stretch of several nucleotides and thus be rendered insusceptible to hydrolysis. The 5′ chain may be peeled back until some point when, for obscure reasons, it competes successfully for the template function by attracting the growing chain to switch templates (as in Fig. 13). Such a sequence of events would produce a covalent fork in the growing chain and has been suggested (27) as part of a mechanism to account for the branched structure and readily renaturable character of DNA synthesized on a helical template primer.

Pyrophosphorolysis and Pyrophosphate Exchange

Pyrophosphorolysis, the capacity of DNA polymerase to degrade DNA chains with PP_i, reaches a steady state when the accumulation of triphosphates supports synthesis at a rate that balances their removal (15). The enzyme also supports the exchange of PP_i into the β, γ groups of a triphosphate (15). This reaction can occur with only a single triphosphate present, but otherwise requires all the primer and template conditions demanded of replication (Fig. 10). All the evidence fails to suggest the formation of a nucleotidyl-enzyme intermediate.

The behavior of chains terminated with a dideoxynucleotide suggests a mechanism of pyrophosphorolysis and PP_i exchange (12, 15). Such chains are relatively insusceptible to degradation by nucleophilic attack of PP_i, just as they are to that of OH^- (nuclease). Furthermore, a triphosphate analog lacking a 3′-OH group supports little, if any, PP_i exchange. Inasmuch as such an analog can be added to a chain, it would seem that PP_i exchange with the triphosphate does not occur entirely in the triphosphate site. Rather, the lack of a 3′-OH group prevents binding of the transition state formed by attack of the primer terminus on the triphosphate, and the terminal nucleotide in this transition state cannot be displaced efficiently by PP_i or by OH^-.

Pyrophosphorolysis is therefore taken to be a reversal of the polymerization step, including the concerted chain movement. Inorganic pyrophosphate exchange appears to be the result of a sequence of a polymerization step and a pyrophosphorolytic step repeated many times over (Fig. 10). Because the rate of PP_i exchange is considerably faster than that of pyrophosphorolysis (15), we suggest that the attack by PP_i occurs at a transition state short of completion of the polymerization step, and that this transition state is attained more readily from the direction of polymerization.

Insights from an Enzyme Modified by Acylation

Chemical alterations of DNA polymerase are beginning to provide important clues about its structure and function (3). Inasmuch as alkylation or Hg substitution of the single sulfhydryl group of DNA polymerase does not alter any of its activities, we assume that this group is relatively remote from the active center. However, acylation of the enzyme with N-carboxymethylisatoic anhydride results in a highly fluorescent derivative with markedly altered functional properties.

A derivative with 11 N-carboxymethylisatoyl groups had only 0.2 percent of the original polymerase activity but had 920 percent of the exonuclease activity when measured with DNA as primer or substrate at pH 7.4. Binding measurements and kinetic studies indicated that a major effect of this modification was a marked reduction in the affinity of deoxyribonucleoside triphosphate substrates for their site. Possibly, one or a few of the 61 lysine residues in the enzyme are located in this site, and their acylation had severely damaged the function of the site.

The remarkable increase in exonuclease activity implies that concomitant changes in the interaction of the enzyme with DNA are probably also involved. It remains to be determined whether one or both of the nuclease activities have been modified and how closer studies of the altered enzyme with various DNA's as substrates may elucidate the nature of the active center.

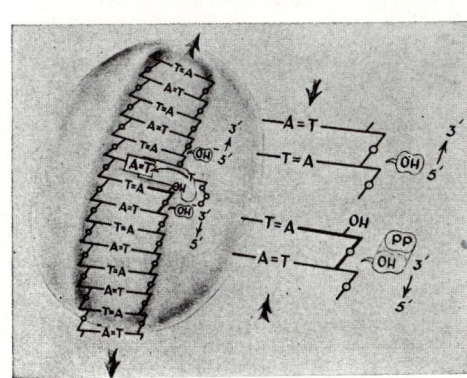

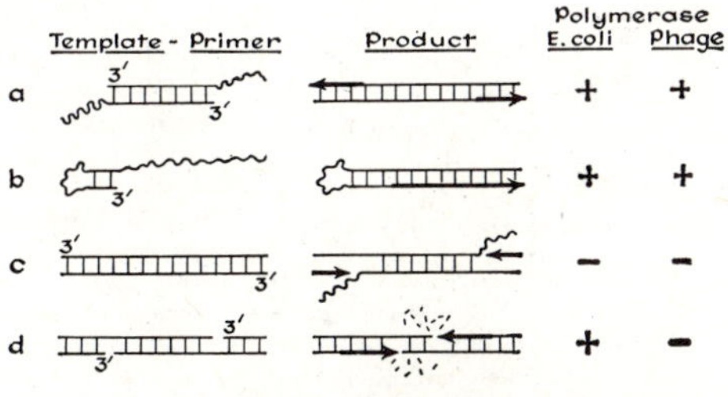

Fig. 11 (left). Model to account for the multiple functions of DNA polymerase within a single active center.

Fig. 12 (above). Template-primer requirements of the *E. coli* and phage-induced DNA polymerases.

Multiple Sites in One Active Center

A recapitulation of the several functional and physicochemical features of DNA polymerase is provided in a model proposed in Fig. 11. The template and primer functions in polymerization, the distinctive $3' \to 5'$ and $5' \to 3'$ exonuclease activities, and the various binding capacities of DNA are oriented in one active center. While alternative models have been considered (28–31), and some of these still have attractive features, the one we offer here goes farthest in explaining available data. A model by Beyersmann and Schramm (28), based exclusively on kinetic data, proposes an identical site for pyrophosphorolysis and hydrolysis and one active center for the degradative and polymerization functions of the enzyme. However, their model also proposes the pyrophosphorolysis of 3'-P-terminated chains and roles for specific inhibitors that are inconsistent with our formulation.

Yet, our model, even if proven correct in basic outline, is still far too simple to account for many of the enzyme's features. For example, the capacity of the enzyme to discriminate secondary structure both in polymerization and in nuclease activities will require a more striking three-dimensional recognition of template and primer strands within the active center. To accommodate the screwlike translation of an essentially helical structure through the active center would demand a surface adapted to one or more turns of the helical duplex. The important influences of specific metal cations, ionic strength, and temperature on the enzyme functions have not even been considered. Finally, it seems likely that enzymes and factors related to the cellular functions of DNA polymerase will interact with it, at or near the active center, in the regulation of these functions.

Comparison of Various DNA Polymerases

Studies analogous to those described here for the *E. coli* enzyme have not yet been carried out on DNA polymerase from other sources. However, a comparison of the template-primer requirements of the *E. coli* enzyme with those of enzymes induced by infection with phages T2 (8), T4 (32), and T5 (33) is of considerable interest

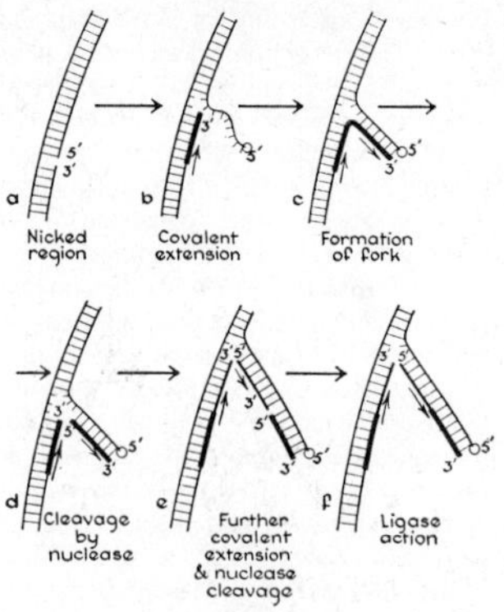

Fig. 13. Speculative scheme for unidirectional replication of a duplex chain.

(Fig. 12). Both the bacterial and the phage polymerases can utilize a double-stranded DNA helix, partially denuded from each 3'-OH end by exonuclease III (Fig. 12a). The helix is restored by replication to its original length by extention of the 3' strands at each end. Similarly, both types of enzyme replicate single-stranded DNA. The latter, upon forming a looped structure with a short 3'-OH end, is converted by replication to a hairpin-like product (Fig. 12b).

The bacterial enzyme was regarded as different from the phage-induced polymerases in its apparent ability to utilize a linear segment of double-stranded helical DNA. This difference had been ascribed to the unique capacity of the *E. coli* enzyme to initiate a DNA strand *de novo* with the 3'-OH strand of the DNA duplex as template (Fig. 12c). It seems likely from the foregoing discussion that this earlier interpretation (34) was mistaken and that neither kind of polymerase can employ an intact linear duplex.

What seems clearer now is that a segment of duplex DNA which serves as template-primer for the *E. coli* enzyme does so by virtue of the nicked region that it contains and not by supporting new chain synthesis at its ends (Fig. 12d). It follows therefore that the phage-induced enzymes cannot exploit such nicked regions in the DNA. In support of this formulation are the long-standing observations that "activation" of duplex DNA by pancreatic deoxyribonuclease (that is, introduction of 3'-OH nicks) increases its template-primer capacity 10- to 20-fold for the *E. coli* enzyme while providing no significant improvement for the phage-induced enzymes (8).

How does one explain the inability of phage-induced enzymes to replicate at a nicked region of a duplex? Let us assume that the $5' \to 3'$ degradation by the $5' \to 3'$ nuclease function of the *E. coli* enzyme in the nicked region clears a path for the advancing synthetic chain, and that this is an essential step in the initiation of replication in vitro. Although the phage-induced polymerases are known to degrade polynucleotides from a $3' \to 5'$ direction, they may not possess the $5' \to 3'$ nuclease activity. Recent tests (26) show that there is, in fact, a striking absence of this $5' \to 3'$ activity in the phage T4-induced enzyme. Perhaps such a nuclease activity is present in the phage-infected cell, but has been eliminated on purification of the polymerase.

Inasmuch as the phage-induced DNA polymerase, as well as that from *E. coli*, has the $3' \to 5'$ nuclease activity, is there some physiological purpose attributable to this function? Again we can only conjecture. It appears that both nuclease activities are favored by some destabilization of the tight helical structure (23). Removal of DNA at or near disordered regions such as those produced by irradiation might then be performed by one of the nuclease activities. There is also the intriguing possibility that $3' \to 5'$ hydrolysis provides an opportunity for double-checking and editing to eliminate any newly polymerized nucleotide member of a poorly stacked or faulty base pair.

A Hypothetical Scheme for Replication Process in vivo

How might DNA polymerase serve a physiological role in replication? What follows is speculative. Bacterial chromosomes are duplex circular structures and, when intact, are inert in replication. Introduction of a nick, possibly at a specific site, starts replication. DNA polymerase binds at the nick and replication proceeds by covalent extension of the 3'-OH end (Fig. 13, a and b). The 5' end may be degraded to some extent by $5' \to 3'$ nuclease action, or, if displaced, is freed from further attack. Fixation to some membrane site (35, 36) may facilitate displacement and preservation of the 5' strand.

Replication proceeds for some distance and then switches to the complementary strand as template to form a fork; the fork is then cleaved by an endonuclease (Fig. 13, c and d). A repetition of this sequence leads to interruptions or small pieces of DNA near the replicating fork (Fig. 13e). Such pieces have been isolated by Okazaki and co-workers (37) at or near the nascent replicating region. These interruptions are sealed by ligase (Fig. 13f).

If this scheme were essentially correct, it would explain how one enzyme, DNA polymerase, replicating exclusively in a $5' \rightarrow 3'$ direction, would copy, almost simultaneously, the two maternal strands of opposite polarity. Examination of dividing bacteria at a gross level by autoradiography (38) or by gene duplication (39) makes it appear that there is a simultaneous, sequential replication of both strands. However, at the nucleotide level, as proposed in this scheme, the replicative action is staggered, alternating from one strand to the other.

Summary

DNA polymerase, a homogeneous protein of molecular weight 109,000, appears to be a single polypeptide chain. The enzyme contains one triphosphate substrate binding site and one site for binding a nicked region of duplex DNA. A model of the active center of the enzyme has been proposed (Fig. 11) in which there are distinctive sites for the template strand, the primer strand, the 3'-hydroxyl primer strand terminus, the triphosphate substrate, and the 5'-phosphate–terminated strand beyond the nick (point of scission). The model attempts to account for the various synthetic and degradative functions within the closely related sites in the active center of the enzyme.

Initiation of replication is favored at a nicked region of a duplex. In the first phase of replication, extension of the 3'-hydroxyl primer strand appears to be related to the $5' \rightarrow 3'$ hydrolytic removal of the 5'-phosphate–terminated strand. The failure of phage-induced DNA polymerases to initiate replication at a nicked region may be due to the lack of a $5' \rightarrow 3'$ nuclease function in the phage enzyme. A speculative model for helix replication, in vivo (Fig. 13), suggests how DNA polymerase, in conjunction with endonuclease and ligase, may achieve the sequential and almost simultaneous replication of both strands of a helix.

References and Notes

1. P. T. Englund, M. P. Deutscher, T. M. Jovin, R. B. Kelly, N. R. Cozzarelli, A. Kornberg, *Cold Spring Harbor Symp. Quant. Biol.*, in press.
2. T. M. Jovin, P. T. Englund, L. L. Bertsch, *J. Biol. Chem.*, in press.
3. T. M. Jovin, P. T. Englund, A. Kornberg, *ibid.*, in press.
4. L. F. Cavalieri and E. Carroll, *Proc. Nat. Acad. Sci. U.S.* **59**, 951 (1968).
5. A. G. Lezius, S. B. Hennig, C. Menzel, E. Metz, *Eur. J. Biochem.* **2**, 90 (1967).
6. P. T. Englund, R. B. Kelly, A. Kornberg, *J. Biol. Chem.*, in press.
7. N. R. Cozzarelli, R. B. Kelly, A. Kornberg, *Proc. Nat. Acad. Sci. U.S.* **60**, 992 (1968).
8. H. V. Aposhian and A. Kornberg, *J. Biol. Chem.* **237**, 519 (1962).
9. C. C. Richardson, C. L. Schildkraut, A. Kornberg, *Cold Spring Harbor Symp. Quant. Biol.* **28**, 9 (1963).
10. P. T. Englund, J. A. Huberman, T. M. Jovin, A. Kornberg, *J. Biol. Chem.*, in press.
11. C. C. Richardson, R. B. Inman, A. Kornberg, *J. Mol. Biol.* **9**, 46 (1964).
12. M. R. Atkinson, M. P. Deutscher, A. Kornberg, A. Russell, J. Moffatt, in preparation; M. R. Atkinson, J. A. Huberman, R. B. Kelly, A. Kornberg, *Fed. Proc.*, in press.
13. L. H. Toji and S. S. Cohen, personal communication.
14. J. A. Huberman, M. R. Atkinson, A. Kornberg, unpublished results.
15. M. P. Deutscher and A. Kornberg, *J. Biol. Chem.*, in press.
16. R. B. Kelly, N. R. Cozzarelli, A. Kornberg. unpublished results.
17. S. Mitra, P. Reichard, R. B. Inman, L. L. Bertsch, A. Kornberg, *J. Mol. Biol.* **24**, 429 (1967).
18. M. Goulian and A. Kornberg, *Proc. Nat. Acad. Sci. U.S.* **58**, 1723 (1967).
19. M. Goulian, *ibid.* **61**, 284 (1968); *Cold Spring Harbor Symp. Quant. Biol.*, in press.
20. ———, A. Kornberg, R. L. Sinsheimer, *Proc. Nat. Acad. Sci.. U.S.* **58**, 2351 (1967).
21. J. Vinograd and J. Lebowitz, *J. Gen. Physiol.* **49** (6), 103 (1966).
22. R. Knippers, T. Komano, R. L. Sinsheimer, *Proc. Nat. Acad. Sci. U.S.* **59**, 577 (1968); T. Komano, R. Knippers, R. L. Sinsheimer, *ibid.*, p. 911.
23. M. P. Deutscher, R. B. Kelly, A. Kornberg, unpublished results.
24. R. P. Klett, A. Cerami, E. Reich, *Proc. Nat. Acad. Sci. U.S.* **60**, 943 (1968).
25. M. P. Deutscher and A. Kornberg, *J. Biol. Chem.*, in press.
26. N. R. Cozzarelli, R. B. Kelly, A. Kornberg, in preparation.
27. C. L. Schildkraut, C. C. Richardson, A. Kornberg, *J. Mol. Biol.* **9**, 24 (1964).
28. D. Beyersmann and G. Schramm, *Biochim. Biophys. Acta* **159**, 64 (1968).
29. H. Jehle, *Proc. Nat. Acad. Sci. U.S.* **53**, 1451 (1965).
30. H. E. Kubitschek and T. R. Henderson, *ibid.* **55**, 512 (1966).
31. A. Kornberg, in *Regulation of Nucleic Acid and Protein Biosvnthesis*, V. V. Koningsberger and L. Bosch, Eds. (Elsevier, Amsterdam, 1967), p. 22.
32. M. Goulian, Z. J. Lucas A. Kornberg, *J. Biol. Chem.* **243**, 627 (1968).
33. C. D. Steuart, S. R. Anand, M. J. Bessman, *ibid.*, p. 5319.
34. S. Mitra and A. Kornberg, *J. Gen. Physiol.* **49** (6), 59 (1966).
35. F. Jacob, S. Brenner, F. Cuzin, *Cold Spring Harbor Symp. Quant. Biol.* **28**, 329 (1963).
36. W. Gilbert and D. Dressler, *ibid.*, in press.
37. R. Okazaki, T. Okazaki, K. Sakabe, K. Sugimoto, A. Sugino, *Proc. Nat. Acad. Sci. U.S.* **59**, 598 (1968).
38. J. Cairns and C. I. Davern, *J. Cell. Physiol.* **70** (Suppl.), 65 (1967).
39. N. Sueoka, *Mol. Genet.* **2**, 1 (1967).
40. The article is adapted from the inaugural Stanhope Bayne-Jones Lecture deliverd at Johns Hopkins University School of Medicine, 19 November 1968. For the recent work which forms the basis of this lecture, I want to acknowledge the contributions of M. R. Atkinson, N. R. Cozzarelli, M. P. Deutscher, P. T. Englund, J. A. Huberman, T. M. Jovin, and R. B. Kelly. I want particularly to cite Jovin's initiative and skill in the physicochemical aspects of the work. I am also appreciative of the skillful efforts of Mrs. L. M. Follett, director of medical illustration at Stanford.

ENZYMATIC SYNTHESIS OF DNA, XXIV.
SYNTHESIS OF INFECTIOUS PHAGE φX174 DNA*

By Mehran Goulian,† Arthur Kornberg, and Robert L. Sinsheimer

DEPARTMENT OF BIOCHEMISTRY, STANFORD UNIVERSITY SCHOOL OF MEDICINE, PALO ALTO, AND DIVISION OF BIOLOGY, CALIFORNIA INSTITUTE OF TECHNOLOGY, PASADENA

Communicated September 25, 1967

Past attempts at *in vitro* replication of transforming factor present in DNA have given negative or inconclusive results.[1-3] Rigid proof was lacking that template material had been excluded from the synthetic product. Even if a rigorous demonstration of net synthesis of transforming factor for a given genetic marker were forthcoming, it would still prove only that some relatively short sequence of nucleotides, sufficient for replacement of the mutant locus, had been synthesized. If enzymatic synthesis of infectious bacteriophage DNA were achieved, it would be made clear at once that relatively few, if any, mistakes had been made in replicating a DNA sequence of several thousand nucleotides.

Escherichia coli DNA polymerase can replicate single-stranded circular DNA from phage M13 or φX174[4] and in conjunction with a polynucleotide-joining enzyme produces a fully covalent duplex circle.[5] Analyses of this product by equilibrium and velocity sedimentation and by electron microscopy have shown it to be indistinguishable, except for supercoiling, from replicative forms (RF)[6] of the viral DNA.[5] By substitution of bromouracil for thymine in the complementary strand ((−) circle), it should be possible on the basis of density difference to isolate this strand from the duplex circle and determine whether it has the infectivity known to reside in (−) circles.[7,8]

This report will describe: (1) the isolation of infective, synthetic (−) circles from the partially synthetic replicative form, (2) the ability of the isolated (−) circles to serve as templates for the production of infective, completely synthetic duplex circles, and (3) the isolation of infective, synthetic (+) circles from the latter.

Thus, DNA polymerase carries out the relatively error-free synthesis of the φX174 genome from the four deoxyribonucleoside triphosphates on direction from phage DNA templates.

Results.[9]—*Isolation of synthetic* (−) *circle and test of infectivity:* A duplex circle was synthesized by replicating H^3-φX174 DNA with DNA polymerase in the presence of a polynucleotide-joining enzyme. Details for the production and isolation of this partially synthetic RF, containing $\overline{BU}$ and P^{32} in the (−) circle, were described in an earlier report.[5] Separation of the synthetic (−) circle from the duplex form followed the plan outlined in Figure 1. The duplex circles were exposed to pancreatic DNase to an extent sufficient to produce a single scission in one of the strands in about half of the molecules. The resulting mixture of intact and nicked molecules was denatured by heating. The mixture, which now contained circular and linear H^3-T (+) strands, and P^{32}-$\overline{BU}$ (−) strands, in addition to intact RF, was fractionated by equilibrium density-gradient sedimentation in CsCl (Fig. 2).[10] Three peaks of radioactivity were evident, corresponding, in order of decreasing density, to single-stranded DNA containing $\overline{BU}$, a duplex hybrid

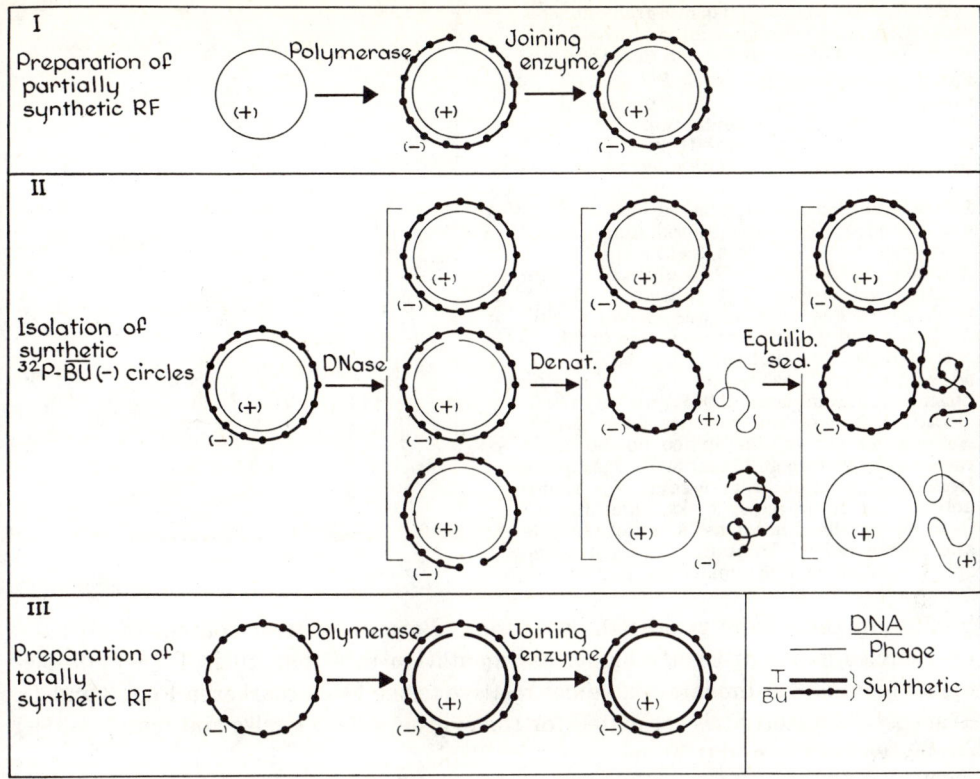

FIG. 1.—Schematic representation of the preparation of synthetic (−) circles and RF. For details see text and Figs. 2 and 6.

of $\overline{BU}$ and T, and single strands containing T, with mean densities of 1.809, 1.747, and 1.722 gm/ml, respectively. These values may be compared to the values of 1.732 and 1.725 previously determined[13] for the native hybrid and T (+) single strands prepared *in vivo* or to the calculated values[14] of 1.815 and 1.753 for the $\overline{BU}$ (−) strands, and the hybrid, respectively. In addition to the three peaks, there was an area on the heavy side of the hybrid zone, which in other experimental trials appeared as a more distinct shoulder and is attributable to some duplex circles that had failed to renature after the heat treatment (Fig. 2).

Inasmuch as the (−) circle is infectious in the spheroplast assay,[7, 8] it was possible to test the enzymatically synthesized material directly for biologic activity. Four peaks of infectivity were found (Fig. 2). One corresponded to the position of heavy, P^{32}-$\overline{BU}$, synthetic (−) single strands and another to that of light, H^3-T (+) single strands. Specific infectivity values for the single-stranded regions could not be determined from these data because there was an unknown quantity of linear strands. The P^{32}-$\overline{BU}$ and H^3-T peaks were therefore each subjected to velocity sedimentation in a neutral, low-salt sucrose gradient to give a partial separation of the circles from linear forms. As seen in Figure 3 for the P^{32}-$\overline{BU}$ (−) strands, and in Figure 4 for the H^3-T (+) strands, the infective material was found, in each case, in the leading shoulder of the peak which contains the circles

FIG. 2.—Equilibrium density-gradient sedimentation analysis of partially synthetic RF after limited DNase action and denaturation. Partially synthetic RF, with P^{32} and $\overline{BU}$ in the synthetic ($-$) strand,[5] was incubated for 20 min at 20° at a concentration of 0.1 mM, in 0.2 ml of 10% glycerol–10 mM Tris HCl (pH 7.6)–2 mM $MgCl_2$–0.25 mμg/ml pancreatic DNase. (The DNase (Worthington 1× recrystallized), 5 mg/ml in 0.01 N HCl, was stored at 0°[11] and diluted immediately before use in 10 mM Tris acetate (pH 5.5)–5 mM $MgCl_2$–0.2 M KCl–50% glycerol.) The reaction was stopped by addition of EDTA to 8 mM. The mixture was heated at 90° for 2 min and adjusted to a volume of 9.8 ml with 0.01 M Tris HCl (pH 7.6); EDTA was added to 1 mM, as well as 1 mg of bovine plasma albumin and 9.961 gm of CsCl. Centrifugation of this mixture ($\rho = 1.750$) was carried out in the Spinco no. 50 angle rotor at 45,000 rpm at 25° for 50 hr. Aliquots from each fraction were assayed for radioactivity on filter paper disks,[5] and for infectivity by the spheroplast assay of Guthrie and Sinsheimer.[12] Inhibition by CsCl in the spheroplast assay was avoided by dilution.

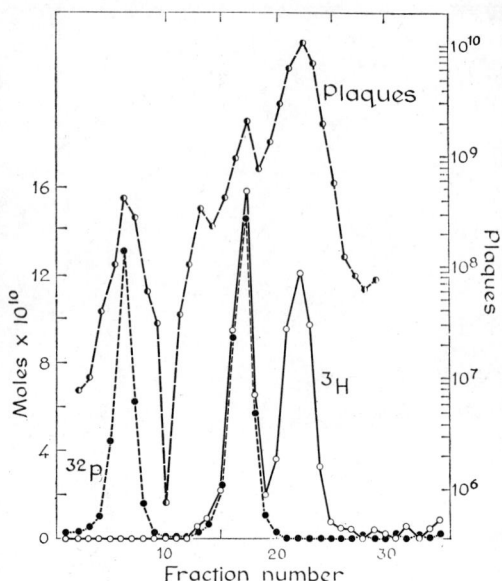

because of their more rapid sedimentation. Because of their content of $\overline{BU}$, the ($-$) circles had a distinctly higher sedimentation rate than their T (+) complements (compare sedimentation values relative to the DNA marker in Figs. 3 and 4). The specific infectivities estimated for the synthetic ($-$) circles and template (+) circles were 0.074 and 0.80, respectively (Table 1).

The other two peaks of infectivity in Figure 2 corresponded to the position of denatured and native forms of duplex hybrid molecules. Their respective specific infectivities were 0.066 and 0.012 (Table 1).

Proof that the infectivity of the P^{32}-$\overline{BU}$ peak resides in the enzymatically synthesized DNA: (1) A peak of infectivity coincides with the P^{32}-$\overline{BU}$ peak[17] in the density gradient (Fig. 2) and is separated from neighboring peaks. (2) Phage (+) circles are absent from the single-stranded, P^{32}-$\overline{BU}$ peak as judged by the absence of detectable H^3-labeled material. In view of the sensitivity of the radioactivity measurements, the *upper* limit for the amount of template material in the synthetic peak is 8 $\mu\mu$moles/ml; this concentration is one-tenth of that necessary to account for the infectivity of the peak. (3) In velocity sedimentation in sucrose gradients, the peak of infectivity corresponds to the position of intact P^{32}-$\overline{BU}$ ($-$) circles, and sediments more rapidly because of the presence of $\overline{BU}$ than the analogous peak of intact H^3-T (+) circles. (4) The photoinactivation of P^{32}-$\overline{BU}$ ($-$) DNA as compared with H^3-T (+) DNA (Fig. 5) demonstrates the more rapid inactivation of most of the infectious particles in the CsCl gradient peak corresponding to P^{32}-$\overline{BU}$ ($-$) strands and is consistent with the known greater photosensitivity of $\overline{BU}$-containing DNA.[18] The presence of approximately 5 per cent of the infectious material displaying an inactivation rate similar to that of T DNA[19] (Fig. 5) indicates the extent of contamination by T (+) circles. Inasmuch as the (+) circles of Figure 2 have about ten times the specific infectivity of these ($-$) circles, the residual content of phage DNA in the $\overline{BU}$ fraction is estimated to be closer to 0.5 per cent than 5 per cent.

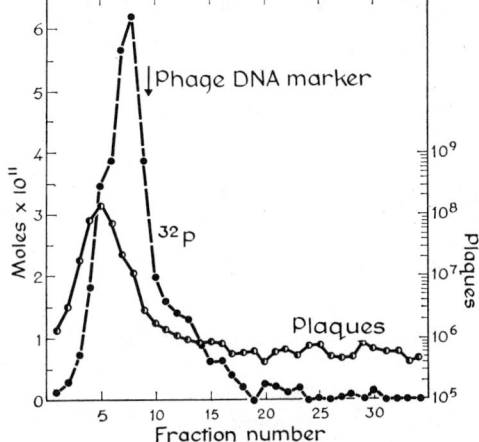
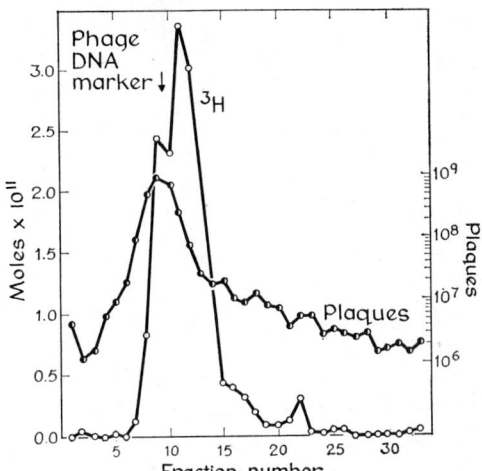

FIG. 3.—Velocity sedimentation of P^{32}-$\overline{BU}$, (−) synthetic DNA derived from partially synthetic RF. The P^{32}-$\overline{BU}$ peak fractions were pooled, dialyzed against 2 mM Tris HCl (pH 7.6)–0.2 mM EDTA and then concentrated five fold to a volume of 0.1 ml by rotary evaporation under reduced pressure. An aliquot of 20 μl was centrifuged in a 5–20% sucrose gradient in 5 mM NaCl–5 mM Tris HCl (pH 7.6)–1 mM EDTA, at 60,000 rpm and 10° for 360 min. H^3 was not detectable (<0.3 μμmole/fraction). The position of ϕX174 DNA was obtained from a separate tube containing this DNA as marker.

FIG. 4.—Velocity sedimentation of H^3-T, (+) phage DNA derived from partially synthetic RF. The H^3-T peak (Fig. 2) was treated as described in Fig. 3 for P^{32}-$\overline{BU}$ except that half as much H^3-T was placed on the sucrose gradient.

Replication of synthetic (−) *circle, isolation of fully synthetic replicative forms, and a test of infectivity:* The synthetic, P^{32}-$\overline{BU}$ (−) circles, separated from phage (+) circles, could now be used as templates for the production of fully synthetic RF (Fig. 1) which proved to be infective. Incubation conditions for synthesis of the RF were as previously employed,[5] except that H^3-dCTP was the labeled sub-

TABLE 1

INFECTIVITIES OF NATURAL AND SYNTHETIC ϕX174 DNA

	Plaques (ml^{-1} × 10^{-8})	DNA (μμmole ml^{-1})	Specific infectivity (plaques/ particle)	Relative infectivity	Ref.
(+) Circle, natural	37	64	0.80*	1.0	Fig. 4
(−) Circle, natural				~0.2	8
" " , synthetic	6.2	150	0.074*	0.09	Fig. 3
RF (native), natural				0.05	15
" " , part. synthetic	6,000	200,000	0.058	0.07	†
" " , part. synthetic	61	9,100	0.012	0.01	Fig. 2
RF (denat.), natural				1.0	16
" " , part. synthetic	9.5	260	0.066	0.06	Fig. 2
" " , fully synthetic	24	120	0.36	0.3	Fig. 6

Specific infectivity was calculated on the basis of 1.1 × 10^8 particles/μμmole of nucleotide residues for single-stranded molecules and half that value for the duplexes. *Relative infectivity* of the natural (+) circle was arbitrarily taken as 1.0 and the other figures adjusted, with inclusion of a correction, for variations between different assays, from phage DNA standards that were run in each assay. Technical difficulties resulting from the low concentrations of DNA have thus far prevented reliable estimates of the specific infectivities of native, fully synthetic RF and synthetic (+) circles.
* Includes a correction for estimated contamination with linear forms.
† Sample assayed prior to exposure to DNase as in Fig. 2.

strate, dTTP replaced dBUTP, and the pooled P^{32}-$\overline{BU}$ (−) peak from the CsCl gradient (Fig. 2) was the template. Evidence that a duplex circle was synthesized was obtained by velocity sedimentation analysis in an alkaline sucrose gradient (Fig. 6). The hybrid peak contained H^3 and P^{32} in approximately equimolar amounts and had the S value expected of a covalent duplex circle in alkali. The infectivity coincided exactly with the radioactivity, and the specific infectivity values were within the range expected for the denatured form of natural RF (Table 1). Additional evidence for the covalent duplex structure was obtained by density-gradient centrifugation in the presence of ethidium bromide (Fig. 7). This analysis was preceded by an initial density-gradient centrifugation with ethidium bromide in which a peak of higher buoyant density was identified as corresponding to the duplex covalent zone by alkaline sucrose gradient analysis of each fraction (legend to Fig. 7).

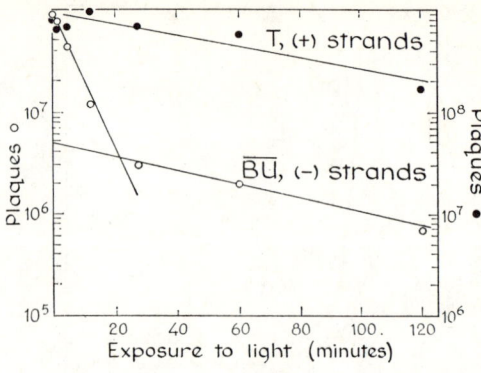

FIG. 5.—Photoinactivation of synthetic P^{32}-$\overline{BU}$ DNA and H^3-T, phage DNA. The P^{32}-$\overline{BU}$ (−) and H^3-T, (+) peaks (Fig. 2), dialyzed and concentrated as described in Fig. 3, were each diluted into 10 mM Tris HCl (pH 7.6)–1 mM EDTA. The diluted DNA's were exposed in identical fashion to a 15-watt daylight fluorescent tube at a distance of 3 cm; 0.05-ml aliquots were placed in the dark at the indicated times and subsequently assayed for infectivity. The ratio of the initial slope for $\overline{BU}$-DNA to that for T-DNA is 12.6.

Isolation of a synthetic (+) circle from the fully synthetic replicative form: A procedure similar to the one employed to separate synthetic (−) circles from partially synthetic RF forms was used (Fig. 1). A limited digestion by pancreatic DNase, followed by alkaline denaturation, achieved the release of H^3-T (+) circles from the fully synthetic RF containing H^3-T (+) and P^{32}-$\overline{BU}$ (−) circles. The mixture was fractionated directly in an alkaline sucrose gradient (Fig. 8). The synthetic H^3-T (+) circles, complementary to the synthetic P^{32}-$\overline{BU}$ (−) circles and now corresponding in structure to the original (+) phage DNA template, were evident as a H^3-labeled shoulder, with corresponding infectivity, partially separated from the slower sedimenting linear forms. Trailing from this infective (+) strand peak obscured the position of the more rapidly sedimenting and less infective $\overline{BU}$ (−) template circles (Fig. 8; also note legend for method of collecting the fractions); all hybrid molecules which had remained were pelleted under the conditions used.

Discussion.—Physical studies on the partially synthetic RF prepared by enzymatic replication of phage DNA showed its structure to be like that of the RF form I isolated from infected cells.[5] The only distinction was the relative absence of supercoiling in the partially synthetic molecule and this can be attributed, at least in part, to the difference between the *in vitro* and *in vivo* conditions of salt and temperature[20] at the time of strand closure to form the circular duplex. The test of infectivity is a more rigorous and meaningful measure of the accuracy of replication and ring closure of phage DNA by polymerase and joining enzyme.

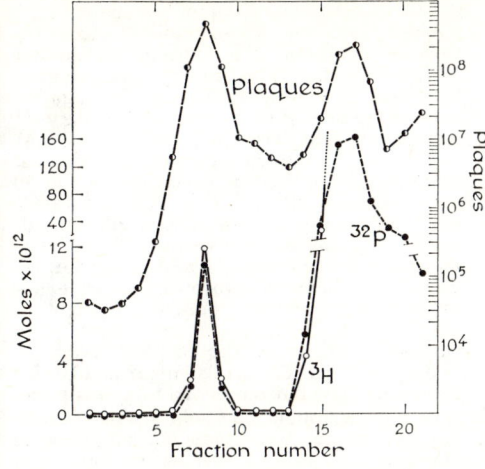
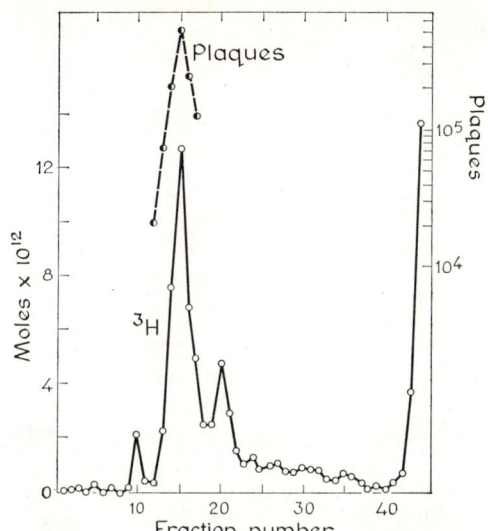

FIG. 6.—Alkaline sucrose-gradient sedimentation of fully synthetic RF. P^{32}-$\overline{BU}$, (−) strands (40 μl of peak sample from Fig. 2, dialyzed and concentrated as described in Fig. 3) were replicated in a volume of 0.1 ml as described previously;[5] the labeled nucleotide was H^3-dCTP (Schwarz BioResearch, 1000 cpm/μμmole), and dTTP rather than d$\overline{BU}$TP was used. After 180 min, the mixture was made 20 mM in EDTA, 0.1 M in NaOH, and centrifuged in a sucrose gradient in 0.2 M NaOH–0.8 M NaCl–1 mM EDTA, at 60,000 rpm and 1° for 100 min. The fractions were neutralized with 1 M Tris citrate (pH 5) before being assayed for radioactivity and infectivity.

FIG. 7.—Density-gradient sedimentation of synthetic RF in the presence of ethidium bromide. The synthetic RF, prepared as described in Fig. 6, was purified by a preliminary density-gradient centrifugation in CsCl-ethidium bromide, as described previously.[5] The covalent duplex zone, identified by alkaline sucrose-gradient sedimentation of aliquots from the fractions, was collected and refractionated in the same type of CsCl-ethidium bromide gradient with results shown above. Fractions were diluted 200-fold for the spheroplast assay but were not otherwise treated to remove CsCl or ethidium bromide. P^{32} in the $\overline{BU}$, (−) template was not measurable due to radioactive decay and low recoveries.

Numerous φX174 mutants are known[21] in which the change of a single nucleotide results in loss of infectivity under the assay conditions employed. The fact that isolated synthetic circles and fully synthetic RF forms made with these circles as templates had specific infectivity values in the range measured for natural forms of viral DNA (Table 1) attests to the precision of the enzymatic operation.

It should now be possible to apply the techniques used in this work to the synthesis of the duplex circular genomes of other viruses, such as phage λ and animal viruses, and DNA molecules of comparable structure from cellular organelles. Such synthetic efforts will permit the insertion of base and nucleoside analogues in a manner and variety not attainable with *in vivo* systems. In addition, base changes generated by replication of the DNA with defective polymerases can now easily be studied in combination with standard genetic tools. It is of interest that DNA of approximately normal specific infectivity has been synthesized here without the use of any methylated nucleotide. This result may be related to the lack of host modification or restriction in the *E. coli* C-K12 pair and might not be applicable to other viral DNA's.

Since the conversion of phage DNA to RF-form I is accomplished *in vivo* by

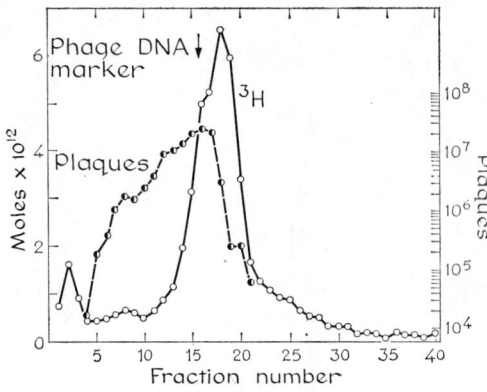

Fig. 8.—Identification of synthetic, (+) circles by alkaline sucrose-gradient sedimentation of synthetic RF exposed to limited DNase action. The rapidly sedimenting fractions of synthetic RF in the alkaline sucrose gradient (Fig. 6) were pooled, dialyzed against 10 mM Tris HCl (pH 7.6)–1 mM EDTA and incubated for 20 min at 20° (final volume of 2 ml) at a concentration of 50 μμmoles/ml in 10 mM Tris HCl (pH 7.6)–5 mM $MgCl_2$–0.1 mg/ml bovine plasma albumin–1.2 mμg/ml pancreatic DNase. The mixture was then made 15 mM in EDTA, reduced in volume to 0.15 ml by rotary evaporation under reduced pressure, brought to pH 12 with NaOH, and centrifuged in a sucrose gradient in 0.2 M NaOH–0.8 M NaCl–1 mM EDTA, at 60,000 rpm and 10° for 240 min. The bottom of the tube was punctured with a hollow needle and the contents were displaced by saturated CsCl solution (containing Blue Dextran from Pharmacia) using a peristaltic pump. The sucrose-gradient fractions were collected from the top of the tube via a fine polyethylene tube in a stopper at the top. The fractions were neutralized (as in Fig. 6) prior to assays. The fractions were numbered in the reverse order of their collection, in order that the direction of sedimentation conform to the illustration of velocity sedimentations in the other figures. P^{32} in the $\overline{BU}$, template strand was not measurable due to radioactive decay and the small amounts of DNA employed.

host enzymes, and since the DNA polymerase and polynucleotide-joining enzyme are so effective in converting phage DNA to RF-form I *in vitro*, it appears likely that these enzymes are used by infected *E. coli* cells to carry out this conversion *in vivo*. Although the predominant pathway of phage replication appears to involve the open RF-form II,[22] the two forms are in fact interconvertible *in vivo*. Questions of the roles of these enzymes and the replicative forms in the production of (+) circles for progeny phage require further study.

The fact that *E. coli* DNA polymerase can synthesize biologically active DNA does not establish its function in the replication of the bacterial chromosome. However, the effectiveness of the combined action of the polymerase and the polynucleotide-joining enzyme in forming infective DNA may have considerable significance for chromosomal replication. In an earlier paper,[4] a mechanism was suggested whereby polymerase, with a then hypothetical polynucleotide-joining enzyme, might function in the simultaneous replication of both strands of helical DNA. The subsequent discovery of this joining enzyme, the requirement for it in phage T4 DNA synthesis,[23] its persistence in the most purified *E. coli* and phage T4 DNA polymerase preparations,[5] as well as the current demonstration of its conjoint action with polymerase, all strengthen the suggestion of this replication mechanism.[4]

Summary.—A partially synthetic, closed replicative form (RF) of φX174 DNA, consisting of phage DNA as the (+) circle and a bromouracil-containing complement synthesized by DNA polymerase as the (−) circle, was used as the source of synthetic (−) circles. The latter were separated from template strands by limited DNase action on the RF followed by denaturation and density-gradient equilibrium sedimentation. The isolated (−) circles were infectious and had the buoyant density, sedimentation velocity, and radiation sensitivity expected for DNA containing bromouracil. These (−) circles served as templates for a second round of replication which produced a fully synthetic RF with the specific infectivity of natural RF. Infective synthetic (+) circles, corresponding to the original phage DNA, were isolated from the synthetic RF after DNase treatment, as in the previous isolation of synthetic (−) circles. These results imply a relatively error-free synthesis of the ϕX174 genome by DNA polymerase.

Note added in proof: A study by Okazaki, R., T. Okazaki, K. Sakabe, and K. Sugimoto (*Jap. J. Med. Sci. Biol.*, **20**, 255 (1967)) of DNA replication in *E. coli* supports a mechanism of discontinuous 5′ → 3′ chain growth on the 5′ template strand (see *Discussion*).

We gratefully acknowledge the expert assistance of Mrs. Gloria Davis of the Division of Biology at the California Institute of Technology in performing the spheroplast assays for infectivity.

* This research was supported by grants from the National Institutes of Health and the National Science Foundation.

† USPHS special fellow. Present address: Department of Medicine, University of Chicago.

[1] Litman, R. M., and W. Szybalski, *Biochem. Biophys. Res. Commun.*, **10**, 473 (1963).

[2] Richardson, C. C., C. L. Schildkraut, H. V. Aposhian, A. Kornberg, W. Bodmer, and J. Lederberg, in *Informational Macromolecules*, ed. H. J. Vogel, V. Bryson, and J. O. Lampen (New York: Academic Press, 1963), p. 13.

[3] Richardson, C. C., R. B. Inman, and A. Kornberg, *J. Mol. Biol.*, **9**, 46 (1964).

[4] Mitra, S., P. Reichard, R. B. Inman, L. L. Bertsch, and A. Kornberg, *J. Mol. Biol.*, **24**, 429 (1967).

[5] Goulian, M., and A. Kornberg, these PROCEEDINGS, **58**, 1723 (1967).

[6] Abbreviations used are: RF for replicative form; T for thymine; $\overline{BU}$ for bromouracil; dCTP, d$\overline{BU}$TP, and dTTP for the deoxyribonucleoside triphosphates of cytosine, $\overline{BU}$, and T, respectively; (+) circle for phage DNA; (−) circle for complementary copy of (+) circle.

[7] Rüst, P., and R. L. Sinsheimer, *J. Mol. Biol.*, **23**, 545 (1967).

[8] Siegel, J. E. D., and M. Hayashi, *J. Mol. Biol.*, **27**, 443 (1967).

[9] Experimental procedures were as described previously[5] or as detailed in the figure legends.

[10] In this figure, and in all succeeding ones, the ordinate values represent the total moles of nucleotide or plaques per fraction. The fractions, except where indicated otherwise, are numbered in the order of their collection from the bottom of the tube.

[11] Elson, E. L., thesis, Stanford University, Stanford (1966).

[12] Guthrie, G. D., and R. L. Sinsheimer, *Biochim. Biophys. Acta*, **72**, 290 (1963).

[13] Denhardt, D. T., and R. L. Sinsheimer, *J. Mol. Biol.*, **12**, 647 (1965).

[14] The ρ values for $\overline{BU}$-containing DNA were calculated from the base composition (Sinsheimer, R. L., *J. Mol. Biol.*, **1**, 43 (1959)), and the figure of 0.2 gm/ml determined by Baldwin and Shooter (*J. Mol. Biol.*, **7**, 511 (1963)) for the difference in ρ between dAT and dA$\overline{BU}$.

[15] Sinsheimer, R. L., M. Lawrence, and C. Nagler, *J. Mol. Biol.*, **14**, 348 (1965).

[16] Burton, A., and R. L. Sinsheimer, *J. Mol. Biol.*, **14**, 327 (1965).

[17] The possibility of a facilitative effect of the synthetic DNA upon the infectivity of a small contaminant of natural DNA was tested by mixing synthetic DNA molecules (P^{32}-$\overline{BU}$; Fig. 2; amber mutant) and natural DNA (γh, temperature-sensitive mutant).[13] The plaque count for the amber mutant was 398 for the $\overline{BU}$ DNA alone and 459 for the mixture, whereas the corresponding figures at similar dilutions for the temperature-sensitive mutant, alone and mixed, were 88 and 126, thus indicating the lack of interaction.

[18] Denhardt, D. T., and R. L. Sinsheimer, *J. Mol. Biol.*, **12**, 674 (1965).

[19] This relatively large amount is unexplained and surprising in view of the low level to which infectivity dips between the P^{32}-$\overline{BU}$ peak and the denatured hybrid duplex region of the CsCl gradient (Fig. 2).

[20] Wang, J. C., D. Baumgarten, and B. M. Olivera, in preparation.

[21] Sinsheimer, R. L., C. Hutchison, and B. H. Lindqvist, in *The Molecular Biology of Viruses* ed. J. S. Colter (New York: Academic Press, in press.)

[22] Lindqvist, B. H., and R. L. Sinsheimer, in preparation.

[23] Fareed, G. C., and C. C. Richardson, these PROCEEDINGS, **58**, 665 (1967).

Enzymatic Synthesis of Deoxyribonucleic Acid

XXXVI. A PROOFREADING FUNCTION FOR THE 3' → 5' EXONUCLEASE ACTIVITY IN DEOXYRIBO-NUCLEIC ACID POLYMERASES*

(Received for publication, July 16, 1971)

Douglas Brutlag‡ and Arthur Kornberg

From the Department of Biochemistry, Stanford University School of Medicine, Stanford, California 94305

SUMMARY

The *Escherichia coli* and T4 DNA polymerases do not extend chains in which the 3'-terminal nucleotide (primer terminus) is not paired with the template. By using synthetic double-stranded polynucleotides, the 3' → 5' exonuclease function of these polymerases was shown to be directed specifically against a mispaired or unpaired primer terminus. Chain extension of such termini begins only after all mispaired nucleotides have been removed and a base-paired terminus is reached. The latter is completely conserved while polymerization is maintained. These results suggest that the function of the 3' → 5' exonuclease activity of DNA polymerases is to remove mispaired nucleotides which have been incorrectly incorporated, thereby increasing the fidelity of template copying. The function of other *E. coli* exonucleases suggested by their specificities on polynucleotide substrates are trimming loose ends of DNA for exonuclease I and enlarging nicks and gaps within helical regions for exonuclease III.

An exonuclease activity which degrades DNA chains in the 3' to 5' direction is a component of DNA polymerase isolated from *Escherichia coli* (1). The 3' → 5' exonuclease is also closely associated with and appears to be part of DNA polymerases induced in *E. coli* by infections with bacteriophages T2, T4, and T5 (2–4). The specificity and possible functions of this nuclease have been obscured by the presence in *E. coli* DNA polymerase of another and even more active exonuclease activity that degrades chains from 5' to 3'. Recently it has become possible to separate and physically isolate these two nuclease activities by proteolytic cleavage of DNA polymerase (5, 6). From the parent polypeptide chain (mol wt 109,000), two active fragments were isolated: the small fragment (mol wt 36,000) contains the 5' → 3' exonuclease activity; the large fragment (mol wt 75,000) contains the 3' → 5' exonuclease activity as well as the polymerizing functions (5, 6).

* This study was supported in part by grants from the National Institutes of Health (United States Public Health Service) and the National Science Foundation. The previous paper in this series is Reference 33.
‡ Predoctoral Fellow of the National Science Foundation.

The studies reported here on the specificity of the 3' → 5' exonuclease support earlier speculations (3, 7, 8) that this activity may be designed to remove mispaired nucleotides which have been incorrectly incorporated at the 3'-hydroxyl end of the double-stranded DNA and thereby serves a corrective function.[1] By the use of specifically labeled homopolymers and block copolymers as substrates we show in this report that under polymerizing conditions a mispaired primer terminus is quantitatively removed by the DNA polymerase whereas a correctly base-paired primer terminus is completely conserved. It thus appears that the 3' → 5' exonuclease may serve a proofreading role in removing mispaired nucleotides at the growing end of a chain and that the polymerase will not extend a chain until the mispaired 3' terminus is removed.

The availability of these polynucleotide substrates has enabled us also to examine the specificities of the *E. coli* exonucleases I and III. These nucleases appear to perform complementary roles in trimming single-stranded ends of DNA chains and in enlarging nicks and gaps within helical regions of double-stranded DNA.

MATERIALS

Nucleotides—Unlabeled deoxyribonucleotides were purchased from P-L Biochemicals except for dUTP which was purchased from Calbiochem. dITP was a gift from Dr. M. Fikus and was prepared by deamination of dATP as described by Inman and Baldwin (9). [^{3}H]dTTP and [^{3}H]dATP were purchased from Schwarz BioResearch. [^{3}H]dUTP was purchased from Amersham-Searle and was purified before use by paper chromatography (Schleicher and Schuell, No. 589 orange ribbon) using an isopropyl alcohol-concentrated NH_4OH-water (7:1:2) solvent. [α-^{32}P]dTTP was purchased from ICN and was checked for radiochemical purity by paper electrophoresis (20 mM sodium citrate buffer, pH 3.5). [^{14}C]dTTP was purchased from New England Nuclear. All specific activities were determined by spectral measurement and radioactivity measured on Whatman GF/C glass filters in a Nuclear Chicago scintillation counter. 2',3'-Dideoxythymidine triphosphate and the 6'-deoxy-6'-homothymidine pyrophosphoryl phosphonate were gifts from Dr. J. G. Moffatt. β,γ-dTTP methylene diphosphonate was purchased from Miles Laboratories.

Enzymes—Micrococcal nuclease and spleen phosphodiesterase were purchased from Worthington Biochemicals. Alkaline

[1] L. E. Orgel and F. H. C. Crick, personal communication.

phosphatase was isolated and purified according to the method of Malamy and Horecker (10). DNA polymerases induced by T4 $amN82$ and T4 $amN82$ $tsL56$, the gifts of Dr. W. M. Huang, had been purified as described (3, 11). $E.\ coli$ DNA polymerase was purified as described (12). The large fragment obtained by proteolytic cleavage of this enzyme appeared in the purification procedure of the native DNA polymerase and was also purified to homogeneity (6). All DNA polymerases used in this study yielded a single protein band on sodium dodecyl sulfate-polyacrylamide gel electrophoresis. $E.\ coli$ exonuclease I was purified through the hydroxylapatite stage (specific activity of 15,000 units per mg) (13). $E.\ coli$ exonuclease III had a specific activity of 130,000 units per mg (12). Terminal deoxynucleotidyltransferase (terminal transferase) purified from calf thymus according to Kato $et\ al.$ (14) was a gift from Dr. F. N. Hayes (Los Alamos Scientific Laboratory, Los Alamos, New Mexico).

METHODS

Polynucleotides—The polynucleotides $d(T)_{260}$ and $d(I)_{265}$ were prepared with terminal transferase as described (15) except that 20 mM potassium phosphate buffer (pH 7.0), 20 mM potassium cacodylate buffer (pH 7.0), and 8 mM $MgCl_2$ were used for the polymerization of dITP. $d(A)_{4000}$ was prepared according to Riley, Maling, and Chamberlin (16) and $d(C)_{1000}$ according to Chamberlin and Patterson (17).

Terminally Labeled Polynucleotides—Polynucleotides, radioactivity labeled at their 3'-termini with various nucleotides ($d(T)_{260}$-[^{3}H]$d(T)_{0.8}$, $d(T)_{260}$-[^{3}H]$d(C)_{1.1}$, $d(T)_{260}$-[^{3}H]$d(A)_{0.9}$, $d(T)_{260}$-[^{3}H]$d(U)_{1.8}$, and $d(I)_{265}$-[^{3}H]$d(T)_{1.6}$)[2] were synthesized by incubating labeled triphosphates with unlabeled polynucleotides and terminal transferase. To add a purine nucleotide, dATP for example, to $d(T)_{260}$, 0.3 mM $d(T)_{260}$[3] was incubated with 5 μM [^{3}H]dATP (17.7 Ci per mmole), and 0.15 mg per ml of terminal transferase in 20 mM potassium phosphate buffer (pH 7.0), 20 mM potassium cacodylate buffer (pH 7.0), 8 mM $MgCl_2$, and 1 mM β-mercaptoethanol at 37° for an amount of time (predetermined by a small scale reaction) sufficient to add about 1 mole of nucleotide per mole of polynucleotide chain. Incorporation was monitored by adsorption to DEAE-paper as described below. To add a pyrimidine nucleotide, 0.1 M potassium phosphate buffer (pH 7.0), 0.1 M potassium cacodylate buffer (pH 7.0), and 0.25 mM $CoCl_2$ were used. The reaction was terminated by cooling and adding 50% KOH to a final concentration of 0.3 M. After 10 min at 0°, the mixture was neutralized with 85% H_3PO_4 and dialyzed extensively against 1 M NaCl, 10 mM Tris buffer (pH 8.0), and then $versus$ 10 mM Tris buffer (pH 8.0) to remove the salt. The final average number of residues added was determined from the specific activity of the added nucleotide, the length of the polynucleotide, and the radioactivity present in the labeled polymer per mole of total nucleotide as determined by ultraviolet absorption.

Determination of Distribution of Labeled Residues—Hayes $et\ al.$ (18) have shown that limited addition of nucleotides to short oligonucleotide primers by terminal transferase gives a Poisson

[2] Polynucleotides labeled at the 3' terminus are abbreviated as $d(T)_{260}$-[^{3}H]$d(C)_{1.1}$, where 260 gives the average number of residues in the polymer of unlabeled thymidylate to which has been added on the average 1.1 moles of ^{3}H-labeled deoxycytidylate per mole of polymer.

[3] Polynucleotide concentrations are expressed in terms of moles of total nucleotide.

distribution of added residues. This conclusion was found to apply also to the longer polynucleotide primers used in this study. This determination was made from an independent measure of the distribution of nucleotides as well as the average number of residues added per chain. Degradation of the labeled polymers by micrococcal nuclease and spleen phosphodiesterase converts all the internal nucleotides to 3'-nucleotides and the 3'-terminal residue to a nucleoside. The ratio of label converted to nucleoside to the total label found in the nucleoside and 3'-nucleotide, combined, gives the fraction of the labeled residues at the 3'-terminal position (F_t). For this analysis, 3 nmoles of each of the terminally labeled polynucleotides were degraded with micrococcal nuclease and spleen phosphodiesterase and the products were separated by chromatography as described by Wu and Kaiser (19). The hydrolysis converted >98% of the total label to nucleoside and nucleotide. Table I summarizes the average number of labeled residues and the fraction of these residues which was terminal. The observed fraction of terminal residues was in good agreement with that predicted by a Poisson distribution. It can be demonstrated that this analysis is formally equivalent to showing that the fraction of chains to which no labeled residues were added conforms to the Poisson distribution.

Determination of Polynucleotide Length—The number-average length of a polynucleotide was determined by end group labeling analysis as described by Weiss, Live, and Richardson (20).

Nuclease Assays—Nuclease assays were performed in 0.2 ml containing 2 nmoles of [^{3}H]$d(T)_{300}$, with or without 5 nmoles of unlabeled $d(A)_{4000}$, 0.05 M N-hydroxyethylpiperazine-N'-ethanesulfonate buffer (pH 7.4), and 5 mM $MgCl_2$. The mixture was heated to the appropriate temperature and a 20-μl sample was taken and applied to a 1.5-cm square of Whatman DE-81 paper. Then 10 μl of enzyme diluted in bovine serum albumin (1 mg per ml), 0.05 M HEPES[4] buffer (pH 7.4), and containing less than 1 pmole of enzyme added to the reaction; 20-μl samples were removed at appropriate intervals and adsorbed onto 1.5 cm squares of DE-81 paper. The squares were washed three times by gentle agitation for 5 min in 100 ml of ammonium formate, 0.3 M (adjusted to pH 7.8), and dehydrated by two washes in 95% ethanol and one in anhydrous ether. The squares were air dried and the amount of radioactive label remaining in polynucleotide determined in a scintillation counter. (Polymerization was also measured with this technique by using labeled deoxyribonucleoside triphosphates.) Nuclease rates were determined from a least squares fit to initial linear points of nucleotide hydrolyzed plotted against time. Rates were determined from only those assays in which >20% of the polynucleotide had been degraded and in which the rate was linear for at least 3 time points.

RESULTS

3' → 5' Exonuclease Rates of DNA Polymerases: Influence of Secondary Structure and Temperature—The nuclease rate of the large fragment (a proteolytic fragment of the $E.\ coli$ enzyme containing only polymerase and 3' → 5' exonuclease activities) on $d(T)_{300}$ was 4-fold greater with the polymer in a

[4] The abbreviations used are: HEPES, N-2-hydroxyethylpiperazine-N'-2-ethanesulfonic acid; ddTTP, 2',3'-dideoxythymidine triphosphate; dTMPPCP, β,γ-dTTP methylene diphosphonate.

TABLE I
Distribution of terminal residues on synthetic polynucleotides

Polynucleotide composition	Fraction of labeled residues at the 3' terminus	
	Observed[a]	Poisson[b]
$d(T)_{260}$-$[^3H]d(T)_{0.85}$	0.54	0.67
$d(T)_{260}$-$[^3H]d(C)_{1.11}$	0.56	0.60
$d(T)_{260}$-$[^3H]d(A)_{0.87}$	0.42	0.67
$d(T)_{260}$-$[^3H]d(U)_{1.81}$	0.50	0.46
$d(I)_{265}$-$[^3H]d(T)_{1.63}$	0.47	0.49

[a] The fraction of labeled residues which are at the 3'-OH terminal position of a chain (F_t) was determined by nuclease degradation. Specific activities of the labeled nucleotides were from 7 to 20 Ci per mmole.

[b] $F_t = (1 - e^{-r})/r$, for a Poisson distribution of labeled residues where r is the average number of labeled residues per chain.

TABLE II
Influence of temperature on $3' \rightarrow 5'$ exonuclease rates of DNA polymerase acting on single- and double-stranded polynucleotides

DNA polymerase	Temperature	Substrate	
		Single-stranded $[^3H]d(T)_{300}$	Double-stranded $[^3H]d(T)_{300}$:$d(A)_{4000}$
		moles nucleotide hydrolyzed/mole enzyme/min	
E. coli large fragment	37°	19	4.8
T4 wild type[a]	37	1100	350
T4 wild type[a]	30	850	27
T4 L56[a]	30	790	11

[a] β-Mercaptoethanol (10 mM) was included in assays of the T4 enzyme.

single-stranded as compared to a double-stranded conformation ($d(T)_{300}$:$d(A)_{4000}$) (Table II). A similar preference for single-stranded $d(T)_{300}$ was observed at 37° with the T4 enzyme which has a much higher nuclease activity than the E. coli enzyme.

When the nuclease rates of the T4 polymerase on single-stranded $d(T)_{300}$ at 30° are compared with the rates at 37° only a slight reduction in rate (23%) was found. However, there was a 13-fold reduction in rate on the double-stranded substrate (Table II). A similar marked temperature dependence of the nuclease rate of the E. coli large fragment on double-stranded substrates has also been observed.[5] This temperature dependence indicates a high energy of activation for some rate-limiting step on double-stranded substrates, possibly local denaturation at the 3' terminus. Thus the $3' \rightarrow 5'$ exonuclease might require that the terminal nucleotides not be base-paired (i.e. frayed).

Nuclease Action of DNA Polymerases on Primer Terminus—Polynucleotides containing on the average about one radioactive nucleotide residue per chain at the 3' terminus were synthesized employing terminal transferase (Table I). The terminal residue was either the same as the rest of the polymer or distinctive from it. The distribution of labeled residues added to the polynucleotide chains approximates a Poisson distribution. For simplicity, these 3' terminally labeled polymers are referred to with a

[5] P. Setlow and A. Kornberg, unpublished results.

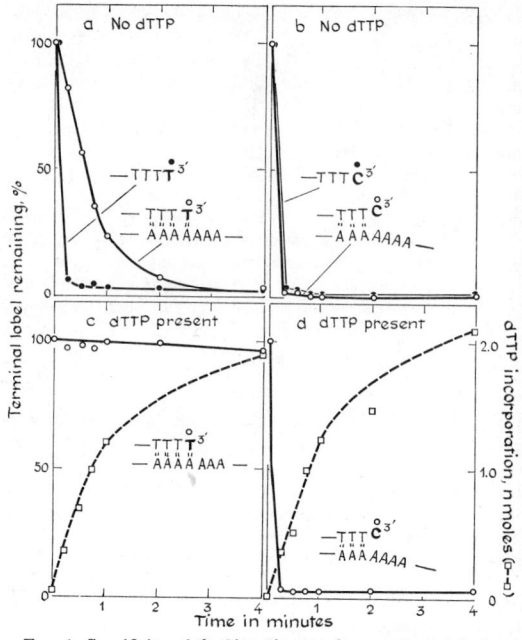

FIG. 1. Specificity of the $3' \rightarrow 5'$ exonuclease action of the large fragment of the Escherichia coli DNA polymerase on terminally labeled polynucleotides. An excess of large fragment (5.4 pmoles) was incubated at 37° under standard assay conditions with the indicated polymers and triphosphates. Samples were taken at indicated times. a, 0.6 nmoles of $d(T)_{260}$-$[^3H]d(T)_1$ with and without 5.4 nmoles of $d(A)_{4000}$; b, 0.66 nmole of $d(T)_{260}$-$[^3H]d(C)_1$ with and without 5.4 nmoles of $d(A)_{4000}$; c, same as a with $d(A)_{4000}$ and with 50 μM [α-^{32}P]dTTP(18,000 cpm per nmole); d, same as b with $d(A)_{4000}$ and with 50 μM [α-^{32}P]dTTP.

subscript "1" even though the actual average number of labeled residues only approximates unity.

The rate of removal of the terminal label from $d(T)_{260}$-$[^3H]d(T)_1$ by the large fragment was too great to permit accurate rate measurements (Fig. 1a). However, the decrease in nuclease rate with the polynucleotide in a double-stranded conformation was large enough to be apparent. Thus the rate of hydrolysis of these terminal residues reflects the preference for single-stranded substrate initially observed with uniformly labeled polynucleotides. The terminal residue of $d(T)_{260}$-$[^3H]d(C)_1$, which does not base pair with $d(A)_{4000}$, was removed at a rapid rate both in the single and the double-stranded conformation (Fig. 1b). Although the relative rates of hydrolysis of dCMP in these two cases cannot be determined, the rate of removal of dCMP from $d(T)_{260}$-$[^3H]d(C)_1$:$d(A)_{4000}$ was far more rapid than that of dTMP from $d(T)_{260}$-$[^3H]d(T)_1$:$d(A)_{4000}$ (compare Fig. 1, a and b). Thus the preference of the nuclease for nucleotides which are not able to base pair with the template is readily apparent.

Removal of Terminal Nucleotides Prior to Polymerization—Polymerization had a dramatic effect on the $3' \rightarrow 5'$ exonuclease activity. When the large fragment and the polymer $d(T)_{260}$-$[^3H]d(T)_1$:$d(A)_{4000}$ were incubated with dTTP so that polymerization could proceed, there was no detectable loss of terminal la-

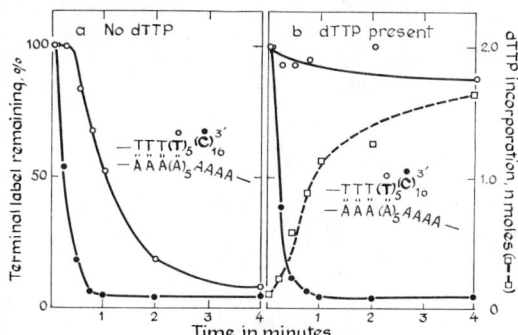

FIG. 2. $3' \rightarrow 5'$ exonuclease action of the large fragment on the terminal residues of $d(T)_{260}$-[^{32}P]d(T)$_5$-[^{3}H]d(C)$_{16}$·d(A)$_{4000}$. The large fragment (5.4 pmoles) was incubated under standard assay conditions at 37° with the following; a, 0.75 nmole of $d(T)_{260}$-[^{32}P]d(T)$_5$-[^{3}H]d(C)$_{16}$ and 2.0 nmoles of $d(A)_{4000}$; b, same as a but with 50 μM[^{14}C]dTTP (10,400 cpm per nmole).

TABLE III

Inhibition of $3' \rightarrow 5'$ exonuclease is specific for deoxyribonucleoside triphosphates which are incorporated

The deoxyribonucleoside triphosphate, 50 μM, was incubated with 0.6 nmole of $d(T)_{260}$-[^{3}H]d(T)$_1$, 2.0 nmoles of $d(A)_{4000}$, and 5.4 pmoles of large fragment at 37° under standard assay conditions for 4 min. Incorporation was followed using ^{14}C- or ^{32}P-labeled triphosphate (10,000 cpm per nmole) except where noted.

Nucleoside triphosphate added	Triphosphate incorporation[a]	Labeled primer terminus remaining[b]
		% of initial
dATP	None	<5
dGTP	None	<5
dCTP	None	<5
Homophosphonate[c]	None[d]	<5
dTMPPCP	None[d]	<5
dTTP	Normal	100
dUTP	Normal	95
ddTTP	One	97

[a] "None" indicates no incorporation detectable (less than residue per chain); "normal" indicates rapid chain extension; "one" indicates a single residue incorporated per chain (22).

[b] Values of <5% indicate that both the rate and the extent of hydrolysis were the same as a control without any deoxyribonucleoside triphosphate added.

[c] 6'-Deoxy-6'-homothymidine pyrophosphoryl phosphonate.

[d] The inability of DNA polymerase to incorporate these analogues was shown by their inability to replace dTTP in a poly-[d(A-T)] primed reaction (Brutlag and Kornberg, unpublished results).

bel during the incorporation of dTTP (Fig. 1c). The incorporation of dTTP completely protected the terminal nucleotides from exonuclease action. In a similar experiment in which $d(T)_{260}$-[^{3}H]d(T)$_1$ was present in a single-stranded conformation, there was no detectable incorporation of dTTP nor did the dTTP protect the terminal nucleotides from nuclease. Thus both the template and the triphosphate which participate in polymerization are required for protection of the terminal residues.

A third requirement for the protection of a terminal residue from exonuclease action is that it be base-paired to the template. When $d(T)_{260}$-[^{3}H]d(C)$_1$:d(A)$_{4000}$ was incubated with polymerase and dTTP, the terminal dCMP residue was removed rapidly after which polymerization proceeded normally with the primer that remained (Fig. 1d). After extensive polymerization little or no [^{3}H]dCMP was detectable in the product, indicating that the DNA polymerase does not polymerize on a mispaired 3' terminus; instead the enzyme first removes the non-base-paired nucleotides by $3' \rightarrow 5'$ exonuclease action.

In order to investigate further how strict is the base pairing specificity of this exonuclease during polymerization, a copolymer was synthesized containing a block of 5 [α-^{32}P]dTMP residues subterminal to a block of 16 [^{3}H]dCMP residues ($d(T)_{260}$-[^{32}P]d(T)$_5$-[^{3}H]d(C)$_{16}$:d(A)$_{4000}$). The fate of the terminal block of 16 non-base-paired dC residues and the subterminal block of base-paired dT residues were both observed. In the absence of dTTP, the enzyme (large fragment) removed the terminal dCMP residues at a maximum rate of 20 nucleotides per min per chain, whereas the dTMP residues were removed only after a significant lag and then only at a maximum rate of 4 residues per min per chain (Fig. 2a). The rate difference is not immediately apparent from Fig. 2a because the plot of the percentage of residues remaining does not reflect that there are initially three times as many dCMP residues as dTMP residues. When dTTP was included so that polymerization could proceed, the terminal dCMP residues were still removed as rapidly but the subterminal, base-paired dTMP residues were retained (Fig. 2b); loss of dTMP was less than 0.5 residue per chain. As expected, the polynucleotide supported synthesis only after a lag, during which the dCMP residues were removed. This experiment indicates not only the specificity of the $3' \rightarrow 5'$ exonuclease for a mispaired

nucleotide, but also the inability of the polymerase to utilize such a nucleotide as a primer terminus for polymerization.

Inhibition of $3' \rightarrow 5'$ Exonuclease by Deoxyribonucleoside Triphosphates—The previous experiments demonstrated that dTTP in an exonuclease assay with the base-paired template primer $d(T)_{260}$-[^{3}H]d(T)$_1$:d(A)$_{4000}$ completely inhibited the $3' \rightarrow 5'$ exonuclease activity of the DNA polymerase on terminal dTMP residues. Incorporation of nucleotides on a terminally labeled polynucleotide covers the labeled residue and protects it from exonuclease action. In order to determine whether incorporation of the triphosphate is required for protection of a terminal nucleotide, other deoxyribonucleoside triphosphates were tested for their ability to inhibit exonuclease (Table III). Nucleotides not complementary to the template $d(A)_{4000}$ (i.e. dATP, dGTP, and dCTP) were not incorporated and also failed to inhibit exonuclease. The phosphonates, dTMPPCP and 6'-deoxy-6'-homothymidine pyrophosphoryl phosphonate, are analogues of dTTP, which bind to the triphosphate-binding site on the enzyme (21),[6] and have the same base pairing specificity as dTTP. Neither analogue is incorporated by DNA polymerase nor does either analogue inhibit exonuclease. Three nucleotides which were incorporated (dTTP, dUTP, and ddTTP) all inhibited exonuclease (Table III).

The behavior of ddTTP is of interest in two respects. Since ddTTP lacks a 3'-hydroxyl group only 1 residue is incorporated per chain (22). Nevertheless, this limited incorporation was sufficient to protect essentially all of the terminal nucleotides from exonuclease. Secondly, ddTTP is incorporated at a rate

[6] D. Brutlag and A. Kornberg, unpublished results.

TABLE IV
Mispairing of 3'-terminal nucleotide as basis of nuclease action by large fragment and T4 DNA polymerase

An excess of DNA polymerase (4 to 5 pmoles) was incubated with 0.6 nmole of primer polynucleotide $d(T)_{260}$ (first three columns) or $d(I)_{265}$ (fourth column) labeled at the 3' terminus with the designated (*) nucleotide and annealed to 5 pmoles of a template ($d(A)_{4000}$ or $d(C)_{1000}$). dTTP or dGTP (50 μM) was present, as indicated. Incubation was 4 min under standard assay conditions except that 10 mM β-mercaptoethanol was included in assays of the T4 polymerase. The temperature was 37° except as noted.

DNA polymerase	--TTT*OH --AAAAA-- dTTP	--TTC*OH --AAAAA-- dTTP	--TTA*OH --AAAAA-- dTTP	--IIT*OH --CCCCC-- dGTP
	% of initial labeled primer terminus remaining			
E. coli large fragment	98	3	2	3
T4 wild type	97	1	1	1[a]
T4 L56	100[a]	1[a]	1[a]	3[a]

[a] These assays were performed at 30°.

which is a 1000-fold less than that of dTTP (22). Although this rate of ddTTP incorporation (0.4 nucleotide per chain per min) is 10-fold slower than that of exonuclease action on the terminal base-paired residues in the absence of a triphosphate, the presence of ddTTP completely inhibited the exonuclease. Thus the ability of a nucleoside triphosphate to bind to the triphosphate binding site on the enzyme and to base pair with the template strand is not sufficient to inhibit exonuclease. It is required that the triphosphate establish and maintain an effective polymerization complex with the template and primer terminus, even if the completion of the phosphodiester bond is exceedingly slow.

Comparison of Specificity of E. coli and T4 DNA Polymerases in 3' → 5' Hydrolysis of Various Mismatched Primer Termini—The T4 polymerase showed the same capacity as the *E. coli* enzyme to preserve the terminal dTMP residue of $d(T)_{260}$-$[^3H]d(T)_1$:$d(A)_{4000}$ during polymerization and to remove the terminal dCMP residue of $d(T)_{260}$-$[^3H]d(C)_1$:$d(A)_{4000}$ (Table IV). These enzymes also effectively removed a purine nucleotide from a purine-purine mismatch ($d(T)_{260}$-$[^3H]d(A)_1$:$d(A)_{4000}$) and a pyrimidine nucleotide from a pyrimidine-pyrimidine mismatch ($d(I)_{265}$-$[^3H]d(T)_1$:$d(C)_{1000}$). With the latter polymer incubated in the presence of dGTP, there was efficient removal of the terminal dTMP residue (Table IV). This primer then supported incorporation of dGTP. The polymer $d(T)_{260}$-$[^3H]d(U)_1$:$d(A)_{4000}$, in which the terminal dUMP residue base pairs with the template, was also tested with the *E. coli* large fragment. The result, as expected, was that more than 96% of the dUMP residue was retained in the presence of either dUTP or dTTP.

It is of interest to compare the base pairing specificity of the 3' → 5' exonuclease activity of the mutant T4-induced DNA polymerase (L56) with that of the wild type enzyme. A mutator role has been proposed for the enzyme in T4 L56, a temperature-sensitive mutant in gene 43, the gene which codes for the DNA polymerase (23, 24). The exonucleases of both T4 DNA polymerases had approximately the same rate at 30° (Table I) and both showed base pairing specificity identical with the *E. coli* large fragment (Table IV). Terminal base-paired residues were retained during polymerization whereas mispaired residues were removed. The base pairing specificity of the mutator

TABLE V
Nature of cleavage of terminally labeled polynucleotides in presence of inorganic pyrophosphate

E. coli large fragment (7.5 pmoles) was incubated with 1 mM pyrophosphate, 1.2 nmoles of $d(T)_{260}$-$[^3H]d(T)_1$ or 1.4 nmoles of $d(T)_{260}$-$[^3H]d(C)_1$, with or without 2.1 nmoles of $d(A)_{4000}$ under standard assay conditions at 37° for 6 min. The products were adsorbed to Norit, eluted, and analyzed by paper electrophoresis (20 mM sodium citrate buffer, pH 3.5). The fraction of the radioactivity migrating as monophosphate ($[^3H]dTMP$ or $[^3H]dCMP$) and as triphosphate ($[^3H]dTTP$ or $[^3H]dCTP$) was determined.

Structure at 3' primer terminus	Products released	
	Mono-phosphate (hydrolysis)	Tri-phosphate (pyrophos-phorolysis)
	% of total	
Single-stranded chain ($d(T)_{260}$-$[^3H]d(T)_1$)	100	<0.5
Double-stranded, base-paired ($d(T)_{260}$-$[^3H]d(T)_1$:$d(A)_{4000}$)	53	47
Single-stranded chain ($d(T)_{260}$-$[^3H]d(C)_1$)	100	<0.5
Double-stranded, mispaired ($d(T)_{260}$-$[^3H]d(C)_1$:$d(A)_{4000}$)	100	<0.5

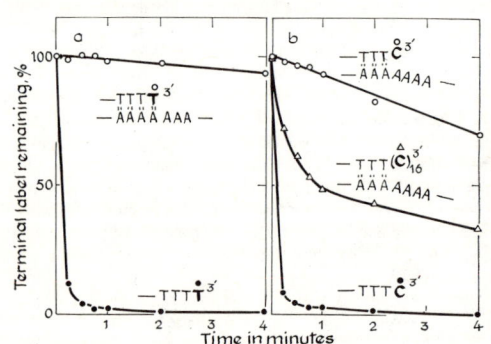

Fig. 3. Exonuclease I action on single strands and on base-paired and mispaired 3' termini. *Escherichia coli* exonuclease I (1.4 units) was incubated under standard assay conditions at 37° with 10 mM β-mercaptoethanol and the following polynucleotides: *a*, 0.6 nmole of $d(T)_{260}$-$[^3H]d(T)_1$ with and without 2.7 nmoles of $d(A)_{4000}$, *b*, 0.66 nmole of $d(T)_{260}$-$[^3H]d(C)_1$ with and without 2.7 nmoles of $d(A)_{4000}$ and 0.54 nmole of $d(T)_{260}$-$[^3H]d(C)_{16}$ annealed to 2.7 nmoles of $d(A)_{4000}$.

DNA polymerase appears at this level of analysis not to be markedly different from the wild type enzyme. However, a 1% failure by the mutant polymerase to remove a mispaired primer terminus would not have been detected above the background in our assay.

Pyrophosphorolysis Is Reversal of Polymerization and Is Distinct from 3' → 5' Exonuclease—Suggestions that the 3' → 5' exonuclease activity of the *E. coli* DNA polymerase is analogous to pyrophosphorolysis (25, 26) appeared unlikely on the basis of later evidence (27). The requirement for all the components of the polymerization reaction argued that pyrophosphorolysis is the reversal of polymerization; the ability of the 3' → 5' exonuclease activity to degrade single-stranded DNA was taken as

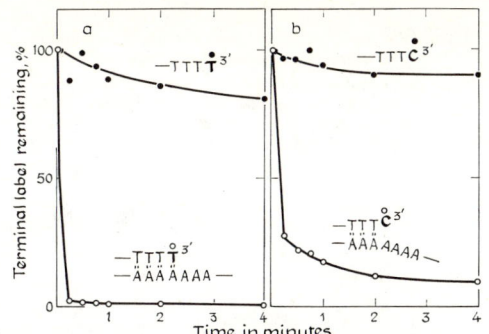

FIG. 4. Exonuclease III action on base-paired and mispaired 3' termini. *Escherichia coli* exonuclease III (19.6 units) was incubated with the same polymers and under the same conditions as in Fig. 3.

evidence that this hydrolytic reaction was distinct from pyrophosphorolysis which requires a double-stranded DNA. The availability of primer templates in which only the terminal nucleotide of the primer is mispaired provides an even more exacting test of the requirement of pyrophosphorolysis for a base-paired 3' terminus.

When 1 mM pyrophosphate was present, the sole radioactive products of degradation of single-stranded $d(T)_{260}$-[^{3}H]d(T)$_1$ and $d(T)_{260}$-[^{3}H]d(C)$_1$ by the large fragment were still monophosphates: [^{3}H]dTMP and [^{3}H]dCMP, respectively (Table V). This result verifies that 3'-terminal nucleotides of single-stranded chains are not removed by pyrophosphorolysis. With the double-stranded polynucleotides $d(T)_{260}$-[^{3}H]d(T)$_1$:d(A)$_{4000}$ and $d(T)_{260}$-[^{3}H]d(C)$_1$:d(A)$_{4000}$ as substrates, 47% of the [^{3}H]dT residues were released by pyrophosphorolysis but there was no detectable pyrophosphorolysis of the non-paired [^{3}H]dC residues (Table V). These results establish that pyrophosphorolysis has the specificity expected for the reversal of polymerization and that its characteristics are distinct from those of 3' → 5' exonuclease action.

Base Pairing Specificities of E. coli Exonuclease I and Exonuclease III—Exonuclease I degrades double-stranded DNA at a 40,000-fold slower rate than denatured DNA (13). The specificity of this enzyme was further analyzed with terminally labeled polynucleotides. With enzyme present in excess over the available 3' termini, the terminal nucleotide was removed rapidly from $d(T)_{260}$-[^{3}H]d(T)$_1$ but only slowly (if at all) from the same polymer annealed to $d(A)_{4000}$ (Fig. 3a). Attack on the mispaired terminal nucleotide of $d(T)_{260}$-[^{3}H]d(C)$_1$ was similar (Fig. 3b). Whereas, removal of the dCMP residues from the single-strand was rapid, removal from $d(T)_{260}$-[^{3}H]d(C)$_1$:d(A)$_{4000}$ was very slow. When the number of terminal dCMP residues was increased to 16, exonuclease I removed up to 60% of the dC residues of $d(T)_{260}$-[^{3}H]d(C)$_{16}$:d(A)$_{4000}$ rapidly, but the remaining nucleotides were removed very slowly. These data indicate that exonuclease I removes unpaired nucleotides that are within 6 to 8 residues of a base-paired region very slowly compared with nucleotides in a single-stranded chain. Thus exonuclease I shows a general specificity for extended single-stranded structures.

Exonuclease III (a 3' → 5' exonuclease which attacks native DNA preferentially (28)) shows a specificity for double-stranded structure. The terminal residues of $d(T)_{260}$-[^{3}H]d(T)$_1$ were hydrolyzed very rapidly by exonuclease III only when the polymer was annealed to $d(A)_{4000}$ (Fig. 4a). Likewise, most of the residues of $d(T)_{260}$-[^{3}H]d(C)$_1$ were sensitive to exonuclease III only when the polymer was in the double-stranded conformation (Fig. 4b). In experiments with polymers that contained blocks of dCMP residues, resistance to hydrolysis increased in proportion to the length of the sequence that did not anneal to the $d(A)_{4000}$. From these results it appeared that exonuclease III hydrolyzes double-stranded polymers which contain one, two, or even three mispaired terminal nucleotides at a rapid rate. About 90% of the terminal residues were removed from $d(T)_{260}$-[^{3}H]d(C)$_1$:d(A)$_{1000}$ in 4 min (Fig. 4b). But polymers with 5 or more residues were attacked very slowly, if at all, as 10% of the terminal residues of $d(T)_{260}$-[^{3}H]d(C)$_1$:d(A)$_{4000}$ remained and a Poisson distribution predicts that 10% of the total label resides in chains with 5 or more labeled residues.

DISCUSSION

Specificity of 3' → 5' Exonuclease Associated with DNA Polymerases—The 3' → 5' exonuclease activity of the *E. coli* and the T4 DNA polymerases degrades both single- and double-stranded DNA. Temperature had no more than the usual influence on degradation of single strands. However, with double-stranded DNA, the influence of temperature was profound. An increase of only 7° (from 30° to 37°) enhances the rate more than 10-fold. This behavior suggests that it is the frayed end of a helix which is required for action of the nuclease, as does an alkaline pH optimum (1), at which melting of the helix is favored. These suggestions, indicating a requirement of the nuclease for an unpaired 3'-hydroxyl terminus, have been confirmed by our studies with synthetic DNA polymers.

In the studies reported here, homopolymer pairs were used as nuclease substrates in which the 3'-hydroxyl end of the primer was either a proper base pair with the template or any of the several kinds of possible mispairings. When the triphosphates appropriate for polymerization were present, then two things were found. A properly base-paired terminus was extended without measurable nuclease action, whereas a mispaired terminus was quantitatively removed before polymerization was initiated. Mispaired termini included purine-purine and pyrimidine-pyrimidine pairs as well as purine-pyrimidine mismatches.

Suppression of 3' → 5' Exonuclease by Polymerization—What components of the polymerization reaction are required to suppress the 3' → 5' exonuclease? We have found that only deoxyribonucleoside triphosphates which match the template and can be incorporated into polymers suppress nuclease action. Phosphonate analogues of dTTP which bind to the triphosphate site and are suitable for base pairing with a poly(dA) template do not serve as substrates and are correspondingly ineffective in suppressing nuclease. Of special interest is the dideoxyribonucleotide analogue (ddTTP) which is incorporated at a rate 1000-fold slower than that of dTTP and to the extent of only 1 nucleotide per chain. Nevertheless ddTTP inhibits exonuclease activity as completely as does dTTP. These results indicate that the primer terminus in its complex with template and triphosphate is invulnerable to nuclease even though the rate-limiting step in polymerization is exceedingly slow.

Pyrophosphorolysis as Reversal of Polymerization—Template-directed chain extension as catalyzed by DNA polymerase results in a properly base-paired 3' terminus which is utilized for subsequent polymerization. Pyrophosphorolysis, as the reversal of polymerization, would be expected to show specificity for the product of polymerization, *i.e.* a base-paired 3' terminus. The experiments described show that only such a terminus is removed by pyrophosphorolysis. A mispaired primer terminus was not attacked by pyrophosphate. If such a terminus can be removed by pyrophosphorolysis, then it must be at a rate very much slower than that of hydrolysis. This specificity clearly distinguishes pyrophosphorolysis which attacks only base-paired termini from hydrolysis in which attack is only on unpaired termini. Thus there may be two forms of an enzyme-DNA complex. One form, with the primer terminus base-paired, would allow incorporation of a triphosphate or attack by pyrophosphate, and the other form, in which the 3' terminus is frayed would be available only for hydrolysis.

Proofreading Role for 3' → 5' Exonuclease—The observed specificity of the 3' → 5' exonuclease during polymerization suggests that this nuclease may act as a proofreading mechanism to remove a mismatched nucleotide. Inasmuch as the polymerase cannot extend a mispaired terminus, its exonuclease would remove a mismatched nucleotide generated by itself as well as those in the synthetic polymers studied here. The fidelity of template copying would be far greater as the result of having two base pairing selection steps: first the selection of a nucleotide for polymerization, and second, a determination that the primer terminus is properly paired before adding the next nucleotide. The presence of the same base pairing specificity in the exonuclease activity of two different DNA polymerases (T4 and *E. coli*) points to the fundamental importance of the exonuclease for accuracy in a template-directed polymerization process. Englund (8) has also suggested a role for the exonuclease of the T4 DNA polymerase in fidelity of template copying based on removal of unpaired nucleotides by this enzyme prior to polymerization. We have recently investigated yet a third DNA polymerase, DNA polymerase II isolated from *E. coli* by Kornberg and Gefter (29), and have shown that this enzyme may also contain a 3' exonuclease activity with this same specificity as greater than 80% of a base-paired terminus was retained during polymerization, while a mismatched terminus was quantitatively removed.

Defect in 3' → 5' Exonuclease May Be Mutagenic—Gene 43 of T4 codes for the DNA polymerase and certain lesions in the gene are known to have a profound effect on the spontaneous mutation rate (24, 30, 31). Allen *et al.* (32) have suggested on genetic grounds that the exonuclease activity of the T4 polymerase is involved in the fidelity of replication. If the proofreading exonuclease function of the polymerase is important for replication accuracy, then an alteration in exonuclease might cause mutagenic effects. A known mutagenic polymerase has been purified, and Hall and Lehman (11) have demonstrated in an *in vitro* experiment with the polymerase induced by T4 L56, a 4-fold increase in misincorporation of dTTP with a poly(dC) template as compared with the wild type T4 polymerase. We tested the L56 polymerase to determine whether it is defective in its ability to remove a mispaired nucleotide. In none of the cases tested was the mutator polymerase significantly different from the wild type polymerase in its exonuclease specificity (Table IV). In an experiment with polyinosinic acid, terminally labeled with dTMP residues and annealed to $d(C)_{1000}$ (the case most closely resembling the conditions of Hall and Lehman), there was less removal of the mispaired primer dT terminus by the mutator polymerase. However, more rigorous and sensitive assays, such as measuring direct transfer of dGTP to the terminus, will have to be performed to establish differences. Studies with other mutator DNA polymerases, as well as more extensive evaluation of the base pairing specificity, including all 12 possible nucleotide mismatches, will help in providing insights into the function of this nuclease.

Coordinated Roles of Exonuclease I and Exonuclease III—The specificities of the DNA polymerase and the other exonucleases of *E. coli* demonstrate that they are capable of coordinated hydrolytic action on a variety of structures at the 3' terminus (Table VI). Displaced single-stranded 3' termini which might result from various processes in repair, recombination, and replication of DNA would be rapidly degraded by either exonuclease I or the 3' → 5' exonuclease of DNA polymerase. Once the length of the displaced strand is reduced to 5 or 6 nucleotides, exonuclease I action becomes very slow, but the specificity of exonuclease III (or again, the DNA polymerase) would allow degradation into the base-paired region (Table VI). The nuclease activity of the DNA polymerase on displaced single strands is similar to the combined action of exonuclease I and exonuclease III. In a double-stranded region of DNA, the DNA polymerase would begin chain extension, whereas exonuclease III would continue hydrolysis leaving a single-stranded region or a gap. Gaps might also be produced by DNA polymerases if the amount of DNA precursors was severely limited. Such gaps might be important in the metabolism of DNA. The redundant action of these enzymes, and also their multifunctional nature indicates that each may be involved in more than one physiological process. Their redundant nature would insure the integrity of the metabolism of DNA and their multifunctional

TABLE VI

Summary of action of E. coli 3' → 5' exonucleases and DNA polymerase on various 3' termini at 37°

The structures depicted are a base-paired 3' terminus, an unpaired 3' terminus (mismatched), and a displaced 3'-terminal strand containing many unpaired nucleotides.

Structure	Action on 3' primer terminus			
	Chain extension by DNA polymerase	3' → 5' Hydrolysis by DNA polymerase	3' → 5' Hydrolysis by exonuclease I	3' → 5' Hydrolysis by exonuclease III
Base-paired terminus	Yes	No[a]	No	Yes
Mismatched terminus	No	Yes	No[b]	Yes
Displaced strand	No	Yes	Yes	No

[a] In the presence of triphosphates.
[b] Hydrolysis is very slow.

character would insure proper coordination of these vital molecular events.

REFERENCES

1. LEHMAN, I. R., AND RICHARDSON, C. C. (1964) *J. Biol. Chem.*, **239**, 233.
2. APOSHIAN, H. V., AND KORNBERG, A. (1962) *J. Biol. Chem.*, **237**, 519.
3. GOULIAN, M., LUCAS, Z. J., AND KORNBERG, A. (1968) *J. Biol. Chem.*, **243**, 627.
4. STEUART, C. D., ANAND, S. R., AND BESSMAN, M. J. (1968) *J. Biol. Chem.*, **243**, 5308.
5. BRUTLAG, D., ATKINSON, M. R., SETLOW, P., AND KORNBERG, A. (1969) *Biochem. Biophys. Res. Commun.*, **37**, 982.
6. SETLOW, P., BRUTLAG, D., AND KORNBERG, A. (1972) *J. Biol. Chem.*, **247**, 224.
7. KORNBERG, A. (1969) *Science*, **163**, 1410.
8. ENGLUND, P. T. (1971) *J. Biol. Chem.*, **246**, 5684.
9. INMAN, R. B., AND BALDWIN, R. L. (1964) *J. Mol. Biol.*, **8**, 452.
10. MALAMY, M. H., AND HORECKER, B. L. (1964) *Biochemistry*, **3**, 1893.
11. HALL, Z. W., AND LEHMAN, I. R. (1968) *J. Mol. Biol.*, **36**, 321.
12. JOVIN, T M., ENGLUND, P. T., AND BERTSCH, L. L. (1969) *J. Biol.. Chem.*, **244**, 2996.
13. LEHMAN, I. R., AND NUSSBAUM, A. L. (1964) *J. Biol. Chem.*, **239**, 2628.
14. KATO, K., GONÇALVES, J. M., HOUTS, G. E., AND BOLLUM, F. J. (1967) *J. Biol. Chem.*, **242**, 2780.
15. KELLY, R. B., COZZARELLI, N. R., DEUTSCHER, M. P., LEHMAN, I. R., AND KORNBERG, A. (1970) *J. Biol. Chem.*, **245**, 39.
16. RILEY, M., MALING, B., AND CHAMBERLIN, M. J. (1966) *J. Mol. Biol.*, **20**, 359.
17. CHAMBERLIN, M. J., AND PATTERSON, D. L. (1965) *J. Mol. Biol.*, **12**, 410.
18. HAYES, F. N., MITCHELL, V. E., RATLIFF, R. L., AND WILLIAMS, D. L. (1967) *Biochemistry*, **6**, 2488.
19. WU, R., AND KAISER, A. D. (1968) *J. Mol. Biol.*, **35**, 523.
20. WEISS, B., LIVE, T. R., AND RICHARDSON, C. C. (1968) *J. Biol. Chem.*, **243**, 4530.
21. ENGLUND, P. T. HUBERMAN, J. A., JOVIN, T. M., AND KORNBERG, A. (1969) *J. Biol. Chem.*, **244**, 3038.
22. ATKINSON, M. R., DEUTSCHER, M. P., KORNBERG, A., RUSSELL A. F., AND MOFFATT, J. G. (1969) *Biochemistry*, **8**, 4897.
23. DEWAARD, A., PAUL, A. V., AND LEHMAN, I. R. (1965) *Proc. Nat. Acad. Sci. U. S. A.*, **54**, 1241.
24. SPEYER, J. F. (1965) *Biochem. Biophys. Res. Commun.*, **21**, 6.
25. ENGLUND, P. T., DEUTSCHER, M. P., JOVIN, T. M., KELLY, R. B., COZZARELLI, N. R., AND KORNBERG, A. (1968) *Cold Spring Harbor Symp. Quant. Biol.*, **33**, 1.
26. BEYERSMANN, D., AND SCHRAMM, G. (1968) *Biochim. Biophys. Acta*, **159**, 64.
27. DEUTSCHER, M. P., AND KORNBERG, A. (1964) *J. Biol. Chem.*, **244**, 3019.
28. RICHARDSON, C. C., LEHMAN, I. R., AND KORNBERG, A. (1964) *J. Biol. Chem.*, **239**, 251.
29. KORNBERG, T., AND GEFTER, M. L. (1971) *Proc. Nat. Acad. Sci. U. S. A.*, **68**, 761.
30. SPEYER, J. F., KARMAN, J. D., AND LENNY, A. B. (1966) *Cold Spring Harbor Symp. Quant. Biol.*, **31**, 693.
31. DRAKE, J. W., ALLEN, E. F., FORSBERG, S. A., PREPARATA, R. M., AND GREENING, E. O. (1969) *Nature*, **221**, 1128.
32. ALLEN, E. F., ALBRECHT, I., AND DRAKE, J. W. (1970) *Genetics*, **65**, 187.
33. HUBERMAN, J. A., AND KORNBERG, A. (1970) *J. Biol. Chem.*, **245**, 5326.

A Possible Role for RNA Polymerase in the Initiation of M13 DNA Synthesis

(DNA replication/chloramphenicol/replicative form/rifampicin/single-stranded DNA)

DOUGLAS BRUTLAG, RANDY SCHEKMAN, AND ARTHUR KORNBERG

Department of Biochemistry, Stanford University School of Medicine, Palo Alto, California 94305

Contributed by Arthur Kornberg, September 16, 1971

ABSTRACT The conversion of single-stranded DNA of bacteriophage M13 to the double-stranded replicative form in *Escherichia coli* is blocked by rifampicin, an antibiotic that specifically inhibits the host-cell RNA polymerase. Chloramphenicol, an inhibitor of protein synthesis, does not block this conversion. The next stage in phage DNA replication, multiplication of the double-stranded forms, is also inhibited by rifampicin; chloramphenicol, although inhibitory, has a much smaller effect. An *E. coli* mutant whose RNA polymerase is resistant to rifampicin action does not show inhibition of M13 DNA replication by rifampicin. These findings indicate that a specific rifampicin–RNA polymerase interaction is responsible for blocking new DNA synthesis. It now seems plausible that RNA polymerase has some direct role in the initiation of DNA replication, perhaps by forming a primer RNA that serves for covalent attachment of the deoxyribonucleotide that starts the new DNA chain.

DNA polymerases from *Escherichia coli* and phage-infected cells extend DNA chains, but as yet they have not been shown to initiate a chain (1, 2). With single-stranded circles of M13 or φX174 phage DNA as templates for *E. coli* DNA polymerase, it appeared that chain initiation did take place (3). Subsequently however, the presence and participation of small, linear DNA fragments as primers to initiate synthesis was recognized (1, 4).

How then is new DNA synthesis initiated? One possibility is that all DNA synthesis takes place by covalent extension of preexisting DNA chains. Thus, oligonucleotide fragments of a DNA chain or ends produced by endonucleolytic scissions of a chain might serve as primers. Experience with available enzymes favors this possibility. An alternative to this possibility is that a new enzyme initiates chains by itself, or in conjunction with one of the known DNA polymerases. Studies with intact cells favor this suggestion, but no enzyme has yet been found to do this job.

Another alternative occurred to us. Since RNA polymerase starts new RNA chains, and DNA polymerase is known to covalently extend a ribonucleotide terminus during DNA synthesis (5), a brief transcriptional operation by RNA polymerase might provide an RNA primer for DNA synthesis. This priming piece of RNA could later be recognized and excised by nuclease action. Thus, a *de novo* initiation event catalyzed by RNA polymerase would be an essential step for the start of DNA synthesis. The synthesis of the double-stranded replicative forms of M13 DNA provided an excellent system

Abbreviations: SS, (phage) single-stranded DNA; RF, (phage) double-stranded replicative form.

in which to test this possibility. The first step in M13 DNA synthesis involves the conversion of the phage single-stranded DNA (SS) to a double-stranded replicative form (RF) (6). This step relies entirely on host-cell enzymes; it does not require the expression of any known viral gene or any new protein synthesis. The next stage, in which the replicative forms multiply (RF→RF), requires, among other things, the product of a viral gene (6).

In this report we show that both of these DNA replication events (SS→RF, RF→RF) are strongly and immediately inhibited by the antibiotic rifampicin, a specific inhibitor of the RNA polymerase initiation step. This rifampicin action is not due to inhibition of protein synthesis because chloramphenicol, a direct inhibitor of protein synthesis, does not prevent the SS→RF conversion and has only a small effect on RF multiplication. Rifampicin does not affect M13 DNA synthesis in an *E. coli* mutant with an RNA polymerase resistant to rifampicin.

MATERIALS AND METHODS

Materials were as follows: rifampicin and chloramphenicol from Calbiochem; unlabeled ribonucleotides from P-L Biochemicals; [^{3}H]CTP (3.2 Ci/mmol) and [^{3}H]uridine (27Ci/mmol) from Schwarz BioResearch Inc.; [^{3}H]thymidine (6.7 Ci/mmol) and [^{3}H]leucine (55 Ci/mmol) from New England Nuclear Corp.

Growth of cells and phage, and [^{3}H]thymidine-labeling of *E. coli* 5274 (F$^+$, T$^-$) and the wild-type M13 strains have been described (7, 8).

Rifampicin-resistant mutants of *E. coli* 5274 were isolated by plating 10^8 cells on agar that contained tryptone, as well as 50 μg/ml of rifampicin. The resistance to rifampicin of each of five independently isolated mutants was demonstrated to be due to an altered RNA polymerase by an *in vitro* assay of the polymerase activity in extracts of the bacteria. Both the resistant and sensitive strains were grown to late-log phase in 10 ml of Hershey broth, collected by centrifugation at 5000 × *g* for 5 min, resuspended in 1 ml of buffer A (see below), disrupted with a 100-W M.S.E. sonicator, and centrifuged at 10,000 × *g* for 10 min. 0.2 Volume of 1% protamine sulfate was added to the supernatant. The resulting precipitate was collected by centrifugation at 10,000 × *g* for 10 min, and suspended in 1 ml of buffer A. This fraction (50 μl) was used in an RNA polymerase assay (9) with [^{3}H]CTP (14,000 cpm/nmol) as the labeled nucleotide and calf-thymus DNA (125 μg/ml) as the template. Potassium phosphate (20 mM, pH 7.4) was present to inhibit polynucleotide phosphorylase. The enzyme fractions were incubated with rifampicin (5 μg/ml) at 25°C for 5 min before the assay. The rifampicin-sensitive strain incorporated 320 cpm in 10 min without rifampicin, but only 5 cpm with rifampicin present (corrected for incorporation in the absence of the three unlabeled triphosphates). The rifampicin-resistant strain incorporated 184 cpm without, and 121 cpm with, rifampicin.

In most experiments, the cells were grown on a glucose–casamino acids medium supplemented with 2 μg/ml of thymine (7). When protein synthesis was measured by the incorporation of [^{3}H]leucine, cells were grown on this medium with the casamino acids replaced by 19 amino acids (excluding leucine) at 5 μg/ml.

[^{3}H]Thymidine-labeled M13 phage was prepared as described (8). The titer was 1.9×10^{12} phage/ml, with a specific activity of 5.8×10^6 cpm/ml.

Intracellular M13 DNA was extracted by a modification of the procedure of Hirt (10). Cells treated with lysozyme were lysed by the addition of sodium dodecyl sulfate to 0.5% and incubated at 37°C for 10 min. A concentrated solution of NaCl was added to a concentration of 1 M, and mixed with the lysate by slow inversion and rolling of the tube. The lysate was chilled to 0°C for 2 hr, centrifuged at $20,000 \times g$ for 15 min, and the supernatant, containing most of the M13 DNA, was removed. In some cases the DNA was precipitated by adding 0.1 volume of 3 M sodium acetate (pH 6.5) and 2 volumes of cold ethanol and storing the mixture at −20°C overnight; the precipitate was collected by centrifugation at $10,000 \times g$ for 10 min and dissolved in 0.01 M Tris·HCl (pH 8.0)–1 mM EDTA.

Buffer A contains 0.05 M Tris·HCl (pH 8.0)–0.01 M MgCl$_2$–0.01 M 2-mercaptoethanol–0.05 mM EDTA. NET buffer (0.1 M) contains 0.1 M NaCl–1 mM EDTA–0.01 M Tris·HCl (pH 8.0). NET (1 M) is the same as NET (0.1 M), except that the NaCl concentration is 1 M.

Sucrose gradients were fractioned by collection of drops from the bottom of the tube directly into scintillation vials. Scintillation fluid(Triton X-100–toluene base; 10 volumes) was added and the samples were counted in a Nuclear Chicago scintillation counter.

RESULTS

Conversion of SS to RF is inhibited by rifampicin but not by chloramphenicol

Upon infection of *E. coli*, the M13 phage DNA (SS) was promptly converted to the double-stranded replicative forms (Fig. 1a). Fully covalent forms (RF I) predominated over nicked ones (RF II). Some of the parental DNA, cosedimenting with free phage*, still remained. With *E. coli* treated with 200 μg/ml of rifampicin for 5 min before infection, there was very little conversion of parental SS to RF; most of the DNA sedimented in the position of intact phage (Fig. 1b). Adsorption of the phage to cells was not affected by rifampicin; about two parental phage equivalents per cell were recovered with or

* This peak still possessed about 5% of the original specific infectivity of the infecting phage, and it also cosedimented with a phage-infectivity marker added to the lysate in other experiments. Uncoating of the viral DNA may be linked to the conversion of SS to RF (A. B. Forsheit and D. S. Ray, personal communication).

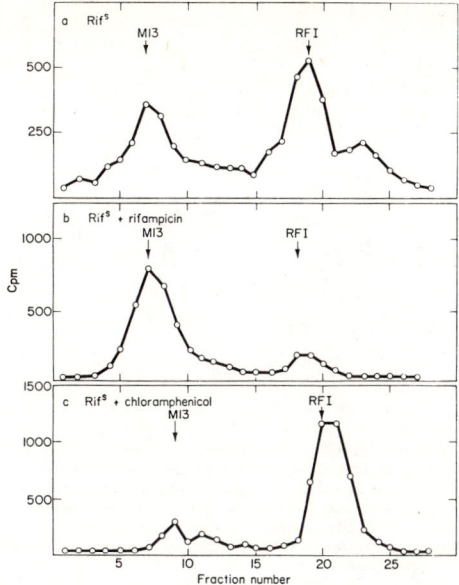

FIG. 1. Rifampicin inhibits the conversion of parental single-stranded DNA to the replicative forms. *E. coli* 5274 grown to 3×10^8/ml in 4 ml of medium at 37°C was either (a) untreated, (b) treated for 5 min with 200 μg/ml of rifampicin before infection, or (c) treated for 5 min with 100 μg/ml of chloramphenicol before infection. Cultures were then infected with M13-containing [^{3}H]DNA at a multiplicity of 100; the infection proceeded for 10 min at 37°C. Cultures were chilled on ice and 0.4 ml of 0.1 M KCN was added. Cells were collected by centrifugation at $5000 \times g$ for 10 min and resuspended with vigorous stirring in 1 ml of NET buffer (0.1 M) containing 10 mM KCN. The cells were washed one more time and resuspended in 1 ml of the same buffer. Lysozyme (0.1 ml of 4 mg/ml) was added and the cells were incubated at 37°C for 10 min. Sarkosyl (50 μl of 10%) was added to lyse the cells and the lysate was chilled and layered directly on a 5–20% sucrose gradient in NET buffer (1 M). After centrifugation for 15 hr at 23,000 rpm at 5°C in the SW27 rotor of the Spinco L2-65B ultracentrifuge, 0.8-ml fractions were collected and counted. Aliquots were titered for infectivity.

without rifampicin present. Uridine incorporation (presumably RNA synthesis) measured during the 10-min period of infection was reduced to 4–9% of control values by rifampicin treatment. Exposure of *E. coli* to 100 μg/ml of chloramphenicol did not inhibit the conversion of SS to RF (Fig. 1c, see also ref. 6).

To insure that the rifampicin inhibition is not a nonspecific effect due to the high concentrations of antibiotic used (and required for inhibition of intracellular RNA initiation, see ref. 11), a rifampicin-resistant mutant was obtained for study. This spontaneous mutant of *E. coli* 5274 possessed a rifampicin-resistant RNA polymerase as determined by assays of the enzyme in extracts (see *Methods*). With this mutant strain, conversion of SS to RF was not affected by

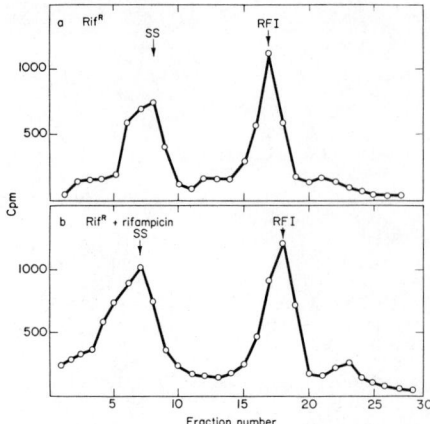

FIG. 2. Rifampicin does not inhibit conversion of parental DNA to RF in a rifampicin-resistant cell. This experiment was performed as described in Fig. 1, except that the *E. coli* 5274, rifampicin-resistant, mutant was used and the cells were lysed with 0.5% sodium dodecyl sulfate instead of Sarkosyl. (a) no treatment; (b) treated with 200 μg/ml of rifampicin for 5 min before infection.

rifampicin (Fig. 2). Thus, the primary effect of rifampicin in its inhibition of SS→RF appears to be on RNA polymerase.

Multiplication to RF is inhibited by rifampicin

Replication of RF requires at least one viral gene product, in addition to host-cell factors. The multiplication of double-stranded circular DNA molecules would appear to resemble chromosomal replication more than does the SS→RF conversion. In order to determine whether rifampicin would inhibit this step as well, *E. coli* 5274 was infected with phage M13 for 5 min, a time at which RF replication is well under way. Rifampicin was added to the culture at this point and the RF replication was observed by pulse-labeling of aliquots with [^{3}H]thymidine and measurement of its incorporation into RF by sucrose-gradient analysis. The rate of RF replication increased exponentially beyond 5 min after infection in the untreated control culture (12), and then leveled off at a constant rate about 5 min later (Fig. 3). Addition of 100 μg/ml of chloramphenicol at 5 min after infection did not inhibit the rate of RF replication for several minutes thereafter, even though the rate of protein synthesis, as judged by [^{3}H]leucine incorporation, had fallen to less than 2% after the first 3 min of treatment (Fig. 4). Thereafter, the rate of RF synthesis did fall, as expected from the requirement for continued protein synthesis during infection. Addition of rifampicin, however, produced an immediate inhibition of the rate of RF synthesis (Fig. 3), which fell exponentially for 20 min. In a separate experiment that measured the rate of [^{3}H]uridine incorporation under these conditions, the value fell to less than 4% within the first 3 min of treatment (Fig. 4). By 15 min after exposure to rifampicin, the rate of RF synthesis was nearly 100-times lower than that of chloramphenicol-treated cells. With the rifampicin-resistant strain, the rate of RF

synthesis continued normally in the presence of rifampicin (Fig. 3).

DISCUSSION

Our findings demonstrate that rifampicin inhibits the conversion of parental single-stranded phage DNA to the double-stranded RF, as well as inhibiting the further multiplication of RF. Because mutants with an altered RNA polymerase resistant to rifampicin also resist these inhibitory effects, we conclude that this inhibition of DNA synthesis is caused by a specific interaction with RNA polymerase. What has not been excluded is the possibility that the rifampicin–RNA polymerase complex inhibits DNA replication by an attachment to DNA that prevents DNA chain initiation at that point, or extension beyond that point. Yet we know that rifampicin does not inhibit ongoing replication of the host DNA, which must also be undergoing transcription.

Should the rifampicin–RNA polymerase interaction prove to be an interruption of an essential function of the enzyme in DNA synthesis, then a transcriptional role would seem most likely. SS→RF conversion is not dependent on protein synthesis, and RF multiplication is nearly 100-times more

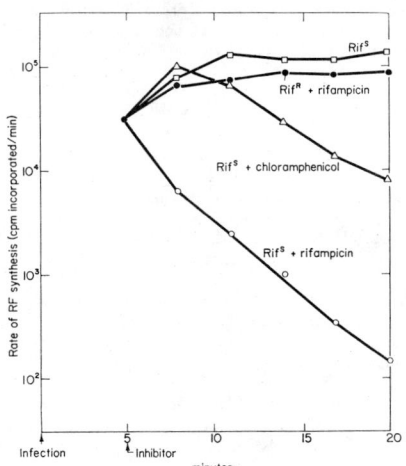

FIG. 3. Rifampicin inhibits the rate of RF synthesis. *E. coli* 5274 (or the rifampicin-resistant mutant) were grown at 37°C to 3×10^8/ml and infected with M13 at a multiplicity of 100. 100 μg/ml of chloramphenicol or 200 μg/ml of rifampicin were added 5 min after infection. At 3-min intervals, 9.5-ml samples were removed and exposed to 20 μCi of [^{3}H]thymidine for 1 min. Incorporation was stopped by pouring samples into 2 volumes of NET buffer (0.1 M)–10 mM KCN at 0°C. The cells were collected, washed, resuspended in 0.7 ml of 0.05 M Tris·HCl (pH 8.0)–1 mM EDTA, and lysed with 100 μg of lysozyme for 1–2 hr at 0°C. The M13 DNA was extracted by the modified Hirt procedure, precipitated by ethanol, and dissolved in 0.01 M Tris·HCl (pH 8.0)–1 mM EDTA. Samples were layered on 5–20% sucrose gradients in NET buffer (1 M) and centrifuged for 3 hr at 5°C in the SW56 rotor; 0.1-ml fractions were collected and counted. Data points are expressed as total counts in replicative forms I and II in each gradient.

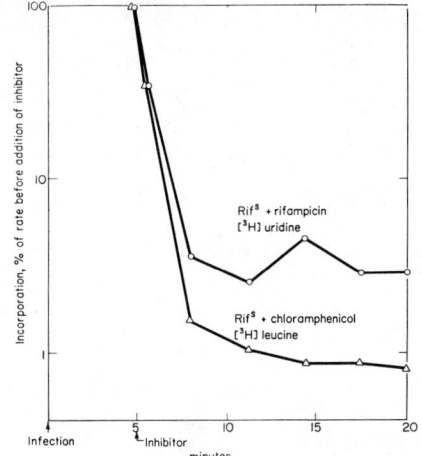

FIG. 4. The effect of chloramphenicol on protein synthesis and rifampicin on RNA synthesis in M13-infected *E. coli*. *E. coli* 5274 was grown to 3×10^8/ml at 37°C. At zero time, M13 was added at a multiplicity of 100. Just before 100 μg/ml of chloramphenicol or 200 μg/ml of rifampicin was added at 5 min after infection, 0.5-ml samples were labeled with 10 μCi of either [³H]uridine or [³H]leucine for 1 min and the labeling was stopped by the addition of cold 10% Cl₃CCOOH. Acid-insoluble material was determined by filtration on a Millipore HAWP filter and was counted in a scintillation counter. At 3-min intervals, samples were again taken and similarly pulse-labeled. The rates of incorporation are expressed relative to the initial rate without inhibitor added.

inhibited by rifampicin than by chloramphenicol. Thus, the RNA chains that are made by RNA polymerase would appear to serve some function other than as messengers for protein synthesis. Although these *in vivo* experiments cannot discriminate between transcription of the M13 DNA or *E. coli* DNA, preliminary experiments with cell-free extracts lacking almost all of the *E. coli* chromosomal material show a conversion of SS to RF that is inhibited by 5 μg/ml of rifampicin. Furthermore, this inhibition can be overcome by the addition of RNA polymerase purified from a rifampicin-resistant strain. These results suggest that RNA synthesis with the M13 viral strand as template is required for replication.

If such RNA, copied from the viral strand, is required for DNA synthesis, then this RNA could serve one of several functions. For example, RNA synthesis at a certain region on the parental DNA might disrupt some short, self-complementary regions that are known to exist in M13 DNA (13). Such disruption of helical structure might be necessary for proper initiation (14). The RNA synthesized might play a role itself, as in some macromolecular complex required for replication of M13 DNA (as has been proposed for the synthesis of *E. coli* DNA, K. G. Lark, personal communication). However, an attractive hypothesis for the function of such an RNA chain is that which originally prompted this investigation: to act as a primer terminus for covalent extension by a DNA polymerase.

We are grateful for the helpful advice of Drs. Costa Georgopoulos, Barry Marrs, and Michael Chamberlin. This work was supported in part by grants from the National Institutes of Health and the National Science Foundation. D. B. is a predoctoral fellow of the National Science Foundation.

1. Goulian, M., and A. Kornberg, *Proc. Nat. Acad. Sci. USA*, **58**, 1723 (1967).
2. Goulian, M., Z. J. Lucas, and A. Kornberg, *J. Biol. Chem.*, **243**, 627 (1968).
3. Mitra, S., P. Reichard, R. B. Inman, L. L. Bertsch, and A. Kornberg, *J. Mol. Biol.*, **24**, 429 (1967).
4. Goulian, M., *Cold Spring Harb. Symp. Quant. Biol.*, **33**, 11 (1968).
5. Berg, P., H. Fancher, and M. Chamberlin, in *Informational Macromolecules*, ed. H. J. Vogel, V. Bryson, and J. O. Lampen (Academic Press, New York, 1963), p. 467.
6. Pratt, D., and W. S. Erdahl, *J. Mol. Biol.*, **37**, 181 (1968).
7. Ray, D. S., and R. W. Schekman, *Biochim. Biophys. Acta*, **179**, 398 (1969).
8. Forsheit, A. B., and D. S. Ray, *Proc. Nat. Acad. Sci. USA*, **67**, 1534 (1970).
9. Berg, D., K. Barrett, and M. Chamberlin, in *Methods in Enzymology*, ed. L. Grossman and K. Moldave (Academic Press, New York, 1971), Vol. 21, p. 506.
10. Hirt, B., *J. Mol. Biol.*, **26**, 365 (1967).
11. Reid, P., and J. Speyer, *J. Bacteriol.*, **104**, 376 (1970).
12. Hohn, B., H. Lechner, and D. A. Marvin, *J. Mol. Biol.*, **56**, 143 (1971).
13. Schaller, H., H. Voss, and S. Gucker, *J. Mol. Biol.*, **44**, 445 (1969).
14. Dove, W. F., H. Inokuchi, and W. F. Stevens, in *The Bacteriophage Lambda*, ed. A. D. Hershey (Cold Spring Harbor Laboratory, New York, 1971), in press.

Isolation of an *E. coli* Strain with a Mutation affecting DNA Polymerase

by

PAULA DE LUCIA
JOHN CAIRNS
Cold Spring Harbor Laboratory,
Cold Spring Harbor,
New York 11724

By testing indiscriminately several thousand colonies of mutagenized *E. coli*, a mutant has been isolated that on extraction proves to have less than 1 per cent of the normal level of DNA polymerase. The mutant multiplies normally but has acquired an increased sensitivity to ultraviolet light.

KORNBERG'S discovery of an enzyme that could faithfully copy DNA *in vitro*[1] was a crucial step in the history of molecular biology because it firmly established the fact that only a small part of a cell's DNA is needed to code for a mechanism that can duplicate the whole. Whether this is the enzyme responsible for DNA duplication *in vivo* was rightly thought, at that time, to be of secondary importance. Since then, however, circumstantial evidence has accumulated suggesting that, at least in bacteria, this particular enzyme is used for the repair of DNA rather than for its duplication. The various mutants of *Escherichia coli* and *Bacillus subtilis* that are unable to duplicate their DNA at high temperature have all been shown to contain normal polymerase and normal deoxyribonucleoside triphosphate pools at the non-permissive temperature[2-6], and at least one of them has been shown to carry out repair synthesis at high temperature[7]. Repair replication and the process of DNA duplication apparently differ in the extent to which they discriminate against 5-bromouracil as an acceptable substitute for thymine, suggesting that the two reactions involve different polymerases[8]. Finally, the 5'-exonucleolytic activity, recently shown to be an intrinsic property of the *E. coli* polymerase[9], is clearly a desirable attribute for an enzyme responsible for excision and repair but is of no obvious advantage for an enzyme carrying out semiconservative replication.

These and other less persuasive arguments prompted us to look for mutants of the polymerase, in the hope that they would either establish a role for the polymerase in DNA duplication or exclude it and, at the same time, provide convenient strains in which to search for the right enzyme. Although we have not succeeded in these more distant objectives, we have isolated such a mutant and here describe the method of isolation and some of its properties. The accompanying article describes a genetic study of the mutation.

The Selective Procedure

The successful isolation of mutants of *E. coli* lacking ribonuclease I[10] demonstrated that it is possible to find the mutant one wants simply by testing individually several hundred colonies grown from a heavily mutagenized stock. Because we wished to avoid having to guess what symptoms, if any, would result from a lack of DNA polymerase, we decided to follow that example and assay the polymerase in clones of a mutagenized stock until we found what we were looking for. We had to allow for the possibility that the mutation we sought might be a conditional lethal, so we began by assaying at 45° extracts made from clones grown at 25° or 30°; later we tested clones grown at 37°, thinking that temperature-sensitive mutants of the polymerase might be more readily detectable if the enzyme had been assembled at a higher temperature. As it turned out, the mutant we eventually isolated would have been found whatever approach had been adopted, and we shall therefore simply give the history of the mutant when we describe its isolation and properties.

Extraction of Polymerase

Because we expected to have to test many hundred colonies, we required a very simple method for preparing extracts. In addition, we needed a procedure which made the bacteria incapable of incorporating deoxyribonucleosides, to ensure that labelled triphosphates could not enter DNA by way of breakdown to nucleosides and incorporation by those few cells that might have survived the extraction procedure. These two requirements were satisfied by the slight modification of a method devised for extracting polysomes, using the non-ionic detergent Brij-58 (ref. 11). *E. coli* is suspended at a concentration of about 3×10^9/ml. in ice cold 10 per cent sucrose 0·1 M Tris (pH 8·5); lysozyme and EDTA are added to final concentrations of 50 µg/ml. and 0·005 M, respectively, and the mixture is kept on ice for 30 min; addition of a mixture of Brij and MgSO$_4$ (at room temperature) to give final concentrations of 5 per cent and 0·05 M, respectively, results in partial clearing; following centrifugation (1,500g for 30 min), the deposit contains 99·9 per cent of the DNA and the supernatant contains the polymerase, which may then be assayed simply by adding sonicated calf thymus DNA (to 50 µg/ml.) and the four deoxyribonucleoside triphosphates (to a final concentration of 4 nmoles/ml. dATP, dGTP, dCTP and 2 nmoles/ml. ³H-TTP).

This extraction procedure demonstrates one point of interest: any method of lysis that liberates fragmented DNA will automatically create sites for the attachment of polymerase and therefore cannot give a true picture of the location of the polymerase *in vivo*[12]. Extraction with Brij yields cells which still contain their DNA but, on resuspension, have little if any ability to incorporate deoxyribonucleoside triphosphates. Because Brij apparently does not dissociate polymerase from its template (the polymerase being assayable in the presence of Brij), we can conclude that most of the polymerase in *E. coli* is normally not attached to DNA but lies free within the cell—as might befit an enzyme awaiting the summons to repair synthesis. This conclusion is supported by the observation that when *E. coli* segregates daughter cells which lack DNA these cells nevertheless retain their full quota of DNA polymerase[13,14].

Isolation of the Mutant

E. coli W3110 thy⁻, growing in minimal medium, was washed and suspended in 0·15 M acetate (pH 5·5), treated with N-methyl-N'-nitro-N-nitrosoguanidine (1 mg/ml.) for 30 min, and then centrifuged and suspended in Penassay broth[15]. Following growth at 25° C for 18 h, the culture was plated; after incubation overnight at 37° C, the colonies were picked into 1 ml. lots of Penassay broth which were incubated overnight at 37° C and then centrifuged and extracted with lysozyme and Brij.

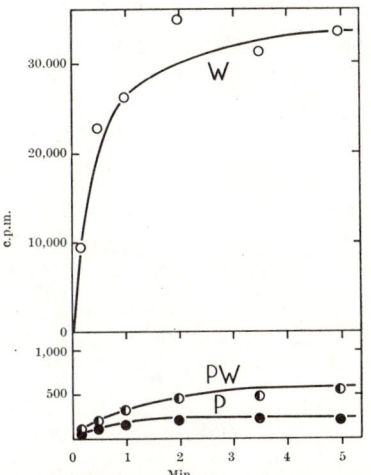

Fig. 1. Triphosphate incorporation by extracts of the parent strain (W), the mutant strain (P) and a mixture of 99 per cent mutant and 1 per cent parent (PW). *E. coli* W3110 thy⁻ and the mutant derivative, p3478, were grown with aeration in Penassay broth at 37° to about 5×10^8/ml. Each culture was then chilled, centrifuged and suspended in 0·1 M Tris–0·01 M MgSO₄ (pH 7·4) at a concentration of 6×10^9/ml. A mixture of 1 per cent parent strain, 99 per cent mutant strain was prepared. This and the two unmixed suspensions were centrifuged and suspended in Tris-Mg²⁺ at a concentration of 1×10^{10}/ml. The three suspensions were disrupted by sonic vibration and mildly centrifuged (1,000g for 10 min). To 0·9 ml. of each supernatant at 25° 0·1 ml. sonicated calf thymus DNA was added (final concentration 20 μg/ml.) and, 5 min later, 0·3 ml. triphosphate solution (final concentrations 100 nmoles/ml. dGTP, dATP, dCTP, and 0·6 nmoles/ml., 2·5 μCi/ml., ³H-TTP). Samples of 0·2 ml. were taken from each reaction mixture into 5 ml. 5 per cent trichloroacetic acid–1 per cent sodium pyrophosphate[16]. These samples were then washed on Whatman GFA filters with 5 per cent trichloroacetic acid and with 5 per cent acetic acid, dried and counted in a scintillation counter.

After testing a few thousand colonies in this way we found a clone, p3478, that appeared to lack polymerase activity. It was therefore tested again using a more conventional method for extracting the enzyme. According to this test (Fig. 1), extracts of the mutant have 0·5–1·0 per cent of the normal activity. This decrease in activity does not seem to arise from the presence of an inhibitor.

Some Properties of the Mutant

As far as we can determine, the mutant multiplies at the same rate as the parent strain, in minimal and complete media, and at temperatures from 25° to 42° C. On plating, it forms slightly smaller colonies than those of the parent strain, and occasionally it seems to have difficulty in getting out of stationary phase, but we have not investigated further either of these phenomena.

Parent and mutant are equally susceptible to infection with T4, T5, T7 and λ bacteriophages. When converted to spheroplasts, they are equally susceptible to φX174 DNA and produce equal yields of phage (personal communication from David Dressler). This finding was somewhat surprising, but it should be remembered that all stages in the replication of φX174 DNA are temperature sensitive in a mutant that is temperature sensitive for normal DNA replication[17] but not for repair synthesis[7].

With regard to host cell reactivation, there is no detectable increase in the rate of inactivation of T7 by ultraviolet light, when the survivors are assayed on the mutant rather than the parent. Thus the mutant is hcr⁺.

The mutant has a marked increase in sensitivity to ultraviolet light. For convenience, this effect will be documented in the following article[18], where the sensitivities of various derivative strains are compared.

The parent strain will form colonies normally in the presence of 0·04 per cent methylmethanesulphonate, whereas the mutant plates with an efficiency of about 10^{-7}. We assume that these rare methylmethanesulphonate-resistant cells are revertants that have either arisen spontaneously or been created by the methylmethanesulphonate. Because every one of twenty such independently arising revertants exhibited normal sensitivity to ultraviolet light and had normal or near-normal levels of polymerase, it is clear that the three basic properties of the mutant (UVˢ, MMSˢ and lack of polymerase) are the result of a single mutational step.

Repair or Replication

The accompanying article[18] demonstrates that we are dealing with an amber mutation which is recessive in partial diploids. We assume that it is in the gene coding for DNA polymerase, although proof will require the demonstration that it—or other similar mutations—results in changes in the polymerase protein. Because the mutation produces an increased sensitivity to ultraviolet light, it seems likely that recovery from the effects of ultraviolet light is partly the responsibility of this polymerase.

Unfortunately, it is not going to be easy, by a study of this or other such mutants, to show that this polymerase plays no part in normal DNA duplication. Because *E. coli* contains several hundred polymerase molecules per bacterium[19], the residual activity found in extracts of our mutant could represent perhaps 5–10 molecules per cell—a number that could well be sufficient for normal duplication. Even if we could somehow prove that the residual activity were entirely that of another enzyme (in other words, that this amber mutation is not measurably leaky), we should still not have proved that duplication is carried out by some other enzyme, for it could readily be argued that those few polymerase molecules concerned with duplication are necessarily incorporated into some larger enzyme complex the activity of which is not assayable *in vitro*. It could even be argued that more of the polymerase molecule must be intact for it to serve as a repair enzyme (and, incidentally, to survive extraction) than for it to act when part of the replicating machinery. We therefore believe that the question will be resolved either by engineering a total deletion of the polymerase gene or by determining, in some direct manner, which enzymes and what precursors are used for normal DNA duplication. It is our hope that each of these exercises will have been made easier now that the polymerase gene has probably been located[18] and a mutant is generally available.

We thank Dr Raymond Gesteland (who pioneered this kind of mutant hunt) for encouragement; Dr David Dressler for testing our mutant with φX174 and for permission to cite his results; and Drs Julian and Marilyn Gross for arranging to stay on at Cold Spring Harbor to conduct most of the experiments reported in the next article.

The work was supported by a grant from the US National Science Foundation.

Received November 26, 1969.

[1] Lehman, I. R., Bessman, M. J., Simms, E. S., and Kornberg, A., *J. Biol. Chem.*, **233**, 163 (1958).
[2] Bonhoeffer, F., *Z. Vererbungslehre*, **98**, 141 (1966).
[3] Buttin, G., and Wright, M., *Cold Spring Harbor Symp. Quant. Biol.*, **33**, 259 (1968).
[4] Fangman, W. L., and Novick, A., *Genetics*, **60**, 1 (1968).
[5] Gross, J. D., Karamata, D., and Hempstead, P. G., *Cold Spring Harbor Symp. Quant. Biol.*, **33**, 307 (1968).
[6] Hirota, Y., Ryter, A., and Jacob, F., *Cold Spring Harbor Symp. Quant. Biol.*, **33**, 677 (1968).
[7] Couch, J., and Hanawalt, P. C., *Biochem. Biophys. Res. Commun.*, **29**, 779 (1967).
[8] Kanner, L., and Hanawalt, P. C., *Biochim. Biophys. Acta*, **157**, 532 (1968).
[9] Kornberg, A., *Science*, **163**, 1410 (1969).
[10] Gesteland, R. F., *J. Mol. Biol.*, **16**, 67 (1966).
[11] Godson, G. N., and Sinsheimer, R. L., *Biochim. Biophys. Acta*, **149**, 476 (1967).
[12] Billen, D., *Biochim. Biophys. Acta*, **68**, 342 (1963).
[13] Cohen, A., Fisher, W. D., Curtiss, R., and Adler, H. I., *Cold Spring Harbor Symp. Quant. Biol.*, **33**, 635 (1968).
[14] Hirota, Y., Jacob, F., Ryter, A., Buttin, G., and Nakai, T., *J. Mol. Biol.*, **35**, 175 (1968).
[15] Adelberg, E. A., Mandel, M., and Chien Ching Chen, G., *Biochem. Biophys. Res. Commun.*, **18**, 788 (1965).
[16] Hurwitz, J., Gold, M., and Anders, M., *J. Biol. Chem.*, **239**, 3462 (1964).
[17] Steinberg, R. A., and Denhardt, D. T., *J. Mol. Biol.*, **37**, 525 (1968).
[18] Gross, J. D., and Gross, M., *Nature*, **224**, 1166 (1969).
[19] Richardson, C. C., Schildkraut, C. L., Aposhian, H. V., and Kornberg, A., *J. Biol. Chem.*, **239**, 222 (1964).

Deoxyribonucleic Acid Replication *in vitro*

Heinz Schaller, Bernd Otto, Volker Nüsslein, Julita Huf,
Richard Herrmann and Friedrich Bonhoeffer

*Friedrich-Miescher-Laboratorium der Max-Planck-Gesellschaft
and Max-Planck-Institut für Virusforschung
74 Tübingen, Germany*

(*Received 2 September 1971*)

We describe an *in vitro* system for DNA replication which uses a highly concentrated lysate of DNA polymerase I deficient *Escherichia coli* bacteria. The DNA synthesis observed *in vitro* proceeds over long periods of time, and the rates of total DNA synthesis and of chain elongation are 10 to 20% of the *in vivo* rate.

This *in vitro* DNA synthesis resembles *in vivo* replication in many aspects. It is semiconservative. Furthermore the DNA is synthesized in small pieces which become joined together upon prolonged incubation. Treatments which specifically inhibit *in vivo* replication affect the rate of *in vitro* synthesis. The rate is much reduced when the cells have been pretreated with ultraviolet light or mitomycin, when cells have been allowed to finish the replication cycle in the absence of further initiation, when synthesis takes place in the presence of nalidixic acid, or when synthesis takes place at non-permissive temperature in lysates of mutants that are temperature-sensitive with respect to DNA replication.

DNA synthesis depends on the presence of soluble macromolecules other than DNA polymerase I or II. These molecules must be present at a concentration comparable to their *in vivo* concentration.

1. Introduction

In order to study the complex enzymology of DNA replication and the function of the various components involved in this process, we will eventually be forced to investigate DNA replication in open (*in vitro*) systems, in which the site of synthesis is freely accessible to added macromolecules instead of being hidden behind the cellular membranes. In such systems we should be able to test single components for their function by artificial addition, removal or inactivation of the specific component.

DNA synthesis has been observed in several *in vitro* systems (Kornberg, A., 1969; Ganesan, 1968; Smith, Schaller & Bonhoeffer, 1970; Knippers & Strätling, 1970; Kornberg, T. & Gefter, 1970; Okazaki, Sugimoto, Okazaki, Imae & Sugino, 1970; Knippers, 1970; Moses & Richardson, 1970a; Ganesan, 1971). Recent studies indicate, however, that DNA synthesis observed *in vitro* is not necessarily related to DNA replication (Cairns, 1970). With the possible exception of Ganesan's system (Ganesan, 1971), DNA synthesis in all the open *in vitro* systems reported differs from replication in one or more respects, as for example in the sensitivity to specific inhibitors, in the formation and joining of Okazaki pieces (Okazaki *et al.*, 1968), in temperature sensitivity (when bacteria are used which are temperature sensitive with respect to DNA replication) in the kinetics of synthesis, in the rate of synthesis, in the rate of chain elongation, etc.

In this paper we describe a new and more complete *in vitro* system which differs from other systems in that it contains all macromolecular components of the bacterial cell at very high but ill-defined concentrations. The DNA synthesis observed resembles DNA replication in almost every respect tested.

2. Materials and Methods

(a) Bacterial strain

A thymine-requiring derivative of *E. coli* H560 *pol* A1$^-$ *endo*I$^-$, kindly provided by Dr H. Hoffman-Berling, was used.

(b) Chemicals

Penassay broth, Bacto agar (Difco Laboratories, Detroit, Michigan). TTP, dATP, dCTP, dGTP, ATP, thymidine, cytidine, thymine (Schwarz Bioresearch Co., Orangeburg, N.Y.)†. Lysozyme, chloramphenicol, mitomycin C, Brij 58 (polyethyleneglycol monostearyl ether) (Serva Feinbiochemica GmbH & Co., Heidelberg). Deoxyribonuclease (Worthing Biochemical Corporation, Freehold, New Jersey). Morpholino propane sulphonic acid, EGTA (ethylene glycol-bis-(2 amino ethylether)-N,N'-tetra-acetic acid), EDTA (Sigma Chemical Co., St. Louis, Missouri). Nalidixic acid (The Bayer Products Co., Surbiton upon Thames). Sarkosyl NL 97 (J. R. Geigy A. G., Basle), NAD (Boehringer GmbH, Mannheim). 5-Bromodeoxyuridine triphosphate was prepared chemically.

(i) *Radiochemicals*

[^{14}C]Thymine, [^{3}H]thymine, [^{3}H]TTP, [^{3}H]dCTP (The Radiochemical Centre, Amersham, Bucks.). [α-^{32}P]TTP was prepared enzymically.

(ii) *Filters*

Nitrocellulose filters, 24 mm diameter, 0·15 μm pore size (Membranfiltergesellschaft, Göttingen).

(iii) *Cellophane membrane*

Kalle Einmach Cellophan^R, 22 μm thick (Kalle, Wiesbaden Biebrich).

(c) Media and buffers

(i) *Growth medium*

Penassay broth supplemented with 2 μg [^{14}C]thymine/ml. (320 or 960 mCi/mole).

(ii) *Minimal medium*

Tris-buffered salt–glucose medium supplemented with 2 μg thymine/ml.

(iii) *Agar plate B*

2% Bacto agar, 20 mM-morpholino propane sulphonic acid, 5 mM-$MgCl_2$, adjusted to pH 7·5 with NaOH.

(iv) *Agar plate A*

Same as B, plus 10 mM-EGTA, 0·33 M-sucrose.

(v) *Incorporation buffer*

20 mM-morpholino propane sulphonic acid, pH 7·2, 100 mM-KCl, 5 mM-$MgCl_2$.

† The incorporation with Sigma deoxyribonucleoside triphosphates is low and not linear. This may be due either to a poisonous substance in the Sigma triphosphate preparations, or to a necessary co-factor which is present in the Schwarz triphosphates.

(vi) *Incorporation mixture*

Incorporation buffer plus 1 mM-ATP 20 μM each of dATP, dCTP, dGTP and TTP, and either tritiated dCTP or TTP (500 Ci/mole), and 170 μM-cytidine or thymidine depending on the radioactive label. For density label, TTP is replaced by 5-bromodeoxyuridine triphosphate.

(d) *Experimental methods*

The rationale of the procedure is to lyse the cells very carefully in order to leave the DNA and any DNA membrane complex intact, and at the same time to keep all the macromolecular components in the lysate at high concentrations comparable to those *in vivo*. To this end, the following procedure is used.

(i) *Preparation of the system*

Bacteria were grown at 37°C in [^{14}C]thymine-containing medium to 2×10^8 cells/ml. The culture was cooled to 0°C. The bacteria were harvested, washed once in ice-cold minimal medium and resuspended at a final concentration of 5×10^{10} cells/ml. in minimal medium containing 0·5% Brij 58. 1 μl. of lysozyme (1 mg/ml.) was spread on a Cellophane membrane disc of 12 mm diameter. This disc had previously been placed on agar plate A at 0° to 4°C (see above). 1 μl. of the cell suspension was spread on to the disc. When more than 5×10^7 bacteria were needed, a correspondingly larger disc area was used. After 20 min, the disc was transferred to agar plate B and incubated for another 10 min in the cold. The cells lysed, due to osmotic shock. The second agar plate was kept uncovered to allow the lysate to dry off.

(ii) *DNA synthesis* in vitro

The disc was transferred to a warm 50-μl. drop of incorporation mixture at 37°C and incubated in a Petri dish which was closed in order to avoid evaporation. The temperature of the lysate was kept close to 35°C, as measured with a thermocouple. Care was taken that the lysate-carrying surface had no direct contact with the liquid. The reaction was stopped by putting the disc into 0·5 ml. of a solution of 0·5 M-NaOH and 1% sodium dodecyl sulphate.

A large amount of non-radioactive thymidine or deoxycytidine was included in the incorporation mixture to ensure that [^{3}H]TTP or [^{3}H]dCTP was not incorporated by any intact cells after degradation to [^{3}H]dT or [^{3}H]dC. Control experiments showed that, in general, this precaution was unnecessary.

(iii) *DNA synthesis* in vivo

Cells were grown as described above. Tritiated thymine (final activity 32 Ci/mole) was added to the culture, or the cells were filtered and resuspended in growth medium containing tritiated thymine (32 Ci/mole) and incubated at 37°C.

(iv) *Determination of incorporated activity*

All samples (discs, fraction of gradients and 0·5-ml. samples of cultures) were added to 0·5 ml. of 0·5 N-NaOH, 1% sodium dodecyl sulphate, 0·5 mg crude herring sperm DNA/ml., 10% saturated sodium pyrophosphate. After heating for 5 min to 80°C and cooling, DNA was precipitated by addition of 1 ml. of 2 M-trichloroacetic acid and collected on nitrocellulose filters. The filters were washed with 15 ml. of 0·05 M-trichloroacetic acid, dried and counted in a liquid-scintillation counter. The amount of incorporated TTP or thymine was calculated from the specific activity, counting efficiency and ^{3}H cts/min corrected for overlap from ^{14}C radioactivity.

The ^{14}C radioactive material incorporated into 10^8 bacteria was determined. This value served to normalize the ^{3}H incorporation values with respect to the number of lysed cells used in the assay. This normalization eliminated any inaccuracy due to handling small volumes.

(v) *Ultraviolet irradiation*

An u.v. lamp (Sterisol F1140 with an Osram 20-W bulb) was used. As measured by the surviving rate of T4v_1 phages (Wulff, 1963), its dosage at a sample distance of 72 cm is 5 erg/mm²/sec. The phage was kindly provided by Dr N. Symonds.

3. Results

(a) Description of the system

The system consists of a highly concentrated bacterial lysate covering a Cellophane membrane disc which is layered on a buffer containing the four deoxyribonucleoside triphosphates. These DNA precursors diffuse through the membrane disc to the lysate and are incorporated into acid-precipitable alkali-stable material (see Materials and Methods). As will be shown in the following sections, the incorporation had many of the specific properties expected of *in vitro* DNA replication. In this section, some less specific properties and dependencies of the system will be discussed.

The kinetics of DNA synthesis is shown in Figure 1. At first there is a brief lag which is probably due to the time needed for the triphosphates to diffuse through the Cellophane membrane to the lysate. After this initial lag, DNA synthesis starts and continues over a long period of time. This is different from the synthesis observed in other *in vitro* systems (Smith *et al.*, 1970; Knippers & Strätling, 1970; Okazaki *et al.*, 1970). The overall rate of synthesis is proportional to the number of lysed bacteria covering the disc, provided the number of bacteria per cm^2 is kept between 10^7 and 10^8. The dependence of the activity on KCl, hydrogen ion, ATP and deoxyribonucleotide concentration is shown in Figures 2, 3, 4 and 5. These curves are similar to those

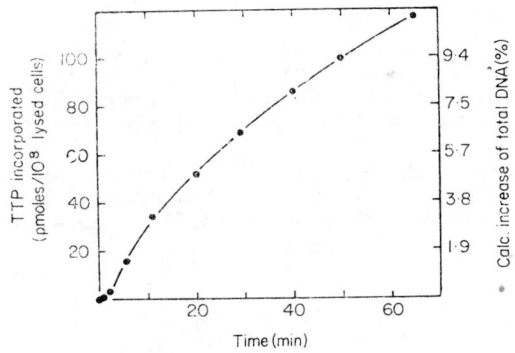

Fig. 1. Kinetics of DNA synthesis. The calculation of the relative increase of DNA during synthesis is based on the DNA-thymine content of 0·134 μg thymine/10^8 cells.

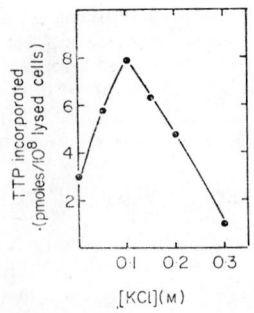

Fig. 2. Dependence of *in vitro* synthesis on KCl concentration. Synthesis is carried out as described in Materials and Methods, except that the incorporation mixture contains various concentrations of KCl. Incorporation time: 5 min.

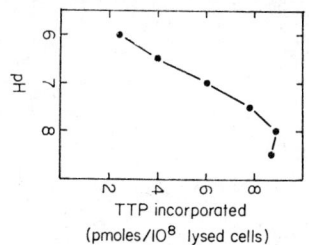

FIG. 3. pH dependence of *in vitro* synthesis. Synthesis is carried out as described in Materials and Methods, except that the incorporation mixture has been adjusted to various pH values. Incorporation time: 5 min.

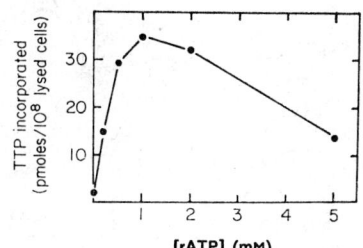

FIG. 4. rATP dependence of *in vitro* synthesis. Synthesis is carried out as described in Materials and Methods, except that the incorporation mixture contains various concentrations of rATP. Incorporation time: 10 min.

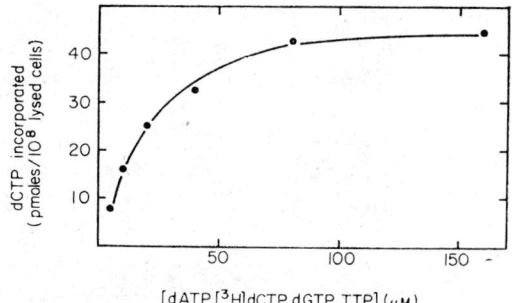

FIG. 5. Dependence of *in vitro* synthesis on the concentration of deoxyribonucleoside triphosphates. Incorporation is carried out as described in Materials and Methods, with the exception that the incorporation mixture contains various concentrations of the four deoxyribonucleoside triphosphates. Incorporation time: 10 min.

observed for DNA synthesis in other *in vitro* systems (Smith *et al.*, 1970; Knippers & Strätling, 1970) and in *E. coli* cells, which have been made permeable to low molecular weight precursors (Vosberg & Hoffmann-Berling, 1971). DNA synthesis is dependent on the presence of ATP; deoxynucleoside triphosphates are utilized with an apparent K_s of 20 μM. The temperature dependence (Fig. 6) is unusual: the rate is maximal below the physiological temperature and a sharp decrease in rate is observed above 40°C.

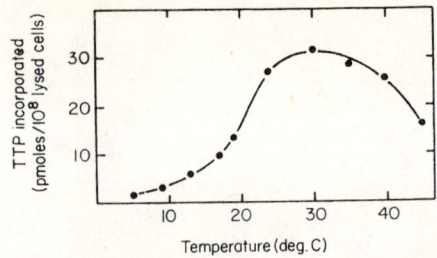

Fig. 6. Temperature dependence of *in vitro* synthesis. Synthesis was carried out as described in Materials and Methods, with the exception that the incorporation takes place at various temperatures. Incorporation time: 8 min.

As can be seen from the preceding paragraph, the standard conditions that we used for *in vitro* synthesis did not give the maximal rate of synthesis. They have rather been chosen to be similar to *in vivo* conditions.

The effects of various poisons on the system is shown in Table 1: *p*-chloromercuribenzoate and the magnesium complexing agent EDTA almost completely destroyed the activity. Azide (1 mM) and cyanide (1 mM) showed no inhibitory effect. More specific inhibitors for DNA replication are discussed in section (f). As expected for an *in vitro* system, practically no incorporation of deoxyribonucleoside triphosphates into acid-precipitable material was observed in the presence of deoxyribonuclease, which degraded 95 to 98% of the parental DNA. It should be mentioned that RNase inactivates the system by a factor of two, when cells are lysed in the presence of EDTA (K. G. Lark, personal communication). As shown in Table 1, DNA synthesis is dependent on the presence of all four deoxyribonucleoside triphosphates. If one or two triphosphates are omitted from the system, synthesis is markedly reduced. Deoxyribonucleoside diphosphates compete with the corresponding triphosphates very efficiently as indicated by the reduction of incorporation of [^{3}H]TTP when TDP is present. This result could be explained by the very rapid interconversion of TDP and TTP in the *in vitro* system, as demonstrated by experiments (not shown) with reaction mixtures containing [^{3}H]TTP and [α-^{32}P]TDP. Therefore, this *in vitro* system does not show

TABLE 1

Properties of the system

Standard incorporation	100%
+ PCMB (20 mM)	< 2%
+ EDTA (20 mM)	< 2%
+ EGTA (20 mM)	60%
+ NaN$_3$ (1 mM)	100%
+ KCN (1 mM)	100%
+ TTP (20 mM)	60%
+ TDP (20 mM)	58%
+ TMP (100 mM)	70%
+ dBrUTP (20 mM)	50%
− dCTP	12%
− dCTP − d-GTP	6%
+ DNase (1 μg/disc)	< 2%
+ RNase (1 μg/disc)	> 90%

whether the immediate precursor for DNA synthesis is the nucleoside di- or triphosphate. Thymidine monophosphate is not used efficiently for DNA synthesis, as shown in Table 1. 5-Bromodeoxyuridine triphosphate competes efficiently with its analogue TTP. As will be shown later, TTP can be completely replaced by 5-bromodeoxyuridine.

(b) *Semi-conservative DNA synthesis*

In order to show that DNA synthesis in the system is semi-conservative (Meselson & Stahl, 1958), newly synthesized DNA was labelled with a density marker and radioactive material. In the incorporation mixture dTTP was replaced by its analogue dBUTP and [^{3}H]dCTP was used for radioactive label. The replacement of dTTP by dBUTP had no influence on the rate of DNA synthesis, as determined by the uptake of [^{3}H]dCTP. Figure 7(a) shows the density distribution of parental ^{14}C and newly synthesized [^{3}H]DNA. Newly synthesized DNA bands at the position of hybrid DNA, which consists of one heavy (BrUra) and one light (thymine) strand. Therefore, some ^{14}C prelabelled DNA would be expected at hybrid density. In this particular experiment, the prelabelling of the parental DNA was too weak to allow the detection of the expected 6% of hybrid [^{14}C]DNA in the presence of ^{32}P marker DNA. In other experiments, with stronger prelabelling, the expected amount of ^{14}C parental DNA was found at hybrid density.

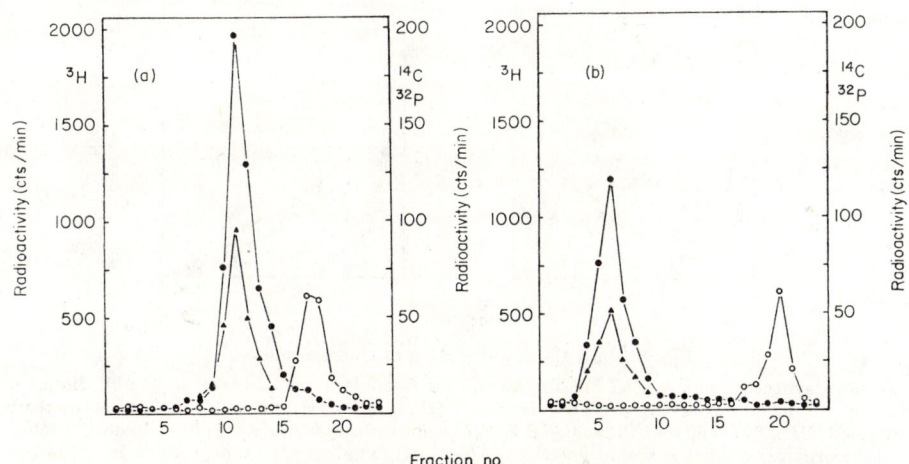

Fig. 7. Semi-conservative DNA synthesis. *In vitro* synthesis was carried out for 30 min on Cellophane membrane of 24 mm diameter floating on 0·2 ml. of incorporation mixture. The mixture contained [^{3}H]dCTP (20 μM, 500 mCi/m-mole) and dBUTP (20 μM). 2·2 pmoles of dCTP were incorporated/10^8 cells/min. DNA preparation for CsCl-gradient centrifugation was carried out as described in Fig. 9, except that only a part of the DNA was denatured (b), whereas the other part remained native (a). 2·65 g of CsCl were dissolved in 2·0 g solution. [^{32}P]BrUra hybrid *E. coli* DNA and single-stranded [^{32}P]BrUra *E. coli* DNA served as density markers. (a) native DNA; (b) denatured DNA.

—●—●—, ^{3}H radioactivity in newly synthesized BrUra labelled DNA. —○—○—, ^{14}C radioactivity prelabelled parental *E. coli* DNA; 33% of the ^{32}P counts are subtracted to account for overlap. —▲—▲—, ^{32}P radioactivity reference.

If this newly synthesized DNA is denatured (Fig. 7(b)), the light and the heavy strand separate and the ^{3}H radioactivity is found at the density of heavy single-stranded DNA. From these results we conclude that the DNA synthesis is semiconservative.

(c) *Formation and joining of Okazaki pieces*

In vivo, at least one of the new DNA strands is synthesized discontinuously (Okazaki *et al.*, 1968); first, molecules consisting of about 2000 nucleotides are formed; subsequently, these are joined together to form high molecular weight DNA. The same process can be observed in our *in vitro* system.

DNA which is synthesized during a short early pulse sediments slowly in alkaline sucrose gradients, with a sedimentation coefficient similar to that of Okazaki pieces. Such DNA molecules could grow either continuously or discontinuously. In order to exclude the continuous growth mechanism, the following two experiments are presented. The first experiment (Fig. 8(a)) involves a late pulse. It shows that even after 10 minutes pre-incubation, most of the ^{3}H pulse-labelled DNA sediments slowly, whereas the DNA synthesized during the 10 minutes immediately preceding the pulse (and labelled with ^{32}P) sediments much faster. Thus, the pulse-labelled DNA is initially not joined to the DNA, which was synthesized during the previous 10 minutes. The second experiment (Fig. 8(b)) is a pulse-chase experiment, which shows that pulse-labelled DNA which

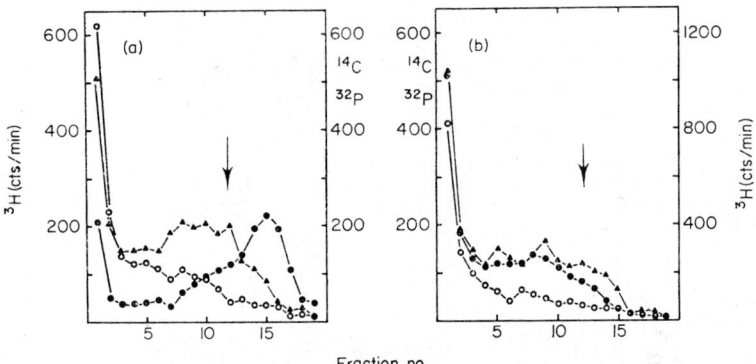

FIG. 8. Formation and joining of Okazaki pieces.

In vitro synthesis was carried out for 30 min on Cellophane membranes of 24 mm diameter floating on 0·2 ml. of incorporation mixture. For pulse-labelling, the incorporation mixture contained [^{3}H]TTP, 2500 mCi/m-mole and for long-time incorporation [α-^{32}P]TTP, 10 mCi/m-mole. Both incorporation mixtures contained NAD (0·2 mM). The reaction was stopped in 0·7 ml. of 0·2 N-NaOH and 1% Sarkosyl.

The DNA was carefully removed from the disc to avoid shearing and the sample was layered on an alkaline sucrose gradient (20 to 5% sucrose, 0·5 M-NaCl, 0·1 mg herring sperm DNA/ml., 10^{-3} M-EDTA, 0·1% Sarkosyl, 0·01 M-Tris adjusted to pH 12 with NaOH). Centrifugation was carried out in a Spinco SW41 Ti rotor for 7 hr at 20°C at 33,000 rev./min. ^{32}P-labelled fd DNA was used as reference in a separate tube.

(a) 10 min *in vitro* synthesis ([α-^{32}P]TTP) is followed by a 40-sec pulse ([^{3}H]TTP).

(b) A 40-sec pulse ([^{3}H]TTP) is chased by 10 min long-time incorporation ([α-^{32}P]TTP). Before the pulse, the disc had been incubated at 4°C for 20 sec in the presence of [^{3}H]TTP.

—●—●—, ^{3}H radioactivity (40-sec pulse); —▲—▲—, ^{32}P radioactivity (10 min incorporation); —○—○—, ^{14}C radioactivity (prelabel of the *E. coli* cells). The arrow shows the position of fd DNA.

was originally low in molecular weight can be chased to high molecular weight DNA during a subsequent 10 minutes incubation period.

Thus, similarly to *in vivo* replication, low molecular weight DNA molecules are intermediates in the *in vitro* synthesis. Whether both or only one strand is synthesized discontinuously awaits further investigation.

(d) *Rate of total DNA synthesis*

The rate of total synthesis can be expressed as the number of thymine nucleotides incorporated per lysed bacterium per second. This may be calculated from the amount of [^{3}H]TTP incorporated into the lysate, the specific activity of [^{3}H]TTP, the length of the incubation period and the number of lysed cells contained in the lysate. The last of these can be obtained from the ^{14}C prelabel in the lysate, if the ^{14}C prelabel per bacterium has been determined. For strain H560, the *in vitro* rates at 37°C vary from 200 to 300 thymine-nucleotides/sec/bacterium. This corresponds roughly to the synthesis of half a chromosome per bacterium per generation (50 min). For certain mutant strains or with optimized conditions, higher rates have been observed.

This rate of incorporation is still low compared to the rate of *in vivo* replication, which is 2100 thymine-nucleotides/sec/bacterium. This number has been determined from the generation time (50 min) and the number of acid-precipitable thymine molecules per bacterium (6.4×10^6, see the legend of Fig. 1). From this number we calculate that, *in vivo* 3·5 chromosomes are synthesized per bacterium per generation. This is at least seven times higher than the *in vitro* synthesis rate. Since, however, the temperature dependence of *in vitro* and *in vivo* synthesis are different (Fig. 6), the rates approach each other with decreasing temperature.

(e) *Rate of chain elongation*

There are two possible causes for the finding that the rate of total DNA synthesis *in vitro* is lower than the *in vivo* rate: conceivably not all the growth points which are active *in vivo* are active *in vitro*; it is also possible that the actual rate of fork movement is reduced *in vitro*. In this section, we present evidence that the latter explanation is correct.

In the following, we have to distinguish between four different terms: rate of total DNA synthesis, rate of polymerization, rate of chain elongation and rate of fork movement. The rate of total DNA synthesis (moles thymine/bacterium/sec) has been discussed in the last section; it is measured in moles of thymine incorporated into the DNA of one bacterium (or the corresponding lysate) during one second. The rate of polymerization (nucleotides/strand/sec) is the rate at which a low molecular weight intermediate (Okazaki piece, approx. 500,000 mol. weight) is synthesized, probably by terminal addition of nucleotides. The rate of chain elongation (nucleotides/strand/sec.) is the rate at which high molecular weight DNA (molecular weight greater than 2×10^6) is growing, possibly by terminal addition of Okazaki pieces. Thus, the rate of chain elongation depends on at least two processes, polymerization and joining. The fork rate, finally, is the rate at which the growth point traverses the chromosome. It is measured in nucleotide pairs per second per fork, and would normally correspond to one chromosome/generation/fork if replication is unidirectional, or to half that value if replication is bidirectional. In healthy bacteria or in well-functioning *in vitro* systems, the rate of chain elongation equals the fork rate. In systems in which the joining activity is inhibited, however, the fork rate may become greater than the rate of chain elongation. In such systems one would expect to find an accumulation of low molecular weight DNA.

The following experiments show that: (a) the rate of chain elongation in this system can be as high as 10 or 20% of the *in vivo* rate; (b) the rate of chain elongation of the average growing molecule is not more than 5 or 10% of the *in vivo* rate; (c) there is no accumulation of small molecular weight DNA and that therefore: (d) the *in vitro* fork-rate cannot be considerably greater than the *in vitro* rate of chain elongation, i.e. 10% of the *in vivo* fork-rate. The fact that in the *in vitro* system the fork-rate is reduced about as much as the total rate of synthesis, suggests that the reduction is not due to a loss of active growth points.

The rate of chain elongation can be determined by measuring the time required for the synthesis of a long continuous stretch of single-stranded DNA of known size (Bonhoeffer & Gierer, 1963). In order to measure the length of the single-stranded DNA synthesized during a given period of time, the newly synthesized DNA was labelled both with radioactivity ([^{3}H]dCMP) and with a density label (dBUMP). The DNA was then extracted and gently sheared to disconnect most of the newly synthesized DNA from most of the pre-existing DNA. The strength of shear treatment and the time of synthesis were chosen so that the size of DNA molecules produced by shearing were of the same order of magnitude as the size of the newly synthesized DNA. After shear treatment the DNA was denatured and analysed in CsCl density-gradients. The density distribution of radioactive DNA was broad, extending from the density of heavy single-stranded DNA to the density of light single-stranded DNA (Fig. 9(a)).

The experiments recorded in Figure 9 show that the size of the largest fragments of completely new (i.e. fully heavy) DNA increases steadily with the time of DNA synthesis. By 8 minutes, newly synthesized pieces of the size of denatured λ DNA (1.5×10^7 daltons) can be observed. Therefore, the rate of chain elongation can be at least as high as 15×10^6 daltons/8 minutes. This is 10% of the *in vivo* fork-rate if replication occurs unidirectionally, or 20% if it is bidirectional.

In the last paragraph we only considered the rate of chain elongation of the fastest-growing molecules. One can also estimate from the same experiment the rate of chain elongation of the average molecule. Figure 9(b) shows clearly that the average molecular weight of the newly synthesized DNA increases with time. A rough estimate of the average molecular weight after 8 or 16 minutes of synthesis leads to an average rate of chain elongation of 10^6 daltons/min/strand. This is 5% of the *in vivo* fork-rate if replication occurs unidirectionally, or 10% if it is bidirectional.

The slow rate of chain elongation is not a consequence of insufficient joining, because Figure 9 does not show any accumulation of DNA of small molecular weight during long periods of synthesis. Therefore, the rate of chain elongation equals the fork-rate. We conclude that the fork-rate is reduced as much as is the total rate of synthesis, and that all the genuine growth points which are active *in vivo* are also active *in vitro*.

There are several possible reasons why the rate of DNA replication is reduced in the *in vitro* system. The reaction conditions are certainly not optimal, the topology of the replication complex may have been changed during lysis, the diffusion of essential components to the site of synthesis may have become rate limiting, or necessary components of low molecular weight may have been lost during the preparation of the lysate, etc.

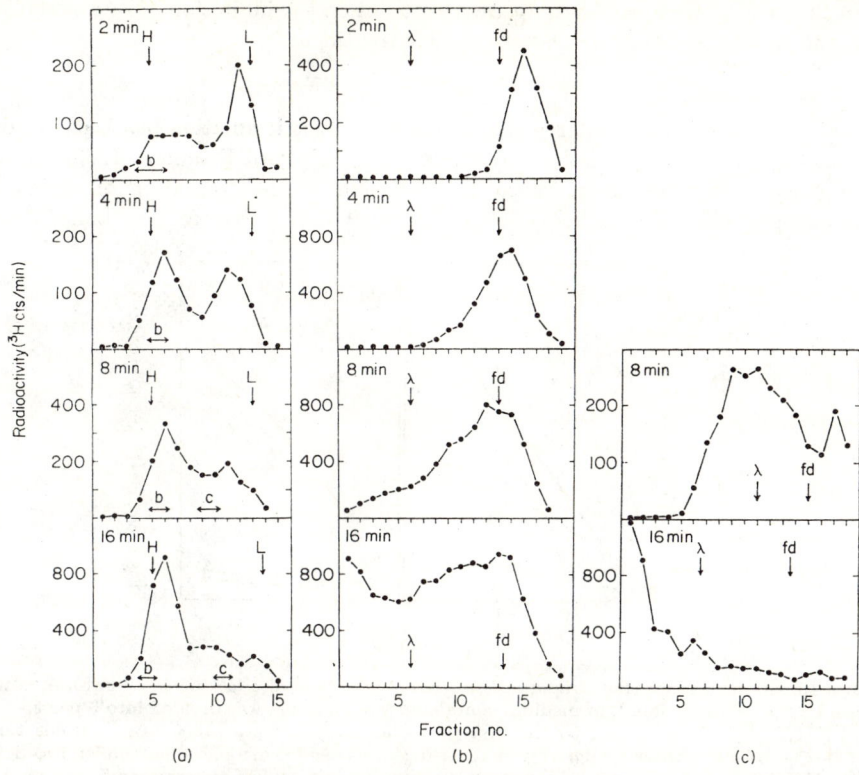

Fig. 9. Density distribution (a) and size distribution (b) of BrUra-labelled denatured DNA.

(a) *In vitro* synthesis was carried out on Cellophane membranes of 34 mm diameter floating on 0·4 ml. of incorporation mixture, which contained [^{3}H]dCTP (20 μM, 500 mCi/m-mole) dBUTP (10 μM) and TTP (0·5 μM). Incorporation was carried out during 2, 4, 8 and 16 min. The reaction was stopped in 1·5 ml. of a solution containing EDTA (0·1 M, pH 7·6) and 1% Sarkosyl. The samples were treated with pronase (1 mg/ml.) for 12 hr. The DNA was removed from the membrane by treatment with a Vortex mixer for 30 sec. 100 μg of crude herring sperm DNA and 3 μg of *E. coli* BrUra DNA was added; then the DNA was fragmented by pressing the solution twice through a needle (0·6 mm × 38 mm) of a motor-driven syringe at constant rate (1·1 ml./3 sec). The DNA was denatured by alkali treatment, the solution was readjusted to pH 8·5 by addition of 0·2 M-Tris (pH 7). 2·70 g CsCl were added to 2 g of DNA solution. The solution was centrifuged for 48 hr at 30,000 rev./min and 20°C in a SW50.1 rotor of a Spinco centrifuge. Thereafter fractions were collected and a part of each was used for radioactivity assay. ^{14}C-pre-labelled DNA and BrUra ^{14}C-labelled *E. coli* DNA served as density markers. Their positions are indicated by arrows.

(b) and (c). Size distribution of BrUra labelled DNA.

Some heavy (b) as well as some lighter (c) fractions of the CsCl gradient which are indicated in Fig. 9(a), were pooled and dialysed against Tris (pH 7·6; 0·01 M) EDTA (10^{-4} M). Then they were analysed in neutral sucrose gradients (20 to 5% sucrose, 0·5 M-NaCl, 0·1 mg herring sperm DNA/ml., 10^{-3} M-EDTA, 0·1% Sarkosyl, 0·01 M-Tris, pH 7·5). ^{32}P-labelled denatured λ DNA and ^{32}P-labelled DNA of phage fd were used as sedimentation markers. Their positions in the gradients are indicated by arrows. All gradients (except the 8-min value of (c)) were run at 33,000 rev./min and 20°C for 2·5 hr in an SW41 rotor; the 8-min value of (c) was run for only 75 min.

(f) *Inhibition of DNA synthesis*

In this section we describe comparative studies on the inhibition of *in vivo* replication and *in vitro* synthesis by certain specific treatments.

(i) *Ultraviolet light*

Inhibition of *in vivo* DNA replication by ultraviolet irradiation has been studied extensively (Swenson & Setlow, 1966; Rupp & Howard-Flanders, 1968). It has been shown that the main cause of inhibition is the formation of dimers of adjacent pyrimidines (Wacker, 1963). Figure 10 shows that *in vivo* and *in vitro*

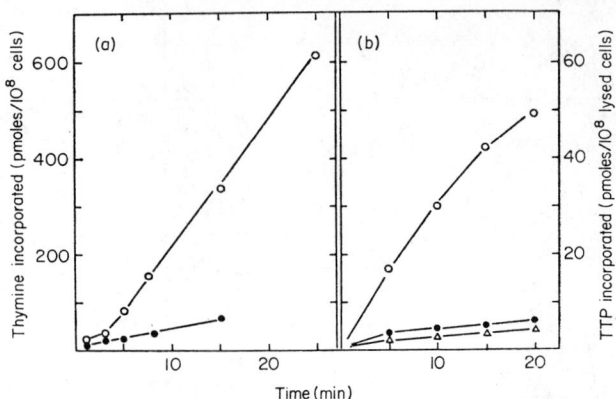

FIG. 10. Effect of ultraviolet irradiation on DNA synthesis *in vivo* (a) and *in vitro* (b). A culture grown to 2×10^8 bacteria/ml. in medium containing [^{14}C]thymine was divided into 3 parts.

(a) [^{3}H]Thymine was added to the first part and samples were taken after various times (—○—○—). Cells of the second part were centrifuged, suspended in a phosphate buffer, irradiated with 1500 erg/mm^2, resuspended in growth medium containing [^{3}H]thymine and incubated as above (—●—●—).

(b) The third part was washed and concentrated for preparation of the *in vitro* system. Normal unirradiated assay (—○—○—). Cells were spread on membrane discs without lysozyme, irradiated with 1500 erg/mm^2, lysozyme was added to the cells, which were lysed as in the untreated system. Incorporation of the irradiated lysed cells was measured over various periods of time (—●—●—). The cells were lysed on the membrane discs as in a normal assay. The lysate was irradiated with 1500 erg/mm^2 just before incorporation (—△—△—).

synthesis of DNA are affected similarly. Synthesis is strongly reduced by an ultraviolet dose of 1500 erg/mm^2. The rate of residual synthesis after such treatment decreases with increasing dose. The effect of ultraviolet light in the *in vitro* system (Fig. 10(b)) is the same whether the cells are irradiated before or after lysis.

When DNA synthesis is due to a repair process, the rate of synthesis would be expected to increase with increasing ultraviolet dosage (Couch & Hanawalt, 1967). The opposite is the case. This observation, together with the quantitative similarity of the *in vivo* and *in vitro* inhibition, can be taken as suggestive evidence that the *in vitro* synthesis reflects DNA replication rather than repair.

(ii) *Mitomycin*

Figure 11 shows that cells treated by mitomycin *in vivo* are inhibited in their DNA synthesis *in vivo* and *in vitro*. This agent attacks DNA directly as a bifunctional alkylating agent building cross-links between the complementary strands (Waring,

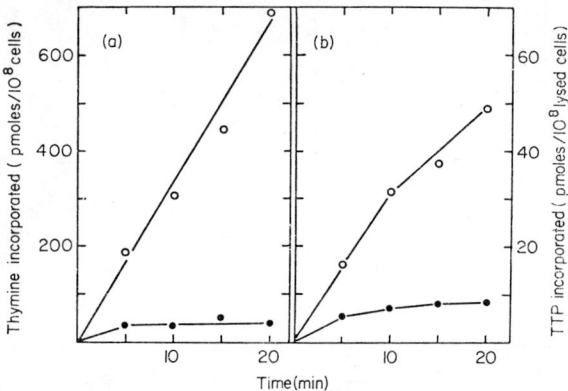

Fig. 11. Effect of mitomycin treatment on DNA synthesis *in vivo* (a) and *in vitro* (b).

—○—○—, *In vivo* and *in vitro* assay with untreated cells as described in Fig. 10.
—●—●—, Cells were harvested and resuspended in minimal medium to a final concentration of 5×10^8 bacteria/ml. Mitomycin (100 µg/ml.) was added and the suspension incubated at 37°C in the dark without aeration for 10 min. The cells were centrifuged, washed and resuspended in [^{3}H]thymine containing growth medium (a), and in minimal medium for the preparation of the *in vitro* system (b).

1968). It is believed that these cross-links prevent the progress of the replication fork and thus account for the blockage of DNA synthesis (Waring, 1968). The similarity between the *in vivo* and *in vitro* inhibition by mitomycin can be taken as another indication that the synthesis observed resembles replication.

(iii) *Chloramphenicol*

Chloramphenicol inhibits protein synthesis (Friedman, Lu & Rich, 1969). Indirectly, it also inhibits DNA synthesis (Lark, 1969), since in the presence of chloramphenicol those proteins that are necessary for the initiation of DNA replication cannot be formed. Thus, cells which are growing in chloramphenicol finish their DNA replication cycle and then discontinue synthesis of DNA. Lysates of cells harvested in the late stage, after addition of chloramphenicol, show a much reduced synthesizing activity (Fig. 12). The same observation has been made with lysates of cells that have been starved for essential amino acids and with lysates of mutants that are temperature-sensitive with respect to DNA initiation and have been grown at non-permissive temperature for 1·5 generations. All these results support the idea that the synthesis observed *in vitro* occurs at the growth points which had been active *in vivo*.

(iv) *Nalidixic acid*

Nalidixic acid rapidly and specifically inhibits DNA replication (see Goulian, 1971). The precise mechanism is still unknown. Figure 13 shows that the incorporation of [^{3}H]thymine *in vivo* and [^{3}H]TTP *in vitro* is inhibited by nalidixic acid. This was not observed in other *in vitro* systems (Boyle, Cook & Goss, 1969; Okazaki *et al.*, 1970). *In vivo* and *in vitro*, there is some residual synthesis, which cannot be significantly reduced by higher concentrations of nalidixic acid.

E. coli cells which had been treated with nalidixic acid *in vivo* show a rate of synthesis about twice as high as that of the control (Boyle, Goss & Cook, 1967). There are indications that by this treatment premature rounds of chromosome replication are

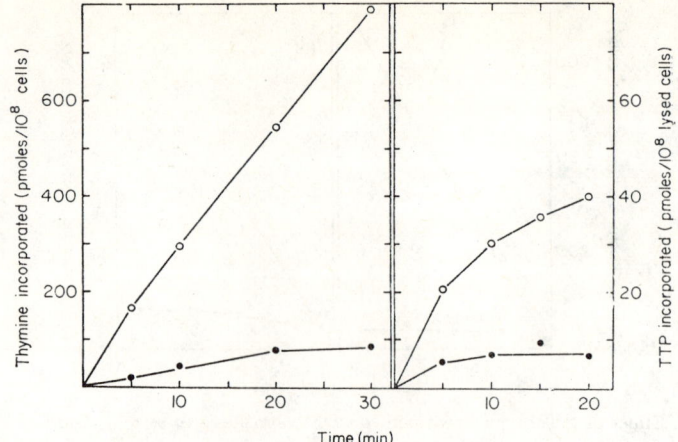

Fig. 12. Effect of inhibition of protein synthesis on DNA synthesis *in vivo* (a) and *in vitro* (b).
—○—○—, *In vivo* and *in vitro* assay with untreated cells. —●—●—, Chloramphenicol (150 μg/ml.) was added to part of a growing culture and the culture was incubated for another 100 min. [³H]Thymine was added to half of the culture and synthesis was measured as described in Materials and Methods (a). The other half of the culture was used for preparation of the *in vitro* system (b).

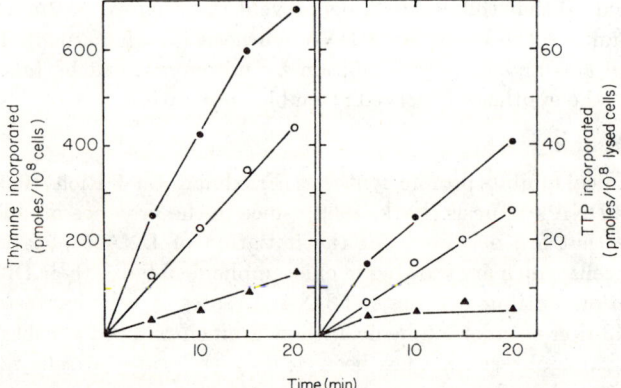

Fig. 13. Nalidixic acid stimulation and inhibition of *in vivo* (a) and *in vitro* (b) DNA synthesis.
—○—○—, Untreated control. —●—●—, Pre-treated with nalidixic acid. Nalidixic acid (20 μg/ml.) was added to a growing culture and the culture was incubated for another 30 min. The bacteria were filtered, washed and resuspended in [³H]thymine containing growth medium (a) and in minimal medium for the preparation of the *in vitro* system (b).
(a) —▲—▲—, Nalidixic acid (20 μg/ml.) and [³H]thymine were added to an exponentially growing culture. The culture was incubated at 37°C and samples were taken after various times.
(b) —▲—▲—, Normal *in vitro* assay was performed with nalidixic acid (100 μg/ml.) in the incorporation mixture.

initiated at the normal origin of vegetative DNA synthesis, in some respects like the premature initiation of chromosomal replication observed after thymine starvation (Pritchard & Lark, 1964). In our experiments, we found such a stimulation by nalidixic acid *in vivo* and *in vitro* (Fig. 13).

We have isolated nalidixic acid-resistant strains. The lysates of these strains are resistant to high levels of nalidixic acid.

(v) *Replication mutants*

By using mutants which are temperature-sensitive with respect to DNA replication, it is possible to inhibit *in vivo* replication very specifically (see Gross, 1971). There are at least two types of such mutants, namely those in which DNA replication is arrested immediately and those which finish the ongoing replication cycle of DNA which has begun at non-permissive temperature before they stop synthesizing DNA. The latter are inhibited as regards initiation of DNA replication and have been discussed earlier. We have isolated and partially characterized some 50 different ts_{DNA} mutants of the first type, using a DNA polymerase I-deficient strain as parental strain (de Lucia & Cairns, 1969). These mutants will be described in a separate paper. In about half of them the DNA synthesizing activity *in vitro* is significantly reduced at non-permissive temperatures. In Figure 14 the temperature-dependence of one of them is given as an

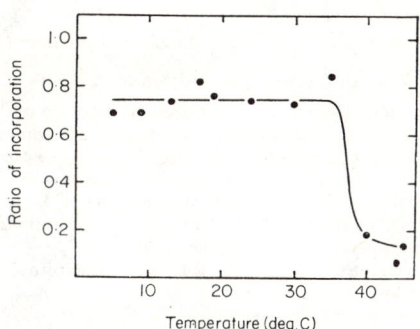

Fig. 14. Temperature-dependence of *in vitro* synthesis in lysates of the temperature-sensitive mutant H560/1040.
DNA synthesis was measured for H560 wild type and for a temperature-sensitive mutant H560/1040 as described in Materials and Methods. The incorporation temperature was varied between 9 and 45°C. The time of incorporation was 8 min. The ratio of the incorporation into the mutant lysate to the incorporation into wild type lysate (see Fig. 6) is plotted against temperature.

example. Probably some of those mutants which are not impaired as regards *in vitro* activity are affected by the pathway of the DNA precursors, which in the *in vitro* system are supplied externally. It should be mentioned that none of the mutants with low *in vitro* activity at the non-permissive temperature shows temperature-sensitive DNA polymerase II activity when assayed in crude cell extracts. It should be mentioned here that, under the various inhibitory conditions used, we often observe *in vitro* a residual DNA synthesis which is higher than the corresponding residual *in vivo* synthesis. It remains an open question whether this residual *in vitro* synthesis is related to true replication or whether it reflects some other kind of DNA synthesis.

(g) *Soluble factors involved in replication*

The Cellophane membrane disc used in the system not only serves as a carrier for the lysate allowing quick dialysis against different buffers. Its main and essential function is to prevent the macromolecular components of the lysate from being diluted by the incorporation mixture. If one allows direct contact between lysate and incorporation mixture, by placing the disc upside down on the drop, the DNA synthesizing capacity is strongly reduced (Fig. 15(a)). Even small dilutions of the lysate reduce the

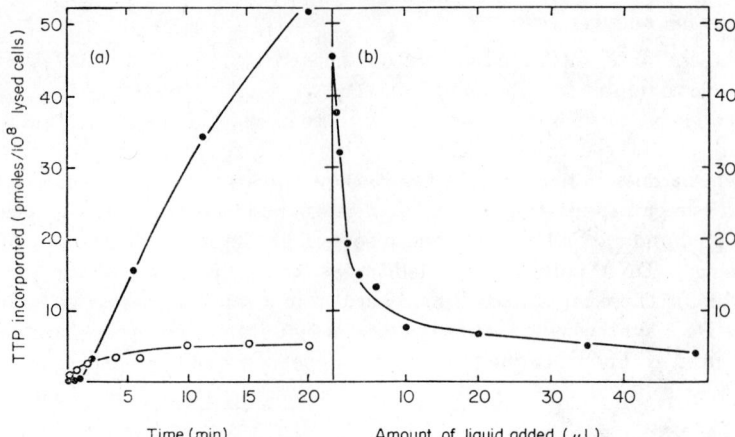

Fig. 15. Effect of lysate concentration on the rate of synthesis.

(a) Kinetics of incorporation. During incorporation, the lysate and the drop of incorporation mixture were separated by the membrane disc and the lysate remained at high concentration (—●—●—). The disc was placed upside down on the drop, so that the lysate and the drop (50 µl.) were in direct contact during incubation, and the lysate was diluted (—○—○—).

(b) Amount of DNA synthesized during a 20-min incubation. During the incubation, the drop and the lysate (about 10^8 cells) were separated by the membrane and various additional amounts of incorporation mixture were pipetted on top of the lysate.

In the upside down experiment and in (b), the total acid-precipitable radioactive material of the disc and the drop was counted.

activity of the system drastically (Fig. 15(b)). These results indicate that one or more macromolecular factors are necessary at high concentrations for the proper functioning of the system, and that they can be removed from the DNA synthesizing complex. It should be possible to divide the system into a soluble and an insoluble fraction and to do complementation and reconstitution experiments. Preliminary results of this type are shown in Table 2; washed disc preparations (B) show much reduced activity. The activity is partly recovered upon adding back the soluble factor(s). Recovery was only 40%, and could not be increased by the addition of more wash.

Among other enzymic activities, a DNA synthesizing activity is observed in the wash preparation. Because of its template specificity (Moses & Richardson, 1970(a); Knippers, 1970), it is probably due to polymerase II. Purified polymerase II (kindly supplied by Dr R. Knippers) did not stimulate DNA synthesis when added to the washed system. Moreover, the stimulation was sensitive to nalidixic acid (Table 2), whereas polymerase II activity is not (R. Knippers, personal communication). These results indicate that the stimulation is due rather to a restoration of the system than to an unspecific synthesis stimulated by polymerase II.

Experiments on the nature and function of the soluble factors necessary for the function of the system are in progress.

4. Discussion

A technique has been developed which allows the preparation of cell lysates at very high concentration. These lysates are prepared on Cellophane membranes. They are capable of incorporating nucleotides into macromolecular DNA, using deoxyribonucleoside triphosphates as substrate. The incorporation continues for a long period of time.

TABLE 2

Separation of soluble and insoluble macromolecular components and reconstitution

Preparation	Amount of TTP (pmoles) incorporated by preparation of 10^8 cells during 20 min	
	No nalidixic acid	Nalidixic acid (300 µg/ml.)
A Normal system	57	11
B Washed system	6	
C Wash	0·2	
D Wash on washed system	22	6

A. Preparation of the normal system: as described in Materials and Methods.
B. Preparation of the washed system. The normal system was prepared as described in Materials and Methods. After lysis of the cells by osmotic shock, the lysate was dried on an agar plate B with a fan. 20-ml. of cold incorporation buffer (see Materials and Methods) were layered over the discs: the liquid was exchanged after 30 min, and after another 30 min the discs were transferred to a second agar plate, B, and dried.

Wash preparation. Three discs with an area of 10 cm² containing 5×10^8 cells each were prepared as described in Materials and Methods and dried on an agar plate, B. These 3 discs were layered one after the other upside down on a 250-µl. drop of incorporation buffer, and remained there for 10 min at 4°C.

Preparation B. 5 µl. of incorporation buffer were spread on a washed disc lying on an agar plate B, then the disc was dried.

Preparation C. 10 µl. of wash preparation was spread on an empty disc lying on an agar plate B, then the disc was dried.

Preparation D. 5 µl. of wash preparation was dried on washed discs as above.

The purpose of this paper is to show that this synthesis is *in vitro* DNA replication. The evidence that the synthesis really occurs *in vitro* rather than in unlysed cells is threefold. First, the deoxyribonucleoside triphosphates are accepted as precursors, whereas neither the doexyribonucleosides nor the monophosphates serve as precursors in this system. Second the template and the product are accessible to externally added macromolecules like DNase. Third, synthesis can be inhibited and then stimulated by the removal and then re-addition of some (unknown) macromolecular components.

The observed synthesis is probably true DNA replication and not some other kind of DNA synthesis. Thus, the DNA is synthesized semi-conservatively and discontinuously, as it is *in vivo*, and the low molecular weight DNA synthesized during pulses is subsequently joined together to form high molecular weight DNA. The overall rate of DNA synthesis and the rate of chain elongation are high, although somewhat lower than the *in vivo* rate. DNA synthesis is equally inhibited *in vivo* and *in vitro* by a pre-treatment of the bacteria with ultraviolet light, mitomycin, or chloramphenicol or by the presence of nalidixic acid, whereas protein and RNA synthesis continues. Lastly, lysates of ts_{DNA} mutants exhibit a strong temperature sensitivity for DNA synthesis.

The only other *in vitro* system in which DNA replication occurs is in cells the membranes of which have become penetrable by low molecular weight components (Hoffmann-Berling, 1968; Vosberg & Hoffmann-Berling, 1971; Moses & Richardson, 1970b; Mordoh, Hirota & Jacob, 1970; Kohiyama & Kolber, 1970). Such systems, however, are not suitable for studying the enzymology of replication, since the site of synthesis is not freely accessible to macromolecular components. In contrast, the azide-poisoned *Bacillus subtilis* cells described by Ganesan (1971) are permeable to

DNase and therefore appear to be a promising system for the study of the function of the macromolecular components involved in DNA replication.

We have started to analyse the various components which take part in replication. Thus, the system can be divided into a soluble fraction and a fraction which contains the DNA, cell membranes and bound proteins. Neither fraction by itself can synthesize DNA over an extended period, whereas they do synthesize DNA when they are combined. It should be possible to purify the soluble factors and to analyse them by standard biochemical and genetical techniques, but it will probably be a difficult task to characterize and isolate those components which are involved in replication and are bound to the replicating complex.

Note added in proof: Recently it has been shown that the mutant H560/1044 is a *dna* E mutant and temperature-sensitive in a soluble DNA polymerase (polymerase III) which is one of the components needed in the *in vitro* system at approximately *in vivo* concentration (Gefter, M. L., Hirota, Y., Kornberg, T. & Wechsler, J. A., manuscript submitted for publication. Nüsslein, V., Otto, B., Bonhoeffer, F. & Schaller, H., manuscript submitted for publication).

REFERENCES

Bonhoeffer, F. & Gierer, A. (1963). *J. Mol. Biol.* **7**, 534.
Boyle, J. V., Goss, W. A. & Cook, T. M. (1967). *J. Bact.* **94**, 1664.
Boyle, J. V., Cook, T. M. & Goss, W. A. (1969). *J. Bact.* **97**, 230.
Cairns, J. (1970). *Harvey Lecture.* New York and London: Academic Press.
Couch, J. T. & Hanawalt, P. (1967). *Biochem. Biophys. Res. Comm.* **29**, 779.
De Lucia, P. & Cairns, J. (1969). *Nature,* **224**, 1164.
Friedman, H., Lu, P. & Rich, A. (1969). *Cold Spr. Harb. Symp. Quant. Biol.* **34**, 255.
Ganesan, A. T. (1968). *Cold Spr. Harb. Symp. Quant. Biol.* **33**, 45.
Ganesan, A. T. (1971). *Proc. Nat. Acad. Sci., Wash.* **68**, 1296.
Goulian, M. (1971). *Ann. Rev. Biochem.* **40**, 855.
Gross, J. D. (1971). In *Current Topics in Microbiology and Immunology.* Berlin, Heidelberg and New York: Springer-Verlag; in the press.
Hoffmann-Berling, H. (1968). *Molecular Genetics,* 38. Berlin, Heidelberg and New York: Springer-Verlag.
Knippers, R. & Strätling, W. (1970). *Nature,* **226**, 713.
Knippers, R. (1970). *Nature,* **228**, 1050.
Kohiyama, M. & Kolber, A. (1970). *Nature,* **228**, 1157.
Kornberg, A. (1969). *Science,* **163**, 1410.
Kornberg, T. & Gefter, M. L. (1970). *Biochem. Biophys. Res. Comm.* **40**, 1348.
Lark, K. G. (1969). *Ann. Rev. Biochem.* **38**, 569.
Meselson, M. & Stahl, F. W. (1958). *Proc. Nat. Acad. Sci., Wash.* **44**, 671.
Mordoh, J., Hirota, Y. & Jacob, F. (1970). *Proc. Nat. Acad. Sci., Wash.* **67**, 773.
Moses, R. E. & Richardson, C. C. (1970a). *Biochem. Biophys. Res. Comm.* **41**, 1557.
Moses, R. E. & Richardson, C. C. (1970b). *Proc. Nat. Acad. Sci., Wash.* **67**, 674.
Okazaki, R., Okazaki, T., Sakabe, K., Sugimoto, K., Kainuma, R., Sugino, A. & Iwatsuki, N. (1968). *Cold Spr. Harb. Symp. Quant. Biol.* **33**, 129.
Okazaki, R., Sugimoto, K., Okazaki, T., Imae, Y. & Sugino, A. (1970). *Nature,* **228**, 223.
Pritchard, R. H. & Lark, K. G. (1964). *J. Mol. Biol.* **9**, 288.
Rupp, W. D. & Howard-Flanders, P. (1968). *J. Mol. Biol.* **31**, 291.
Smith, D. W., Schaller, H. & Bonhoeffer, F. (1970). *Nature,* **226**, 711.
Swenson, P. A. & Setlow, R. B. (1966). *J. Mol. Biol.* **15**, 201.
Vosberg, H-P. & Hoffmann-Berling, H. (1971). *J. Mol. Biol.* **58**, 739.
Wacker, A. (1963). *Progr. Nucleic Acid Res.* **1**, 369.
Waring, M. J. (1968). *Nature,* **219**, 1320.
Wulff, D. L. (1963). *J. Mol. Biol.* **7**, 431.

Analysis of DNA Polymerases II and III in Mutants of *Escherichia coli* Thermosensitive for DNA Synthesis

(*pol*A1 mutants/phosphocellulose chromatography/*dna*E locus)

MALCOLM L. GEFTER, YUKINORI HIROTA*, THOMAS KORNBERG, JAMES A. WECHSLER, AND C. BARNOUX*

Department of Biological Sciences, Columbia University, New York, N.Y. 10027; and * Service de Génétique Cellulaire de l'Institut Pasteur, Paris

Communicated by Cyrus Levinthal, October 18, 1971

ABSTRACT A series of double mutants carrying one of the thermosensitive mutations for DNA synthesis (*dna*A, B, C, D, E, F, and G) and the *pol*A1 mutation of DeLucia and Cairns, were constructed. Enzyme activities of DNA Polymerases II and III were measured in each mutant. DNA Polymerase II activity was normal in all strains tested. DNA Polymerase III activity is thermosensitive specifically in those strains having thermosensitive mutations at the *dna*E locus. From these results we conclude that DNA Polymerases II and III are independent enzymes and that DNA Polymerase III is an enzyme required for DNA replication in *Escherichia coli*.

The isolation by DeLucia and Cairns (1) of an *Escherichia coli* mutant that lacks DNA Polymerase I activity (*pol*A1) has prompted many investigations into the nature of the DNA synthetic capacity of such strains. The purification and characterization of DNA Polymerase II has been reported by ourselves (2) and others (3, 4). In addition, we have reported the existence of a third DNA polymerase in *E. coli* (DNA Polymerase III) (2). A physiological function for these enzymes has not been determined.

The viability of cells devoid of measurable DNA Polymerase I activity suggests that this enzyme is not an obligatory component of the DNA replication machinery of *E. coli*. To determine whether polymerases II and III are essential for replication, we examined the DNA polymerases of *E. coli* mutants that were temperature-sensitive for DNA replication in an attempt to correlate the genetic lesions with altered DNA polymerase activity *in vitro*. We will present evidence indicating that DNA Polymerase III is the product of an essential gene mapping at the *dna*E locus.

MATERIALS AND METHODS

The following bacterial strains were used†:

(1) CRT4637:   F⁻ *thr⁻ leu⁻ his⁻ str*ʳ *malA mtl⁻ thi⁻ dna*AT46
(2) CRT2667:   F⁻ *his⁻ str*ʳ *malA thi⁻ pol*A1 *sup⁻ dna*BT266
(3) BT1029:   H560 *thy⁻ endo*I⁻ *pol*A1 *dna*B
(4) PC22:   F⁻ *his⁻ str*ʳ *malA xyl⁻ arg⁻ mtl⁻ thi⁻ pol*A1 *sup⁻ dna*C2

† In the text, these strains will be referred to by their number in the above list, followed by the *dna* mutation designation in parenthesis.

(5)   PC79:   F⁻ *his⁻ str*ʳ *malA xyl⁻ mtl⁻ thi⁻ pol*A1 *sup⁻ dna*D7
(6)   E5111:   F⁻ *his⁻ str*ʳ *malA xyl⁻ mtl⁻ arg⁻ thi⁻ sup⁻ pol*A1 *dna*E511
(7)   E4860:   F⁻ *his⁻ str*ʳ *malA xyl⁻ mtl⁻ arg⁻ thi⁻ sup⁻ dna*E486
(8)   E4868:   F⁻ *his⁻ str*ʳ *malA xyl⁻ mtl⁻ arg⁻ thi⁻ sup⁻ pol*A1 *dna*E486
(9)   BT1026: H560 *thy⁻ endo*I⁻ *pol*A1 *dna*E
(10) BT1040: H560 *thy⁻ endo*I⁻ *pol*A1 *dna*E
(11) E1011:   F⁻ *his⁻ str*ʳ *malA xyl⁻ mtl⁻ arg⁻ thi⁻ sup⁻ pol*A1 *dna*F101
(12) JW207:   *thy⁻ rha⁻ str*ʳ *pol*A1 *dna*F101
(13) NY73:   *leu⁻ thy⁻ metE rif*ʳ *str*ʳ *pol*A1 *dna*G3
(14) CRT2668: F⁻ B1⁻ *his⁻ malA str*ʳ *sup⁻ pol*A1 *dna*⁺
(15) JG112:   W3110 *thy⁻ rha⁻ lac⁻ sup⁻ pol*A1 *dna*⁺

The isolation of the double mutant *dna*B *pol*A1, was described (5). A further series of *dna*–*pol*A1 double mutants, *dna*C, D, E, and F, with *pol*A1, were constructed through two successive steps. (i) Each thermosensitive mutation was introduced into an Hfr strain (HfrP4x8: an Hfr that injects its chromosome in the order, O-*proA-leu-lac*-F, or Hfr-Cavalli: an Hfr that injects its chromosome in the order, O-*lac-leu—gal*-F) by crossing the Hfr with an F-strain carrying a thermosensitive mutation affecting DNA synthesis. *lac*y (The site of F integration of HfrP4x8 is near *lac*y) (6) and *gal* (the site of F integration of Hfr Cavalli is near *gal*) (6) are used for selection. (ii) Each thermosensitive Hfr strain isolated was then crossed with an F⁻ strain, PA33612; F⁻ *arg⁻ his⁻ thi⁻ leu⁻ malA⁻ xyl⁻ mtl⁻ pol*A1⁻ *sup⁻*, using a closely linked marker (*leu* for *dna*E, C, and D and *his* for *dna*F) for selection. Recombinants were then tested for both *pol*A1, *dna*, and *sup*.

As controls, thermoresistant *dna*⁺*pol*A1⁻ were constructed by selection at a high temperature, either after P1-transduction of the thermosensitive allele, *dna*⁺, or spontaneous occurrence of revertants from the double mutants. Strain JW 207 was isolated after bacteriophage P1 transduction of *dna*F101 from strain E101 (7) into W3110 *thy⁻ rha⁻ pol*A1 *purF*, by selection for *purF*⁺. Strain NY73 was isolated by introduction of *pol*A1 into PC3 with JG78 (Hfr *R1*, *metE rha*⁺ *pol*A1 Rif) (Peacey, M., and J. D. Gross, unpublished data).

3150

The isolation and mapping of thermosensitive mutants have been reported by others (5, 7–14). Strains PC2:dnaC2 and PCF dnaD7 were a gift from Dr. P. Carl. Strains 1026, 1040 (dnaE), and 1029 (dnaB) were a gift from Dr. F. Bonhoeffer (classification of dna lesion was by co-transduction, ref. 11).

The materials used for purification and assay of DNA Polymerases II and III were described. [³H]TTP (2×10^5 cpm/nmol) was used throughout to assay DNA polymerase activity.

Cells were grown in three-times concentrated L.B. broth (Difco) (15) at 25°C with aeration, and harvested in mid-log phase at $3-4 \times 10^9$ cells/ml. Cell-free extracts (10 g of cells) and the S100 fraction were prepared as described (2). Separation of DNA Polymerases II and III (see Fig. 1A) was also described, except that all volumes and column dimensions were scaled down 10-fold and preliminary dialysis and batch elution from phosphocellulose (step II) were omitted. For the addition of large amounts of DNA Polymerase III to reaction mixtures, the enzyme activity that eluted from phosphocellulose was concentrated 10-fold by precipitation with ammonium sulfate.

Assays of rates of reaction at 30 and 45°C were done by first equilibrating the assay mixture (0.9 ml) at the appropriate temperature. The reaction was begun by the addition of enzyme. 0.2-ml aliquots were withdrawn at various times and pipetted into 1.0 ml of 5% trichloroacetic acid. The acid-insoluble material was collected on a Millipore filter and the radioactivity was determined in a liquid scintillation counter.

RESULTS

The results of a typical isolation of DNA Polymerase II and III are shown in Fig. 1A. Polymerases II and III are distinguished on the basis of chromatographic behavior; Polymerase III elutes at 0.1 M PO_4^{3-} (fraction 17) and Polymerase II at 0.2 M PO_4^{3-} (fraction 37). The two enzymes can further be distinguished on the basis of their response to ionic strength (2). In two instances (strains 1 and 7), Polymerases II and III were isolated from cells containing the normal amount of DNA Polymerase I. The result of phosphocellulose chromatography of extracts from such cells is shown in Fig. 1B. Although DNA Polymerase III is not completely resolved from Polymerase I activity (fraction 20), the activity of Polymerase III can be uniquely determined by assay of column fractions in the presence of either N-ethylmaleimide (dotted line, Fig. 1B) or antiserum to DNA Polymerase I. Since Polymerase III activity is completely abolished in the presence of N-ethylmaleimide, and is unaffected by antiserum directed against DNA Polymerase I (2), it is possible to obtain preparations active only due to Polymerase III despite the presence of Polymerase I. DNA Polymerase II is obtained in normal yield from pol+ strains; it is completely resolved from Polymerase I by phosphocellulose chromatography.

The peak fraction of each polymerase activity was used directly for measurements of the rate of synthesis at 30 and 45°C. The rate of reaction catalyzed by Polymerase II was 1.8-times faster at 45°C than at 30°C; the rate of the Polymerase III reaction was 1.5-times faster at 45°C than at 30°C. DNA Polymerase III activity is not linear with time after 5 min at 45°C and, therefore, relative rates were calculated only from the initial slopes.

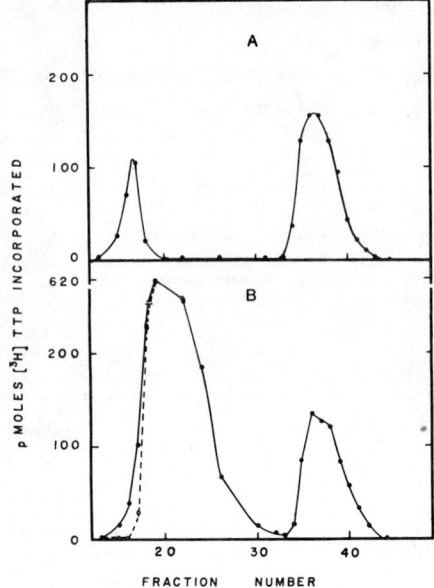

FIG. 1. Separation of DNA Polymerases II and III by phosphocellulose chromatography. Cell-free extracts from polA1 strains (A) and pol+ strains (B) were prepared and subjected to phosphocellulose chromatography. Each fraction was assayed for DNA polymerase activity. Results obtained by assay of fractions (12–19) in the presence of N-ethylmaleimide are shown by the dotted line. Polymerase II and III activities elute at fractions 35–45 and 12–20, respectively.

The results of a typical analysis of DNA Polymerase II activity at 30 and 45°C are shown in Fig. 2. Fig. 2A represents the Polymerase II activity isolated from strain 13 (dnaG) and 2B the results from strain 9 (dnaE). On the basis of these analyses, DNA Polymerase II activity appeared normal in all strains tested. The results of these relative rate measurements are summarized in Table 1.

The results of a typical analysis of wild-type DNA Polymerase III activity are shown in Fig. 3A. These results were obtained with DNA Polymerase III isolated from strain 13 (dnaG) and are representative of all enzyme preparations tested except for those isolated from strains 6, 9, and 10 (carrying dnaE mutations). These results are also summarized in Table 1.

The results obtained for DNA Polymerase III activity isolated from strain 9 (dnaE) are shown in Fig. 3B. (Polymerase III activity isolated from strain 10 (dnaE) behaves essentially the same way.) In contrast to the rate observed with a normal enzyme, the rate of synthesis at 45°C with enzyme preparations from the dnaE mutants 9 and 10 was undetectable. To rule out the possibility of a temperature-dependent inhibitor present in these preparations, concentrated Polymerase III preparations (see Methods) from strains 9 (dnaE) and 13 (dnaG) were mixed and the rates at

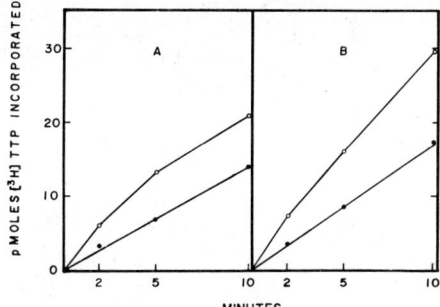

FIG. 2. Effect of temperature on Polymerase II-catalyzed synthesis. The rate of polymerization at 30°C (—●—) and 45°C (—○—) was measured. The results obtained for enzymes isolated from strain 13 (*dna*G) and strain 9 (*dna*E) are shown in parts *A* and *B*, respectively.

TABLE 1. *Effect of temperature on the rate of reaction*

Strain		Reaction rate $\frac{45°C}{30°C}$	
		DNA Polymerase II	DNA Polymerase III
1	(*dna*A)	1.85	1.45
2	(*dna*B)	1.73	1.74
3	(*dna*B)	1.95	1.55
4	(*dna*C)	1.86	1.20
5	(*dna*D)	1.99	1.47
6	(*dna*E)	1.87	1.0
7	(*dna*E)	1.70	—*
8	(*dna*E)	2.00	—*
9	(*dna*E)	1.95	<0.1
10	(*dna*E)	1.70	<0.1
11	(*dna*F)	1.55	1.33
12	(*dna*F)	1.77	1.43
13	(*dna*G)	1.75	1.57
14	(*dna*+)	1.92	1.70
15	(*dna*+)	1.90	1.55

* A dash indicates that the enzyme activity was not detectable at 30°C.

30 and 45°C were determined. The presence of Polymerase III from strain 9 does not render the wild-type enzyme temperature sensitive (Fig. 3C).

In order to obtain further evidence that the temperature-sensitive character of Polymerase III, derived from strains 9 or 10, was due to a specific alteration of the enzyme, Polymerase III from strain 9 was further purified. The procedures used (T. K. and M. G., manuscript in preparation) are sufficient to purify the wild-type enzyme 2000-fold with respect to the S100 fraction. The properties of such an enzyme preparation are identical to those described for the enzyme activity eluted from the phosphocellulose column.

The specific activity (assayed at 30°C) of preparations from strains 9 and 10 are 10% that of the wild-type enzyme throughout the purification procedure. In all cases "mutant" Polymerase III is totally inactive when assayed at 45°C.

Polymerase III activity could not be detected in extracts from strains 7 and 8 (*dna*E486). The DNA Polymerase III isolated from strain 6 (*dna*E511) was only marginally temperature sensitive.

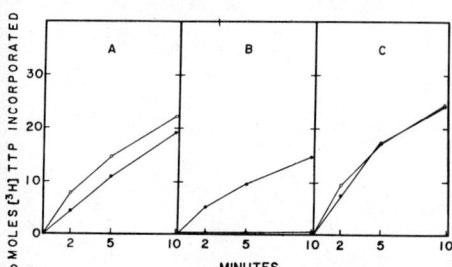

FIG. 3. Effect of temperature on Polymerase III-catalyzed synthesis. The rate of polymerization at 30°C (—●—) and 45°C (—○—) was measured. The results obtained for enzymes isolated from strain 13 (*dna*G) and strain 9 (*dna*E) are shown in Parts *A* and *B*, respectively. The results obtained with a mixture of equal amounts of enzyme from strain 13 and strain 9 are shown in part *C*.

DISCUSSION

The enzymes DNA Polymerase II and III have been analyzed in several mutants thermosensitive for DNA synthesis. DNA Polymerase II activity appears to be normal in all strains tested, including a strain carrying a thermosensitive *rec*A gene (data not presented). The failure to associate Polymerase II activity with any *dna* locus tested does not prove that this enzyme is dispensable; isolation of a mutant defective in Polymerase II will be required to resolve this question. That Polymerase II is not thermolabile in *dna*E mutants indicates that this enzyme is not related to Polymerase III, as was previously suggested (2).

DNA Polymerase III activity appears to be normal in strains carrying mutations at the *dna*A, B, C, D, F, and G loci. Of four independently isolated *dna*E mutants, all had altered DNA Polymerase III activity *in vitro*. DNA Polymerase III activity could not be detected in extracts of *dna*E486 mutants, either in the presence or absence of DNA Polymerase I. We believe that this result is due to instability *in vitro* of the Polymerase III in these strains. DNA Polymerase III in the *dna*E511 mutant is only slightly temperature sensitive, which perhaps reflects the slow cessation of *in vivo* DNA synthesis that this mutant displays at the restrictive temperature (11). Polymerase III activity in strains 9 and 10 (*dna*E) is thermolabile.

Our results suggest that the structural gene for DNA Polymerase III is located at the *dna*E locus. Furthermore, since *dna*E mutants fail to replicate their DNA at 42°C *in vivo* (10) and contain a thermolabile Polymerase III *in vitro*, we conclude that this enzyme is required for DNA replication in *E. coli*.

NOTE ADDED IN PROOF

Drs. H. Schaller, B. Otto, V. Nüsslein, J. Huf, R. Herrmann, and F. Bonhoeffer have isolated 50 independent

dna^{ts} mutants of *E. coli*. Of these, 20 were shown to be temperature-sensitive in highly concentrated lysates designed to measure DNA replication *in vitro*. Four such mutants that include strains 9 and 10, were shown not to complement each other but were complemented by a soluble factor isolated from dna^+ cells. Drs. B. Otto and V. Nüsslein have purified the soluble factor (E-protein) 1,000-fold using a complementation assay, and have independently shown that its properties (polymerase activity, stability, response to ionic conditions, and molecular weight) are in keeping with those determined for DNA polymerase III. According to the above criteria, DNA polymerase III and the E-protein appear to be the same.

We thank Miss C. Ganier, Mrs. S. Yang, and Mr. G. Grandusky for excellent technical assistance and Dr. M. Oishi for helpful discussions. Dr. H. Ogawa performed enzyme isolations from the $recA^{ts}$ strain. We are grateful to Dr. F. Bonhoeffer for making strains 3, 9, and 10 available before publication of their isolation. This work was supported by a grant from the Délégation Générale a la Recherche Scientifique et Technique to Y. H.; by Public Health Service Research Grant CA12590-01 to J. A. W., and by Grant no. NP-6A from the American Cancer Society to M. G.

1. DeLucia, P., and J. Cairns, *Nature*, **224**, 1164 (1969).
2. Kornberg, T., and M. L. Gefter, *Proc. Nat. Acad. Sci. USA*, **68**, 761 (1971).
3. Knippers, R., *Nature*, **228**, 1050 (1970).
4. Moses, R., and C. C. Richardson, *Biochem. Biophys. Res. Commun.*, **41**, 1565 (1970).
5. Mordoh, J., Y. Hirota, and F. Jacob, *Proc. Nat. Acad. Sci. USA*, **67**, 773 (1970).
6. Jacob, F., and E. Wollman, *Sexuality and Genetics of Bacteria* (Academic Press, Inc., New York, 1961).
7. Carl, P., *Mol. Gen. Genet.*, **109**, 107 (1970).
8. Ricard, M., and Y. Hirota, *C.R.-Acad. Sci.*, **268**, 1335, (1969).
9. Bonhoeffer, F., *Z. Vererbungslehre.*, **98**, 141 (1966).
10. Gross, J. D., *Current Topics in Microbiology and Immunology* (Berlin, Springer-Verlag, 1971).
11. Wechsler, J. A., and J. D. Gross, *Mol. Gen. Genet.*, in press.
12. Kohiyama, M., D. Cousin, A. Ryter, and F. Jacob, *Ann. Inst. Pasteur*, **110**, 465 (1966).
13. Hirota, Y., A. Ryter, and F. Jacob, *Cold Spring Harbor Symp. Quant. Biol*, **33**, 677 (1968).
14. Hirota, Y., J. Mordoh, and F. Jacob, *J. Mol. Biol*, **53**, 369 (1970).
15. Bertani, G., *J. Bacteriol.* **62**, 293 (1951).

Replication and Repair of DNA in Cells of *Escherichia coli* Treated with Toluene*

Robb E. Moses† and Charles C. Richardson‡

DEPARTMENT OF BIOLOGICAL CHEMISTRY, HARVARD MEDICAL SCHOOL, BOSTON, MASS. 02115

Communicated by Eugene P. Kennedy, July 13, 1970

Abstract. DNA synthesis has been studied in *Escherichia coli* cells made permeable to nucleotides by treatment with toluene. Replicative synthesis, as distinguished from repair synthesis, occurs at a rate comparable to that observed *in vivo*; it is dependent on the presence of all four deoxyribonucleoside triphosphates, but does not require exogenous DNA; and it is stimulated by ATP. Furthermore, replicative synthesis can be abolished at the restrictive temperature in DNA temperature-sensitive mutants. N-ethylmaleimide completely inhibits this type of synthesis, whereas it does not inhibit repair synthesis. Repair synthesis further differs from replicative synthesis in the following points: it does not require ATP; it persists at the restrictive temperature in DNA temperature-sensitive mutants; it can be induced by endogenous or exogenous nuclease activity; and its demonstration requires a Pol+ strain.

The bacterial chromosome replicates semiconservatively. Synthesis starts at a fixed point and proceeds in a linear, sequential fashion.[1-3] Attempts to duplicate this process *in vitro* using purified DNA polymerases have failed thus far. While DNA polymerase can accurately repair regions of single-stranded DNA, when it is presented with a duplex DNA the product is biologically inactive and contains structural aberrations.[4-6]

Several attempts have been made to isolate the replicating complex while maintaining its integrity. On the hypothesis that this complex may involve the cell membrane, some investigators have attempted to isolate membrane fractions capable of synthesizing DNA.[7-9] In another approach, Smith *et al.*[10] have utilized gentle lysis of cells imbedded in agar. Both approaches have yielded a system capable of carrying out semiconservative DNA synthesis for short periods of time at rates comparable to those observed *in vivo*.[9,10] Buttin and Kornberg[11] have described a technique for measuring intracellular DNA synthesis. Cells treated with EDTA-Tris become permeable to deoxynucleoside triphosphates, but the synthesis observed is mainly of the repair type.[12]

This report describes DNA synthesis in *Escherichia coli* cells treated with toluene. Such cells maintain many of their physiological functions,[13] but have become permeable to compounds of low molecular weight, including deoxynucleoside triphosphates. Although these cells are no longer viable, it has been possible to obtain extended semiconservative replication, and to distinguish this process from a repair type synthesis. In this report the DNA synthesis induced by nuclease action is termed "repair synthesis." Ether-treated cells have been used to study intracellular φX174 DNA replication.[14]

Materials. Unlabeled deoxynucleotides, ATP, [³H]dATP, and [¹⁴C]dTDP were obtained from Schwarz BioResearch. [α-³²P]dTTP and [α-³²P]dATP were obtained from International Chemical and Nuclear Corporation, and their purity was established enzymatically and chromatographically. Deoxybromouridine triphosphate (BrdUTP) was a gift from Dr. A. Kornberg. Crystalline pancreatic DNase and snake venom phosphodiesterase were purchased from Worthington. Antibody to purified *E. coli* DNA polymerase was the gift of Dr. I. R. Lehman.

Bacterial strains: *E. coli* W 3110 (pol^+endI^+), a K12 derivative, was provided by Dr. J. Cairns, as was *P3478* (pol^-endI^+), a derivative of *W3110* lacking DNA polymerase activity in extracts. In the original description[15,16] *P3478* is indicated as *polA1*, but since it is the only polymerase-negative mutant used in this study, pol^- will here indicate a mutation at the *polA1* locus. Both strains require thymine. *E. coli ER22* is a B derivative lacking endonuclease I (pol^+endI^-). *D110* is an endonuclease I negative strain (pol^-endI^-) derived by us from *P3478* using the method of Dürwald and Hoffmann-Berling.[17] Extracts of this strain contain less than 2% of wild type endonuclease I activity. *E. coli* strains *CR266-26* and *CRT26-43* were provided by Dr. G. Buttin. $recB_{21}$, a strain lacking Rec B exonuclease, was obtained from Dr. C. A. Thomas. Strains were routinely grown in tryptone broth supplemented with thymine (10 μg/ml) or in M-9 medium[18] supplemented with casamino acids (2 mg/ml) and thymine (10 μg/ml).

Toluene treatment was by a modification of the method of Levin et al.[19] Cells in the log phase of growth were harvested by centrifugation at 4°C, resuspended in 0.05 M potassium phosphate buffer (pH 7.4), the suspension was made 1% in toluene and shaken at 25° or 37°C for 10 min unless otherwise noted. After toluene treatment, survival was less than 10^{-8}, as measured by ability to grow and form colonies on agar plates. Toluene-treated cells, washed once with buffer and frozen at $-60°C$, maintain their activity for at least 1 month.

Results. Characteristics of the reaction: (a) *Requirements:* Synthesis of DNA in toluene-treated cells has been followed by incorporation of [α-³²P]-dTTP, [α-³²P]dATP, [³H]dATP, or [³H]dCTP into acid-precipitable material. The reaction requires the presence of all four deoxynucleoside triphosphates and Mg^{++} (Table 1). Mn^{++} is 50% as effective as Mg^{++}.

(b) *Stimulation by ATP:* The rate of reaction is markedly stimulated by ATP; Mg^{++} is present in the reaction mixture at ten times the concentration of ATP (Table 1). The optimal concentration of ATP for freshly prepared cell

TABLE 1. *Characteristics of DNA synthesis in toluene-treated E. coli.*

System	Activity (%)	System	Activity (%)
Complete	100	−dTTP	8
+DNA	104	−dCTP	11
−ATP	9	−4 dXTP	0.3
−Mg⁺⁺	5	−4 dXTP, + 4 dXMP	1
−dGTP	0.4	−4 dXTP, + 4 dX	1

W3110 cells were grown to 7×10^8 cells/ml, concentrated 5-fold in 0.05 M potassium phosphate buffer (pH 7.4), and agitated 10 min at 37°C with 1% toluene. The reaction mixture (0.3 ml) contained 70 mM potassium phosphate buffer (pH 7.4), 13 mM Mg^{++}, 1.3 mM ATP, 33 μM [³H]-dATP, dGTP, dTTP, dCTP, and 1.5×10^8 toluene-treated cells. Salmon sperm DNA was present at 20 μg/0.3 ml reaction where indicated. After incubation for 30 min at 37°C, the reaction was stopped by the addition of cold 10% TCA–0.1 M PP$_i$. After mixing, each sample was filtered through a Whatman GF/C glass filter (2.4 cm) and washed three times with 3 ml of cold TCA–PP$_i$, followed by three washes of 3 ml each with cold 0.01 M HCl. The filters were dried and the radioactivity was measured. dX, dXMP, and dXTP stand for deoxynucleoside, 5'-deoxynucleoside monophosphate, and 5'-deoxynucleoside triphosphate, respectively.

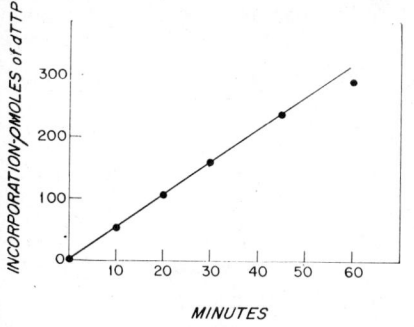

FIG. 1. Rate of DNA synthesis in toluene-treated cells. *D110* cells were grown to a concentration of 6×10^8 cells/ml, harvested by centrifugation at 4°C, concentrated 50-fold in 0.05 M potassium phosphate buffer (pH 7.4), and shaken for 10 min in 1% toluene at 37°C before freezing at $-60°C$. The reaction mixture was as in Table 1, except that $[\alpha^{-32}P]dTTP$ was used in place of $[^3H]dATP$. The reaction was initiated by the addition of 0.14 ml of thawed cell suspension to 2.1 ml of reaction mixture. At each time point 0.3 ml of the reaction mixture was withdrawn, the reaction stopped, and the product assayed as in Table 1.

suspensions is 1.0–2.0 mM. The stimulation does not appear to be due to the *recB* gene product, an ATP-requiring exonuclease,[20] since it is also observed in toluene-treated $recB_{21}$ cells.

(c) *Time course of reaction:* The incorporation of labeled dTTP into DNA in toluene-treated cells is linear for 1 hr (Fig. 1). The extent of the reaction is directly proportional to the concentration of the cell suspension from 10^7 to 10^9 cells per reaction mixture. With freshly prepared cell suspensions of *pol⁻ endI⁻* cells and with optimal ATP concentrations, a rate of 1.5×10^2 nucleotides cells^{-1} sec^{-1} can be achieved at 35°C, comparable to that observed *in vivo*.

This ATP-stimulated synthesis represents semiconservative replication (see below). However, nuclease action can induce a repair-type synthesis that can achieve a rate several times that observed during replicative synthesis.

Toluene treatment: (a) *Time course:* The optimal time of exposure of the cells to toluene for assays performed in the presence of ATP is 10 min (Fig. 2, left column); the stimulation of synthesis produced by ATP can be as much as 20-fold. After 20 min the amount of ATP-stimulated synthesis decreases. Polymerase-negative strains (*P3478* and *D110*) show normal levels of ATP-stimulated activity.

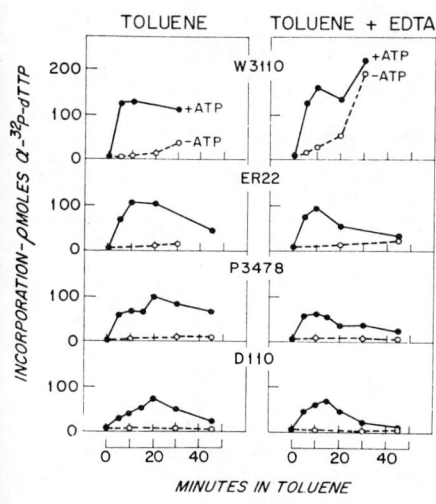

FIG. 2. Time course of toluene treatment. Cells were grown to a concentration of $7-9 \times 10^8$ cells/ml, collected, concentrated 5-fold in 0.05 M potassium phosphate buffer (pH 7.4), exposed to 1% toluene or to 1% toluene–2 mM EDTA, and placed on a rotary shaker at 37°C. At various times, aliquots were withdrawn and assayed directly for 30 min at 37°C as in Table 1, with or without ATP. $[\alpha^{-32}P]$-dTTP was present in the reaction.

(b) *Effect of EDTA on toluene treatment:* If EDTA is present during toluene treatment, a stimulation of DNA synthesis is observed in pol^+endI^+ (W3110) cells between 20 and 30 min of treatment (Fig. 2, right column). The enhanced synthesis does not depend on ATP.

A possible explanation, which might be applicable to Tris–EDTA-treated cells as well,[14] is that EDTA produces ribosomal breakdown, release of RNase I, and unmasking of endonuclease I activity. In support of this hypothesis is the observation that pol^+endI^- (ER22) cells lacking endonuclease I do not show increased activity when EDTA is present during toluene treatment (Fig. 2). Neither $endI^+$ nor $endI^-$ strains of pol^- (P3478, D110) show increased synthesis under these conditions. In *E. coli* pol^-endI^+ this suggests a defect in the ability to repair damage induced by endonuclease I.

Characterization of the DNA product: (a) *Product of hydrolysis:* The radioactive product synthesized in toluene-treated cells is acid-precipitable and alkali-resistant. It can be degraded to acid-soluble material by pancreatic DNase and snake venom phosphodiesterase. When the product of this enzymatic digestion, after incorporation of $[\alpha\text{-}^{32}P]dTTP$, is analyzed by chromatography, the label is associated with dTMP. We conclude that the isolated product is DNA synthesized from deoxynucleoside triphosphate precursors.

(b) *Sedimentation analysis:* *E. coli* was grown for several generations in the presence of $[^3H]dT$. The cells were then treated with toluene and incubated in the standard DNA-synthesizing system containing $[\alpha\text{-}^{32}P]dTTP$. As shown in Fig. 3, the newly synthesized DNA is distributed over the same size range as the prelabeled DNA. This is true for both pol^+ (W3110) and pol^- (P3478), but pol^- shows more pieces of smaller size. These findings are in contrast to those observed with Tris–EDTA-treated cells, where newly synthesized product was found in pieces smaller than the DNA synthesized before treatment.

(c) *Pycnographic analysis:* Pycnographic analysis (Fig. 4) of newly syn-

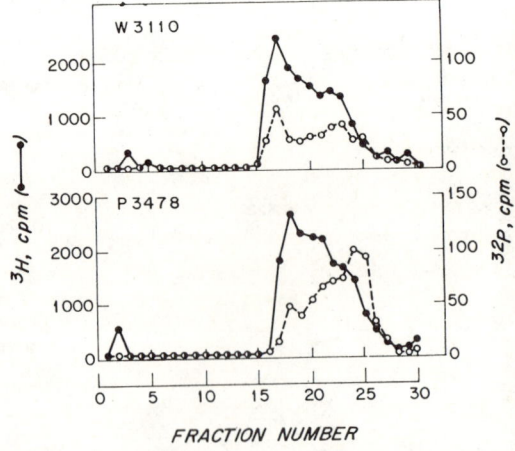

FIG. 3. Alkaline sedimentation analyis of DNA synthesized in toluene-treated cells. Cells were grown in $[^3H]$thymidine for two and one-half generations, harvested, concentrated, and treated with toluene as in Fig. 4. Cells were added to a reaction mixture as in Table 1, with $[\alpha\text{-}^{32}P]dTTP$. Reaction was continued for 40 min at 37°C. The cells were then chilled to 4°C, harvested by centrifugation, resuspended in 0.5 M KOH, and incubated 20 min at 37°C for lysis. The sample was centrifuged to remove debris. An aliquot of the supernatant was placed on a linear 0.7 M NaCl–0.3 M NaOH–1 mM EDTA 30–70% sucrose gradient, and centrifuged for 3.5 hr at 35,000 rpm in the Spinco type SW50.1 rotor at 4°C. 3-drop fractions were collected from the bottom of the tube and treated with 10% TCA in the presence of 25 μg of carrier DNA. The precipitates were washed 3 times with 3 ml of 0.01 M HCl over Whatman GF/C filters (2.4 cm), dried, and counted in a toluene scintillator.

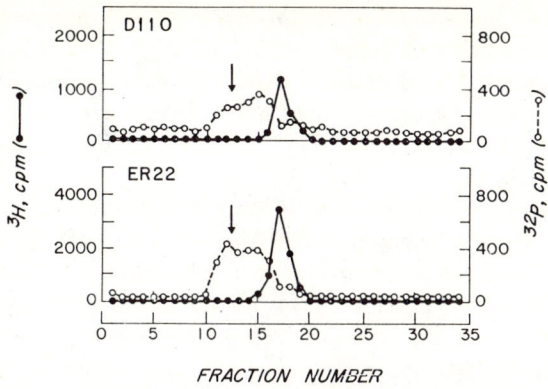

FIG. 4. Equilibrium sedimentation of BrdUTP-containing DNA synthesized in toluene-treated cells. Ten ml of cells were grown to a concentration of 8×10^8 cells/ml in [^{3}H]thymidine, harvested, and suspended in 0.5 ml of 0.05 M potassium phosphate buffer (pH 7.4). After shaking for 10 min in 1% toluene at 37°C, the cells were added to a reaction mixture as in Table 1, but containing in addition 0.13 mM NAD, α[^{32}P]dATP (specific activity of 1.0 Ci/mmole), and 33 μM BrdUTP in place of dTTP. The reaction mixture was incubated for 20 min at 37°C. Cells were lysed and DNA was extracted.[21] DNA was sedimented in neutral CsCl[22] for 50 hr at 22°C at 30,000 rpm in the IEC SB-405 rotor. Two-drop fractions were collected, precipitated with acid, and the precipitates were collected on Whatman GF/C filters, dried, and counted in toluene-based scintillator. The arrows indicate the position of hybrid DNA.

thesized density-labeled DNA demonstrates that the synthesis occurring in toluene-treated cells is semi-conservative. *E. coli* cells, grown in the presence of [^{3}H]dT, were treated with toluene and incubated in the presence of BrdUTP, [α-^{32}P]dATP, dGTP, and dCMP. After sedimentation to equilibrium in CsCl, a major portion of the newly synthesized DNA is found at a position characteristic of molecules of hybrid density, thereby indicating semiconservative synthesis. When this DNA was sedimented to equilibrium in alkaline CsCl, all of the newly synthesized DNA was found at the position of heavy single strands.

Effect of DNase on the reaction: Repair synthesis can be stimulated not only by activation of endogenous endonuclease as described above, but by exogenous DNase as well. Hoffmann-Berling[14] has demonstrated an increased rate of DNA synthesis in ether-treated cells exposed to pancreatic DNase. As shown in Table 2, addition of pancreatic DNase to toluene-treated *pol*$^+$*endI*$^-$ (*ER22*) cells leads to a 3-fold stimulation of DNA synthesis. DNA synthesis in toluene-treated *pol*$^-$ *endI*$^-$ (*D110*) cells (lacking DNA polymerase) is inhibited rather than enhanced by DNase treatment (Table 2), suggesting that the stimulation in *pol*$^+$ cells is due to the repair activity of polymerase at sites created by the DNase. This conclusion is further supported by the finding that the DNase stimulated synthesis does not depend on ATP. Furthermore, when pycnographic analysis, as described in Fig. 4, was performed on the product of the DNase-stimulated reaction in *pol*$^+$ cells, the newly synthesized DNA was found at a position corresponding to molecules of light density. With *pol*$^-$ cells the

TABLE 2. *Effect of pancreatic DNase on synthesis.*

Strain	—Activity—	
	−DNase (nmol)	+DNase (nmol)
ER22 (*pol*$^+$*endI*$^-$)	0.11	0.27
D110 (*pol*$^-$*endI*$^-$)	0.06	0.004

The reaction was as in Table 1, with 0.3 μg/ml of pancreatic DNase present where indicated. ATP was omitted from the reactions with DNase present.

newly synthesized DNA, although reduced in amount, appears at hybrid density. This result suggests a defect in repair synthesis in pol^- cells.

DNA synthesis in DNA thermosensitive mutants (DNA$_{ts}$): Replicative and repair synthesis can be differentiated in toluene-treated mutant cells which are unable to replicate DNA at elevated temperatures *in vivo* (Table 3). Two such mutants, *CR266-26* and *CRT26-43*, rapidly cease DNA synthesis at 41°C, although extracts of these cells possess normal DNA polymerase activity, even at the restrictive temperature.[12] Replicative synthesis (ATP-stimulated DNA synthesis) after toluene treatment in these mutants is 10 times lower at 43°C as compared to only 2 times lower in wild-type cells. However, normal levels of repair synthesis (DNase-stimulated activity) are present in the temperature-sensitive strains at the restrictive temperature.

Differential inhibition: DNA synthesis in toluene-treated cells is not inhibited by 5 mM cyanide or 10 mM azide. ATP-stimulated synthesis is inhibited by sulfhydryl compounds or sulfhydryl-blocking agents (Table 4). However, DNase-stimulated activity is not inhibited by sulfhydryl blockers, but is inhibited by antibody to *E. coli* DNA polymerase.

Discussion. Two types of DNA synthesis have been identified *in vivo*: that responsible for replication of the chromosome, and that occurring during repair of DNA.[6] In fact, mutants of *E. coli* have been isolated which can carry out

TABLE 3. *Effect of temperature on DNA synthesis in toluene-treated DNA$_{ts}$ mutants.*

	DNA synthesis			
	Assayed at 33°C		Assayed at 43°C	
Mutant	+ATP (%)	+DNase (%)	+ATP (%)	+DNase (%)
ER22	100	150	51	230
D110	100	...	52	...
CR266-26	100	300	12	300
CRT26-43	100	94	8	170

Cells were prepared as in Table 1, but with growth at 28°C and toluene treatment at 29°C. 50-μl aliquots of treated cells were added directly to warmed reaction mixtures, and incubation continued for 30 min, then terminated as in Table 1. Pancreatic DNase was present at 0.3 μg/ml in the reaction. ATP was omitted from reactions with DNase. Activity for each strain is expressed as percent of activity at 33°C with ATP.

TABLE 4. *Inhibition of DNA synthesis.*

	Activity	
Additions	D110 (pol^-endI^-) %	ER22 (pol^+endI^-) %
Control	100	100
+2-Mercaptoethanol	45	52
+NEM	3	2
+Antibody	95	97
+DNase, +NEM	...	135
+DNase, +Antibody	...	33

Cells were prepared and the reaction was performed as in Table 1, but with [α-^{32}P]dTTP in place of [^{3}H]dATP. 2-Mercaptoethanol was present at 13 mM. When N-ethylmaleimide (NEM) was present (1.5 mM), the cells were first incubated for 2 min at 0°C with 1 mM 2-mercaptoethanol. Antibody was present at 20 μl of a 1:10 dilution in 0.15 M NaCl/0.3 ml of reaction mixture. The antibody was heated at 70°C for 20 min before use to destroy nucleases. ATP was omitted from the reactions with DNase present. Pancreatic DNase was present at 0.3 μg/ml when indicated.

DNA repair but not replication under nonpermissive conditions. Both repair and replication of DNA can be observed in toluene-treated *E. coli* cells. The permeability of toluene-treated cells to compounds of low molecular weight permits further distinction between the two types of DNA synthesis.

DNA synthesis in toluene-treated cells resembles replication *in vivo* in its rate, the size of the product, a requirement for all four deoxyribonucleotides, and its semiconservative nature. The phase of DNA synthesis we believe to represent replication occurs even in toluene-treated pol^- cells at normal levels, as replication must. Toluene-treated DNA_{ts} mutants lack the capacity for replicative synthesis at the restrictive temperature, as do untreated mutants *in vivo*. Freedom from a requirement for an exogenous template in the toluene-treated cells implicates the bacterial chromosome as the site of synthetic activity.

Our studies with toluene-treated cells significantly add to observations made *in vivo* in revealing that replicative synthesis depends, not only on the presence of all four deoxyribonucleoside triphosphates, but also on the presence of ATP. The ATP stimulation of DNA synthesis is not highly specific, since other nucleoside triphosphates can partially substitute for ATP. The effect of ATP could be due to protection of the precursors against a phosphatase, or it may serve either as an energy source or as an enzyme cofactor. ATP appears to be required only for replicative synthesis, without effect on repair synthesis.

Toluene-treated cells furnish a new tool for distinguishing repair from replication with endogenous nuclease and polymerase activities. *PolAl* strains show a normal capacity for replicative synthesis, but a markedly reduced capacity for repair synthesis as evidenced by the results in $endI^-$ cells of pol^- compared to $endI^-pol^+$. These observations agree with current concepts of a role for DNA polymerase in repair synthesis.[23-25] Although replicative synthesis appears to proceed normally in the toluene-treated pol^- mutant, sedimentation analysis suggests a large proportion of low molecular DNA, as would be expected if repair activity were impaired and/or degradation of DNA were taking place.

DNase-stimulated synthesis in toluene-treated cells furnishes readily controllable, easily induced levels of repair synthesis in cells which are $endI^-$. We believe that the DNase-stimulated synthesis is similar to that observed during repair of ultraviolet damage of DNA *in vivo*, but this has not yet been determined. Pol^- cells do not respond to stimulation by DNase, a finding compatible with the lack of DNA polymerase activity in extracts of this mutant. Treatment of the toluene-treated cells with antibody to DNA polymerase inhibits the DNase-stimulated synthesis, thus further implicating the polymerase in repair synthesis. The results with DNA_{ts} mutants extend the differentiation between replicative and repair synthesis further: repair synthesis can be demonstrated in DNA_{ts} strains under conditions where replicative synthesis, as well as replication *in vivo* has ceased. Studies with sulfhydryl blocking agents provide separate documentation for the duality of DNA synthetic processes; they demonstrate inhibition of replicative synthesis under conditions not affecting repair synthesis as demonstrated here, or DNA polymerase *in vitro*.[26]

It is clear that toluene-treated cells differ from the Tris–EDTA cells described by Buttin and Kornberg.[11] These appear to be dependent upon the presence of an endogenous nuclease or on the activation of a nucleolytic activity. In the Tris–EDTA system pol^- strains show lower synthetic activity than do pol^+ strains, supporting this interpretation.

Our results demonstrate a dichotomy of events in DNA synthesis: repair and replication. Repair synthesis seems to require an activity similar to, or identical with, the known functions of DNA polymerase. Our data do not exclude DNA polymerase from functioning in replication. However, the observed differences in the effect of inhibitors and ATP on repair and replicative synthesis imply a necessity for steps or structures in addition to those already known.

Abbreviations: DNA_{ts}, a mutant with a block in DNA synthesis at elevated temperatures; NEM, N-ethylmaleimide; TCA, trichloroacetic acid; BrdUTP, bromodeoxyuridine triphosphate.

* The work described in this paper was sponsored in part by research grants from the National Institutes of Health, U.S. Public Health Service, No. AI-06045, and from the American Cancer Society, No. P-486.

† Supported by Public Health Service Fellowship, 5-FO3-GM42, 968-02 from General Medical Sciences.

‡ Recipient of a Public Health Service Research Career Program Award, GM-13,634.

[1] Cairns, J., *J. Mol. Biol.*, **6**, 208 (1963).
[2] Nagata, T., *Proc. Nat. Acad. Sci. USA*, **49**, 551 (1963).
[3] Yoshikawa, H., and N. Sueoka, *Proc. Nat. Acad. Sci. USA*, **49**, 559 (1963).
[4] Schildkraut, C. L., C. C. Richardson, and A. Kornberg, *J. Mol. Biol.*, **9**, 24 (1964).
[5] Richardson, C. C., R. B. Inman, and A. Kornberg, *J. Mol. Biol.*, **9**, 46 (1964).
[6] Richardson, C. C., *Ann. Rev. Biochem.*, **38**, 795 (1969).
[7] Frankel, F. R., C. Majumdar, S. Weintraub, and D. M. Frankel, *Cold Spring Harbor Symp. Quant. Biol.*, **33**, 495 (1968).
[8] Ganesan, A. T., *Cold Spring Harbor Symp. Quant. Biol.*, **33**, 45 (1968).
[9] Knippers, R., and W. Strätling, *Nature*, **226**, 713 (1970).
[10] Smith, D. W., H. Schaller, and F. J. Bonhoeffer, *Nature*, **226**, 711 (1970).
[11] Buttin, G., and A. Kornberg, *J. Biol. Chem.*, **241**, 5419 (1966).
[12] Buttin, G., and M. Wright, *Cold Spring Harbor Symp. Quant. Biol.*, **33**, 259 (1968).
[13] Jackson, R. W., and J. A. DeMoss, *J. Bacteriol.*, **90**, 1420 (1965).
[14] Hoffmann-Berling, H., in *Molecular Genetics*, eds. Wittmann, H. C., and M. Schuster (Berlin: Springer, 1968), p. 38.
[15] DeLucia, R., and J. Cairns, *Nature*, **224**, 1164 (1969).
[16] Gross, J., and M. Gross, *Nature*, **224**, 1166 (1969).
[17] Dürwald, H., and H. Hoffmann-Berling, *J. Mol. Biol.*, **34**, 331 (1968).
[18] Anderson, E. H., *Proc. Nat. Acad. Sci. USA*, **32**, 120 (1946).
[19] Levin, D. H., M. N. Thang, and M. Grunberg-Manago, *Biochim. Biophys. Acta*, **76**, 558 (1963).
[20] Oishi, M., *Proc. Nat. Acad. Sci. USA*, **64**, 1292 (1969).
[21] Thomas, C. A., Jr., K. I. Berns, and T. J. Kelley, Jr., in *Procedures in Nucleic Acid Research*, eds. Cantoni, G. L., and D. R. Davies, (New York: Harper and Row, 1966), p. 535.
[22] Pettijohn, D. E., and P. C. Hanawalt, *J. Mol. Biol.*, **9**, 395 (1964).
[23] Kelly, R. B., N. R. Cozzarelli, M. P. Deuscher, I. R. Lehman, and A. Kornberg, *J. Biol. Chem.*, **245**, 39 (1970).
[24] Kanner, L., and P. Hanawalt, *Biochem. Biophys. Res. Commun.*, **39**, 149 (1970).
[25] Boyle, J. M., M. C. Patterson, and R. B. Setlow, *Nature*, **226**, 708 (1970).
[26] Englund, P. T., J. A. Huberman, T. M. Jovin, and A. Kornberg, *J. Biol. Chem.*, **244**, 3038 (1969).

Joining of DNA Strands by DNA Ligase of *E. coli*

MARTIN GELLERT, JOHN W. LITTLE, CAROL K. OSHINSKY,
AND STEVEN B. ZIMMERMAN

*Laboratory of Molecular Biology, National Institute of Arthritis and Metabolic Diseases,
National Institutes of Health, Bethesda, Maryland*

Enzymes which covalently join DNA strands have been isolated in the past two years from uninfected *E. coli* (Gellert, 1967; Olivera and Lehman, 1967a; Zimmerman et al., 1967; Gefter et al., 1967) and from *E. coli* infected with phage T4 (Weiss and Richardson, 1967a; Cozzarelli et al., 1967; Becker et al., 1967). It is likely that these DNA ligases have a role in DNA repair and recombination. They have also recently been implicated in the process of DNA replication (Okazaki et al., 1968; Newman and Hanawalt, 1968).

In this paper we discuss some of our studies on the DNA ligase of *E. coli*, first summarizing and commenting on our previously published results and then sketching more recent lines of work.

Assay of DNA Ligase

Assays for joining of DNA strands have been based on one of two characteristics of the product: (1) Resistance of the joined structure to dissociation by DNA-denaturing conditions (Gellert, 1967; Zimmerman et al., 1967; Cozzarelli et al., 1967; Bautz, 1967), and (2) incorporation of the 5'-terminal phosphate of a DNA chain into a phosphatase-resistant linkage (Weiss and Richardson, 1967a; Olivera and Lehman, 1967a). We have devised two assays of the first type, using λ DNA as substrate.

Since it was found that the DNA of phage λ is intracellularly converted to a covalently circular form (Young and Sinsheimer, 1964; Bode and Kaiser, 1965), the cohesive ends of λ DNA have offered a promising basis for an assay of DNA-joining activity. Suitable incubation of λ DNA yields a cohered structure (Hershey et al., 1963) which aligns the DNA termini in a reasonable configuration to allow end-to-end joining of the DNA strands. By varying the DNA concentration in the cohesion step, formation either of hydrogen-bonded circles or intermolecular aggregates can be favored (Hershey et al., 1963). While covalent closure of hydrogen-bonded circles, as detected by the rapid alkaline sedimentation of covalent circles, has been used by ourselves and others as a measure of joining activity (Gellert, 1967; Olivera and Lehman, 1967a; Gefter et al., 1967), it suffers some disadvantages as a routine assay: (1) It is extremely sensitive to endonucleolytic attack on the DNA, because the rapid sedimentation depends on both strands being covalently continuous. (2) It is exceedingly laborious, since each assay requires analysis of one tube from a sucrose gradient centrifugation.

We thus developed another assay which, while still using joining at the ends of λ DNA, does not suffer from these problems. The assay measures the covalent joining of two types of λ DNA which differ in their response to denaturation (Fig. 1). One DNA species has been cross-linked by reaction with a bifunctional alkylating agent. After alkaline denaturation, it rapidly renatures on return to neutral pH. The other DNA species contains ^{3}H-thymidine but, since it has no cross-links, is irreversibly denatured by alkali.

A mixture of these two DNAs is incubated at 50° to form hydrogen-bonded aggregates, concentrations being adjusted so that almost all the tritiated DNA is bound to cross-linked DNA rather than to itself. After incubation with DNA ligase, the DNA is denatured in alkali and then brought back to neutral pH (in the presence of formaldehyde,

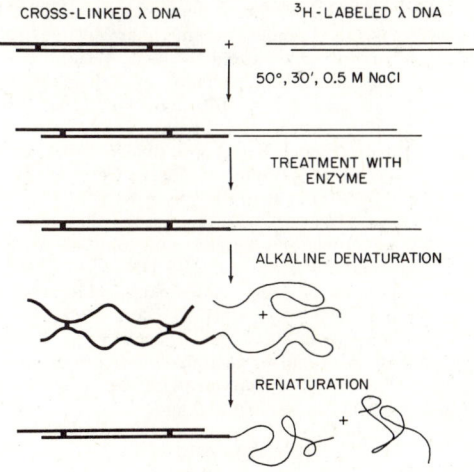

FIGURE 1. Assay of DNA ligase (Zimmerman et al., 1967).

to prevent any spontaneous renaturation of tritiated DNA). By these means, any ^{3}H-labeled DNA which has been covalently joined to cross-linked DNA will remain bound to a native DNA 'handle', while the unjoined fraction of ^{3}H-labeled DNA will be irreversibly denatured and completely disaggregated from the cross-linked DNA.

Denatured and native (or partly native) DNA are then separated by adsorption on hydroxylapatite. At a suitable salt concentration, native DNA adsorbs to hydroxylapatite while denatured DNA does not (Bernardi, 1962; Miyazawa and Thomas, 1965). Thus when the DNA reaction product is mixed with a hydroxylapatite slurry, the cross-linked DNA 'handle' is specifically adsorbed together with any covalently joined ^{3}H-labeled DNA. The bound ^{3}H is thus a measure of DNA-joining activity. Details of this assay have been published previously (Zimmerman et al., 1967).

With this assay, a DNA-joining activity was found in extracts of *E. coli* B and was purified 200-fold.

PROPERTIES OF THE DNA-JOINING REACTION

In addition to joining the ends of λ DNA, DNA ligase can rejoin internal breaks in a DNA duplex structure. Thus breaks made by limited digestion with pancreatic DNase are efficiently sealed, as shown both by an increase in single-strand molecular weight and by the disappearance of phosphatase-sensitive terminal phosphate groups (Zimmerman et al., 1967). Entirely parallel results are found with the enzyme from T4-infected cells (Weiss and Richardson, 1967a). To be sealed, the DNA strands must terminate in 5'-phosphoryl and 3'-hydroxyl groups. Thus dephosphorylation of λ DNA termini renders them insusceptible to joining; activity is restored if the 5' termini are rephosphorylated (Zimmerman et al., 1967). Similarly, Becker et al. (1967) have shown that the T4-induced ligase can join breaks in DNA introduced by micrococcal nuclease (which yields 3'-phosphoryl termini) only if these are first dephosphorylated and then phosphorylated at the 5' end. The strands to be joined must also be held in the correct hydrogen-bonded conformation with adjoining ends: denatured nicked DNA, or λ DNA with separated cohesive ends, cannot be joined (Weiss and Richardson, 1967a; Gellert, 1967).

The strands become joined through a 3'-5' phosphodiester bond; thus a phosphate originating on a 5' end can be shown, after joining, also to be bonded in a 3'-phosphoryl linkage.

DPN AS COFACTOR

The DNA ligase of *E. coli* uses DPN as cofactor (Zimmerman et al., 1967; Olivera and Lehman, 1967b). The K_m for DPN is remarkably low, half-maximal activity being achieved with 3×10^{-8} M DPN. The DPN requirement is highly specific; of the analogs which we have tested, only thionicotinamide-DPN and 3-acetylpyridine-DPN have partial activity, while a variety of related compounds including DPNH, TPN, TPNH and nucleoside di- and triphosphates are totally inactive (Zimmerman et al., 1967).

To investigate how DPN is used in the DNA-joining reaction, we prepared two DPN species, one labeled in the adenylate half of the molecule with ^{32}P (schematized AR^{32}PPRN), and the other labeled in the nicotinamide moiety with ^{3}H (ARPPRN*). When a mixture of these DPN species was incubated with DNA ligase and an excess of nicked DNA, the DPN was entirely broken down, with the ^{3}H now appearing in NMN and the ^{32}P mainly in AMP (some inorganic phosphate was liberated from AMP by a phosphatase contaminant of the DNA ligase) (Fig. 2B). In the absence of DNA, very little breakdown of DPN occurred (Fig. 2A). The number of DNA breaks sealed in this experiment indicated that one molecule of DPN is broken down to NMN and AMP for each joining event. Thus the overall stoichiometry of the reaction can be written as follows:

$$\underset{\text{OH P}}{\rule{2cm}{0.4pt}}\;+\;\text{DPN}\;\longrightarrow\;\underset{\text{P}}{\rule{2cm}{0.4pt}}\;+\;\text{AMP}\;+\;\text{NMN}$$

ENZYME-ADENYLATE INTERMEDIATE

When DNA ligase was incubated with the mixture of labeled DPN species in the experiment of Fig. 2A (note the absence of DNA), a small amount of ^{32}P but no ^{3}H became acid-insoluble. This result suggested the formation of an enzyme-adenylate complex; subsequent results have confirmed this inference. First, we looked for the exchange reaction predicted by this formulation. Indeed, if DNA ligase was incubated with DPN and radioactive NMN, radioactivity was rapidly incorporated into DPN; no such incorporation of radioactive AMP occurred (Little et al., 1967). It thus appeared likely that an enzyme-adenylate was being formed in the reversible reaction:

$$\text{enzyme} + \text{DPN} \rightleftarrows \text{enzyme-AMP} + \text{NMN}.$$

It proved possible to isolate a ligase-adenylate complex directly. Figure 3 shows the result of an experiment in which DNA ligase was incubated with the mixture of labeled DPN species described above. The product was passed through a Sephadex G-25 column to separate enzyme from unattached small molecules. It can be seen that a large fraction of the ^{32}P (coming from the AMP moiety of DPN) appears with the ligase activity; the lack of ^{3}H in

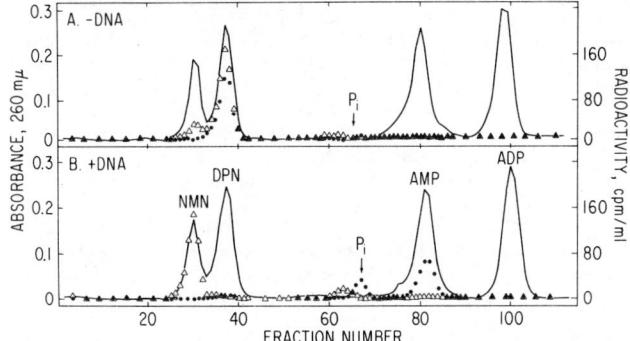

FIGURE 2. DNA-dependent breakdown of DPN. A mixture of ^{3}H-nicotinamide-labeled DPN and ^{32}P-AMP-labeled DPN was incubated with DNA ligase in the absence (A) or presence (B) of DNase-treated DNA. The acid-soluble products were chromatographed on DEAE-cellulose. (For details see Zimmerman et al., 1967.) △, ^{3}H; ●, ^{32}P; ———, absorbance at 260 mμ.

the enzyme region indicates that DPN *per se* is not being bound. (In other experiments with the AMP moiety of DPN labeled in different positions, it was shown that the entire AMP residue is bound.)

The AMP is complexed very tightly and presumably is covalently bound to the enzyme. It is not removed on precipitation of the enzyme with cold perchloric acid, or by brief boiling at pH 7 or 13.

The enzyme-adenylate complex has properties to be expected of an intermediate in the DNA-joining reaction:

(1) Incubation with NMN liberates free enzyme and DPN.

(2) The bound AMP is held at a free energy level not far from that of the pyrophosphoryl linkage in DPN, since incubation with a hundred-fold excess of NMN over DPN releases half the bound AMP.

(3) Incubation with native DNA carrying single-strand breaks liberates free enzyme and AMP.

(4) A variety of other nucleotides and of nucleic acid species (e.g. denatured DNA, tRNA) fail to release AMP from the complex.

(5) Even in the absence of added DPN, enzyme-adenylate is capable of carrying out the DNA-joining reaction, as tested either in our standard assay or by the increase in single-strand molecular weight of DNase-treated λ DNA (Little et al., 1967).

Thus the second half of the overall reaction can be written:

enzyme-AMP + nicked DNA →
 enzyme + AMP + joined DNA.

The equilibrium of this reaction strongly favors the joining process. It has not been possible to introduce single-strand breaks into DNA by incubation with DNA ligase and AMP.

FURTHER PURIFICATION OF DNA LIGASE

The existence of the ligase-adenylate complex has simplified further purification of DNA ligase in two ways.

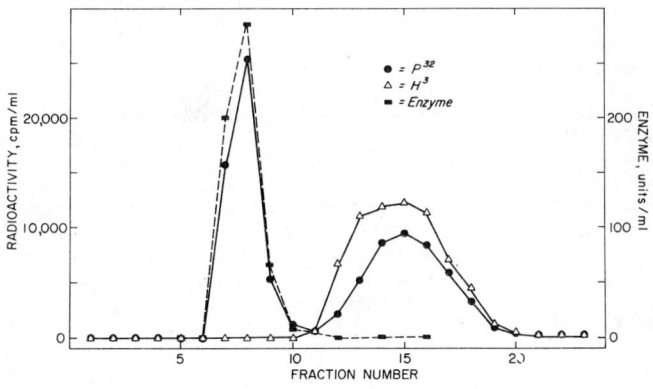

FIGURE 3. Isolation of enzyme-adenylate specifically labeled from the ^{32}P-AMP moiety of DPN. After incubation of DNA ligase with a mixture of ^{3}H-nicotin labeled DPN and ^{32}P-AMP-labeled DPN, the sample was applied to a Sephadex G-25 column. (For details see Little et al., 1967.)

First, it provides an extremely rapid and easy assay for DNA ligase. Since with excess DPN the formation of enzyme-adenylate rapidly goes to completion, the acid-precipitable radioactivity obtained after incubation of ligase with AMP-labeled DPN is a measure of the ligase concentration in the sample. Even at early stages in the purification, there seems to be no contaminating activity with similar properties.

Second, in at least one step of the purification, chromatography of DNA ligase as the adenylate improves the purification and recovery.

The current purification follows the one previously described (Zimmerman et al., 1967) through Fraction IV. The enzyme is then concentrated by ammonium sulfate precipitation, converted to the adenylate, and chromatographed on DEAE-cellulose with a potassium phosphate gradient at pH 8.0. The peak is fractionated with ammonium sulfate, the adenylate removed by treatment with excess NMN, and the stripped enzyme chromatographed on phosphocellulose (potassium phosphate gradient, pH 6.8). The pooled phosphocellulose fraction is purified about 1000-fold.

An approximate molecular weight of the enzyme-adenylate formed from the phosphocellulose fraction has been determined by use of a sizing column (Sephadex G-200). Using the method of Andrews (1965), with several protein markers, a value of 1.0×10^5 was estimated.

Knowing the molecular weight and the specific activity of the phosphocellulose fraction (1400 $\mu\mu$ moles AMP bound per mg protein), one can conclude that this enzyme fraction is at best 15% pure.

Multiple Forms of DNA Ligase

In the DEAE-cellulose gradient chromatography step of the purification, the bulk of enzyme-adenylate is eluted at 0.14 M potassium phosphate. Similarly, if ligase which has been purified through the phosphocellulose step is adenylated and rechromatographed on DEAE-cellulose, it is eluted at the same salt concentration. If, however, the enzyme-adenylate is stored for a day in low salt (e.g. 0.02 M potassium phosphate, pH 8.0, 5°) before rechromatography on DEAE-cellulose, then about 40% of the enzyme-adenylate elutes as a discrete peak at 0.11 M potassium phosphate, the remainder eluting at 0.14 M. Longer storage in low salt increases the fraction eluting early. Storage in 0.10 M potassium phosphate effectively prevents this conversion.

As compared with the 'normal' form of the enzyme-adenylate, this altered form has the following properties:

(1) The bound AMP can also be removed by incubation with NMN or nicked DNA.

(2) The stripped enzyme can be re-adenylated to the same extent with DPN.

(3) By chromatography on Sephadex G-200, both forms have the same molecular weight.

(4) The discharge of AMP from the '0.11 M' form by moderate concentrations of nicked DNA is significantly slower.

(5) The DNA-joining activity in our standard assay is at least tenfold lower.

Thus this form of DNA ligase appears to have undergone a specific structural transition which renders it inactive in the DNA-joining step of the overall reaction. It has not so far been possible to reverse this transition.

Activity of DNA Ligase in Various *E. coli* Strains

In order to understand the physiological role of DNA ligase, it would clearly be desirable to find mutant *E. coli* strains with altered ligase activity. In particular, one might suppose that changes in ligase activity would be associated with some of the known mutations affecting DNA repair or recombination.

Accordingly, extracts from a number of mutant strains were prepared (as described by Gellert, 1967) and assayed for DNA ligase. Among the strains tested (genotype in parenthesis) were: AB 1886 (*uvr* A), AB 1885 (*uvr* B), AB 1884 (*uvr* C) (Howard-Flanders et al., 1966); AB 2463 (*rec* A), AB 2470 (*rec* B), (Howard-Flanders and Theriot, 1966); KMBL 90, 91, 99, 100, 101 (all *uvr*) (Van de Putte et al., 1965); KMBL 239, 240, 241, 243 (all *rec*) (Van de Putte et al., 1966); KMBL 237 (*exr*), KMBL 102 (*dir*) (Rörsch et al., 1966); B_{s-1} (*exr uvr*), B_{s-2} (*exr*) (Hill, 1961; Mattern et al., 1966) and B/r (Witkin, 1947). The specific activity of DNA ligase in all these extracts was the same, within a factor of two, as in extracts of representative *E. coli* strains with normal repair and recombination proficiency, e.g., *E. coli* B, C600, or AB 1157, which all fall in the range of 120–160 units/mg. Thus the deficiencies in DNA repair and recombination which characterize these mutants cannot be explained as resulting from changes in DNA ligase activity. (It should be noted that our results are in marked disagreement with those of Gefter et al. [1967], who reported a 30-fold variation in DNA ligase activity among a set of mutant strains included in the above list, and proposed a plausible correlation between ligase activity and DNA repair proficiency. The explanation of the discrepancy may be in the different assays used; Gefter et al. measured formation of covalent circles of λ DNA which, as mentioned above, is an assay exceptionally sensitive to endonucleases.)

In screening more strains for alterations in DNA ligase activity, we have proceeded similarly to Fareed and Richardson (1967), who showed that an altered form of the T4-induced ligase was produced by mutation in a gene known to affect T4 DNA synthesis. We have thus assayed DNA ligase in crude extracts of temperature-sensitive *E. coli* mutants which fail to synthesize DNA at high temperature (42°). Twenty-two such mutants, isolated by the procedure of Bonhoeffer and Schaller (1965), were kindly sent to us by Dr. Bonhoeffer. Extracts were prepared from cells grown at 30°, and from cells incubated at 43° for 30 min after growth at 30°. In none of the strains was there a significant decrease in ligase activity following the high-temperature incubation.

Since these mutants were obtained by a very special selection procedure (which would exclude, for example, all strains which die at high temperature), a screening of other mutants with temperature-sensitive DNA synthesis may still reveal strains with altered ligase activity.

DISCUSSION

A unique feature of *E. coli* DNA ligase is the participation of DPN, usually considered as a coenzyme of oxidative reactions, in a synthetic reaction; DNA phosphodiester bond synthesis proceeds at the expense of the pyrophosphoryl linkage of DPN. The highly specific requirement for DPN is indicated by the lack of activity of structurally or functionally related compounds, and by the remarkably low K_m (ca. 3×10^{-8} M), which may reflect the importance to the cell of a reaction maintaining the integrity of its genetic complement. The intracellular concentrations of DPN and NMN are such that a large fraction of the enzyme probably exists in vivo in the adenylated form required for rejoining breaks in DNA.

One may speculate whether the internal pyrophosphoryl linkage of DPN or related coenzymes is used as a means of activation in other synthetic reactions. Without prior knowledge of the synthetic reaction, such a system might appear simply to hydrolyze the coenzyme. Thus, lacking a measure of the joining of DNA, one might have described DNA ligase as a DNA-dependent DPNase.

Except for the cofactor requirement, the *E. coli* DNA ligase and the T4-induced ligase have very similar properties. Starting from ATP, the T4 enzyme also forms an enzyme-adenylate complex which is then functional in the DNA-joining reaction (Weiss and Richardson, 1967b; Becker et al., 1967). Both enzymes require similar structural properties of the DNA substrate. The physiological significance of the shift in cofactor requirement on phage infection is not clear.

ACKNOWLEDGMENTS

We are indebted to Drs. F. Bonhoeffer, P. Howard-Flanders, A. Rörsch, and E. M. Witkin for their generosity in sending us numerous mutant strains of *E. coli*.

REFERENCES

Andrews, P. 1965. The gel-filtration behaviour of proteins related to their molecular weights over a wide range. Biochem. J. *96:* 595.

Bautz, E. K. F. 1967. A biological assay for polynucleotide ligase: recovery of marker activity in DNA-transformation. Biochem. Biophys. Res. Commun. *28:* 641.

Becker, A., G. Lyn, M. Gefter, and J. Hurwitz. 1967. The enzymatic repair of DNA, II. Characterization of phage-induced sealase. Proc. Nat. Acad. Sci. *58:* 1996.

Bernardi, G. 1962. Chromatography of denatured deoxyribonucleic acid on calcium phosphate. Biochem. J. *83:* 32.

Bode, V. C., and A. D. Kaiser. 1965. Changes in the structure and activity of λ DNA in a superinfected immune bacterium. J. Mol. Biol. *14:* 399.

Bonhoeffer, F., and H. Schaller. 1965. A method for selective enrichment of mutants based on the high UV sensitivity of DNA containing 5-bromouracil. Biochem. Biophys. Res. Commun. *20:* 93.

Cozzarelli, N. R., N. E. Melechen, T. M. Jovin, and A. Kornberg. 1967. Polynucleotide cellulose as a substrate for a polynucleotide ligase induced by phage T4. Biochem. Biophys. Res. Commun. *28:* 578.

Fareed, G. C., and C. C. Richardson. 1967. Enzymatic breakage and joining of deoxyribonucleic acid. II. The structural gene for polynucleotide ligase in bacteriophage T4. Proc. Nat. Acad. Sci. *58:* 665.

Gefter, M. L., A. Becker, and J. Hurwitz. 1967. The enzymatic repair of DNA, I. Formation of circular λ DNA. Proc. Nat. Acad. Sci. *58:* 240.

Gellert, M. 1967. Formation of covalent circles of λ DNA by *E. coli* extracts. Proc. Nat. Acad. Sci. *57:* 148.

Hershey, A. D., E. Burgi, and L. Ingraham, 1963. Cohesion of DNA molecules isolated from phage lambda. Proc. Nat. Acad. Sci. *49:* 748.

Hill, R. F., and E. Simson. 1961. A study of radiation sensitive and radiation resistant mutants of *E. coli* B. J. Gen. Microbiol. *24:* 1.

Howard-Flanders, P., R. P. Boyce, and L. Theriot. 1966. Three loci in *Escherichia coli* K-12 that control the excision of pyrimidine dimers and certain other mutagen products from DNA. Genetics. *53:* 1119.

Howard-Flanders, P., and L. Theriot. 1966. Mutants of *Escherichia coli* K-12 defective in DNA repair and in genetic recombination. Genetics. *53:* 1137.

Little, J. W., S. B. Zimmerman, C. K. Oshinsky, and M. Gellert. 1967. Enzymatic joining of DNA strands, II. An enzyme-adenylate intermediate in the DPN-dependent DNA ligase reaction. Proc. Nat. Acad. Sci. *58:* 2004.

Mattern, I. E., H. Zwenk, and A. Rörsch. 1966. The genetic constitution of the radiation-sensitive mutant *Escherichia coli* B$_{s-1}$. Mutation Res. *3:* 374.

Miyazawa, Y., and C. A. Thomas, Jr. 1965. Nucleotide composition of short segments of DNA molecules. J. Mol. Biol. *11:* 223.

NEWMAN, J., and P. HANAWALT. 1968. Role of polynucleotide ligase in T4 DNA replication. J. Mol. Biol. *35:* 639.

OKAZAKI, R., T. OKAZAKI, K. SAKABE, K. SUGIMOTO, and A. SUGINO. 1968. Mechanism of DNA chain growth, I. Possible discontinuity and unusual secondary structure of newly synthesized chains. Proc. Nat. Acad. Sci. *59:* 598.

OLIVERA, B. M., and I. R. LEHMAN. 1967a. Linkage of polynucleotides through phosphodiester bonds by an enzyme from *Escherichia coli*. Proc. Nat. Acad. Sci. *57:* 1426.

———, ———. 1967b. Diphosphopyridine nucleotide: a cofactor for the polynucleotide-joining enzyme from *Escherichia coli*. Proc. Nat. Acad. Sci. *57:* 1700.

RÖRSCH, A., P. VAN DE PUTTE, I. E. MATTERN, and H. ZWENK. 1966. Genetic and enzymic control of radiation sensitivity in *Escherichia coli*, p. 105–128. *In* Genetical Aspects of Radiosensitivity: Mechanisms of Repair. International Atomic Energy Agency, Vienna.

VAN DE PUTTE, P., C. A. VAN SLUIS, J. VAN DILLEWIJN, and A. RÖRSCH. 1965. The location of genes controlling radiation sensitivity in *Escherichia coli*. Mutation Res. *2:* 97.

VAN DE PUTTE, P., H. ZWENK, and A. RÖRSCH. 1966. Properties of four mutants of *Escherichia coli* defective in genetic recombination. Mutation Res. *3:* 381.

WEISS, B., and C. C. RICHARDSON. 1967a. Enzymatic breakage and joining of deoxyribonucleic acid, I. Repair of single-strand breaks in DNA by an enzyme system from *Escherichia coli* infected with T4 bacteriophage. Proc. Nat. Acad. Sci. *57:* 1021.

———, ———. 1967b. Enzymatic breakage and joining of deoxyribonucleic acid, III. An enzyme-adenylate intermediate in the polynucleotide ligase reaction. J. Biol. Chem. *242:* 4270.

WITKIN, E. M. 1947. Genetics of resistance to radiation in *Escherichia coli*. Genetics *32:* 221.

YOUNG, E. T., II, and R. L. SINSHEIMER. 1964. Novel intra-cellular forms of lambda DNA. J. Mol. Biol. *10:* 562.

ZIMMERMAN, S. B., J. W. LITTLE, C. K. OSHINSKY, and M. GELLERT. 1967. Enzymatic joining of DNA strands: a novel reaction of diphosphopyridine nucleotide. Proc. Nat. Acad. Sci. *57:* 1841.

T4 Bacteriophage Gene 32: A Structural Protein in the Replication and Recombination of DNA

by

BRUCE M. ALBERTS
LINDA FREY
Department of Biochemical Sciences,
Frick Chemical Laboratory,
Princeton University,
Princeton, New Jersey

A new type of protein essential for DNA replication and genetic recombination has been isolated from T4 bacteriophage-infected cells of E. coli. This protein binds cooperatively to single-stranded DNA, and it catalyzes DNA denaturation and renaturation under physiological conditions in vitro.

GENETIC recombination involves the precise breakage and reunion of "mating" double-stranded DNA molecules at points of mutual sequence homology[1-3]. Recombinant DNA molecules have been shown to contain a heterozygous region, which seems to be formed during the fundamental event in the recombination process[2,4]. Although the actual mechanism of genetic recombination is unknown, several relatively simple models have been proposed[5-7]; these assume an unusual fluidity of DNA structure within the cell which allows efficient testing for complementary base pairings between strands of randomly colliding DNA molecules. For example, in the scheme proposed by Holliday[5], local DNA denaturation is invoked to open mating DNA helices at homologous regions, followed by DNA renaturation between single strands thereby exposed on opposite molecules. In vitro, however, the DNA double-helix is overwhelmingly stable relative to the single strands in physiological conditions[8], and locally denatured regions more than a few base-pairs long should consequently occur only very rarely. This expectation is borne out by experimental studies of the stability of the short helix formed by the "cohesive ends" of isolated bacteriophage lambda DNA, for which, even in low [Na+] (0·033 M), transient melting of twelve contiguous base-pairs occurs with a relaxation time of about 7 days at 37° C (ref. 9). By contrast, at 37° C within the cell, the average T4 DNA molecule participates in more than one recombination exchange every 10 min[10], and it must unwind completely during each round of DNA replication.

Single-stranded regions of DNA postulated as intermediates in various models for the recombination process could easily be generated by the action of exonuclease[7,11,12] or DNA polymerase[6] within the cell, if not by denaturation. It might be expected that the complementary single-stranded regions so generated would rapidly pair by renaturation because of the overwhelming stability of the double-helical conformation at 37° C. Single-stranded DNA folds on itself, however, to create imperfectly hydrogen-bonded, intra-strand helices under physiological conditions *in vitro*[13]. These folds make the DNA bases relatively inaccessible, and thereby prevent complementary single strands from finding satisfactory pairings[14,15]. As a result, raising the temperature from 37° C to 68° C increases renaturation rates as much as 1,000-fold. This is purely a kinetic effect, for the equilibrium stability of the double-helix relative to single strands is greater at the lower temperature.

We have discovered a DNA-binding protein in the T4 bacteriophage system the properties of which suggest a solution to this problem of DNA mechanics. The protein is the product of T4 gene 32. The "32-protein" is required for the genetic recombination of T4 bacteriophage DNA[16]; in addition, it is one of several gene products known to be essential for T4 DNA replication[17].

Biological Role of T4 Gene 32

The product of T4 gene 32 is required for T4 DNA replication throughout the infectious cycle: an amber mutant in this gene requires 40 min at 37° C to approximate even a single round of replication[18], whereas temperature-sensitive mutants which are allowed to begin synthesizing DNA at 25° C stop replication when shifted to a non-permissive temperature[19,20]. (For one particular mutant, ts P7, all replication ceases within 1 min after a shift to 42° C: S. Riva, A. Cascino, and E. P. Geiduschek, manuscript submitted for publication; unpublished results of M. Curtis and B. M. A.) Moreover, gene dosage experiments show that gene 32 is unique among the T4 genes known to affect DNA metabolism in that its product is required stoichiometrically rather than catalytically; that is, as for structural proteins of the phage particle, the quantity of 32-protein synthesized in the infected cell directly limits the number of progeny phage produced[21]. This finding must be reconciled with the fact that about 10,000 molecules of 32-protein are made in a normal infection, and very few, if any, are used up in the construction of mature phage particles[20]. It therefore seems that 32-protein plays a structural part in the replication of T4 DNA.

Another important biological observation concerning gene 32 is that, as first shown by Tomizawa and co-workers, its function is necessary for the formation of the hydrogen-bonded joint DNA molecules believed to be the initial products of genetic recombination[2,16,18]. In this connexion, it should be noted that DNA replication does not seem to be required for recombination of T4 DNA[16], and that recombination-deficient mutants in another bacteriophage system (phage λ) replicate their DNA normally[11]. It is therefore likely that 32-protein functions directly in both of these genetic processes.

Properties of Purified 32-Protein and its Binding to DNA

We have previously reported that at least twenty different DNA-binding proteins are synthesized after T4 bacteriophage infection of *E. coli*, as judged by DNA-cellulose chromatography[20]. One of the principal DNA-binding proteins was identified as the product of T4 gene 32, for it is altered after infection with bacteriophages carrying amber and temperature-sensitive mutations in this gene[19,20]. The 32-protein is made in large quantities at both early and late times of infection, about 10,000 molecules accumulating per infected cell.

In the absence of a direct assay for 32-protein, the course of its purification was originally monitored by polyacrylamide gel electrophoresis. In the work to be described stepwise elution from a single-stranded DNA–cellulose column followed by DEAE-cellulose chromatography has been used to prepare 32-protein which is electrophoretically homogeneous.

In spite of its tight binding to polyanionic DNA, 32-protein carries a net negative charge at pH 7. As estimated from a combination of sedimentation and gel filtration data, the molecular weight of the native protein is 35,000, and the axial ratio for an equivalent prolate ellipsoid is about 4 (ref. 19). Because the same molecular weight is obtained for denatured, reduced 32-protein in sodium dodecyl sulphate (SDS)-containing polyacrylamide gels[22], the native protein seems to consist of a single polypeptide chain[20].

Purified 32-protein binds strongly to single-stranded DNA, as seen by the co-sedimentation of ^{3}H-leucine-labelled protein with such DNA through stabilizing sucrose gradients. The affinity of 32-protein for DNA decreases gradually as the salt concentration is increased from 0·15 to 0·60 M, suggesting the importance of electrostatic forces in the binding. It seems likely, therefore, that the region of polypeptide chain in direct contact with the DNA includes a concentration of positively charged residues spaced so as to interact with the DNA phosphates, even though the protein as a whole carries a net negative charge.

The stoichiometry of the tight complex which 32-protein forms with single-stranded DNA at low salt concentrations has been examined by sucrose gradient sedimentation of a fixed quantity of labelled 32-protein in the presence of varying amounts of the circular, single-stranded DNA from bacteriophage fd[23]. At lower concentrations of DNA, two distinct peaks of radioactive protein are seen; one sediments rapidly with the DNA, the second at the slow rate characteristic of the free protein. The free protein peak is absent above a weight ratio of DNA to protein of 1 : 12. The complex therefore contains about one protein molecule of 35,000 molecular weight for every ten single-stranded DNA nucleotides. Because ten nucleotides can span a distance of not more than 70 Å, whereas 32-protein may be as much as 120 Å long, adjacent molecules of 32-protein could overlap in the complex. Consistent with this expectation, it was shown previously that in crude extracts 32-protein binds cooperatively to single-stranded DNA-cellulose[19].

To determine whether the purified 32-protein also binds cooperatively to DNA, two different concentrations of 32-protein (containing the same amount of tritiated 32-protein) were mixed with a constant amount (large excess) of fd DNA at an elevated salt concentration where the complex is only marginally stable. The results of sucrose gradient sedimentation analyses are shown in Fig. 1. It is clear that a 14-fold increase in 32-protein concentration dramatically increases its DNA affinity. This result requires that 32-protein molecules interact with each other in the complex (see caption to Fig. 1), and suggests a model in which there are two types of binding sites for 32-protein on single-stranded DNA: ten nucleotides adjacent to a previously bound molecule of 32-protein ("contiguous" site) and ten not adjacent to any previously bound molecule ("isolated site"). If this model is correct, the affinity of 32-protein for a contiguous site must be at least eighty times greater than its affinity for an isolated site, for strong cooperative binding is observed even in conditions where the number of isolated sites available exceeds the number of contiguous sites by at least this factor (Fig. 1).

The highly cooperative nature of the DNA affinity of 32-protein should cause it to bind to DNA in long clusters even in conditions of large DNA excess. Direct evidence for such clustered binding is obtained when labelled 32-protein is mixed with a large excess of fd DNA and sedimented through sucrose gradients at low salt concentrations. In this experiment, a larger portion of the 32-protein sediments ahead of the main DNA peak, being tightly bound to a small fraction of the DNA molecules. The mean size of a 32-protein cluster in these conditions must therefore be an appreciable fraction of the length of an fd DNA molecule (6,600 nucleotides)[23]. Clustered binding to poly dA can be detected by this method, indicating that the cooperativity observed is the result of direct stabilizing interactions between adjacent 32-protein monomers. This view is also supported by our finding (unpublished) that 32-protein self-aggregates in the absence of DNA at a concentration of 0.5 mg/ml. or higher.

double-stranded DNA which occurs in its presence should be accompanied by the large hyperchromic change that is characteristic of this helix–coil transition. By this criterion, double-stranded T4 DNA is not denatured in the presence of excess 32-protein in a variety of ionic conditions at temperatures up to 37° C. By contrast, poly dAT, which normally has a T_m about 16° C lower than T4 DNA (65° C as against 81° C in 0.01 M KCl–0.01 M $MgSO_4$), is readily denatured by 32-protein even at 25° C. Typical kinetics for poly dAT denaturation by 32-protein are shown in Fig. 2. In the presence of 0.01 M Mg^{2+}, half-denaturation of poly dAT by 32-protein is attained in about 20 min in the conditions used; this denaturation is reversible, for the absorbance at 260 mµ can be restored to its original value either by addition of NaCl to 0.5 M at 25° C (to dissociate 32-protein), or by direct cooling to 4° C. The initial rate of denaturation at 25° C is reduced with increasing [Mg^{2+}], decreasing at least 15-fold when [Mg^{2+}] is increased from 0.01 M to 0.04 M, and increasing about three-fold when all Mg^{2+} is removed.

To denature poly dAT at 25° C, ΔG for the coil→complex reaction (single-stranded DNA coil + 32-protein → DNA – protein complex) must be sufficiently negative to make $\Delta G_{helix \to coil} + \Delta G_{coil \to complex} < 0$ at that temperature. The $\Delta G_{helix \to coil}$ should be about +1.0 kcalories/mole base-pairs for poly dAT in 0.01 M Mg^{2+} at 25° C ($T_m = 65°$ C) (ref. 28). Consequently, to obtain denaturation, $\Delta G_{coil \to complex}$ will need to be < −5 kcalories/mole of binding sites (ten single-stranded DNA nucleotides) because $K_{dissociation} = \exp(-\Delta G/RT) \frac{\text{(free protein) (free sites)}}{\text{(bound protein)}}$, half-denaturation of poly dAT in 0.01 Mg^{2+} at 25° C with 170 µg/ml. of 32-protein (Fig. 2) will require an effective dissociation constant for 32-protein of $< 1.1 \times 10^{-9}$ M. This is in agreement with direct measurements in sucrose gradients under similar conditions, which yield a dissociation constant (averaged for cooperactivity) of $< 10^{-9}$ M for the 32-protein complex with fd DNA.

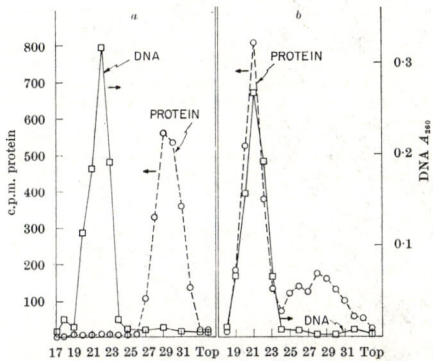

Fig. 1. Cooperative binding of 32-protein to single-stranded DNA. Purified fd DNA (10 µg) was mixed with about 0.5 µg (a) or 7 µg (b) of [3H]-labelled 32-protein in 0.2 ml. of 0.02 M Tris–HCl (pH 8.1)—0.5 mM Na_2EDTA—0.30 M NaCl—100 µg/ml. bovine serum albumin (BSA)—10 per cent glycerol—1 mM β-mercaptoethanol at 4° C. After 20 min, the mixture was layered at 4° C onto a 5 ml, 5–30 per cent sucrose gradient prepared in the same buffer. Following centrifugation for 2 h at 46,000 r.p.m. in the Spinco 'SW50' rotor, 0.15 ml. fractions were collected and monitored for radioactivity by standard techniques. Recoveries of [3H]-protein added averaged about 75 per cent. Concentrations of 32-protein ($A_{280}=1.1$ mg/ml.) and fd DNA ($A_{260}=23.8$ mg/ml. in 0.15 M NaCl—0.015 M sodium citrate, pH 7)[24] were determined by absorbance measurements. Note that, for ordinary binding, the protein distribution would have been identical in the above two experiments, for (free protein)/(bound protein)=K/(free DNA sites), and the concentration of free DNA sites was held essentially constant.

Although 32-protein binds very tightly to all single-stranded DNAs tested, including the synthetic polynucleotide poly dA, no binding of the purified protein to double-stranded DNAs or to R17 RNA could be detected by sucrose gradient sedimentation at 4° C.

Denaturation of DNA with 32-Protein

Histones and polyamines bind to the double-helical form of DNA more tightly than to single strands and thereby raise the temperature required for DNA denaturation[25,26]. Conversely, the strong selective affinity of 32-protein for single-stranded DNA should lower the thermal denaturation temperature of double-stranded DNA. A precedent for such an effect is the destabilization of DNA observed in the presence of pancreatic ribonuclease which likewise preferentially binds to DNA single strands[27].

Because single-stranded DNA is fully hyperchromic when complexed with 32-protein, any denaturation of

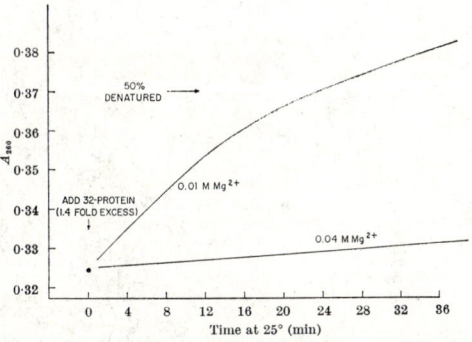

Fig. 2. Poly dAT denaturation catalyzed by 32-protein at 25° C. In addition to the concentration of $MgSO_4$ indicated, each sample contained 10 µg/ml. of poly dAT (from A. Kornberg, $s_{20,w}=29$ in 0.1 M NaOH—0.9 M NaCl), 170 µg/ml. 32-protein, 0.01 M KCl, 2 mM Tris-HCl (pH 8.1), 1 mM β-mercaptoethanol, 0.1 mM Na_2EDTA, and 2 per cent glycerol. Samples were placed in cuvettes in the thermostated compartment of a Gilford spectrophotometer, so that the absorbance could be monitored automatically. At time zero, concentrated 32-protein was added to start the reaction. Similar results have been obtained at a KCl concentration of 0.12 M.

Renaturation of DNA with 32-Protein

As already noted, the renaturation of purified DNA in physiological conditions *in vitro* is an extremely slow process, because of the intrastrand folding in denatured DNA. Our results imply, however, that in the T4 system single-stranded DNA does not exist as such *in vivo*, but is instead always present as a tight complex with 32-

protein. We find that fd DNA saturated with 32-protein sediments only about 1·3 times faster than the free DNA, although its mass is thirteen times greater. This means that the frictional coefficient of fd DNA increases about six-fold in the complex. Because frictional coefficients only double when single-stranded DNA is unfolded in alkali[29], complexed DNA must be held in a highly expanded conformation by 32-protein. (This expansion can be seen directly by electron microscopy; personal communication of H. Delius.) DNA in such a conformation might be expected to renature much more rapidly than free denatured DNA at low temperatures.

The rate of renaturation of DNA covered with 32-protein was measured by an absorbance assay similar to

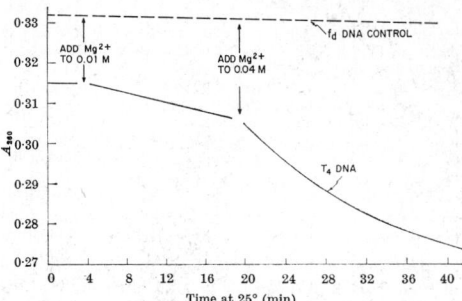

Fig. 3. DNA renaturation catalyzed by 32-protein at 25° C. Single-stranded T4 DNA was prepared by alkaline denaturation and dialysis into low salt buffer as described by Studier[12], except that the DNA was sheared in alkali to molecular weight 5.4×10^6 before the dialysis. The fd DNA serves as a control, for it is not self-complementary and therefore cannot renature. At zero time, 10 µg/ml. of each DNA was mixed in separate cuvettes with 170 µg/ml. of 32-protein in the buffer used in Fig. 2. At the times indicated, MgSO$_4$ was added from a 1 M stock to both fd and T4 DNA reactions. A decrease in absorbance of about 0·072 units is expected for full renaturation of the T4 DNA.

that used to monitor denaturation in Fig. 2. In this case, a decrease in the absorbance of single-stranded DNA is expected proportional to the amount of reformed double-helix. Typical results obtained for single-stranded T4 DNA in the presence of excess 32-protein at 25° C are presented in Fig. 3. It can be seen that a rapid decrease in absorbance is observed in 0·04 M Mg^{2+}, representing a more than 1,000-fold acceleration of the renaturation rate without 32-protein. The dependence of this reaction on [Mg^{2+}] is the reverse of that found for denaturation in Fig. 2: the renaturation rate drops about four-fold in 0·01 M Mg^{2+}, whereas no renaturation is detected without Mg^{2+}.

A more sensitive measure of the course of renaturation is obtained from CsCl gradients, where renatured DNA has a lower buoyant density than single strands[14]. This assay

Table 1. RENATURATION RATES FOR T$_4$ DNA SINGLE STRANDS OF MOLECULAR WEIGHT 5.4×10^6

32-Protein	DNA concentration (µg/ml.)	Ionic composition	Temperature	K_2 (l. mole^{-1} s^{-1})
—	15	1·0 M NaCl	68°	270
—	76	0·04 M MgSO$_4$ 0·01 M KCl	37°	<0·2
+	15	0·04 M MgSO$_4$ 0·01 M KCl	37°	300
+	72	0·011 M MgSO$_4$ 0·12 M KCl	37°	22

For the 32-protein catalyzed reaction: (1) renaturation rates decrease with storage of the protein, so that the maximum rates are probably greater than those listed; (2) between 0·01 M and 0·04 M Mg^{2+}, the rate at 37° C is roughly proportional to [Mg^{2+}]; it falls drastically at lower [Mg^{2+}] levels; (3) renaturation rates are reduced as [KCl] is increased, being severely affected above 0·15 M; and (4) addition of spermidine (0·001 M) is without significant effect. The rates shown were measured by CsCl banding after incubation at pH 7·6 (the catalyzed reaction has a broad optimum between pH 7 and pH 8). Enough 32-protein was used to saturate fully the DNA.

can be used for kinetic analyses, for the addition of concentrated CsCl dissociates 32-protein from the DNA and prevents further renaturation. By this technique, it has been found that the rate of renaturation of T4 DNA in the presence of excess 32-protein is proportional to the square of the DNA concentration, showing that, as in the uncatalyzed reaction, the rate measured is that for the nucleation of complementary pairings. Some second order rate constants for the renaturation of T4 DNA single strands of molecular weight 5.4×10^6 are listed in Table 1. Note that the rate of renaturation catalyzed by 32-protein at 37° C can exceed the uncatalyzed rate observed in standard conditions (68° C in 1 M NaCl).

If 32-protein accelerates DNA renaturation by imparting favourable conformation to the single strands, the bases of which would otherwise be inaccessible on highly folded chains, the dependence of renaturation rate on the ratio of 32-protein to DNA should be unusual. Below the saturating protein : DNA ratio of 12 : 1, the rate of renaturation should be drastically lowered, initial rates being proportional to at least the square of the amount of 32-protein added. Results of renaturation assays performed at sub-saturating 32-protein levels are shown in Fig. 4, where it is seen that a four-fold drop in the concentration of 32-protein decreases the rate of renaturation of T4 DNA at least twenty-five-fold, as expected. A second expectation is that above a protein : DNA ratio of 12 : 1, additional 32-protein should not further increase renaturation rates. This prediction has also been confirmed (experiment not shown).

On the basis of these results, we conclude that 32-protein accelerates renaturation in physiological conditions by forcing DNA single strands into an unfolded conformation which leaves their bases available for pairing during chance collisions between complementary strands. It seems likely that the bound 32-protein is rapidly displaced from the rewinding single strands as the double-helix forms (see Fig. 3).

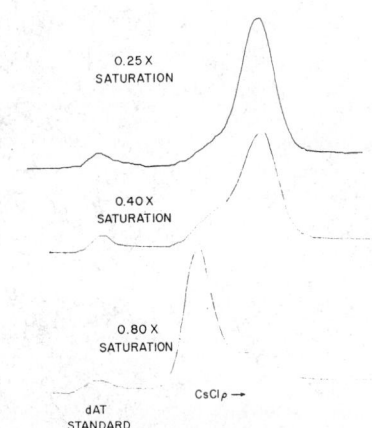

Fig. 4. T4 DNA renaturation as a function of 32-protein : DNA ratio. Renaturation was carried out with T4 DNA single strands (Fig. 3) at 72 µg/ml., and the quantity of 32-protein added was varied as indicated. Incubation was at 37° C in buffer containing 0·12 M KCl and 0·011 M MgCl$_2$. After 30 min, the reaction was quenched by addition of concentrated CsCl (to ϱ=1·700) plus 50 µg of sodium dodecyl sulphate. The band patterns shown are tracings of photographs taken after centrifugation for 20 h at 44,770 r.p.m. in the Spinco model E analytical ultracentrifuge. The renatured band has shifted to a buoyant density representing molecules which are about two-thirds double-helical, as expected from the pairing of strands randomly cut from a longer sequence[30]. As controls, no density shift was observed when fd DNA was processed in an identical manner, whereas fragments of single-stranded T7 DNA renatured rapidly in the presence of 32-protein.

Function of 32-Protein in Genetic Recombination

The denaturation of poly dAT by 32-protein at 25° C indicates that within the cell, the cooperative action of 32-protein should generate fluctuating regions of local denaturation in T4 DNA. The requirement for 32-protein in early steps of genetic recombination can be explained by this ability to open up local regions of native DNA, while simultaneously facilitating helix formation between matching, complexed single strands. Both of these functions are probably necessary for efficient testing of complementary pairings between double-stranded DNA molecules. In addition, experiments with infected cells have suggested that the formation of single-strand breaks ("nicks") in double-stranded intracellular T4 DNA is a prerequisite for the initiation of strand exchanges[31]. Order-of-magnitude calculations, based on the treatment of DNA renaturation formulated by Wetmur and Davidson[30], suggest that random collisions between specifically nicked, double-stranded T4 DNA molecules (see ref. 5) might be efficient enough in the presence of 32-protein to account for the high recombination rates observed in the T4 system (unpublished results). This, however, merely indicates that the mechanism of genetic recombination may be relatively simple, for the properties of 32-protein are also compatible with most other types of models suggested for the recombination process.

Function of 32-Protein in DNA Replication

The genetic results which indicate a structural role for 32-protein in DNA replication suggest that local unwinding by 32-protein might be required in the replication fork in order for productive replication to proceed. In agreement with this role, in vitro experiments have revealed that T4 DNA polymerase[32] uses single-stranded DNA templates much more rapidly in the presence of 32-protein than in its absence (unpublished results of Huberman, Kornberg and B. M. A.). This stimulation is probably the result of favourable template alignment by 32-protein. Because stimulation is not observed in similar experiments with E. coli DNA polymerase, a direct interaction of T4 polymerase with 32-protein may also be involved.

In a normal T4 bacteriophage infection, the number of replication forks present per cell increases linearly until 30 min after infection (25° C)[33]. Because the rate of polymerization observed at each fork is constant throughout this period[33], this rate must be independent of the level of 32-protein, inasmuch as this increases continuously during infection. Yet the gene dosage experiments, which reveal a direct proportionality between the quantity of phage progeny produced and the quantity of 32-protein present, seem to demand that the overall rate of DNA synthesis be proportional to the amount of 32-protein made[21]. To account for these facts, we propose that a functioning replication fork has a unique tertiary structure that contains a fixed number of 32-protein molecules. (About sixty new replication forks are eventually generated in a normal T4-infected cell, so that each fork could incorporate no more than 170 molecules of 32-protein.) In our view, the amount of 32-protein determines the quantity of DNA made, for a new replication fork can be formed only as fast as a threshold level of free 32-protein becomes available.

If each cycle of DNA replication begins at a special point on the T4 genome[34], new replication forks must be generated only at a unique nucleotide sequence. Both during this process and as the replication fork travels, 32-protein may interact with other proteins in addition to T4 DNA polymerase. Likely candidates for such proteins include the products of T4 genes 41, 44, 45, 59 and 62, all of which have as yet unidentified functions essential for T4 DNA replication[17]. Further studies involving 32-protein may therefore provide a fresh insight concerning the unknown mechanism by which DNA is replicated in biological systems.

Preparation of Homogeneous 32-Protein

An E. coli culture, grown to 5×10^8 cells per ml. at 32° C in M-9 minimal media containing 0.3 per cent casein hydrolysate plus 1 per cent glucose, was infected twice at 10 min intervals with a total m.o.i. of ten T4e bacteriophages (lysozyme$^-$). The cells were harvested and washed after 90 min of aeration at 32° C and stored at $-20°$ C. Cells (50 g) were broken by sonication after resuspension in 200 ml. of 0.02 M Tris-HCl (pH 8.1)—0.01 M $MgCl_2$—2 mM $CaCl_2$—1 mM β-mercaptoethanol—1 mM Na_3EDTA containing 20 µg/ml. pancreatic deoxyribonuclease I (Worthington). After incubation for 90 min at 10° C, the extract was centrifuged at low speed to remove cell debris and then clarified at 30,000 r.p.m. for 3 h in the Spinco 30 rotor. The supernatant was dialysed for 24 h against several changes of 0.02 M Tris-HCl (pH 8.1)—0.05 M NaCl—5 mM Na_3EDTA—1 mM β-mercaptoethanol (buffer A) to remove the divalent cations necessary for the activity of deoxyribonuclease I. After centrifugation to remove a light precipitate, the dialysed extract was made 10 per cent in glycerol and forced at 100 ml./h through a column containing 20 ml. packed volume of denatured calf thymus DNA–cellulose (approximately 1 mg of DNA per ml.). The DNA–cellulose (7 cm × 3 cm²) had been equilibrated with a buffer consisting of 10 per cent glycerol in buffer A, and this basic buffer was used for an 80 ml. rinse and for elutions in which increasing concentrations of NaCl were added. The column was eluted at 20 ml./h, in 40 ml. steps of 0.15, 0.40, 0.60 and 2.0 M NaCl. The peak 2.0 M NaCl eluting fraction contained 32-protein as its principal component. This fraction (8 ml.) was dialysed against 0.02 M Tris-HCl (pH 8.1)—10 per cent glycerol—1 mM Na_3EDTA—1 mM β-mercaptoethanol (buffer B) and applied to a 7 cm × 0.8 cm² column of DEAE–cellulose (Whatman DE32). The column was washed with 5 ml. of buffer B and then eluted with a 30 ml. linear gradient of 0–0.5 M NaCl in this buffer. Fractions of 1.3 ml. were collected every 20 min. The 32-protein (A_{280}/A_{260} absorbance ratio of 1.7) appeared in three adjacent fractions with a mean NaCl concentration of 0.20 M. These fractions were either used directly for the studies to be described, or concentrated further by vacuum dialysis against buffer B containing 0.05 M KCl. Approximately 8 mg of electrophoretically homogeneous 32-protein is obtained by this procedure. As determined by the subsequent recovery of purified ^{3}H-labelled 32-protein added to crude extracts, this represents about a 65 per cent yield. An identical procedure was used on a smaller scale for preparation of ^{3}H-labelled 32-protein, except that the cells were grown at 25° C and labelled with 500 µC of ^{3}H-leucine after 35 min of infection. Unless otherwise stated, all operations were carried out at 4° C; at this temperature, concentrated solutions of 32-protein (> 500 µg/ml), may be kept for several weeks. At $-80°$ C, 32-protein has been stored for up to 8 months without a noticeable change in its DNA affinity.

This work was supported by grants from the US National Institutes of Health and the American Cancer Society, and in part by the Eugene Higgins Trust Fund, and facilities made available by the Whitehall and John A. Hartford Foundation to the Department of Biology, Princeton University. We thank Walter Kauzmann, Noboru Sueoka, and Jacques Fresco for discussions, and Barbara Bamman for skilfully performing the analytical ultracentrifugations.

Received June 11; revised July 7, 1970.

[1] Meselson, M., in Heritage from Mendel (edit. by Brink, R. A.), (Univ. Wisconsin Press, 1967).
[2] Tomizawa, J., J. Cell. Physiol., 70, Suppl. 1, 201 (1967).
[3] Shahn, E., and Kozinski, A., Virology, 30, 455 (1966).
[4] Hershey, A. D., and Chase, M., Cold Spring Harbor Symp. Quant. Biol., 16, 471 (1951).
[5] Holliday, R., Genet. Res., 5, 282 (1964).
[6] Whitehouse, H. L. K., Nature, 199, 1034 (1963).
[7] Thomas, C. A., Prog. Nucleic Acid Res. Mol. Biol., 5, 315 (1966).
[8] Marmur, J., and Doty, P., J. Mol. Biol., 5, 109 (1962).
[9] Wang, J. C., and Davidson, N., Cold Spring Harbor Symp. Quant. Biol., 33, 409 (1968).
[10] Doermann, A. H., and Parma, D. H., J. Cell. Physiol., 70, Suppl. 1, 147 (1967).
[11] Signer, E., Echols, H., Weil, J., Radding, C. M., Shulman, M., Moore, L., and Manly, K., Cold Spring Harbor Symp. Quant. Biol., 33, 711 (1968).
[12] Buttin, G., and Wright, M., Cold Spring Harbor Symp. Quant. Biol., 33, 259 (1968).
[13] Doty, P., Boedtker, H., Fresco, J. R., Haselkorn, R., and Litt, M., Proc. US Nat. Acad. Sci., 45, 482 (1959).
[14] Doty, P., Marmur, J., Eigner, J., and Schildkraut, C., Proc. US Nat. Acad. Sci., 46, 461 (1960).

[15] Studier, F. W., *J. Mol. Biol.*, **41**, 199 (1969).
[16] Tomizawa, J., Anraku, N., and Iwama, Y., *J. Mol. Biol.*, **21**, 247 (1966).
[17] Epstein, R. H., Bolle, A., Steinberg, C. M., Kellenberger, E., Boy de la Tour, E., Chevalley, R., Edgar, R. S., Susman, M., Denhardt, G. H., and Lielausis, A., *Cold Spring Harbor Symp. Quant. Biol.*, **28**, 375 (1963).
[18] Kozinski, A., and Felgenhauer, Z. Z., *J. Virol.*, **1**, 1193 (1967).
[19] Alberts, B. M., Amodio, F. J., Jenkins, M., Gutmann, E. D., and Ferris, F. L., *Cold Spring Harbor Symp. Quant. Biol.*, **33**, 289 (1968).
[20] Alberts, B. M., *Fed. Proc.*, **29**, 1154 (1970).
[21] Snustad, D. P., *Virology*, **35**, 550 (1968).
[22] Shapiro, A. L., Vinuela, E., and Maizel, J. V., *Biochem. Biophys. Res. Commun.*, **28**, 815 (1967).
[23] Marvin, D. A., and Hohn, B., *Bact. Rev.*, **33**, 172 (1969).
[24] Knippers, R., and Hoffmann-Berling, H., *J. Mol. Biol.*, **21**, 293 (1966).
[25] Akinrimisi, E. O., Bonner, J., and Tso, P. O. P., *J. Mol. Biol.*, **11**, 128 (1965).
[26] Mahler, H. R., and Mehrotra, B. D., *Biochim. Biophys. Acta*, **68**, 211 (1963).
[27] Felsenfeld, G., Sandeen, G., and von Hippel, P. H., *Proc. US Nat. Acad. Sci.*, **50**, 644 (1963).
[28] Scheffler, I. E., Elson, E. L., and Baldwin, R. L., *J. Mol. Biol.*, **48**, 145 (1970).
[29] Studier, F. W., *J. Mol. Biol.*, **11**, 373 (1965).
[30] Wetmur, J. G., and Davidson, N., *J. Mol. Biol.*, **31**, 349 (1968).
[31] Kozinski, A. W., Kozinski, P. B., and James, R., *J. Virol.*, **1**, 758 (1967).
[32] Goulian, M., Lucas, Z. J., and Kornberg, A., *J. Biol. Chem.*, **243**, 627 (1968).
[33] Werner, R., *Cold Spring Harbor Symp. Quant. Biol.*, **33**, 501 (1968).
[34] Mosig, G., and Werner, R., *Proc. US Nat. Acad. Sci.*, **64**, 747 (1969).

Structure of Branch Points in Replicating DNA: Presence of Single-stranded Connections in λ DNA Branch Points

Ross B. Inman and Maria Schnös

*Biophysics Laboratory and Biochemistry Department
University of Wisconsin, Madison, Wis. 53706, U.S.A.*

(*Received 16 July 1970, and in revised form 26 October 1970*)

Replicating λ DNA molecules can be isolated from *Escherichia coli* shortly after phage infection. During the first round of replication λ DNA has a double-branched circular structure and frequently single-strand connections are observed at the branch points.

A study has been made of the single-strand connections to branch points to obtain information about the molecular fine structure at the growing point during replication. The results are consistent with models which allow replication along both DNA strands but which require DNA synthesis only in the 5′ to 3′ direction. A model is presented which requires two sites of DNA synthesis to be associated with each branch point.

1. Introduction

Replicating λ DNA molecules can be isolated from infected *Escherichia coli* (growing in a medium containing ^{15}N and D) by cesium chloride density-gradient centrifugation (Ogawa, Tomizawa & Fuke, 1968; Tomizawa & Ogawa, 1968). A recent study of replicating λ DNA indicated that double-branched circular molecules† often exhibit a single-strand region at the connection between a DNA segment and a branch point (Schnös & Inman, 1970). This phenomenon has now been examined in more detail to obtain information about the molecular fine structure at branch points.

Previously it was shown that branch points in double-branch circles were of two types; either they represented a growing point, or they marked the position of the starting point of replication. Within the group of double-branch molecules studied, the growing points greatly outnumbered the starting points (89 and 11% respectively) because usually replication proceeded in both directions from a starting point. In a double-branch molecule, both branch points were therefore very often found to be growing points.

In the present study an attempt is made to characterize the single-strand connections often observed at branch points (Plate I and Schnös & Inman, 1970). The determination of whether a branch point represents a growing point or a starting point requires a knowledge of the denaturation map of that molecule. Partial denaturation will complicate the study of pre-existing single strands; therefore, in the present study, we rely on the previous experimental finding that most branch points are in fact growing points and assume that frequently the single-stranded connections are associated with growing points.

† A double-branched circular molecule is defined as a circular molecule containing an additional DNA segment extending between two branch points situated on the circle.

The assumption is also made that the single-strand segments existed *in vivo* and that they were not introduced as a result of the isolation procedure. In defense of this assumption, it is found that carefully prepared mature λ DNA rarely exhibits single-strand regions when prepared for electron microscopy by the method used in this study.

2. Materials and Methods

(a) *Preparation of intracellular λ DNA*

E. coli K12 strain 594 (W3350 $sm^r\ su^-$) was grown in a heavy medium containing ^{15}N and D and infected with [^{3}H]thymidine-labeled $\lambda cIII_{67}cII_{68}$. The intracellular DNA was isolated 10 min after infection, by the procedure already described (Schnös & Inman, 1970), and purified by two CsCl density-gradient centrifugations.

(b) *Electron microscopy*

DNA from the second CsCl density-gradient fractionation (containing ^{3}H activity with buoyant density intermediate between LL and HL) was diluted to $\text{O.D.}_{260} = 0.015$ with CsCl ($\rho = 1.67$). A portion of the DNA solution (7 μl.) was then mixed with 3 μl. of 0.068 M-Na_2CO_3, 34% HCHO and 0.0107 M-EDTA (final pH 9.9) followed by the addition of 10 μl. of formamide and 2 μl. of 0.1% cytochrome *c*. Samples were then prepared for electron microscopy by a modification of the protein film technique (Kleinschmidt, Lang, Jacherts & Zahn, 1962) which consisted of spreading onto drops of water rather than onto water contained in a trough (Inman & Schnös, 1970).

Electron micrographs were recorded on 70-mm roll film in a Philips EM300 at a magnification of about 10,000. Magnification was determined from electron micrographs of a ruled grating (54,800 lines/in., E. Fullam Inc., New York). DNA length measurements were made from the electron micrographs with a curve length integrating device which automatically computes, and prints out, the distance through which a hand-held pen is moved.

3. Results and Discussion

(a) *Proportion of branch points with single-strand connections*

Altogether 96 double-branch circular molecules were photographed but not all were analyzed due to ambiguity in length measurements. Of the 82 molecules analyzed, 29% did not appear to have single-strand material associated with either branch point, 39% were found to have a single-strand connection at one of the two branches (Fig. 1(a)), 31% were found to have single-stranded sections at both branch points (Fig. 1(b)) and 1%† had two single-stranded regions associated with one branch point (Fig. 1(c)). Out of the 164 branch points examined, 51% had a single-strand connection.

(b) *Are single-strand connections associated with parental or daughter segments?*

Circular replicating molecules should consist of an unreplicated parental segment and two newly formed daughter strands. Within any one molecule, daughter segments should be recognized by their similarity in length. Length measurements of the three DNA segments in a double-branch molecule will therefore reveal if a particular segment is parental or daughter. (This will not be true if replication has proceeded to exactly 50% because then the three segments will have an equal length.)

Do the single-strand branch connections occur on daughter or parental segments or both? Length measurements were made on 58 molecules which exhibited single-strand branch connections (these consisted of the types already discussed, Fig. 1(a), (b) and

† Molecular types which occur at low frequency can very well be artifacts produced by the electron microscopic preparation technique.

(c)) and it was found that all but one contained single-strand connections associated with daughter segments; out of 84 single-strand connections, 99% involved daughter strands. Growing points therefore often exhibit single-strand connections which are almost always associated with daughter strands.

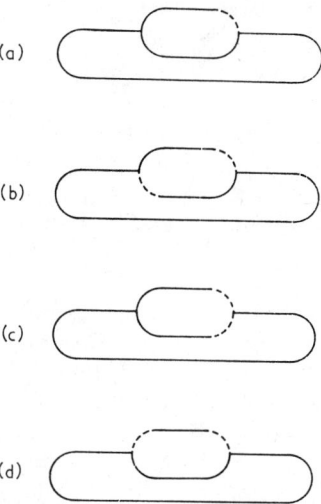

FIG. 1. Representations of the single-strand branch connection observed in double-branch circular molecules. The solid line represents double-stranded DNA and the dashed line indicates the single-strand branch connections. The length of the single-strand sections is greatly exaggerated.

(c) *Observations on molecules possessing two single-strand branch connections*

The 58 molecules discussed above consisted, in part, of 26 molecules which had two single-strand connections. These were almost always (96%) of the type shown in Figure 1(b), where the single strands are on opposite ends of the two daughter arms. Only one molecule had two single strands connected to one branch point (Fig. 1(c)). No molecules were observed to have single-stranded connections at both ends of one daughter segment (Fig. 1(d)). If there are two single-stranded connections to growing points on a molecule, then they are usually deployed such that there is one on each daughter and situated at opposite ends of daughter segments as shown in Figure 1(b).

(d) *Number of bases in single-strand branch connections*

The electron microscopic length of the single-strand connections is shown in Figure 2; the distribution of length is broad, ranging from a maximum of $0.40\,\mu$ down to $0.04\,\mu$ (and possibly below $0.04\,\mu$). The average length is $0.18\,\mu$. At the present time it is not possible to make an exact conversion of electron microscopic length of single-stranded DNA into number of bases present. The mass per unit length of single-stranded DNA is highly variable and presumably depends on the spreading conditions used during preparation for electron microscopy.

ϕX174 DNA is about 6% shorter than ϕX174 replicative form under somewhat similar spreading conditions (Freifelder, Kleinschmidt & Sinsheimer, 1964; Chandler, Hayashi, Hayashi & Spiegelman, 1964). However, Follett & Crawford (1967), find that

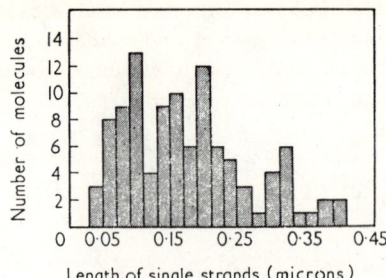

Fig. 2. Histogram showing the length distribution of single-strand branch connections. Apparent single-strand lengths of 300 Å (or less) were ignored. Samples were prepared for electron microscopy by examination of molecules from a CsCl density-gradient centrifugation (see Materials and Methods section (b)).

single-stranded papilloma virus DNA is 40% shorter than the native material. Ionic strength has been found to influence single-strand lengths when prepared for electron microscopy by the diffusion method (Bujard, 1970). We have also encountered, what is presumed to be, variations in mass per unit length of single-stranded DNA. Partial denaturation of P2 DNA by heat leads to a decrease in total length (when spread from 10% HCHO onto 0·5% HCHO in water (Inman & Bertani, 1969)) whereas, the opposite is true if P2 DNA is partially denatured in alkali and spread from 10% HCHO and 50% formamide onto water (unpublished results). Although the molecular weight of the single-stranded DNA represented in Figure 2 is not at all well defined, molecular weight limits of ± 50% given by half the ratio of mass per unit length of double-stranded DNA should be appropriate. (It is assumed therefore that 17·4 μ, the observed length of native DNA under the present spreading conditions, represent the length of 47,000 (± 50%) bases in the single strands.) The number of bases present in the average single-strand connection is therefore between 240 and 730 with a range extending up to 540 to 1600 bases and down to 54 to 160 bases (and possibly less).

(e) *Comparison of results with the Okazaki model*

Figure 3 shows a growing-point model based on the principles described by Okazaki *et al.* (1968), Mitra, Reichard, Inman, Bertsch & Kornberg (1967) and Richardson (1969), which allows replication to proceed along both DNA strands and yet requires DNA synthesis only in the 5′ to 3′ direction. We define a "replication unit" (Okazaki *et al.*, 1968) as the amount of replicated material that will result from the events pictured in Figure 3 ((a) to (e)). The length of a replication unit may be quite variable: it could be determined by the particular bases encountered by the growing point in Figure 3(c) or may result from some other event. The successful completion of synthesis within a replication unit requires at least two modes of action. First, the synthesis of DNA will be required to proceed with a growing point that constantly encounters parental DNA in the form of a double helix (Fig. 3(b)). Second, after the switch to the alternate parental template, the synthesis of DNA can proceed with a growing point that encounters only a single-stranded template (Fig. 3(d)). One might suspect that the two phases of the complete excursion occur at quite different rates.

According to the model in Figure 3, a single-stranded region that exists between a daughter segment and a branch point can only result if a molecule is isolated at a time

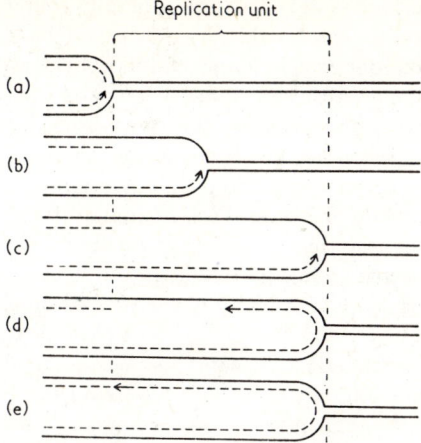

Fig. 3. Model for growing point motion resulting in the synthesis of one replication unit of DNA (see text for details on (a), (b), (c), (d) and (e)).

corresponding to the first half of a complete excursion (Fig. 3(b) or (c)). These types of molecules were observed (Fig. 1(a) and (b)) and give a very approximate measure of the size of an average replication unit (540 to 1600 base pairs or $0\cdot 4\,\mu$ of single-strand DNA). However, according to the model, the second phase of synthesis (Fig. 3(d)) will also result in single strands but now separated from the branch point by a short section of double-stranded DNA. This type of single strand was not usually observed; out of the 164 branch points examined, less than 2% were of the above type (apparent double-stranded sections less than 300 Å were ignored) whereas 51% had a single-strand connection between a daughter segment and the branch point. One possibility is that the second half of the synthetic excursion (Fig. 3(d)) occurs very much faster than the first (for the reasons discussed above) and that, at any point in time, one observes either temporarily inactive growing points (Fig. 3 (a) or (e), branch points will appear to be completely double-stranded) or a growing point executing an outward movement (Fig. 3(b)). The low frequency of single strands of the type depicted in Figure 3(d) is therefore not necessarily inconsistent with the growing point model (see also section (f) below).

The two findings which are quite consistent with the model are: (a) single-strand connections are almost always (99%) associated with daughter segments and (b), in the case of molecules possessing two single-strand connections, these are usually deployed (96%) in the way shown in Figure 1(b). It should be noted that the observed single-strand connections of the type shown in Figure 1(b) are quite consistent with the previous finding of bi-directional replication in λ phage DNA (Schnös & Inman, 1970).

(f) *Single-strand regions not physically associated with growing points*

An electron microscopic examination of carefully prepared mature λ DNA reveals that single-strand regions can be observed only rarely. It is therefore surprising to find that double-branch circular molecules often have short single-strand regions at positions other than those involving a connection to a branch point. Out of 82 molecules examined 23 were found to have short single-strand regions not directly connected to a branch point. The following rules were generally found to be true: (a) single-stranded segments were always on daughter arms (100%); (b) they were almost always situated on a daughter arm which possessed a single-strand branch connection (87%); (c) they were almost always situated very close to the single-strand branch connection. Daughter arms which display the above property would, on the average, consist of $0{\cdot}18\,\mu$ of single-strand DNA connected to the branch point, $0{\cdot}17\,\mu$ of double-strand DNA, $0{\cdot}17\,\mu$ of single-strand DNA followed by double-strand DNA extending to the second branch point. Two examples of single-strand segments are given in Plate I. It can be seen that according to (b) above, the single-strand segments are not of the type expected from Figure 3(d).

A model which is consistent with many of the present observations is presented in Figure 4. The model is simply an extension of the events pictured in Figure 3, except that now a new forward synthetic cycle can take place (in replication unit 2) while the backward cycle (in replication unit 1) is being executed (Fig. 4(e)). At least two different series of events which take place at the terminus of a replication unit can be imagined. A terminus could be associated with the switch from one template to another (as shown previously in Fig. 3(d)) followed immediately by the creation of a nick (and a new growing point) at the position of the switch (Fig. 4(d)). The resulting small DNA segment, containing the old growing point, must be short because it is not observed in the electron microscope. After these two events there will be two

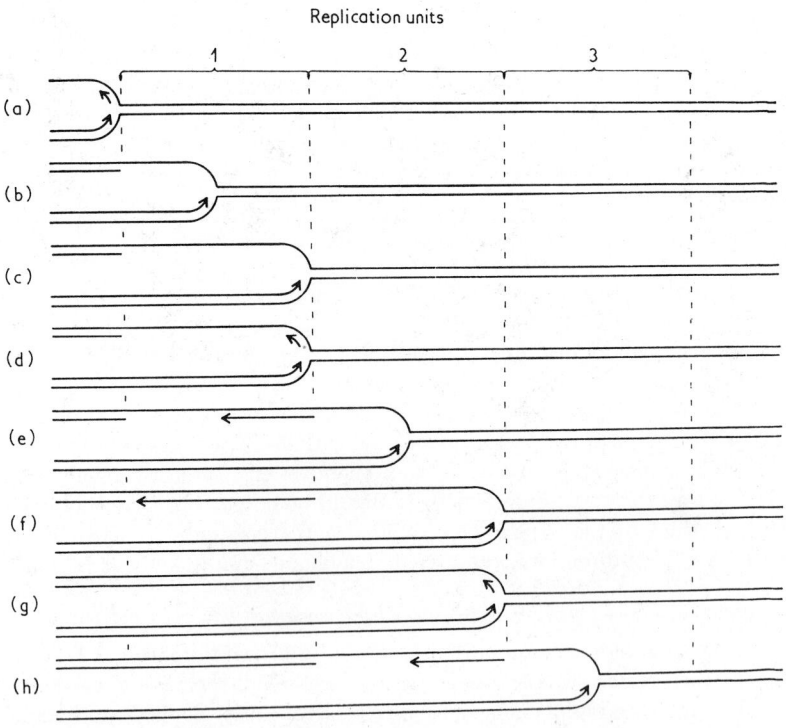

FIG. 4. Model for growing point motion involving simultaneous DNA synthesis within two adjacent replication units.

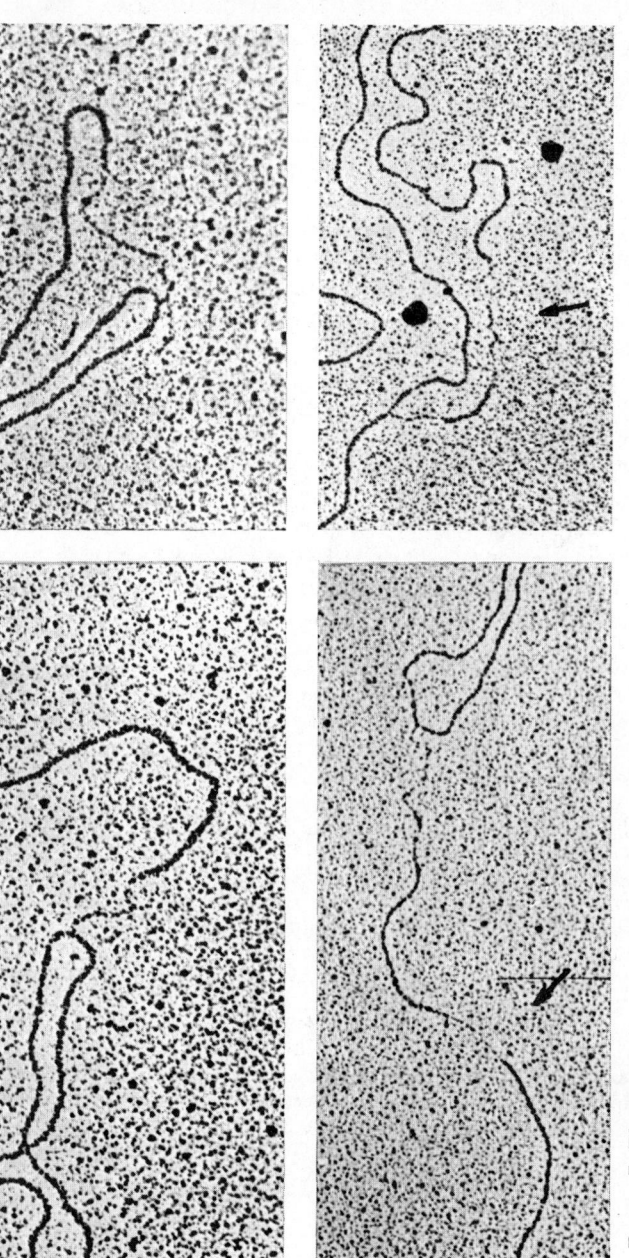

PLATE I. Electron micrographs of single-strand connections and segments. Top micrographs (scale 0.5 μ = 6.1 cm); examples of single-strand connections between daughter segments and branch points. Bottom micrographs (scale 0.5 μ = 3.9 cm); examples of single-strand segments (arrows) on strands containing single-strand connections.

[*facing p. 324*

growing points associated with a branch point; the old growing point executing a backward motion and the new growing point moving forward (Fig. 4(e)).

A somewhat simpler series of events can be imagined to take place at the terminus of a replication unit. We will differentiate between the two parental strands by calling one a (+)-strand (the upper strand in Fig. 4(c)) and the other a (−)-strand (the lower strand in Fig. 4(c)). Assume that the terminus of replication unit 1 in Figure 4 is marked, not by a switch from one template to the other, but by an event leading to the creation of a new growing point (whose template is the (+)-strand). The new growing point, which will execute a backward synthetic movement, is presumed to form as soon as the (+)-template becomes single-stranded due to the passage of the original growing point past this position. According to this alternate view, we can consider that the original growing point moves continuously along the (−)-template and is not subject to any control by the events associated with the terminus of a replication unit. The terminus of a replication unit may therefore simply be a position from which a new initiation can take place once the (+)-template becomes single-stranded (Fig. 4(d)).

Although the model does explain many of the electron microscopic observations, it is also necessary to assume that the events shown in Figure 4 do not occur with strict synchronization because one often observes completely double-stranded branch points (Fig. 3(a)) which according to the model (Fig. 4) would not occur.

We thank Drs R. L. Baldwin, W. F. Dove and A. D. Kaiser for their thoughtful criticism.

This research was supported by grants from the United States Public Health Service, National Institutes of Health, the American Cancer Society and the Graduate School, University of Wisconsin.

REFERENCES

Bujard, H. (1970). *J. Mol. Biol.* **49**, 125.
Chandler, B., Hayashi, M., Hayashi, M. N. & Spiegelman, S. (1964). *Science*, **143**, 47.
Follett, E. A. C. & Crawford, L. V. (1967). *J. Mol. Biol.* **28**, 461.
Freifelder, D., Kleinschmidt, A. K. & Sinsheimer, R. L. (1964). *Science*, **146**, 254.
Inman, R. B. & Bertani, B. (1969). *J. Mol. Biol.* **44**, 533.
Inman, R. B. & Schnös, M. (1970). *J. Mol. Biol.* **49**, 93.
Kleinschmidt, A. K., Lang, D., Jacherts, D. & Zahn, R. K. (1962). *Biochim. biophys. Acta*, **61**, 857.
Mitra, S., Reichard, P. Inman, R. B., Bertsch, L. L. & Kornberg, A. (1967). *J. Mol. Biol.* **24**, 429.
Ogawa, T., Tomizawa, J. & Fuke, M. (1968). *Proc. Nat. Acad. Sci., Wash.* **60**, 861.
Okazaki, R., Okazaki, T., Sakabe, K., Sugimoto, K., Kainuma, R., Sugnino, A. & Iwatsuki, N. (1968). *Cold Spr. Harb. Symp. Quant. Biol.* **33**, 129.
Richardson, C. C. (1969). *Ann. Rev. Biochem.* **38**, 795.
Schnös, M. & Inman, R. B. (1970). *J. Mol. Biol.* **51**, 61.
Tomizawa, J. & Ogawa, T. (1968). *Cold Spr. Harb. Symp. Quant. Biol.* **33**, 533.

The Rolling Circle for φX DNA Replication, II. Synthesis of Single-Stranded Circles*

David Dressler

THE BIOLOGICAL LABORATORIES, HARVARD UNIVERSITY, CAMBRIDGE, MASSACHUSETTS 02138

Communicated by James D. Watson, September 25, 1970

Abstract. φX-infected cells have been allowed to incorporate tritiated thymidine late in the phage life cycle when single-stranded circles are the product of DNA synthesis. Virtually all of the radioactivity is recovered in a continuum of actively replicating viral DNA molecules. These molecules are termed rolling circle intermediates because they are characterized by three structural properties. They possess positive strands that are longer than the length of a mature viral genome, and negative strands that are covalently closed single-stranded circles. The 3′ termini of the long positive strands lie upon the template rings, while the 5′ ends are free in solution.

From these experimental data, the basic mode of synthesis is deduced to involve the continuous elongation of the open positive strand by endless copying around the circular negative strand template. As new bases are added to the template-bound (3′) end of the positive strand, the distal (5′) end is displaced from the template ring as a single-stranded tail of increasing length. It is the tail which serves as the source of material for progeny chromosomes.

These data confirm our characterization of this φX intermediate, which initially was based only on the possession of long positive strands, and extend this characterization to include experimental statements about the circular nature of the template DNA strand, and the 5′ to 3′ direction of polynucleotide chain growth within the intermediate. Moreover, the description can now be applied to all of the molecules which acquire label during a pulse.

The replication of φX DNA involves a period of double-stranded circle synthesis followed by a period of single-stranded circle synthesis.[1] In the rolling circle model[2] (Fig. 1) an attempt has been made to explain both types of φX-circle synthesis under one unified mechanism. For each synthesis, the replicating DNA molecule is characterized by three properties: (1) the possession of one copy of the genetic information (+) in the form of a longer-than-unit-length polynucleotide strand, (2) the maintenance of the other copy of the genetic information (−) in the form of a covalently closed single-stranded circle, to be used as an endless template, and (3) the positioning of the long DNA strand so that its 3′-OH terminus lies upon tne template ring where it may be endlessly elongated by the Kornberg polymerase, or another enzyme with analogous properties.

This paper presents a set of experiments which show that φX single-stranded circles are in fact made by a replicating DNA molecule which has the three

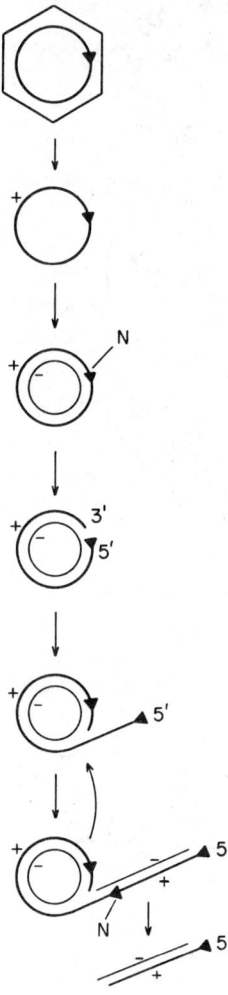

FIG. 1. The rolling circle intermediate making φX double-stranded and single-stranded circles.

The positive strand circle from the infecting φX particle is operated upon sequentially by the DNA polymerase and the ligase which put the viral chromosome into the form of a covalently closed duplex ring. The replication of this duplex ring then begins as one strand (say the positive strand) is nicked by a sequence-recognizing endonuclease (N). The 5' terminus of the open positive strand (*triangle*) is then peeled back, and nucleotide triphosphates are condensed upon the 3'-OH terminus.

The actively replicating DNA molecule which is thus formed is characterized by three experimentally testable properties: (1) One copy of the genetic information (+) is present as a polynucleotide strand which is longer than the length of a mature viral genome, (2) The other copy of the genetic information (−) is maintained in the form of a covalently closed single-stranded circle, to be used as an endless template, and (3) The 3'-OH end of the long strand remains upon the template ring and is thus in a position to be elongated by a polymerase of the known type that is, a polymerase which is able to extend a DNA strand by the condensation of a nucleotide triphosphate upon a 3'-OH terminus.

The result of the rolling circle intermediate is to generate a linear tail that is longer than the length of a complete chromosome. Then, the terminal redundancy of the tail can be used to support either a generalized or a site-specific recombination process leading to the excision of a lambda-type DNA rod. This type of DNA rod is intrinsically capable of circularization.[3,4]

The intermediate can produce either double-stranded or single-stranded circles with equal facility; to make a single-stranded circle, the tail of the intermediate is simply prevented from becoming duplex *except* for the limited region that is involved in the site-specific recombination process.

structural properties expected for a rolling circle intermediate. These data extend our characterization of this replicating intermediate, which initially was based on its possession of long polynucleotide strands, to include statements about the nature of the template strand, and the direction of polynucleotide chain growth.

Labeling of φX Replicating Intermediates. The replication of φX DNA involves a period of double-stranded circle synthesis followed by a period of single-stranded circle synthesis.[1] We have studied the way in which φX single-stranded circles are made.

Cells that were accumulating ϱX at a normal rate following synchronized infection were allowed to incorporate [³H]dT for 50 sec during the period of single-stranded circle synthesis.

After the pulse, the infected complexes were harvested and broken open with lysozyme and detergent. The unfractionated cell lysate was sedimented through a neutral velocity gradient, and the pulse-labeled DNA forms were recovered in a continuum of structures (Figure 2, fractions 18–42). The pulse-labeled DNA sedimented heterogeneously from 16 S, (the velocity of unit φX duplex rings) up to about 30 S (the velocity expected for relatively massive replicating intermediates).

That the pulse-label is contained in *viral* DNA forms is demonstrated by the DNA–DNA hybridization study of Fig. 2. The data show that the [³H]dT

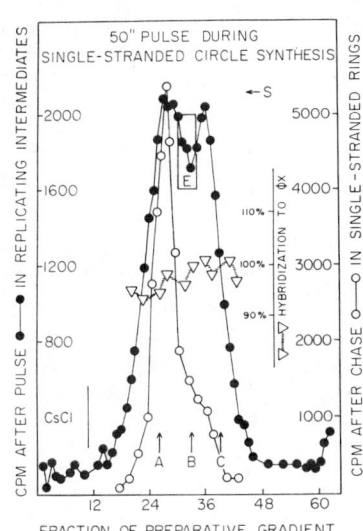

FIG. 2. Infection and pulse-labeling of intermediates. Cells were infected with φX and exposed to tritiated thymidine briefly during the period of single-stranded circle synthesis: *E. coli* strain HF 4704 (hcr^- thy^-) was grown to a titer of 3×10^8/ml at 28°C in 100 ml of mT3XD medium.[5] During the final 20 min of growth, mitomycin C (Calbiochem) was present at 80 μg/ml. This antibiotic selectively inhibits host DNA synthesis while allowing a normal φX life cycle.[5,11] The cells were washed, resuspended in 25 ml of holding buffer[12] containing mitomycin, starved for 40 min, and then infected at a multipilcity of 2 with φX am3, a mutant that cannot lyse the host cell. After 15 min, when more than 99% of the phage had eclipsed, an equal volume of double-strength m3XD (containing 1 μg/ml of thymine) was added to initiate phage growth.

50 min later the infected cells were maturing 4 φX/cell per min, indicating a normal infection. At this time, [³H]dT was added to the culture (Schwartz Bioresearch, Orangberg, N. Y., 5 mCi, 16 Ci/mmole). 50 sec after the addition of label, half of the culture was pipetted directly into an equal volume of acetone at −70°C to stop incorporation. The other half of the culture was allowed to continue incorporation for an additional 10 min in the presence of a thousand-fold excess of nonradioactive thymidine.

The cultures were harvested by centrifugation, washed, and resuspended at 4×10^9/ml in lysis buffer. (100 mM Tris(pH8)– 100 mM NaCl–10 mM KCN–10 mM Iodoacetate–1 mM EDTA). The infected complexes were then broken open with lysozyme (400 μg/ml, 37°C, 20 min) and detergent (2% sarkosyl, 65°C, 20 min) and, lastly, exposed to self-digested pronase (1 mg/ml, 4 hr, 37°C).

The unfractionated cell lysates were sedimented through preparative neutral velocity gradients (5–20% sucrose, 0.5 M NaCl, 1 mM EDTA, 0.1% sodium lauroyl sarcosinate in 0.05 M Tris, pH 8; underlaid with a saturated CsCl–sucrose cushion). Centrifugation was at 24,000 rpm for 17 hr at 8°C in the Beckman SW-25.1 rotor.

The solid dots show the distribution of [³H]dT after the pulse, the open dots show the distribution after the chase. Although part of the pulse-label is sedimenting in the 27 S position characteristic of single-stranded circles, this label is actually present in rolling circle intermediates with long tails (see Fig. 3A).

To show that the pulse-label is present in φX positive strand base sequences, aliquots of each fraction were denatured, reneutralized, and hybridized[5] to membrane filters containing 1) immobilized positive and negative strands from φX duplex rings, or 2) only φX positive strands. The pulse-labeled strands hybridized with almost 100% efficiency to the filters which contained both φX positive and negative strands (*triangles*), but not at all (< 2%) to the filters which contained only positive strands (not shown).

has been incorporated entirely into φX DNA, not *Escherichia coli* DNA. And, as expected since labeling was carried out during the period of single-stranded circle synthesis, the replicating structures have incorporated radioactivity almost exclusively into positive strand base sequences.

The acceptance of the pulse-labeled structures as replicating *intermediates* is based on the observation that when the 50-sec pulse was followed by a 10-min exposure to nonradioactive thymidine, label was chased out of the heterogenously-sedimenting structures and quantitatively reappeared in the 27 S position characteristic of single-stranded circles (Fig. 2, fractions 24–30).

Long Positive Strands. Fig. 3 shows that the replicating intermediates contain, as one component, positive strands that are longer than unit length. Intermediates of various sizes were recovered from areas *A*, *B*, and *C* of Fig. 2, and denatured into their component polynucleotide strands with alkali. The released strands were then sedimented through secondary neutral velocity gradients, together with a marker for the position of unit length genomes.

Fig. 3C shows that the slowest sedimenting intermediates, upon alkaline denaturation, yielded pulse-labeled positive strands that were near unit length or slightly longer. In contrast the faster-sedimenting intermediates, representing over 90% of the pulse-labeled structures of Fig. 2, released radioactive positive strands which sedimented in advance of the marker (fractions 28–30) and were thus judged to be longer than unit length.

The lengths of the longest pulse-labeled positive strands can be estimated from the empirical finding of Studier[6] that a flexible polynucleotide chain in a

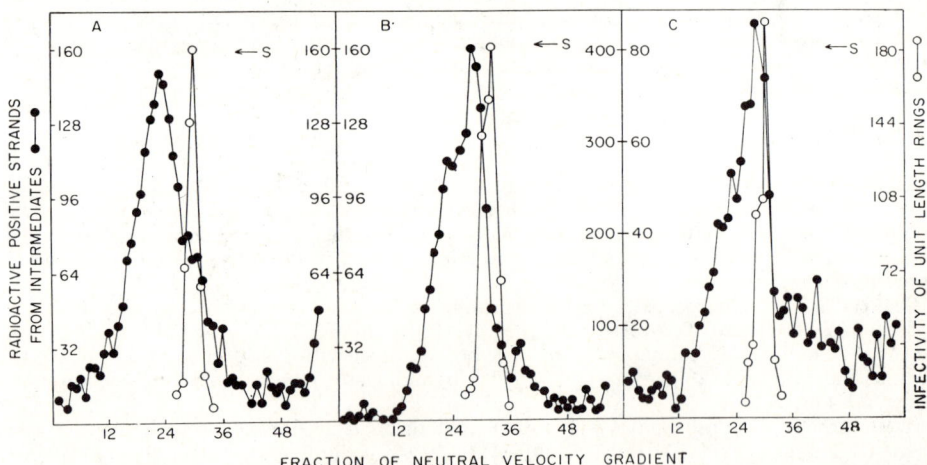

Fig. 3. Long positive strands. Pulse-labeled replicating intermediates were recovered from areas *A*, *B*, and *C* of the gradient shown in Fig. 2, and denatured into their component polynucleotide strands with 0.25 M NaOH. After 4 min at 37°C the solutions were reneutralized with HCl. The single strands from the intermediates were then sedimented through secondary neutral velocity gradients (10–30% sucrose, 1 M NaCl, 1 mM EDTA, 0.1% sodium lauroyl sarcosinate; for 2.5 hr. 64,000 rpm, 8°C, in the Beckman SW-65 rotor).

Alternate fractions of the gradients were assayed for radioactivity (representing the pulse-labeled positive strands from the intermediates) and for infectivity to spheroplasts[7] of marker φXh4 positive strand circles (which had been added prior to alkaline denaturation to define the sedimentation velocity of unit-length φX strands).

neutral pH, high ionic strength gradient will sediment 49% faster as its mass increases from m to 2 m. Thus, while a unit length single-stranded rod (or ring) sediments at 27 S, a double-length single strand is expected to sediment at 40 S. 40 S, in fact, is the velocity of the fastest sedimenting strands from the most massive replicating intermediates (Fig. 3A, fractions 12–15).

Circular negative strand templates: The long positive strands of the intermediates were readily seen (Fig. 3) because they acquired label during a pulse. However, the negative strand templates from which they were synthesized are unlabeled and thus somewhat more difficult to detect. To account for the synthesis of the long positive strands, we would expect that the negative templates must be redundant; but, this could be achieved either with DNA strands that are longer-than-unit-length or circular. In the event that the negative templates are circular, they should be visible by virtue of their infectivity to spheroplasts.

Pulse-labeled φX replicating intermediates were obtained from region B of Fig. 2 and, after repeated purification by sedimentation, denatured into their component polynucleotide strands with alkali. Fig. 4A displays the sedimentation profiles of both the radioactive *and* the infective strands from the replicating

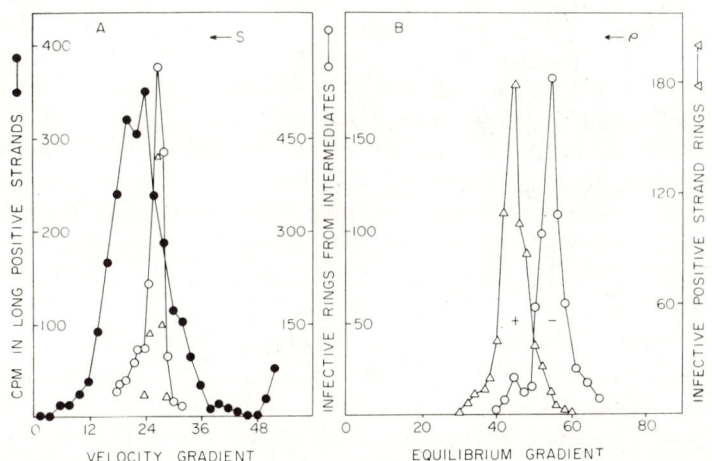

FIG. 4. Circular negative strands. (A) Replicating intermediates were recovered from area B of Fig. 2 and denatured with alkali. The released single strands were then sedimented through a neutral velocity gradient. The pulse-labeled positive strands (solid dots) are seen sedimenting ahead of unit-length φX strands. The unit length position is marked by two kinds of infective single-stranded circles: marker φXh4 positive strand rings, and φX am-3 rings, derived from the intermediates. (B) The infective single-stranded circles (both those from the intermediates and also the marker positive strand rings) were recovered from fractions 25–28 (A) and centrifuged to equilibrium in alkaline CsCl. The material was taken up to 2500 μl with Na$_3$PO$_4$ containing 1 mM EDTA and 5 μg of denatured lambda phage DNA. 3.325 g of CsCl was added and the solution was centrifuged at 40,000 rpm for 60 hr at 17°C in the angle 60 rotor of the IEC B60 centrifuge.

Gradient fractions were assayed for the ability to infect spheroplasts and produce either am-3 phage (representing the rings from the replicating intermediates) or φXh4 phage (representing the marker positive strand circles). The rings from the intermediates separated from the positive strand circles and came to equilibrium in the negative strand position, (B, fractions 50–60).

intermediates. In addition to the long pulse-labeled positive strands (Fig. 4A, fractions 12–24) the denatured intermediates have indeed released an infective component; this is responsible for the ability of fractions 25–28 to generate φX am3 phage particles upon incubation with *E. coli* spheroplasts. The infective component from the intermediate is taken to be a single-stranded ring because it sediments with precisely the same velocity as the marker positive strand circles (which are identified in the spheroplast assay by their production of φX h4 phage).

The infective single-stranded circles were recovered from fractions 25–28 of Fig. 4A. They were then centrifuged to equilibrium in alkaline CsCl to determine whether those corresponding to the genotype of the replicating intermediates would form a single infectivity peak at the negative strand position.

In alkaline CsCl, as determined by Vinograd, Morris, Davidson, and Dove, φX positive and negative strands separate from each other and band at two different densities,[8] since the φX positive strand contains more thymine and guanine residues than the negative; above pH 12.5 the ring protons of these residues are titrated off and replaced by density-enhancing Cs^+ ions.

When the fractions of the alkaline CsCl gradient were assayed with spheroplasts for the ability to produce phage of the marker and experimental genotypes, it was found that the single-stranded circles from the intermediates had in fact separated from the marker positive strand circles. The infectivity corresponding to the genotype of the replicating intermediates had peaked in the negative-strand position (Fig. 4B, fractions 50–60).

In several experiments, the number of long positive strands (as judged by the radioactivity content of fractions 12–24 of Fig. 4A) was compared to the number of negative strand circles (as judged by the amount of infectivity, in fractions 25–28 of Fig. 4A). In each case, the replicating structures contained equal numbers of long positive strands and negative template rings within a factor of two.

The 3′-OH end of the long strand is the growing end: The data of the previous two sections has shown that the replicating intermediate contains two components: a long positive strand and a circular negative strand. If the long positive strand is growing by chain elongation around the negative template ring, then replication must proceed with the displacement of one end of the long strand from the template. Which end of the long stand is displaced as a result of synthesis, and which retains its association with the template ring, can be determined by exposing the pulse-labeled replicating intermediates to exonuclease III.

Exonuclease III, as characterized by Richardson, Lehman, and Kornberg, is a highly specific nuclease that will depolymerize a polynucleotide strand from its 3′ terminus if, and only if, that terminus is double-stranded.[9] Thus, only if the long, pulse-labeled, positive strand of the intermediate has its 3′ terminus positioned upon the negative template ring will the enzyme be able to shorten the strand and release acid-soluble mononucleotides.

Fig. 5 shows the results of exposing pulse-labeled replicating intermediates to exonuclease III. The native intermediates were incubated with exonuclease

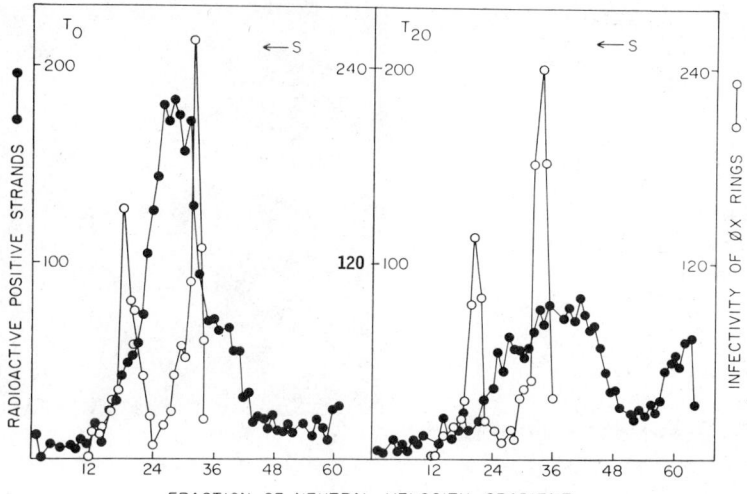

FIG. 5. The 3' end of the long strand lies upon the template ring. Pulse-labeled intermediates were obtained from a preparative velocity gradient, identical to that shown in Fig. 2, and were diluted into a buffer for treatment with exonuclease III (70 mM Tris, (pH 8)–0.7 mM $MgCl_2$–10 mM mercaptoethanol).

φX positive strand rings and supercoils were added to the solution, then exonuclease III (the phosphocellulose fraction, from Drs. William Reznikoff and Charles A. Thomas, Jr.). The mixture was incubated at 37°C; aliquots were withdrawn at 0 and 20 min.

Enzymic digestion was stopped by the addition of sodium dodecyl sulfate to 0.5%, followed by alkaline denaturation of the sample (0.25 M NaOH). After reneutralization, the remaining DNA strands were sedimented through neutral velocity gradients.

Solid dots: radioactive positive strands from the pulse-labeled intermediates before digestion (*A*) and after digestion (*B*).

Open dots: infective supercoils (fractions 18–21) and single-stranded circles (fractions 32–35) present before and after exonuclease digestion.

for either 0 or 20 min, and then denatured. Their remaining polynucleotide strands were sedimented through neutral velocity gradients. Fig. 5*A* shows the sedimentation profile of the denatured reaction mixture after *zero minutes* of enzyme treatment. Pulse-labeled positive strands sediment in a distribution between unit length and twice-unit-length rods, the positions of which are defined by the infectivity to spheroplasts of marker single-stranded circles (27 S) and supercoils (40 S) present in the reaction mixture. Fig. 5*B* shows the effect of a limited (20 min) exposure of the intermediates to the enzyme: the long positive strands have been shortened, partially or completely. This result indicates that the long positive strands of the intermediates do in fact have their 3' termini positioned upon the negative ring templates.

An important control is represented by the equal number of infective single-stranded circles and supercoils present at the beginning of enzyme digestion (Fig. 5*A*, *open dots*) and at the end (Fig. 5*B*, *open dots*). The fact that these infective forms were not destroyed indicates that during the exonuclease III digestion there was no significant nicking activity directed against either single-stranded or double-stranded DNA. Such nicks would lead to shortened positive strands, or to the creation of 3'-OH targets for exonuclease III attack in substrates that would otherwise be inert.

Electron microscopy: A collaboration was undertaken with Dr. Lorne MacHattie to process for the electron microscope several of the same preparations of replicating intermediates which had been analyzed by physcial chemical methods. Since the replicating intermediates were expected to contain single-stranded regions, the pulse-labeled structures were processed by the Westmoreland adaptation[10] of the Kleinschmidt technique. This procedure renders single-stranded DNA visible, though thinner and less rigid than duplex DNA.

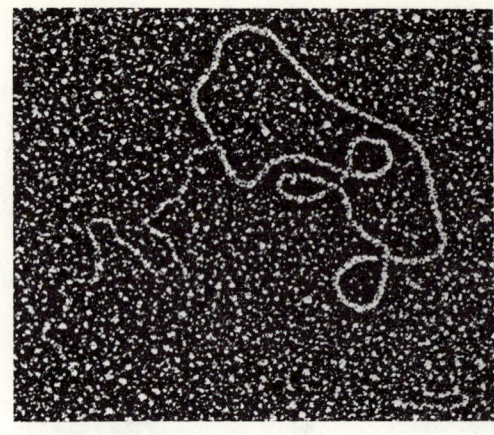

FIG. 6. A φX rolling circle.

In preparations of pulse-labeled intermediates, double-stranded circles with single-stranded tails were often seen (Fig. 6). Characteristically, about 20–40% of the viral structures isolated from the replicating intermediate region of a preparative velocity gradient (such as Fig. 2, area E) had this configuration. The remaining viral structures were nicked or supercoiled φX duplex rings, present in essentially equal numbers.

Discussion. This paper shows that nascent positive strand material for progeny φX chromosomes first appears in greater-than-genome-length polynucleotide strands. These strands are associated with, and presumably generated by, continuous copying around circular negative strand templates. In this synthesis, the 5′ terminus of the positive strand is displaced from the template ring as the 3′ end is simultaneously elongated. These results correspond to the rolling circle description of φX DNA synthesis. The properties of this actively replicating DNA molecule have also been studied by Knippers, Komano, Razin, Davis, and Sinsheimer for φX[13,14] and by Ray[15] and Wirtz and Hofschneider[16] for M13. Their findings are in close agreement with ours.

The DNA synthesis which occurs during bacterial mating provides another instance of replication that can be explained in terms of the rolling circle intermediate. When bacteria mate, a single preexisting strand of the male chromosome is detached and transferred, 5′-end first, to the female recipient. Simultaneously, a new copy of the transferred strand is synthesized and retained in the male. Because males can, during prolonged mating, transfer a second, linked set of markers to the female, it appears likely that the synthesis involves continuous copying around a circular template. These experiments and concepts have been established by Rupp, Ihler, Ohki, Tomizawa, and Fulton.[17–19]

Several systems which make double stranded DNA also appear to involve a rolling circle intermediate. Schnöss and Inman[20] have obtained electron microscopic evidence that the replication of phage P2 DNA proceeds via an intermediate which consists of a duplex ring with an attached double-stranded tail. Furthermore, studies of the late life cycle of phage lambda by Smith and Skalka[21] and Kiger and Sinsheimer[22] have yielded evidence for catomeric replicating structures which may be duplex rings with double-stranded tails. Moreover, rolling circle structures have occasionally been observed among the replicating DNA of polyoma,[23] colicin factors,[24] and plasmids.[25]

For several other organisms studied thus far, however, the more frequently observed conformation for the replicating chromosome is the double-forked circle of the type originally found for *E. coli* by Cairns.[26] Hopefully, these two types of configurations for replicating DNA, and the different modes of replication which they imply, may be reconciled within a larger framework, but this remains for future work.

It is a privilege to acknowledge the collaboration of Walter Gilbert in this work, also, I am grateful to Dr. Lorne MacHattie for the application of his skill in electron microscopy. John Cairns and James Watson offered thoughtful readings of the manuscript, after John Wolfson had provided persistent encouragement for its preparation.

This work was aided by grants from the American Cancer Society (E-592) and from the National Institutes of Health (GM 17088). DHD is a fellow of the Helen Hay Whitney Foundation.

* Paper I in this series, Gilbert and Dressler,[2] discusses the general properties and applicability of the rolling circle model. Paper III is Dressler and Wolfson (*Proc. Nat. Acad. Sci. USA*, **67**, 456 (1970)) and presents experiments in support of the rolling circle mechanism for the synthesis of ϕX *duplex* rings.

[1] Sinsheimer, R., B. Linqvist, and C. Hutchinson, *The Molecular Biology of Viruses*, eds. J. Colter and W. Paranchych (New York: Academic Press, 1967).
[2] Gilbert, W., and D. Dressler, *Cold Spring Harbor Symp. Quant. Biol.*, **33**, 473 (1968).
[3] Hershey, A., E. Burgi, and L. Ingraham, *Proc. Nat. Acad. Sci. USA*, **49**, 748 (1963).
[4] Strack, H., and A. Kaiser, *J. Mol. Biol.*, **12**, 36 (1965).
[5] Dressler, D., and D. Denhardt, *Nature*, **219**, 346 (1968).
[6] Studier, F., *J. Mol. Biol.*, **11**, 373 (1965).
[7] Guthrie, G., and R. Sinsheimer, *Biochim. Biophys. Acta*, **72**, 290 (1963).
[8] Vinograd, J., J. Morris, N. Davidson, and W. Dove, *Proc. Nat. Acad. Sci. USA*, **49**, 12 (1963).
[9] Richardson, C., I. Lehman, and A. Kornberg, *J. Biol. Chem.*, **239**, 251 (1964).
[10] Westmoreland, B., *Science*, **163**, 1343 (1969).
[11] Linqvist, B., and R. Sinsheimer, *Fed. Proc.*, **25**, 651 (1967).
[12] Denhardt, D., and R. Sinsheimer, *J. Mol. Biol.*, **12**, 641 (1965).
[13] Sinsheimer, R., R. Knippers, and T. Komano, *Cold Spring Harbor Symp. Quant. Biol.*, **33**, 443 (1968).
[14] Knippers, R., A. Razin, R. Davis, and R. Sinsheimer, *J. Mol. Biol.*, **45**, 237 (1969).
[15] Ray, D., *J. Mol. Biol.*, **43**, 631 (1969).
[16] Wirtz, A., and P. H. Hofschneider, *Biochim. Biophys. Acta*, in press.
[17] Rupp, D., and G. Ihler, *Cold Spring Harbor Symp. Quant. Biol.*, **33**, 647 (1968).
[18] Ohki, M., and J. Tomizawa, *Cold Spring Harbor Symp. Quant. Biol.*, **33**, 651 (1968).
[19] Fulton, C., *Genetics*, **52**, 55 (1965).
[20] Schnös, M., and R. Inman, *J. Mol. Biol.*, (1970) in press.
[21] Smith, M., and A. Skalka, *J. Gen. Physiol.*, **49**, 127 (1966).
[22] Kiger, J., and R. Sinsheimer, *J. Mol. Biol.*, **40**, 467 (1969).
[23] Hirt, B., *J. Mol. Biol.*, **40**, 141 (1969).
[24] Inselburg, J., and M. Fuke, *Science*, **169**, 590 (1970).
[25] Lee, C., and N. Davidson, *Biochim. Biophys. Acta*, **204**, 285 (1970).
[26] Cairns, J., *Cold Spring Harbor Symp. Quant. Biol.*, **28**, 43 (1963).

DNA Restriction Enzyme from E. coli

by
MATTHEW MESELSON
ROBERT YUAN*

The Biological Laboratories,
Harvard University,
Cambridge, Massachusetts

An endonuclease which degrades foreign DNA has been isolated. The enzyme requires S-adenosylmethionine, ATP and Mg^{++}.

MANY strains of *E. coli* can recognize and degrade DNA from foreign *E. coli* strains. Whether a foreign DNA molecule will be rejected can depend on non-heritable characteristics imparted to it by the cell from which it is obtained. Such characteristics are called host-controlled modifications[1-3]. For example, the ability of λ and several other bacteriophages to multiply on *E. coli* strain K depends on the bacterial host in which the phages were last grown. Phages grown in bacteria possessing the modification allele m_K multiply well, but phages grown in bacteria lacking m_K do not. Instead, their DNA is quickly degraded on entering cells of strain K. The ability of strain K to reject or "restrict" DNA from cells lacking m_K is itself under genetic control, the responsible allele being designated r_K (refs. 4 and 5).

More generally, cells with the restriction allele r_i can degrade DNA from cells lacking the corresponding modification allele m_i. Several different modification and restriction alleles are known[6,7]. As well as certain phage DNAs, bacterial DNA transferred between cells by conjugation or transduction is subject to host-controlled modification and restriction, suggesting that these phenomena play a part in regulating the flow of genetic information between bacteria.

There is evidence that the modification character of a DNA molecule is determined by its pattern of methylation[8,9]. The simplest hypothesis for the biochemical basis of restriction is that each restriction allele directs the formation of a nuclease specific for DNA lacking the corresponding modification character. We have detected, isolated and characterized such an enzyme; it is an endonuclease present in strain K that is specifically active against λ DNA from strains lacking m_K.

Restriction by bacterial extracts. Restriction by extracts of r_K bacteria was discovered as an ATP-dependent nucleolytic activity specific for DNA from λ grown on a m_K^- host. Fig. 1 shows the sedimentation profiles of ^3H λ·K DNA and ^{32}P λ·C DNA incubated without extract (a), with extract of the λ strain *E. coli* 1100 (b), and with extract of its r_K^- derivative, strain 1100.293 (c). The sedimentation of λ·K DNA is seen not to be affected by incubation with either extract. In contrast, the sedimentation rate of λ·C DNA is specifically decreased by incubation with the r_K extract. We conclude that the activity is specific for unmodified DNA and occurs only in the restricting strain.

Purification of the enzyme. Purification of the restriction activity was at first frustrated by its apparent loss on dialysis of bacterial extracts. The difficulty was overcome when we found that both ATP and S-adenosylmethionine (SAM) are needed for restriction to occur. A role for SAM was suggested by the report that methionine, a biosynthetic precursor of SAM, is necessary for optimal restriction

* Fellow of the Damon Runyon Memorial Foundation.

in vivo by methionineless bacteria (unpublished observations of W. B. Wood, cited in ref. 8).

Following the establishment of suitable assay conditions, the activity was purified as follows: *E. coli* 1100 was grown to 6×10^8 cells/ml. at 37° C in tryptone broth with 1 µg/ml. of thiamine hydrochloride. The cells were sedimented and stored at $-20°$ C. 120 g of cells was suspended in 0·01 M *tris* pH 7·4, 2×10^{-4} M $MgCl_2$, 10^{-4} M EDTA, 2×10^{-3} M mercaptoethanol. All subsequent operations were conducted at approximately 4° C. The cells were extracted by blending with glass beads and the extract was centrifuged at low speed and then at 35,000 r.p.m. for two hours in the International *A*-170 rotor. The sediment possessed very little restriction activity and was discarded.

The high speed supernatant (300 ml.) was fractionally precipitated by adding dry $(NH_4)_2SO_4$. The material precipitating between 35 and 55 per cent saturation was dissolved in 100 ml. of 0·02 M potassium phosphate, pH 7·0, 10^{-4} M EDTA, 2×10^{-3} M mercaptoethanol (PEM), dialysed against the same buffer, and applied to a column of Whatman *DE* 52 DEAE cellulose 6 cm in diameter and 11 cm long that had been equilibrated with buffer. The column was washed with 600 ml. of buffer and eluted with a 1,000 ml. linear gradient of PEM running from 0·06 to 0·30 M potassium phosphate and collected in 50 ml. fractions. The restriction activity was found in two adjacent fractions with a mean phosphate concentration of 0·15 M.

The active DEAE fractions were combined, dialysed against PEM containing 0·02 M potassium phosphate, and applied to a 1×15 cm column of Whatman *P*11 phosphocellulose equilibrated with the same buffer. The column was washed with 40 ml. of buffer and eluted in 9 ml. fractions with a 300 ml. linear gradient running from 0·05 to 0·3 M phosphate. The restriction activity emerged in three adjacent fractions with a mean phosphate concentration of 0·16 M.

The pooled active phosphocellulose fractions were dialysed against 0·01 M potassium phosphate, pH 7·5, 10^{-4} M EDTA, and 5×10^{-4} M dithiothreitol, concentrated to 7·5 ml. by dialysis against dry 'Sephadex *G*-200', and dialysed again. The concentrate was applied in 2·5 ml. portions to 30 ml., 10–25 per cent glycerol gradients made up in the dithiothreitol buffer. After centrifugation at 25,000 r.p.m. for 20 h at 2° C in an International *SB*101 rotor, the activity was found in three adjacent 1 ml. fractions at a position corresponding to a sedimentation coefficient of approximately 12*S*. These were stored at $-10°$ C and have been kept for nearly a year without noticeable inactivation. Table 1 summarizes the purification of the restriction enzyme.

Assays with a limiting amount of pure enzyme were conducted with and without the addition of extract of

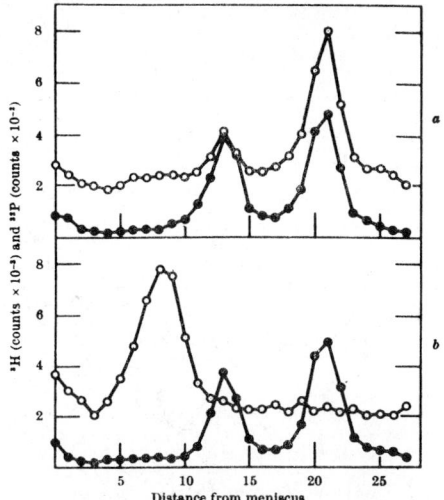

Fig. 3. Sedimentation profiles showing the action of the restriction enzyme on twisted circular λ DNA molecules. (a) A mixture of ³H λ·K and ³²P λ·C DNA consisting mostly of twisted circular molecules but also containing the less rapidly sedimenting non-twisted circular form. (b) The same mixture after 10 min of incubation with enzyme. Samples were centrifuged for 1 h under the conditions given in the legend for Fig. 1. ——○——, ³H; ——●——, ³²P.

are plotted as fractions of the weight of the intact duplex or single chain. The near superposability of the two distributions shows that at least most of the limit product consists of duplexes without single chain breaks. Only if located near the ends of the duplexes would any substantial number of single chain breaks have gone undetected by this test.

Single chain scission precedes duplex cleavage. Prolonged action by the restriction enzyme breaks the DNA duplex into several pieces. What is the timing with which individual chains of the duplex are broken? Single and double chain scissions may be sensitively detected and

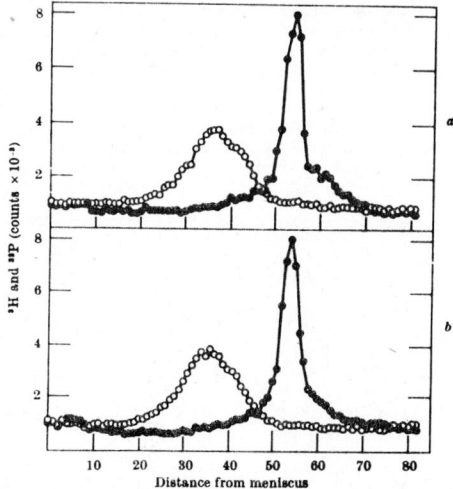

Fig. 4. Sedimentation profiles of a mixture of ³H λ·K DNA and ³²P λ·C DNA incubated with enzyme (a) once or (b) twice. See text for details. Samples of 0·4 ml. were layered on 12 ml. 15–28 per cent sucrose gradients containing 0·01 M tris pH 8·0, 0·001 M EDTA, 0·04 per cent sodium dodecyl sulphate and centrifuged for 17 h at 35,000 r.p.m. in an International SB283 rotor at 2° C. ——●——, ³H; ——○——, ³²P.

distinguished by their effects on the twisted circular form of λ DNA (ref. 18). Single chain scission converts twisted circles to the more slowly sedimenting non-twisted circular configuration. Double strand scission converts them to even more slowly sedimenting non-circular molecules. Fig. 6 shows the result of incubating a mixture of ³²P λ·C twisted circles and ³H λ·K non-circular molecules with enzyme for 30, 60 and 120 s; after an initial lag, untwisted circles are produced before non-circular molecules appear. Thus single strand scission precedes cleavage of the duplex. The same conclusion has been reached by a comparison of neutral and alkali sedimentation analyses performed in parallel on samples taken at various times in the course of the reaction. The occurrence of single chain scission early in the reaction, taken together with the paucity of single chain breaks in the limit product, indicate that the enzyme first breaks only one chain and then, a few seconds later, breaks the complementary chain at a point directly or nearly opposite. Our results do not reveal whether a given enzyme molecule remains bound to catalyse the breaks on both chains or whether the two chains are attacked independently.

The infectivity assay employed to determine the requirements of the reaction is probably insensitive to single strand scissions, for λ DNA can be rather extensively nicked by pancreatic DNase without loss of infectivity[19]. We therefore reinvestigated the requirements for ATP, Mg++ and SAM by sedimentation analysis with ³²P λ·C twisted circles as substrate. We found that neither double nor single chain scission occurs if any one of the three components is omitted from the reaction mixture.

Duplexes with only one modified chain are not attacked. Separated polynucleotide chains of ³H λ·K DNA and ³²P λ·C DNA were mixed in appropriate complementary combinations and annealed to obtain the two possible heteroduplexes (one chain modified, the other unmodified) and both homoduplexes. Incubation with enzyme followed by sedimentation analyses on neutral and alkali gradients showed that unmodified homoduplexes are cleaved but that the enzyme does not act on modified homoduplexes

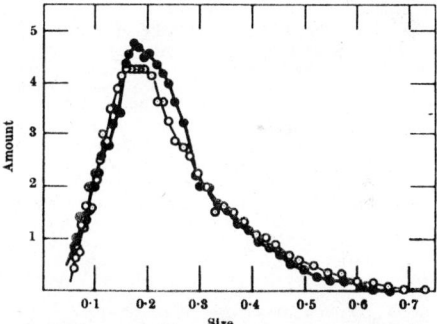

Fig. 5. Molecular size distributions of the reaction product before (——○——) and after (——●——) strand separation. Each point represents a fraction collected from the corresponding gradient. A mixture of ³H λ·K DNA and ³²P λ·C DNA was extensively incubated with enzyme and then analysed on 11·2 ml. D₂O–H₂O gradients containing 0·9 M NaCl, 0·001 M EDTA and either 0·04 per cent sarcosyl NL97, 0·01 M tris pH 8·0 or 0·1 M NaOH. The gradients were centrifuged in the SB283 rotor at 41,000 r.p.m. for 3·3 h at 20° C, which moves the intact ³H reference DNA about 85 per cent of the way down the gradient. Sedimentation coefficients were computed for each fraction, taking into account the variation of acceleration, solution density, and viscosity along the gradient. Molecular size relative to intact ³H λ·K duplexes or single chains taken as unity was obtained from the sedimentation coefficients[17]. The ordinate for each point is the ³²P counting rate corrected for background and normalized to give the amount of DNA per unit molecular size. No correction has been made for the effect of finite band width. Parallel centrifugations of the DNA mixture incubated without enzyme showed the ³²P DNA to sediment in both neutral and alkali solution as a band with a half-width of three fractions at half its maximum concentration. In the neutral gradient, the trailing edge was sharp while in alkali it displayed a tail indicating that approximately 1/4 of the polynucleotide chains were broken. In neutral solution 10 per cent of the material sedimented slightly ahead of the leading edge while in alkali the leading edge was sharp.

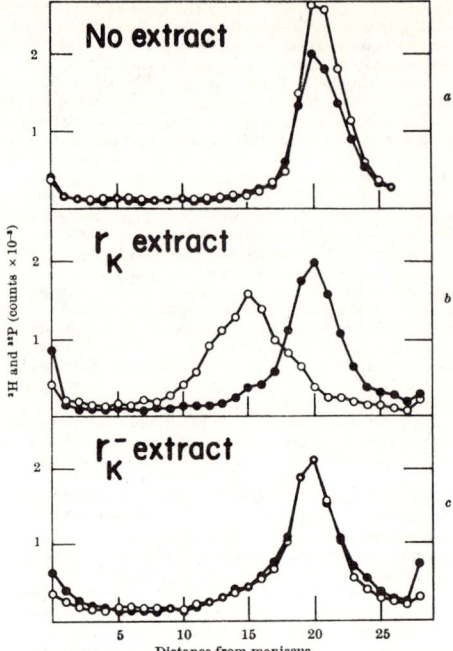

Fig. 1. Sedimentation profiles of a mixture of ^{3}H λ·K DNA and ^{32}P λ·C DNA incubated for 10 min (a) without extract, (b) with extract of *E. coli* strain 1100, or (c) with extract of strain 1100.293. Incubation mixtures (b) and (c) contained 1 mg/ml. bacterial protein. A 0·25 ml. sample of each mixture was layered on a 3·5 ml. 6–20 per cent sucrose gradient containing 0·01 M *tris* pH 8, 0·001 M EDTA, 0·04 per cent sodium dodecyl sulphate and centrifuged for 1·8 h at 55,000 r.p.m. in an International SB405 rotor at 20° C. Fractions were collected and counted by standard techniques. ——●——, ^{3}H; ——○——, ^{32}P.

the r_K^- strain 1100.293. The extract had no effect on the extent of the reaction, showing that the activity of r_K^- extract can be directly compared with that of the purified enzyme in computing the degree of purification achieved. The failure of the r_K^- extract to inhibit the reaction also shows that the r_K^- phenotype results from a reduction in the amount or activity of the restriction enzyme rather than from the presence of an inhibitor.

Requirements of the reaction. Table 2 shows the effects of omitting various components of the reaction mixture. Restriction is measured as the inactivation of infectious λ·C DNA. Enzyme, ATP, Mg^{++} and SAM are all seen to be essential while gelatine, EDTA and mercaptoethanol are not. Fig. 2 shows the effect of lowering the concentration of ATP or SAM. The requirement for SAM cannot be satisfied by 1-methionine, 5′-thiomethyladenosine, S-adenosyl-*dl*-homocysteine or S-adenosylethionine. ATP can be replaced by dATP, although at limiting concentrations approximately twice as much must be used for the same degree of inactivation. ATP cannot be replaced by GTP, ADP, AMP, adenosine, pyrophosphate or phosphate. The optimal pH for the restriction reaction lies between 7·5 and 8·0.

The restriction enzyme is an endonuclease. The action of the enzyme on ^{32}P λ·C DNA liberates little or no non-sedimenting ^{32}P, showing that degradation is endo rather than exonucleolytic. The action of the enzyme on λ twisted circles, DNA molecules without ends, confirms this. Sedimentation analyses of a mixture of ^{3}H λ·K and ^{32}P λ·C twisted circles incubated with and without enzyme (Fig. 3) prove that the enzyme readily degrades λ·C twisted circles. The enzyme, however, has no effect on λ·K twisted circles; neither double nor single chain scissions are produced.

The limit product is duplex DNA. A mixture of ^{3}H λ·K DNA and ^{32}P λ·C DNA was incubated with enzyme, re-isolated by phenol extraction, dialysed against TNE, and again incubated with enzyme. Fig. 4 shows that the second incubation caused little or no additional degradation. A control incubation proved that intact ^{32}P λ·C DNA added to the second incubation mixture was extensively degraded. We conclude that the sedimentation profiles seen in the figure represent the limit product of the restriction enzyme under our reaction conditions. The maxima of these profiles occur close to the position to which λ quarter molecules would sediment. The sedimentation distribution of the limit product in 0·1 M salt was compared with that in 0·01 M salt. No difference was observed. Single DNA chains sediment about three times faster in the stronger salt solution, whereas duplexes sediment at the same rate in both solvents. We conclude that the limit product consists of duplexes containing little or no single stranded DNA.

The limit product is without single-chain breaks. Clearly the enzyme is able to break the λ·C duplex into a number of segments. Conceivably, there are sites within these segments where the enzyme has broken only one of the two polynucleotide chains. Such single-chain breaks would have little effect on the sedimentation behaviour of DNA duplexes. They would, however, become apparent from a comparison of the molecular weight distribution of the reaction product before and after chain separation. A mixture of ^{32}P λ·C DNA and ^{3}H λ·K DNA was extensively incubated with enzyme and then sedimented at 20° C on neutral and on alkali D$_2$C-H$_2$O gradients containing 0·9 M NaCl. The alkali solvent dissociates duplexes into their component single chains. Fig. 5 shows the molecular weight distributions calculated from the sedimentation profiles in the two solvents. Molecular weights

Table 1. PURIFICATION OF THE ENZYME

	Total protein (mg)	Activity
Low speed supernatant	8,000	~1
High speed supernatant	5,700	
(NH$_4$)$_2$SO$_4$ precipitate	3,400	
DEAE eluate	260	
Phosphocellulose eluate	2	
Glycerol gradient fraction	≤0·5	≳5,000

Activity is expressed as the reciprocal of the mg/ml. of protein required to break 90 per cent of λ·C duplexes during a 10 min incubation under the assay conditions described in the text.

Table 2. REQUIREMENTS FOR REACTION

	log$_{10}$ λ·K/λ·C
Complete system	2·0
−EDTA	2·0
−Gelatine	2·0
−Mercaptoethanol	2·0
−MgCl$_2$	0·0
−SAM	0·0
−ATP	0·0

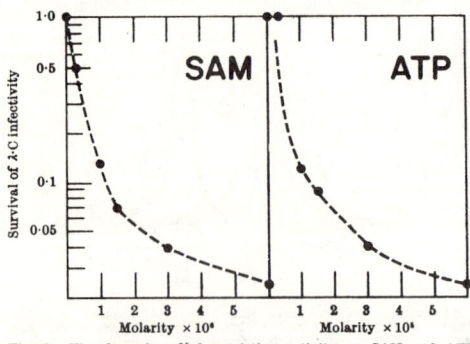

Fig. 2. The dependence of restriction activity on SAM and ATP concentration.

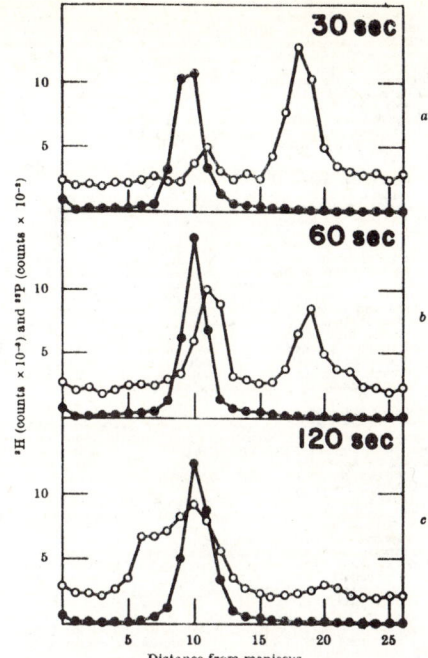

Fig. 6. Sedimentation distributions of a mixture of ³H λ·K non-circular DNA and ³²P λ·C twisted circular DNA after incubation with enzyme for (a) 30, (b) 60 and (c) 120 s. The distributions of the untreated DNA mixture and of the mixture incubated with enzyme in a reaction mixture lacking ATP and SAM were essentially the same as that seen in (a). Samples were centrifuged for 1·1 h under the conditions given in the legend for Fig. 1. ——●——, ³H; ——○——, ³²P.

or on either of the two heteroduplexes (Fig. 7). Thus heteroduplexes are resistant even to single chain scissions.

A restriction enzyme from P1 lysogenized cells. We have partly purified a restriction activity from cells lysogenic for phage P1 which specifically breaks DNA duplexes of phage λ grown on strains not lysogenic for P1. Like the restriction enzyme from strain K, it requires Mg^{++}, ATP and SAM. The product of its action also consists mainly or entirely of relatively large fragments of the λ duplex. We propose that these enzymes be designated *E. coli* endonuclease III·K and III·P, respectively.

Discussion. The degree of purification achieved and the diversity of the fractionation procedures employed make it most unlikely that the restriction activity we have isolated corresponds to more than a single molecular species, although it may well be a stable aggregate of two or more sub-units. Indeed, the enzyme's high sedimentation coefficient, the complexity of its action and co-factor requirements suggest an aggregate.

The ATP and SAM requirements of endonucleases III·K and III·P are unprecedented and we do not yet know whether either compound is consumed in the reaction or acts only catalytically, possibly as an allosteric effector. The requirement for SAM is especially interesting because of the likelihood that modification corresponds to the specific methylation of DNA[8,9], presumably using SAM as a methyl donor. It appears that the enzyme interacts directly with ATP, for it is rapidly inactivated when they are incubated together if SAM is absent.

Takano, Watanabe and Fukasawa[20] have described a specific inactivation of unmodified infectious λ DNA by extracts of bacteria carrying certain restrictive episomes. No ATP was added to their reaction mixtures. In our experiments and in an earlier investigation (unpublished results of S. Lederberg and M. Meselson) using the λ infectivity assay system, neither K nor P restriction activity was detected in bacterial extracts without added ATP; the addition of SAM was unnecessary in our crude extracts. It remains to be seen whether the enzymes associated with the restrictive episomes studied by Takano *et al.* differ in their co-factor requirements from the K and P enzymes or whether there was adequate ATP (and SAM) naturally present in their extracts.

The size distribution of the DNA fragments in the limit digest of the restriction enzyme poses several problems. The very limited action of the enzyme *in vitro* contrasts sharply with the nearly complete and very rapid breakdown of unmodified DNA to acid soluble products seen *in vivo*[21,22]. Perhaps the production of an end, or a certain type of end, quite generally exposes a DNA duplex to such breakdown within the cell, even though it is not clear that the known nucleases of *E. coli* would be sufficient for this purpose[23]. Although it seems very likely that the restriction enzyme attacks DNA at fixed sites, no clear indication of discrete species was seen in the sedimentation profiles of the limit product. Depending on

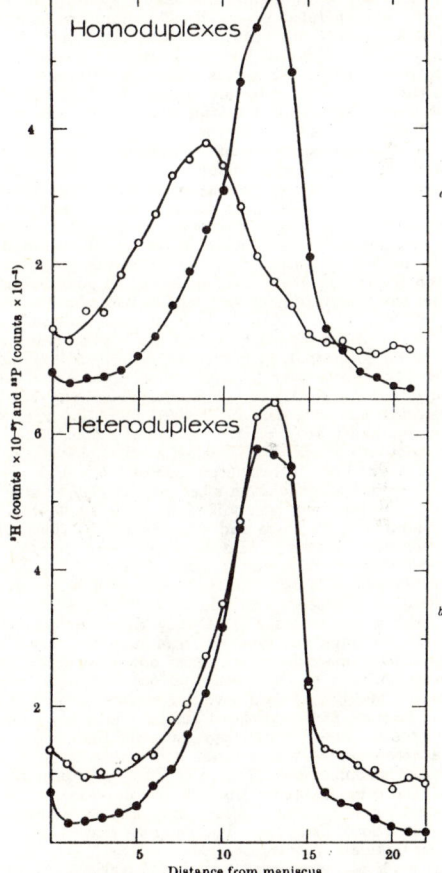

Fig. 7. Alkali sedimentation profiles of (a) reconstituted ³H λ·K and ³²P λ·C homoduplexes and (b) the two possible heteroduplexes after incubation with enzyme. Samples of 0·2 ml. were layered onto 3·5 ml. 6–20 per cent sucrose gradients containing 0·005 M EDTA and 0·3 M NaOH and centrifuged 1·8 h at 55,000 r.p.m. in the SB405 rotor at 20° C. ——●——, ³H; ——○——, ³²P.

the distribution of susceptible sites, however, it is possible that different classes of fragments would not have been resolved under our conditions of analysis.

The molecular size distribution of the limit product indicates that not all λ chromosomes are broken to the same degree. For example, although there are segments of approximately 40 per cent length, they comprise much less than 40 per cent of the digest, as though an appreciable number of chromosomes did not react to give such segments. If Fig. 5 accurately portrays the molecular size distribution of the limit product, it may be that the initial population of DNA molecules was heterogeneous; some may already have been modified at one or a few sites. Alternatively, the restriction enzyme itself may occasionally render some sites resistant to cleavage. Conceivably the restriction enzyme is also the modifying enzyme but with its modifying activity only weakly manifested in our assay system. The presence of SAM in the incubation mixture, the close genetic linkage between r^- and m^- mutations, and the fact that a large proportion of r^- mutants are also m^- (refs. 4, 5 and 24), are consistent with this interpretation, although other explanations can be given.

It is definitely known that one, and somewhat less well established that both, of the two possible λ chromosomes with only one modified polynucleotide chain are able to infect and multiply normally in a restricting host[2]. Our finding that both types of λ heteroduplex are resistant to the restriction enzyme confirms these conclusions. This shows that both polynucleotide chains of λ·K DNA are modified at every site of potential attack. The fact that heteroduplexes are resistant even to single chain scission indicates that the enzyme somehow takes account of the structure of both chains before cutting any. The resistance of heteroduplexes may serve to protect newly replicated bacterial DNA from attack by the cells' own restriction enzyme, allowing time for modification of the newly synthesized chain. Indeed, if restriction and modification are accomplished by the same enzyme, the choice between the two reactions may normally be governed by whether the substrate is an unmodified homoduplex or a heteroduplex, respectively. The extreme destructive potential of the restriction system may have enforced selection for additional mechanisms to ensure that cellular DNA newly synthesized by replication or repair is modified before it can be restricted. Various protective control mechanisms can obviously be devised involving the cofactor requirements for restriction and the sensitivity of the restriction enzyme to ATP when SAM is absent.

If DNA synthesis in restricting bacteria could be continued beyond one generation after methylation has been blocked, the resulting unmodified homoduplexes should be vulnerable to cleavage and degradation by the cells' own restriction system. Having made this prediction, we were struck by the report of Lark[25] that DNA synthesis in methionineless *E. coli* 15T- halts one generation after methionine is replaced with ethionine or norleucine and that much of the cellular DNA may then become acid soluble. Although the level of SAM may be very low during methionine deprivation, even one cleavage of the bacterial chromosome by restriction enzyme might stop replication and trigger extensive degradation. According to this picture, DNA synthesis should continue beyond one generation and without degradation if Lark's experiments were repeated with a strain deficient in restriction.

Finally, endonucleases III·K and III·P may provide a model for other systems that cleave duplexes or cut single chains at specific locations, not only in connexion with restriction phenomena, but possibly also in replication, recombination or transcription.

This work was supported by the US National Science Foundation. We thank Miss Ann Hablanian for technical assistance and Martin Oster and Fred. Blattner for their help. Samples of S-adenosylhomocysteine and 5'-thiomethyladenosine were provided by Dr S. Harvey Mudd.

Materials and Methods. *Bacteria and bacteriophages.* Strain C600 (ref. 10), its derivative CR34 (thy (ref. 11) and the DNA endonuclease I deficient strain 1100 (ref. 12) have the restriction and modification properties $r_K m_K$. Strain 1100.293 is an $r_K^- m_K^-$ mutant of strain 1100. Strains C600.4 and C600.5 are $r_K^- m_K^-$. Bacteriophage λ is the wild type of Kaiser[14] and λc is a clear plaque forming mutant. Phages grown on m_K^- strains are designated λ·C, following the custom based on the fact that *E. coli* strain C is m_K^-. Taking the plating efficiency on strain C600.4 as unity, that on all other sensitive r_K^- strains employed was 1·0 and on r_K^- strains was 1·0 for λ·K and 10⁻⁴ for λ·C.

DNA preparations. Stocks of λ·K and λc·C were made on strains C600 and C600.4, respectively, by the agar layer method. Stocks of ³H-thymidine λ·K and ³²P λ·C were produced by ultraviolet light induction of CR34 (λ) and C600.4 (λ) in radioactive media. Phages were purified by two cycles of differential centrifugation followed by equilibrium sedimentation in CsCl solution. DNA was extracted with water-saturated phenol, extensively dialysed against 0·01 M tris, 0·01 M NaCl, 2 × 10⁻⁴ M EDTA, pH 7·4 (TNE) and stored at 0° C. Before use for analysis, DNA solutions were kept at 55° C for 5 min in order to disassociate end-to-end aggregates. Twisted circular λ DNA was prepared by infection of *E. coli* C600.4 (λ) or C600 (λ) with 10 ³²P λ·C or ³H λ·K per cell, respectively. After incubation for 30 min at 37° C in tryptone broth, the infected cells were sedimented, washed with 0·08 M tris pH 8·0, 0·01 M NaCl, and resuspended in the same medium at 5 × 10⁹ cells/ml. The suspension was made 0·02 per cent in lysozyme, incubated for 1 min at 37° C, made 0·01 M in EDTA, incubated for an additional minute, and lysed by the addition of 10 mg/ml. sodium dodecyl sulphate. The lysate was sheared by vigorous passage through a No. 26 hypodermic needle, extracted with water-saturated phenol, dialysed against TNE, and sedimented through a neutral sucrose gradient containing 10⁻³ M EDTA. After shear treatment, host cell DNA sediments much more slowly than λ twisted circles, which are relatively insensitive to shearing. The fraction of the gradient containing twisted circles was dialysed against TNE and stored at 0° C.

Polynucleotide chain separation and annealing. Separated polynucleotide chains were prepared according to Hradecna and Szybalski[15], using poly UG rather than poly IG. Broken chains and poly UG were removed by sedimentation through an alkali D_2O-H_2O gradient. Complementary mixtures containing approximately 10 μg/ml. of purified single chains were annealed by overnight dialysis at 37° C against 0·001 M EDTA, 0·1 M NaCl, NaOH pH 10·5 followed by dialysis against TNE (private communication from W. Doerfler and D. Hogness). More than 80 per cent of the annealed material sedimented at the same rate as native DNA duplexes in neutral 0·01 M salt. Sedimentation analysis in alkali showed that most of the annealed duplexes were free of single chain breaks.

Enzyme assays. Restriction activity was assayed in a mixture containing, per ml.: 50 μmoles tris, 5 μmoles MgCl₂, 0·2 μmoles EDTA, 5 μmoles mercaptoethanol, 100 μg gelatine, 0·2 μmoles ATP, 0·02 μmoles S-adenosylmethionine (SAM), approximately 10¹⁰ phage units of ³²P λ·C DNA and ³H λ·K DNA, and the enzyme sample to be assayed. SAM was purified from the commercial product by elution from 'Biorex 70' with acetic acid. The pH of the reaction mixture was 7·7 at 20° C. After incubation for the desired time at 37° C, 1/10 volume of 0·4 M EDTA pH 9·5 was added and the mixture was subjected to sedimentation analysis.

In some assays, radioactive DNA was replaced with approximately 10⁶ phage units of a mixture of λc·C DNA and λ·K DNA. After incubation, the reaction mixture was diluted 1 : 100 in TNE and assayed for infectivity. Restriction was measured as a decrease in the ratio of clear plaques to turbid plaques, relative to the ratio for the untreated DNA mixture. λ·K DNA is not inactivated by the purified enzyme. The infectivity assay was that of Kaiser and Hogness[18] except for the following: strain C600.5 was used as recipient and C600.4 (λ¹⁸⁵⁷) as indicator. Recipient cells were prepared by 4·5 h growth at 37° C from a 1 : 3 dilution of an overnight culture. Glycerol rather than glucose was used as carbon source. The bacteria were sedimented, resuspended at 2 × 10⁹/ml. in 0·01 M MgSO₄, infected with 5 λ¹⁸⁵⁷ phages per cell, incubated for 7 min at 37° C, diluted 1 : 1 with iced 0·01 M tris pH 8·0, 0·01 M CaCl₂, 0·01 M MgSO₄, sedimented, and resuspended at 2 × 10⁹/ml. in the same solution. The cells were kept at 0° C for 2–12 h before use. The efficiency of the assay was approximately 0·01 plaques per phage unit of λ DNA. Infectivity assays were useful for detecting restriction activity only after the phosphocellulose chromatography step in the purification of the enzyme. Less purified enzyme preparations caused severe non-specific inactivation of infectivity.

Received March 11, 1968.

[1] Luria, S. E., *Cold Spring Harbor Symp. Quant. Biol.*, 18, 237 (1953).
[2] Bertani, G., and Weigle, J. J., *J. Bacteriol.*, 65, 113 (1953).
[3] Arber, W., *Ann. Rev. Microbiol.*, 365 (1965).
[4] Colson, C., Glover, S. W., Symonds, N., and Stacey, K. A., *Genetics*, 52, 1043 (1965).
[5] Wood, W. B., *J. Mol. Biol.*, 16, 118 (1966).
[6] Eskridge, R. W., Weingeld, H., and Paigen, K., *J. Bacteriol.*, 93, 835 (1967).
[7] Kerszman, G., Glover, S. W., and Avonovitch, J., *J. Gen. Virol.*, 1, 333 (1967).
[8] Arber, W., *J. Mol. Biol.*, 11, 247 (1965).
[9] Klein, A., and Sauerbier, W., *Biochem. Biophys. Res. Commun.*, 18, 440 (1965).
[10] Appleyard, R. K., *Genetics*, 39, 440 (1954).
[11] Okada, T., Homma, J., and Sonohara, H., *J. Bact.*, 84, 602 (1962).
[12] Durwald, H., and Hoffman-Berling, H., *J. Mol. Biol.* (in the press).
[13] Meselson, M., *J. Mol. Biol.*, 9, 734 (1964).
[14] Kaiser, A. D., *Virology*, 3, 42 (1957).
[15] Hradecna, S., and Szybalski, W., *Virology*, 32, 633 (1967).
[16] Kaiser, A. D., and Hogness, D. S., *J. Mol. Biol.*, 2, 392 (1960).
[17] Studier, F. W., *J. Mol. Biol.*, 11, 373 (1965).
[18] Bode, V. C., and Kaiser, A. D., *J. Mol. Biol.*, 14, 399 (1965).
[19] Heinemann, S., dissert., Harvard Univ. (1967).
[20] Takano, T., Watanabe, T., and Fukasawa, T., *Biochem. Biophys. Res. Commun.*, 25, 192 (1966).
[21] Arber, W., Hattman, S., and Dussoix, D., *Virology*, 21, 30 (1963).
[22] Lederberg, S., and Meselson, M., *J. Mol. Biol.*, 8, 623 (1964).
[23] Lehman, I. R., *Ann. Rev. Biochem.*, 36, 645 (1967).
[24] Boyer, H., *J. Bact.*, 88, 1652 (1964).
[25] Lark, C., *J. Mol. Biol.*, 31, 401 (1968).

Integrative and Excisive Recombination by Bacteriophage λ: Evidence for an Excision-specific Recombination Protein

H. Echols†

*Department of Biochemistry, University of Wisconsin
Madison, Wisc. 53706, U.S.A.*

(Received 4 August 1969, and in revised form 20 October 1969)

The nature of the site-specific recombination used by bacteriophage λ to integrate into and excise from the host chromosome has been studied. The specificity of these integrative and excisive recombination events was analyzed through measurements of lytic recombination between phage DNA molecules which contain the structural determinants appropriate to the two types of recombination. The results of these experiments show that an active *int* protein is required for both integrative and excisive recombination, but an active *xis* protein is probably essential only for the latter event. One type of site-specific recombination was ineffective even if active *int* and *xis* proteins were present: λ*gal* with a phage presumed to carry the host recognition region.

1. Introduction

Previous experiments have demonstrated that bacteriophage λ integration involves a site-specific recombination event, catalyzed at least in part by the protein product of the *int* gene (Weil & Signer, 1968; Echols, Gingery & Moore, 1968). These experiments entailed measurement of recombination between phage during lytic growth under conditions where the normally available general recombination systems were blocked by mutation. Integrative recombination was recognized by its site-specificity: substantial lytic recombination was observed only between λ genes on opposite sides of the integration site (e.g. *J* and *N* of Fig. 1).

With the discovery of the *xis* gene, which appears to have a critical role only in excision (Guarneros & Echols, 1970), the question obviously arises as to whether the *xis* protein product is an excision-specific recombination enzyme. In order to study the excision event ("excisive recombination") by lytic crosses, one needs phage which carry the recombinant recognition regions found at the prophage ends (see Fig. 1). Such phage are available in the form of galactose- and biotin-transducing phages: λ*gal* has acquired the "left" recombinant recognition region and the *gal* genes; λ*bio* has acquired the "right" recombinant recognition region and the *bio* genes. Both λ*gal* and λ*bio* typically lose phage genetic material from the prophage end opposite the addition and thus are deletion–addition mutations (see Fig. 1).

This report describes experiments designed to study the specificities of the *int* and *xis* proteins in integrative and excisive recombination, using λ, λ*gal* and λ*bio* derivatives to provide the appropriate structural determinants. The results indicate

† Present address: Department of Molecular Biology, University of California, Berkeley, Calif., U.S.A.

strongly that the *xis* protein is essential only for excisive recombination, whereas the *int* protein is essential for both normal integrative and excisive recombination, as well as for various "hybrid" events.

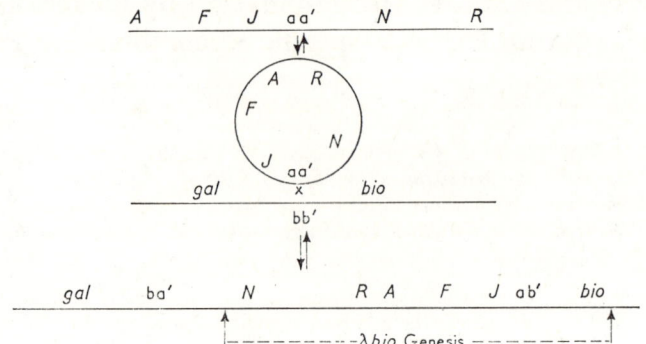

FIG. 1. The Campbell (1962) model for integration and excision and the origin of deletion-addition transducing phages.

A few λ genes (*A*, *F*, *J*, *N*, *R*) are indicated for reference. The linear λ DNA molecule (top) forms a circle by joining ends, and a reciprocal recombination between phage and host "recognition regions" (aa′ and bb′, respectively) provides for linear insertion of the λ DNA between the *gal* and *bio* gene clusters of the host DNA (bottom). Recombinant recognition regions (ba′ and ab′) are produced at the ends of the prophage DNA. Excision involves a reciprocal recombination event between prophage recombinant recognition regions to regenerate free phage DNA and the original phage and host recognition regions.

Transducing phages are presumed to arise through rare aberrant excision events in which recombination may join a point within the prophage and a point within the host DNA (outside of the normal recognition region). A typical galactose-transducing (λ*gal*) phage acquires host genetic material near the left prophage end (including the *gal* genes) in an event which deletes some λ genetic material at the right prophage end; a typical biotin-transducing (λ*bio*) phage acquires host genetic material near the right prophage end (including the *bio* genes) in an event which deletes some λ genetic material at the left prophage end. Regions of DNA which might be joined to create a λ*bio* are indicated by arrows on the Figure. λ*gal* phage will carry the "left" prophage recognition region (ba′) and λ*bio* phage will carry the "right" prophage recognition region (ab′).

2. Materials and Methods

(a) *Bacteriophage and bacteria*

The *Escherichia coli* strains used in this investigation and their relevant genetic characteristics are as follows: C600 su^+ and MS0 su^-—permissive and non-permissive hosts for the λ*sus* nonsense mutants (Campbell, 1961); 152 rec^-su^-—recombination defective strain used for phage crosses; QR48 rec^-su^+—recombination defective su^+ strain used to score the *bio10* deletion, which leads to failure to make plaques on $recA^-$ hosts (Manly, Signer & Radding, 1969); W3104 gal^-su^+ and B9 gal^-su^+—galactose non-fermenting (gal^-) strains used to score for the presence of the *gal* genes in λ*gal* derivatives.

The λ mutations employed in various combinations were the following: λ*sus* conditional-lethal nonsense mutations A32, J6 and P80 (Campbell, 1961); the general recombination mutation *redB114* (Echols & Gingery, 1968; Signer *et al.*, 1968); the *int* gene mutation *int2* (Gingery & Echols, 1967; Guarneros & Echols, 1970); the *xis* gene mutation *xis1* (Guarneros & Echols, 1970); the *bio10* deletion–addition mutation, which has deleted *int* through *red* (Signer, Manly & Brunstetter, 1969) and therefore is $int^- xis^- red^-$ (Guarneros & Echols, 1970); and the *gal8* deletion–addition mutation (Adhya & Court, manuscript in preparation). The λ*gal8* strain forms plaques, in contrast to previously studied galactose-transducing phages (λ*dgal*), which are defective in lytic growth because of a loss of essential genes in the A–J region (Fig. 1—see Campbell, 1961). This strain was obtained by "moving" gal close to λ through a chlD deletion between gal and λ prophage (Adhya, Cleary & Campbell, 1968). Although $\lambda gal8$ will form plaques, it resembles the "classical" $\lambda dgal$ in its inability to integrate effectively. $\lambda gal8$ has a buoyant density in CsCl which is 0·008 g/cm^3 less than normal λ (Adhya & Court, manuscript in preparation); $\lambda gal8$ must therefore have lost a considerable amount of phage genetic material from the right prophage end—presumably including the right recombinant recognition region (ab' of Fig. 1).

Phage strains carrying various combinations of these mutations were constructed by standard crosses, except for $susP\ red^-\ gal$ derivatives. Scoring for int^-, red^-, and xis^- has been described previously (Gingery & Echols, 1967; Echols & Gingery, 1968; Guarneros & Echols, 1970). The $susP\ red^-gal$ derivatives were prepared by crosses of red^-galc^+ with $susPint^{\pm}xis^{\pm}red^-cI857$ in the rec^- host 152; under these conditions (with general recombination blocked), $galcI857$ derivatives mainly have the int and xis allele of the parental $susP$ (see discussion of Table 3 and Fig. 4). For gal derivatives, it was necessary to check $int^{\pm}$ by complementation for excision with an int^- lysogen (Guarneros & Echols, 1970).

(b) *Media and bacteriophage crosses*

The media used are described in the previous paper (Guarneros & Echols, 1970), except that T-broth was supplemented with 0·2% maltose and 0·01% yeast extract for phage crosses. Phage crosses were carried out by lytic growth in strain 152 rec^-, after adsorption in 0·01 M-MgSO$_4$ (20 min at 25°C). A multiplicity of infection of 5 (approx. 2×10^9/ml.) for each parental phage was used. After phage adsorption, the infected cells were diluted 1:500 into supplemented T-broth, incubated for 80 min at 37°C, treated with CHCl$_3$, and plated on C600 su^+ and MSO su^- to measure sus^+ recombination frequency. The frequency of sus^+ revertants in stocks of sus mutants was less than 10^{-5} in every stock used for a cross.

3. Results

(a) *Specificity of* xis *and* int *for vegetative bacteriophage recognition regions*

As a first step in exploring the recombinational specificity of the xis and int proteins, we examined the effect of xis^- and int^- mutations on recombination between phage with the integrative recognition region of the normal vegetative phage (Fig. 2). The results are presented in Table 1. Part of Table 1 simply repeats previous data (Weil & Signer, 1968; Echols et al., 1968) which showed that: (1) there is substantial residual recombination across the recognition region even if all general recombination is blocked by the use of recombination-defective (red^-) phage and a recombination-defective (rec^-) host (line 2); (2) nearly all of this residual recombination is eliminated by an additional int^- mutation in the parental phage (line 3). When the effect of a xis^- defect was studied (line 4), there was at most a small effect on the site-specific

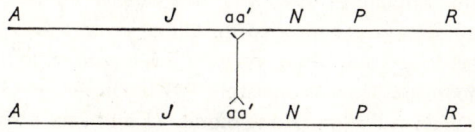

Fig. 2. Schematic representation of recombination between two phage recognition regions. This site-specific recombination becomes readily observable during lytic growth under conditions where general recombination is blocked by mutation. Substantial recombination is found between phage genes on opposite sides of the aa' recognition region (e.g. J and N or J and P), but not between phage genes on the same side of the aa' recognition region (e.g. A and J or P and R) (Weil & Signer, 1968; Echols et al., 1968).

Table 1

Effect of xis *and* int *mutations on recombination between phage with vegetative "recognition regions"*

Cross	% Recombination
$J^- \times P^-$	8·2
$J^-red^- \times P^-red^-$	1·1
$J^-red^-int^- \times P^-red^-int^-$	0·04
$J^-red^-xis^- \times P^-red^-xis^-$	0·6

Parental phage in each case carried the *susJ6* or *susP80* nonsense mutation. The red^-, int^-, and xis^- mutations employed were *red114*, *int2*, and *xis1*. Phage crosses were carried out by lytic growth in 152 rec^-. Total progeny phage were scored by plating on C600 su^+, and sus^+ recombinants were scored by plating on MSO su^-. The numbers in the Table are plaques on MSO/plaques on C600 × 100. Details of cross procedures and strains used are given in Materials and Methods. The data represent the average of three separate experiments; the maximum variability encountered was approximately ±50% of the numbers in the Table.

recombination. Thus we conclude that the *xis* and *int* proteins probably have different cellular functions.

(b) *The role of* xis *and* int *in excisive recombination*

Although the experiments above show that the *xis* and *int* genes probably have different functional roles, they provide no direct evidence that the *xis* protein is involved in site-specific recombination. To investigate this question, it is necessary to study recombination between phage with the recognition regions of the prophage ends—λ*gal* and λ*bio* (see Introduction and Fig. 3).

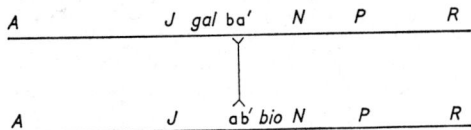

Fig. 3. Schematic representation of recombination between phage DNA molecules carrying the prophage recognition regions—the recombination event characteristic of excision. The left prophage recognition region (ba') is carried by λ*gal*; the right prophage recognition region (ab') by λ*bio*.

The results of such crosses (Table 2) show that an active *xis* gene is essential for the excision type of recombination event, as well as an active *int* gene. Thus both the *int* and *xis* proteins appear to have a critical role in excisive recombination, but the *xis* protein is likely to have the interesting role of an *excision-specific* recombination protein. It is also interesting to note the very high recombination frequency associated with the excisive recombination—consistent with the very efficient excision of a prophage following release of repression. Signer (1969) has also observed very high recombination in this type of cross.

(c) *Specificity of* xis *and* int *for mixed recognition regions*

The experiments of section (b) demonstrate that the *xis* protein is an essential element of the excisive recombination system, presumably having a critical role in

EXCISIVE RECOMBINATION BY λ

TABLE 2

Effect of xis *and* int *mutations on recombination between bacteriophage with recombinant prophage recognition regions*

Cross	% Recombination
A⁻bio × P⁻gal	9·3
A⁻bio × P⁻gal red⁻	10
A⁻bio × P⁻gal red⁻int⁻	0·05
A⁻bio × P⁻gal red⁻xis⁻	0·03

Parental phage in each case carried the *susA32* or the *susP80* nonsense mutation and the *bio10* or *gal8* deletion–addition mutation. The deletion of *bio10* renders the phage *int⁻xis⁻red⁻*. The A⁻*bio10* phage would be expected to have only the right recombinant recognition region and the P⁻*gal8* would be expected to have only the left recombinant recognition region. The *red⁻*, *int⁻*, and *xis⁻* point mutations, the cross conditions, and the scoring of recombinants were identical to those of Table 1. The *sus⁺* recombinants in line 2 were checked for the expected presence of both *gal* and *bio*; greater than 99% were indeed *gal bio* phenotype.

providing for recombination between the left and right prophage ends (ba' and ab, of Fig. 1). We wanted to know whether the role of the *xis* protein involved pairwise recognition (e.g. ba' *and* ab') or perhaps just one of the two prophage ends. Therefore, we carried out crosses between phage with normal vegetative recognition regions and phage with one or the other of the prophage recognition regions (Fig. 4).

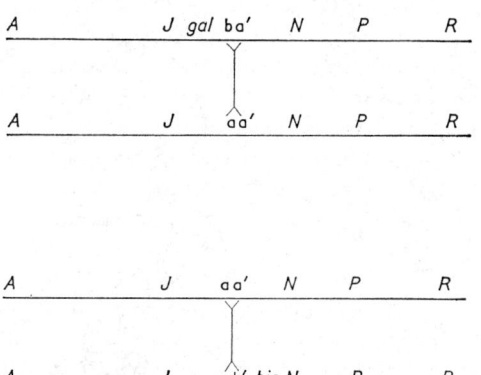

FIG. 4. Schematic representation of recombination between one DNA molecule with a phage recognition region and another DNA molecule with a prophage recognition region. The left prophage recognition region (ba') is carried by λ*gal* (upper Figure); the right prophage recognition region (ab') is carried by λ*bio* (lower Figure).

The results (Table 3) show that an active *int* gene is again essential for the site-specific recombination, whereas an active *xis* gene is not essential (although there may be a small effect of *xis⁻*). Thus these results are consistent with the concept that the *xis* protein has a critical role in site-specific recombination only if both prophage ends are present; however, since all possibilities have not yet been tested (e.g. ba' × ba'), further experiments will be required to establish this point.

The factor of ten difference in site-specific recombination between λ*gal*-λ and

Table 3

Effect of xis *and* int *mutations on recombination between one bacteriophage with a vegetative recognition region and another bacteriophage with a recombinant recognition region*

Cross	% Recombination
$J^- \times P^- gal$	17
$J^- red^- \times P^- gal\ red^-$	10
$J^- red^- int^- \times P^- gal\ red^- int^-$	0·05
$J^- red^- xis^- \times P^- gal\ red^- xis^-$	4·4
$A^- bio \times P^-$	6·3
$A^- bio \times P^- red^-$	1·0
$A^- bio \times P^- red^- int^-$	0·07
$A^- bio \times P^- red^- xis^-$	0·4

Parental phage, cross conditions and scoring of recombinants were those of Table 1 and Table 2.

λbio-λ (Table 3, lines 2 and 6), is presumably understandable in terms of the role of "structural features" of the recognition regions in determining the efficiency of the recombination event; similar quantitative differences have been observed by Signer, Weil & Kimball (1969) in a comparison of λb2-λ and λbio-λ crosses. However, there is not at present any clear reason why "left" (ba′) is better than "right" (ab′).

(d) *The role of* xis *and* int *in integrative recombination*

The experiments presented so far demonstrate a clear difference between *xis* and *int* in recombinational specificity; however, they have not examined a true integration type of recombination—that is, a recombination event involving the recognition regions of a normal vegetative phage and a normal non-lysogenic host. Guerrini (personal communication) has pointed out that the excisive recombination between λ*gal* and λ*bio* (Fig. 3) should generate a λ*gal bio* recombinant phage with the *host* recognition region (bb′). This follows from the fact that the reciprocal products of excisive recombination are phage (aa′) and host (*gal* bb′ *bio*) (see Figs 1 and 3).

To investigate the role of *xis* and *int* in integrative recombination between phage and host recognition regions, we studied recombination between λ and λ*gal bio* (Fig. 5). The results (Table 4) show that an active *int* gene is essential to integrative recombination, but an active *xis* gene is not. The experiments also indicate that integrative recombination may be more efficient for a phage–host pairing than for phage–phage (Table 1).

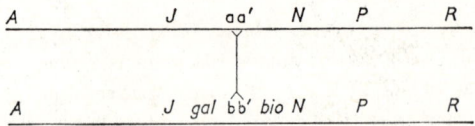

Fig. 5. Schematic representation of recombination between phage DNA molecules carrying phage and host recognition regions—the recombination event characteristic of integration. The host recognition region (bb′) is carried by λ*gal bio* and is derived from the excisive recombination λ*gal* × λ*bio* (ba′ × ab′).

Table 4

Effect of xis *and* int *mutations on recombination between bacteriophages with host and bacteriophage recognition regions*

Cross	% Recombination
A$^-$gal bio × P$^-$	6·7
A$^-$gal bio × P$^-$red$^-$	2·9
A$^-$gal bio × P$^-$red$^-$int$^-$	0·01
A$^-$gal bio × P$^-$red$^-$xis$^-$	1·9

Parental phage in each case carried the *susA32* or *susP80* nonsense mutation; the A$^-$ phage carried both the *gal8* and *bio10* deletion–addition mutations, and would be expected to have the host recognition region. The *red$^-$*, *int$^-$* and *xis$^-$* point mutations, the cross conditions, and the scoring of recombinants were identical to those of Tables 1 to 3.

(e) *The structural defect of* λgal—*failure of site-specific recombination*

One classical property of the λ*gal* transducing phage is its inability to integrate effectively; thus *gal* transduction is typically carried out in the presence of a normal "helper" phage (Campbell, 1957). The failure of λ*gal* to integrate does not result from a failure to replicate (Joyner, Isaacs, Echols & Sly, 1966), nor from an inability to provide *int* protein (Gingery & Echols, 1968; Weisberg & Gottesman, 1969; see also Table 2). It has been suggested that the integration defect of λ*gal* results from a structural defect in the integrative recombination between λ*gal* and the host (Guerrini, personal communication 1966 and 1969; Gingery & Echols, 1968; Weisberg & Gottesman, 1969).

To investigate this question in our recombination system, we studied recombination between λ*gal* and host recognition regions. The results show that recombination is very inefficient, even if active *int* and *xis* genes are present (Table 5, line 2). Thus the integration defect of λ*gal* does seem to involve a failure of the site-specific recombination which inserts the prophage. An understanding of the nature of the structural defect may have to await biochemical knowledge of the specificity of the *int* and *xis* proteins.

Table 5

Structural defect of λgal—*recombination between a* gal *transducing bacteriophage and a bacteriophage with the host recognition region*

Cross	% Recombination
A$^-$gal bio × P$^-$gal	5·3
A$^-$gal bio × P$^-$gal red$^-$	0·07
A$^-$gal bio × P$^-$gal red$^-$int$^-$	0·03
A$^-$gal bio × P$^-$gal red$^-$xis$^-$	0·05

Parental phage in each case carried the *susA32* or *susP80* nonsense mutations. The A$^-$ phage carried both the *gal8* and *bio10* deletion–addition mutations, and would be expected to have the host recognition region; the P$^-$ parent carried the *gal8* deletion–addition mutation, and would be expected to have only the left recombinant recognition region. The *red$^-$*, *int$^-$* and *xis$^-$* point mutations, the cross conditions, and the scoring of recombinants were identical to those of Tables 1 to 4.

4. Discussion

(a) *Catalytic features of integration and excision:* int *and* xis *proteins*

The experiments reported here amplify previous work on the role of the *int* protein (Weil & Signer, 1968; Echols *et al.*, 1968; Signer, Weil & Kimball, 1969) and provide evidence for a new site-specific recombination protein—the product of the *xis* gene. The *int* protein appears to be essential for both integrative and excisive recombination —the *xis* protein for excisive recombination only. One possible mechanism would involve the functioning of an *int–xis* protein complex in the excision reaction and the *int* protein only in the integration reaction. The *int–xis* protein complex might provide for a new reaction pathway which leads to a reversal of the "monomer" *int* protein pathway. The "decision" whether to integrate or to excise would depend on the level of "free" *int* protein and *int–xis* complex. Mechanisms by which the presence of the *xis* protein might control the equilibrium are considered in the accompanying paper by Dove (1970).

One potentially interesting feature of the excisive recombination is the possibility that there may be an obligatory pairwise recognition (both prophage ends) for activity (see Results, section (c)); this might be a general feature of all recombination events, but only readily observable for this special type of structural recognition.

(b) *Structural features of integration and excision*

The fact that an active *xis* gene is essential only for excisive recombination demonstrates that the vegetative phage and host structural recognition regions (which provide integrative recognition) cannot be identical to the prophage ends (which provide excisive recognition); this implies also that the vegetative phage and host recognition regions are not identical (i.e. aa' ≠ bb' in Fig. 1). The latter point is supported experimentally by the very high recombination between λ*gal* and normal λ (Table 3), compared to the very low recombination between λ*gal* and "host" (Table 5). A considerable amount of previous evidence also supports the concept that the recognition regions of phage, host, and prophage ends are not identical (see Signer, 1968).

The evolutionary advantage of such structural differences presumably resides in the resultant possibility for regulation of the integration–excision reaction, so that both integration and excision can proceed very efficiently under the cellular conditions appropriate to each process. To choose the simple example mechanism discussed in section (a) above, under repressed conditions (low concentration of *int* and *xis* proteins), most *int* protein may be free, and integration thereby is favored; after effective derepression (high concentration of *int* and *xis* proteins) mostly *int–xis* complex may exist, and excision thereby is favored.

It is considerably easier to say that structural differences exist between the recognition regions and to speculate as to why they exist in evolutionary terms than it is to say *what* the structural differences really are. This may have to await biochemical analysis of proteins and sites.

We thank Sankar Adhya, Allan Campbell, Kenneth Manly and Ethan Signer for phage stocks and bacterial strains, and Sankar Adhya and Ethan Signer for the communication of unpublished results. This research was supported in part by U.S. Public Health Service grant GM08407.

REFERENCES

Adhya, S., Cleary, P. & Campbell, A. (1968). *Proc. Nat. Acad. Sci., Wash.* **61**, 956.
Campbell, A. (1957). *Virology*, **4**, 366.
Campbell, A. (1961). *Virology*, **14**, 22.
Campbell, A. (1962). *Advanc. Genetics*, **11**, 101.
Dove, W. F. (1970). *J. Mol. Biol.* **47**, 585.
Echols, H. & Gingery, R. (1968). *J. Mol. Biol.* **34**, 239.
Echols, H., Gingery, R. & Moore, L. (1968). *J. Mol. Biol.* **34**, 251.
Gingery, R. & Echols, H. (1967). *Proc. Nat. Acad. Sci., Wash.* **58**, 1507.
Gingery, R. & Echols, H. (1968). *Cold Spr. Harb. Symp. Quant. Biol.* **33**, 721.
Guarneros, G. & Echols, H. (1970). *J. Mol. Biol.* **47**, 565.
Guerrini, F. (1969). *J. Mol. Biol.* **46**, 523.
Joyner, A., Isaacs, L. N., Echols, H. & Sly, W. S. (1966). *J. Mol. Biol.* **19**, 174.
Manly, K. F., Signer, E. R. & Radding, C. M. (1969). *Virology*, **37**, 177.
Signer, E. R. (1968). *Ann. Rev. Microbiol.* **22**, 451.
Signer, E. R. (1969). *Virology*, in the press.
Signer, E. R., Echols, H., Weil, J., Radding, C., Shulman, M., Moore, L. & Manly, K. (1968). *Cold Spr. Harb. Symp. Quant. Biol.* **33**, 711.
Signer, E. R., Manly, K. F. & Brunstetter, M. (1969). *Virology*, **39**, 137.
Signer, E. R., Weil, J. & Kimball, P. C. (1969). *J. Mol. Biol.* **46**, 543.
Weil, J. & Signer, E. R. (1968). *J. Mol. Biol.* **34**, 273.
Weisberg, R. A. & Gottesman, M. (1969). *J. Mol. Biol.* **46**, 565.

Mechanism for the Action of λ Exonuclease in Genetic Recombination

by

ERA CASSUTO & CHARLES M. RADDING

Departments of Internal Medicine and Molecular Biophysics and Biochemistry, Yale University

---

Lambda exonuclease eliminates *in vitro* redundant regions in DNA which probably occur during genetic recombination.

A ROLE of the exonuclease and β protein of bacteriophage λ in genetic recombination has been shown by the discovery that the mutations (*red*) that affect general recombination of λ are located in the structural genes of the exonuclease and its associated β protein[1-3]. Studies *in vitro* have been designed to seek clues to the role of the enzyme *in vivo* by examining its action and binding at terminal and internal sites in DNA (ref. 4 and unpublished observations of C. M. R. and D. M. Carter). λ exonuclease acts at the 5′ ends of double stranded molecules of DNA[5], and binds weakly at nicks[4], although it does not readily initiate digestion at nicks[4]. We wish to describe the synthesis of a new substrate and show that λ exonuclease acts at an internal site in DNA where there is a redundant single strand.

We were led to synthesize and test this substrate by a specific hypothesis that incorporated these observations: λ exonuclease acts at the site of a redundant single strand (structure III, Fig. 1), digesting a strand in the duplex and making way for the single strand to be assimilated into the duplex (step *c*, Fig. 1). When all the redundant strands were assimilated, the enzyme would find itself at a site indistinguishable from a nick where it would stop.

The proposed mechanism, called single strand assimilation, provides a precise and self-controlled way of trimming the products of a cruder kind of joining, and might be of more general significance in genetic recombination. Our observations also provide a more rational basis for the *in vitro* simulation of recombination.

We refer to a molecule such as structure III in Fig. 1 as a "redundant joint molecule". It contains two redundant joints. Structures that contain a single redundant joint are shown in Fig. 2, *b* and *c*. This terminology is an extension of that used by Tomizawa[6] to describe recombinant molecules of DNA in which the two parental moieties are held together only by hydrogen bonds. Accordingly, joint molecules of DNA from which nucleotides are missing on either side of the hydrogen-bonded biparental region might be called incomplete joint molecules.

Substrate with a Redundant Joint

A substrate was constructed from λ DNA in which: (1) a 3′ terminal fragment of one of the two strands was labelled with ^{32}P (Fig. 2*a*), (2) a redundant joint was located at the 5′ end of the ^{32}P-labelled fragment (Fig. 2*b*), and (3) the normal homologous ends were protected by annealing to form hydrogen-bonded joints as in a Hershey circle (Fig. 2*c*). (Such joints are resistant to λ exonuclease[4].) The two strands of a tritiated preparation of λ DNA were isolated by complexing denatured DNA with poly UG according to Hradecna and Szybalski[7]. The left hand fragment of the *r* strand was isolated in the same way from a preparation of ^{32}P-DNA broken into halves by hydrodynamic shear (see legend to Fig. 3*a*). This fragment was annealed with the ^{3}H-*l* strand to produce the intermediate structure shown in Fig. 2*a*. Annealing of the intact tritiated *r* strand to the intermediate structure (Fig. 2*a* and *b*), followed by annealing of the complementary ends of the λ DNA, formed a modified Hershey circle with a single redundant joint (Fig. 2*c*). When used as substrate for λ exonuclease, this material will be referred to as the preparation of redundant joints.

Our conclusions about the nature of the substrate which we prepared are based on the following characterization of intermediates and product. (a) ^{32}P-fragment: the isolated left hand fragment originated from the r strand, for it annealed with the l strand in the final substrate (Fig. 3c and 4a), but not with the r strand (Fig. 3b). It was shown to be approximately half as long as an intact strand of λ DNA by centrifugation in alkaline sucrose (Fig. 3a)[8]. (b) Intermediate structure (Fig. 2a): we have already found that λ exonuclease is unable to re-initiate digestion of a molecule of DNA that has been partially degraded from the 5' ends by prior digestion with λ exonuclease (unpublished observations of D. M. Carter and C. M. R.). The ^{32}P-DNA in an intermediate which has the structure shown in Fig. 2 should therefore be resistant to λ exonuclease. An amount of exonuclease sufficient to digest 21% of intact DNA from bacteriophage P22 released no ^{32}P counts from the intermediate. (At this concentration of λ exonuclease, the other single stranded 5' terminus of the intermediate or of intact λ DNA was also resistant.) Fifty times as much λ exonuclease hydrolysed most (37%) of the susceptible part of the ^{3}H strand, but only 15% of the ^{32}P-labelled material. According to the properties of λ exonuclease, the relative release of ^{3}H and ^{32}P would have been reversed had the ^{32}P come from the right hand end of the r strand. As will be seen, these results add to the significance of the experiments on redundant joints. (c) Substrate with a redundant joint (Fig. 2c): after the addition of the r strand and the final annealing, a structure was produced which sedimented more rapidly than linear λ DNA (Fig. 3c) and contained a constant ratio of ^{3}H-labelled and ^{32}P-labelled DNA (Fig. 4a). The observed ratio 3.7 : 1 agrees with the expected ratio of approximately 4 : 1 (Fig. 2a and b). Like authentic Hershey circles (unpublished observations of D. M. Carter and C. M. R.), the product continued to sediment in a discrete peak after prolonged treatment with a large excess of λ exonuclease (Fig. 4). Conditions of heating and quenching which caused Hershey circles to open also caused the product to sediment more slowly than a circular marker (Fig. 3d). These properties support the conclusion that the product is a hydrogen bonded circle, like the Hershey circle. Before the annealing treatment, the separated l and r strands were shown to be intact by sedimentation in alkaline sucrose, so that any hydrogen bonded circle containing these strands must have a redundant joint. Furthermore, exonuclease I of *E. coli* degraded about 20% of the ^{3}H-DNA, but none of the ^{32}P (Table 2).

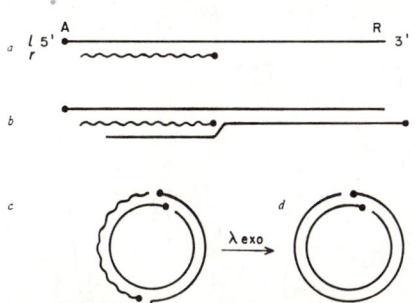

Fig. 2 Diagram for the preparation of a Hershey circle with a single stranded branch proximal to the 5' end (●—) of a specifically labelled ^{32}P-fragment (a redundant joint, b and c). The interpretation of the action of λ exonuclease is shown in d. ———, ^{3}H; ∼∼∼∼, ^{32}P.

Table 1 Action of λ Exonuclease on Redundant Joints

| | Units of exonuclease | Incubation time | DNA made acid soluble | | | |
| | | | ^{3}H | | ^{32}P | |
			nmol	%	nmol	%
A	1	30'	0		0.022	7
	2	1.7	30'	0.03	1.8	0.073	23
B	1	0	30'	0		0	
	2	3.3	30'	0.056	3.5	0.12	39
	3	3.3	60'	0.11	7	0.17	56
	4	2.5	120'	0.19	8.1	0.44	95

Each incubation mixture contained, in a total volume of 0.15 ml.: 1.9 to 2.8 nmol of substrate, 10 μmol of glycine buffer (pH 9.6), 0.3 μmol of MgCl$_2$, and 1.5 nmol of sRNA. Incubation was at 37° C. Acid soluble radioactivity was measured by scintillation counting. One unit of λ exonuclease digests 10 nmol of DNA in the conditions of the standard assay[19].

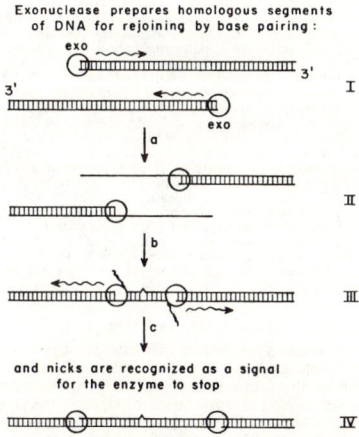

Fig. 1 A model for the action of λ exonuclease in general genetic recombination. (a) λ exonuclease prepares homologous segments of DNA for base pairing; (b) a redundant joint molecule (III) is formed by pairing; and (c) exonuclease continues to cleave nucleotides from the duplex, making way for the assimilation of the redundant strand behind nicks. Complete assimilation is associated with a specific mechanism that stops the enzyme at a site that is the same as a nick. The caret in structures III and IV represents a mismatched base pair in the heterozygous region. This model is a particular variant of a general model described and analysed by Thomas *et al.*[21].

λ Exonuclease and Redundant Joints

When the preparation of redundant joints was treated with λ exonuclease*, the ^{32}P was specifically degraded to acid-soluble products, while the physical integrity of the circular structure was maintained (Fig. 4 and Table 1). An amount of enzyme that released 7% of the ^{32}P from the preparation of redundant joints (Table 1, line 1) released no ^{32}P from the intermediate structure already described. With more enzyme and longer incubation, all the ^{32}P was made acid-soluble (Fig. 4 and Table 1). Only a small fraction of the ^{3}H was released. Because, as we have described, the ^{3}H-labelled circles did not break, we suppose that the released ^{3}H did not

* The λ exonuclease used in these experiments was complexed with β protein[9]. Gel electrophoresis of this preparation of enzyme yielded one band corresponding to λ exonuclease and another band corresponding to β protein; other bands were either faint or absent.

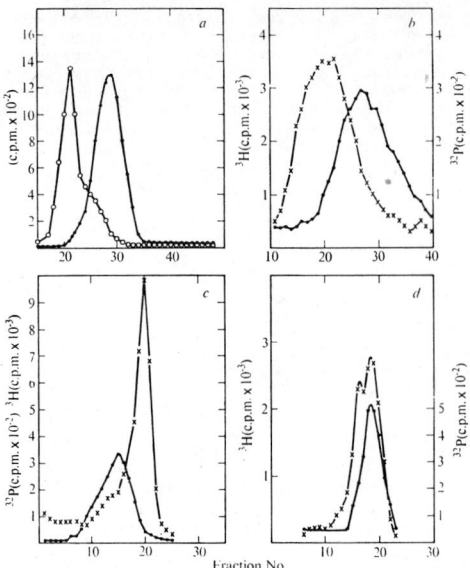

Fig. 3 Characterization of redundant joint substrate and intermediates. All centrifugations were carried out in a Spinco SW 50.1 rotor at 50,000 r.p.m. and 4° C for 2 h. Forty-six to forty-eight fractions were collected from preformed gradients of 5–20% sucrose. *a*, Sedimentation in alkaline sucrose (0.1 M NaCl, 0.05 M K$_2$HPO$_4$, 0.05 M EDTA, pH 12) of intact λ DNA (○) and of the left hand fragment of the ^{32}P-r strand (●). This fragment was isolated by shearing of ^{32}P–λ DNA[18] (25 μg/ml. in 0.01 M MgSO$_4$, 0.015 M NaCl, 0.01 M Tris, pH 6.7) 30 min at 1,500 r.p.m. and 0° C, followed by centrifugation in CsCl in the presence of poly UG[7] (23 h at 42,000 r.p.m. in a Spinco No. 50 fixed angle rotor). *b*, Characterization of the left hand fragment of the *r* strand (^{32}P): sedimentation in neutral sucrose (0.01 M Tris, pH 7.5, 0.001 M EDTA) of the fragment heated with the intact *r* strand (^{3}H) in annealing conditions (15 h at 65° C in 4 SSC). ●, ^{32}P; ×, ^{3}H. *c*, Sedimentation of the preparation of redundant joints in neutral sucrose. The slow moving peak corresponds to linear ^{3}H-λ DNA used as a marker (see Fig. 4*a* where no marker was present). The shoulder moving ahead of this peak is due to ^{3}H associated with the ^{32}P-*r* strand fragment (see Fig. 4*a*). Symbols are as in *b*. *d*, Sedimentation in neutral sucrose of the final product heated 2 min at 75° C and quenched (Fig. 2*b*). The fast moving peak corresponds to authentic Hershey circles used as a marker. Symbols are as in *b*.

come from our constructed substrate, but from some contaminating material such as that which sedimented more rapidly than the major peak of DNA (Fig. 4*a*).

Substrate, Products and Exonuclease I

The model predicts that the redundant single strand becomes hydrogen bonded into the duplex structure of DNA as a result of the action of λ exonuclease which excises nucleotides ahead of the branch point (Fig. 1). Thus, the removal of ^{32}P from the redundant joint by λ exonuclease should be accompanied by a corresponding reduction in the amount of single stranded ^{3}H-DNA (3' OH terminated) that is susceptible to digestion by exonuclease I. Fulfilment of this prediction (Table 2) strongly supports the model. Before treatment with λ exonuclease, exonuclease I released 20% of the ^{3}H; the predicted amount is 25%. After treatment with λ exonuclease, which digested most of the ^{32}P and 4.5% of the ^{3}H (Table 2, line 2), exonuclease I digested only a further 2% of the ^{3}H (Table 2, line 3).

Arrest of Exonuclease

In conditions that removed most of the ^{32}P from the preparation of redundant joints but left most of the ^{3}H intact (Table 1 and Fig. 4*c*), linear λ DNA was 31% degraded. Even longer incubation (Fig. 4*d*) removed all of the ^{32}P but left ^{3}H-DNA intact. After degrading all the ^{32}P-labelled *r* fragment, λ exonuclease was unable to continue. The removal of twelve nucleotides from the 5' end of the *l* strand would have broken the Hershey circle and would have led to extensive digestion of the ^{3}H-DNA (Fig. 2*c*). Unless most of the ^{32}P-*r* fragments had lost nucleotides from their 3' ends, we can conclude that exonuclease stopped at a nick.

Exonuclease in Genetic Recombination

The action of λ exonuclease on redundant joints is a new exonucleolytic mechanism, the biological significance of which is suggested by the participation of this enzyme in genetic recombination[1–3] and by the plausibility of redundant joint molecules as intermediates in recombination. Fig. 1 illustrates how exonuclease could play a role in forming joint molecules, and unifies present knowledge of what the enzyme can do in a simple model for its role in recombination.

Table 2 Susceptibility of the Preparation of Redundant Joints to Exonuclease I Before and After Treatment with λ Exonuclease

	Treatment	Incubation time	DNA made acid soluble			
			^{3}H		^{32}P	
			nmol	%	nmol	%
1	exo I 2.9 u	60'	0.94	20	0	
2	λ exo 5.0 u	60'	0.21	4.5	0.76	79
3	exo I 2.9 u after λ exo (line 2)	60'	0.31	6.5	0.69	72

Incubation mixture No. 1 contained, in 0.15 ml: 5.7 nmol of redundant joint substrate, 10 μmol of glycine buffer (pH 9.6), 1 μmol of MgCl$_2$, 0.2 μmol of 2 mercaptoethanol, and 2.9 units of exonuclease I. Incubation mixture No. 2 contained, in 0.3 ml: 11.4 nmol of substrate, 20 μmol of glycine buffer (pH 9.6), 1.5 μmol of MgCl$_2$, 3 nmol of sRNA, and 5 U of λ exonuclease. After incubation for 60 min, an aliquot of 0.15 ml. was withdrawn and acid soluble radioactivity was measured. The remaining 0.15 ml. (incubation mixture No. 3) was subjected to a temperature of 75° C for 10 min to inactive λ exonuclease, and 0.7 μmol of MgCl$_2$, 0.2 μmol of 2 mercaptoethanol, and 2.89 U of exonuclease I were added. All incubations were at 37° C. One unit of exonuclease I produces 10 nmol of acid soluble nucleotides in 30 min at 37° C[20]. Control experiments showed that the presence of λ exonuclease did not inhibit the action of exonuclease I on denatured DNA, and that denatured DNA heated in the presence of λ exonuclease was unchanged in its susceptibility to exonuclease I. In the conditions of these experiments denatured DNA was less than 3% digested by λ exonuclease in 60 min. (Exonuclease I was given by Dr R. S. Samaha.)

By constructing a substrate that contains a redundant joint, we were able to observe that as λ exonuclease excised one strand from the double helix there was a corresponding reduction in the amount of single stranded DNA in the preparation. After removing all of the ^{32}P-labelled strand from the duplex, the enzyme stopped as predicted at a point which was probably a nick. Our hypothesis, however, proposes that the enzyme stops at a nick created by its own action (Fig. 1); the enzyme allegedly displaces a single interruption in the DNA by converting a redundant joint into a nick. In the substrate that we made, a nick existed at the 3' end of the ^{32}P-labelled fragment and the enzyme might have been stopped by a different, though similar, mechanism. To test our idea about the mechanism of arrest we need to show that the length of the redundant strand determines the extent of exonuclease action on the duplex strand and that polynucleotide ligase can seal the joint after λ exonuclease has acted.

In other experiments, we have found that λ exonuclease keeps a tenacious hold on any molecule of DNA on which it starts; the enzyme moves down the molecule releasing about 700 mononucleotides per min at 37° C and pH 7·5 (unpublished observations of D. M. Carter and C. M. R.). An arrest mechanism, like that we have proposed, would preserve the integrity of the recombining molecules of DNA. Similarly, the resistance of DNA at a gap (D. M. Carter and C. M. R., unpublished) would prevent exonuclease from acting in conditions in which it might serve only a degradative function.

Information about the substrate containing redundant joints comes from previous knowledge about the structure of λ DNA[7,10] and from physical and enzymatic characterization at each step in the synthesis. The diagram in Fig. 2c represents the probable structure of the product. But one might wonder about the exact structure of the 5' end of the ^{32}P-labelled fragment. It seems likely that rotation of the helix partially unwinds the 5' end of the ^{32}P-fragment and winds a corresponding length of the redundant strand in its place. A displacement, to and fro[11], of this joint might contribute to the relative slowness with which λ exonuclease removes the ^{32}P-labelled fragment. If we assume that all of the potential sites were saturated by λ exonuclease, ^{32}P-labelled nucleotides were excised from redundant joints at a rate estimated to be 0.1 to 0.2 of the rate of release from a double stranded end of DNA. On the basis of a few observations, the rate of digestion of redundant joints appears to be at least ten times greater than the rate at which ^{32}P-nucleotides were released from the intermediate structure (Fig. 2a) before the redundant joint was made.

The role of the redundant single strand in the enzymatic mechanism is of particular interest. We have stressed the possible effect of the single strand on the mechanism of arrest. In addition, the length of this strand might influence the rate of initiation of enzyme action. Recently we have observed that denatured DNA, which is a poor substrate, binds λ exonuclease (unpublished observations of J. Rosenzweig, E. C. and C. M. R.). This binding could either inhibit or facilitate the entry of the enzyme to the substrate site on the duplex strand.

The mechanism of exonuclease action indicated by these experiments is formally similar to the action of *E. coli* DNA polymerase in which the enzyme excises nucleotides ahead of itself and synthesizes a new complementary strand behind itself[12]. The details of the polymerase and λ exonuclease mechanisms, however, are quite different.

We still lack positive information about the function of β protein in genetic recombination (unpublished observations of C. M. R., J. Rosenzweig, E. C. and F. F. Richards). All the experiments described here were done with exonuclease complexed with β protein (the αβ complex)[9]. One experiment with the uncomplexed enzyme showed no difference in the release of ^{32}P from the preparation of redundant joints.

As a speculative extension of our observations, we might suggest that exonucleolytic action at a redundant joint would be ideally suited for what happens in bacterial transformation. A single strand is extracted from a fragment of duplex donor of DNA and inserted into the recipient duplex in place of a similar sequence[13]. Once an initial complex of donor and recipient were formed in which there is a nick in the recipient strand, an exonuclease might, by a similar mechanism to that described here, make way in the recipient duplex, lay down a donor strand behind itself, and automatically unwind the donor helix.

Rec B and C mutants of *E. coli* lack an ATP-dependent nuclease activity[14-16]. Low has shown that the recombination defect of *rec B* and *C* mutants can be corrected by the λ *red* system, suggesting that the *rec* and *red* systems may operate in a similar way (personal communication from B. Low). The nuclease activity associated with the *rec* system might operate

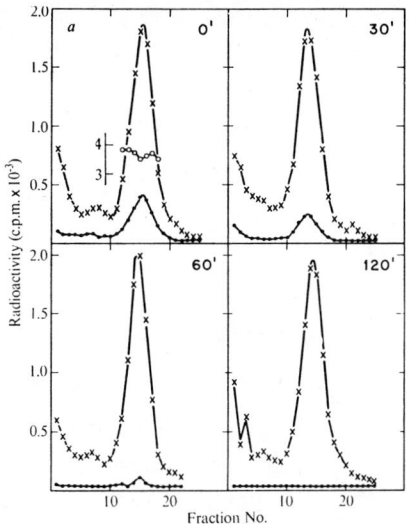

Fig. 4 Action of λ exonuclease on the preparation of redundant joints (Fig. 2c): release of the ^{32}P. The composition and treatment of the reaction mixtures are shown in Table 1B. Centrifugation in neutral sucrose was as described in the legend to Fig. 3. ●, ^{32}P; ×, ^{3}H. The inset (○) in *a* represents the ratio of DNA labelled with ^{3}H to DNA labelled with ^{32}P.

in a way similar to λ exonuclease. On the other hand, the exonuclease model is probably not general for non-reciprocal recombination. The intermediates in T4 recombination are incomplete joint molecules from which about 300 nucleotides are missing[17]; the origin and repair of such intermediates seem to be quite different.

The action of λ exonuclease on redundant joints and the scheme shown in Fig. 1 point the way for simulating genetic recombination *in vitro* and exploring other steps in the overall process.

This work was supported by a US Public Health Service grant. We thank Dr Paul Howard-Flanders and Dr Brooks Low for discussion and criticism.

Received August 3; revised November 10, 1970.

[1] Signer, E. R., Echols, H., Weil, J., Radding, C. M., Shulman, M., Moore, L., and Manly, K., *Cold Spring Harbor Symp. Quant. Biol.*, **33**, 711 (1968).
[2] Shulman, M. J., Hallick, L. M., Echols, H., and Signer, E. R., *J. Mol. Biol.*, **52**, 501 (1970).
[3] Radding, C. M., *J. Mol. Biol.*, **52**, 491 (1970).
[4] Carter, D. M., and Radding, C. M., *Fed. Proc.*, **29**, 405 (1970).
[5] Little, J. W., *J. Biol. Chem.*, **242**, 679 (1967).
[6] Tomizawa, J., *J. Cell. Physiol.*, **70**, Suppl. 1, 201 (1967).
[7] Hradecna, Z., and Szybalski, W., *Virology*, **32**, 633 (1967).
[8] Abelson, J., and Thomas, C. A., *J. Mol. Biol.*, **18**, 262 (1966).
[9] Radding, C. M., and Shreffler, D. C., *J. Mol. Biol.*, **18**, 251 (1966).
[10] Wu, R., and Kaiser, A. D., *J. Mol. Biol.*, **35**, 523 (1968).
[11] Lee, C. S., Davis, R. W., and Davidson, N., *J. Mol. Biol.*, **48**, 1 (1970).
[12] Kelly, R. B., Atkinson, M. R., Huberman, J. A., and Kornberg, A., *Nature*, **224**, 495 (1969).
[13] Gurney, jun., T., and Fox, M. S., *J. Mol. Biol.*, **32**, 83 (1968).
[14] Barbour, S. D., and Clark, A. J., *Proc. US Nat. Acad. Sci.*, **65**, 955 (1970).
[15] Wright, M., and Buttin, G., *Bull. Soc. Chim. Biol.*, **51**, 1373 (1969).
[16] Oishi, M., *Proc. US Nat. Acad. Sci.*, **64**, 1292 (1969).
[17] Anraku, N., Anraku, Y., and Lehman, I. R., *J. Mol. Biol.*, **46**, 481 (1969).
[18] Hogness, D. S., and Simmons, J. R., *J. Mol. Biol.*, **9**, 411 (1964).
[19] Radding, C. M., *J. Mol. Biol.*, **18**, 251 (1966).
[20] Lehman, I. R., *J. Biol. Chem.*, **235**, 1479 (1960).
[21] Watson, G. S., Smith, W. K., and Thomas C. A., jun., *Progr. Nucleic Acid Res. and Mol Biol.*, **5**, 338 (1966).

On the Mechanism of Integration of Transforming Deoxyribonucleate

MAURICE S. FOX

From the Department of Biology, Massachusetts Institute of Technology, Cambridge

ABSTRACT The characteristics of the intermediates in the reaction, between DNA and pneumococcus, that results in genetic transformation are described in so far as they have been characterized. Transformation with DNA isolated from bacteria carrying in addition to genetic markers ^{32}P as a radioactive label and ^{2}H and ^{15}N as density labels has permitted the characterization of the product of recombination between the newly introduced DNA and the DNA of a recipient bacterium. The evidence for a single strand displacement mechanism producing a hybrid structure in the DNA of the recipient bacteria is presented. Progeny of single transformants have been examined. The results of these segregation studies permit the further characterization of this hybrid product of transformation as a genetically heterozygous structure.

The process whereby bacterial transforming deoxyribonucleate (DNA) is utilized by transformable bacteria to effect a genetic transformation has been examined in three organisms, pneumococcus (1), *B. subtilis* (2, 3), and *H. influenzae* (4, 5). For each organism there is now evidence that transforming DNA is physically integrated into the genomes of the recipient bacteria, but it is certainly clear that the mechanisms differ, at least, in detail.

Since other participants in this symposium will be discussing the other two systems, I will limit my discussion to what is known about the integration process as it occurs in pneumococcus with only limited reference to other organisms.

Formally the over-all reaction between bacteria and transforming DNA can be described in terms of two steps. The first is a rapidly equilibrating reversible reaction, whose rate is probably diffusion-limited, to form a complex. The second is a slow irreversible process, whereby the DNA is fixed in the sense that it cannot be removed by exhaustive washing nor is it any longer sensitive to deoxyribonuclease (DNAase) added to the medium (6).

The stoichiometry of the over-all reaction has been determined (7, 8). There is a strict linear correspondence between the number of bacteria transformed with respect to a single marker and the quantity of DNA fixed by the treated competent population. With the most active DNA preparations, the number of bacterial equivalents of donor DNA fixed, per bacterium transformed, is less than two and may approach one for some single markers. Since the population fixes fragments of the transforming DNA at random (9), and only occasionally does the fragment fixed carry the marker under examination, the process appears to be remarkably efficient.

Little is known about the state of the DNA that is presumed to be reversibly bound in the form of the complex. It might be bound to the bacterial surface but this is by no means the only possibility. The high efficiency of the over-all process suggests an alternative explanation. Material in the form of the complex may include elements of DNA that have reversibly penetrated various permeability barriers and are therefore able to "search," by diffusion, for homologous regions of the recipient genome. "Finding" the homologous region, perhaps by complementary pairing, might initiate the irreversible reaction and thus account for the remarkable efficiency of the process. This hypothesis is further suggested by the long time constant (about 180 min) (1) for the fixation step and by the rapidity and certainty with which the succeeding steps occur (10). In addition, this proposal is consistent with the failure to observe, in recently transformed pneumococcus, any DNAase-resistant donor DNA retaining its original double-stranded integrity.

In contrast to what is observed in pneumococcus, recently transformed populations of *H. influenzae* have, in fact, been shown to contain intact double-stranded donor DNA which cannot be removed by washing or by treatment with DNAase. It is likely that the complex that has been described includes a sequence of intermediate steps. All these intermediates must be considered to be reversible in the case of pneumococcus. In the case of *H. influenzae*, however, the passage of DNA into one of these intermediate states might be irreversible. On further incubation the donor DNA that has been fixed, intact, might then pass slowly through the various further steps that are involved in integration. Such a proposal could perhaps resolve the differences between the observations describing the integration processes in these two organisms (1, 4, 5, 11).

We return to the case of pneumococcus, and the DNA that has been irreversibly fixed. Total DNA, extracted immediately following a short exposure of bacteria to transforming DNA, manifests very little of the transforming activity of the newly fixed DNA present. During further incubation at 37°C the transforming activity, as measured in extracts, of the newly fixed DNA is recovered with a half-time of 3 min (10). Following this recovery and during subsequent growth of the transformed population the extracted DNA contains a constant ratio of donor to recipient transforming activities. It can be therefore concluded that following escape from eclipse the newly introduced transforming DNA multiplies in synchrony with the DNA of the recipient bacteria (12).

Certain pairs of markers are said to be genetically linked since they are frequently carried on single fragments of transforming DNA (9). Linkage provides a biological criterion for examing the establishment of molecular associations that occur as a consequence of the recombination events in transformed bacteria. DNA extracted from transformed bacterial populations can exhibit activity characteristic of newly created linkage pairs, one member of which has been provided by the donor DNA and the other by the DNA recipient bacterium.

The formation of this new association and hence the recombination process itself occurs with a half-time of about 6 min at 37°C. This reaction has been shown to be distinct from that involved in the recovery from eclipse (10). Both these reactions can be completed at their maximum rates even under conditions wherein there has been a less than 5% and probably less than 1% net increase in DNA in the transformed bacterial population (13).

Incorporation into transforming DNA of radioactive phosphorous (^{32}P) at high specific activities leads to destruction of the transforming activity as a primary consequence of the ^{32}P harbored in individual molecules (14). This observation has been used to examine the degree to which the integrity of the newly introduced DNA is retained through the process of fixation, recovery from eclipse, and establishment of linkage (15). Disintegration of ^{32}P destroys the biological activity with respect to a single marker present in the heavily labeled transforming DNA at a rate consistent with a minimum sensitive target of about 600,000 daltons (14).

Heavily labeled DNA that was about to lose transforming activity as a consequence of ^{32}P disintegration was used to transform cold bacteria. DNA was isolated from the treated bacteria after allowing sufficient time for the newly introduced DNA to escape from eclipse and nearly complete the establishment of linkage. The transforming activity of a marker present in the donor DNA was determined after allowing various amounts of ^{32}P disintegration to occur. The biological activity of the newly introduced marker decayed at the same rate as did the same marker in the original transforming DNA. It follows therefore, that the elements of DNA manifesting a newly introduced biological activity, have the same arrangement of ^{32}P atoms and must have retained their integrity over a molecular region of at least 600,000 daltons.

Given the demonstration that elements of transforming DNA are themselves genetically integrated into the genomes of the recipient bacteria, it now becomes possible to inquire into the physical structure of the integrated product. This inquiry has been accomplished by the use of isotopic density labels coupled with density gradient centrifugation. The technique permits the separation and identification of different species of DNA molecules by virtue of their different densities.

The use of such methods has made it possible to demonstrate that the product of the reaction between transforming DNA and the DNA of the recipient bacterium is physically a hybrid. The hybrid, composed of one strand of donor and one strand of host DNA, extends over a region of about 2 or 3 thousand nucleotide pairs. Furthermore, the terminals of the newly intro-

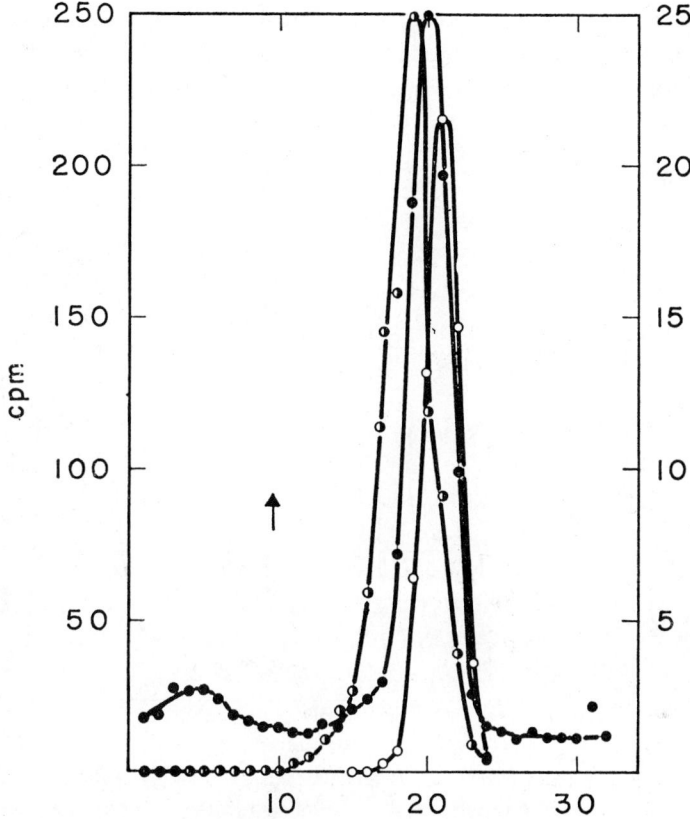

FIGURE 1. Equilibrium density gradient of DNA isolated from light cold bacteria that had been treated with triply labeled transforming DNA for 15 min at 30°C, followed by 3 min at 37°C. The arrow indicates the gradient position to be expected for heavy DNA. The distribution shows the radioactivity (●) and transforming activity (◐) of the donor DNA and the transforming activity of the recipient DNA (o). The scale on the right is in transformants per ml $\times 5 \times 10^{-3}$ for the donor marker and $\times 10^{-5}$ for the recipient marker.

duced polynucleotide chain are covalently bound to elements of the host DNA.

This conclusion is reached by an examination of the DNA extracted from unlabeled bacteria that had been transformed with triply labeled DNA carrying the density labels ^{2}H and ^{15}N and the radioactive label, ^{32}P. DNA was isolated from recipient bacteria after a 15 min exposure to the transforming

DNA, at 30°C and 3 min at 37° to allow escape from eclipse. This DNA was allowed to come to equilibrium in a CsCl density gradient centrifuged at 140,000 × g. The distribution of the various components is described in Fig. 1.

All the biological activity of the newly introduced DNA and the bulk of its ^{32}P are found at a density position very near that of the host DNA. There is none to be found at the position characteristic of the donor DNA. The covalent linkage between the donor DNA and the recipient DNA is demon-

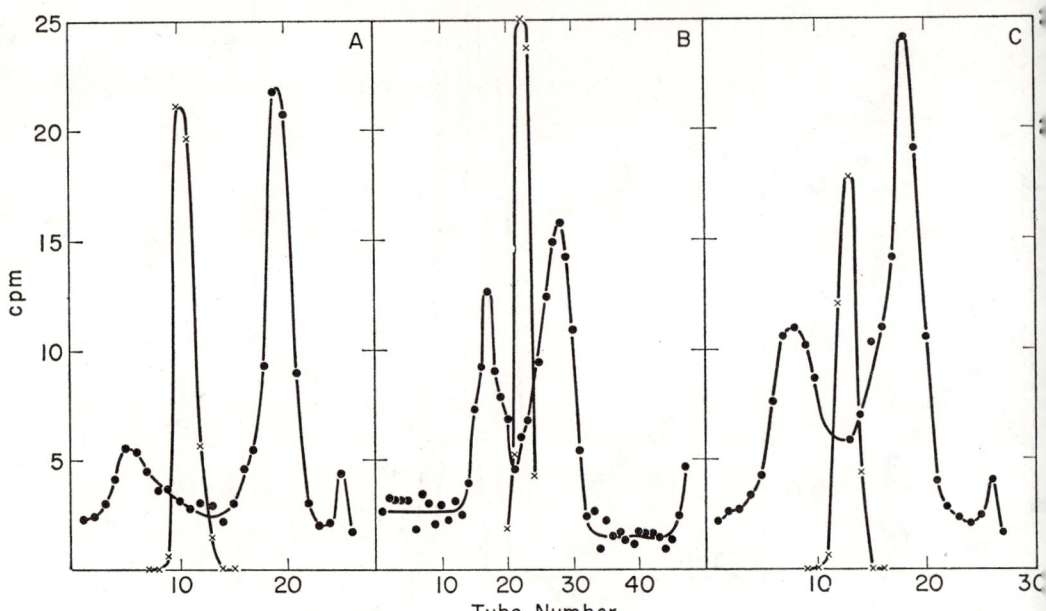

FIGURE 2. The extract from transformed bacteria showing the distribution of radioactivity (●) present in the donor DNA. The extract was (A) untreated, (B) heat-denatured, and (C) alkali-denatured. The biological activity of added heavy DNA (X) constitutes a position marker.

strated by the persistence of their association following either thermal or alkali denaturation, as indicated in Fig. 2.

The hybrid structure of the intermediate was demonstrated by shearing the DNA extracts prior to CsCl centrifugation. Shearing by sonication was used to reduce the molecular weight of the DNA in the extract from 20 million to about 1 to 2 million. The effect of this shearing is described in Fig. 3A. The reduction in molecular weight results in shifting of the buoyant density of the DNA carrying the newly introduced marker from a position near that of the light host DNA to a position nearly halfway between the buoyant densities of the heavy donor and light recipient DNA's. A substantial fraction of radioactivity, perhaps half, also assumes the hybrid density.

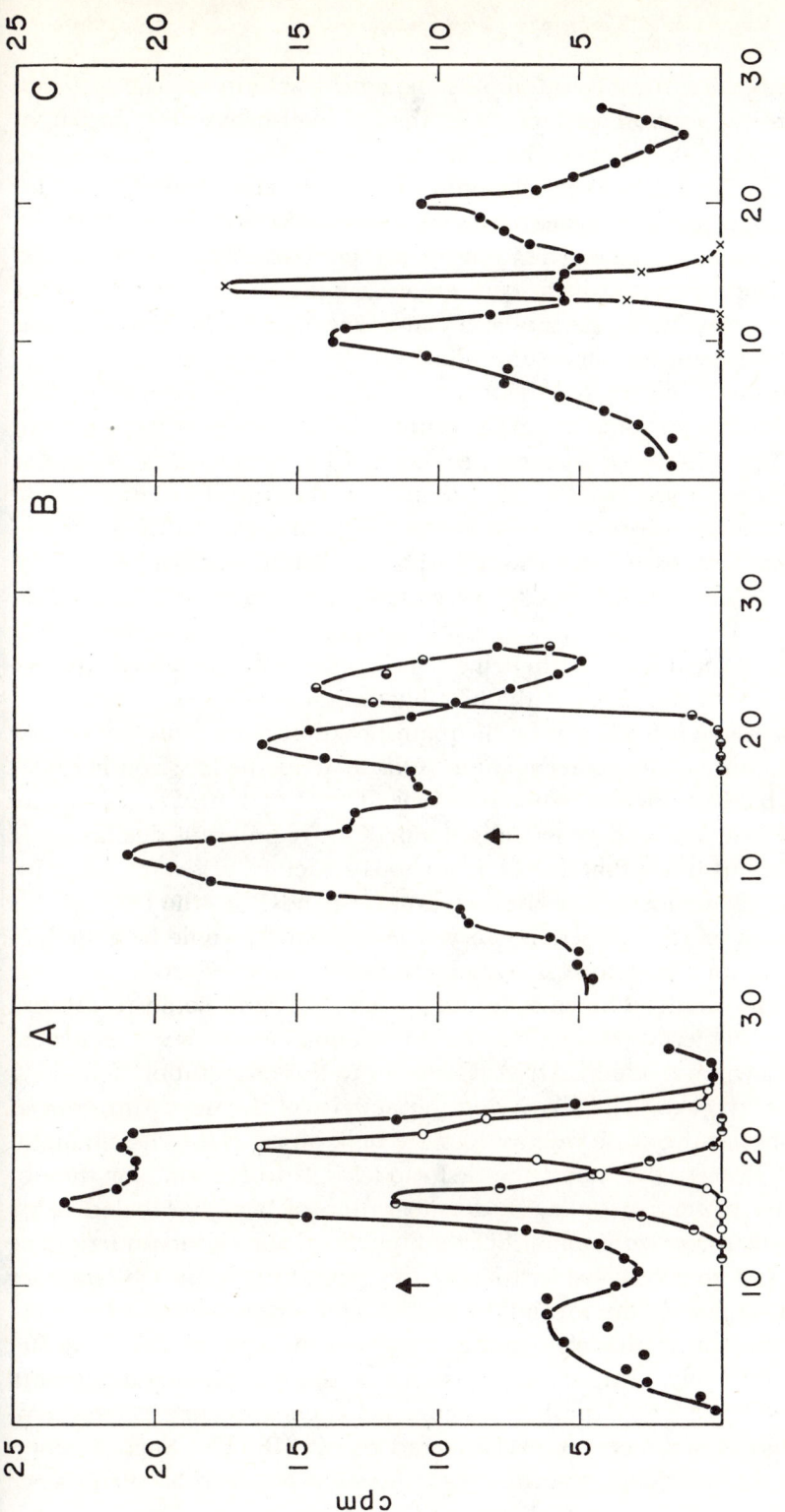

FIGURE 3. Density gradient distributions of an extract from transformed bacteria that had been sonicated (*A*), then (*B*) heat-denatured, or (*C*) alkali-denatured. The arrows indicate the position that native heavy DNA would be expected to assume. The distributions show the radioactivity (●) and biological activity (◐) of the donor DNA and the biological activity of the recipient DNA (○), as well as the biological activity of a light position marker (◑) in (*B*) and a heavy position marker (*X*) in (*C*). The scale on the right is in transformants per ml × 1.25 for the donor marker, × 10^{-3} for the resident marker, × 2 × 10^{-2} for the heavy position marker, and × 5 × 10^{-5} for the light position marker.

Further shearing fails to move either the biological activity or the radioactivity to a density position greater than that of half-heavy and half-light DNA.

In contrast to what is observed as a consequence of denaturation of the intact extract, denaturation of a sheared DNA isolate, described in Fig. 3B and C, results in the bulk of the ^{32}P banding in the position of heavy denatured DNA as if most of it had lost its association with light DNA. Thus the DNA which had become hybrid in density as a consequence of shearing was hybrid because it was heavy in one strand and light in the other.

The first intermediate carrying the transforming activity of the donor DNA can now be described in some detail. A segment of the DNA of the recipient bacterium has been displaced by a single strand of the donor DNA. A similar structure has been described as the product of the transformation of *B. subtilis* (2, 3). The demonstration of this intermediate confirms the displacement mechanism proposed by Lacks (16) and by Guild and Robison (17). In pneumococcus, this region is 2 to 3 thousand nucleotides in length, substantially shorter than the usual length of molecules of transforming DNA. Furthermore, the terminals of this single polynucleotide sequence are covalently coupled with the DNA of the host bacterium.

Four possible naive models can be distinguished that permit further definition of the structure of this intermediate. One pair of models assumes that the physical hybrid is genetically heterozygous: The light strand is genetically "host" and the heavy strand genetically donor. One member of this first pair of models further assumes that the heterozygous structure arises from the introduction of a particular one of the two donor strands. On the basis of this model, the product of the first semiconservative replication would be a double-stranded homozygous structure in which the newly synthesized strand was inactive in transformation. Since only the particular strand could transform, the first round of bacterial replication would yield no increase in the biological activity of the newly introduced DNA. Contrary to this expectation, following recovery from eclipse, the DNA carrying the activity of the newly introduced marker multiplies at the same rate as does the bulk host DNA, with no lag in its replication (12). Furthermore, the first model fails to account for the observed reduction of sensitivity to ^{32}P disintegration of heavily labeled DNA reisolated from transformed populations that had been allowed to undergo one doubling. We will reserve consideration of the second model in which either strand can participate in the formation of the heterozygous structure.

An additional pair of models assumes that the physical hybrid is in fact genetically homozygous. Rapid "conversion" of the complementary strand of the recipient DNA could perhaps occur as the consequence of new synthesis or some kind of process of excision and repair (18, 19). Such a genetically homozygous physically hybrid structure might occur as a consequence of the introduction of either of the two strands of a fragment of donor DNA. This model holds that a heavily ^{32}P-labeled element of newly integrated DNA would rapidly create its "converted" complement. The complement would be biologically active, harbor no ^{32}P atoms, and therefore be much less sensitive or insensitive to subsequent ^{32}P disintegration. Contrary to this expectation, heavily labeled transforming DNA, reisolated from transformed bacteria manifests little, if any, loss in its sensitivity to subseqeunt ^{32}P disintegration (15).

The remaining possibility may be considered single strand displacement, with a particular one of the two complementary strands participating, followed by conversion. This model is to be contrasted with the more appealing model of single strand displacement, with either strand participating and resulting in the formation of a genetic heterozygote.

Recent studies in this laboratory by Guerrini (20) on the segregation of progeny of single transformed bacteria have ruled out simple models involving conversion. The conversion models predict that all the progeny of a single newly transformed bacterium be transformant. The heterozygote model predicts that a single bacterium transformed at a given locus will give rise to a mixed clone. Each clone would be expected to contain bacteria transformed at the indicated locus as well as bacteria that manifest the recipient genotype at that locus.

Guerrini used a pair of allelic markers concerned with sulfanilamide resistance (d, d+) that have been described by Hotchkiss and Evans (21). Either allelic form of the pair may be detected by a selective procedure. Mutant bacteria, carrying the genotype d, are sensitive to *p*-nitrobenzoic acid, but are able to grow in the presence of sulfanilamide, whereas wild type bacteria, d+, are sensitive to sulfanilamide and may be identified by their capacity to grow in the presence of *p*-nitrobenzoic acid.

Guerrini has distributed newly transformed populations in an agar growth medium containing a small amount of sulfanilamide which limits the growth of wild type bacteria to small colonies, while permitting normal growth for transformants. Total colonies of transformants were picked, resuspended, and grown in a nonselective medium. Each clone was tested for the presence of bacteria carrying the d marker and also for bacteria carrying the d+ marker.

The colony-forming units in a transformable population of pneumococcus are chains containing on the average about four bacteria. As a consequence the clones isolated in the manner described above would all be expected to be mixed and among about 200 clones that have been examined all are indeed mixed. To reduce the number of bacteria that participate in the formation of a colony, the newly transformed bacteria were heat-killed to the point where only about 5% of the colony formers survived. Assuming that thermal inactivation occurs at random, these colony-forming units would only rarely be expected to include more than one bacterium. Clones of transformed bacteria were isolated after the treatment and characterized with regard to their composition. Twenty-seven of the twenty-eight clones analyzed were mixed. In these mixed clones the bacteria carrying the recipient genotype occurred with frequencies between 0.1 and 10%.

To confirm this observation, freshly transformed populations were exposed to ultrasonic agitation, so as to allow only about 10^{-3} to 10^{-4} survivors, prior to plating on low levels of sulfanilamide. Here again random inactivation would be expected to yield colony formers containing more than one bacterium only rarely. Among the thirty-five colonies that have been examined, again all were mixed.

The following corollary experiment justifies the assumption that the clones that were examined indeed arose from single bacteria and confirms the heterozygote model. A transformed population was allowed to increase 40% in the number of bacteria as observed by counting in a Petroff-Hauser chamber. The culture was then sonicated and plated as above in the presence of a low concentration of sulfanilamide. Of 44 clones examined, 36 were still mixed. The remaining 8 clones contained only transformants. Of 78 clones isolated as above from a transformed population that had been allowed one doubling (Petroff-Hauser count) before sonication and plating, 71 clones proved to be pure transformant and only 7 were still mixed. The segregation of almost all the transformed bacteria after one doubling justifies the assumption that the colony-forming units after sonication are single bacteria. Even more interesting, this observation excludes most models involving heterozygosity arising as the consequence of multiple nuclei.

Guerrini's results mean that transformants are genetically heterozygous, and that segregation is practically complete by the first bacterial doubling. This observation excludes conversion models and, barring some rather unlikely assumptions, itself suggests single strand displacement as the mechanism of integration of transforming DNA.

We can summarize the major conclusions:
1. Recently integrated transforming DNA exists as a physical hybrid with its complementary segment in the recipient bacterium.
2. The fragment simultaneously carries the genetic information of both participating DNA's.
3. Neither conversion nor double strand events are involved in a significant fraction of transformation events.

A speculative cartoon describing the local structure of the donor and recipient DNA molecules is shown as structure I in Fig. 4. We know that the molecules of transforming DNA are substantially larger than the average element integrated, and that multiple exchanges can occur between a single donor molecule and the recipient genome (22, 23). These concepts are incorporated in structures II or III in Fig. 4. A single transformation event might have as its consequence any of the three possibilities.

The cartoons significantly avoid indicating strand-switching events in which first a length of one strand of donor DNA is incorporated and then a

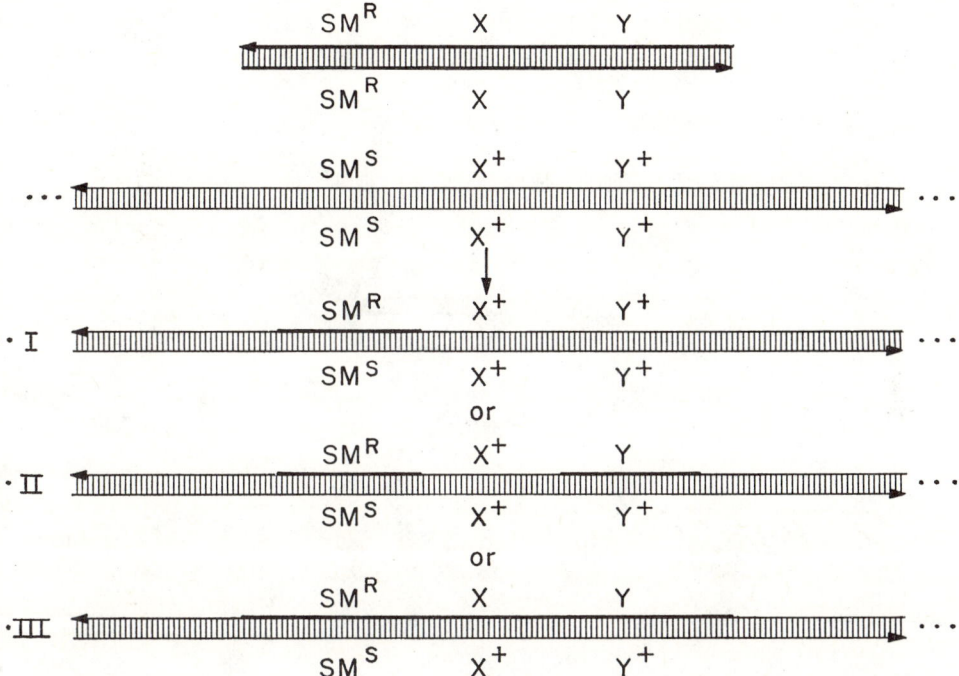

FIGURE 4. Cartoon describing several kinds of products of a single integration event. Light horizontal lines represent the strands of the host DNA and heavy horizontal lines represent the strands of donor DNA. As an example SM^R might represent the region of the DNA responsible for resistance to streptomycin and SM^S the homologous region in sensitive cells. The X^+ and Y^+ are meant to designate the recipient genotype and the X and Y other adjacent genetic elements present in the donor DNA and linked to SM^R.

length of the other. The reason for their omission is that such events occur only at the low frequency to be expected for random participation of two independent elements of transforming DNA (3).

The cartoons and the accompanying discussion are largely speculative. In this realm it seems worthwhile to discuss the possible mechanism of formation of the hybrid heterozygous structures. Some of the newly introduced DNA can, on reisolation, be characterized as denatured (16, 1, 3). Although this denatured DNA has been proposed as an intermediate in the integration process (16), it might equally well be a by-product of the reaction.

Still in the realm of cartoons, Fig. 5 illustrates possible intermediates in the formation of the recombinant structure. Since integration must involve breakage of the recipient genome and reunion with the donor DNA, a start can be made by breaking at position A. The process of trading strand partners would release one strand of the transforming DNA, as denatured material. A second break at position B would yield the structure illustrated as pathway I in Fig.

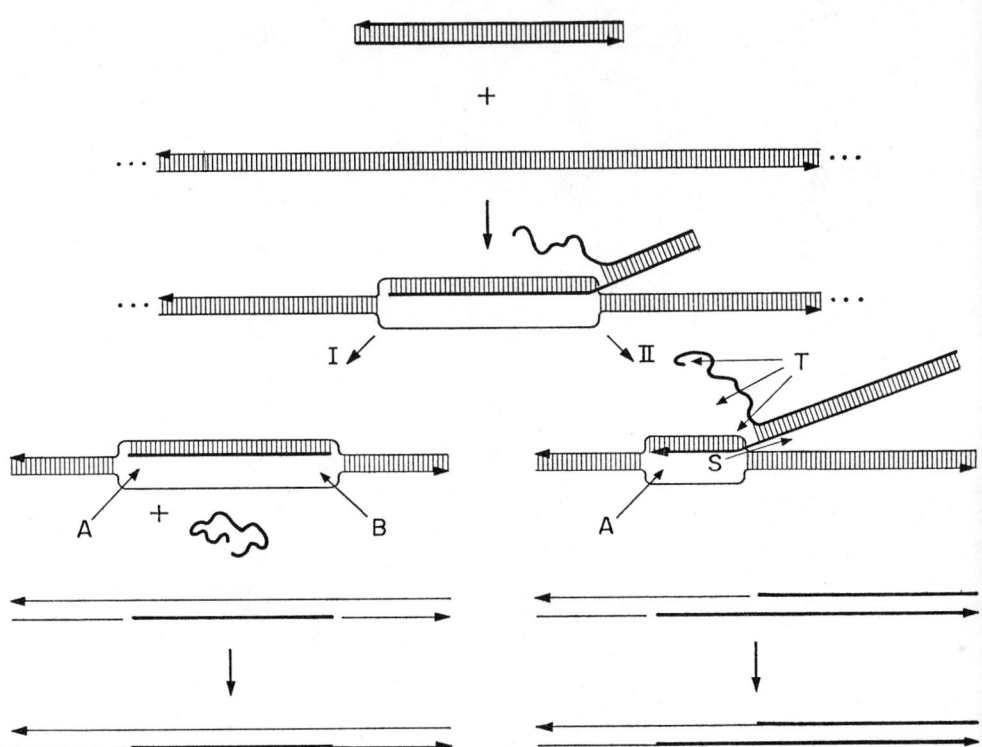

FIGURE 5. Cartoon describing a possible intermediate in the production of the fina integrated structure. Arrows are meant to designate points of breakage of polynucleotide chains. Light lines again represent strands of the host DNA and heavy lines, strands of the donor DNA.

5. Repair processes complete the reaction. Breaks of the kind illustrated in pathway II at A and S with hydrolysis of the single-stranded region at T are apparently excluded in pneumococcus. The product, however, is identical with that proposed to describe bacteriophage T2 heterozygotes (24) and bacteriophage λ recombinants (25).

The principal argument against the product of pathway I as a frequent occurrence in bacteriophage T2 or T4 recombination is the demonstration, by Levinthal (24), that most T2 heterozygotes are recombinant for outside markers. We now know that about one-third of the heterozygotes were probably of the terminally redundant type and would therefore be expected to be frequently recombinant (26). If the remaining heterozygotes were of the overlap variety a substantial fraction of them could be of the type described as the product of pathway I. This consideration could still yield the result that most T2 heterozygotes were recombinant for the outside markers.

Heterozygotes that are parental for outside markers have been reported in λ (27) as well as in T4 (28, 29), the latter occurring in substantial frequency. In both cases the heterozygotes have been accounted for by structures of the kind proposed as the products of pathway I (27, 29).

In three point crosses involving closely linked bacteriophage markers double cross-overs occur with an unexpectedly high frequency (high negative interference) (30, 31). Hotchkiss has suggested that this observation might be accounted for by assuming that a substantial fraction of recombination events were double events (32). The product of pathway I could be such a double event and in moderate frequency could account for high negative interference.

Both pathways may be considered possible as alternative products of a single recombination event in bacteriophage. Most recombinants for markers distant with respect to the average length of the heterozygous overlap, would therefore be expected to be products of pathway II. On the other hand, recombination events between closely linked markers might be produced frequently by either pathway.

It is obvious that the cartoon is incomplete, and that the completed product might be unrecognizable in these speculations. The fact is, a single strand of the donor transforming DNA displaces a segment of the DNA of the recipient bacterium to create a hybrid, heterozygous product.

The interesting questions that now arise are concerned with how such a product is created, and to what extent it might occur as a product of recombination in other organisms.

The author wishes to express his gratitude to Dr. H. Crespi and the Argonne National Laboratory for providing the ^{2}H- and ^{15}N-labeled substrates used to label transforming DNA.
This work was in part supported by Grant No. AI 05388 from the National Institutes of Health and Grant No. GB-644 from the National Science Foundation.

REFERENCES

1. Fox, M. S., and ALLEN, M. K., On the mechanism of deoxyribonucleate integration in pneumococcal transformation, *Proc. Nat. Acad. Sc.*, 1964, **52**, 412.
2. BODMER, W. F., and GANESAN, A. T., Biochemical and genetic studies of integration and recombination in *B. subtilis* transformation, *Genetics*, 1964, **50**, 717.
3. BUC, H., On the mechanism of DNA integration in *Bacillus subtilis* transformation, Abstracts, 9th Annual Meeting of the Biophysical Society, 1965, 47.
4. STUY, J. H., Fate of transforming DNA in the *Haemophilus influenzae* transformation system, *J. Mol. Biol.*, 1965, **13**, 554.

5. Notani, N. K., and Goodgal, S. H., Decrease in integration of transforming DNA of *Hemophilus influenzae* following ultraviolet irradiation, *J. Mol. Biol.*, 1965, **13**, 611.
6. Fox, M. S., and Hotchkiss, R. D., Initiation of bacterial transformation, *Nature*, 1957, **179**, 1322.
7. Fox, M. S., Deoxyribonucleic acid incorporated by transformed bacteria, *Biochim. et Biophysica Acta*, 1957, **26**, 83.
8. Lerman, L. S., and Tolmach, L. J., Cellular incorporation of DNA accompanying transformation in pneumococcus, *Biochim. et Biophysica Acta*, 1957, **26**, 68.
9. Hotchkiss, R. D., and Marmur, J., Double marker transformation as evidence of linked factors in desoxyribonucleate transforming agents, *Proc. Nat. Acad. Sc.*, 1954, **40**, 55.
10. Fox, M. S., Fate of transforming deoxyribonucleate following fixation by transformable bacteria. II, *Nature*, 1960, **187**, 1004.
11. Voll, M. J., and Goodgal, S. H., Loss of activity of transforming deoxyribonucleic acid after uptake by *Haemophilus influenzae*, *J. Bact.*, 1965, **90**, 873.
12. Fox, M. S., and Hotchkiss, R. D., Fate of transforming deoxyribonucleate following fixation by transformable bacteria. II, *Nature*, 1960, **187**, 1002.
13. Fox, M. S., unpublished observation.
14. Fox, M. S., Biological effects of the decay of incorporated radioactive phosphorus in transforming deoxyribonucleate, *J. Mol. Biol.*, 1963, **6**, 85.
15. Fox, M. S., The fate of transforming deoxyribonucleate following fixation by transformable bacteria. III, *Proc. Nat. Acad. Sc.*, 1962, **48**, 1043.
16. Lacks, S., Molecular fate of DNA in genetic transformation of pneumococcus, *J. Mol. Biol.*, 1962, **5**, 119.
17. Guild, W. F., and Robison, M., Evidence for message reading from a unique strand of pneumococcal DNA, *Proc. Nat. Acad. Sc.*, 1963, **50**, 106.
18. Setlow, R. B., and Carrier, W. L., The disappearance of thymine dimers from DNA: An error-correcting mechanism, *Proc. Nat. Acad. Sc.*, 1964, **51**, 226.
19. Boyce, R. P., and Howard-Flanders, P., Release of ultraviolet light–induced thymine dimers from DNA in *E. coli* K-12, *Proc. Nat. Acad. Sc.*, 1964, **51**, 293.
20. Guerrini, F., in preparation.
21. Hotchkiss, R. D., and Evans, A. H., Analysis of the complex sulfonamide resistance locus of pneumococcus, *Cold Spring Harbor Symp. Quant. Biol.*, 1958, **23**, 85.
22. Nester, E. W., Schafer, M., and Lederberg, J., Gene linkage in DNA transfer: A cluster of genes concerned with aromatic biosynthesis in *Bacillus subtilis*, *Genetics*, 1963, **48**, 529.
23. Kent, J. L., and Hotchkiss, R. D., Kinetic analysis of multiple, linked recombinations in pneumococcal transformation, *J. Mol. Biol.*, 1964, **9**, 308.
24. Levinthal, C., Recombination in phage T2: Its relationship to heterozygosis and growth, *Genetics*, 1954, **39**, 169.
25. Meselson, M., and Weigle, J. J., Chromosome breakage accompanying genetic recombination in bacteriophage, *Proc. Nat. Acad. Sc.*, 1961, **47**, 857.
26. Sechaud, J., Streisinger, G., Emrich, J., Newton, J., Lanford, H., Reinhold,

H., and STAHL, M. M., Chromosome structure in phage T4. II. Terminal redundancy and heterozygosis, *Proc. Nat. Acad. Sc.*, 1965, **54**, 1332.
27. KELLENBERGER, G., ZICHICHI, M. L., and EPSTEIN, H. T., Heterozygosis and recombination of bacteriophage, *Virology*, 1962, **17**, 44.
28. DOERMANN, A. H., and BOEHNER, L.. An experimental analysis of bacteriophage T4 heterozygotes. I. Mottled plaques from crosses involving six r11 loci, *Virology*, 1963, **21**, 551.
29. BERGER, H., Genetic analysis of T4D phage heterozygotes produced in the presence of 5-fluorodeoxyuridine, *Genetics*, 1965, **52**, 729.
30. CHASE, M., and DOERMANN, A. H., High negative interference over short segments of the genetic structure of bacteriophage T4, *Genetics*, 1958, **43**, 332.
31. AMATI, P., and MESELSON, M., Localized negative interference in bacteriophage, *Genetics*, 1965, **51**, 369.
32. HOTCHKISS, R. D., Size limitations governing the incorporation of genetic material in the bacterial transformations and other non-reciprocal recombinations, *Symp. Soc. Exp. Biol.*, 1958, **13**, 49.

Discussion

Dr. Walter R. Guild: Are there any strong positive arguments against the participation of a single-stranded intermediate?

Dr. Fox: Since it is clear that the transformation process involves the integration of only one strand of DNA, it becomes necessary in the case of transformation by double-stranded DNA to discard the complementary strand. Although this does not really constitute an argument, I think that there are no compelling reasons for excluding either possibility, a single strand as an intermediate or a single strand as a product, in the normal reaction. Even the clear demonstration of transforming activity of single-stranded DNA might not necessarily distinguish between these possibilities.

Dr. Guild: I could comment simply that we can fractionate strands far better than anything reported. Materials with density differences as large as 0.011 g/cm^3, which do not renature appreciably, unless mixed, have been separated from each other. It is quite clear that in this case material single-stranded in the region of the marker itself can constitute an intermediate. Now, whether it does, of course, in the normal case is another matter.

DNA Replication and Recombination after UV Irradiation

Paul Howard-Flanders, W. Dean Rupp, Brian M. Wilkins, and Ronald S. Cole

Departments of Radiology and Molecular Biophysics, Yale University, New Haven, Connecticut

Recent research on the replication of the DNA of bacteria and bacteriophages has been presented in a number of papers in this volume (Gellert et al.; Olivera et al.; Ganesan; Okazaki et al.; Sadowski et al.; Sinsheimer et al.; and Huberman), and these provide an introduction to the subject of this paper—the replication of DNA containing damaged bases.

When cells are exposed to radiations and mutagenic agents, they suffer structural changes and damaged bases in their DNA. Many damaged bases may be removed by excision enzymes, so that the DNA can be repaired before it acts as template for replication. However, other damaged bases may remain in the DNA at the time it undergoes replication and affect its performance as template in DNA synthesis.

Among the many different types of structural defects that can be produced in DNA by mutagens, the most favorable for these studies have proved to be the ultraviolet (UV) photoproducts, of which pyrimidine dimers are the most stable and abundant (Setlow, 1966). They are formed between adjacent pyrimidine bases in the same single strand in known yield, and they interfere with normal replication and transcription (Swenson and Setlow, 1966; Rupp and Howard-Flanders, 1968; Pardee and Prestidge, 1967). Strains of *E. coli* K12 that carry mutations in one of the three loci, $uvrA$, B, or C, are very UV-sensitive and are unable to excise pyrimidine dimers (Setlow and Carrier, 1964; Boyce and Howard-Flanders, 1964; Howard-Flanders et al., 1966). Because of the stability of UV-induced pyrimidine dimers in excision-defective mutants, it is possible to investigate the replication of the bacterial DNA containing a known number of damaged bases. We therefore set out to measure the molecular weight of the DNA synthesized upon damaged templates in UV-irradiated bacteria.

If the molecular weight of the newly synthesized DNA is to be limited only by the spacing between the damaged bases, the template DNA must remain of high molecular weight after UV irradiation. To verify this, the cells were prelabeled by growth for two hours in ^{14}C-thymidine prior to UV irradiation, lysed in the surface layer of an alkaline sucrose gradient, and sedimented under conditions in which shear degradation is reduced to a minimum (McGrath and Williams, 1966). Fortunately, as can be seen from the almost identical sedimentation patterns obtained in alkaline sucrose for the ^{14}C-labeled DNA in Fig. 1, and from the relative positions at which bacteriophage λ and bacterial DNA sediment, the ^{14}C-DNA has a molecular weight of 100 or 200 million daltons. This confirms the expectation that the DNA, containing pyrimidine dimers, is not being cut at a significant rate by excision in these excision-defective mutants, and is of sufficiently high molecular weight to serve as a satisfactory template in the following experiments.

DNA synthesized during a ten-minute pulse in UV-irradiated cells is of lower molecular weight than the template upon which it is synthesized (Rupp and Howard-Flanders, 1968). Figure 1 shows typical distributions for the ^{3}H-radioactivity in alkaline sucrose gradients of the DNA from cells that were labeled for 10 min after exposure to 0 or 60 ergs/mm^2. In Fig. 1A, the DNA synthesized during a 10-min ^{3}H-labeling pulse in the untreated cells is seen to sediment nearly as fast as the material uniformly prelabeled with ^{14}C. In contrast, DNA synthesized after UV irradiation sediments more slowly than the uniformly labeled template DNA, and is seen on the right in Fig. 1B. Because of the reduced rate of DNA replication following exposure to UV light, the same chromosome containing UV photoproducts is still undergoing its first round of replication at 50 to 60 min after irradiation, and the DNA synthesized during this interval is also of low molecular weight, as seen on the right in Fig. 1C.

As the DNA synthesized upon a UV-irradiated template is of lower than normal molecular weight, we may ask whether the length of this newly synthesized material is comparable to the distance between dimers. A UV dose of 1 erg/mm^2 of 2537 Å UV light produces 1.3 pyrimidine dimers per million base pairs, so that for a UV dose of 60 ergs/mm^2, the mean molecular weight between dimers is about nine million daltons (Setlow et al., 1963; Wulff, 1963; Boyce and Howard-Flanders, 1964; Setlow and Carrier, 1966; Rupp and Howard-Flanders, 1968). For UV doses ranging from 10 to 100 ergs/mm^2, the molecular weight distributions

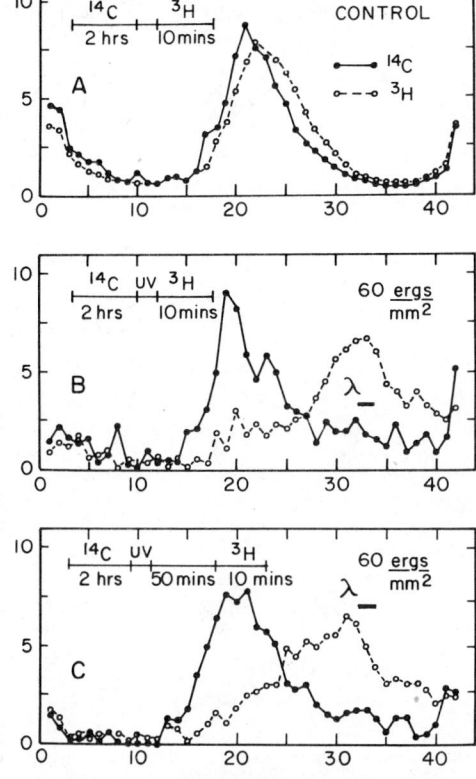

FIGURE 1. The distribution of radioactivity in fractions collected from alkaline sucrose gradients of labeled DNA synthesized in normal (A) and UV-irradiated (B, C) *E. coli*, showing that material (^{3}H in B and C) synthesized on a template containing UV photoproducts sediments more slowly than normal (^{3}H in A). Sedimentation is from right to left, and the fast sedimenting ^{14}C peaks are the uniformly prelabeled DNA.

An excision-defective, thymine requiring strain, *E. coli* K12 AB2500, was uniformly labeled by growth at 37° with aeration in a glucose-salts-casamino acids medium with ^{14}C-thymidine for 2 hr. The culture was divided into three aliquots. One (A) was further labeled by growth with ^{3}H-thymidine for 10 min; the second (B) was exposed to 60 ergs/mm^2 of 2537 A UV light from a low pressure mercury discharge lamp and labeled by growth with ^{3}H-thymidine for 10 min; the third (C) was exposed to 60 ergs/mm^2 of UV light, incubated for 50 min and then transferred to ^{3}H-thymidine medium for 10 min. All three labeled samples were then treated with lysozyme and about 10^7 of the spheroplasts so obtained were lysed in 0.1 ml of 0.1 N NaOH in the surface layer of a 5–20% sucrose gradient at pH 12 (McGrath and Williams, 1966; Rupp and Howard-Flanders, 1968). The gradients were centrifuged for 60 min at 42,500 rpm at 20°C, and fractions were collected from the bottom of the tube onto filter discs, washed in cold trichloroacetic acid and counted in toluene-ethanol scintillator fluid in a scintillation counter. The figures also show the position at which intact phage λ DNA sediments under identical conditions. AB2500 was described by Howard-Flanders, Boyce, and Theriot (1966).

determined by sedimentation in alkaline sucrose are in reasonably good agreement with the calculated distribution of the molecular weight of the DNA between randomly spaced pyrimidine dimers (Rupp and Howard-Flanders, 1968; Howard-Flanders, Rupp, and Wilkins, 1968).

The amount of DNA synthesized in the irradiated cells during the ten-minute labeling period is between 30 and 100 million daltons. Since the average number of pyrimidine dimers in the DNA which undergoes replication during this time is between 15 and 20, it is likely that 7 to 10 of these random length molecules are synthesized upon each of the two template strands containing pyrimidine dimers.

The slow sedimenting material described in this paper should not be confused with the DNA fragments selectively labeled by a radioactive pulse lasting only a few seconds, which are 10 or 100 times smaller and short-lived (Sakabe and Okazaki, 1966; Okazaki et al., this volume).

Although the DNA synthesized upon a template containing pyrimidine dimers sediments more slowly than normal in alkaline sucrose, it is possible that this material is of higher molecular weight, but is joined together by alkali-labile bonds that are disrupted in the alkaline sucrose gradients. To investigate this question, control and UV-irradiated cells that had been uniformly labeled by growth for several generations in ^{14}C-thymidine were exposed to 0 or 120 ergs/mm^2 of UV light and then labeled for 10 min with ^{3}H-thymidine. The DNA was extracted with phenol and denatured by heating for 5 min to 100°C. When sedimented in neutral sucrose, the 10-min pulse-labeled DNA from the untreated cells sedimented almost as rapidly as the prelabeled DNA, as seen in Fig. 2A. However, the 10-min pulse-labeled DNA from the UV-irradiated cells sedimented more slowly than the prelabeled DNA. To judge from the position at which heat-denatured phage λ DNA sediments, the molecular weight of the polynucleotide chains in this newly synthesized material is about one quarter of that of one strand of phage λ DNA, and is comparable to the mean molecular weight of the DNA between pyrimidine dimers, which is 4.5 million daltons for a dose of 120 ergs/mm^2. Although experiments in neutral sucrose are less reliable than those carried out in alkali, it is clear that material synthesized on a UV-irradiated template is of relatively low molecular weight and can be detected in the absence of alkali treatment. Thus it is unlikely that the low molecular weight chains synthesized upon the damaged template are joined by alkali-labile bonds.

The lifetime of the low molecular weight material synthesized on the damaged template can be investigated by labeling UV-irradiated cells as before, and then incubating with a cold thymidine chase before sedimentation in alkaline sucrose. As previously seen, slow sedimenting material is formed during the 10-min pulse in UV-irradiated cells (Fig. 3B). With 60 min incubation in cold media, however, the slow sedimenting material is converted to a fast sedimenting form, as seen in Fig. 3C, a change which occurs mainly between 15 and 40 min. This increase in molecular weight

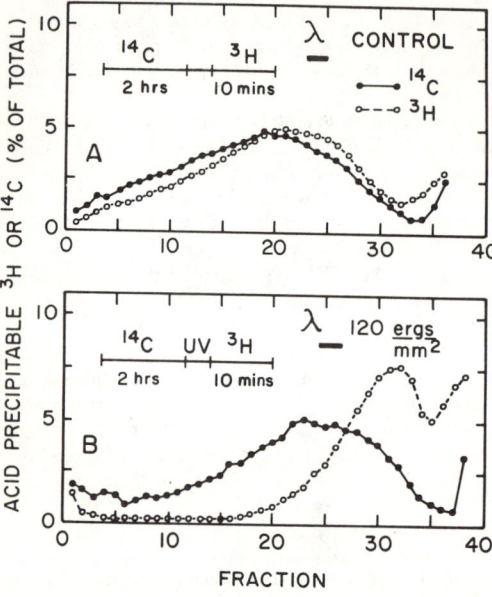

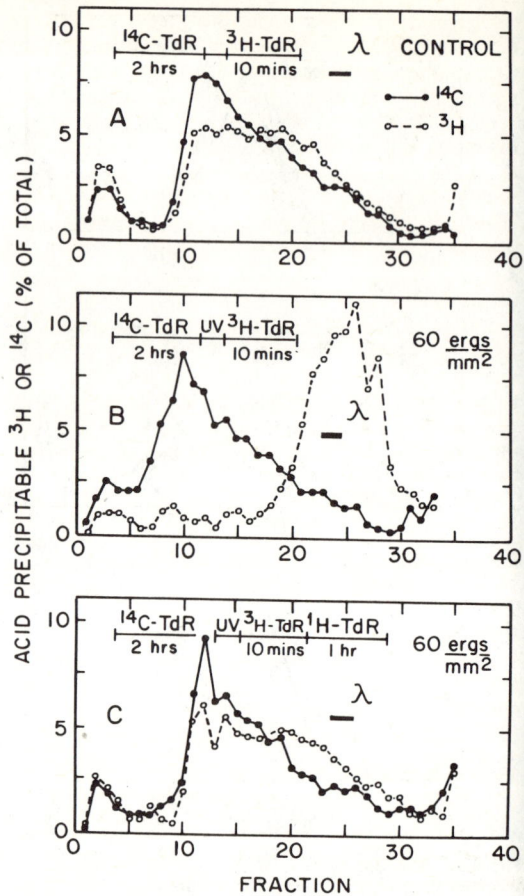

FIGURE 2. The distribution of radioactivity in fractions collected from neutral sucrose gradients containing heat denatured labeled DNA from normal (A), and UV-irradiated (B) *E. coli*, showing that material (^{3}H in B) synthesized on a template containing UV photoproducts sediments more slowly than normal (^{3}H in A) at neutral pH. Sedimentation is from right to left and the fast sedimenting ^{14}C peaks are the uniformly prelabeled DNA.

An excision-defective, thymine requiring, endonuclease I deficient strain of *E. coli* K12 AB3048 was grown in a glucose-salts-casamino acids medium with ^{14}C-thymidine at 37°C with aeration for 2 hr. Half the labeled cells (A) were kept as controls and half (B) were exposed to 120 ergs/mm² of UV light. They were incubated for 10 and 15 min respectively in growth medium containing ^{3}H-thymidine; resuspended in 0.03 M Tris at pH 8.1 and 30% sucrose; treated with lysozyme, EDTA, and trypsin for 5 min; and lysed with sodium dodecylsulfate. The DNA was twice extracted by shaking with buffer-saturated phenol, and the phenol was extracted with ethyl ether and blown out with nitrogen. The DNA solutions were twice dialyzed with 1/10 SSC (Marmur, 1961). The DNA in 1/10 SSC was heated to 100°C for 5 min, chilled to 0°C and gently layered on 5–20% sucrose, 0.7 M NaCl gradients at neutral pH. These were centrifuged at 60,000 rpm for 15 min at 20°C. Samples were collected from the bottom of the tube and counted in liquid scintillator containing toluene, Triton X 100 detergent, water, and scintillator. AB3048 is an Endonuclease I deficient strain derived from strain 1100 (Durwald and Hoffmann-Berling, 1968) and carries *xyl*, *thyA*, *thyR*, *uvrB*, and *Str^s*.

FIGURE 3. The distribution of radioactivity in fractions collected from alkaline sucrose gradients containing labeled DNA synthesized in normal (A) and UV-irradiated (B, C) *E. coli* AB3048, showing that material (^{3}H in B) synthesized on a template containing UV-photoproducts sediments more slowly than normal (^{3}H in A), but with further incubation sediments rapidly (^{3}H in C). Sedimentation is from right to left and the fast sedimenting ^{14}C peak is the uniformly prelabeled DNA.

The experiment was performed as in the legend for Fig. 1. Samples (B, C) were exposed to 60 ergs/mm² of UV light and incubated for 10 min with ^{3}H-thymidine. Sample (B) was held at 0°C while sample (C) was incubated in nonradioactive medium for 60 min. The cultures were then lysed and centrifuged.

could not be due to degradation and reutilization of label for the synthesis of high molecular weight material, since newly synthesized material is always of lower molecular weight, as seen in Fig. 1B and 1C. One interpretation of these observations is that the lower molecular weight material is enzymatically joined during incubation into high molecular weight DNA. This change may reflect the action of a genetic recovery mechanism and will be discussed again later.

Other types of damaged bases may cause the newly synthesized strands to be discontinuous. Cells treated with the monofunctional alkylating agent methyl methane sulfonate (MMS) undergo strand cutting during incubation and presumably excise the alkylated bases from their DNA (Reiter et al., 1967; Friedberg and Goldthwaite, this volume; Boyce and Tepper, 1968). At 0.04 M, this agent induces only a slight change in the molecular weight of the template strand, but nevertheless causes a tenfold reduction in the molecular weight of the newly synthesized material, as judged by sedimentation in alkaline sucrose. The amount of DNA synthesized upon each template strand during a 20-min labeling period is sufficient for only two or three low molecular weight fragments, and the effect is therefore less impressive than that obtained with UV light. It is interesting, however, that an improvement in the quality of the DNA as a template for replication can be observed as the cells are incubated and the alkylated bases are presumably removed. The mean molecular weight of newly synthesized DNA is at first 1 to 2×10^7 daltons in MMS-treated cells, but increases with time after alkylation until the recovery of the template is complete and the molecular weight of the newly synthesized DNA approaches that in control untreated cells, about 2×10^8 daltons (Finesilver and Howard-Flanders, in prep.).

Transfer of DNA from UV-Irradiated Male Bacteria

Further evidence for the existence of discontinuities or gaps in the DNA synthesized in UV-irradiated bacteria, and an indication of their positions relative to the pyrimidine dimers in the template strand, can be sought in experiments with mating bacteria. The advantage of a conjugating

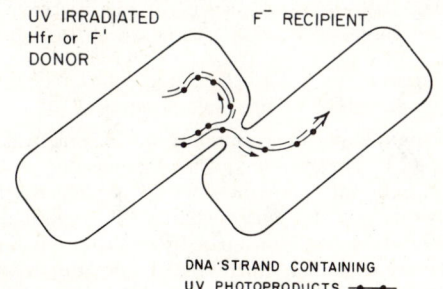

FIGURE 4. Hypothetical diagram of the transfer of DNA containing photoproducts from a UV-irradiated Hfr, or F episome donor to a recipient during mating. The UV photoproducts remain in the DNA in the excision-defective donor so that the DNA strand presumably contains photoproducts as it is replicated during transfer to the recipient cell. The possibility that the newly synthesized strands are discontinuous is discussed in the following sections.

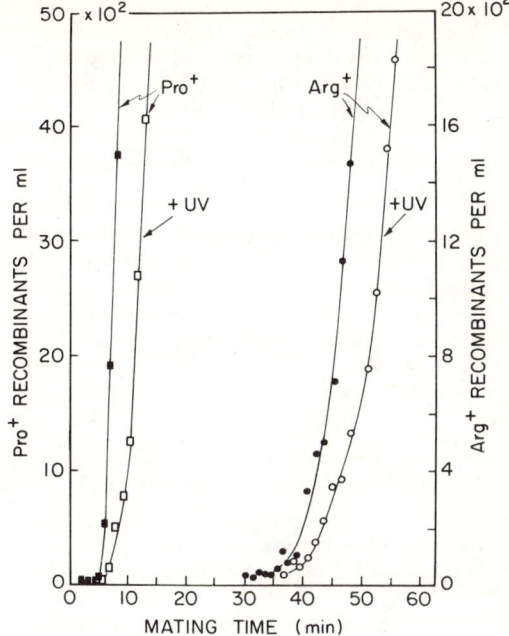

FIGURE 5. The effect of UV-irradiating excision-defective Hfr bacteria on the kinetics of chromosome transfer during mating.

An exponentially growing culture of *E. coli* AB3108 Hfr J2 *uvrB5 his* T6^s was irradiated with 200 ergs/mm² of UV light before mating at 37° with AB 3105 F⁻ *proA argE* T6^R. Mating was interrupted at various times by vortex blending and the addition of excess phage T6, and the suspension was plated on selective agar. The number of Pro⁺ His⁺ T6^R (□) and Arg⁺ His⁺ T6^R (○) recombinants formed on selective agar per ml of mating mixture is plotted against the mating time. The numbers of recombinants formed in the unirradiated control experiment are shown by the closed symbols. Exposure to 200 ergs/mm² delays the entry of *proA*⁺ and *argE*⁺ by no more than 3 min.

system is that the DNA appears to be replicated during transfer from the donor (Jacob et al., 1963; Gross and Caro, 1966; Bonhoeffer, 1966; Bonhoeffer et al., 1967). Moreover, the action of repair enzymes in the recipient can be controlled by mating the male bacteria with suitable female mutants. If mating experiments with UV-irradiated donors are to be useful, it is necessary to test whether bacterial chromosomes carrying UV photoproducts are transmitted from male to female cells during conjugation and, if so, whether the strand synthesized during transfer is discontinuous at each photoproduct, as illustrated in Fig. 4. After exposure to UV irradiation, excision-defective Hfr cells were mated with genetically marked F⁻ recipients, and the times of entry of genetic markers were investigated by interrupting the mating at intervals and plating the cells on selective agar. As seen in Fig. 5, exposure of the Hfr bacteria to 200 ergs/mm² causes no more than a two or three minute delay in the times of entry of the two male genes, *proA* and *argE*, which control proline and arginine independence, and are separated by about one fifth of the genetic map of the bacterial chromosome. This delay is small considering that over 150 pyrimidine dimers must be present in the single strand of the Hfr chromosome between these two genetic markers, and that this UV dose reduces the colony-forming ability of the excision-defective Hfr strain by a factor of about 10^6. Evidently, the presence of a few hundred pyrimidine dimers in the Hfr chromosome does not prevent DNA transfer between conjugating cells, and delays the process by no more than one or two seconds per transferred dimer.

Are the Discontinuities in a Newly Synthesized Strand Opposite Pyrimidine Dimers?

If discontinuities are formed in the DNA synthesized during the transfer of a chromosome or episome from a UV-irradiated donor, they may be opposite the pyrimidine dimers, or may be displaced from them, as shown in Fig. 6. A distinction between these two possible configurations can be made by investigating their susceptibility to various repair processes. The repair of DNA containing pyrimidine dimers by excision depends upon the presence of an intact complementary strand to serve as template during repair synthesis (Setlow and Carrier, 1964; Boyce and Howard-Flanders, 1964; Yarus and Sinsheimer, 1964), and will presumably be blocked if the newly synthesized strand is discontinuous opposite each pyrimidine dimer, as in the upper line of Fig. 6. Single-strand cuts induced by X-rays in DNA can be repaired within a few minutes (Boyce and Tepper, 1968). However, gaps may remain open for longer periods if positioned opposite pyrimidine dimers which may block repair enzymes. Thus, the dimer may be stabilized by the opposing gap, while the gap may be long-lived because of the dimer. A dimer which is opposite a gap should still be subject to photoreactivation, since this process involves the splitting of pyrimidine dimers *in situ* without need for the opposite strand as template, and because UV-irradiated phage φX can be photoreactivated

	Repair by Excision	Strand cut next to Dimer	Repair of Gap	Photoreactivation
──•──	−	−	−	+
── ──	+	+	+	+

FIGURE 6. Two models for the positions of the gaps relative to the pyrimidine dimers in the DNA transferred from a UV-irradiated donor during mating.

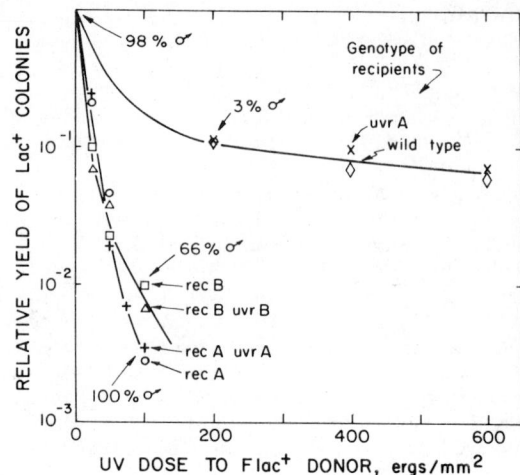

FIGURE 7. The relative yield of Lac⁺ colonies obtained as a function of the UV dose to an excision-defective donor F *lac⁺* when mated with Lac⁻ recipients of various genotypes: at UV doses above 100 ergs/mm², most of the Lac⁺ colonies formed in Rec⁺ cells are recombinants.
The % ♂ in the figure denotes the fraction of Lac⁺ colonies found to act as secondary F *lac⁺* donors and is always 100% in *recA* recipients. The strains employed and pertinent genetic characteristics are:
Donor: GY854 *uvrB ura*/F *lac⁺*
Recipients: AB1157 +; AB1886 *uvrA6*; AB2463 *recA13*; AB3114 *recA13 uvr*; AB3071 *recB21*; AB3072 *recB21 uvrB*.

to a small extent (Setlow, 1961). If, however, the situation is as depicted in the lower line of Fig. 6, and the single-strand gaps are displaced from the pyrimidine dimers, then both the dimers and the gaps should be subject to the repair processes which depend upon the presence of an intact complementary strand.

The susceptibility to repair of DNA from a UV-irradiated donor can conveniently be tested in a mating system with a donor carrying an F episome.

Properties of F *lac⁺* Episomes Transferred from UV-Irradiated Donors

The next experiments employed a mating system with a UV-irradiated excision-defective F *lac⁺* donor, and were concerned with the properties of the system and the susceptibility of the damaged transferred episome to repair by excision or photoreactivation. When untreated F *lac⁺* donors are mated with suitable Lac⁻ females, the episomes are transferred and many of the females become secondary F *lac⁺* donors. The F *lac⁺* episomes appear to be transferred efficiently after exposing the excision-defective donor cells to UV radiation, as the yield of Lac⁺ colonies obtained decreases slowly with increasing dose, and 5 to 10% of the normal yield is obtained at 600 ergs/mm² (Fig. 7).

The susceptibility to repair of the episomes transferred from UV-irradiated donors can be investigated by studying the colonies formed in these mating experiments when UV-sensitive bacteria are used as recipients. Suitable recipient bacteria include excision-defective strains carrying *uvrA*, *uvrB*, or *uvrC* mutations, recombination-deficient mutants, of which different types have been recognized and found to carry *recA*, *recB* or *recC* mutations (Clark and Margulies, 1965; Howard-Flanders and Boyce, 1966; Clark, 1967; Emmerson and Howard-Flanders, 1967; Howard-Flanders, 1968; Willetts and Mount, in prep.), and double mutants with both *uvr* and *rec* mutations. When a UV-irradiated F *lac*+ donor is mated with recombination-deficient recipients that carry *recA*, the yield of Lac+ colonies decreases rapidly with increasing UV dose, as seen in Fig. 7. All Lac+ colonies formed from *recA* recipients consist of secondary F *lac*+ donors and harbor intact episomes, so that this system provides a convenient means for measuring the number of intact episomes transferred, and permits a sensitive test of the susceptibility of F *lac*+ episomes from UV-irradiated donors to repair by excision in the recipient. As seen in Fig. 7, there is no significant difference between the yields of Lac+ colonies in the recipient strains that are either normal or excision-defective. Moreover, this is true for normal, as well as for recombination-deficient mutants carrying either *recA* or *recB*. If these damaged episomes transferred from UV-irradiated donors were subject to repair by excision in the recipient, a higher yield of Lac+ colonies would be obtained in normal (carrying *uvr*+) than in excision-defective (carrying *uvr*−) recipient strains. The failure to detect such a difference in crosses with any of these strains indicates that DNA transferred from a UV-irradiated donor either is not subject to repair by excision, or is repaired too slowly to affect the yield of Lac+ colonies. Other failures to detect any repair by excision of the genetic properties of DNA transferred from UV-irradiated donors have been reported in several different systems (Devoret et al., 1965; Mattern et al., 1965; Devoret and George, 1967; Pardee and Prestidge, 1967).

PHOTOREACTIVATION BY EXPOSURE TO VISIBLE LIGHT AFTER MATING

A proportion of the damaged episomes transferred from a UV-irradiated donor can be reactivated by exposing the recipients to visible light after mating has been interrupted, as may be seen in Fig. 8. Photoreactivation can most readily be detected in a cross with a *recA* recipient, as only intact F *lac*+ episomes are able to form Lac+ colonies. The effect of the visible light is to convert

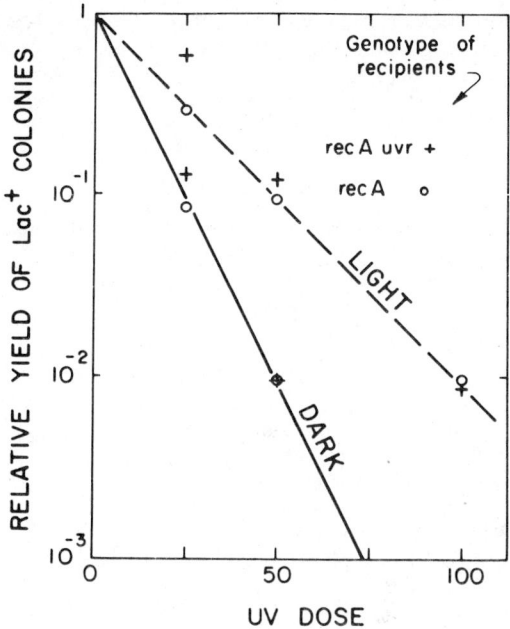

FIGURE 8. The relative yield of Lac+ recombinants is plotted as a function of UV dose to excision-defective *ura* T6^S F *lac*+ donors when mated with T6^R *lac* recipients that carry either *recA13* or *recA13 uvr*: the UV-damaged episomes can be photoreactivated in the recipient after transfer.

Mating was interrupted after 30 min by vortex agitation, and incubation with excess phage T6 for five min. The mated bacteria were then plated on agar selective for Lac+ Ura+ colonies and incubated. The plates were kept in the dark or under white fluorescent lights for the first hour. The strains used were: GY854 T6^S *ura uvrB*/F *lac*+; AB2463 T6^R *lac recA13*. At 100 ergs/mm² and after photoreactivation, 20/20 of the Lac+ colonies proved to be secondary F *lac*+ donors.

damaged to intact episomes, as shown by the observation that when the donor culture was exposed to 100 ergs/mm², 20/20 of the Lac+ colonies obtained in increased yield after photoreactivation proved to be secondary F *lac*+ donors.

Since damaged F episomes transferred from a UV-irradiated donor during conjugation can be repaired in the recipient by exposure to visible light though not through the action of excision enzymes present in normal cells but not in *uvr* mutants, it seems likely that the discontinuities in the newly synthesized strand are opposite the pyrimidine dimers, as in the upper structure of Fig. 6. Thus, these experiments in which excision is genetically controlled may be regarded as an in vivo test of the structure in the region of each transferred pyrimidine dimer. No indication is obtained, however, as to whether the number of bases missing in the discontinuity is small or large.

These damaged F episomes transferred from a

UV-irradiated donor are remarkably stable in recA recipients, and lose their sensitivity to photoreactivation by visible light with a long half-life (approximately 1.5 hours) in either recA or recA uvrA recipients. As this result could not be obtained if excision enzymes in the recipient cut the transferred DNA chain containing the photoproducts, it appears unlikely that excision enzymes act upon DNA containing pyrimidine dimers opposite discontinuities. This observation is hard to reconcile with an earlier report that UV-irradiated single strand DNA of phage ϕX is cut by excision enzymes (Ono and Shimazu, 1967). Further work will be required to resolve this apparent conflict.

It is also of interest that a dose of 13 ergs/mm^2 is required to reduce by one natural logarithm, the number of Lac$^+$ colonies formed from recA recipients (Fig. 7). The mean dimer spacing for this dose is about 40 million daltons, which is to be compared with 25 to 37 million daltons for one single strand of the F lac$^+$ episome, as estimated by Freifelder (1968; Freifelder et al., 1968; this volume). These results indicate that approximately one pyrimidine dimer in the transferred strand of an episome renders it defective. In crosses with Rec$^+$ recipients, most of the F lac$^+$ episomes transferred from donors exposed to UV doses in excess of 100 ergs/mm^2 are unable to convert these recipients to secondary F lac$^+$ donors, and instead produce sectored female Lac$^+$ colonies, which contain F$^-$ lac$^+$ recombinants and cells that form sectored colonies, giving rise to Lac$^+$, Lac$^-$ cells and to a minority of cells that form further sectored colonies (R. Devoret, J. George, M. Thompson, and P. Howard-Flanders, in prep.).

RECOMBINATION BETWEEN DNA TRANSFERRED FROM A UV-IRRADIATED DONOR AND THE CHROMOSOME IN THE RECIPIENT CELL

The chromosomes transferred from UV-irradiated Hfr donors may be unusual in that they have defects in both strands at each pyrimidine dimer. The presence of these defects involving both strands influences the yield and genetic constitution of the recombinants obtained. When excision-defective Hfr bacteria are UV-irradiated and mated with genetically marked females, the yield of recombinants carrying a single marker from the male is depressed to between 20 and 50% of the normal value for UV doses of up to 200 ergs/mm^2, even though only one in a million of these excision-defective Hfr bacteria will survive this UV dose. An examination of the recombinants from such a cross, for each of four linked genes in the thr-leu region, has given the results summarized in Table 1. In crosses with unirradiated Hfr bacteria, the recombinants selected for one male character have been found to carry male markers at all four loci in about 70% of recombinants, indicating a tendency for the Hfr chromosome to be integrated in lengths greater than the distance spanning these markers. However, when the male is exposed to a UV dose of 200 ergs/mm^2, this class of recombinants is reduced to 1%. Male markers are then integrated less frequently at unselected loci. The predominant class contains female characters at all unselected loci, but the proportion of recombinants with two or more exchanges in this region is increased from 4 to 16% (Wilkins and Howard-Flanders, 1968). These results show a loss of continuity in the integration of characters from the Hfr chromosome into recombinants following irradiation of the male. In addition, the increase in the proportion of recombinants that either carry unselected female characters or show multiple genetic exchanges between the four linked markers suggests that there may be many sites of genetic exchange and that only short lengths of male genome are integrated in these crosses. The length of Hfr chromosome that can be incorporated into viable recombinants may therefore be limited to the lengths of chromosome between pyrimidine dimers and opposing defects.

Crosses with F lac$^+$ donors also give results of interest. When UV-irradiated, excision-defective F lac$^+$ donors are mated with an equal number of suitable females, the fraction of F$^-$ cells giving

TABLE 1. THE ANALYSIS OF RECOMBINANTS FORMED BY MATING NORMAL OR UV-IRRADIATED EXCISION-DEFECTIVE Hfr CELLS WITH A GENETICALLY MARKED RECIPIENT: EFFECTS ON THE FREQUENCY OF INCLUSION OF PROXIMAL MALE MARKERS IN RECOMBINANTS

Recombinants carrying male (m) or female (−) markers at:				Frequency per 100 selected recombinants	
thyR	thr	ara	leu	Unirradiated control	After 200 ergs/mm^2 to Hfr uvrB5
m	m	m	m	69.0	1.5
m	−	−	−	6.5	56.0
m	−	m	−		
m	−	−	m	4.0	15.5
m	m	−	m		
m	−	m	m		
m	m	m	−	20.5	27.0
m	m	−	−		

Control and UV-irradiated cultures of AB3108 Hfr J2 uvrB5 thyR his T6^S were mated at 37° with AB 3105 F$^-$ leu ara thr thyA T6^R. Mating was stopped after 60 min by the addition of T6 phage, and the cells were incubated in broth +30 μg/ml of thymine for 3 hr to allow the recombinants to segregate. The bacteria were then plated on an appropriate minimal medium containing 2 μg/ml of thymine to select recombinants carrying thyA thyR his$^+$. These were then tested for the presence of the proximal male markers thr$^+$, ara$^+$ and leu$^+$. Very similar frequencies were obtained when the recipient strain carried the uvrB5 mutation.

colonies containing Lac$^+$ recombinants may be as high as 10%. This is a high proportion of recombinants to obtain in the absence of any selection, and indicates a tendency of the defective episome to recombine with the bacterial DNA.

This tendency to recombine is an important characteristic of DNA transferred from UV-irradiated donor bacteria. It is seen in the crosses with Hfr as well as with F′ donors and may be due to dimers and opposing gaps. The high efficiency of these dimers and gaps may be due to their relative longevity in comparison with other types of defect which are subject to removal by repair enzymes. It is possible that the internal free ends initiate recombination, but there is nothing in these experiments to indicate that it is the free ends rather than the single stranded regions which carry the dimers that initiate recombination.

Recovery of F lac$^+$ Episomes in Excision-Defective Cells Prior to Transfer

Mating systems involving the transfer of F lac$^+$ episomes provide a favorable method for investigating the recovery after UV irradiation of these episomes in the irradiated excision-defective donor cells. To test whether recovery takes place in recA uvrB or in rec$^+$ uvrA cells, UV-irradiated lac$^+$ donors carrying these mutations were mated with recA recipients. It is convenient to employ recA recipients, as the only Lac$^+$ colonies obtained are those due to the transfer of an intact episome. The results in Fig. 9 show that, if the rec$^+$ uvrB donors are UV irradiated with 100 ergs/mm^2 and incubated prior to mating, there is a substantial increase in the number of intact episomes transferred from these cells, even though only 0.1% of donor cells survive exposure to this UV dose. No such recovery of the episome can be detected in the recombination-deficient donor. Evidently, the recA$^+$ gene is required for recovery of the episome in the donor.

A diagram summarizing a possible interpretation of these experiments is shown in Fig. 10, with events in the donor cell indicated on the left-hand side of the barrier, while events in the recipient are drawn on the right. As shown at the top, a single strand of the episome containing photoproducts is transferred from the UV-irradiated donor and is replicated during transfer. The defective episome so formed is subject to repair only by photoreactivation. Alternatively, if the donor is incubated after UV irradiation so that the episome can replicate before transfer, there is an opportunity for sister exchanges to occur in normal, but not in recombination-deficient donor cells. The increase of intact episomes transferred may reflect the successful reconstruction of the

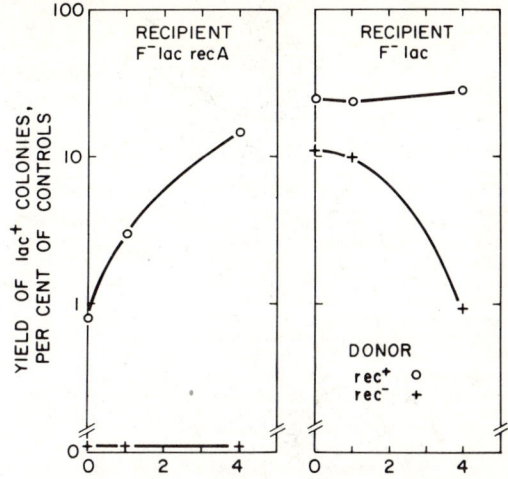

FIGURE 9. The yield of Lac$^+$ colonies from normal and recombination-deficient F lac$^+$ donors mated with Lac$^-$ recipients is plotted against the time between exposing the donor to 100 ergs/mm^2 of UV light and mating to recipients; recovery of the F lac$^+$ episome occurs in the excision-defective donor cells prior to transfer, but is blocked if the donor also carries recA.

Exponentially growing cultures of the F lac$^+$ donors GY854 uvrB ura/F lac$^+$, and GY567 uvrB ura recA13/F lac$^+$ were exposed to 100 ergs/mm^2 of UV light and incubated at 37° for a time ranging from 0 to 4 hr before mating with the AB2463 F$^-$ lac recA13, or AB1157 F$^-$ lac recipients. Cells were mated for 30 min and plated on agar selective for Lac$^+$ Ura$^+$ colonies. Only intact F lac$^+$ episomes form Lac$^+$ colonies in the recA recipient and, as seen on the left, the number of episomes transferred from the donor increases with incubation, even though cell division is blocked by the UV irradiation. Evidently, recovery before transfer occurs in this uvrB donor. As no comparable recovery occurs in the uvrB recA13 donor, it appears that recA$^+$ is required for recovery in the absence of excision.

episome from the two defective structures formed by replication.

Genetic Recovery after DNA Replication

If newly synthesized strands of DNA are discontinuous at each defective base in the template strand, the structure obtained may be as indicated in the second line on the right of Fig. 11. At any point where there is a damaged base and opposing gap in one duplex of this structure, the sister duplex is generally intact and contains the correct sequence of base pairs. Thus, the complete genetic code in this length of the chromosome may be present in the form of overlapping fragments in the two duplexes. Genetic exchanges between the sister duplexes, perhaps initiated by the free ends at the gaps and opposing dimers, could lead to the reconstruction of the complete genome from the overlapping fragments.

FIGURE 10. A hypothetical diagram of the transfer and recovery of F lac^+ episomes in a mating system with an excision-defective UV-irradiated donor.

In the top line, the F lac^+ episome containing UV photoproducts in both strands is represented on the left. Only one strand is transferred and produces a damaged episome containing pyrimidine dimers in one strand, while the other is discontinuous. The effect of exposing the recipient to light is to monomerize the dimers *in situ* by photoreactivation, so that the strand containing dimers is effectively repaired and an intact episome is formed. If the UV-irradiated donor is incubated, no dimer excision occurs as the cell is excision-defective. However, after replication, there is an opportunity for sister exchanges to occur in a Rec+ cell, and intact episomes to be formed which can be detected by testing for the transfer of infective F lac^+ episomes by the formation of Lac+ colonies by recipients carrying *lac recA*.

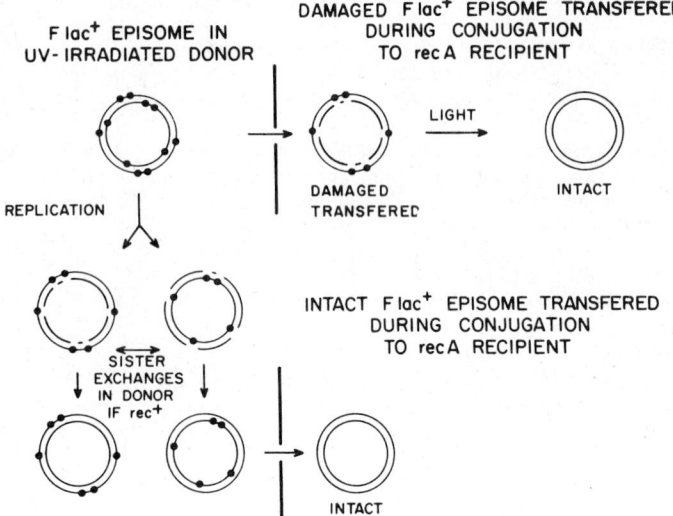

FIGURE 11. Hypothetical diagram of two mechanisms for recovery from UV-induced photoproducts in the DNA of bacteria. Two-strand DNA molecules are represented by double lines, and pyrimidine dimers are represented by black circles.

In normal cells, UV-induced pyrimidine dimers may be removed from the DNA by excision, following which the single strand gaps are repaired as on the left (Setlow and Carrier, 1964; Boyce and Howard-Flanders, 1964).

In excision-defective cells, however, about 50 pyrimidine dimers are formed in the DNA of the bacterial genome per lethal event. The effects of the great majority of dimers may be obviated by a second recovery mechanism which may operate after DNA replication, as seen on the right. If DNA containing pyrimidine dimers is replicated, the newly synthesized strands (thin lines) are discontinuous opposite each dimer. Recovery may be due to the reconstruction of an intact genome with the correct base sequence by means of sister exchanges between the two DNA duplexes, each of which contains fragments of the complete base pair sequence (Rupp and Howard-Flanders, 1968).

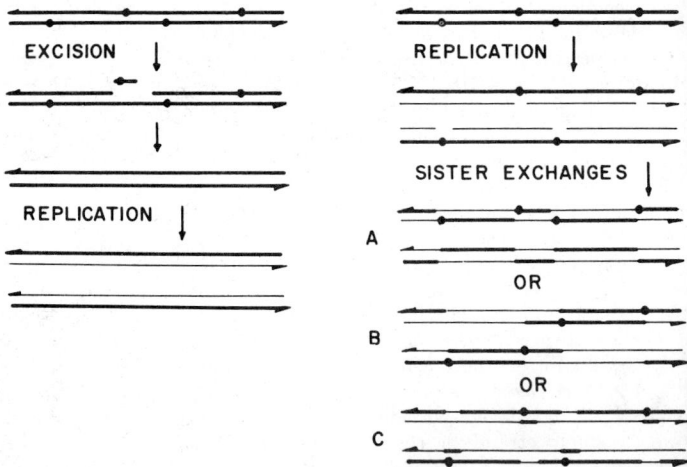

Possible structures formed by different mechanisms are shown. (A) pyrimidine dimers induce two-strand crossing-over in such a way that the dimers tend to accumulate in one twin helix: (B) dimers cause single strand exchanges in the opposite strands. Both conserved and semiconserved twin helices are formed: (C) dimers induce short lengths of single strand exchange. Predominantly semiconserved material is formed.

There are a number of reasons for believing that these overlapping fragments may be utilized during recovery from radiation or mutagen-induced damage by a mechanism which promotes the survival of colony-forming ability.

Recombination appears to be required for efficient recovery from radiation damage, since recombination-deficient mutants carrying recA, recB, or recC are abnormally sensitive to radiations and to mutagenic agents.

If the complete genome were reconstructed by sister exchanges, the low molecular weight material formed in irradiated cells would subsequently be increased in size until it became comparable to that of undamaged DNA. Just such an increase in molecular weight of the newly synthesized DNA in UV-irradiated cells was seen in Fig. 3C (Rupp and Howard-Flanders, 1968). Furthermore, sister exchanges occur following DNA replication in UV-irradiated cells, and can be detected by means of density labels, as a disturbance of normal semiconservative replication. Experimental evidence for exchanges, possibly similar to those shown in Fig. 11 lower right, will be described in a later publication.

Although excision-defective bacteria are sensitive to UV light, about 50 pyrimidine dimers are required to produce a lethal event in these cells. These irradiated cells produce colonies which are frequently indistinguishable from the original strain, and must contain the complete bacterial genome with no more than unimportant alterations in base sequence. Thus, if damaged bases are not removed from DNA by excision, their effect may be offset by another genetic recovery process which operates after DNA replication and depends upon sister exchanges for utilizing regions containing overlapping fragments of the correct base sequence, to reconstruct an undamaged genome.

SUMMARY AND CONCLUSIONS

Pyrimidine dimers are formed in the DNA of UV-irradiated *E. coli* and remain *in situ* in excision-defective mutants. When DNA containing UV-induced pyrimidine dimers is replicated, the newly synthesized strands are of low molecular weight, as if formed with gaps at a spacing approximately equal to that between the pyrimidine dimers.

Since DNA is also replicated during transfer between mating bacteria, it is possible to use mating systems involving UV-irradiated donor cells to test whether the newly synthesized strand also contains gaps or discontinuities and, if so, whether these are directly opposite the pyrimidine dimers. When excision-defective male bacteria are exposed to UV light and mated with suitable females, DNA is transferred at almost the normal rate, even after UV doses of several hundred ergs/mm^2. When UV-irradiated F *lac*$^+$ donors are mated with suitable recipients, damaged episomes are transferred, but fail to form secondary F *lac*$^+$ donors. These damaged episomes can be photoreactivated in the recipient by exposure to visible light and restored to a fully active form. They are not, however, subject to repair by excision after transfer. Since an intact complementary strand is required for repair by excision, the lack of detectable repair of the transferred episome by this process suggests that there may be gaps or defects of some kind in the strand synthesized during transfer, and that these are directly opposite the dimers.

An increased frequency of genetic exchanges occurs in crosses with UV-irradiated donors. This is seen in crosses with Hfr strains as a loss of recombinants containing long pieces of integrated Hfr DNA, and as an increase in the fraction of recombinants with multiple exchanges between linked markers. It can also be seen in crosses between UV-irradiated F *lac*$^+$ donors and Lac$^-$ recipients as a tendency to produce Lac$^+$ recombinants in which the *lac*$^+$ gene of the episome is integrated into the bacterial chromosome.

UV-induced recombination appears to be initiated with high efficiency after replication, when the DNA contains dimers and opposing gaps. The efficiency of these dimers and gaps may be due to their relative longevity in comparison with other types of defects which are subject to removal by repair enzymes. Although there is nothing in these experiments to suggest that recombination is due to the strands with internal free ends rather than to the single stranded regions containing the dimers, it is possible that the strands with free ends at the gaps are able to uncoil for a limited distance and are free for sufficient time to initiate recombination with homologous DNA.

Acknowledgments

We would like to thank Mrs. Donna Reno and Miss Margaret Thompson for performing certain of these experiments.

This work was supported by United States Public Health Service Grants CA 06519 and AMK 69397.

REFERENCES

Bonhoeffer, F., 1966. DNA transfer and DNA synthesis during bacterial conjugation. Z. Vererbungslehre 98: 141.

Bonhoeffer, F., R. Hosselbarth, and K. Lehmann, 1967. Dependence of the conjugational DNA transfer on DNA synthesis. J. Mol. Biol. 29: 539.

Boyce R. P., and P. Howard-Flanders, 1964. Release of ultraviolet light-induced thymine dimers from DNA in *E. coli* K-12. Proc. Nat. Acad. Sci. 51: 293.

Boyce, R. P., and M. Tepper, 1968. X-ray-induced single-strand breaks and joining of broken strands in superinfecting λ DNA in *Escherichia coli* lysogenic for λ. Virology 34: 344.

Clark, A. J., 1967. The beginning of a genetic analysis of recombination proficiency. J. Cellular Physiol. 70: 165.

Clark, A. J., and A. D. Margulies, 1965. Isolation and characterization of recombination deficient mutants of *Escherichia coli* K-12. Proc. Nat. Acad. Sci. 53: 451.

Devoret, R., and J. George, 1967. Induction indirecte du prophage λ par le rayonnement ultraviolet. Mutation Res. 4: 713.

Devoret, R., M. Monk, and J. George, 1965. Indirect ultraviolet induction of prophage λ and Colicin I factor. Zentralblatt fur Bakteriologie 196: 193.

Durwald, H., and H. Hoffmann-Berling. 1968. Endonuclease I-deficient and ribonuclease I-deficient *E. coli* mutants. J. Mol. Biol. 34: 331.

Emmerson, P. T., and P. Howard-Flanders, 1967. Cotransduction with *thy* of a gene required for genetic recombination in *Escherichia coli*. J. Bacteriol. 93: 1729.

Freifelder, D. R., 1968. Studies on *Escherichia coli* sex factors. IV. Molecular weights of the DNA of several F' elements. J. Mol. Biol. 35: 95.

Freifelder, D. R., and D. Freifelder, 1968. Studies on *Escherichia coli* sex factors. II. Some physical properties of F' Lac and F DNA. J. Mol. Biol. 32: 25.

Gross, J. D., and L. G. Caro, 1966. DNA transfer in bacterial conjugation. J. Mol. Biol. 16: 269.

Howard-Flanders, P., 1968. Genes that control DNA repair and genetic recombination in *Escherichia coli*. Adv. Biol. Med. Phys. 12: 299.

Howard-Flanders, P., and R. P. Boyce, 1966. DNA repair and genetic recombination: Studies on mutants of *Escherichia coli* defective in these processes. Radiation Res. Suppl. 6: 156.

Howard-Flanders, P., R. P. Boyce, and L. Theriot, 1966. Three loci in *Escherichia coli* K-12 that control the excision of pyrimidine dimers and certain other mutagen products from DNA. Genetics 53: 1119.

Howard-Flanders, P., W. D. Rupp, and B. M. Wilkins, 1968. The replication of DNA containing photoproducts and UV-induced genetic recombination. *In* W. J. Peacock [ed.] Replication and Recombination of Genetic Material. Australian Acad. Sci. Canberra, Australia, p. 142.

Jacob, F., S. Brenner, and F. Cuzin, 1963. On the regulation of DNA replication in bacteria. Cold Spring Harbor Symp. Quant. Biol. 28: 329.

Marmur, J., 1961. A procedure for the isolation of deoxyribonucleic acid from micro-organisms. J. Mol. Biol. 3: 208.

Mattern, I. E., M. P. van Winden, and A. Rorsch, 1965. The range of action of genes controlling radiation sensitivity in *E. coli*. Mutation Res. 2: 111.

McGrath, R. A., and R. W. Williams, 1966. Reconstruction *in vivo* of irradiated *E. coli* deoxyribonucleic acid: the rejoining of broken pieces. Nature 212: 534.

Ono, J., and Y. Shimazu, 1967. *In vivo* cleavage of a circular, single-stranded DNA of bacteriophage φR irradiated with ultraviolet light. J. Mol. Biol. 24: 491.

Pardee, A. B., and L. S. Prestidge, 1967. Ultraviolet-sensitive targets in the enzyme-synthesizing apparatus of *Escherichia coli*. J. Bacteriol. 93: 1210.

Reiter, H., J. Strauss, M. Robbins, and R. Marone, 1967. Nature of the repair of methyl methane sulfonate-induced damage in *B. subtilis*. J. Bacteriol. 93: 1056.

Rupp, W. D., and P. Howard-Flanders, 1968. Discontinuities in the DNA synthesized in an excision-defective strain of *Escherichia coli* following ultraviolet irradiation. J. Mol. Biol. 31: 291.

Sakabe, K., and R. Okazaki, 1966. A unique property of the replicating region of chromosomal DNA. Biochim. Biophys. Acta 129: 651.

Setlow, R. B., 1961. Macromolecular Radiosensitivity. Brookhaven Symp. Biol. 14: 1.

———. 1966. Cyclobutane-type pyrimidine dimers in polynucleotides. Science 153: 379.

Setlow, R. B., and W. L. Carrier, 1964. The disappearance of thymine dimers from DNA: An error-correcting mechanism. Proc. Nat. Acad. Sci. 51: 226.

———, ———. 1966. Pyrimidine dimers in ultraviolet-irradiated DNAs. J. Mol. Biol. 17: 237.

Setlow, R. B., P. A. Swenson, and W. L. Carrier, 1963. Thymine dimers and inhibition of DNA synthesis by ultraviolet irradiation of cells. Science 142: 1464.

Swenson, P. A., and R. B. Setlow, 1966. Effects of ultraviolet radiation on macromolecular synthesis in *Escherichia coli*. J. Mol. Biol. 15: 201.

Wilkins, B. M., and P. Howard-Flanders, 1968. The genetic properties of DNA transferred from ultraviolet-irradiated Hfr cells of *Escherichia coli K-12* during mating. Genetics. In press.

Wulff, D. L., 1963. Kinetics of thymine photodimerization in DNA. Biophysical J. 3: 355.

Yarus, M., and R. L. Sinsheimer, 1964. The ultraviolet resistance of double-stranded φX174 DNA. J. Mol. Biol. 8: 614.

DISCUSSION

RESCUE OF DNA FROM UV-IRRADIATED T4 PHAGE

E. Shahn

Department of Biological Sciences
Hunter College of The City University of New York

The results of the following two experiments, performed several years ago, on the rescue of DNA from UV-irradiated T4 phage, are in substantial agreement with the conclusions of the experiments reported by Howard-Flanders regarding the frequency and enzymatic requirements of genetic exchanges following UV irradiation of bacteria. Specifically, with increasing doses of UV, the size of the parental contributions decreases, the number of such contributions per phage increases, and the enzymes necessary for rescue are other than those required merely for replication of DNA.

In both experiments, heavy (BU-labeled) bacteria in heavy medium were simultaneously infected with light, ^{32}P-labeled, irradiated phage and light, nonlabeled, nonirradiated rescuing phage. Replication and the extent of material recombination (i.e., dispersion) of the irradiated phage DNA could be detected by the observed increase in density of the ^{32}P label in CsCl density gradients.

The first experiment (Fig. 1) examined the effect of different shearing conditions on the DNA extracted from the resulting progeny phage. UV

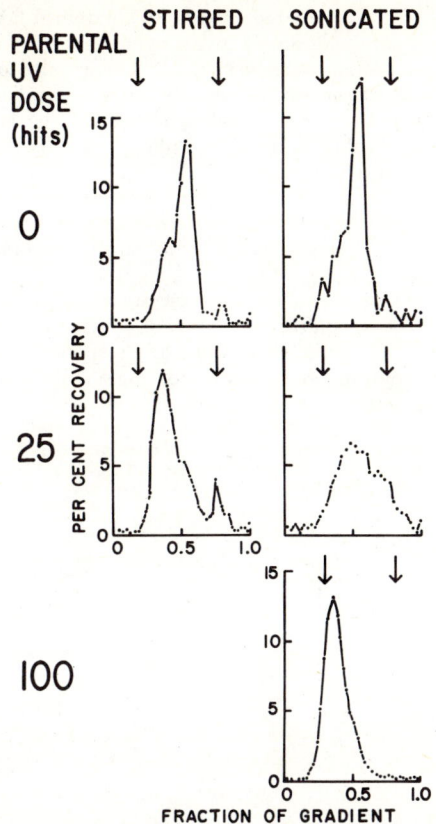

FIGURE 1. *E. coli* B23 were grown for two generations in a heavy medium (Kozinski and Kozinski, Virology 20: 213, 1963) and infected with light, ^{32}P-labeled (3 mc/mg P), UV-irradiated phage T4BOr[1] and nonirradiated phage, moi = 3 each. Irradiation doses were determined by extrapolation of the inactivation curve obtained with an unfiltered germicidal lamp. Progeny phage were purified by filtration and differential centrifugation. DNA was extracted with phenol. Samples were stirred in a Virtis homogenizer or sonicated in a Raytheon ultrasonic oscillator. Arrows give locations of reference DNA. The parental density contributions in the second row (25 hits) most likely come from uninjected phage.

doses corresponding to 0, 25, and 100 phage-lethal hits were used. The shearing procedures used were vigorous stirring and sonication. With no shearing, the parental label banded close to the density characteristic of the progeny DNA for all doses—these results are not shown. Sucrose gradient analysis of the stirred DNA indicated that a homogeneous population of fragments of approximately 2–5% of a complete phage molecule had been produced. Analysis of this material in CsCl gradients (Fig. 1, first column) shows the hybrid (light parental-heavy progeny) structure containing the parental contribution from nonirradiated phage,

but not from irradiated phage. Sonication, resulting in roughly ½ million Dalton fragments, succeeds in revealing the hybrid nature of the 25-hit parental contribution (column 2). However, no such configuration can be seen in the progeny DNA containing contributions from parental phages exposed to 100 hits.

These results clearly suggest that the size of the parental contribution to progeny phages decreases with increasing doses of UV irradiation administered to the parental phages.

It has previously been shown (Shahn and Kozinski, Virology 30: 455, 1966) that the parental contribution from nonirradiated phages consists of one continuous single-stranded piece. It is also known that in experiments similar to the one described both (1) the total amount of parental DNA appearing in the progeny population, and (2) the amount of DNA per progeny phage, decreases with increasing doses of UV (Shahn, dissertation, Univ. of Penna. 1965). From these facts one can calculate that the number of parental pieces per recipient progeny phage increases with increasing doses of UV administered to the parental phage.

In the second experiment, chloramphenicol (CM) was added to samples of the infected bacteria at 5 and 7 min after infection. Incubation continued for 20 and 45 min, at which times intracellular DNA was extracted and analyzed in CsCl density gradients. Similar analyses were also made on DNA

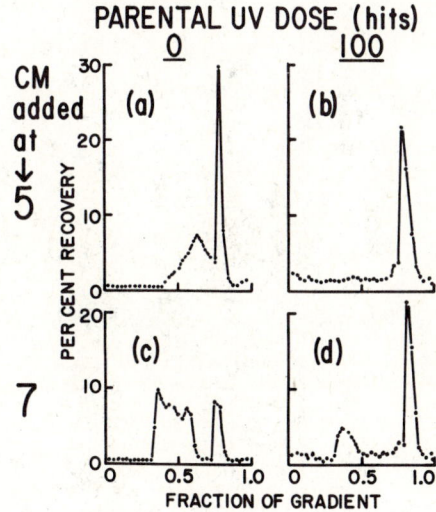

FIGURE 2. Chloramphenicol was added to samples of infected bacteria to a final concentration of 100 μg/ml. Intracellular DNA was extracted with SDS and phenol. Light, ^{3}H-labeled DNA was added as a reference prior to analysis in CsCl. The ^{3}H peaks coincided with the parental density ^{32}P peaks and have not been included.

samples extracted at several times from nontreated infected bacteria. UV doses of zero and 100 hits were used. The results of the 45-min samples are shown in Fig. 2.

When CM is added to the nonirradiated system at 5 min, semiconservative replication occurs, some of the label appearing at the hybrid density, (a). No similar replication of irradiated-phage DNA is seen in (b). Thus, UV-irradiated DNA cannot serve as a template in an in vivo system which can replicate DNA. When CM is added to the nonirradiated system at 7 min, (c), the density of a part of the parental label increases beyond hybrid, indicating that the phage enzymes responsible for recombination have by this time been synthesized (Kozinski, Kozinski and Shannon, PNAS 50: 746, 1963). In the irradiated system, (d), parental label is also seen at a density greater than hybrid, and in a somewhat narrower band. The results obtained at 20 min were essentially the same as these. The earliest time at which a significant peak of parental label from irradiated phage was observed at an increased density was 12 min. At this time it was as far removed from the parental density as it ever got. There was no indication at any time of label from irradiated DNA at an intermediate density.

It is concluded that rescue of irradiated phage DNA requires an enzyme or enzymes which, in the nonirradiated system, are responsible for recombination and repair, or are synthesized concurrently.

Enzymatic Repair of Deoxyribonucleic Acid. IV. Mechanism of Photoproduct Excision[*]

Sidney R. Kushner,[†] Joan C. Kaplan,[‡] Howard Ono, and Lawrence Grossman[§]

ABSTRACT: The enzymatic mechanisms controlling the incision and excision of UV-irradiated *Escherichia coli* DNA have been examined. The UV-endonuclease, which catalyzes the incision step in repair, hydrolyzes a single phosphodiester bond, 5' to the photoproduct region. This step is associated with the formation of a 3'-phosphoryl group on the DNA and a 5'-hydroxyl group either at, or one nucleotide 5', to the photoproduct. The release of either nucleotides or photoproducts does not occur to any significant extent under these circumstances. The excision of nucleotides, and those containing photoproducts, from incised DNA is dependent on the unique exonucleolytic properties of a second enzyme, the UV-exonuclease. This enzyme catalyzes the release of approximately seven nucleotides for each phosphodiester bond hydrolyzed during the incision of irradiated *E. coli* DNA. There are ten nucleotides released during the excision for each incision event of irradiated *Micrococcus luteus* DNA. The cavity of the excised *E. coli* DNA is terminated by two phosphoryl groups. A 3'-phosphoryl group arises from incision and after the excision event a 5'-phosphoryl group remains. Chromatographic analyses of the products released during the initial phases of excision revealed that oligonucleotides were liberated and subsequently degraded exonucleolytically by the same enzyme. It is concluded that the primary hydrolytic event during excision is endonucleolytic in nature. Moreover, the distribution of nucleotides 3' to the photoproduct in the excised fragment from irradiated *E. coli* DNA is enriched in adenine and thymine containing deoxynucleoside 5'-monophosphates. As the irradiation is intensified this distribution is modified to include cytosine-containing deoxynucleotides. The significance of these findings is discussed.

The recovery of organisms from the effects of ultraviolet irradiation during dark repair is mediated by a multistep enzymatic process, which is presumably initiated by the excision of pyrimidine dimers. Setlow and Carrier (1964) and Boyce and Howard-Flanders (1964) investigated this mechanism and showed that pyrimidine dimers were selectively removed from irradiated DNA during *in vivo* dark repair in *Escherichia coli*. Dimer excision has now been observed in UV-resistant strains of many bacteria including: *Hemophilus influenzae* (Setlow et al., 1968), *Bacillus subtilis* (Strauss et al., 1966), and *Micrococcus luteus* (Kaplan et al., 1969). In the human disease xeroderma pigmentosum sensitivity to sunlight, which is expressed as a high incidence of skin lesions and cancer, has been correlated with a defect in one of the early steps of dark repair (Cleaver, 1969). In fact, Setlow has shown that these cells excise pyrimidine dimers with difficulty (Setlow et al., 1969).

Both of the current models which explain excision-repair (Howard-Flanders and Boyce, 1966; Haynes, 1966) involve incision of the damaged DNA strand by an endonuclease and either immediate ("cut and patch") or delayed excision of the photoproduct-containing regions ("patch and cut").

We have shown previously that the excision of pyrimidine photoproducts *in vitro* requires two enzymes from *M. luteus* (Kaplan et al., 1969; Grossman et al., 1968). The properties of the UV-endonuclease and UV-exonuclease have been described in a separate paper (Kaplan et al., 1971). In this communication we describe the enzyme-catalyzed *M. luteus* photoproduct excision system and its relation to overall excision-repair mechanisms.

Experimental Section

Materials

The UV-endonuclease and UV-exonuclease from *M. luteus* are prepared as described previously (Kaplan et al., 1971). Polynucleotide kinase (kinase) was the generous gift of Dr. C. C. Richardson. Yeast-photoreactivating enzyme was kindly donated by Dr. J. K. Setlow. *E. coli* DNA polymerase was generously provided by Dr. A. Kornberg. Crystalline pancreatic DNase I, micrococcal nuclease, snake venom phosphodiesterase, calf spleen phosphodiesterase, and bacterial alkaline phosphatase were all products of the Worthington Biochemical Corp.

^{32}P-Labeled DNA is prepared from either *E. coli* B or *M. luteus* by the method of Grossman (1967). [^{3}H]Thymine-labeled DNA is obtained from *E. coli* B/r thymine$^-$ by the method of Mahler (1967). UV-irradiated double-stranded DNA and heat-denatured DNA are prepared as previously described (Kaplan et al., 1971). [^{3}H]Thymine-labeled poly-(dA:dT) is synthesized enzymatically (Riley et al., 1966).

Methods

Enzyme Assays. UV-Endonuclease and UV-exonuclease are assayed as described previously (Kaplan et al., 1971). Pancreatic DNase I is assayed with native ^{32}P-labeled *E. coli*

---

[*] Contribution 788 from the Graduate Department of Biochemistry, Brandeis University, Waltham, Massachusetts 02154. *Received April 28, 1971.* This work was supported by an American Cancer Society Research Grant P-573A, Research Contract AT (30-1)3449 from the Atomic Energy Commission, Research Grant GM 15881 from the National Institute of General Medical Sciences, and Research Grant GB6208 from the National Science Foundation. In addition L. G. has been supported by a Research Career Development Award K03-GM 04845 from the National Institute of General Medical Sciences.
[†] Present address: Department of Biochemistry, Stanford University Medical Center, Stanford, California 94305.
[‡] Present address: Infectious Disease Unit, Massachusetts General Hospital, Boston, Mass. 02114.
[§] To whom correspondence should be addressed.

DNA as a substrate in which the reaction course is followed by the standard UV-endonuclease assay (Kaplan *et al.*, 1971). The incubation mixture contains 7.7 nmoles of ^{32}P-labeled native DNA (specific activity 20–100 cpm/pmole), 3 μmoles of sodium acetate buffer (pH 5), and 1.5 μmoles of MgSO$_4$ in 0.3-ml volume. *One unit of activity is defined as that amount of enzyme catalyzing the release of 10 pmoles of [^{32}P]P$_i$ phosphomonoester groups by bacterial alkaline phosphates In 30 min at 37°.* Micrococcal nuclease is assayed in the same manner. Each reaction mixture includes 10 μmoles of sodium borate buffer (pH 8.8), 0.5 μmole of CaCl$_2$, and 7.7 mμmoles of ^{32}P-labeled native *E. coli* DNA (specific activity 20–100 cpm/pmole of nucleotide equivalent). The enzyme is diluted immediately before use in a solution of bovine serum albumin (1 mg/ml in H$_2$O). One unit is defined in the same manner as described for pancreatic DNase I.

Polynucleotide kinase is assayed according to the method of Richardson (1965). The yeast-photoreactivating enzyme is assayed using the procedure developed for the UV-endonuclease. The incubation mixture (0.5 ml) contains 6 mμmoles of UV-irradiated ^{32}P-labeled native *E. coli* DNA (specific activity 20–40 cpm/pmole of nucleotide equivalent, total incident UV dose = 1.3×10^5 ergs/mm^2 (precalibrated 15-W GE germicidal lamp)), 25 μmoles of Tris-HCl buffer (pH 7.5), 5 μmoles of MgCl$_2$, and 1 μmole of EDTA. Following a 10-min incubation in the dark at 37°, the solution is incubated for 2 hr at 37° in front of a 150-W General Electric photoflood lamp. The photoreactivated DNA is then examined under the standard UV-endonuclease reaction conditions to measure the disappearance of UV-endonuclease-susceptible pyrimidine–pyrimidine dimers. *One unit of activity is defined as that amount of enzyme catalyzing the disappearance of 10 pmoles of UV-endonuclease-susceptible photoproducts under standard assay conditions.*

^{32}*P-Labeled DNA Incised with Pancreatic DNase I.* Pancreatic DNase I catalyzes the formation of single-strand breaks in DNA producing 5′-phosphoryl and 3′-hydroxyl end groups (Laskowski, 1961). The incubation mixture (0.3 ml) contains 7.7 nmoles of native ^{32}P-labeled *E. coli* DNA (specific activity 50–80 cpm/pmole), 3 μmoles of sodium acetate buffer (pH 5.0), 1.5 μmoles of MgSO$_4$, and 8 units of pancreatic DNase I. The enzyme is diluted immediately before use in distilled water. After incubation for 30 min at 37° the reaction is stopped by adding 0.2 ml of calf thymus DNA (2.5 mg/ml) and 0.3 ml of 7% perchloric acid. The precipitated DNA is collected by centrifugation at 12,500g for 10 min and washed with cold distilled water. For determination of the number of single-strand breaks, the DNA is dissolved in 0.5 ml of 0.1 N NaOH and assayed using the second part of the standard UV-endonuclease assay (Kaplan *et al.*, 1971).

^{32}*P-Labeled DNA Incised with Micrococcal Nuclease.* Micrococcal nuclease cleaves phosphodiester bond to yield 3′-phosphoryl and 5′-hydroxyl groups (Cunningham *et al.*, 1956). The reaction mixture is similar to that described for the micrococcal nuclease assay described above and includes 8 units of enzyme. A 30-min incubation at 37° is followed by treatment as described in the previous section.

UV-Irradiated ^{32}P-Labeled DNA Incised with UV-Endonuclease. The incubation mixture (0.3 ml) contains 7.7 nmoles of UV-irradiated ^{32}P-labeled DNA (specific activity 50–80 cpm/pmole of nucleotide equiv), 5 μmoles of potassium phosphate buffer (pH 7.5), 3 μmoles of MgCl$_2$, 5 μmoles of 2-mercaptoethanol, and 4–10 units of UV-endonucleaes. After incubation at 37° for 30–45 min, the enzymatic reaction is terminated by heating at 65° for 15 min. Large batches of incised UV-irradiated DNA are prepared by pooling individual tubes. In some cases the DNA was precipitated by procedures already described for pancreatic DNase I.

UV-Irradiated ^{32}P-Labeled Excised DNA. Excised DNA refers to UV-irradiated ^{32}P-labeled DNA which is first treated with UV-endonuclease and then with UV-exonuclease. Sufficient time is permitted for complete removal of those photoproduct regions recognized by the UV-endonuclease. The initial incubation is identical with that described for incised UV-irradiated DNA. After the UV-endonuclease activity is terminated by heating at 65° for 15 min and the reaction tubes chilled, 10.0 units of UV-exonuclease are added, and the mixture incubated for 2 hr at 37°.

The isolation of excision products is accomplished in one of three ways. The reaction sequences may be terminated by perchloric acid treatment followed by solubilization arising from neutralization with alkali. The acid-soluble fractions are chromatographed directly. In some cases the reaction mixtures are chromatographed directly without acid precipitation, or the reactions are stopped with 2 M NaCl and ethanol in the presence of carrier DNA. In all cases the nature and distribution of excision products are chromatographically similar.

Analysis of Excision Products from UV-Irradiated DNA. Two types of chromatographic systems were employed for measuring levels of oligonucleotides and mononucleotides. Mononucleotide resolution is best obtained by spotting reaction mixtures on Whatman No. 3MM chromatography paper which is developed for 24 hr in a descending direction in isobutyric acid–NH$_4$OH–H$_2$O (99:1:33, v/v) (Carrier and Setlow, 1966). Resolution of the oligonucleotides according to chain length is achieved in a two-step DEAE-chromatography system developed by Kelley *et al.* (1969, 1970).

Excision products are categorized into two major classes according to their chromatographic mobility with standard oligonucleotide markers. Mononucleotide excision products (MEP)[1] consist of the four deoxynucleoside monophosphates and oligonucleotide excision products (OEP) consist of resolvable nucleotides of chain length (pX)$_2$–(pX)$_7$. Large quantities of excision products are obtained by combining individaul reaction mixtures.

Digestion of Incised ^{32}P-Labeled DNA with Snake Venom Phosphodiesterase. Incised ^{32}P-labeled *E. coli* DNA, which is prepared with pancreatic DNase I, micrococcal nuclease, or UV-endonuclease, is dissolved with gentle vortex mixing in 0.01 N NaOH following perchloric acid precipitation. These DNA samples are then brought to 100° for 10 min and rapidly chilled. The pH is adjusted to 8.8 by adding 6 μmoles of Tris-HCl (pH 6.45). The reaction mixture (0.6 ml) also contains 3 μmoles of MgCl$_2$, 1 μmole of 2-mercaptoethanol, and 1 unit of snake venom phosphodiesterase. The reaction is stopped at various times with perchloric acid precipitation as described for incision with pancreatic DNase I. The supernatant fluid is counted in Bray's–dioxane solution.

Digestion of Incised ^{32}P-Labeled DNA with Calf Spleen Phosphodiesterase. Incjsed ^{32}P-labeled *E. coli* DNA is prepared as described previously and dissolved with gentle vortex mixing in 0.01 N NaOH. The pH is adjusted to 6.5 by adding 2.5 μmoles of succinic acid and 12.5 μmoles of sodium succinate, (pH 6.5). The incised DNA is denatured by heating. The final mixture (0.605 ml) also contains 3 μmoles of MgCl$_2$, 10 μmoles of 2-mercaptoethanol, and 1.0 unit of calf spleen

---

[1] Abbreviations used are: MEP, mononucleotide excision products; OEP, oligonucleotide excision products.

phosphodiesterase. The reaction is terminated as described in the previous section.

Digestion of Oligonucleotide Excision Products with Snake Venom Phosphodiesterase. Each reaction mixture (0.1 ml) contains 3 μmoles of Tris-HCl buffer (pH 8.8), 0.5 μmole of $MgCl_2$, 0.2 μmole of 2-mercaptoethanol, 20–100 nmoles of OEP, and 0.5 μg of snake venom phosphodiesterase. The solution is incubated for 4 hr at 37° and then spotted directly onto either DEAE- or Whatman chromatography paper.

Digestion of OEP with Calf Spleen Phosphodiesterase. The reaction mixture (0.1 ml) contains 5 μmoles of potassium phosphate buffer (pH 7.0), 0.5 μmole of $MgCl_2$, 0.2 μmole of 2-mercaptoethanol, 20–100 nmoles of OEP, and 30 μg of calf spleen phosphodiesterase. The solution is incubated for 4 hr at 37° and then spotted directly onto DEAE- or Whatman No. 1 chromatography paper.

Norit Assay. Following a 2-hr excision of UV-irradiated DNA as described previously, 60 μmoles of Tris-HCl buffer (pH 8.0) and 2.0 units of bacterial alkaline phosphatase are added to each reaction mixture. After a 30-min incubation at 45°, 0.1 ml of 1 N HCl, 0.2 ml of 20% Norit, and 0.12 ml of H_2O are added. The resulting mixture is shaken vigorously for 20 min. At this point 0.2 ml of bovine serum albumin (5.0 mg/ml) is added and the solution (1.0 ml) shaken for another 10 min. The supernatant fraction obtained from a 10-min centrifugation is counted in Bray's–dioxane scintillation fluid.

Dephosphorylation of Excision Products. The reaction mixture contains 10 μmoles of Tris-HCl buffer (pH 8.0), 20–100 nmoles of excision products, and 0.2 unit of bacterial alkaline phosphatase. After an incubation of 3 hr at 45°, the reaction is terminated by spotting onto either DEAE- or Whatman chromatography paper.

Results

Site of Initial Phosphodiester-Bond Breakage. SPLEEN PHOSPHODIESTERASE SENSITIVITY. The introduction of single phosphodiester-bond breaks either 3' or 5' to the photoproduct by the UV-endonuclease is presumably the first step in dark repair. These two possibilities were examined by measuring the rate of calf spleen phosphodiesterase catalyzed hydrolysis. This enzyme digests single-stranded DNA exonucleolytically initiating hydrolysis from the 5'-hydroxyl terminus liberating deoxynucleoside 3'-monophosphates (Razzell and Khorana, 1958). The course of its exonucleolytic action is inhibited by the presence of photoproducts (Pearson and Johns, 1966). The action of the UV-endonuclease results in 5'-hydroxyl and 3'-phosphoryl groups during incision (described in a subsequent section). It is to be expected, therefore, that the spleen phosphodiesterase should catalyze at a normal, or unimpaired, initial rate of hydrolysis on denatured incised UV-irradiated DNA if the initial break is 3' to the photoproduct. On the other hand, the reverse situation involving a break 5' to the photoproduct would be expected to interfere with the enzymatic reaction. In Figure 1 it can be seen that spleen phosphodiesterase was in fact unable to digest incised irradiated DNA after it had been denatured, although it readily hydrolyzed denatured DNA bearing 5'-hydroxyl and 3'-phosphoryl groups at its termini following micrococcal nuclease treatment. The inhibition of spleen phosphodiesterase activity indicated, therefore, that the initial incision was not only 5' to the photoproduct but was fairly close to its site.

The lack of a 5'-phosphoryl group on the damaged portion of the DNA seems not to be responsible for the retardation of exonucleolytic rates by the spleen phosphodiesterase. The

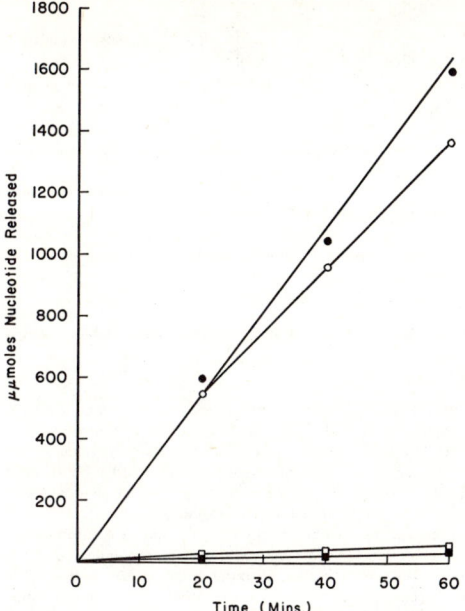

FIGURE 1: Digestion of incised ^{32}P-labeled DNA with spleen phosphodiesterase. 5'-Hydroxyl and incised irradiated DNA were prepared according to procedures outlined in the Methods section; similarly, the specific requirements for spleen phosphodiesterase activity are listed in the same section. The samples to be photoreactivated were treated with 150 μmoles of Tris-HCl buffer (pH 7.4); 7.5 μmoles of EDTA and 3.5 units of yeast-photoreactivating enzyme (PR) are added in a final volume of 3.0 ml. The conditions in photoreactivation are described in the legend to Table II. 5'-Hydroxyl-terminated DNA (●), incised irradiated DNA (■), incised irradiated DNA treated with photoreactivating enzyme in the presence of visible light (○), and incised irradiated DNA treated with photoreactivating enzyme in the absence of light (□).

block to hydrolysis if due exclusively to the presence of thymine-containing dimers should be relieved by enzymatic photoreactivation of the lesion prior to the action of spleen phosphodiesterase. Incised DNA was subjected to enzymatic photoreversal of the thymine-containing dimers prior to hydrolysis by the spleen enzyme. Enzymatic photoreactivation, requiring visible light, produced an active substrate for the spleen enzyme, whereas an equivalent dark incubation (in the presence of the photoreactivating enzyme but in the dark) did not provide a sensitive substrate.

PHOSPHOMONOESTER STOICHIOMETRY. Further proof concerning the position of the photoproduct relative to the initial break was obtained from the stoichiometry of [^{32}P]P_i release by bacterial alkaline phosphatase following incision and then excision. UV-endonuclease action produces one phosphomonoester group which is sensitive to bacterial alkaline phosphatase. Since the UV-exonuclease produces 5'-mononucleotides (Kaplan *et al.*, 1971), a second bacterial alkaline phosphatase susceptible phosphomonoester group should, in fact, appear after excision. If the initial phosphodiester break by

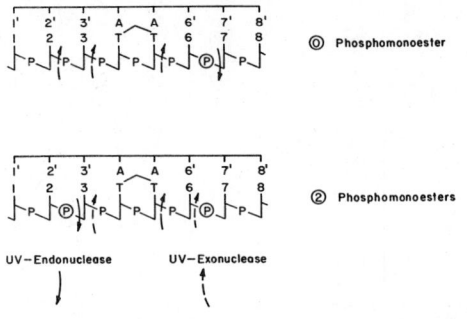

FIGURE 2: The structural basis for phosphomonoester stoichiometry.

the UV-endonuclease was 3' to the photoproduct, then excision would result in a quantitative loss of phosphomonoester groups from the incised DNA (Figure 2). Experiments which determined the number of phosphodiester bonds after incision alone and incision plus excision clearly demonstrated that there was a stoichiometric appearance of a second phosphomonoester group (Table I). This apparent doubling of phosphomonoester groups could only be accounted for if the initial break was, in fact, 5' to the site of damage in the DNA.

The incision process, therefore, includes an initial phosphodiester-bond break 5' to the photoproduct. Since a negligable number of nucleotides were released by spleen phosphoesterase, the initial break may be adjacent to or possibly one nucleotide removed from the dimer region.

LOCATION OF PHOSPHOMONOESTER GROUP AFTER INCISION. Polynucleotide kinase catalyzes the phosphorylation of the γ-phosphorous from [γ-^{32}P]ATP specifically into 5'-hydroxyl groups of DNA (Richardson, 1965). Incorporation of labeled phosphorous into incised DNA would suggest phosphodiester cleavage resulted in the formation of 5'-hydroxyl groups. However, if pretreatment with bacterial alkaline phosphatase is required in order to achieve quantitative labeling, then it may be concluded that the original incision produced a 5'-phosphomonoester. When UV-irradiated incised DNA was treated with polynucleotide kinase under conditions in which pancreatic DNase I treated DNA was quantitatively phosphorylated after treatment with bacterial

TABLE I: Effect of UV-Exonuclease on the Apparent Number of Phosphodiester Bonds Broken.[a]

	Phosphomonoester Groups Detected after:	
Incubn Time (min)	Incision (pmoles)	Excision (pmoles)
90	22.3	43.6
150	36.2	68.9
270	58.0	112.0

[a] Incised and excised DNA were prepared as described in Methods. The number of phosphomonoester groups was determined as described in the previous publication (Kaplan et al., 1971).

TABLE II: Location of Phosphomonoester Group after Incision.[a]

Exptl Condn	Phosphodiester Bonds Broken (pmoles)	^{32}P-Labeled 5'-OH Groups (pmoles)
A. Native *E. coli* DNA		
1. Pancreatic DNase I	21.2	
2. (1) + kinase[b]		0.4
3. (1) + BAP[c] + kinase		17.0
B. UV-Irradiated *E. coli* DNA		
1. UV-Endonuclease	234.0	
2. Photoreactivation + UV-endonuclease	20.0	
3. (1) ± BAP + kinase		25.0
4. (1) + photoreactivation ± BAP + kinase		206.4

[a] DNA substrates were treated sequentially with the listed enzymes. The incubation mixture in part A (0.3 ml) contained 20 μmoles of Tris-HCl buffer (pH 8.0), 1.5 μmoles of MgCl$_2$, 78.6 nmoles of native *E. coli* DNA, and 1.0 unit of pancreatic DNase I. After a 30-min incubation at 25°, the DNA was collected by adding 0.3 ml of 1.0 M NaCl and 1.2 ml of 95% EtOH, and centrifuging at 12,500g for 20 min. The precipitate was dissolved in 0.1 ml of 0.1 N NaOH and brought to pH 8.0 with 20 μmoles of Tris-HCl buffer (pH 6.45). For dephosphorylation 0.2 unit of bacterial alkaline phosphatase was added and the mixture incubated for 30 min at 45°. For reaction with polynucleotide kinase the above solution was chilled and 1.0 μmole of potassium phosphate buffer (pH 7.5), 3 μmoles of MgCl$_2$, 5 μmoles of 2-mercaptoethanol, 20 μmoles of γ-^{32}P-labeled ATP (specific activity 2.2 × 10^3 cpm/nmole), and 2.0 units of enzyme were added. The incubation at 37° for 60 min was terminated by adding 3.0 ml of cold 5% trichloroacetic acid and collecting the DNA on GF/C glass filters. The reaction mixture in part B (3.0 ml) contained 5.0 μmoles of Tris-HCl buffer (pH 7.5), 3.0 μmoles of MgCl$_2$, 5.9 μmoles of 2-mercaptoethanol, 83.4 nmoles of UV-irradiated *E. coli* DNA (1.3 × 10^5 ergs/mm^2), and 23 units of UV-endonuclease. After an incubation at 37° for 30 min, the reaction was stopped by heating at 65° for 15 min. To the chilled solutions were then added 150 μmoles of Tris-HCl buffer (pH 7.4), 7.5 μmoles of EDTA, and 3.5 units of yeast-photoreactivating enzyme. An initial 15-min incubation at 37° in the dark was followed by 2.5 hr at 37° in front of a 150-W photoflood lamp. The DNA was collected by alcohol precipitation as described in part A. The dephosphorylation and polynucleotide kinase labeling procedures were identical with those described in part A. The number of phosphodiester bonds broken was determined with ^{32}P-labeled *E. coli* DNA as described previously (Kaplan et al., 1971). [b] Polynucleotide kinase. [c] Bacterial alkaline phosphatase.

alkaline phosphatase, incorporation of the terminal phosphate from [γ-^{32}P]ATP into internal breaks was not observed. These negative findings were inconsistent with the single-stranded nature of the breaks produced during incision (Grossman et al., 1968; Carrier and Setlow, 1970).

The possibility that thymine-containing photoproducts

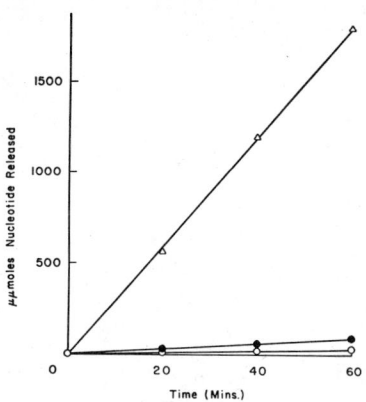

FIGURE 3: Digestion of incised DNA with venom phosphodiesterase. 3′-Hydroxyl-terminated, 3′-phosphonyl-terminated, and incised UV-irradiated *E. coli* DNA were prepared according to the procedures described in the Methods section. The number of phosphodiester bonds hydrolyzed in the various substrates were as follows: 3′-hydroxyl-terminated DNA (250 pmoles), 3′-phosphoryl-terminated DNA (140.2 pmoles), and incised UV-irradiated DNA (171.1 pmoles). Digestion of the heat-denatured incised DNAs with venom phosphodiesterase was carried out according to the procedures described in the Methods section. The figure illustrates the kinetics of digestion of 3′-hydroxyl-terminated DNA (△), 3′-phosphoryl-terminated DNA (○), and incised UV-irradiated DNA (●). Nucleotide release was corrected for self-absorption.

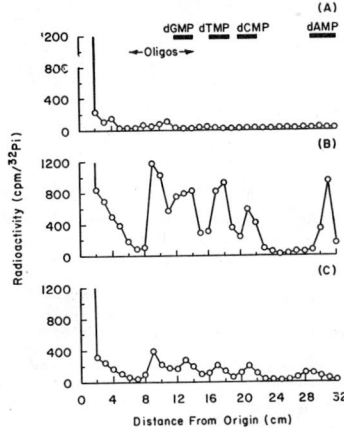

FIGURE 4: Excision as a two-step process. Incised and excised ^{32}P-labeled UV-irradiated *E. coli* DNA were prepared as described in Methods. Prior to precipitation with perchloric acid, the reaction mixtures (0.3 ml) were spotted directly on Whatman No. 3MM paper and developed as described in Methods. Radioactivity was determined as described in Methods. The bars on the graph represent reference compounds. The figure illustrates irradiated DNA plus UV-endonuclease (A), irradiated DNA plus UV-endonuclease and UV-exonuclease (B), and irradiated DNA plus UV-exonuclease (C).

might be interfering with the action of the polynucleotide kinase was investigated. Incised DNA was incubated with the yeast-photoreactivating enzyme and visible light—conditions which monomerize thymine dimers *in situ* (Cook and McGrath, 1967). Under these circumstances stoichiometric phosphorylation was obtained with the polynucleotide kinase independent of bacterial alkaline phosphatase pretreatment (Table II). These results clearly indicated that the endonucleolytic incision produced a 5′-hydroxyl group. Surprisingly, however, the action of the relatively nonspecific polynucleotide kinase was sensitive to the presence of pyrimidine photoproducts. It can be concluded, furthermore, that the photoproduct must be fairly close to the site of the initial phosphodiester bond hydrolyzed by the UV-endonuclease to impose this kind of restriction on the course of action of the polynucleotide kinase.

The presence of a 3′-phosphomonoester group was identified by examining the effects of dephosphorylation on the rates of exonucleolytic hydrolysis by snake venom phosphodiesterase. This enzyme initiates hydrolysis of denatured DNA from the 3′ terminus liberating 5′-deoxynucleoside monophosphates. In Figure 3 it can be seen that the enzyme degrades 3′-hydroxyl-terminated DNA at a rate faster than 3′-phosphoryl-terminated substrates. When incised DNA was exposed to the action of venom enzyme, this rate of hydrolysis coincided with that of the 3′-phosphoryl-terminated denatured DNA. Incised UV-irradiated DNA after dephosphorylation with bacterial alkaline phosphatase was digested normally by the enzyme. Since it has been shown that the initial break is 5′ to the photoproduct, this result is in direct support for a phosphomonoester group at the 3′ terminus. Other indirect support for the presence of a phosphomonoester located at this position is derived from the phosphomonoester stoichiometry in Table II.

Mechanism of Excision. DEMONSTRATION OF A TWO-STEP EXCISION PROCESS. The initial steps of the dark repair process involve the enzymatic recognition of DNA which contains photoproducts followed by the removal of pyrimidine dimers from such molecules. The primary incision by the UV-endonuclease does not result in the release of photoproducts (Kaplan *et al.*, 1971). However a second enzyme, the UV-exonuclease, can excise the photoproduct-containing region from the DNA molecule once the initial break has been made by the UV-endonuclease (Kaplan *et al.*, 1969, 1971). A chromatographic analysis of the two-step excision process is shown in Figure 4. When UV-irradiated DNA was treated sequentially, first with UV-endonuclease and then UV-exonuclease, extensive release of mono- and oligonucleotides occurred (Figure 4B). As expected from previously reported results (Kaplan *et al.*, 1971) very small amounts of radioactivity appeared in the oligonucleotide region when UV-irradiated *E. coli* DNA was reacted with UV-endonuclease alone (Figure 4A). At the dosage of irradiation employed, some pyrimidine dimers may have been sufficiently close so that successive single-strand breaks could release some oligonucleotides. For example, irradiation of [^{3}H]thymine-labeled poly(dA:dT) leads to extensive degradation by the UV-endonuclease under similar conditions (unpublished results).

Products obtained from an identical incubation of DNA with UV-exonuclease alone contained small amounts of mono- and DNA, probably arising from digestion of DNA termini (Figure 4C). The excision system, therefore, consists of two distinct steps, incision by UV-endonuclease and excision by UV-exonuclease. Further demonstration of this process is discussed and seen in Figure 5.

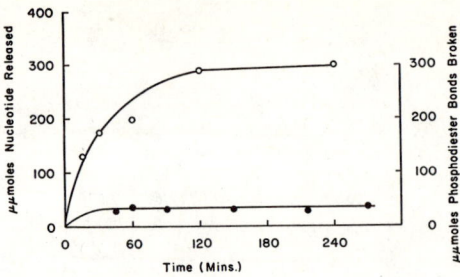

FIGURE 5: Excision of nucleotides from UV-irradiated ^{32}P-labeled native *M. luteus* DNA. The number of phosphodiester bonds broken and the amount of nucleotides released were measured by the standard UV-endonuclease assay (see Methods). The figure illustrates the number of phosphodiester bonds broken by the UV-endonuclease (●) and the amount of nucleotide released resulting from UV-exonuclease action (○).

EXTENT OF DNA EXCISION. The ability of the UV-exonuclease to degrade substrates containing photoproducts makes it uniquely suited as an excision enzyme. After UV-irradiated ^{32}P-labeled *E. coli* DNA is incised with the UV-endonuclease, treatment with the exonuclease produces a rapid but limited release of ^{32}P-labeled nucleotides (Kaplan *et al.*, 1969). Similar results are obtained when UV-irradiated *M. luteus* DNA is exposed to the two enzymes (Figure 5). A size estimate of the excised region was obtained by comparing the ratio of ^{32}P released to the number of phosphodiester bonds broken by the UV-endonuclease. A ratio of 6 is obtained for *E. coli* DNA at all doses of irradiation between 1.0×10^4 and 1.5×10^5 ergs/mm^2 (Table III). In the case of *M. luteus* DNA a significant increase to a ratio of 10 was obtained (Figure 5).

The limited degradation of incised DNA by the UV-exonuclease is attributable to its ability to act only on denatured DNA. This implies that the excisable region is of limited size and single-stranded in character. In addition, preliminary evidence suggests that the UV-exonuclease does not cleave the phosphodiester bond between the pyrimidine dimer moieties. These findings imply that the exonuclease can in fact act endonucleolytically under certain structurally constrained conditions.

POSITION OF THE SECOND PHOSPHOMONOESTER GROUP ON EXCISED DNA. The UV-exonuclease hydrolyzes both unirradiated and irradiated denatured DNA substrates to produce 5′-mononucleotides (Kaplan *et al.*, 1969). Similarly nucleoside 5′-monophosphates are released during excision of incised UV-irradiated native DNA. It is expected as a result that a

TABLE III: Effect of Ultraviolet Dose on the Size of Excised Regions.

UV Dose (ergs/mm^2 × 10^{-3})	Phosphodiester Bonds Broken (pmoles) Incision	[^{32}P]Nucleotides Released (pmoles) Excision	Released Nucleotides: Phosphodiester Bonds
9	19.5	120.1	6.15
36	63.0	359.2	5.7
108	85.1	471.7	5.5

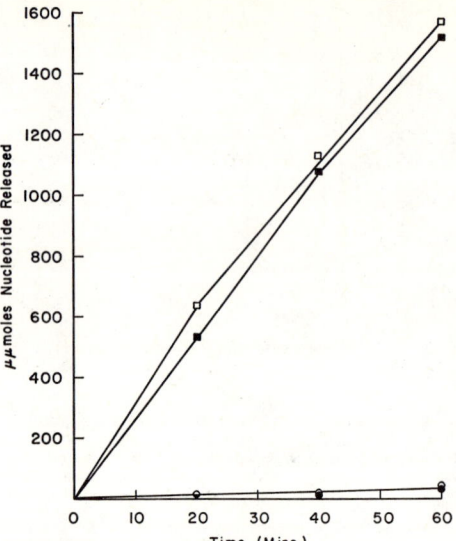

FIGURE 6: Digestion of excised ^{32}P-labeled DNA with spleen phosphodiesterase. 5′-Hydroxyl, 5′-phosphorylated, and excised DNAs were prepared according to the procedures described in the Methods section. Conditions describing the techniques for dephosphorylation of excised DNA are the same as those described in the legend associated with Table II. The experimental procedures followed for spleen phosphodiesterase digestion are described in the Methods section of this paper. The figure illustrates the kinetics of digestion of 5′-hydroxyl-terminated denatured DNA (□), 5′-phosphorylated denatured DNA (○), excised denatured DNA (●), and dephosphorylated excised denatured DNA (■).

5′-phosphomonoester should remain on the DNA after excision by this same enzyme.

The incision step results in the formation of a 3′-phosphomonoester group, consequently the appearance of a second phosphomonoester group as a consequence of the excision step implies that its specific location is likely to be at the 5′ position. Verification of the 5′-phosphoryl group in the cavity of excised DNA is provided from exonucleolytic hydrolysis rates catalyzed by the spleen phosphodiesterase which is sensitive to the presence of a 5′-phosphomonoester at the site of its initiation. The resistance of excised ^{32}P-excised DNA to hydrolysis by the spleen phosphodiesterase before treatment with bacterial alkaline phosphatase implicates the presence of a phosphomonoester at this specific site on excised DNA (Figure 6).

NATURE OF THE PRIMARY EXCISION PRODUCTS. Chromatographic analyses of the excision products released as a function of time revealed that the initial events of the excision process appear to be endonucleolytic in nature (Figure 7). The larger oligonucleotide excision products are liberated from the irradiated DNA substrate and can be identified directly from the reaction milieu—from either an alcohol-soluble fraction of the reaction or from an acid-soluble supernatant fraction arising from termination of the reaction with 7% perchloric acid. As the DNA becomes further damaged by UV-irradiation, it can be assumed that the possibility exists for damaged

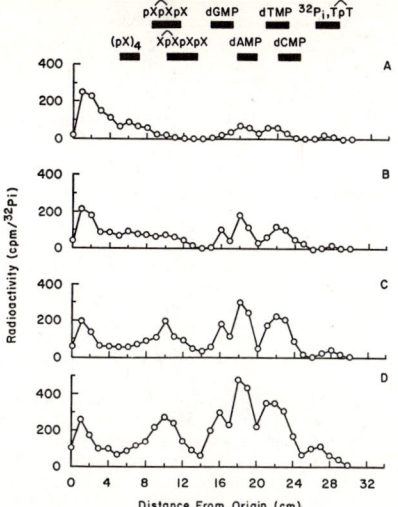

FIGURE 7: Composition of excision products as a function of time. Excision products from UV-irradiated E. coli DNA (3.6 × 10⁴ ergs/mm²) were isolated as described in Methods. The figure illustrates excision products at 5 min (A), 10 min (B), 40 min, (C), and 130 min (D).

TABLE IV: Size Distribution of Oligonucleotide Excision Products.[a]

Excision Time (min)	Pentanucleotide or Longer (%)	Tetranucleotide or Pentanucleoside Tetraphosphate (%)	Trinucleotide or Tetranucleoside Triphosphate (%)
5	68	23	9
10	54	24	22
40	39	11	49
130	33	7	60

[a] The size estimates were obtained from the data shown in Figure 6. Size assignments based on chromatographic mobility were as follows: pentanucleotides, 0–4 cm; tetranucleotides, 5–7 cm; trinucleotides, 8–13 cm. Corrections were made for overlapping of peaks. The radioactivity appearing in strips 0–13 was taken as the total amount of oligonucleotide excision products.

regions becoming correspondingly closer exists. If this phenomenon is involved then acid denaturation of excised DNAs should liberate acid-soluble fragments between and including excised regions. If this in fact were the case, a rise in the ratio of acid-soluble nucleotides released per phosphodiester bond broken by the endonuclease might be observed as a function of dose. This is not the case, however, as seen in Table III in which this ratio remains relatively constant over a 10-fold increase in the dose range of ultraviolet light.

During the course of excision, the large oligonucleotides are degraded to smaller fragments. At 5 min, 68% of the excised oligonucleotides were at least as large as a pentanucleotide. Only 9% appeared as either a trinucleotide or a tetranucleoside triphosphate (Table IV). As the time course of excision increases, a marked shift in the size of the OEP shifted to shorter chain lengths. For example, at 130 min of excision 60% of the original radioactivity in the oligonucleotide region was converted into trinucleotides or tetranucleoside triphosphates. Shorter oligonucleotides such as XpXpX and pXpX were indistinguishable from the mononucleotides.

The UV-exonuclease appears, therefore, to initially act endonucleolytically on the incised DNA molecule. The released fragment is subsequently degraded exonucleolytically by the same enzyme. The ability of the UV-exonuclease to hydrolyze the single-stranded DNA and oligothymidylates exonucleolytically has been previously discussed (Grossman et al., 1968; Kaplan et al., 1971).

The 5′→3′-exonuclease activity associated with the homogeneous E. coli DNA polymerase also acts endonucleolytically on nicked irradiated poly(dA:dT) (Kelly et al., 1969).

OLIGONUCLEOTIDE:MONONUCLEOTIDE RATIOS. The ratio of phosphodiester bonds to phosphomonoester groups in excision products is presented in Figure 8. At early times of excision, 81% of the excised material was insensitive to bacterial alkaline phosphatase, while at 90 min this percentage had dropped to 53% (Table V). The UV-exonuclease was therefore degrading the fragments, but the extent of hydrolysis appeared to decrease with the size of the fragments.

Results from chromatographic analyses of the oligonucleotide to mononucleotide ratio are summarized in Table VI. Because small oligonucleotides such as XpXpX or pXpX migrated with the mononucleotides, the level of mononucleo-

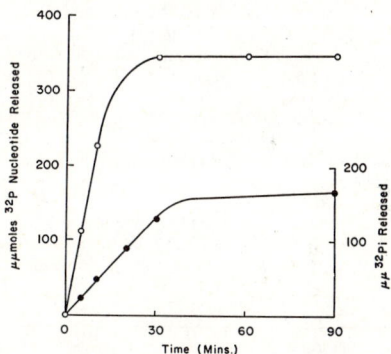

FIGURE 8: ³²P-Labeled nucleotide release vs. [³²P]Pi release by bacterial alkaline phosphatase from excision products: a kinetic comparison. Incised ³²P-labeled UV-irradiated E. coli DNA was prepared as described in Methods. To each tube was added 4.0 units of UV-exonuclease and excision allowed to take place at 37°. At specified times tubes were removed and heated at 100° for 5 min. After chilling, one-half of the tubes were precipitated as described in Methods and the supernatant fluid counted in Bray's–dioxane scintillation fluid. The other tubes were treated with the Norit assay (see Methods). All values were corrected for self-absorption. The figure illustrates ³²P-labeled nucleotide release (○) and [³²P]Pi release by bacterial alkaline phosphatase from the ³²P-labeled excision products (●).

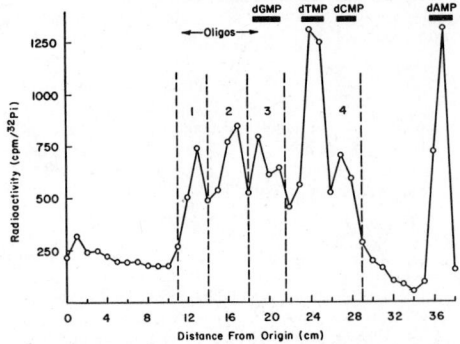

FIGURE 9: Chromatographic analysis of excision products. Excision products from ^{32}P-labeled UV-irradiated *E. coli* DNA (3.6 × 10^4 ergs/mm^2) were prepared as described in Methods and chromatographed as described in Figure 1. The dotted vertical lines indicate those regions which were eluted for further analysis.

TABLE VI: Distribution of Excision Products.

Excision Time (min)	A (pX)$_6$– (pX)$_8$ (%)	% Change in Concn	B (pX)$_3$– (pX)$_5$ (%)	C (pX)$_1$ (%)	%^{a} Change in Concn of B + C
5	51.8	—	18.0	27.4	—
10	29.6	−22.2	24.1	43.7	+22.4
40	16.2	−13.4	26.1	53.7	+12.0
130	12.3	−4.1	24.6	57.0	+3.3

^{a} This is an incremental change calculated as a sum of differences B + C from one time period to the next later period.

tides arising from extended excision times may have been artifically increased. However, even after extended periods of excision the limit digest product, accounting for 40–45% of the total oligonucleotide excised, is not larger than a di- or trinucleotide. Even at 130 min not all of the fragments had been digested. The level of 40–45% for oligonucleotide excision product concentration, which correlates well with the data presented in Figure 7, was therefore taken to mean that at 130 min the average chain length of the photoproduct containing fragments was between two and three nucleotides in length.

IDENTIFICATION OF RELEASED MONONUCLEOTIDES. A comparison was made of the mononucleotides released by the UV-endonuclease from DNA irradiated as a function of dose. At a dose of 9 × 10^3 ergs/mm^2, 46.8% of all the nucleotides released were dTMP-5, indicating that initial dimer formation occurred in thymine-rich regions (Table VII). The second most prevalent nucleotide was dAMP-5'. As the dose was increased to 108 × 10^3 ergs/mm^2, the level of dTMP dropped significantly to 30.7% while dAMP-5' and dCMP-5' increased to 35.8 and 24.2%, respectively. At the higher doses, detectable dimer formation occurred in less homogeneous areas of the DNA molecule.

A similar compositional analysis of nucleotides released from *M. luteus* DNA (GC:AT = 2.7) showed a different spectrum of excised nucleotides. At a comparable dose the excised region of *M. luteus* DNA is more pyrimidine rich than is the same area from *E. coli* DNA. Moreover, the region appears to be less GC rich than the overall base composition of this DNA.

PERCENTAGE OF THYMINE IN THE EXCISED REGION. We have shown previously that approximately three thymine-containing nucleotides are released for every single-strand break (Kaplan *et al.*, 1969). Accordingly, if a total of 7 nucleosides (molar ratio of 6 phosphates:phosphodiester bond break) were released, 43% of them would have been thymine. When excision products of [^{3}H]thymine-labeled *E. coli* DNA were examined chromatographically, approximately 45% of the total released nucleotides appeared as dTMP-5' with the remainder as oligonucleotides. Chromatographic analysis of ^{32}P-labeled excision products, such as shown in Figure 9, showed an overall composition of the excision products to be: oligonucleotide, 45.2%; dGMP-5', 7.2%; dAMP-5', 16.1%; dCMP-5', 10.8%; and dTMP-5', 20.7%. Since this amount of dTMP-5' represented 45% of the thymine released, the total amount of thymine excised was equivalent to 46% of the entire excised regions, a value which was in excellent agreement with the percentage calculated previously.

NUCLEOTIDES 3' TO PYRIMIDINE DIMER. The excised fragment (Figure 10A) was isolated and analyzed for the nucleotides 3' to the photoproduct region by means of snake venom phosphodiesterase, which initiates hydrolysis from 3'-hydroxy termini liberating 5'-deoxynucleoside monophosphates. As shown in Figure 10B, digestion resulted in the appearance of large amounts of mononucleotides as expected from the 5' position of the dimer in the excised fragment. Moreover, the

TABLE V: Susceptibility of Excision Products to Phosphomonoesterase.

Excision Time (min)	Extent of Excision (%)	Phosphomono- esterase Insensitive: Sensitive (Ratio)
5	32.5	5.4
10	65.0	4.7
20	93.0	3.6
30	100	2.6
90	100	2.1

TABLE VII: Excised Deoxynucleotides Located 3' to Photoproducts.

UV Dose (ergs/mm^2 × 10^{-3})	% Distribution of			
	TMP-5'	dAMP-5'	dCMP-5'	dGMP-5'
A. *E. coli* DNA (GC:AT = 1.0)				
9	46.8	23.7	19.9	9.6
36	39.0	30.9	17.4	12.6
108	30.7	35.8	24.2	9.2
B. *M. luteus* DNA (GC:AT = 2.7)				
108 × 10^3	35.1	18.1	34.3	12.4

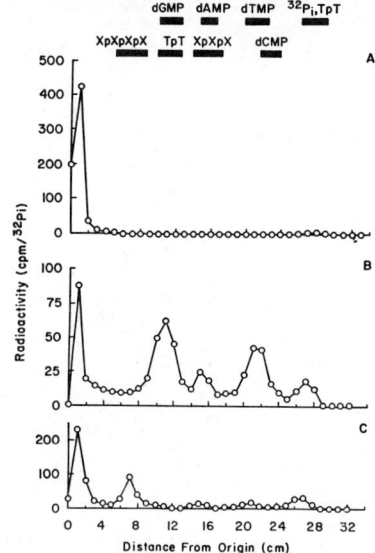

FIGURE 10: Digestion of early excision products with snake venom phosphodiesterase or calf spleen phosphodiesterase. Excision products from a 5-min excision of incised ^{32}P-labeled UV-irradiated *E. coli* DNA (3.6 × 10^4 ergs/mm^2) were prepared and isolated as described in Methods. All radioactivity which appeared prior to the region 1 of Figure 8 was eluted with H$_2$O and digested with snake venom phosphodiesterase or calf spleen phosphodiesterase (see Methods). The reaction mixtures (0.1 ml) were spotted on DEAE-cellulose chromatography paper and developed as described in Figure 9. The figure illustrates undigested excision products (A), snake venom phosphodiesterase digestion (B), and calf spleen phosphodiesterase digestion (C).

distribution of nucleotides was similar to that released by the UV-exonuclease during excision (Table VII). The apparent release of small amounts of radioactivity from the excised fragment by spleen phosphodiesterase (Figure 10C) indicates the occasional occurrence of nucleotides 5′ to the photoproduct. This possibility is presently under investigation.

Discussion

The excision of pyrimidine dimers occurs in several stages. As shown in Figure 11, the UV-endonuclease catalyzes a single phosphodiester-bond break proximate to the photoproduct. Because of the ready reversibility of lesions catalyzed by the photoreactivating enzyme, it is assumed that the primary photoproduct formed at these doses is probably a thymine dimer. The incision which is 5′ to this dimer leaves a 5′-hydroxyl and a 3′-phosphomonoester group. In the second step, catalyzed by the UV-exonuclease, initial cleavage results in the release of a large photoproduct-containing fragment. This process leaves a 5′-phosphomonoester on the DNA molecule and a 3′-hydroxyl group on the fragment. The large fragment is subsequently degraded by the UV-exonuclease. The size of the limit product is dependent on the time of the excision reaction. After 2 hr the major products are trinucleotide or tetranucleoside triphosphates and deoxynucleoside

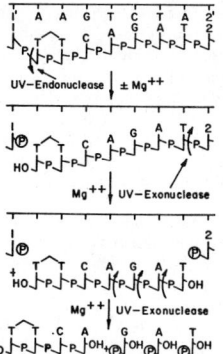

FIGURE 11: Model of initial steps in the enzymatic repair of UV-irradiated DNA.

5′-monophosphates. A summary of this repair cycle is depicted in Figure 11.

The proposed model for the enzymatic excision of photoproducts is consistent with the "cut-and-patch" mechanism currently proposed to explain dark repair (Howard-Flanders and Boyce, 1966). The UV-endonuclease and UV-exonuclease would catalyze the first two steps. The presence, however, of a 3′-phosphomonoester blocking a potential nucleophilic site for DNA polymerase initiation may require the participation of either a polynucleotide phosphomonoesterase or an exonuclease-III-type digestion in preparation for polymerization by the *M. luteus* polymerase. Resynthesis would then be finalized by a joining reaction mediated by DNA ligase to complete the repair process.

However, the recent work by Kelley and coworkers (1970) indicates that the "patch-and-cut" model (Haynes, 1966) may be a more accurate description of the repair process. The ability of the 5′→3′-exonuclease activity of the *E. coli* DNA polymerase to excise photoproduct-containing regions suggests that the DNA polymerase complex may start polymerization immediately after the initial incision has been made, excising the photoproduct after resynthesis. While an enzyme resembling the UV-endonuclease has not been isolated from uninfected *E. coli*, such an activity has been detected in bacteriophage T$_4$ infected *E. coli* (Friedberg and King, 1969; Yasuda and Sekuguchi, 1970).

This same type of patch-and-cut system may function in *M. luteus*, but under slightly different conditions.

A mechanism of polymerization as proposed by Kelly *et al.*, (1970) requires the participation of an enzyme possessing 5′→3′-exonuclease activity. The UV-exonuclease could fulfill such a role although it has been shown to prefer denatured DNA while the nuclease activity associated with the polymerase from *E. coli* is specific for native DNA. It might also be noted that the UV-exonuclease could be an integral part of the DNA polymerase *in vivo*, but appears as an independent activity *in vitro* presumably due to limited proteolysis during purification from this organism. In short, the model presented in Figure 11 could also fit a patch-and-cut system by which polymerization would start immediately following incision.

Both the UV-exonuclease and the 5′→3′-exonuclease activity of the DNA polymerase represent a third class of deoxyribonuclease activity since they act either as endo- or exonucleases under certain conditions. The nature of the enzymatic attack is a function of the substrate and its structure. With incised UV-irradiated DNA the UV-exonuclease acts as an endonuclease, while with denatured DNA it functions exclusively as an exonuclease. It appears that a redefinition of nuclease classification is needed in order to better describe these multifunctional enzymes. Thus, an exonuclease is a phosphodiesterase requiring a terminus for its action whereas an endonuclease does not and can therefore initiate hydrolysis randomly.

The studies on the composition of the excised region indicate that the phenomenon of UV resistance in certain organisms is in part a function of DNA base composition. Fewer dimers are recognized in *M. luteus* than in *E. coli* DNA under equivalent conditions. Furthermore, dimer formation which is recognized by the excision–repair system occurs in thymine-rich regions. Since the UV-endonuclease action releases some oligonucleotides even at low doses of ultraviolet irradiation, some of the dimers must be fairly close to each other (Kaplan *et al.*, 1971). Regions must exist in the *E. coli* chromosome which are highly susceptible to ultraviolet damage. Also it is likely that only those photoproducts which result in large distortions of the DNA helix, that is, those formed in AT-rich regions are recognized by the UV-endonuclease. Further work is now in progress to determine whether the excision system of *M. luteus* is capable of acting on lesions other than those produced by ultraviolet light. These studies will help determine the generality of the cellular excision–repair system.

References

Boyce, R. P., and Howard-Flanders, P. (1964), *Proc. Nat. Acad. Sci. U. S. 51*, 293.
Carrier, W. L., and Setlow, R. B. (1966), *Biochim. Biophys. Acta 129*, 318.
Cleaver, J. E. (1969), *Proc. Nat. Acad. Sci. U. S. 63*, 428.
Cook, J. S., and McGrath, J. R. (1967), *Proc. Nat. Acad. Sci. U. S. 58*, 1359.
Cunningham, L., Cutlin, B. W., and Privat de Garihle, M. J. (1956), *J. Amer. Chem. Soc. 78*, 4642.
Friedberg, E. C., and King, J. J. (1969), *Biochem. Biophys. Res. Commun. 37*, 646.
Grossman, L. (1967), *Methods Enzymol. 12*, 700.
Grossman, L., Kaplan, J., Kushner, S. R., and Mahler, I. (1968), *Cold Spring Harbor Symp. Quant. Biol. 33*, 229.
Haynes, R. H. (1966), *Radiat. Res., Suppl. 6*, 8.
Howard-Flanders, P., and Boyce, R. P. (1966), *Radiat. Res., Suppl. 6*, 156.
Kaplan, J. C., Kushner, S. R., and Grossman, L. (1969), *Proc. Nat. Acad. Sci. U. S. 63*, 144.
Kaplan, J. C., Kushner, S. R., and Grossman, L. (1971), *Biochem. J.* (in press).
Kelley, R. B., Atkinson, M. R., Huberman, J. A., and Kornberg, A. (1969), *Nature (London) 224*, 495.
Kelley, R. B., Cozzarelli, N. R., Deutcher, M. P., Lehman, I. R., and Kornberg, A. (1970), *J. Biol. Chem. 245*, 39.
Laskowski, M., Sr. (1961), *Enzymes 5*, 123.
Mahler, I. (1967), *Methods Enzymol. 12*, 693.
Pearson, M., and Johns, H. E. (1966), *J. Mol. Biol. 19*, 303.
Razzell, W. E., and Khorana, H. G. (1958), *J. Amer. Chem. Soc. 80*, 1770.
Richardson, C. C. (1965), *Proc. Nat. Acad. Sci. U. S. 54*, 158.
Riley, M., Maling, B., and Chamberlin, M. (1966), *J. Mol. Biol. 20*, 359.
Setlow, J. K. (1966), *in* Current Topics in Radiation Research, Ebert, M., and Howard, A., Ed., Vol. II, Amsterdam, North Holland Publishing Co., p 195.
Setlow, R. B., Brown, D. C., Boling, M. E., Mattingly, A., and Gordon, M. P. (1968), *J. Bacteriol. 95*, 546.
Setlow, R. B., and Carrier, W. L. (1964), *Proc. Nat. Acad. Sci. U. S. 51*, 226.
Setlow, R. B., Regan, J. D., German, J., and Carrier, W. L. (1969), *Proc. Nat. Acad. Sci. U. S. 64*, 1035.
Strauss, B. S., Searashi, T., and Robbins, M. (1966), *Proc. Nat. Acad. Sci. U. S. 56*, 932.
Yasuda, S., and Sekiguchi, M. (1970), *Proc. Nat. Acad. Sci. U. S. 67*, 1839.

A Determination of Mutagen Specificity in Bacteria using Nonsense Mutants of Bacteriophage T4

Mary Osborn, Stanley Person, Stephen Phillips and Fred Funk

Biophysics Department, College of Science
The Pennsylvania State University
University Park, Pennsylvania, U.S.A.

(*Received 19 October 1966, and in revised form 4 April 1967*)

We present a method for the determination of mutagen specificity in bacteria, using an *Escherichia coli* strain that is mutant because of an amber triplet in a gene necessary for arginine biosynthesis. A large number of revertants occurring spontaneously and after treatment with 2-aminopurine, [5-^{3}H]uracil radioactive decay, ethyl methanesulfonate, 5-bromodeoxyuridine, hydroxylamine and ultraviolet light were tested for their ability to support the growth of 24 amber mutants and one ochre mutant of bacteriophage T4. Regardless of the mutagen used to produce the revertants, only 6 patterns, called classes, of phage growth were obtained. Revertants of classes 1, 2 and 3 contained amber suppressors, revertants of classes 4 and 5 contained ochre suppressors and those of class 6, since they suppressed none of the mutant phages used, were assumed to be structural gene revertants. By comparing the patterns of phage growth obtained to those for bacterial strains with characterized suppressors, the amino acid inserted by a revertant of class 1, 2 or 3 was inferred. Strikingly different distributions into the classes were observed for some of the mutagens. 2-Aminopurine, ethyl methanesulfonate, and [5-^{3}H]uracil radioactive decay were found to be very specific in their action. Assuming that reversion occurred by single base changes in DNA specifying either the amber codon or certain sRNA anticodons, it was also possible to infer the base changes produced by the mutagens. The sRNA anticodons assumed to be altered are those that can become the inverse complements of either UAG or UAA by single base changes.

1. Introduction

The existence of nonsense codons was implicit in the work of Benzer & Champe (1962) and Garen & Siddiqi (1962). Recent studies on nonsense codons and their suppression indicate that (1) the amber nonsense codon is UAG (Brenner, Stretton & Kaplan, 1965; Weigert & Garen, 1965a) and the ochre nonsense codon is UAA (Brenner *et al.*, 1965); (2) the presence of an amber codon results in polypeptide chain termination (Sarabhai, Stretton, Brenner & Bolle, 1964) and the presence of an ochre codon probably also causes chain termination (Brenner & Beckwith, 1965); (3) the chain terminating effect can be suppressed as a result of a mutation in the protein synthesizing machinery of the cell, and in at least one case as a result of production of an altered sRNA (Capecchi & Gussin, 1965; Engelhardt, Webster, Wilhelm & Zinder, 1965); (4) the different amber suppressors which have been characterized each inserts a characteristic amino acid at the position of the amber codon (Weigert & Garen, 1965b; Stretton & Brenner, 1965; Weigert, Lanka & Garen, 1965; Kaplan, Stretton & Brenner, 1965; Notani, Engelhardt, Konigsberg & Zinder, 1965).

The fact that a bacterium containing a suppressor can cause the translation of its own nonsense codon and one in phage as well (Notani et al., 1965) is very useful for reversion studies on bacterial mutants containing nonsense codons. If a sufficient number of amber and ochre mutant phages are used one can detect those revertants that contain amber and ochre suppressors. One can also infer the amino acids inserted by some of the suppressors by comparing the patterns of phage growth obtained to those for bacterial strains containing suppressors that cause the insertion of known amino acids. In this paper we have used nonsense mutant phages to characterize a large number of revertants, from an *E. coli* mutant, occurring spontaneously and after treatment with the six mutagenic agents, 2-aminopurine, [5-^{3}H]uracil radioactive decay, ethyl methanesulfonate, 5-bromodeoxyuridine, hydroxylamine and ultraviolet light. This characterization, along with an assumed model for suppressor mutation, allows us to infer DNA base changes leading to reversion by the different mutagens.

2. Materials and Methods

(a) *Bacterial strains*

E. coli WWU, also called TAU-bar, a polyauxotroph requiring thymidine, uridine, proline, arginine, methionine and tryptophan has been described by Person & Bockrath (1964). Revertants of the arginine locus were characterized. The following *E. coli* K12 bacteria were given to us by Dr Alan Garen, Yale University: S26, an amber mutant of the alkaline phosphatase gene, S26R1e, S26R1d and H12R8a, which have been shown by Weigert & Garen (1965b) and Weigert et al. (1965) to contain amber suppressors inserting serine, glutamine and tyrosine, respectively.

(b) *Phage stocks*

Wild-type T4 and T4 amber mutants *55*, *N122*, *B22*, *A453*, *N82* and *120* were obtained from Dr William Ginoza of our department. Mutants *55* and *120* were originally from the collection of Dr S. E. Luria, Massachusetts Institute of Technology. Amber *B17* and ochre mutant *oc427* were obtained from Dr Jonathan Beckwith, Harvard Medical School. All other amber mutants listed in Table 2 were from the collection of Dr R. S. Edgar, California Institute of Technology and have been described by Epstein et al. (1963) and by Edgar & Wood (1966). Stocks of the mutants were grown on *E. coli* CR63 and contained less than one revertant in 5×10^4 amber mutants.

(c) *Phage assay plates*

Bottom agar (1·2% contained Tryptone (10 g/l.), yeast extracts (5 g/l.), sodium chloride (5·8 g/l.) and thymidine (40 mg/l.)). Top agar (0·4% for plating, 0·6% for spot printing) contained nutrient broth (8 g/l.), sodium chloride (5·8 g/l.) and thymidine (40 mg/l.). The addition of thymidine prevents viability decline (Bockrath, 1965), a phenomenon that occurs in some suppressed mutants of WWU.

(d) *Spot printer*

The printer consists of a brass plate to which are soldered 30 brass prongs $\frac{7}{64}$ in. in diameter and 1¾ in. long arranged in a concentric circular array. The tube holder, made of thick aluminum plate, holds 30 small tubes in a matching array. Each tube contains a suspension of a different amber phage (3×10^7/ml.). Upon withdrawing the printer from the tubes the flat end of each prong retains a drop of phage suspension. The printer is inverted and the drops are transferred, in a manner analogous to replica plating, to a plate already containing bacteria as hosts. A similar printer has been devised independently by Hill & Stent (1965).

(e) *Isolation of revertants*

WWU was grown in supplemented minimal medium as described previously (Person & Bockrath, 1964). Unless otherwise noted, all treatments were at 37°C with aeration on

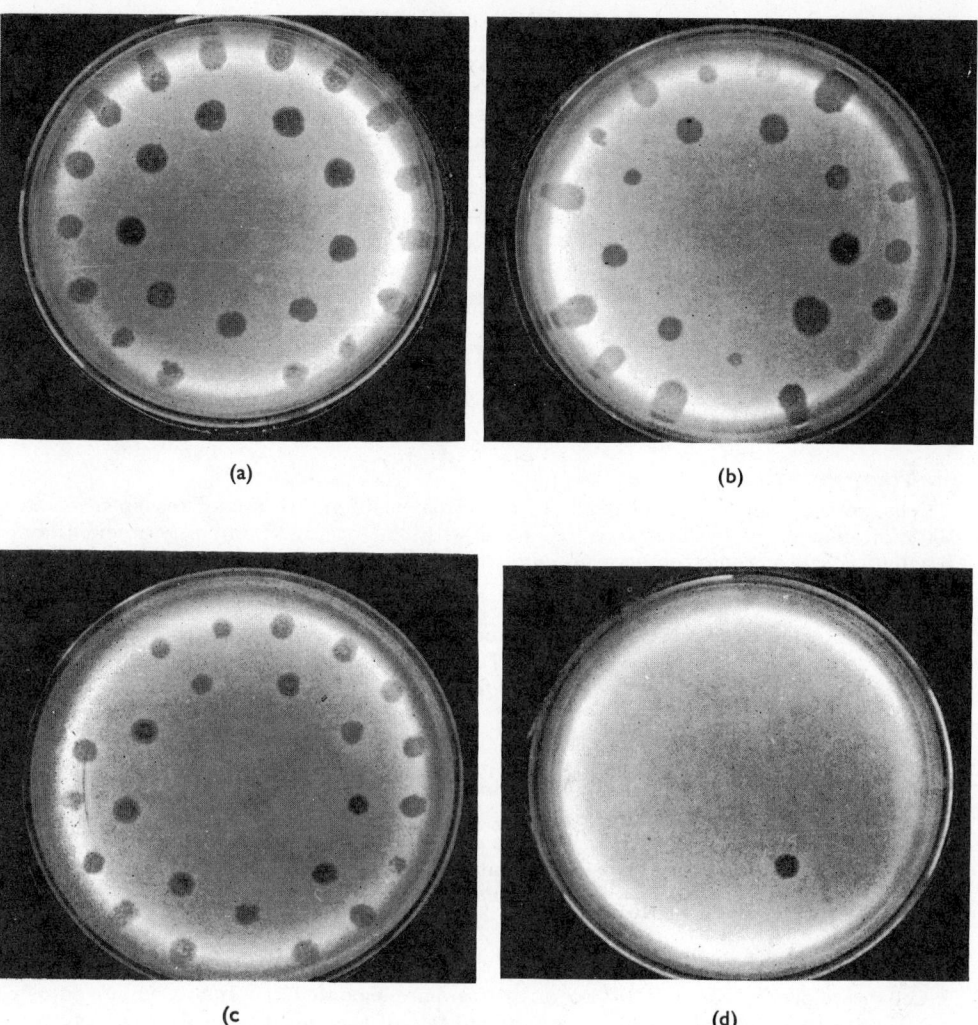

PLATE I. Spot prints of WWU revertants characteristic of classes 1, 2 and 3 and WWU are shown in (a), (b), (c) and (d), respectively. Classes 1, 2 and 3 have a pattern of suppression similar to Su_1^+, Su_2^+ and Su_3^+ shown by Garen and his associates to insert serine, glutamine and tyrosine, respectively. A clear or cloudy area in the region of the prong print is taken as evidence of phage growth and, hence, suppression. Amber 55 is immediately to the left of the blank at the bottom of each photograph. The pattern of suppression can be ascertained by reading the individual prong prints beginning with amber 55, proceeding clockwise around the outer circle of prong prints and then switching to amber *N133* (above the blank spot on the outer circle) and continuing in a clockwise direction around the inner circle, finishing with T4. The mutants in Table 1 are arranged in a corresponding order beginning from the top of the Table. The ochre phage is not in the photographs.

[facing p. 439

cells at 4 to 6×10^8 grown from an inoculum of 3×10^7. After the mutagenic treatments described below, the cells were filtered and washed to remove the mutagen and growth supplements, and resuspended in minimal medium. 0·2 ml. of the resuspension was plated to isolate revertants of the arginine locus. Arginine at 3·5 μg/ml. replaced nutrient broth at 0·2 mg/ml. in the plating medium used by Person & Bockrath (1964). These concentrations of arginine and nutrient broth allow equivalent mutation expression. After 36 hr at 37°C, revertant colonies were individually streaked on a similar medium lacking arginine and regrown. The streaks were kept at 4°C. An inoculum from each streak was grown at 37°C with aeration for 4 to 12 hr in nutrient broth (8 g/l.) containing thymidine (40 mg/l.) before use as host cells in the spot-print test.

(f) *Mutagenic treatments*

(i) *Spontaneous*

Cells which had been resuspended in minimal medium and plated directly served as controls for all treatments except hydroxylamine and [5-³H]uracil decay.

(ii) *Ultraviolet light*

Cells were filtered and resuspended in minimal medium. 10 ml. of resuspension in a 100-mm diameter Petri dish were irradiated with 300 ergs/mm² from a germicidal lamp. Irradiation was at room temperature and the cells were plated without further filtration.

(iii) *Ethyl methanesulfonate*

Cells at 6 to 12×10^8 were filtered and concentrated $2 \times$ by resuspending in minimal medium. 2 ml. was added to 7·6 ml. of 1 M-Tris, pH 7·4, containing 0·4 ml. EMS† (Eastman Organic Chemicals) and shaken vigorously for 2·5 to 5·0 min at 37°C.

(iv) *2-Aminopurine*

Two methods of treatment were used. In the first, 2AP was added to cells at 4×10^8 in supplemented minimal medium to a final concentration of 10^{-2} M and the cultures incubated for 35 min at 37°C. In the second method, 0·1 ml. of 2AP at 40 mg/ml. was added to ½-in. diameter filter paper in the center of a revertant isolation plate on which cells, filtered to remove growth supplements, had just been plated. Approximately half of the colonies were isolated by each method and gave similar percentages of suppressed mutants.

(v) *Hydroxylamine*

Cells were grown from 3×10^7 to 4×10^8 in supplemented minimal medium adjusted to pH 6·0. HA, adjusted to pH 6·0, was added to half the culture to a final concentration of 3·5 mg/ml. and the cultures incubated for 40 min at 37°C. The untreated portion of the culture was used to isolate control revertants for this mutagen.

(vi) *[5-³H]Uracil decay*

Revertants produced by the decay of incorporated [5-³H]uracil and control revertants for this mutagen were isolated as in Person & Bockrath (1964).

(vii) *5-Bromodeoxyuridine*

5-BUdR was added to cells at 4×10^8 in supplemented minimal medium to a final concentration of 10^{-2} M and the cultures incubated for 65 to 75 min at 37°C.

3. Results

From the results of spot printing one ochre and 24 amber mutant phages we divided more than 2000 revertants, produced spontaneously and by six mutagens, into six classes. The characteristics of each class, in terms of phage growth, are shown in Table 1(a). Photographs of the spot prints for revertants representative of classes 1, 2 and 3 are shown in Plate I(a), (b) and (c). There is no overlap in the phage mutants

† Abbreviations used: EMS, ethyl methanesulfonate; 2AP, 2-aminopurine; HA, hydroxylamine; [5-³H]U, [5-³H]uracil; 5BUdR, 5-bromodeoxyuridine; UV, ultraviolet light.

TABLE 1

Response of revertants of WWU, WWU and bacteria containing known amber suppressors to nonsense mutants of T4

Phage Mutants (gene number)	Table (a) Class of WWU revertants						WWU	Table (b) Alkaline phosphatase suppressed mutants		
	1	2	3	4	5	6		S26R1e (serine)	S26R1d (glutamine)	H12R8a (tyrosine)
Wild type T4	+	+	+	+	+	+	+	+	+	+
Amber 55	+	+	+	W	—	—	—	+	+	+
N122 (42)	+	+	+	W	—	—	—	+	+	+
B22 (43)	+	+	+	+	W	—	—	+	+	+
A453 (32)	+	P	+	—	—	—	—	+	W	+
N82 (44)	+	+	+	W	—	—	—	+	+	+
B17 (23)	+	+	P	—	—	—	—	+	+	P
120	+	+	+	+	—	—	—	+	+	+
N135 (5)	+	+	+	W	X	—	—	+	+	+
N102 (6)	+	W	+	X	—	—	—	+	W	+
B16 (7)	+	+	+	W	X	—	—	+	+	+
N132 (8)	+	W	+	—	—	—	—	+	W	+
B255 (10)	+	+	+	X	X	—	—	+	+	+
N93 (11)	+	+	+	—	—	—	—	+	+	+
N69 (12)	+	+	W	—	—	—	—	+	+	W
E609 (13)	+	W	+	X	—	—	—	+	W	+
B20 (14)	+	+	+	X	—	—	—	+	+	+
N133 (15)	+	W	+	—	—	—	—	+	P*	+
N66 (16)	+	+	+	—	—	—	—	+	+	+
B25 (34)	+	+	+	—	—	—	—	+	+	+
A455 (34)	+	+	+	—	—	—	—	+	+	+
N58 (34)	+	+	+	—	—	—	—	+	+	+
N52 (37)	+	+	+	—	—	—	—	+	+	+
H28 (53)	+	+	+	X	—	—	—	+	+	+
E1102 (64)	+	+	+	X	—	—	—	+	+	+
Ochre 427†	—	—	—	+	W	—	—	—	—	—

Table (a) shows the response of WWU revertants in classes 1 to 6, and WWU, on the 24 amber mutants and one ochre mutant of phage T4. Table (b) shows the response of 3 alkaline phosphatase-suppressed mutants on the same mutant phages. S26R1e has been shown to insert serine, S26R1d, glutamine and H12R8a, tyrosine by Garen and his associates. By comparing the data in (a) and (b) we conclude that WWU mutants of class 1 insert serine, class 2 insert glutamine and class 3 insert tyrosine.

+, Cloudy or clear lysis in the area of the prong print; W, incomplete lysis, corresponding to a lower plating efficiency; X, further work is required to know if *all* suppressed mutants in these classes cause suppression of these phages; P, test is mostly negative by printing test but corresponds to a very low efficiency of suppression by plating (plaques develop and are amber); P*, N133 plates with very low efficiency on S26R1d. However, plaques that do develop are amber as shown by testing on CR63 and B. After passage through S26R1d it plates with equal efficiency on S26R1d, class 2 and class 4 WWU suppressed mutants. However, the original N133 stock was used for all spot printing.

† Only about half of the total number of WWU revertants isolated were tested with the ochre phage. All revertants tested gave the response indicated.

used to distinguish between classes 1, 2 and 3. Classes 4 and 5 are readily distinguished from classes 1, 2 and 3 as they caused suppression of a minor number of amber mutants and also of the ochre mutant. They are also separable from each other. However, classes 4 and 5 probably are not unique, and subclasses representing a minor fraction of each may exist within each one. Revertants of class 6 are unable to suppress any of the mutant phages used and give a response identical to the parent cell, WWU, shown in Plate I(d). Assuming that these phages are sufficient to detect all suppressors, class 6 revertants correspond to structural gene alterations. Revertants could be assigned unambiguously to a particular class and results obtained using the printer were confirmed by plating all of the mutant phages on revertants representative of each of the six classes.

To determine whether a particular pattern of suppression represents a particular amino acid insertion we compared the patterns for the various classes of suppressed mutants with those for the three characterized amber suppressors, Su_1^+, which inserts serine (S26R1e), Su_2^+ which inserts glutamine (S26R1d), and Su_3^+ which inserts tyrosine (H12R8a) (Weigert & Garen, 1965a; Weigert et al., 1965). The patterns of phage growth for these bacteria are shown in Table 1(b). Examination of Table 1(a) and (b) shows that the pattern for class 1 revertants is the same as that for S26R1e, that for class 3 revertants is the same as that for H12R8a, and that for class 2 revertants is the same as that for S26R1d except for a variation in the degree of response to the amber mutants $N133$ and $A453$ (see Table 1). Spot printing at 25°C and at 42°C did not alter the comparison. We infer that the class 1 suppressor inserts serine, the class 2 suppressor inserts glutamine and the class 3 suppressor inserts tyrosine. Since revertants contain amber suppressors, it follows that the parent cell, WWU, contains an amber triplet in an arginine gene.

The number of independent isolations of revertants and the total and net numbers of revertants spot printed are shown in Table 2.

The percentage distributions of revertants into classes 1 to 6 produced spontaneously and by the six mutagens used are shown in Table 3. The percentages refer to the net number tested (Table 1). For example, 203 of 226, or 90%, of EMS-produced revertants are in class 2. It is clear that the percentage of revertants that are sup-

TABLE 2

Number of arginine revertants characterized by spot printing for each mutagen

Mutagen	Number of independent isolations of revertants	Number of revertants isolated after treatment with mutagen and spot printed	Net number† of revertants spot printed from all isolations
2AP	4	185	124·3
5BUdR	3	210	154·5
HA	4	309	261·9
UV	5	273	249·5
EMS	4	334	225·8
[5-³H]U decay	5	370	331·9
Spontaneous	15	792	792

† The net number is the number spot printed less those revertants that were of spontaneous origin. The fraction of the total that are spontaneous in origin was determined for each experiment on a per plate basis using several plates.

TABLE 3

Percentage distributions into classes 1 to 6 of WWU revertants occurring spontaneously and by the mutagens listed

Mutagen	Amber suppressors			Ochre suppressors		Structural gene revertants
	Class 1 (serine)	Class 2 (glutamine)	Class 3 (tyrosine)	Class 4	Class 5	Class 6 (su^-)
2AP	0·5	1·9	1·6	−1·8	3·0	94·7
5BUdR	0·6	3·2	0·5	−3·8	16·2	83·4
HA	0·9	32·3	0·8	3·6	4·1	58·4
UV	6·4	41·3	0·9	18·6	23·8	8·8
EMS	2·2	90·0	−0·3	9·4	−6·2†	5·0
[5-³H]U decay	0·7	81·0	0·7	15·4	0·1	2·2
Spontaneous	4·4	19·1	1·5	10·5	15·0	49·5

The net numbers of revertants in classes 1 to 5 were determined by subtracting the number of revertants in a particular class that can be attributed to the control, from the total number of revertants assigned to that class. The net numbers of revertants in each class were converted to percentages using as 100% the net number tested for a particular mutagen shown in Table 2. This correction for revertants of spontaneous origin is the source of negative and non-integral percentages. [5-³H]Uracil decay produced revertants were corrected by using the control percentages in classes 1 to 5 averaged for many experiments.

† Probably percentages of this magnitude are only significant for spontaneous revertants. Therefore, except for spontaneous revertants, percentages ≤ 6·4% were taken as 0 in constructing Table 5.

pressed mutants (sum of classes 1 to 5) depends strikingly on the mutagen (5·2% for 2AP to 98% for [5-³H]U decay). In addition some mutagens produce revertants that are predominantly in one of the six classes. For example 2AP produces class 6 revertants almost exclusively, and 5BUdR produces revertants that are predominantly in class 6. In contrast almost all the revertants produced by [5-³H]U decay and EMS contain suppressors, the large majority of which are in class 2. The remaining mutagens are less specific. The distribution of spontaneous revertants among the six classes is shown in the bottom line of Table 3. The ability of a mutagen to produce revertants in particular classes is indicative of its specificity.

4. Discussion

(a) *Relation of amino acid inserted to base change*

Each pattern of suppression is presumably representative of a particular amino acid insertion. Since the genetic code is known, perhaps the DNA base changes causing the amino acid insertions could be determined if we knew the mechanism by which suppressor mutation occurs. There is direct evidence that a pre-existing seryl-sRNA is involved in Su_I^+ suppression (Bergquist & Capecchi, 1966). Also, it is a striking finding that the amino acids, serine, glutamine and tyrosine, inserted by the three characterized amber suppressors, have a codon assignment differing from the amber codon, UAG, by a single base. These findings have reinforced the idea discussed by some workers (Kaplan *et al.*, 1965; Capecchi & Gussin, 1965) that amber suppressors might result from changing the anticodons of these sRNA's so they become the inverse complement of UAG. Since it is only the anticodon that is altered by mutation, the sRNA is still charged by the same amino acid, and the sRNA translates the amber codon by inserting this amino acid. We assume this mechanism for the production of amber suppressors. We can use the finding of Brenner & Beckwith (1965) that amber suppressors recognize the codon UAG but not UAA and Crick's hypothesis concerning recognition sets (Crick, 1966) to identify the anticodons that would be altered. From Crick's hypothesis the anticodon of an sRNA involved in amber suppression, since it recognizes codons of the type XYG but not XYA, has a C at the 5′ end and is CUA. A change in the anticodon, from CGA to CUA, of an sRNA that formerly recognized the serine codon UCG would mean that this sRNA now recognizes UAG and would cause serine insertion by an amber suppressor. Similarly an anticodon change from CUG to CUA of an sRNA that formerly recognized the glutamine codon CAG would cause glutamine insertion by an amber suppressor. An anticodon change from AUA to CUA would cause tyrosine insertion by an amber suppressor if the tyrosyl-sRNA formerly recognized UAU. Alternatively an anticodon change from GUA to CUA would cause tyrosine insertion by an amber suppressor if the tyrosyl-sRNA formerly recognized both UAU and UAC. The DNA base changes leading to these anticodon changes are shown in Table 4 along with single base changes that lead to the production of all other possible amber suppressors (column 3). We assume in Table 4 that alterations of either strand of DNA can lead to reversion.

Ochre suppressors as well as amber suppressors might also arise by alterations of certain sRNA anticodons. Ochre suppressors recognize both UAA and UAG (Brenner & Beckwith, 1965). From Crick's hypothesis the anticodon of an sRNA involved in ochre suppression, since it recognizes codons of the type XYA and XYG, has a U at the 5′ and is UUA. Thus, an anticodon change from UUG to UUA in an sRNA that formerly recognized the glutamine codon CAA would mean that this sRNA now recognizes UAA and would cause glutamine insertion by an ochre suppressor. All ochre suppressors produced by anticodon changes are shown in column 4 of Table 4.

Since this model for suppressor mutation involves the alteration of an anticodon of a pre-existing sRNA, suppression by anticodon changes is possible only if another sRNA exists that will preserve the reading of codons for that amino acid (Kaplan *et al.*, 1965). For all amber suppressors, except those inserting tyrosine, recognition of codons of the type XYG could be preserved either by an sRNA charged with the same amino acid that has a 5′ U in its anticodon, or alternatively by having multiple copies of the genetic information for the particular altered sRNA. For all ochre suppressors except those inserting tyrosine, recognition of codons of the type XYA can only be preserved by having multiple genetic copies of the particular sRNA that was altered.

We also show in Table 4 the amino acid insertions that would result from single base changes in the DNA specifying the amber codon. These changes result in the replacement of the amber codon with one that codes for an amino acid and correspond to the events leading to structural gene reversion.

(b) *Assignment of base changes*

The classes of revertants can be related to DNA base changes by reference to Table 4. Class 1 revertants inserting serine by an amber suppressor arise from the transversions GC → TA. Class 2 revertants inserting glutamine by an amber suppressor

Table 4

Relation of amino acid insertions at the site of the amber codon to DNA base changes

Base changes	Structural gene reversion	Amber suppressors	Ochre suppressors
GC → AT	none	tryptophan† glutamine	glutamine
AT → GC	tryptophan glutamine	none	none
GC → TA	tyrosine	serine glutamic acid	serine glutamic acid tyrosine
AT → CG	serine glutamic acid	tyrosine	none
GC → CG	tyrosine	tyrosine	none
AT → TA	leucine lysine	leucine lysine	leucine lysine tyrosine

Mutation is assumed to be due to single base changes in DNA specifying either the amber codon (structural gene reversion) or certain sRNA anticodons. The sRNA anticodons assumed to be altered are those that become the inverse complements of UAG (amber suppressors) or UAA (ochre suppressors) by single base changes.

† There is only one codon for tryptophan and if there is only one tryptophanyl-sRNA that translates this codon, an amber suppressor inserting tryptophan could only exist if there were multiple genetic copies for this sRNA. Experimentally an amber suppressor for tryptophan has not been reported.

arise from the transitions GC → AT. Class 3 revertants inserting tyrosine by an amber suppressor may arise from either of the transversions AT → GC or GC → CG.

In assigning base changes to revertants of classes 4 and 5 uncertainties exist because more than one mechanism can give rise to ochre suppressors, the amino acids inserted by ochre suppressors have not been identified, and classes 4 and 5 may contain subclasses. If, however, the ochre suppressors of either class arise by any of the base changes leading to classes 1, 2 or 3 they should consistently occur with one of the classes. Table 3 shows that class 4 revertants consistently occur with class 2 revertants and therefore may be caused by the base changes GC → AT. The ochre suppressor in class 4 revertants would then result from glutamine insertion (Table 4), and this would imply that multiple copies of this sRNA would exist. This assignment is consistent with the observation that, when revertants of a cell containing an ochre codon, isolated after treatment with [5^3H]U decay or EMS, were spot printed, all revertants had a pattern of suppression similar to that for class 4 revertants. Class 5 revertants do not consistently occur with revertants of classes 1, 2 or 3 and this class is therefore unlikely to arise by any of the base changes leading to revertants in these classes.

Thus, if class 5 revertants arise through anticodon changes they could only be caused by the base changes that are not yet assigned, AT→TA. However, we note that when the parent su⁻ cell, WWU, was spot printed in the presence of streptomycin at 1 μg/ml. a pattern of phage growth similar to that for class 5 revertants was obtained. If class 5 revertants are caused by a mechanism other than altered anticodons, they could be due also to other changes, most likely AT→GC. The assignment of base changes to this class must await further chemical and genetic evidence.

Class 6 revertants arise from changes in the DNA specifying the amber codon. Table 4 predicts that a mutagen producing exclusively class 6 revertants would be caused by TC→AG changes. For mutagens not producing class 6 revertants exclusively, class 6 is heterogeneous because structural gene revertants can arise from several base changes. For example, Table 4 predicts that the base changes that produce amber suppressors leading to class 1 and class 3 revertants will also produce structural gene revertants.

(c) *Mutagen specificity*

Table 4 suggests that a mutagen giving 100% suppressed mutants produces GC→AT transitions exclusively since other base changes would give some structural gene revertants. Two of the mutagens tested, EMS and [5-³H]U decay, produced suppressed mutants of which 90 and 81%, respectively, are class 2 revertants (glutamine insertion by an amber suppressor). From Table 4 glutamine insertion by an amber suppressor (class 2 revertants) should be produced by mutagens causing GC→AT transitions. In addition, both [5-³H]U decay and EMS produce a small percentage of revertants containing the class 4 ochre suppressor and which, as already noted, might also arise from the same base changes.

Table 4 also suggests that a mutagen giving 0% suppressed mutants produces AT→GC transitions exclusively since other base changes would give some suppressed mutants. 2AP produced almost all structural gene revertants and therefore probably produces AT→GC changes. To summarize, as shown in Table 5 for the three mutagens found to be the most specific:

$$GC \xrightleftharpoons[\text{2AP (95\%)}]{\substack{\text{[5-}^3\text{H]U decay (81 to 96\%)} \\ \text{EMS (90 to 99\%)}}} AT$$

Ultraviolet, 5BUdR and HA produced revertants that could be assigned to more than one class. The base change assignments for these mutagens are also assigned in Table 5. The base change assignments for all the mutagens used are in general agreement with those assigned previously for these mutagens (for review see Adelberg, 1966). It is clear from Table 5 that while none of the mutagens used produces a significant percentage of the six transversions listed, some spontaneous revertants are produced by these transversions.

The base changes in Table 4 are listed as base-pair changes. Additional information is needed to decide which member of the base pair is altered. For example, [5-³H]U decay, the mutagen we have studied most, might produce one or both of the changes C→T or G→A. We know that a molecular rearrangement occurring at the site of the decay is important for reversion production (Person & Bockrath, 1965). Metabolic conversion of uracil and cytosine results in the incorporation of approximately 6% of the label in the DNA (Person & Bockrath, 1965) as [5-³H]cytosine, and the mutagenic effect of [5-³H]U decay is probably due to decays that originate from the

TABLE 5

Assignment of base changes† to mutagens using the mutation model of Table 4 and the distribution of revertants into classes in Table 3

	AT → CG GC → CG	GC → TA	GC → AT‡	AT → GC
2AP	0	0	0	95
5BUdR	0	0	0	83
HA	0	0	32	58
UV	0	0	41–60	9
EMS	0	0	90–99	0
[5-^{3}H]U decay	0	0	81–96	0
Spontaneous§	3	≥9	19–30	42

† The percentage of class 5 revertants in Table 3 could not be assigned to particular base changes (see text) and were therefore omitted from Table 5. If these revertants are due to AT → TA changes, the percentages listed above for AT → GC changes (5BUdR, u.v. and spontaneous) may be lowered slightly.

‡ The lower limit for base changes GC → AT is the percentage of class 2 revertants and the upper limit is the sum of class 2 and class 4 revertants (see text).

§ The inequality sign for GC → TA changes reflects the uncertainty of the production of the ochre suppressors or of the amber suppressor for glutamic acid for these base changes (Table 4). It was assumed for AT → CG and GC → CG changes that glutamic acid inserted by structural gene reversion did not give rise to functional protein.

[5-^{3}H]cytosine in the DNA (Osborn, unpublished results). Finally, [5-^{3}H]cytosine efficiently induces reversion of an S13 bacteriophage (Funk, unpublished results) shown by Tessman, Poddar & Kumar (1964) to revert by a C→T transition. These observations suggest that for organisms labeled with [5-^{3}H]uracil, decays causing mutation originate as [5-^{3}H]cytosine and cause C→T transitions. Such transitions might occur by the formation of a cytosine hydration product that spontaneously deaminates to uracil in a manner analogous to that shown for cytosine following ultraviolet irradiation (Johns, LeBlanc & Freeman, 1965). EMS also produces either or both of the transitions G→A or C→T. The chemical analysis of Lawley & Brookes (1963) and the mutagenic data of Bautz & Freese (1960), Freese (1961) and Tessman et al. (1964) suggest that EMS produces G→A transitions. If the anticodon model is correct, the C→T change inferred for [5-^{3}H]U decay and the G→A change inferred for EMS affect different bases of the same C–G base pair in the DNA triplet specifying the anticodon for a glutamyl-sRNA. This implies that alteration of either strand of the DNA can result in mutation.

The method of analysis presented here is useful in investigating mutagen specificities for cells containing nonsense codons. In principle, the use of an extensive collection of amber and ochre mutant phages allows one to divide revertants into classes on the basis of single amino acid insertions except for structural gene revertants. In addition a large number of revertants can be characterized, in this manner, in a short time. This characterization, which might be useful on a comparative basis for different mutagens, is independent of the model used to relate amino acid insertions by the various suppressors to base changes in the DNA. If the anticodon model for suppressors is substantiated, the method would be of increased value because, given a complete knowledge of the genetic code, the base change assignments may be made without reference to the chemical action of a mutagen and also without knowledge of the action of other mutagens. We suggest that the agreement between the base change assignments in Table 5 and those assigned previously for the same mutagens implies that some suppressors arise through altered anticodons.

This work was made possible by grants from the National Science Foundation (GB-4485) and the National Aeronautics and Space Administration (NsG-324) and by the kindness of Drs Alan Garen, R. S. Edgar and J. R. Beckwith in supplying bacterial and phage stocks. We thank Dr Ernest Pollard and Dr Evelyn Witkin for several stimulating exchanges of ideas. We also acknowledge discussion of our data with Dr Alan Garen, Dr William Ginoza and Morton Sclair and thank Dr Sydney Brenner for his work on the manuscript. Finally, we acknowledge that this study might never have materialized without earlier studies on tritium decay, pursued so enthusiastically by Dr Richard C. Bockrath.

REFERENCES

Adelberg, E. A. (1966). *Papers on Bacterial Genetics*, p. 1. Boston: Little Brown & Co.
Bautz, E. & Freese, E. (1960). *Proc. Nat. Acad. Sci., Wash.* **46**, 1585.
Benzer, S. & Champe, S. P. (1962). *Proc. Nat. Acad. Sci., Wash.* **48**, 1114.
Bergquist, P. L. & Capecchi, M. R. (1966). *J. Mol. Biol.* **19**, 202.
Bockrath, R. C., Jr. (1965). Ph. D. thesis, The Pennsylvania State University.
Brenner, S. & Beckwith, J. R. (1965). *J. Mol. Biol.* **13**, 629.
Brenner, S., Stretton, A. O. W. & Kaplan, S. (1965). *Nature*, **206**, 994.
Capecchi, M. R. & Gussin, G. N. (1965). *Science*, **149**, 417.
Crick, F. H. C. (1966). *J. Mol. Biol.* **19**, 548.
Edgar, R. S. & Wood, W. B. (1966). *Proc. Nat. Acad. Sci., Wash.* **55**, 498.
Engelhardt, D. L., Webster, R. E., Wilhelm, R. C. & Zinder, N. D. (1965). *Proc. Nat. Acad. Sci., Wash.* **54**, 1791.
Epstein, R. H., Bolle, A., Steinberg, C. M., Kellenberger, E., Boy de la Tour, E., Chevalley, R., Edgar, R. S., Sussman, M., Denhardt, G. & Lielausis, A. (1963). *Cold Spr. Harb. Symp. Quant. Biol.* **28**, 375.
Freese, E. B. (1961). *Proc. Nat. Acad. Sci., Wash.* **47**, 540.
Garen, A. & Siddiqi, O. (1962). *Proc. Nat. Acad. Sci., Wash.* **48**, 1121.
Hill, R. & Stent, G. S. (1965). *Biochem. Biophys. Res. Comm.* **18**, 757.
Johns, H. E., LeBlanc, J. C. & Freeman, K. B. (1965). *J. Mol. Biol.* **13**, 849.
Kaplan, S., Stretton, A. O. W. & Brenner, S. (1965). *J. Mol. Biol.* **14**, 528.
Lawley, P. D. & Brookes, P. (1963). *Biochem. J.* **89**, 127.
Notani, G. W., Engelhardt, D. L., Konigsberg, W. & Zinder, N. D. (1965). *J. Mol. Biol.* **12**, 439.
Person, S. & Bockrath, R. C., Jr. (1964). *Biophys. J.* **4**, 355.
Person, S. & Bockrath, R. C., Jr. (1965). *J. Mol. Biol.* **13**, 600.
Sarabhai, A. S., Stretton, A. O. W., Brenner, S. & Bolle, A. (1964). *Nature*, **201**, 13.
Stretton, A. O. W. & Brenner, S. (1965). *J. Mol. Biol.* **12**, 456.
Tessman, I., Poddar, R. K. & Kumar, S. (1964). *J. Mol. Biol.* **9**, 352.
Weigert, M. G. & Garen, A. (1965*a*). *Nature*, **206**, 992.
Weigert, M. G. & Garen, A. (1965*b*). *J. Mol. Biol.* **12**, 448.
Weigert, M. G., Lanka, E. & Garen, A. (1965). *J. Mol. Biol.* **14**, 522.

GENERAL NATURE OF THE GENETIC CODE FOR PROTEINS

By Dr. F. H. C. CRICK, F.R.S., LESLIE BARNETT,
Dr. S. BRENNER and Dr. R. J. WATTS-TOBIN

Medical Research Council Unit for Molecular Biology,
Cavendish Laboratory, Cambridge

THERE is now a mass of indirect evidence which suggests that the amino-acid sequence along the polypeptide chain of a protein is determined by the sequence of the bases along some particular part of the nucleic acid of the genetic material. Since there are twenty common amino-acids found throughout Nature, but only four common bases, it has often been surmised that the sequence of the four bases is in some way a code for the sequence of the amino-acids. In this article we report genetic experiments which, together with the work of others, suggest that the genetic code is of the following general type:

(a) A group of three bases (or, less likely, a multiple of three bases) codes one amino-acid.

(b) The code is not of the overlapping type (see Fig. 1).

(c) The sequence of the bases is read from a fixed starting point. This determines how the long sequences of bases are to be correctly read off as triplets. There are no special 'commas' to show how to select the right triplets. If the starting point is displaced by one base, then the reading into triplets is displaced, and thus becomes incorrect.

(d) The code is probably 'degenerate'; that is, in general, one particular amino-acid can be coded by one of several triplets of bases.

The Reading of the Code

The evidence that the genetic code is not overlapping (see Fig. 1) does not come from our work, but from that of Wittmann[1] and of Tsugita and Fraenkel-Conrat[2] on the mutants of tobacco mosaic virus produced by nitrous acid. In an overlapping triplet code, an alteration to one base will in general change three adjacent amino-acids in the polypeptide chain. Their work on the alterations produced in the protein of the virus show that usually only one amino-acid at a time is changed as a result of treating the ribonucleic acid (RNA) of the virus with nitrous acid. In the rarer cases where two amino-acids are altered (owing presumably to two separate deaminations by the nitrous acid on one piece of RNA), the altered amino-acids are not in adjacent positions in the polypeptide chain.

Brenner[3] had previously shown that, if the code were universal (that is, the same throughout Nature), then all overlapping triplet codes were impossible. Moreover, all the abnormal human hæmoglobins studied in detail[4] show only single amino-acid changes. The newer experimental results essentially rule out all simple codes of the overlapping type.

If the code is not overlapping, then there must be some arrangement to show how to select the correct triplets (or quadruplets, or whatever it may be) along the continuous sequence of bases. One obvious suggestion is that, say, every fourth base is a 'comma'. Another idea is that certain triplets make 'sense', whereas others make 'nonsense', as in the comma-free codes of Crick, Griffith and Orgel[5]. Alternatively, the correct choice may be made by starting at a fixed point and working along the sequence of bases three (or four, or whatever) at a time. It is this possibility which we now favour.

Experimental Results

Our genetic experiments have been carried out on the B cistron of the r_{II} region of the bacteriophage $T4$, which attacks strains of *Escherichia coli*. This is the system so brilliantly exploited by Benzer[6,7]. The r_{II} region consists of two adjacent genes, or 'cistrons', called cistron A and cistron B. The wild-type phage will grow on both *E. coli* B (here called B) and on *E. coli* $K12$ (λ) (here called K), but a phage which has lost the function of either gene will not grow on K. Such a phage produces an r plaque on B. Many point mutations of the genes are known which behave in this way. Deletions of part of the region are also found. Other mutations, known as 'leaky', show partial function; that is, they will grow on K but their plaque-type on B is not truly wild. We report here our work on the mutant $P\ 13$ (now re-named $FC\ 0$) in the $B1$ segment of the B cistron. This mutant was originally produced by the action of proflavin[8].

We[9] have previously argued that acridines such as proflavin act as mutagens because they add or delete a base or bases. The most striking evidence in favour of this is that mutants produced by acridines are seldom 'leaky'; they are almost always completely lacking in the function of the gene. Since our note was published, experimental data from two sources have been added to our previous evidence: (1) we have examined a set of 126 r_{II} mutants made with

2

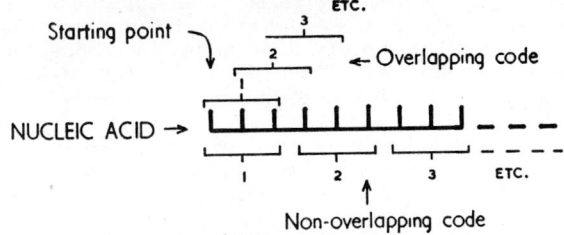

Fig. 1. To show the difference between an overlapping code and a non-overlapping code. The short vertical lines represent the bases of the nucleic acid. The case illustrated is for a triplet code

acridine yellow; of these only 6 are leaky (typically about half the mutants made with base analogues are leaky); (2) Streisinger[10] has found that whereas mutants of the lysozyme of phage $T4$ produced by base-analogues are usually leaky, all lysozyme mutants produced by proflavin are negative, that is, the function is completely lacking.

If an acridine mutant is produced by, say, adding a base, it should revert to 'wild-type' by deleting a base. Our work on revertants of $FC\ 0$ shows that it usually reverts not by reversing the original mutation but by producing a second mutation at a nearby point on the genetic map. That is, by a 'suppressor' in the same gene. In one case (or possibly two cases) it may have reverted back to true wild, but in at least 18 other cases the 'wild type' produced was really a double mutant with a 'wild' phenotype. Other workers[11] have found a similar phenomenon with r_{II} mutants, and Jinks[12] has made a detailed analysis of suppressors in the h_{III} gene.

The genetic map of these 18 suppressors of $FC\ 0$ is shown in Fig. 2, line a. It will be seen that they all fall in the $B1$ segment of the gene, though not all of them are very close to $FC\ 0$. They scatter over a region about, say, one-tenth the size of the B cistron. Not all are at different sites. We have found eight sites in all, but most of them fall into or near two close clusters of sites.

In all cases the suppressor was a non-leaky r. That is, it gave an r plaque on B and would not grow on K. This is the phenotype shown by a complete deletion of the gene, and shows that the function is lacking. The only possible exception was one case where the suppressor appeared to back-mutate so fast that we could not study it.

Each suppressor, as we have said, fails to grow on K. Reversion of each can therefore be studied by the same procedure used for $FC\ 0$. In a few cases these mutants apparently revert to the original wild-

3

type, but usually they revert by forming a double mutant. Fig. 2, lines b–g, shows the mutants produced as suppressors of these suppressors. Again all these new suppressors are non-leaky r mutants, and all map within the $B1$ segment for one site in the $B2$ segment.

Once again we have repeated the process on two of the new suppressors, with the same general results, as shown in Fig. 2, lines i and j.

All these mutants, except the original FC 0, occurred spontaneously. We have, however, produced one set (as suppressors of FC 7) using acridine yellow as a mutagen. The spectrum of suppressors we get (see Fig. 2, line h) is crudely similar to the spontaneous spectrum, and all the mutants are non-leaky r's. We have also tested a (small) selection of all our mutants and shown that their reversion-rates are increased by acridine yellow.

Thus in all we have about eighty independent r mutants, all suppressors of FC 0, or suppressors of suppressors, or suppressors of suppressors of suppressors. They all fall within a limited region of the gene and they are all non-leaky r mutants.

The double mutants (which contain a mutation plus its suppressor) which plate on K have a variety of plaque types on B. Some are indistinguishable from wild, some can be distinguished from wild with difficulty, while others are easily distinguishable and produce plaques rather like r.

We have checked in a few cases that the phenomenon is quite distinct from 'complementation', since the two mutants which separately are phenotypically r, and together are wild or pseudo-wild, must be put together in the same piece of genetic material. A simultaneous infection of K by the two mutants in separate viruses will not do.

The Explanation in Outline

Our explanation of all these facts is based on the theory set out at the beginning of this article. Although we have no direct evidence that the B cistron produces a polypeptide chain (probably through an RNA intermediate), in what follows we shall assume this to be so. To fix ideas, we imagine that the string of nucleotide bases is read, triplet by triplet, from a starting point on the left of the B cistron. We now suppose that, for example, the mutant FC 0 was produced by the insertion of an additional base in the wild-type sequence. Then this addition of a base at the FC 0 site will mean that the reading of all the triplets to the right of FC 0 will be shifted along one base, and will therefore be incorrect. Thus the amino-acid sequence of the protein

4

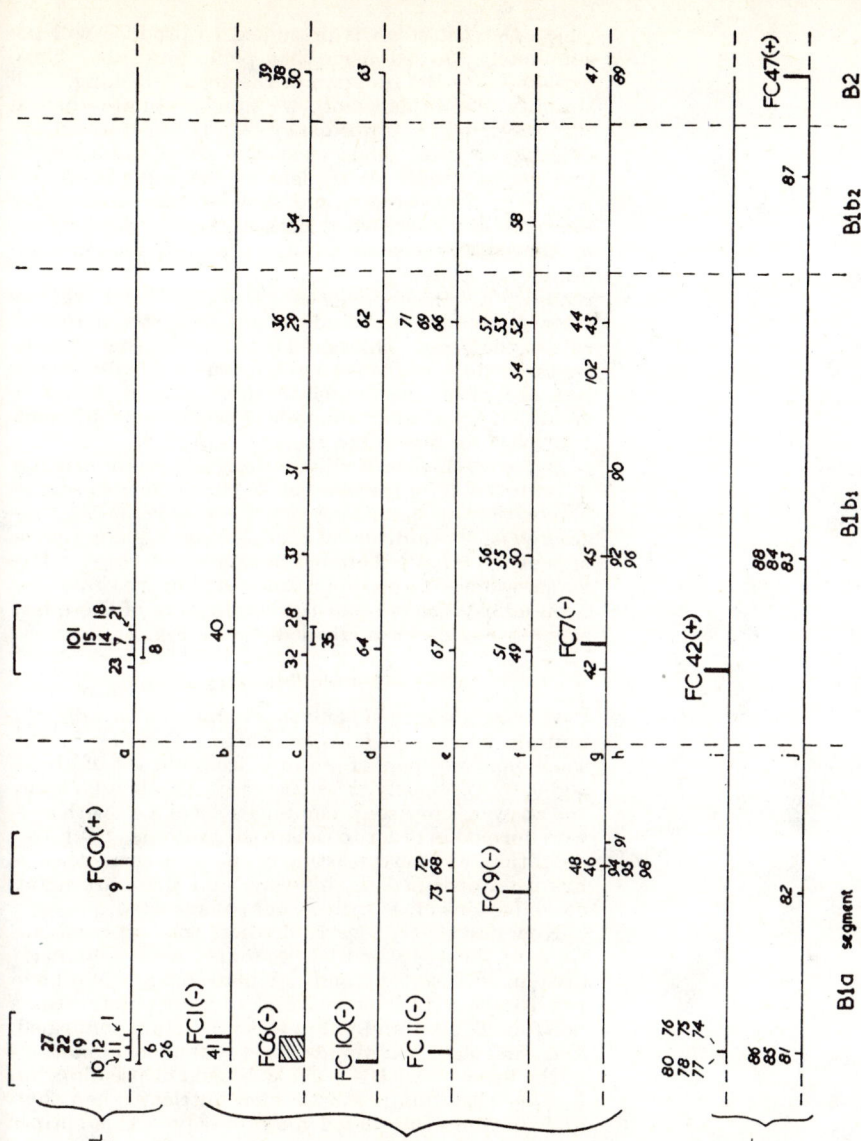

Fig. 2. A tentative map—only very roughly to scale—of the left-hand end of the *B* cistron, showing the position of the *FC* family of mutants. The order of sites within the regions covered by brackets (at the top of the figure) is not known. Mutants in italics have only been located approximately. Each line represents the suppressors picked up from one mutant, namely, that marked on the line in bold figures

which the B cistron is presumed to produce will be completely altered from that point onwards. This explains why the function of the gene is lacking. To simplify the explanation, we now postulate that a suppressor of FC 0 (for example, FC 1) is formed by deleting a base. Thus when the FC 1 mutation is present by itself, all triplets to the right of FC 1 will be read incorrectly and thus the function will be absent. However, when both mutations are present in the same piece of DNA, as in the pseudo-wild double mutant FC (0 + 1), then although the reading of triplets between FC 0 and FC 1 will be altered, the original reading will be restored to the rest of the gene. This could explain why such double mutants do not always have a true wild phenotype but are often pseudo-wild, since on our theory a small length of their amino-acid sequence is different from that of the wild-type.

For convenience we have designated our original mutant FC 0 by the symbol + (this choice is a pure convention at this stage) which we have so far considered as the addition of a single base. The suppressors of FC 0 have therefore been designated − . The suppressors of these suppressors have in the same way been labelled as +, and the suppressors of these last sets have again been labelled − (see Fig. 2).

Double Mutants

We can now ask: What is the character of any double mutant we like to form by putting together in the same gene any pair of mutants from our set of about eighty? Obviously, in some cases we already know the answer, since some combinations of a + with a − were formed in order to isolate the mutants. But, by definition, no pair consisting of one + with another + has been obtained in this way, and there are many combinations of + with − not so far tested.

Now our theory clearly predicts that all combinations of the type + with + (or − with −) should give an r phenotype and not plate on K. We have put together 14 such pairs of mutants in the cases listed in Table 1 and found this prediction confirmed.

At first sight one would expect that all combinations of the type (+ with −) would be wild or pseudo-wild, but the situation is a little more intricate than that, and must be considered more closely. This springs

Table 1. DOUBLE MUTANTS HAVING THE r PHENOTYPE

− With −	+ With +	
FC (1 + 21)	FC (0 + 58)	FC (40 + 57)
FC (23 + 21)	FC (0 + 38)	FC (40 + 58)
FC (1 + 23)	FC (0 + 40)	FC (40 + 55)
FC (1 + 9)	FC (0 + 55)	FC (40 + 54)
	FC (0 + 54)	FC (40 + 38)

6

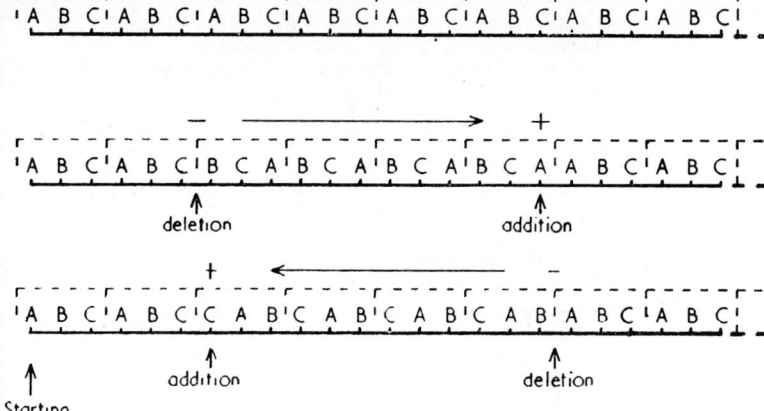

Fig. 3. To show that our convention for arrows is consistent. The letters, A, B and C each represent a different base of the nucleic acid. For simplicity a repeating sequence of bases, ABC, is shown. (This would code for a polypeptide for which every amino-acid was the same.) A triplet code is assumed. The dotted lines represent the imaginary 'reading frame' implying that the sequence is read in sets of three starting on the left

from the obvious fact that if the code is made of triplets, any long sequence of bases can be read correctly in one way, but incorrectly (by starting at the wrong point) in two different ways, depending whether the 'reading frame' is shifted one place to the right or one place to the left.

If we symbolize a shift, by one place, of the reading frame in one direction by → and in the opposite direction by ←, then we can establish the convention that our + is always at the head of the arrow, and our − at the tail. This is illustrated in Fig. 3.

We must now ask: Why do our suppressors not extend over the whole of the gene? The simplest postulate to make is that the shift of the reading frame produces some triplets the reading of which is 'unacceptable'; for example, they may be 'nonsense', or stand for 'end the chain', or be unacceptable in some other way due to the complications of protein structure. This means that a suppressor of, say, $FC\ 0$ must be within a region such that no 'unacceptable' triplet is produced by the shift in the reading frame between $FC\ 0$ and its suppressor. But, clearly, since for any sequence there are *two* possible misreadings, we might expect that the 'unacceptable' triplets produced by a → shift would occur in different places on the map from those produced by a ← shift.

Examination of the spectra of suppressors (in each

7

case putting in the arrows → or ←) suggests that while the → shift is acceptable anywhere within our region (though not outside it) the shift ←, starting from points near FC 0, is acceptable over only a more limited stretch. This is shown in Fig. 4. Somewhere in the left part of our region, between FC 0 or FC 9 and the FC 1 group, there must be one or more unacceptable triplets when a ← shift is made; similarly for the region to the right of the FC 21 cluster.

Thus we predict that a combination of a + with a − will be wild or pseudo-wild if it involves a → shift, but that such pairs involving a ← shift will be phenotypically r if the arrow crosses one or more of the forbidden places, since then an unacceptable triplet will be produced.

Table 2. Double Mutants of the Type (+ with −)

+ \ −	FC 41	FC 0	FC 40	FC 42	FC 58*	FC 63	FC 38
FC 1	W	W	W		W		W
FC 86		W	W	W	W	W	
FC 9	r	W	W	W	W		W
FC 82	r		W	W	W	W	
FC 21	r	W			W		W
FC 88	r	r			W	W	
FC 87	r	r	r	r			W

W, wild or pseudo-wild phenotype; W, wild or pseudo-wild combination used to isolate the suppressor; r, r phenotype.
* Double mutants formed with FC 58 (or with FC 34) give sharp plaques on K.

We have tested this prediction in the 28 cases shown in Table 2. We expected 19 of these to be wild, or pseudo-wild, and 9 of them to have the r phenotype. In all cases our prediction was correct. We regard this as a striking confirmation of our theory. It may be of interest that the theory was constructed before these particular experimental results were obtained.

Rigorous Statement of the Theory

So far we have spoken as if the evidence supported a triplet code, but this was simply for illustration. Exactly the same results would be obtained if the code operated with groups of, say, 5 bases. Moreover, our symbols + and − must not be taken to mean literally the addition or subtraction of a single base.

It is easy to see that our symbolism is more exactly as follows:

$$+ \text{ represents } + m, \text{ modulo } n$$
$$- \text{ represents } - m, \text{ modulo } n$$

8

where n (a positive integer) is the coding ratio (that is, the number of bases which code one amino-acid) and m is any integral number of bases, positive or negative.

It can also be seen that our choice of reading direction is arbitrary, and that the same results (to a first approximation) would be obtained in whichever direction the genetic material was read, that is, whether the starting point is on the right or the left of the gene, as conventionally drawn.

Triple Mutants and the Coding Ratio

The somewhat abstract description given above is necessary for generality, but fortunately we have convincing evidence that the coding ratio is in fact 3 or a multiple of 3.

This we have obtained by constructing triple mutants of the form (+ with + with +) or (− with − with −). One must be careful not to make shifts

Table 3. TRIPLE MUTANTS HAVING A WILD OR PSEUDO-WILD PHENOTYPE

$$FC\,(0 + 40 + 38)$$
$$FC\,(0 + 40 + 58)$$
$$FC\,(0 + 40 + 57)$$
$$FC\,(0 + 40 + 54)$$
$$FC\,(0 + 40 + 55)$$
$$FC\,(1 + 21 + 23)$$

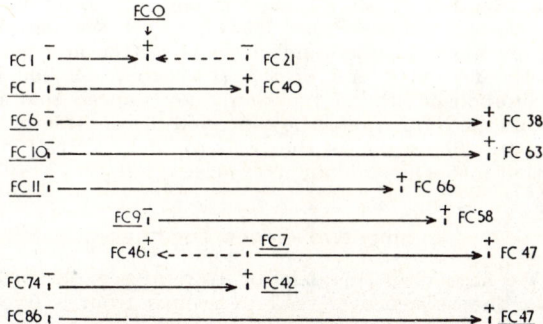

Fig. 4. A simplified version of the genetic map of Fig. 2. Each line corresponds to the suppressor from one mutant, here underlined. The arrows show the range over which suppressors have so far been found, the extreme mutants being named on the map. Arrows to the right are shown solid, arrows to the left dotted.

across the 'unacceptable' regions for the ← shifts, but these we can avoid by a proper choice of mutants.

We have so far examined the six cases listed in Table 3 and in all cases the triples are wild or pseudo-wild.

The rather striking nature of this result can be

9

seen by considering one of them, for example, the triple (FC 0 with FC 40 with FC 38). These three mutants are, by themselves, all of like type ($+$). We can say this not merely from the way in which they were obtained, but because each of them, when combined with our mutant FC 9 ($-$), gives the wild, or pseudo-wild phenotype. However, either singly or together in pairs they have an r phenotype, and will not grow on K. That is, the function of the gene is absent. Nevertheless, the combination of all three in the same gene partly restores the function and produces a pseudo-wild phage which grows on K. This is exactly what one would expect, in favourable cases, if the coding ratio were 3 or a multiple of 3.

Our ability to find the coding ratio thus depends on the fact that, in at least one of our composite mutants which are 'wild', at least one amino-acid must have been added to or deleted from the polypeptide chain without disturbing the function of the gene-product too greatly.

This is a very fortunate situation. The fact that we can make these changes and can study so large a region probably comes about because this part of the protein is not essential for its function. That this is so has already been suggested by Champe and Benzer[13] in their work on complementation in the r_{II} region. By a special test (combined infection on K, followed by plating on B) it is possible to examine the function of the A cistron and the B cistron separately. A particular deletion, 1589 (see Fig. 5) covers the right-hand end of the A cistron and part of the left-hand end of the B cistron. Although 1589 abolishes the A function, they showed that it allows the B function to be expressed to a considerable extent. The region of the B cistron deleted by 1589 is that into which all our FC mutants fall.

Joining two Genes Together

We have used this deletion to reinforce our idea that the sequence is read in groups from a fixed starting point. Normally, an alteration confined to the A cistron (be it a deletion, an acridine mutant, or any other mutant) does not prevent the expression of the B cistron. Conversely, no alteration within the B cistron prevents the function of the A cistron. This implies that there may be a region between the two cistrons which separates them and allows their functions to be expressed individually.

We argued that the deletion 1589 will have lost this separating region and that therefore the two (partly damaged) cistrons should have been joined together. Experiments show this to be the case,

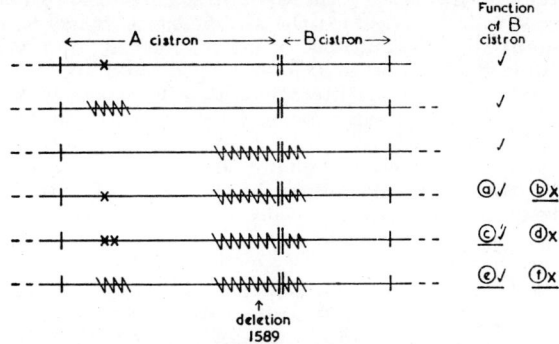

Fig. 5. Summary of the results with deletion 1589. The first two lines show that without 1589 a mutation or a deletion in the A cistron does not prevent the B cistron from functioning. Deletion 1589 (line 3) also allows the B cistron to function. The other cases, in some of which an alteration in the A cistron prevents the function of the B cistron (when 1589 is also present), are discussed in the text. They have been labelled cases (*a*), (*b*), etc., for convenience of reference, although cases (*a*) and (*d*) are not discussed in this paper. √ implies function; × implies no function

for now an alteration to the left-hand end of the A cistron, if combined with deletion 1589, can prevent the B function from appearing. This is shown in Fig. 5. Either the mutant $P43$ or $X142$ (both of which revert strongly with acridines) will prevent the B function when the two cistrons are joined, although both of these mutants are in the A cistron. This is also true of $X142\,S1$, a suppressor of $X\,142$ (Fig. 5, case *b*). However, the double mutant ($X142$ with $X142\,S1$), of the type (+ with −), which by itself is pseudo-wild, still has the B function when combined with 1589 (Fig. 5, case *c*). We have also tested in this way the 10 deletions listed by Benzer[7], which fall wholly to the left of 1589. Of these, three (386, 168 and 221) prevent the B function (Fig. 5, case *f*), whereas the other seven show it (Fig. 5, case *e*). We surmise that each of these seven has lost a number of bases which is a multiple of 3. There are theoretical reasons for expecting that deletions may not be random in length, but will more often have lost a number of bases equal to an integral multiple of the coding ratio.

It would not surprise us if it were eventually shown that deletion 1589 produces a protein which consists of part of the protein from the A cistron and part of that from the B cistron, joined together in the same polypeptide chain, and having to some extent the function of the undamaged B protein.

Is the Coding Ratio 3 or 6?

It remains to show that the coding ratio is probably 3, rather than a multiple of 3. Previous rather rough estimates[10,14] of the coding ratio (which are admittedly very unreliable) might suggest that the coding ratio is not far from 6. This would imply, on our theory, that the alteration in FC 0 was not to one base, but to two bases (or, more correctly, to an even number of bases).

We have some additional evidence which suggests that this is unlikely. First, in our set of 126 mutants produced by acridine yellow (referred to earlier) we have four independent mutants which fall at or

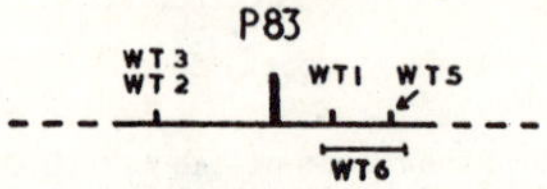

Fig. 6. Genetic map of P 83 and its suppressors, WT 1, etc. The region falls within segment B 9a near the right-hand end of the B cistron. It is not yet known which way round the map is in relation to the other figures

close to the FC 9 site. By a suitable choice of partners, we have been able to show that two are $+$ and two are $-$. Secondly, we have two mutants ($X146$ and $X225$), produced by hydrazine[15], which fall on or near the site FC 30. These we have been able to show are both of type $-$.

Thus unless both acridines and hydrazine usually delete (or add) an even number of bases, this evidence supports a coding ratio of 3. However, as the action of these mutagens is not understood in detail, we cannot be certain that the coding ratio is not 6, although 3 seems more likely.

We have preliminary results which show that other acridine mutants often revert by means of close suppressors, but it is too sketchy to report here. A tentative map of some suppressors of P 83, a mutant at the other end of the B cistron, in segment B 9a, is shown in Fig. 6. They occur within a shorter region than the suppressors of FC 0, covering a distance of about one-twentieth of the B cistron. The double mutant WT (2 + 5) has the r phenotype, as expected.

Is the Code Degenerate?

If the code is a triplet code, there are 64 (4 × 4 × 4) possible triplets. Our results suggest that it is unlikely that only 20 of these represent the 20 amino-acids and that the remaining 44 are nonsense. If this were the case, the region over which suppressors of the FC 0 family occur (perhaps a quarter of the B cistron) should be very much smaller than we observe, since a shift of frame should then, by chance, pro-

12

duce a nonsense reading at a much closer distance. This argument depends on the size of the protein which we have assumed the B cistron to produce. We do not know this, but the length of the cistron suggests that the protein may contain about 200 amino-acids. Thus the code is probably 'degenerate', that is, in general more than one triplet codes for each amino-acid. It is well known that if this were so, one could also account for the major dilemma of the coding problem, namely, that while the base composition of the DNA can be very different in different micro-organisms, the amino-acid composition of their proteins only changes by a moderate amount[16]. However, exactly how many triplets code amino-acids and how many have other functions we are unable to say.

Future Developments

Our theory leads to one very clear prediction. Suppose one could examine the amino-acid sequence of the 'pseudo-wild' protein produced by one of our double mutants of the (+ with −) type. Conventional theory suggests that since the gene is only altered in two places, only two amino-acids would be changed. Our theory, on the other hand, predicts that a string of amino-acids would be altered, covering the region of the polypeptide chain corresponding to the region on the gene between the two mutants. A good protein on which to test this hypothesis is the lysozyme of the phage, at present being studied chemically by Dreyer[17] and genetically by Streisinger[10].

At the recent Biochemical Congress at Moscow, the audience of Symposium I was startled by the announcement of Nirenberg that he and Matthaei[18] had produced polyphenylalanine (that is, a polypeptide all the residues of which are phenylalanine) by adding polyuridylic acid (that is, an RNA the bases of which are all uracil) to a cell-free system which can synthesize protein. This implies that a sequence of uracils codes for phenylalanine, and our work suggests that it is probably a triplet of uracils.

It is possible by various devices, either chemical or enzymatic, to synthesize polyribonucleotides with defined or partly defined sequences. If these, too, will produce specific polypeptides, the coding problem is wide open for experimental attack, and in fact many laboratories, including our own, are already working on the problem. If the coding ratio is indeed 3, as our results suggest, and if the code is the same throughout Nature, then the genetic code may well be solved within a year.

13

We thank Dr. Alice Orgel for certain mutants and for the use of data from her thesis, Dr. Leslie Orgel for many useful discussions, and Dr. Seymour Benzer for supplying us with certain deletions. We are particularly grateful to Prof. C. F. A. Pantin for allowing us to use a room in the Zoological Museum, Cambridge, in which the bulk of this work was done.

[1] Wittman, H. G., Symp. 1, Fifth Intern. Cong. Biochem., 1961, for refs. (in the press).
[2] Tsugita, A., and Fraenkel-Conrat, H., *Proc. U.S. Nat. Acad. Sci.*, **46**, 636 (1960); *J. Mol. Biol.* (in the press).
[3] Brenner, S., *Proc. U.S. Nat. Acad. Sci.*, **43**, 687 (1957).
[4] For refs. see Watson, H. C., and Kendrew, J. C., *Nature*, **190**, 670 (1961).
[5] Crick, F. H. C., Griffith, J. S., and Orgel, L. E., *Proc. U.S. Nat. Acad. Sci.*, **43**, 416 (1957).
[6] Benzer, S., *Proc. U.S. Nat. Acad. Sci.*, **45**, 1607 (1959), for refs. to earlier papers.
[7] Benzer, S., *Proc. U.S. Nat. Acad. Sci.*, **47**, 403 (1961); see his Fig. 3.
[8] Brenner, S., Benzer, S., and Barnett, L., *Nature*, **182**, 983 (1958).
[9] Brenner, S., Barnett, L., Crick, F. H. C., and Orgel, A., *J. Mol. Biol.*, **3**, 121 (1961).
[10] Streisinger, G. (personal communication and in the press).
[11] Feynman, R. P.; Benzer, S.; Freese, E. (all personal communications).
[12] Jinks, J. L., *Heredity*, **16**, 153, 241 (1961).
[13] Champe, S., and Benzer, S. (personal communication and in preparation).
[14] Jacob, F., and Wollman, E. L., *Sexuality and the Genetics of Bacteria* (Academic Press, New York, 1961). ‡Levinthal, C. (personal communication).
[15] Orgel, A., and Brenner, S. (in preparation).
[16] Sueoka, N., *Cold Spring Harb. Symp. Quant. Biol.* (in the press).
[17] Dreyer, W. J., Symp. 1, Fifth Intern. Cong. Biochem., 1961 (in the press).
[18] Nirenberg, M. W., and Matthaei, J. H., *Proc. U.S. Nat. Acad. Sci.*, **47**, 1588 (1961).

Mapping of Deletions and Substitutions in Heteroduplex DNA Molecules of Bacteriophage Lambda by Electron Microscopy

Abstract: Electron microscopy of heteroduplex DNA molecules, composed of one strand of Escherichia coli phage λ^+ DNA annealed to the complementary DNA strand of a λ deletion or substitution mutant, permits visualization, as well as precise measurements and mapping, of the unpaired single-stranded regions of nonhomology in the otherwise double-stranded molecules. In the $\lambda b2$ mutant, the central segment (13 percent) of the λ^+ DNA molecule is shown to be deleted. In the hybrid phages λi^{434} and λi^{21} a segment of the right arm of the λ^+ genome (5.5 or 7.6 to 9 percent) is replaced by the corresponding immunity regions of phage 434 (3.3 percent) or phage 21 (4 percent) DNA. The b5 region in the $\lambda b5$ mutant appears to be identical to the i^{21} segment. From these data it is possible to estimate the size and position of those λ genes which are replaced by the i^{434} and i^{21} segments. The method permits preparing complete physical maps of viral genomes with a precision heretofore unattainable.

One of the aims of classical cytogenetics is to relate the chromosome dimensions and general morphology to the genetic maps deduced from recombination experiments. This approach has been only partially successful because of the complex and largely unknown fine structure of the chromosomes in higher organisms. The genome of the bacteriophage λ is much simpler, consisting of a single linear DNA molecule, which can be seen in the electron microscope. Furthermore, there are numerous mutants of λ readily available, in which practically any region of the genome has been deleted or replaced by nonhomologous DNA derived from the host or other phages. These may be classified into three most common types: (i) deletions within the central region of the λ DNA molecule, as exemplified by the b2 deletion; (ii) hybrids between λ and λ-related phages, including the special case of substitutions in the immunity region, such as λi^{434}, λi^{21}, and $\lambda b5$; and (iii) substitutions consisting of *Escherichia coli* DNA within the left ($\lambda dgal$) or right arm (λbio and $\lambda dbio$) of the λ molecule (*1*). The position of all these deletions can be mapped by standard genetic crosses.

We now show that very precise physical maps of these deletions can be prepared by combining electron microscopy and DNA-DNA hybridization techniques. The precision of the mapping is much higher than that for any other known method; the standard deviation is frequently less than 2 percent. The principle of the technique is as follows. The complementary DNA strands of wild-type λ (λ^+) and of a deletion or substitution mutant are preparatively separated (*2*). One strand of λ^+ DNA is hybridized with the *complementary* strand derived from the mutant strain DNA. The resulting *heteroduplex* is a double-stranded molecule with the exception of the regions of nonhomology. For a simple deletion, such as that of region b2+, the heteroduplex between the λ^+ and $\lambda b2$ strands of DNA should appear as an uninterrupted double-stranded molecule of $\lambda b2$ length with a single-stranded loop formed from the unpaired segment of the λ^+ strand, as schematically depicted in Fig. 1. On the other hand, substitution by a nonhomologous region in one strand of the heteroduplex should be seen as a stretch of two unpaired single DNA strands (such as region i^λ/i^{21} in Fig. 1) bridging a gap between the double-stranded regions. These predictions were fully confirmed by preparing electron micrographs of heteroduplex molecules in which the single-stranded DNA segments are well extended and could be readily identified and measured; their exact position was accurately determined with respect to the termini of the molecule and to other single-stranded regions (see Figs. 3 to 5). Independently, Davis and Davidson (*3*) have used a somewhat similar approach, although the single DNA strands were collapsed ("bushes") in their electron microscopic preparations; thus, they were unable to determine the length and fine structure of the unpaired single-stranded regions. They did not fractionate the complementary DNA strands, and their method of renaturation of the mixture of denatured DNA's differed from ours.

The separated l and r DNA strands of λc_{72}, $\lambda cb2$, $\lambda cb2b5$, λi^{434}, λi^{21} and $\lambda b2i^{21}$ (*1*) were isolated as described by Hradecna and Szybalski (*2*), with the use of equilibrium centrifugation in CsCl gradients containing poly(U,G), a copolymer of guanylic and uridylic acids. Heteroduplexes were prepared by the following procedure, adapted from Subirana (*4*). To 1 ml of 50 percent formamide solution (1 volume of formamide and 1 volume of 0.01M NaHCO$_3$, pH 8.6) was added 10 μl from each of two freshly prepared 6M CsCl gradient fractions, one containing the separated strands l or r of λ DNA and the other containing the complementary strands of mutant λ DNA. In a similar manner, homoduplex controls were prepared. The final concentrations were 1 to 4 μg of DNA per milliliter of 0.12M CsCl. Annealing was carried out at 4°C for 5 days or longer.

The DNA was prepared for electron microscopy by a modification of the basic-protein film technique of Kleinschmidt and Zahn (*5*). The cytochrome c solution was filtered through 0.2-μ Flotronics membranes (FM-13; Selas), whereas all other solutions were filtered

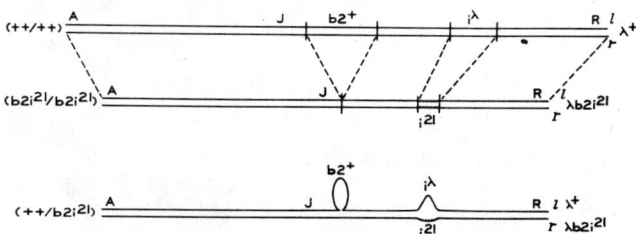

Fig. 1. Representation of a heteroduplex DNA molecule (++/b2i^{21}; bottom drawing) formed by annealing the l strand (++) of λ^- with the r strand (b2i^{21}) of $\lambda b2i^{21}$ (*1*). The corresponding homoduplexes are represented by the two upper diagrams. In $\lambda b2i^{21}$ the b2+ region is deleted, whereas region i^λ is replaced by a shorter nonhomologous region i^{21} (knobby line).

through 0.22-μ Millipore membranes to eliminate the particulate matter, which contributes to the clumping of the DNA molecules. A mixture composed of 0.1 ml of the DNA solution in 50 percent formamide and 0.01 ml of 0.1 percent cytochrome c (Nutritional Biochemicals) was spread on a surface of twice-distilled water from an inclined, acid-cleaned, glass slide. The monolayer containing DNA was then transferred to specimen grids and dehydrated for 5 seconds each in absolute ethanol, amyl acetate, and 2-methylbutane. The most satisfactory results were obtained with platinum specimen grids (19 drilled holes; Siemens America) coated with Formvar membranes. A fresh solution of 0.3 percent Formvar (Shawinigan Resins) in 1,2-dichloroethane was dried on a glass cover slip. The film was then cast onto distilled water in a sintered-glass funnel and drawn down onto platinum specimen grids, lightly coated with carbon, and used the same day.

The DNA was rendered visible by shadow-casting with uranium oxide. Approximately half of the nitrocellulose binder was removed from commercial uranium oxide (E. F. Fullam) by extracting with acetone, discarding half of the supernatant from the sediment, and evaporating the solvent remaining with the uranium oxide sediment, while mixing to obtain an even distribution of binder. This uranium oxide (12 mg) was evaporated from a tungsten wire basket at an angle of 6° to 7°, with a distance of 11 cm between the basket and rotating specimen (10 rev/min). The nitrocellulose was first removed by slowly heating at a pressure less than 0.1 μ-Hg. The uranium oxide was evaporated by rapid heating for 5 to 10 seconds at a pressure lower than 0.03 μ-Hg (Kinney evaporator SC-3 with a liquid nitrogen trap).

Electron micrographs (Kodak electron image plates) were taken with a Siemens Elmiskop I (intermediate lens off, projector polepiece III, double condenser illumination, 10,700×, 60 kv, 15-μ objective aperture). Magnification was calibrated with a carbon replica of a diffraction grating (54,864 lines per inch; E. F. Fullam). The DNA molecules were measured as described by Ris and Chandler (6). To minimize variation in magnification all the micrographs from one grid were taken without realigning the microscope or repositioning the grid. The illuminated area on the viewing screen was ad-

Table 1. Length comparisons between homoduplexes or homoduplex regions of DNA. Each set of two was placed on the same grid.

Homoduplex (or homoduplex region)	Length (μ)	Standard deviation* (%)
$\lambda b2/\lambda b2$	14.8	1.6
λ^+/λ^+	17.0	0.9
$\lambda b2/\lambda b2$	14.8	2.1
$\lambda b2/\lambda^+$	14.7	1.5
$\lambda b2b5/\lambda b2b5$	14.2	1.2
λ^+/λ^+	17.0	1.4
$\lambda i^{434}/\lambda i^{434}$	16.6	2.1
$\lambda b2/\lambda^+$	14.8	2.0

* Percent of indicated length, for example, 14.8 ± 0.24 (1.6 percent).

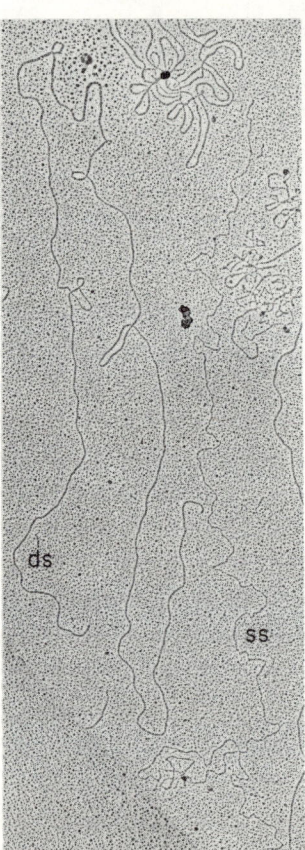

Fig. 2. Electron micrograph of a double-stranded (ds) homoduplex of λ^+ DNA, prepared by annealing the separated l and r strands, and of an unpaired single-stranded λ^+ DNA molecule (ss).

justed to the same size before each photograph. One hour was allowed after the lenses were turned on before the micrographs were taken, and the lenses were turned off and on several times to reduce hysteresis. Where comparison of the lengths of two different molecules was needed, a mixture of both duplexes was placed on the same grid. Each value represents 20 to 40 measurements on nonambiguous configurations. The standard deviations were about 2 percent for double-stranded DNA and about 5 percent for single-stranded DNA.

As evident in Fig. 2, the annealing of two complementary strands of λ DNA results in the formation of uniformly double-stranded homoduplexes (ds). The number of such duplexes formed depends on the time allowed for annealing (80 percent after 2 weeks). Some circular duplexes (6) (Fig. 4A) were present in all preparations, an indication that the conditions of annealing permit pairing of homologous terminal regions of λ as short as only 20 base pairs. Under the described conditions of annealing, spreading, and shadowing, the homoduplexes (ds) appear somewhat thicker and more rigid than the single DNA strands (ss), an example of which also appears in Fig. 2. Moreover, the single strands, although somewhat kinky in appearance, are reasonably well extended, although their length is variable, being frequently 5 to 10 percent longer than the corresponding double strands (Tables 1 and 2). Thus, by this technique double-stranded and single-stranded molecules can be easily distinguished.

Heteroduplexes (++/b2b5) composed of an l strand of λ^+ DNA (++) and an r strand of $\lambda b2b5$ (b2b5) are shown in Fig. 3, A–D. The molecules appear double-stranded except for two regions: a single-stranded loop (b2+), which must correspond to the b2+ region comprising the central sector (13 percent) of the λ^+ strand and deleted in the $\lambda b2$ mutant (7), and an unpaired region with one single-stranded segment longer than the other. An enlargement of a b2+ loop is shown in Fig. 3C. As expected for a simple deletion (Fig. 1), no apparent interruption in the double-stranded configuration occurs at the point where the single-stranded loop emanates from the double helix (Fig. 3, A–C). Figure 3D shows in greater detail the b5/i^λ (b5/+) nonhomology region. It is obvious from this micrograph and Fig.

3A that there is a discontinuity in the double-stranded structure, which bifurcates into two single-stranded regions before rejoining. The shorter single-stranded segment must correspond to the b5 region, since the λb5 (b5/b5) DNA molecule is shorter than λ+ (+/+) DNA (*1, 2, 7* and Tables 1, 2, and 3). Practically identical results, well within the standard experimental error (Table 2), were obtained when one strand of the heteroduplex was derived from λ+ and the other from the λb2i^{21} hybrid, in which the λ genes N, *rex*, c_I, *x*, *y*, and c_{II} have been replaced by the analogous segment of the phage 21 genome (*1, 8, 9, 10*). This similarity suggests that the i^{21} region is identical with the b5 region. More recently, corroboration of this notion has been provided by studies carried out in cooperation with Dr. Z. Hradecna, in which it was shown that heteroduplexes between λb2b5 and λb2i^{21} appear as perfectly double-stranded molecules, free of any readily discernible single-stranded regions.

Another example of an unpaired single-stranded region is provided by a heteroduplex between strand *l* of λb2 and strand *r* of λi^{434} (*7, 11*). In the latter phage, λ genes *rex*, c_I, and *x* have been replaced by a corresponding but somewhat shorter segment [as judged by buoyant density data and length measurements; see (*1*) and Tables 1, 2,

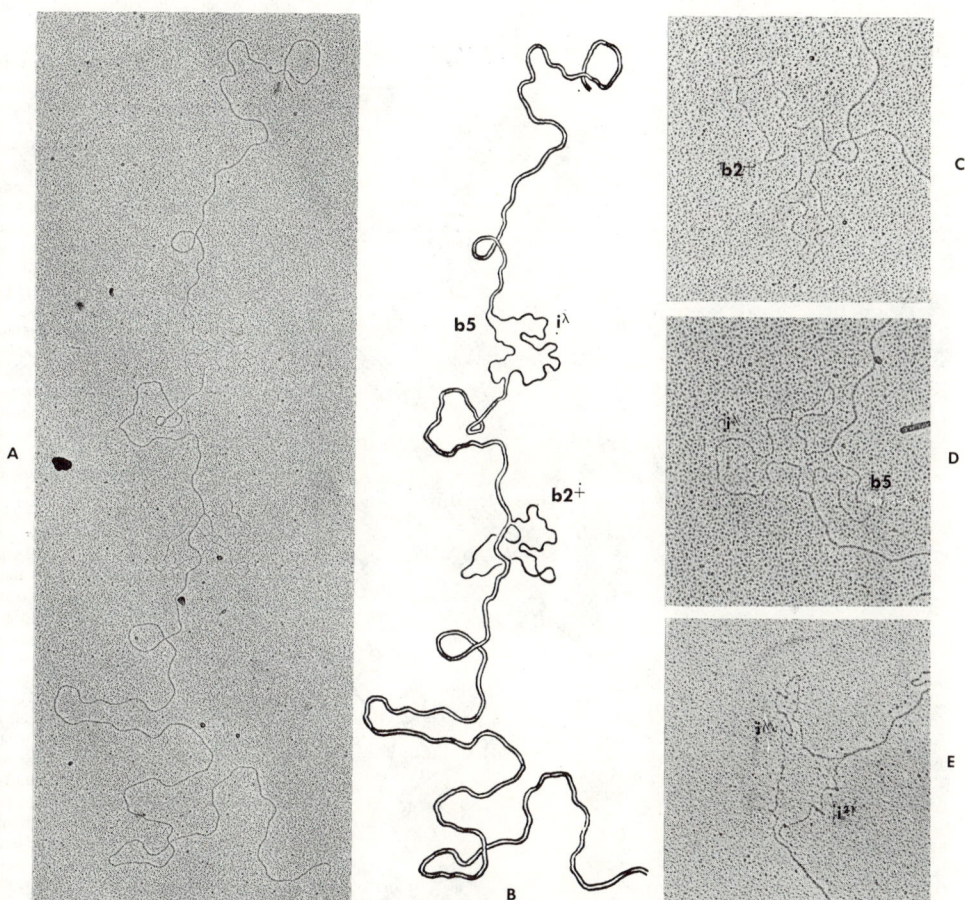

Fig. 3. Electron micrographs of heteroduplex λ DNA molecules. (A) Heteroduplex between strand *l* of λ+ and strand *r* of λb2b5 (++/b2b5). (B) An interpretive drawing of the ++/b2b5 heteroduplex, including the single-stranded b2+ loop and the unpaired segment $i^λ$/b5 (+/b5). (C) Enlargement of another b2+ loop. (D) Enlargement of another $i^λ$/b5 (+/b5) unpaired region. (E) Enlargement of the $i^λ$/i^{21} (+/i^{21}) unpaired region.

231

Table 2. Length of single-stranded (ss) and double-stranded (ds) regions in heteroduplexes of λ DNA.

Segment of λ	Length (μ)	S.D.* (%)	Length (% of λ+)
λi^{434}/λb2			
Left λ terminus to b2+ loop (ds)	7.53	2.3	44.3
b2+ loop (ss)	2.33	5.9	13.0†
b2+ loop to left bifurcation of i^{434}/i^λ (ds)	2.78	4.3	16.3
i^λ (ss)	0.99	5.1	5.5†
i^{434} (ss)	0.58	4.6	3.3†
Right bifurcation of i^{434}/i^λ to right λ terminus (ds)	3.54	2.9	20.9
λb2b5/λ+			
Left λ terminus to b2+ loop (ds)	7.53	1.9	44.3
b2+ loop (ss)	2.48	5.9	13.0†
b2+ loop to left bifurcation of b5/i^λ (ds)	2.30	7.3	13.5
i^λ (ss)	1.71	3.4	9.0†
b5 (ss)	0.76	5.5	4.0†
Right bifurcation of b5/i^λ to right λ terminus (ds)	3.43	1.8	20.2
λb2i^{21}/λ+‡			
Left λ terminus to b2+ loop (ds)	7.53	4.5	44.3
b2+ loop (ss)	1.78	17.2	13 †
b2+ loop to left bifurcation of i^{21}/i^λ (ds)	2.37	9.4	13.8
i^λ (ss)	1.17	11.5	8.4†
i^{21} (ss)	0.62	13.6	4.4†
Right bifurcation of i^{21}/i^λ to right λ terminus (ds)	3.52	4.9	20.5

* Percent standard deviation for each region, for example, 7.53 ± 0.17 (2.3 percent).   † Corrected to equivalent of double-stranded length, taking the b2+ loop as equal to 13 percent of the λ+ length. ‡ Preliminary measurements for 20 heteroduplex molecules, with the distance from the left λ terminus to the b2+ loop used as a standard (7.53 μ).

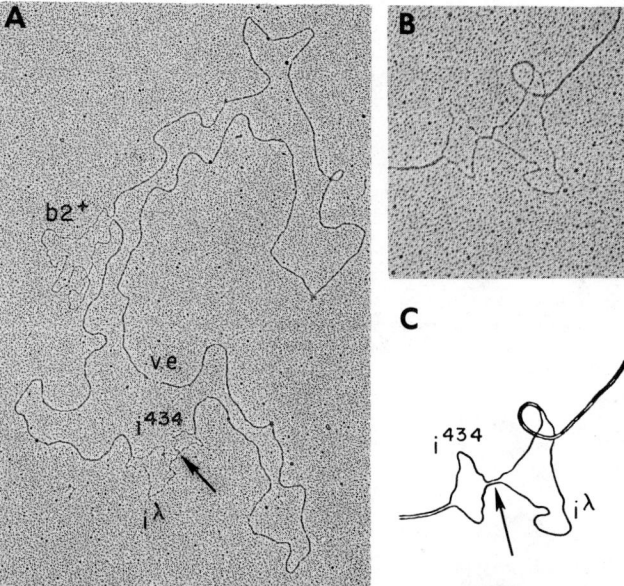

Fig. 4. Electron micrographs of heteroduplex molecules between the *l* strand of λb2 and the *r* strand of λi^{434} DNA. (A) Intact circular heteroduplex molecule. It appears that, due to tensions generated during drying of the DNA heteroduplex on the Formvar film, the vegetative (cohesive) ends (*v.e.*) have become disengaged and the *l* strand of the λb2 DNA has broken at the point where the single-stranded b2+ loop emerges. (B) Enlargement of another unpaired segment i^λ/i^{434}, and its interpretative drawing (C).

and 3] carrying the immunity region i^{434} of phage 434 (8, 9, 11). As shown in Figs. 4 and 5, the region of nonhomology between λi^{434} and λ+ is shorter than that observed for the i^{21} or b5 substitutions. A further complexity, which was not observed in the b5/+ or i^{21}/+ nonhomology regions, is indicated by the arrows in Fig. 4. In most of the unpaired i^{434}/i^λ regions the single strands meet (pair?) at a specific location (Fig. 5B). This probable homology region corresponds to less than 0.3 percent (150 base pairs) of the λ+ DNA length but may actually be much shorter, since homologies consisting of only 20 nucleotides can be visualized by our technique, as pointed out for circular λ molecules.

Our results have been compared with molecular and genetic data obtained by other investigators using different or similar approaches. The figure of 13 percent, representing the length of the b2 deletion, compares favorably with recent sedimentation data which indicate that 13.6 percent of λ+ DNA is deleted in the λb2 mutant (12). These figures are somewhat lower than those reported earlier and summarized in Table 3. The only lower value (11.2 percent) for the length of the b2 deletion was computed indirectly by comparing the electron micrographic lengths of λ+, λb2b5, and λb5 DNA (3), although more recent measurements performed in the same laboratory indicate a b2+ length of 12.6 percent (13). On the basis of genetic data (14) and the nucleotide distribution in λ+ and +λb2 DNA (15), it was previously concluded that the b2+ region occupies the central portion of the λ+ DNA molecule. In our micrographs of the b2/+ heteroduplex the b2+ loop emanates at a point 50.8 percent from the left and 49.2 percent from the right end of the double-stranded structure. There is less than 1 percent difference between this result and other electron micrographic measurements (3, 13).

The immunity region of λ, which is within the unpaired segments i^{434}/+, and i^{21}/+ (1, 8-11), is located near the middle of the right arm of the λ+ DNA molecule. Comparison of the electron micrographic measurements (Fig. 5, A and B) with the genetic data permits assignment of lengths and positions to the λ genes located within the region replaced by the i^{434} and i^{21} segments and in their immediate vicinity. The approximate position of the left end of gene N, as based on the sedimentation data for infectious DNA fragments [73 percent from the left λ^+ terminus (16)] agrees well with its placement between the left ends of the i^{434} and i^{21} substitution and with our measurements of the respective distances from the left terminus amounting to 73.6 percent and 70.8 to 72.2 percent of the λ^+ molecule (16a). The distance measured between the left ends of the i^{21} and i^{434} substitutions permits assignment of 1.4 to 2.8 percent (see 16a) to that region of the λ genome which contains mutations sus7 and sus53 (see 9, 17) in gene N (Fig. 5C). However, the size of that portion of gene N which lies between the i^{21} and i^{434} substitutions appears much smaller, when one compares the difference in length between heterologous regions i^{21} and i^{434}, which is only 0.7 percent of the λ^+ length. To make the latter comparison, one must assume that the gene supplied by phage 21 in the λi^{21} hybrid corresponds to gene N of λ, both in function and in length. Since the only known difference between the i^{21} and i^λ regions is the absence of the rex function, the length of the i^{21} substitution (4 percent of the λ^+ length) must accommodate the functions corresponding to the following λ genes and controlling elements: N, promoter (p_L) and operator (o_L) for the N operon, repressor gene c_I, promoter (p_R) and operator (o_R) for the x-O-P operon, regions x and y, and gene c_{II} (1, 9, 10, 17).

Gene c_I, whose length should correspond to approximately 2 percent of the λ^+ DNA molecule, as judged from the molecular weight of the c_I protein (18), should be located to the right of gene rex within the region defined by the i^{434} substitution (Fig. 5C). This result, which places the right end of c_I at least 76 percent from the left λ^+ terminus, indicates that the λ region which is characterized by 43 percent guanine + cytosine (G+C) content must extend more than 5 percent beyond its assigned 71 percent right boundary (15), because c_I-specific messenger RNA hybridizes with those DNA fragments which contain 43 percent G+C (19).

The size of the λ immunity region, as defined by the i^{434} substitution, is 5.5 percent of the λ^+ DNA length, which should be ample for genes c_I (2 percent) and rex, together with the two c_I-controlled operators, o_L (left) and o_R (right), the latter in region x (8, 17, 18). The analogous functions, with the exception of gene rex, must occupy only 3.3 percent in phage 434

Table 3. Various estimates of the net loss in the λ^+ DNA length (percent) as a result of the b2 deletion and the b5, i^{21} and i^{434} substitutions.

Measurement	Phage							Reference
	λ^+	λb2	λi^{434}	λi^{21}	λb5	λb2i^{21}	λb2b5	
Electron microscopy								
Percent deleted	0						23	(23)
Percent deleted	0	17.8						(24)
Percent deleted	0	(11.2)*			5.3		16.5	(3)
Percent deleted	0	13.0	2.2†	4.0† 3.6†	5.0† 3.5*		16.5	(Present data) (16a)
Density of heterodimers								
Percent deleted	0	15.3						(25)
Zone sedimentation								
Percent deleted	0	15.1			7.5		19.0–20.7	(26)
Percent deleted	0	13.6						(12)
Buoyant density of phage in CsCl gradient								
Percent deleted	0	14.7	2.8	5.9	5.9	20.5	20.5	(1, 2)‡
Density (g/cm³)	1.508	1.491	1.505	1.501s	1.501s	1.483s	1.483s	
Percent deleted	0	14.7			6.4		20.9	(7)‡
Density (g/cm³)	1.508	1.491			1.501		1.483	
Percent deleted	0				5.5		17.9	(3)‡
Density (g/cm³)	1.508				1.502		1.487	

* Value obtained indirectly from the difference between measurements of double-stranded DNA. † Difference between the lengths of the single DNA strands in unpaired regions (corrected to equivalent of double-stranded length). ‡ Percent deleted was calculated (or presently recalculated) from the formula in (3).

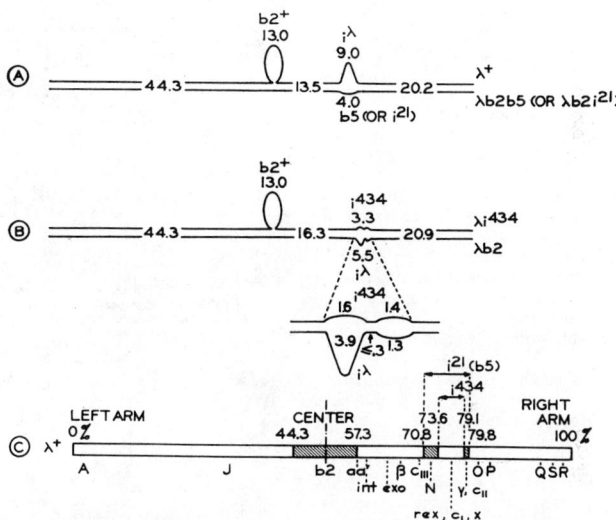

Fig. 5. Representation of heteroduplex λ DNA molecules and the physical map of λ^+ DNA derived from measurements of the double-stranded and unpaired regions in the heteroduplexes. (A) Linear dimensions of a heteroduplex formed by annealing one strand of λ^+ with the complementary strand of λb2b5. Practically identical results were obtained when one strand of λ^+ was annealed with the complementary strand of λb2i^{21} DNA (Fig. 3E and Table 2; see also 16a). (B) Linear dimensions of a heteroduplex formed by annealing one strand of λi^{434} with the complementary strand of λb2. The enlarged section of the map represents the fine structure of the unpaired i^λ/i^{434} region including the short segment of apparent homology. (C) Physical map of λ^+ DNA with the position of the deleted or substituted regions indicated as the distance from the left terminus, expressed in percent of the total length of the λ^+ DNA molecule. The location of the various genes is indicated on this physical map. Site aa' (57.3 percent) is the crossover point between λ and E. coli DNA in int-mediated λ integration or excision (16a, 17). [For a description of the genes see (2, 8, 9, 17); for more recent measurements of the position of the left end of the i^{21} substitution (72.2 percent instead of 70.8 percent) see (16a).]

233

and less than 4 percent in phage 21. The c_I gene of λ should be similar in size to the repressor gene of phage 434, since the molecular weights are similar (20). At present it is difficult to speculate on the meaning of the short region of apparent homology within the unpaired $i^{434}/+$ segment, probably within gene c_I (Fig. 5B).

The distance between the right ends of the i^{434} and i^{21} substitutions permits assignment of 0.7 percent of the $λ^+$ length to region y and gene c_{II}. There is excellent agreement between the earlier mapping of the left end of gene O [79 to 81 percent from the left terminus (16)] and the position of the right end of the i^{21} substitution (79.8 percent from the left terminus), which is located between genes c_{II} and O (Fig. 5C).

Another conclusion to be derived from our data is the identity or near identity of the so-called b5 region in the λb5 "mutant" and the i^{21} non-homology region in $λi^{21}$ (1, 7). Phage λb5, which has a density identical to that of phage $λi^{21}$ and which has the immunity of phage 21 (1, 7, 21), is a recombinant between $λ^+$ and the descendents of a plaque which appeared as a fortuitous contaminant during crossing of λ mutants (22). Our data on the position of the b5 region agree well with the results of Davis and Davidson (3), although the length computed by them for the corresponding deletion in the λ genome is substantially lower (5.3 percent) than the value we obtained by direct measurement of the homoduplexes (Table 1 and 3) and the single-stranded λ DNA [7.6 to 9 percent (16a)] within the unpaired b5/+ region (Fig. 5A).

The lack of pairing between particular regions of the l and r strands in a heteroduplex could be caused either by substitution or by inversion of a segment of the genome. In the case of inversion, the base sequences of the single DNA strands in the unpaired region would be identical instead of complementary. The differences in length between the single-stranded $i^λ$ segment on one hand, and the corresponding i^{434}, i^{21}, or b5 segments on the other, provide an argument against inversion; the absence of any observable homologies between the r strands of $λ^+$ and the r strands of the λb5 mutant is also an argument against inversion.

It can be concluded that the method described, whereby both the single-stranded and double-stranded regions of various heteroduplexes of viral DNA can be accurately measured, permits the construction of precise molecular maps, including the assignment of both position and size to various genes (28).

BARBARA C. WESTMORELAND
WACLAW SZYBALSKI, HANS RIS
Department of Zoology and McArdle Laboratory, University of Wisconsin, Madison 53706

References and Notes

1. The strains used are listed in Ref. 2. Phage $λc_{72}$ will be referred to as $λ^+$ or (+) in our report, since the electron microscopic appearance of $λc_{72}$ DNA is indistinguishable from that of the parental ("wild type") strain $λ_{papa}$ (9–11), from which it differs by a point mutation in gene c_I. In phages λb2 and λcb2, the central b2+ region has been deleted (7), whereas the symbols i^{21} and i^{434} indicate that a segment of the λ genome, including the so-called immunity region $i^λ$, has been deleted and replaced by analogous regions of the λ-related phages 21 or 434, respectively. Strains $λi^{434}$ and $λi^{21}$ were originally described as 434hy and 21hyl, respectively (10, 11). As evident in this communication and from immunity studies (9, 21, 22), region b5 in phages λb5 or λb2b5 (7) is probably identical to i^{21}. The buoyant densities of these phages, as measured by Hradecna and Szybalski (2), are 1,508 g/cm ($λc_{72}$ or $λ_{papa}$), 1.491 (λb2 or λcb2), 1.483 to 1.484 (λb2b5), 1.505 ($λi^{434}$), 1.501 ($λi^{21}$), and 1.483 to 1.484 ($λb2i^{21}$); see also Table 1. The symbols λdgal, λbio, and λdbio designate defective (d) or plaque-forming phages in which some λ genes were replaced by *E. coli* DNA (9, 27). For definition of DNA strands l(=W) and r(=C) see (2) and (17).
2. Z. Hradecna and W. Szybalski, *Virology* **32**, 633 (1967); unpublished data.
3. R. W. Davis and N. Davidson, *Proc. Nat. Acad. Sci. U.S.* **60**, 243 (1968).
4. J. A. Subirana, *Biopolymers* **4**, 189 (1965).
5. A. K. Kleinschmidt and R. K. Zahn, *Z. Naturforsch.* **14b**, 770 (1959).
6. H. Ris and B. L. Chandler, *Cold Spring Harbor Symp. Quant. Biol.* **28**, 1 (1963).
7. G. Kellenberger, M. L. Zichichi, J. Weigle, *Nature* **187**, 161 (1960).
8. H. A. Eisen, C. R. Fuerst, L. Siminovitch, R. Thomas, L. Lambert, L. Pereira Da Silva, F. Jacob, *Virology* **30**, 224 (1966); λ genes N and c_{II} are also deleted in phage λb5 (W. F. Dove, personal communication).
9. W. F. Dove, *Ann. Rev. Gen.* **2**, 305 (1968).
10. M. Liedke-Kulke and A. D. Kaiser, *Virology* **32**, 475 (1967).
11. A. D. Kaiser, *ibid.* **3**, 42 (1957); ———— and F. Jacob, *ibid.* **4**, 509 (1957).
12. E. Burgi, personal communication.
13. J. S. Parkinson and R. W. Davis, personal communication.
14. E. Jordan, *J. Mol. Biol.* **10**, 341 (1964).
15. A. Skalka, E. Burgi, A. D. Hershey, *ibid.* **34**, 1 (1968).
16. D. S. Hogness, W. Doerfler, J. B. Egan, L. W. Black, *Cold Spring Harbor Symp. Quant. Biol.* **31**, 129 (1966).
16a. Recent electron micrographic measurements of heteroduplexes formed between l strands of $λi^{21}$ and $λi^{434}$ DNA and r strands of λbio phages (1), in which most of the λ DNA between the right end of b2+ and the left end of the i^{21} region has been replaced by *E. coli* DNA, indicate that the left end of the i^{21} substitution is located 72.2 percent from the left $λ^+$ terminus, that is 1.4 percent farther to the right than the distance of 70.8 percent shown in Fig. 5c (Z. Hradecna and W. Szybalski, *Virology*, in press; D. M. Zuhse, M. J. Fiandt, Z. Hradecna, W. Szybalski, unpublished). The $i^λ$ region in the $λ^+/λb2i^{21}$ heteroduplex (Fig. 5A) thus would have to be shortened to 7.6 percent of the $λ^+$ length. The difference between the lengths of the $i^λ$ and i^{21} regions would then become 3.6 percent (7.6 minus 4.0), which value is in excellent agreement with the computed difference of 3.5 percent between the total lengths of the λb2/λb5 and λb2b5/λb2b5 homoduplexes [100 × (14.8 − 14.2)/17.0 = 3.5; Tables 1 and 3]. Other measurements summarized in Fig. 5C remain unaltered by the aforementioned recent studies.
17. W. Szybalski, *Can. Cancer Conf.*, in press.
18. M. Ptashne, *Nature* **214**, 232 (1967); ———— and N. Hopkins, *Proc. Nat. Acad. Sci. U.S.* **60**, 1282 (1968); mutation v_2 identifies the o_L operator and mutations v_1v_3 identify the o_R operator (17).
19. P. D. Bear and A. Skalka, *Carnegie Inst. Wash. Year B.* **67**, 557 (1968); personal communication.
20. V. Pirrotta, personal communication.
21. S. Brenner, personal communication.
22. J. Weigle, personal communication.
23. L. A. MacHattie and C. A. Thomas, Jr., *Science* **144**, 1142 (1964).
24. L. G. Caro, *Virology* **25**, 226 (1965).
25. R. L. Baldwin, P. Barrand, A. Fritsch, D. A. Goldthwait, F. Jacob, *J. Mol. Biol.* **17**, 343 (1966).
26. E. Burgi, *Proc. Nat. Acad. Sci. U.S.* **49**, 151 (1963).
27. B. Westmoreland, thesis, University of Wisconsin (1968).
28. Supported in part by NSF grants RG4748 and GB-2096 and by NIH grant GM 4738. We thank Drs. Z. Hradecna and A. Guha for the separated strands of λ DNA and Miss Elaine C. Babitz for measuring the contour length of the DNA molecules. We also thank Drs. J. Weigle, N. Davidson, and A. D. Hershey for helpful information and for reading this manuscript. Dr. N. Davidson pointed out to us that, in 1961, M. Nomura and S. Benzer [*J. Mol. Biol.* **3**, 684 (1961)] proposed the mapping of deletions by the electron microscopy of heteroduplexes, whereas our study was initiated only in 1962 by B.W. (27).

2 December 1968

SECTION II

RNA Synthesis

Polynucleotide Phosphorylase

The first enzyme capable of catalyzing the synthesis of nucleic acid polymers, polynucleotide phosphorylase, was discovered by Grunberg-Manago and Ochoa (*J. Am. Chem. Soc.* 1955, 77, 3165). The enzyme has now been isolated from a wide variety of bacteria as well as from eukaryotes. The enzyme uses nucleotide 5'-diphosphates as substrates and requires divalent Mg^{++} and/or Mn^{++}. Although the presence of preformed polymer often accelerates the rate of synthesis, the effects are complex and there is little to suggest that the preformed polymer primer is acting as a template. More likely it often functions as a primer by supplying a 3'-OH end on which growth can take place (Equation 1). Polynucleotide phosphorylase can also degrade polymer by phosphorolysis. Indeed, the polymerization and phosphorylytic reaction can be said to be in equilibrium as suggested by Equation 2.

(1) pApA + nUDP $\longrightarrow$ pApA (pU)n
(2) pApApU + Pi $\rightleftharpoons$ pApA + UDP

According to the mass action law high phosphate concentration should favor degradation, which in fact it does. Polynucleotide phosphorylase has been extremely useful in the production of homopolymers of all types and many heteropolymers, and they have been found to be most useful in a variety of physicochemical (Felsenfeld and Miles, *Ann. Rev. Biochem.* 1967, 36, 407) and biochemical investigations. Although this was the first nucleic acid polymerase to be discovered, its biological role is still not clear.

DNA Primed RNA Synthesis

The successful search for an enzyme which synthesized DNA using preformed DNA as a template led others to look for a similar enzyme that would synthesize RNA using preformed DNA as template (Sethi, *Prog. Biophys. Mol. Biol.* 1971, 23, 67). It seemed likely that this was the main way in which the genes communicated their information for use in the cell. Weiss (*Proc. Nat. Acad. Sci. U. S.* 1960, 46, 1020) and Hurwitz (see *paper 28*) and their coworkers detected DNA-directed RNA polymerase in mammalian and bacterial sources, the latter containing only a single species of RNA polymerase. The RNA polymerases of eukaryotic cells, each of which contains several enzyme species that appear to have separate functions, is currently being studied very extensively (e.g. see *Cold Spring Harbor Symp. Quant. Biol. 1970*, 35, and Jacob, *Prog. Nuc. Acid Res. Mol. Biol.* 1973, 13, 93). One eukaryotic RNA polymerase appears to be specifically involved in transcribing the ribosomal RNA genes.

Most of the work leading to an understanding of the mechanism of transcription has been done with bacterial enzymes. The most extensively studied bacterial RNA polymerases (from *E. coli* and *B. subtilis*) synthesize RNA in the presence of DNA and all four triphosphates plus Mn^{++} or Mg^{++}. The single-stranded RNA product may be separated from the double-stranded DNA primer by CsCl density gradient centrifugation because of its higher buoyant density. The RNA has a base composition complementary to the strand of the DNA that is being transcribed. A more accurate and yet simpler measure of the sequence complementarity of the RNA is available through the technique of DNA-RNA hybrid formation. It has been shown that RNA will only anneal with high fidelity with single-stranded DNA derived from the double-stranded DNA that was used as a template to prime its synthesis; the same is true of *in vivo* synthesized RNA. The identity or percent relatedness of RNA transcripts synthesized *in vivo* and *in vitro* can be determined by competition hybridization. Sites on the DNA that would be occupied by the complementary, labeled RNA species can be competed for by cold RNA to test for relatedness between the cold and labeled RNA. The treatment for annealing DNA to RNA is similar to the recipe used for annealing complementary strands of DNA (for example, see *papers 1 and 31*, as well as Gillespie, *Methods in Enzymology*, 1968, 12B, 641.)

The DNA-directed RNA polymerase reaction may be analysed by considering three distinct steps: initiation, propagation and termination.

The chromosome of bacteria and phage are organized in such a manner that meaningful transcription occurs *in vivo* from a unique DNA strand, thus requiring the existence of special recognition sites at the beginning of genes or, in some cases, groups of such recognition sites (*papers 25, 26, 59 and 60*). Syzbalski and his coworkers (*Cold Spring Harbor Symp. Quant. Biol.* 1970, 35, 341) as well as Losick (*Ann. Rev. Biochem.* 1972, 41, 409) have defined a unit of transcription ("transcripton" or "scripton") as "a cluster of genes" whose transcription requires initiation at an autonomous promotor site (that is a promotor site which is not under the *cis* dominant control of another promotor). The intricacies involved in the control of RNA synthesis are examined in Section IV where regulation *per se* is discussed. The *core* RNA polymerase molecule of *E. coli*, which is a tetramer consisting of two α (M. W. 45,000), and one β (M. W. 150,000) and one β' (M. W. 165,000) polypeptide chain with a combined molecular weight of about 400,000, is incapable by itself of selecting normal *in vivo* initiation sites on the DNA template (*paper 24*). An additional protein component, found in *E. coli* called σ (M. W. 90,000), forms a moderately stable, noncovalent complex with the core polymerase (holoenzyme). The sigma subunit of RNA polymerases is involved in the specific initiation of transcription. In its presence T4 DNA is transcribed much more efficiently; the holoenzyme binds more tightly to DNA than to the core enzyme; and finally, the sigma factor acts catalytically in transcription, being released from the polymerase-DNA complex soon after initiation. Once the complex has been formed at the promotor (or entry) site and the nucleotide triphosphates are added, synthesis is triggered at the start site by addition of the divalent cation in an environment containing the appropriate pH and ionic strength. The entry of the RNA polymerase and the start of the synthesis of the RNA chain are two separate functions (*paper 26*). That the initiation complex is formed by DNA and σ-core alone has been confirmed by inhibition studies with the antibiotic rifampicin. This drug specifically inhibits the initiation of RNA synthesis. If rifampicin is added to the σ-core complex before the DNA is added, it prevents subsequent initiation; if rifampicin is added to the σ-core complex after the DNA is added, its effectiveness is considerably reduced, at least on the first round of initiation. It seems likely that other classes of proteins controlling gene expression on both viral and bacterial genomes must exist. This is a very active area of research, aimed at isolating RNA polymerase stimulatory or regulatory factors, some of which may be associated with the polymerase. It also appears that the components of the RNA polymerase holoenzyme may be chemically modified, especially during bacteriophage infection and bacterial differentiation, to change their initiation site specificity (Linn *et al.*, *Proc. Nat. Acad. Sci. U. S.* 1973. 70, 1865).

By using γ-^{32}P labeled nucleotides, Maitra and Hurwitz (*paper 27*) were able to show that RNA molecules are synthesized (*propagated*) in the same directional sense as DNA molecules, namely from the 5' and to the 3' terminus. When using purified enzyme, the γ-^{32}P labeled triphosphate is left intact at the 5' terminus of the polymers. Such studies have shown that there is a strong bias favoring purine nucleotides at the 5' terminus of the polymer with a particular RNA containing one or the other purine exclusively. The rate of RNA chain elongation by *E. coli* RNA polymerase, *in vivo* and *in vitro* is about 20 nucleotides/sec.; the rate is faster with the RNA polymerase coded for by phage T7.

When initiation and propagation of RNA synthesis takes place the DNA is most likely melted out locally (von Hippel and McGhee, *Ann. Rev. Biochem.* 1972 41, 231; Bick *et al.*, *J. Mol. Biol.* 1972, 71, 1) so that one strand of the DNA may serve as template. As the RNA is synthesized, the locally denatured DNA region would presumably refold and adjacent regions unfold and refold successively until RNA synthesis is complete. *In vitro* either double helix DNA or single-stranded DNA can be used as primer; when the former is used the RNA product is single-stranded, but associated with the DNA by way of the polymerase. In contrast, when the latter is used as primer, the RNA is attached to the DNA as a DNA-RNA hybrid double helix. Of particular importance is the fact that a primer is not essential for the initiation of RNA chain growth. Thus, the initiation of DNA synthesis, which requires both template and primer (which may be either a deoxy-or ribo-oligonucleotide) may take advantage of this fact by having RNA synthesis precede DNA synthesis. It is not clear at the present time whether the RNA polymerase which synthesizes all the cellular species of RNA is also involved in the initiation of DNA synthesis as well.

Under special *in vitro* conditions E. coli RNA polymerase does not need to initiate with a triphosphate, but can extend a pre-existing RNA chain. There are two distinct forms of RNA-primed RNA synthesis. Downey and So have shown that the dinucleotides UpA and ApU stimulate transcription of poly dAT (*Biochem.* 1970, 9, 2520). Subsequent work has shown that the dinucleotides are incorporated into the 5' end of the RNA chains. With T7 DNA as a template, specific dinucleotides stimulate the synthesis of RNA from selected initiation sites (Minkley and Pribnou, *J. Mol. Biol.* 1973, 77, 255). It appears likely that the dinucleotides cannot act by base pairing to the template alone for such structures would be unstable at room temperature (in addition, there are too many complementary deoxydinucleotides in the template) but rather must interact directly with the RNA polymerase before or after it has bound to the appropriate initiation site.

Termination of RNA synthesis on some DNA templates (such as T4 and T7) is probably due in part to the recognition by the polymerase of a non-transcribable base sequence. At this point propagation of RNA synthesis will stop generating transcripts of discrete size, terminating with U_6AOH as the 3'OH end (papers 25 and 26). In other termination studies, it was found by Roberts (paper 28) that a protein factor isolated from *E. coli*, called ρ, is essential for proper termination of so-called early RNA from bacteriophage λ DNA. The ρ molecule is a tetramer with a molecular weight of about 200,000. In the absence of ρ factor the transcribed RNA becomes indefinitely large and remains firmly attached in a DNA-polymerase complex. It appears that the mechanism of action of ρ is that it interacts with *E. coli* RNA polymerase at specific DNA sites and thus leads to the termination of RNA transcription to generate chains of defined length (Goldberg and Hurwitz, *J, Biol. Chem.* 1972, 247, 5637).

Little is known about termination of other genes or the existence of alternative methods of termination. Since ρ factor is present to uninfected bacteria, it is possible that it is a termination factor for some, if not all, bacterial genes.

Certain viral DNAs do not appear to require ρ for termination and release of RNA, at least *in vitro*. The meaning of the *in vitro* biochemical results using ρ factor should be interpreted cautiously since they are without a genetic correlate.

Inhibitors of RNA Synthesis

Three principal types of RNA inhibitors exist: substrate analogues of RNA precursors such as cordycepin triphosphate, those that interact with DNA and are thus relatively non-specific, and those that interact with the RNA polymerase. The last type are the most interesting since they can be used to isolate antibiotic resistant mutants with altered RNA polymerases. Actinomycin D has been a useful antibiotic for studying the effects of arresting RNA synthesis. This compound intercalates into the DNA, specifically binding to the deoxyguanosine residues (Sobel and Jain, *J. Mol. Biol.* 1972, 68, 21; Sobell, *Prog. Nuc. Acid Res, Mol. Biol.*, 1973, 13, 153) inhibiting the propagation step in RNA synthesis. Because of the higher GC content of rRNA compared to mRNA, actinomycin D may inhibit the rRNA species more efficiently than the mRNA. It is very important to realize this fact when using this inhibitor in *in vivo* studies and one must be cautious in interpreting results if the inhibition of mRNA synthesis is not shown. Other efficient RNA inhibitors of bacterial and eukaryotic RNA synthesis that act by binding to the template, but which may inhibit only *in vitro* because of permeability barriers, include anthracyclines, mithramycin and distamycin (Hash, *Ann. Rev. Pharmacol.* 1972, 12, 35; Goldberg and Freedman, *Ann. Rev. Biochem.* 1971, 40, 775).

Those inhibitors of RNA synthesis that act by binding to RNA polymerase subunits include the rifamycins (Wehrli and Staehelin, *Bact. Revs.* 1971, 35, 290) which block initiation of RNA synthesis in prokaryotic cells (paper 33), streptolydigin and α-amantin (Blatte et al., *Cold Spring Harb. Symp. Quant. Biol.* 1970, 35, 649). The latter blocks nonribosomal RNA synthesis in eukaryotic cells both *in vivo* and *in vitro*.

Synthesis and Degradation of Messenger RNA

Analysis of the steady state composition of the RNA in *E. coli* indicates that about 3 percent is messenger RNA (Kennell, *Prog. Nuc. Acid Res. Mol. Biol.* 1971, 11, 259). In spite of this, the vast majority of the DNA (90 percent) encodes messenger RNA. However, at any particular time only about 10 percent of the genes are expressed and the lifetime of messenger RNA is very limited (about 2—3 minutes). In prokaryotic cells, such as *E. coli* which have no well defined nucleus or nuclear membrane, the processes of transcription and translation are usually coupled so that translation begins before transcription is completed. As the first part of the messenger RNA peels off the DNA template, it becomes associated with the ribosomal translation apparatus. A necessary corollary of this coupling process is that transcription and translation both begin from the $5'$ terminus of the messenger RNA. Before the synthesis of messenger RNA is complete, its degradation from the $5'$ end has already begun (*paper 29*). Because of different size transcriptional units, the range of sizes of messenger RNA is very great; in some cases the messenger RNA reflects the gene for only one polypeptide chain, in other cases (polycistronic messenger RNA) as many as eight polypeptide chains, usually with related function, may be represented in a single messenger. In the case of T7 phage DNA transcription by host RNA polymerase it appears that the polymerase attaches to one of three promotors generating a large transcript that encompasses the entire early region. An endonuclease then cleaves the long RNA at five specific sites (Dunn and Studier, *Proc. Nat. Acad. Sci. U.S.* 1973, 70, 1559).

Since eukaryotic cells have defined nuclei enclosed in nuclear membranes, the processes of transcription and translation are separated. RNA synthesized in the nucleus can be very large (called heterogeneous nuclear RNA) and is frequently modified post-transcriptionally by the addition of a poly A segment 50 to 100 residues in length to the $3'$ end. The nuclear RNA is processed to a smaller size and is transported by an unknown mechanism to the cytoplasm where translation takes place (*paper 30*). Mitochondrial DNA transcripts also have poly A segments on their $3'$ ends. In lower eukaryotes, such as slime molds and yeast (Reed and Wintersberger, *FEBS Letters*, 1973, 32, 213), there appears to be no large heterogeneous nuclear RNA; however, most of the nuclear and cytoplasmic messengers do have poly A sequences at the $3'$ ends. The lifetime of different messengers in higher eukaryotes, even within the same cell, is extremely varied, spanning periods of hours to days. The variation depends upon cell types as well as the specific messenger RNA; presumably this variation is due to the relative susceptibility of different messenger RNAs to some cytoplasmic nuclease.

Synthesis of Ribosomal RNA

Ribosomal RNA (rRNA) cistrons comprise only 0.2 to 0.4 percent of the genome of *E. coli* as measured by saturation hybridization experiments (*paper 31*). Bacterial ribosomal cistrons can be isolated free of other parts of the genome; the bacterial DNA is first sheared to a small size, denatured, then hybridized to rRNA. The non-base paired regions are then removed by exposure to a nuclease which acts only on single stranded nucleic acids. The DNA-RNA hybrids, composed of equal length DNA and complementary rRNA can be separated from either DNA or RNA by a variety of techniques, including density gradient centrifugation in concentrated cesium salt solutions. Even though there are only a few ribosomal cistrons, rRNA synthesis can account for up to 40 percent of the instantaneous rate of total RNA synthesis. This high rate of synthesis per rRNA cistron is much greater than that of most mRNA species and suggests that there may be, some fundamental difference (for example, efficient ribosomal gene promotors) in the transcription processes.

In *E. coli* current evidence favors the notion that the three RNA molecules found in ribosomes (23S, 16S and 5S with molecular weights of 1.2×10^6, 9.6×10^6 and 0.4×10^4 respectively) are made as separate molecules (*paper 32*; Brownlee, Sanger and Barrell, *J. Mol. Biol.* 1968, 34, 379) but are derived from larger precursors. Whether any RNA polymerase stimulatory factors specify the transcription of ribosomal cistrons is not clear. Eukaryotes initially transcribe the ribosomal cistrons as a single RNA molecule which is partly degraded after synthesis to yield the two large RNA molecules found in the ribosomal subunits. Using electron microscopy and partial digestion with a $3'$-exonuclease, Wellauer and Dawid (*Proc. Nat. Acad. Sci. U.S.*, in the press, 1973) have elegantly shown that the the 45S ribosomal RNA precursor of HeLa cells has the 28S

RNA region at the 5′-end, followed by a spacer, the 18S RNA region and another spacer at the 3′-end. The 5S species is synthesized from a locus adjacent to the larger ribosomal cistrons in *E. coli* and *Saccharomyces cerevisiae*, but not in *Drosophila* or *Xenopus*. Some modification of the individual bases of ribosomal RNA, mainly by endonucleases and methylating enzymes, occurs during or after synthesis of the main chain. Except under starvation conditions the ribosomal RNA is very long-lived.

In *E. coli*, about 6 to 10 ribosomal genes are present per genome. A low level of heterogeneity is seen in the primary sequence of 16S RNA, indicative of a small degree of cistron heterogeneity (Fellner *et al.*, *Nature New Biology* 1972, **239**, 1). In higher forms several hundred copies exist which are usually closely associated with the nucleolus of the cell. It has been shown that ribosomal RNA (except for the 5S RNA molecule) is synthesized in close association with the nucleolus, by an RNA polymerase molecule distinguishable from the main RNA polymerase of the nucleus. These two polymerases are readily distinguishable *in vitro* by their physical properties and by the susceptibility of the main chromatin polymerase to inhibition by the antibiotic α-amanitin (Blatti *et al.*, *Cold Spring Harb. Symp. Quant. Biol.* 1970, **35**, 649).

Direct visualization of nucleolar genes by electron microscopy (Miller and Beatty, *Science*, 1969, **164**, 955) shows that each rRNA cistron is actively engaged in the simultaneous synthesis of about 100 rRNA molecules. Significant segments of inactive (spacer) DNA are visible between each active rRNA cistron.

Other means of determining the location of the rRNA genes, which are clustered, can be shown by *in situ* hybridization in animal cells (Pardue *et al.*, *Chromosoma*, 1970, **29**, 268) or in bacteria (*E. coli*) by genetic mapping (Matsubara *et al.*, *Mol. Gen. Genet.* 1972, **117**, 311), as well as by DNA-rRNA hybridization using as a source of DNA selected segments of the chromosome isolated in density shift experiments (in *B. subtilis*) of synchronized cell populations (Smith *et al.*, *J. Mol. Biol.* 1967, **45**, 137).

Transfer RNA Synthesis
(papers 33 and 34)

Very little is known about transfer RNA (tRNA) synthesis. *E. coli* probably carries only a single gene for each tRNA species, and mapping studies with suppressor tRNA mutants show some clustering of tRNA genes around the *E. coli* chromosome (Smith, *Ann Rev. Genet.* 1972, **6**, 235). Some characteristic tRNA genes are also carried by bacteriophages; some of them can be mutated so that their altered tRNA product can act to suppress nonsense mutants (Wilson and Kells, *J. Mol. Biol.* 1972, **69**, 39). However, in the case of wild type T-even phages, deletion of the tRNA genes does not affect phage viability on the usual laboratory hosts. Like rRNA the tRNA is generally long-lived; also the genes for tRNA, which account for about 10 percent of the total cellular RNA but only about 0.02 percent of the total genome (Giacomone and Spiegelman, *Science* 1962, **138**, 1328) are extremely active.

The transcribed tRNA precursor is about 50 percent larger than the mature tRNA and must be cleaved by a specific nuclease after transcription. Extensive modification of many of the bases in tRNA occurs after transcription. These post-transcriptional modifications include a variety of methylations, pseudouridine or thiouridine formation from uridine, deamination of adenine and base modification such as the addition of an isopentyl group to particular bases. In addition to these post-transcriptional modifications of tRNA an enzyme system exists which adds a common ending of $-CCA-3′OH$ on all the tRNAs. This is a stepwise process with individual nucleotide triphosphates serving as the substrates (Deutscher, *Prog. Nuc. Acid Res. Mol. Biol.* 1973, **13**, 5).

The rRNA and tRNA genes have conserved their identity during evolution and have resisted extensive base sequence changes. Thus, while mRNA from heterologous species within a fairly closely related bacterial or eukaryotic group will hybridize poorly with the DNA of one member of the group, rRNA and tRNA will hybridize with the heterologous DNA almost as well as the corresponding RNA from a homologous organism.

Phage Specified RNA Polymerase
(paper 35)

Although the replication of phages T4 or λ in *E. coli* or SP01 in *B. subtilis* is sensitive to rifampicin, T7 replication is insensitive to this antibiotic early after infection. This is due to the fact that phage T7 codes for a totally new RNA polymerase. Early in the infection, the host RNA polymerase transcribes a small portion (about 20 percent) of the T7 genome; one of the gene products translated from this early RNA, coded for by gene 1, is an RNA polymerase differing in size, antigenicity, specificity and antibiotic sensitivity from the host RNA polymerase.

RNA-Directed RNA Synthesis
(papers 36, 37, 38)

Cellular RNA polymerase which is normally directed by DNA templates to synthesize polyribonucleotides *in vitro* can utilize RNA templates with considerably reduced efficiency. The biological significance of this is unknown. There are, however, new polymerases formed in RNA phage infected cells that have now been thoroughly studied and it appears that these RNA phages code for their own polymerase that is part of the machinery necessary for multiplication of the viral RNA (Stavis and August, *Ann. Rev. Biochem.* 1970, 39, 527). The RNA containing *E. coli* bacteriophages f2 and Qβ have been most extensively studied. These viruses in the native state contain a single, linear RNA genome, with a molecular weight of about 1×10^6 daltons. The polymerase of the Qβ phage—as well as of f2 (Fedoroff and Zinder, *Nature New Biology*, 1973, 241, 105)—is a multimer with dissimilar subunits; only one of the subunits is coded for by the phage genome (*paper 38*). The other three subunits are endogenous to uninfected *E. coli* but are not found in the usual DNA-dependent *E. coli* RNA polymerase described above: two of these subunits correspond to elongation factors, EFT_u and EFT_s, involved in protein synthesis (see Section II, and Blumenthal, Landers and Weber, *Proc. Nat. Acad. Sci. U. S.* 1972, 69, 1313); the third subunit corresponds to the i factors believed to be involved in the control of translation (see Section III, and Scheps, *et al.*, *Nature New Biology* 1972, 239, 19).

The essential features of the replication of the RNA phage in infected *E. coli* are as follows. The phage genome, called the plus strand, which is also a messenger RNA, must direct the synthesis of an RNA-dependent RNA polymerase. This polymerase is very specific in its template requirements and cannot replicate the RNA of cellular RNA or of closely related RNA phages. The phage specific RNA polymerase synthesizes a free minus strand, which in turn acts as a template for the synthesis of progeny plus strands. All RNA synthesis proceeds in the $5'$ to $3'$ direction, as in the synthesis of RNA by the DNA-dependent RNA polymerase. These steps can be mimicked *in vitro* with a high degree of fidelity since the progeny plus strands are infectious.

Spiegelman and his colleagues (*Science*, 1973, 180, 916) have been able to isolate "variants" during the *in vitro* replication of Qβ RNA in which dispensable segments of the viral genome have been eliminated. The base sequence of one such variant, containing 218 nucleotides and that can still be replicated *in vitro*, has been determined.

RNA-Directed DNA Synthesis
(paper 39)

While the usual flow of genetic information is from DNA to RNA and thence to protein, there is no inherent structural or informational reason why RNA could not direct the synthesis of DNA. The combination of a proper enzyme system and template are all that should be required. This type of synthesis has been found by Temin and by Baltimore (*paper 39*; for a recent review of this subject see Temin and Baltimore, *Adv. Virus Research* 1972, 17, 129.) and their associates for certain RNA tumor producing viruses. The exact function of the DNA synthesis step is unclear but it appears to be necessary for virus multiplication. Recently, such RNA-dependent DNA synthetases have also been reported to exist in a large number of normal cell types. However, there is a fundamental difference between the reverse transcriptases from tumor producing RNA viruses and those from normal cells. The synthetase from the virus has a strong preference for an RNA template while those from normal cells prefer DNA templates for synthesizing DNA.

Factor Stimulating Transcription by RNA Polymerase

by
RICHARD R. BURGESS
ANDREW A. TRAVERS
Biological Laboratories,
Harvard University

JOHN J. DUNN
EKKEHARD K. F. BAUTZ
Institute of Microbiology,
Rutgers University

A protein component usually associated with RNA polymerase can be separated from the enzyme by chromatography on phosphocellulose. The polymerase is unable to transcribe T4 DNA unless this factor is added back.

IN *E. coli* the synthesis of all types of cellular RNA is thought to be mediated by a single enzyme, DNA-dependent RNA polymerase. The highly purified enzyme[1-6] can catalyse the synthesis of RNA *in vitro* in the presence of DNA and the ribonucleoside triphosphates. When an intact double helical DNA is used as template, transcription *in vitro* is asymmetric—at any given region along the DNA only one of the complementary DNA strands is transcribed[7-10]. This is also characteristic of transcription *in vivo*[11-13]. Moreover, the selective transcription *in vitro* of certain regions of T4 and λ DNA[10,14-17], coupled with studies on the binding of RNA polymerase to DNA[18-22], suggests that the polymerase initiates RNA synthesis at specific sites on the DNA. The state of aggregation of the enzyme is strongly influenced by ionic strength[19,21,23,24], substrate[25], and possibly by enzyme concentration and temperature. Most investigators agree that the molecular weight of the active enzyme is in the range of 350,000–700,000 daltons. This large size is consistent with the observation that the enzyme is composed of several different polypeptide chains[26,27]. Furthermore, it has been assumed that "highly purified" RNA polymerase is a protein entity from which nothing can be further removed without destroying its enzyme activity. We report here, however, the separation of polymerase into two components. One contains enzyme activity, but its ability to transcribe certain DNA templates is greatly reduced; the other is a factor able to stimulate RNA synthesis on these restrictive templates to normal levels.

It was independently observed at Harvard and Rutgers that when RNA polymerase was purified by chromatography on a phosphocellulose column, the enzyme obtained, although able to transcribe calf thymus DNA almost normally, was much less active when assayed with T4 DNA as template. Enzyme purified by an alternative procedure, however, was almost equally active on both templates. This suggested that some component necessary for the transcription of T4 DNA was separated from RNA polymerase by the phosphocellulose column. Furthermore, the activity of the phosphocellulose-purified enzyme on T4 DNA could be greatly enhanced by the addition of another fraction from the phosphocellulose column. This fraction lacked significant RNA polymerase activity of its own. We describe here the identification and some properties of this stimulating component.

Isolation of RNA Polymerase

The RNA polymerase we used was purified as outlined in Fig. 1 from *E. coli* K12. Three methods were used to achieve the final purification of the polymerase. Enzyme purified with phosphocellulose (PC enzyme) and with glycerol gradient (GG enzyme) was prepared by the method of Burgess (manuscript in preparation). Using

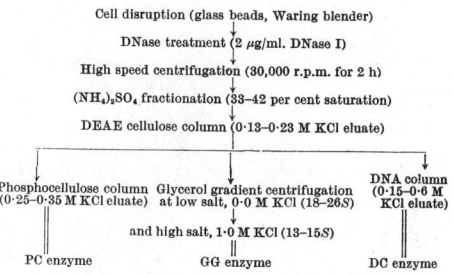

Fig. 1. Outline of enzyme purification.

the same method through the DEAE cellulose step, J. J. D. and E. K. F. B., from Rutgers, purified the enzyme by binding it to a cellulose column containing immobilized T4 DNA and eluting the bound enzyme with high salt[28]. DNA column enzyme (DC enzyme) obtained in this way and GG enzyme are essentially identical and do not differ significantly from enzymes purified by the method of Chamberlin and Berg[1]. Unless otherwise stated, the work described here was carried out with GG enzyme, but similar results were obtained with DC enzyme.

Identification of the Component Stimulating Transcription of T4 DNA

To isolate the component necessary for the transcription of T4 DNA, GG enzyme was chromatographed on a

Table 1. ASSAY OF STIMULATING ACTIVITY OF PEAKS A, B AND C

Sample assayed for stimulation	μμmole AMP incorporated/min		Stimulation ratio
	No PC enzyme added	4 μg PC enzyme added	
PC enzyme (4·0)	3	9	2
GG enzyme (1·1)	57	170	38
Peak A (1·0)	13	114	34
Peak B (1·3)	66	203	46
Peak C (1·9)	2	8	2

The assay mixture (0·25 ml.) contained 0·04 M tris-HCl buffer, pH 7·9, at 25° C, 0·01 M MgCl₂, 0·0001 M EDTA, 0·0058 M 2-mercaptoethanol, 0·15 M KCl, 0·5 mg/ml. bovine serum albumin (Calbiochem, crystalline), 0·15 mM CTP, GTP and UTP, 0·15 mM ¹⁴C-ATP (1 mCi/mmole), 20 μg/ml. T4 phage DNA, and varying amounts of enzyme and stimulating factor. The mixture was incubated for 10 min at 37° C, chilled, precipitated with 5 per cent TCA, and filtered on 'Millipore' filters. Samples were counted on an end-window gas-flow counter. The amount of incorporation is expressed as μμmoles of AMP incorporated per min of incubation. To obtain a measure of the stimulating activity in a particular sample, that sample was assayed in the absence and presence of PC enzyme. The additional incorporation in the presence of PC enzyme was due to the stimulation of this added PC enzyme by the factor in the sample. When this additional incorporation is divided by the small amount of incorporation obtained with PC enzyme alone, the stimulation ratio is obtained. The values in parentheses indicate the amount of protein, in μg, added to the assay mixture.

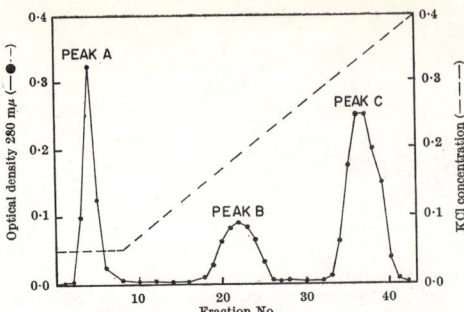

Fig. 2. Phosphocellulose column profile. Phosphocellulose (Whatman P11, 7·4 mequiv./g) was washed with base and acid and titrated to pH 7·9 at 25° C with KOH. This material was placed in a column (0·9 × 10 cm) and equilibrated extensively with a buffer containing 0·05 M tris-HCl, pH 7·9, at 25° C, 0·05 M KCl, 0·0001 M EDTA, 0·0001 M dithiothreitol and 5 per cent glycerol. The pH of the outflow was 8·1 ± 0·05 at 4° C. A sample containing 9 mg of GG enzyme in 4 ml. of this buffer was applied to the column. The column was then washed with 5 ml. of buffer and eluted with a 100 ml. gradient from 0·05 M to 0·4 M KCl. The flow rate was 0·5 ml./min and 2·2 ml. fractions were collected. Tubes 4, 22 and 36 were taken as representative of peaks A, B and C respectively. Assuming a value of $E^{1\%}_{280 m\mu} = 6\cdot6$ for all fractions, these tubes contain 0·5, 0·14 and 0·4 mg of protein per ml. respectively. A small amount of polymerase activity was present in the flow-through (peak A). If the sample is applied to the column too rapidly (at a rate greater than one column volume per 30 min), then much more enzyme will be found in the flow-through.

phosphocellulose column. This column separated the material present in GG enzyme into three peaks, A, B and C (Fig. 2). Each peak was assayed for ability to promote the transcription of T4 DNA in the presence and absence of PC enzyme. Peak A contains stimulating activity, peak B is a mixture of PC enzyme and stimulating activity, while peak C is identical to normal PC enzyme. The results (Table 1) show that, in the absence of PC enzyme, peak B can use T4 DNA as a template for RNA synthesis while peaks A and C have little activity. The addition of peak A material to PC enzyme, however, results in a stimulation of RNA synthesis which increases with increasing amounts of peak A, until a rate of RNA synthesis comparable with that obtained with GG enzyme is reached. Although peak B is able to stimulate PC enzyme, it can itself be stimulated by the addition of peak A, and thus is a mixture of PC enzyme and stimulating component.

Each of the fractions was analysed by polyacrylamide gel electrophoresis in various conditions to determine the species of protein present. These gels were run in the presence of either 8 M urea or 0·1 per cent sodium dodecyl sulphate (SDS), both of which can cause the dissociation of oligomeric proteins into single polypeptide chains. The protein bands observed in this analysis are shown in Fig. 3.

Previous studies[27] have shown that phosphocellulose enzyme is composed of two chief types of polypeptide chains, which we here designate α and β. These chains are present in equimolar amounts and have molecular weights of $\sim 40{,}000$ and $\sim 160{,}000$ respectively. α and β are present in all fractions but only in very small amounts in peak A. GG enzyme contains, in addition, two extra bands which we shall designate σ and τ. These bands are present in peak A in amounts greatly in excess of the amounts of α and β present. Peak B contains bands α, β and σ, but lacks τ. These patterns are observed on both the 8 M urea gels and the 0·1 per cent SDS gels. These data are consistent with band σ being the component responsible for the stimulating activity, but they do not exclude the possibility that band τ or even some other material might also stimulate.

The stimulating component was identified as σ by the zone sedimentation of peak A material on a glycerol density gradient (Fig. 4). The factor required for T4 transcription was found to sediment at $\sim 5S$. Analysis of the gradient fractions by electrophoresis on 8 M urea gels showed that band σ sedimented identically to the stimulating activity whereas band τ sedimented at about $8S$ in a region with no stimulating activity.

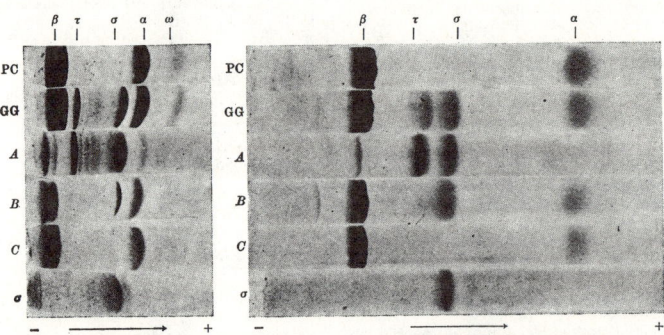

Fig. 3. Polyacrylamide gel electrophoresis patterns of different preparations of RNA polymerase and of purified factor. From top to bottom: PC enzyme (20 μg), GG enzyme (20 μg), peak A (10 μg), peak B (7 μg), peak C (8 μg) and purified factor (2 μg) from tube 17 of the glycerol gradient shown in Fig. 4. 8 M urea gels were prepared and run according to the general method described by Davis[24]. These gels contained 7·5 per cent acrylamide. 0·1 per cent SDS gels were run according to the procedure of Shapiro et al.[25] and contain 5 per cent acrylamide. In both cases the gels were stained by immersing them in a 0·25 per cent solution of Coomassie brilliant blue in methanol : acetic acid : water (5 : 1 : 5 v/v/v) for at least 2 h. The gels were then soaked in 7·5 per cent acetic acid, 5 per cent methanol for 0·5 h and finally destained electrophoretically in this same solvent. The bands observed are designated β, τ, σ, α and ω; α, β and ω are the bands normally seen in PC enzyme. The molecular weights of α and β are $\sim 40{,}000$ and $\sim 160{,}000$ respectively. The molecular weight of β was previously reported to be 110,000 (ref. 27), but recent measurements (Burgess, in preparation) indicate that 160,000 is a more accurate value. The band β as seen on 0·1 per cent SDS gels appears as two closely spaced bands of equal intensity which probably correspond to two different polypeptide chains, with molecular weights of about 155,000 and 165,000. For simplicity these are both called β in the text. In addition, a small polypeptide, ω, with a molecular weight of about 10,000 can be seen moving ahead of α on the 8 M urea gels. It is present in GG enzyme, PC enzyme and DC enzyme but it is not yet known whether it is a component of RNA polymerase or merely a tightly binding impurity. The amount of band τ observed in GG enzyme and also in peak A is variable. DC enzyme (not shown) contains only traces of τ and thus it is probably an impurity. From the SDS gels it is estimated that 45 per cent of peak A protein is σ. Purified factor is estimated to be about 80 per cent pure and is completely free of τ. The right hand two-thirds of the urea gels which contain no bands are not shown.

Table 2. HEAT INACTIVATION OF THE STIMULATING FACTOR

Treatment of factor before assay	μμmole AMP incorporated/min	Per cent inactivation of stimulating activity
5 min at 0° C	294	0
5 min at 37° C	295	0
5 min at 45° C	154	52
5 min at 50° C	31	98

Peak A protein (0·9 μg) was added to each assay tube which also contained assay solution lacking the four triphosphates and T4 DNA. Separate tubes were incubated at the indicated temperatures for 5 min and then chilled to 0° C. The triphosphates, T4 DNA, and PC enzyme (6 μg) were added and the assay performed as described in the legend to Table 1. With no added factor, PC enzyme incorporated 17 μμmoles of AMP. Factor alone incorporated 9 μμmoles of AMP. The percentage inactivation at a given temperature was calculated by subtracting 26 μμmoles from the total incorporation observed after treatment at that temperature and then setting the incorporation of the unheated sample equal to 100 per cent activity.

Stimulating Factor is a Protein

The association of the stimulating activity with a specific band on polyacrylamide gels suggests that the factor is a protein. To confirm this, we determined the heat stability of the stimulating activity (Table 2). Incubation of peak A material for 5 min at 45° C and 50° C resulted in an inactivation of the stimulating activity of 52 per cent and 98 per cent respectively. Furthermore, the activity was sensitive to trypsin. We conclude therefore that the factor is a heat labile protein, although it is still possible that the factor is associated with a nucleic acid component of low molecular weight.

The molecular weight of a polypeptide chain can be estimated from its mobility on 0·1 per cent SDS polyacrylamide gels[29]. Using this procedure with β-galactosidase, bovine serum albumin and ovalbumin, with molecular weights of 130,000, 67,000 and 45,000 respectively, as molecular weight markers, the molecular weight of σ was calculated to be about $95,000 \pm 5,000$. This is consistent with the S value of about 5 obtained from the glycerol gradient.

Factor Requirements on Various DNA Templates

RNA synthesis in the presence and absence of factor was measured for several different DNA templates (Table 3). The greatest stimulation was observed with native T4 DNA, where the presence of factor increased the amount of synthesis seventy-five-fold. With all other templates tested the stimulation was considerably lower. This variation may be ascribed to two factors. First, the fully stimulated levels of RNA synthesis vary according to the template used: this has been observed by several investigators[1-3]. Second, in the absence of factor, different DNA templates direct the synthesis of differing amounts of RNA. From analytical gels we estimate that PC enzyme contains less than 2 per cent as much factor as GG enzyme. Thus it is possible that the very small amount of RNA synthesis off T4 DNA with PC enzyme is due to traces of remaining factor. With calf thymus DNA, however, this could not easily explain the results. A more likely possibility is that there are some sites at which initiation can occur in the absence of factor. It is clear that such sites are not merely fully denatured regions, for denaturation of T4 DNA does not result in increased synthesis with PC enzyme. Furthermore, φX-174 DNA is also a poor template. The behaviour of PC enzyme on

Table 3. STIMULATION ON VARIOUS DNA TEMPLATES

DNA template	mμmoles AMP incorporated/min/mg enzyme		
	PC enzyme	PC enzyme + factor	GG enzyme
T4—native	0·5	33·0	37·5
T4—denatured	0·5	6·1	3·0
Calf thymus—native	14·2	32·8	30·5
Calf thymus—denatured	3·3	14·5	10·7
φX-174	0·9	6·2	4·9

RNA synthesis in the presence and absence of factor was assayed as described in the legend to Table 1, except that the DNA concentration was 40 μg/ml. DNA was denatured by adding 1/10 volume of 2 N NaOH to a 20 μg/ml. DNA solution. After standing at 25° C for 10 min, the solution was neutralized. Almost identical results were obtained if the DNA was denatured by heating at 95° C for 10 min, and then rapidly chilling in ice. The concentrations of PC enzyme and GG enzyme in the reaction mixture were both 4 μg/ml. Peak A protein was present, where indicated, at a concentration of 4 μg/ml. This corresponds to 2 μg/ml. of factor. The ratio of factor to enzyme in the mixture of PC enzyme and factor was about twice that normally occurring in GG enzyme. Incorporation by factor alone was negligible for all types of DNA tested. In all cases saturating amounts of DNA were used. The φX-174 phage DNA was a gift from Dr D. T. Denhardt.

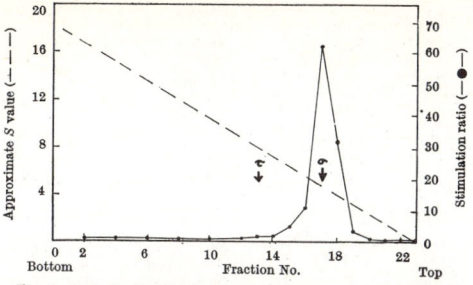

Fig. 4. Analysis of peak A by zone centrifugation. A 0·2 ml. sample containing 100 μg of peak A protein was layered on a 4·8 ml. linear 10–30 per cent glycerol density gradient containing 0·01 M tris-HCl buffer, pH 7·9, 0·01 M MgCl, 0·15 M KCl, 0·0001 M EDTA, and 0·0001 M dithiothreitol. The gradient was centrifuged for 11 h at 60,000 r.p.m. in a Spinco SW65 rotor at 4° C. 0·22 ml. fractions were collected. Each fraction was assayed for stimulating activity as described in the legend to Table 1. The proteins present in each fraction were analysed by electrophoresis on 8 M urea polyacrylamide gels. Band σ peaked in tube 17, band τ in tube 13. Molecular weight markers (E. coli β-galactosidase, human gamma-globulin, and egg lysozyme) were centrifuged on a parallel gradient. The sedimentation coefficients of σ and τ were estimated to be about 5S and 8S, respectively.

T4 DNA and calf thymus DNA is remarkably similar to that of the RNA polymerase isolated from T4 infected cells by Walter et al.[30]

Factor is not a Nuclease

It could be argued that σ is a type of nuclease which makes single-stranded breaks in the DNA, thus "activating" the DNA. One prediction of this hypothesis is that DNA which has been used as a template for transcription in the presence of both enzyme and factor should then be capable of supporting the initiation of RNA chains in the absence of factor. This is not the case. The experiment shown in Table 4 shows that DNA which has been used as a template for factor-stimulated transcription exhibits virtually the same unstimulated and stimulated levels of transcription on re-use as DNA which has not previously been incubated with factor. In addition, evidence obtained by several investigators[19-21,31] argues that DNA used for transcription by polymerase purified by the methods of Chamberlin[1] and Furth[2] remains intact. Because enzyme prepared in this way contains σ, it seems unlikely that the stimulation observed is due to nuclease action.

Formation of a Factor–Polymerase Complex

Several lines of evidence suggest that the factor can exist in a complex with RNA polymerase. First, the factor purifies with polymerase through steps involving protamine sulphate precipitation, ammonium sulphate fractionation and DEAE–cellulose chromatography. Furthermore, it remains with polymerase during low and high salt glycerol gradient centrifugation ($\Gamma/2 = 0.04$ and 1.0, respectively) where the polymerase sediments at 24S and 14S. The free factor sediments at about 5S, so it must be tightly bound to polymerase in all these conditions. Second, from the molecular weights of α, β and σ, and from the intensity of their stained bands on polyacrylamide gels, it is possible to make an estimate of the relative amounts of each band present in GG enzyme and also in peak B. In both, such an estimate yields a very approximate molar ratio of $\alpha : \beta : \sigma$ of 2 : 2 : 1. Third, complex formation between PC enzyme and factor can be demonstrated by running standard pH 8·7 polyacrylamide gels in the absence of dissociating agents (Fig. 5). In these conditions PC enzyme is resolved into several bands, which probably represent aggregates. Peak B, which contains the complex of α, β and σ, migrates as a single band which moves ahead of the PC enzyme bands. Purified σ moves

Table 4. EFFECT OF PREINCUBATION WITH FACTOR AND PC ENZYME ON THE ABILITY OF T4 DNA TO DIRECT FACTOR-DEPENDENT TRANSCRIPTION

Material present during preincubation of DNA	$\mu\mu$moles AMP incorporated/min	
	PC enzyme alone	PC enzyme + factor
No additions	8	264
PC enzyme (10 μg)	10	282
Factor (1·5 μg)	17	238
PC enzyme (10 μg) + factor (1·5 μg)	15	259
No preincubation	12	259

Four reaction mixtures (0·25 ml. each) for RNA synthesis were set up as described in the legend to Table 1, with the modification that BSA was omitted and non-radioactive ATP replaced ^{14}C-ATP. PC enzyme and factor were added as indicated. Each reaction mixture was incubated for 10 min at 37° C. They were then diluted to 1 ml. with distilled water and extracted with 1 ml. of water-saturated phenol. The DNA was precipitated from the aqueous phase by the addition of 2 ml. ethanol. The precipitates were washed twice with 2 ml. of ethanol, collected by centrifugation and dried *in vacuo* to remove all traces of ethanol. Finally the DNA samples were redissolved in 0·2 ml. 0·01 M *tris* buffer pH 7·9. The ability of this DNA to direct RNA synthesis by PC enzyme in the presence and absence of factor was then assayed as described in the legend to Table 1 with the modification that the reaction volume was 0·10 ml. and BSA was omitted. Each tube contained 4 μg of PC enzyme. 0·6 μg factor (1·3 μg of peak *A* material) was also added where indicated.

faster than the complex. If σ and PC enzyme are mixed in approximately equivalent amounts and then subjected to gel electrophoresis, a single band corresponding to the complex is seen.

Even though PC enzyme and σ form a complex, for example as in peak *B*, the addition of either factor or PC enzyme to this complex results in stimulation. We are investigating whether the enzyme and factor are in rapid equilibrium with complex and whether factor is released for re-use in other complexes by the act of initiation.

Enzyme and complex elute at different ionic strengths from a phosphocellulose column, so it seems possible that they would elute from a DNA cellulose column at different ionic strengths. We found, however, that they behave identically on such a column. This provides some indication that the factor does not function merely by increasing the affinity of the polymerase for DNA. Furthermore, the free factor was not retained by the DNA cellulose column at 0·1 M KCl. This suggests that the free factor does not bind to DNA in conditions where the basic enzyme and the complex do so.

The Function of σ

The results clearly show that PC enzyme can by itself initiate the synthesis of RNA chains, and can catalyse chain elongation. Thus it is possible that this enzyme is the fundamental RNA polymerase. The presence of the stimulating factor, σ, greatly enhances the amount of RNA synthesis, the degree of enhancement being dependent on the DNA template used. Several possible modes of action of σ can be proposed. It could stimulate initiation, increase the rate of polymerization or prevent unusually early cessation of chain growth. Preliminary evidence (Travers and Burgess, manuscript in preparation) indicates that σ markedly increases the number of RNA chains initiated. This suggests that σ acts at the level of initiation.

We can thus pose the question: is the additional initiation observed in the presence of σ merely due to an increase in the rate of initiation at sites poorly utilized in its absence, or does this initiation occur at sites which absolutely require σ for their expression? The first possibility implies that the PC enzyme determines the specificity of initiation and that σ may have some other function in the process of initiation. If σ itself determines the specificity of initiation, however, the interesting possibility arises that several similar factors could exist, each with a specificity for a different type of initiation site. This latter idea is attractive, for recent studies using the antibiotic rifamycin suggest that *in vivo* only one kind of RNA polymerase exists[32]. Yet there is also much evidence to indicate that *in vivo* the control of *m*RNA, *t*RNA and *r*RNA synthesis is not coordinate[33]. σ and similar factors could then act as positive control elements regulating the amount of synthesis of different classes of RNA, including the late RNA of certain bacteriophages.

We thank Professor J. D. Watson for his enthusiastic interest and support, Professor K. Weber, Mr J. Roberts and Dr J. Tkacz for helpful discussions and suggestions, and Miss Anne-Marie Piret and Mrs Christine Roberts for technical assistance. This work was supported by grants from the US National Science Foundation and the US Public Health Service. One of us (A. A. T.) is a postdoctoral fellow of the Damon Runyon Memorial Fund for Cancer Research, and R. R. B. is a National Science Foundation predoctoral fellow.

Received December 2, 1968.

[1] Chamberlin, M., and Berg, P., *Proc. US Nat. Acad. Sci.*, **48**, 81 (1962).
[2] Furth, J. J., Hurwitz, J., and Anders, M., *J. Biol. Chem.*, **237**, 2611 (1962).
[3] Stevens, A., and Henry, J., *J. Biol. Chem.*, **239**, 196 (1964).
[4] Fuchs, E., Zillig, W., Hofschneider, P. H., and Preuss, D., *J. Mol. Biol.*, **10**, 546 (1964).
[5] Richardson, J. P., *Proc. US Nat. Acad. Sci.*, **55**, 1616 (1966).
[6] Babinet, C., *Biochem. Biophys. Res. Commun.*, **26**, 639 (1967).
[7] Hayashi, M., Hayashi, M. N., and Spiegelman, S., *Proc. US Nat. Acad. Sci.*, **51**, 351 (1964).
[8] Geiduschek, E. P., Tocchini-Valentini, G. P., and Sarnat, M., *Proc. US Nat. Acad. Sci.*, **52**, 486 ( ).
[9] Green, M., *Proc. US Nat. Acad. Sci.*, **52**, 1388 (1964).
[10] Luria, S. E., *Biochem. Biophys. Res. Commun.*, **18**, 735 (1965).
[11] Hayashi, M., Hayashi, M. N., and Spiegelman, S., *Proc. US Nat. Acad. Sci.*, **50**, 664 (1963).
[12] Tocchini-Valentini, G. P., Stodolsky, M., Sarnat, M., Aurisicchio, A., Graziosi, F., Weiss, S. B., and Geiduschek, E. P., *Proc. US Nat. Acad. Sci.*, **50**, 935 (1963).
[13] Marmur, J., and Greenspan, C. M., *Science*, **142**, 387 (1963).
[14] Khesin, R. B., Shemyakin, M. F., Gorlenko, Zh. M., Bogdanova, S. L., and Afanaseva, T. P., *Biokhimiya*, **27**, 1092 (1962).
[15] Geiduschek, E. P., Snyder, L., Colvill, A. J. E., and Sarnat, M., *J. Mol. Biol.*, **19**, 541 (1966).
[16] Naono, S., and Gros, F., *Cold Spring Harbor Symp. Quant. Biol.*, **31**, 363 (1966).
[17] Cohen, S. N., Maitra, U., and Hurwitz, J., *J. Mol. Biol.*, **26**, 19 (1967).
[18] Crawford, L. V., Crawford, E. M., Richardson, J. P., and Slayter, H. S., *J. Mol. Biol.*, **14**, 593 (1965).
[19] Richardson, J. P., *J. Mol. Biol.*, **21**, 83 (1966).
[20] Jones, O. W., and Berg, P., *J. Mol. Biol.*, **22**, 199 (1966).
[21] Pettijohn, D., and Kamiya, T., *J. Mol. Biol.*, **29**, 275 (1967).
[22] Sentenac, A., Ruet, A., and Fromageot, P., *Europ. J. Biochem.*, **5**, 385 (1968).
[23] Priess, H., and Zillig, W., *Biochim. Biophys. Acta*, **140**, 540 (1967).
[24] Stevens, A., Emery, jun., A. J., and Sternberger, N., *Biochem. Biophys. Res. Commun.*, **24**, 929 (1966).
[25] Smith, D. A., Martinez, A. M., Ratliff, R. L., Williams, D. L., and Hayes F. N., *Biochemistry*, **6**, 3057 (1967).
[26] Zillig, W., Fuchs, E., and Millette, R., in *Proc. Nucleic Acid Res.* (edit. by Cantoni, G. L., and Davies, D. R.), 323 (Academic Press, New York 1966).
[27] Burgess, R. R., *Fed. Proc.*, **27**, 295 (1968).
[28] Alberts, B. M., Amodio, F. J., Jenkins, M., Gutman, E. D., and Ferris, F. L., *Cold Spring Harbor Symp. Quant. Biol.*, **33** (in the press).
[29] Shapiro, A., Vinuela, E., and Maizel, J. V., *Biochem. Biophys. Res. Commun.*, **28**, 815 (1967).
[30] Walter, G., Seifert, W., and Zillig, W., *Biochem. Biophys. Res. Commun.*, **30**, 240 (1968).
[31] Maitra, U., and Hurwitz, J., *J. Biol. Chem.*, **242**, 4897 (1967).
[32] Tocchini-Valentini, G. P., Marino, P., and Colvill, A. J., *Nature*, **220**, 27 (1968).
[33] Maaløe, O., and Kjeldgaard, N. O., *Control of Macromolecular Synthesis* (Benjamin, New York, 1966).
[34] Davis, B. J., *Ann. NY Acad. Sci.*, **121**, 406 (1964).

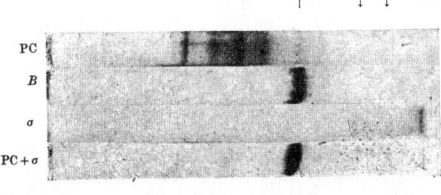

Fig. 5. Reconstitution of the enzyme-factor complex as demonstrated by electrophoresis on polyacrylamide gels. From top to bottom PC enzyme (4 μg), peak *B* (4 μg), purified factor (1 μg) from tube 17 of the glycerol gradient shown in Fig. 4, and a mixture of PC enzyme and purified factor. Polyacrylamide gels, pH 8·7, 4 per cent acrylamide, were prepared and run as described by Davis[34], and stained as described in the legend to Fig. 3. σ appears as two very faint bands which move about 20 and 30 per cent faster than the complex and are indicated by arrows. The marker dye, seen very near the right end of the third gel from the top, just ran off of the other three gels.

Printed in Great Britain by Fisher, Knight & Co., Ltd., St. Albans.

F. R. Blattner and J. E. Dahlberg, 1972. RNA synthesis startpoints in bacteriophage λ: Are the promoter and operator transcribed? Nature New Biology 237 (1972) 227-232.

F. R. Blattner, J. E. Dahlberg, J. K. Boettiger, M. Fiandt and W. Szybalski, 1972. Distance from a promoter mutation to an RNA synthesis startpoint on bacteriophage λ DNA. Nature New Biology 237 (1972) 232-236.

RNA Synthesis Startpoints in Bacteriophage λ: Are the Promoter and Operator Transcribed?

FREDERICK R. BLATTNER
McArdle Laboratory for Cancer Research, University of Wisconsin, Madison, Wisconsin 53706

JAMES E. DAHLBERG
Department of Physiological Chemistry, University of Wisconsin, Madison, Wisconsin 53706

Hybridization and fingerprint analysis of *in vitro* synthesized λ RNA shows that four chains are initiated at sites corresponding to those seen *in vivo*, and that each molecule starts with a specific sequence. In one case examined, the major leftward operon, the promoter and operator are not transcribed into RNA.

THE chromosomes of bacteria and bacteriophages are organized into transcriptional units called operons[1], providing a basis for physiologically regulated gene expression. Initiation, termination and regulation of transcription by RNA polymerase involve specific regions of DNA. Therefore, a thorough understanding of DNA to RNA transcription will require knowledge of the nucleotide sequences in these regions, particularly at the beginnings and ends of specific transcriptional units. In addition, it will be necessary to identify the points in these sequences where RNA synthesis starts and the regions to which RNA polymerase and control proteins bind.

We are studying this problem by analysing initial RNA sequences, that is those immediately downstream from the points where RNA synthesis starts, in specific operons of bacteriophage λ (ref. 2). Very short RNA products were synthesized *in vitro* by limiting the period of synthesis. These were analysed by DNA-RNA hybridization and fingerprinting. Four RNA molecules, each with a unique sequence, were synthesized. The points where synthesis starts have been located on the λ chromosome with respect to genetic markers including promoter and operator mutants (see the following article[3]). Unexpectedly we found that the promoter–operator region of the major leftward operon in λ is not transcribed into RNA. (Interpretations of this finding will be discussed in the following article[3].)

The DNA of λcb2 was used as the template for *E. coli* RNA polymerase in order to eliminate transcription from the b2 region[4]. The kinetics of incorporation using two methods for starting the reaction are shown in Fig. 1. We found a lag in incorporation when the reaction was started by addition of triphosphates. However, when the reaction was started by addition of magnesium ions[5], the lag was substantially reduced, and at 25° C RNA chains grew to approximately 175 nucleotides in 0.4 min. This latter procedure was used for all subsequent experiments.

Characterization of the Products

Four RNA species were produced in the synthesis reaction, corresponding to four startpoints on the λ chromosome. These have been located by hybridization of the RNA to

separated strands of deletion and substitution mutants of λ as shown in Fig. 2.

More than 85% of the RNA originated from the divergently oriented promoters p_L and p_R, which are responsible for RNA synthesis in the early phase of lytic growth[6]. We shall denote these species p_L and p_R RNA. From the genetic map of λ, these RNAs probably correspond to the beginnings of the messengers for genes N and *tof*, respectively[4]. A small amount of RNA was also observed originating from two other promoters, one directed leftward and located to the right of the

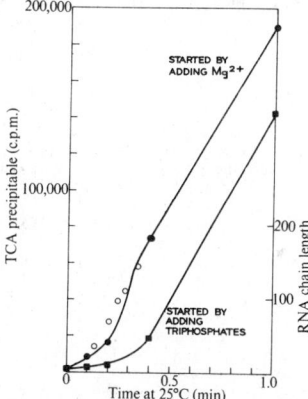

Fig. 1 Kinetics of *in vitro* transcription. *E. coli* RNA polymerase was prepared by the Chamberlin and Berg[34] procedure followed by a low salt glycerol gradient. The enzyme preparation contained sigma factor[35] but no rho factor[4], as assayed by polyacrylamide gel electrophoresis. λb2cI72 DNA was extracted by the phenol procedure. Unlabelled nucleoside triphosphates ATP, GTP, CTP and UTP were purchased from Schwarz. α-^{32}P-labelled triphosphates were prepared by our modification of the method of Symons[36]. The standard preincubation mixture contained 200 μg of RNA polymerase, 200 μg of DNA, 50 pmol of each nucleoside triphosphate (one labelled in the α position with 1–10 mCi of ^{32}P), 4 μmol of EDTA, 2 μmol of MgCl$_2$, 25 μmol of Tris, pH 7.9, and 85 μmol of KCl, all in a volume of 0.5 ml. After 2 min at 25° C the reaction was started by adding 20 μmol of MgCl$_2$ mixed with 5 μg of rifampin in 0.05 ml of H$_2$O. The reaction was stopped by adding 50 μmol of EDTA. Incorporation was measured by precipitation with TCA (●) and by the progressive appearance of oligonucleotides in p_L RNA (○). Although RNA chains commenced growing at time zero, they did not become fully TCA precipitable until the chain length reached about 50 nucleotides. When the reaction mixture was preincubated with MgCl$_2$ and started by adding the 4 triphosphates without rifampin there was a delay of 0.3 min before the RNA chains began to grow as assayed by TCA precipitation (■).

immunity region*, and the other located near the right end of the chromosome and directed rightward. Although the functions of these two minor RNAs are not clear, the leftward one may be involved with initiation of DNA replication and/or establishment of lysogeny[7–9]; the rightward one may be involved in initiation of late transcription[2,10]. Transcription from the *l* strand in the c*I*-*rex* region[11], which codes for the λ repressor and *rex* product, is apparently absent (less than 1% of the total). *In vitro* transcription on the left arm of the chromosome is also absent, although this is the region where poly (U, G) binding sites, once thought to correspond to promoter sites, are predominantly clustered[6,11–13].

To determine the 5′ terminal nucleotide of each RNA species[14], RNA was labelled with γ-^{32}P-ATP and γ-^{32}P-GTP. Hybridization patterns of these products are presented in Table 1. Each of the four RNA molecules is labelled at its 5′ end with a specific nucleoside triphosphate. The minor leftward RNA starts with GTP; the others start with ATP.

The sequence of each RNA molecule has been studied using the fingerprinting techniques of Sanger et al.[15] and Billeter et al.[16]. Each of the four molecules was purified from the others by a series of preparative hybridizations to separated strands of deletion mutants of λ. After elution of each RNA from the DNA–RNA hybrid it was digested either with pancreatic or T$_1$ ribonuclease, and fingerprints were prepared. Fig. 3 shows T$_1$ RNAase fingerprints of each of the RNAs, labelled with α-^{32}P-ATP. The first few nucleotides of each sequence are indicated in Fig. 2. The complete sequences of these products will be published later.

It is instructive to compare the fingerprints in Fig. 3 with two λ RNA fingerprints already published. Lebowitz et al.[17] have determined the sequence of an RNA synthesized *in vitro* using λb2*imm*21 DNA as template. Their fingerprint is almost identical with the fingerprint of the minor rightward molecule. This identifies the RNA molecule sequenced by these authors as originating from the *r* strand on the right arm within 10% of the right end of the λ chromosome. This region of the genome includes the late promoter p'_R which normally is under the control of a positive transcriptional regulator, the product of gene *Q*. As Lebowitz et al.[17] point out, this RNA chain terminates at the sequence UUUUUUA in the absence of rho factor. Thus if the minor rightward RNA in fact originates from p'_R, we might propose that gene *Q* product activates late transcription by overcoming the transcriptional block at its 3′ end.

A second λ RNA has been described by Marcaud et al.[18]. This 4.5S RNA fragment called l_1 was isolated after induction of a λ lysogen. Comparison of their fingerprints with those in Fig. 3 shows that the oligonucleotides in the l_1 fragment are a subset of those present in the *in vitro* p_L-RNA molecule. This identifies the l_1 fragment as originating somewhere near the beginning of the leftward operon, in agreement with the result of Lozeron et al.[19]. From our unpublished sequence of p_L RNA, we estimate that l_1 in fact begins about 90 nucleotides downstream from the 5′ terminus.

Lozeron et al. (ref. 20 and manuscript in preparation) have also purified ^{32}P-labelled λ RNA synthesized *in vivo* using techniques that minimize mRNA breakdown. In various experiments they found RNA species which have fingerprints identical to the *in vitro* synthesized p_L, p_R, and minor leftward

---

* The detailed mapping of the minor leftward RNA presents a unique problem. The RNA represents a single species, as determined by sequence analysis. It hybridizes well to *l* strands of λ*imm*434, λ*nin*5, φ80*att*80*imm*λ and λp4[39] (not shown in Fig. 2), so it must map between *imm*434 and *nin*5. It seems, however, to hybridize variably to the *l* strands of λ*imm*21. This may be because λ*imm*21 is not completely isogenic with λ in this region. Fingerprints of the RNA templated from λb2*imm*21 show that the sequence is basically very similar to that of λ, so it probably maps to the right of *imm*21. Since it does not hybridize to λ*att*80*imm*80, it probably maps outside of any λ-φ80 homology regions[39]. Thus, we provisionally place this RNA between the immunity region of λ*imm*21 and the *O–P* homology[39] of λ*att*80*imm*80, in an area near the origin (*ori*) of λ replication[44,47], rather than in the *y-cII* region.

RNA molecules. In particular the 5′ ends of the *in vivo* molecules were identical to the 5′ ends of the corresponding *in vitro* molecules. In addition, all internal oligonucleotides in T$_1$ and pancreatic RNAase digests were also present. Thus, they have concluded that RNA synthesis starts at the same point both *in vivo* and *in vitro*.

Precise Mapping of Startpoint s_L

Each of the four RNA chains we are studying starts exclusively with its unique nucleotide sequence. Thus, RNA synthesis starts at precise points on the chromosome. We have adopted the term startpoint, *s*, to denote the exact nucleotide on a template where an RNA chain is started. Taking advantage of opportunities afforded by the λ system, we have succeeded in precisely mapping the startpoint s_L for the leftward operon.

As shown in Fig. 2, the two strains λ*bio*N2-1*nin*5 and λ*imm*434 share a short overlapping region of λ homology located to the left of p_L. The length of this overlap is about 215 nucleotide pairs as determined by the heteroduplex electron microscopic measurements[3,21]. On analytical hybridization the p_L RNA molecule hybridized 100% efficiently to *l* strands of each of these strains relative to the *l* strands of λ$^+$. Therefore, the p_L RNA chain, whose length is about 175 nucleotides, must hybridize within the 215 nucleotide overlap region. Strain λ*bio*30-7d*vnin*5 has a shorter overlap with λ*imm*434 measuring about 120 nucleotides[3]. By comparison, p_L RNA hybridized only 60% efficiently to *l* strands of this strain. This indicates that the λ*bio*30-7d*vnin*5 deletion endpoint is located near the middle of the *in vitro* molecule, and the RNA startpoint is close to the right end of the overlap region.

To determine the location of the startpoint more exactly, we have performed two RNAase digestion experiments, taking advantage of the two deletions that cut near the startpoint from each side. In the first experiment, a purified preparation of *in vitro* synthesized p_L RNA was hybridized to *l* strands of λ*bio*30-7d*vnin*5. The DNA–RNA hybrid was treated with pancreatic RNAase to remove any RNA not protected by hybridization to DNA, and the protected RNA was eluted. In Fig. 4, the pancreatic RNAase fingerprints of this RNA, and the untreated control, are presented. As we suspected, certain oligonucleotides were selectively eliminated from the fingerprint by the digestion step although all oligonucleotides near the 5′ end were retained. Interestingly, all the oligonucleotides missing from Fig. 4*b* are present in the l_1 fingerprints published by Marcaud et al.[18]. This confirms that l_1 is at the 3′ end of the *in vitro* molecule.

The data shown in Fig. 4*b* allow us to estimate the distance from s_L to the dv endpoint. All oligonucleotides labelled with UTP, which occur before position 120 in the sequence, are present in this fingerprint and also in the T$_1$ RNAase fingerprint of the same material (not shown). All those beyond this position are absent. Thus, the chain length of the dv portion of p_L RNA is 120 nucleotides assuming that RNAase can digest the free RNA precisely at the left end of the dv insertion. Since the overlap of λ*bio*30⅞d*vnin*5 with λ*imm*434 is 120 ± 20 nucleotides[3], the startpoint must be within 20 nucleotides of the left end of the λ*imm*434 nonhomology region.

In the second digestion experiment, γ-^{32}P-ATP was the only radioactive precursor so that all *l* strand specific radioactivity was incorporated into the 5′ nucleotide of the p_L RNA chain. This RNA was hybridized with *l* strands of λ*imm*434 and again the effect of pancreatic ribonuclease on the hybrid was examined. If the RNA startpoint were within the region of nonhomology, there would be an unpaired segment at the 5′ end. If this were sufficiently long, all the radioactivity would be removed by RNAase (the second and third nucleotides of this molecule are pyrimidines). As Table 1 shows, the radioactivity was completely stable to pancreatic RNAase, indicating that the startpoint is to the left of the λ*imm*434 nonhomology region, or at most a few nucleotides to the right of it.

Proof that RNA synthesis in fact starts to the left of this region comes from our RNA sequence work. If the p_L RNA chain actually began within the region of nonhomology, we would expect RNA transcribed from λb2imm434 DNA to have a region at the 5′ end with different sequence from the RNA of λb2. As we discuss below, however, this RNA sequence was identical for at least the first eighteen nucleotides at the 5′ end with that of λb2.

Therefore we conclude that p_L RNA synthesis starts at a startpoint s_L located to the left of the region of nonhomology between λ and λimm434 but within 20 nucleotides of it. Thus the distance from λimm434 to s_L is 0 to 20 nucleotide pairs.

Promoter and Operator are not Transcribed

Precise localization of the RNA startpoint allows us to examine the spatial relationship between genetically identified promoter and operator regions of the λ leftward operon and the physically identified point of RNA chain initiation.

The promoter and operator in this case can be defined by the point mutations sex1 (ref. 22 and M. Gottesman, personal communication) and $v2^{1,23,24}$ and the substitution mutant λimm434[26]. The p_L mutant λsex1 is characterized by a sub-

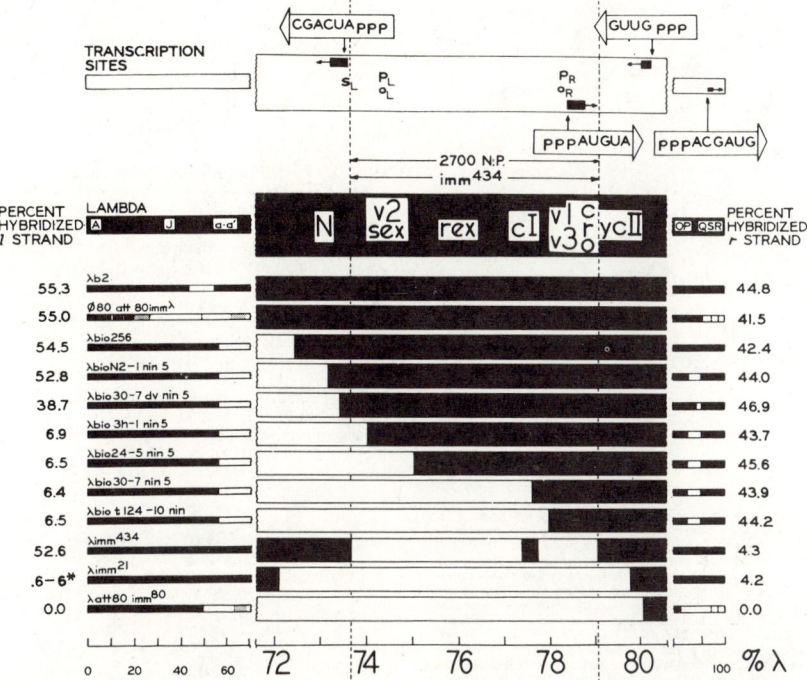

Fig. 2 Hybridization pattern of *in vitro* synthesized λ RNA. λ RNA, synthesized as described in the legend to Fig. 1, was purified from unincorporated nucleoside triphosphates by passage of the reaction mixture over a 'Sephadex G-50' column equilibrated with 0.2 M NH$_4$-acetate, pH 7.9. The front peak fractions were adjusted to 0.05 M MgCl$_2$ and 0.005 M CaCl$_2$, 100 μg of tRNA was added as a carrier, and the DNA was digested with 200 μg/ml. of DNAase for 10 min at 25° C. To remove protein, EDTA (0.06 M), SDS (0.05%), and diethyl pyrocarbonate (0.5%) were added and the mixture was incubated for 10 min at 37° C. About 1/10 volume of 4 M KCl was added, dissolved CO$_2$ was removed by vacuum, protein-SDS precipitate was removed by centrifugation at low speed and RNA was precipitated with 80% ethanol and then resuspended in water. About 5% of the RNA was used for analytical hybridization to separated strands of deletion and substitution mutants of λ. The remainder was used in preparative hybridization followed by fingerprinting. Hybridization techniques were based on the method of Nygaard and Hall[37] developed by Lozeron (see ref. 38) for separated DNA strands. Each hybridization mixture contained at least a 100-fold molar excess of DNA strands over RNA, which resulted in 80–90% hybridization efficiency. The homology structure of each strain is shown relative to a genetic map in which the centrally located immunity region has been expanded twenty-fold. Regions of DNA homologous to λ as judged in the electron microscope are in black, and deletions or heterologous substitutions are in white[3,31,39]. A few areas showing partial or variable homology in electron micrographs are cross hatched. The numbers on the left and right sides of the figure are percentages of ^{32}P-RNA hybridizing to *l* and *r* strands respectively. Percentages are normalized to the total RNA hybridizable to λb2 strands *l* and *r*. Since hybridization to the *l* strand of λimm21 appears to be variable, the range of 15 experiments is given (see footnote to text *). Locations and terminal sequences of the four RNA molecules made in this reaction are shown at the top of the figure. Note that the leftwardly transcribed molecules are shown with the 5′ end to the right. The origins of the strains are as follows: λimm434, λimm21, φ80att80immλ, λatt80imm80, λbio30-7nin5, λbio256 and λbio30-7nin5 are the same as mapped by Westmoreland et al.[31] and Fiandt et al.[39]. The nin5 derivatives[40] of λbio strains N2-1, M3h-1, and R24-5 (refs. 41, 42) were prepared by G. Kayajanian or by us under his supervision. Their buoyant densities are 1.495, 1.495 and 1.492 g/ml. respectively. The nin derivative of strain λbiot124-10 (ref. 43) was obtained from C. R. Fuerst and A. Guha. Strain λbio30-7dvnin5 was isolated by S. Kumar as a virulent derivative of λbio30-7nin5 after growth in a cell carrying the plasmid λdv, and was designated as λdb30-7v1v3nin by Kumar and Szybalski[27]. Recognition and proof of its structure as an insertion of the plasmid into the λbio30-7nin5 genome were through the efforts of S. Kumar, W. Szybalski, G. Kellenberger-Gujer, S. Hayes, H. Lozeron, and ourselves. Its buoyant density is 1.506 g/ml. For the designation of the genes and mutations and for the electron micrographic maps, see Szybalski et al.[2] and Fiandt et al.[39].

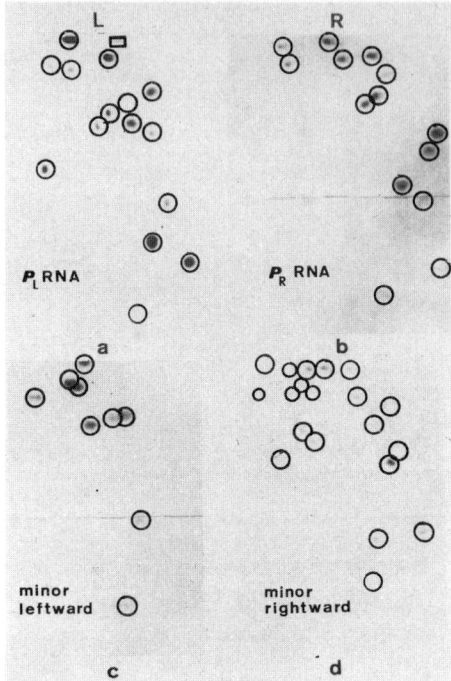

Table 1 Hybridization of Terminally Labelled λb2 RNA

Hybridized to DNA of.	Strand	RNAase	Percentage of γ-^{32}P-labelled RNA hybridized	
			ATP	GTP
λb2	l	+	40.7	88.1
	l	−	44.1	
λbioN2-1nin5	l	+	42.1	98.6
λbio3h-1nin5	l	+	0.8	88.9
λimm434	l	+	40.2	71.3
	l	−	40.5	
λb2	r	+	59.6	11.9
λimm434	r	+	8.6	2.2

RNA was synthesized using either γ-^{32}P-labelled ATP or GTP as the radioactive precursors and λb2 DNA as the template as described in the legend to Fig. 1. The RNA was purified and hybridized as shown in the legend to Fig. 2. The numbers given are percentages of total γ-^{32}P-RNA hybridizable to l and r strands of λb2, using RNAase treatment. The deletion endpoints of the various strains are shown in Fig. 2.

some of λ. In the electron microscope, heteroduplexes between λ and λimm434 show a block of nonhomology where the strands fail to form duplexes[31] (Fig. 2). This DNA is the source of the nonhomology region to which reference was made above in connexion with hybridization experiments. Strain λimm434 does not contain a λ specific operator o_L or promoter p_L, but it does contain an operator that is specific for phage 434 repressor[26,32], as well as a promoter that has been defined genetically by sex-type mutations (D. Friedman, personal communication). Neither λv2 nor λsex1 can be crossed with λimm434 to yield a wild type recombinant[22,33]. Conversely, the λimm434sex mutants cannot be rescued from wild type λ (D. Friedman, personal communication). Therefore, it can be concluded that λimm434 does not contain DNA sequences corresponding to p_L and o_L elements of λ. Instead, it contains phage 434 DNA of analogous function but different specificity and nucleotide sequence.

Our finding that the startpoint s_L is outside (to the left)

Fig. 3 T$_1$ RNAase fingerprints of the λ RNAs synthesized in 0.4 min in vitro. The fingerprints are designated: a, p_L RNA; b, p_R RNA; c, minor leftward RNA; d, minor rightward RNA. RNA was synthesized and purified as described in the legends to Figs. 1 and 2, using λb2 DNA as template and α-^{32}P-labelled ATP (specific activity 100 mCi/μmol). The RNA molecules were separated from one another by a series of preparative-scale hybridizations to the following sequence of DNA strands: (1) l strands of λbio30-7nin5; (2) l of λb2; (3) r of λimm434; (4) r of λb2. After each step the hybrid was collected by filtration, and RNA that did not hybridize was collected by ethanol precipitation of the filtrate for use in the next step. RNA was eluted from each filter by placing it in 4.0 ml. H$_2$O at 85° C for 4 min. Thus, each of the four RNA species was eluted in pure form from one of the filters for fingerprinting. 100 μg of carrier RNA and 0.5 ml. of 2 M ammonium acetate, pH 6.5, were added to each eluate, followed by 25 ml. of absolute ethanol. The precipitates were collected by centrifugation, and dissolved in 25 μl. of water. Half of each preparation was used for T$_1$ RNAase fingerprints (shown here) and half for pancreatic RNAase fingerprints (not shown). Enzymatic digestion was in a total volume of 3 μl. of 1 mg/ml. T$_1$ RNAase (Sankyo) in 0.01 M Tris-Cl, pH 7.6, for 30 min at 37° C. The digests were fingerprinted and autoradiographed according to the procedures of Sanger et al.[15]. Electrophoresis on 90 × 25 cm cellulose acetate strip in pH 3.5 pyridine acetate plus 7 M urea was for 3 h at 5 kV (right to left). Ionophoresis on DEAE cellulose paper (80 × 40 cm) in 7% formic acid was for 16 h at 0.95 kV, 170 mA (top to bottom).

stantial reduction in transcription to the left, both in vivo[6,25] and in vitro[4]. The o_L mutant λv2 has a mutant operator with reduced affinity for the λ repressor. This has also been demonstrated both in vivo[24,27] and in vitro[28,29].

Strain λimm434 was derived from a series of crosses between λ and phage 434[26] in which two recombinational events took place within short regions of interspecies homology[30] that flank the immunity regions[31] of the parental phages. As a result, the immunity region of 434 and short segments of the flanking homologous regions were substituted into the chromo-

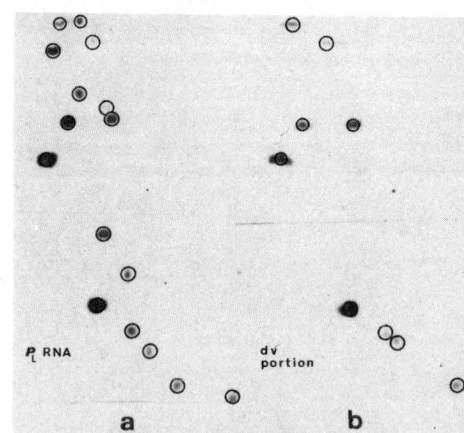

Fig. 4 Pancreatic RNAase fingerprints of whole p_L RNA (a) and of the λdv complementary segment (b). The p_L RNA, labelled with α-^{32}P-UTP, was prepared and purified as described in the legends to Figs. 1, 2 and 3. It was either fingerprinted directly (a) or first hybridized to the l strands of λbio30-7dvnin5 DNA (b). The hybrid was collected by filtration and the nitrocellulose filter was treated with pancreatic RNAase for 1 h at 25° C to destroy any RNA that was not in DNA–RNA hybrid form. After inactivation of RNAase[38], the RNA was eluted, digested with pancreatic RNAase and fingerprinted as described in the legend to Fig. 3.

Note: Spots in Figs. 3–5 are circled to compensate for the loss of contrast in printing the λimm434 nonhomology region demonstrates that the promoter and operator are upstream of s_L and therefore are not transcribed into RNA in this operon. This conclusion was verified by the use of λb2sex1 and λb2imm434 DNAs as templates. If the sex1 mutation occurred in a region that is transcribed into RNA, we would expect to find a nucleotide change in the sequence. Moreover, for the λb2imm434 template we would expect more extensive alterations sufficient to change the specificity of both the operator and promoter from that of λ to that of 434.

In our $in\ vitro$ system λb2sex1 DNA gave 90% reduction in the level of p_L RNA relative to p_R RNA. A T_1 RNAase fingerprint of α-GTP-labelled RNA made from λb2sex1 DNA is shown in Fig. 5b. The fingerprint is identical to that of RNA from wild type DNA (Fig. 5a). This experiment has also been done with α-^{32}P-CTP as the precursor. Both products were also fingerprinted after pancreatic RNAase digestion. In all cases every oligonucleotide was analysed for molar yield and nearest neighbour phosphate transfer with the same result: the sequences of p_L RNA made from λb2 DNA and λb2sex1 DNA are identical over their entire length. Therefore, we conclude that at least that part of the DNA which is directly affected by the sex1 mutation is not transcribed into RNA. Furthermore, since the same RNA chain length was reached in 0.4 min, we conclude the sex1 mutation acts by reducing the frequency of initiation, rather than by introducing a premature termination signal[4,29].

When λb2imm434 DNA was used as template, the yield of p_L RNA was even more severely reduced than with λsex1, confirming the result of Steinberg and Ptashne[29] that leftward promoter of λimm434 is functionally different from p_L of λ. The T_1 RNAase fingerprints of α-^{32}P-GTP labelled products from the two templates are shown in Figs. 5a and 5c and summarized in Fig. 5d. Although these molecules are not identical, they are extremely similar as evidenced by the fingerprint patterns. The differences have been examined in the GTP-labelled products and by repeating the experiment with labelled ATP and UTP. Each product was fingerprinted after digestion with pancreatic as well as T_1 RNAase, and the eluted oligonucleotides were further digested and analysed. All the oligonucleotides that comprise the first eighteen residues of the p_L RNA from λ are present in the digests of the molecules from λimm434. We conclude that the two molecules have the same sequence up to position of at least 18. Beyond that position there are four widely spaced base changes at positions 19 or 22, 34, 124 and 152. It is possible that in the original λ × 434 crosses this area of the chromosome originated from phage 434 rather than λ. These few changes might reflect the lack of perfect isogenicity between N regions of phages λ and λimm434; mutational differences between genes N of these two phages have been reported[45]. Moreover, several of these sequence changes have also been observed when comparing several strains having λ immunity (J. E. D. and H. A. Lozeron, unpublished observation). The important fact, however, is that the two RNAs have a very high degree of homology and are actually identical for a large region at the 5′ end, whereas substantial changes would have been predicted if promoter and operator were transcribed. We therefore conclude that the two genetically defined elements p_L and o_L which control the expression of this operon are both located upstream from the point where the RNA chain starts and neither of them is transcribed into RNA.

In summary, DNA-RNA hybridization and RNA fingerprint analysis of the initial sequences of RNA synthesized $in\ vitro$ have been used to map and characterize the RNA startpoints of bacteriophage λ. These studies also relate to the mechanism of chain initiation by RNA polymerase. In particular, all molecules start with their own specific sequence, indicating that the point on the DNA where synthesis starts is very precise. The lag that we observe when synthesis is started by the addition of nucleoside triphosphates must also be explained in any model of events that occur at initiation. The molecules that

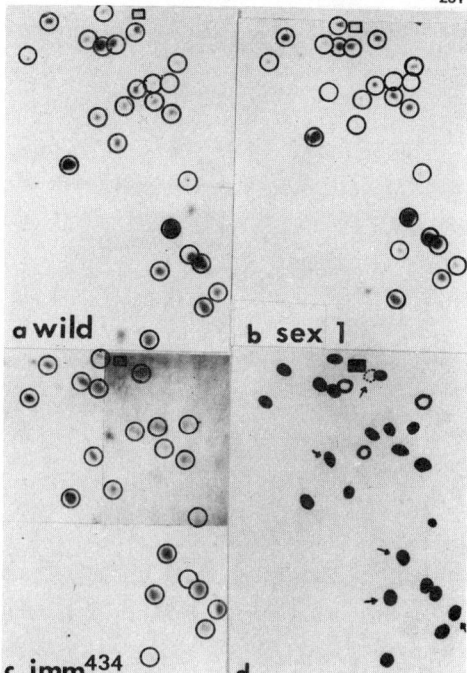

Fig. 5 Effect of alterations in the p_Lo_L region on the T_1 RNAase fingerprints of p_L RNA. RNAs were synthesized from λb2 (a) λb2sex1 (b), or λb2imm434 (c) DNA, as described in the legends to Figs. 1 and 2, in the presence of α-^{32}P-GTP. Separation of each RNA species was as described in the legend to Fig. 3 except that it included two hybridizations to l strands λbio30-7nin5 DNA in order to remove the minor leftward RNA more thoroughly. This was followed by hybridization to l strands of λbioN2-1nin5, from which the RNA was eluted for fingerprinting. Digestion and the conditions of electrophoresis are as described in the legend to Fig. 3. Diagram d summarizes differences between fingerprints a and c. Oligonucleotides which are in common are shown in solid black and those missing in fingerprint c are circled with a line. Arrows indicate oligonucleotides that are present (dotted) or augmented only in fingerprint c. The 5′ oligonucleotide is shown as a rectangle. Oligonucleotides of fingerprint c that are not in diagram d represent contamination by a small amount of the minor leftward RNA (Fig. 3c).

are synthesized $in\ vitro$ correspond very well to the initial regions of $in\ vivo$ transcriptional units. Our finding that the startpoint is not identical with the promoter p_L, and that neither p_L nor o_L is transcribed into RNA, will be extended and discussed in the following paper. A preliminary account of this work has been reported already[46].

Much of this work was done in the laboratory of Dr W. Szybalski where F. R. B. is a postdoctoral fellow. We thank him for interest and support. J. Boettiger, B. Stark, W. Szybalski, E. Thiede, P. Twose and W. Wacker provided technical assistance. We also thank M. Fiandt, D. Friedman, C. R. Fuerst, M. Gottesman, S. Hayes, T. Ikemura, G. Kayajanian and H. Lozeron for various things. This work was supported by grants from the National Institutes of Health and the National Science Foundation to J. E. D. and W. Szybalski. J. E. D. holds an NIH research career development award.

$Note\ added\ in\ proof$. After this manuscript was submitted for publication, we fingerprinted p_L RNA and p_R RNA synthesized on λv2v1v3 DNA which contains mutations in both the o_L operator ($v2$) and the o_R operator ($v1v3$). These RNAs have fingerprints identical to those of the wild type λ. This confirms our conclusion that o_L is not transcribed into RNA and extends this conclusion to the major rightward operon.

Received April 21; revised May 17, 1972.

[1] Jacob, F., and Monod, J., *J. Mol. Biol.*, **3**, 318 (1961).
[2] Szybalski, W., Bøvre, K., Fiandt, M., Hayes, M., Hradecna, Z., Kumar, S., Lozeron, H. A., Nijkamp, H. J. J., and Stevens, W. F., *Cold Spring Harbor Symp. Quant. Biol.*, **25**, 341 (1970).
[3] Blattner, F. R., Dahlberg, J. E., Boettiger, J. K., Fiandt, M., and Szybalski, W., *Nature New Biology*, **237**, 232 (1972).
[4] Roberts, J. W., *Nature*, **224**, 1168 (1969).
[5] Davis, R. W., and Hyman, R. W., *Cold Spring Harbor Symp. Quant. Biol.*, **35**, 269 (1970).
[6] Szybalski, W., Bøvre, K., Fiandt, M., Guha, A., Hradecna, Z., Kumar, S., Lozeron, H. A., Maher, V. M., Nijkamp, K. J. J., Summers, W. C., and Taylor, K., *J. Cell. Physiol.*, **74** (Suppl. 1), 33 (1969).
[7] Champoux, J. J., *Cold Spring Harbor Symp. Quant. Biol.*, **35**, 319 (1970).
[8] Hayes, S. J., *Fed. Proc.*, **31**, 444, Abstr. (1972).
[9] Reichardt, L., and Kaiser, A. D., *Proc. US Nat. Acad. Sci.*, **68**, 2185 (1971).
[10] Herskowitz, I., and Singer, E., *Cold Spring Harbor Symp. Quant. Biol.*, **35**, 355 (1970).
[11] Taylor, K., Hradecna, Z., and Szybalski, W., *Proc. US Nat. Acad. Sci.*, **57**, 1618 (1967).
[12] Szybalski, W., Kubinski, H., and Sheldrick, P., *Cold Spring Harbor Symp. Quant. Biol.*, **31**, 123 (1966).
[13] Champoux, J. J., thesis, Stanford Univ. (1969).
[14] Takanami, M., Okamoto, T., and Sugiura, M., *Cold Spring Harbor Symp. Quant. Biol.*, **35**, 179 (1970).
[15] Sanger, F., Brownlee, G. G., and Barrell, B. G., *J. Mol. Biol.*, **13**, 373 (1965).
[16] Billeter, M. A., Dahlberg, J. E., Goodman, H. M., Hindley, J., and Weissmann, C., *Cold Spring Harbor Symp. Quant. Biol.*, **34**, 635 (1969).
[17] Lebowitz, P., Weissman, S. M., and Radding, C. M., *J. Biol. Chem.*, **246**, 5120 (1971).
[18] Marcaud, L., Portier, M., Kourilsky, P., Barrell, B. G., and Gros, F., *J. Mol. Biol.*, **57**, 247 (1971).
[19] Lozeron, H. A., Funderburgh, M. L., and Szybalski, W., *Fed. Proc.*, **30**, 1315, Abstr. (1971).
[20] Lozeron, H. A., Funderburgh, M. L., Dahlberg, J. E., Stark, B. P., and Szybalski, W., *Abstracts Ann. Meet. Amer. Soc. Microbiol.*, 237 (1972).
[21] Blattner, F. R., Boettiger, J. K., Fiandt, M., Lozeron, H. A., Funderburgh, M. L., Dahlberg, J. E., and Szybalski, W., *Fed. Proc.*, **31**, 472, Abstr. (1972).
[22] Gottesman, M. E., and Weisberg, R. A., in *The Bacteriophage Lambda* (edit. by Hershey, A. D.), 113 (Cold Spring Harbor Laboratory, New York, 1971).
[23] Jacob, F., and Wollman, E. L., *Ann. Inst. Pasteur*, **87**, 653 (1954).
[24] Ptashne, M., in *The Bacteriophage Lambda* (edit. by Hershey, A. D.), 221 (Cold Spring Harbor Laboratory, New York, 1971).
[25] Nijkamp, H. J. J., Bøvre, K., and Szybalski, W., *J. Mol. Biol.*, **54**, 599 (1970).
[26] Kaiser, A. D., and Jacob, F., *Virology*, **4**, 504 (1957).
[27] Kumar, S., and Szybalski, W., *Virology*, **47**, 665 (1970).
[28] Chadwick, P., Pirrotta, V., Steinberg, R., Hopkins, N., and Ptashne, M., *Cold Spring Harbor Symp. Quant. Biol.*, **35**, 283 (1970).
[29] Steinberg, R. A., and Ptashne, M., *Nature New Biology*, **230**, 76 (1971).
[30] Simon, M. N., Davis, R. W., and Davidson, N., in *The Bacteriophage Lambda* (edit. by Hershey, A. D.), 313 (Cold Spring Harbor Laboratory, New York, 1971).
[31] Westmoreland, B. C., Szybalski, W., and Ris, H., *Science*, **163**, 1343 (1969).
[32] Pirrotta, V., and Ptashne, M., *Nature*, **222**, 541 (1969).
[33] Pereira da Silva, L., and Jacob, F., *Ann. Inst. Pasteur*, **115**, 145 (1968).
[34] Chamberlin, M., and Berg, P., *Proc. US Nat. Acad. Sci.*, **48**, 81 (1962).
[35] Burgess, R. R., Travers, A. A., Dunn, J. J., and Bautz, E. K. F., *Nature*, **221**, 43 (1969).
[36] Symons, R. H., *Biochim. Biophys. Acta*, **190**, 548 (1969).
[37] Nygaard, A. P., and Hall, B. D., *J. Mol. Biol.*, **9**, 125 (1964).
[38] Bøvre, K., Lozeron, H. A., and Szybalski, W., in *Methods in Virology* (edit. by Maramorosch, K., and Koprowski, H.), **5**, 271 (Academic Press, New York, 1971).
[39] Fiandt, M., Hradecna, Z., Lozeron, H. A., and Szybalski, W., in *The Bacteriophage Lambda* (edit. by Hershey, A. D.), 329 (Cold Spring Harbor Laboratory, New York, 1971).
[40] Court, D., and Sato, K., *Virology*, **39**, 348 (1969).
[41] Kayajanian, G., *Virology*, **41**, 170 (1970).
[42] Kayajanian, G., *Virology*, **36**, 30 (1968).
[43] Guha, A., Saturen, Y., and Szybalski, W., *J. Mol. Biol.*, **56**, 53 (1971).
[44] Stevens, W. F., Adhya, S., and Szybalski, W., in *The Bacteriophage Lambda* (edit. by Hershey, A. D.), 515 (Cold Spring Harbor Laboratory, New York, 1971).
[45] Ghysen, A., and Pirono, M., *J. Mol. Biol.* (in the press).
[46] Blattner, F. R., Boettiger, J. K., and Dahlberg, J. E., *Fed. Proc.*, **30**, 1315, Abstr. (1971).
[47] Schnös, M., and Inman, R. B., *J. Mol. Biol.*, **55**, 316 (1970).

Distance from a Promoter Mutation to an RNA Synthesis Startpoint on Bacteriophage λ DNA

FREDERICK R. BLATTNER, JAMES E. DAHLBERG, JULIE K. BOETTIGER, MICHAEL FIANDT & WACLAW SZYBALSKI

McArdle Laboratory and Department of Physiological Chemistry, University of Wisconsin, Madison, Wisconsin 53706

Sites of promoter and operator mutations are separated from the RNA synthesis startpoint by approximately 200 nucleotide pairs. Several models could explain how RNA polymerase can interact with such widely separated sites without transcribing the intervening region into RNA.

ONE of the chief conclusions of the preceding article[1] was that the promoter–operator region of the major leftward operon is not transcribed into RNA[2]. In this connexion, the promoter site p_L was defined by the position of known mutations, such as $sex1$, which are located upstream from the first protein-coding sequence and which modify the rate of transcription initiation. Operator site o_L was defined by mutations, such as $v2$, which affect the binding of repressor to DNA. A new transcription element, denoted startpoint s_L, was recognized and defined as the nucleotide pair in DNA at which the RNA chain starts.

The purpose of this article is to determine the physical

distance between the startpoint and the mutation sites defining the promoter–operator region. To do this we used several deletion–substitution λ mutants which cut within or near the s_L–$p_L o_L$ interval (Fig. 2 of the preceding article and Figs. 1–3). This interval was divided into three segments, each of which was measured by somewhat different combinations of physical and genetic methods.

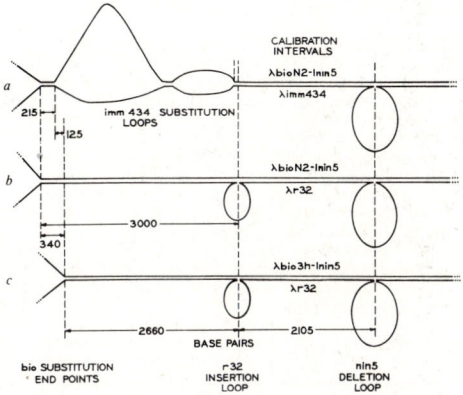

Fig. 1 Electron micrographic measurements of *bio* insertion endpoints relative to the left end of *imm*434 substitution. Heteroduplexes as shown were prepared by the technique of Westmoreland *et al.*[4] and examined in the electron microscope. Double-stranded regions are shown to scale. In determining the positions of the *bio* endpoints, the interval between *r*32 insertion loop[42] and the *nin*5 deletion loop[43,44] served for calibration. The ratios between the *bio*–*r*32 intervals and this calibration interval were measured on molecules containing both intervals intact (*b*: 28 molecules, and *c*: 16 molecules). This ratio is insensitive to those sources of variation which tend to affect molecules as a whole such as magnification fluctuations. Thus, the distance between the *bio* endpoints could be determined with high precision in terms of the difference of the ratios multiplied by the length of the calibration interval. The length of this *r*32–*nin*5 interval was measured in 103 molecules by Dr E. H. Szybalski and found to be 4.53% of λ length (2,105 base pairs), as compared with the left arm of the λ DNA (*m-A-J-att* interval) corresponding to 57.3% of λ length[4]. The distance from *bio*N2-1 to the *imm*434 substitution was measured on 14 molecules relative to a slightly longer calibration interval (2,220 base pairs) existing between the right end of the *imm*434 substitution and the *nin*5 loop (*a*). Since the *bio*N2-1–*imm*434 distance is so short (215 base pairs), this measurement is inherently accurate. Six heteroduplexes between λdv and λ*imm*434 (not shown here) were measured in a similar manner giving 120 nucleotide pairs for the overlap (see Fig. 3). The mean size of each interval is shown in base pairs on the figure. This assumes that 1% of λ DNA length corresponds to 465 nucleotide pairs[45]. The measurements have been analysed statistically and the error flags are presented in Fig. 3. Strains employed in this study are described in the legend to Fig. 2 in the preceding article[1].

We are indebted to Dr. P. Brachet for λr32.

(i) Segment s_L–*imm*434 was measured by subtracting the distance between s_L and the λdv left endpoint, determined by RNA sequence data in the preceding paper[1], from the distance between λdv and *imm*434 determined by electron microscopy (Figs. 1 and 3). (ii) Segment *imm*434–*bio*3h-1 was determined by electron micrographic measurements on appropriate heteroduplexes (Fig. 1). (iii) Segment *bio*3h-1–$p_L o_L$ was measured by recombinational analysis employing a physically known distance between two nearby deletion endpoints as a calibration standard.

The total length of the s_L–$p_L o_L$ interval is the sum of the three segments and measures about 200 nucleotide pairs. This corresponds to several diameters of the RNA polymerase molecule. This unexpected finding will be considered in relation to several models. A preliminary account of this study has appeared elsewhere[3].

Electron Micrographic Mapping

Since the electron micrograph measurements enter into size determination of each of the segments in the s_L–$p_L o_L$ interval (Fig. 3), they are discussed in the first section of this article. The principle of the electron micrographic mapping, based on the work of Westmoreland *et al.*[4], is illustrated in Fig. 1 and explained in the legend to this figure.

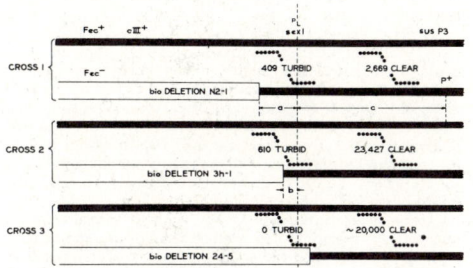

Fig. 2 Mapping of the p_L promoter mutant *sex*1. Three λ*bio* deletion strains, obtained from Dr G. Kayajanian, whose endpoints have been precisely measured in the electron microscope and which are drawn to scale, were each crossed with the lysogen W3102(λ*sex*1*cI*857*sus*P3), obtained from Dr M. Gottesman. For each cross, the cells growing exponentially at 30°C were concentrated and infected at multiplicities of 3 with one of the plaque-forming λ*bionin*5 phages and a recombinational helper λ*imm*21*sus*P3. Without this helper no recombination was observed due to the Red⁻ Gam⁺ phenotype of λ*sex*1, which shuts off the host recombination system[46], and the Red⁻ phenotypes of the *bio* strains. After 90 min of growth at 37°C, the crosses were plated at 30°C on lysogenic strain N100*rec*A⁻ *su*⁻(λ*imm*21). On this strain neither the *bio* parent (Fec⁻)[47], nor the *sex* parent (*susP*), nor the helper (*imm*21) can plate. The P⁺ Fec⁺ recombinants, however, do plate, and these are of two types: *sex*⁺ which is turbid, and *sex*⁻ which is clear due to the deficiency of cIII product, whose synthesis requires an active p_L promoter. It should be noted that λ*sex*1 is genotypically cIII⁺ Fec⁺, although phenotypically it is cIII⁻ Fec⁻. The numbers of turbid and clear recombinants scored in each cross are shown. All crosses were also plated on the permissive strain C600 and on C600(λ*imm*21) to assess the level of recombination. In each cross recombinants were found at about 0.1% of the *bio* parent output. Control crosses were also performed, with each *bio* phage alone, with helper alone, and with bacteria alone.

The distance between the *bio*N2-1 endpoint and the left end of *imm*434 was determined by a direct measurement as shown in Fig. 1*a* and found to be about 215 ± 10 base pairs. The distance between the λdv endpoint and *imm*434 was determined in an analogous way using λdv/λ*imm*434 heteroduplexes (not shown in Fig. 1). This distance corresponded to 120 ± 20 nucleotide pairs.

The distance between the left end of *imm*434 and the *bio*3h-1 cut (segment ii) was calculated from a series of direct measurements illustrated in Fig. 1. Although this calculation involves subtraction of two relatively large numbers, the precision of the measurements is quite high because each measured molecule contained its own calibration interval (Fig. 1). First, the distance between *bio*N2-1 and *bio*3h-1 was calculated by subtraction of the intervals between the *r*32 insertion and the two *bio* endpoints (Fig. 1*b* and *c*). This distance was found to be 340 ± 70 nucleotide pairs. From this, the $bioN2\text{-}1\text{-}imm434$ interval was subtracted leaving a distance of 125 ± 75 nucleotide pairs for the $imm434\text{-}bio3h\text{-}1$ segment. These results constitute a part of the predoctoral work of M. F.

Mapping of Promoter and Operator Mutations

The position of the $sex1$ mutation site relative to the $bio3h\text{-}1$ endpoint (segment iii) was determined by a somewhat novel mapping method combining genetic and physical techniques, as illustrated in Fig. 2. Three plaque-forming λbio deletion phage strains were crossed with the lysogen W3102($\lambda sex1\text{-}cI857susP3$). The bio endpoints of all three λbio strains have been precisely measured by electron microscopy (Figs. 1–3) and the distance between $bioN2\text{-}1$ and $bio3h\text{-}1$ served as a physical calibration standard. The details of the cross and the peculiarities of $bio \times sex1$ recombination are specified in the legend to Fig. 2.

As Fig. 2 shows, no turbid recombinants were observed in cross 3, although they were found in cross 2. This indicates qualitatively that the $sex1$ mutation maps between $bio3h\text{-}1$ and $bio24\text{-}5$. A quantitative measurement of how far to the right of $bio3h\text{-}1$ the $sex1$ mutation maps was obtained by combining electron micrographic and genetic data. In cross 1, the map interval in which recombination gives rise to turbid plaques (interval a in Fig. 2) is longer than the corresponding interval b for cross 2. The ratio of genetic map distances a/b was determined by dividing the ratio of turbid to clear plaques in cross 1 by the analogous ratio in cross 2. This measurement is unaffected by variations in overall recombination rates between the two crosses, since the common denominator, interval c, is the same length in both crosses. From the data in Fig. 2, $a/b = 5.89 \pm 0.39$. The 90% confidence limits are based on plaque counting error. This was combined with the electron micrographically measured difference $(a-b) = 340 \pm 70$ nucleotide pairs (see Figs. 1 and 3). The relevant formulae are

$$b = \frac{D}{R-1} \text{ and } a = \frac{D}{1-1/R}$$

where D is the difference $a-b$ measured electron micrographically and R is the ratio a/b measured genetically. Thus $a = 410 \pm 90$ and $b = 70 \pm 15$ nucleotide pairs. The 90% confidence limits in this case include both genetic and electron micrographic sources of error.

The quantitative accuracy of this method of mapping is based on several assumptions, one of which is that the bio substitutions do not interfere with recombination occurring very near to the bio endpoint. But if there is such an interference[5,6] and it affects each bio endpoint similarly, our technique underestimates the size of the $bio3h\text{-}1\text{-}p_L$ interval. Thus, the $s_L\text{-}p_L$ interval would be even greater than our measurements indicate. On the other hand, if such interference affects each bio endpoint differently, or if recombinational hot spots are present, the results might be somewhat influenced in either direction, but only within the limits of the qualitative results of the deletion mapping.

Using analogous genetic crosses between the three bio deletion phages and $\lambda v2$, Kayajanian[7] found that the $v2$ operator mutation also maps between the $bio3h\text{-}1$ and $bio24\text{-}5$ endpoints. We have confirmed and extended this observation. Quantitatively, the $v2$ mutation maps within 75 nucleotides of the $sex1$ mutation. A detailed account of the mapping of p_L and o_L mutations using the present technique developed by F. R. B. and J. K. B. will be published later.

Size of the $s_L\text{-}p_L o_L$ Interval

The $s_L\text{-}p_L o_L$ interval is composed of three segments. Segment $s_L\text{-}imm434$ is 0 to 20 nucleotide pairs long, as discussed in the preceding paper[1]. This was calculated by subtracting the chain length of the dv portion of p_L RNA (120 nucleotides) from the electron micrographic distance between λdv and $imm434$ (120 ± 20 nucleotides) and excluding the possibility that s_L is to the right of $imm434$. Segments $imm434\text{-}bio3h\text{-}1$ and $bio3h\text{-}1\text{-}sex1$ contain 125 ± 75 and 70 ± 15 nucleotides, respectively, so the $s_L\text{-}sex1$ interval is at least 195 ± 80 nucleotide pairs. Thus, the length of the interval (680 ± 270 Å) which is not transcribed into RNA corresponds to several diameters of RNA polymerase if the latter is a globular protein with a diameter of 75–105 Å[8,9]. These results are summarized in Fig. 3.

Models of Transcription Initiation

The present study in conjunction with the preceding article[1] indicates that two well separated regions of DNA, namely the $sex1$ mutation site and the RNA startpoint s_L, both interact with RNA polymerase during initiation of p_L RNA synthesis. Since correct starting and the effect of the $sex1$ mutation are seen both *in vivo* and *in vitro*, we conclude that the purified polymerase system contains all components that are necessary for these interactions. In other words, both interactions probably occur directly between DNA and RNA polymerase.

Since the separation between $sex1$ and s_L measures several polymerase diameters, one might suppose that either (1) the DNA is condensed in some way to bring the sites closer together, or (2) the polymerase drifts from one site to the other so that interaction at the two separated sites would be sequential rather than simultaneous.

On the first model, DNA might be condensed by formation of symmetrical base-paired hairpin branches analogous to those proposed by Gierer[10]. The branches might be stabilized by binding of polymerase, whereas linear unbranched DNA might bind repressor. This model would require that the sequence of the $sex1\text{-}s_L$ DNA be able to form specific branches. An alternative form of the condensation model assumes that the intervening DNA is simply looped out. This is somewhat analogous to a model postulated for replication of phage Qβ RNA[11,12]. The sequence of the loop region need not be capable of forming any special secondary structures, and the region could be of any length. Branch or loop formation could occur before, during or after polymerase binding. If it occurred after binding, this model could formally be considered as a special case of the "drift model", which will be discussed below.

The Drift Model

In the second or drift model, we propose that the beginning of a transcriptional unit comprises a polymerase entry region, a drift region, a start site and a start point, and that RNA polymerase can drift from one end to the other without producing RNA chains. The "entry region"* might correspond to (a) a single polymerase entry site, (b) a series of such sites or (c) an extended region of DNA which binds RNA polymerase rather non-specifically. The latter two arrangements can be called "antennae" because they might serve as collectors of polymerase molecules. The "drift region" is defined as the area between entry site(s) and start site, where DNA-bound polymerase could move without producing RNA chains. As will be discussed below, polymerase drift might also occur in other regions of DNA. The "start site" is defined as a nucleotide sequence which triggers RNA synthesis at the "startpoint" nucleotide. If such a specific sequence exists, it is likely to be just upstream from the startpoint since the 5′ terminal sequences are different for each of the four λ RNAs[1]. However, the possibility of self-complementary sequences flanking the startpoint should also be considered.

RNA polymerase drift may have several underlying mechanisms. (i) The enzyme may move from one site to the other

---

* We use the term "entry region" to denote a site where a functional interaction between DNA and polymerase first occurs. We consider this term to be more precise than the alternative term, "binding site"[13,14], since the latter might be construed as any region of DNA where bound polymerase can be found.

in a process that is essentially diffusional, perhaps with a barrier to prevent upstream drift or with a gradient of affinity leading the enzyme toward downstream sequences. (ii) Alternatively, there may be migration involving energy release, which we call "pseudo-transcription", in which nucleoside triphosphates are hydrolysed but not polymerized. (iii) A third possibility, in which the RNA chain in the p_L–s_L region is degraded immediately after synthesis, is unlikely, since the p_L RNA chain starts with a triphosphate in the *in vitro* enzymatic system[1].

example, they were to block drift. Furthermore, if an operator site were located in the drift region, promoter and operator mutations could be interspersed or coincident, as has been observed[20-24].

Extensions of the Drift Model

The drift model provides a framework for visualizing some transcriptional controls. For example, if an operator were located in the drift region, as discussed above, bound repressor would prevent initiation of RNA chains by blocking access to

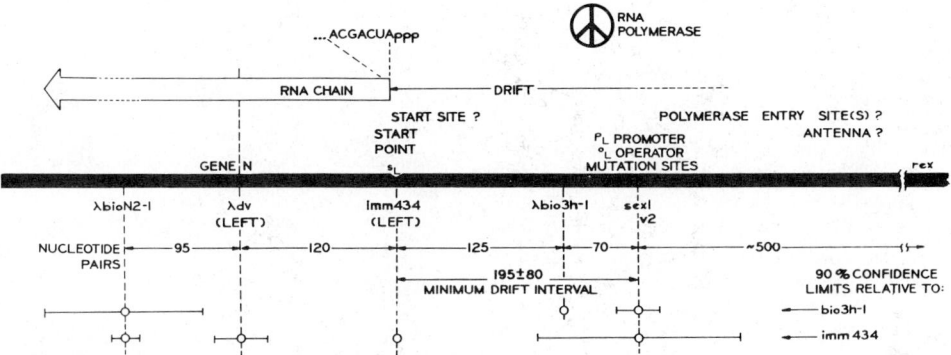

Fig. 3 Physical map of the beginning of leftward operon of λ including hypothetical entities involved in the initiation of transcription. The measurements of Fig. 1 and Fig. 2 are combined to provide a physical and genetic map of this region. Error flags representing 90% confidence limits for these measurements were calculated relative to *imm*434 endpoint (bottom line) and *bio*3h-1 endpoint (second line from the bottom) from the random fluctuations in the genetic and electron micrographic data, using t distribution. According to the "drift model", RNA polymerase attaches to the DNA and drifts leftward for at least 200 nucleotide pairs before reaching the RNA startpoint, s_L defined by the deoxynucleotide pair coding for the 5′ terminal nucleoside triphosphate in the RNA chain. The startpoint s_L is located less than 20 nucleotides to the left of the *imm*434 substitution. A hypothetical start site, comprising a region of the chromosome near the startpoint, triggers the polymerase to start RNA synthesis. The *sex*1 mutation may lie within a specific polymerase entry site. Alternatively, entry may occur at a site further upstream or on an extended "antenna", in which case the *sex*1 mutation might function by blocking the drift of the RNA polymerase. The open arrow indicates the location and orientation of mRNA which codes for gene *N*. The 5′ terminal nucleotide sequence is specified. The size of RNA polymerase, as drawn, corresponds to a diameter of about 100 Å, with the sectors representing the respective molecular weights of the β, β′ and α subunits.

One prediction of the drift model is that a period of time would elapse while a polymerase moved from entry site to startpoint. The kinetic experiment using different conditions of preincubation (Fig. 1 of the preceding article[1]) demonstrates that nucleoside triphosphates but not magnesium ions are required for a period of time before RNA synthesis actually starts. This time might be correlated with polymerase drift, whereas the triphosphate requirement could indicate a need for energy. Alternatively, triphosphates and time might be necessary to induce a change in the polymerase or secondarily in DNA. Sigma factor[15] could conceivably act at any or all of the sites that we have discussed, although its release might occur subsequently[16].

Another prediction of the drift model is that a number of polymerase molecules might line up in queue in the drift region. This could explain observations that substantially more polymerase molecules can bind to DNA than would be predicted from the number of startpoints[14,17-19].

Each of the models we proposed predicts that "promoter mutations" should be found at several sites of polymerase-DNA recognition, such as the entry site and start site. The models also raise the possibility that some parts of the promoter region, such as the drift region (Fig. 3), might be dispensable, for example, by deletion. Nevertheless, point mutations in such dispensable regions might inactivate promoter function if, for

the start site. This region could also accommodate various positive control sites including those which might interact with specific proteins, like CAP (CRP), psi or sigma-like factors[25,26], possibly to overcome pre-existing drift barriers. In fact, several positive or negative control sites might be clustered in a modular arrangement. In this way, the various subregions would be free to evolve and function independently. In addition, a single entry site feeding two start sites might explain some cases of divergent patterns of transcription[23,24].

The idea of polymerase drift may also have application to areas other than promoter function. Szybalski *et al.*[24] have proposed the "dry run" in which polymerase molecules could move through a structural gene without net RNA synthesis. That provides a mechanism for obtaining higher transcription in promoter-distal regions of operons. Further control possibilities can also be imagined if RNA polymerase were to drift through a region of DNA between a termination signal, where one RNA chain would be released, and a following start site, where a new RNA chain could start without necessity for a new polymerase entry region. Operator-like sites in such a drift area could be targets of termination and anti-termination factors[27,28].

Miller *et al.*[9,29] have observed what seem to be RNA polymerase molecules on the spacer DNA between repeated regions coding for the ribosomal RNA precursor in *Xenopus*.

This may indicate that intergenic drift occurs in higher organisms.

Another interesting application of this model might be in the synthesis of a single polypeptide chain from two distant genes, as in the case of antibodies containing a variable and a constant segment[30-32]. The polymerase might start by transcribing a variable segment. If the polymerization were terminated but the nascent RNA chain were not released at the site marking the end of this segment, the enzyme carrying this unfinished chain might drift to a restart point, corresponding to the beginning of the constant region, where this chain would be further extended.

What is a Promoter?

A promoter was originally described as a region of DNA which is "an indispensable initiator element"[33] and subsequently as a "site which serves to initiate transcription of an operon"[34]. Operationally it could be defined by *cis*-acting mutations located upstream from the first protein-coding sequence and affecting both the *in vivo* and *in vitro* rates of transcription initiation. Promoters have often been visualized as a small polymerase binding and initiation site. This picture, however, seems to be inconsistent with the large minimum size of this region we have measured here in relation to the RNA polymerase diameter. Actually, a large promoter–operator region with modular construction might help to explain the complex controls postulated in some well studied operons[24,26,34-38], as well as the series of steps by which RNA polymerase initiates transcription[39,40].

If transcriptional initiation requires various subregions of DNA, possibly interspersed by control sites and dispensable sequences, a question arises as to how the term promoter should be applied. For the present it might be most appropriate to apply the term "promoter region" for the extended meaning of promoter which includes the polymerase entry, drift and start sites. If this region also includes the operator site, as might be the case in the p_L operon of λ, the term promoter region would be equivalent to the original dual meaning of the word "operator" as defined in 1961[41]. Assignment of the term promoter to one specific site, perhaps corresponding to the very beginning of the transcriptional unit or some other crucial site of transcription initiation, should await the time when a large number of promoter-type mutations have been isolated and thoroughly studied.

In summary, we have found that at the beginning of the major leftward λ operon there exists a region which is not transcribed into RNA, although RNA polymerase must traverse it. This region commences with the area characterized by the *sex*1 promoter mutation, which represents some site of interaction between polymerase and the λ DNA template, possibly includes the o_L operator site, and ends with the start site which triggers polymerization of RNA chains at the s_L startpoint. The length of this nontranscribed region is at least 195 ± 80 nucleotide pairs, which corresponds to several diameters of RNA polymerase. To interpret the initiation of transcription several possible mechanisms were explored. In the "drift" model, polymerase moves along the *sex*1–s_L region with or without energy input, but no RNA chains are synthesized. In the "DNA condensation" model the promoter mutation site and startpoint are brought physically together. The data presented could be interpreted in terms of both models.

We thank P. Twose and Dr E. H. Szybalski for their help, and H. Echols, M. Gottesman and G. Kayajanian for discussion. This work was supported by grants from the National Institutes of Health and National Science Foundation to J. E. D. and W. S.; J. E. D. holds an NIH research career development award.

Received April 21; revised May 17, 1972.

[1] Blattner, F. R., and Dahlberg, J. E., *Nature New Biology*, **237**, 227 (1972).
[2] Blattner, F. R., Boettiger, J. K., and Dahlberg, J. E., *Fed. Proc.*, **30**, 1315, Abstr. (1971).
[3] Blattner, F. R., Boettiger, J. K., Fiandt, M., Lozeron, H. A., Funderburgh, M. L., Dahlberg, J. E., and Szybalski, W., *Fed. Proc.*, **31**, 472, Abstr. (1972).
[4] Westmoreland, B. C., Szybalski, W., and Ris, H., *Science*, **163**, 1343 (1969).
[5] Bode, W., *Z. Vererbungslehre*, **94**, 190 (1963).
[6] Matsubara, K., and Kaiser, A. D., *Cold Spring Harbor Symp. Quant. Biol.*, **33**, 769 (1968).
[7] Kayajanian, G., *Virology*, **41**, 170 (1970).
[8] Lubin, M., *J. Mol. Biol.*, **39**, 219 (1969).
[9] Miller, jun., O. L., Hamkalo, B. A., and Thomas, jun., C. A., *Science*, **169**, 392 (1970).
[10] Gierer, A., *Nature*, **212**, 1480 (1966).
[11] Haruna, I., and Spiegelman, S., *Proc. US Nat. Acad. Sci.*, **54**, 189 (1965).
[12] Kolakofsky, D., and Weissmann, C., *Nature New Biology*, **231**, 42 (1971).
[13] Le Talaer, J.-Y., and Jeanteur, P., *Proc. US Nat. Acad. Sci.*, **68**, 3211 (1971).
[14] Blattner, F. R., thesis, Johns Hopkins Univ. (1968).
[15] Burgess, R. R., Travers, A. A., Dunn, J. J., and Bautz, E. K. F., *Nature*, **221**, 43 (1969).
[16] Krakow, J. S., and von der Helm, K., *Cold Spring Harbor Symp. Quant. Biol.*, **35**, 73 (1970).
[17] Hinkle, D. C., and Chamberlin, M., *Cold Spring Harbor Symp. Quant. Biol.*, **35**, 73 (1970).
[18] Ihler, G., thesis, Harvard Univ. (1967).
[19] Richardson, J. P., *J. Mol. Biol.*, **21**, 83 (1966).
[20] Smith, T. F., and Sadler, J. R., *J. Mol. Biol.*, **59**, 273 (1971).
[21] Ordal, G. W., in *The Bacteriophage Lambda* (edit. by Hershey, A. D.), 565 (Cold Spring Harbor Laboratory, New York, 1971).
[22] Fankhauser, D. B., Ely, B., and Hartman, P. E., *Genetics*, **68**, 518 (1971).
[23] Guha, A., Saturen, Y., and Szybalski, W., *J. Mol. Biol.*, **56**, 53 (1970).
[24] Szybalski, W., Bøvre, K., Fiandt, M., Hayes, M., Hradecna, Z., Kumar, S., Lozeron, H. A., Nijkamp, H. J. J., and Stevens, W. F., *Cold Spring Harbor Symp. Quant. Biol.*, **35**, 341 (1970).
[25] Burgess, R. R., *Ann. Rev. Biochem.*, **40**, 711 (1971).
[26] Losick, R., *Ann. Rev. Biochem.*, **41** (in the press).
[27] Roberts, J. W., *Cold Spring Harbor Symp. Quant. Biol.*, **35**, 121 (1970).
[28] Szybalski, W., Bøvre, K., Fiandt, M., Guha, A., Hradecna, Z., Kumar, S., Lozeron, H. A., Maher, V. M., Nijkamp, H. J. J., Summers, W. C., and Taylor, K., *J. Cell. Physiol.*, **74**, Suppl. 1, 33 (1969).
[29] Miller, O. L., Beatty, B. R., Hamkalo, B. A., and Thomas, jun., C. A., *Cold Spring Harbor Symp. Quant. Biol.*, **35**, 505 (1970).
[30] Hood, L. E., *Fed. Proc.*, **31**, 177 (1972).
[31] Smith, G. P., Hood, L., and Fitch, W. M., *Ann. Rev. Biochem.*, **40**, 969 (1971).
[32] Edelman, G. M., and Gall, W. E., *Ann. Rev. Biochem.*, **38**, 415 (1969).
[33] Jacob, F., Ullman, A., and Monod, J., *CR Acad. Sci.*, **258**, 3125 (1964).
[34] Epstein, W., and Beckwith, J. R., *Ann. Rev. Biochem.*, **34**, 411 (1968).
[35] Reznikoff, W. S., *Ann. Rev. Genet.*, **6** (in the press).
[36] Brenner, M., and Ames, B. N., in *Metabolic Regulation, Metabolic Pathways* (edit. by Vogel, H. J.), **5**, 349 (Academic Press, New York, 1971).
[37] Englesberg, E., in *Metabolic Regulation, Metabolic Pathways* (edit. by Vogel, H. J.), **5**, 257 (Academic Press, New York, 1971).
[38] Hiraga, S., *J. Mol. Biol.*, **39**, 159 (1969).
[39] Anthony, D. D., Zeszotek, E., and Goldthwait, D. A., *Proc. US Nat. Acad. Sci.*, **56**, 1026 (1966).
[40] Zillig, W., Zechel, K., Rabussay, D., Schachner, M., Sethi, V. S., Palm, P., Heil, A., and Seifert, W., *Cold Spring Harbor Symp. Quant. Biol.*, **35**, 47 (1970).
[41] Jacob, F., and Monod, J., *J. Mol. Biol.*, **3**, 318 (1961).
[42] Brachet, P., Eisen, H., and Rambach, A., *Mol. Gen. Genet.*, **108**, 266 (1970).
[43] Court, D., and Sato, K., *Virology*, **39**, 348 (1969).
[44] Fiandt, M., Hradecna, Z., Lozeron, H. A., and Szybalski, W., in *The Bacteriophage Lambda* (edit. by Hershey, A. D.), 329 (Cold Spring Harbor Laboratory, New York, 1971).
[45] Davidson, N., and Szybalski, W., in *The Bacteriophage Lambda* (edit. by Hershey, A. D.), 45 (Cold Spring Harbor Laboratory, New York, 1971).
[46] Ungar, R., Echols, H., and Clark, A. J., *J. Mol. Biol.* (in the press).
[47] Zissler, J., Signer, E., and Schaefer, F., in *The Bacteriophage Lambda* (edit. by Hershey, A. D.), 455 (Cold Spring Harbor Laboratory, New York, 1971).

THE ROLE OF DNA IN RNA SYNTHESIS, IX. NUCLEOSIDE TRIPHOSPHATE TERMINI IN RNA POLYMERASE PRODUCTS*

By Umadas Maitra† and Jerard Hurwitz

DEPARTMENT OF MOLECULAR BIOLOGY, ALBERT EINSTEIN COLLEGE OF MEDICINE, BRONX, NEW YORK

Communicated by A. D. Hershey, July 22, 1965

It has been shown previously that in RNA polymerase reactions primed with a variety of DNA preparations there is incorporation of P^{32} from $\beta\gamma$-labeled ATP into an acid-insoluble product and that triphosphate groups are present at the ends of the RNA chains formed during the reaction.[1] Two schemes for initiation of the chains can thus be envisaged. In one scheme, the initial nucleotide incorporated into RNA would retain its triphosphate end, while the growing end of the molecule would be a nucleoside. In the second scheme, the situation is reversed: The nucleoside end would be the initiation point, and the triphosphate, the growing site of the molecule. It is clear that these two schemes of initiation of RNA synthesis differ specifically in that the first would result in the initial nucleotide retaining the β and γ phosphate groups, whereas in the second scheme the last entering nucleotide would contain the β and γ phosphate group.

In the present communication, evidence will be presented that (1) initiation and subsequent elongation of RNA chains formed by RNA polymerase under the direction of a DNA template occur by a mechanism in which the first nucleotide incorporated into the RNA chain retains its triphosphate moiety, and (2) adenosine and guanosine triphosphate ends are preferentially formed.

Materials and Methods.—γ-P^{32}-GTP and UTP were prepared by photophosphorylation of the corresponding nucleoside diphosphates with $^{32}P_i$ and spinach chloroplasts by a modification of the procedure of Avron.[2] γ-P^{32}-CTP was prepared by the action of nucleoside diphosphokinase[3] on $\beta\gamma$-P^{32}-ATP and CDP in the presence of an excess of myokinase. $\beta\gamma$-P^{32}-ATP was prepared as described by Penefsky and Racker[4] in the presence of excess myokinase. These P^{32}-labeled compounds were purified by chromatography on Dowex-1-Cl$^-$ and were free of P^{32} in the α-phosphate position. The methods of preparation of other materials, including the DNA-dependent RNA polymerase of *Escherichia coli*, have been previously described.[1, 5] Calf thymus DNA was obtained from General Biochemicals.

Enzyme assay: The presence of a triphosphate terminus in the RNA formed in an RNA polymerase reaction was measured by the incorporation of γ-P^{32} ribonucleoside triphosphate into an acid-insoluble product, and RNA synthesis was measured by the incorporation of α-P^{32} or C^{14}-labeled ribonucleoside triphosphate. Reaction mixtures (0.50 ml) contained Tris buffer, pH 8.0, 25 μmoles; 2-mercaptoethanol, 4 μmoles; $MnCl_2$, 0.5 μmole; $MgCl_2$, 2.5 μmoles; DNA, 25 mμmoles; ATP, UTP, GTP, and CTP, 10 mμmoles each, one labeled with P^{32} in the γ-phosphate group (containing 1–2 × 10^9 cpm/μmole) for measurement of triphosphate termini or with C^{14}-ATP or GTP (containing 2–5 × 10^6 cpm/μmole) for measurement of the total amount of RNA formed in the reaction. In either case, the reaction was initiated by the addition of enzyme.[6] After incubation at 37° for the desired time, the reaction mixture was chilled in ice, and 0.1 ml of a bovine plasma albumin solution (5 mg/ml) was added, followed by 0.3 ml of 7% $HClO_4$. The resulting precipitate was centrifuged for 5 min at 15,000 × g, and the pellet dissolved in 0.2 ml of ice-cold 0.2 N NaOH. This was followed by the addition of 0.1 ml of nonradioactive triphosphate (25 μmoles/ml) corresponding to the γ-P^{32}-labeled triphosphate used in the reaction mixture and 5 ml of cold 5% TCA solution containing 0.01 M sodium pyrophosphate. After the reaction mixture had stood in ice for 5 min, the acid-insoluble material was collected by centrifugation, and the washing procedure was repeated two more times. The final pellet was dissolved in 1.5 ml of 0.2 N NH$_4$OH, transferred to an aluminum planchet, and after drying, counted in a windowless gas-flow counter. The high specific radioactivities of the γ-P^{32}-labeled substrates necessitated the washing procedure to obtain consistently low blanks. When the total amount of RNA synthesized from C^{14}-nucleoside triphosphate (specific radioactivity 10^6–10^7 cpm/μmole) was measured, the washing procedure was not necessary, and the acid-insoluble RNA product was isolated by filtration on membrane filters as described previously.[1]

Results.—*Incorporation of γ-P^{32}-labeled nucleoside triphosphates into RNA polymerase products:* RNA polymerase, primed with dAT copolymer, catalyzes the incorporation of $\beta\gamma$-P^{32}-labeled ATP into an acid-insoluble polyribonucleotide product.[1] The incorporation of P^{32} is dependent on the presence of UTP, dAT copolymer, and RNA polymerase. The omission of any of these components, or the addition of RNase or DNase, results in a marked decrease in P^{32}-ATP incorporation. In contrast, similar experiments carried out with γ-P^{32}-UTP do not result in significant incorporation of P^{32}, as shown in Table 1. The low level of incorporation observed is not dependent on the simultaneous presence of ATP.[7]

TABLE 1

REQUIREMENTS FOR THE INCORPORATION OF $\beta\gamma$-P^{32}-LABELED ATP AND γ-P^{32}-UTP IN dAT-PRIMED POLY·rAU SYNTHESIS

Additions	$\beta\gamma$-P^{32}-ATP incorporated ($\mu\mu$moles)	γ-P^{32}-UTP incorporated ($\mu\mu$moles)
1. Complete system	4.40	0.40
2. Omit UTP	0.37	—
3. Omit ATP	—	0.40
4. Omit enzyme	0.13	0.11
5. Omit dAT copolymer	0.29	0.28
6. Complete + DNase (1 μg)	0.30	0.24
7. Complete + RNase (1 μg)	0.30	0.24
8. Complete with C^{14}-ATP in place of $\beta\gamma$-P^{32}-ATP	7200 (poly rAU)	

The conditions of the experiment were as described under *Enzyme Assay*, except that GTP and CTP were omitted and 10 mμmoles of dAT copolymer replaced DNA. $\beta\gamma$-P^{32}-ATP incorporation and γ-P^{32}-UTP incorporation were measured in separate mixtures. In the complete system, the total amount of nucleotide incorporated was calculated from the amount of C^{14}-AMP incorporated into the poly rAU product. Mixtures were incubated for 40 min and contained 3 units of RNA polymerase.

The finding that poly rAU chains contain ATP ends and very few UTP ends prompted an examination of the relative incorporation of P^{32} from each of the four γ-P^{32}-labeled nucleoside triphosphates with different DNA templates of varying base composition. The results, presented in Table 2, show that RNA chains are

TABLE 2

INCORPORATION OF γ-P^{32}-NUCLEOSIDE TRIPHOSPHATES WITH DIFFERENT DNA PREPARATIONS

DNA primer	RNA synthesis ($\mu\mu$moles)	γ-P^{32}-Nucleotide Incorporated ($\mu\mu$moles)			
		ATP	GTP	UTP	CTP
T2	4800	2.40	1.2	0.12	0.10
T5	4000	1.80	1.4	0.41	0.23
SP3	5480	1.25	1.0	0.39	0.12
Cl. perfringens	2800	1.60	2.1	0.28	0.25
E. coli	2660	0.43	1.4	0.13	0.10
M. lysodeikticus	2560	0.36	2.5	0.10	0.12
Calf thymus	3560	0.77	1.3	0.33	0.18
dAT copolymer	rAU = 7200	4.40	—	0.20	—
dGC homopolymer	rG = 1350 rC = 120	—	4.8	—	0.30

The conditions of the assay were as described under *Methods*. In each reaction mixture, only one of the four nucleoside triphosphates was labeled with P^{32} in the γ-phosphate group; the other three were nonradioactive. Incubation was for 40 min at 37° with 2 units of enzyme. Where indicated, 7.5 mμmoles of dGdC homopolymer and 10 mμmoles of dAT copolymer were added. Controls without enzyme and without DNA were included. In these controls incorporation was 0.2–0.3 $\mu\mu$mole of nucleotide, and the higher value was subtracted from the results listed above.

formed predominantly with ATP and GTP ends, whereas few chains are initiated with UTP and CTP. The relative number of triphosphate ends beginning with adenosine or guanosine varied with the DNA used to direct the reaction (Table 2). With various bacterial DNA's and calf thymus DNA, GTP ends predominated, although significant ATP incorporation also was observed. With DNA preparations from phages T2, T5, and SP3 (which have an A + T/G + C ratio > 1), ATP ends occurred slightly more often than GTP ends. Denaturation of DNA had marked effects on both RNA synthesis and the incorporation of P^{32} from γ-P^{32}-labeled nucleoside triphosphates (Table 3). These effects can be summarized as

TABLE 3

Effect of Denaturation of DNA on the Incorporation of γ-P^{32}-Labeled Nucleoside Triphosphates

DNA primer	RNA synthesis ($\mu\mu$moles)	γ-P^{32}-Nucleotide Incorporated ($\mu\mu$moles)			
		ATP	GTP	UTP	CTP
T2	4800	2.3	1.2	0.13	0.10
T2 heat-denatured	1000	3.6	5.1	0.88	0.40
T2 alkali-denatured	1050	3.5	4.7	0.91	0.45
Calf thymus	5700	1.0	1.8	0.33	0.20
Calf thymus heat-denatured	2000	3.8	10.1	0.82	0.66
E. coli	2000	0.6	1.5	0.13	0.10
E. coli heat-denatured	1300	2.4	8.1	0.56	0.44

The condition of the assay was as described under *Methods*, with 20 mμmoles of each of the various DNA preparations, 2 units of enzyme, and 40 min of incubation at 37°. RNA synthesis was followed by incorporation of both C^{14}-AMP and C^{14}-GMP, and total RNA synthesis was calculated by multiplying the sum of these values by two.

follows: (1) RNA synthesis was markedly inhibited, (2) the incorporation of all four γ-P^{32} nucleoside triphosphates increased severalfold, and (3) there was an increase in the ratio of GTP to ATP termini as well as a significant though small number of RNA chains containing UTP and CTP ends.

Identification of guanosine triphosphate ends of RNA formed in the RNA polymerase reaction: As with $\beta\gamma$-P^{32}-labeled ATP,[1] the incorporation of γ-P^{32}-GTP into an acid-insoluble material had the same requirements found for RNA synthesis.

For identification of the site of GTP incorporation, an RNA product containing γ-P^{32}-GTP was prepared using denatured thymus DNA as template. Subjecting the P^{32}-labeled RNA product to the action of alkaline phosphatase or to acid hydrolysis (1 N HCl for 10 min at 100°) rendered the P^{32} acid-soluble and Norit nonadsorbable. The P^{32} present in the RNA product was not converted to P_i by the action of prostatic phosphomonoesterase.[8] The P^{32} product was insensitive to pancreatic DNase, since it remained acid-insoluble, whereas pancreatic RNase and alkaline hydrolysis released all the P^{32} into an acid-soluble but Norit-adsorbable form. These results indicate that the P^{32} incorporated from γ-P^{32}-GTP was in a terminal portion of the RNA structure, presumably $\overset{*}{p}ppGpXpYpZ$----, and not in an internucleotide link.

The expected products of alkaline hydrolysis of polynucleotides with the structure $\overset{*}{p}ppGpXpYpZ$---- are 2'(3')-nucleoside monophosphates and nucleoside tetraphosphates. Radioactivity from γ-P^{32}-GTP should be found only in guanosine tetraphosphate ($\overset{*}{p}ppGp$). The prediction was tested as follows: An alkaline hydrolysate (0.3 N KOH at 37° for 18 hr) of the labeled product was neutralized with Dowex-50 (H$^+$), and an aliquot of the solution containing 30,000 cpm was mixed with 2 μmoles each of GMP, GDP, GTP, ATP, and guanosine-5'-tetraphosphate.[9] The mixture was added to a column (1 × 12 cm) of Dowex-1-Cl$^-$ (100-

200 mesh, 2% cross-linked), and the nucleotides were eluted as follows: (a) 150 ml of 0.01 M HCl + 0.05 M LiCl (GMP); (b) 150 ml of 0.01 M HCl + 0.1 M LiCl (GDP followed by ATP); (c) 150 ml of 0.01 M HCl + 0.2 M LiCl (GTP); (d) 150 ml of 0.05 M HCl + 0.2 M LiCl, which eluted guanosine 5'-tetraphosphate. The elution profile and identification of each of the nucleotides were determined by measuring the optical densities of the effluents at 260 and 280 mμ. More than 90 per cent of the P^{32} added to the column was eluted as a sharp symmetrical peak in the last solvent with guanosine 5'-tetraphosphate, whereas 8 per cent of the added radioactivity was eluted in the GTP region. No P^{32} was detected in the other regions of the chromatogram. These results are consistent with the presence of $\overset{*}{p}ppGp$. To characterize further the P^{32}-labeled product in the alkaline hydrolysate, another aliquot of the alkaline hydrolysate (containing 30,000 cpm) was incubated with 3.5 units[10] of prostatic phosphomonoesterase at pH 5.0 at 37° for 30 min. The mixture was then chromatographed on Dowex-1-Cl$^-$ under the conditions described above. Approximately 75 per cent of the added radioactivity now chromatographed with GTP and the remainder with guanosine 5'-tetraphosphate. These results are consistent with the structure $\overset{*}{p}ppGpXpYpZ$----, i.e., γ-P^{32}-GTP is incorporated as such at the end of RNA chains.

Kinetics of nucleoside triphosphate incorporation: The rates of RNA synthesis and $\beta\gamma$-P^{32}-ATP incorporation were compared (Table 4). Whereas T2 DNA-directed

TABLE 4

Kinetics of RNA Synthesis versus $\beta\gamma$-P^{32}-ATP Incorporation

Time of incubation (min)	RNA synthesis ($\mu\mu$moles)	$\beta\gamma$-P^{32}-ATP incorporated ($\mu\mu$moles)	Ratio
1	300	1.2	250
2	600	1.6	375
5	1500	2.4	630
10	2700	2.8	960
20	4500	3.2	1410
40	6100	3.7	1695
60	8400	3.7	2270

The assay was performed as described under *Methods*, except that 5 μmoles of MgCl$_2$ replaced MnCl$_2$, 2 units of RNA polymerase and 20 mμmoles of each of the nucleoside triphosphate were added, and the incubation was carried out at 25°. This procedure permitted slow growth of the RNA chains with T2 DNA as primer.

RNA synthesis continued during the entire experiment, 65 per cent of the total amount of $\beta\gamma$-P^{32}-ATP was incorporated by 5 min. The ratio of $\beta\gamma$-P^{32}-ATP ends to total nucleotide incorporation decreased progressively with time from a ratio of 1 ATP terminus per 250 RNA nucleotides during the first minute to 1 ATP terminus per 2270 nucleotides after 60 min.

A comparison of the kinetics of P^{32} incorporation from γ-P^{32}-GTP with native and with heat-denatured T2 DNA as template is summarized in Table 5. With native T2 DNA as template the incorporation of P^{32} was virtually complete in 20 min, although RNA synthesis continued throughout the incubation period. The ratio of GTP ends to RNA formed decreased progressively with time from a ratio of 1 GTP terminus per 1300 nucleotides in the first 5 min to 1 GTP terminus per 5100 nucleotides after 90 min. Under the same conditions, experiments with $\beta\gamma$-P^{32}-ATP indicated the presence of 1 ATP terminus per 600 nucleotides during the first 5 min of incubation and 1 ATP terminus per 3100 nucleotides after 90 min. In contrast, with denatured DNA as template, the incorporation of γ-P^{32}-GTP was severalfold faster and continued during the entire period of RNA synthesis. Similar results were obtained with $\beta\gamma$-P^{32}-ATP. In this case, with denatured DNA as template, the ratio of ATP ends to RNA formed was lower than that found with GTP (1/160 after 5 min of incubation and 1/540 after 60 min), since with denatured DNA more γ-P^{32}-GTP than $\beta\gamma$-P^{32}-ATP was incorporated. However, with either of these two triphosphates, the ratio of triphosphate ends formed to RNA synthesized was considerably smaller than that obtained with the corresponding native DNA as template. The ratios obtained in these experiments can be used as a measure of the length of the RNA product formed. Wood and Berg[11] found that RNA products formed with denatured DNA as template were smaller than those obtained with native DNA. The results summarized in Table 5 are in agreement with their findings.

TABLE 5

Kinetics of RNA Synthesis and γ-P^{32}-GTP Incorporation with Native and Denatured T2 DNA as Templates

Experiment no.	Time of incubation (min)	RNA synthesized ($\mu\mu$moles)	γ-P^{32}-GTP incorporated ($\mu\mu$moles)	Ratio
I	5	1500	1.1	1360
	10	2700	1.4	1800
	20	5000	1.7	2900
	40	6200	1.9	3200
	60	8300	1.9	4150
	90	10400	1.9	5200
II	5	150	3.2	47
	10	300	4.8	62
	20	600	6.6	90
	40	1140	7.7	143
	60	1500	8.1	180
	90	2100	10.1	210

The conditions of the experiment were as described under *Methods*, except that 20 mμmoles each of γ-P^{32}-GTP, UTP, CTP, and ATP were added. Twenty-four mμmoles of native T2 DNA (approximate size = 35S) were added in experiment I, and an equimolar amount of heat-denatured T2 DNA was added in experiment II. Incubation was at 37°.

Direction of growth of RNA chains: In order to determine whether incorporation of the triphosphate terminus in RNA occurs according to the scheme in which the first nucleotide incorporated retains the triphosphate end, or by a mechanism in which the triphosphate moiety is at the growing point of the RNA molecule, the following experiment was performed. γ-P^{32}-GTP was incorporated into RNA for 5 min, and then a large excess of cold GTP was added to reduce the specific activity of the labeled nucleotide. The fate of P^{32} already incorporated at the ends of the RNA chains was then followed during subsequent RNA synthesis. If the initial nucleotide incorporated was present as the triphosphate end, subsequent synthesis should have no effect on the P^{32} already incorporated. In contrast, if the last entering nucleotide existed as the triphosphate terminus, subsequent synthesis should release the previously incorporated P^{32}. The fate of γ-P^{32}-GTP ends under the conditions described above is summarized in Figure 1. As shown, the addition of unlabeled GTP halted γ-P^{32} uptake immediately, and continuing RNA synthesis did not diminish the amount of P^{32} already incorporated.

Proof that the labeled chains actually increased in length after the addition of unlabeled substrates was obtained by the following experiment. RNA synthesis was carried out in the presence of γ-P^{32}-GTP and $\beta\gamma$-P^{32}-ATP. After the reaction

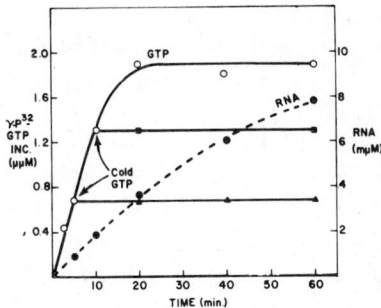

FIG. 1.—Effect of dilution on γ-P^{32}-GTP incorporation. Reaction mixtures were as described under *Methods*, with the exception that 20 mμmoles of each of the four nucleoside triphosphates were added, γ-P^{32}-GTP (1.1 × 10^9 cpm/μmole) was used, and incubation was carried out at 37°. In two separate reaction mixtures, one after 5 min and the other after 10 min, a 30-fold excess of unlabeled GTP was added. RNA synthesis was measured in separate reaction mixtures in which C^{14}-GTP was used as the only radioactive substrate, with all other additions as above. ▲, Incorporation of γ-P^{32}-GTP in the reaction mixture diluted with unlabeled GTP after 5 min; ■, incorporation in the reaction mixture diluted after 10 min.

had been allowed to proceed for 4 min, a large excess of unlabeled GTP and ATP was added. One sample was removed at this time, and a second after an additional 8 min of incubation. The sedimentation of the labeled RNA is illustrated in Figure 2, which shows that the sedimentation rate increased from about 6S to about 20S after the addition of unlabeled substrates. An increase in size of RNA products with time was also noted by Bremer and Konrad.[12]

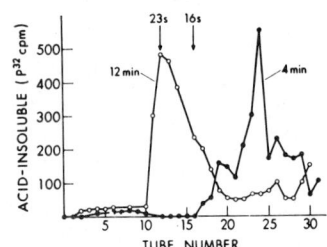

FIG. 2.—Zone sedimentation analysis of RNA. Two 0.5-ml reaction mixtures containing Tris buffer, pH 8.0, 25 μmoles; 2-mercaptoethanol, 4 μmoles; MgCl$_2$, 5 μmoles; ADP, 5 mμmoles; T5 DNA, 25 mμmoles; UTP and CTP, 25 mμmoles each; $\beta\gamma$-P^{32}-ATP and γ-P^{32}-GTP, specific activity 1 × 10^9 cpm/μmole, 25 mμmoles each; and 3 units of enzyme were incubated for 4 min at 25°. After that time, 1.5 μmoles each of nonradioactive ATP and GTP were added to each tube. One reaction was stopped immediately by the addition of sodium dodecyl sulfate and EDTA (0.5% and 0.01 M final concentration, respectively). The other tube was treated in the same manner after 12 min at 25°. Both samples were diluted to 1 ml with 0.5% sodium dodecyl sulfate, layered on 30 ml of a 15–30% sucrose solution gradient containing 0.05 M Tris buffer, pH 8.0, 0.1 M NaCl, and 0.2% sodium dodecyl sulfate, and centrifuged for 15 hr at 25,000 rpm in an SW 25.1 Spinco rotor at approximately 25°. Ribosomal RNA was used as an optical density marker. Fractions (1 ml) were collected through a hole punched in the bottom of the tube. The fractions were scanned through a Gilford recording spectrophotometer to locate the position of 23S and 16S ribosomal markers and assayed for P^{32}.

These results show that nucleoside triphosphates are incorporated at the point of initiation and not at the growing end of RNA chains, and, therefore, that chain growth occurs by the addition of nucleotides to the 3′-hydroxyl end.

Discussion.—The above results clearly show that RNA chains are initiated with ribonucleoside triphosphates, principally or exclusively ATP and GTP. Upon denaturation of DNA by heat or alkali, the incorporation of nucleoside triphosphates into terminal positions increases. Thus, RNA polymerase finds more and different initiation sites for RNA synthesis on denatured DNA than on double-stranded DNA. This conclusion is also supported by the observation that single-stranded and denatured DNA saturate RNA polymerase more effectively than native DNA.[13, 14]

The finding that the pyrimidine sites of DNA, especially when native, are preferentially utilized as initiation points for RNA synthesis was totally unexpected. In fact, the pyrimidine nucleoside triphosphate ends found in small numbers in the RNA may reflect the presence of small amounts of denatured DNA in all the primers used. The reason for this specificity is unknown, but it probably results from the manner in which the enzyme interacts with DNA. It is probably not due to the selective binding of ATP and GTP versus UTP and CTP to the enzyme, since there is no difference in affinity constant of these nucleotides for RNA polymerase.[5] The selective copying of the pyrimidine-rich strand of the DNA of *Bacillus subtilis* phages SP8 and 2C and *Bacillus megatherium* phage α *in vivo*[15, 16] and *in vitro*[17, 18] may be related to the preferential initiation of RNA chains with purine nucleotides. The selection of one DNA strand over another may thus be governed by runs of pyrimidine bases in native DNA. The loss of asymmetric copying of DNA in RNA synthesis upon denaturation of the DNA[17] may be related to our finding that denaturation uncovers new sites in the DNA at which RNA chains can be started.

In vivo, RNA synthesis (i.e., gene expression) must begin at particular sites on DNA. Since DNA of *E. coli* (and others) is uninterrupted,[19] there must be a high degree of specificity for initiation of RNA chains within the DNA duplex. The results presented above suggest that these sites in DNA may be pyrimidine bases. In accord with this idea is the finding that sRNA molecules contain considerable amounts of guanine and adenine at the 5'-phosphate end.[20, 21] How RNA polymerase can specifically recognize initiation points on DNA, and what factor controls the accessibility of the enzyme to such sites for RNA synthesis, are problems intimately involved in the mechanism of gene control.

Summary.—In the DNA-dependent RNA polymerase reaction, the RNA chains formed contain ribonucleoside triphosphates at their starting points and grow by the subsequent addition of ribonucleotides to the 3'-hydroxyl group of the ribonucleoside end. Purine nucleoside triphosphates are preferentially found at the triphosphate end. When denatured DNA is used as a template, there are an increase in the number and a change in the kind of starting points.

* This research was supported by grants from the National Institutes of Health, the National Science Foundation, and the New York City Public Health Research Council. Paper VIII of this series was concerned with the inhibition of RNA polymerase by histones (Skalka, A., A. Fowler, and J. Hurwitz, *J. Biol. Chem.*, in press). Communication no. 40 of the Joan and Lester Avnet Institute for Molecular Biology.

† Postdoctoral fellow of the Jane Coffin Childs Memorial Fund for Medical Research.

[1] Maitra, U., A. Novogrodsky, D. Baltimore, and J. Hurwitz, *Biochem. Biophys. Res. Commun.*, **18**, 801 (1965).

[2] Avron, M., *Anal. Biochem.*, **2**, 535 (1961).

[3] Berg, P., and W. K. Joklik, *J. Biol. Chem.*, **210**, 657 (1954).

[4] Penefsky, H., and E. Racker, *J. Biol. Chem.*, **235**, 3330 (1960).

[5] Furth, J. J., J. Hurwitz, and M. Anders, *J. Biol. Chem.*, **237**, 2611 (1962).

[6] In all reaction mixtures containing $\beta\gamma$-P^{32}-ATP, 2 mμmoles of ADP per 10 mμmoles of ATP were included to suppress the action of any possible contaminating polyphosphate forming enzyme [Kornberg, A., S. R. Kornberg, and E. S. Simms, *Biochim. Biophys. Acta*, **20**, 235 (1956)].

[7] In other experiments the incorporation of γ-P^{32}-UTP was <0.2 $\mu\mu$moles. Evidence that this low incorporation was not due to the presence of an inhibitor in the P^{32}-UTP preparation was obtained by the finding that γ-P^{32}-UTP supported both C^{14}-ATP and $\beta\gamma$-P^{32}-ATP incorporation with dAT copolymer as primer. The other γ-P^{32}-nucleoside triphosphates also supported RNA synthesis, and the omission of a single triphosphate resulted in a marked decrease in both RNA synthesis and chain initiation.

[8] Ostrowski, W., and A. Tsugita, *Arch. Biochem. Biophys.*, **94**, 68 (1961).

[9] Gardner, J. A. A., and M. B. Hoagland, *J. Biol. Chem.*, **240**, 1244 (1965). We are indebted to Dr. M. Hoagland for a gift of guanosine 5'-tetraphosphate.

[10] One unit of enzyme will cleave 1 μmole of 0-nitrophenylphosphate per minute at 37°.

[11] Wood, W. B., and P. Berg, *J. Mol. Biol.*, **9**, 452 (1964).

[12] Bremer, H., and M. W. Konrad, these PROCEEDINGS, **51,** 801 (1964).
[13] Hurwitz, J., J. J. Furth, M. Anders, and A. Evans, *J. Biol. Chem.,* **237,** 3752 (1962).
[14] Berg, P., R. D. Kornberg, H. Fancher, and M. Dieckmann, *Biochem. Biophys. Res. Commun.,* **18,** 932 (1965).
[15] Marmur, J., and C. M. Greenspan, *Science,* **142,** 387 (1963).
[16] Tocchini-Valentini, G. P., M. Stodolsky, M. Sarnat, A. Aurisicchio, F. Graziosi, S. B. Weiss, and E. P. Geiduschek, these PROCEEDINGS, **50,** 935 (1963).
[17] Colvill, A. J. E., L. C. Kanner, G. P. Tocchini-Valentini, M. T. Sarnat, and E. P. Geiduschek, these PROCEEDINGS, **53,** 1140 (1965).
[18] Fowler, A. V., J. Marmur, and J. Hurwitz, unpublished observations.
[19] Cairns, J., *J. Mol. Biol.,* **6,** 208 (1963).
[20] Ralph, R. K., R. J. Young, and H. G. Khorana, *J. Am. Chem. Soc.,* **85,** 2002 (1963).
[21] Bell, D., R. V. Tomlinson, and G. M. Tener, *Biochem. Biophys. Res. Commun.,* **10,** 304 (1963).

Termination Factor for RNA Synthesis

by
JEFFREY W. ROBERTS
The Biological Laboratories,
Harvard University

A new protein has been isolated from *E. coli* which causes specific termination and release of RNA during synthesis *in vitro*. It has been given the name ρ-factor.

The synthesis of a messenger RNA molecule requires that the enzyme RNA polymerase initiate and terminate transcription at appropriate sites on the DNA template. RNA polymerase does frequently initiate RNA synthesis at the correct sites *in vitro*[1-5], a process which requires a protein factor (the sigma factor) that is normally bound in a tight complex to purified RNA polymerase[6,7]. The usual observation, however, has been that RNA synthesis *in vitro* does not end at unique sites, but instead terminates at random to produce RNA molecules of various, and often clearly excessive, lengths (refs. 8, 9 and personal communication from M. Green). Furthermore, these RNA molecules are not released from the enzyme, but remain complexed with DNA and RNA polymerase when synthesis is completed[8,10]. I now report the isolation and purification of a protein factor from *Escherichia coli* which causes the termination and release of RNA molecules in an *in vitro* reaction using bacteriophage λ DNA as a template. The product contains several discrete RNA species which may correspond to the natural early messenger RNA of lambda.

The termination protein, which is named the ρ-factor (ρ for release), was discovered during a search for factors in a crude extract of *E. coli* which might increase the overall accuracy of *in vitro* RNA synthesis. The criterion of accuracy was that RNA synthesis initiated at the c_{17} promoter of the DNA of bacteriophage $λc_{17}$ should dominate the *in vitro* transcription from $λc_{17}$ DNA; a discussion of the reasoning which suggested this approach is given at the end of this article, where the exact relation of the termination factor to the c_{17} mutation is explained. This criterion provided an assay which allowed a partial purification of the ρ-factor, after which a convenient activity of the ρ-factor became evident: it depresses net RNA synthesis in a standard *in vitro* synthesis reaction. This depression is the basis for a much simpler assay which has allowed purification of the ρ-factor to homogeneity; a summary of the purification procedure is given at the end of this text. The experiments presented below demonstrate that the ρ-factor depresses RNA synthesis from λ DNA template by terminating the synthesis of RNA molecules at specific sites on the template.

ρ-Factor affects Chain Propagation

Fig. 1 illustrates an experiment which shows two properties of the depression activity: only a limited fraction of *in vitro* transcription is depressed by the ρ-factor, and the propagation rather than initiation of RNA chains is affected. Both total RNA synthesis and the number of RNA chains initiated with GTP are measured with increasing concentrations of ρ-factor in a standard reaction mixture, which contains purified λ DNA, purified RNA polymerase, nucleotide triphosphate precursors and the appropriate ions. Total RNA synthesis is determined as the incorporation of ^{3}H-UTP into acid insoluble material; the number of RNA chains initiated with GTP is indicated by the incorporation of ^{32}P from GTP labelled in the γ position with $^{32}PO_4$, because only the γ phosphate from the 5′ terminal nucleotide is retained in the product after polymerization. Total RNA synthesis is depressed by ρ-factor to a plateau value, whereas the number of molecules initiated with GTP remains constant or increases slightly as a saturating amount of ρ-factor is added. An identical result is found if initiation with ATP is measured. Because essentially no initiation occurs with pyrimidine nucleotide triphos-

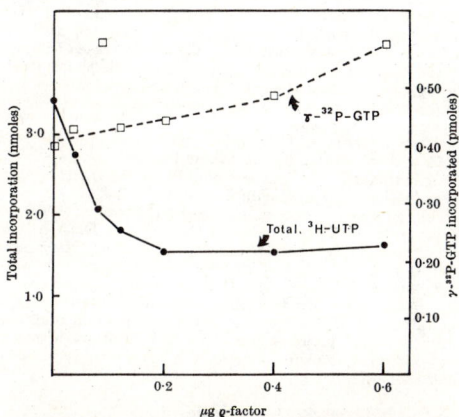

Fig. 1. Effect of ρ-factor on total RNA synthesis and initiation of RNA synthesis with GTP. Each reaction mixture contained in 100 μl.: 0·02 M Tris (pH 7·9); 0·10 M KCl; 0·010 M magnesium acetate; 0·0001 M dithiothreitol; 0·15 mM CTP; 0·05 mM ATP; 0·05 mM ^{3}H-UTP at 80 μCi/μmole; 0·05 mM γ-^{32}P-GTP at 2·1 mCi/μmole; 1·2 μg λ DNA; 3 μg RNA polymerase; 25 μg bovine serum albumin; and ρ-factor as indicated. The amount of ρ-factor was estimated by assuming an absorbance at 280 nm for the purified protein of 0·5/mg/ml. After incubation for 20 min at 37°, the reaction was stopped by addition of 3 ml. 6 per cent trichloroacetic acid (TCA) containing 0·01 M sodium pyrophosphate and the RNA was filtered on a Millipore filter. The filters were washed on the filtration apparatus with 20–30 ml. of 2 per cent TCA with pyrophosphate, followed by soaking for 1 h in 2 per cent TCA containing 1·0 M KCl and pyrophosphate. They were dried and counted in toluene-liquifluor on a liquid scintillation spectrometer. The growth and purification of phage and preparation of DNA were described previously[1]. *E. coli* RNA polymerase was purified to greater than 90 per cent purity by the glycerol gradient technique[11]. ^{3}H-UTP was purchased from Schwarz Bioresearch and γ-^{32}P-GTP was made by the procedure of Glynn and Chappell[12].

phates[4], this experiment shows that the ρ-factor does not interfere with initiation. The depression activity must therefore be the result of some later effect on chain growth. Depression of RNA synthesis by ρ-factor also occurs with DNA templates from bacteriophages T4 and T7, so that the effect does not depend on any peculiar property of phage λ DNA.

ρ-Factor produces Shorter RNA Molecules

The most direct demonstration that the ρ-factor terminates RNA synthesis and produces shorter RNA molecules is provided by examining the size of RNA synthesized in the presence and absence of the factor; in order to make such an experiment meaningful, it is necessary to establish rigorously that the ρ-factor preparation contains no ribonuclease activity which could produce fragments from longer RNA molecules. The following experiment achieves this, and also reveals that the ρ-factor does cause shorter RNA chains to be synthesized. RNA labelled with ^{3}H-UTP was synthesized in a standard reaction with no ρ-factor present; it was purified by phenol extraction and equal portions were added to two new RNA synthesis reactions incubated in the presence and absence of ρ-factor. The new RNA synthesized in the presence of ρ-factor was labelled with ^{14}C-ATP, and no further label was added to the second incubation mixture without ρ-factor. After the second incubation period each reaction was treated with sodium dodecyl sulphate (SDS) and the sizes of the ^{3}H-RNA and ^{14}C-RNA were estimated by sucrose gradient centrifugation. Fig. 2a reveals that ^{3}H-RNA synthesized in the absence of ρ-factor has a broad distribution of sizes from greater than 35S to less than 5S; this appearance is typical of RNA synthesized in the conditions used. By contrast, the ^{14}C-RNA synthesized in the presence of ρ-factor is restricted to sizes less than 15S (Fig. 2b). It will be demonstrated that this peak contains two RNA species which are not resolved here. Because the ^{3}H-RNA present in the synthesis reaction with ρ-factor has essentially the same size distribution as that in the reaction without ρ-factor, I conclude that the ρ-factor decreases the size of the product by directly interacting with the synthesizing complex, and that the factor preparation does not fragment RNA under the conditions of RNA synthesis.

Other less rigorous tests confirm the absence of ribonuclease activity. Incubation of bacteriophage R17 RNA with purified ρ-factor does not reduce its infectivity to spheroplasts, nor does incubation of factor with ^{3}H-poly U release acid soluble radioactivity.

The ρ-factor is free of deoxyribonuclease activity to the extent that it can be incubated with λ DNA in the conditions of the standard assay (including Mg^{2+} and nucleotide triphosphates) without introducing detectable single strand breaks in the DNA. Thus the decrease in size of the RNA product cannot be attributed to fragmentation of the template. This absence of nuclease activity might be expected from the final purification step, for most known nucleases are considerably smaller than the ρ-factor.

ρ-Factor releases RNA Molecules

Besides restricting the size of RNA synthesized *in vitro*, the ρ-factor causes it to be released from the complex of DNA and RNA polymerase. After a normal *in vitro* synthesis of RNA with purified RNA polymerase, most of the RNA is found in a fast sedimenting complex with DNA and RNA polymerase[8,10]. This is revealed in the experiment of Fig. 3a, in which an *in vitro* reaction using λ DNA was stopped by addition of EDTA and chilling and centrifuged on a glycerol gradient without being subjected to conditions which denature proteins. Most of the RNA and template DNA sediment faster than free DNA, at approximately 50S. Fig. 3b presents a companion experiment for which ρ-factor was present during the

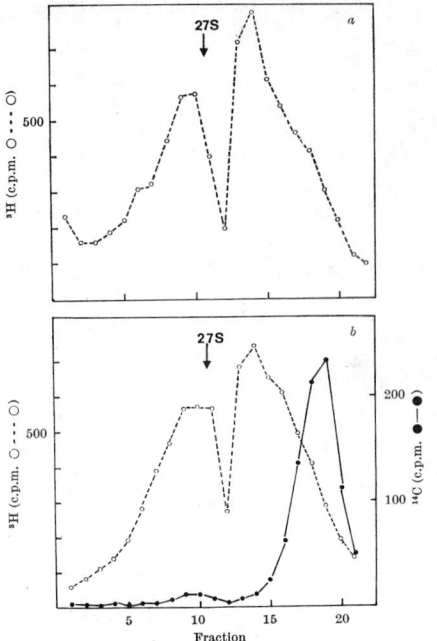

Fig. 2. Effect of ρ-factor on pre-synthesized and newly synthesized RNA. ^{3}H-RNA was synthesized without ρ-factor in a reaction mixture containing in 100 μl.; 0·02 M Tris (pH 7·9); 0·10 M KCl; 0·010 M magnesium acetate; 0·0001 M dithiothreitol; 0·15 mM ATP; 0·15 mM GTP; 0·15 mM CTP; 0·05 mM ^{3}H-UTP at 2·0 mCi/μmole; 1·2 μg λ DNA; and 2 μg RNA polymerase. After incubation for 20 min at 37°, the reaction mixture was chilled and extracted with 50 μl. phenol; the aqueous phase was dialysed against 0·01 M Tris (pH 7·4), 0·0001 M EDTA, and 1·0 M KCl, and finally against 0·01 M Tris (pH 7·4) and 0·0001 M EDTA. Equal portions of this ^{3}H-RNA were added to two further reaction mixtures: (a) as above, except with 1·0 μg RNA polymerase and 0·15 mM UTP (unlabelled), and (b) as above, except with 1·0 μg RNA polymerase, 0·15 mM UTP (unlabelled), 0·15 mM ^{14}C-ATP at 40 μCi/μmole, and 1·0 μg ρ-factor. These were incubated 20 min at 37°, followed by addition of 50 μl. 0·3 per cent sodium dodecyl sulphate (SDS); each was mixed and chilled. Precipitated SDS was removed by centrifugation, and each reaction mixture was layered on a 5 ml. 5–20 per cent sucrose gradient in SSC (0·15 M NaCl; 0·015 M sodium citrate), followed by centrifugation for 1·8 h at 50,000 r.p.m. and 4° in the Spinco SW50 rotor. Twenty-one fractions were collected of each gradient, and the fractions were precipitated with 6 per cent TCA and 0·01 M sodium pyrophosphate and filtered on Millipore filters. Distribution of ^{3}H and ^{14}C was determined by counting in toluene-liquifluor on a liquid scintillation spectrometer. R17 RNA was centrifuged in a separate gradient as a marker at 27S. ○---○, Radioactivity (c.p.m.) in ^{3}H; ●—●, in ^{14}C.

incubation. In this case most of the DNA cosediments with free λ DNA (32S), and the peak of DNA is only slightly skewed toward the higher S values. Most of the RNA sediments well behind the DNA so that clearly it has been released from the DNA. Because the previous experiment ensures that the factor contains no nuclease activity which could cut RNA chains away from the complex, the ρ-factor must free RNA chains from the complex of DNA and RNA polymerase, an event which is probably part of the act of termination.

RNA Product consists of Discrete Sizes

If the ρ-factor terminates at specific sites on the DNA template, the resulting RNA molecules should be of discrete sizes. It will be shown that the RNA product from λ DNA does consist of two discrete sizes, each initiated at a single site on the DNA; thus the ρ-factor causes specific rather than random termination of RNA synthesis.

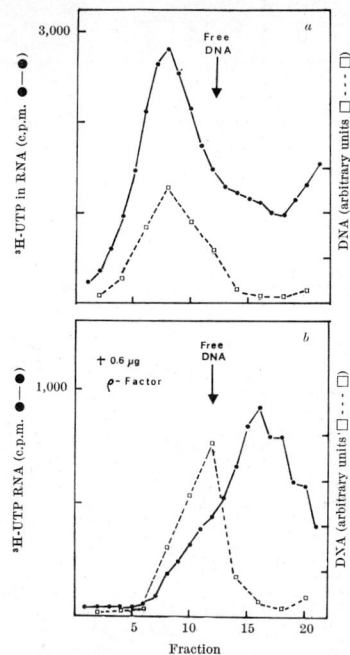

Fig. 3. Release of RNA from complex by ϱ-factor. The conditions of synthesis were as in Fig. 2, except with 0·05 mM ³H-UTP at 1·0 mCi/μmole, 1·8 μg RNA polymerase, and 25 μg bovine serum albumin per 100 μl. reaction mixture. The reaction mixture of 3b contained in addition 0·6 μg ϱ-factor. After incubation for 20 min at 37°, synthesis was stopped by chilling and addition of EDTA to 0·02 M, and each was layered on a 5 ml. 10–30 per cent glycerol gradient in 0·05 M Tris (pH 7·4), 0·1 M KCl, 0·01 M magnesium acetate, 0·0001 M EDTA and 0·0001 M dithiothreitol. The gradients were centrifuged 1·6 h at 64,000 r.p.m. in the Spinco SW65 rotor at 4°, and twenty-one fractions were collected. To determine the distribution of RNA, 50 μl. portions of each fraction were precipitated with TCA, filtered on to Millipore filters, and counted as before. The distribution of DNA in the gradients was determined by assaying portions of each fraction for template activity, measured as incorporation of ¹⁴C-ATP into RNA under conditions of excess polymerase. The position of free DNA was determined in a parallel gradient. ●—●, Radioactivity (c.p.m.) in ³H-RNA; □---□, DNA, arbitrary units. a, No ϱ-factor; b, with 0·6 μg ϱ-factor.

What parts of the λ genome does one expect to be transcribed by RNA polymerase in the absence of phage specific functions ? Fig. 4 is a map of λ which shows the positions of the promoters for early transcription and the directions of RNA synthesis. If phage repressor is absent or is¹ removed, synthesis of messenger RNA for the N gene is initiated at a promoter close to the beginning of the N gene, and proceeds to the end of the gene. Unless the N protein is made from this messenger, there is no transcription of genes to the left of N (refs. 13 and 14). In an in vitro RNA synthesis reaction which lacks the N protein, one therefore would expect transcription to the left only of the N gene itself. (The function of the N gene product will be discussed below.) RNA synthesis to the right will be initiated at a promoter which is likely to be in region x (refs. 15 and 16). This transcription also appears to be dependent on the N gene product, although some genes to the right are expressed at least slightly in its absence[17,18]. Thus if both initiation and termination occur at the correct sites in vitro, the RNA product should include discrete species initiated at the leftward and possibly the rightward promoter.

Fig. 5a shows the sedimentation profile of RNA synthesized by RNA polymerase in vitro from λb_2 template in the presence of ϱ-factor. Two peaks of RNA are apparent, sedimenting at approximately 12S and 7S. If RNA is synthesized in the absence of ϱ-factor and treated identically, there is little RNA in this area of the gradient (see also Fig. 2) and there is no trace of these peaks. The following experiments establish that these peaks represent distinct RNA species by showing (a) that they are synthesized from small regions of the genome adjacent to the promoters; and (b) that they are initiated at the rightward and leftward promoters.

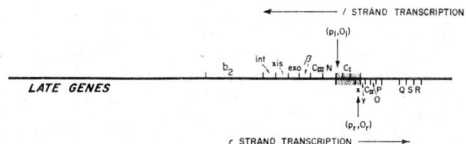

Fig. 4. A genetic and physical map of bacteriophage λ. The shaded zone is the immunity region, which includes the C_1 (repressor) gene and the promoters and operators of early transcription[19,20]. Messenger RNA synthesis toward the right from the immunity region is initiated at a promoter p_R located in region x, and the RNA is homologous to the r strand of DNA[15,16]. Transcription toward the left of messenger for gene N (and probably genes c_{III}–int as well) is initiated from a promoter p_L at the beginning of the N gene, and the RNA is homologous to the l strand[16,20].

To determine the regions of the template to which the 12S and 7S RNA is homologous, portions of each fraction of the gradient of Fig. 5a were hybridized to separated strands of λb_2 or wild type, λi^{434}, and $\lambda b_2 i^{21}$ (see Fig. 6). RNA homologous to l strands DNA is synthesized toward the left of the standard λ map (see Fig. 4), and RNA homologous to r strand DNA is synthesized toward the right[14,16]. The λi^{434} DNA effectively deletes the immunity region for hybridization purposes, and the $\lambda b_2 i^{21}$ DNA effectively deletes the immunity region and the N gene. The presence of the b_2 region in the λi^{434} DNA is irrelevant because the template does not contain the b_2 region. Fig. 5b reveals that the 12S peak is homologous to the l strands of both λb_2 and λi^{434} DNA, but not to the l strand of $\lambda b_2 i^{21}$ DNA; thus the 12S RNA is synthesized to the left and from a region in the vicinity of the N gene. The 7S RNA is homologous to the r strand of λb_2 DNA, but not to r strands of either λi^{434} or $\lambda b_2 i^{21}$ DNA (Fig. 5c). The 7S RNA is therefore synthesized towards the right and from an area within the immunity region. Both RNA peaks arise from highly limited regions of the template DNA.

The 12S and 7S RNA combined constitute only approximately 50 per cent of the total product. The remaining background of RNA is transcribed from both strands and is not found in definite peaks, but instead exhibits a wide range of sizes. This RNA might result from incorrect initiation by RNA polymerase at unknown locations on the template, followed by termination events which produce RNA molecules of random length; such incorrect initiation is known to occur (see ref. 1). Substantially more of the background RNA is synthesized from λ DNA template which contains the b_2 region, so that λb_2 DNA (that is, DNA deleted for the b_2 region) has been used for these experiments whenever possible.

12S and 7S RNA initiated at Known λ Promoters

How can it be shown that each of the 7S and 12S RNA species is initiated at a unique locus on the template ? Mutants of λ exist which abolish or greatly reduce transcription in vivo either toward the right or toward the left from the immunity region. These are good candidates to be mutations in the promoters which prevent the normal interaction between RNA polymerase and DNA to initiate RNA synthesis. The experiments below demonstrate that DNA which carries either of these mutations is not a template for *in vitro* transcription of the RNA species synthesized in the direction corresponding to the *in vivo* defect; thus the two RNA species are initiated at the sites of the mutations, and these sites are in fact the promoters.

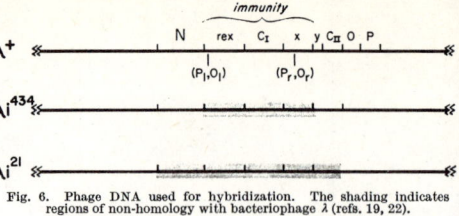

Fig. 6. Phage DNA used for hybridization. The shading indicates regions of non-homology with bacteriophage λ (refs. 19, 22).

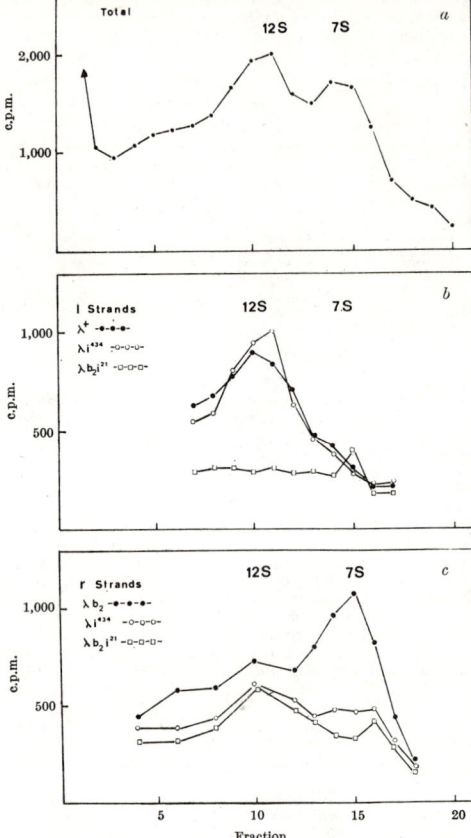

Fig. 5. Sucrose gradient analysis of RNA synthesized from λb_2 DNA template in the presence of ρ-factor. The conditions of synthesis were as in Fig. 2, except that the 200 μl. reaction mixture contained 2·4 μg λb_2 DNA, 4·0 μg RNA polymerase, 0·05 mM ³H-UTP at 2·0 mCi/μmole, and 1·5 μg ρ-factor. After incubation for 20 min in SDS, mixed, chilled and centrifuged to remove precipitated SDS. It was layered on to a 5 ml. 5–20 per cent sucrose gradient in SSC and centrifuged 4 h at 50,000 r.p.m. and 4° in the Spinco SW50 rotor. Twenty-one fractions were collected. The sedimentation constants were determined by centrifuging 16S ribosomal RNA and 4S tRNA as markers in a parallel gradient. *a*, Thirty μl. portions of each fraction were precipitated with TCA, filtered on to Millipore filters, and counted as before. *b*, Thirty μl. portions were annealed to *l* strand DNA from wild type λ, λi^{434}, and $\lambda b_2 i^{21}$ phage. The growth of phage and preparation of separated strands were described previously¹. For hybridization, each RNA portion was combined with 200 μl. 2×SSC containing 0·4–0·5 μg DNA and phenol at 25 per cent saturation. The mixtures were annealed for 5 h at 67°, chilled, diluted to 1·0 ml. with 2×SSC containing 10 μg ribonuclease A, incubated 30 min at 37°, and diluted to 4 ml. with 0·01 M Tris (pH 7·4) and 0·50 M KCl (ref. 21). They were filtered on to Schleicher and Schuell B6 filters, rinsed with 30–40 ml. of Tris-KCl, dried, and counted in toluene-liquifluor. It was previously determined that 0·4–0·5 μg DNA is a saturating amount. RNA homologous to wild type λ *l* strand DNA (●—●); to λi^{434} *l* strand DNA (○—○); and $\lambda b_2 i^{21}$ *l* strand DNA, (□—□). *c*, Thirty μl. portions were hybridized to 0·3–0·5 μg *r* strand DNA of λb_2 (●—●); λi^{434} (○—○); and $\lambda b_2 i^{21}$ (□—□).

The mutant defective in transcription toward the left is λ_{sex}. This mutant was isolated by M. Gottesman, who showed that it greatly reduces the expression of several gene products in the region to the left of gene *N*. The sex mutation diminishes *in vivo* RNA synthesis throughout the whole p_L-N-*int* region by approximately a factor of ten (H. J. J. Nijkamp and W. Szybalski, in preparation), and thus differs from N⁻ mutations, which do not affect the transcription of the p_L-N region but abolish RNA synthesis in the region to the left of *N* (refs. 14, 23). Lambda$_{sex}$ maps as indicated in Fig. 9. These facts strongly suggest that it affects the promoter at the beginning of the *N* gene.

To test this mutant, RNA was synthesized from λb_2 and λb_2 $_{sex}$ DNA templates in the usual *in vitro* system in the presence of ρ-factor. RNA from λb_2 template was labelled with ³H-ATP, and RNA from λb_2 $_{sex}$ template was labelled with ¹⁴C-ATP; the RNA preparations were mixed and analysed by sucrose gradient centrifugation as before. Fig. 7 shows that the 7S *r* strand RNA is synthesized from both templates, whereas the 12S *l* strand peak is not synthesized in significant amount from the λ_{sex} DNA. It might be argued that the 12S RNA is not seen because the sex mutation has created immediately adjacent to the promoter a new termination site which the ρ-factor recognizes; this is not true, however, because I have observed that the sex mutation also decreases net *l* strand RNA synthesis in an *in vitro* reaction without ρ-factor. The sex mutation must therefore affect RNA synthesis by impairing the interaction between RNA polymerase and DNA by which synthesis is initiated, and the 12S RNA must be initiated at the site affected in λ_{sex}. I conclude that the 12S RNA is initiated at the leftward promoter at the beginning of gene *N*.

The mutation which has been shown to abolish rightward transcription in λ is t_{11} (ref. 16). Lambda t_{11} maps in region *x* (see Fig. 9), as do numerous other mutants which are deficient in expression of phage genes on the right[15,16]. The x mutant examined here for its effect on *in vitro* RNA synthesis is λx_{13}, which was chosen because it could be grown in sufficient quantities to provide template DNA (see legend to Fig. 9); it is presumed to be similar to t_{11}. For this experiment, ³H-RNA synthesized *in vitro* from wild type λ template in the presence of ρ-factor was mixed with ¹⁴C-RNA synthesized similarly from λx_{13} template and the combined RNA was analysed by sucrose gradient. This experiment is complicated by the fact that neither template contains the b_2 deletion, and that, as mentioned above, there is RNA synthesized *in vitro* from the b_2 region which somewhat obscures the sucrose gradient pattern of RNA seen in the previous figures. Nevertheless, it is evident from Fig. 8*a* that considerably less RNA in the 7S region is synthesized from λx_{13} DNA than from wild type λ DNA.

The 7S RNA can be examined without interference from *r* strand RNA synthesized on other regions of the template by measuring RNA which anneals to the *r* strand of φ80*iλ* DNA; this phage is a hybrid of λ and φ80 which is homologous to λ only to the right of some point probably located in the exonuclease gene[24,25], so that RNA synthesized from the b_2 region or farther to the left will no

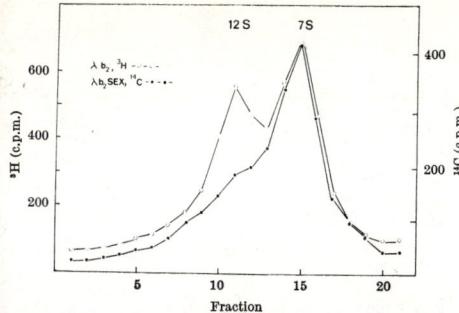

Fig. 7. Sucrose gradient analysis of RNA products from λb₂ and λb₂sex DNA templates. Synthesis reactions were as in Fig. 2, except that 100 μl. reaction mixtures contained 0·05 mM UTP (unlabelled), 1·5 μg DNA template, 1·3 μg RNA polymerase, and 0·8 μg ρ-factor. RNA synthesized from λb₂ DNA was labelled with ³H-ATP (0·15 mM at 330 μCi/μmole), and RNA from λb₂sex DNA was labelled with ¹⁴C-ATP (0·15 mM at 67 μCi/μmole). After a 20 min incubation at 37°, the reaction mixtures were made 0·1 per cent in SDS, chilled, combined and centrifuged to remove precipitated SDS. The RNA was layered on to a sucrose gradient as before and centrifuged 4 h at 65,000 r.p.m. and 5° in a Spinco SW 65 rotor. Twenty-one fractions were collected; each was precipitated with TCA, filtered on to Millipore filters, dried and counted in toluene-liquifluor on a liquid scintillation spectrometer. ○, Radioactivity (c.p.m.) in ³H; ●, in ¹⁴C.

Table 1. TRANSCRIPTION FROM WILD TYPE λ AND λc₁₇ TEMPLATES IN THE PRESENCE OF ρ-FACTOR

Annealed to	Template	
	λc₁₇	λ⁺
λ⁺ l	697	704
λ⁺ r	1,847	947
φ80iλ r	1,164	297

RNA synthesis in 200 μl. reaction mixtures was as in Fig. 2, except with 0·05 mM ³H-UTP at 1·0 mCi/μmole, 0·6 μg template DNA, 0·9 μg RNA polymerase, 50 μg bovine serum albumin, and 1 μg ρ-factor. Incubation was for 20 min at 37° C. Aliquots of each reaction mixture were hybridized to separated DNA strands by a method described previously[1]. Figures represent radioactive material (c.p.m.) annealed to a saturating amount of DNA, and are the averages of duplicate samples; a background of 307 c.p.m. was subtracted.

Conclusion and Discussion

Fig. 9 summarizes the effect of ρ-factor on *in vitro* transcription from bacteriophage λ DNA. This work has shown that a large fraction of *in vitro* RNA synthesis is initiated at promoters p_L and p_R, as defined experimentally by the mutations λsex and λx₁₃. When ρ-factor is present, RNA synthesis is terminated at points indicated by triangles to produce a 12S RNA species from the *l* strand and a 7S RNA species from the *r* strand. Is this pattern of RNA synthesis reasonable? One can estimate that the molecular weight of the 12S RNA is approximately 250,000 (ref. 28), so that as a messenger RNA it could encode a protein of approximately 25,000–30,000 molecular weight. The size of the N gene protein is unknown, but

hybridize to its DNA. Fig. 8b shows the result of annealing RNA from the gradient of Fig. 8a to φ80iλ DNA: it is apparent that the 7S RNA is synthesized from wild type DNA and is not synthesized from λx₁₃ DNA. Hybridization of portions of the gradient to *l* strand DNA of λb₂ demonstrates that the 12S *l* strand RNA is synthesized from both templates. The x₁₃ mutation also reduces *r* strand specific RNA synthesis in an *in vitro* reaction without ρ-factor, so that, as in the case of λsex, the mutation cannot function by interacting with ρ-factor to terminate synthesis of the 7S RNA prematurely. It appears very likely that the 7S RNA is initiated at the site of mutation x₁₃, and that this site is the rightward promoter.

More Accurate RNA Synthesis with ρ-Factor

An example of the increased overall accuracy of *in vitro* RNA synthesis which the ρ-factor produces is provided by the mutant λc₁₇. The c₁₇ mutation creates a new promoter which allows constitutive synthesis of messenger RNA from the *x–O–P* region, even in the presence of phage repressor and the absence of the N gene product[1,26,27]. In an *in vitro* system using RNA polymerase and no ρ-factor it was found that approximately 60 per cent more RNA homologous to φ80iλ *r* strand DNA (which includes the *x–O–P* region) is synthesized from λc₁₇ DNA than from wild type DNA (ref. 1). It was argued that the effect of the c₁₇ mutation on mRNA synthesis *in vivo* is likely to be much greater than this, for the mutation has a profound effect on the behaviour of the phage *in vivo*. A comparison of *in vitro* RNA synthesis from wild type λ and λc₁₇ templates in the presence of ρ-factor is presented in Table 1. One measure of the *in vitro* effect of the c₁₇ mutation is the ratio of φ80iλ *r* strand specific RNA synthesis from λc₁₇ and λ wild type templates under conditions which produce identical *l* strand specific RNA synthesis; with ρ-factor this ratio is 3 to 4, whereas without ρ-factor it was approximately 1·6 (ref. 1). The apparent excess RNA specific to φ80iλ *r* DNA synthesized from λc₁₇ DNA is increased five-fold by ρ-factor. Because RNA homologous to φ80iλ *r* strand DNA includes the immunity specific 7S RNA which is synthesized from both wild type and λc₁₇, the actual excess synthesized from the *x–O–P* region of λc₁₇ DNA may be closer to ten times that from the wild type template.

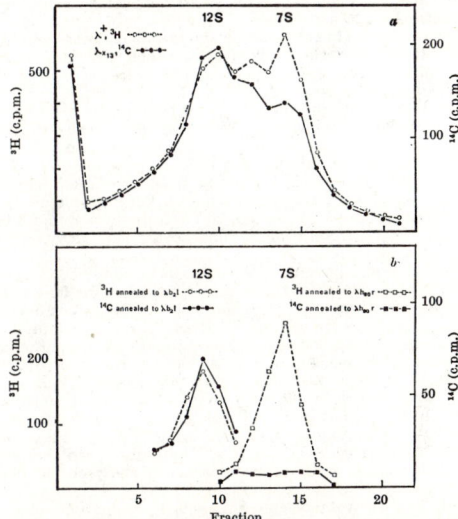

Fig. 8. Comparison of RNA product from λx₁₃ and wild type λ templates. Conditions of synthesis were as in Fig. 2, except that 190 μl. reaction mixtures contained 0·05 mM UTP (unlabelled), 2·4 μg DNA template, 2·5 μg RNA polymerase, and 1·5 μg ρ-factor. The mutant λx₁₃ was isolated by H. Eisen. λx₁₃ phage were prepared by mitomycin induction of a tandem double lysogen of λi⁴³⁴ii₂ and λc₂₆₇x₁₃ constructed by M. Gottesman. The phage burst contains about 90 per cent λx₁₃ and 10 per cent λi⁴³⁴ (unpublished observation and personal communication from M. Gottesman). The λx₁₃ were separated from the λi⁴³⁴ by banding the phage to equilibrium in a CsCl density gradient. RNA synthesized from wild type λ DNA was labelled with ³H-ATP (0·15 mM at 0·9 mCi/μmole), and RNA synthesized from λx₁₃ template was labelled with ¹⁴C-ATP (0·15 mM at 110 μCi/μmole). The reaction mixtures were incubated, combined and centrifuged on a sucrose gradient as in Fig. 7, except that the centrifugation time was 5 h. Twenty-one fractions were collected. *a*, 50 μl. portions of each fraction were precipitated with TCA, filtered and counted as before. ○ - - - ○, Radioactivity (c.p.m.) ³H; ● — ●, in ¹⁴C. *b*, 50 μl. portions of each fraction were annealed to *r* strand DNA of φ80iλ or to *l* strand DNA of λb₂, as described in the legend to Fig. 5. ○ - - - ○, Radioactivity (c.p.m.) in ³H annealed to λb₂ *l* DNA; ● — ●, in ¹⁴C annealed to λb₂ *l* DNA; □ - - - □, in ³H annealed to φ80iλ *r* DNA; ■ — ■, in ¹⁴C annealed to φ80iλ *r* DNA,

the size ascribed to the mRNA made in vivo for the N gene region is 225,000 (Kourilsky, P., Marcaud, L., Portier, M. M., aud Gros, F., manuscript in preparation). It remains to be determined if the species made in vivo and in vitro are indeed identical. The $7S$ RNA is estimated to have a molecular weight in the vicinity of 100,000, which might encode a protein of approximately 10,000 molecular weight. This protein may correspond to the turn-off function (tof) which limits expression of genes to the left of N after early stages of infection, and, like the $7S$ in vitro RNA, is both encoded by the immunity region and under the control of promoter p_R (ref. 30). Either the $7S$ RNA or tof, or both, may correspond to CRN, a gene which prevents repressor synthesis during lytic growth and is also synthesized from promoter p_R in the immunity region (H. Eisen, manuscript in preparation). An RNA species with the properties of the $7S$ RNA recently has been detected in vivo (S. Heinemann, in preparation), and it has been shown that most in vivo r strand transcription in the absence of the N function is restricted to the immunity region (S. Kumar and W. Szybalski, personal communication).

When ρ-factor is not present during in vitro synthesis, the $12S$ and $7S$ RNA species are not found and most of the RNA product is larger. Because the ρ-factor was shown to affect propagation rather than initiation of RNA chains, I infer that in its absence RNA synthesis is not terminated at these points, but instead proceeds into the areas indicated in Fig. 9 by the dotted lines. This model explains both the change in size of the RNA product and the decrease in net RNA synthesis which the ρ-factor produces.

The existence of the ρ-factor suggests a possible mechanism of action of the λ N protein, which also might be relevant to other systems such as the arabinose operon[31] which are under positive control. The N function greatly stimulates transcription in the regions to the left and right of the termination points in Fig. 9. Because these regions are directly adjacent to the termination points, the N protein might function simply by preventing the termination process and allowing RNA synthesis to proceed beyond the termination signals. For example, if the ρ-factor functions by binding to DNA at the termination point, the N protein might bind to the same site more tightly and thereby deny ρ-factor access to it. (The N function would have to be specific to the phage DNA and not to the ρ-factor or another host function, for different phages which grow in the same host are known to have incompatible N functions—H. Eisen, personal communication.) This "conditional termination" hypothesis requires that RNA synthesized under stimulation by N consists of molecules initiated at promoters p_L and p_R, and there is strong evidence that this is true (D. Luzatti, in preparation; J. Pero, in preparation; and refs. 16, 32). A reasonable explanation for the slight synthesis of messenger for genes O and P which occurs in the absence of N function is that termination on the right is not completely efficient, but instead occasionally allows an RNA polymerase molecule to pass the termination points. The conditional termination hypothesis makes several predictions: (1) Messenger RNA molecules for genes beyond the termination points should include the sequences of the $12S$ and $7S$ RNA species at their 5′ ends. (2) There should exist mutations which allow λ to grow in the absence of N function by destroying termination signals against which the N protein acts. N-independent mutants of λ exist[33–35], but their mechanism is unknown. (3) If the $7S$ RNA corresponds to the tof function, then the tof function should not be under control of the N gene product.

Why is the fraction of RNA specific to the region affected by the c_{17} mutation greater in the presence of ρ-factor? The c_{17} mutation is located to the right of the immunity region: it can be crossed into phage 434hy (ref. 26). Because the $7S$ RNA is synthesized within the im-

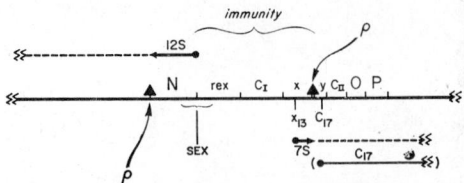

Fig. 9. Model of in vitro RNA synthesis from λ DNA template. RNA synthesis in the presence of ρ-factor is indicated by solid arrows, and the extension of this synthesis in the absence of ρ-factor by dotted lines. The proposed sites of action of the ρ-factor are indicated by triangles. For further details, see the text.

munity region, the factor must terminate RNA synthesis at a point within the immunity region or very close to its right boundary. Thus it is likely that the c_{17} mutation is located outside the site at which the ρ-factor acts, so that even in the presence of ρ-factor, RNA polymerase molecules which initiate at the c_{17} promoter can proceed through the region of the O and P genes. Transcription initiated at p_R does not continue beyond the termination site in the presence of ρ-factor. Whereas in the absence of factor the c_{17} promoter simply adds to transcription initiated at p_R and continuing through the O–P region, in the presence of factor it is essentially the only source of transcription of this region.

It seems safe to infer that the ρ-factor is responsible for termination events in the synthesis of early messenger RNA of λ in infected cells, and that it has the same function in bacterial RNA synthesis. It might, of course, be only one component of the complete termination system of E. coli: there could exist in addition several factors like ρ which recognize distinct classes of termination signals, or possibly several distinct termination processes. I have observed that at least one termination event occurs with λ DNA template in the absence of ρ-factor to produce an RNA species of approximately $22S$, although its physiological significance is unknown. There is little information about the molecular mechanism of termination by the ρ-factor. It might bind to DNA and act when it is met by an RNA polymerase molecule progressing along the template; alternatively, it might form a complex with RNA polymerase and function when the complex reaches a termination point. A third possibility is that the configuration of RNA polymerase changes at the termination site, allowing ρ-factor to interact with it and remove it from the DNA. Whatever the mechanism, there must exist specific DNA sequences which function as termination signals in the presence of the ρ-factor.

Purification and Properties of ρ-Factor

All steps are performed at 0°–4° C. Frozen E. coli cells (50 g) are ground with 135 g alumina; to the paste is added 140 ml. of a buffer containing 0·05 M Tris (pH 7·4), 0·20 M KCl, 0·01 M $MgCl_2$, 0·0001 M dithiothreitol, 0·0001 M ethylenediaminetetraacetate, and 5 per cent glycerol. Electrophoretically pure deoxyribonuclease (600 μg; Worthington) is added and the mixture is allowed to stand for 15 min. Alumina and débris are removed by low speed centrifugation and ribosomes are removed by centrifugation for 2 h at 100,000g. To the supernatant is added 32 g ammonium sulphate per 100 ml., and precipitated proteins are dissolved in 0·05 M potassium phosphate buffer (pH 7·5), containing 0·0001 M dithiothreitol and 5 per cent glycerol. (The molar ratio of K_2HPO_4 to KH_2PO_4 in the buffer is 5·2.) The proteins are dialysed against this buffer, loaded on a 100 ml. phospho-cellulose column in the same buffer, and eluted with a gradient from 0·05 M to 0·50 M phosphate. Portions (5 μl.) of column fractions are assayed in a reaction mixture similar to that described in the legend to Fig. 1. Fractions around the peak of depression at approximately 0·16 M phosphate are pooled, protein is precipitated with ammonium sulphate at 60 per cent saturation, and the precipitated protein is dissolved in 0·02 M potassium phosphate buffer (pH 7·5), containing 0·0001 M dithiothreitol and 5 per cent glycerol. The protein is dialysed into this buffer and loaded on a 20 ml. DEAE cellulose column equilibrated in the same buffer. Protein is eluted with a salt gradient from 0·02 M to 0·20 M phosphate, and 3 µl. of each fraction is assayed as before. Fractions around the peak of depression eluted at approximately 0·10 M phosphate are pooled, concentrated by ammonium sulphate precipitation as before, and dissolved in 400 µl. of a buffer containing 0·05 M Tris (pH 7·4), 0·10 M KCl, 0·0001 M dithiothreitol, 0·0001 M ethylenediaminetetraacetate, and 5 per cent glycerol. After a brief dialysis against this buffer, the sample is layered on to two 5 ml. glycerol gradients (10–30 per cent, by volume) in the same buffer and centrifuged approximately 12 h at 65,000 r.p.m. in the Spinco SW 65 rotor. Twenty-two fractions of each gradient are collected; the $A_{280\ nm}$ value of each fraction is determined and 2 µl. portions of fractions are assayed as before.

Fig. 10 illustrates the sedimentation profile obtained from the glycerol gradient step. The depression activity clearly coincides with a peak in $A_{280\ nm}$ which sediments at 8–10S and is well separated from the bulk of contaminating material. The assay is not linear over most of the concentration range in the peak, so that the activity appears to trail well beyond the absorbance peak. Because the factor cannot be assayed at the early steps of purification, and the assay has not been quantitated adequately for any step, the recovery of ϱ-factor cannot be calculated; I suspect that it is small.

The ϱ-factor purified to this extent appears to be a pure protein. Its $A_{280\ nm}$ to $A_{260\ nm}$ ratio is 1·5–2·0, which is characteristic of a protein and implies that very little nucleic acid is present. Heating the factor for 5 min at 50° C destroys the depression activity. It adheres to both DEAE-cellulose and phospho-cellulose at low ionic strength, so that it contains regions of both positive and negative charge. The sedimentation constant of 8–10S suggests a molecular weight in the vicinity of 200,000, assuming an average configuration. When fractions across the 8–10S peak of Fig. 10 are examined by polyacrylamide gel electrophoresis in SDS, one predominant band is found; its maximum intensity coincides with the absorbance peak. A gel resolution of purified ϱ-factor is shown in Fig. 11; its purity is estimated to be greater than 95 per cent. The molecular weight of the polypeptide chain is approximately 50,000 measured on polyacrylamide gels in SDS by the method of Shapiro et al.[36], using the subunits of RNA polymerase as markers[11]. These molecular weights suggest that the 8–10S protein may be a tetramer.

I thank Professor Walter Gilbert, who suggested and guided these experiments; Professor J. D. Watson, for support and advice; Professor Klaus Weber, for helpful advice; Dr Max Gottesman, for providing phage strains and for suggesting that λ_sex might be a promoter mutant; Drs R. Burgess, M. Ptashne, M. Schwartz and A. Travers, and R. Hendrix, N. Hopkins, R. Kamen and J. Pero for helpful conversations; Mrs Ann Soderquist for technical assistance; and Mrs C. Roberts for figure drawings. I was supported by a predoctoral fellowship from the US National Science Foundation and by grants to the laboratory from the US National Science Foundation and the US National Institutes of Health.

Received November 11, 1969.

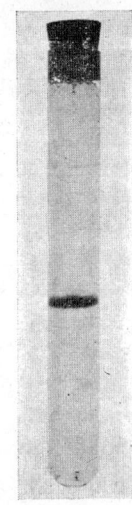

Fig. 11. Polyacrylamide gel electrophoresis of purified ϱ-factor. The electrophoresis and staining procedure were as described in ref. 11; the gel contained 5 per cent acrylamide and 0·1 per cent SDS.

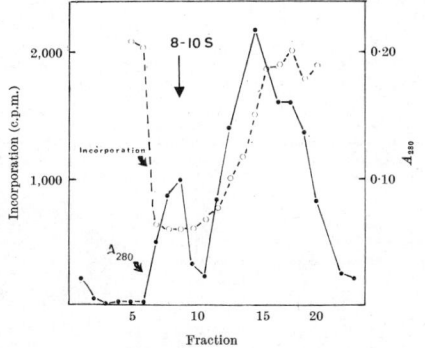

Fig. 10. Glycerol gradient centrifugation of ϱ-factor. Details are given in the text. The direction of centrifugation is from right to left. ○---○, Radioactivity (c.p.m.) in ³H; ●—●, $A_{280\ nm}$.

[1] Roberts, J. W., Nature, 223, 480 (1969).
[2] Milanesi, G., Brody, E. N., and Geiduschek, E. P., Nature, 221, 1015 (1969).
[3] Naono, S., and Gros, F., Cold Spring Harbor Symp. Quant. Biol., 31, 361 (1966).
[4] Cohen, S. N., Maitra, U., and Hurwitz, J., J. Mol. Biol., 26, 19 (1967).
[5] Summers, W. C., and Siegel, R. B., Nature, 223, 1111 (1969).
[6] Travers, A. A., and Burgess, R. R., Nature, 222, 537 (1969).
[7] Bautz, E. K. F., Bautz, F. A., and Dunn, J. J., Nature, 223, 1022 (1969).
[8] Richardson, J. P., J. Mol. Biol., 21, 115 (1966).
[9] Geiduschek, E. P., and Haselkorn, R., Ann. Rev. Biochem., 38, 647 (1969).
[10] Bremer, H., and Konrad, M. W., Proc. US Nat. Acad. Sci., 51, 801 (1964).
[11] Burgess, R. R., Travers, A. A., Dunn, J. J., and Bautz, E. K. F., Nature, 221, 43 (1969).
[12] Glynn, I. M., and Chappell, J. B., Biochem. J., 90, 147 (1964).
[13] Radding, C., and Echols, H., Proc. US Nat. Acad. Sci., 60, 707 (1968).
[14] Kumar, S., Bøvre, K., Guha, A., Hradecna, Z., Maher, Sr. V. M., and Szybalski, W., Nature, 221, 823 (1969).
[15] Eisen, H., Fuerst, C., Siminovitch, L., Thomas, R., Lambert, L., Pereira da Silva, L., and Jacob, F., Virology, 30, 224 (1966).
[16] Taylor, K., Hradecna, Z., and Szybalski, W., Proc. US Nat. Acad. Sci., 57, 1618 (1967).
[17] Pereira da Silva, L. H., and Jacob, F., Virology, 33, 618 (1967).
[18] Ogawa, T., and Tomizawa, J., J. Mol. Biol., 38, 217 (1968).
[19] Westmoreland, B. C., Szybalski, W., and Ris, H., Science, 163, 1343 (1969).
[20] Ptashne, M., and Hopkins, N., Proc. US Nat. Acad. Sci., 60, 1282 (1968).
[21] Bolle, A., Epstein, R. H., Salser, W., and Geiduschek, E. P., J. Mol. Biol., 31, 325 (1968).
[22] Signer, E. R., Manly, K. F., and Brunstetter, M., Virology, 39, 137 (1969).
[23] Szybalski, W., Bøvre, K., Fiandt, M., Guha, A., Hradecna, Z., Kumar, S., Lozeron, H. A., Maher, Sr. V. M., Nijkamp, H. J. J., Summers, W. C., and Taylor, K., J. Cell Physiol., 74, suppl. 1, 33 (1969).
[24] Szpirer, J., Thomas, R., and Radding, C. M., Virology, 37, 585 (1969).
[25] Lozeron, H. A., and Szybalski, W., Virology, 39, 373 (1969).
[26] Pereira da Silva, L., and Jacob, F., Ann. Inst. Pasteur, 115, 145 (1968).
[27] Packman, S., and Sly, W. S., Virology, 34, 778 (1968).
[28] Spirin, A. S., Biokhimiya, 26, 511 (1961).
[29] Kourilsky, P., Marcaud, L., Sheldrick, P., Luzzati, D., and Gros, F., Proc. US Nat. Acad. Sci., 61, 1013 (1968).
[30] Pero, J., Virology (in the press).
[31] Sheppard, D., and Englesberg, E., Cold Spring Harbor Symp. Quant. Biol., 31, 345 (1966).
[32] Schwartz, M., Virology (in the press).
[33] Hopkins, N., Virology (in the press).
[34] Butler, B., and Echols, H., Virology (in the press).
[35] Court, D., and Sato, K., Virology, 39, 348 (1969).
[36] Shapiro, A., Viñuela, E., and Maizel, J. V., Biochem. Biophys. Res. Commun., 30, 240 (1967).

Printed in Great Britain by Fisher, Knight & Co., Ltd., St. Albans.

Degradation of Tryptophan Messenger

Contrary to a previous report, messenger RNA for the tryptophan operon is degraded in the 5′—3′ direction. The following two articles describe how pulse labelling and hybridization experiments established this. In some conditions, the degradation of individual messengers starts before their synthesis is complete.

On the Degradation of Messenger RNA for the Tryptophan Operon in *Escherichia coli*

IN spite of the attention which has been paid to the short lived messenger RNA (mRNA) fractions in bacteria, very little is known about the mechanism by which individual mRNA molecules are broken down or the factors regulating the turnover of mRNA *in vivo*. We report here experiments which demonstrate that tryptophan mRNA molecules are degraded from the E gene end (5′ end of the molecule) toward the A gene end (3′ end). This conclusion is in conflict with that presented by Baker and Yanofsky[1]. (See also the following paper[2].)

The tryptophan operon of *E. coli* consists of at least five contiguous genes[3] (Fig. 1) with an operator region at the E gene end[4]. Following a shift from repression to derepression, the synthesis of polycistronic mRNA is initiated at the E gene end and proceeds to the A gene end, taking about 8 min to complete a round of transcription in the conditions used[5]. After 8 min, the rate of transcription of the operon reaches steady state.

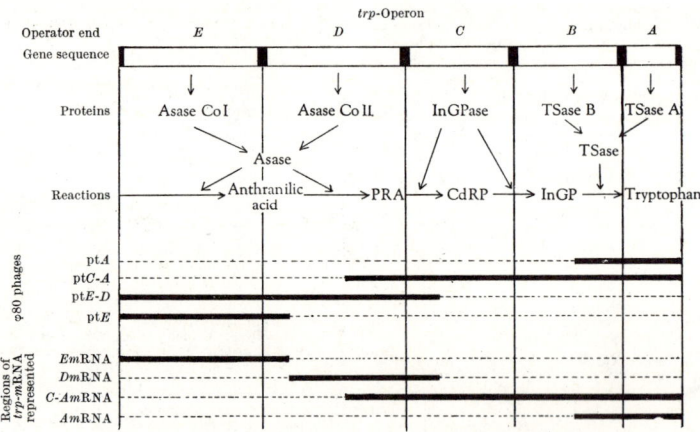

Fig. 1. *Trp*-operon segments carried by φ80 pt phages. The region indicated by the solid line is carried by each φ80 pt. The relative size of the *trp*-genes corresponds to the molecular weights of the corresponding polypeptides. Location of the termini in phages pt$E$ and pt$C$-A is taken from the data of Imamoto[9] and Deeb et al.[8], respectively, and those in pt$E$-D and pt$A$ are only approximate. *Trp*-mRNA hybridizable with pt$E$, pt$E$-D, pt$C$-A and pt$A$ is designated EmRNA, E-DmRNA, C-AmRNA and AmRNA, respectively.

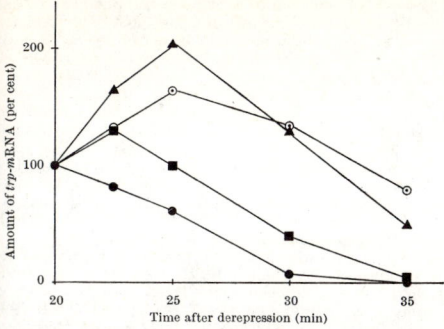

Fig. 2. Time course of degradation of trp-mRNA molecules. Hybridization values with ³H-RNA pulse-labelled for 30 s at the 19·5th min after derepression were 1,520, 1,060, 910 and 653 c.p.m./100 μg of RNA for E, D, C-A and AmRNA, respectively. The ordinate is the relative value (per cent) calculated by dividing the amount of each trp-mRNA segment remaining at the indicated time by the hybridization values given here. ●, Hybrid values with ptE DNA; ■, difference between hybrid values with ptE-D DNA and ptE DNA; ▲, hybrid values with ptC-A DNA; ○, hybrid values with ptA DNA.

We grew strain W3110, wild type K12, in repressed conditions (excess tryptophan) at 37° C with vigorous shaking. The exponentially growing culture was cooled rapidly and centrifuged in the cold. Cells were washed with cold minimal medium⁶ and transferred at 2×10^9 cells/ml. to prewarmed minimal medium containing 1·0 per cent glucose and a mixture of nineteen amino-acids at 0·5 mM each, but without L-tryptophan. Cells were shaken vigorously in a water bath at 30° C. RNA was pulse-labelled by exposure of the cells to ³H-uridine (20 μCi/ml., 18 Ci/mmole) for 30 s, beginning 19·5 min after derepression. The pulse-labelling was followed by dilution with unlabelled uridine, added to a final concentration of 1 mg/ml. At various times afterwards the cultures (5 ml. each) were poured onto 30 ml. of crushed frozen medium containing 1×10^{-2} M tris HCl buffer at pH 7·3, 5×10^{-3} M MgCl$_2$ and 1×10^{-2} M NaN$_3$. RNA was prepared by phenol extraction as described previously⁷. Trp-mRNA was detected by specific hybrid formation between the DNA of phage φ80 carrying different trp-regions of the E. coli chromosome (Fig. 1) and labelled mRNA⁷. The results are presented as c.p.m. of ³H/100 μg total RNA hybridized with 5 μg (an excess amount) of φ80 pt DNA. The counts trapped by φ80 DNA are subtracted in each case. The averages of duplicate determinations are presented.

The experiment presented in Fig. 2 was performed to determine the orientation of trp-mRNA degradation in vivo. A short pulse of ³H-uridine is given during steady state transcription to label all the segments of the trp-mRNA. The addition of unlabelled uridine then rapidly dilutes the radioactivity of the uridine derivatives in the RNA precursor pools. The rate of appearance of label in trp-mRNA declines rapidly after the "chase" with unlabelled uridine, becoming undetectable after 5 min. Such lingering incorporation of label can reasonably be assumed to be proportionate for all segments of the trp-mRNA. Fig. 2 shows that labelled trp-mRNA corresponding to the E gene (EmRNA) began disappearing relatively quickly (on the order of 1 min) after the pulse label was diluted by unlabelled uridine. The messenger segments corresponding to the D gene (DmRNA) and C-A genes (C-AmRNA) started to disappear 2 to 3 and 5 min, respectively, after the label was diluted. Almost all label was removed from the trp-mRNA molecules by 10 min for EmRNA, 15 min for DmRNA and 25 min for C-AmRNA. The obvious conclusion from this experiment is that degradation of the polycistronic trp-mRNA molecules proceeds in a sequential fashion from the E gene to the A gene.

The time lag in the start of disappearance of various trp-mRNA segments shown in the previous experiment indicates an inequality in the rates of synthesis and degradation of the messenger. In our experiments about 8 min are required to synthesize a complete trp-mRNA molecule⁵. As indicated, total disappearance of C-AmRNA lags 10 min or so behind that of E-DmRNA. It therefore appears that in the conditions used the degradation of the trp-mRNA is slower by a factor of about 2 than the synthesis of the messenger. This relatively slow degradation of trp-mRNA was confirmed in experiments presented in Fig. 3, in which the mode of accumulation of the trp-mRNA segments was investigated during steady state derepression. An excess of ³H-uridine (100 μCi/ml., 1·22 μCi/mmole) was added to a culture 20 min after derepression and samples were removed at the times indicated. RNA was then isolated from each sample of cells and hybridized with φ80 ptE, φ80 ptE-D, φ80 ptC-A and φ80 ptA DNAs. Values are expressed as c.p.m. of hybrid/100 μg RNA (Fig. 3a). These data can be plotted as the ratio of the amounts accumulated to the amounts of the corresponding molecules synthesized during a short period of steady state transcription (Fig. 3b). As can be seen in Fig. 3b, the amount of EmRNA labelled increases for 5 min after the addition of ³H-uridine, at which time it reaches a steady state. The amounts of DmRNA and C-AmRNA reach steady state levels at 10 min and 15 min, respectively. This indicates that degradation of different segments of the trp-mRNA molecules reaches a balance with synthesis at different times. The lifetime of the different trp-mRNA segments is longer in the operator-distal portion than in the operator-proximal portion. The data presented in Fig. 3b also show that the final levels of trp-mRNA accumulation are higher in the order

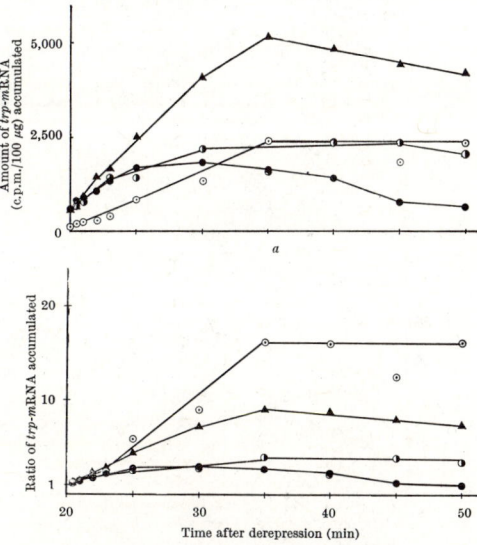

Fig. 3. Time course of accumulation of trp-mRNA molecules. a, Absolute amounts observed during continued exposure of derepressed cells to ³H-uridine; b, ratio of the total accumulated to that seen following 30 s pulse-labelling (see text). Hybridization values were 603, 587, 557 and 148 c.p.m./100 μg for E, D, C-A and AmRNA, respectively, when hybridization was carried out with ³H-RNA prepared from the same bacteria pulse-labelled for 30 s. ●, EmRNA; ◐, DmRNA; ▲, C-AmRNA; ○, AmRNA.

AmRNA, C-AmRNA, DmRNA and EmRNA. Because the values are related to the length of survival of the trp-mRNA molecules, the higher level of accumulation indicates a longer lifetime of the messenger molecules. Thus following the synthesis of trp-mRNA segments corresponding to the E, D and C-A genes, degradation begins at the E gene end. Because degradation is slower than synthesis, the operator-distal segments of the trp-mRNA molecules remain intact for a longer period than the operator-proximal segments. Thus we again estimate that the average rate of degradation of trp-mRNA in these conditions is about half the rate of synthesis. More trp-mRNA segments from operator-distal genes are present at steady state than segments from operator-proximal genes. Hence the time lag in the disappearance of different trp-mRNA segments (Fig. 2).

We have found that degradation of trp-mRNA progresses sequentially from the E gene to the A gene in various physiological conditions (to be published); we have studied the cases of degradation of trp-mRNA synthesized during a single round of transcription induced by ordinary derepression for a brief period, of trp-mRNA synthesized following addition of the tryptophan analogue, indole-3-propionic acid[1], and of trp-mRNA degraded in the presence of inhibitors such as actinomycin D and puromycin. We observed that the rate of degradation was significantly stimulated in the presence of tryptophan. This agrees with the findings of Morse, Mosteller, Baker and Yanofsky (see following paper[2]), where degradation of trp-mRNA is considerably faster than in our experiments.

The foregoing experiments suggest an exonucleolytic mechanism of degradation of trp-mRNA. This possibility was examined in sedimentation studies using trp-mRNA pulse-labelled with ^{3}H-uridine (20 μCi/ml., 18 Ci/mmole). The conditions of labelling were as in Fig. 2.

Fig. 4a shows sedimentation profiles for trp-mRNA from bacteria pulse-labelled in the steady derepressed state and collected immediately. Most of the trp-mRNA molecules corresponding to the C-A region (C-AmRNA) sediment faster than 23S RNA. It is known that the C-A region of mRNA is at the operator-distal portion of the polycistronic trp-mRNA molecule[10]. As expected, many of the trp-mRNA molecules corresponding to the

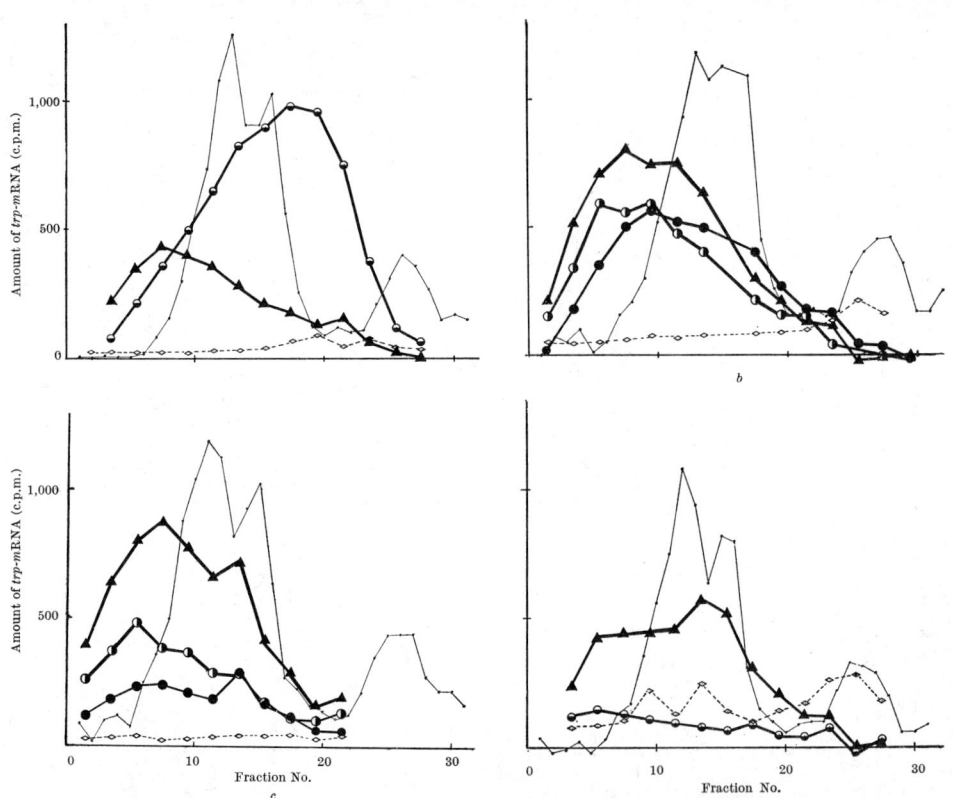

Fig. 4. Sedimentation profiles of trp-mRNA molecules in the process of degradation. The conditions of pulse-labelling and preparation of ^{3}H-RNA are the same as in Fig. 2. 600–700 μg of RNA was sedimented in each case in a 5 to 30 per cent linear sucrose gradient containing 0·01 M $tris$-HCl buffer at pH 7·3 and 0·05 M KCl, for 5 h at 36,000 r.p.m. in the cold in an SW39. After centrifugation the bottom of the tube was punctured and fractions were collected. After determining the absorbance in each fraction, two neighbouring fractions were combined and a portion hybridized with 5 μg of heat-denatured phage DNA in a 0·25 ml. reaction mixture. The time of harvest of cells for a, b, c and d is 0, 2·5, 5·0 and 10·0 min after pulse-labelling, respectively. The values presented were obtained with a one-fourth portion of each sample. ●, EmRNA; ◐, DmRNA; ◓, E-DmRNA; ▲, C-AmRNA; ◇----◇, hybridization value with φ80 DNA (background value); ●—●, optical density at 260 nm.

E-D region (E-DmRNA) sediment more slowly than those containing the C-A region. The sedimentation profile of pulse-labelled RNA prepared from derepressed wild-type cells generally exhibits this kind of heterogenous distribution[10], presumably because of the varied sizes of trp-mRNA molecules in the process of synthesis.

Fig. 4b–d presents the sedimentation profiles of trp-mRNA molecules pulse-labelled as in Fig. 4a, followed by a further incubation for 2·5, 5·0 and 10·0 min, respectively, in the presence of unlabelled uridine. In Fig. 4b the distribution pattern is shifted uniformly to the faster sedimenting region of the profile. This indicates that most of the pulse-labelled molecules continued to grow following the addition of unlabelled uridine. Because in the conditions used the dilution of pulse-labelling by unlabelled uridine is complete after about 5 min (Fig. 2), the effect of degradation of trp-mRNA molecules would be expected to be visible in the sedimentation profiles shown in Fig. 4c, d. When the sedimentation profiles of Fig. 4b, c are compared, it is evident that the trp-mRNA segments disappear more rapidly in the order EmRNA, DmRNA and C-AmRNA. It is also clearly seen in Fig. 4d that a significant amount of C-AmRNA remains even when the bulk of the E-DmRNA has disappeared. It has been estimated previously that the length of C-A region is one-half that of the whole operon, on the basis of the molecular weight of the polypeptides specified by the trp genes[11]. It has also been shown that the trp-mRNA segments corresponding to the E gene, which region corresponds to one-fourth the length of the operon, sediment at a site near 16S RNA[11]. From the data presented in Fig. 4c, d, it is evident that trp-mRNA fragments corresponding to the operator-distal portion, where the operator-proximal portion has already been degraded, have sedimentation profiles indicating a fairly large size. This suggests that the trp-mRNA molecules are degraded by exonucleolytic action. An alternative possibility, which cannot be ruled out at present, is that the trp-mRNA molecules are first segmented in a sequential fashion from the E gene end to the A gene end by endonucleolytic action, then the small fragments are digested exonucleoytically from either end.

Potassium-activated RNase II is the enzyme thought most likely to be responsible for degradation of mRNA in vivo[12]. Evidence for the possible role of this exonuclease in mRNA degradation has been reviewed recently[13]. It has recently been reported, however, that this enzyme digests RNA in the 3′→5′ direction[12]. Another ribonuclease, polynucleotide phosphorylase, also has 3′-exonuclease activity. At present no bacterial ribonuclease with 5′-exonucleolytic activity has been found. It therefore seems plausible that degradation of mRNA in the 5′→3′ direction results from the cooperation of an endonuclease and an exonuclease with 3′-exonucleolytic activity.

The mechanism, and regulation, of messenger degradation remain a fascinating problem in molecular biology. Our recent finding that degradation of trp-mRNA molecules is specifically stimulated by tryptophan (to be published elsewhere) may offer an approach to the problem. Studies of this type are currently in progress.

We thank Dr Irving P. Crawford for his helpful suggestions during the preparation of the manuscript.

NOBUKO MORIKAWA
FUMIO IMAMOTO

Department of Microbial Genetics,
The Research Institute for Microbial Diseases,
Osaka University,
Yamada-kami, Suita, Osaka, Japan.

Received February 4, 1969.

[1] Baker, R. F., and Yanofsky, C., *Nature*, **219**, 26 (1968).
[2] Morse, D. E., Mosteller, R., Baker, R. F., and Yanofsky, C., *Nature*, **223**, 40 (1969) (following article).
[3] Yanofsky, C., and Ito, J., *J. Mol. Biol.*, **21**, 313 (1966).
[4] Matsushiro, A., Sato, K., Ito, J., Kida, S., and Imamoto, F., *J. Mol. Biol.*, **11**, 54 (1965).
[5] Ito, J., and Imamoto, F., *Nature*, **220**, 441 (1968).
[6] Vogel, H., and Bonner, D. M., *J. Biol. Chem.*, **218**, 97 (1956).
[7] Imamoto, F., Morikawa, N., Sato, K., Mishima, S., Nishimura, T., and Matsushiro, A., *J. Mol. Biol.*, **13**, 157 (1965).
[8] Deeb, S. S., Okamoto, K., and Hall, B. D., *Virology*, **31**, 289 (1967).
[9] Imamoto, F., and Ito, J., *Nature*, **220**, 27 (1968).
[10] Imamoto, F., Morikawa, N., and Sato, K., *J. Mol. Biol.*, **13**, 169 (1965).
[11] Imamoto, F., and Yanofsky, C., *J. Mol. Biol.*, **28**, 1 (1967).
[12] Nossal, N. G., and Singer, M. F., *J. Biol. Chem.*, **243**, 913 (1968).
[13] Singer, M. F., and Leder, P., *Ann. Rev. Biochem.*, **35**, 195 (1966).

Polyadenylic Acid Sequences: Role in Conversion of Nuclear RNA into Messenger RNA

Abstract. Polyadenylic acid [poly(A)] segments containing 150 to 250 nucleotides appear to be covalently linked to heterogeneous nuclear RNA (HnRNA) and messenger RNA (mRNA) in eucaryotic cells. The poly(A) is synthesized in the nucleus, and is probably linked initially to HnRNA that is ultimately transported as mRNA to the cytoplasm. Studies with inhibitors of RNA or poly(A) synthesis indicate that synthesis of poly(A) segments is independent of transcription. The poly(A) marker may prove useful to elucidate mRNA modification and transport in eucaryotic cells.

Rapidly labeled nuclear RNA from mammalian cells can be divided into two major classes (1). In the nucleolus there are large ribosomal precursor RNA molecules of two uniform sizes ($45S$ and $32S$ which undergo specific cleavage to yield the RNA eventually found in cytoplasmic ribosomes (2, 3). The second type of nuclear RNA, found outside the nucleolus, consists of molecules ranging in size up to 20,000 nucleotides. The base composition of

Table 1. Kinetics of labeling of poly(A) in nuclear and polysomal RNA of HeLa cells. Three cultures (each 5×10^7 cells in 20 ml) of HeLa cells were incubated for 25 minutes with 0.05 µg of actinomycin D per milliliter and subsequently labeled, in the continued presence of actinomycin, for 15, 25, and 150 minutes, with 1 mc of [^{3}H]adenosine. The HnRNA was extracted from detergent-cleaned nuclei without deoxyribonuclease digestion and precipitated twice with $2M$ LiCl (21). The RNA was then analyzed for radioactivity and poly(A). Polysomal DNA refers to alkali-labile radioactivity released by EDTA from polysomes prepared as in Fig. 2.

Time of labeling (min)	Radioactivity (10^4 count/min)					
	Nuclear			Polysomal		
	RNA	Poly(A)	Percent in poly(A)	RNA	Poly(A)	Percent in poly(A)
15	580	4.8	0.9	18	2.5	14
25	650	11	1.7	40	7	17
150	300	3.7	1.2	250	27	11

these molecules resembles that of DNA. This class of RNA is called heterogeneous nuclear RNA (HnRNA).

With the knowledge that mammalian cells derive ribosomal RNA from nucleolar RNA of high molecular weight (2–7), the possibility of a similar relationship between HnRNA and polysomal messenger RNA (mRNA) was raised by two findings. (i) Both HnRNA and mRNA have base compositions resembling DNA. (ii) Both are rapidly labeled after exposure of cells to radioactive nucleosides (2–5). Because both HnRNA and mRNA presumably are mixtures of molecules with different base sequences, it has not been possible to test whether HnRNA contains a precursor of mRNA. The majority of HnRNA molecules turn over in the nucleus and thus are not precursors of mRNA (5, 7). However, this finding does not necessarily exclude the possibility that mRNA molecules are derived from a fraction of metabolically active HnRNA. A satisfactory test of this possibility requires some chemical tracer, more precise than average base composition, to identify the same sequences in the two classes of RNA. Such a chemical tracer is available in the form of virus-specific nuclear and cytoplasmic RNA sequences. Cells transformed by small DNA viruses contain viral DNA, covalently integrated into cellular DNA, which is the template for synthesis of virus-specific mRNA (8, 9). In the nucleus, the virus-specific RNA is part of large HnRNA molecules, but in the cytoplasm, the molecules are smaller and uniform in size (10). These results suggest that virus-specific mRNA in transformed cells may derive from HnRNA.

Workers in many laboratories (11–13) are studying regions of mammalian RNA that are rich in polyadenylic acid [poly(A)] (14). The question of whether other nucleotides are present in the poly(A) regions has not been conclusively answered. Since these distinctive sequences are covalently linked to both HnRNA and mRNA, they might also serve as a chemical tracer to study the relationship between HnRNA and mRNA. In this paper we show that nuclear RNA has a higher content of poly(A) than does cytoplasmic mRNA after brief labeling periods, whereas after longer labeling periods the reverse is true. In addition, experiments with drugs that affect RNA metabolism indicate that poly(A) is added after transcription, and that this modification enables certain mRNA molecules to reach the cytoplasm. These results suggest that HnRNA can be specifically converted into mRNA.

Resistance to digestion by ribonuclease is the basis for different assays for poly(A) sequences in various RNA species (11–13). To facilitate a wide variety of experiments, we needed an assay that would accurately determine the size and base composition of the ribonuclease-resistant polynucleotides in large numbers of unpurified samples. Therefore, ribonuclease digests of the total HnRNA and mRNA from HeLa cells were examined directly by polyacrylamide gel electrophoresis. Figure 1 shows that digests of both types of RNA labeled with ^{32}P contain a major

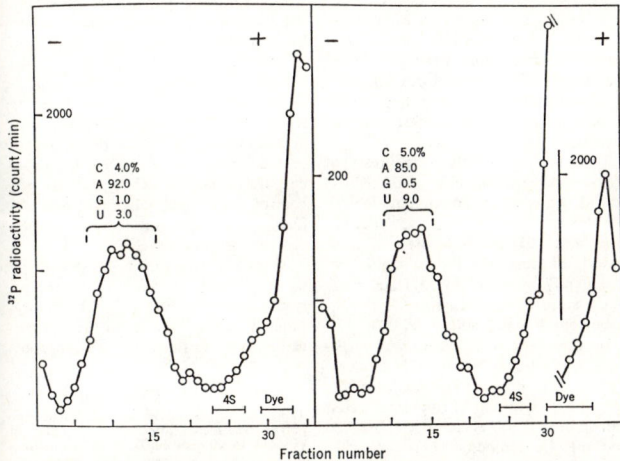

Fig. 1. Polyacrylamide gel electrophoresis of HnRNA (left) and mRNA (right) after ribonuclease treatment. HeLa cells (10^9) were labeled for 4 hours with ^{32}P in low phosphate medium, and HnRNA and mRNA were prepared (19, 20). Samples of RNA were digested, with 100 µg per milliliter of deoxyribonuclease in a solution containing $0.1M$ NaCl, $0.01M$ tris [tris(hydroxymethyl)-aminomethane], pH 7.4, and $0.001M$ MgCl$_2$. After 30 minutes at 37°C, EDTA (ethylenediaminetetraacetic acid) was added to $0.01M$, 2 µg of pancreatic ribonuclease per milliliter and 5 units of ribonuclease T1 were added, and the incubation was continued for 30 minutes at 37°C. The reaction was halted by addition of SDS (sodium dodecyl sulfate) to 0.5 percent, and the RNA was precipitated along with 100 µg yeast RNA per milliliter by addition of 2 volumes of ethanol, chilling for 1 hour at −20°C, and centrifugation at 20,000g for 45 minutes. Above the peaks are the base compositions of pooled peak fractions. Abbreviations are as follows: C, cytidylate; A, adenylate; G, guanylate; U, uridylate. The dye marker indicated was bromphenol blue.

ribonuclease-resistant fraction which migrates in the region expected for RNA molecules 150 to 250 nucleotides long (15). Polynucleotides resistant to ribonuclease and with the same electrophoretic mobility have previously been purified by chromatography on polythymidylate-cellulose columns (11). The base composition of the peaks obtained by gel electrophoresis of both HnRNA and mRNA shows that the molecules are more than 85 percent poly(A) (Fig. 1) and that prior cellulose chromotography is not necessary to obtain homogeneous peaks by gel analysis.

Using this assay, we performed several studies to elucidate, for nuclear and for cytoplasmic poly(A), the pathway of synthesis and its sensitivity to drugs. Cells labeled with [³H]adenosine for 15, 25, and 150 minutes were fractionated, and poly(A) content of HnRNA and mRNA was measured. Table 1 shows that the ratio of radioactivity in nuclear poly(A) to that in cytoplasmic poly(A) was 1.9 at 15 minutes and 1.6 at 25 minutes, and then fell to 0.14 after 150 minutes of labeling. Thus, poly(A) in the nucleus becomes labeled prior to that in polyribosomes, a result consistent with (but not proof of) the derivation of cytoplasmic from nuclear poly(A) sequences. A much higher percentage of adenosine label from polyribosomes is recoverable as poly(A) than is label

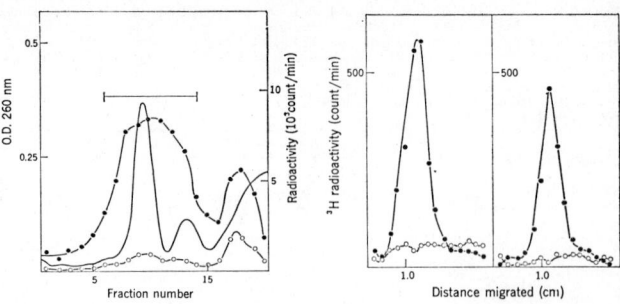

Fig. 2 (left). Isolation of labeled mRNA from cells treated with a high dose of actinomycin D. Polyribosomes labeled with [³H]adenosine, prepared from the cultures of the experiment described in Table 2, were treated with EDTA buffer (0.01M EDTA, 0.01M NaCl, 0.01M tris) and resedimented through sucrose gradients containing EDTA (20). The optical density at 260 nm was the same for samples from both cultures and is given by the solid line. The alkali-labile radioactivity for cells treated before labeling with a high level of actinomycin D (5 μg/ml) is shown by open circles; and radioactivity for control cells is shown by solid circles. The RNA isolated from the region indicated by the enclosed bar was considered to represent mRNA. Fig. 3 (right). Polyacrylamide gel pattern of poly(A) in HnRNA (left) and mRNA (right) from cells treated with cordycepin (open circles) and control cells (solid circles). The nuclease digests were adsorbed and eluted from Millipore filters according to Lee et al. (12) and subjected to electrophoresis as in Fig. 1. Only the region of the poly(A) peaks is shown.

from HnRNA. The radioactivity in these studies remains almost exclusively in adenylate; for example, only 15 percent of the total label in nuclear RNA was in guanylic acid after 150 minutes of labeling, and 5 percent or less was in guanylic acid after the shorter labeling periods.

To determine whether the presence of poly(A) in HnRNA and mRNA was dependent on continuous DNA-dependent transcription, we labeled cells with [³H]adenosine after various periods of treatment with a high dose (5 μg/ml) of actinomycin D, a drug which stops almost instantly all DNA-dependent RNA synthesis (2, 16).

In the first such experiment cells were treated initially for 25 minutes with a low dose of actinomycin D, which stops synthesis of ribosomal RNA precursors but allows continued HnRNA and mRNA synthesis (17). The culture was then divided, half was treated with a high dose of actinomycin D for 1.5 minutes, and both portions were then labeled for 25 minutes with [³H]-adenosine. Polyribosomal mRNA and HnRNA were prepared and analyzed for total radioactivity and incorporation into poly(A). In this experiment labeled mRNA was the radioactivity associated with polysomes that could be released by EDTA (ethylenediaminetetraacetic acid) disruption of polysomes (Fig. 2). Most of the labeled RNA

after short labeling periods is present in the nucleus (12), and Table 2 shows that HnRNA synthesis was suppressed by 96 percent in the culture treated with high actinomycin D. The total incorporation into mRNA was reduced by 93 percent. Synthesis of poly(A) was not suppressed nearly so much. In the presence of actinomycin D poly(A) incorporated into HnRNA was 40 percent of the control value, and poly(A) incorporated into mRNA was 25 percent of the control value.

A second experiment was performed to determine the effects of longer periods of exposure to high concentrations of actinomycin D on the incorporation of adenosine into poly(A). Table

Table 2. Effect of actinomycin D on the incorporation of adenosine into HnRNA, mRNA, and poly(A). Two cultures of HeLa cells (7.5 × 10⁷ in 20 ml) were treated with low doses of actinomycin D (0.05 μg/ml) for 25 minutes. Actinomycin D at a high level (5 μg/ml) was added to one culture 1.5 minutes before a 25-minute exposure to [³H]adenosine and the other culture was maintained in low actinomycin for the labeling period. The cells were fractionated as described in Table 1. Because actinomycin inhibits incorporation into RNA more than into DNA, a larger fraction of the extracted radioactivity is in DNA in such samples. Therefore, the gel analysis of poly(A) was carried out after the poly(A) Millipore purification technique of Lee et al. (12). This technique allows total recovery of poly(A) and eliminates other ribonuclease-resistant material (primarily DNA).

Total radioactivity (10³ count/min)		Poly(A) radioactivity (10³ count/min)	
Control	Actinomycin	Control	Actinomycin
Heterogeneous nuclear RNA			
6800	260	56.69	22.13
Messenger RNA			
58.0	4.4	7.42	1.66

Table 3. Incorporation of adenosine into poly(A) segments after different times of exposure to actinomycin D. HeLa cell cultures (10⁸ cells) were treated with actinomycin D (5 μg/ml) for 1, 5, and 15 minutes, respectively. Each sample was then labeled, in the continued presence of actinomycin, with 1 mc of [³H]adenosine for 40 minutes, and nuclear and polysomal RNA were prepared and analyzed as described in Tables 1 and 2 and Fig. 1.

Actinomycin D pre-treatment (min)	Radioactivity (10² count/min)	
	Nuclear poly(A)	Polysomal poly(A)
1	410	79.0
5	217	20.5
15	31	3.0

Table 4. Differential effect of actinomycin D on labeling of total HnRNA and of poly(A) in HnRNA. Two HeLa cell cultures (1.5 × 10⁸ cells each) were incubated in low actinomycin D (0.05 µg/ml) for 25 minutes. Half of each culture was treated with high actinomycin D (7.5 µg/ml) for 1.5 minutes and both portions of each culture were labeled for 2 or 4 minutes with 1.5 mc of [³H]adenosine in the continued presence of actinomycin (7.5 µg/ml in actinomycin cells, 0.05 µg/ml in control cells). The HnRNA was extracted and purified by sucrose gradient sedimentation, and total incorporation into alkali-labile acid-precipitable material and into poly(A) was measured.

Culture	Radioactivity (10³ count/min)	
	Total HnRNA	Poly(A) in HnRNA
2-minute label		
Control	320	2.70
Actinomycin D	36	1.35
4-minute label		
Control	830	9.05
Actinomycin D	120	4.18

3 shows that cells treated for 5 minutes with high actinomycin D incorporate adenosine into poly(A), but incorporation is almost completely stopped by a 15-minute treatment with high actinomycin D.

In the two experiments described above, we used long labeling periods (25 to 40 minutes) after addition of actinomycin, so that newly labeled molecules would be given time to appear in polyribosomes. However, for the most sensitive test of the relation between transcription and synthesis of poly(A) in RNA, we should examine HnRNA synthesis during very short labeling periods. Therefore, cells were labeled for 2 or 4 minutes with and without a 1.5-minute exposure to a high dose of actinomycin D. The HnRNA was extracted, and total radioactivity and radioactivity in poly(A) were measured. In these very brief labeling periods actinomycin D caused a 7- to 9-fold reduction in the total incorporation but only about a 2- to 2.5-fold reduction in incorporation into the poly(A) of the HnRNA (Table 4).

Thus, actinomycin D treatment has a much greater effect on continued transcription than on poly(A) synthesis, results which suggest that the poly(A) segment may be added after transcription is completed. In addition, it appears that HnRNA, which turns over at a very high rate, is only available for a brief time after transcription for the addition of poly(A).

Penman, Rosbach, and Penman (*18*) have described a reduction in the appearance of mRNA labeled with uridine in polysomes of cells treated with cordycepin (3'-deoxyadenosine), although this drug had little effect on incorporation of uridine into HnRNA. Since cordycepin is an analog of adenosine, it seemed possible that its effect on labeling of mRNA was caused by interference with posttranscriptional addition of poly(A) to preformed chains of HnRNA. To investigate this question, we labeled cells with [³H]-adenosine for 30 minutes with or without a 10-minute prior treatment with cordycepin. The HnRNA and mRNA were prepared and assayed for poly(A). Table 5 and Fig. 3 show a tenfold reduction, in the presence cordycepin, of labeled poly(A) in HnRNA and in mRNA. There was, however, only a twofold reduction in label in the total HnRNA and a threefold reduction in mRNA. In the HnRNA of cells treated with cordycepin there was a reasonable amount of ribonuclease-resistant material, but it did not migrate as a discrete species.

These experiments confirm that the poly(A) in mRNA and HnRNA is of similar size and composition (*9*). The poly(A) in HnRNA becomes labeled before that in mRNA, and this labeling appears on recently formed molecules without further DNA-mediated transcription. Moreover, the blockage of poly(A) addition by cordycepin may explain the decrease in synthesis of mRNA caused by this drug.

Most HnRNA is not a precursor of mRNA but is synthesized and degraded within the cell nucleus (*2, 4, 7*). Any scheme for the use of a portion of the HnRNA to form a portion of the mRNA must distinguish between HnRNA destined for mRNA and the bulk of HnRNA. The attachment of poly(A) could be part of such a selective process.

The following scheme of mRNA biosynthesis in HeLa cells seems justified by these experiments. After transcription, HnRNA molecules are selected by some unspecified mechanism for conversion to mRNA. A poly(A) segment is attached, probably at a terminus of the HnRNA molecule. One or more nucleases might first recognize and cleave off the mRNA plus poly(A) unit and then destroy the remainder of the HnRNA, or might simply destroy all the HnRNA except for this unit. This posttranscriptional modification is necessary for mRNA molecules to reach the cytoplasm.

J. E. DARNELL, L. PHILIPSON
R. WALL, M. ADESNIK
Department of Biological Sciences,
Columbia University, New York 10027

Table 5. Effect of cordycepin on the incorporation of adenosine into HnRNA and mRNA. Two cultures (10⁶ cells each) were incubated with a low dose of actinomycin D (0.05 µg/ml) for 25 minutes. One culture was treated for the final 10 minutes with cordycepin (50 µg/ml). Both cultures were then labeled with [³H]adenosine for 30 minutes, and the total nuclear and polysomal RNA was isolated as described in Table 1.

	Radioactivity (10³ count/min)	
	Total	Poly(A)
Heterogeneous nuclear RNA		
Control	11,000	77.60
Cordycepin	6,700	11.70
Messenger RNA		
Control	46	0.46
Cordycepin	12	0.46

References and Notes

1. J. E. Darnell, *Bacteriol. Rev.* **32**, 262 (1968).
2. "Section on eucaryotic RNA," *Cold Spring Harbor Symp. Quant. Biol.* **35**, 504 (1970).
3. K. Scherrer, H. Latham, J. E. Darnell, *Proc. Natl. Acad. Sci. U.S.* **49**, 240 (1963).
4. S. Penman, K. Scherrer, Y. Becker, J. E. Darnell, *ibid.*, p. 654.
5. K. Scherrer and L. Marcaud, *Bull. Soc. Chim. Biol.* **47**, 1697 (1965).
6. J. F. Houssais and G. Attardi, *Proc. Natl. Acad. Sci. U.S.* **56**, 616 (1966); R. Soeiro, H. C. Birnboim, J. E. Darnell, *J. Mol. Biol.* **19**, 362 (1966).
7. R. Soeiro, M. H. Vaughan, J. R. Warner, J. E. Darnell, *J. Cell Biol.* **39**, 112 (1968).
8. J. Sambrook, H. Westphal, P. R. Srinivasan, R. Dulbecco, *Proc. Natl. Acad. Sci. U.S.* **60**, 1288 (1968).
9. J. Benjamin, *J. Mol. Biol.* **16**, 539 (1966).
10. U. Lindberg and J. E. Darnell, *Proc. Natl. Acad. Sci. U.S.* **65**, 1089 (1970); G. Winocur, R. A. Weinberg, M. Herzberg, in *The Biology of Oncogenic Viruses*, L. Silverstein, Ed. (North-Holland, Amsterdam, 1971).
11. M. Edmonds and M. G. Caramela, *J. Biol. Chem.* **244**, 1314 (1969); M. Edmonds, M. H. Vaughan, H. Nakazato, *Proc. Natl. Acad. Sci. U.S.* **68**, 1336 (1971).
12. Y. Lee, J. Mendecki, G. Brawerman, *Proc. Natl. Acad. Sci. U.S.* **68**, 1331 (1971).
13. J. E. Darnell, R. Wall, R. J. Tushinski, *ibid.* **68**, 1321 (1971); R. Lim and E. S. Canellakis, *Nature* **227**, 710 (1970).
14. We refer to these RNA regions as poly(A) since over 90 percent of the constituent residues are adenylic acid [Fig. 1 and (*11*)].
15. U. E. Loening, *Biochem. J.* **113**, 131 (1969).
16. E. Reich, R. M. Franklin, A. Shatkin, E. H. Tatum, *Science* **134**, 556 (1961).
17. R. Perry, *Natl. Cancer Inst. Monogr.* **14**, 73 (1964).
18. S. Penman, M. Rosbach, M. Penman, *Proc. Natl. Acad. Sci. U.S.* **67**, 1678 (1970).
19. R. Soeiro and J. E. Darnell, *J. Mol. Biol.* **44**, 557 (1969); J. E. Darnell and A. Balint, *J. Cell Physiol.* **76**, 349 (1970).
20. S. Penman, *J. Mol. Biol.* **17**, 117 (1966); D. Baltimore and M. Girard, *Proc. Natl. Acad. Sci. U.S.* **56**, 741 (1967).
21. Supported by grants from NIH (CA 11159-03), NSF (GB 8497), and the American Cancer Society (410-3070-2372). J.E.D. is a career research scientist of the City of New York. L.P. holds an Eleanor Roosevelt International Cancer Fellowship awarded by the International Union of Cancer. R.W. and M.A. are Damon Runyon fellows. We thank R. Tushinski and Z. Davidovits for excellent technical assistance.

25 June 1971

Physical Linkage between 5 s, 16 s and 23 s Ribosomal RNA Genes in *Bacillus subtilis*

WALTER COLLI†, ISSAR SMITH AND MICHIO OISHI

*The Public Health Research Institute of the City of New York, Inc.
455 First Avenue, New York, N.Y. 10016, U.S.A.*

(*Received 8 June 1970*)

The arrangement of 5 s, 16 s and 23 s ribosomal RNA genes on the *Bacillus subtilis* chromosome has been studied by means of Cs_2SO_4–$HgCl_2$ density gradient centrifugation of rRNA–DNA hybrids. The DNA base sequences complementary to the three rRNA species are completely linked in fragments of DNA with a single strand molecular weight of 2 to 4×10^6. In DNA fragments possessing an average single strand molecular weight of 1×10^6, 5 s and 23 s rRNA genes showed complete linkage to each other and these genes showed partial linkage to the genetic determinants for 16 s rRNA. Complete linkage of 5 s RNA genes to 23 s RNA genes was still found when DNA with a single strand molecular weight of 3 to 5×10^5 was used, but little or no linkage between 5 s RNA genes and 16 s RNA genes was observed with DNA of this molecular weight. On the basis of these results, we conclude that the 5 s and 23 s rRNA genes are much more closely linked to each other than to the 16 s rRNA genes and that the 5 s and 16 s rRNA genes must be separated by a DNA "space".

1. Introduction

Using a technique of synchronous gene replication, the chromosomal location of structural genes for ribosomal RNA in *Bacillus subtilis* has been studied (Oishi & Sueoka, 1965; Dubnau, Smith & Marmur, 1965; Smith, Dubnau, Morell & Marmur, 1968). However, these experiments provided little information concerning the fine structural organization of these genes. Recently, by utilizing a method involving Cs_2SO_4–$HgCl_2$ isopycnic centrifugation (Nandi, Wang & Davidson, 1965), of rRNA–DNA hybrids, we have concluded that the genes coding for 16 s and 23 s rRNA are linked in DNA fragments with an average single strand molecular weight of approximately 2 million (Colli & Oishi, 1969). This result suggested that in bacteria the redundant genes coding for both rRNA species (Oishi & Sueoka, 1965; Smith *et al.*, 1968) alternate with one another on the chromosome in sets, each set containing one 16 s and one 23 s rRNA gene. Since it was previously shown that the structural genes for 5 s rRNA, a component of the 50 s ribosome, map close to the other rRNA genes (Smith *et al.*, 1968), an effort was made to study the fine structure relationships between the genes for 5 s RNA and those for the other rRNA species.

In this paper we would like to report on the nature of the physical linkage of the 5 s rRNA genes to 16 s and 23 s rRNA genes elucidated by Cs_2SO_4–$HgCl_2$ density gradient centrifugation and to discuss the possible arrangement of the rRNA genes in bacteria.

† Present address: Conjunto das Quimicas, Cidade Universitaria, Bloco 10, Sao Paulo, Brazil.

2. Materials and Methods

(a) Bacterial strains

Bacillus subtilis 168W was used to prepare 5 s rRNA, while 168 *ura,his* was the source of 16 s and 23 s rRNA. DNA was isolated from 168 *thy, ind, str, ery*.

(b) Preparation of radioactive RNA

For ^{32}P-labeled 5 s rRNA, 168W was grown in 150 ml. of a low phosphate medium (1 liter contained: 3 g NaCl; 0·4 g $MgSO_4 \cdot 7H_2O$; 4 mg $CaCl_2 \cdot 2H_2O$; 2 g sodium lactate; 0·4 mg $MnSO_4$; 100 mg arginine; 5 g Bacto-peptone; 5 g glucose, and was buffered with 25 mM-Tris, pH 7·4) in the presence of 330 μc of ^{32}P/ml. at 34°C. When the culture grew to a level of 4×10^8 cells/ml., it was diluted with an equal volume of high phosphate medium (the same medium described above plus 7 g K_2HPO_4 and 3 g KH_2PO_4/l.) and the cells were allowed to grow for an additional 1 hr after which they were chilled and immediately centrifuged. The purification of ^{32}P-labeled 5 s RNA, including Sephadex G100 and MAK† column chromatography, was as described (Morell, Smith, Dubnau & Marmur, 1967). The specific activity of the ^{32}P-labeled 5 s RNA preparation ranged from 1 to 2×10^6 cts/min/μg. ^{3}H-labeled 16 s and 23 s RNA were prepared as reported previously (Oishi & Sueoka, 1965).

(c) Isolation, shearing and denaturation of DNA

DNA was isolated from 168, *thy, ind, str, ery* as previously described (Colli & Oishi, 1969). The DNA (50 μg/ml.) in 0·1 × SSC (SSC is 0·15 M-NaCl + 15 mM-sodium citrate) was sheared in a Sorvall Omnimixer (30 min at 41,000 rev./min; 0°C). The sheared DNA was denatured by alkali and neutralized by NaH_2PO_4 as previously described (Colli & Oishi, 1969).

(d) Separation of DNA complementary strands by methylated albumin kieselguhr column and hydroxylapatite chromatography

After denaturation, the DNA complementary strands were separated on methylated albumin kieselguhr columns as previously described (Colli & Oishi, 1969). The fractions containing hybridizing activity toward rRNA were pooled and were extensively dialyzed against 5 mM-Na_2SO_4 + 5 mM-sodium borate buffer, pH 8·0. The single strand DNA samples of different molecular weights after self-annealing (see below) were passed through hydroxylapatite columns (Colli & Oishi, 1969, 1970) to separate single-stranded DNA from renatured molecules.

(e) Preparation of single-strand DNA–ribosomal RNA hybrids with different molecular weights

Three DNA preparations were used in the experiments presented here. We have previously found that reannealing of single-strand DNA after MAK column chromatography and hybridization with rRNA at 68°C frequently causes single-strand scission, resulting in DNA fragments with molecular weights lower than 5×10^5 (see experiment 3, Table 1, Colli & Oishi, 1969). To obtain DNA samples with molecular weights greater than 1×10^6 reproducibly, reannealing and RNA hybridizations were performed at 37°C in the presence of 30% formamide (Bonner, Kung & Bekhor, 1967; Colli & Oishi, 1970).

(i) 2 to 4×10^6 molecular weight DNA

The pooled fraction from the MAK column (fractions 64 to 70, Fig. 1) was dialyzed against 5 mM-Na_2SO_4 + 5 mM-sodium borate buffer, pH 8·0 and was then incubated in the presence of 0·3 M-NaCl, 10 mM-sodium phosphate, pH 7·0, and 30% formamide at 37°C for 24 hr. The sample was then dialyzed against 50 mM-sodium phosphate, pH 7·0, and renatured DNA was removed from single strands by hydroxylapatite column chromatography (Colli & Oishi, 1969, 1970). The purified single-stranded DNA was dialyzed against 12 l. of 10 mM-Na_2SO_4 + 10 mM-sodium borate buffer, pH 8·0. After dialysis, portions of the single-

† Abbreviation used: MAK, methylated albumin kieselguhr.

stranded DNA sample (13·6 μg) were incubated with either 1·7 μg of ³H-labeled 16 s RNA (spec. act. = 1·27 × 10⁵ cts/min/μg) or 2·4 μg of ³H-labeled 23 s RNA (spec. act. = 1·30 × 10⁵ cts/min/μg) in the presence of 0·3 M-Na₂SO₄, 10 mM-sodium borate buffer, pH 8·0, and 30% formamide (total volumes 2·7 ml.), at 37°C for 24 hr. After the incubation, the samples were dialyzed for 24 hr against 2 × 3 l. of 10 mM-Na₂SO₄+10 mM-sodium borate buffer, pH 8·0. The dialyzed samples were then incubated with pancreatic RNase (4 μg/ml.) and RNase T₁ (5 units/ml.) at 37°C for 15 min in order to destroy excess RNA. Immediately after RNase treatment, the samples were centrifuged in Cs₂SO₄–HgCl₂, as described below.

(ii) *1 × 10⁶ molecular weight DNA*

A portion of the dialyzed pooled fraction in 5 mM-Na₂SO₄+5 mM-sodium borate buffer, pH 8·0 (tubes 64 to 70, Fig. 1) was further sheared in the Sorvall Omnimixer to give a molecular weight of 1·0 × 10⁶. This sample was reannealed at 37°C, in the presence of 30% formamide and portions were hybridized with 16 s rRNA and 23 s rRNA, as described above for the 2 to 4 × 10⁶ mol. wt DNA sample.

(iii) *3 to 5 × 10⁵ molecular weight DNA*

This DNA was prepared as described previously (Colli & Oishi, 1969) including dialysis of the pooled MAK fractions against 0·1 × SSC, reannealing at 68°C in the presence of 0·3 M-NaCl–10 mM-sodium phosphate, pH 7·0, hydroxylapatite column chromatography and hybridization with 16 s or 23 s RNA at 68°C in the presence of 0·15 M-Na₂SO₄ and 10 mM-sodium borate buffer, pH 8·0. Unhybridized RNA was removed as described above.

(f) *Cs₂SO₄(Hg²⁺) isopycnic centrifugation of the DNA–RNA hybrid*

This method was essentially the same as described by Colli & Oishi (1969). After hybridization with rRNA and RNase digestion, the samples were mixed with 30 μg of native DNA (double-stranded marker), sodium borate buffer, final concentration 20 mM, pH 9·0, solid Cs₂SO₄ and HgCl₂ in this order to give a final density of 1·580 g/cm³ and an R_F (Hg²⁺/phosphorus) = 0·25 to 0·30. The mixtures (usually 3·8 to 4·0 ml.) were centrifuged at 4°C for 40 hr at 36,000 rev./min in the Spinco SW50 rotor. After centrifugation, 5-drop fractions were collected and mixed with 0·3 ml. of 15 mM-NaCl and 1 mM-EDTA. The absorbance and trichloroacetic acid-precipitable radioactivity of portions of each fraction were then monitored.

(g) *Determination of molecular weights*

The molecular weights of the larger DNA samples (annealed and hybridized in the presence of formamide) were determined by centrifugation in sucrose gradients (5 to 20%) containing 0·15 M-NaCl, 15 mM-sodium citrate, pH 7·0. The equation used for the molecular weight calculation was $S_0 = 0.022 M^{0.48}$, as suggested by Eigner & Doty (1965). The standard DNA used as a reference for S values was fd phage DNA which under the conditions used had a $S_0 = 24.4$ s (Marvin & Hoffman-Berling, 1963). It was found that the once-sheared sample had an average molecular weight of 3.0×10^6 ($S_0 = 28.6$ s) with some heterogeneity (2 to 4 × 10⁶). The other sample (twice sheared) had an average molecular weight of 1.0×10^6 ($S_0 = 17$ s). The molecular weight of the smallest DNA sample, 3 to 5 × 10⁵, was determined by alkaline sucrose gradient centrifugation (5 to 20% sucrose, 0·9 M-NaCl, 0·1 M-NaOH, 1 mM-EDTA) using sheared coliphage T4 DNA as a standard. (The standard had a molecular weight of 1·5 × 10⁶, as measured in the Spinco model E analytical centrifuge.)

(h) *RNA hybridization*

Hybridizations of MAK column and alkali treated Cs₂SO₄ fractions with radioactive rRNA were performed as described previously (Smith *et al.*, 1968; Colli & Oishi, 1969, 1970).

(i) *Reagents*

All reagents were prepared and all chemicals were obtained as previously described (Smith *et al.*, 1968; Colli & Oishi, 1969, 1970).

3. Results

(a) Methylated albumin kieselguhr column chromatography of 5 s RNA genes

It was previously shown that 16 s and 23 s rRNA in *B. subtilis* are transcribed from one of the two complementary DNA strands (H strand) (Oishi, 1969; Margulies, Rameza & Rudner, 1970). We have recently shown that 16 s and 23 s rRNA hybridize with the same DNA fractions eluted from MAK columns (Colli & Oishi, 1969). Figure 1 illustrates that the single-stranded DNA fragments complementary to 5 s RNA show an MAK elution pattern identical to that of the DNA sequences for 16 s and 23 s rRNA.

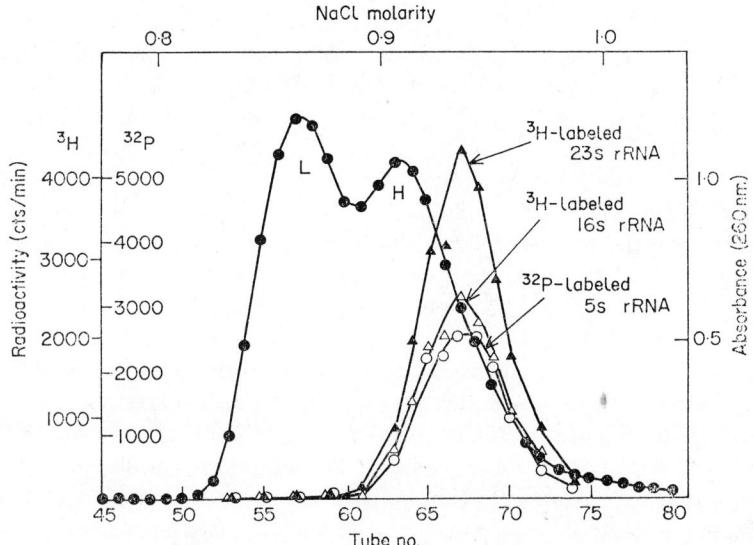

FIG. 1. Distribution of the DNA complementary to 5 s rRNA on MAK columns.

DNA isolated from *B. subtilis* (W168) was dissolved in $0.1 \times$ SSC to a final concentration of 50 μg/ml. and sheared by a Sorvall mechanical mixer to an average molecular weight of 7×10^6. The sheared DNA was denatured by alkali and neutralized (mol. wt 2 to 4×10^5). The MAK column was prepared according to Mandel & Hershey (1960), except that the quantities of all components were increased 1.5 times. Approximately 5 mg of denatured DNA were applied. The elution was performed by a linear gradient of NaCl (0.6 to 1.2 M, total 400 ml.) containing 0.05 M-sodium phosphate, pH 6.7. Total recovery was 80%. Fractions were collected (6 ml.) and the absorbance at 260 nm (—●—●—) was measured. From each indicated fraction, 0.1-ml. samples were taken for hybridization with ^{32}P-labeled 5 s rRNA (—○—○—), ^{3}H-labeled 16 s rRNA (—△—△—) and ^{3}H-labeled 23 s rRNA (—▲—▲—), under the conditions previously described (Smith *et al.*, 1968; Colli & Oishi, 1969).

This result encouraged us to study the physical linkage between 5 s RNA genes and the other two genes coding for rRNA. The approach used for this purpose, previously described (Colli & Oishi, 1969), consists of hybridizing the single-stranded DNA obtained in the MAK column with one of the species of rRNA followed by density gradient centrifugation in Cs_2SO_4–$HgCl_2$ and analysis of the displacement of the genes coding for the other two ribosomal RNA species. In the present work, DNA of three sizes were analyzed. The pooled fraction from the MAK column (tubes 64 to 70, Fig. 1) had a molecular weight of 2 to 4×10^6, which is slightly greater than the combined molecular weight of the three rRNA genes (1.7×10^6). In order to obtain smaller molecular weight DNA a portion of the pooled fraction was further sheared to a molecular weight of 1.0×10^6. Both DNA fractions were then processed in the same way which included reannealing and hydroxylapatite chromatography to eliminate any renatured DNA (Colli & Oishi, 1969). To avoid scission of the DNA strands, the reannealing was carried out by incubating the samples in the presence of 30% formamide at 37°C for 24 hours. The third DNA sample, with a single-strand molecular weight of 3 to 5×10^5, was obtained as previously described (Colli & Oishi, 1969).

(b) *Linkage of 5 s RNA genes with 16 s and 23 s RNA genes*

Single-stranded DNA binds mercuric ions more effectively than double-stranded DNA (Nandi *et al.*, 1965), and this differential binding can be used to separate these two species of DNA by Cs_2SO_4 isopycnic centrifugation as the single-stranded DNA–Hg^{2+} complex bands at a higher density than the double-stranded DNA–Hg^{2+} complex. It was previously demonstrated that DNA–RNA hybrids also band at a lower density than single-stranded DNA in the presence of mercuric ions (Colli & Oishi, 1969, 1970). This property of the DNA–RNA hybrids was used to demonstrate linkage between a 23 s RNA gene and a 16 s RNA gene (Colli & Oishi, 1969) and to purify the rRNA genes from *B. subtilis* (Colli & Oishi, 1970). The same method was applied to resolve the topological relationships between the 5 s RNA genes and the genes coding for the other two species of rRNA. The outline of the experiment is as follows: single-stranded DNA, after purification by hydroxylapatite chromatography, is hybridized with ^{3}H-labeled 23 s rRNA or ^{3}H-labeled 16 s rRNA and subjected to Cs_2SO_4 density gradient centrifugation in the presence of mercuric ion. After equilibrium, fractions are collected and the absorbance and radioactivities are measured in order to ascertain the position of the DNA–RNA hybrid relative to that of the single-stranded DNA. Portions of these fractions are subjected to alkaline hydrolysis to destroy hybridized RNA and are then *re*-hybridized with ^{32}P-labeled 5 s rRNA and the other species of rRNA which was not used for the initial hybridization. In Figures 2 and 3, single-stranded DNA with a molecular weight ranging between 2 to 4×10^6 was hybridized with ^{3}H-labeled 23 s rRNA and ^{3}H-labeled 16 s rRNA, respectively. After centrifugation, hybridization with ^{3}H-labeled 16 s rRNA in Figure 2 and ^{3}H-labeled 23 s rRNA in Figure 3 and with ^{32}P-labeled 5 s rRNA in both experiments showed that the DNA sequences homologous to the rRNA's used for *re*-hybridization were displaced together with the hybrids prepared with either 23 s or 16 s rRNA prior to centrifugation. As we have shown previously, the 23 s rRNA–DNA hybrid is shifted more than the 16 s rRNA–DNA hybrid, reflecting the difference in the size of the hybrid region on the DNA fragments. These results confirm the previous conclusion that in fragments with a molecular weight above 2×10^6 the genes coding for 16 s and 23 s rRNA are completely linked (Colli & Oishi, 1969) and furthermore indicate that with this DNA molecular weight the 5 s rRNA genes are linked to the genes coding for the other two species of rRNA.

When the same experiment was performed with a single-stranded DNA with an average molecular weight of 1.0×10^6, a different result was obtained. Figure 4 shows the pattern obtained by hybridization with ^{3}H-labeled 23 s rRNA before centrifugation followed by *re*-hybridization with ^{3}H-labeled 16 s rRNA and ^{32}P-labeled 5 s rRNA. As can be seen, approximately 50% of the 16 s rRNA sequences were displaced with 23 s rRNA sequences. This means that when the DNA fragments have a molecular size below the size of the sum of both genes, some bearing 16 s rRNA sequences contain no sequences homologous to 23 s rRNA. Therefore, prior hybridization with

23 s rRNA did not displace the former. It is apparent, however, that all the 5 s rRNA sequences were displaced together with the 23 s rRNA genes. In Figure 5, the reciprocal experiment was attempted. The low molecular weight single-stranded DNA was hybridized with ^{3}H-labeled 16 s RNA before centrifugation after which the fractions collected were *re*-hybridized with ^{3}H-labeled 23 s rRNA and ^{32}P-labeled 5 s rRNA. Although some sequences homologous to 23 s rRNA accompanied the ^{3}H-labeled 16 s rRNA–DNA hybrid displacement, others remained with the single-stranded DNA peak. *Re*-hybridization with ^{32}P-labeled 5 s rRNA produced a pattern similar to that of the ^{3}H-labeled 23 s rRNA which differed from the ^{3}H-labeled 16 s rRNA pattern.

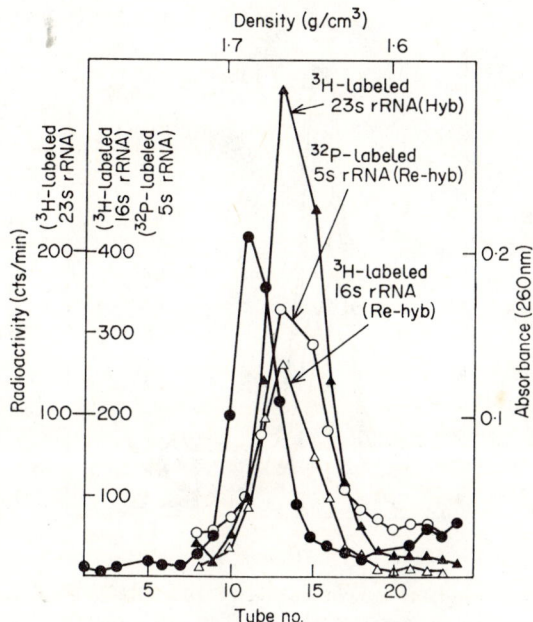

Fig. 2. Cs_2SO_4–Hg^{2+} density gradient centrifugation of 23 s rRNA–DNA hybrid (DNA mol. wt 2 to 4×10^6).

The pooled single-stranded DNA (fractions 64 to 70 from the MAK column illustrated in Fig. 1, mol. wt 2 to 4×10^6) after hydroxylapatite chromatography was dialyzed extensively against 10 mM-Na_2SO_4 + 10 mM-sodium borate buffer, pH 8·0, and a portion (13·6 µg of DNA) was incubated with 2·4 µg of ^{3}H-labeled 23 s rRNA (spec. act. $1·30 \times 10^5$ cts/min/µg) in the presence of 0·3 M-Na_2SO_4, 10 mM-sodium borate (pH 8·0) and 30% formamide (final volume 2·7 ml.) at 37°C for 24 hr. After incubation, the sample was dialyzed for 24 hr against 2×3 l. of 10 mM-Na_2SO_4 + 10 mM-sodium borate, pH 8·0. The dialyzed samples were then incubated with pancreatic RNase (4 µg/ml.) and RNase T_1 (5 units/ml.) at 37°C for 15 min to destroy unhybridized RNA. Immediately after RNase treatment, the sample was mixed with 23 µg of native DNA (reference marker), sodium borate buffer (final concentration, 20 mM, pH 9·0), solid Cs_2SO_4 and $HgCl_2$ in this order to give a final density of 1·580 g/cm³ and an R_F (molar ratio of total Hg^{2+} to total DNA phosphate) of 0·27. The mixture was centrifuged at 36,000 rev./min at 4°C for 40 hr in the Spinco SW50 rotor. After centrifugation, fractions were collected and mixed with 0·3 ml. of 15 mM-NaCl and 1 mM-EDTA. After the absorbance (—●—●—) was monitored, 0·5 ml. of the same buffer was added and each fraction was divided in 3 portions. One of the portions of each fraction was used to monitor the trichloroacetic acid-precipitable radioactivity due to ^{3}H-labeled 23 s rRNA (—▲—▲—). The other two portions from each fraction were treated overnight at room temperature with 0·1 N-NaOH, neutralized with NaH_2PO_4, and *re*-hybridized with ^{3}H-labeled 16 s rRNA (—△—△—) and ^{32}P-labeled 5 s rRNA (—○—○—) as previously published (Colli & Oishi, 1969; Smith *et al.*, 1968). The reference double-stranded DNA banded at a lighter density (approx. 1·650) and its absorbance profile was omitted for clarity. The ^{32}P-labeled 5 s rRNA had a specific activity of $1·0 \times 10^6$ cts/min/µg.

The larger shifts in density of the rRNA–DNA hybrids in this experiment (Figs 4 and 5) reflect the smaller size of the DNA and the proportionally larger hybrid region when compared to the previous experiment (Figs 3 and 4).

When DNA with a molecular weight of 3 to 5×10^5 was hybridized with 23 s rRNA before centrifugation, and then fractions were *re*-hybridized with 5 s rRNA, the results illustrated in Figure 6 were obtained. Essentially all of the sequences complementary

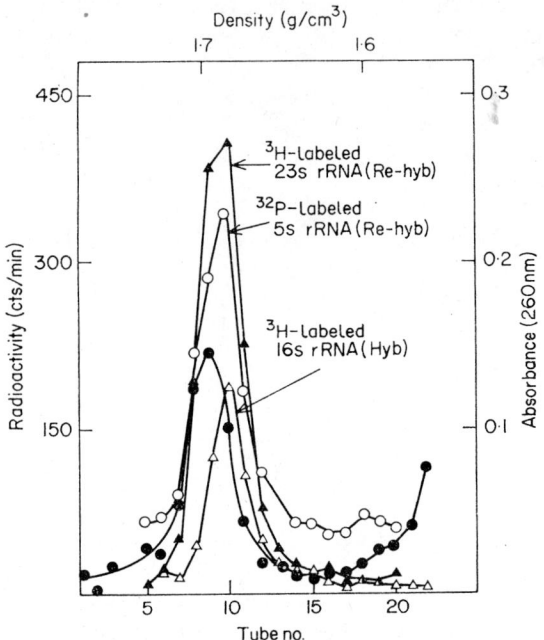

Fig. 3. Cs_2SO_4 density gradient centrifugation of 16 s rRNA–DNA hybrid (DNA mol. wt 2 to 4×10^6).

The conditions are the same as described in Fig. 2, except that the single-stranded DNA (13·6 μg) was hybridized with 1·7 μg of ³H-labeled 16 s rRNA (—△—△—) with a specific activity of $1·27 \times 10^5$ cts/min/μg before centrifugation. *Re*-hybridization was performed with ³H-labeled 23 s rRNA (—▲—▲—) and ³²P-labeled 5 s rRNA (—○—○—). Absorbance, —●—●—.

to 5 s rRNA were shifted to the position of the 23 s rRNA–DNA hybrid. On the other hand, when 16 s rRNA was hybridized with the low molecular weight DNA sample, very few of the sequences complementary to 5 s rRNA were displaced from the single-strand DNA peak (Fig. 7). As expected, the density shifts of the rRNA–DNA hybrids in these experiments (Figs 6 and 7) were greater than the previous series and reflected the low molecular weight of the DNA (only ³²P-labeled 5 s rRNA was used for *re*-hybridization in these experiments because small amounts of DNA were recovered from the gradients).

4. Discussion

Although a considerable amount of information pertaining to the genes coding for rRNA in *B. subtilis* has been obtained, no data were yet available concerning the fine structural relationships among them. Previously we have demonstrated physical linkage between the 16 s and the 23 s rRNA genes on DNA fragments with an average

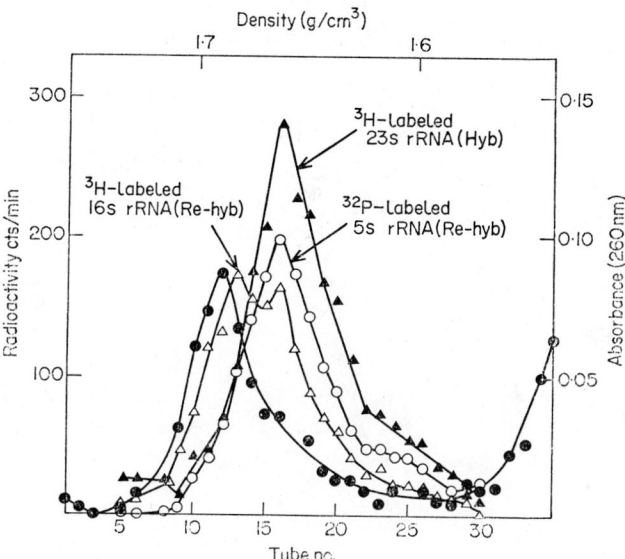

FIG. 4. Cs_2SO_4 density gradient centrifugation of 23 s rRNA–DNA hybrid (DNA mol. wt 1×10^6).
The conditions and symbols are the same as described in Fig. 2, but DNA with an average single-stranded molecular weight of 1×10^6 was used and the volume for the hybridization was 1·5 ml. This DNA sample was obtained by further shearing a portion of the pooled fractions (tubes 64 to 70) from the MAK column illustrated in Fig. 1.

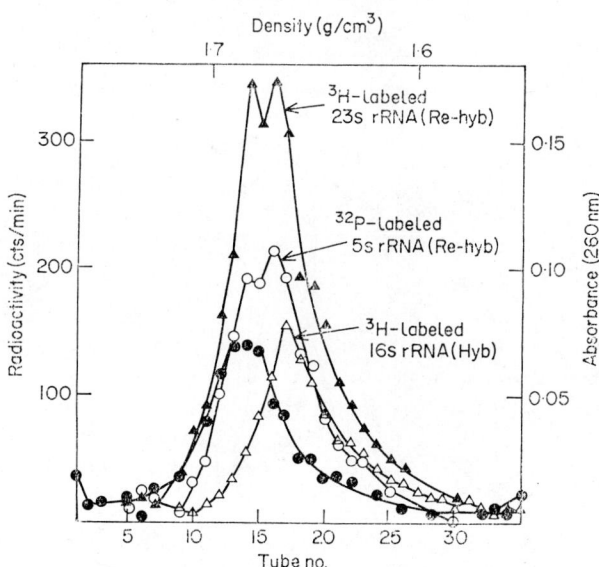

FIG. 5. Cs_2SO_4 density gradient centrifugation of 16 s rRNA–DNA hybrid (DNA mol. wt 1×10^6). The conditions and symbols are the same as described in Fig. 3, but DNA with a molecular weight of 1×10^6 was employed and the volume for the hybridization was 1·5 ml.

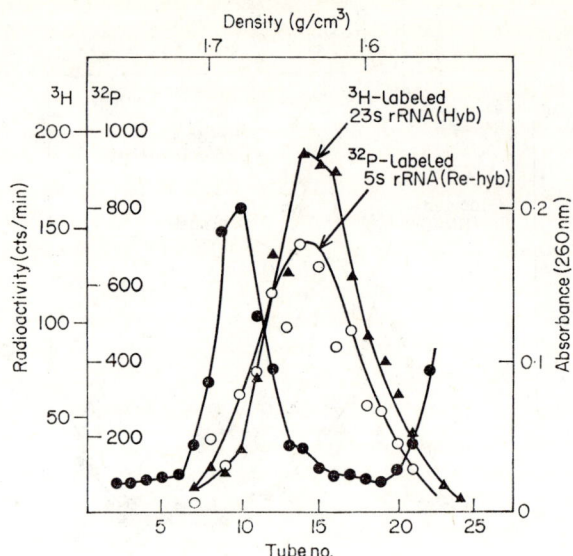

Fig. 6. Cs_2SO_4 density gradient of 23 s rRNA–DNA hybrid (DNA mol. wt 3 to 5×10^5). The conditions and symbols were as described in Fig. 2, but the single-stranded rRNA–DNA hybrid was prepared by reannealing and hybridization with 23 s rRNA at 68°C. 15 μg of DNA were hybridized with 1·5 μg of ^{3}H-labeled 23 s rRNA. Because of the low amounts of DNA recovered from the gradient, only 5 s rRNA was used for rehybridization. The ^{32}P-labeled 5 s rRNA had a specific activity of 2×10^6 cts/min/μg.

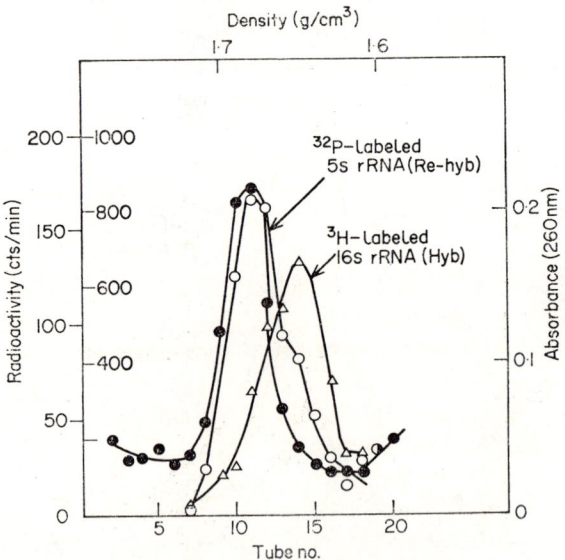

Fig. 7. Cs_2SO_4 density gradient centrifugation of 16 s rRNA–DNA hybrid (DNA mol. wt 3 to 5×10^5).

The conditions were as described in Fig. 6, but 0·8 μg of ^{3}H-labeled 16 s rRNA was used to form the hybrid.

molecular weight of 2×10^6 (Colli & Oishi, 1969). Since the expected molecular weight of both structural genes combined is 1.66×10^6, our results strongly suggested that these redundant genes for both rRNA species were interspersed among each other, forming sets, each set containing one gene for each species (Oishi & Sueoka, 1965; Smith *et al.*, 1968). Such tandem genes for ribosomal RNA and evidence for spacing between these sets have been described in higher organisms (Quagliorotti & Ritossa, 1968; Birnstiel, Spiers, Purdom, Jones & Loening, 1968; Brown & Weber, 1968*a,b*; Miller & Beatty, 1969). More recently it has been shown that in *Proteus mirabilis* the ribosomal tandem genes are widely spaced (Purdom, Bishop & Birnstiel, 1970). The results shown in the present paper show that in DNA fragments with a molecular weight 2 to 4×10^6, the 5 s rRNA genes are also linked to the genes coding for the other two species (Figs 2 and 3). However, when the DNA fragments have a molecular weight of 1×10^6, a breakdown of linkage between the 16 s and the 23 s genes is observed. Under this condition the 5 s genes remain completely linked to the 23 s genes but are approximately 50% linked to the 16 s genes. With DNA fragments of 3 to 5×10^5 molecular weight, 5 s and 23 s rRNA genes are still completely linked, while little, if any, linkage is observed between 5 s and 16 s rRNA genes. This indicates that the 5 s rRNA and the 23 s rRNA genes are more closely linked to each other than they are to 16 s rRNA genes. If these genes form sets which contain one gene for each rRNA species, it must be concluded that a 5 s gene is not contiguous to a 16 s gene. The DNA "spacer" separating the genes for these RNA species probably has a molecular weight of 1×10^6 or less (since the 5 s and 16 s rRNA genes are more than 50% linked in DNA fragments with a molecular weight of 1×10^6) but greater than 3 to 5×10^5 (little or no linkage is observed between 5 s and 16 s rRNA genes with DNA fragments of 3 to 5×10^5).

Some possibilities for the order of the rRNA genes on the chromosome are shown in Figure 8. Before deciding on which model is most likely, the problem of contiguity of adjacent rRNA sets must be considered. If the rRNA sets are separated on the chromosome so that linkage between two would not occur with DNA of the molecular weight utilized in these experiments, the third model is favored. The first and second possibilities would be unlikely because the 5 s and 16 s rRNA are separated by a "space" with a molecular weight greater than 1×10^6. As shown above (Fig. 5), the genes for these two species are linked in excess of 50% at this molecular weight. The "space" in the third model would have a molecular weight of less than 1×10^6 but greater than $\sim 4 \times 10^5$. If, however, adjacent rRNA gene sets are linked in DNA fragments of the size utilized in these experiments, the first and second models would be possible. The only requirement would be that the 16 s rRNA gene in the adjacent set would be close enough to the 5 s rRNA gene in the original set. Thus, until more is known about the spacing of adjacent rRNA gene sets on the *B. subtilis* chromosome, it is difficult to favor any of these models.

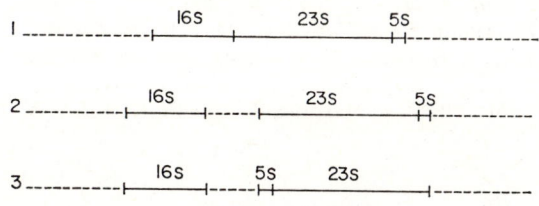

FIG. 8. Possible orders for rRNA genes on the *B. subtilis* chromosome.

From the results presented here, it is tempting to speculate on some problems of biosynthesis of these RNA species *in vivo*. First of all, the evidence of the close linkage of 5 s and 23 s rRNA genes reported here may indicate that both genes form an operon and that these RNA molecules, components of 50 s ribosomes, are transcribed under a common regulatory mechanism. Possibly 5 s and 23 s rRNA are transcribed at first as a contiguous molecule which is then split into two species before or at the time of 50 s ribosome assembly. Using a kinetic analysis, others have also postulated a common 5 s–23 s rRNA operon in *B. subtilis* (Bleyman, Kondo, Hecht & Woese, 1969). In this respect, it is noteworthy that in higher organisms, i.e. *Xenopus*, genes for 5 s rRNA are not linked to the 18 s and 28 s rRNA genes (Brown & Weber, 1968a), unlike the situation in *B. subtilis*. As discussed above, the technique used here could not resolve the question of the existence of a small space between 16 s rRNA genes and 5 s–23 s rRNA genes as suggested for 18 s and 28 s rRNA genes in *Xenopus* (Brown & Weber, 1968b). One of the possible approaches to elucidate the fine structure of the rRNA genes in bacteria, including the problem of such a space, would be the direct observation of the rRNA–DNA hybrid under an electron microscope, after extensive purification of the hybrid which has been recently achieved in this laboratory (Colli & Oishi, 1970).

We thank Misses Gloria Slywka and Sharon Pifko for expert technical assistance. Dr Loren Day for a gift of fd bacteriophage, Drs David Dubnau, Paul Margolin and Leonard Mindich for helpful discussions, and Mrs Annabel Howard for invaluable secretarial aid.

This work was supported by a National Science Foundation grant no. GB14313, awarded to one of us (M. O.), and by a National Science Foundation grant no. GB16782 and American Cancer Society grant no. E380C awarded to another of the authors (I. S.). One of the authors (W. C.) was the recipient of a travel grant from Fundação de Amparo à Pesquisa do Estado de São Paulo.

REFERENCES

Birnstiel, M. L., Spiers, J., Purdom, I., Jones, K. & Loening, U. E. (1968). *Nature*, **219**, 454.
Bleyman, M., Kondo, M., Hecht, N. & Woese, C. (1969). *J. Bact.* **99**, 535.
Bonner, J., Kung, G. & Bekhor, I. (1967). *Biochemistry*, **6**, 3650.
Brown, D. D. & Weber, C. W. (1968a). *J. Mol. Biol.* **34**, 661.
Brown, D. D. & Weber, C. W. (1968b). *J. Mol. Biol.* **34**, 681.
Colli, W. & Oishi, M. (1969). *Proc. Nat. Acad. Sci., Wash.* **64**, 642.
Colli, W. & Oishi, M. (1970). *J. Mol. Biol.* **51**, 657.
Dubnau, D., Smith, I. & Marmur, J. (1965). *Proc. Nat. Acad. Sci., Wash.* **54**, 724.
Eigner, J. & Doty, P. (1965). *J. Mol. Biol.* **12**, 549.
Mandel, J. D. & Hershey, A. D. (1960). *Analyt. Biochem.* **1**, 66.
Margulies, L., Remeza, V. R. & Rudner, R. (1970). *J. Bact.* **103**, 560.
Marvin, D. A. & Hoffman-Berling, H. (1963). *Z. Naturf.* **18b**, 884.
Miller, O. L., Jr. & Beatty, B. R. (1969). *Science*, **164**, 955.
Morell, P., Smith, I., Dubnau, D. & Marmur, J. (1967). *Biochemistry*, **6**, 258.
Nandi, O. S., Wang, J. C. & Davidson, N. (1965). *Biochemistry*, **4**, 1687.
Oishi, M. (1969). *Proc. Nat. Acad. Sci., Wash.* **62**, 256.
Oishi, M. & Sueoka, N. (1965). *Proc. Nat. Acad. Sci., Wash.* **54**, 483.
Purdom, I., Bishop, J. O. & Birnstiel, M. L. (1970). *Nature*, **227**, 239.
Quagliorotti, G. & Ritossa, F. M. (1968). *J. Mol. Biol.* **36**, 57.
Smith, I., Dubnau, D., Morell, P. & Marmur, J. (1968). *J. Mol. Biol.* **33**, 724.

Submitted to:
Proc. Int. Symp. Protein Synthesis and Nucleic Acids (XI Latin American Symposium, La Plata, 1971)

Visualization of Genetic Transcription*

BARBARA A. HAMKALO AND O. L. MILLER, JR.

From the Biology Division, Oak Ridge National Laboratory,

Oak Ridge, Tennessee 37830

U.S.A.

---

*Research sponsored by the U. S. Atomic Energy Commission under contract with the Union Carbide Corporation.

SUMMARY

Electron microscopic observations of the extruded contents of osmotically shocked bacterial cultures permit the direct study of microbial genetic activity. The bacterial chromosome, ~40 Å in diameter, is stained by a protein-specific stain, suggesting that the genome is a deoxyribonucleoprotein complex. Transcription-translation complexes at sites of structural gene activity are seen as irregularly spaced chains of polyribosomes attached to the bacterial genome by RNA polymerase molecules. Putative RNA polymerases, granules ~80 Å in diameter, appear to be bound also to genetically "silent" regions of the genome.

Bacterial ribosomal RNA genes are identified as segments of the chromosome about 1.3 μ in length, with ribonucleoprotein fibrils rather than polyribosomes attached. Under rapid growth conditions, these ribonucleoprotein fibrils are closely spaced and number 60 to 80 per region. Although the ribonucleoprotein fibrils form two contiguous gradients, corresponding to the 16S and 23S ribosomal RNA cistrons, respectively, there is a single initiation site for the transcription of these two linked cistrons. There are several 16S-23S doublets per bacterial chromosome; however, unlike eukaryotic ribosomal RNA genes, these regions are separated by chromosomal segments that contain structural genes.

INTRODUCTION

Transcription of structural genes by RNA polymerase to produce messenger RNA's (mRNA's) and the translation of such messengers by polyribosomes to produce proteins are intimately coupled processes in bacterial cells (1). In fact, it is possible to reconstruct coupled transcription and translation systems <u>in vitro</u> from separated bacterial components (2,3). Using techniques developed by Miller and Beatty (4) to prepare the nuclear contents of amphibian oocytes for electron microscopy, we can now directly visualize genetically active bacterial chromosomes. In these studies, we have identified both structural genes and ribosomal RNA (rRNA) genes (5,6).

TECHNIQUE

Bacterial cultures in log phase are rendered osmotically sensitive by a brief treatment with T4 lysozyme in the presence of sucrose at 4°; the protoplasts are then osmotically shocked by rapid dilution into distilled water; and a sample of burst cells is deposited onto a carbon-coated electron microscope grid by low-speed centrifugation. Preparations are stained with phosphotungstic acid (PTA) under conditions that stain net positive groups of proteins (7). Therefore, all the structures observed have protein associated with them.

STRUCTURAL GENES

At low magnification, the extruded contents of a shocked bacterial cell appear as a network of deoxyribonuclease-digestible fibers with attached ribonuclease-sensi polyribosomes (Fig. 1). The fibers are stained with PTA and measure approximately 40 Å in diameter, twice the diameter of duplex DNA. Thus, the bacterial chromosor appears to be uniformly coated with protein. The protein may become associated with the DNA during isolation; or the DNA may, in fact, exist as a deoxyribonucleoprotein complex in the cell, as suggested by the experiments of Zubay and Watson (8). In this volume, August et al. (9) describe the isolation of several classes of histone-like proteins that could associate with the bacterial chromosome.

Figs. 2 and 3 show relatively long regions of the genome exhibiting short-to-lor polyribosome gradients. Since the length of genetically active DNA in both figure is sufficient to code for several proteins,[1] we conclude that these active segments are polycistronic operons exhibiting intimately coupled transcription and translation. The absence of free polyribosomes on grids or in the supernatant fraction after centrifugation and the presence of short polyribosomes distal to the longest polyribosome of a gradient (Fig. 3) strongly suggest that mRNA's are degraded while attached to the DNA. There is biochemical evidence (10-12) substantiating the degradation of bacterial messengers from the $5'-PO_4$ to the $3'-OH$ end, as woul be required in this case.

---

[1] Lactose operon (3 genes), ~1.4 μ; tryptophan operon (5 genes), ~2.3 μ.

Polyribosomes on active loci typically are unequally spaced, possibly because the initiation is aperiodic. If initiation were periodic, however, polyribosomes could become irregularly spaced during transcription, provided that individual polymerases transcribe at different rates.

A comparison of the number of ribosomes within a polyribosome with its relative distance from the initiation site of the locus permits an estimate of the average length of mRNA associated with a ribosome. This estimate takes into consideration the fact that, when transcribed, 1 μ of duplex DNA in the B conformation gives rise to a single-stranded RNA molecule which, if fully extended, would measure 2 μ. Based on several measurements, approximately 1000 to 1500 Å (150 to 200 nucleotides) of messenger is associated with each ribosome. Since the diameter of a ribosome would protect only about 1/20 of this amount of extended mRNA, the messenger must be rather extensively coiled in the normal polyribosome configuration. In fact, we have measured lengths of mRNA up to 2000 Å between adjacent ribosomes within polyribosomes that have been stretched during preparation.

In unstretched polyribosomes, the first ribosome attached to the mRNA typically is closely apposed to the bacterial chromosome. However, when polyribosomes are slightly stretched during preparation, a granule about 80 Å in diameter can be seen at each site of polyribosome attachment to the genome (Figs. 2 and 3). These granules undoubtedly are RNA polymerases that were actively transcribing at the time of isolation. Fig. 4 shows, at higher magnification, a portion of a negatively stained preparation. The RNA polymerases can be seen as lightly stained granules smaller than individual ribosomes. In some cases, the individual 30S and 50S subunits of a single ribosome can be resolved. The mRNA strand is seen associated primarily with the 30S subunits, in agreement with biochemical data (13).

Granules about the same size as active RNA polymerases also are bound to genetically "silent" regions of the genome, as defined by the absence of polyribosom This suggests that inactive portions of the genome may have RNA polymerases associated with promoter sites while awaiting the proper initiation signals. Althoug only qualitative estimates can be made from micrographs of burst cells, it appears that a relatively small percentage of the bacterial chromosome is active at any one instant. This observation is in agreement with a similar conclusion by Kennel (14), based on DNA-RNA hybridization with Escherichia coli nucleic acids.

RIBOSOMAL RNA GENES

Hybridization data indicate that bacterial rRNA genes reside in DNA segments containing three contiguous cistrons: 16S–23S–5S (15). There are about six such segments per E. coli chromosome (16), and it has been calculated (17, 18) that under optimal growth conditions 80 to 90 RNA polymerases must simultaneously transcribe each region in order to meet cellular requirements for ribosomes. The molecular weights of the three rRNA molecules are: 16S, 0.55×10^6; 23S, 1.1×10^6; and 5S, 0.04×10^6 daltons (19,20). Since 1 μ of duplex DNA codes for a 1×10^6-dalton single-stranded polyribonucleotide, the length of B-conformati

DNA required to accomodate the three cistrons is ~1.7 μ. These considerations can be used to predict the structure of active rRNA genes — i.e. chromosomal segments ~1.7 μ long, which under optimal growth conditions are transcribed simultaneously by a large number of closely spaced RNA polymerases, and which can be distinguished from structural genes by attached ribonucleoprotein (RNP) fibrils rather than polyribosomes.

The extruded contents of the burst cell shown in Fig. 1 exhibit a number of such regions. Positive identification of these regions as rRNA genes rests on the use of a temperature-sensitive mutant of E. coli isolated and characterized by Atherly (unpublished data). At the permissive temperature the mutant synthesizes normal amounts of rRNA; however, after a shift to the nonpermissive temperature no rRNA synthesis can be detected biochemically. The rRNA loci in preparations of mutant cultures grown at 30° are normal in appearance, but at 42° no rRNA segments are visible, although structural gene activity appears to be normal.

Fig. 5A shows a single rRNA segment. The length of such regions is about 1.3 μ, shorter than the DNA length expected from the combined molecular weights of the products. A large number of RNA polymerases transcribing adjacent to one another might foreshorten the DNA relative to its B-conformation length, because of extensive local denaturation at each polymerase site.

Each rRNA segment is composed of 60 to 70 RNP fibrils, which form two contiguous short-to-long fibril gradients. The first gradient is half as long as the second, in agreement with expectations based on the molecular-weight differences between the 16S and 23S rRNA's. A few polymerases are sometimes seen (Fig. 5B) beyond the 23S gradient; biochemical data suggest that these regions could be active 5S rRNA cistrons (15).

When cellular growth rate is reduced, the rRNA segments become increasingly difficult to visualize, because the number of RNP fibrils decreases (Fig. 5C). If one assumes that all polymerases travel at approximately the same speed, the initiation of transcription of these genes must be aperiodic, since the fibrils on these loci are unevenly spaced.

The fact that the rRNA segments are composed of two fibril gradients (Fig. 5A) suggests the occurrence of two RNA polymerase initiation sites per region but does not exclude the existence of only one initiation site. Rifampin, a drug that inhibits initiation but not elongation by RNA polymerase (21), can be used to distinguish between these two alternatives. Fig. 6 schematically depicts the configuration of rRNA genes after addition of the drug based on one and on two polymerase binding sites. With appropriate sampling times it can be seen that the 16S fibril gradient disappears first, and only then does the 23S gradient begin to disappear. In Fig. 7a complete 23S cistron is evident, although the entire 16S cistron has been cleared of fibrils. Thus, there is a single initiation site for the transcription of the linked rRNA cistrons. These observations agree with indirect data based on the relative amounts of labeled uridine in the two rRNA molecules after addition of rifamycin (22) and the appearance of label first in the 16S rRNA when amino-acid-starved cells synthesizing no rRNA (under stringent RNA synthesis control) are supplied with the required amino acid (23).

The relative physical location of bacterial rRNA segments on the chromosome is unclear. Data of Yu et al. (24) suggest that all regions are virtually contiguous, while Birnbaum and Kaplan (25) localize only half of the genes in a region about 30 μ long, and Spadari and Ritossa (26) conclude that there is some nonribosomal DNA between bacterial rRNA segments. Fig. 8 illustrates a single rRNA region bracketed by structural gene activity. Similar regions over 10 μ long have been seen, an observation which excludes the possibility that bacterial rRNA genes are closely adjacent, as are rRNA genes of most eukaryotic cells.

BACTERIOPHAGE TRANSCRIPTION AND TRANSLATION

Biochemical studies of the genetic activity of E. coli cells infected with bacteriophage T7 suggest that it should be relatively simple to identify active viral genomes. First, host transcription ceases after infection, due to a viral function (27); and secondly, the phage-induced RNA polymerase, unlike the host enzyme, is totally insensitive to inhibition by rifamycin and its derivatives, such as rifampin (28). Therefore, any transcription occurring early in the infection cycle in the presence of the drug, and all transcription later in the cycle, must be directed by the viral enzyme.

Preliminary observations indicate that our isolation procedures do permit visualization of phage genetic activity and maturation. Fig. 9 shows the extruded contents of a cell late in the infection cycle. Large quantities of genetically silent DNA, with phage-sized particles in the midst of this DNA, are visible.

The densely stained particles are mature phage, and the paler particles most probably are phage heads being filled with DNA. Unlike the extruded contents of uninfected cells, nearly all polyribosomes in these preparations appear to be associated with membrane fragments. Although no direct attachment between membrane and polyribosomes has been seen, there are several possible explanations for such a configuration: (1) the polyribosomes of infected cells associate with the membrane artifactually, during isolation; (2) the phage DNA, a molecule only 12 µ long, is attached to the bacterial membrane, and consequently structures associated with the viral DNA are found in close proximity to the membrane; or (3) T7 mRNA's are translated on membrane-associated polyribosomes. No evidence that favors any of these possibilities is available at present. However, this unique prokaryotic polyribosome configuration may be relevant to the long half-life (~ 20 min) of T7 mRNA's (29), at the same time that at least some host messengers decay with their normal half-lives (~ 3 min) (30).

CHLOROPLAST GENETIC ACTIVITY

Based on the differences in drug sensitivities between the RNA- and protein- synthesizing systems of chloroplasts and the corresponding nuclear and cytoplasmic systems (31, 32), as well as the size of chloroplast rRNA's and ribosomes (33), it has been suggested that genetic activity in these organelles is prokaryotic in nature — *i.e.*, transcription and translation are closely coupled. Opposing evidence, however, is also available (34). Fig. 10 shows the contents extruded from a Euglena gracilis chloroplast by osmotic shock. The extruded fibers are portions of the chloroplast genome, and structures closely resembling polyribosomes appear to be attached to the genome. Although preliminary in nature, these observations are direct visual support for the suggestion that transcription and translation are coupled in this system.

CONCLUSIONS

Development of rapid and simple preparative techniques have made it possible to visualize directly the structure of active genes in both eukaryotic and prokaryotic chromosomes (5). The vast amount of indirect data that has been accumulated in biochemical and genetic studies of prokaryotic chromosomes provides a background of information that can be complemented by observations of the structure of active loci, in order to analyze problems of the coordination and control of transcription and translation under different growth conditions or in strains which harbor mutations altering the coupling between these processes (e.g. polar mutants). With this multidisciplinary approach, the mechanism of action of drugs affecting transcription and translation can also be elucidated.

Finally, refinement of our preparative methods should allow visualization of the mechanism of integration of viral DNA into the bacterial genome. Such a study in the simple microbial system may provide guidelines that will be helpful in the visualization of the structural aspects of the integration of viral nucleic acid into human chromosomes.

REFERENCES

1. STENT, G. S., Science, 144, 816 (1964).
2. BYRNE, R., LEVIN, G., BLADEN, H. A., AND NIRENBERG, M. W., Proc. Nat. Acad. Sci. U.S.A., 52, 140 (1964).
3. ZUBAY, G., AND CHAMBERS, D. A., Cold Spring Harbor Symp. Quant. Biol., 34, 753 (1969).
4. MILLER, O. L., JR., AND BEATTY, B. R., Science, 164, 955 (1969).
5. MILLER, O. L., JR., HAMKALO, B. A., AND THOMAS, C. A., JR., Science, 169, 392 (1970).
6. MILLER, O. L., JR., BEATTY, B. R., HAMKALO, B. A., AND THOMAS, C. A., JR., Cold Spring Harbor Symp. Quant. Biol., 35, 505 (1970).
7. SILVERMAN, L., AND GLICK, D., J. Cell. Biol., 40, 761 (1969).
8. ZUBAY, G., AND WATSON, M. R., J. Biophys. Biochem. Cytol., 5, 51 (1959).
9. AUGUST, J. T., this symposium.
10. KUWANO, M., KWAN, C. N., APIRION, D., AND SCHLESSINGER, D., Lepetit Colloq. Biol. Med., 1, 222 (1969).
11. MORIKAWA, N., AND IMAMOTO, F., Nature, 223, 37 (1969).
12. MORSE, D., MOSTELLER, R., BAKER, R., AND YANOFSKY, C., Nature, 223, 40 (1969).
13. MOORE, P., J. Mol. Biol., 22, 145 (1966).
14. KENNEL, D., J. Mol. Biol., 34, 85 (1968).

15. DOOLITTLE, W. F., AND PACE, N. R., Proc. Nat. Acad. Sci. U.S.A., 68, 1786 (1971).

16. PURDOM, I., BISHOP, J. O., AND BIRNSTIEL, M. L., Nature, 227, 239 (1970).

17. BREMER, H., AND YUAN, D., J. Mol. Biol., 38, 163 (1968).

18. MANOR, H., GOODMAN, D., AND STENT, G. S., J. Mol. Biol., 39, (1969).

19. KURLAND, C. G., J. Mol. Biol., 2, 83 (1960).

20. SMITH, I., DUBNAU, D., MORELL, P., AND MARMUR, J., J. Mol. Biol., 33, 123 (1968).

21. LILL, H., LILL, U., SIPPEL, A., AND HARTMANN, G., Lepetit Colloq. Biol. Med., 1, 55 (1969).

22. PATO, M., AND VON MEYENBURG, K., Cold Spring Harbor Symp. Quant. Biol., 35, 497 (1970).

23. KOSSMAN, C. R., STAMATO, T. D., AND PETTIJOHN, D. E., Nature New Biol., 234, 102 (1971).

24. YU, M. T., VERMEULEN, C. W., AND ATWOOD, K. C., Proc. Nat. Acad. Sci. U.S.A., 67, 26 (1970).

25. BIRNBAUM, L. S., AND KAPLAN, S., Proc. Nat. Acad. Sci. U.S.A., 68, 925 (1971).

26. SPADARI, S., AND RITOSSA, F., in Proc. 7th FEBS Symposium, in press.

27. SUMMERS, W. C., AND SIEGEL, R. B., Cold Spring Harbor Symp. Quant. Biol., 35, 253 (1970).

28. CHAMBERLIN, M., MCGRATH, J. M., AND WASKELL, L., Nature, 228, 227 (1970).

29. SUMMERS, W. C., J. Mol. Biol., 51, 671 (1970).

30. MARRS, B. L., AND YANOFSKY, C., Nature New Biol., 234, 168 (1971).

31. HAWLEY, E. S., AND GREENAWALT, J. W., J. Biol. Chem., 245, 3573 (1970).

32. ELLIS, R. J., AND HARTLEY, M. R., Nature New Biol., 233, 193 (1971).

33. ASHWELL, M., AND WORK, T. S., Ann. Rev. Biochem., 39, 251 (1970).

34. SUMMERS, D. F., MAIZEL, J. V., AND DARNELL, J. E., Proc. Nat. Acad. Sci. U.S.A., 54, 505 (1965).

FIGURE LEGENDS

FIG. 1. Low magnification of the extruded contents of an osmotically shocked Salmonella typhimurium cell from a log-phase, broth-grown culture. Arrow designates one of the 19 rRNA loci in the field.

FIGS. 2 and 3. Genetically active segments of E. coli chromosome with attached polyribosomes. The arrows indicate RNA polymerase molecules on or very near the initiation sites for transcription.

FIG. 4. Negatively stained (uranyl acetate) portion of E. coli genome showing polyribosomes attached to chromosome by RNA polymerase molecules (arrow).

FIG. 5. A, an rRNA locus showing activity of 16S (short fibril gradient) and 23S (longer gradient) in log-phase, broth-grown E. coli. B, an rRNA locus showing RNA polymerases (arrow) distal to the 23S cistron in E. coli. Linkage data indicate that this site may be the location of the 5S rRNA cistron. C, activity of an rRNA locus from E. coli grown under suboptimal conditions (synthetic medium with glycerol and 0.05% Casamino acids).

FIG. 6. Schematic representation of rifampin readouts of rRNA cistrons based on one and on two RNA polymerase binding sites per 16S–23S doublet.

FIG. 7. An E. coli rRNA locus 40 sec after rifampin (200 µg/ml) treatment. The 16S cistron is essentially cleared, whereas the 23S cistron shows a normal complement of fibrils.

FIG. 8. An rRNA locus bracketed by polyribosomes attached to active structural genes in E. coli.

FIG. 9. Extruded contents from an E. coli cell infected with bacteriophage T7, with rifampin (200 µg/ml) added 2 min after infection to inhibit bacterial RNA polymerase. The preparation was made 8 min after infection at 37°; lysis occurs at 12 min.

FIG. 10. Portion of an osmotically shocked chloroplast, showing extruded genome with associated structures resembling polyribosomes (arrow) in Euglena gracil

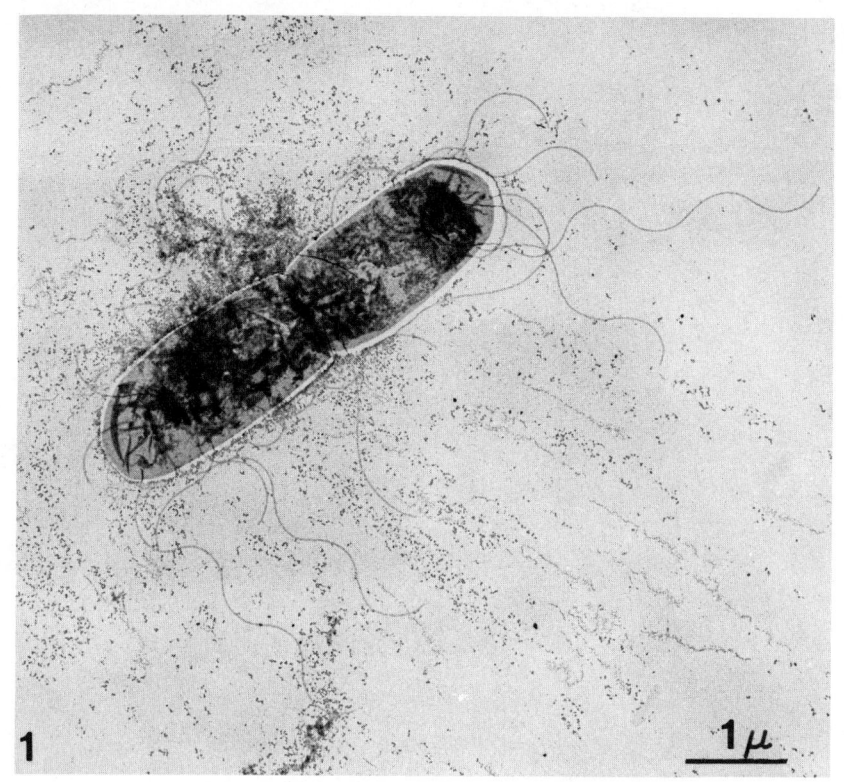

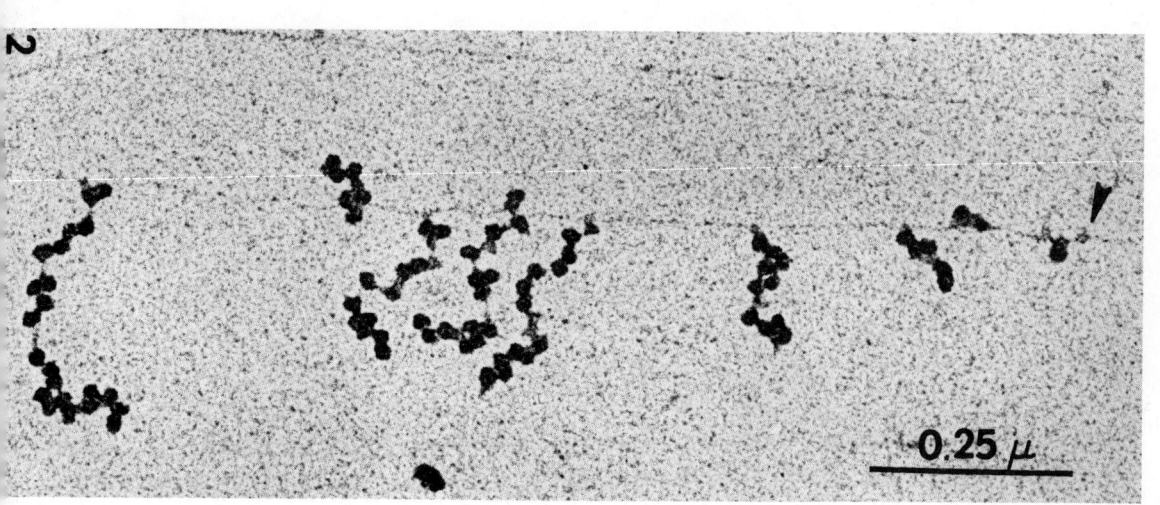

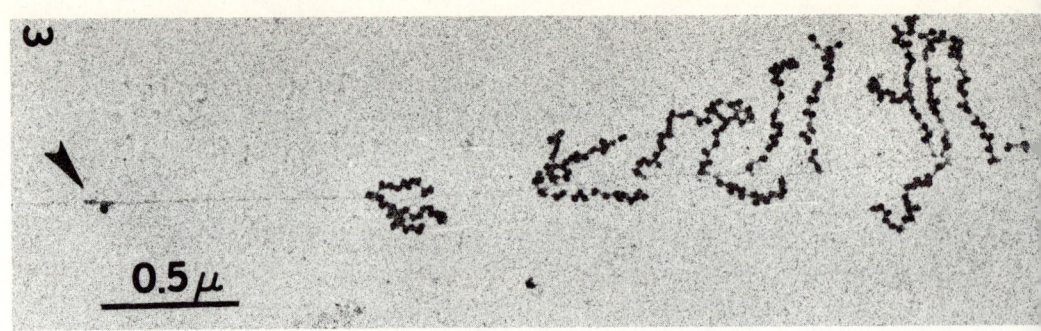

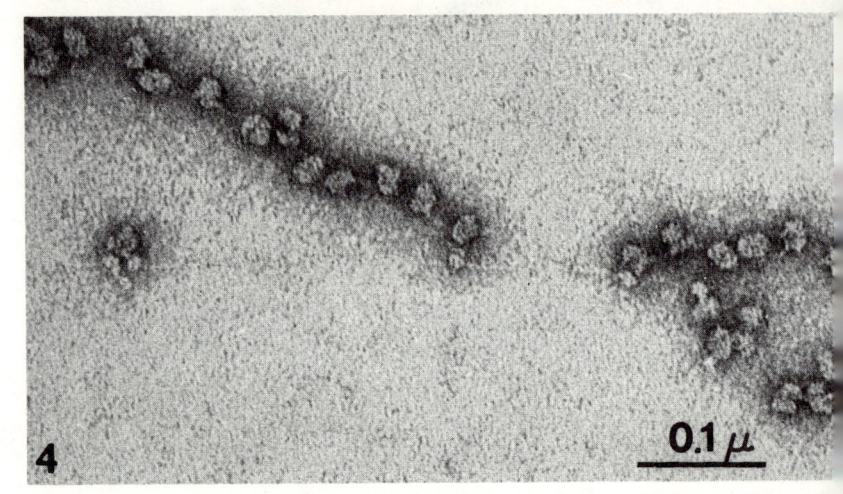

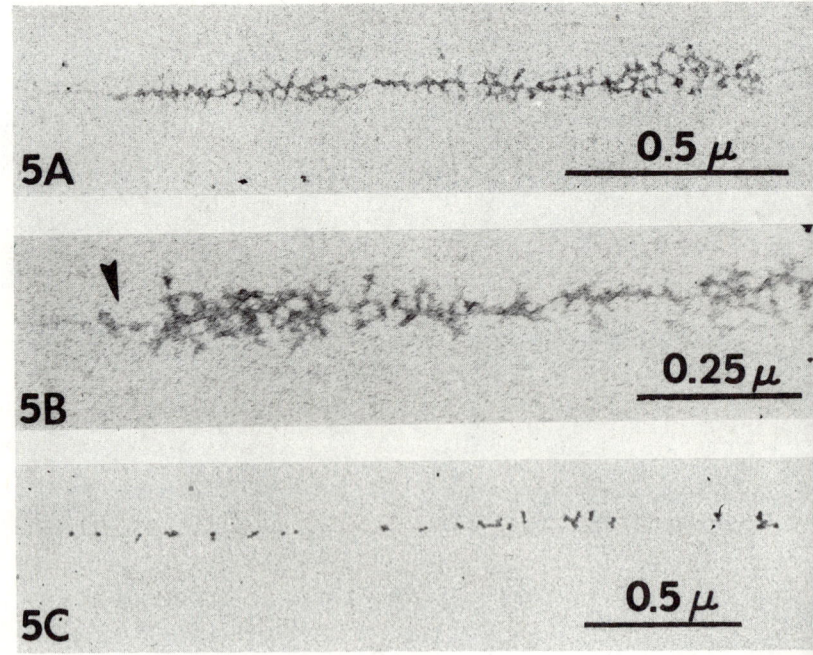

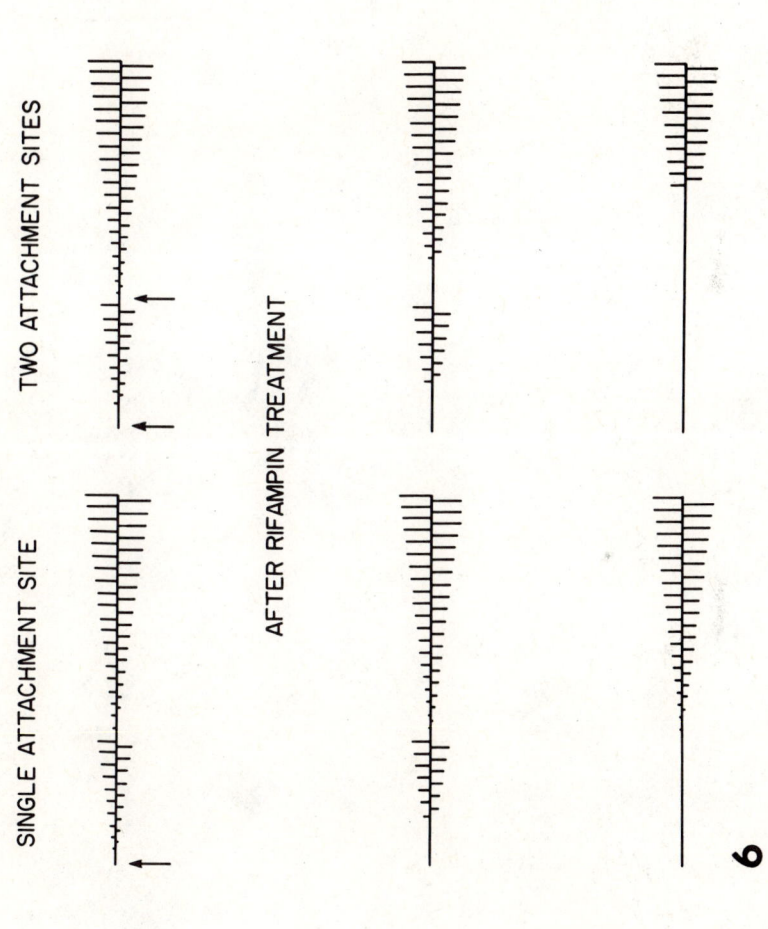

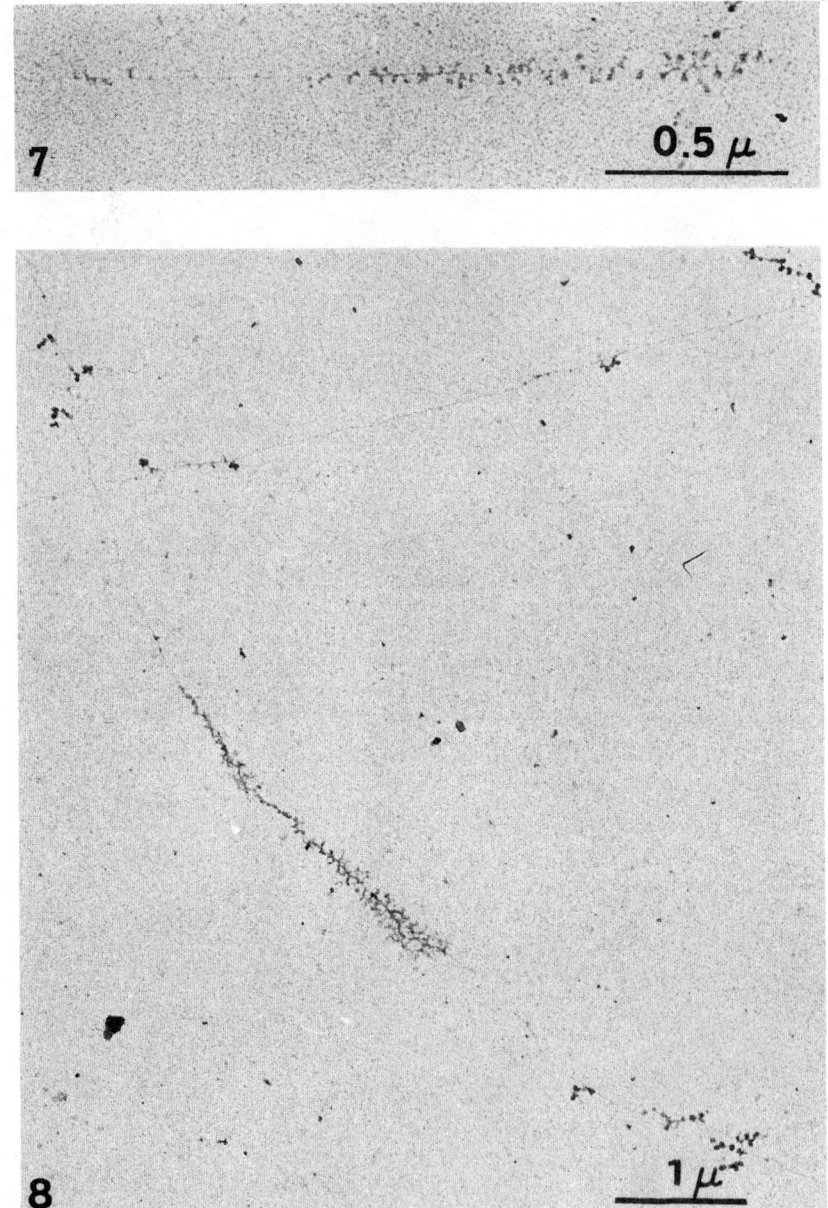

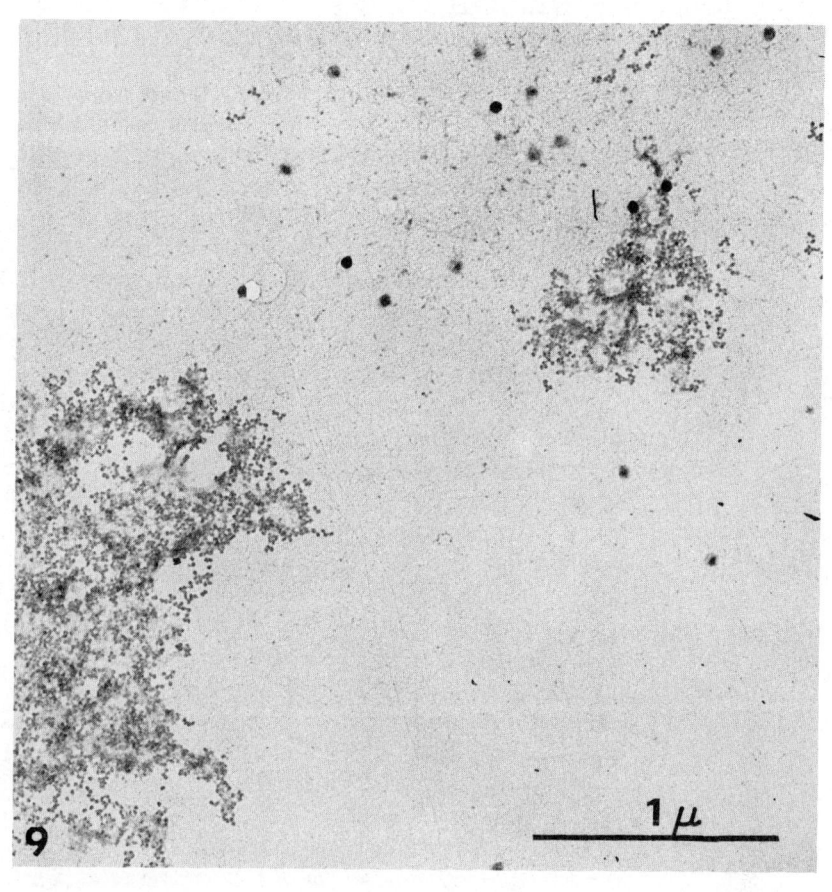

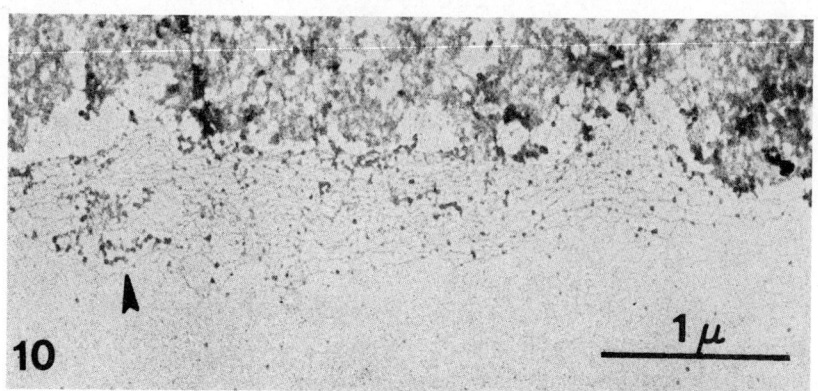

Isolation of Tyrosine tRNA Precursor Molecules

MOLECULES of tRNA are probably produced by a series of alterations of the original transcription product of tRNA cistrons. The nucleotide modifications occur on an unmodified precursor[1-8], and there is also suggestive evidence that the common 3′ –CCA$_{OH}$ end may be added to a precursor, for this sequence seems to be absent from tRNA cistrons[9] and there are enzymes which can bring about the addition[8]. Precursor molecules also are likely to have short lifetimes *in vivo*.

This communication describes the isolation of short lived RNA molecules with those properties expected for a precursor of tyrosine tRNA. The molecules contain at least thirty-five to forty-five extra nucleotides; they have different 5′ and 3′ ends and lack some of the usual nucleotide modifications. Nevertheless, they can be unequivocally identified as tRNA precursors because they contain a nucleotide sequence appropriate for the gene involved. There have been reports of the isolation of proposed precursors of tRNA based chiefly on the kinetics of labelling and modification of unfractionated low molecular weight RNA in a mammalian system[10,11], and the ability of an *in vitro* transcription product of a bacterial extract to compete with tyrosine tRNA in DNA–RNA hybridization experiments[12]. These identifications of precursor are compatible with the properties of the precursor to tyrosine tRNA described here.

When *Escherichia coli* is infected at 37° C with the transducing bacteriophage φ 80 carrying the gene for certain uniquely useful mutants of tyrosine tRNA, labelled briefly with ^{32}P-phosphate, and extracted quickly with phenol, examination of the resulting labelled material in acrylamide gels[13] reveals a hitherto unidentified and prominent band of radioactive material which is either barely or not at all observed after prolonged labelling. These molecules are very labile and the critical step in their isolation is the rapid phenol extraction (see legend to Fig. 1): even if the cells are collected in the cold by centrifugation before phenol extraction, much if not all of the material becomes degraded. Fig. 1, which is a composite illustration of two experiments, shows the usual bands of 4S and 5S RNA in each extract. A band of RNA moving slightly faster than 4S RNA, characteristic of infected cells, is present in each of those extracts (personal communication from M. L. Gefter and B. G. Barrell). Large quantities of tyrosine tRNA are made in cells infected with the phages carrying either the su_o^- or su_{III}^+ genes[14] whereas little or no tyrosine tRNA with varying amounts of a nearby contaminant is present in the other cell extracts. Cells infected with mutant 9313, 9311 or A15, however, contain a higher molecular weight material not previously seen.

Mutant 9313 has the su^- phenotype and does not make RNA with the same electrophoretic mobility of tyrosine tRNA. But it reverts to the su^+ genotype with a 4% frequency, and consequently was thought to have a partial gene duplication which might be transcribed *in vivo* (unpublished observations of S. A.). The mobility of its new band, labelled X, indicates a composition of about 200 nucleotides. (Mutant 9311 differs from 9313 only in its reversion properties and makes X band material as well as some lower molecular weight fragments labelled Z.) Analysis by the two dimensional fingerprinting technique of Sanger and his colleagues[15,16] of the material in the X band revealed at least one complete tyrosine tRNA sequence, aside from the terminal nucleotides[17]. The two spots containing the usual 5′ (pGp) and 3′ (AAUCCUU-CCCCCACCACCA$_{OH}$) ends were missing from the fingerprint of the T_1 ribonuclease digest, showing that the ends were in some way altered. On the other hand, several new spots appeared and one of these has been tentatively identified as the usual 3′ end lacking the terminal –CCA$_{OH}$. Large alterations in the structure of the 9313 product were expected because of the complex genetic structure of the mutant. The interesting result, however, is that the Y band, made by mutant A15 (ref. 18), a previously characterized single base mutant, and the W band made by the wild type phage (discussed later) also had the expected tyrosine tRNA sequence, the absence of the usual 5′ and 3′ ends and a subset of the same new spots which appeared in the X band fingerprint. The extra spots common to all new bands can account for approximately forty more nucleotides than the usual tyrosine tRNA complement, and suggest that the same biosynthetic mechanism produces precursor in each case. The fingerprints also show that at least three of the nucleotides normally found modified in tyrosine tRNA[17] (2′O-methyl G, modified A next to the anti-codon triplet and a nearby pseudo-uracil) are not modified in the precursor. A complete description of the sequences and extent of modification of the precursor will be published later.

Neither the wild type gene nor that carrying the amber suppressor mutant gave a new band in the experiments shown in Fig. 1. These latter two phages make large amounts of tRNA and the other mutants little or none, and so the ability to make tRNA efficiently seemed to be an indication of a very short half life for the precursor (that is, the necessary

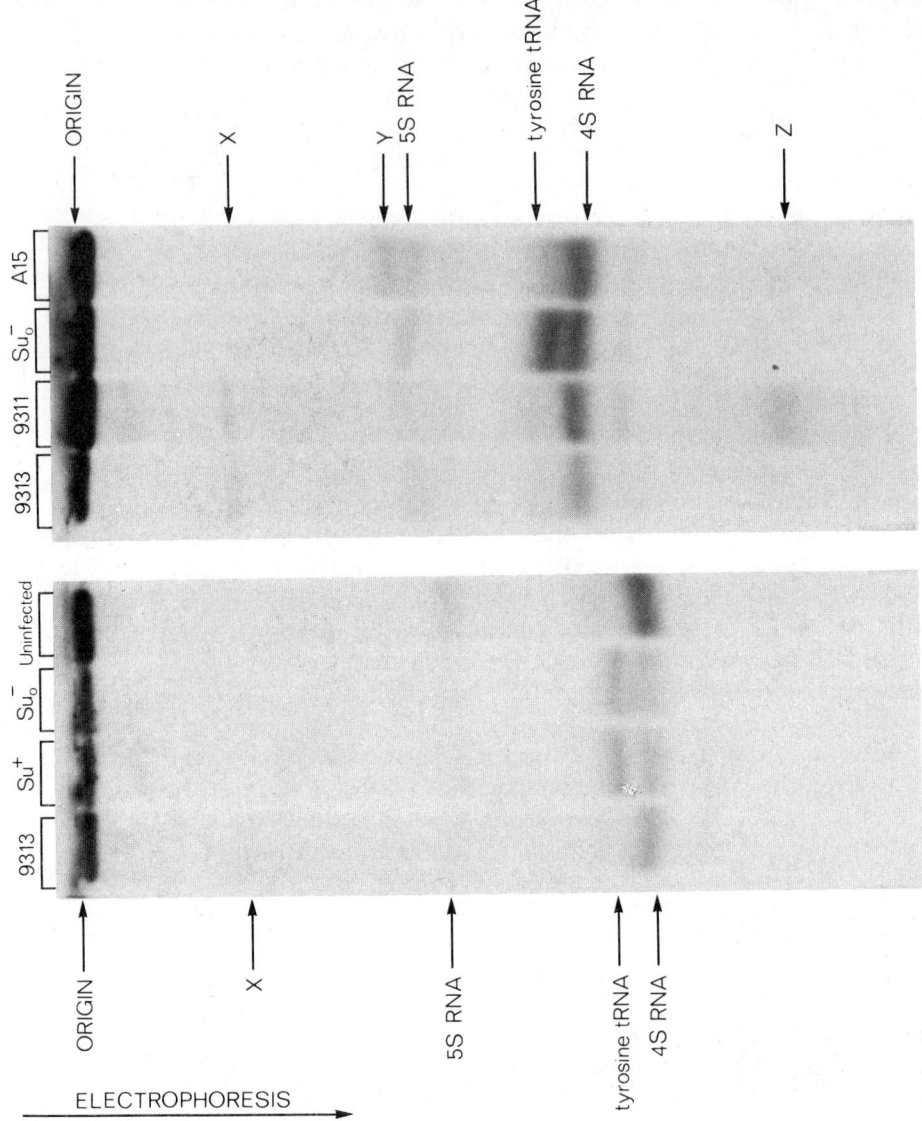

Fig. 1 Separation of labelled RNA from infected cells by electrophoretic mobility in acrylamide gels. The procedure for infection of cells and gel electrophoresis is generally as described in ref. 13. The host strain used was *E. coli* MB931, an su⁻ derivative of strain MB93 (F⁻(ara⁻amber T6R)lac₁₂₅ amber gal⁻U42 kinase amber gal⁻ epimerase su⁺III (singlet)tryp⁻amber/F8 gal⁻U42 kinase amber gal⁻ epimerase⁻) from the Cambridge collection. ³²P-phosphate (2 mCi) was added to 10 ml. of infected cell culture 30 min after the start of infection; 3.5 min later the growing, radioactive culture was mixed quickly with an equal volume of redistilled, neutralized phenol and shaken vigorously for 15 min at room temperature. In most experiments, about 40 μg of carrier *E. coli* tRNA was then added per ml. of aqueous phase from the phenol extract. This was made 0.2 M in sodium acetate, pH 5; two volumes of ethanol were added and the RNA precipitated at –20° C. Another precipitation was carried out from 1 ml. of the resuspended pellet in the standard way[13], after which the labelled material was prepared for gel electrophoresis as previously described[13]. Each column in the gel patterns is titled according to the tyrosine tRNA gene used in infection. The su⁺ phage used is a revertant of 9313 and has the normal suⅡ sequence. The positions of the familiar and prominent new bands are marked. The two groups of gel patterns represent runs in single slab gels from separate experiments.

tailoring enzymes worked very well on these substrates). Indeed, similar pulse-labelling experiments carried out at 25° C yielded a new prominent band (W) in gels of material from su_o^--infected cells (Fig. 2), indicating that metabolism of the precursor was retarded to a greater extent than the incorporation of ^{32}P at the lower temperature. No such

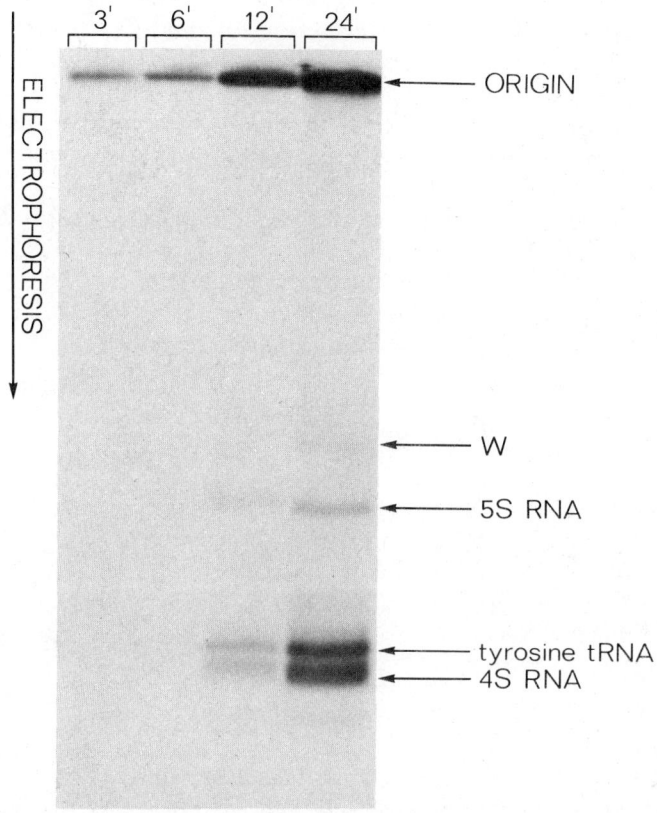

Fig. 2 Separation of labelled RNA from su_o^--infected cells pulse labelled at 25° C. Conditions are as described in the legend to Fig. 1, except that adsorption of phages was carried out at 25° C for 30 min and then aeration proceeded for 90 min at 25° C before 4 mCi of radioactive phosphate was added. Samples (2.5 ml.) were taken at the times after ^{32}P-addition indicated at the top of the gel picture, and were treated as described in the legend to Fig. 1.

prominent band was observed in uninfected cells, although a minor band is seen at the same position in gels of both uninfected and su_o^--infected cells which have undergone prolonged labelling. Preliminary pulse chase experiments indicate that the half life of the mutant precursors is about 3 min at 37° C and that of the wild type precursor is similar at 25° C. It is evident, considering these relative half lives, that even a single base change quite far from the ends of the tyrosine sequence, can radically alter the specificity of the enzymes which tailor the ends of the molecule.

The precise nature of the new RNA segments found in the precursor molecules is not yet known. It is possible that complete promoter or terminate signals will be found, but the molecules described here may not necessarily be the first product of tRNA cistron transcription: they may already be a further step in the production of mature, functional tRNA. Nor is it certain that the "signals" themselves are transcribed *in vivo*.

The fast extraction technique described here may be useful in the isolation of other tRNA precursors or unstable messenger RNA.

I thank Drs S. Brenner and F. Crick for hospitality, encouragement and criticism. During this work I was supported by a fellowship from the Anna Fuller Fund. I benefited greatly from discussions with Drs J. D. Smith, K. W. Anderson, A. R. Cashmore, U. K. Laemmli and W. H. McClain.

SIDNEY ALTMAN

MRC Laboratory of Molecular Biology,
Hills Road, Cambridge

Received September 28, 1970.

[1] Fleissner E., and Borek, E., *Proc. US Nat. Acad. Sci.*, **48**, 1199 (1962).
[2] Hayward, R. S., and Weiss, S. B., *Proc. US Nat. Acad. Sci.*, **55**, 1161 (1966).
[3] Heinrikson, R. L., and Goldwasser, E., *J. Biol. Chem.*, **239**, 1177 (1964).
[4] Weiss, S. B., and Legault-Demare, J., *Science*, **149**, 429 (1965).
[5] Moshowitz, D. B., *J. Mol. Biol.*, **50**, 143 (1970).
[7] Burdon, R. H., and Glason, A. E., *J. Mol. Biol.*, **39**, 113 (1969).
[6] Bernhardt, D., and Darnell, J., *J. Mol. Biol.*, **42**, 43 (1969).
[8] Zachau, H. G., *Angew. Chem.*, intern. edit., **8**, 711 (1969).
[9] Daniel, V., Sarid, S., and Littauer, U. Z., *Science*, **167**, 1682 (1970).
[10] Burdon, R. H., Martin, B. T., and Lal, B. M., *J. Mol. Biol.*, **28**, 357 (1967).
[11] Smillie, E. J., and Burdon, R. H., *Biochim. Biophys. Acta*, **213**, 248 (1970).
[12] Daniel, V., Sarid, S., Beckmann, J. S., and Littauer, U. Z., *Proc. US Nat. Acad. Sci.*, **66**, 1260 (1970).
[13] Smith, J. D., Barnett, L., Brenner, S., and Russell, R. L., *J. Mol. Biol.*, **54**, 1 (1970).
[14] Goodman, H. M., Abelson, J., Landy, A., Brenner, S., and Smith, J. D., *Nature*, **217**, 1019 (1968).
[15] Sanger, F., Brownlee, G. G., and Barrell, B. G., *J. Mol. Biol.*, **13**, 373 (1965).
[16] Brownlee, G. G., and Sanger, F., *J. Mol. Biol.*, **23**, 338 (1967).
[17] Goodman, H. M., Abelson, J., Landy, A., Zadrazil, S., and Smith, J. D., *Europ. J. Biochem.*, **13**, 461 (1970).
[18] Abelson, J. N., Gefter, M. L., Barnett, L., Landy, A., Russell, R. L., and Smith, J. D., *J. Mol. Biol.*, **47**, 15 (1970).

Printed in Great Britain by Fisher, Knight & Co., Ltd., St. Albans.

DNA-Directed Cell-Free Synthesis of Biologically Active Transfer RNA: su^+_{III} Tyrosyl-tRNA

(E. coli/β-galactosidase/bacteriophage)

GEOFFREY ZUBAY, LORETTA CHEONG, AND MALCOLM GEFTER

Department of Biological Sciences, Columbia University, New York, N.Y. 10027

Communicated by Cyrus Levinthal, July 8, 1971

ABSTRACT Biologically active su^+_{III} tyrosyl-tRNA has been synthesized in a DNA-directed cell-free system from *Escerichia coli*. Such a system should be most useful for studying the mechanism of tRNA synthesis. This tRNA is capable of suppressing amber mutations in the gene coding for β-galactosidase (EC 3.2.1.23) and, therefore, must be capable of being charged and transferring amino acids. A 4-fold stimulation in the activity of the tRNA formed *de novo* is obtained with isopentenyl pyrophosphate, a compound involved in the post-transcriptional acylation of an adenine base adjacent to the anticodon. It has been suggested elsewhere that formation of RNA subject to stringent control may be inhibited by guanosine tetraphosphate (ppGpp). However, guanosine tetraphosphate did not affect the synthesis of su^+_{III} tyrosyl-tRNA, even though the synthesis of this tRNA is subject to stringent control.

DNA-directed cell-free systems are being used with increasing frequency to study mRNA and protein synthesis (1). Recently, attempts have been made to use such systems for the study of rRNA and tRNA synthesis. The rRNA studies have been done with *Escherichia coli* DNA, which contains about 10–12 genes for rRNA, which constitute about 0.5% of the *E. coli* chromosome (2). A protein factor called ψ has been partially purified. This factor is believed to be necessary for initiation of ribosomal gene transcription; this initiation is inhibited by guanosine tetraphosphate (ppGpp) *in vitro* (3, 4). DNA-directed cell-free studies of tRNA synthesis have been done with bacteriophage φ80 psu^+_{III} DNA (5), a viral DNA that contains a single bacterial suppressor tRNA that constitutes about 0.2% of the viral chromosome (6). The tRNA synthesized is much larger than mature tRNA, with a sedimentation coefficient of about 8 S. It has been suggested elsewhere that this is the immediate transcriptional product of the tRNA gene, and that partial degradation of this product in a subsequent maturation reaction is necessary. In the reported RNA synthesis studies, no evidence was presented for the synthesis of biologically competent RNA. The main technique for characterization of the cell-free synthesized RNAs has been hybridization to the transcriptional site on the DNA template.

Possibly a more meaningful approach would be the synthesis of a biologically competent RNA, coupled with analysis of the components essential at the various stages of formation, these stages being initiation, propagation, and termination of transcription, as well as post-transcriptional modification. We have been successful in synthesizing a biologically active su^+_{III} tyrosyl-tRNA and defining some of the requirements and conditions for its synthesis.

METHODS

Bacterial and viral strains

λ*dlac* was isolated from a doubly lysogenic strain containing λhφ80 and λhφ80*dlac*.

λ*dlac 545* was isolated from a doubly lysogenic strain containing λhφ80 and λhφ80*dlac* 545. The λCI857t68h80*dlacz*545 used to make the double lysogen was constructed by L. Caine and J. Beckwith, by the following technique: Strain X-7060 (F$^-$, lac-deletion Sm^S) was transduced to *lac*$^+$ at low multiplicity (about 1 phage particle/200 bacteria) with a lysate containing both λCI 857t68h80 and λCI857t68 h80*dlac*$^+$. Lac$^+$ transductant clones were purified and tested to see whether they gave rise to transducing lysates without helper phage. A strain (XS-7001) that was presumed to be a defective lysogen by this test was used for further experiments. Strain E-7119 (F'*lac* z545/*lac* z545*trp*$^-$) was mated with XS-7001 at 30°C and clones of XS-7001 were picked from glucose–minimal media to detect those that had received the F-*lac* episome. These clones were mated with strain M-7025 (F$^-$*lacz* 90 Sm^R), and the mating mixture was spread on lactose minimal agar to look for *lac*$^+$ recombinants. Those clones that gave rise to recombinants should have the structure F-*lac* z545/λCI857h80*dlac*$^+$. These clones were grown up in culture and aliquots were spread on lactose–tetrazolium agar to look for *lac*$^-$/*lac*$^-$ homogenotes. A red colony on such medium that still carries phage immunity was presumed to have incorporated the *lac* z545 mutation onto the defective phage. Mating techniques are described by Miller *et al.* (7).

φ80*psu*$^+_{III}$ was grown and isolated as described (6).

Strain 514 was used to make all cell-free extracts. This strain is F$^-$Δ*lac*, *trp*$^-$, T6^r, Sm^r.

Preparation of DNA has been described elsewhere (8). The double lysogen is grown to a suitable titer and induced to make viruses. The defective virus is purified and DNA is made from it. The DNA is stored in a solution of 0.01 M Tris acetate, pH 8.2, over chloroform at 5°C.

Conditions for cell-free synthesis and assay

Synthesis and Assay of β-galactosidase (EC 3.2.1.23). Except for slight modifications described herein, all procedures used for synthesis, enzyme assay, and preparation of bacterial extracts and DNA have been described in detail (8). The procedures for synthesis and assay were: the incubation mixture contains per ml: 44 μmol of Tris acetate (pH 8.2); 1.37 μmol dithiothreitol; 55 μmol KAc; 27 μmol NH$_4$AC; 14.7 μmol

MgAc$_2$; 7.4 μmol CaCl$_2$; 0.22 μmol amino acids; 2.2 μmol ATP; 0.55 μmol (each) of GTP, CTP, and UTP; 21 μmol phosphoenolpyruvic acid; 0.5 μmol cyclic AMP 100 μg of tRNA; 27 μg of pyridoxine·HCl; 27 μg TPN; 27 μg FAD; 11 μg of p-aminobenzoic acid. The above ingredients are incubated for 3 min at 37°C with 50 μg/ml of $\lambda dlac$ DNA, with shaking, before 6.5 mg of S-30 extract protein is added. Incubations with shaking are allowed to continue for 60 min at 37°C. After the synthesis has been completed, a 0.2-ml aliquot is removed and mixed with 1.5 ml of assay buffer containing 0.53 mg of O-nitrophenyl-β-D-galactoside–0.1 M sodium phosphate (pH 7.3)–0.14 M 2-mercaptoethanol. After a suitable length of time (1–40 hr), the mixture is treated with one drop of glacial acetic acid, chilled, and centrifuged to remove the precipitate. The supernatant is mixed with an equal volume of 1 M Na$_2$CO$_3$ and read against water in a 1-cm quartz cell at a wavelength of 420 nm. A zero-time value of 0.035 is subtracted from all readings. Duplicate analyses usually agree within 2%.

Synthesis of su^+_{III} Tyrosyl-tRNA. As above, except for the use of $\phi 80 psu^+_{III}$ DNA instead of $\lambda dlac$ DNA and addition of 10^{-4} M isopentenylpyrophosphate (9).

Partial purification of su^+_{III} tyrosyl-RNA made in the cell-free system

After synthesis, the 5-ml incubation mixture contained a precipitate consisting of most of the DNA, presumably in an mRNA–ribosome complex. This precipitate is removed by centrifugation at 1000 $\times$ g for 5 min. The resulting supernatant is vigorously agitated with an equal volume of 88% phenol for 15 min. The resulting emulsion was centrifuged at 13,000 $\times$ g for 1 min, and the upper aqueous-layer was collected and precipitated by the addition of 0.1 volumes of 5 M KAc and 2.0 volumes of ethanol. The precipitate is centrifuged as above, redissolved in 3 ml of H$_2$O, and precipitated again by the addition of 0.3 ml of 5 M KAc and 6 ml of EtOH. The resulting precipitate is redissolved in 1.5 ml of H$_2$O. Then, 1.6 ml of 2 M NaCl is added and the solution allowed to stand at 0°C for 20 min. The precipitate is removed by centrifugation at 15,000 $\times$ g for 1 hr. The resulting supernatant is precipitated by the addition of 2 volumes of ethanol. The precipitate is collected by centrifugation at 10,000 $\times$ g for 1 min, dissolved in 1.5 ml of H$_2$O, and precipitated by the addition of 0.15 ml of 5 M KAc and 3 ml of EtOH. The precipitate is collected, dissolved, and precipitated as described above; the final precipitate is dissolved in 0.5 ml of H$_2$O. This solution is stored at -75°C until ready for use.

Preparation of su^+_{III} tyrosyl-tRNA made *in vivo*

Suppressor tyrosyl-tRNA was isolated and purified from $\phi 80 psu^+_{III}$-infected cells essentially as described (6). For these studies, the partially modified tRNA (isopentenyl adenosine adjacent to the anticodon) was used throughout. Moles of tyrosyl-tRNA added are calculated on the basis of tyrosine-acceptor capacity.

RESULTS AND DISCUSSION

The most crucial aspect of our approach was to develop a suitably sensitive and selective assay for a biologically competent tRNA. Advantage was taken of the specific suppression properties of the mutant su^+_{III} tyrosyl-tRNA not normally found in *E. coli*. The assay used was the su^+_{III} tyrosyl-tRNA stimulation of DNA-directed β-galactosidase synthesis using DNA with an amber triplet in the β-galactosidase gene. Any tRNA competent in such a process must be capable of amino-acid acylation, as well as transfer of amino acid to the correct position on the growing polypeptide chain in protein synthesis. We shall first describe the DNA-directed system for β-galactosidase synthesis, then the suppression assay, and finally the requirements for the cell-free synthesis of the suppressor tRNA.

Comparison of β-galactosidase synthesized in the DNA-directed cell-free system with normal (λdlac) and mutated (λdlac 545) DNA

A DNA-directed cell-free system for β-galactosidase synthesis has been used in gene regulation studies for several years. This system comprises a cell-free extract of *E. coli* (called S-30 extract), DNA from the defective transducing virus $\lambda dlac$, and the cofactors and substrates necessary for RNA and protein synthesis. The amount of β-galactosidase synthesized in the cell-free system is proportional to the amount of active DNA containing the *lac* operon present.

In order to adapt this system for suppression studies, a mutant DNA (mutant 545) containing an amber triplet in the

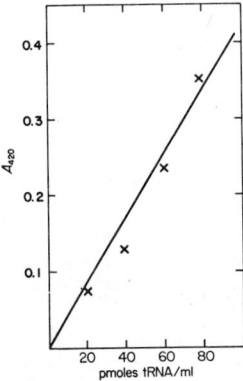

FIG. 1. Synthesis of β-galactosidase as a function of added su^+_{III} tyrosyl-tRNA with amber-suppressible *lac* DNA in the cell-free system (see text for explanation).

TABLE 1. *Amounts of su^+_{III} tyrosyl-tRNA synthesized under various conditions*

System	su^+_{III} Tyrosyl-tRNA synthesized/ml (pmoles)
Complete	40
+ zero time for synthesis	<1
− $\phi 80 psu^+_{III}$ DNA	<1
+ $\phi 80$ DNA instead of $\phi 80 psu^+_{III}$ DNA	<1
+ 2 μg/ml of rifampicin	<1
+ 5 $\times$ 10^{-4} M cyclic AMP	43
+ 50 μM ppGpp	44
− isopentenyl pyrophosphate	10
+ 10^{-4} M S-adenosylmethionine	38

Each value is the average of two determinations. The average error of a determination is about $\pm 10\%$.

early region of the β-galactosidase gene was constructed; this DNA, in the form of λdlac 545, was substituted for the normal λdlac DNA. The effect of the 545 mutation has been described by Zipser (10); very low levels of β-galactosidase are made *in vivo* because of the nontranslatability of the amber triplet. In the cell-free system, this altered DNA gives about 1/4000th of the normal level of β-galactosidase. Under standard assay conditions (see *methods*), this means a ΔA_{420} in a 1-cm tube of 0.005 after 20 hr of incubation. Thus, the single amber triplet introduces a severe block for β-galactosidase synthesis both in whole cells and in the cell-free system.

The amber-mutated β-galactosidase gene can be suppressed by the addition of su^+_{III} tyrosyl-tRNA to the cell-free system

The su^+_{III} tyrosyl-tRNA is a highly effective amber suppressor that inserts tyrosine into the position corresponding to the amber triplet. As high as 70% effective suppression has been obtained when this tRNA is present *in vivo*. When low levels of purified su^+_{III} tyrosyl-tRNA are added to the cell-free system, no effect on the amount of β-galactosidase synthesized is observed as long as normal λdlac DNA is used. However, when the mutant λdlac DNA 545 is used, there is a stimulation of β-galactosidase synthesis that is directly proportional to the amount of tRNA added (see Fig. 1). As little as 10 pmol of tRNA gives a readily detectable increment in enzyme activity.

With minor modifications, the DNA-directed system for β-galactosidase synthesis can be used for tRNA synthesis

The same system used for β-galactosidase synthesis can be used for tRNA synthesis; only minor modifications are necessary. The λdlac DNA is replaced by $\phi80psu^+_{III}$ DNA and 10^{-4} M isopentenylpyrophosphate is added. After the synthesis step, the resulting tRNA is partially purified, concentrated (see *Methods*), and assayed by its ability to stimulate β-galactosidase synthesis in the suppressible system described above. The amount of tRNA is estimated by comparison with the stimulation produced by a known amount of su^+_{III} tyrosyl-tRNA (see Fig. 1). The estimated amounts of tRNA made in the cell-free system under various conditions are indicated in Table 1. Under optimum conditions, about 40 pmol of active tRNA are synthesized per ml of incubation mixture. This low level can be easily detected because of the sensitive suppression assay and because it can be concentrated before assay. The gross amount of ribopolynucleotide synthesis in the cell-free system (measured by cold-acid-precipitation of [^{14}C]CTP labeled RNA) exceeds that required to account for the biologically mature tRNA by about 20-fold. Since the tRNA gene only comprises about 0.2% of the viral DNA used, it is obvious that the bulk of the RNA made must be viral in origin. It is also likely that more tRNA is transcribed than reaches maturation and, therefore, escapes detection by the suppression assay. Comparison of the complete system (Table 1, line 1) with the controls (lines 2–5) shows that tRNA activity requires incubation with DNA containing the tRNA gene and rifampicin-sensitive initiation of transcription. The compounds adenosine-3'5'-cyclic monophosphate (cAMP) and guanosine tetraphosphate (ppGpp) have no significant effect on the synthesis. Cyclic AMP is a gene coactivator for catabolite-sensitive genes, such as the lactose and arabinose operons, but it has no known effect on other types of genes. ppGpp stimulates the synthesis of catabolite-sensitive gene products, at least *in vitro* (12). Before this fact was known, it was suggested that ppGpp functions as an inhibitor of rRNA synthesis in the stringent response (4). However, under the cell-free conditions used here, ppGpp greatly stimulates catabolite-sensitive genes but has a negligible effect on tRNA synthesis. This is quite interesting, since the tyrosyl-tRNA gene is known to be subject to stringent control (4) and others (2, 3) have suggested that ppGpp inhibits initiation of genes subject to stringent control. The precise role of ppGpp, as well as its mechanism of action, is clearly an important matter for future study.

Post-transcriptional modification is necessary for tRNA activity

Mature tRNA is probably smaller than the immediate transcriptional product (11). It also must be modified by attachment of the trinucleotide pCpCpA to its 3'-hydroxy end and chemical alteration of a number of bases. Some bases are methylated in mature tRNA, and it is known that the active methylating agent is S-adenosylmethionine (AdMet). Addition of AdMet to the cell-free system has no effect on the amount of active tRNA synthesized. However, since this system contains methionine and ATP, it is possible that saturating amounts of AdMet are present endogenously. Alternatively, methylation may not be required for activity of this particular tRNA. Further experiments are necessary to resolve this question.

Isopentenyl pyrophosphate (IPP) functions as a precursor of an acetylated adenine base adjacent to the anticodon. Modification of this base is required for optimum activity of su^+_{III} tyrosyl-tRNA in protein synthesis (9). Consistent with this, it is observed that IPP stimulates accumulation of active tRNA by about 4-fold (see Table 1), but has no direct effect on β-galactosidase synthesis. Although much work remains to be done, we believe that the system described here represents a new, fruitful approach for probing the mechanisms of tRNA synthesis.

We thank Mr. Tettah Blankson for his assistance in the growth of viruses and cells and Mrs. S. Yang for assistance in the preparation of tRNAs. This work was supported by a grant from the National Institutes of Health (5-RO1-GM-16648-03) and the National Science Foundation (BO1 8733-000) to G. Zubay, and by grant E-561 from the American Cancer Society and project no. 273 from the Jane Coffin Childs Memorial Fund for Medical Research to M. Gefter.

1. *Cold Spring Harbor Symp. Quant. Biol.*, **35** (1970).
2. Travers, A., R. Kamen, and M. Cashel, *Cold Spring Harbor Symp. Quant. Biol.*, **35**, 415 (1970).
3. Cashel, M., *Cold Spring Harbor Symp. Quant. Biol.*, **35**, 407 (1970).
4. Primakoff, P., and P. Berg, *Cold Spring Harbor Symp. Quant. Biol.*, **35**, 391 (1970).
5. Daniel, V., S. Sarid, J. S. Beckman, and U. Z. Littauer, *Proc. Nat. Acad. Sci. USA*, **66**, 1260 (1970).
6. Russell, R. L., J. N. Abelson, A. Landy, M. L. Gefter, S. Brenner, and J. D. Smith, *J. Mol. Biol.*, **47**, 1 (1970).
7. Miller, J., K. Ippen, J. Scaife, and J. R. Beckwith, *J. Mol. Biol.*, **38**, 413 (1968).
8. Chambers, D., L. Cheong, and G. Zubay, *The Lac Operon*, ed. D. Zipser and J. Beckwith (Cold Spring Harbor Laboratory of Quantitative Biology, Cold Spring Harbor, N.Y. 1970).
9. Gefter, M. L., and R. L. Russell, *J. Mol. Biol.*, **39**, 145 (1969).
10. See Zipser, D., in *The Lac Operon*, ed. J. R. Beckwith and D. Zipser (Cold Spring Harbor Laboratory, Cold Spring Harbor, 1970), p. 226.
11. Altman, S., *Nature*, **229**, 19 (1970).
12. Zubay, G., L. Gielow, and E. Englesberg, *Nature*, in press.

New RNA Polymerase from *Escherichia coli* infected with Bacteriophage T7

by
MICHAEL CHAMBERLIN
JANET McGRATH
LUCY WASKELL

Department of Molecular Biology
and Virus Laboratory,
University of California,
Berkeley, California 94720

A new T7-specific RNA polymerase is found in T7 phage-infected cells. It is the product of T7 gene I and is physically and biochemically distinct from the host cell RNA polymerase. The synthesis of T7 RNA polymerase provides a direct explanation for the pleiotropic control of late T7 functions exerted by gene I.

THE growth of a bacteriophage in a bacterial cell provides a useful system for the study of the regulation of gene expression. In the case of several bacteriophages of *Escherichia coli*, regulation of gene expression seems to occur at the level of transcription; only messenger RNA species containing information for early bacteriophage proteins are formed early in infection. At the appropriate time, one or more "switching" events occur which trigger the synthesis of messenger RNA for late bacteriophage genes (for reviews, see refs. 1, 2) and which terminate the transcription of some early bacteriophage genes. It has been proposed that the change in transcription specificity leading to synthesis of late messenger RNA is provided in bacteriophages T4[3] and T7[4] by synthesis of a new protein component, which functions in place of the sigma subunit of *E. coli* RNA polymerase and which allows the host cell RNA polymerase to initiate RNA synthesis at promoter sites for late bacteriophage genes, where it otherwise would not function. We show here that bacteriophage T7 induces the formation of a new RNA polymerase activity in infected *E. coli*. The T7 RNA polymerase is distinct from *E. coli* RNA polymerase and seems to be the protein product of gene 1 of T7 bacteriophage. Synthesis of messenger RNA for late bacteriophage functions seems to be mediated by this phage-induced RNA polymerase and not by an alteration of the host RNA polymerase.

Procedures

[3]H-labelled and unlabelled nucleotides were obtained from Schwarz Bioresearch; α-^{32}P-ATP was prepared by the method of Symons[5]. DNA from T2, T7 and λ bacteriophages was prepared according to Thomas and Abelson[6]. dAT copolymer and dG·dC homopolymer pairs were prepared according to Schachman et al.[7] and Radding et al.[8], respectively. DNA concentrations are expressed in terms of nucleotide equivalents. Rifamycin was the gift of Ciba Pharmaceuticals, and streptolydigin was donated by the Upjohn Co. Gel electrophoresis was carried out according to Shapiro et al.[9] using *E. coli* RNA polymerase to provide molecular weight standards[10]. *E. coli* RNA polymerase was purified by the procedure of Berg et al.[11] and was fraction V enzyme. Strains of T7 bacteriophage containing amber mutations and *E. coli* strains B and O11' were obtained from Dr F. W. Studier and were grown according to Studier[12]. T7 342-15, which contains a mutation in gene 1 rendering the growth of the phage temperature sensitive, was obtained from Dr W. C. Summers[4]. *E. coli* infected with T7 bacteriophage were grown in a "Biogen" (American Sterilizer Co.). *E. coli* B/1 were grown to a density of 3×10^9 cells per ml. with maximum aeration using M9 medium[12] supplemented with additional ammonium chloride and 5 g/l. of glucose. T7 bacteriophage were added to give a multiplicity of 3 and after 12·5 min the cells were rapidly chilled and harvested by continuous flow centrifugation. Infected cells were stored at $-15°$ C until use.

RNA polymerase was purified from T7-infected cells by two methods. Procedure I was carried out essentially as described for isolation of *E. coli* RNA polymerase from uninfected cells[11] using T7-infected cells. RNA polymerase assays were done during this procedure were done as described by Berg et al.[11] using T7 DNA as template. Procedure II was devised to obtain T7 specific RNA polymerase (Table 1). RNA polymerase assays for this enzyme

Table 1. SUMMARY OF PURIFICATION PROCEDURE II

Fraction	Volume (ml.)	Total activity (units × 10⁻³)	Protein (mg/ml.)	Specific activity (U/mg)	Percentage total activity
A. Extract	47	570	20	600	100
B. Streptomycin supernatant fluid	40	370	15	630	65
C. (NH₄)₂SO₄ extract fraction	9	202	14·5	1,500	35
D. DEAE-cellulose pooled fractions	90	200	0·25	8,800	35
E. Phosphocellulose pooled fractions	33	125	(0·03)	(120,000)	22
F. Concentrated phosphocellulose fractions	3·1	87	0·23	120,000	15

All operations were carried out at 0°–4° C and all centrifugations were carried out for 30 min at 15,000 r.p.m. in the 'SS 34' rotor of the Sorvall 'RC IIB' centrifuge except where specified otherwise. T7-infected cells (13 g) were mixed with 26 g of cold, acid-washed alumina powder and 0·4 ml. of 0·05 M Tris-HCl (pH 8)–0·01 M MgCl₂–0·0001 M dithiothreitol (DTT)–5 per cent glycerol (buffer A) and the mixture was ground in a mortar for 12 min. Buffer A (30 ml.) was added and the viscous mixture was mixed thoroughly and was centrifuged for 1 h. The supernatant fluid was diluted to a protein concentration of 20 mg/ml. (fraction A, 47 ml.). To fraction A was added 3·3 ml. of 10 per cent streptomycin sulphate and the mixture was stirred for 15 min and then centrifuged. The supernatant fluid (fraction B, 39 ml.) was decanted and 10·8 g of ammonium sulphate added. After 30 min of gently stirring the precipitate was collected by centrifugation and extracted in a 50 ml. centrifuge tube with 8 ml. of 1·2 M ammonium sulphate in 0·01 M Tris-HCl (pH 8)–0·0001 M dithiothreitol–DTT 5 per cent glycerol (buffer I), using a tightly fitting pestle. After 30 min the suspension was centrifuged. The supernatant fluid was saved and the pellet extracted again exactly as before to give a second supernatant fluid. The two supernatant fluid fractions were pooled (volume 16 ml.) and 2·06 g of ammonium sulphate added. After 30 min the precipitate was collected by centrifugation and dissolved in buffer I (fraction C, 9 ml.). Fraction C was dialysed for 3 h against two changes of 500 ml. each of buffer I in a rapid dialysis apparatus[13] and then diluted to 3 mg/ml. protein concentration. The enzyme was adsorbed to a column of DEAE-cellulose (20 cm length × 3 cm² cross section) equilibrated with buffer I. Elution was carried out with a linear gradient from 0·05–0·4 M KCl in buffer I (total volume 400 ml.). Fractions containing a high specific activity of T7 RNA polymerase were pooled (fraction D, 90 ml.) and immediately passed through a column of phosphocellulose (Whatman 'P-11'; 20 cm length × 3 cm²) previously equilibrated with buffer I containing 0·15 M KCl. A flow rate of about 1 ml./min is maintained. The enzyme is eluted from the column with a linear gradient from 0·15–1·0 M KCl in buffer I (total volume 300 ml.). Fractions containing a high specific activity of T7 RNA polymerase and showing only a single protein band on sodium dodecyl sulphate SDS-acrylamide gel electrophoresis were pooled (fraction E, 22 ml.) and concentrated by pervaporation[14] against buffer I to give a final protein concentration of more than 0·25 mg/ml. (fraction F, 3·1 ml.). These fractions were stored at 0° C or, after addition of an equal volume of glycerol, at −15° C.

were carried out as in procedure I with the following modifications. The enzyme was diluted in a solution containing 0·05 M Tris-HCl (pH 8), 0·01 M 2-mercaptoethanol and 2 mg/ml. of bovine serum albumin (BSA). Reaction mixtures contained 4 μmoles of Tris-HCl (pH 8), 2 μmoles of $MgCl_2$, 1 nmole of 2-mercaptoethanol, 100 μg of BSA, 36 nmole of T7 DNA, 40 nmole of ATP, CTP, GTP and UTP, and 0·5–10 units of enzyme in a final volume of 0·1 ml. One of the nucleoside triphosphates was labelled with α-^{32}P or ^{3}H at a specific activity of 2,000–20,000 c.p.m. per nmole. The rate of nucleotide incorporation with T7 RNA polymerase was constant for at least 20 min at 37° C with infected cell extracts or with the purified enzyme. Standard assays were terminated after 10 min incubation at 37° C and the amount of acid-insoluble radioactivity was determined as before. The amount of nucleotide incorporation was directly proportional to the amount of enzyme added to the reaction over the range of activity given above. One unit of activity is the amount of enzyme which gives a rate of 1 nmole of AMP incorporation in 1 h at 37° C in the standard assay. The activity of *E. coli* RNA polymerase holoenzyme is essentially the same in the two different assay procedures.

Two RNA Polymerase Activities

As a result of the studies of Summers and Siegel[4] it seemed likely that *E. coli* RNA polymerase in cells infected with T7 would be modified by replacement of the sigma subunit. To study the properties of such a modified polymerase we chose procedure I, which should allow isolation of the virtually intact enzyme. During the purification there were extensive losses in RNA polymerase activity. The missing activity did not appear in discarded fractions, so it was assumed that the activity was unstable, although the *E. coli* RNA polymerase is quite stable in similar conditions. Gradient elution of the enzymatic activity from DEAE-cellulose resolved two separate peaks of RNA polymerase activity, one in the range of 0·15 M KCl (fraction 5A), and the second in the range of 0·25 M KCl (fraction 5B). RNA polymerase in uninfected extracts elutes at about 0·25 M KCl, in a position similar to fraction 5B. Fractions 5A and 5B contained 1·5 and 3 per cent, respectively, of the initial RNA polymerase activity measured in the original extract.

Characterization of these two fractions revealed that fraction 5A was highly active with T7 DNA as template, but did not use dAT copolymer whereas fraction 5B was active with both templates but most active with dAT copolymer (Table 2). The template specificity of fraction 5A was remarkable, for *E. coli* RNA polymerase is highly active with dAT copolymer as template whether or not it contains the sigma subunit (see Table 2). From the specific activities of the two fractions and the known specific activity of the *E. coli* RNA polymerase it was estimated that the fractions were probably about 30 per cent pure, that is, one-third of the protein in each fraction would be RNA polymerase. Analysis of the two fractions by electrophoresis in SDS-acrylamide gels indicated that fraction 5B contained an appreciable amount of a protein component having the size of the subunits β' and β of the *E. coli* RNA polymerase[15,10]. Fraction 5A contained no detectable β bands, however, suggesting that the enzymatic activity was not associated with these subunits of the host RNA polymerase. We tested this possibility further by measuring the sensitivity of the two fractions to inhibitors of *E. coli* RNA polymerase: rifamycin[17], streptolydigin[18], and a purified γ-globulin fraction prepared from a high titre antiserum to *E. coli* RNA polymerase (Table 3). It is clear that fraction 5B behaves as would be expected if most of its RNA polymerase activity were *E. coli* polymerase*. By contrast, fraction 5A is insensitive to both drugs and its activity is not affected by antibody to *E. coli* RNA polymerase. This suggested that T7 bacteriophage might induce a new RNA polymerase protein in infected cells.

Table 3. INHIBITION OF T7 RNA POLYMERASE FRACTIONS BY RIFAMYCIN, STREPTOLYDIGIN AND ANTIBODY TO *E. coli* RNA POLYMERASE

Enzyme	Additions	Activity for T7 RNA synthesis (per cent)
E. coli B RNA polymerase	—	100
	Streptolydigin	4
	Rifamycin	< 1
	Antibody	10
T7 RNA polymerase, fraction 5A	—	100
	Streptolydigin	100
	Rifamycin	100
	Antibody	91
T7 RNA polymerase, fraction 5B	—	100
	Streptolydigin	22 (7)
	Rifamycin	18 (3)
	Antibody	40 (31)

Incorporation of ^{14}C-AMP into T7 RNA was followed as before with the addition of 2×10^{-4} M streptolydigin, 20 μg/ml. rifamycin or 0·02 A_{280} unit of purified antibody to *E. coli* RNA polymerase where noted. With *E. coli* RNA polymerase, T7 RNA polymerase fraction 5A, and 5B, 1·2, 1·1 and 1·5 nmoles of ^{14}C-AMP were incorporated, respectively, in the absence of inhibitors. Values given in parentheses for T7 RNA polymerase fraction 5B are those obtained with dAT copolymer as template.

Purification and Characterization of T7 RNA Polymerase

Preliminary characterization of the T7 RNA polymerase using fraction 5A led to the modification of the assay conditions. In particular, the activity—both in extracts and in purified fractions—is lost if the enzyme is diluted for assay in solutions without high concentrations of BSA. Considering other differences between the T7 RNA polymerase and the *E. coli* enzyme, we believe that this factor accounts for the extensive losses of enzyme activity incurred during purification by procedure I and also for the failure of previous workers to identify T7 RNA polymerase. Using the modified RNA polymerase assay it was found that RNA polymerase activity increases at least thirty-fold after infection by T7. This activity is stable in cell-free extracts for at least 1 week when the extracts are kept at 0° C. A purification for the T7 RNA polymerase was devised using the assay described and adding 20 μg/ml. of rifamycin to all assays. In these conditions the *E. coli* RNA polymerase is not active and one can fractionate selectively for the T7 RNA polymerase activity. Procedure II gives about a 130-fold purification of the initial T7 RNA polymerase activity with a reasonable recovery of total activity. Analysis

Table 2. ACTIVITIES OF RNA POLYMERASE FRACTIONS FROM T7-INFECTED *E. coli*

Fraction	Source	Specific activity		Ratio T7/dAT
		T7 DNA	dAT copolymer (U/mg)	
5A	T7-infected *E. coli*	2,500	64	39
5B	T7-infected *E. coli*	1,700	4,000	0·4
RNA polymerase holoenzyme	*E. coli*	8,100	20,000	0·4
Core polymerase	*E. coli*	880	16,000	0·06

Data for *E. coli* polymerases are from Berg *et al.*[11]. Where indicated, 20 nmoles of dAT copolymer replaced T7 DNA as template.

* The residual activity seen with fraction 5B in the presence of rifamycin or streptolydigin arises from trace contamination of this fraction by T7 RNA polymerase as shown by the lack of rifamycin resistant activity with dAT as template. The activity of fraction 5B with dAT was also more than 90 per cent sensitive to anti-*E. coli* RNA polymerase antibody when larger amounts of antibody were added.

of the enzyme fractions obtained after chromatography on phosphocellulose using SDS-acrylamide gel electrophoresis (Fig. 1) shows that tubes with the highest enzyme activity contain a single major protein component which has a molecular weight of about 107,000 (Fig. 1). When large amounts of protein from T7 polymerase fraction F are analysed, traces of other proteins are detected (Fig. 1). But fraction F T7 RNA polymerase consists of at least 80 per cent of this single protein component. Protein components which could be subunits of E. coli RNA polymerase are present in only trace amounts if at all. The final specific activity of T7 RNA polymerase is extremely high. Although protein determinations by the Lowry procedure are made difficult by the low concentrations of protein in fraction E and by the fact that buffer I interferes with the determination, it is clear that the specific activity of these fractions is about fifteen times higher than that of the E. coli RNA polymerase acting with T7 DNA template.

As found for fraction 5A, the purified T7 RNA polymerase is not inhibited by high concentrations of rifamycin (100 μg/ml.), streptolydigin (2×10^{-3} M), or of antibody to E. coli RNA polymerase. The enzyme requires Mg^{2+} as a divalent metal ion and is not active when Mn^{2+} replaces Mg^{2+}. T7 RNA polymerase shows optimal activity at 2×10^{-2} M Mg^{2+} as compared with 8×10^{-3} M Mg^{2+} for the E. coli enzyme[19]. T7 RNA polymerase is strongly inhibited by KCl at concentrations exceeding 0.05 M and is essentially inactive at 0.15 M KCl. The T7 enzyme shows a remarkable template specificity. Whereas E. coli RNA polymerase shows equivalent activities with T7 and T2 DNAs and is highly active with dAT copolymer, T7 RNA polymerase is active only with T7 DNA (Table 4). Denatured T7 DNA is active but

SDS-ACRYLAMIDE GEL ANALYSIS OF T7 RNA POLYMERASE FRACTIONS

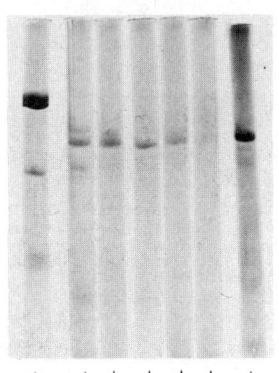

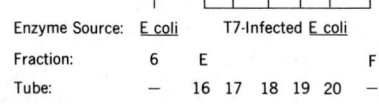

Enzyme Source:	E coli	T7-Infected E coli
Fraction:	6	E F
Tube:	—	16 17 18 19 20 —

Fig. 1. Analysis of T7 RNA polymerase fractions in SDS-acrylamide gels. Samples contained 8 μg of E. coli RNA polymerase fraction 6, 1.5–3 μg of T7 RNA polymerase fractions from phosphocellulose chromatography (fractions E 16–20) or 11 μg of T7 RNA polymerase fraction F. The protein was analysed by gel electrophoresis in SDS-acrylamide gels and was stained with coomassie blue. The peak of T7 RNA polymerase activity eluting from phosphocellulose was found in fractions 17–19 which contained enzyme with a constant specific activity of 120,000 U/mg. Protein in these fractions was concentrated for gel analysis by precipitation with 10 per cent trichloroacetic acid.

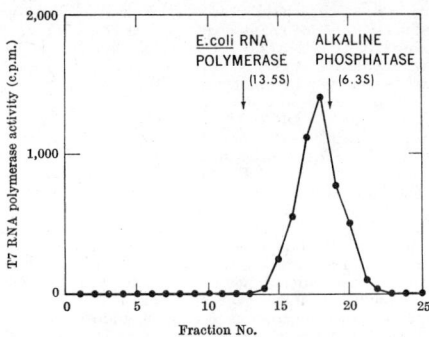

Fig. 2. Zone sedimentation of T7 RNA polymerase in a glycerol gradient. T7 RNA polymerase (33 U), fraction 5A, was layered on a 10–30 per cent glycerol gradient and sedimented at 40,000 r.p.m. for 11.5 h. All solutions contained 1 mg/ml. BSA, 0.2 M KCl and buffer I. Fractions were collected and were analysed for T7 RNA polymerase activity. Sedimentation standards, E. coli RNA polymerase fraction 5 and E. coli alkaline phosphatase, were run in a duplicate tube during the same run. The arrows indicate the peak position of each.

much less so, and T2 DNA, λ DNA and dAT copolymer are almost inactive as templates. The homopolymer pair dG·dC is highly active for poly rG synthesis but does not direct poly rC synthesis at all with T7 RNA polymerase. This suggests that the T7 enzyme is highly specific in its recognition of initiation sites on DNA; although such sites occur on T7 DNA, they are lacking on any of the other DNAs we have tested and are not replaced by dAT copolymer. The activity of T7 RNA polymerase on dG·dC suggests that the promoter sites recognized by this enzyme may be C-rich regions of the DNA. Such regions are known to occur in T7 DNA[20].

Table 4. COMPARISON OF ACTIVITIES OF E. coli RNA POLYMERASE AND T7 RNA POLYMERASE WITH DIFFERENT DNA TEMPLATES

	Specific activity (U/mg)	
Template	E. coli polymerase	T7 polymerase
dAT copolymer	15,000	140
dG·dC	17,000 (^{3}H-GTP)	40,000 (^{3}H-GTP)
	8,700 (^{3}H-CTP)	< 90 (^{3}H-CTP)
T7 DNA	7,700	68,000
Denatured T7 DNA	950	24,000
T2 DNA	6,000	< 60
λ DNA	1,700	300

Where indicated, 40 nmoles of the indicated DNA or 20 nmoles of dAT or dG·dC replaced T7 DNA in the standard assay for T7 RNA polymerase. When dG·dC was used as template, only ^{3}H-CTP and GTP or ^{3}H-GTP and CTP were added as nucleotide substrates. Assays were carried out using 0.15 μg of fraction F T7 RNA polymerase or 0.14 μg of fraction 5 E. coli RNA polymerase.

To test the specificity of the T7 RNA polymerase further we determined the strand specificity of the T7 RNA formed by T7 RNA polymerase with native T7 DNA as template. In vivo, all T7 RNA seems to be transcribed from the r-strand of T7 DNA[21], so that synthesis of l-strand specific RNA would signal that non-specific transcription was occurring. More than 95 per cent of the T7 RNA formed by T7 RNA polymerase is r-strand specific, indicating that T7 RNA polymerase does initiate RNA chains at specific sites on the T7 template (Table 5).

Analysis of T7 RNA polymerase by sedimentation through a glycerol gradient showed a single peak of enzymatic activity (Fig. 2). Calibration of the gradient[22], using E. coli RNA polymerase ($s_{20,w} = 13.5S$[10]) and E. coli alkaline phosphatase ($s_{20,w} = 6.3S$[23]) as sedimentation standards, gave an estimate of $s_{20,w} = 7.2S$ for the

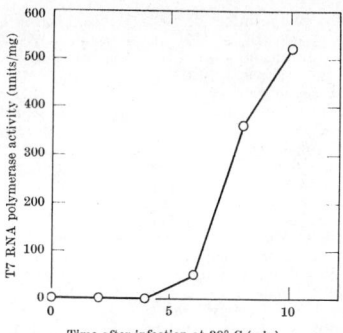

Fig. 3. Induction of T7 RNA polymerase activity after T7 bacteriophage infection. *E. coli* B/1 were grown at 37° C in 1·5 l. of M9 medium containing 5 g/l. of glucose. When cell density reached 1×10^9 cells/ml. the cells were chilled and harvested and were suspended in 28 ml. of M9 medium. An aliquot (2·5 ml.) was taken as an uninfected sample and $7·5 \times 10^{13}$ purified T7 phage were added in 0·3 ml. After 10 min at 0° C the cells were diluted into 475 ml. of prewarmed M9 medium which contained 5 g/l. glucose and the culture was shaken vigorously at 30° C. Aliquots (100 ml.) were taken at 2 min intervals and were chilled rapidly by pouring the solution on ice at −15° C. The cells were collected by centrifugation, were suspended in 1 ml. of buffer A, and were disrupted by sonic vibration for 2·5 min at 50 V using a Bronwill sonic vibrator. The extracts were centrifuged and the supernatant fluid was assayed for protein and for T7 RNA polymerase activity. Assay reactions contained 20 μg/ml. of rifamycin and 4×10^{-4} M streptolydigin and used ^{3}H-CTP as the labelled nucleotide. In these conditions *E. coli* RNA polymerase is not active. Uninfected *E. coli* extracts contain about 25 units/mg of RNA polymerase activity with T7 DNA as template, when assayed in similar conditions in the absence of rifamycin.

T7 RNA polymerase. If one assumes that the protein is globular and that $\bar{v} = 0·73$, this would correspond to a molecular weight of about 100,000 (ref. 24). This result, taken with the results of SDS-gel analysis, suggests that the native T7 RNA polymerase consists of a single polypeptide chain of molecular weight 100,000–110,000.

Table 5. STRAND SPECIFICITY OF T7 RNA SYNTHESIZED USING T7 RNA POLYMERASE

Enzyme	RNA annealing to T7 DNA (c.p.m.)		Percentage r-strand
	r-strand	l-strand	RNA
E. coli RNA polymerase	1,159	282	80
T7 RNA polymerase	1,045	63	94

T7 RNA was synthesized in a standard T7 RNA polymerase assay scaled up to 1·0 ml. using either 18 μg of fraction 6 *E. coli* RNA polymerase (2 min reaction, 28 nmoles RNA formed) or 4 μg of fraction F T7 RNA polymerase (10 min reaction, 70 nmoles RNA formed). ATP was labelled with ^{32}P at a specific activity of 100,000 c.p.m./nmole. The reaction was terminated with 10 μg of pancreatic deoxyribonuclease, and after 10 min at 37° C the RNA was extracted three times with phenol. Aliquots of the aqueous phase were assayed for acid-insoluble radioactivity and for their ability to anneal to the separated strands of T7 DNA[21] using a modification of the method of Gillespie and Spiegelman[27]. The total amount of RNA annealing to r and l strands was about 30 per cent of that added to the annealing mixture. Similar efficiencies were obtained when annealing either RNA product to T7 DNA which had not been separated into r and l strands.

T7 RNA polymerase did not interact physically with *E. coli* RNA polymerase holoenzyme or core polymerase when the two purified enzymes were mixed and sedimented together in a glycerol gradient. Furthermore, T7 polymerase did not stimulate the synthetic activity of *E. coli* core RNA polymerase on T7 DNA when the two were mixed in an assay. *E. coli* polymerase holoenzyme did not stimulate synthesis by T7 RNA polymerase with T7 DNA as template. Thus T7 RNA polymerase seems to be a complete polymerase *per se* and neither replaces the *E. coli* RNA polymerase sigma component nor binds to the *E. coli* core polymerase in its stead.

Appearance of T7 RNA Polymerase in Infected Cells

Studier and Maizel[25] have shown that there are two principal classes of T7 induced proteins; early proteins which begin to appear 5–7·5 min after infection at 30° C, and late proteins after about 10 min. To determine to which of these classes the T7 RNA polymerase belongs, *E. coli* B/1 cells were grown in minimal medium with glucose and were infected at 0° C to allow synchronous infection[21]. The infected cells were then diluted into prewarmed medium and samples were taken at intervals to follow the appearance of T7 RNA polymerase activity in extracts. The results (Fig. 3) show that T7 RNA polymerase activity appears 4–6 min after infection. Thus T7 RNA polymerase is an early T7 protein.

Structural Gene for T7 RNA Polymerase

Because only three early genes are currently known for T7 phage[25], it was of interest to determine if any of these might represent the structural gene for T7 RNA polymerase. The size of the T7 polymerase and its ability specifically to transcribe T7 DNA suggested that T7 gene 1 was a likely candidate, because it specifies a protein of about 100,000 molecular weight, the protein is required for the synthesis of late mRNA in T7-infected cells[25,26], and gene 1 mutants are defective in all known late T7 functions. In fact, no other protein of this size is synthesized in T7-infected cells[25]. Because of the availability of a variety of mutant T7 phages[12,4], this possibility was easily tested. Table 6 records the activity of T7 RNA polymerase in extracts of *E. coli* B/1 infected with amber mutants for T7 genes 1, 2 and 3, and with a T7 mutant which contains a temperature-sensitive mutation in gene 1. Whereas extracts of cells infected with amber mutants for genes 2 and 3 have about half of the normal specific activity of T7 RNA polymerase, the extract from cells infected with an amber mutation in gene 1 has no detectable T7 RNA polymerase activity. When extracts are prepared from cells infected with T7 tsl (342-15) the resulting extract (tsl extract) contains detectable but reduced levels of T7 RNA polymerase. By contrast with the T7 RNA polymerase activity obtained with extracts prepared from cells infected with wild-type T7 (T7+ extract), the activity in a tsl extract was unstable and 50 per cent of the activity was lost after overnight storage. Preliminary experiments showed that the rate of inactivation of T7 RNA polymerase activity in a tsl extract at either 37° C or 45° C was similar to the rate of inactivation found for RNA polymerase from a T7+ extract. The enzymatic activity of T7 RNA polymerase in a tsl extract was, however, strikingly reduced when assays were carried out at 45° C (Table 6). By contrast, the activity of T7 RNA polymerase in a T7+ extract was slightly enhanced when assayed at 45° C relative to 37° C. Thus a temperature-sensitive mutation in gene 1 affects the T7 RNA polymerase protein and seems to lead to reduced enzymatic activity at elevated temperatures, but not to an increased sensitivity to thermal denaturation. These results, taken with the lower initial activity of the enzyme in cells infected with T7 tsl and with the physical and biochemical properties of the T7 RNA polymerase, indicate that the structural gene for T7 RNA polymerase is gene 1.

Table 6. ACTIVITY OF T7 POLYMERASE IN EXTRACTS OF CELLS INFECTED WITH MUTANT T7 PHAGES

Phage	Genotype	T7 polymerase activity (U/mg)	
		37° C	45° C
T7+	Wild type	520	570
T7 tsl (342-15)	Temperature-sensitive, gene 1	48	18
T7 am 23	Gene 1 amber	<2	—
T7 am 64	Gene 2 amber	290	—
T7 am 29	Gene 3 amber	260	—

100 ml. aliquots of *E. coli* B/1 (a nonpermissive strain for T7 amber mutants) were grown at 30° C in M9 medium containing 5 g/l. glucose to a density of 1×10^9 cells/l. A lysate of T7 phage was added to give a final multiplicity of five and growth was continued for 15 min. Infected cells were harvested, and extracts were prepared and assayed as described in Fig. 3.

Model for Regulation

The identification of T7 gene 1 protein as a new RNA polymerase provides a clear model for the positive control of transcription by T7 phage: the infecting T7 DNA is transcribed initially by the host RNA polymerase giving rise to early T7 mRNA. Translation of this early mRNA leads to T7 RNA polymerase along with two other early proteins of unknown function[25]. The T7 RNA polymerase then carries out the transcription of late T7 mRNA. The question remains how negative control of early T7 genes might occur. Synthesis of gene 1 protein seems to cease about 10 min after infection[25], at a time when late particle proteins are just beginning to appear. Thus T7 may have a mechanism to turn off transcription of early genes. Such a negative control of early transcription could result from modification leading to inactivation of the *E. coli* RNA polymerase by a late phage function, for the host enzyme is no longer needed after synthesis of the T7 RNA polymerase*. Extensive modification of *E. coli* RNA polymerase occurs after infection by T4 phage[28], and may serve a negative control function in the growth of that phage as well.

What implications does the identification of a T7-specific RNA polymerase have for our understanding of the control of transcription in other bacteriophage systems? At least three possible mechanisms have been proposed by which a bacteriophage could initiate transcription of a set of genes which it had not previously transcribed: (1) alteration of the initiation specificity of the host RNA polymerase by modification of the polymerase protein or by the synthesis of a new subunit such as sigma subunit[3]; (2) *de novo* synthesis of a new RNA polymerase having an altered initiation specificity, and (3) synthesis of a protein factor (anti-termination factor) which prevents normal termination of RNA chains and so allows the host RNA polymerase to "read through" to distal genes[29,30]. These mechanisms are not mutually exclusive; in the case of complex bacteriophages such as λ phage and T4 phage, at least three classes of transcriptional units are known, and more than one positive control mechanism may well be used.

Alteration of transcriptional specificity of RNA polymerase by replacement of sigma subunit with a subunit carrying a new initiation specificity has been proposed to occur during growth of T4 and λ phages and during sporulation of *B. subtilis*[31]. Although there is good evidence for modification of bacterial RNA polymerases in some of these systems, there is still no evidence that this modification confers a new transcriptional specificity on the RNA polymerase *in vivo*. In particular, as discussed, negative control mechanisms through which certain genes are turned off are present in all of these systems. It is possible that most modifications of bacterial RNA polymerase are directed to such an end and not toward activation of the enzyme for a new function.

In the case of bacteriophages λ and T4, it seems likely that synthesis of early and delayed early classes of mRNA is mediated directly by the host cell RNA polymerase by a mechanism such as the first or third and not by phage-directed synthesis of a new RNA polymerase. Activation of late transcriptional units in λ and T4 infected cells is, however, more complex. At least one new phage protein is directly required for late transcription in each case, T4 gene 55 (ref. 32) and λ gene Q[33]. Evidence that the host RNA polymerase itself is responsible for late transcription in T4 and λ-infected cells is based primarily on studies which show that phage growth and also mRNA synthesis remain sensitive to rifamycin throughout phage infection[34,35]. By contrast with these results, the growth of T7 phage becomes resistant to rifamycin inhibition after 5 min, the time at which the T7 RNA polymerase appears[4]. One interpretation of the continued sensitivity of late transcription to rifamycin in λ and T4-infected cells is that the host cell RNA polymerase is directly involved in late transcription as proposed by mechanism 1 or 3. Alternatively, our results suggest that late transcription might be carried out by a new phase-specific polymerase as in mechanism 2. Such a late RNA polymerase could be specified by gene 55 and T4 and by gene Q in λ phage. By this hypothesis, the sensitivity of late transcription to rifamycin would be the result of a continued requirement for host cell RNA polymerase for some other phage function. This would result, for example, if a small amount of early or delayed early transcription were continually required for late transcription to continue. In the case of phage λ, some evidence supports this latter possibility, for both N gene function and an active, early r-strand promoter site seem to be continually required for late transcription[36,37], and the N gene product seems to be unstable during growth[38,39]. Thus it is possible that the initiation of transcription of late phage genes by *de novo* synthesis of a bacteriophage specific RNA polymerase may prove to be a general mechanism for the activation of late bacteriophage transcription.

This investigation was supported by a US Public Health Service research grant and training grant from the Institute of General Medical Sciences. We would like to thank Drs Paul Berg, Richard Calendar and Harrison Echols for criticism and discussion.

Received July 24, 1970.

[1] Geiduschek, E. P., and Haselkorn, R., *Ann. Rev. Biochem.*, **38**, 647 (1969).
[2] Calendar, R., *Ann. Rev. Microbiol.*, **24** (in the press).
[3] Travers, A. A., *Nature*, **223**, 1107 (1969); **225**, 1009 (1970).
[4] Summers, W. C., and Siegel, R. B., *Nature*, **223**, 1111 (1969).
[5] Symons, R. H., *Biochim. Biophys. Acta*, **155**, 609 (1968).
[6] Thomas, jun., C. A., and Abelson, J., in *Proc. Nucleic Acid Res.* (edit. by Cantoni, G. L., and Davies, D. R.), 553 (Harper and Row, New York, 1966).
[7] Schachman, H. K., Adler, J., Radding, C. M., Lehman, I. R., and Kornberg, A., *J. Biol. Chem.*, **235**, 3242 (1960).
[8] Radding, C. M., Josse, J., and Kornberg, A., *J. Biol. Chem.*, **237**, 2869 (1962).
[9] Shapiro, A. L., Viñuela, E., and Maizel, jun., J. V., *Biochem. Biophys. Res. Commun.*, **28**, 815 (1967).
[10] Berg, D., and Chamberlin, M., *Biochemistry* (in the press).
[11] Berg, D., Barrett, K., and Chamberlin, M., *Methods Enzymol.*, **13** (in the press).
[12] Studier, F. W., *Virology*, **39**, 562 (1969).
[13] Englander, S. W., and Crowe, D., *Anal. Biochem.*, **12**, 579 (1965).
[14] Richardson, C. C., and Kornberg, A., *J. Biol. Chem.*, **239**, 242 (1964).
[15] Burgess, R. R., *J. Biol. Chem.*, **244**, 6168 (1969).
[16] Wehrli, W., Nüesch, J., Knüsel, F., and Staehelin, M., *Biochim. Biophys. Acta*, **157**, 215 (1968).
[17] Tocchini-Valentini, G., Marino, P., and Colvill, A. J., *Nature*, **220**, 275 (1968).
[18] Siddhikol, C., Erbstoeszer, J., and Weisblum, B., *J. Bact.*, **99**, 151 (1969).
[19] Chamberlin, M., and Berg, P., *Proc. US Nat. Acad. Sci.*, **48**, 81 (1962).
[20] Summers, W. C., and Szybalski, W., *Biochim. Biophys. Acta*, **166**, 371 (1968).
[21] Summers, W. C., and Szybalski, W., *Virology*, **34**, 9 (1968).
[22] Martin, R. G., and Ames, B. N., *J. Biol. Chem.*, **236**, 1372 (1961).
[23] Garen, A., and Levinthal, C., *Biochim. Biophys. Acta*, **38**, 470 (1960).
[24] Schachman, H. K., *Ultracentrifugation in Biochemistry*, 247 (Academic Press, New York, 1959).
[25] Studier, F. W., and Maizel, jun., J. V., *Virology*, **39**, 575 (1969).
[26] Siegel, R. B., and Summers, W. C., *J. Mol. Biol.*, **49**, 115 (1970).
[27] Gillespie, D., and Spiegelman, S., *J. Mol. Biol.*, **12**, 829 (1965).
[28] Walter, G., Seifert, W., and Zillig, W., *Biochem. Biophys. Res. Commun.*, **30**, 240 (1968).
[29] Roberts, J., *Nature*, **223**, 480 (1969).
[30] Salser, W., Bolle, A., and Epstein, R., *J. Mol. Biol.*, **49**, 271 (1970).
[31] Losick, R., and Sonenshein, A. L., *Nature*, **224**, 35 (1969).
[32] Pulitzer, J. F., and Geiduschek, E. P., *J. Mol. Biol.*, **49**, 489 (1970).
[33] Skalka, A., Butler, B., and Echols, H., *Proc. US Nat. Acad. Sci.*, **58**, 576 (1967).
[34] Takeda, Y., Oyama, Y., Nakajima, K., and Yura, T., *Biochem. Biophys. Res. Commun.*, **36**, 533 (1969).
[35] Haselkorn, R., Vogel, M., and Brown, R., *Nature*, **221**, 836 (1969).
[36] Butler, B., and Echols, H., *Virology*, **40**, 212 (1970).
[37] Rabovsky, D., and Konrad, M., *Virology*, **40**, 10 (1970).
[38] Konrad, M., *Proc. US Nat. Acad. Sci.*, **59**, 171 (1968).
[39] Schwartz, M., *Virology*, **40**, 23 (1970).

---

* T7 RNA polymerase fraction 5B from procedure I seems to be composed of host cell enzyme subunits and is active as an RNA polymerase on both T7 DNA and T2 DNA templates. This does not rule out the possibility that the *E. coli* RNA polymerase is modified or inactivated in T7-infected cells, however, because we cannot yet exclude the possibility that the active polymerase in this fraction is derived entirely from uninfected cells present, in the infected cell population.

In Vitro Synthesis of an Infectious Mutant RNA with a Normal RNA Replicase

N. R. PACE AND S. SPIEGELMAN

Abstract. *When purified Qβ-RNA replicase is presented alternately with two genetically different Qβ-RNA molecules, the RNA synthesized is identical to the initiating template. The results establish that the RNA is the instructive agent in the replicative process and hence that it satisfies the operational definition of a self-duplicating entity. The data also eliminate alternative explanations which do not involve self-propagation of the input RNA. An opportunity is now provided for studying the genetics and evolution of a self-duplicating nucleic acid molecule under conditions permitting detailed control of environmental parameters and chemical components.*

Previous experiments with two serologically distinct [1, 2] RNA coliphages (MS-2 and Qβ) established [3, 4] that each induces in *Escherichia coli* a replicase (RNA replicating enzyme) which exhibits a unique requirement for intact [5] homologous RNA as a template. Further studies with purified Qβ replicase showed that the RNA synthesized is physically [6] and chemically [7] indistinguishable from the strands found in the Qβ virus.

The ability of the synthetic RNA to program the synthesis of complete virus particles was examined by protoplast infection in the course of a serial transfer [8]. In these experiments, the products of a reaction initiated with Qβ-RNA was serially diluted to prime successive reactions until the original RNA was reduced to less than one strand per reaction mixture. The final tube contained new radioactive RNA which, when assayed in a bacterial protoplast system [9, 10], displayed the same ability to generate virus particles as the viral RNA used to start the reaction in the first tube. These experiments indicated that the synthesis of a self-propagating and infectious entity had been achieved in a simple system of known components.

The significance and potential usefulness of the finding encouraged further efforts at purifying the enzyme. A procedure for more extensive purification was developed [11] involving equilibrium banding in CsCl followed by zonal centrifugation in linear gradients of sucrose. Here, advantage was taken of expected disparities in size and density between the replicase protein and unwanted impurities. The resulting preparation was effectively free of residual virus particles, permitting direct assay for infectivity and thereby obviating the laborious purification of the RNA product required in the earlier [8] study. The concomitant removal of polynucleotide contaminants did not decrease, qualitatively or quantitatively, the ability of the replicase to respond to added Qβ-RNA by synthesizing infectious copies. This latter finding makes even more implausible arguments which would explain the increase in infectious units in terms of an "activation" of RNA preexistent in the enzyme by an unknown reaction which requires both added template and new RNA synthesis.

We now come to the central issue of the present communication which stems from the fact that two informed components are present in the reaction mixture, replicase and RNA template. None of the experiments thus far described proved that the RNA synthesized in this system is, in fact, a self-duplicating entity—that is, one which contains the requisite information and directs its own synthesis. What is required is a rigorous demonstration that the RNA, and not the replicase, is the instructive agent in the replicative process. A definitive decision would be provided by an experimental answer to the following question: if the replicase is provided alternatively with two distinguishable RNA molecules, is the product produced always identical with the initiating template?

A positive outcome would establish that the RNA is directing its own synthesis and simultaneously completely eliminate any remaining possibility of "activation" of preexisting RNA. Our data establish that the RNA synthesized is a self-duplicating entity. The discriminating selectivity of the replicase for its own genome as template makes it impossible to employ heterologous RNA in the test experiments and recourse was, therefore, had to mutants. For ease in isolation and simplicity in distinguishing between mutant and wild type, temperature sensitive (ts) mutants were chosen. Their diagnostic phenotype is poor growth at 41°C as compared with that at 34°C. The wild type grows equally well at both temperatures.

Temperature-sensitive mutants of Qβ were isolated by a modification of the method described by Davern [12]. *Escherichia coli* K-38 [13] was grown in a rotary shaker at 34°C in modified 3XD medium [14] to an optical density (660 mμ) of 0.15. Qβ bacteriophage was added to a multiplicity of 5, and the suspension was mixed and allowed to stand for adsorption of virus at 34°C for 5 minutes. Shaking was reinstituted for 10 minutes, whereupon 20 μg of 5-fluorouracil was added per milliliter of culture, and the incubation was continued for 2 hours. The resulting lysate was cleared by low-speed centrifugation and plated for plaques arising at 34°C. Isolated plaques were stabbed with a needle and suspended in 1 ml of water. A small loopful of the suspension was transferred to each of two plates seeded with *E. coli* K-38, and respective plates were incubated at 34° or 41°C. Plaques arising only at 34°C were picked for further testing, and those which retained the ts phenotype were chosen. Mutant virus particles isolated in this manner are quite stable to passage and

Table 1. Relative efficiency of plating at 34° and 41°C. Dilutions were plated with *E. coli* K-38 as the indicator organism, and duplicate plating series were incubated at 34° and 41°C. The relative efficiency of plating (REOP) of 100 is defined relative to the plaque forming units (PFU) observed at 34°C.

	34°C	41°C
	Virus Qβ	
REOP	100	100
PFU	1.14×10^{13}/ml	1.16×10^{13}/ml
	Virus ts Qβ	
REOP	100	2.5×10^{-5}
PFU	4.4×10^{7}/ml	1.1×10^{4}/ml

Table 2. Efficiency of infection of protoplasts by three RNA preparations. Infectious RNA assays were carried out on Qβ RNA, synthetic Qβ RNA and ts RNA. Duplicate pairs were incubated at 34° and 41°C. Efficiencies at 34° are defined as 100. The synthetic Qβ-RNA was the result of a 20-fold synthesis carried out by Qβ replicase purified through CsCl and sucrose centrifugation; 0.1 μg of Qβ RNA was used to initiate the standard reaction.

	34°C	41°C
	Natural Qβ-RNA	
REOP	100	93
PFU	4.56×10^{5}/ml	4.24×10^{5}/ml
	Synthetic Qβ-RNA	
REOP	100	92
PFU	2.90×10^{6}/ml	2.66×10^{6}/ml
	Natural ts-Qβ-RNA	
REOP	100	1.5
PFU	1.86×10^{6}/ml	2.75×10^{4}/ml

possess low efficiencies of plating at 41°C (Table 1). To provide a supply of mutant RNA, large lysates were prepared from plaque inocula of the ts-$Q\beta$ and RNA was isolated from the virus as previously described (5).

The ts phenotype is easily recognized by parallel platings of intact virus particles at 34° and 41°C on receptor cells (Table 1). It remained, however, to be seen whether this difference would be retained when the corresponding purified mutant RNA preparations were assayed for infectivity in the protoplast system. This check is particularly necessary since one of the steps requires a 10-minute incubation of the infected protoplasts at 35°C. During this interval, "revertants" could be produced and contribute to the background of plaques developing at 41°C. In addition, it was necessary to establish that the synthetic product of the replicase, primed by a normal $Q\beta$-RNA, behaves like the natural viral RNA in its behavior at 41°C (Table 2). It is evident that the synthetic wild type $Q\beta$-RNA behaves exactly like its natural counterpart at the two temperatures. On the other hand, the ts-$Q\beta$-RNA again shows the lower efficiency at 41°C, although it will be noted that the background at 41°C is higher than in the intact cell assay (Table 1), as expected. The 65-fold difference at the two temperatures is, however, more than adequate for a clear diagnosis.

It is evident that the system available will permit us to determine whether the product produced by a normal replicase primed with ts-$Q\beta$-RNA is mutant or wild type. As in previous investigations, this is best done by a serial transfer experiment to avoid the ambiguity of examining reactions containing significant quantities of the initiating RNA. Accordingly, seven standard reaction mixtures (0.25 ml) were prepared, each containing 60 μg of $Q\beta$ replicase isolated from cells infected with normal virus and purified through the CsCl banding sucrose sedimentation steps (9). To the first reaction mixture was added 0.2 μg of RNA, and synthesis was allowed to proceed at 35°C. After a suitable interval, one-tenth of this reaction mixture was used to initiate a second reaction which, in turn, was diluted into a third reaction mixture, and so on for seven transfers. A control series was carried out in a manner identical to that just described, save that no RNA was added to the first tube.

Portions from each reaction mixture

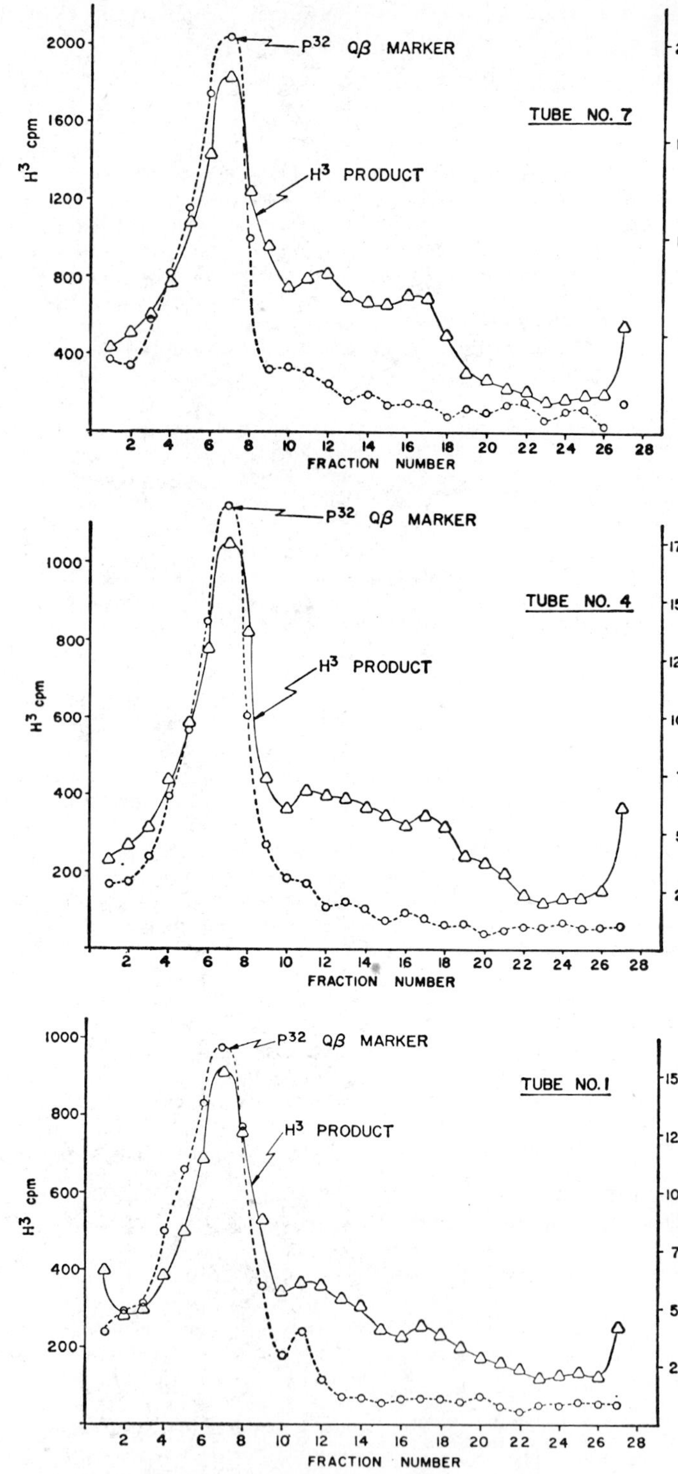

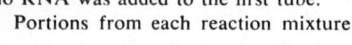
1 JULY 1966

Fig. 1. Sedimentation analysis of products. A portion (0.04 ml) from reaction mixtures 1, 4, and 7 (see Table 3) were each mixed with 0.01 ml of P^{32}-$Q\beta$-RNA, 0.01 ml 20 percent sodium dodecyl sulphate, and 0.20 ml TM (tris magnesium solution), and layered onto linear gradients of 2.5 to 15 percent sucrose in a solution of 0.01M tris, pH 7.4; 0.005M $MgCl_2$; and 0.1M NaCl. Gradients were centrifuged at 10°C for 14 hours in the Spinco SW-25 rotor. Fractions were collected and analyzed for radioactivity (count/min) as described previously (6).

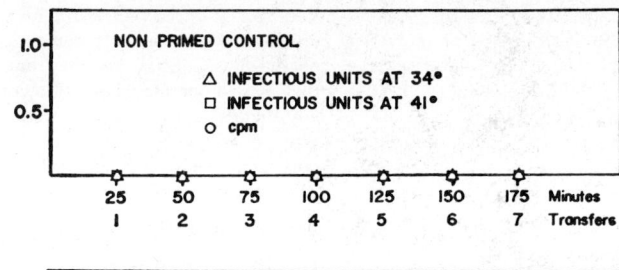

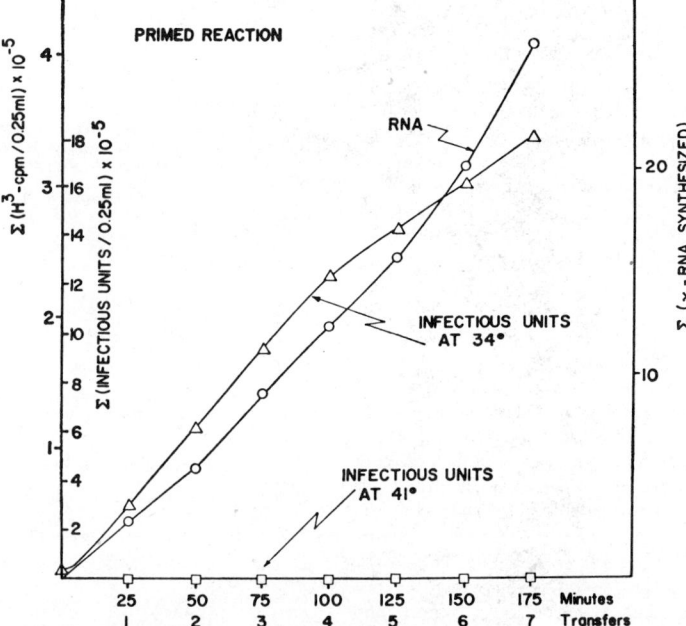

Fig. 2. RNA synthesis and formation of infectious units. The data are from the serial transfer experiment of Table 3.

were examined for radioactivity in material precipitable by trichloroacetic acid (TCA) and assayed for infectious RNA at 34° and 41°C. In addition, samples from reactions 1, 4, and 7 were examined for physical similarity to the input RNA by sedimentation through sucrose gradients. As shown in Fig. 1, the bulk of material synthesized is similar in sedimentation characteristics to ts-$Q\beta$-RNA derived from virus particles.

Table 3 is a record of such a serial transfer experiment. If we first focus attention of the RNA formation in the experimental series (columns 3 to 6), it is evident that ts-$Q\beta$-RNA serves as an excellent initiator for the normal replicase. Included also is the radioactivity (count/min) observed in the nonprimed control series (columns 7 and 8). No detectable synthesis occurs in the first three tubes although a few counts accumulate near the end which are, however, negligible from the point of view of the chemical amounts of RNA synthesized. Though quantitatively insignificant, this "long-term background" is persistently observed with some enzyme preparations.

Columns 9 and 10 of Table 3 give the actual number of plaques counted in the assay for infectious units at each transfer, the numbers representing the average of two duplicate plates. Comparisons of columns 9 and 10 reveal that the relative number of plaque formers at the two temperatures agree with those obtained with the original ts-$Q\beta$-

Table 3. *Transfer Experiment with ts-$Q\beta$-RNA.* Each 0.25 ml of standard reaction mixture (4) contained 60 µg of $Q\beta$ replicase purified through CsCl and sucrose centrifugation, and H^3-CTP (cytidine triphosphate) at a specific activity such that 15,600 count/min signifies 1 µg of synthesized RNA. The first reaction was initiated by addition of 0.2 µg of temperature sensitive infectious RNA. Each reaction was carried out at 35°C for 25 minutes, whereupon 0.02 ml was drawn for counting, and 0.025 ml was used to prime the next reaction. All samples were stored frozen at −70°C until infectivity assays were carried out. Dilutions for infectious RNA assays were made into a solution of 0.01M tris, pH 7.4, and 0.005M $MgCl_2$, and used immediately. Columns 1 and 2 give the reaction number and total time elapsed during the experiment. Column 3 lists acid-precipitable radioactivity (count/min) found in each 0.25 ml reaction mixture and column 4 lists the corresponding sum. Similarly, columns 5 and 6 list the RNA formation during each reaction and their cumulative amounts. Columns 7 and 8 present radioactivity incorporated in the control transfer without added RNA. Columns 9 and 10 are the averages of plaques observed on duplicate plates in the assays for infectious RNA, on plates incubated at 34° and 41°C. In all cases, reaction products were diluted 1.6×10^{-3} during the course of the assay. Column 11 presents the actual number of infectious units appearing in each reaction tube, and column 12 is the sum of infectious units appearing at 34°C.

		Formation of RNA						Formation of infectious units			
Transfer No.	Time (Min)	With RNA				Without RNA Radioactivity (count/min × 10⁻⁵)		With RNA			
		(count/min × 10⁻³) Radioactivity		Amt RNA				PFU observed		Infectious units × 10⁻⁵ at 34°C	
		Each	Sum	Each (µg)	Sum (µg)	Each	Sum	34°C	41°C	Each	Sum
1	2	3	4	5	6	7	8	9	10	11	12
1	25	0.446	0.446	2.86	2.86	0	0	487	9	3.04	3.04
2	50	.418	.864	2.68	5.54	0	0	486	10	3.04	6.08
3	75	.560	1.424	3.59	9.13	0	0	500	12	3.12	9.20
4	100	.508	1.932	3.26	12.39	0.002	0.002	464	4	2.90	12.10
5	125	.527	2.459	3.38	15.77	.012	.014	299	6	1.87	13.97
6	150	.685	3.149	4.39	20.16	.0007	.014	295	5	1.85	15.82
7	175	.927	4.071	5.94	26.10	.004	.018	289	2	1.81	17.63

RNA (Table 2) in the protoplast assay. The proportions of plaques seen at 41°C (Table 3, column 10) is not significantly different from the expected 1 to 2 percent of the numbers developing at 34°C. Thus the ts phenotype of the initiating ts-$Q\beta$ is faithfully inherited. Column 11 gives the number of ts-infectious units per reaction mixture calculated from the dilution used; column 12 lists the corresponding cumulative sums. No evidence of the synthesis of infectious RNA which could produce plaques at either 34° or 41°C appeared in the control nonprimed reaction. The corresponding negative columns are therefore omitted from Table 3.

The average infective efficiency of the RNA in the protoplast assay is 2×10^{-7}. The initial input in tube 1 was 0.2 μg corresponding to 1.2×10^{11} strands and 2.4×10^4 plaque forming units. Since each transfer involves a 1 to 10 dilution, it is clear that less than one of the 1.87×10^5 plaque formers observed in the 5th tube can be ascribed to the initiating ts-$Q\beta$-RNA. Finally, by tube 7 which contains 3.6×10^{12} new strands, the number of plaque formers (1.8×10^5) exceeds in absolute terms the number (1.2×10^4) of old strands present. It is clear that the serial dilution experiment has demonstrated the appearance of newly synthesized infectious RNA possessing the temperature-sensitive phenotype.

In the lower portion of Fig. 2 the outcome of the experiment shown in Table 3 is summarized by plotting against time the cumulative sums of the RNA synthesized (column 6) and plaque formers at 34°C (column 12). The fact that the plaque formers at 41°C are not statistically above the background of the assay of ts-$Q\beta$-RNA means that no detectable wild type $Q\beta$-RNA has been produced, a fact indicated by the open squares. For comparison the control reaction of Table 3, in which the initiating RNA was omitted, is similarly plotted on the same scale in the upper part of Fig. 2. No significant synthesis of either RNA or infectious units were observed.

It is apparent from the experiments described that one and the same normal replicase can produce distinguishably different but genetically related RNA molecules. The genetic type produced is completely determined by the RNA used to start the reaction and is always identical to it. The following two conclusions would appear to be inescapable from these findings; (i) the RNA is **the instructive agent in the replicating process and therefore satisfies the operational definition of a self-duplicating entity**; (ii) **it is not some cryptic contaminant of the enzyme but rather the input RNA which multiplies**.

The experiments described generate an opportunity for studying the genetics and evolution of a self-replicating nucleic acid molecule in a simple and chemically controllable medium. Of particular interest is the fact that such studies can be carried out under conditions in which the only demand made on the molecules is that they multiply; they can be liberated from all secondary requirements (for example, coding for coat protein, and so forth) which serve only the needs and purposes of the complete organism.

N. R. PACE
S. SPIEGELMAN

Department of Microbiology,
University of Illinois, Urbana

References and Notes

1. I. Watanabe, *Nihon Rinsho* 22, 243 (1964).
2. L. R. Overby, G. H. Barlow, R. H. Doi, M. Jacob, S. Spiegelman, *J. Bacteriol.* 91, 442 (1966).
3. I. Haruna, K. Nozu, Y. Ohtaka, S. Spiegelman, *Proc. Nat. Acad. Sci. U.S.* 50, 905 (1963).
4. I. Haruna and S. Spiegelman, *ibid.* 54, 579 (1965).
5. ———, *ibid.*, p. 1189.
6. ———, *Science* 150, 884 (1965).
7. ———, *Proc. Nat. Acad. Sci. U.S.*, in press (1966).
8. S. Spiegelman, I. Haruna, I. B. Holland, G. Beaudreau, D. Mills, *ibid.* 54, 919 (1965).
9. G. D. Guthrie and R. L. Sinsheimer, *Biochem. Biophys. Acta* 72, 290 (1963).
10. J. A. Strauss, Jr., *J. Mol. Biol.* 10, 422 (1963).
11. N. Pace and S. Spiegelman, *Proc. Nat. Acad. Sci. U.S.* (in press).
12. C. I. Davern, *Australian J. Biol. Sci.* 17, 726 (1964).
13. Supplied by N. Zinder, Rockefeller Institute.
14. D. Fraser and E. A. Jerrel, *J. Biol. Chem.* 205, 291 (1953).
15. Supported by PHS research grant No. CA-01094 and research grant No. GB-2169 from NSF. N.R.P. is a predoctoral trainee in microbial and molecular genetics (grant no. USPH 5-T1-GM-319).

25 April 1966

REPLICATION OF VIRAL RNA, XVI. ENZYMATIC SYNTHESIS OF INFECTIOUS VIRAL RNA WITH NONINFECTIOUS Q_β MINUS STRANDS AS TEMPLATE*

BY GÜNTER FEIX, ROBERT POLLET, AND CHARLES WEISSMANN†

DEPARTMENT OF BIOCHEMISTRY, NEW YORK UNIVERSITY SCHOOL OF MEDICINE, NEW YORK, NEW YORK

Communicated by Severo Ochoa, October 16, 1967

Earlier studies on the replication of the RNA[1] of RNA phages, both *in vivo* and *in vitro*, provided evidence that a viral minus strand, i.e., an RNA strand complementary to the one found in the phage particle (the plus strand) was formed early in replication and served as template for the synthesis of progeny RNA (for references, see ref. 2). Spiegelman and his colleagues[3, 4] purified a viral RNA polymerase from *E. coli* infected with the RNA phage Q_β and showed that this enzyme preparation (Q_β replicase) utilized infectious Q_β RNA as template for the synthesis of more infectious viral RNA.[4, 5] On the basis of their studies they suggested that minus strands were neither formed nor required in this reaction.[6, 7] However, a reinvestigation of the Q_β replicase system in this laboratory[8] clearly showed that in the early phase of the *in vitro* reaction, minus strands were synthesized almost exclusively, while plus strands were detected only after several minutes of incubation. More recent experiments by Spiegelman and his collaborators led these authors to similar conclusions.[9, 10]

While it is commonly held that minus strands occur as part of a double-stranded helix, hydrogen bonded to plus strands, we have suggested that the replicating complex may, in its native state, contain the template and the nascent product in a predominantly single-stranded[11] state (i.e., not extensively hydrogen-bonded to each other), possibly held together by the enzyme or by short hydrogen bonded regions or both. Following treatment with phenol or sodium dodecylsulfate, the structure apparently rearranges to become largely double-helical.[12, 13] It is relevant to note in this connection that, whereas species of double-stranded Q_β RNA ("replicative form," "replicative intermediate") extracted from Q_β-infected *E. coli* are not able to prime Q_β replicase *in vitro*,[9, 14] the same material after heat denaturation is an excellent template for the enzyme.[14] Under these conditions, synthesis of the product is directed preferentially by the minus strand in the first minutes of incubation.

The above observations made it desirable to study the behavior of isolated minus strands, and we therefore undertook the isolation and purification of Q_β minus strands.[15] We now show that, on incubation with Q_β replicase and nucleoside triphosphates, noninfectious minus strands direct a rapid synthesis of infectious viral RNA in substantial excess over the added template.

Materials and Methods.—Q_β *replicase:* The enzyme was prepared essentially according to Spiegelman and collaborators,[3, 4] including the CsCl and -sucrose-density gradient centrifugations. The latter was carried out as detailed earlier.[13] The specific activity of the enzyme used in these experiments was 15 (mμmoles UMP incorporated/mg/min). No nuclease activity was detected using the criteria previously described.[13] Contamination of the enzyme preparation by virus was less than 2500 PFU/mg.

145

TABLE 1. *Infectivity of different viral RNA preparations before and after heating.*

RNA	Infectious units (PFU $\times 10^{-6}/\mu g$)	
	Not heated	Heated 90 sec at 100°
1. Q_β plus strands	1650	1200
2. Q_β minus strands (step 6)	<0.001	0.19
3. Double-stranded Q_β RNA	<0.001	130

Aliquots containing about 0.06 μg of RNA in 0.2 ml of 0.003 M EDTA were assayed for infectivity either directly or after heating.

Phage RNA: Labeled and unlabeled phage were prepared[8, 16] and the RNA was extracted as described elsewhere.[8, 17]

Partially double-stranded viral RNA: Partially double-stranded MS2 and Q_β RNA were prepared as described earlier.[14] These preparations contained both "replicative form," presumed to be a double-helix containing a plus and a minus strand,[16, 18-20] and "replicative intermediate," described as a double-stranded core with tails of nascent plus strands attached.[18, 21] They will be designated as "double-stranded RNA" for the sake of simplicity. After digestion with RNase under standard conditions,[22] 78% of the Q_β preparation and 67% of the MS2 preparation remained acid-insoluble. After heat denaturation, more than half of the MS2 and one third of the Q_β preparation sedimented as expected for full-length viral RNA (27S and 30S, respectively).

Q_β minus strands: Minus strands were prepared as previously described,[15] with addition of one further purification step. A step 5 preparation (52 μg RNA) was dialyzed against 0.5 mM EDTA–0.1% sodium dodecylsulfate (SDS) for 12 hr at 4°, concentrated to about 0.15 ml by lyophilization, and centrifuged through a 5–23% linear sucrose gradient in 0.05 M Tris-HCl (pH 7.6)–0.1% SDS for 120 min, at 4° and 65,000 rpm in the Spinco SW65 rotor. The leading fractions of the 30S band containing the single-stranded RNA were pooled, dialyzed for 16 hr against 3 changes of 0.5 M EDTA, pH 7, and concentrated by lyophilization (step 6 preparation). No infectious RNA was found in this preparation (Table 1, expt. 2; see also Table 3), although a specific infectivity 10^{-5} that of Q_β plus strands could have been detected. However, since residual plus strands were separated from single minus strands (steps 5 and 6) after conversion into double-stranded form through self-annealing, some double-stranded RNA containing infectious plus strands might still contaminate the minus strand preparation. This was indeed the case but the contamination was small. After heating samples under conditions known to denature double-stranded Q_β RNA without substantial loss of infectivity of the plus strands (Table 1, expts. 1 and 3), some infectivity (about 10^{-4} that of plus strands) appeared in the step 6 preparation (Table 1, expt. 2).

Incubations with Q_β replicase: Incubations were carried out at 37°, with $MgCl_2$, 12.8 mM; Tris-HCl buffer, pH 7.4, 84 mM; ATP, UTP, GTP, and CTP (one of which had P^{32} label), each 0.8 mM; replicase 0.1 mg/ml, and template, added as specified. Acid-insoluble radioactive RNA and RNase-resistant labeled RNA were determined as described earlier.[22, 23]

Determination of infectious RNA: Samples (15 μl) were mixed with 1 μl of SDS (1.5%) and 1 μl of pronase (previously incubated for 2 hr at 37°; 7.5 mg/ml in 0.15 M EDTA, pH 7). An aliquot (10 μl) was drawn into a 10-μl micropipette and incubated for 30 min at 35°.[4] The pronase-digested sample (containing 0.05–0.1 μg of RNA) was mixed with 3 mM EDTA, pH 7.0 (0.55 ml) and then added to 0.55 ml of *E. coli* K12W6 spheroplasts. Purified RNA preparations were assayed without the pronase digestion. The preparation of spheroplasts and the assay of RNA infectivity was carried out as described by Strauss.[24] One μg of Q_β RNA gave rise to 10^8–10^9 PFU.

Determination of base composition: P^{32}-labeled RNA (20 μg, 20,000 cpm) and 0.25 mg of ribosomal RNA, added as carrier, were digested for 20 hr at 37° in 0.2 M acetate buffer, pH 4.5 (0.2 ml) with heat-treated Takadiastase[25] (15 units of RNase activity). After adding 10 μl of 1.5 N HCl, aliquots (4000 cpm) were spotted on Whatman 3 MM paper strips (4 $\times$ 96 cm) and the nucleotides were separated by electrophoresis (3 hr at

42 v/cm) in a mixture of acetic acid, pyridine, and H_2O (10:1:89), pH 3.5. The spots were cut out and the radioactivity was determined by scintillation counting in Liquifluor (Packard Instrument Co.).

Other materials: Heated Takadiastase was a gift of Mr. H. Schwam of this department. Other reagents were obtained from the sources indicated earlier.[13, 14, 23]

Results.—Properties of purified minus strands: The electrophoretic pattern, the sedimentation profile, and the annealing behavior of purified minus strands have been described.[15] With H^3-labeled MS2 RNA as a density marker (ρ = 1.626 gm/cm^3),[26] Q_β plus strands were estimated to have a buoyant density of 1.627, and Q_β minus strands 1.617 gm/cm^3. The base composition of Q_β plus and minus strands, given in Table 2, is in good agreement with the assumption that the complementarity relationship postulated for the two strands of a DNA double-helix is also applicable to the two viral strands capable of forming a hydrogen-bonded RNA double-helix.[27]

TABLE 2. *Nucleotide composition of P^{32}-labeled Q_β plus and Q_β minus strands.*

Nucleotide	P^{32}-Q_β plus strand (Moles %)	P^{32}-Q_β minus strand
AMP	22.9 ± 0.2*	29.1 ± 0.1
UMP	29.2 ± 0.1	22.8 ± 0.1
GMP	24.1 ± 0.0	23.9 ± 0.1
CMP	23.8 ± 0.1	24.2 ± 0.1

P^{32}-labeled Q_β minus strands (step 6) were prepared as described, from *E. coli* infected with Q_β and labeled from 0 to 40 min with P^{32}-phosphate (40 mc/l).[24]

* Standard error of the mean $\bar{x} = \sqrt{\frac{\Sigma(x - \bar{x})^2}{N(N - 1)}}$. Three analyses were carried out on each preparation.

Minus strands as template for Q_β replicase: Purified noninfectious Q_β minus strands promoted vigorous synthesis of RNA from the onset of incubation and infectious units appeared within two minutes (Table 3, expt. 1, see also Fig. 2). After four minutes, the amount of RNA synthesized (Table 3, col. 3) was equivalent to three times the input. The infectious units increased from 0 (<60) to 2.7×10^6 and 18.3×10^6 PFU, respectively, after 4 and 15 minutes. No infectivity appeared when nucleoside triphosphates or minus strands were omitted (expts. 2 and 3). Assuming that the specific infectivity of newly synthesized Q_β RNA and natural Q_β RNA is similar, the infectivities in experiment 1 (calculated from Table 3, expt. 4, time 0) correspond to 0.014 µg of Q_β RNA, i.e., an amount equivalent to about half the template added, after 4 minutes, and to 0.095 µg, i.e., 3 times the input after 15 minutes of incubation. The greater discrepancy between total and infectious RNA at earlier than at later times of incubation may be due to the presence of a relatively larger proportion of incomplete strands early in incubation.

In marked contrast with the above results, the incorporation of nucleotides directed by plus strands (Table 3, expt. 4) lagged behind that promoted by minus strands. Only a fraction (about one fifth) of the amount of RNA added as template was synthesized during the first two minutes. Moreover, as noted by Mills *et al.*,[9] in the early phase of incubation there was a disappearance rather than an increase of infectivity, possibly due to involvement of the added template

TABLE 3. *Synthesis of infectious RNA by Q_β replicase with either Q_β plus or Q_β minus strands as template.*

	(1) Time of incubation (min)	(2) P^{32}-UMP incorporated (cpm/10 μl)	(3) RNA synthesized (μg/10 μl)	(4) Infectious Units (PFU × 10^{-6}/10 μl) Total	Net synthesis
1. Q_β minus strands (step 6) (0.03 μg/10 μl)	0	0	0	0*	0
	2	2900	0.064	1.6	1.6
	4	4020	0.089	2.7	2.7
	6	4850	0.11	3.2	3.2
	10	8600	0.19	11.4	11.4
	15	13,000	0.29	18.3	18.3
2. Q_β minus strands (step 6) (0.03 μg/10 μl). No nucleoside triphosphates	0	—	—	(0.001)	—
	15	—	—	0†	—
3. No template	0	—	—	(0.003)	—
	6	—	—	(0.0005)	—
	15	—	—	0†	—
4. Q_β plus strands (0.038 μg/10 μl)	0	0	0	7.3	0
	2	280	0.0064	5.0	−2.3
	4	900	0.020	3.7	−3.6
	6	2470	0.057	6.2	−1.1
	10	5100	0.11	13.0	5.7
	15	9200	0.20	24.0	16.7

The standard reaction mixtures (0.12 ml) contained, per 10 μl, 1 μg of Q_β replicase. UTP was P^{32}-labeled (spec. radioactivity, 54,000 cpm/mμmole). At the times indicated, aliquots (15 μl) were withdrawn for the determination of acid-insoluble radioactivity and infectious RNA. Each value in col. 2 was corrected for the blank (25 cpm). One mμmole of UMP corresponds to 1.16 μg of Q_β plus strands or 1.54 μg of Q_β minus strands (free acid). A value of 1.2 μg RNA per mμmole of UMP was used throughout to calculate the amount of RNA synthesized (col. 3), since in all but the early incubations with Q_β plus strands the product consisted predominantly of plus strands.[14] All values were recalculated for 10-μl aliquots. Expts. 1 and 4 were carried out in parallel. The values in parentheses were too low to be statistically significant.

* <60 PFU.
† <100 PFU.

in a replicating complex. Even after 15 minutes less RNA and fewer infectious units were synthesized with plus than with minus strands.

The template activity of our minus strand preparation is not due to either (*a*) Q_β plus strand or (*b*) double-stranded Q_β RNA contaminants. Besides the fact that contamination by infectious plus strands, either in a single or a double-stranded form, is extremely low (specific infectivity relative to that of plus strands, less than 10^{-5} and 10^{-4}, respectively), possibility (*a*) is ruled out by the finding (Table 3) that with plus strands there was an increase of infectivity only after six minutes of incubation. Possibility (*b*) is ruled out by our previous observation[14] that double-stranded Q_β RNA does not stimulate nucleotide incorporation by Q_β replicase during the first 30 minutes of incubation and by the present finding (Table 4) that only a minute amount, if any, of infectious units was produced (expt. 2) unless the double-stranded RNA was heat-denatured (expt. 3). MS2 minus strands apparently do not serve as template for Q_β replicase since denatured, double-stranded MS2 RNA did not promote the formation of infectious RNA (Table 4, expt. 4).

Dependence of RNA synthesis on template concentration: As seen in Figure 1,

TABLE 4. *Template activity of different RNA preparations in the Q_β replicase system.*

	Time of incubation (min)	UMP incorporated (cpm/10 µl)	Total infectious units (PFU × 10^{-6}/10 µl)
1. Q_β minus strands (step 6) (0.03 µg/10 µl)	0	0	0*
	5	4250	2.8
	15	10,050	19.0
2. Double-stranded Q_β RNA (0.04 µg/10 µl)	0	0	(0.0003)
	5	0	(0.001)
	15	0	(0.03)
3. Denatured double-stranded Q_β RNA (0.04 µg/10 µl)	0	0	0.39
	15	7500	18.2
4. Denatured double-stranded MS2 RNA (0.04 µg/10 µl)	0	0	3.2
	15	324	2.6

The incubation mixtures (0.08 ml) had the composition indicated in the *Methods* section. Template was added as indicated. Double-stranded RNA dissolved in 0.5 mM EDTA, pH 7, was denatured by heating at 100° for 90 sec. At the times indicated, aliquots were removed for the determination of infectious units (15 µl) and acid-insoluble radioactivity (5 µl). A blank (90 cpm) was subtracted from each value in col. 2. One-thousand cpm are equivalent to 0.035 µg RNA. One µg of Q_β RNA gave 3.3 × 10^8 PFU in the infectivity assay. The values in parentheses were too low to be statistically significant.

* <100 PFU.

minus strands were a more effective template than plus strands at any concentration tested. Moreover, whereas the enzyme system was saturated by plus strands at a level of about 0.11 µg of RNA/µg of protein, there was no indication of even incipient saturation with 0.13 µg of minus strands/µg of protein irrespective of the assay used, whether RNA synthesis or appearance of infectious units. At high template concentrations, incorporation was about seven times greater with minus than with plus strands. Since saturation was not reached with minus strands, the actual difference in V_{max} may be even greater. An approximate calculation shows that the number of template molecules (either plus or minus strands) added to the enzyme incubation was less than the number of enzyme molecules estimated to be present.[28]

The results of Figure 1 are compatible with the assumption that the enzyme preparation has a small number of sites specific for plus, and a larger number specific for minus strands, perhaps associated with two different polypeptide chains. If this were true, the incorporation resulting from short-time incubations with both plus and minus strands (the former at saturating concentrations) should be additive. Table 5 shows that this was indeed the case.

Time required for the synthesis of a plus strand: The *in vitro* system directed by minus strands is ideally suited to determine the time of synthesis of an infectious RNA strand. As the initial infectivity is nil, the completion of a

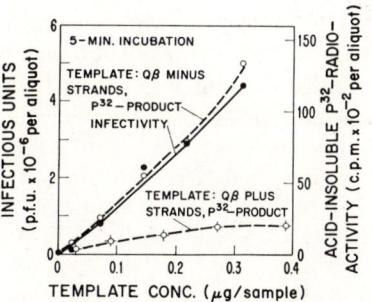

FIG. 1.—Template activity of Q_β plus and minus strands at different concentrations. Standard incubation mixtures (24 µl), containing the amount of template indicated, were incubated for 5 min at 37°. The specific radioactivity of the P^{32}-UTP was 39,500 cpm/mµmole.

TABLE 5. *Additive effect of Q_β plus and Q_β minus strands in promoting nucleotide incorporation by Q_β replicase.*

Template Added		
Plus strand (µg)	Minus strand (µg)	UMP incorporated (cpm)
0.5	—	373
0.75	—	425
—	0.024	231
—	0.072	720
0.5	0.024	657
0.5	0.072	1071

Standard assay mixtures (24 µl) contained the amount of template indicated. After a 5-min incubation at 37°, the acid-insoluble radioactivity was determined. The results are the average of duplicate determinations. The specific radioactivity of the GTP-C^{14} was 5000 cpm/mµmole.

small number of RNA molecules is readily detectable. In Figure 2 the synthesis of infectious units at 37° is plotted as a function of the incubation time. Whereas no infectious units were present up to 90 seconds, infectious RNA equivalent to about 10^7 PFU was generated within the next 20 seconds. This marks the completion of the first crop of progeny RNA molecules. Assuming that the association of enzyme and template was largely completed during the two-minute preincubation in the absence of three nucleoside triphosphates, the time required for the completion of a plus strand (3000 nucleotides) is about 100 seconds and phosphodiester bonds would be formed at the rate of 3000/100 = 30 per second at 37°. The value greatly exceeds that observed for *in vitro* RNA synthesis by RNA polymerase directed by T4 or T7 DNA (2.5 and 7 phosphodiester bonds/sec, respectively)[29, 30] but may still be less than that occurring *in vivo*.

Discussion.—Since the single minus strand[11] efficiently promotes the final step in viral RNA replication, i.e., the synthesis of infectious plus strands, it meets an important requirement for a functional intermediate in the process. Such a requirement is not met by species of double-stranded RNA isolated from infected cells.

The finding that purified minus strands are biologically competent and yet not infectious shows that lack of infectivity is an intrinsic property of minus strands and not a consequence of damage suffered during purification or even during synthesis and sojourn in the host. A reason for the lack of infectivity can easily be given. Initiation of infection in the host requires a virus-specific

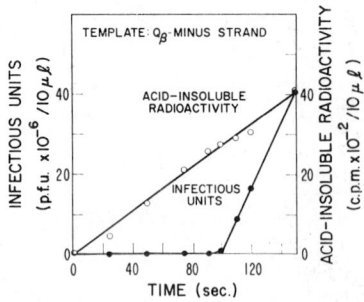

FIG. 2.—Determination of the time required for completion of an infectious RNA strand. Q_β minus strands (0.96 µg), Q_β replicase (15 µg), ATP (0.12 µmoles), and $MgCl_2$ (1.92 µmoles) in 0.1 ml of 0.126 M Tris-HCl, pH 7.6, were incubated for 2 min at 37°. The reaction was initiated by the addition of GTP, CTP, and C^{14}-UTP (0.12 µmoles each, in 50 µl). The final reaction mixture had the usual composition. At the times indicated, 15 µl of the mixture were pipetted into tubes containing 1 µl of 1.5% SDS, mixed, and cooled. Infectivity and acid-insoluble radioactivity were determined as described in the *Methods* section. The specific activity of the C^{14}-UTP was 34,000 cpm/mµmole. One µg of Q_β RNA gave 7.5×10 PFU.

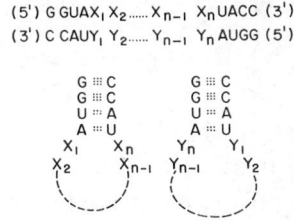

FIG. 3.—Two hypothetical RNA strands which (a) are commentary to each other and (b) share common sequences at the ends. Each of these strands can assume the "amphora" configuration shown at the bottom of the figure. (X,Y), any set of complementary bases.

RNA polymerase (synthetase or replicase) coded for by the plus [31–33] but not by the minus strand. The minus strand in fact appears to be incapable of directing the *in vitro* synthesis of acid-insoluble polypeptides.[34] The fact that the minus strands were obtained from *E. coli* infected with Q_β phage proves the assumption that biologically competent minus strands (either free or complexed) are formed during replication *in vivo*. The remarkable template specificity of Q_β replicase,[3] thought to recognize only Q_β plus strands among natural RNA's, has now been extended to a second species of RNA, the Q_β minus strand. If plus and minus strands had the end sequences in common, as in the model of Figure 3, Q_β replicase could recognize both strands with one recognition site. Such a relationship between the two strands implies terminal self-complementarity in each plus and minus strand, as in the "amphora" model proposed for other reasons for the Q_β plus strand.[35] We have looked for double-helical regions in purified Q_β RNA subjected to annealing conditions but have failed to find significant amounts of RNase-resistant RNA (less than 0.2%),[36] and we feel that if such regions existed they would not comprise more than ten base pairs. However, since ten nucleotides might be sufficient to specify a recognition sequence we cannot eliminate the model on the basis of our findings. On the other hand, we have presented evidence which is compatible with the assumption that the recognition sequences of plus and minus strands differ and that the replicase preparation contains different amounts of two specific binding sites, a small number for plus and a large number for minus strands. These sites may be associated with different proteins or protein subunits. The finding of Lodish and Zinder[37] that synthesis of plus and minus strands can be dissociated in certain temperature-sensitive mutants of phage f2 can be explained within the framework of this hypothesis.

Summary.—Purified, single Q_β minus strands are inherently noninfectious and effectively promote the synthesis of infectious viral RNA by Q_β replicase *in vitro*.

We thank Mr. Morton C. Schnieder and Mr. Winston Burrell for excellent technical assistance. We are indebted to Dr. Severo Ochoa for his constant support and valuable advice.

* Aided by grants AM-01845, AM-08953, and FR-05399 from the National Institutes of Health, U.S. Public Health Service, and E. I. du Pont de Nemours and Company, Inc.

† Present address: Institut für Molekularbiologie der Universität Zürich, Zürich, Switzerland.

[1] Definitions and abbreviations are as in previous work.[13]

[2] Weissmann, C., and S. Ochoa, in *Progress in Nucleic Acid Research and Molecular Biology*, ed. J. N. Davidson and W. E. Cohn (New York: Academic Press, 1967), vol. 6, p. 353.

[3] Haruna, I., and S. Spiegelman, these PROCEEDINGS, **54**, 579 (1965).
[4] Pace, N. R., and S. Spiegelman, these PROCEEDINGS, **55**, 1608 (1966).
[5] Spiegelman, S., I. Haruna, I. B. Holland, G. Beaudreau, and D. R. Mills, these PROCEEDINGS, **54**, 919 (1965).
[6] Spiegelman, S., and I. Haruna, *J. Gen. Physiol.*, **49**, 263 (1966).
[7] Haruna, I., and S. Spiegelman, these PROCEEDINGS, **55**, 1256 (1966).
[8] Weissmann, C., and G. Feix, these PROCEEDINGS, **55**, 1264 (1966).
[9] Mills, D. R., N. R. Pace, and S. Spiegelman, these PROCEEDINGS, **56**, 1778 (1966).
[10] Pace, N. R., D. H. L. Bishop, and S. Spiegelman, these PROCEEDINGS, **58**, 711 (1967).
[11] We designate as "single" an RNA strand which is RNase-sensitive under standard assay conditions and is therefore not part of a double helix.
[12] Borst, P., and C. Weissmann, these PROCEEDINGS, **54**, 982 (1965).
[13] Feix, G., H. Slor, and C. Weissmann, these PROCEEDINGS, **57**, 1401 (1967).
[14] Weissmann, C., G. Feix, H. Slor, and R. Pollet, these PROCEEDINGS, **57**, 1870 (1967).
[15] Pollet, R., P. Knolle, and C. Weissmann, these PROCEEDINGS, **58**, 766 (1967).
[16] Weissmann, C., P. Borst, R. H. Burdon, M. A. Billeter, and S. Ochoa, these PROCEEDINGS, **51**, 682 (1964).
[17] Weissmann, C., L. Colthart, and M. Libonati, *Biochemistry*, in press.
[18] Franklin, R. M., these PROCEEDINGS, **55**, 1504 (1966).
[19] Ammann, J., H. Delius, and P. H. Hofschneider, *J. Mol. Biol.*, **10**, 557 (1964).
[20] Francke, B., and P. H. Hofschneider, these PROCEEDINGS, **56**, 1883 (1966).
[21] Fenwick, M. L., R. L. Erikson, and R. M. Franklin, *Science*, **146**, 527 (1964).
[22] Billeter, M. A., and C. Weissmann, in *Procedures in Nucleic Acid Research*, ed. G. L Cantoni and D. R. Davies (New York: Harper and Row, 1966), p. 498.
[23] Weissmann, C., these PROCEEDINGS, **54**, 202 (1965).
[24] Strauss, J. H., Jr., *J. Mol. Biol.*, **10**, 422 (1964).
[25] Himamasu, M., T. Uchida, and F. Egami, *Anal. Biochem.*, **17**, 135 (1966).
[26] Billeter, M. A., C. Weissmann, and R. C. Warner, *J. Mol. Biol.*, **17**, 145 (1966).
[27] Langridge, R., M. A. Billeter, P. Borst, R. H. Burdon, and C. Weissmann, these PROCEEDINGS, **52**, 114 (1964).
[28] The specific activity of our Q_β replicase preparation, purified according to Pace and Spiegelman,[4] was 60 (mµmoles of total nucleotide incorporated per min at 37°), i.e., somewhat higher than that reported from Spiegelman's laboratory.[4, 9, 35] The electrophoretically homogeneous Q_β enzyme described by August and his colleagues (personal communication) has a corresponding value of 1000. We may therefore estimate our preparation to be about 5% pure. Assuming that the enzyme has a molecular weight[4] of about 10^5, 2.5 µg of the preparation would contain $\frac{0.05 \times 2.5 \times 10^{-6} \times 6 \times 10^{23}}{10^5} = 7.5 \times 10^{11}$ molecules of enzyme. This amount of enzyme is not saturated by the addition of $\frac{0.3 \times 10^{-6} \times 6 \times 10^{23}}{10^6} = 1.8 \times 10^1$ molecules of minus strands.
[29] Bremer, H., and M. W. Konrad, these PROCEEDINGS, **51**, 801 (1964).
[30] Richardson, J. P., *J. Mol. Biol.*, **21**, 115 (1966).
[31] Lodish, H. F., S. Cooper, and N. D. Zinder, *Virology*, **24**, 60 (1964).
[32] Nathans, D., M. P. Oeschger, K. Eggen, and Y. Shimura, these PROCEEDINGS, **56**, 184 (1966).
[33] Viñuela, E., I. D. Algranati, and S. Ochoa, *Europ. J. Biochem.*, **1**, 3 (1967).
[34] Iwasaki, K., personal communication.
[35] Haruna, I., and S. Spiegelman, these PROCEEDINGS, **56**, 1333 (1966).
[36] Weissmann, C., in *Perspectives in Virology*, ed. E. Pollard (New York: Academic Press 1967), p. 1.
[37] Lodish, H., and N. D. Zinder, *Science*, **152**, 372 (1966).

Subunit Structure of Qβ Replicase

by
M. KONDO
R. GALLERANI
C. WEISSMANN

Institut für Molekularbiologie
der Universität Zürich

Qβ replicase comprises four different polypeptides, only one of which is coded for by the phage genome.

THERE is currently considerable interest in the structure of the RNA of phage Qβ and its relationship to the virus-specified proteins[1-3]. The Qβ phage particle comprises two minor proteins (A_1 and A_2 (ref. 4; unpublished results of Farron and C. W.)), besides the coat protein. Labelling experiments on actinomycin-inhibited *Escherichia coli* infected with Qβ revealed a fourth phage specified protein, tentatively identified as a Qβ replicase subunit[4]. We have examined highly purified Qβ replicase preparations and have found them to contain four polypeptides of different electrophoretic mobility, only one of which, with a molecular weight of about 69,000, is Qβ specific. The functional and structural relationship of the three host-specific polypeptides to the Qβ replicase molecule is unclear. Because the molecular weight of Qβ replicase is 130,000 or greater, the enzyme must consist of more than one subunit. We estimate that the four polypeptides thought to be specified by the Qβ genome, A_1, A_2, coat and replicase β subunits, comprise about 1,500 amino-acids; on the other hand, the number of nucleotides available for coding in Qβ RNA corresponds to not more than 1,150 codons[2].

Qβ replicase was purified by a slightly modified version of the method of Eoyang and August[5] (see Table 1) to a specific activity of 7,000 units/mg, the highest value reported so far. It was dependent on host-specific factor for activity with Qβ RNA as template[6]. The sedimentation coefficient of the enzyme was determined by preparative zonal sedimentation through a sucrose gradient using human haemoglobin ($s_{20,w} = 4·31$)[7], rabbit muscle aldolase ($s_{20,w} = 7·35$)[8,9] and beef liver catalase as reference proteins ($s_{20,w} = 11·1$)[10]. The $s_{20,w}$ value of Qβ replicase was interpolated to be 6·7 from the calibration curve shown in Fig. 1. Assuming the relationship between sedimentation coefficient and molecular weight to be the same for replicase as for an average globular protein, an approximate molecular weight of 130,000 can be estimated. There was no significant effect on the relative sedimentation rate by varying the ammonium sulphate concentration between 0·02 and 0·8 M although, at low ionic strength (0·01 M Tris, 0·005 M Mg^{2+}), aggregation occurs[11].

Highly purified Qβ replicase was dissociated with sodium dodecyl sulphate (SDS) and mercaptoethanol and analysed by polyacrylamide gel electrophoresis in the presence of 0·1 per cent SDS. Four bands labelled α, β, γ and δ were found in amounts that varied from one preparation to another. The β band was usually the strongest component (Figs. 2 and 3). Similar results were obtained at higher concentrations of SDS and in the presence of 0·5 M urea. To determine which of the polypeptides was specified by the virus we did the following experiment. Two batches of *E. coli* labelled with ^{35}S

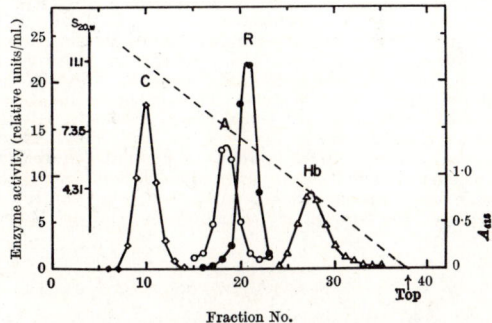

Fig. 1. Determination of the sedimentation coefficient of Qβ replicase. Qβ replicase (specific activity, 4,800 units/mg, about 150 units), rabbit muscle aldolase (about 200 μg, Boehringer) and beef liver catalase (about 300 μg, Boehringer) were mixed and dialysed against a buffer containing 50 mM Tris (pH 7·8), 5 mM $MgCl_2$, 1 mM EDTA, 5 mM β-mercaptoethanol and 0·1 M ammonium sulphate for 2 h. Human haemoglobin (about 200 μg) was added to the mixture which was centrifuged through a 5–23 per cent sucrose gradient in the same buffer as that used for dialysis, at 40,000 r.p.m. and 4° C for 24 h in the Spinco 'SW 41' rotor. Qβ replicase (R, ●) activity was assayed as usual[5]; the value 10 on the ordinate equals 5 replicase units. Catalase (C, ◊) was assayed according to the method of Beers and Sizer[12]; the value 10 on the ordinate corresponds to 50 catalase units. Aldolase (A, ○) was assayed according to Jagannathan *et al.*[13]; the value 10 on the ordinate corresponds to 5 aldolase units. Haemoglobin (Hb, △) was measured spectrophotometrically at 415 nm.

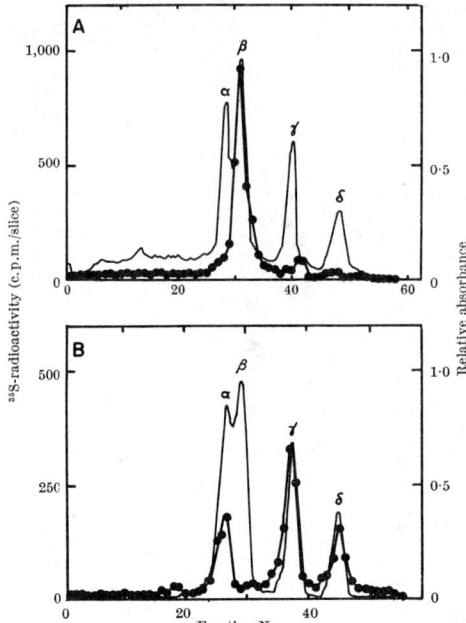

Fig. 2. Analysis of ³⁵S-labelled polypeptides of Qβ replicase preparations by gel electrophoresis. ³⁵S-labelled Qβ replicase (step 5, about 500 units) was incubated for 2 h at 37° C in a buffer containing 50 mM Tris (pH 7·8), 5 mM MgCl₂, 1 mM EDTA, 2·0 per cent β-mercaptoethanol, 0·22 per cent SDS and 3·3 per cent sucrose. The samples (about 75 μl.) were mixed with 5 μl. of bromphenol blue (0·25 per cent) in 30 per cent sucrose, layered on 10 per cent polyacrylamide gels¹³,¹⁷ (7 × 100 mm) and run at 8 mA per tube for 6 h. The gels were stained with 0·25 per cent Coomassie brilliant blue in 50 per cent methanol–7 per cent acetic acid overnight, destained with the same solvent and scanned with the Joyce–Loebl 'Chromoscan' at 620 nm. After soaking in 0·1 per cent SDS for several hours, the gels were frozen and cut into 1 mm slices. The slices were digested with H₂O₂ and alcoholic hyamine hydroxide and counted in Bray's solution²⁰. The general methods and buffers have been described¹³,¹⁷. A, Qβ replicase purified from a mixture of ³⁵S-labelled Qβ-infected cells (1 g) and unlabelled Qβ-infected cells (100 g); 6 μg (9,600 c.p.m. of ³⁵S) of the step 5 preparation was applied to the gel. B, Qβ replicase purified from a mixture of ³⁵S-labelled uninfected cells (1 g) and unlabelled Qβ infected cells (100 g); 88 μg (4,800 c.p.m. cf ³⁵S of the step 5 preparation was used.

stained and scanned to locate the four bands; the gels were then sectioned and the radioactivities were determined. The distribution of (unlabelled) protein among the four bands was similar in both preparations (Fig. 2), but in the sample containing ³⁵S-labelled protein from infected cells virtually all radioactivity was in band β, whereas in the sample containing ³⁵S-labelled proteins from uninfected cells bands α, γ and δ but not β were labelled. Thus β corresponds to the virus-specified polypeptide. The same conclusion has been reached independently by R. Kamen (personal communication and following article¹²).

The molecular weight of the virus-specified replicase subunit was determined in an SDS-containing polyacrylamide gel system¹³,¹⁴, by comparing its mobility with that of aldolase, catalase, bovine serum albumin and phosphorylase a (all dissociated in mercaptoethanol and sodium dodecyl sulphate) side by side in split gels or gel slabs as well as by co-electrophoresis. Fig. 3 shows that the mobility of the β band is slightly less than that of bovine serum albumin (68,000)¹⁵. The mobility of the reference proteins was plotted against log molecular weight¹³,¹⁴ and the molecular weight of the β band was interpolated to be 69,000. The values for the α, γ and δ

were prepared, one uninfected and labelled with ³⁵S throughout log phase growth, the other labelled with ³⁵S after infection with Qβ. About 1 g of each preparation was mixed with 100 g of Qβ-infected E. coli and Qβ replicase was purified as usual. Table 1 shows that after the final purification step (zonal centrifugation) the ratio of ³⁵S to enzymatic activity was ten times lower when the radioactive proteins were from non-infected rather than from infected cells. After dissociation with SDS and mercaptoethanol both preparations were subjected to polyacrylamide gel electrophoresis. The gels were

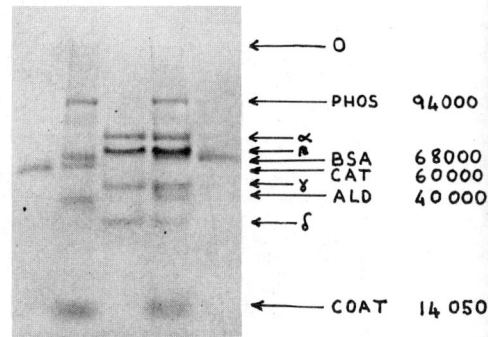

Fig. 3. Polyacrylamide gel electrophoresis of Qβ replicase and reference proteins. The conditions for electrophoresis are described in the legend to Table 2. 1, Catalase; 2, phosphorylase a, bovine serum albumin, catalase, aldolase, Qβ coat protein; 3, Qβ replicase; 4, Qβ replicase, phosphorylase a, bovine serum albumin, aldolase, Qβ coat protein. 15 μ of Qβ replicase, 8 μg Qβ coat protein and 4 μg each of the other reference proteins were run in one lane.

Table 1. PURIFICATION OF Qβ REPLICASE FROM ³⁵S-LABELLED E. coli

Step	Replicase activity (units) A	B	³⁵S Radioactivity (c.p.m.) A	B	Specific activity (units/mg) A	B	³⁵S Radioactivity/replicase activity (c.p.m./units) A	B
1 (NH₄)₂SO₄	6,700	5,300	2·8 × 10⁷	9·2 × 10⁷	9·1	6·4	4,180 (1,000)*	17,360 (1,000)
2 DEAE-cellulose	8,200	9,400	2·4 × 10⁶	3·6 × 10⁶	119	162	293 (70)	383 (22)
3 Hypatite	5,400	5,400	1·1 × 10⁶	2·5 × 10⁶	—	—	204 (49)	463 (27)
4 'Biogel'	12,000	5,000	3·0 × 10⁵	1·3 × 10⁵	4,690	3,270	25 (6)	26 (1·5)
5 Sucrose gradient	6,200	4,600	1·2 × 10⁵	3·2 × 10⁴	7,210	6,770	19 (4·5)	7 (0·4)

* Normalized to an initial value of 1,000 c.p.m./unit.

(A) Starting material was a mixture of 100 g of unlabelled, Qβ-infected E. coli Q13 and 1 g of Qβ-infected E. coli Q13 labelled with ³⁵S (7 × 10⁸ c.p.m. after infection. For the labelled preparation, cells were grown in 1 l. of medium containing 0·5 g NaCl, 3·0 g KCl, 1·1 g NH₄Cl, 0·2 g MgCl₂·6H₂O, 12·1 Tris, 0·023 g KH₂PO₄, 0·294 g CaCl₂·2H₂O, 50 μg FeSO₄·7H₂O, 0·01 g sodium citrate, 1·6 g casamino-acids, 10 g glycerin, 10 mg thiamine, 100 μmole (NH₄)₂SO₄. When the cell number reached 1·6 × 10⁹ cells/ml., the bacteria were harvested, suspended in 1 l. of the above medium lacking sulphate and glycerin After 20 min at 37° C 10 g of glycerin and 20 mCi of ³⁵S-SO₄²⁻ were added to give a final SO₄²⁻ concentration of 5 μM. After 10 min, Qβ was added at multiplicity of infection of 10. The infected cells were harvested after 40 min at 37° C. (B) Starting material was a mixture of 100 g of Qβ-infected E. coli Q13 and 1 g of uninfected E. coli Q13 labelled with ³⁵S (10⁹ c.p.m.). The cells were grown in the medium described above except that ³⁵S-SO₄²⁻ (80 μM, 0·25 mCi/μmole) was present throughout growth; the cells were harvested in log phase (3 × 10⁸ cells/ml.). Qβ replicase was purified as described by Eoyar and August⁵, except that (a) elution from the DEAE-cellulose column was with an NaCl gradient (0·1 to 0·3 M), (b) chromatography on 0·5 M 'Biogel' was substituted for that on 'Sephadex G-200', and (c) zonal centrifugation was carried out as described by Feix et al.¹¹.

Table 2. MOLECULAR WEIGHTS OF POLYPEPTIDE FROM Qβ REPLICASE PREPARATIONS AND Qβ PARTICLES

1) Polypeptides from Qβ replicase preparations	Mol. wt.	(2) Polypeptides from Qβ particles	Mol. wt.
α	74,000	A_1	38,000
β	69,000	A_2	44,000
γ	47,000	Qβ coat	14,050*
δ	33,000		

* From the amino-acid composition given by Konigsberg et al.[16]. Molecular weights were estimated by the general procedure of Shapiro et al.[13] and Weber and Osborn[14]. All proteins were incubated for 2 h at 37° C with 0.1 per cent mercaptoethanol and 0.1 per cent SDS before electrophoresis. Samples were either run side by side with a mixture of reference proteins in split gels (0.7 × 12 cm tubes filled with 3 ml. of gel for 6 h at 10 mA/gel) or in adjacent slots in gel slabs (6 × 13 cm, 4 mm thickness, 2 h at 30–35 mA), or co-electrophoresis from the same slot of a gel slab flanked by unmixed samples (see Fig. 3). Buffers and gels were prepared according to Viñuela et al.[17]. All gels contained 10 per cent acrylamide and 0.27 per cent bis-acrylamide; gel slabs contained 0.5 M urea in addition. The same relative mobilities were found whether or not urea was present. The molecular weights were obtained from a plot of log (molecular weight of the reference proteins) against distance travelled.

The values in column 1 were obtained from eleven independent experiments; those in column 2 from two. The standard error of the mean was less than 5 per cent. The reference proteins[14] were phosphorylase a (94,000), bovine serum albumin (68,000)[15], beef liver catalase (60,000), rabbit muscle aldolase (40,000) and Qβ coat protein (14,050)[16].

bands are given in Table 2. None of the components occurring in Qβ replicase preparations has the same mobility as any of the subunits of DNA-dependent RNA polymerase. It is not immediately evident which, if any, of the subunits besides β constitute the native enzyme. Two observations are of interest in this connexion. First, a Qβ replicase preparation derived from a mixture of labelled uninfected and unlabelled infected cells contains radioactivity in the α, γ and δ bands. This means that the labelled polypeptides either exchange with the putative subunits of replicase, or that they are present in the preparation as independent protein molecules. Because Qβ replicase prepared from a mixture of a small amount of E. coli labelled with ^{35}S after infection and a large excess of infected unlabelled cells contains more than ten times more radioactivity in the β than in all the other bands, it seems that synthesis of α, γ and δ is strongly depressed after infection. Second, if a Qβ replicase preparation is incubated with Qβ RNA and the resulting 30S binding complex[11] is isolated, only the α and β polypeptides are recovered. This may mean either that α and β are associated as subunits of one protein, or that they constitute two different proteins both of which have the capacity to bind to Qβ RNA. If all four polypeptides are part of a structural unit, it is necessary to account for the fact that their added molecular weights (223,000) greatly exceed the value of 130,000 estimated from the sedimentation properties of the enzyme. Such an apparent discrepancy could arise if the enzyme were highly asymmetric or if it were in an association–dissociation equilibrium with its subunits. The latter possibility could also account for the finding that the four polypeptides occur in non-stoichiometric and variable amounts within the sharply sedimenting enzyme band.

Assuming an average molecular weight of 110 for an amino-acid, the added molecular weights of the polypeptides thought to be specified by the Qβ genome correspond to 1,500 amino-acids (Table 2). Qβ RNA comprises 3,500 nucleotides of which not more than about 3,400 are used for coding[2,3]; it can therefore specify only 1,150 amino-acids. Beyond assuming errors in the molecular weight determinations of the RNA or the polypeptides the discrepancy could be explained in several ways. First, only one of the A proteins is virus-specified while the other is derived from the host. Neither the data of Garwes et al.[4] nor labelling experiments done in our laboratory encourage this view, but it cannot be ruled out. Second, the virus preparations consist of a mixture of two kinds of particles, containing different A proteins. Third, the smaller of the A proteins is derived from the larger by proteolytic cleavage. Comparison of the fingerprints of A_1 and A_2 may provide the answer.

We thank Dr R. Kamen and Dr H. Schachman for valuable discussions and Mr Eduard Peier and Mr Werner Römer for their technical assistance. This project is supported by grants from the Schweizerische Nationalfonds and the Jane Coffin Childs Fund.

Received August 11, 1970.

[1] Billeter, M. A., Dahlberg, J. E., Goodman, H. M., Hindley, J., and Weissmann, C., Nature, 224, 1083 (1969).
[2] Hindley, J., Billeter, M. A., and Weissmann, C., Adv. Microbiol. (in the press).
[3] Goodman, H. M., Billeter, M. A., Hindley, J., and Weissmann, C., Proc. US Nat. Acad. Sci. (in the press).
[4] Garwes, D., Sillero, A., and Ochoa, S., Biochim. Biophys. Acta, 186, 166 (1969).
[5] Eoyang, L., and August, J. T., in Methods in Enzymology, 12, B (edit. by Grossman, L., and Moldave, K.), 530 (Academic Press, New York, 1968).
[6] Franze de Fernandez, M. T., Eoyang, L., and August, J. T., Nature, 219, 588 (1968).
[7] Braunitzer, G., Hilse, K., Rudloff, V., and Hildschmann, N., Adv. Protein Chem., 19, 1 (1964).
[8] Taylor, J. F., and Lowry, C., Biochim. Biophys. Acta, 20, 109 (1956).
[9] Hass, L. F., Biochemistry, 3, 535 (1964).
[10] Samejima, T., and Shibata, K., Arch. Biochem. Biophys., 93, 407 (1961).
[11] Feix, G., Slor, H., and Weissmann, C., Proc. US Nat. Acad. Sci., 57, 1401 (1967).
[12] Kamen, R., Nature, 228, 527 (1970) (following article).
[13] Shapiro, A. L., Viñuela, E., and Maizel, J. V., Biochem. Biophys. Res. Commun., 28, 815 (1967).
[14] Weber, K., and Osborn, M., J. Biol. Chem., 244, 4406 (1969).
[15] Tanford, C., Kawahara, K., and Lapanje, S., J. Amer. Chem. Soc., 89, 729 (1967).
[16] Konigsberg, W., Maita, T., Katze, J., and Weber, K., Nature, 227, 27 (1970).
[17] Viñuela, E., Algranati, I. D., and Ochoa, S., Europ. J. Biochem., 1, 3 (1967).
[18] Beers, jun., R. F., and Sizer, I. W., J. Biol. Chem., 195, 133 (1952).
[19] Jagannathan, V., Singh, K., and Damodaran, M., Biochem. J., 63, 94 (1956).
[20] Bishop, D. H. L., Claybrook, J. R., and Spiegelman, S., J. Mol. Biol., 26, 373 (1967).

Covalently Linked RNA–DNA Molecule as Initial Product of RNA Tumour Virus DNA Polymerase

INDER M. VERMA, NORA L. MEUTH,
ESTHER BROMFELD, KENNETH F. MANLY &
DAVID BALTIMORE

Department of Biology, Massachusetts Institute of Technology, 77 Massachusetts Avenue, Cambridge, Massachusetts 02139

> Reverse transcriptase from avian myeloblastosis virus initially synthesizes from endogenous RNA template a covalently linked DNA–RNA species, which Verma et al. suggest contains a small RNA primer molecule.

THE DNA polymerase found in virions of the RNA tumour viruses[1,2] can be assayed in two ways. If disrupted virions are incubated without addition of a template, the endogenous viral RNA is copied by the DNA polymerase[3-5]. If exogenous templates are provided, often these are copied at a much higher rate than the endogenous RNA[6-12]. With exogenous templates, however, the DNA polymerase requires the presence of a homologous polynucleotide primer to initiate polymerization of nucleotides[10]. The primer, which can be as short as a tetranucleotide[11,12], is physically incorporated into the product

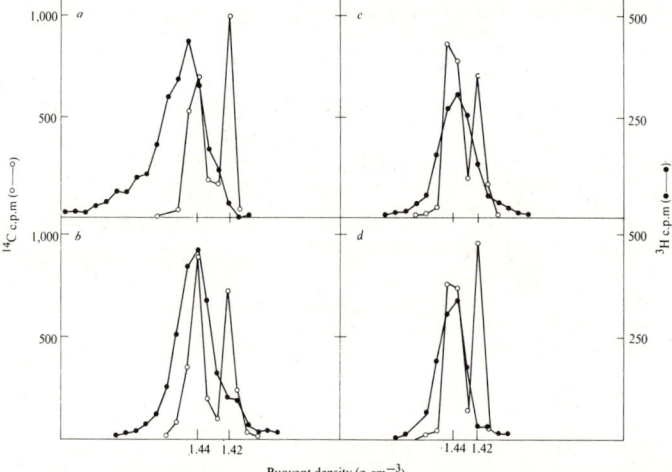

Fig. 1 Analysis in Cs_2SO_4 gradients of the product of the mouse leukaemia virus DNA polymerase. A 20 min reaction was carried out in standard conditions[13] with virions of Moloney mouse leukaemia virus and ^{3}H-TTP. The fast sedimenting fraction was separated on sucrose gradients, collected by precipitation with ethanol, and dissolved in 0.01 M Tris-HCl (pH 7.6), 0.01 M NaCl. Portions were placed in a boiling water bath for 5 min and chilled. Parts of the heated samples were treated for 10 min at 37° C in 0.4 ml. with 16 µg of pancreatic ribonuclease (Worthington) plus 3 µg of T_1 ribonuclease (Calbiochem, 5,000 U/mg) contained either in 0.01 M Tris-HCl (pH 7.6), 0.01 M NaCl (low salt) or 0.3 M NaCl, 0.03 M Na citrate (high salt). A separate portion of the product was adjusted to 0.3 M NaOH and placed in a boiling water bath for 5 min, chilled and neutralized with HCl. For analysis, the treated samples were mixed with 1.40 ml. of 0.01 M Tris-HCl (pH 7.4), 0.001 M EDTA saturated at 24° C with Cs_2SO_4 and enough Tris-EDTA buffer to make 3.0 ml. The samples were placed in polyallomer tubes, a cushion of 0.15 ml. of saturated Cs_2SO_4 was carefully placed at the bottom and the top was covered with 'Nujol'. The solutions were centrifuged for 65–70 h at 33,000 r.p.m. and 23° C in a Spinco SW 50.1 rotor in a model L ultracentrifuge. All the gradients contained ^{14}C-labelled heat-denatured and native P22 DNA as standard markers with buoyant densities of 1.44 and 1.42 respectively[13]. Three to five drop fractions were collected from the bottom of the tube, acid-precipitated, collected on filters as previously described[13] and counted in a mixture of 5 g PPO and 100 g naphthalene in 1 l. of p-dioxane. *a*, Boiled product; *b*, alkali-treated product; *c*, boiled product after ribonuclease treatment in low salt; *d*, boiled product after ribonuclease treatment in high salt.

(Smoler, Molineaux and Baltimore, submitted for publication). The primer requirement of the enzyme, demonstrated with exogenous polynucleotide templates, suggests that in the absence of such templates, when the enzyme copies the endogenous 60–70S viral RNA, a primer might also be present to initiate polymerization.

The initial reaction product formed when the virion DNA polymerase copies the endogenous viral RNA consists of small pieces of DNA attached to the 60–70S RNA[3,4,13–15]. Analysis by buoyant density in Cs_2SO_4 indicated that the DNA product could be released from the viral RNA by procedures which disrupt hydrogen bonds[3].

Further analysis of the product released from the viral RNA by heat treatment has now revealed that the material is not free DNA. After 30 min or less of reaction the product contains molecules which behave like DNA–RNA duplexes even after heat denaturation. One interpretation of this result is that the initial product of the reaction might be a covalently linked DNA–RNA molecule and that the primer for initiation of synthesis might be an RNA species.

The initial observation was with Moloney mouse leukaemia virus (MLV). After 20 min of incubation, the product of the virion DNA polymerase reaction was extracted with 'Sarkosyl', purified on a 'Sephadex G-50' column, boiled to denature the secondary structure and centrifuged to equilibrium in a Cs_2SO_4 gradient. The principal component of the boiled product banded at a density slightly greater than denatured P22 phage DNA (Fig. 1a). Some material of greater density was also evident. After treatment with alkali (Fig. 1b), ribonuclease in low salt (Fig. 1c) or ribonuclease in 0.3 M NaCl plus 0.3 M Na citrate (Fig. 1d), the product DNA banded coincidentally with the denatured P22 DNA although the polymerase product formed a much wider band than the marker. The effects of alkali and ribonuclease indicate that RNA was attached to the product DNA even after boiling and was causing the DNA to band at a higher density than the marker phage DNA.

To confirm and extend these data the reaction product of the avian myeloblastosis virus (AMV) DNA polymerase was investigated because larger amounts of this virus were available. The endogenous product formed after 2, 5, 10, 15 and 30 min of reaction was banded in Cs_2SO_4 in its native state, after boiling and after alkali treatment.

The native products consisted of some DNA banding coincidentally with an RNA marker, at a density of 1.65 (Fig. 2a). Increasing amounts of DNA banding at lower density appeared as incubation time increased. These complexes have been described previously[3,4,13–15].

The heat-denatured product after 2 min of incubation (Fig. 2b) banded heterogeneously with most of the DNA heavier than 1.5 g/ml. which is the density of a 1 : 1 DNA · RNA hydrogen-bonded duplex[16]. With increasing time of incubation, the heterogeneous band moved to lighter positions in the gradient. By 30 min the product resembled that seen after 20 min of reaction by the MLV DNA polymerase.

Treatment of the DNA polymerase products with alkali in conditions which hydrolyse all RNA caused them to band in Cs_2SO_4 at a lower average density than the boiled product (Fig. 2c). The bands were quite broad, indicating a low molecular weight. The samples from later times of incubation gave a sharper band which centred at a density slightly lighter than the denatured P22 phage DNA. The earliest sample gave an especially broad band but still the alkali-treated product was on the average less dense than the boiled product.

To show that the boiled product did not spontaneously renature, the effect of the single-strand-specific nuclease from

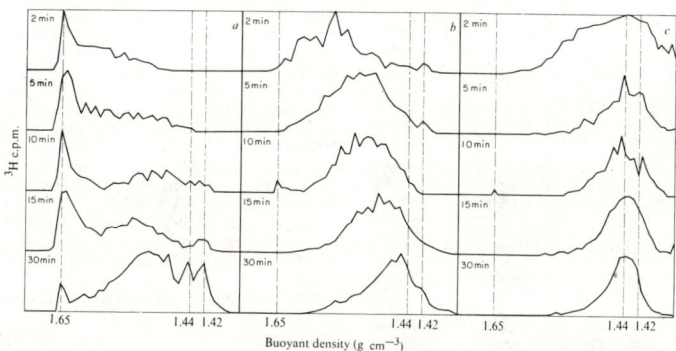

Fig. 2 Analysis in Cs_2SO_4 gradients of the product of the avian myeloblastosis virus DNA polymerase after various times of reaction. The products were prepared using the following 0.2 ml. reaction mixture: 50 mM Tris-HCl (pH 8.3), 6 mM magnesium acetate, 20 mM dithiothreitol, NaCl, 1 mM dATP, 1 mM dGTP, 1 mM dCTP, 6.5 µM ^{3}H-dTTP (6,000 c.p.m./pmol), 0.2% 'Nonidet P-40' and 1–2 mg of AMV protein in purified virions[10]. Samples were incubated at 37° C for the given length of time and the reaction terminated by the addition of an equal volume of 3% sodium dodecyl sarcosinate. The samples were further incubated for 5 min at 37° C. Portions were withdrawn from the samples to measure acid-insoluble material. The 2, 5, 10, 15 and 30 min products in a typical experiment had 51,000, 165,000, 280,000, 400,000 and 500,000 total c.p.m. respectively. The products were then separated from low molecular weight material by chromatography on 'Sephadex G-50' columns using 0.1 M NH_4HCO_3 buffer, pH 8.0, for elution. The recovery of the purified product was over 70%. For heat denaturation, 0.01–0.05 ml. of the product was diluted to 0.5 ml. with 0.01 M Tris-HCl, pH 7.6, and 0.01 M NaCl, placed in a boiling water bath for 5 min and cooled rapidly. For alkaline hydrolysis, portions in the same buffer containing 0.3 M NaOH were placed in a boiling water bath for 5 min, chilled and neutralized with HCl to pH 7.0. Cs_2SO_4 gradients containing 8,000–12,000 c.p.m. of products were prepared and analysed as in Fig. 1. ^{14}C-18S ribosomal RNA was also included as a marker. Recoveries from the gradients containing native material were 20–50% and from the other gradients were 55–80%. Drops from the gradient tubes were collected directly into scintillation fluid. a, Native products; b, boiled products; c, alkali-treated products.

Neurospora* was investigated. A portion of the boiled product from a 15 min incubation of AMV was treated with *Neurospora* nuclease in previously specified conditions which only degrade single stranded regions of nucleic acids[13]. The treated material was then centrifuged to equilibrium in Cs_2SO_4 as was an equivalent amount of boiled but undigested material. The nuclease-treated product contained only 4% of the radioactivity of the untreated product and this was distributed heterogeneously. Therefore, after the boiling treatment, almost all of the DNA remains in a single stranded state.

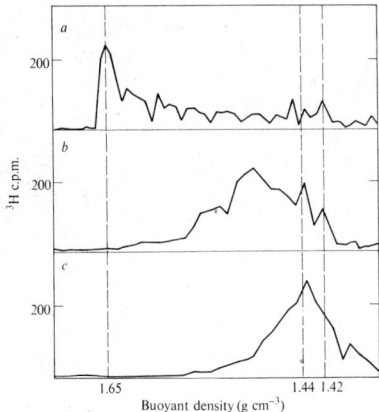

Fig. 3 Analysis in Cs_2SO_4 gradients of the product of purified avian myeloblastosis virus DNA polymerase copying 60–70S viral RNA. The polymerase was purified by disruption of the virions with 0.2% 'Nonidet-P40', chromatography on 'DEAE-Sephadex' and then chromatography on phosphocellulose. A yield of 44% of the initial activity was obtained using poly(C)·(dG)$_{14}$ as template-primer[10] and a purification of fifty times was achieved (I. M. V. and D. B., unpublished results). The 60–70S RNA was isolated by centrifugation through a sucrose gradient of purified avian myeloblastosis virus[10] disrupted with sodium dodecyl sulphate. The RNA was identified by absorbance at 260 nm and ethanol precipitated. Two μg of DNA polymerase and 5 μg of viral RNA were incubated at 37° C for 15 min in the reaction mixture described in Fig. 2, lacking detergent and virus. The product was purified as in Fig. 2 and contained 25,000 c.p.m. of ^{3}H label. Portions were treated as in Fig. 2 and analysed as in Fig. 1. *a*, Native product; *b*, heat-denatured product; *c*, alkali-treated product.

To see if the 60–70S viral RNA contained an RNA primer, the reaction of purified AMV DNA polymerase with isolated 60–70S AMV RNA was studied. A reaction was carried out for 15 min and portions of the product were centrifuged to equilibrium in Cs_2SO_4 either without any treatment, after boiling, or after alkaline hydrolysis. As with the early endogenous product, most of this product banded with RNA at $\rho = 1.65$ g cm^{-3} before treatment (Fig. 3*a*). Boiling converted it to material banding between RNA and DNA (Fig. 3*b*) and alkaline hydrolysis caused it to form a wide band centring slightly lighter than the P22 denatured DNA (Fig. 3*c*). The product formed when purified enzyme is allowed to copy 60–70S viral RNA therefore behaves similarly to the product of the endogenous reaction, supporting the idea that these two systems are comparable.

These data indicate that the initial product of the endogenous DNA polymerase reaction of both MLV and AMV and of the reaction of purified AMV polymerase with AMV 60–70S RNA is a covalently bonded DNA–RNA molecule. Covalent bonding is indicated by the resistance of the product to boiling. The sensitivity of the boiled product to *Neurospora* nuclease and the change of density of the product after treatment with ribonuclease in 0.3 M NaCl indicate that little or no RNA remains hydrogen-bonded to the product DNA after heat denaturation. The ability of alkali and ribonuclease to convert the boiled product to material banding coincidentally with DNA indicates that RNA and not some other material was responsible for the density of the boiled product being greater than that of free DNA. The decreasing density of the boiled product with time of incubation (Fig. 2*b*) coupled with the increasing size of the DNA product indicated by its decreasing band width (Fig. 2*c*) suggest that a constant size piece of RNA is attached to a growing DNA strand.

The RNA in the product presumably acts as a primer for the endogenous reaction. Experiments utilizing the inhibition of the polymerase reaction by dideoxythymidine triphosphate (Smoler, Molineaux and Baltimore, submitted for publication) indicate that the DNA is attached to the 3'-OH of the primer RNA. Experiments are in progress to determine whether unique nucleotides in the DNA and RNA are linked.

The nature of the attached RNA is under investigation. Two models seem most likely. One is that the 60–70S RNA bends back on itself to form a "hair pin" and that DNA synthesis is initiated on an end of 60–70S RNA. The sensitivity of boiled product to the *Neurospora* nuclease argues against this model because a "hair pin" structure should renature spontaneously. The second model invokes a small RNA primer and we favour this idea at present. The existence of transfer RNA species in the 60–70S RNA complex[17] suggests that they might serve this function. Previous evidence that multiple DNA molecules are formed from a single region of viral RNA[13] argues that multiple primer molecules may exist attached to the viral RNA.

The use of an RNA primer for initiating DNA synthesis might occur in systems other than the RNA tumour viruses. For example, an RNA molecule might serve the function of the postulated initiator in bacterial DNA synthesis[18,19]. Such an initiator molecule could be synthesized from a DNA template or could be formed by specific enzymes in the absence of a template.

This work was supported by a grant from the American Cancer Society and a contract from the Special Virus Cancer Program of the National Cancer Institute. I. M. V. was a fellow of the Jane Coffin Childs Memorial Fund for Medical Research, K. F. M. was a fellow of, and D. B. was a faculty research awardee of, the American Cancer Society.

Received August 19, 1971.

* In a previous paper[13] this enzyme was called the *Neurospora* endonuclease and was thought to be identical to the enzyme prepared by Linn and Lehman[20]. Dr M. Fraser, who prepared the nuclease used in the previous studies, has informed us that the nuclease is a different enzyme from the Linn and Lehman enzyme. It is an exonuclease combined with some endonuclease activity but has very high specificity for single stranded regions of nucleic acid like the Linn and Lehman enzyme. Using a modification of the procedure of Rabin, Preiss and Fraser[21], we have isolated the enzyme used in the present experiments and have confirmed its high specificity for single stranded nucleic acid as previously reported[13].

[1] Baltimore, D., *Nature*, **226**, 1209 (1970).
[2] Temin, H., and Mizutani, S., *Nature*, **226**, 1211 (1970).
[3] Spiegelman, S., Burny, A., Das, M. R., Keydar, J., Schlom, J., Travnicek, M., and Watson, K., *Nature*, **227**, 563 (1970).
[4] Rokutanda, M., Rokutanda, H., Green, M., Fujinaga, K., Ray, R. K., and Gurgo, C., *Nature*, **227**, 1026 (1970).
[5] Duesberg, P. H., and Canaani, E., *Virology*, **42**, 783 (1970).
[6] Spiegelman, S., Burny, A., Das, M. R., Keydar, J., Schlom, J., Travnicek, M., and Watson, K., *Nature*, **227**, 1029 (1970).
[7] Spiegelman, S., Burny, A., Das, M. R., Keydar, J., Schlom, J., Travnicek, M., and Watson, K., *Nature*, **228**, 430 (1970).
[8] Mizutani, S., Boettiger, D., and Temin, H. M., *Nature*, **228**, 424 (1970).
[9] Riman, J., and Beaudreau, G. S., *Nature*, **228**, 427 (1970).

[10] Baltimore, D., and Smoler, D., *Proc. US Nat. Acad. Sci.*, **68**, 1507 (1971).
[11] Baltimore, D., and Smoler, D., *Proc. Third Annual Miami Winter Biochemistry Symp.* (in the press).
[12] Baltimore, D., Smoler, D., Manly, K. F., and Bromfeld, E., in Ciba Foundation Symp. *The Strategy of the Viral Genome* (Churchill, London, in the press).
[13] Manly, K., Smoler, D. F., Bromfeld, E., and Baltimore, D., *J. Virol.*, **7**, 106 (1971).
[14] Fujinaga, K., Parsons, J. T., Beard, J. W., Beard, D., and Green, M., *Proc. US Nat. Acad. Sci.*, **67**, 1432 (1970).
[15] Garapin, A.-C., McDonnell, J. P., Levinson, W. E., Quintrell, N., Fanshier, L., and Bishop, J. M., *J. Virol.*, **6**, 589 (1970).
[16] Chamberlain, M., and Berg, P., *J. Mol. Biol.*, **8**, 297 (1964).
[17] Erikson, E., and Erikson, R. L., *J. Virol.* (in the press).
[18] Jacob, F., and Brenner, S., *CR Acad. Sci.*, **256**, 298 (1963).
[19] Jacob, F., Brenner, S., and Cuzin, F., *Cold Spring Harbor Symp. Quant. Biol.*, **28**, 329 (1963).
[20] Linn, S., and Lehman, I. R., *J. Biol. Chem.*, **240**, 1287 and 1294 (1965).
[21] Rabin, Preiss and Fraser, *Preparative Biochemistry* (in the press).

SECTION III

Protein Synthesis

Proteins are the most complex biopolymers, containing as many as 20 different amino acid residues, and in some cases a total of more than 1,000 residues in a single polypeptide chain. Further complexity is introduced by reaction of the amino acids after their polymerization, as in occasional disulfide formation of 2 cysteine residues or in the conversion of proline to hydroxyproline. Accompanying this complexity of the individual protein there is an enormous variety in the number of proteins in existence. In the bacterium *E. coli* it has been estimated that there may be up to 5,000 different proteins present. In higher forms, where much more genetic information is available, it seems possible that even a much greater variety of protein species will be found, but it is too early to make a reasonable guess as to what the number will be.

Early Work on Mechanism of Protein Synthesis Done in Zamecnik's Laboratory (paper 40)

An intricate biochemical machinery exists for the ordering and polymerization of twenty types of α-amino acids into the polypeptide chains which are the precursors of proteins. The requirements for protein synthesis include 40 to 60 transfer RNA molecules, 20 or more amino acyl transfer RNA enzymes, special enzymes for initiation, propagation and termination of peptide synthesis, ribosomes containing at least three different structural RNAs and as many as 50 different structural proteins, messenger RNA, ATP, GTP and Mg^{++}. Much of the early work pinpointing the importance of ribonucleoprotein particles as the site of peptide synthesis, and of transfer RNA as the agent that transports the amino acid to the ribosome was done in Zamecnik's laboratory in the 1950s (*paper 41*). The biochemical pathway followed by an amino acid was traced by injecting it in a radioactive form into a whole animal and subsequently fractionating the subcellular components for their content of the radioactive amino acid at varying times thereafter. In this way, it was shown the amino acids first become linked to a low molecular weight RNA fraction which then migrates to the ribosome where the peptide linkage is formed. The crucial biochemical steps in protein synthesis discovered in the Zamecnik laboratory consist of amino acid activation by formation of an amino acid-AMP, transfer of this activated amino acid to an amino acyl-transfer RNA complex, and finally formation of peptide linkages on the ribosome.

The Adapter Hypothesis and tRNA-Ribosome Interaction

Gamov and Crick provided the theoretical framework for understanding the template process involved in protein synthesis. From the fact that there are 20 commonly occurring amino acids in proteins and 4 commonly occurring bases in either RNA or DNA, Gamov reasoned that the simplest nucleic acid representation that could encode 20 amino acids would be a sequence of 3 nucleotides on the DNA.

Thus, if one arranges 4 different bases of a nucleic acid in all possible combinations, there are 16 possible sequences of 2 bases and 64 possible sequences of 3 bases. Since RNA is transcribed from one DNA strand the same coding relationship should exist for the RNA. The interaction of amino acids with codons on the nucleic acid was explained by Crick's adapter hypothesis which states that amino acids become linked to specific adapter-RNA molecules, presumably by specific enzymes; these become transferred to the ribosome where the adapter RNAs attach by hydrogen bonds to complementary sites on an RNA template (the latter to be known as messenger RNA). The complementary base pair interactions are due to the formation of Watson-Crick type base pairs. Amino acids placed in the proper order then become linked together by peptide synthesis. Crick originally thought the adapter molecule might consist of trinucleotides containing only the anti-codon for adsorption to the template; in fact, the transfer RNAs which serve this adapter function are considerably larger, containing, in addition to the anti-codon region, about 70 nucleotides in a single polynucleotide chain with a complex secondary and tertiary structure. It is clear that the adapter RNA, or transfer RNA (tRNA) as it is more commonly called, must contain, in addition to the anti-codon triplet for binding to the template, a specific site which insures that only one amino acid becomes linked to the adapter by a specific transfer enzyme. It must also contain one or more sites for transient attachment to the ribosome, and it is clear that a certain number of nucleotides neighboring each of these sites are necessary to insure a stable configuration.

The exact three-dimensional configuration of the anti-codon loop of tRNA (discussed in *paper 50*; also see Kim et al. *Proc., Nat. Acad. Sci. U.S.* 1972, 69, 3746) and its manner of interaction with the messenger RNA template is not known. Sequence analyses of a number of tRNA molecules have led to the "cloverleaf" model where some bases are invariably hydrogen bonded, some are invariably not hydrogen bonded, and some are in loop regions. Two bases on either side of the anti-codon are left unpaired as indicated in the structure of tRNA shown in the paper by Fuller and Hodgson (*paper 50*).

A Detailed Analysis of the Steps in Protein Synthesis*

The formation of the protein polypeptide chain from amino acids may be conceptually divided into three phases: initiation, elongation, and termination. Most of the conclusions concerning these three phases have resulted from a detailed study of subcellular, fractionated systems of *E. coli*.

Initiation (papers 42 and 43) The first step

---

*Kerwar and Weissbach, *Ann. Rev. Biochem.* 22, in press, 1973.

of protein synthesis involves formation of an enzyme-bound amino acyl-adenylate complex in which the amino acid carboxyl forms an anhydride with the phosphate of AMP (Loftfield, *Prog. Nuc. Acid Res. Mol. Biol.* 1972, 12, 87). This is followed by the transfer of the aminoacyl moiety to a specific tRNA; the carboxyl group of the amino acid becomes linked by an ester bond to the 3'-hydroxyl group of the ribose of a terminal adenylic acid residue of the tRNA. Both steps are catalyzed by a single enzyme, a Mg^{++}-dependent aminoacyl synthetase, which is specific for ATP, a particular amino acid, and all tRNAs that normally accept the amino acid. Although there is probably only one synthetase for most amino acids, the possibility that there is more than one for some amino acids cannot be excluded. The region on the tRNA which results in the specific interaction between the tRNA and its synthetase is unknown although it is known that the anti-codon region does not participate in determining this specificity. There are indications that this region is not the same for every tRNA molecule.

All polypeptide chain synthesis in *E. coli* (*paper 42*), and probably in most mitochondria (Raff and Mahler, *Science*, 1972, 177, 575) and chloroplasts, is initiated with N-formylmethionine. In most eukaryotes, however, methionine *without* the N-formyl group is used (Lucas-Lenard and Lipmann, *Ann. Rev. Biochem.* 1971, 40, 441). In completed protein this fact is often disguised because the formyl group is invariably hydrolyzed from the completed product *in vivo* and frequently one or more of the amino acids from the NH_2-terminal end of the polypeptide chain is also removed. Thus amino terminal group analysis indicates that only 30 to 40 percent of the N-terminal amino acids in total *E. coli* protein is methionine. There is only one codon, AUG, for methionine. In spite of this there exist two tRNA molecules which can accept methionine. These are referred to as $tRNA^{met}$ and $tRNA^{met}_f$. Both of these tRNAs carry the anti-codon CAU. The $tRNA^{met}_f$ interacts with the protein initiator codon AUG only, while $tRNA^{met}$ interacts only with the AUG internal methionine codon. It is believed that protein initiation factors form a complex with $tRNA^{met}$ which determines this positional specificity for the initiator transfer RNA. The process of initiation begins with the activation of methionine and the formation of met-$tRNA^{met}_f$ (that is, the methionine charged tRNA$_f^{met}$). This moiety reacts with N^{10}-formyltetrahydrofolate-met-tRNA$_f$-transformylase with the formation of fmet-tRNA$_f^{met}$. Of all the amino acyl tRNAs, the transformylase will only react with met-tRNA$_f^{met}$. The details of initiation are far from clear and the concept of what takes place in initiation could change drastically in the next few years. The formation of the initiation complex including the mRNA, fmet-tRNA$_f^{met}$, and 30S ribosomal subunit requires GTP and three proteins, called initiation factors (IFs), IF_1, IF_2, and IF_3. These initiation factors can be bound to free 30S ribosomal subunits, but are displaced after reaction with the 50S ribosomal subunit. At the same time GDP is released and the messenger RNA fmet-tRNA$_f^{met}$ is translocated.

The binding of the initial tRNA to the ribosome requires GTP but without any attendant hydrolysis of the triphosphate. The GTP hydrolysis is required later, however, during the translocation step. This conclusion is partly based on the fact that the GTP analog, guanylyl-5'-methylenediphosphonate (GMPPCP), in which the β-γ-pyrophosphate bond is resistant to enzymatic hydrolysis, will substitute for GTP in the binding step but not in subsequent steps (Skoultchi et al., Biochem. 1970, 9, 508).

There are two tRNA sites on the ribosome, operationally called P and A (paper 44). All tRNAs react initially at the A site and are transferred to the P site. The mRNA shifts coordinately, so that the codon-anti-codon complex is kept intact during translocation. The translocation step requires a GTP $\longrightarrow$ GDP + P cleavage. The ribosomal complex is now ready to accept another amino acyl-tRNA molecule.

Initiation factor IF_3 has been found to influence the specificity of mRNA-ribosome interaction. IF_3 has been separated into several fractions which show various specificities for different mRNA cistrons (Lee-Huang and Ochoa, Nature New Biology, 1971, 234, 236). An important problem is the possibility of intracellular changes in IF_3 activity. A protein has been isolated from E. coli which changes the specificity of IF_3 towards different mRNAs; this protein, which binds to IF_3 is called interference factor i (Groner et al., Nature New Biology, 1972, 239, 16).

Elongation (paper 44) Three protein elongation factors, S_1, S_2, and S_3, (or Ts, G, and Tu, respectively) are required for this second phase of protein synthesis, called *elongation*. A charged tRNA reacts with S_3 and GTP. This complex is transferred to the A site on the ribosome, complementary pairing taking place between the second site on the mRNA and the tRNA anti-codon; then an S_3-GDP complex and P_i are released as peptide polymerase, an enzyme bound to the 50S ribosomal subunit, makes the first peptide linkage. This peptide linkage is made with the α-carboxyl group of N-formyl methionine and requires prior simultaneous cleavage of the ester linkage of this amino acid to the tRNA$_f^{met}$. The system is now ready for translocation, which requires the S_2 elongation factor and another GTP $\longrightarrow$ GDP + Pi cleavage. The precise way in which the S_2 factor works is not understood. Translocation involves the displacement of the discharged tRNA$_f^{met}$ and the coordinate movement of the peptidyl tRNA-mRNA complex from the A site to the P site on the ribosome. The process of adding another amino acid to the carboxyl end of the growing chain can now be repeated and the elongation steps continue until the penultimate codon in the cistron message is reached. The S_1 elongation factor catalyzes the reutilization of the S_3-GDP complex for reuse in elongation.

The energetics of peptide synthesis are interesting. One ATP is hydrolyzed to AMP and PPi and 2GTPs are hydrolyzed to GDP and Pi. This is considerably more energy than the estimated bond energy of the peptide linkage (7 kcal) and is probably necessitated by the energy required for the ordering process in positioning the amino acids.

Termination There are three triplets, UAA, UAG and UGA, which do not code for any amino acids. These three triplets signal *termination* of a polypeptide chain. Nirenberg and coworkers (paper 45) found two protein release factors R_1 and R_2 which recognize the triplets UAA and UAG, and the triplets UAA and UGA respectively. Thus, the insertion of a nontranslatable terminator triplet by itself in the mRNA is not sufficient for release. This merely stops the process of translation; a protein factor which recognizes the terminator triplet is required. As the nascent peptidyl tRNA is released from the ribosome, the linkage between the peptides and the tRNA is broken. The messenger RNA is also released from the ribosome and the ribosomal subunits may dissociate. In the case of most polycistronic messengers it seems unlikely that the ribosomal subunits invariably dissociate after the reading of each cistron.

The Determination of Codon Assignments*

Whereas Zamecnik's laboratory had established that peptide linkages are formed on cytoplasmic ribonucleoprotein particles, in this early work there was no way of distinguishing between template and ribosome. The earlier concept that ribosomes themselves contained the message for the proper alignment of amino acids had to be discarded when it was shown that phage infected cells were able to synthesize completely new (phage specific) proteins without synthesizing new ribosomes (Brenner, Jacob and Meselson, *Nature*, 1961, **190**, 576). It turned out that the phage specific messages, transcribed from phage DNA, directed the synthesis of phage protein, after attachment to *preexisting* ribosomes.

The detailed analysis of the code was initiated by Nirenberg and Matthaei (*paper 46*) who showed that both natural and synthetic polyribonucleotides could be added to a cell-free system (containing ribosomes, tRNA, and the other compounds necessary for protein synthesis) in such a way as to become adsorbed to the ribosome to direct peptide bond formation. The first break in the coding problem came when Nirenberg and Matthaei showed that polyuridylic acid stimulated polyphenylalanine synthesis. In line with the Gamov-Crick ideas discussed earlier, this suggested that a sequence of three U residues was the codon for phenylalanine. For three years after Nirenberg's important basic discovery, a number of laboratories prepared and tested a wide variety synthetic messenger RNAs to determine the stimulatory effect of various synthetic heteropolynucleotide messengers on the extent and sequence of amino acid polymerization. This work established a correlation between the amino acids incorporated into peptide linkage and the base composition of the added messenger. The weakness of the early approach was that it was not possible to obtain a messenger RNA with a defined sequence of bases. Consequently, only rough correlations were possible between messenger base sequence and the amino acid being coded for. The next major advance in decoding came again from Nirenberg's laboratory (Nirenberg and Leder, *paper 47*) where it was shown that particular trinucleotides when added to the ribosome stimulated the binding of particular amino acid-tRNAs to the ribosome. This turned out to be due to a specific affinity between codon and anti-codon. By studying the binding of labelled amino acid-tRNAs, stimulated by all 64 possible trinucleotides, the sense and nonsense codon assignments have been made.

Khorana and his coworkers employed another method for determining codon assignments (Jones, Nishimura and Khorana, *J. Mol. Biol.* 1966, **16**, 454; *paper 48*). This consisted of using synthetically produced polyribonucleotides of well-defined sequence as messengers and studying the sequence of the polymerized amino acids. Because of the limitation of synthetic techniques, such studies have thus far been confined to polynucleotides with repeating sequences such as poly UC or poly AAG. Poly UC stimulates the synthesis of ser-thr- alternating copolymers, as would be expected on the basis of the Nirenberg triplet binding studies. Poly AAG stimulates the synthesis of three homopolymers: polylysine, polyarginine, and polyglutamate. Correlated tRNA binding studies show that GpApA stimulates glutamyl-tRNA absorption to ribosome, ApApG stimulates lysyl-tRNA binding to ribosome, but that ApGpA, surprisingly, does not stimulate arginyl-tRNA binding.

Both Nirenberg's and Khorana's methods have been valuable tools in determining codon assignments. However, the methods may suffer from one major defect—all studies are done in cell-free systems where the choice of chemical conditions is somewhat arbitrary. For example, the magnesium ion concentration has to be decided upon arbitrarily, and yet the exact concentration may be decisive in determining exactly which triplets will be specifically effective in coding for a particular amino acid. Ultimately natural nucleotide sequences in naturally occurring messenger will have to be correlated with naturally occurring peptide sequences in the corresponding proteins. Studies of this type are underway and will be discussed below. It is worth mentioning that in studies thus far carried out the *in vitro* determined code is similar to that which has been determined from *in vivo* studies.

Nirenberg and coworkers studied the question of codon universality by examining the recognition capacity of nucleotide triplets of

---

*Ninio, *Prog. Nuc. Acid Res. Mol. Biol.* 1973, **13**, 301.

amino acid-tRNAs from different organisms—bacterial, amphibian and mammalian (*paper 48*). Insofar as this question has been investigated, codon assignments appear to be universal although the extent to which different species use the various synonym codons differs appreciably. Evolutionary studies have estimated that the divergence of vertebrates and bacteria occurred about 50 million years ago, yet, similiarity of the codes used by bacteria and vertebrates suggests that the code as it is now known has been in use for at least that period of time. The specificity between tRNA and activating enzyme is much more species dependent and there is no indication of universality at this level of protein synthesis.

Several characteristics of the genetic code are apparent from inspection of *paper 48*. First, except for three nonsense triplets, all the other triplets code for an amino acid. Since there are 64 triplets and only 20 amino acids the code is highly degenerate, that is, most amino acids are coded for by more than one triplet (or codon). There is a pattern to the degeneracy such that the first two bases in the triplet are usually the same but the third may vary.

The degeneracy among the codons immediately suggests a corresponding ambiguity in anti-codon on tRNAs. In *E. coli* several tRNAs for leucine and other amino acids have, in fact, been found. There are at least forty different tRNA species in *E. coli*, that is, more than one tRNA species for a single amino acid. The ambiguity among the tRNAs helps explain how different codons are utilized, but it does not explain the frequently observed ability of one tRNA to interact with more than one codon. The tRNA-codon-ribosome binding experiments of Nirenberg and his coworkers has already shown the different tRNA species, charged with different labeled amino acids could be made to bind to the ribosome by the same triplet.

The Wobble Hypothesis

Crick suggested an amendment to his adapter hypothesis to account for the fact that some tRNA molecules appear to be capable of favorable interaction with more than one codon (*paper 49*) or that several tRNAs can bind to a single codon. The available evidence suggests that standard Watson-Crick base pairs may be used rather strictly in the first 2 positions of the triplet, and that the "wobble" is confined to the third base. The structural basis for the pairing allowable by wobble is that, by a slight alteration in the position of the base in the double helix, base pairs other than the standard A:U or G:C base pairs become possible (see also *paper 50*).

The *in vivo* Code (papers 51 and 52)

Most of the information on the coding properties of codons and on initiation and termination signals has come from studies of trinucleotides or synthetically constructed messengers. It would obviously be a step forward if one could correlate the sequence of nucleotides in a natural messenger with the sequence of amino acids in the corresponding protein as well as to ascertain the *in vivo* termination triplets. Certain single-stranded *E. coli* RNA bacteriophages, R17, f2, and Qβ code for only three gene products. The parental viral RNA serves both as the genome as well as the single, polygenic (or polycistronic) messenger RNA molecule. In order to isolate pure viral messenger RNA one has merely to isolate the virus and prepare the RNA from it. The latter is accomplished by phenol extraction and precipitation of the RNA with ethanol. This is far easier than trying to isolate a cellular messenger from among the wide array of messengers contained in the whole cell. Once the RNA is purified from phage preparations it can be used as a messenger RNA to synthesize proteins *in vitro* (*paper 51*).

The R17 RNA molecule contains about 3500 nucleotides in a single chain; much is known about its base sequence and the total sequence should soon be known (*paper 52*). The steps involved in this base sequence determination involve growth of R17 RNA on ^{32}P-labeled medium and isolation of ^{32}P-labeled viral RNA (Hindley, *Prog. Biophys. Mol. Biol.* 1973, 26, 271). The production of nuclease generated fragments is followed by their fractionation on columns and by two-dimensional paper chromatography. The methods of overlapping sequences for sequence determination is being used in much the same way for the determination of the sequence of very large RNA molecules as it was for amino acid sequences in proteins. It is remarkable that all the necessary fragments have been produced by two enzymes with limited specificity: pancreatic RNase which cleaves on the 3' side of pyrimidines, and takadiastase T_1 which cleaves on the 3' side of guanines. After exhaustive digestion these enzymes break all linkages of the above specified type. Under limiting conditions—low temperature, limited time of digestion, and high magnesium ion concentration—only the more exposed linkages are cleaved. By using hydrolytic conditions varying between limiting and exhaustive, fragments of any desired size range may be obtained. The order of genes of the RNA genome has been determined (*paper 52*). Starting about 300 nucleotides from the 5' end, there are about 300 nucleotides before the first cistron, which codes for the maturation (or A) protein, found in small concentration in the native virus. The coat protein constitutes the middle cistron and the RNA synthetase cistron is located near the 3'-OH proximal end. There is a sequence of over 200 nucleotides between the end of the RNA synthetase cistron and the 3'-OH end of the RNA. The sequence of nucleotides in the region of the initiation sites for the three cistrons has been determined by taking advantage of the fact that in the absence of translation, only the initiation sites bind firmly to the ribosome. This binding protects the region of the initiation site of the messenger RNA from pancreatic RNase digestion. Large ^{32}P-labeled fragments containing the initiation sites are first complexed with ribosomes, then exhaustively treated with pancreatic RNase to remove portions of RNA in direct contact with the ribosome. The remaining undigested RNA is then sequenced. The sequence of bases in the region of the initiation sites is indicated by comparing it with the known amino acid sequences for the corresponding polypeptide chains. All initiation sites begin, as predicted, with AUG and continue with predetermined codons for the amino acids in the three viral specific proteins. In an even more recent *tour de force*, Fiers and his colleagues (*Nature* 1972, **237**, 82) have established the entire nucleotide sequence for the gene of the RNA phage MS2 that codes for the coat protein. The codons specifying the 129 amino acids of the coat polypeptide are in very good agreement with the *in vitro* determined code determined by Nirenberg and Khorona and their coworkers.

The translation of the polycistronic messenger RNA of animal RNA viruses such as polio differs from that of the RNA bacterial virus in that in the former case the entire messenger RNA is translated into a large protein and is then cleaved to smaller proteins, presumably by specific proteases (Jacobson and Baltimore, *J. Mol. Biol.* 1970, 49, 657).

Colinearity of the Gene and Polypeptide Chain

The seemingly endless variety of phenotypic effects resulting from gene mutation is beginning to be understood in molecular terms. This understanding is being reached by tracing the effect of single base-pair changes, involving substitutions as well as deletions or additions, in the DNA. Fortunately, there are a number of chemical reagents (discussed in Section I) that are known to produce single base-pair alterations with a high frequency. Since most permutations of three bases code for some amino acid but only a few code for no amino acid, it is likely that the most probable effect of a single base-pair change in the DNA would be the substitution of one amino acid for another in the proteins; a single base-pair change might also result in no change in amino acid if the resulting triplet codes for the same amino acid. In a small number of cases where the resulting triplet does not code for any amino acid we might expect premature termination of the protein polypeptide chain. These three kinds of mutation are referred to as missense, sense, and nonsense, respectively (*paper 56*). Yanofsky and his coworkers (*Proc. Nat. Acad. Sci. U.S.* 1964, **51**, 266) in studies of the A subunit of the trytophan synthetase protein, and of the gene of *E. coli* that codes for it, have found a colinear relationship between the position of a missense mutation in the gene and the position of the amino acid alteration in the polypeptide chain. In an equally informative study Sarabhai and his colleagues (*Nature* 1964, **201**, 13) found that the position of the nonsense mutation in the gene for the head protein of T2 bacteriophage was related in a linear fashion to the point of premature termination in the polypeptide chain of the head protein. More recently, Streisinger and coworkers (*paper 53*) have used frameshift mutagens and analysis of the resultant mutant proteins to determine the direction of translation of a number of T4 bacteriophage genes relative to the T4 linkage map.

Suppression (paper 54)

In certain cases the effects of missense and nonsense mutations, like those studied by Yanofsky, and by Sarabhai, and their coworkers, can be suppressed by inducing second point mutations elsewhere on the bacterial chromosome. Observations made in Benzer's laboratory were fundamental in bringing clarity to the subject of suppression at the molecular level. Benzer and Champe (*Proc. Nat. Acad. Sci. U.S.* 1961, 47, 1025) observed that certain mutations in the rII cistron of T4 bacteriophage were ambivalent in the sense that the mutants could grow on one strain of *E. coli*, but not on another strain. The inability to replicate on the one *E. coli* strain could be overcome or suppressed either by mutation of the host or by adding 5-fluorouracil (5-FU) during virus growth. Since bacterial mutation and 5-FU have parallel effects on the ambivalent viral mutants, and since 5-FU is incorporated exclusively into the RNA, it was reasoned that the effect of the suppression was to produce some alteration in translation of genetic information. A specific suppressor mutation might be expected to produce a general change in translation which would tend to reduce the activity of most enzymes to some degree, but it could also greatly increase the activity of a mutant enzyme by overcoming the deleterious effects of missense or nonsense mutations. A large number of missense suppressors have been found and further analysis has supported the explanation for the phenomenon given here. For example, Gupta and Khorana (*Proc. Nat. Acad. Sci. U.S.* 1966, 56, 772) showed that altered tRNAs are involved in different suppressor mutations of trytophan synthetase. Two kinds of alteration in the tRNA could explain the observed suppressor effects. Either the tRNA accepts a different amino acid or it tends to pair with a different codon after accepting the right amino acid. The former explanation probably applies to the case investigated by Gupta and Khorana.

There appear to be three classes of nonsense suppressors depending upon which of the three existing nonsense triplets is being suppressed (*paper 54*). The three nonsense triplets are UAG or amber, UAA or ochre and UGA or opal. Studies by Capecchi and Gussin, (*Science*, 1965, 149, 417) and by Engelhardt and his associates (*Proc. Nat. Acad. Sci. U.S.* 1965, 54, 1791) have demonstrated that altered tRNAs are the responsible agents in the suppression of the amber-associated nonsense triplet. In several cases where the nonsense suppressor mutation has been shown to be due to an altered tRNA, base sequence analyses have shown that the change in the tRNA resulted from a single base change in its anti-codon or elsewhere on the anticodon loop (Goodman *et al.*, *Europ. J. Biochem.* 1970, 13, 461; Anderson and Smith, *J. Mol. Biol.* 1972, 69, 349). Presumably one or two (in tandem) of the nonsense triplets is used to signal *in vivo* termination of the polypeptide chain. Nonsense triplets may be required for purposes other than the termination of a polypeptide chain. Special sequences of bases may also be required to serve as "spacers" on polycistronic messengers. More extensive studies on the base sequence analyses of RNA phages and other messenger RNAs should clear up this point.

Translation level suppressors of a different type operate by modifying the reading of messenger RNA on the ribosomes. Streptomycin binds to the small subunit of bacterial ribosomes (more so when it is combined with the larger subunit) in such a way that it can modify translation and in some cases overcome the otherwise lethal effects of some mutations. Gorini and his coworkers (Davies, Gilbert and Gorini, *Proc. Nat. Acad. Sci. U.S.* 1964, 51, 883) have extensively studied this type of antibiotic-induced suppression as well as certain mutations (*ram*) that affect the 30S ribosomal subunit of *E. coli*, mimicking in some ways the effects of streptomycin (Rosset and Gorini, *J. Mol. Biol.* 1969, 39, 95).

Complementation

Genetic complementation is a more complex topic than translational suppression since it depends not only upon the type or position of the mutation but upon the secondary and tertiary structure of the gene-encoded protein as well. Two chromosomes (or two replicons), each defective in a given function, are said to complement one another if the defect is remedied without recombination when the two mutant chromosomes are present in the same cell. Two fundamentally different types of complementation must be distinguished. When the complementing chromosomes have their defects in *different* functional units or cistrons, the intercistronic complementation process is invariably successful. If the complementing chromosomes have their defects in the *same* cistrons, successful complementation process depends upon which mutant pairs are combined, and usually the mutant defects are only partially corrected by the complementation process. Various explanations have been put forth to explain this phenomenon of intracistronic complementation. The working hypothesis used by Schlessinger and Levinthal, (*J. Mol. Biol.* 1963, 7, 1) and by Garen and Garen (*J. Mol. Biol.* 1963, 7, 13) was that *intracis-*

tronic complementation can be observed for a protein which normally contains two or more identical subunits. When the mutant protein subunits from two mutant protein subunits have defects of the appropriate kind, they can sometimes interact to produce a hybrid protein with some activity. The above authors have provided parallel *in vitro* and *in vivo* observations on alkaline phosphatase mutants of *E. coli* which conform to this hypothesis. Crick and Orgel (*J. Mol. Biol.* 1964, 8, 161) have further examined why subunit interaction should lead to correction of the mutant defects. They speculate that the complementation correctable defects are probably the result of abnormal folding of the polypeptide chain. Effective association between appropriately chosen protein monomers, they believe, stabilizes the normal folded structure of the protein.

Other results indicate that there may be additional types of intracistronic complementation. The pertinent complementation studies have been done mostly with mutants of β-galactosidase which normally occurs as a tetramer containing four identical monomers. Complementation, it has been found, can occur between mutants in the β-galactosidase structural gene having large complementary deletions. It seems likely that the complementary *fragments* of intact polypeptide chains interact to form effective subunits that may or may not undergo conversion to multimers (Ullmann, Jacob and Monod, *J. Mol. Biol.* 1968, 32, 11). In support of this notion it has been found that the defective enzyme containing a deletion of a certain peptide segment exists as a dimer rather than as the normally occurring tetramer. Complementation with a low molecular weight fragment containing all or a portion of the missing monomer peptide segment produces active enzyme without necessarily producing the transition from dimer to tetramer. Physical-chemical data indicate that one complementing peptide fragment per dimer is sufficient to produce active enzyme. If complementation between polypeptide fragments is common, making it possible to assay for terminal fragments, there might be ways of generating the necessary N- and C-terminal peptide fragments (by arresting translation, for example) other than by genetic deletions. Zipser and Perrin (*Cold Spring Harbor Symp. Quant. Biol.* 1963, 28, 533) have shown that complementation can occur on the ribosome between a nascent polypeptide chain and its soluble complementing fragment. If complementation on ribosomes is sufficiently rapid it could provide a means of detecting N-terminal fragments, since the growing polypeptide chain could interact with its C-terminal complement when it reached the appropriate length rather than after normal termination.

Differences between Translation in Prokaryotes and Eukaryotes

Eukaryotes have well-defined nuclei, several sets of chromosomes, can be haploid or diploid, and contain a nuclear membrane. On the other hand, prokaryotes have diffuse nuclear bodies, called nucleoids, often have a single, usually circular, chromosome, are haploid, and have no nuclear membrane. All eukaryotes possess mitochondria and sometimes chloroplasts, while prokaryotes do not possess such organelles.

Much more is known about the mechanism of translation in prokaryotes than eukaryotes. Much of what is known about translation in eukaryotes is in substantial agreement with what happens in lower forms. Thus tRNA binding studies indicate universality of the code. Initiation begins with methionine instead of N-formyl methionine in most cases, and ribosomal subunit dissociation and reassociation seems to accompany the translation cycle (Lucas-Lenard and Lipmann, *Ann. Rev. Biochem.* 1971, 40, 409). A major difference between prokaryotes and eukaryotes is in the degree of separation of the processes of transcription and translation. In prokaryotes, such as *E. coli*, the processes are believed to be coupled; soon after the initiation of transcription one observes that the messenger becomes associated with ribosomes and translation begins. This is possible since transcription progresses from the $5'$ to the $3'$ end of the messenger RNA as does translation, that is, the processes are colinear. As the messenger RNA is being synthesized the number of ribosomes on it increases so that the resultant polysomes are involved in many translations along the length of the messenger before transcription is completed.

In eukaryotes the processes of transcription and translation are clearly separated in both space and time. Transcription occurs in the nucleus; the completed heterogeneous nuclear RNA undergoes appreciable post-transcriptional modification prior to being transported to the cytoplasm where its translation takes place. There are two types of ribosomes in the cytoplasm: unattached ribosomes which probably make proteins for use in the cell and ribosomes attached to endoplasmic reticulum which probably make proteins destined for transport outside the cell.

Inhibitors of Protein Synthesis and Their Mechanism of Action

A number of highly specific inhibitors of protein synthesis have been found (Pestka, *Ann. Rev. Biochem.* 1971, 40, 697). As the steps in protein synthesis become better understood, the site of action of those antibiotics that interfere with protein synthesis can be better pinpointed. Aurintricarboxylic acid interferes with the binding of mRNA to the 30S ribosome (Grollman and Huang, *Fed. Proc.* 1973, 32, 1673). The mode of action of puromycin, which mimics in part an amino acid-tRNA, is probably the best understood of all antibiotics. It serves as an acceptor for the growing peptidyl-tRNA complex on the ribosome, thereby terminating the growing chain as a peptidyl-puromycin. Puromycin is an effective protein inhibitor in both microorganisms and higher forms. Some inhibitors work preferentially on one or the other. Thus, chloramphenicol and cycloheximide both inhibit protein synthesis by binding to the larger subunit of the ribosome. However, chloramphenicol is effective against bacterial and mitochondrial ribosomes, while cyclohexamide is effective against the cytoplasmic ribosomes of eukaryotes.

Streptomycin interferes with bacterial ribosomes by specifically binding, at low concentrations to the smaller ribosomal subunit (more so when it is in the 70S state than when it is free), causing distortion of the subunit, misreading messenger during translation as well as inhibition of peptide bond formation. This was shown by examination of the ribosomal subunits from normal, streptomycin sensitive bacteria and streptomycin resistant mutants. The 70S ribosomes of *E. coli* are first dissociated into their 50 and 30S component subunits at low magnesium ion concentration, and then separated by fractional ultracentrifugation. Purified 50S and 30S subunits are recombined making all possible combinations of 70S ribosomes from the subunit of normal and mutant cells. Polyuridylic acid stimulated cell-free polyphenylalanine synthesis is studied by the Nirenberg technique (described earlier) using normal, mutant and hybrid combinations of ribosomal subunits. Polypeptide synthesis is severely inhibited by streptomycin except when the 30S ribosomal subunit is derived from a streptomycin resistant cell. Nomura and his coworkers (*paper 55*) have taken the analysis of the streptomycin action one step further by separating all twenty of the proteins on the 30S subunit, and reconstituting the 16S RNA with different combinations of normal and mutant ribosomal proteins (Kurland, *Ann. Rev. Biochem.* 1972, 41, 377). It was possible to replace one of the twenty proteins on the sensitive 30S subunit by a protein from a resistant 30S subunit and thereby confer resistance. Thus, the resistance in streptomycin resistant bacteria is due to the alteration of amino acid sequence in one of the 20 proteins of the 30S ribosome so that translation in the presence of the antibiotic is again normal and the reconstituted ribosomal subunit will not bind streptomycin at low antibiotic concentrations.

Another interesting antibiotic, fusidic acid, interacts specifically with the translocase S_2 factor (Haenni and Lucas-Lenard, *Proc. Nat. Acad. Sci. U.S.* 1968, 61, 1363). In cell-free extracts it has been found that fusidic acid interfers specifically at the level of tripeptide formation, demonstrating that the S_2 translocase factor is not needed until this stage. Evidently the initiation tRNA, N-formylmethionyl $tRNA_f^{met}$, can be translocated without this particular translocase factor. Erythromycin inhibits protein synthesis in bacteria by binding to the 50S ribosomal subunit. 5-Methyltryptophan, an amino acid analogue of tryptophan, has found widespread use as an inhibitor of protein synthesis. Its effectiveness is due to the competitive inhibitory action it has on the formation of trp-tRNA. The amino alcohol derivative of the amino acids appear to have similar effects. Pactamycin appears to inhibit the initiation of polypeptide synthesis in eukaryotes at low concentrations, whereas sparsomycin inhibits the function of peptidyl transferase on the larger ribosomal subunit (Goldberg et al., *Fed. Proc.* 1973, 32, 1688).

Mutants resistant to antibiotics that inhibit protein synthesis have proven to be very useful in genetic mapping studies, particularly to show clustering and possible organization of genes coding for ribosomal proteins into operons

(Nomura and Engback, *Proc. Nat. Acad. Sci. U.S.* 1972, **69**, 1526). Such mutants have also been useful in pinpointing the activity of particular proteins in the overall protein synthesizing complex.

Some Aspects of Protein Degradation

Studies on bacteria have generally involved the use of rapidly growing cultures, and since protein degradation is minimal in such conditions, the amount of enzyme degradation is very small compared to the amount of enzyme synthesis. This is not true in starved bacterial cultures, nor is it true with animal cells where protein degradation is a continual process, even at the fastest growth rates.

It is very likely that specific intracellular proteolytic enzymes play an important regulatory role, particularly during microbial development and phage infection. It has been shown, for instance, that during sporulation of *B. cereus* the vegetative aldolase is specifically cleaved by protease action (which can be duplicated *in vitro*) to yield an altered enzyme which is incorporated into the spore. RNA polymerase may undergo similar changes (see Section IV), leading to an altered specificity and altered transcription products during and just after the commitment to sporulation.

The cleavage of bacteriophage proteins is also of interest, even if the reason for the proteolysis is not known. In studies with bacteriophage T4 it has been shown that the gene 23 protein, the major head structural protein, is first synthesized as a molecule with a molecular weight of 56,000 daltons. This protein and at least three other T4 proteins are cleaved during assembly and some host enzymes are altered as well. Proteases may serve a monitoring function by degrading altered or mutant proteins. For instance, Zipser and his colleagues (Morrison and Zipser, *J. Mol. Biol.* 1970, **50**, 359) have shown that protein fragments, synthesized as a result of a nonsense mutation in the structural gene of β-galactosidase, are difficult to recover intact due to their degradation by intracellular proteases (also see Goldberg, *Proc. Nat. Acad. Sci. U.S.* 1972, **69**, 422).

A SOLUBLE RIBONUCLEIC ACID INTERMEDIATE IN PROTEIN SYNTHESIS*†

By MAHLON B. HOAGLAND,‡ MARY LOUISE STEPHENSON, JESSE F. SCOTT, LISELOTTE I. HECHT, AND PAUL C. ZAMECNIK

(From the John Collins Warren Laboratories of the Huntington Memorial Hospital of Harvard University at the Massachusetts General Hospital, Boston, Massachusetts)

(Received for publication, September 27, 1957)

The cell-free rat liver system in which C^{14}-amino acids are incorporated irreversibly into α-peptide linkage in protein has been used in our laboratories for a number of years as a measure of protein synthesis. The essential components of this system are the microsomal ribonucleoprotein particles, certain enzymes derived from the soluble protein fraction, adenosine triphosphate, guanosine di- or triphosphate, and a nucleoside triphosphate-generating system (1–3). The ribonucleoprotein particles of the microsomes appear to be the actual site of peptide condensation. The soluble enzymes and ATP[1] have been found to effect the initial carboxyl activation of the amino acids (4). The role of GTP is not yet understood, although the present paper sheds light on its probable locus of action.

Much evidence has accumulated in the past 8 years, beginning with the studies of Caspersson (5) and Brachet (6), implicating a role for cellular RNA in protein synthesis. The intermediate stages between amino acid activation and final incorporation into protein in the rat liver *in vitro* system offered us unexplored regions in which to seek more direct evidence for a chemical association of RNA and amino acids. A preliminary report of such an association has recently been presented by us (7). There it was shown that the RNA of a particular fraction of the cytoplasm hitherto uncharacterized became labeled with C^{14}-amino acids in the presence of ATP and the amino acid-activating enzymes, and that this labeled RNA subsequently was able to transfer the amino acid to microsomal protein

---

* This work was supported by grants from the United States Public Health Service, the American Cancer Society, and the United States Atomic Energy Commission.

† This is publication No. 916 of the Cancer Commission of Harvard University.

‡ Scholar in Cancer Research of the American Cancer Society. Present address, Cavendish Laboratory, Cambridge University.

[1] The abbreviations used in this paper are as follows: RNA, ribonucleic acid; pH 5 RNA, ribonucleic acid derived from the enzyme pH 5 fraction; AMP, adenosine 5′-phosphate; ATP, GTP, CTP, UTP, the triphosphates of adenosine, guanosine, cytosine, and uridine; PP, inorganic pyrophosphate; PPase, inorganic pyrophosphatase; PEP, phosphoenol pyruvate; Tris, tris(hydroxymethyl)aminomethane; and ECTEOLA, cellulose treated with epichlorohydrin and triethanolamine.

241

in the presence of GTP and a nucleoside triphosphate-generating system. This paper is a more definitive report on these studies.

Materials and Methods

Cellular fractions (microsomes and pH 5 enzymes) of rat liver and mouse Ehrlich ascites tumor were prepared by methods previously described (2, 3). Microsomes were generally sedimented at 105,000 × g for 90 to 120 minutes instead of the usual 60 minutes in order to insure more complete sedimentation of microsome-like particles. pH 5 enzymes were precipitated from the resulting supernatant fraction by adjusting the pH to 5.2.

Preparation of Labeled pH 5 Enzyme Fraction—The labeling of the pH 5 enzyme fraction was carried out by incubating 10 ml. of pH 5 enzyme preparation (containing 100 to 200 mg. of protein) dissolved in buffered medium (2) with 4.0 μmoles of C^{14}-L-leucine (containing 7.2 × 10^6 c.p.m.) and 200 μmoles of ATP in a final volume of 20 ml. for 10 minutes at 37°. The reaction mixture was then chilled to 0°, diluted 3-fold with cold water, and the enzyme precipitated by addition of 1.0 M acetic acid to bring the pH to 5.2. The precipitate was redissolved in 5 to 10 ml. of buffered medium, diluted again (to 60 ml.) with water, and the enzyme reprecipitated at pH 5.2 with M acetic acid. This final precipitate was washed with water and dissolved in 5.0 to 10.0 ml. of the cold buffered medium.

Isolation of pH 5 RNA—Isolation of pH 5 RNA was carried out by a minor modification of the method of Gierer and Schramm (8) and Kirby (9). The labeled pH 5 enzyme solution as prepared above was shaken in a mechanical shaker at room temperature for 1 hour with an equal volume of 90 per cent phenol, followed by centrifugation at 15,000 × g for 10 minutes. The top aqueous layer containing the RNA-leucine-C^{14} was removed with a syringe, more water was added, and, after thorough mixing, the centrifugation and withdrawal of the aqueous solution were repeated. Phenol was removed from the pooled aqueous solutions by three successive ether extractions. 0.1 volume of 20 per cent potassium acetate (pH 5) was then added, and the RNA was precipitated with 60 per cent ethanol at $-10°$, redissolved in water, and again precipitated from 60 per cent ethanol. The final precipitate was dissolved in a small volume of water and dialyzed against water for 4 hours in the cold. This method of extraction was used as a preparative procedure and yielded 50 to 70 per cent of the RNA initially present in the enzyme preparation, and was also used to prepare microsomal and unlabeled pH 5 RNA.

For analysis of pH 5 RNA-leucine-C^{14} in smaller incubations, NaCl was used to extract the RNA. To the incubation mixture (usually a volume of 2.0 ml.), 10 volumes of cold 0.4 N perchloric acid were added. The resulting acid-insoluble precipitate, containing RNA and protein, was washed four times with cold 0.2 N perchloric acid, once with 5:1 ethanol-0.2 N perchloric acid, once with ethanol in the cold, and once with 3:1 ethanol-ether at 50°. The RNA was then extracted with 10 per cent NaCl at 100° for 30 minutes. (During this extraction, the pH drops to around 2 to 3 and it is essential to permit this to occur; if the pH is held above 6, the isolated RNA contains little or no radioactivity.) The RNA was precipitated from the NaCl extract with 60 per cent ethanol at $-10°$, and was dissolved in water and again precipitated with ethanol. The final ethanol suspension was filtered by suction onto disks of No. 50 Whatman paper. The dried RNA was counted by using a Nuclear micromil window gas flow counter, was then eluted from the paper with 0.005 N alkali, and the concentration determined by measuring the absorption at 260 mμ in a Beckman spectrophotometer by using an extinction coefficient of 34.2 mg.$^{-1}$ cm.^{2} (10). This extraction procedure yielded 30 to 35 per cent of the RNA originally present in the incubation mixture. In experiments in which total counts are recorded, the specific activity of this NaCl-extracted RNA was multiplied by the total quantity of RNA initially added as determined by the method of Scott *et al.* (10). This was based on the assumption that the RNA extracted was a representative sample of the total.

For the determination of the specific activity of the protein, the methods described previously (1, 2) were employed.

The nucleoside triphosphate preparations, the triphosphate-generating system, and the C^{14}-amino acids used in these studies were the same as those used in other recent work reported from this laboratory (2). 1 μmole of Mg^{++} was added per micromole of nucleoside triphosphate in all cases.

Results

Labeling of RNA Cellular Fractions with Amino Acids—In the complete system required for incorporation of C^{14}-amino acids into protein (microsomes, pH 5 enzymes, ATP, GTP, nucleoside triphosphate-generating system, and C^{14}-amino acids), the RNA subsequently isolated was found to be labeled with C^{14}-amino acids. Incubation of the (pH 5) enzyme fraction without microsomes under these conditions resulted in substantially more RNA labeling than in the complete system. Little labeling of RNA was observed when microsomes were incubated alone under the above conditions. Further analysis of the requirements for labeling of pH 5 RNA revealed that ATP alone was sufficient and that GTP and the generating system were not necessary. A survey of the extent of labeling of the RNA of various isolated liver cellular fractions with leucine is shown in Fig. 1, which shows that pH 5 RNA has the highest specific activity.

Fig. 2 shows the dependence of labeling of pH 5 RNA upon leucine concentration. Glycine-C^{14}, valine-C^{14}, or alanine-C^{14} gave about the same

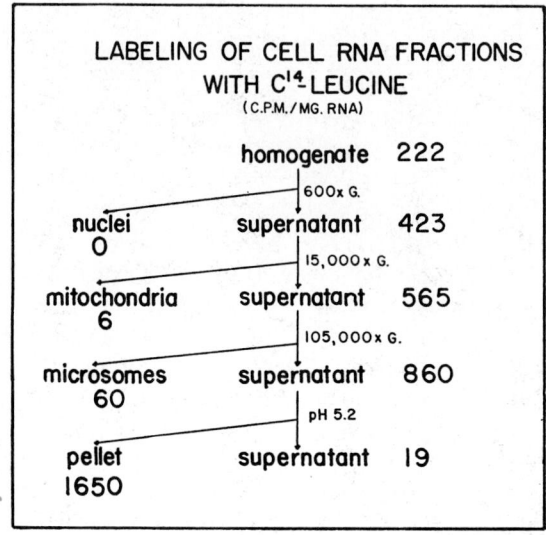
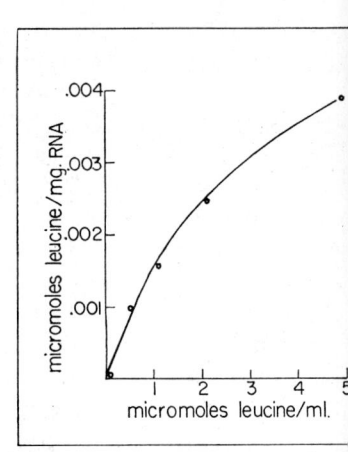

FIG. 1. FIG. 2.

FIG. 1. Cellular fractions were prepared from rat liver and resuspended or dissolved in buffered medium (2). 0.75 ml. of each fraction was then incubated at 37° for 10 minutes with 10 μmoles of ATP, 0.1 μmole of L-leucine-C^{14} containing 180,000 c.p.m., 5 μmoles of PEP, and 0.04 mg. of pyruvate kinase in a final volume of 1.0 ml. 4 μmoles of potassium fumarate, 10 μmoles of potassium glutamate, and 10.0 μmoles of orthophosphate were added to the incubation mixtures containing the original homogenate, the 600 × g supernatant fraction, and the mitochondrial fraction, and these were shaken in an atmosphere of 95 per cent oxygen-5 per cent CO_2 during the incubation.

FIG. 2. Leucine concentration curve for labeling of pH 5 RNA. 1.2 ml. of rat liver pH 5 enzyme preparation (approximately 13 mg. of protein), in buffered medium were incubated for 10 minutes at 37° with 20 μmoles of ATP, L-leucine-C^{14} containing 3.6×10^5 c.p.m. at the concentrations indicated, in a final volume of 2.0 ml.

extent of labeling as leucine-C^{14} when each was present at a concentration of 0.007 M. When these amino acids were combined (0.007 M each), the labeling was approximately additive. The addition of a mixture containing fifteen C^{12}-amino acids (lacking leucine) did not affect the extent of labeling with leucine-C^{14}. Maximal labeling of 0.04 μmole of leucine per mg. of RNA was attained with the most active liver preparations by using 0.005 M leucine and 0.01 M ATP.

ATP was necessary for the labeling of the RNA with amino acids, and the extent of labeling depended upon the concentration of ATP (Fig. 3).

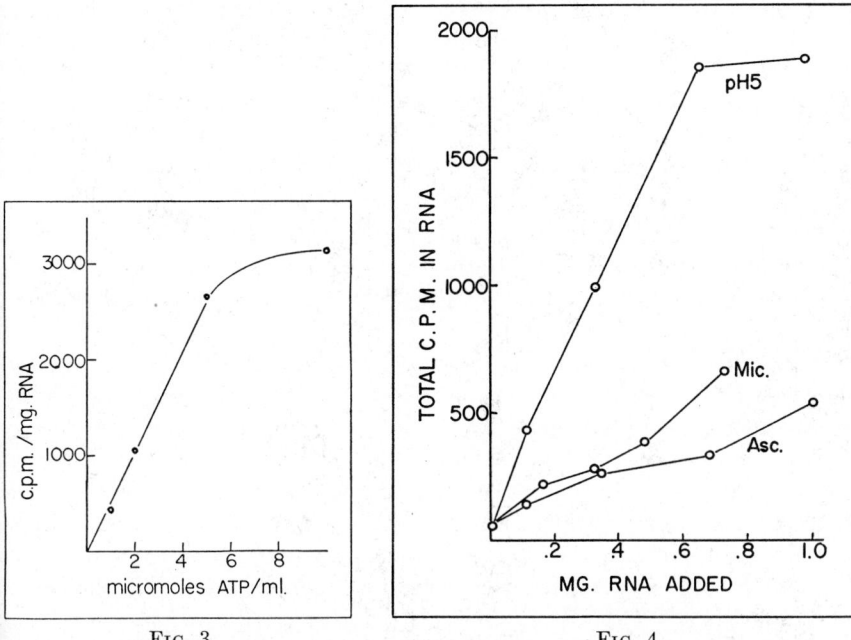

Fig. 3.   Fig. 4.

Fig. 3. ATP concentration curve for labeling of pH 5 RNA. 1.8 ml. of rat liver pH 5 enzyme preparation (20 mg. of protein) in buffered medium were incubated for 20 minutes at 37° with 0.2 μmole of L-leucine-C^{14} containing 3.6 × 10^5 c.p.m. and ATP at the concentrations indicated, in a final volume of 2.0 ml.

Fig. 4. Effect of addition of certain RNA preparations upon the labeling of pH 5 RNA with leucine-C^{14}. RNA was prepared from the pH 5 enzyme (pH 5) and microsome (Mic.) fractions of rat liver and from whole ascites cells (Asc.) by the phenol method. The quantities of RNA indicated, dissolved in 0.02 M Tris buffer, pH 7.6 (0.5 μmole of $MgCl_2$ added per mg. of RNA), were incubated at 37° for 10 minutes in 1.0 ml. volumes with pH 5 enzyme obtained from Ehrlich ascites cells (containing 0.12 mg. of RNA), and the following final concentrations of additions: 0.01 M ATP; 0.01 M PEP; 0.0038 M $MgCl_2$; 0.038 M Tris, pH 7.6; 0.018 M KCl; and 0.04 mg. of pyruvate kinase. The total radioactivity in RNA plotted was calculated as the product of the specific activity of the RNA isolated by extraction by the NaCl method and the total quantity of RNA present at the end of the incubation period. Addition of liver pH 5 RNA to the liver pH 5 enzyme fraction produces a similar enhancement of incorporation to that shown here with the tumor enzyme.

Maximal incorporation was reached at about 0.01 M. A similarly shaped ATP concentration curve had earlier been found for the amino acid activation reaction, as measured by hydroxamic acid formation (4). Also, as in the activation reaction, the requirement for ATP was specific since GTP, CTP, and UTP did not replace this nucleotide. GTP, which is necessary for the over-all incorporation of amino acids into protein, did not affect the rate of labeling of pH 5 RNA by the pH 5 enzyme fraction, in the presence or absence of a mixture containing fifteen C^{12}-amino acids (lacking leucine).

The amino acid labeling of RNA was sensitive to ribonuclease. 10 γ per ml. of Worthington crystalline ribonuclease gave 40 per cent inhibition, and 40 γ gave 90 per cent inhibition in the presence of 10 mg. of enzyme protein per ml. This inhibition was accompanied by a smaller loss of absorbance at 260 mμ in the acid-precipitable fraction. It is worth recalling at this point that the activation reaction, as measured by PP^{32}-ATP exchange and hydroxamic acid formation, is *not* affected by ribonuclease (4).

The extent of incorporation of leucine-C^{14} into pH 5 RNA was markedly stimulated by the addition of isolated pH 5 RNA as may be seen in Fig. 4. The enhancement of labeling was relatively specific for this particular RNA, rat liver microsomal RNA and mouse ascites whole cell RNA being of low activity. The small amount of stimulation by microsomal RNA shown in Fig. 4 may well be due to contamination of microsomes with the supernatant fraction, since the microsomes were centrifuged from the undiluted 15,000 × g supernatant fluid of a concentrated (30 per cent) homogenate.

The labeling reaction proceeded linearly with time for 3 minutes and was complete in 10 minutes. In those preparations in which precaution was taken to minimize contamination with microsomes (by preparing pH 5 enzymes from a 105,000 × g supernatant fraction obtained after a centrifugation for 90 to 120 minutes), there was no loss of leucine-C^{14} for a period of 20 minutes after maximal labeling had been reached. Slight microsomal contamination, however, resulted in a loss of leucine from RNA after maximal labeling had been reached.

After incubation of the (pH 5) enzyme fraction with leucine-C^{14} and ATP, these latter compounds could be largely removed by reprecipitation of the enzymes at pH 5 from dilute solution, as described. Upon subsequent incubation of this reprecipitated fraction, the leucine label was rapidly lost from the RNA unless ATP was added (Table I). The equivalent effect of a nucleoside triphosphate-generating system (PEP and pyruvate kinase), also shown in Table I, was probably mediated through the presence of very small amounts of adenylates which coprecipitate with the pH 5 enzyme. PP, on the other hand, increased the extent of loss of label from the RNA. These findings suggested that the labeling process might be *reversible*. This possibility was rendered more probable by the finding that, in the presence of added ATP, the addition of leucine-C^{12} produced a dilution of the leucine-C^{14} labeling, as shown in Experiment 2, Table I. This would be expected if the following reactions were occurring:

$$ATP + \text{leucine-}C^{14} + E \rightleftharpoons E(AMP \sim \text{leucine-}C^{14}) + PP \qquad (1)$$

$$E(AMP \sim \text{leucine-}C^{14}) + RNA \rightleftharpoons RNA \sim \text{leucine-}C^{14} + E + (AMP) \qquad (2)$$

The loss of label in the absence of added ATP would depend upon the presence of small amounts of indigenous PP. The failure of leucine-C^{12}

TABLE I
Effect of Various Additions upon Loss of Leucine-C^{14} from Labeled pH 5 Enzyme

Experiment No.	Addition	Amount	Per cent initial specific activity lost
		M	
1	None	0	75
	ATP	0.001	44
	"	0.005	35
	" + AMP	0.005 each	37
	PP	0.01	96
	PEP	0.01	24
2	None	0	79
	Leucine-C^{12}	0.01	84
	ATP	0.01	39
	" + leucine-C^{12}	0.01 each	68

L-Leucine-C^{14}-labeled pH 5 enzyme (0.4 ml.) was incubated at 37° for 7 minutes in a volume of 2.0 ml. with the concentrations of additions indicated. A concentration of $MgCl_2$ equal to that of PP was added with the latter. Pyruvate kinase (0.04 mg. per ml.) was added with the PEP. The initial specific activities of the RNA which were isolated from the pH 5 enzymes labeled during the preincubation were: Experiment 1, 770 c.p.m. per mg.; Experiment 2, 440 c.p.m. per mg.

to effect a dilution in the absence of added ATP would be anticipated since, due to the high ATPase activity of the preparation and the absence of a generating system, the ATP concentration would be effectively zero and the reaction would proceed rapidly to the left. It is of interest in this connection that Holley (11) has described an alanine-dependent, ribonuclease-sensitive incorporation of C^{14}-AMP into ATP catalyzed by the pH 5 enzyme preparation. This would suggest a reversal of an ATP-dependent reaction between alanine and RNA. However, other amino acids have not been found to stimulate such an exchange, suggesting that AMP is generally not a free product of reaction (2). The possibility must still be entertained, however, that ATP has some stabilizing effect upon the pH 5 RNA-amino acid bond not related to mass action.

A high concentration of NH_2OH such as 1.2 M, which was used to obtain amino acid hydroxamic acid formation with this preparation (4), also inhibits (90 per cent) the labeling of RNA with leucine.

Some Properties of pH 5 RNA-Leucine-C^{14}—The RNA of the enzyme (pH 5) fraction of rat liver represents 2 per cent of the total RNA of the cell and only 20 per cent of the RNA of the 105,000 × g 2 hour supernatant fraction. It is present in a concentration of 3 mg. of RNA per 100 mg. of protein. In the mouse ascites tumor essentially all the RNA of the 105,000 × g supernatant fraction precipitates at pH 5.2 and amounts to 20 per cent of the total RNA of the cell.

The active component of the pH 5 enzyme fraction does not sediment at 105,000 × g in 3 hours. If one compares the activity and RNA content of the pH 5 enzyme prepared from a supernatant fraction obtained after 1 hour or 3 hour centrifugations at 105,000 × g, one finds that the latter preparation contains only 50 per cent as much RNA as the former. The amount of leucine incorporated into the RNA of both preparations is, however, the same, suggesting that the RNA sedimented during the additional centrifugation time is not active.

RNA-leucine-C^{14} gave a mean sedimentation constant of 1.85 $s_{20,w}$ at a concentration of 0.003 per cent in 0.15 M NaCl, 0.015 M citrate, pH 6.8.[2] Preliminary studies indicate that this value is lower when effort is made to remove magnesium ion first by dialysis against citrate buffer. The material does not appear homogeneous, however, and probably represents a range of molecular sizes. Preliminary results with paper electrophoresis suggest at least two major components.

A sample of pH 5 RNA-leucine-C^{14} extracted by the phenol method was fractionated on ECTEOLA (12). 1 mg. of RNA, dissolved in 0.01 M phosphate buffer at pH 7 and containing 4040 c.p.m. as leucine-C^{14}, was placed on a column 0.2 cm. in diameter containing 50 mg. of ECTEOLA-SF (0.16 meq. of N per gm.).[3] Elution was carried out with a gradient of NaCl in 0.01 M phosphate buffer at pH 7, which was established by feeding buffer containing 2.5 M NaCl into a 500 ml. mixing flask. 1.5 ml. fractions were collected at a flow rate of 1.8 ml. per hour. The NaCl gradient was continued until the molarity of the effluent was about 2. In accordance with the general procedure of Bradley and Rich (12), the gradient was discontinued, the column washed with water, and 10 ml. of 1 N NaOH were run through. Three fractions emerged: Fraction 1 failed to adhere to the exchanger and contained 14 per cent of the ultraviolet absorbance and 8 per cent of the radioactivity (*free* leucine, if present

[2] We wish to thank Dr. J. Fresco and Dr. P. Doty of Harvard University for performing these analyses.

[3] Kindly furnished by Dr. Alexander Rich.

would have been found in this fraction); Fraction 2 emerged at a mean molarity of 0.15 NaCl and contained 48 per cent of the absorbance and 2 per cent of the radioactivity; and Fraction 3 was eluted with NaOH and contained 36 per cent of the absorbance and 68 per cent of the radioactivity. The final recovery amounted to 98 per cent of the ultraviolet absorbance and 78 per cent of the radioactivity. The low recovery of the radioactivity is most likely due to self-absorption in the NaOH-eluted fractions when plated for counting. These results, compared with those published by Bradley and Rich, suggest that at least 68 per cent of the leucine is bound to 36 per cent of the RNA of high sedimentation coefficient relative to that of the bulk of the sample.

pH 5 RNA-leucine-C^{14} isolated from the pH 5 enzyme fraction by both the phenol and NaCl methods was readily bound by Dowex 1 and charcoal at neutral pH value. However, when the RNA-leucine-C^{14} was associated with pH 5 enzyme protein in the natural state, these agents did not take up the RNA. The isolated RNA-leucine-C^{14} was non-dialyzable and stable against water, 10 per cent NaCl, or 8 M urea. There was no detectable acid-precipitable protein in the RNA extracted by the phenol method (1 per cent contamination could have been detected (13)).

The leucine was completely released from the pH 5 RNA by 0.01 N KOH in 20 minutes at room temperature. At pH 4 to 6, it was relatively stable and the labeled material as prepared by the phenol method could be kept some weeks in the frozen state. The leucine appeared to be covalently linked to the RNA, as judged from the following indirect evidence. Treatment of the RNA-leucine-C^{14} with the ninhydrin reagent indicated the absence of free leucine, although leucine is slowly released from the RNA during the course of the ninhydrin procedure. Treatment with anhydrous hydroxylamine, followed by chromatography of the products on paper (75 per cent secondary butanol, 15 per cent formic acid, 10 per cent water), resulted in a spot corresponding to leucine hydroxamic acid which contained all the radioactivity originally bound to the RNA. (A control of this experiment, in which the RNA-leucine-C^{14} bond was first hydrolyzed in 0.01 N alkali, gave no radioactivity associated with the leucine hydroxamic acid spot.)

Labeling of RNA with Leucine-C^{14} in Intact Cell—If pH 5 RNA were on the pathway of protein synthesis, it would be reasonable to expect that in the intact cell it would become labeled with leucine-C^{14} earlier than microsome protein. Previous studies in this laboratory by Littlefield and Keller (3) had shown that treatment of mouse ascites tumor microsomes with 0.5 M sodium chloride facilitates the centrifugal separation of ribonucleoprotein particles rich in RNA (about 50 per cent RNA, 50 per cent protein). The protein moiety of these "sodium chloride-insoluble" particles was found to be the most highly labeled protein fraction after incorporation of leucine-C^{14} by intact cells. A preliminary experiment in the rat showed that, at the earliest time point which it was possible to obtain after injection of leucine-C^{14} (1 minute), both RNA of the pH 5 fraction and the protein of the ribonucleoprotein particles of the microsomes were already maximally labeled. By use of mouse ascites tumor cells, it was possible to slow down the reaction by reducing the temperature of incubation. After incubation of these cells with leucine-C^{14} at 25°, the cells were washed and lysed, and concentrated solutions were added to give a final concentration of 0.5 M NaCl, 0.005 M $MgCl_2$, and 0.01 M Tris buffer, pH 7.6 (3). "NaCl-insoluble" (NaCl particles) and "soluble" fractions of a 10 minute 15,000 × g supernatant fraction were separated by centrifugation at 78,000 × g for 2 hours. Both the protein and RNA were isolated. Since almost all of the RNA present in the soluble fraction of the ascites cells precipitates at pH 5, the RNA of this fraction may be considered pH 5 RNA. The proteins of the soluble fraction represent the proteins of the NaCl-soluble components of the microsomes and the soluble cell proteins. Littlefield and Keller (3) have shown that these two fractions become labeled at a slow rate and therefore they were not separated. The results of this experiment are shown in Fig. 5. Soluble and particle RNA became labeled maximally in 2 minutes and remained so as if a steady state had been reached, while the protein of the ribonucleoprotein particles continued to acquire new amino acid content throughout the incubation period. Incorporation into the other cell proteins started after an initial lag period and proceeded at the slowest rate. The rate of labeling of the pH 5 RNA is so rapid that it occurs to some extent at 0° and no satisfactory rate curve for this labeling process could be obtained, since the reaction is proceeding even during centrifugal separation of the fractions. Similar results were obtained when the cell fractions were prepared from a sucrose homogenate and the pH 5 RNA and the protein of the deoxycholate-soluble and -insoluble fractions of the microsomes were isolated. These data suggest that the pH 5 RNA-amino acid compound *could* be an intermediate in the incorporation of amino acids into the proteins of the ribonucleoprotein particles of the microsomes.

Transfer of Leucine-C^{14} from Labeled pH 5 Enzyme Fraction to Microsomal Protein—We have reported (7) that the leucine-C^{14}-labeled enzyme fraction at pH 5, freed from ATP and leucine-C^{14} by reprecipitation at pH 5.1 from dilute solution, will transfer the RNA-bound leucine-C^{14} to microsomal protein upon subsequent incubation with microsomes, a nucleoside triphosphate-generating system, and GTP (Table II). The other nucleoside triphosphates, including ATP, would not replace GTP in this reaction; ATP also failed to stimulate the transfer in the presence of GTP.

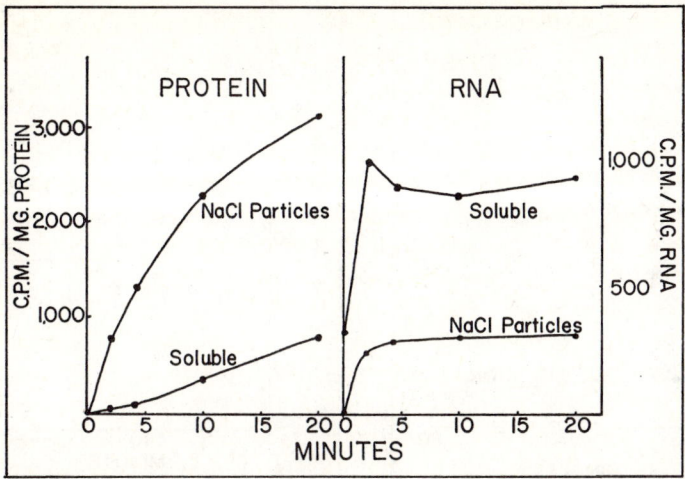

FIG. 5. Time-curve of incorporation of L-leucine-C^{14} into the RNA and protein of the ribonucleoprotein particles and the soluble fraction in intact ascites cells. Ascites tumor cells (approximately 10 gm. of packed cells) were incubated at 25° in 50 ml. of their own ascitic fluid fortified with glucose (0.04 M), Tris buffer, pH 7.6 (0.02 M), and containing 3 μmoles of L-leucine (3.5×10^6 c.p.m. per μmole). Aliquots were taken at the time points shown; NaCl-insoluble and -soluble fractions were prepared from the $15,000 \times g$ supernatant fraction. The specific activities of the RNA and protein of these fractions are shown.

TABLE II

Transfer of Leucine-C^{14} from Labeled pH 5 Enzyme Fraction to Microsome Protein

	Total c.p.m. in	
	RNA	Protein
Before incubation: complete system	478	22
After " : " "	182	433
Complete system minus GTP	116	67
" " " generating system	62	101
" " " " " and minus GTP	29	23
" " " " " but with 5 × GTP	176	91
Complete system CTP replacing GTP	98	79
" " UTP " "	117	100
" " plus 0.005 M leucine-C^{12}	178	371

The results shown are averaged values from two experiments. In each experiment 0.6 ml. of a microsome suspension containing about 15 mg. of protein and 0.4 ml. of a pH 5 enzyme fraction prelabeled with leucine-C^{14}, containing about 5 mg. of protein, were incubated for 15 minutes at 37° with the nucleoside triphosphates (0.0005 M), PEP (0.01 M), and pyruvate kinase (0.04 mg.) as indicated, in a final volume of 2.0 ml.

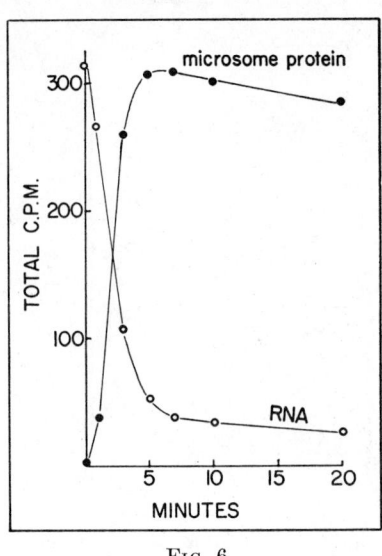
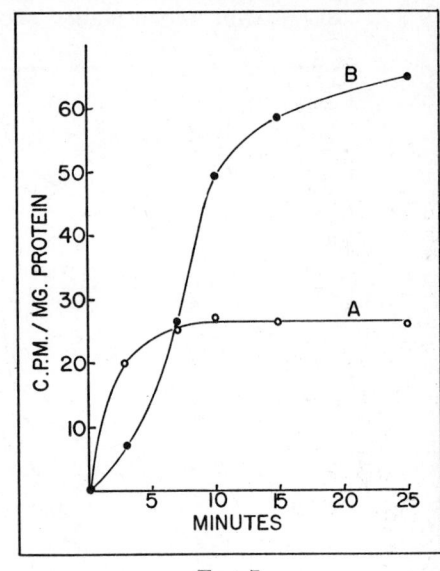

Fig. 6. Fig. 7.

Fig. 6. Time-curve of transfer of leucine-C^{14} from prelabeled pH 5 enzyme fraction to microsome protein. 4.0 ml. of a leucine-C^{14}-labeled pH 5 enzyme preparation (containing 2.6 mg. of RNA), and 6.0 ml. of microsomes (21 mg. of RNA) were incubated at 37° with 20 μmoles of GTP, 200 μmoles of PEP, and 0.8 mg. of pyruvate kinase in a volume of 20.0 ml. 2.5 ml. aliquots were taken at the time points shown. These were chilled, diluted to 12.5 ml., and centrifuged at 105,000 × g for 60 minutes in the cold. The RNA and protein of the supernatant fluid and of the microsomes were separated. The total counts per minute in the pH 5 RNA (○) and in the microsomal protein (●) is plotted. Since there is about 50 per cent enzymatic loss of leucine-C^{14} from pH 5 RNA during the hour's centrifugation (determined directly by centrifuging an aliquot of labeled enzyme, pH 5, of known specific activity under the same conditions), a correction for this loss was applied to the specific activity of RNA to give the final figures used.

Fig. 7. A comparison of the rates of the over-all incorporation reaction and the incorporation when starting with labeled pH 5 enzyme fraction. 40 ml. of an enzyme preparation at pH 5 were incubated for 10 minutes at 37° with 100 μmoles of ATP and 2 μmoles of leucine-C^{14} (3.6 × 10^6 c.p.m.) in a volume of 10 ml. An equal aliquot of the same enzyme was incubated identically with ATP and leucine-C^{12} Both enzymes were then precipitated twice at pH 5.1 from dilute solution. An aliquot of the labeled enzyme was taken for determination of RNA content and another for determination of specific activity of RNA-leucine-C^{14}. 1.8 ml. of leucine-labeled enzyme, dissolved in buffered medium and containing 1540 c.p.m. of bound leucine-C^{14} were then incubated with 1.8 ml. of microsomes, 3 μmoles of GTP, 60 μmoles of PEP and 0.24 mg. of pyruvate kinase in a volume of 6.0 ml. The same volume and amount of unlabeled enzyme were incubated with the same quantity of microsomes, GTP PEP, and pyruvate kinase, plus 3 μmoles of ATP and 0.6 μmole of leucine-C^{14} containing 1.1 × 10^6 c.p.m. The incubation mixtures were equilibrated at 30° for 1 minute before addition of microsomes and incubation was carried out at 30°. 1.0 ml. aliquots of each incubation were taken, each containing approximately 9 mg. of protein, at the times indicated and the protein of the samples was precipitated washed, plated, and counted. Curve A, reaction with prelabeled pH 5 enzyme Curve B, over-all reaction.

252

In the absence of GTP there was an equally rapid microsome-dependent loss of leucine-C^{14} from the intermediate, without concomitant appearance of amino acid in protein. (Rat liver microsomes contain considerable "ATPase" activity. Whether the loss of label is due to destruction of ATP still present in the system, thus permitting reversal of the reaction, or a manifestation of an uncoupling of the basic mechanism for trans-

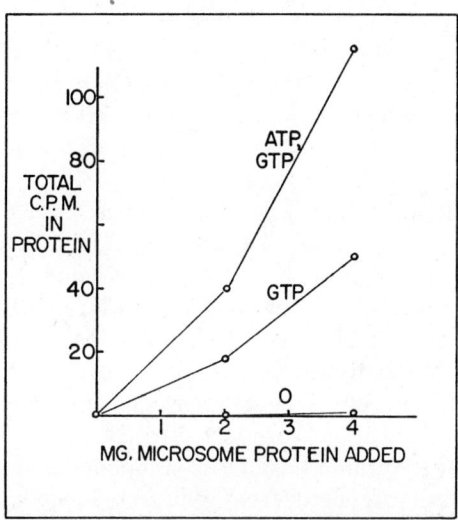

Fig. 8. Transfer of leucine-C^{14} from isolated pH 5 RNA-leucine-C^{14} to microsomal protein. 0.40 mg. of RNA, prepared by the phenol method, containing 600 c.p.m. of bound leucine-C^{14}, was incubated at 37° for 15 minutes in buffered medium (0.5 ml.) with the amount of microsomes indicated with added PEP (0.01 M), pyruvate kinase (0.04 mg.) in a volume of 1.0 ml. 0.5 μmole of ATP and GTP was added as indicated (0 = no addition of nucleotide). ATP alone gave the same activity as with no nucleotide additions. The microsomes used were sedimented from a 15,000 × g supernatant fraction which was diluted 3.5-fold before centrifugation at 105,000 × g.

ferring amino acid to microsomes is not known.) GTP had no effect on the considerably slower rate of loss of label from enzyme at pH 5 in the absence of microsomes. The addition of a relatively high concentration of leucine-C^{12} did not appreciably dilute the radioactivity transferred to protein. Fig. 6 shows a time-curve for the transfer reaction and emphasizes the high efficiency of the transfer as well as its relatively rapid rate. Fig. 7 presents a comparison of the rates of the transfer of leucine-C^{14} to microsomal protein from free leucine-C^{14} and ATP (Curve B) and from the prelabeled intermediate in the absence of free leucine and ATP (Curve A). When starting with the labeled intermediate, the lag in the initial rate of the over-all reaction was absent, the efficiency of transfer was much greater, and the reaction attained completion at an earlier time.

Transfer of Leucine-C^{14} from Isolated Labeled pH 5 RNA to Microsomal Protein—pH 5 RNA-leucine-C^{14}, extracted from labeled enzyme at pH 5 by the phenol method, precipitated twice from ethanol and dialyzed against water, will, upon incubation with microsomes, transfer leucine-C^{14} to microsomal protein. In seven experiments of this type, an average of 20 per cent of the leucine was transferred to protein (25 per cent maximum). In every case, pretreatment of the RNA-leucine-C^{14} with 0.01 N KOH at room temperature for 10 to 20 minutes resulted in lack of transfer of the leucine. Again, leucine-C^{12} did not inhibit the transfer. pH 5 RNA-leucine-C^{14} extracted by the NaCl method was consistently found to be inactive.

GTP was again found to be necessary for this transfer, and there was no transfer in the absence of a nucleoside triphosphate-generating system. Furthermore, a partial requirement for ATP became apparent with this simplified system as shown in Fig. 8. The failure to elicit an ATP requirement for the transfer of amino acid from labeled pH 5 enzyme fraction to microsomes (previous section) was apparently due to the presence of ATP not washed free from the enzyme when reprecipitated at pH 5.

Microsomes alone appear not to react directly with pH 5 RNA-leucine but to require the mediation of enzymatic components of the pH 5 enzyme fraction. Microsomes prepared from dilute homogenates (to minimize contamination with pH 5 enzymes) were low in activity when incubated with pH 5 RNA-leucine-C^{14}, ATP, GTP, and the generating system but activity could be restored by addition of the pH 5 enzyme fraction.

DISCUSSION

The evidence presented supports the conclusion that there occur ATP-dependent enzymatic reactions between ribonucleic acid and amino acids. These reactions are catalyzed by an enzyme preparation which is known to activate the carboxyl groups of amino acids in the presence of ATP. The product formed, an RNA or ribonucleoprotein to which amino acids are apparently covalently linked, is capable of interacting with enzymatic components of the activating enzyme preparation and with microsomes to effect the transfer of the amino acid to peptide linkage in protein. It is therefore suggested that this particular RNA fraction functions as an intermediate carrier of amino acids in protein synthesis. A growing body of evidence from other laboratories also suggests the presence of an intermediate similar to the one herein described (14, 15, 11, 16).

Since the amino acid activation reaction is insensitive to ribonuclease and since an activating enzyme has been isolated relatively free from RNA (17), it is still necessary to invoke an initial enzymatic activation reaction as originally postulated (4), followed by a transfer of amino acid to linkage on RNA. Because of the impurity of the enzyme system at pH 5, it cannot be stated that pH 5 RNA is naturally linked to amino acid-activating enzymes or that other enzymatic steps intervene between activation and linkage to RNA. The relative specificity of the reaction of the pH 5 enzyme fraction with pH 5 RNA shown in Fig. 4, does, however, emphasize the uniqueness of this particular RNA fraction in regard to ATP-dependent amino acid binding.

The present data suggest that the pH 5 RNA molecules, when associated with protein in the natural state, are considerably lower in average sedimentation coefficient than are the ribonucleoprotein particles of the microsomes. The latter are probably of the order of 80 S (18), while the former appear to be much lower. Furthermore, the results of other experiments from this laboratory, in which pH 5 RNA is enzymatically terminally labeled with the nucleoside monophosphate moieties of nucleoside triphosphates (19), suggest that the average molecular weight of the RNA is not likely to exceed 20,000 (based on maximal labeling, and assuming no branching). The sedimentation constant of 1.85 would be consistent with a molecular weight considerably lower than this.

Thus far we cannot assign a specific structure to the amino acid-RNA linkage. An attractive possibility is an acyl anhydride involving internucleotide phosphate groups or a terminal nucleotide residue. The acid stability and alkali lability of the linkage, qualitatively similar to the behavior of the synthetic amino acyl adenylates (20, 21), the formation of a hydroxamic acid, and the relative high energy of the linkage suggested by the possible reversibility of the reaction would support this type of anhydride linkage. The linkage would appear, however, to be more stable than a phosphate diester anhydride might be expected to be. We have also given thought to the possibility that internucleotide P—O bonds may be opened by reaction with an amino acyl adenylate, with resulting attachment of the amino acyl group to one of the opened ends of the nucleoside chain and adenylate to the other. Other possible linkages to be considered are carboxyl bonding to 2'-OH on ribose and bonding involving groups on the nucleotide bases themselves. It is, nevertheless, likely that, regardless of the type of bonding, amino acids are individually linked to pH 5 RNA and do not *condense* at this stage, for the amino acid may be recovered as the specific hydroxamic acid upon treatment with hydroxylamine.

The high efficiency of the GTP-dependent transfer of amino acid from intermediate to microsome protein is striking. There is no evidence that GTP is concerned either in the activation step or in the transfer of amino acid to pH 5 RNA. Its locus of action is thus narrowed down to the area of interaction between pH 5 RNA-amino acid and microsomes. The fact that enzymatic components of the pH 5 fraction are still required for the transfer from pH 5 RNA-leucine to microsomes could mean either that a new transfer enzyme is required or that reassociation of intermediate with activating enzymes is necessary. If this latter is the case, the possibility that pH 5 RNA acts simply as a storage site for activated amino acids must be considered.

Other studies in this laboratory to be reported have shown that the same pH 5 enzyme fraction also catalyzes a rapid incorporation of the nucleotide monophosphate moieties of ATP, CTP, and UTP into pH 5 RNA. The appearance of these reactions in the same fraction which catalyzes the amino acid binding to RNA is intriguing, but thus far it has not been possible to obtain evidence for any clear direct link between the two reactions.

We have suggested elsewhere (22) a hypothetical reaction sequence for protein synthesis which accounts for the findings presented in this paper. Its central idea is that pH 5 RNA molecules, each charged with amino acids in characteristic sequence, polymerize in microsomes (in specific order determined by the complementary structure of microsomal RNA) to higher molecular weight units with resultant configurational changes which permit peptide condensation between contiguous amino acids. This working hypothesis will form the basis for further studies in these laboratories on the mechanism of protein synthesis.

SUMMARY

Evidence is presented that a soluble ribonucleic acid, residing in the same cellular fraction which activates amino acids, binds amino acids in the presence of adenosine triphosphate. Indirect evidence indicates that this reaction may be reversible. The amino acids so bound to ribonucleic acid are subsequently transferred to microsomal protein, and this transfer is dependent upon guanosine triphosphate.

The authors wish to thank Dr. Robert B. Loftfield for the radioactive amino acids used as well as for helpful criticism.

BIBLIOGRAPHY

1. Zamecnik, P. C., and Keller, E. B., *J. Biol. Chem.*, **209**, 337 (1954).
2. Keller, E. B., and Zamecnik, P. C., *J. Biol. Chem.*, **221**, 45 (1956).
3. Littlefield, J. W., and Keller, E. B., *J. Biol. Chem.*, **224**, 13 (1957).
4. Hoagland, M. B., Keller, E. B., and Zamecnik, P. C., *J. Biol. Chem.*, **218**, 34 (1956).
5. Caspersson, T. O., Cell growth and cell function, New York (1950).

6. Brachet, J., in Chargaff, E., and Davidson, J. E, The nucleic acids, New York, **2** (1955).
7. Hoagland, M. B., Zamecnik, P. C., and Stephenson, M. L., *Biochim. et biophys. acta*, **24**, 215 (1957).
8. Gierer, A., and Schramm, G., *Nature*, **177**, 702 (1956).
9. Kirby, K. S., *Biochem. J.*, **64**, 405 (1956).
10. Scott, J. F., Fraccastoro, A. P., and Taft, E. B., *J. Histochem. and Cytochem.*, **4**, 1 (1956).
11. Holley, R., *J. Am. Chem. Soc.*, **79**, 658 (1957).
12. Bradley, D. F., and Rich, A., *J. Am. Chem. Soc.*, **78**, 5898 (1956).
13. Nayyar, S. N., and Glick, D., *J. Histochem. and Cytochem.*, **2**, 282 (1954).
14. Hultin, T., *Exp. Cell Res.*, **11**, 222 (1956).
15. Hultin, T., von der Decken, A., and Beskow, G., *Exp. Cell Res.*, **12**, 675 (1957).
16. Koningsberger, V. V., van der Grinten, C. O., and Overbeek, J. T., *Ned. Akad. Wetnsch, Proc.*, series B, **60**, 144 (1957).
17. Davie, E. W., Koningsberger, V. V., and Lipmann, F., *Arch. Biochem. and Biophys.*, **65**, 21 (1956).
18. Petermann, M. L., and Hamilton, M. G., *J. Biol. Chem.*, **224**, 725 (1957).
19. Zamecnik, P. C., Stephenson, M. L., Scott, J. F., and Hoagland, M. B., *Federation Proc.*, **16**, 275 (1957).
20. DeMoss, J. A., Genuth, S. M., and Novelli, G. D., *Proc. Nat. Acad. Sc.*, **42**, 325 (1956).
21. Berg, P., *Federation Proc.*, **16**, 152 (1957).
22. Hoagland, M. B., Zamecnik, P. C., and Stephenson, M. L., Current activities in molecular biology, in press.

ERRATUM:

In all cases where the terms "enzyme at pH 5" or "(pH 5) enzyme" are used, the term "pH 5 enzyme" should be substituted.

*ASSEMBLY OF THE PEPTIDE CHAINS OF HEMOGLOBIN**

BY HOWARD M. DINTZIS

DEPARTMENT OF BIOLOGY, MASSACHUSETTS INSTITUTE OF TECHNOLOGY

Communicated by John T. Edsall, January 16, 1961

The mechanism by which proteins are synthesized has been a matter of intense speculation in recent years.[1,2] Some published speculative models propose simultaneous bond formation between all neighboring activated amino acids on a preloaded template (a sort of stamping machine operation). Others suggest various forms of sequential addition of amino acids to a steadily growing polypeptide chain. In addition there have been hypothesized all degrees of exchange between amino acids already incorporated into growing peptide chains on the template and various classes of precursor "activated" amino acids in solution.[3]

A common concept of how peptide chains may grow is based on their linear chemical nature and assumes serial addition of amino acids, starting at one end of the chain and progressing steadily to the other end. A less orderly picture involves peptide sections growing randomly here and there on the template and finally coalescing into a single chain. Since we know very little about the geometric nature of the templates upon which protein synthesis occurs,[4] we cannot a priori rule out all manner of complex growth mechanisms. For example, if the substructure of the template is folded or coiled in a regular manner it is possible that short, evenly spaced bits of peptide chain are made first on those parts of the template most accessible to the external solution and that the intervening bits are added later at a slower rate. Also since nothing is known about the type of bonds holding the activated amino acids to the template just prior to peptide bond formation, we cannot assume that chain growth is necessarily unidirectional. It is possible that chain growth is initiated at both the amino end and the carboxyl end and progresses towards the middle, or conversely, begins in the middle and proceeds toward both ends.

It is apparent that there exists no shortage of hypothetical models of protein chain growth. The difficulty lies in finding an analytical technique capable of yielding enough information to eliminate conclusively most wrong models and, if possible, to narrow the choice to a single correct one.

Data concerning the actual mechanism of protein assembly should in principle be obtainable by studying both the newly formed protein molecules and the ribosome templates on which they are supposedly formed. However, no method exists for fractionating from a cellular extract all ribosomes engaged in the production of a single type of protein molecule. If a type of cell could be found which is engaged solely in the synthesis of a single kind of protein molecule, then presumably all ribosomes in such a cell would contain incomplete bits of that kind of protein molecule and no others.

Fortunately, a close approximation to this highly desirable situation exists in the case of immature mammalian blood cells producing hemoglobin. These cells, reticulocytes, account for 80 to 90 per cent of the red cells present in the blood of rabbits made anemic by daily injection of phenylhydrazine. The cells may be isolated from the blood and placed in an incubation medium where they will continue producing hemoglobin for many hours.[5, 6] During such an incubation over 90 per cent of the soluble protein produced appears as hemoglobin. It is therefore reasonable to expect most of the growing peptide chains present in the ribosome fraction of such cells to represent incomplete hemoglobin molecules.

If we have available a technique for splitting the peptide chains of both completed and incompleted hemoglobin molecules at a definite number of specific sites, we should be in a position to test which one, if any, of the above hypothetical mechanisms of protein assembly is correct. This is so because each model of protein assembly leads to a definite prediction as to the time and space distribution of newly added amino acids in short sections of peptide chain, both in finished hemoglobin and in the ribosomal particles. Amino acids labeled with radioactive isotopes provide a means of detecting newly added amino acids. In living reticulocyte cells there exist a very large number of finished hemoglobin molecules (10–20% of the cell by weight) and in addition a large number of ribosomal particles, supposedly containing unfinished hemoglobin molecules in different stages of completion. If, at a given moment, we add a radioactively labeled amino acid to the incubation medium containing reticulocytes, then we expect polypeptide produced hereafter to be labeled with radioactive amino acid.

The data to be presented in this paper strongly support a model of protein synthesis involving growth by some kind of sequential addition of amino acids. In Figure 1 are shown some of the predicted consequences of this type of model. For the purposes of illustration we have chosen a model involving chain initiation at one end of the polypeptide followed by sequential addition of one amino acid after another until the other end is reached. We shall assume that some digestion technique can be used to split each polypeptide chain at a definite number of specific sites, yielding the set of peptides $a, b, c, \ldots g$, and that furthermore, the set can be separated and the amount of newly added amino acid present in each member $a, \ldots g$ determined quantitatively.

In the finished hemoglobin at short times, we would then expect a steep gradient of radioactive label through the peptides, with only a few peptides labeled at very short times. At longer times the gradient of radioactivity along the peptide chain should become shallower as more and more completely labeled molecules are produced. At all times, the peptide g, closest to the finish line, should have the most radioactive label, and the peptide a, closest to the starting line, should have the least radioactive label.

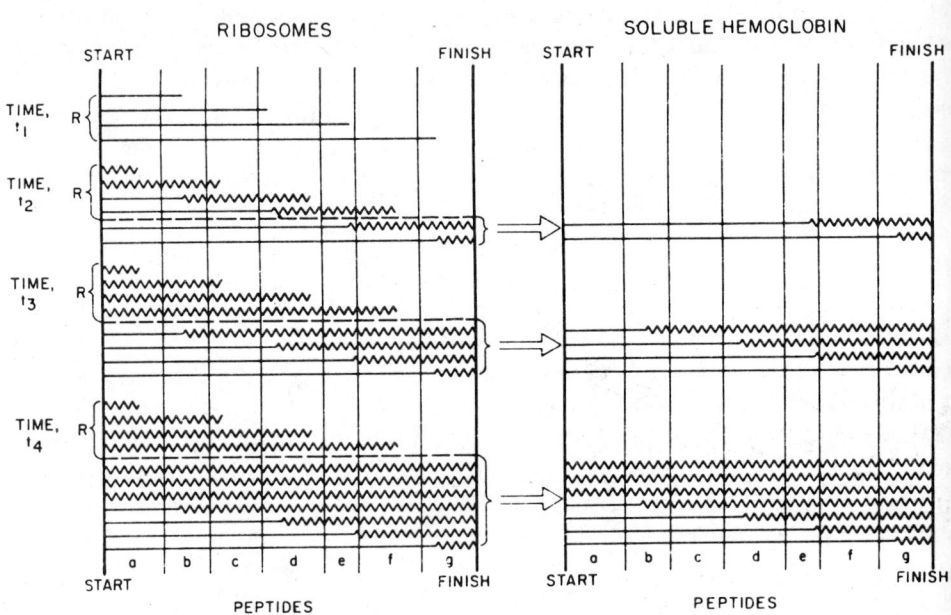

FIG. 1.—Model of sequential chain growth. The straight lines represent unlabeled polypeptide chain. The zigzag lines represent radioactively labeled polypeptide chain formed after the addition of radioactive amino acid at time t_1. The groups of peptides labelled R are those unfinished bits attached to the ribosomes at each time; the rest, having reached the finish line, are assumed to be present in the soluble hemoglobin. In the ribosome at time t_2, the top two completely zigzag lines represent peptide chains formed completely from amino acids during the time interval between t_1 and t_2. The middle two lines represent chains which have grown during the time interval but have not reached the finish line and are therefore still attached to the ribosomes. The bottom two chains represent those which have crossed the finish line, left the ribosomes, and are to be found mixed with other molecules of soluble hemoglobin.

On the other hand, in the ribosomes at very short times we would expect an almost uniform distribution of total label among the various peptides since each growing chain will have added only a small radioactive section (Fig. 1). After times long enough to flush out the nonradioactive bits of growing chain, there should be a gradient of total radioactivity from the initial peptide a, with the most, to the final peptide g, with the least. For this model this is so because there exist in a population of ribosomes at any moment, more sections of peptide a than b, more b than c, and so on. Thus, the expected gradient of label in the ribosomes is opposite to that in the finished hemoglobin, both in space direction and in time development.

General Experimental Considerations.—The technique used for forming and separating a reproducible set of peptides was a modification of the method involving a combination of paper electrophoresis and chromatography, at right angles, used by Ingram for human hemoglobin,[7] and termed "fingerprinting" by him. The enzyme trypsin, which splits polypeptides with high specificity wherever the amino acids lysine and arginine occur, furnishes the means of splitting at a definite number of sites. For various reasons, many details of Ingram's procedure for tryptic digestion, paper electrophoresis, and chromatography were modified in the present study.

The problem of obtaining quantitative data on the amount of radioactivity in each peptide was solved by the use of two different isotopic labels. Short incubations were done with H^3-leucine, and very long incubations with C^{14}-leucine. The very long incubations were assumed to give hemoglobin of uniform specific activity in each leucine position. The H^3- and C^{14}-labeled preparations were mixed and carried through the stages of digestion, electrophoresis and chromatography together. The ratio of H^3 to C^{14} was taken as a measure of the amount of label in each peptide obtained from the short time incubations. This method gave an internal standardization automatically correcting both for the differential losses and for the different number of leucine residues in the peptides.

In order to slow the rate of hemoglobin synthesis to the point where samples could be handled with convenience, incubations were tried at various temperatures below body temperature. It was found that the rate of incorporation of C^{14}-leucine into hemoglobin fell slowly with temperature until a point about 10° was reached, whereupon incorporation abruptly stopped. Incorporation of labeled amino acid was found to proceed smoothly at 15° at approximately $1/4$ of the rate at 37° (Table 1) and this temperature was routinely used for all short-time experiments.

TABLE 1
Incorporation of C^{14} Leucine into Rabbit Hemoglobin at Various Temperatures of Incubation

Temperature of incubation	Experiment 1		Experiment 2	
	Cpm/mg	% of 37° value	Cpm/mg	% of 37° value
0	0	0.00	...	...
5	14	0.22	...	...
10	280	4.3	...	...
15	2,230	34	8,150	17
20	...	...	17,700	38
25	...	...	52,600	110
30	...	...	45,000	95
37	6,500	100	47,000	100

Hemoglobin was dialyzed for 5 days against water, precipitated with trichloroacetic acid, dissolved in dilute NaOH, reprecipitated with trichloroacetic acid, washed with acetone and ether, and then plated in thin layers containing approximately 20 mg. Counting was done using a Nuclear Chicago end window gas flow counter, the results corrected to zero thickness.

It has been previously shown that the structural protein of ribosomes is not appreciably labeled at short times of incubation.[8] It was therefore assumed that the labeled peptides resulting from a digest of ribosomes with ribonuclease and trypsin represent growing hemoglobin chains and not ribosomal structural proteins. On tryptic digestion the ribosome structural protein did yield a large number of ninhydrin staining peptides which were distinct from those of hemoglobin but, as expected, they did not contain radioactive label.

Incubation of cells: Rabbit reticulocytes prepared from phenylhydrazine-treated animals were washed and incubated according to the procedures of Borsook et al.[6] The cells were incubated at 37° for 15 min to allow them to renew metabolites, then at 15° for 5 min. To 1.8 ml cells in a total volume of 4 ml incubation mixture was added 0.24 mg 4, 5 H^3-DL-leucine (5 mc, New England Nuclear Corporation, 3.6 mc per μmole) and the incubation continued at 15°. At various intervals aliquots of approximately 1 ml were removed with a pipet and quickly placed in precooled vials surrounded by solid carbon dioxide.

Uniformly labeled C^{14} leucine hemoglobin was prepared in a similar manner from approximately 1 mg of L-leucine, uniform C^{14}, (50μc Nuclear Chicago), which was incubated with 10 ml of sterile whole blood at 37° for 5 or 24 hr. During 24-hr incubations a significant amount of cell lysis occurred, partly offsetting the approximately 50 per cent higher specific activity obtainable. Typical incubations with C^{14} L-leucine of specific activity 6–8 c/millimole gave hemoglobin of approximate activity 1×10^5 dpm/mg.

Preparation of hemoglobin and ribosomes for tryptic digestion: Samples containing approximately 0.45 ml cells were thawed and the broken cells diluted to a volume of 7 ml with cold solution containing 0.14 M KCl, 0.001 M MgCl$_2$ and 0.01 M Tris-Cl pH 7.3. Solution of this composition had been previously shown to stabilize rabbit reticulocyte ribosomes[8] and was used in all operations where ribosomes were present. The solutions were then centrifuged at 20,000 g for 10 min to remove cell walls and debris, and then at 130,000 g for $1^1/_2$ hr to remove ribosomes.

Hemoglobin: The ribosome-free supernatant was dialyzed for 5–7 days in the cold against daily changes of 5×10^{-4} M KH$_2$PO$_4$, 5×10^{-4} M K$_2$HPO$_4$ saturated with toluene to prevent bacterial growth. The slight precipitate which formed was centrifuged off and the supernatant hemoglobin frozen until used

Short time labeled H^3-leucine hemoglobin (4–60 min at 15°) solution was mixed with long time labeled C^{14}-leucine hemoglobin (5–24 hr at 37°) solution in a ratio such that both the H^3 and C^{14} could be counted with good accuracy. This ratio was usually near 10 dpm H^3 per dpm C^{14}. The combined hemoglobin solution was then used to prepare globin by acid acetone precipitation.[4]

Ribosomes: The ribosome pellets were dissolved in 7 ml stabilizing buffer at 0°C, centrifuged for 5 min at 20,000 g to remove denatured protein and then reprecipitated by centrifuging at 130,000 g for $1^1/_2$ hr. The ribosomes were redissolved and recentrifuged three times to remove

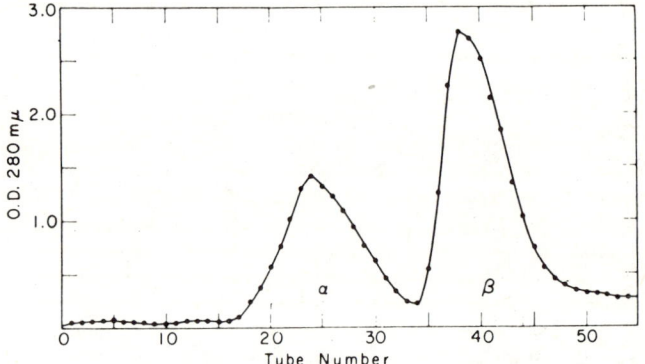

FIG. 2.—Separation of α- and β-chains of rabbit hemoglobin on carboxymethylcellulose column.

free leucine and hemoglobin. The final ribosome pellet was a very light yellowish color and completely transparent.

Separation of peptide chains of hemoglobin: The α- and β-chains of rabbit globin were separated on carboxymethyl cellulose using a linear concentration gradient of buffer between 0.2 M formic acid—0.02 M pyridine and 2 M formic acid—0.02 M pyridine (Fig. 2). Two samples of carboxymethyl cellulose were found to give good results: a preparation of 0.47 meq/g (Brown Co., Berlin N. H.) and a preparation of 0.06 meq/g (Serva, Heidelberg, Germany). Several preparations of higher capacity from various companies did not give as good results. Solutions of separated chains were dried under vacuum in the presence of sulfuric acid and soda lime.

Tryptic digestion of hemoglobin samples: Autotitrator: Dried samples were dissolved in water to a concentration of 10–20 mg/ml. The pH was adjusted to 9.5 with 0.10 N NaOH from an initial value between 3 and 4. Dense precipitation occurred near neutral pH but the solution became clear again at pH 9.5. 0.01 ml of 1% trypsin (Worthington 2x crystallized, salt-free in 10^{-3} M HCl) was added for each ml of solution and the digestion was allowed to proceed at 37° until definite evidence of a plateau in base uptake was obtained (approximately $1^1/_2$ hr).

Buffer: 10 mg of dried sample was dissolved in 0.5 ml water, 0.015 ml 0.5 M NH$_4$OH was added, followed by 0.01 ml 1% trypsin and 0.025 ml buffer made of 1 M NH$_4$OAc + NH$_4$OH to pH 9.75. Digestion proceeded for 4 hours at room temperature.

In all cases digestion was stopped by the addition of several drops of glacial acetic acid. The samples were then dried under high vacuum in the presence of sulfuric acid and soda lime and then dissolved at a concentration of 100 mg/ml in 0.4% acetic acid-0.1% pyridine, giving a preparation which was often clear, but sometimes had slight to medium turbidity.

Tryptic digestion of ribosome samples: The ribosome pellet from 0.45 ml cells, approximately 3 mg dry weight, was dissolved in 1 ml water. 10–20 mg uniformly labeled C^{14} leucine globin was dissolved in 1 ml water. The two solutions were mixed and adjusted to pH 8.5 in an autotitrator at 37°. 0.02 ml 1% ribonuclease (Worthington crystalline) was added, followed, after 15 min, by 0.02 ml 1% trypsin. The digestion was followed in the autotitrator for 15 min, then the pH was raised to 9.5 and the digestion followed for approximately $1^1/_2$ hr until a plateau was reached. The samples were acidified and dried as in the case of hemoglobin digestion.

Paper electrophoresis: Electrophoresis was carried out on a water-cooled metal plate insulated with a thin sheet of polyethylene. Strips of Whatman No. 3MM paper 12 in. wide and 37 in. long were wet with buffer of pH 4.5 (2.5% pyridine, 2.5% acetic acid, 5% n-butanol, all concentrations v/v) and blotted. Eight-inch wicks made of 4 thicknesses of the same paper were overlapped at each end, and 0.02 ml of solution containing 2 mg of sample was applied at the origin. The paper was then covered with polyethylene sheeting pressed flat by weights applied over a sponge rubber pad. Electrophoresis was carried out at 2,000 volts and approximately 100 ma, for 16 hr, after which the paper was dried.

Chromatography: The dried papers were trimmed to a length of 33 in. and stapled into cylinders 12 in. high. Chromatography was then conducted at room temperature in glass jars 12 in. wide and 24 in. high, using a mixture of 42.5 vols n-butanol, 27.5 vols pyridine, 30 vols water. Occasionally it was necessary to increase the chromatographic resolution by sewing a 4-in. strip of paper to the top of the sample sheet before stapling into a circle.

Isolation and counting of peptides: The dried chromatograms were dipped in 0.25 per cent ninhydrin in acetone, dried, and heated at 90° for 5 min. The resulting blue paper spots were cut out, placed in 20 ml counting vials and 5 ml of water was added to each. The vials were then heated in an oven at 90° for 30 min to extract the peptides from the paper, after which time the paper was removed from the vial with a tweezer and the solution evaporated to dryness overnight in an oven at 90°. 0.20 ml of 0.01 HCl was added to each vial, followed by 20 ml of scintillator solution made up of three parts toluene, one part absolute ethanol, and containing 1% phenylbiphenyloxadiazole-1,3,4(PBD) and 0.05% p-bis [2-(5-phenyloxazolyl)]-benzene (POPOP). The resulting solutions were measured for C^{14} and H^3 activity simultaneously using a TriCarb scintillation counter equipped with split channel operation so that the lower voltage channel counted both C^{14} and H^3 while the upper voltage channel counted mainly C^{14}. The recovery of radioactivity from eluted peptides of hemoglobin amounted to approximately 50 per cent of the amount applied at the origin spot for paper electrophoresis.

The TriCarb scintillation counter was run with 1040 volts on the photomultiplier tubes. The lower pulse height discriminator was set to register pulses between 10 and 50 volts, giving an efficiency of 6.5% for H^3 and 20% for C^{14} with a background of 40 cpm. The upper pulse height discriminator was set to register pulses of 100 volts or higher, giving an efficiency of 0.14% for H^3 and 37% for C^{14} with a background of 60 cpm. Mixtures of isotopes ranging from 2 dpm H^3/dpm C^{14} to 40 dpm H^3/dpm C^{14} were used.

Figure 3 shows a rather typical peptide separation. To improve photographic reproduction, the ninhydrin staining was done with twice the usual concentration of ninhydrin. The result shows more clearly than usual the presence of "ghost" spots, which are defined as weak spots sometimes present but usually absent or barely detectable. The spots which are always or almost always present have been numbered arbitrarily from left to right.

Peptide 31 is the leucine-containing peptide farthest from the origin as determined by radioactivity count on peptides made from uniformly labeled hemoglobin. There are approximately four ninhydrin staining spots farther from the origin than peptide 31, but since they were not labeled by leucine, they were routinely removed from the paper by electrophoresis, to increase the separation of the remaining peptides. The total number of peptides found with reasonable reproducibility is thus about 35, appreciably above the number 26 reported in human hemoglobin by Ingram.[7] It should be noted that a number of peptides, e.g., 2, 7, 16, 19, 23, stain quite weakly and may represent products of incomplete tryptic digestion, or partial digestion by other enzymes such as chymotrypsin which may be present as trace impurities in the trypsin.

The separation and identification of peptides was not uniformly good. In some runs spots were either missing or badly smeared into other spots. Consequently it was necessary to eliminate

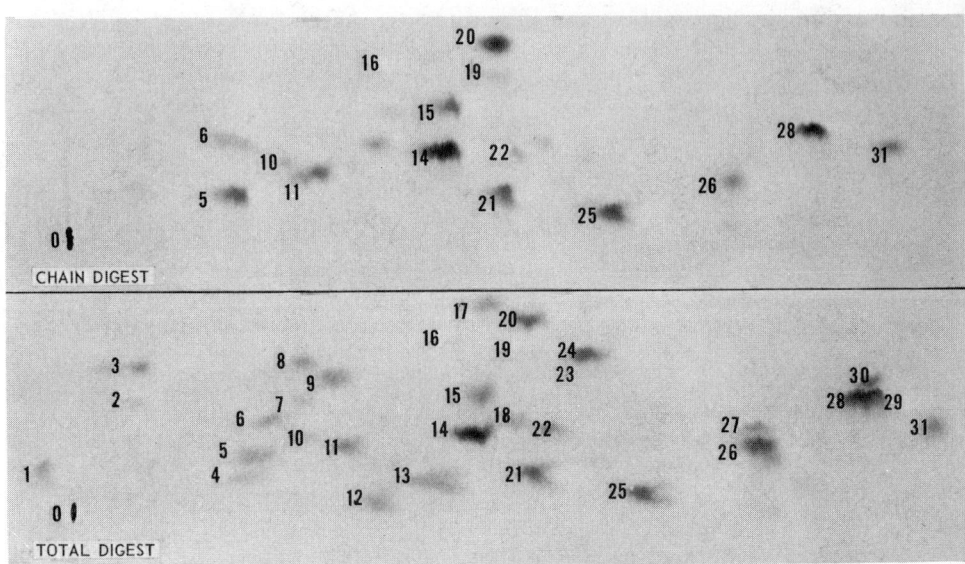

FIG. 3.—Peptide maps of tryptic digest of total rabbit globin (bottom) and column separated α-chain (top). The point of application of digest to the paper is marked by O. The positive electrode is to the left.

from elution and counting in each run those peptides which could not be identified with certainty or which badly overlapped with neighbors known to contain leucine.

A total of 18 peptides, 9 from the α-chain and 9 from the β-chain, were found reproducibly to contain leucine in significant quantity. The average relative yield of C^{14} in these peptides, obtained from digests of uniformly labeled single chains, is shown in Table 2. In addition, smaller

TABLE 2

RELATIVE YIELD OF C^{14} LEUCINE FROM TRYPTIC PEPTIDES

α-Chain Peptides		β-Chain Peptides	
Peptide number	Average relative yield	Peptide number	Average relative yield
10	0.4	1	1.2
11	0.7	3	0.8
14	4.5	9	2.6
16	0.3	12	0.5
20	1.0	13	1.0
21	0.4	17	1.2
22	0.2	18	0.5
25	0.6	24	1.0
31	0.6	27	0.6

amounts of label were found in peptides 2 (β), 6 (α), 19 (α), and 28 (β), but since their yields were quite small and variable, no attempt was made to do quantitative measurements on them (except for a few studies on peptide 28, reported below). This study, which is dependent on the use of radioactive leucine, is therefore based on slightly over half of the total number of recognizable peptides produced from hemoglobin by trypsin digestion. Extension to the remaining peptides awaits the availability of other amino acids, preferably lysine and arginine, of very high H^3 specific activity.

Results.—After 7 min of incubation at 15° in the presence of H^3 leucine, a marked difference in the relative amount of tritium contained could be found in the peptides of both the α- and β-chains. The peptides could be arranged in a more or less definite order of increasing tritium content (Fig. 4), such that only the relative order of nearest neighbors was in doubt.

At different times of incubation the same relative order of the peptides was maintained (Fig. 5). The shape of the curves indicates extreme nonuniformity of labeling at 4 min of incubation, with a number of peptides containing no detectable H^3 leucine. By 60 min of incubation the gradient of radioactivity has been largely, but not entirely, eliminated.

To check the significance of the varying amounts of H^3 leucine found in different peptides two types of control experiments were made (Table 3). First, hemoglobins made by incubation for 5 hr at 37° with H^3 leucine and with C^{14} leucine were mixed, digested and counted for H^3 and C^{14}. These samples gave a uniform ratio within experimental error; see Table 3, column (*a*). Next, samples from a 7-min incubation at 15° giving marked nonuniform labeling with H^3 leucine (Table 3, column (*b*)) were checked to see if any systematic counting error was involved. To each sample

TABLE 3
CONTROLS ON COUNTING ACCURACY

Peptide number	(*a*) Long-time incubation	(*b*) Short-time incubation	(*c*) Increment ratio
α-Chain			
21	1.01	0.08	1.02
10	0.94	0.08	0.98
20	1.06	...	...
25	1.04	0.36	1.03
11	1.04	0.38	1.05
14	1.07	0.69	1.05
31	1.02	...	...
22	1.02	0.84	1.00
16	0.88	1.06	1.02
β-Chain			
13	...	0.05	0.98
24	0.93	0.10	1.02
1	1.01	0.16	1.03
17	0.94	0.23	1.02
3	0.94	0.34	0.88
9	0.99	0.54	1.02
18	0.97	0.59	0.89
12	1.05	0.70	1.00
27	0.86	...	...

(*a*) 5-hr incubation H^3 leucine, 5-hr incubation C^{14} leucine, relative amount of tritium.
(*b*) 7-min incubation H^3 leucine, 30-hr incubation C^{14} leucine, relative amount of tritium.
(*c*) Ratio of increases in H^3 to C^{14} after adding H^3 leucine and C^{14} leucine to each counting vial of (*b*).

vial a constant amount of H^3 standard and C^{14} standard were added, and the radioactivity redetermined. The measured increments in H^3 and C^{14} activity were constant within experimental error and the normalized ratio of increments $\Delta H^3/\Delta C^{14}$ was also constant (Table 3, column (*c*)).

The results obtained by digesting ribosomes were less reproducible for a number of reasons. First, the α- and β-chains could not be separated, since by definition we were looking for incomplete chain fragments in the ribosomes, and thus did not dare lose fragments in an attempt at fractionation. Secondly, the over-all background of radioactivity between ninhydrin staining spots was much higher in the ribosomes. This is perhaps to be expected from the model in Figure 1, where there

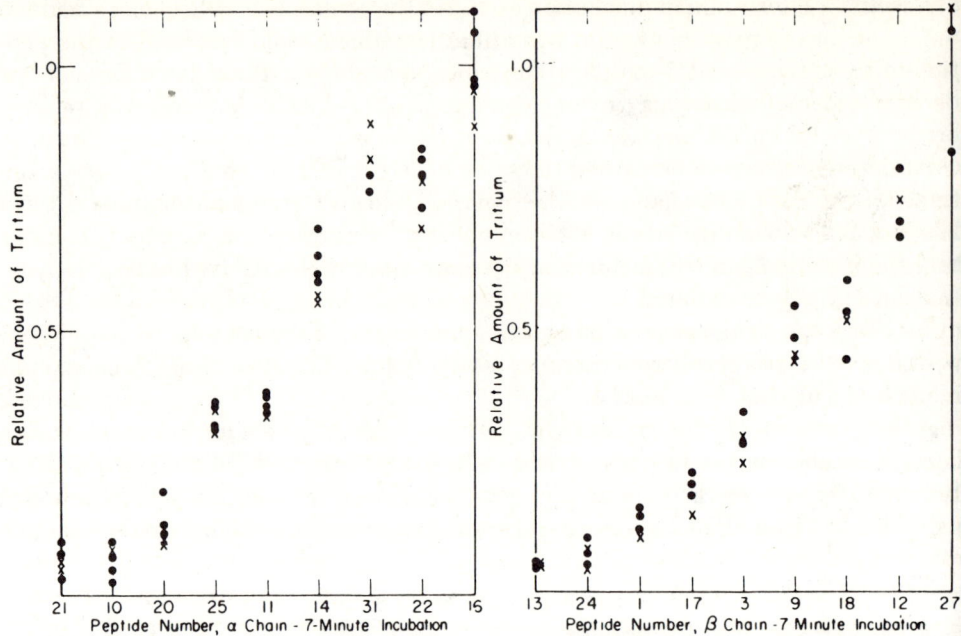

Fig. 4.—Distribution of H³-leucine among tryptic peptides of soluble rabbit hemoglobin. Peptides produced by tryptic digestion in an autotitrater are indicated by ●. Peptides produced from a separate incubation by tryptic digestion in buffer are indicated by X.

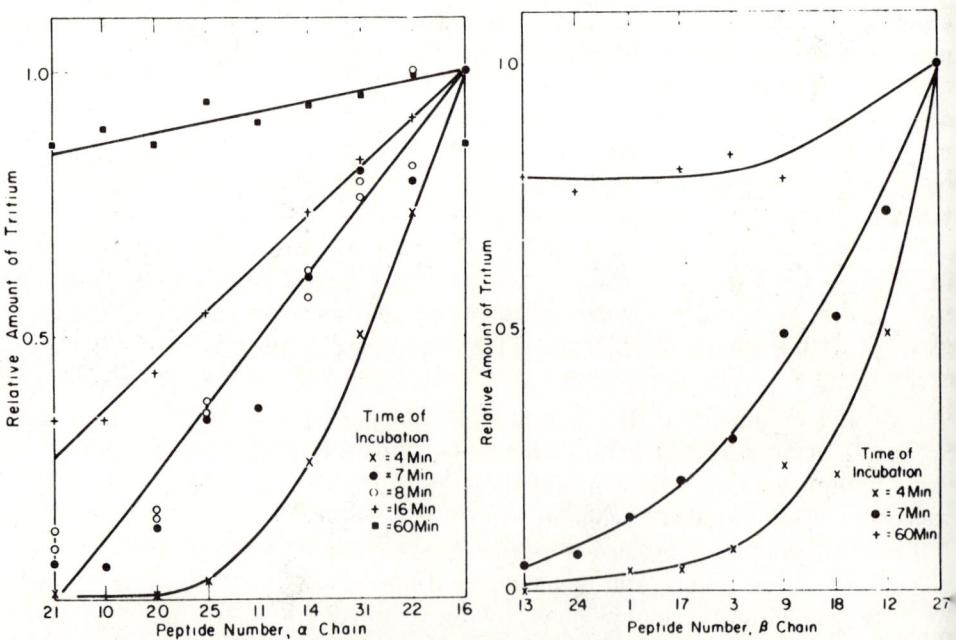

Fig. 5.—Distribution of H³-leucine among tryptic peptides of soluble rabbit hemoglobin after various times of incubation at 15°C. The points indicated for 7 minutes (●) are the result of averaging all points shown in Figure 4.

are shown end bits of growing chain which do not span vertical lines. Such bits would not correspond to tryptic peptides from hemoglobin and would not be expected to separate with the known peptides; hence they would contribute to the background of radioactivity.

Figure 6 shows the data obtained from ribosomes of cells which had been incubated for short periods (4 to 7 min) at 15°. It is hard to see any significant trend to the data, with the possible exception that the terminal peptides (16 and 27) seem lower than the rest. It thus appears that at these short times of incubation the hemoglobin peptides in ribosomes are labeled almost uniformly.

Figure 7 shows results from ribosomes of cells which had been incubated 60 min at 15°. In this case there is a clear trend visible in the peptides of the α-chain with a less definite result in the case of the β-chain. The gradient of radioactivity is opposite to that in Figure 4.

After 7 min of incubation with H^3 leucine at 15° the hemoglobin peptides isolated from soluble hemoglobin (Fig. 4) had an average specific activity of 1.2×10^5 dpm H^3 per mg. The average specific activity of the hemoglobin peptides prepared from the ribosomes isolated from the same cells may be calculated if one can make an estimate of the weight fraction of ribosomal particles which is present as growing hemoglobin chains. If we make the extreme assumption that the purified ribosomes are pure hemoglobin, then the specific activity of the average peptide in the ribosome (Fig. 6) is 7×10^6 dpm H^3 per mg, or 60 times that of the average peptide in soluble hemoglobin. If we take as more likely the previously reported[8] estimate that growing peptide chains amount to approximately 0.1 per cent of the ribosomal mass, then the ratio of peptide specific activity in ribosomes to that in soluble hemoglobin becomes 60,000. This latter assumption also leads to the conclusion that the specific activity of the H^3-leucine in the ribosomal hemoglobin peptides is approximately 1.5 times that of the H^3-leucine used for the incubation, a result obviously too high but within the combined errors of experiment and assumptions. These results indicate conclusively that the tryptic peptide fragments of hemoglobin isolated from ribosomal particles are precursors of finished hemoglobin molecules and do not represent contamination of the ribosomal particles by completed molecules from the soluble pool.

The results given in Figures 4, 5, 6, and 7 are in agreement with the model shown in Figure 1 in all particulars. The predicted gradient of radioactivity in the peptides of soluble hemoglobin, becoming less pronounced with time, and the inverse gradient in the peptides of the ribosomes, becoming more pronounced with time, are both found. The development of a gradient of radioactivity in the ribosomal hemoglobin peptides at long times is perhaps the most direct proof to date that ribosomes contain incomplete growing peptides. A gradient might be expected at short times in the ribosomes due to contamination from nonuniformly labeled molecules produced elsewhere, but it is hard to see how a gradient could develop with increasing time except by means of the mechanism shown in Figure 1.

It must be stressed that the data given thus far do not constitute proof of the correctness of the particular model in Figure 1, although they are in complete agreement with it. This is the case because all the data presented above are limited to time measurements. To test the model completely, the sequence of amino acid residues along the peptide chain must also be known. Specifically it must be

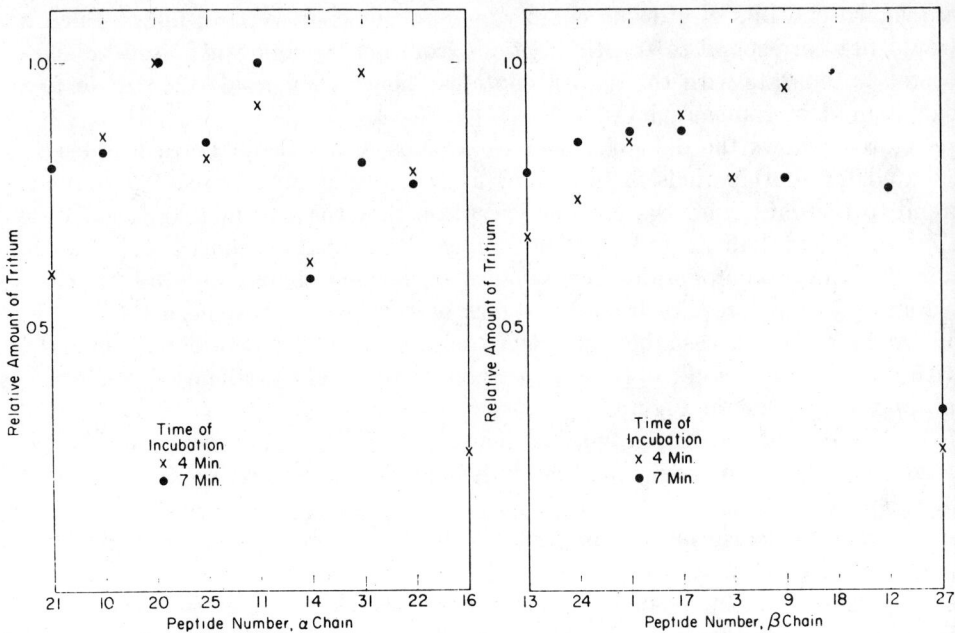

Fig. 6.—Distribution of H³-leucine among tryptic peptides of ribosomes after short incubations.

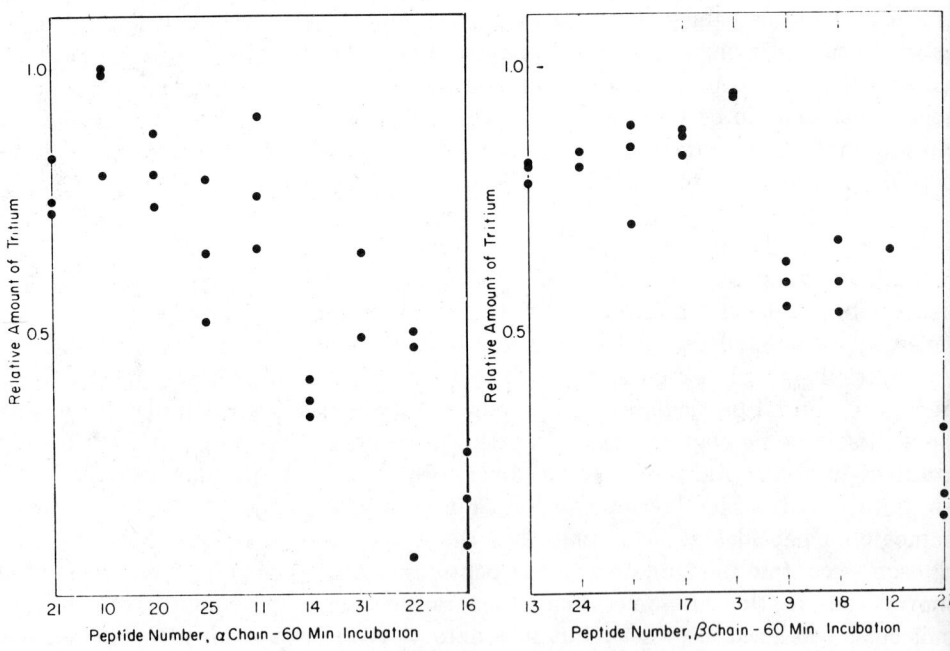

Fig. 7.—Distribution of H³-leucine among tryptic peptides of ribosomes after long incubations

shown that peptides at the end of the time sequence, e.g., 21 and 31, are nearest to the end of the polypeptide chain, and furthermore that peptides which are neighbors in the time sequence of labeling with H³ leucine, e.g., 14 and 31, are also neighbors on the polypeptide chain.

An attempt has been made to identify the tryptic peptide nearest to the free carboxyl end of each hemoglobin chain. Guidotti has reported that by using a mixture of carboxypeptidases A and B he was able to remove sequentially approximately a dozen amino acids from the carboxyl end of the α- and β-chains of human globin.[10] In the case of human globin a leucine residue was one of those removed from each chain. It was therefore reasonable to assume that a similar operation on rabbit globin could remove a leucine residue from the peptide nearest to the carboxyl end.

Uniformly labeled C^{14} leucine globin was incubated according to the conditions of Guidotti with carboxypeptidase A (Worthington, DFP treated) and carboxypeptidase B (kindly donated by Dr. Martha Ludwig). After digestion it was heated at 100° for 15 min to denature the enzyme, and then mixed with undigested uniformly labeled H^3 leucine globin which had received the same treatment, except that no carboxypeptidase had been added to it. The mixture of C^{14} leucine globin and H^3 leucine globin was then digested with trypsin and the peptides were separated and counted as described above. If the carboxypeptidase had no effect on any recognizable leucine-containing peptide, then we would expect to obtain a constant ratio of H^3 to C^{14} in each resulting tryptic peptide e.g., see Table 3, column (a)). If, however, C^{14}-leucine were removed from a peptide by the action of carboxypeptidase, we would expect a decrease in C^{14} leucine content in that peptide with a corresponding increase in the ratio of H^3 to C^{14}. The experiment was also done in reverse, with the H^3 leucine globin being digested with carboxypeptidase. As a final control both C^{14} leucine globin and H^3 leucine globin were carried through all operations except that no carboxypeptidase was added to either. The results are indicated in Table 4.

TABLE 4

Relative Tritium Content of Tryptic Peptides following Carboxypeptidase Action

Peptide number, α-chain	(a) C^{14} Globin Digested with Carboxypeptidases			(b) H^3 Globin Digested with Carboxypeptidases		(c) No Digestion with Carboxypeptidases	
	Digest 1		Digest 2				
10	1.0	1.0	1.2	1.0	0.8	1.3	1.1
11	1.0	1.0	1.2	...	1.0	1.1	1.3
14	1.0	1.0	1.1	1.1	1.0	1.0	1.1
16	11.0	20.0	2.7	0.07	0.0	0.3	0.4
20	1.0	1.0	0.7	1.0	1.0	1.0	1.0
21	1.0	1.0	1.0	1.0	1.1	1.0	1.0
22	0.9	0.9	0.9	1.2	1.6	1.0	1.1
25	1.0	1.0	1.2	1.1	1.1	1.0	1.1
31	1.0	1.1	1.0	1.2	1.0	1.2	1.0
Peptide number, β-chain							
1	1.0	1.0	1.0	0.9	1.0	0.9	1.0
3	0.9	1.0	1.0	0.9	1.0	1.0	1.0
9	0.9	0.9	0.9	0.9	1.0	0.9	0.9
12	1.4	0.8	1.2	1.1	1.1	1.0	0.9
13	1.2	0.8	1.0	0.9	0.9	0.9	1.2
17	1.0	1.0	1.0	1.0	1.0	1.0	1.0
18	0.8	0.9	1.0	0.9	0.9	2.0	0.9
24	1.0	1.0	1.1	0.9	0.9	1.0	1.0
27	1.6	1.4	0.9	0.8	0.8	0.8	0.9
28	30.0	8.0	13.0	0.07	0.01	0.8	0.9

The only α-chain peptide which shows significant deviation from constant ratio in the expected direction is peptide 16. Unfortunately, the control experiment shows some ratio deviation in peptide 16 (column (c)) but not enough to upset the conclusion that peptide 16 is near the carboxyl end of the α-chain. The ratio variation of peptide 16 in the control experiment (Table 4, column (c)) may be due to the fact that the proteins were heated at 100° for 15 min to inactivate carboxypeptidase, whereas in previous experiments (Table 3, column (a)) this was not done.

In the β-chain none of the major yield peptides showed a significant ratio change, but peptide 28, which was previously sometimes present in minor yield, was present in good yield and clearly showed the behavior expected of a peptide near the carboxyl end of the β-chain. On re-examination of the data from 4- and 7-min H^3-leucine incubations, four clear cases were found where peptide 28 had been present but in low yield. The average tritium content of peptide 28 in these four runs was found to be 1.06 ± 0.27 times the tritium content of peptide 27. Although the average yield of peptide 28 was only 0.18 ± 0.04 times the yield of peptide 27, it is tempting to conclude that peptide 28 is closely related to peptide 27 in the time sequence of labeling with H^3 leucine.

It would thus appear that in both the α- and β-chains those leucine-containing peptides which are the first to be labeled with H^3-leucine in the soluble hemoglobin are nearest to the free carboxyl end of the chain. According to the model shown in Figure 1, this implies that chain growth terminates at or near the free carboxyl end of the molecule.

Discussion.—The NH_2-terminal amino acid of both the α- and β-chains of rabbit hemoglobin is valine.[11] Attempts have been reported to find the rate of short time radioactive labeling of the NH_2-terminal valine relative to the average of all other valines in the hemoglobin molecule. Using whole rabbit reticulocytes, Loftfield[2] reported results indicating that the NH_2-terminal valine is labeled last. On the other hand, Bishop et al.,[12] using a cell-free system from rabbit reticulocytes, reported results indicating that the NH_2-terminal valine is labeled first. Reports on other protein-synthesizing systems are equally conflicting. Thus the work of Yoshida and Tobita[13] on bacterial amylase indicates that synthesis proceeds from the amino-terminal toward the carboxyl-terminal end. Complications are present in the interpretation of their work because of the very long times of incubation involved and the presence of various protein precursor pools. Shimura et al.[14] using the fibroin synthesizing gland of the silk worm obtained results indicating that the NH_2-terminal glycine is added last.

Muir et al.[15] reported finding uniform labeling in hemoglobin labeled *in vivo*. This is to be expected from the results reported in this paper. Thus Figure 5 shows that labeling is uniform within 20 per cent after 60 min of incubation at 15°, corresponding to 15 min of incubation at 37°. Kruh et al.[16] have reported nonuniform labeling in hemoglobin after very long *in vivo* experiments. This result is not consistent with the data reported in this paper and possibly represents phenomena different from the original synthesis.

A different approach was used by Loftfield and Eigner who reported kinetics of amino acid incorporation into ferritin[17] and hemoglobin[2] after short times of incubation. Their data indicate that for the first few minutes of labeling the specific activity of newly formed protein increases as the square of the time, becoming linear only after several minutes. From these data they concluded that a scheme essentially the same as that of Figure 1 is indicated. However, this result cannot distinguish between a random and a sequential process of attaching amino acids to the template.

It is perhaps worth noting that if the model shown in Figure 1 is finally proved to be correct, then the experimental technique described in this paper could be useful for structure determination. Thus, it should be possible to determine the spatial sequence of tryptic peptides in proteins of unknown structure by determining the time order of labeling.

It has previously been reported[8] that to account for the production of new hemoglobin in living rabbit reticulocytes, each ribosomal particle must, on the average, make one polypeptide chain of hemoglobin in 1.5 min. That result was obtained by dividing the total rate of hemoglobin synthesis by the total number of ribosomal particles. From Figure 5 it may be seen that the last peptide on each chain to be labeled receives its label at some time between 4 and 7 min of incubation at 15°. Since the rate of labeling was found to be approximately $1/4$ as great at 15° as at 37° (Table 1), this implies that the total time of assembly of each polypeptide chain at 37° is approximately 1.5 min. The agreement between the average rate of synthesis, 1.5 min, and the individual rate of chain synthesis, also 1.5 min, strongly implies that most of the ribosomal particles present in rabbit reticulocytes are, in fact, producing hemoglobin. Since there are approximately 150 amino acid residues in each chain, the average rate of growth is close to two amino acids added per second.

In all of the above discussion a number of possible complications have been ignored because of insufficient data to evaluate their effects. Thus we have ignored the effects of both delay time and dilution of specific activity suffered by labeled leucine during its passage into the cells and subsequent reactions prior to actual peptide bond formation. The fact that we have not needed to invoke these processes to explain the results suggests that the effects are small. Likewise we have ignored the possible existence of hemoglobin in transitory forms between completed polypeptide chains and final soluble hemoglobins. We might imagine, for example, that α-chains and β-chains are produced on separate ribosomal particles and that furthermore single α- and β-chains are insoluble and stay on the ribosomes, while α_2 and β_2 dimers are soluble. This leads to the notion of a small pool of completed chain attached to the ribosome, which would change slightly the results expected in Figure 1, and would lead to a less steep predicted slope in Figure 7.

The figures for this paper have been drawn with uniform spacing between adjacent peptides. This, of course, does not imply that the labelled amino acids are uniformly spaced along the actual polypeptide chain. When the actual sequence of the peptide chains is determined we shall be in a position to plot the relative amount of labeling in each amino acid against its position in the chain. Only when that is done will it be worthwhile to consider the detailed shape of the curves for evidence concerning uniformity of growth rate along the polypeptide chain.

In summary it may be concluded that the growth of the peptide chains of hemoglobin is not a random process but a steady sequential addition of amino acids to growing chains at the rate of approximately two amino acids per second. The number of initiation points per chain is, at most, very small and most likely only one.

The chain growth terminates near or at the free carboxyl end. Taken together, these conclusions indicate that chain growth proceeds steadily from the free amino end toward the free carboxyl end in rabbit hemoglobin.

The author wishes to acknowledge the expert technical assistance of Miss Judith Karossa and Mrs. Ruth Langridge in the early and later parts, respectively, of this investigation.

* This work was supported by a grant from the National Institutes of Health.
[1] Steinberg, D., M. Vaughan, and G. B. Anfinsen, *Science*, **124**, 389 (1956).
[2] Loftfield, R., *Proc. 4th Intern. Congr. Biochem.*, **8**, 222 (1960).
[3] Borsook, H., *Proc. 3rd Intern. Congr. Biochem.*, 92 (1956).
[4] Dibble, W. E., and H. M. Dintzis, *Biochim. et Biophys. Acta*, **37**, 152 (1960).
[5] Kruh, J., and H. Borsook, *J. Biol. Chem.*, **220**, 905 (1956).
[6] Borsook, H., E. H. Fischer, and G. Keighley, *J. Biol. Chem.*, **229**, 1059 (1957).
[7] Ingram, V. M., *Biochim. et Biophys. Acta*, **28**, 539 (1958).
[8] Dintzis, H., H. Borsook, and J. Vinograd, in *Microsomal Particles and Protein Synthesis*, ed. R. B. Roberts (New York: Pergamon Press, 1958), p. 95.
[9] Wilson, S., and D. B. Smith, *Can. J. Biochem. and Physiol.*, **37**, 405 (1959).
[10] Guidotti, G., *Biochim. et Biophys. Acta.* **42**, 177 (1960).
[11] Osawa, H., and K. Satake, *J. Biochem. (Tokyo)*, **42**, 905 (1956).
[12] Bishop, J., J. Leahy, and R. Schweet, these PROCEEDINGS, **46**, 1030 (1960).
[13] Yoshida, A., and T. Tobita, *Biochim. et Biophys. Acta*, **37**, 513 (1960).
[14] Shimura, K., H. Fukai, J. Sato, and R. Saeki, *J. Biochem. (Tokyo)*, **43**, 101 (1956).
[15] Muir, H., A. Neuberger, and J. Perrone, *Biochem. J.*, **52**, 87 (1952).
[16] Kruh, J., J. Dreyfus, and G. Schapira, *J. Biol. Chem.*, **235**, 1075 (1960).
[17] Loftfield, R. B., and E. A. Eigner, *J. Biol. Chem.*, **231**, 925 (1958).

INITIATION OF E. COLI PROTEINS*

By Mario R. Capecchi

THE BIOLOGICAL LABORATORIES, HARVARD UNIVERSITY

Communicated by J. D. Watson, April 18, 1966

Recent experiments and theoretical arguments suggest that formylmethionyl sRNA is employed as an initiator of protein synthesis.[1-6] For example, R17 and f2 bacteriophage coat protein was found to have the amino terminal sequence

formyl met ala ser $AspNH_2$ phe thr...

when synthesized in a cell-free extract programed with viral RNA.[1,2] The known amino-terminal sequence of R17 and f2 coat protein from intact phage (ala ser $AspNH_2$ phe thr...) must be generated by specific cleavage of the formylmethionyl residue from the nascent polypeptide chain. These studies also indicated that other phage proteins synthesized in the *in vitro* system were initiated with formylmethionine.[1] Thus we have an example of an amino acid incorporating system directed by a polycistronic messenger in which the different proteins are initiated by the same mechanism to the extent that the same initiator, formylmethionine, is used. These observations provided a basis for believing that formylmethionine could be the unique initiator of protein synthesis. A necessary consequence of having a unique initiator, as in the bacteriophage system, is the existence of an enzymatic apparatus for the removal of the formylmethionyl residue exposing other known amino-terminal amino acids. The finding by Clark and Marcker[3] and by Nakamoto and Kolakofsky[4] that proteins synthesized under the direction of synthetic messengers such as poly UAG and poly UG are initiated with formylmethionine also supports the unique role of that sRNA species as the initiator of protein synthesis.

One would like to extend the above results to see whether *E. coli* proteins are initiated in a similar manner. To answer this question we have studied *in vitro* protein synthesis programed with endogenous *E. coli* messenger RNA. A number of models can be envisioned which employ formylmethionyl sRNA as the initiator of *E. coli* protein synthesis. For example, let us discuss two models. As stated above, the *in vitro* amino terminal sequence of R17 coat protein, which may be fortuitous, has the tantalizing sequence: formyl met ala ser AspNH$_2$.... When this sequence is coupled with the composition of amino-terminal amino acids of *E. coli* proteins found by Waller:[7]

45% met
30% ala
15% ser,

a simple model emerges. All *E. coli* proteins could be initiated with the sequence F met ala ser.... After synthesis of the protein, an enzyme specifically cleaves after the formyl group, or after methionine, or after alanine, etc. The position of cleavage could be controlled by the protein's three-dimensional conformation. To obtain Waller's results, one need only stipulate a decreasing probability of cleavage further down the polypeptide chain. This model carries with it the implication that the initiation signal for protein synthesis is a longer sequence than the formylmethionyl sRNA codon.

In a second model, *E. coli* proteins could start with the sequences

F met met...
F met ala...
F met ser...,

and so on. After synthesis of the protein, an enzyme specifically removes just the formylmethionyl residue.

These two models and a number of variants can be distinguished by seeing which formylated dipeptides exist in a pool of newly synthesized *E. coli* proteins. Thus, if the first model is correct, the only formylated dipeptide which we should find in *E. coli* protein is F met ala. The second model predicts the existence of F met met, F met ala, F met ser, and so on. Further, the frequency of the above dipeptides should reflect Waller's distribution of amino-terminal amino acids.

Materials and Methods.—(a) *E. coli extracts:* The *E. coli* extracts were prepared by grinding frozen cells with 2.5 gm of alumina and 3.0 ml Tris-Mg buffer per gram wet weight of cells. The extract was clarified by two centrifugations at 15,000 rpm for 15 min each in a Servall SS2 centrifuge. The resulting supernatant was made 0.006 M in β-mercaptoethanol and dialyzed for 6 hr at 4°C against the following buffer: 0.01 M Tris pH 7.8, 0.01 M Mg acetate, 0.03 M KCl, and 0.006 M β-mercaptoethanol. After dialysis the extract was stored at −120°C until used.

(b) *Incorporation experiments:* For the amino acid incorporation experiments the reaction mixture contained per ml: 0.25 ml *E. coli* extract, 0.0075 mmole Mg, 0.06 mmole NH$_4$Cl, 0.003 mmole ATP, 0.0002 mmole GTP, 0.005 mmole PEP, 20 μg pyruvate kinase, 4 × 10^{-5} mmoles of each amino acid except those used as C^{14} or H^3 label, 0.01 mmole glutathione, and 0.05 mmole Tris, pH 7.8.

(c) *H^3-formyltetrahydrofolate (H^3-FTHFA):* H^3-formate (sp. act. 1.92 c/mmole) was activated using formyltetrahydrofolate synthetase from *Clostridium cylindrosporum* (sp. act. 14,000 units/mg).[8] The crystalline enzyme preparation was kindly given by J. C. Rabinowitz. The reaction was stopped by adding 1/6 vol of 1 N HCl.

(d) *Preparation of N-formylated peptide standards:* L-methionylmethionine (Cyclo Chem), L-methionylserine (Mann Research), and L-methionylalanine (Mann Research) were formylated in the presence of acetic anhydride.[9] The N-formylated dipeptides were separated from unreacted peptides by eluting the product from a Dowex-50 column (hydrogen form) with H$_2$O.

(e) *Pronase digestion of E. coli proteins synthesized in a cell-free system:* A 1-ml reaction mixture was incubated with H^3-FTHFA and a C^{14}-amino acid for 12 min at 36°C. The reaction mixture was rapidly chilled to 0°C, and 25 μg of DNase and RNase were added. After 5 min, the reaction mixture was made 0.02 M in EDTA to allow autodigestion of the ribosomes. The protein was precipitated and washed four times in cold 6% TCA. To destroy any aminoacyl sRNA not already digested by the RNase, the reaction mixture was treated with base (pH 12.0 for 10 min at 0°C), followed by TCA precipitation. After removal of residual TCA with ether-ethanol and ether washes, the dried protein was suspended in 200 μl of 0.05 M NH_4HCO_3 and 0.001 M thiodiglycol, pH 7.9. At this point in some experiments, 50 μg of N-formylmethionyl-methionine, N-formylmethionylalanine, and N-formylmethionylserine were added to the protein. Pronase (0.3 mg) was added and hydrolysis was carried out at 0°C for 12 hr. The pronase digest was then put on a 6-ml Dowex-50 column (hydrogen form) in order to separate formylated amino acids, di- and tripeptides, from unblocked amino acids, di- and tripeptides. The formylated peptides were eluted with mild HCl (pH 3.5) and analyzed by paper electrophoresis and chromatography.

(f) *Chromatography and electrophoresis:* Descending chromatography was done with a pyridine, isobutanol, and H_2O (35:35:30) solvent at 20°C for 24 hr. High-voltage electrophoresis was done on a cooled plate (10°C) at 28 v/cm for 8 hr. The electrophoresis buffer contained per liter 25 ml of glacial acetic acid and 30 ml of pyridine.

(g) *Chemicals:* d,l tetrahydrofolate (sealed under nitrogen), Sigma; pronase, Calbiochem; pancreatic RNase, 5× crystallized, Sigma; DNase, Worthington; H^3-formate (1.92 c/mmole), Tracer Lab; and C^{14}-alanine (123 mc/mmole), C^{14}-serine (115 mc/mmole), and C^{14}-methionine (198 mc/mmole), New England Nuclear.

Results.—*Amino acid incorporation mediated by endogenous E. coli messenger RNA:* Incubation of an *E. coli* extract with H^3-formylmethionyl sRNA or H^3-formyl-tetrahydrofolic acid (H^3-FTHFA) results in the incorporation of H^3-formylmethionine into alkali-resistant, TCA-precipitable product. Addition of H^3-formylmethionyl sRNA or H^3-FTHFA to the reaction mixture are equivalent experiments, since (1) it was previously demonstrated that FTHFA is the formyl donor in the

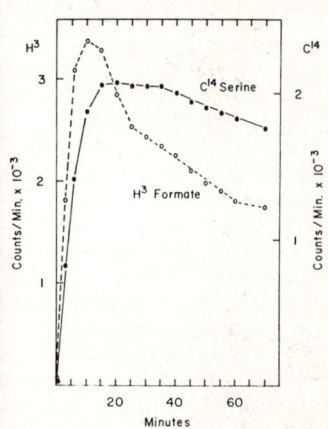

FIG. 1.—Kinetics of *in vitro* incorporation of H^3-formate and C^{14}-serine into *E. coli* proteins. The 1.5-ml reaction mixture contained: 375 μl of *E. coli* extract, 6 × 10⁶ cpm of H^3-formyl THFA (250 cpm/$\mu\mu$mole), and 3.3 × 10⁵ cpm C^{14}-serine (11 cpm/$\mu\mu$mole). At the designated times, 50-μl aliquots were removed and precipitated with cold 7% TCA. The precipitate was incubated at pH 12 for 15 min at 0°C and reprecipitated with cold 7% TCA onto Millipore filters.

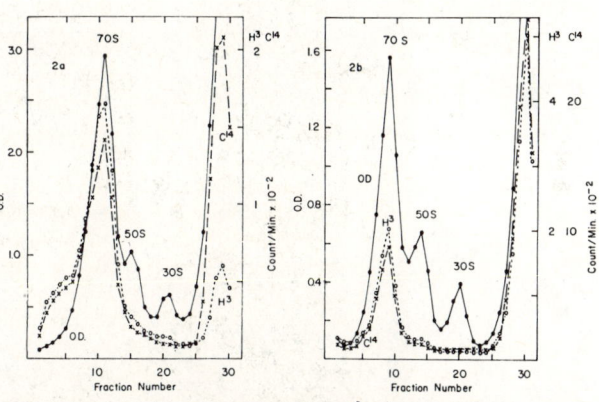

FIG. 2.—Sucrose gradient analysis of the total amino acid incorporation mixture after 15 min incubation with H^3-formyl THFA and C^{14}-alanine, employing *E. coli* extract (*a*) pretreated with DNase, and (*b*) not pretreated with DNase. The 1.5-ml reaction mixtures contained 375 μl of *E. coli* extract, 6 × 10⁶ cpm H^3-formyl THFA, and (*a*) 3.3 × 10⁵ cpm C^{14}-alanine (9.7 cpm/$\mu\mu$mole) and (*b*) 3.3 × 10⁵ cpm C^{14}-alanine (6.1 cpm/$\mu\mu$mole). After 15 min incubation, an aliquot of each reaction mixture was given a mild DNase and RNase treatment prior to layering 225 μl and 150 μl of the respective reaction mixture onto a 5-ml sucrose gradient. Processing of the gradients has been described previously.[1] Digestion with an excess of RNase (50 μg/ml) was used to destroy the aminoacyl sRNA in each gradient fraction.

synthesis of N-formylmethionyl sRNA, and (2) the only N-formylaminoacyl sRNA which can be detected in *E. coli* extracts is N-formylmethionyl sRNA.[1, 10] The magnesium ion dependence of H^3-formylmethionine incorporation is identical to endogenous protein synthesis with a maximum at 7.5 mM Mg.

The high level of formylmethionine incorporation in these *E. coli* extracts is consistent with formylmethionine being the initiator of all *E. coli* protein. At optimal conditions, approximately 11.5 serine and 12 alanine residues per ribosome are incorporated into *E. coli* protein. This corresponds to incorporating approximately 150 amino acids per ribosome. Under the same conditions, we observe incorporation of one formylmethionyl residue per ribosome.

The subsequent enzymatic removal of the formyl group from the newly synthesized polypeptide chain can be followed by examining the kinetics of H^3-formate and

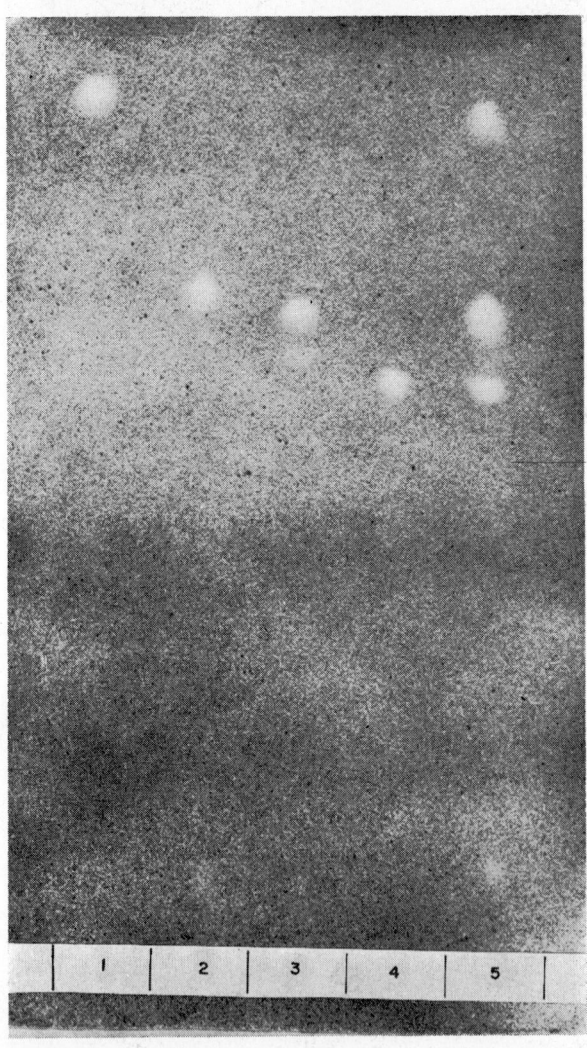

FIG. 3.—An electropherogram of a series of N-formylated dipeptides and formylmethionine. The standards are: *1*, F met; *2*, F met ser; *3*, F met ala; *4*, F met met; and *5*, all four. These standards were detected by the platinic iodide reagent.[11]

C^{14}-serine incorporation in these *E. coli* extracts (see Fig. 1). After 10 min of incubation at 36°C, there is a marked preferential release of H^3-formate from *E. coli* proteins. Whether the H^3-formate is released as free formate or formylmethionine has not yet been directly determined.

For these experiments the *E. coli* extract was not pretreated with DNase, because treatment with DNase (Worthington, electrophoretically pure) reduces the total level of amino acid incorporation by a factor of five. In addition, the level of formylmethionine incorporation is reduced approximately by a factor of 15. This suggests that in a DNase-treated extract, initiation of new polypeptide chains is preferentially reduced and the majority of the synthesis results from completion of pre-existing polypeptide chains. Consistent with this interpretation are the results shown in Figure 2. Figures 2a and 2b are sucrose gradients of the total reaction mixtures after 15 min of incubation with H^3-FTHFA and C^{14}-alanine employing (a) DNase-treated *E. coli* extract, and (b) *E. coli* extract not pretreated with DNase. If one compares the H^3/C^{14} ratio of the released protein in Figures 2a and 2b, it is clear that the majority of proteins released from the ribosomes in a DNase-treated extract do not contain labeled formylmethionine; that is, the majority of the released proteins were not initiated during the incubation period. However, when we employ an *E. coli* extract not pretreated with DNase, we observe that a large proportion of the released proteins were initiated during the incubation period.

Pronase digest of in vitro synthesized E. coli protein: To test which N-formylated amino acid occur in *in vitro* synthesized *E. coli* proteins, and to determine which amino acids are found adjacent to the formylated amino acid, a series of incubations with different pairs of C^{14} and H^3 labels were done. The reaction mixtures were chilled to 0°C after 12 min of incubation and prepared for pronase digestion as described in *Materials and Methods*. The conditions of pronase digestion were empirically selected so as not to drive the hydrolysis to completion. Since we are working with a pool of *E. coli* proteins, separation of the different radioactive-labeled di-, tri-, and tetrapeptides would be difficult. However, we are only interested in the formylated amino acids and the formylated di- and tripeptides. These can be readily separated from unblocked amino acids and unblocked di- and tri-peptides by passing the pronase digest through a Dowex-50 column (hydrogen form). For example, F met, F ala, and F met ala can be nearly quantitatively eluted from the Dowex column (i.e., greater than 90%) under conditions where only 0.5 per cent of the aspartic acid put on the column comes off. The radioactive

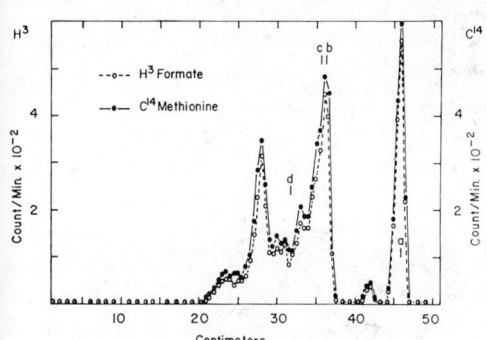

FIG. 4.—Electrophoretic analysis of the pronase digest of *in vitro* synthesized *E. coli* protein labeled with H^3-formate and C^{14}-methionine. The 1.0-ml reaction mixture contained 6×10^6 cpm of H^3-formyl THFA (51 cpm/μμmole) and 6.6×10^5 cpm of C^{14}-methionine (45 cpm/μμmole). Included with each electropherogram (Figs. 4–6) was a set of nonradioactive standards which were detected after radioactive analysis by the platinic iodide reagent.[11] The mid-positions of these standards are designated by the letters a–d: a, formylmethionine; b, formylmethionylserine; c, formylmethionylalanine; and d, formylmethionylmethionine. The origin was at 0 cm.

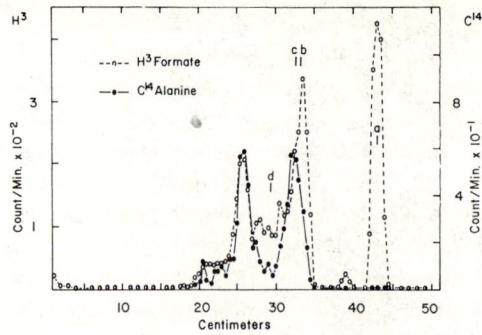

FIG. 5.—Electrophoretic analysis of the pronase digest of *in vitro* synthesized *E. coli* protein labeled with H^3-formate and C^{14}-alanine. The 1.0-ml reaction mixture contained 6×10^6 cpm of H^3-formyl THFA (51 cpm/$\mu\mu$mole) and 2×10^6 cpm of C^{14}-alanine (40 cpm/$\mu\mu$mole).

labeled peptides eluted from the column were analyzed by paper electrophoresis and paper chromatography.

Figure 3 shows an electropherogram of a number of nonradioactive N-formyl peptide standards. The features to note are that this procedure does not resolve F met ser from F met ala (although F met ser does move a little further than F met ala), and that F met ser and F met ala are easily separated from F met met. On this electropherogram, formylalanine and formylserine would have traveled just ahead of, but readily resolved from, formylmethionine.

Figure 4 shows an electropherogram of the pronase hydrolyzate of *in vitro* synthesized *E. coli* protein labeled with H^3-formate and C^{14}-methionine. With each electropherogram (Figs. 4–6) was included a set of nonradioactive formyl peptide standards, *a–d*, which could be detected after radioactive analysis. The presence or absence of these cold markers does not affect the radioactive pattern observed. The standards whose mid-positions are designated by the letters *a* through *d* are: F met, F met ser, F met ala, and F met met, respectively.

The fastest-moving radioactive peak has been identified both by coelectrophoresis and cochromatography with the standard as F met. Throughout these experiments no evidence was found supporting the existence of any other formylated amino acid. For example, if formylalanine and formylserine were incorporated into *E. coli* protein, then we would have picked them up in the electropherograms shown in Figures 5 and 6 in the 44–48-cm region.

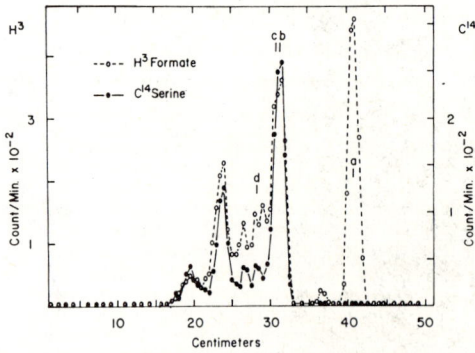

FIG. 6.—Electrophoretic analysis of the pronase digest of *in vitro* synthesized *E. coli* protein labeled with H^3-formate and C^{14}-serine. The 1.0-ml reaction mixture contained 6×10^6 cpm of H^3-formyl THFA and 6.6×10^5 cpm of C^{14}-serine (42 cpm/$\mu\mu$mole).

It is observed in Figure 4 that the ratio of C^{14}-methionine to H^3-formate is fairly constant all along the electropherogram. If there existed detectable amounts of F met met, we would expect the C^{14} to H^3 ratio to double at position *d*. One could argue that F met met is particularly susceptible to pronase and therefore the inability to detect F met met merely indicates that we have destroyed this dipeptide. However, if one adds to the reaction mixture cold F met met prior to pronase digestion (as in the experiment shown), one still cannot detect radioactive F met met in spite of the fact that a large proportion of the cold F met met was recovered after pronase digestion and elution from the Dowex column.

The skewed radioactive peak (Fig. 4), which coelectrophoresed with the standards F met ser and F met ala, was shown to contain both alanine and serine by analysis of the *E. coli* proteins labeled with H^3-formate and C^{14}-alanine (Fig. 5) and H^3-formate and C^{14}-serine (Fig. 6). To rule out the possibility that the skewed peak was a tripeptide such as F met ala ser, this peak was eluted from a number of different electropherograms in which the *E. coli* protein had been labeled with different pairs of radioactive compounds. The eluted peptides were then chromatographed in a system which readily separates the dipeptide F met ala from F met ser. Four such experiments are shown in Figure 7. Each paper chromatogram also included nonradioactive standards (*a*) F met ser and (*b*) F met ala, which were detected after radioactive analysis. It is observed that all four labeling combinations (*I–IV*) are consistent with the skewed peak in the electropherograms (Figs. 4–6) containing two formyl dipeptides, rather than a tripeptide, these dipeptides being F met ala and F met ser.

Discussion.—These studies with *in vitro* protein synthesis directed by endogenous *E. coli* messenger RNA have brought out several points. First, the high level of formylmethionine incorporation into *E. coli* protein is consistent with formylmethionyl sRNA being the initiator of most, if not all, *E. coli* proteins; that is, in a cell-free extract not pretreated with DNase, we are incorporating into *E. coli* protein approximately one formylmethionyl residue per 150 amino acids. Why

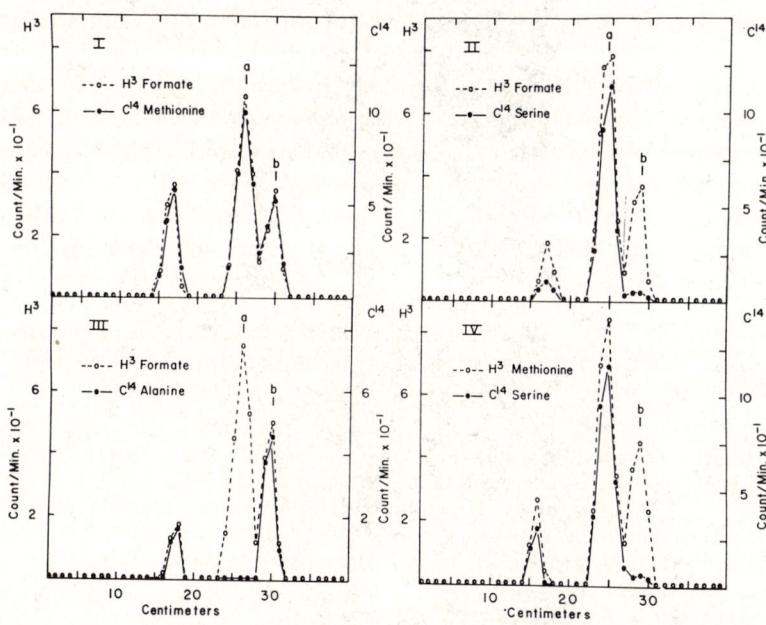

Fig. 7.—Four separate paper chromatographic analyses of the skewed radioactive peak which coelectrophoresed with formylmethionylserine and formylmethionylalanine (see Figs. 4–6). Included with each chromatogram were the nonradioactive standards (*a*) formylmethionylserine, and (*b*) formylmethionylalanine. These standards were detected after radioactive analysis by the chlorination technique.[12] To eliminate problems associated with partial autooxidation, the samples and standards were oxidized with performic acid prior to chromatographic analysis. The combinations of H^3 and C^{14} labels employed in each experiment (*I–IV*) are given with each figure.

DNase drastically reduces the level of amino acid incorporation and, in particular, the level of formylmethionine incorporation, is most simply explained by assuming that in the presence of DNA, new messenger RNA is synthesized which directs the synthesis of new, complete polypeptide chains.

Second, the kinetics of H^3-formate and C^{14}-serine incorporation into *E. coli* protein shows a marked preferential release of H^3-formate from the *E. coli* protein after 10 min of incubation (see Fig. 1). The preferential release of H^3-formate is an indication that the enzymatic hydrolysis of the formyl group from the newly synthesized polypeptide chain can be measured and suggests possible assays for the cleavage enzyme or enzymes.

Third, the position adjacent to formylmethionine on the polypeptide chain is not unique but can be alanine or serine or, to a smaller extent, other amino acids. Note, for example, the small peak which moves behind formylmethionine in Figures 4–6. Our finding that the second position is not unique rules out the first model discussed in the introduction.

Fourth, no F met met could be detected. This implies that the second model which included one cleavage enzyme specific for the hydrolysis of the formyl-methionyl residue from the nascent polypeptide chain is also ruled out. To account for Waller's observation that 45 per cent of *E. coli* proteins have methionine as the amino terminal amino acid, our nondetection of F met met predicts the existence of at least one cleavage enzyme whose specificity is the removal of the formyl group from the polypeptide chain.

E. coli proteins are known to have alanine, serine, and other amino acids as the amino-terminal residue; therefore, the methionyl residue frequently must be also removed. This could be accomplished by the same enzyme that removes the formyl group or a second enzyme specific for formylmethionine or methionine. The question then arises whether cleavage proceeds beyond methionine, presumably controlled by the polypeptides' three-dimensional conformation. Our finding that the amino acid adjacent to formylmethionine is not unique suggests that hydrolysis beyond the methionyl residue need not take place but cannot say whether it does or does not take place.

The author is grateful to Prof. J. D. Watson for his support while pursuing this study, and for his and Prof. W. Gilbert's valuable criticism during the preparation of this manuscript. It again gives the author great pleasure to acknowledge the expert technical assistance of Mrs. Nancy Capecchi.

* This work was supported by grants from the National Institutes of Health RG-9541.
[1] Adams, J. M., and M. R. Capecchi, these PROCEEDINGS, **55**, 147 (1966).
[2] Webster, W., D. Engelhardt, and N. Zinder, these PROCEEDINGS, **55**, 155 (1966).
[3] Clark, B. F. C., and K. A. Marcker, *J. Mol. Biol.*, in press.
[4] Nakamoto, T., and D. Kolakofsky, these PROCEEDINGS, **55**, 606 (1966).
[5] Noll, H., *Science*, **151**, 1241 (1966).
[6] Sundararajan, T. A., and R. E. Thach, *J. Mol. Biol.*, in press.
[7] Waller, J. P., *J. Mol. Biol.*, **7**, 483 (1963).
[8] Rabinowitz, J. C., and W. E. Pricer, Jr., *J. Biol. Chem.*, **237**, 2898 (1962).
[9] Sheehan, J. C., and D. M. Yang, *J. Am. Chem. Soc.*, **80**, 1154 (1958).
[10] Marcker, K., *J. Mol. Biol.*, **14**, 63 (1966).
[11] Toennies, G., and J. J. Kolb, *Anal. Chem.*, **23**, 823 (1951).
[12] Mazur, R. H., B. W. Ellis, and P. S. Cammarata, *J. Biol. Chem.*, **237**, 1619 (1962).

Mechanism and Control of Initiation in the Translation of R17 RNA

MARKUS NOLL & HANS NOLL
Department of Biological Sciences, Northwestern University, Evanston, Illinois 60201

A general model of initiation of protein synthesis based on experiments with R17 RNA as messenger proposes a two-stage mechanism, the first stage requiring dissociation into subunits and initiation factors, the second involving direct attachment of ribosomes.

IN prokaryotes initiation of protein synthesis is thought to require three initiation factors (IF-1, IF-2 and IF-3) and the formation of a 30S subunit·mRNA·fMet-tRNA complex that is converted to a 70S initiation complex by the addition of a 50S subunit. However, an obligatory dissociation step[1] is not proven[2] and the role of the initiation factors is unclear[3]. Initiation can also occur without initiation factors by direct attachment of messenger to 70S ribosomes[4]. We propose below a model for the mechanism and control of initiation and translation of polycistronic messengers based on experiments with R17 RNA.

Two Mechanisms of Initiation

Two mechanisms of initiation are postulated: (i) dissociation of the ribosome into subunits and participation of initiation factors IF-2 and IF-3, and (ii) direct attachment of 70S ribosomes to the messenger without initiation factors (Fig. 1). Which of the two mechanisms applies is governed by two rules. (a) Ribosomes will attach with low efficiency and without dissociation to any open single stranded stretch of mRNA and slide along it until they hit an AUG codon and are fixed by the binding of fMet-tRNA. Initiation at other triplets by binding of the cognate aminoacyl-tRNA leads to relatively unstable complexes[5] that cannot compete with AUG-dependent starts. Direct attachment of 70S ribosomes to mRNA does not require initiation factors and is the general mechanism encountered in initiation with synthetic messengers. (b) Ribosomes attach preferentially to AUG starting triplets at the end of a loop. Recognition and opening of the loop require IF-3, the action of which resembles that of the σ factor. Dissociation of 70S ribosomes is necessary because the loop might not fit into the cleft between the subunits.

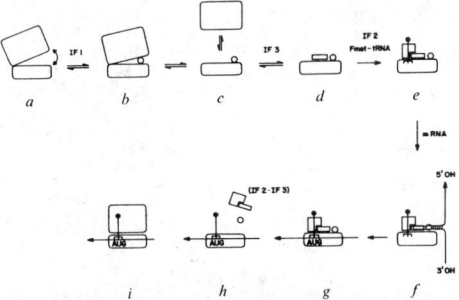

Fig. 2 Detailed model for mechanism of factor-dependent initiation at coat protein initiation site of R17 RNA.

Application of these rules to the translation of R17 RNA is illustrated in Fig. 1. In the native messenger (Fig. 1a) the secondary structure conceals the initiation sites of the synthetase (P) and A-protein (A) as well as all internal AUG triplets[6]. IF-3-dependent translation starts with the attachment of a 30S subunit to the coat protein initiation site C. Ribosome movement successively opens the synthetase and A-protein sites, neither of which is recognized by IF-3 as they are not at the end of a loop. Initiation at these sites is less efficient and occurs by direct binding of 70S ribosomes without initiation factors.

Mechanism of Factor-dependent Initiation

Our model for the mechanism of factor-dependent initiation is shown in Fig. 2. At the Mg concentration* optimal for factor-dependent initiation (3 mM), vacant 70S ribosomes[7,8] are mostly in the associated state, although small changes in the binding energy will produce relatively large shifts in the position of the equilibrium. This labile state may be visualized as a dynamic opening and closing of the subunits as if joined by a hinge (a). Binding of IF-1 to the 30S subunit of the 70S ribosomes (b) enhances their rate of thermal dissociation (c). Binding of IF-3 to 30S subunits (d) shifts the equilibrium toward dissociation. All steps so far are reversible and the equilibrium favours the vacant ribosome couples at 5 mM Mg. Subsequent attachment of the complex [IF-2·fMet-tRNA] to

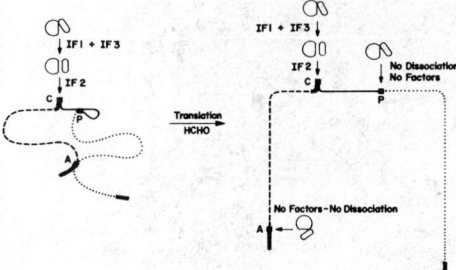

Fig. 1 Overall scheme for mechanism and control of initiation in the translation of RNA from the R17 family.

* In all these experiments the monovalent ion is NH_4^+ at a concentration of 50 mM.

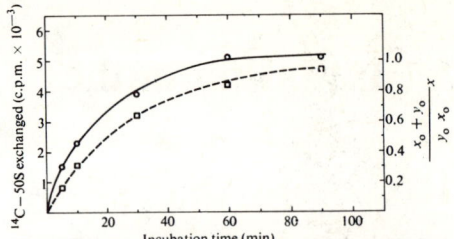

Fig. 3 Rate of exchange of subunits by pure vacant 70S ribosomes at 37° C and 5 mM Mg in the presence (——) and absence (- - -) of crude initiation factors. The incubation mixtures did not contain R17 RNA or fMet-tRNA. The units on the right hand ordinate indicate the extent of exchange (100% exchange = 1.0) obtained from the solution of the differential equation for exchange represented by the theoretical curve (- - -) showing the best fit. Each point was obtained from analysis of a sucrose gradient. ○—○, $k_1 \approx 1.1 \times 10^{-3}$ S^{-1} (5.1 mM Mg^{2+}); □- - -□, $k_1 \approx 3.5 \times 10^{-4}$ S^{-1} (5.1 mM Mg^{2+}).

the [30S·IF-1·IF-3] complex (e) strengthens the binding of IF-3 and hence shifts the equilibrium greatly toward the formation of this complex that now acts as a trap for the messenger. Interaction with the messenger in the following step (f) involves recognition and opening of the double stranded loop at the coat protein initiation site as a result of the cooperative effects of IF-3 and tRNA^{fMet} (g). The initiation factors now leave the 30S subunit in form of an [IF-2·IF-3] complex[9] and the 50S subunit joins the 30S complex in a reaction that requires GTP (i).

We now present the experimental evidence in support of this model. This is described in greater detail in ref. 2 and forthcoming publications from this laboratory.

Obligatory Dissociation for Initiation at Coat Protein Site

Since initiation via 30S subunit is a well established mechanism, the question to be settled is to what extent direct binding of 70S ribosomes to mRNA is possible as well. Thus, a mixed model (a), in which initiation via 30S complex competes (at a rate k_i) with the direct conversion of vacant 70S particles into an initiation complex at a rate k_d, is to be compared with

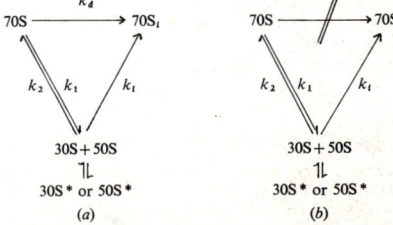

model (b) in which direct conversion is prohibited. An important assumption in these models is that vacant 70S couples are in equilibrium with their subunits

$$70S \underset{k_2}{\overset{k_1}{\rightleftharpoons}} 50S + 30S \qquad (1)$$

and that the rate constants k_1, k_2 have measurable values.

Exchange experiments involving incubation of pure vacant 70S ribosomes with an excess of either kind of subunit (marked with an asterisk) of a different density would detect direct initiation by 70S particles qualitatively by the binding of radio-

active fMet to non-hybrid ribosome couples. Because of the serious technical complications of this method[2,10], we used radioactive 50S particles instead, although discrimination by this approach must rely on quantitative arguments. The strongest prediction distinguishing the two models is that in model (b) any increase in the rate of dissociation (k_1) should stimulate the rate of initiation by the same factor (if $k_i > k_1$), whereas in model (a) the stimulation of the rate of initiation should be less, depending on the relative magnitude of k_d.

To obtain meaningful measurements, it had to be shown that (i) the test particles (vacant 70S couples, 30S and 50S subunits) could be prepared free of detectable contamination with complexed 70S ribosomes[8] or any of the other particles, (ii) all particles were fully active in initiation and (iii) subject to the equilibrium in equation (1). In the experiments summarized below, these conditions were met. At excess fMet-tRNA and a ratio of two messengers per ribosome, more than 85% of vacant 70S couples were converted into a [70S·R17 RNA·fMet-tRNA] complex[2]. The dissociation k_1/k_2 was measured over the range of 1–5 mM Mg^{2+}. At 37° C most particles dissociated between 1.5 and 2.5 mM Mg^{2+}, with less than 1% dissociation at 5 mM and more than 95% at 1 mM

The rate of spontaneous dissociation, k_1, was determined at 37° C and 5 mM Mg^{2+} from the initial rate at which pure labelled 50S particles appeared in the 70S ribosomes after they were mixed with pure non-radioactive vacant couples. The extent of exchange in Fig. 3 follows closely the theoretical curve (broken line) given by the solution of the differential equation describing the exchange process[2]. The rate of spontaneous dissociation is strongly dependent on the Mg concentration and increases about fifty-fold as the Mg concentration is reduced from 5 to 2 mM. The rate is not affected, however, by any components of the initiation system except IF-1 that produces a three-fold stimulation at saturating concentrations.

To distinguish between the two models, the time course of incorporation of both ^{14}C-50S particles and f-^{3}H-Met-tRNA into ribosome couples (vacant couples + initiation complex) was followed in the presence and absence of initiation factors. In the absence of initiation factors, no initiation was observed, and hence the initial rate of exchange reflects the rate of spontaneous dissociation (broken line in Fig. 3). In the presence of initiation factors, the initial rate of exchange was increased about three-fold and exactly equimolar amounts of f-^{3}H-Met-tRNA and ^{14}C-50S subunits were incorporated into ribosome couples (Fig. 4). The three-fold increase of the exchange rate is not a consequence of initiation because it could be produced solely by incubation of 70S

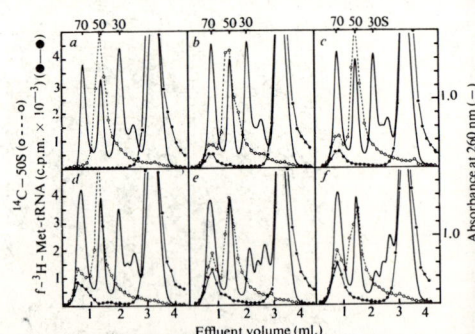

Fig. 4 Time course of initiation and subunit exchange by pure vacant 70S ribosomes in the presence of R17 RNA, f-^{3}H-Met-tRNA, ^{14}C-50S subunits and crude initiation factors at 5 mM Mg^{2+} and 37° C. Techniques and conditions of sedimentation analysis in isokinetic sucrose gradients are described in refs. 2 and 19. a, 0 min; b, 2.5 min; c, 5 min; d, 10 min; e, 20 min; f, 30 min.

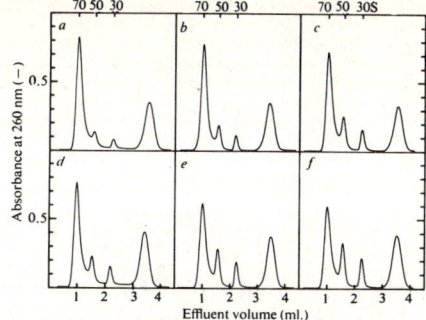

Fig. 5 Initial rates of dissociation of pure vacant 70S ribosomes in the presence of stoichiometric amounts of IF-3 (*a–c*) or IF-1 plus IF-3 (*d–f*) at 37° C and 4.3 mM Mg^{2+}. For conditions of sedimentation analysis see Fig. 4. *a* and *d*, 2.5 min; *b* and *e*, 5 min; *c* and *f*, 10 min.

ribosomes with either crude initiation factors (solid line in Fig. 3) or with stoichiometric amounts of IF-1. Similarly, if IF-1 was omitted from an otherwise complete initiation mixture, initiation occurred at an initial rate corresponding to that of spontaneous dissociation. It follows (i) that IF-1 is not involved in initiation *per se*, and (ii) that it is the only factor capable of stimulating the rate of dissociation (k_1) and hence must act by direct attack on the 70S particles. Finally, the finding that the number of initiations equals the number of dissociation events both before and after stimulation with IF-1 is only compatible with model *b*.

Role of IF-1 and IF-3 in Dissociation

No net dissociation is observed on incubation of vacant 70S couples with IF-1 at 5 mM Mg^{2+} because the equilibrium is so far toward association that the change in dissociation (k_1/k_2) produced by a three-fold increase in k_1 (and presumably a similar decrease in k_2) would hardly be measurable. Apparently, IF-1 bound to the 30S subunit in a 70S particle lowers the interaction between the subunits by reducing the area available for Mg-bridges. On dissociation, however, IF-1 remains firmly bound to the 30S particles but does not prevent them from joining with 50S subunits, although this now occurs at a somewhat lower efficiency.

Incubation of vacant 70S couples with IF-3, on the other hand, causes net dissociation (Fig. 5*a–c*), although the rate of equilibration is identical to the spontaneous exchange rate k_1. By contrast, with both IF-1 and IF-3 present, the initial rate of net dissociation is increased about three-fold (Fig. 5 *d–f*)[11]. Evidently, IF-3 can only bind to free 30S subunits[12] and acts as anti-association factor[13] by preventing them from combining with 50S particles. Because of the much greater affinity of the 50S particles for the 30S subunits, however, a large molar excess of IF-3 is required to keep a significant fraction of the ribosomes dissociated. This also explains the need for IF-2 discussed below.

Role of IF-2 and IF-3 in Initiation

With 30S subunits and fMet-tRNA, undegraded R17 RNA forms a 40S initiation complex that can be well resolved from the unreacted components (Fig. 6*a*). Formation of this complex required IF-3 but neither IF-1, IF-2 or GTP (*b–f, h*). Addition of 50S particles converted the 40S into a 76S initiation complex in a reaction that required GTP but no initiation factors. The 76S initiation complex was fully functional because (*a*) it could be quantitatively converted into the "release complex" carrying the coat protein hexapeptide if the position 6 amber mutant RNA was used, and (*b*) all the fMet could be released with puromycin[2].

When the 50S particles were present in the initiation mixture from the beginning, very little 76S initiation complex was formed unless IF-2 was added in addition to IF-3. We conclude that IF-2, most likely in combination with fMet-tRNA, greatly enhances the affinity of IF-3 for the 30S subunits and thus prevents the latter from joining with 50S particles before the messenger has been inserted. Thus, in contrast to current views, our model stipulates that fMet-tRNA binding precedes, rather than follows, the binding of the messenger.

Initiation by 70S Ribosomes without Dissociation

We confirmed that formaldehyde treatment of R17 RNA exposed three or four additional ribosome binding sites[14], which with excess 30S and in the absence of 50S subunits are revealed by the formation of 30S polysomes (Fig. 6*g*). Binding to these sites did not require any factors[15]. When IF-3 was included, binding increased only at the coat protein site. Neither did initiation at these factor-independent sites require dissociation into subunits. Thus, a similar experiment as in Fig. 4, but carried out at 10 mM Mg^{2+} to suppress spontaneous exchange, showed that initiation occurred without exchange (Fig. 7*a*). The absence of exchange is not due to inactivation of the ^{13}C-50S subunits with respect to either initiation or the ability to form couples because in the control containing an excess of washed 30S subunits instead of 70S ribosomes (*d, e*) nearly all the ^{14}C-50S ribosomes added were converted into 70S particles (*d*) and the extent of initiation (1.2 pmol) was comparable with that obtained with 70S ribosomes (*a*) after subtraction of background (*b*).

Ribosome Cycle

Experiments with purified release complex labelled in either R17 RNA or hexapeptide showed that RF-1 stimulated the release of the hexapeptide but promoted neither the release of messenger nor the exchange of subunits. Nor did RF-1 in combination with any of the three initiation factors stimulate messenger release. Evidently, an additional factor is needed, and the S-factor has been suggested for this role[16]. A dissociation step resulting from the intervention of S-factor could explain the rapid subunit exchange found to be associated with chain termination in mammalian polysomes[17].

Thus, in conditions of active protein synthesis, with most ribosomes in polysomes, dissociation of ribosomes associated

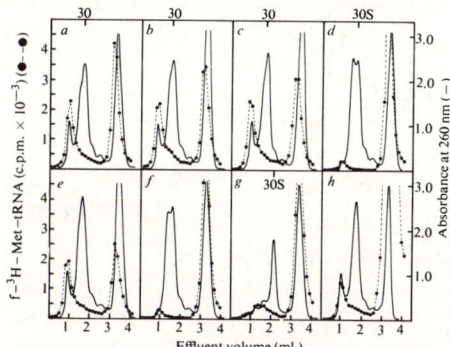

Fig. 6 Requirement for initiation factors and GTP for the formation of 40S initiation complex with native and formaldehyde-treated R17 RNA. Reaction mixtures containing 1 M NH$_4$Cl washed 30S subunits (18), R17 RNA, f-^{3}H-Met-tRNA and pure initiation factors as indicated were incubated at 37° C for 10 min and analysed on sucrose gradients for 1.7 h at 60,000 r.p.m. (1.2 h in *g*). *a*, All factors; *b*, −IF-1; *c*, −IF-2; *d*, −IF-3; *e*, +IF-3; *f*, no factors; *g*, HCHO, no factors; *h*, +IF-3, −GTP.

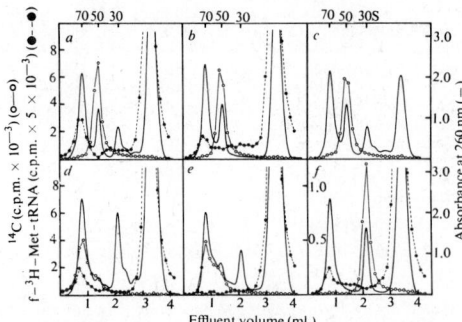

Fig. 7 Formation of 70S initiation complex without subunit exchange after unfolding R17 RNA with formaldehyde. Pure ^{14}C-50S ribosomes were incubated with pure vacant 70S ribosomes, f-^{3}H-Met-tRNA and formaldehyde-treated R17 RNA (14) in the absence of initiation factors at 10 mM Mg^{2+} and 37° C for 20 min. Controls b, e, f lacked R17 RNA, in c fMet-tRNA was omitted; f was as b except that ^{14}C-30S subunits were used to show absence of exchange under conditions of optimal peak separation. In d and e ^{14}C-50S subunits were mixed with 30S subunits instead of pure vacant 70S ribosomes. The reaction mixtures were analysed on sucrose gradients as in Fig. 4.

with mRNA release would feed a major fraction of the 30S subunits from the polysome directly into the initiation complex. If chain initiation becomes limiting, however, vacant ribosomes will accumulate as couples because of the high affinity of the subunits for each other and the high initiation factor to ribosome ratio required to keep them apart. Thus, we can say that most free or native 30S particles contain initiation factors, whereas most 70S particles do not (except for the small fraction containing IF-1). The two forms are freely interconvertible, however, and not functionally different in any fundamental way.

H. N. is a lifetime career professor of the American Cancer Society. This work was supported by research grants from the American Cancer Society and the US Public Health Service.

Received June 19, 1972.

[1] Guthrie, C., and Nomura, M., *Nature*, **219**, 232 (1968).
[2] Noll, L. M., thesis, Northwestern Univ. (1972).
[3] Lucas-Lennart, J., and Lipmann, F., *Ann. Rev. Biochem.*, **40**, 409 (1971).
[4] Falvey, A. K., and Staehelin, T., *J. Mol. Biol.*, **53**, 21 (1970).
[5] Schreier, M. H., and Noll, H., *Proc. US Nat. Acad. Sci.*, **68**, 805 (1971).
[6] Min Jou, W., Haegeman, G., Ysebaert, M., and Fiers, W., *Nature*, **237**, 82 (1972).
[7] Staehelin, T., Brinton, C. C., Wettstein, F. O., and Noll, H., *Nature*, **199**, 865 (1963).
[8] Noll, L. M., Hapke, B., Schreier, M. H., and Noll, H., *J. Mol. Biol.* (in the press).
[9] Groner, Y., and Revel, M., *Europ. J. Biochem.*, **22**, 144 (1971).
[10] Subramanian, A. R., and Davis, B. D., *Proc. US Nat. Acad. Sci.*, **68**, 2453 (1971).
[11] Noll, M., and Noll, H., *Fed. Proc.*, Abstr., **31**, 890 (1972).
[12] Sabol, S., and Ochoa, S., *Nature New Biology*, **234**, 233 (1971).
[13] Kaempfer, R., *Proc. US Nat. Acad. Sci.*, **68**, 2458 (1971).
[14] Lodish, H. F., *J. Mol. Biol.*, **50**, 689 (1970).
[15] Berissi, H., Groner, Y., and Revel, M., *Nature New Biology*, **234**, 44 (1971).
[16] Goldstein, J. L., and Caskey, C. T., *Proc. US Nat. Acad. Sci.*, **67**, 537 (1970).
[17] Falvey, A. K., and Staehelin, T., *J. Mol. Biol.*, **53**, 119 (1970).
[18] Kurland, C. G., *J. Mol. Biol.*, **18**, 90 (1966).
[19] Noll, L. M., Noll, H., and Lingrel, J. B., *Proc. US Nat. Acad. Sci.*, **69**, 1843 (1972).

The S₁ Factor in Peptide Chain Elongation

by
J. WATERSON
G. BEAUD
P. LENGYEL
Department of Molecular Biophysics
and Biochemistry,
Yale University,
New Haven, Connecticut

Aminoacyl-tRNA becomes attached to the ribosome as part of an S_3 factor–GTP–aminoacyl-tRNA complex. GTP is then cleaved and an S_3 factor–GDP complex is released. The S_1 factor promotes the re-formation of the S_3 factor–GTP–aminoacyl-tRNA complex from the S_3 factor–GDP complex, aminoacyl-tRNA and GTP.

A CONVENIENT model for studying peptide chain elongation is a cell-free system in which polyuridylic acid (poly U) directs the synthesis of polyphenylalanyl-tRNA (poly-Phe-tRNA) from phenylalanyl-tRNA (Phe-tRNA)[1,2], in the presence of salts, GTP, ribosomes from *Escherichia coli*, and the amino-acid polymerization factors, S_1, S_2 and S_3, from *Bacillus stearothermophilus*[3]. These factors are analogous to factors from *E. coli* and *Pseudomonas fluorescens*: S_1 corresponds to T_s and FI_s, S_2 to G and FII, and S_3 to T_u and FI_u[3-5]. Subsequently when discussing a factor from these two microorganisms, we shall note in parentheses the designation of the factor from *B. stearothermophilus*.

The addition of each phenylalanine residue to the growing polyPhe-tRNA chain is a cyclic process[1,2]. The first composite step in the cycle is the binding of an S_3–GTP–Phe-tRNA complex[3,5,6] (complex II) to the ribosome–poly U–acetyl-Phe-tRNA complex[7-9]. (Acetyl-Phe-tRNA serves as a simple analogue of polyPhe-tRNA in this complex[9-11].) The GTP from complex II is then cleaved into GDP and P_i and an S_3–GDP complex (complex III) and P_i are released from the ribosome[5,7,9,11,12]. The Phe-tRNA from complex II reacts with the acetyl-Phe-tRNA yielding acetyl-diPhe-tRNA[11,13] which is located at the aminoacyl-tRNA binding site (or A site) of the ribosome[11,13]. Finally, the discharged tRNA is released from the ribosome and the acetyl-diPhe-tRNA is translocated to the peptidyl-tRNA binding site (or P site) of the ribosome, a step catalysed by S_2 and dependent on the cleavage of further GTP into GDP and P_i[7,13-20]. Another phenylalanine residue can then be added.

Previous investigations with S_1 (and with the corresponding factor from *E. coli*) revealed that this factor is involved (together with S_3 and GTP) in the binding of aminoacyl-tRNA to ribosomes[3,13,21]: S_1 catalyses the formation of complex II, which is the source of the aminoacyl-tRNA bound to the ribosome[7,8]. Complex II contains equimolar amounts of S_3, GTP and aminoacyl-tRNA[22]; S_1 is probably not part of the complex[5].

We wondered how the S_3 moiety of complex III is reutilized in chain elongation. The possibility of an involvement of S_1 in this reutilization was indicated by the observation that the GDP in our complex III was exchangeable with GTP or GDP at 0° C (ref. 9), whereas that in the complex III described by Shorey *et al.*[5] was not. The latter preparation presumably contained little or no FI_s (S_1)[5]; our complex III was prepared by a different procedure and did contain S_1[9].

We first established that it was S_1 that made the GDP moiety of our complex III preparation exchangeable. Subsequently we demonstrated that S_1 is required in the conversion of complex III into complex II: S_1 is needed for the repeated utilization of S_3 in peptide chain elongation.

Use of Antiserum against S_1

Anti-S_1 was required for freeing the complex III preparation from S_1 activity. Anti-S_1 was prepared by injecting rabbits with purified S_1. The effect of incubating a constant amount of anti-S_1 with varying amounts of one of the amino-acid polymerization factors was tested by determining the activity of the treated factors in the synthesis of polyPhe-tRNA. The results indicate that the anti-S_1 inactivates S_1 (Fig. 1A), but has no effect on either S_3 (Fig. 1B) or S_2 (data not shown).

If one requires complex III and not acetyl-diPhe-tRNA as a product of the reaction of complex II with a ribosome–poly U-acetyl-Phe-tRNA complex[5,11,13], a ribosome–poly U complex can be substituted for the ribosome–poly U-acetyl-Phe-tRNA complex.

We prepared complex II by incubating S_3, Phe-tRNA, GTP and S_1. To remove GTP not bound in complex II, we fractionated the reaction mixture on 'Sephadex G-25'[7]. The excluded fraction containing complex II, S_1 and Phe-tRNA (but no free GTP) was incubated with anti-S_1 to inactivate the S_1 present. The reaction mixture was then supplemented and incubated with a preformed (ribosome–poly U) complex to allow the binding of Phe-tRNA (from complex II) to the ribosome with the concomitant GTP cleavage and formation of complex III[7,9,11]. The incubated reaction mixture was fractionated on 'Sephadex G-150' (Fig. 2). The use of three differently labelled compounds (^{14}C-Phe-tRNA, ^{3}H-GTP, and γ-^{32}P-GTP) facilitates identification. The excluded material (effluent volume 8–11 ml.) consists of ribosomes with bound ^{14}C-Phe-tRNA (as well as anti-S_1, S_1 complexed to anti-S_1, and other γ-globulins). The next peak (effluent volume 12–18 ml.) contains complex III (as well as 0·11

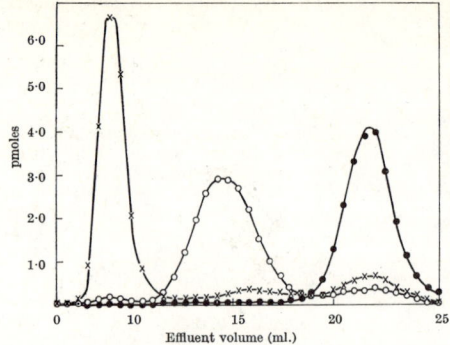

Fig. 2. Isolation of complex III free of S_1 activity. Complex II was prepared in buffer I (0·4 ml.) containing S_1, 5·3 μg; S_3, 60 μg; ^{14}C-Phe-tRNA (specific activity 120 Ci/mole) carrying 500 pmoles of phenylalanine residues (37 $A_{260\,nm}$ units); and a mixture of γ-^{32}P-GTP (125 Ci/mole) and ^{3}H-GTP (1,050 Ci/mole), 2,500 pmoles. The reaction mixture was incubated at 30° C for 5 min, then cooled to 0° C and applied to a 'Sephadex G-25' column (4·5 ml.) which had been equilibrated in 0·16 M NH$_4$Cl–0·04 M Tris-HCl (pH 7·4)–0·01 M magnesium acetate–0·001 M dithiothreitol (buffer II). The column was eluted at 4° C; 0·35 ml. fractions were collected, and 0·01 ml. aliquots of each were counted in Bray's scintillation fluid[24]. The excluded fractions (containing complex II, free Phe-tRNA and S_1) were pooled (0·7 ml.) and incubated with 0·67 mg of anti-S_1 at 0° C for 10 min to inactivate S_1. The solution was supplemented with 0·5 ml. of solution B; incubated at 30° C for 10 min and then cooled to 0° C (solution C). (Solution B (0·5 ml.) consisted of buffer I; ribosomes, 290 $A_{260\,nm}$ units; and poly U, 0·8 mg. It had been incubated at 30° C for 10 min to allow the binding of poly U to ribosomes.) Solution C (1·1 ml.), which included anti-S_1, ribosome bound ^{14}C-Phe-tRNA, complex III, ^{14}C-Phe-tRNA, and P$_i$) was fractionated by gel filtration on a 20 ml. 'Sephadex G-150' column. Fractions (0·5 ml.) were collected; 0·04 ml. aliquots of each were counted in Bray's scintillation fluid. ×, pmoles of ^{14}C-Phe; ○, pmoles of ^{3}H-GTP or GDP; ●, pmoles in γ-^{32}P-GTP or ^{32}P$_i$ (the amounts present in 0·04 ml. aliquots are shown). As expected, the amounts of Phe-tRNA bound to ribosomes, complex III formed and P$_i$ released are close to equimolar[7].

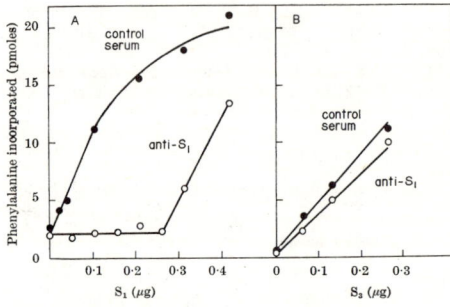

Fig. 1. Specificity of anti-S_1. Different amounts of S_1 (A) or S_2 (B) were incubated in a 0·01 M Tris-HCl (pH 7·4)–0·001 M dithiothreitol solution with either 12·5 μg anti-S_1 (○) or control serum containing 80 μg protein (●) in 0·04 ml. at 0° C for 10 min. The incubated solution was supplemented with the components required for polyPhe-tRNA synthesis: S_3, 0·16 μg; ribosomes, 1·8 $A_{260\,nm}$ units; ^{14}C-Phe-tRNA (specific activity 120 Ci/mole) carrying 33 pmoles of phenylalanine residues, 1·8 $A_{260\,nm}$ units; GTP, 0·25 μmoles; 0·16 M NH$_4$Cl–0·04 M Tris-HCl (pH 7·4)–0·01 M magnesium acetate–0·01 M dithiothreitol (buffer I) and either 0·35 μg S_2 (A) or 0·53 μg S_1 (B). The reaction mixture (0·125 ml.) was incubated at 30° C for 10 min. Subsequently, the amount of ^{14}C-phenylalanine incorporated into hot acid insoluble material was determined[3]. The preparation of thoroughly washed ribosomes and labelled Phe-tRNA from E. coli B, and amino-acid polymerization factors (S_2 purified 70-fold and S_3 100-fold) from B. stearothermophilus was described previously[3]. The S_1 preparation used was purified first according to the published procedure[3] and further purified by chromatography on a DEAE-'Sephadex' column at pH 6·4. The column was eluted with a linear KCl gradient (0·15–0·55 M) in 0·01 M potassium cacodylate. The active fraction was eluted at 0·2 M KCl. The purified S_1 and S_3 preparations moved as single bands on acrylamide gel electrophoresis in 8 M urea at pH 8·9 and pH 4·5. Antiserum against S_1 was obtained from a rabbit injected once with 200 μg and 5 weeks later with 30 μg purified S_1, mixed with Freund's adjuvant. The γ-globulin fraction was isolated by (NH$_4$)$_2$SO$_4$ precipitation and chromatography on a DEAE-cellulose column[24]. Protein was determined according to ref. 25.

mole of free Phe-tRNA per mole of complex III). The third peak (effluent volume 19–25 ml.) contains P$_i$, some free phenylalanine, and some GTP and GDP. The fractions eluted between 13 and 17 ml. were combined and used as complex III. The complex III preparation was tested for S_1 activity by determining the effect of S_1 on the rate of polyPhe-tRNA synthesis in reaction mixtures containing complex III as well as S_2, poly U, ribosomes, Phe-tRNA and GTP. The curves in Fig. 3 reveal that added S_1 increases the rate of polyPhe-tRNA synthesis 10-fold, although polyPhe-tRNA is formed slowly even in the absence of added S_1. It is not yet known whether this arises because of residual S_1 in the complex III which had been treated with anti-S_1 or whether it indicates that polyPhe-tRNA synthesis can occur slowly without S_1.

Exchangeability of GDP

We used the complex III preparation, which was free of, or at least low in, S_1 activity, to study the effect of S_1 on the exchangeability of the GDP moiety of the complex. The assay was based on the fact that complex III is retained on a 'Millipore' filter, whereas free GDP is not[5]. The complex III preparation containing ^{3}H-GDP was incubated at 0° C and the amount of ^{3}H-GDP retained in the complex was determined by filtration through a 'Millipore' filter and counting of radioactivity on the filter. The results in Fig. 4 indicate that no ^{3}H-GDP is released from complex III by incubation at 0° C without further additions. Addition of a large excess of unlabelled GDP results in slow release of ^{3}H-GDP; S_1 greatly increases the rate of release in the presence of unlabelled GDP. Thus S_1 facilitates the exchange between the GDP moiety of complex III and free GDP, although it is to be noted that addition of S_1 alone (without unlabelled GDP) also results in the partial disappearance of ^3H-GDP from complex III.

Formation of Complex II from Complex III

The assay used to test the effect of S_1 on the formation of complex II from complex III, GTP and Phe-tRNA was based on the following considerations: complex II and complex III are excluded from 'Sephadex G-25' whereas GDP and GTP are not; the complex III preparation used contains ^3H-GDP. Consequently, the amount of ^3H radioactivity in the excluded fraction is a measure of complex III. One mole of complex II contains one mole of GTP and one mole of Phe-tRNA[23]. (The amount of Phe-tRNA in the excluded fraction is, however, not a measure of complex II because Phe-tRNA itself is excluded from the gel.) The GTP with which complex III was

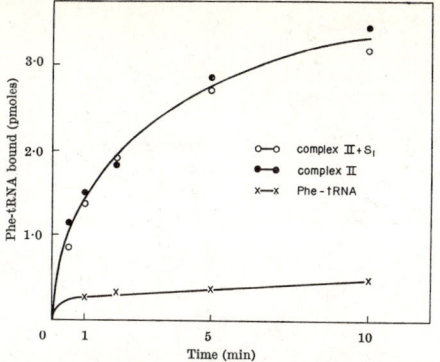

Fig. 5. Lack of effect of S_1 on the rate of binding of Phe-tRNA from complex II to the ribosome. Complex II was prepared, separated from unbound GTP, and treated with anti-S_1 as described in the legend to Fig. 2. The ribosome–poly U-acetylPhe-tRNA complex (RPA complex) was prepared by incubating in buffer II (1·0 ml.): ribosomes, 57·6 $A_{260\,nm}$ units; poly U, 0·20 mg; ^3H-acetylPhe-tRNA[27] carrying 40 pmoles of acetylphenylalanine residues (5·1 $A_{260\,nm}$ units); at 30° C for 1 h. Aliquots (50 μl.) of the solution containing the RPA complex were pipetted into 1·5 ml. aliquots of ice-cold buffer I. The reaction mixtures were supplemented further with either ^{14}C-Phe-tRNA carrying 21 pmoles of phenylalanine residues (×) or complex II preparation containing 8 pmoles of ^3H-GTP and 18 pmoles of ^{14}C-Phe-tRNA (●) (presumably 8 pmoles of this Phe-tRNA were in complex II and 10 pmoles were present as free Phe-tRNA), or complex II preparation in the same amount and 0·42 μg S_1 (○). Incubation was at 0° C. At the times indicated, the reaction mixtures were filtered on 'Millipore'. The filter was then washed with 10 ml. of ice-cold buffer II and dried. The amount of ^{14}C-Phe-tRNA retained on the filter was determined by counting in a toluene based scintillation fluid.

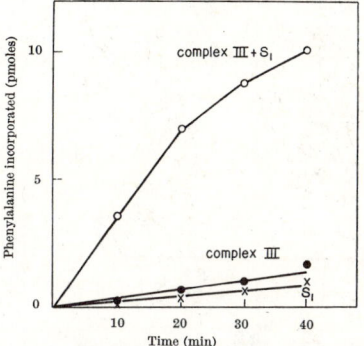

Fig. 3. Dependence on S_1 of polyPhe-tRNA synthesis in the presence of complex III free of S_1 activity. PolyPhe-tRNA synthesis was assayed in buffer I (0·5 ml.) containing: S_2, 0·2 μg; ribosomes, 7·2 $A_{260\,nm}$ units; ^{14}C-Phe-tRNA (120 Ci/mole) carrying 100 pmoles of phenylalanine residues (7·4 $A_{260\,nm}$ units); GTP, 1 μmole; and—as indicated in the figure—either complex III containing 6·6 pmoles of ^3H-GDP (●) or 0·32 μg S_1 (×) or both complex III and S_1 in these quantities (○). The incubation was at 30° C. At the indicated times 0·1 ml. aliquots from the reaction mixture were added to 2·5 ml. of cold 5 per cent trichloroacetic acid. The amount of ^{14}C-phenylalanine residues incorporated into hot acid insoluble material was determined.

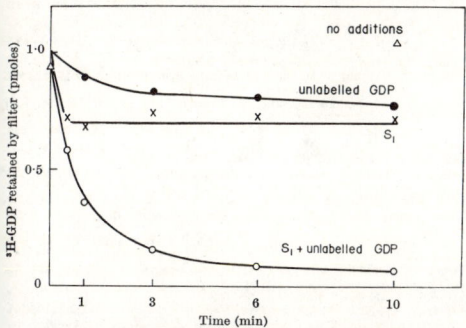

Fig. 4. S_1 increases the rate of exchange between free GDP and the GDP moiety in complex III. Buffer I (0·1 ml.) was supplemented with complex III containing 2·4 pmoles of ^3H-GDP and (as indicated) with either GDP (unlabelled), 1,250 pmoles (●—●) or 0·21 μg S_1 (×—×), both (○—○), or neither (△). The molar ratio of S_3 (in complex III) to S_1 added is about 1:2. The reaction was started by adding complex III. The incubation was at 0° C for the times indicated. The reaction was stopped by the addition of 1·5 ml. of buffer II and the reaction mixture was then immediately filtered through a 'Millipore' filter. The filter was washed with 10 ml. of buffer II, dried, and counted in a toluene based scintillation fluid. Though we added 2·4 pmoles of complex II, only 1·0 pmole of ^3H-GDP was retained by the filter. This loss may have resulted from the dissociation of complex III on dilution.

incubated (in order to convert it into complex II) was γ-^{32}P labelled so that the amount of ^{32}P in the excluded fraction could be taken as a tentative measure of complex II. To put this assumption on a firmer basis, however, it has to be shown that the amount of ^{32}P in the excluded fraction depends, within limits, on the amount of Phe-tRNA in the reaction mixture.

From the results in Table 1 it can be seen that complex III itself is rather labile in the conditions of the experiment: less than 50 per cent is recovered. S_1 (incubated with complex III, GTP and Phe-tRNA) increases the extent of complex II formation nine-fold and results apparently in the conversion of all complex III into complex II. The amount of complex II formed (or more precisely the amount of ^{32}P-GTP in the excluded fraction) is increased over three-fold by added Phe-tRNA. Thus at least 70 per cent of the ^{32}P-GTP in the excluded fraction is present in the form of complex II. Some ^{32}P-GTP appears in the excluded fraction when no Phe-tRNA is added to the reaction mixture. One part of this may be in the form of complex II, for the complex III preparation itself contains some Phe-tRNA (see the legend to Table 1). Another part of the excluded ^{32}P-GTP, however, is present in the form of an S_3-GTP complex (unpublished

Table 1. FORMATION OF COMPLEX II FROM COMPLEX III: DEPENDENCE ON S_1

	Guanine nucleotides in excluded fraction (pmoles)	
	γ-^{32}P-GTP	^3H-GDP
Complete system	10·5	0·2
$-S_1$	1·2	3·3
$-$ Phe-tRNA	3·2	0·6
$-S_1$, $-$ Phe-tRNA	1·2	3·8
$-S_1$, $-$ Phe-tRNA, $-$ GTP	—	4·9

The complete system contained, in 0·25 ml. of buffer I, the following components: complex III, with 11 pmoles of ^3H-GDP (this amount of complex III preparation also contained 1·2 pmoles of Phe-tRNA, see Fig. 2); S_1, 0·21 μg; Phe-tRNA, 25 pmoles; γ-^{32}P-GTP (specific activity 1,250 Ci/mole), 500 pmoles. The components omitted from the various mixtures are indicated. The reaction mixtures were incubated at 30° C for 5 min, cooled to 0° C, and fractionated by gel filtration on a 'Sephadex G-25' column (5·5 ml.) which had been equilibrated with buffer II. Fractions (0·25 ml.) were collected and 0·2 ml. aliquots of each were counted in Bray's scintillation fluid[28]. The table shows the number of pmoles of each isotope in the excluded volume.

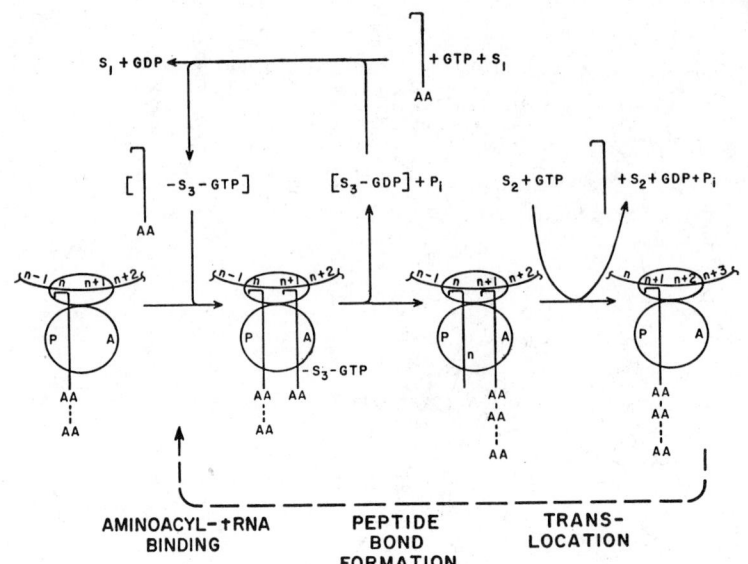

Fig. 6. Scheme of peptide chain elongation. The small oval shape represents the 30S ribosomal subunit, the large oval shape the 50S subunit. A and P indicated in the 50S subunit stand for the hypothetical tRNA binding sites. The symbols n−1, n, n+1, n+2, n+3 represent a series of adjacent codons in a segment of mRNA. The gallows shape stands for tRNA. AA stands for aminoacyl. The scheme of steps shown in the figure is described in the text for the case of the addition of one phenylalanine residue to acetyl Phe-tRNA.

results of G. B.). Clearly S_1 promotes the formation of complex II from complex III.

We proceeded to test if S_1 also affects the subsequent step in chain elongation—the binding of complex II to the ribosome–poly U-acetyl-Phe-tRNA complex[9]—and used a complex II preparation which had been treated with anti-S_1. The results in Fig. 5 indicate that S_1 does not change either the rate or the extent of the binding. The experiments were performed with a complex II preparation containing excess free Phe-tRNA. The same conclusion was drawn from experiments in which no free Phe-tRNA was present.

The Role of S_1

We have demonstrated that S_1 promotes the conversion of complex III to complex II in the presence of aminoacyl-tRNA and GTP. The binding of complex II to the ribosome does not seem to be affected by S_1. Moreover, according to preliminary results, S_1 does not seem to increase the rate of polyPhe-tRNA synthesis from complex II. If this can be proved, it would indicate that the conversion of complex III to complex II is the only step in which S_1 participates. A scheme of the steps in peptide chain elongation, including the conversion of complex III to complex II, is shown in Fig. 6. The need for S_1 in complex II formation has been known for some time. In earlier investigations complex II formation from GTP, aminoacyl-tRNA, S_1 and S_3 was examined[3,21]; we have substituted complex III for S_3 in the reaction mixture. Recent results raise the interesting possibility that the component in the early studies which was presumed to be S_3 may actually have been complex III. Weissbach et al.[23] found that purified T_u (S_3) contains bound GDP, and that it is the presence of this bound GDP which makes the formation of complex II dependent on S_1. Probably the T_u (S_3) containing bound GDP which is obtained by purification is identical to complex III.

This study was supported by a grant from the National Institutes of Health, US Public Health Service. G. B. is a fellow of the Centre National de Recherche Scientifique, France. We thank Y. Ono and A. Skoultchi for discussion and H. Weissbach et al. for sending us their manuscript.

Received April 1, 1970.

[1] Lipmann, F., *Science*, **164**, 1024 (1969).
[2] Lengyel, P., and Söll, D., *Bact. Rev.*, **33**, 264 (1969).
[3] Skoultchi, A., Ono, Y., Moon, H. M., and Lengyel, P., *Proc. US Nat. Acad. Sci.*, **60**, 675 (1968).
[4] Lucas-Lenard, J., and Lipmann, F., *Proc. US Nat. Acad. Sci.*, **55**, 1562 (1966).
[5] Shorey, R., Ravel, J., Garner, G. W., and Shive, W., *J. Biol. Chem.*, **244**, 4555 (1969).
[6] Gordon, J., *Proc. US Nat. Acad. Sci.*, **59**, 179 (1968).
[7] Ono, Y., Skoultchi, A., Waterson, J., and Lengyel, P., *Nature*, **223**, 697 (1969).
[8] Lucas-Lenard, J., and Haenni, A., *Proc. US Nat. Acad. Sci.*, **59**, 554 (1968).
[9] Skoultchi, A., Ono, Y., Waterson, J., and Lengyel, P., *Biochemistry*, **9**, 508 (1970).
[10] Lucas-Lenard, J., and Lipmann, F., *Proc. US Nat. Acad. Sci.*, **57**, 1050 (1967).
[11] Ono, Y., Skoultchi, A., Waterson, J., and Lengyel, P., *Nature*, **222**, 645 (1969).
[12] Gordon, J., *J. Biol. Chem.*, **244**, 5680 (1969).
[13] Haenni, A., and Lenard, J. L., *Proc. US Nat. Acad. Sci.*, **61**, 1363 (1968).
[14] Kuriki, Y., and Kaji, A., *Proc. US Nat. Acad. Sci.*, **61**, 1399 (1968).
[15] Lucas-Lenard, J., and Haenni, A., *Proc. US Nat. Acad. Sci.*, **63**, 93 (1969).
[16] Traut, R., and Monro, R., *J. Mol. Biol.*, **10**, 63 (1964).
[17] Brot, N., Ertel, R., and Weissbach, H., *Biochem. Biophys. Res. Commun.*, **31**, 563 (1968).
[18] Ertel, R., Redfield, B., Brot, N., and Weissbach, H., *Arch. Biochem. Biophys.*, **128**, 331 (1968).
[19] Pestka, S., *Proc. US Nat. Acad. Sci.*, **61**, 726 (1968).
[20] Erbe, R., Nau, M., and Leder, P., *J. Mol. Biol.*, **39**, 441 (1969).
[21] Ertel, R., Brot, N., Redfield, B., Allende, J., and Weissbach, H., *Proc. US Nat. Acad. Sci.*, **59**, 861 (1968).
[22] Ono, Y., Skoultchi, A., Klein, A., and Lengyel, P., *Nature*, **220**, 1304 (1968).
[23] Weissbach, H., Miller, O., and Hachmann, J., *Arch. Biochem. Biophys.*, **137**, 262 (1970).
[24] Calendar, R., and Berg, P., *Biochemistry*, **5**, 1681 (1966).
[25] Lowry, O., Rosebrough, N., Farr, A., and Randall, R., *J. Biol. Chem.*, **193**, 265 (1951).
[26] Bray, G., *Anal. Biochem.*, **1**, 279 (1960).
[27] Haenni, A., and Chapeville, F., *Biochim. Biophys. Acta*, **114**, 135 (1966).

RELEASE FACTORS DIFFERING IN SPECIFICITY FOR TERMINATOR CODONS

By E. Scolnick, R. Tompkins, T. Caskey, and M. Nirenberg

LABORATORY OF BIOCHEMICAL GENETICS, NATIONAL HEART INSTITUTE,
NATIONAL INSTITUTES OF HEALTH, BETHESDA, MARYLAND

Communicated August 16, 1968

In a previous paper, initiator and terminator trinucleotides were shown to stimulate sequentially the binding of f-Met-tRNA to ribosomes and the release of free f-methionine from the ribosomal intermediate.[1] The formation of f-methionine was dependent upon a terminator codon, such as UGA, UAA, or UAG, and the release factor R discovered by Capecchi.[2]

The separation of R into two components is described in this report. R1 corresponds to the codons UAA and UAG; R2, to UAA and UGA.

Methods.—R assay: The termination assay is described elsewhere.[1] Each reaction contained the following components in a final volume of 50 μl: 0.05 M Tris-acetate, pH 7.2; 0.03 M magnesium acetate; 0.05 M potassium acetate; 4–6 $\mu\mu$moles of the (f[^{3}H]-Met-tRNA^f··AUG··Ribosome) complex (4.0–6.7 $\mu\mu$moles of TCA-precipitable f[^{3}H]-Met-tRNA^f, 0.02 A^{260} unit was present/reaction); 0.96 A^{260} unit of *E. coli* B ribosomes (washed with 0.5 M ammonium chloride); 0.17 mμmole of AUG; and, where indicated, 7–7.7 mμmoles of terminator trinucleotide, and an R preparation. Reactions were incubated for 16 min at 30° unless otherwise stated; hence, the rate of f[^{3}H]-methionine formation was determined in all experiments.

The preparation of tRNA^{fMet} (*E. coli* B), fractionated by benzoylated-DEAE-cellulose column chromatography,[3] was the gift of Dr. Michael Wilcox. The tRNA^{fMet} was acylated with [^{3}H]-methyl-methionine (3.1 c/mmole, Schwarz BioResearch Corp.) and then converted to f[^{3}H]-Met-tRNA^{fMet}.[1] Ribosomes were obtained from *E. coli* B as described by Lucas-Lenard and Lipmann,[4] except that seven, rather than five, 0.5 M ammonium chloride washes were employed.

R fractionation: *Escherichia coli* B supernatant and ribosome fractions, treated with DNase but not "preincubated," were prepared as described previously,[5] except that cells were lysed with a French pressure cell at 18,000 psi, 0.005 M DTT replaced β-mercaptoethanol, and extracts were centrifuged for 5 hr at 137,000 × g.

Release factors (prep. A, Table 1) were obtained as follows: R was precipitated by the addition of 116 gm of ammonium sulfate to 359 ml of the S-137 fraction; the pH was maintained at 7.8 by the addition of 0.5 N ammonium hydroxide. All steps were performed at 4°. The precipitate was collected by centrifugation at 30,000 × g for 15 min, dissolved in 139 ml of buffer A (0.05 M Tris-chloride, pH 8.0; 0.15 M potassium chloride; 0.001 M EDTA, adjusted to pH 7 with NaOH; and 0.003 M DTT) and dialyzed against 4 liters of buffer A for 4 hr (0–55% (NH$_4$)$_2$SO$_4$ fraction of Table 1).

DEAE-Sephadex column: The 0–55% (NH$_4$)$_2$SO$_4$ fraction (138 ml, 2200 mg protein) was diluted to 275 ml with buffer A and applied to a DEAE-Sephadex A-50 column (112 × 2.8 cm) equilibrated with buffer A at a flow rate of 0.5 ml/min. Protein was eluted with 3200 ml of buffer A that contained a linear KCl gradient (0.15–0.70 M).

R fractions (prep. B) illustrated in Figures 1 and 2 were obtained by a slightly different procedure.[1] The 0–35% (NH$_4$)$_2$SO$_4$ fraction (450 mg protein)–dissolved in 0.01 M Tris-chloride, pH 8.0; 0.001 M EDTA; 0.001 M DTT; and 0.2 M KCl; (buffer B)—was applied to a 36 × 2.8-cm DEAE-Sephadex A-50 column. Protein was eluted with 1300 ml of buffer B that contained a linear KCl gradient (0.2–0.6 M) at a flow rate of 0.5 ml/min. Each fraction contained 11 ml.

Sephadex G-100 column (Fig. 2): Portions of R1 or R2 (prep. B, DEAE-Sephadex fractions) were applied to a column (85 × 1.8 cm) packed with Sephadex G-100 (bead form) equilibrated with buffer C (0.01 M Tris-chloride, pH 8.0; 0.15 M potassium chloride; 0.001 M EDTA, and 0.001 M DTT). R fractions were eluted with buffer C at a flow rate of 0.22 ml/min. The amounts of R applied and recovered are stated in Table 1.

Pooled R fractions were concentrated by pressure filtration (Amicon Co., Lexington, Mass.) and stored at $-170°$.

G, T_u, and T_s assays: Factor G was determined by the method of Conway and Lipmann;[6] γ-[^{32}P]-GTP (4 c/mmole) was the gift of Dr. H. Weissbach. T_u and T_s were assayed by retention of [^{14}C]-GTP to Millipore filters, as described by Allende and Weissbach;[7] [$^{14}_s$C]-GTP (30 μc/μmole) was obtained from Schwarz BioResearch. Protein was determined by a modification of the Lowry method.[8]

Results.—Release factors and polymerization factors were fractionated by DEAE-Sephadex column chromatography[2] as shown in Figure 1. R was resolved

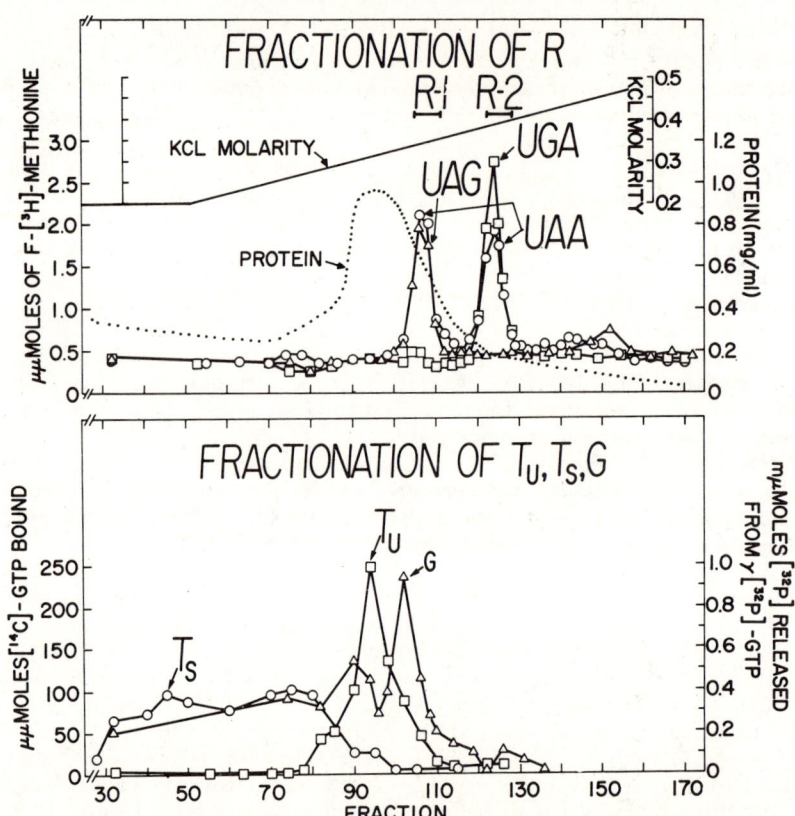

Fig. 1.—Fractionation of R factors and the polymerization factors, T_s, T_u, and G by DEAE Sephadex column chromatography. The conditions of chromatography, the R preparation fractionated (prep. B, 0–35% (NH$_4$)$_2$SO$_4$), and the assay of R are described in *Methods*. Each reaction for the assay of R contained the components described in *Methods*, including 4.3 $\mu\mu$moles of the (f[^{3}H]-Met-tRNA^f··AUG··Ribosome) complex (a total of 4.60 $\mu\mu$moles of TCA-precipitable f[^{3}H]-Met-tRNA^f per reaction), 10 μl of a column fraction, and the trinucleotide preparation indicated. Reactions were incubated for 25 min. The amount of f[^{3}H]-methionine present at zero time (0.30 $\mu\mu$mole) was subtracted from each value.

Fractionation of T_s, T_u, and G is shown in the lower figure. The left ordinate ([^{14}C]-GTP bound) corresponds to T_s and T_u; the right ordinate, to G. Fifteen, 100, and 50 μl of each column fraction indicated were assayed for T_s, T_u, and G, respectively, as described in *Methods*.

into two well-separated peaks, approximately equal in size; R1 corresponds to UAA and UAG, R2 to UAA and UGA. Approximately 60 per cent of the total UAA-dependent R activity was found with R1, 40 per cent with R2. R1 was approximately as active with UAG as with UAA; R2 was more active with UGA than with UAA. Further studies are required to determine whether fractions 140–155 contain additional R components.

Column fractions were also assayed for T_s, T_u, and G (Fig. 1, lower half). The chromatographic mobilities of R factors differ from those of T_s, T_u, and G; also, T_s and T_u were not detected in the pooled R2 fraction. Hence R factors differ from polymerization factors.

R1 and R2 fractions were purified further by Sephadex G-100 column chromatography as shown in Figure 2. No difference was found in the chromatographic mobility of R1 dependent upon UAA or UAG; similarly, the mobilities of UAA- and UGA-dependent R2 were identical. The relative activities of R1 and R2 with appropriate codons (UAG/UAA and UGA/UAA ratios, respectively) were relatively constant throughout each peak. The ratios derived from Figures 1 and 2 differ somewhat because trinucleotide concentrations used for these experiments varied slightly. The limited availability of trinucleotides precluded the addition of excess concentrations of trinucleotides to reactions; however, f[^{3}H]-methionine formation is proportional to the concentration of R at uniform trinucleotide concentrations.[1]

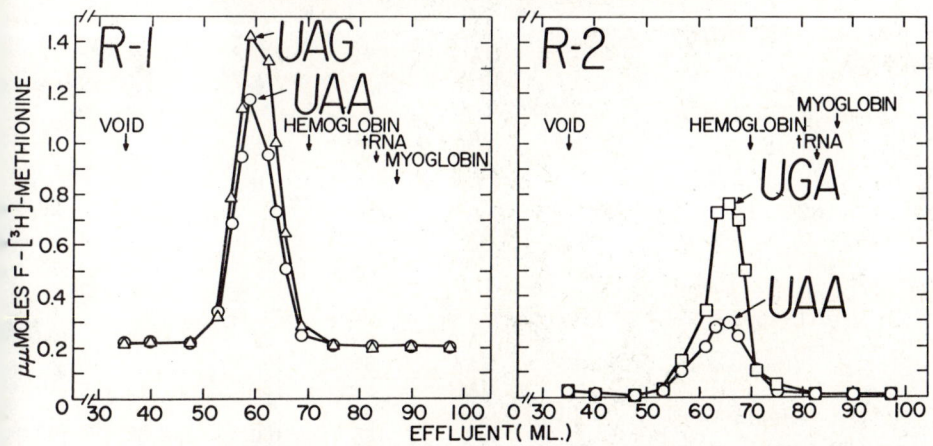

FIG. 2.—Fractionation of R1 and R2 by Sephadex G-100 column chromatography. The symbols represent the following trinucleotides: ○, UAA; △, UAG; □, UGA. Portions of pooled R1 and R2 fractions (prep. B, DEAE-Sephadex column shown in Fig. 1) were fractionated separately as described in *Methods*. R1 and R2 were fractionated in the absence of hemoglobin, tRNA, and myoglobin standards; hence only approximate estimates can be made of relative chromatographic mobilities.

Reactions contained the components described in *Methods*, including 4.36 μμmoles of (f[^{3}H]-Met-tRNA^f..AUG..Ribosome) complex, trinucleotides as indicated and 10 μl of each 1.8-ml fraction. Reactions were incubated 20 min. The amount of f[^{3}H]-methionine present at zero time (0.3 μμmole) was subtracted from each value. The results are summarized in Table 1 (step 4); fractions pooled were 54–65 ml for R1, and 58–70 ml for R2.

R fractionation steps are summarized in Table 1. R1 and R2 fractionated by precipitation with $(NH_4)_2SO_4$ and by DEAE-Sephadex column chromatography were purified more than 15- and 55-fold, respectively, with an over-all recovery of 74 and 45 per cent. Sephadex G-100 column chromatography yielded UAG-dependent R1 and UGA-dependent R2 purified 45- and >200-fold, respectively.

In other studies, R1 and R2 were found to be inactivated 50 per cent by incubation for four minutes at 54° and 59°, respectively. No change in the UAG/UAA or UGA/UAA ratios was detected with partially inactivated R preparations. These results and the fractionation studies suggest that both R1 and R2 are single molecular species which differ in codon specificity.

As shown in Table 2, R1 activity is not stimulated by R2; similarly, R2 is not stimulated by R1. It remains to be determined whether the small inhibition observed when R1 and R2 were assayed together is due to competition between R factors for binding sites on ribosomes. However, the results suggest that termination is dependent upon either R1 or R2, but not upon both factors simultaneously.

The responses of R1 and R2 to terminator codons at 0.012–0.030 M Mg^{++} are

TABLE 1. *Summary of R purification.*

R Prep.	Fraction	Protein (mg)	Codon	μUnits* of R Total	Per mg protein	Purification	Recovery (%)
(A) 1.	Supernatant	4400	UAA	4000	0.91	1	100
			UAG	2300	0.52	1	100
			UGA	3600	0.82	1	100
2.	0–55% $(NH_4)_2SO_4$	2200	UAA	4850	2.2	2	120
			UAG	1750	0.8	2	77
			UGA	3750	1.7	2	104
3.	DEAE-Sephadex						
	R1	184	UAA	2050	11		51
			UAG	1700	9.2	18	74
			UGA	0	0		
	R2	35	UAA	1150	33		29
			UAG	0	0		
			UGA	1600	47	57	45
(B) 4.	Sephadex G-100						
	R1 (applied)	2.51	UAA	77	31		
			UAG	85	34	13	100
	(Recovered)	0.78	UAA	85	109		
			UAG	94	120	45	110
	R2 (applied)	1.21	UAA	94	78		
			UGA	188	156	~38	100
	(Recovered)	0.08 (±0.02)	UAA	45	~560		
			UGA	90	~1120	~200	48

The procedures employed for fractionation of R are described under *Methods*. R (prep. A) was used for fractionation steps 1–3; R (prep. B, shown in Figs. 1 and 2) was used for step 4. The supernatant fraction of prep. B, contained 4.1, 2.7, and 4.1 μunits of R/mg of protein for UAA, UAG and UGA, respectively.

Each reaction for the assay of R contained the components described under *Methods*, including 6.05 μμmoles of (f[^{3}H]-Met-tRNA^f··AUG··Ribosome) complex (6.65 μμmoles of TCA-precipitable f[^{3}H]-Met-tRNA^f per reaction) and the indicated trinucleotide.

* One R unit is defined as the amount of R required for the formation of 1.0 μmole of f[^{3}H]-methionine at 30°C/min due to the addition of 1.5×10^{-4} M of the terminator trinucleotide indicated (f[^{3}H]-methionine formed in the absence of trinucleotide is subtracted from each value).

TABLE 2. *Activity of separated and combined R factors.*

R (μg protein)	μμMoles f[³H]-Methionine Formed Due to Codon	
	+UAG	+UGA
2.5 R1	1.10	−0.05
0.8 R2	0.03	1.31
2.5 R1 + 0.8 R2	0.90	1.19

Each reaction contained the components described in *Methods*, including 6.05 μμmoles of (f[³H]-Met-tRNA^f··AUG··ribosome) complex (6.65 μμmoles of TCA-precipitable f[³H]-Met-tRNA^f were present per reaction); trinucleotides, and R1 and R2 fractions (prep. B, DEAE-Sephadex fractions of Fig. 1) where indicated. The amount of f[³H]-methionine formed in the absence of terminator trinucleotides, 0.08, and 0.03 or 0.02 μμmole, for R1, R2, and R1 + R2, respectively, were subtracted from the values shown.

shown in Table 3. N-Formyl-[³H]-Met-tRNA dissociates from ribosomes rapidly at the lower Mg^{++} concentration (see Table 3 legend); in consequence, the rate of f[³H]-methionine formation was lower at 0.012 than at 0.030 M Mg^{++}. However, the Mg^{++} concentrations tested did not influence the specificity of R for terminator codons.

The effect of antibiotics is shown in Table 4. The formation of f-methionine was inhibited 97, 62, 62, and 20 per cent by tetracycline, streptomycin, sparsomycin, and chloramphenicol, respectively. No effect was observed with fusidic acid at $3 \times 10^{-4} M$ or, in other studies, at $2 \times 10^{-3} M$.

Discussion.—Fractionation of R preparations resulted in the separation of two R factors; R1, purified 45-fold, corresponds to UAA and UAG, and R2, purified approximately 200-fold, corresponds to UAA and UGA. Both fractionation and temperature sensitivity data suggest that R1 and R2 are single molecular species which differ in specificity for codons.

Since R1 and R2 correspond to different sets of codons, terminator codons must be translated in the correct phase by specific molecules. Thus, R-dependent termination is a consequence of codon translation rather than the *absence* of translation. On the basis of these data, a class of suppressors can be predicted in which molecules that normally translate terminator codons erroneously recognize codons for amino acids.

TABLE 3. *Response of R to codons at 0.012–0.03 M Mg^{++}.*

	Addition	μμMoles f[³H]-Methionine Formed (Mg^{++} Molarity)	
		0.012	0.030
R1	None	0.12	0.15
	UAA	0.47	0.86
	UAG	0.53	1.20
	UGA	0.08	0.10
R2	None	0.02	0.14
	UAA	0.22	0.77
	UAG	0.02	0.14
	UGA	0.33	1.08

Each reaction contained the components described under *Methods* (unless indicated otherwise), including 5.70 μμmoles of (f[³H]-Met-tRNA^f··AUG··Ribosome) complex (6.9 μμmoles of TCA-precipitable f[³H]-Met-tRNA^f per reaction); trinucleotides; and 8.0 or 1.5 μg of R1 or R2 protein, respectively (prep. A, DEAE-Sephadex fractions) where indicated. The amount of f[³H]-methionine present at zero time (0.30 μμmole) was subtracted from each value. Control reactions incubated at 0.012 or 0.030 M Mg^{++} contained 2.7 or 6.0 μμmoles of ribosomal-bound f[³H]-Met-tRNA^f, respectively, as determined by the AA-tRNA binding assay.[7]

TABLE 4. *Effect of antibiotics.*

Addition	Termination assay (UAA-dependent f[^{3}H]-methionine formed)	Binding assay (f[^{3}H]-MET-tRNA^f· ribosome complex)
	($\mu\mu$moles)	
None	0.74	5.36
+ Tetracycline	0.02	5.08
+ Streptomycin	0.28	5.36
+ Sparsomycin	0.28	5.36
+ Chloramphenicol	0.59	5.26
+ Fusidic acid	0.74	5.45

Each termination reaction contained the components described under *Methods*, including 5.52 $\mu\mu$moles of (f[^{3}H]-Met-tRNA^f··AUG··Ribosome) complex (6.35 $\mu\mu$moles of TCA-precipitable f[^{3}H]-Met-tRNA^f per reaction); 1.5 μg of R1 protein (prep. B, DEAE-Sephadex fraction shown in Fig. 1); 0.300 A^{260} units of UAA; and 3×10^{-4} M of the indicated antibiotic. The amount of f[^{3}H]-methionine present at zero time (0.33 $\mu\mu$mole) was subtracted from each value.

The effect of antibiotics upon the dissociation of the (f[^{3}H]-Met-tRNA^f··AUG··Ribosome) complex was determined after incubation of reactions identical to those described above (except that R was omitted) by the AA-tRNA binding assay.[17]

The pattern of codon degeneracy found with R1 resembles that found with some species of AA-tRNA; i.e., the molecule interacting with A at the *third* base position of mRNA codons also interacts with G.[9, 10] Alternate recognition of A and G in the *first* position has been found with initiator codons translated by f-Met-tRNA.[11] However, the degeneracy pattern observed with terminator codons for R2 (UAA and UGA) differs markedly from patterns found with AA-tRNA. An equivalence of A and G at the *second* base position of codons, and recognition only of A at the third position of codons, have not been found previously.

R factors are nondialyzable; they are inactivated by incubation with trypsin, N-ethylmaleimide, or at 55–65°, but not by incubation with diisopropylfluorophosphate, T1 RNase,[1] pancreatic RNase A,[1, 2] or by periodate oxidation.[2] Thus, R factors resemble proteins with free sulfhydryl groups. Possible roles for R factors in termination include translation of terminator codons, deacylation of peptidyl-tRNA, and conversion of 70S ribosomes to 30S and 50S subunits.[12]

N-Formyl-methionine release from ribosomes was inhibited completely by tetracycline, and partially by streptomycin and sparsomycin, but not by fusidic acid. Tetracycline also completely inhibits (*a*) initiation factor and GTP-dependent binding of f-Met-tRNA to 30S ribosomes in response to initiator codons,[13] (*b*) NH$_4$$^+$-dependent binding of Phe-tRNA to ribosomes in response to Phe-codons,[14] and (*c*) release of f-methionine from ribosomes in response to terminator codons; hence, these codons may be translated at the same ribosomal site, located at least partially on the 30S ribosome. Streptomycin also interacts with the 30S ribosome, thereby altering the fidelity of codon recognition. The puromycin-dependent deacylation of peptidyl-tRNA is *competitively* inhibited by sparsomycin;[15] inhibition was observed with 50S ribosomes alone (cited in ref. 15). Fusidic acid inhibits GTP hydrolysis catalyzed by G.[16] These results suggest (*a*) that termination requires both 30S and 50S ribosomal subunits; (*b*) that initiator, amino acid, and terminator codons may be recognized at the same ribosomal site; (*c*) that the effect of streptomycin may induce errors in the translation of terminator codons; and (d) that translocation is not required for the release of f[³H]-methionine from ribosomes.

It is clear that the specificity of terminator-codon translation is dependent upon R factors. The simplest explanation is that terminator codons are recognized on ribosomes by corresponding R factors. Alternatively, terminator codons may be translated by other molecules, such as RNA, which interact with appropriate species of R.

Summary.— Release factor preparations were separated into two components; release factor 1, purified 45-fold, corresponds to the terminator codons, UAA and UAG; release factor 2, purified 200-fold, corresponds to UAA and UGA.

The authors take great pleasure in acknowledging the aid of Dr. Bruce Schrier, Gregory Milman, and Joseph Goldstein, and the excellent technical assistance of Mrs. Theresa Caryk and Mr. Wayne Kemper.

The following abbreviations were used: f-Met-tRNA, formyl-methionyl-tRNA; DTT, dithiothreitol; tRNA, transfer RNA; Tris, tris(hydroxymethyl)aminomethane; TCA, trichloroacetic acid; DEAE-cellulose, O-(diethylaminoethyl) cellulose; EDTA, ethylenediaminetetraacetate; GTP, guanosine 5'-triphosphate.

[1] Caskey, C. T., R. Tompkins, E. Scolnick, and M. Nirenberg, *Science*, in press.
[2] Capecchi, M. R., these PROCEEDINGS, **58**, 1144 (1967).
[3] Gillam, I., S. Millward, O. Blew, M. von Tigerstrom, E. Wimmer, and G. M. Tener, *Biochemistry*, **6**, 3043 (1967).
[4] Lucas-Lenard, J., and F. Lipmann, these PROCEEDINGS, **57**, 1050 (1967).
[5] Nirenberg, M. W., in *Methods in Enzymology*, ed. S. P. Colowick and N. O. Kaplan (New York: Academic Press, 1963), vol. 6, p. 17.
[6] Conway, T. W., and F. Lipmann, these PROCEEDINGS, **52**, 1462 (1964).
[7] Allende, J. E., and H. Weissbach, *Biochem. Biophys. Res. Commun.*, **28**, 82 (1967).
[8] Lowry, O. H., N. J. Rosebrough, A. L. Farr, and R. J. Randall, *J. Biol. Chem.*, **193**, 265 (1951).
[9] Nirenberg, M. W., P. Leder, M. Bernfield, R. Brimacombe, F. Rottman, and C. O'Neal, these PROCEEDINGS, **53**, 1161 (1965).
[10] Crick, F. H. C., *J. Mol. Biol.*, **19**, 548 (1966).
[11] Clark, B. F. C., and K. A. Marker, *Nature*, **211**, 378 (1966).
[12] Mangiarotti, G., and D. Schlessinger, *J. Mol. Biol.*, **20**, 123 (1966).
[13] Sarkar, S., *Federation Proc.* (Abstracts), **27**, 398 (1968).
[14] Seeds, N. W., J. A. Retsema, and T. W. Conway, *J. Mol. Biol.*, **27**, 421 (1967).
[15] Jayaraman, J., and I. H. Goldberg, *Biochemistry*, **7**, 418 (1968).
[16] Tanaka, N., T. Kinoshita, and H. Masukawa, *Biochem. Biophys. Res. Commun.*, **20**, 278 (1968).
[17] Nirenberg, M. W., and P. Leder, *Science*, **145**, 1399 (1964).

THE DEPENDENCE OF CELL- FREE PROTEIN SYNTHESIS IN E. COLI UPON NATURALLY OCCURRING OR SYNTHETIC POLYRIBONUCLEOTIDES

BY MARSHALL W. NIRENBERG AND J. HEINRICH MATTHAEI*

NATIONAL INSTITUTES OF HEALTH, BETHESDA, MARYLAND

Communicated by Joseph E. Smadel, August 3, 1961

A stable cell-free system has been obtained from *E. coli* which incorporates C^{14}-valine into protein at a rapid rate. It was shown that this apparent protein synthesis was energy-dependent, was stimulated by a mixture of L-amino acids, and was markedly inhibited by RNAase, puromycin, and chloramphenicol.[1] The present communication describes a novel characteristic of the system, that is, a requirement for template RNA, needed for amino acid incorporation even in the presence of soluble RNA and ribosomes. It will also be shown that the amino acid incorporation stimulated by the addition of template RNA has many properties expected of *de novo* protein synthesis. Naturally occurring RNA as well as a synthetic polynucleotide were active in this system. The synthetic polynucleotide appears to contain the code for the synthesis of a "protein" containing only one amino acid. Part of these data have been presented in preliminary reports.[2, 3]

Methods and Materials.—The preparation of enzyme extracts was modified in certain respects from the procedure previously presented.[1] *E. coli* W3100 cells harvested in early log phase were washed and were disrupted by grinding with alumina (twice the weight of washed cells) at 5° for 5 min as described previously.[1] The alumina was extracted with an equivalent weight of buffer containing 0.01 M Tris(hydroxymethyl)aminomethane, pH 7.8, 0.01 M magnesium acetate, 0.06 M KCl, 0.006 M mercaptoethanol (standard buffer). Alumina and intact cells were removed by centrifugation at 20,000 × g for 20 min. The supernatant fluid was decanted, and 3 μg DNAase per ml (Worthington Biochemical Co.) were added, rapidly reducing the viscosity of the suspension, which was then centrifuged again at 20,000 × g for 20 min. The supernatant fluid was aspirated and was centrifuged at 30,000 × g for 30 min to clear the extract of remaining debris. The liquid layer was aspirated (S-30) and was centrifuged at 105,000 × g for 2 hr to sediment the ribosomes. The supernatant solution (S-100) was aspirated, and the solution just above the pellet was decanted and discarded. The ribosomes were washed by resuspension in the standard buffer and centrifugation again at 105,000 × g for 2 hr. Supernatant fluid was discarded and the ribosomes were suspended in standard buffer (W-Rib). Fractions S-30, S-100, and W-Rib were dialyzed against 60 volumes of standard buffer overnight at 5° and were divided into aliquots for storage at −15°.

In some cases, fresh S-30 was incubated for 40 min at 35°. The reaction mixture components in μmoles per ml were as follows: 80 Tris, pH 7.8; 8 magnesium acetate; 50 KCl; 9 mercaptoethanol; 0.075 each of 20 amino acids;[1] 2.5 ATP, K salt; 2.5 PEP, K salt; 15 μg PEP kinase (Boehringen & Sons, Mannheim, Germany). After incubation, the reaction mixture was dialyzed at 5° for 10 hr against 60 volumes of standard buffer, changed once during the course of dialysis. The incubated S-30 fraction was stored in aliquots at −15° until needed (Incubated-S-30).

RNA fractions were prepared by phenol extraction using freshly distilled phenol. Ribosomal RNA was prepared from fresh, washed ribosomes obtained by the method given above. In later RNA preparations, a 0.2% solution of sodium dodecyl sulfate recrystallized by the method of Crestfield *et al.*[4] was added to the suspension of ribosomes before phenol treatment. The suspension was shaken at room temperature for 5 min. Higher yields of RNA appeared to be obtained when the sodium dodecyl sulfate step was used; however, good RNA preparations were also obtained when this step was omitted. An equal volume of H_2O-saturated phenol was added to ribosomes suspended in standard buffer after treatment with sodium dodecyl sulfate, and the suspension was shaken vigorously at room temperature for 8–10 min. The aqueous phase was aspirated from the phenol phase after centrifugation at 1,450 × g for 15 min. The aqueous layer was extracted two more times in the same manner, using 1/2 volume of H_2O-saturated phenol in each case. The final aqueous phase was chilled to 5° and NaCl was added to a final concentration of 0.1%. Two volumes of ethyl alcohol at −20° were added with stirring to precipitate the RNA. The suspension was centrifuged at 20,000 × g for 15 min and the supernatant solution was decanted and discarded. The RNA pellet was dissolved in minimal concentrations of standard buffer (minus mercaptoethanol) by gentle homogenization in a glass Potter-Elvehjem homogenizer (usually the volume of buffer used was about 1/3 the volume of the original ribosome suspension). The opalescent solution of RNA was dialyzed for 18 hr against 100 volumes of standard buffer (minus mercaptoethanol) at 5°. The dialyzing buffer was changed once. After dialysis, the RNA solution was centrifuged at 20,000 × g for 15 min and the pellet was discarded. The RNA solution, which contained less than 1% protein, was divided into aliquots and was stored at −15° until needed.

Soluble RNA was prepared from 105,000 × g supernatant solution by the phenol extraction method described above. Soluble RNA was also stored at −15°. Alkali-degraded RNA was prepared by incubating RNA samples with 0.3 M KOH at 35° for 18 hr. The solutions then were neutralized and dialyzed against standard buffer (minus mercaptoethanol). RNAase-digested samples of RNA were prepared by incubating RNA with 2 μg per ml of crystalline

RNAase (Worthington Biochemical Company) at 35° for 60 min. RNAase was destroyed by four phenol extractions performed as given above. After the last phenol extraction, the samples were dialyzed against standard buffer minus mercaptoethanol. RNA samples were treated with trypsin by incubation with 20 μg per ml of twice recrystallized trypsin (Worthington Biochemical Company) at 35° for 60 min. The solution was treated four times with phenol and was dialyzed in the same manner.

The radioactive amino acids used, their source, and their respective specific activities are as follows: U-C^{14}-glycine, U-C^{14}-L-isoleucine, U-C^{14}-L-tyrosine, U-C^{14}-L-leucine, U-C^{14}-L-proline, L-histidine-2(ring)-C^{14}, U-C^{14}-L-phenylalanine, U-C^{14}-L-threonine, L-methionine (methyl-C^{14}), U-C^{14}-L-arginine, and U-C^{14}-L-lysine obtained from Nuclear-Chicago Corporation, 5.8, 6.2, 5.95, 6.25, 10.5, 3.96, 10.3, 3.9, 6.5, 5.8, 8.3 mC/mM, respectively; C^{14}-L-aspartic acid, C^{14}-L-glutamic acid, C^{14}-L-alanine, obtained from Volk, 1.04, 1.18, 0.75 m C/mM, respectively; D-L-tryptophan-3 C^{14}, obtained from New England Nuclear Corporation, 2.5 mC/mM; S^{35}-L-cystine obtained from the Abbott Laboratories, 2.4 mC/mM; U-C^{14}-L-serine obtained from the Nuclear-Chicago Corporation, 0.2 mC/mM. Other materials and methods used in this study are described in the accompanying paper.[1] All assays were performed in duplicate.

Results.—Stimulation by ribosomal RNA: In the previous paper,[1] it was shown that DNAase markedly decreased amino acid incorporation in this system after 20 min. For the purpose of this investigation, 30,000 × *g* supernatant fluid fractions previously incubated with DNAase and other components of the reaction mixtures (Incubated-S-30 fractions) were used for many of the experiments.

Figure 1 shows that incorporation of C^{14}-L-valine into protein by Incubated-S-30 fraction was stimulated by the addition of purified *E. coli* soluble RNA. Maximal stimulation was obtained with approximately 1 mg soluble RNA. In some experiments, increasing the concentration 5-fold did not further stimulate the system. Soluble RNA was added to all reaction mixtures unless otherwise specified.

Figure 2 demonstrates that *E. coli* ribosomal RNA preparations markedly stimu-

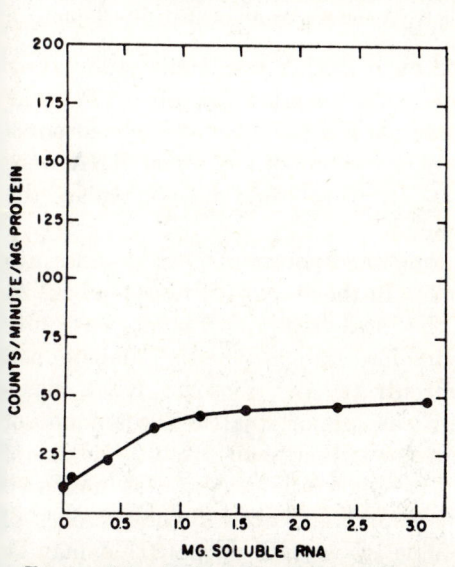

Fig 1.—Stimulation of amino acid incorporation into protein by *E. coli* soluble RNA. Composition of reaction mixtures is specified in Table 1. Samples were incubated at 35° for 20 min. Reaction mixtures contained 4.4 mg. of Incubated-S-30 protein.

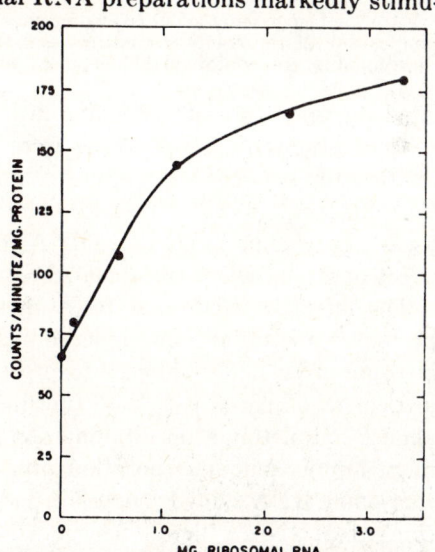

Fig. 2.—Stimulation of amino acid incorporation into protein by *E. coli* ribosomal RNA in the presence of soluble RNA. Composition of reaction mixtures is specified in Table 1. Samples were incubated at 35° for 20 min. Reaction mixtures contained 4.4 mg of Incubated-S-30 protein and 1.0 mg *E. coli* soluble RNA.

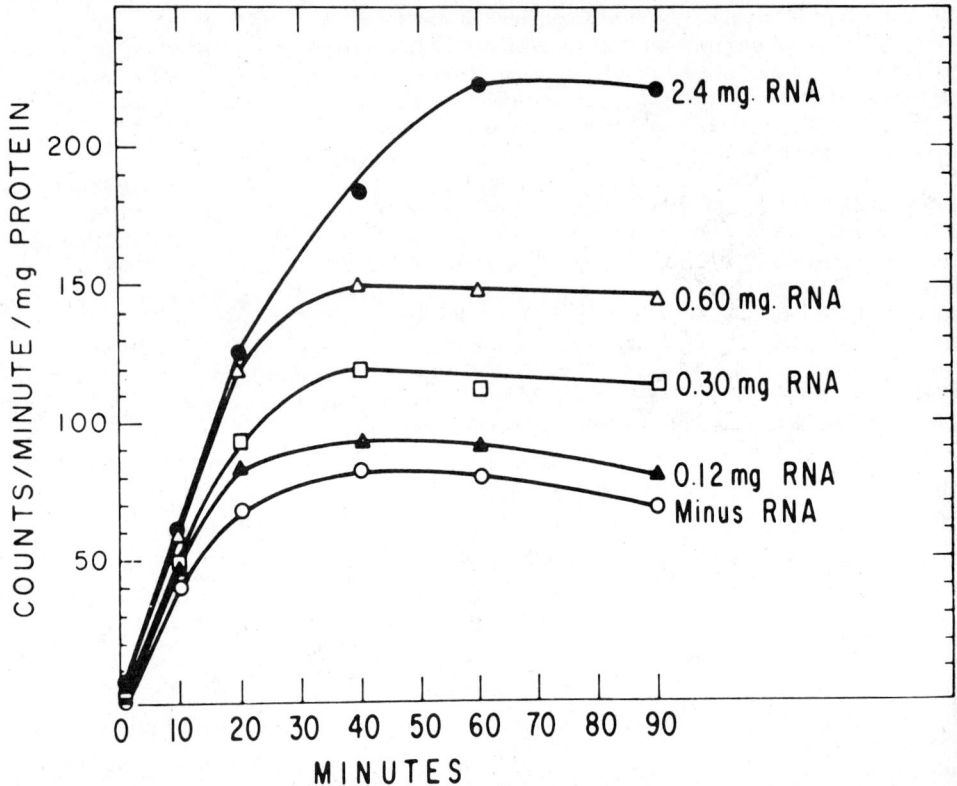

FIG. 3.—Dependence of C^{14}-L-valine incorporation into protein upon ribosomal RNA. The composition of the reaction mixtures and the incubation conditions are presented in Table 1. Reaction mixtures contained 0.98 mg of *E. coli* soluble RNA and 4.4 mg of Incubated-S-30-protein.

lated incorporation of C^{14}-valine into protein even though maximally stimulating concentrations of soluble RNA were present in the reaction mixtures. A linear relationship between the concentration of ribosomal RNA and C^{14}-valine incorporation into protein was obtained when low concentrations of ribosomal RNA were used. Increasing the soluble RNA concentration up to 3-fold did not replace the effect observed when ribosomal RNA was added.

The effect of ribosomal RNA in stimulating incorporation of C^{14}-valine into protein is presented in more detail in Figure 3. In the absence of ribosomal RNA, incorporation of C^{14}-valine into protein by the incubated-S-30 fraction was quite low when compared with S-30 (not incubated before storage at $-15°$) and stopped almost completely after 30 min. At low concentrations of ribosomal RNA, maximum amino acid incorporation into protein was proportional to the amount of ribosomal RNA added, suggesting stoichiometric rather than catalytic action of ribosomal RNA. Total incorporation of C^{14}-valine into protein was increased more than 3-fold by ribosomal RNA in this experiment even in the presence of maximally stimulating concentrations of soluble RNA. Ribosomal RNA may be added at any time during the course of the reaction, and, after further incubation, an increase in incorporation of C^{14}-valine into protein will result.

Characteristics of amino acid incorporation stimulated by ribosomal RNA: In Table 1 are presented the characteristics of C^{14}-L-valine incorporation into protein

TABLE 1
Characteristics of C^{14}-L-Valine Incorporation into Protein

Experiment no.	Addition			Counts/min/mg protein
1	− Ribosomal RNA			42
	+ " "			204
	+ " "	+ 0.15 μmole Chloramphenicol		58
	+ " "	+ 0.20 μmole Puromycin		7
	+ " "	deproteinized at zero time		8
2	− Ribosomal RNA			35
	+ " "			101
	+ " "	− ATP, PEP, PEP kinase		7
	+ " "	+ 10 μg RNAase		6
	+ " "	+ 10 μg DNAase		110
	+ Boiled Ribosomal RNA			127
	+ Ribosomal RNA, deproteinized at zero time			8
3	− Ribosomal RNA			34
	− " "	− 20 L amino acids		21
	+ " "			99
	+ " "	− 20-L-amino acids		52

The reaction mixtures contained the following in μmole/ml: 100 Tris(hydroxymethyl) aminomethane, pH 7.8; 10 magnesium acetate; 50 KCl; 6.0 mercaptoethanol; 1.0 ATP; 5.0 phosphoenolpyruvate, K salt; 20 μg phosphoenolpyruvate kinase, crystalline; 0.05 each of 20 L-amino acids minus valine; 0.03 each of GTP, CTP, and UTP; 0.015 C^{14}-L-valine (∼70,000 counts); 3.1 mg. *E. coli* ribosomal RNA where indicated, and 1.0 mg *E. coli* soluble RNA; 3.2, 3.2, and 1.4 mg of incubated-S-30 protein were present in Experiments 1, 2, and 3, respectively. In addition 4.4 mg protein of W-Rib were added in Experiment 3. Total volume was 1.0 ml. Samples were incubated at 35° for 20 min, were deproteinized with 10 per cent trichloroacetic acid, and the precipitates were washed and counted by the method of Siekevitz.[22]

stimulated by the addition of ribosomal RNA. Amino acid incorporation was strongly inhibited by 0.15 μmoles of chloramphenicol and 0.20 μmoles/ml reaction mixture of puromycin. Furthermore, the incorporation was completely dependent upon the addition of ATP and an ATP-generating system and was totally inhibited by 10 μg/ml RNAase. Equivalent amounts of DNAase had no effect upon the incorporation stimulated by the addition of ribosomal RNA. Placing a ribosomal RNA preparation in a boiling water bath for 10 min did not destroy its C^{14}-valine incorporation activity; instead, a slight increase in activity was consistently observed. However, when these RNA preparations were placed in a boiling water bath, a copious, white precipitate resulted. Upon cooling the suspension in an ice bath, the precipitate immediately dissolved.

The data of Table 1 also demonstrate that the incorporation of amino acids into protein in the presence of ribosomal RNA was further stimulated by the addition of a mixture of 20 L-amino acids, suggesting cell-free protein synthesis.

C- and N-terminal analyses of the ribosomal RNA-dependent product of the reaction were performed with carboxypeptidase and 1-fluoro-2,4-dinitrobenzene respectively (Dr. Frank Tietze kindly performed these analyses). Four per cent of the radioactivity was released from the C-terminal end and 1% was associated with the N-terminal end. The remainder of the C^{14}-label was internal. Similar results were obtained when reactions were performed using S-30 enzyme fractions which had not been treated with DNAase. Protein precipitates isolated from reaction mixtures after incubation were completely hydrolyzed with HCl, and the C^{14}-label incorporated into protein was demonstrated to be valine by paper chromatography.

Many of the experiments presented in this paper were performed with enzyme fractions prepared with DNAase added to reduce their viscosity. Ribosomal

RNA also stimulated C^{14}-valine incorporation when enzyme extracts prepared in the absence of DNAase were used.

To be effective in stimulating amino acid incorporation into protein, the ribosomal RNA *required the presence of washed ribosomes.* The data of Table 2 show

TABLE 2

The Ineffectiveness of Ribosomal RNA in Stimulating C^{14}-L-Valine Incorporation into Protein in the Presence of Ribosomes or 105,000 × g Supernatant Solutions Alone

Additions	Counts/min
Complete	51
" + 2.1 mg Ribosomal RNA	202
" − Ribosomes	17
" − Ribosomes + 2.1 mg Ribosomal RNA	20
" − Supernatant solution	36
" − Supernatant solution + 2.1 mg Ribosomal RNA	45
" Deproteinized at zero time	25

The components of the reaction mixtures and the incubation conditions are presented in Table 1. 0.86 and 3.3 mg protein were present in the ribosome (W-Rib) and 105,000 × g supernatant (S-100) fractions, respectively.

that both ribosomes and 105,000 × g supernatant solution were necessary for ribosomal RNA-dependent amino acid incorporation. No incorporation of amino acids into protein occurred when the 105,000 × g supernatant solution alone was added to ribosomal RNA preparations, demonstrating that ribosomal RNA preparations were not contaminated with intact ribosomes. This conclusion also was substantiated by showing that the activities of ribosomal RNA preparations were not destroyed by boiling, although the activities of the ribosomes were destroyed by such treatment.

The effect of ribosomal RNA upon the incorporation of seven different amino acids is presented in Table 3. The addition of ribosomal RNA increased the incorporation of every amino acid tested.

The effect shown by ribosomal RNA was not observed when other polyanions were used, such as polyadenylic acid, highly polymerized salmon sperm DNA, or a high-molecular-weight polymer of glucose carboxylic acid (Table 4). Pretreatment of ribosomal RNA with trypsin did not affect its biological activity. However, treatment of the ribosomal RNA with either RNAase or alkali resulted in a complete loss of stimulating activity. The active principle, therefore, appears to be RNA.

The sedimentation characteristics of the ribosomal RNA preparations were examined in the Spinco Model E ultracentrifuge (Fig. 4A). Particles having the characteristics of S-30, S-50, or S-70 ribosomes were not observed in these preparations. The S_{20}^W of the first peak was 23, that of the second peak 16, and that of the third, small peak, 4. Pretreatment with trypsin did not affect the S_{20}^W values of the peaks appreciably (Fig. 4C); however, treatment with RNAase completely destroyed the peaks (Fig. 4B), confirming the ancillary evidence which had suggested that the major component was high-molecular-weight RNA.

Preliminary attempts at fractionation of the ribosomal RNA were performed by means of density-gradient centrifugation employing a linear sucrose gradient. The results of one such experiment are presented in Figure 5. Amino acid incorporation activity of the RNA did not follow absorbancy at 260 mμ; instead, the activity seemed to be concentrated around fraction No. 5, which was approximately one-third of the distance from the bottom of the tube. These results again

TABLE 3

SPECIFICITY OF AMINO ACID INCORPORATION STIMULATED BY RIBOSOMAL RNA

C^{14}-Amino Acid	Addition	Counts/min/mg protein
C^{14}-L-Valine	Complete	25
"	" + Ribosomal RNA	137
C^{14}-L-Threonine	"	31
"	" + Ribosomal RNA	121
C^{14}-L-Methionine	"	121
"	" + Ribosomal RNA	177
C^{14}-L-Arginine	"	49
"	" + Ribosomal RNA	224
C^{14}-L-Phenylalanine	"	77
"	" + Ribosomal RNA	147
C^{14}-L-Lysine	"	36
"	" + Ribosomal RNA	175
C^{14}-L-Leucine	"	134
"	" + Ribosomal RNA	272
"	" Deproteinized at zero time	6

The composition of the reaction mixtures are presented in Table 1. The mixture of 20 L-amino acids included all amino acids except the C^{14}-amino acid added to one reaction mixture. Reaction mixtures contained 4.4 mg Incubated-S-30 protein. Samples were incubated at 35° for 60 min. 2.1 mg ribosomal RNA were added where indicated.

TABLE 4

RIBOSOMAL RNA CONTROL EXPERIMENTS DESCRIBED IN TEXT

Experiment No.	Addition	Counts/min/mg protein
1	Complete	54
"	+ 2.4 mg Ribosomal RNA	144
"	+ 2.0 mg Polyadenylic acid	10
"	+ 2.0 mg Salmon sperm DNA	41
"	+ 2.0 mg Polyglucose carboxylic acid	49
"	+ 2.4 mg Ribosomal RNA, deproteinized at zero time	7
2	Complete	39
"	+ 2.0 mg Ribosomal RNA*	150
"	+ 2.1 mg Ribosomal RNA preincubated with trypsin*	166
"	+ 2.0 mg Ribosomal RNA preincubated with RNAase*,†	47
"	Deproteinized at zero time	8
3	Complete	20
"	+ 1.2 mg Ribosomal RNA	82
"	+ 1.2 mg Alkali degraded ribosomal RNA†	21
"	Deproteinized at zero time	7

The composition of the reaction mixtures and the incubation conditions are given in Table 1. 4.4, 3.2, and 4.4 mg Incubated-S-30 protein were present in Experiments 1, 2, and 3, respectively. 2.4, 0.98, and 0 mg *E. coli* soluble RNA were present in Experiments 1, 2, and 3, respectively.
* Ribosomal RNA preparations were deproteinized by phenol extraction after enzymatic digestion as specified under *Methods and Materials*.
† mg Ribosomal RNA refers to RNA concentration before digestion.

demonstrate that the activity was not associated with a soluble RNA fraction, present in maximum concentration in fraction No. 11, near the top of the tube. In addition, all amino acid incorporation analyses were performed in the presence of added soluble RNA, and the addition of more soluble RNA would not stimulate C^{14}-L-valine incorporation into protein.

Effects of RNA obtained from different species: The data of Table 5 demonstrate that RNA from different sources stimulates C^{14}-valine incorporation into protein. Yeast ribosomal RNA prepared by the method of Crestfield et al.[4] was considerably more effective in stimulating incorporation than equivalent amounts of *E. coli* ribosomal RNA. Yeast ribosomal RNA prepared by this method has little or no amino acid acceptor activity and has a molecular weight of about 29,000.[7] Tobacco mosaic virus RNA prepared by phenol extraction and having a molecular weight of

TABLE 5
Stimulation of Amino Acid Incorporation by RNA Fractions Prepared from Different Species

Additions	Counts/min/mg protein
None	42
+ 0.5 mg E. coli ribosomal RNA	75
+ 0.5 mg Yeast ribosomal RNA	430
+ 0.5 mg Tobacco mosaic virus RNA	872
+ 0.5 mg Ehrlich ascites tumor microsomal RNA	65

The components of the reaction mixtures and the incubation conditions are presented in Table 1. Reaction samples contained 1.9 mg Incubated-S-30 protein.

approximately 1,700,000† stimulated amino acid incorporation strongly. Marked stimulation due to tobacco mosaic virus RNA was observed also with *E. coli* enzyme extracts which had not been treated with DNAase. More complete details of this work will be presented in a later publication.

Stimulation of amino acid incorporation by synthetic polynucleotides: The data of Figure 6 show that the addition of 10 µg of polyuridylic acid‡ per ml of reaction mixture resulted in a remarkable stimulation of C^{14}-L-phenylalanine incorporation. Phenylalanine incorporation was almost completely dependent upon the addition of polyuridylic acid, and incorporation proceeded, after a slight lag period, at a linear rate for approximately 30 min.

The data of Table 6 demonstrate that no other polynucleotide tested could replace polyuridylic acid. The absolute specificity of polyuridylic acid was con-

TABLE 6
Polynucleotide Specificity for Phenylalanine Incorporation

Experiment no.	Additions	Counts/min/mg protein
1	None	44
	+ 10 µg Polyuridylic acid	39,800
	+ 10 µg Polyadenylic acid	50
	+ 10 µg Polycytidylic acid	38
	+ 10 µg Polyinosinic acid	57
	+ 10 µg Polyadenylic-uridylic acid (2/1 ratio)	53
	+ 10 µg Polyuridylic acid + 20 µg polyadenylic acid	60
	Deproteinized at zero time	17
2	None	75
	+ 10 µg UMP	81
	+ 10 µg UDP	77
	+ 10 µg UTP	72
	Deproteinized at zero time	6

Components of the reaction mixtures are presented in Table 1. Reaction mixtures contained 2.3 mg Incubated-S-30 protein. 0.02 µmoles U-C^{14}-L-phenylalanine (~125,000 counts/minute) was added to each reaction mixture. Samples were incubated at 35° for 60 min.

firmed by demonstrating that randomly mixed polymers of adenylic and uridylic acid‡ (Poly A-U, 2/1 ratio and 4/1 ratio) were inactive in this system. A solution of polyuridylic acid and polyadenylic acid (which forms triple-stranded helices) had no activity whatsoever, suggesting that single-strandedness is a necessary requisite for activity. Experiment 2 in Table 6 demonstrates that UMP, UDP, or UTP were unable to stimulate phenylalanine incorporation.

The data of Table 7 demonstrate that both ribosomes and $100,000 \times g$ supernatant solution, as well as ATP and an ATP-generating system, were required for the polyuridylic acid–dependent incorporation of phenylalanine. Incorporation was inhibited by puromycin, chloramphenicol, and RNAase. The incorpora-

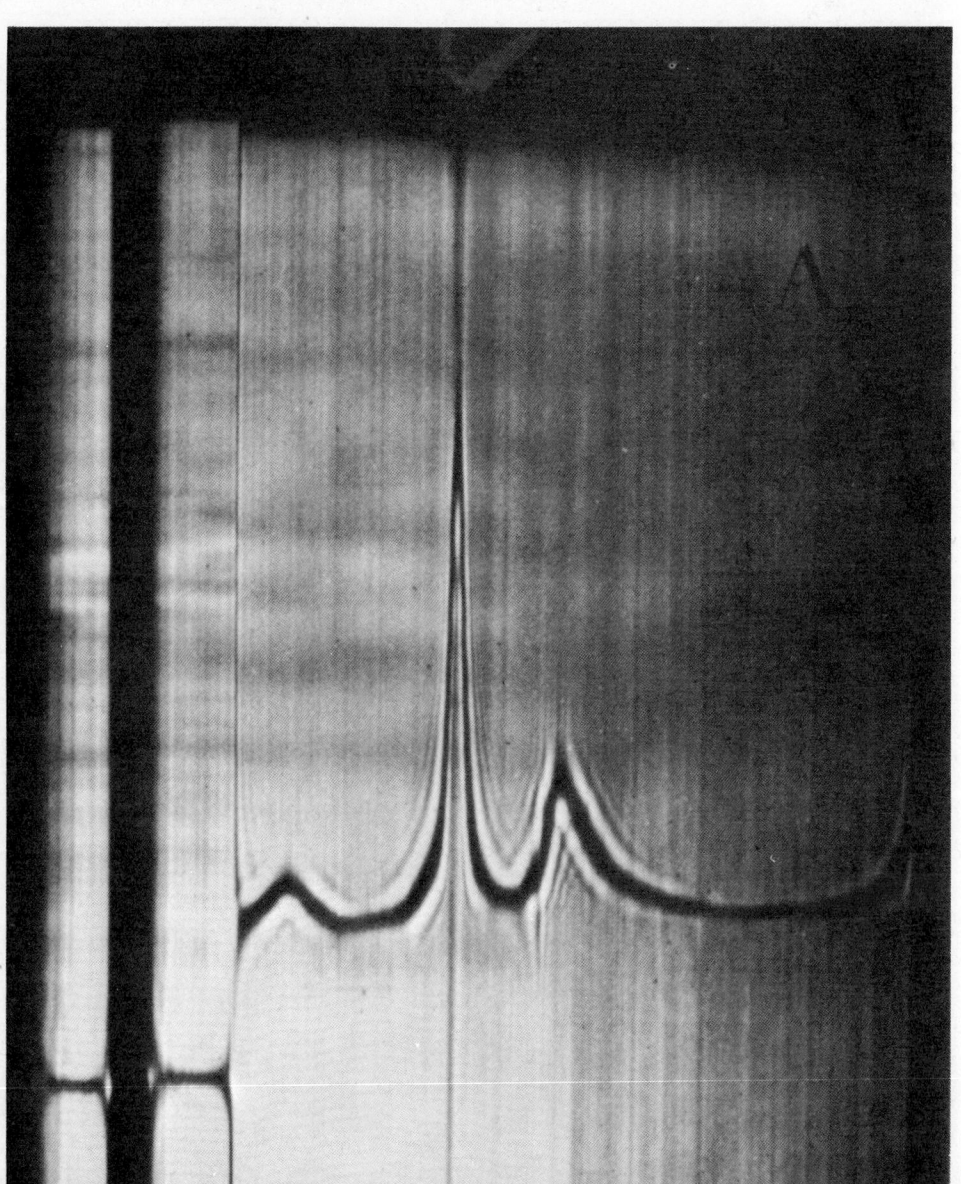

FIG. 4.—*E. coli* ribosomal RNA preparations. (*A*) Untreated (above). (*B*) digested with RNAase; (*C*) digested with trypsin. Preparation and digestion of samples presented under *Methods and Materials*. 9.8 and 10.5 mg/ml RNA were present in *A* and *C*. 11.5 mg/ml RNA was present in *B*
(*Continued on facing page*)

tion was not inhibited by addition of DNAase. Omitting a mixture of 19 L-amino acids did not inhibit phenylalanine incorporation, suggesting that polyuridylic acid stimulated the incorporation of L-phenylalanine alone. This conclusion was substantiated by the data presented in Table 8. Polyuridylic acid had little effect in stimulating the incorporation of 17 other radioactive amino acids. Each labeled amino acid was tested individually, and these data, corroborating the results given in Table 8, will be presented in a subsequent publication.

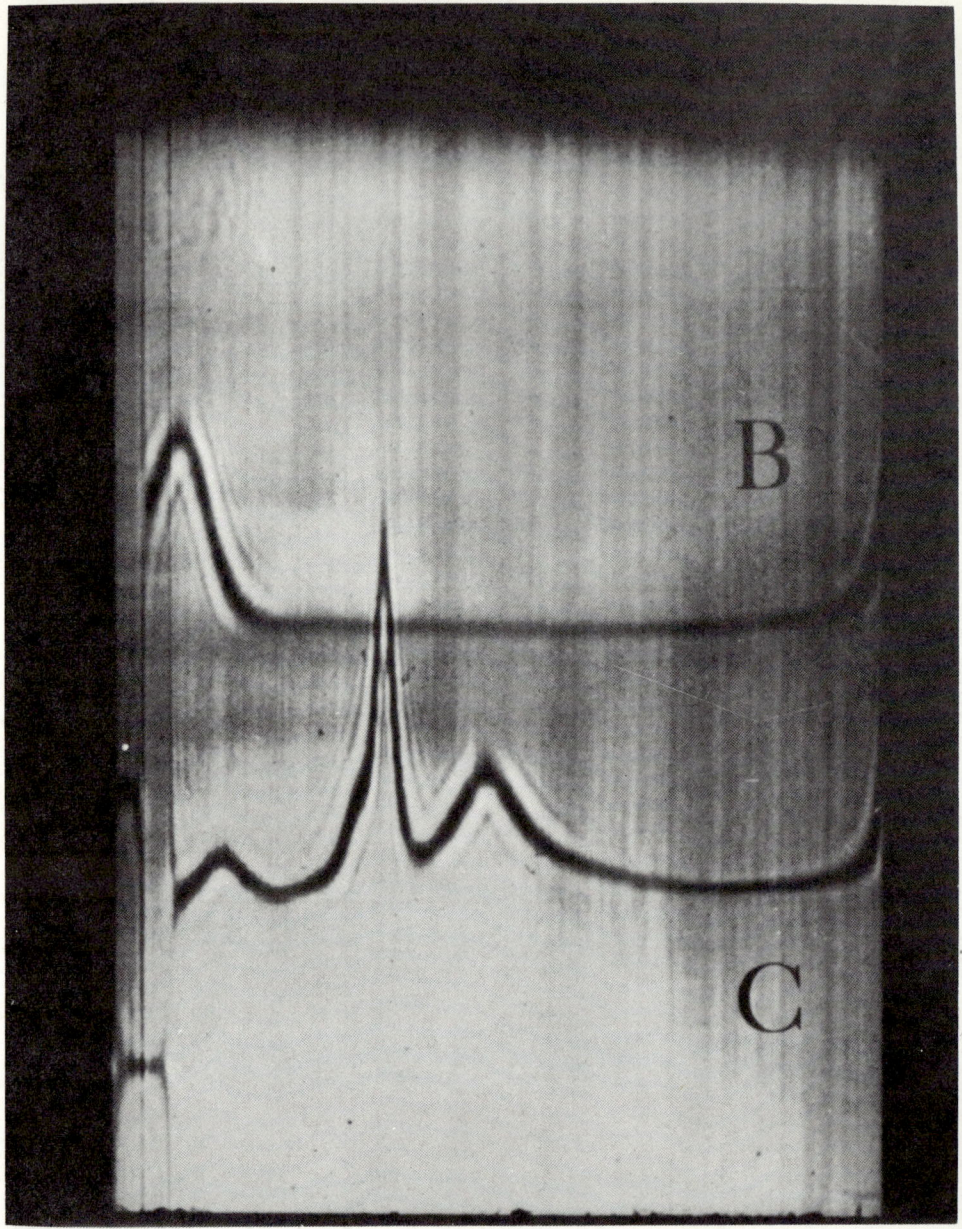

(*Fig. 4—continued*)
before digestion. Photographs were taken in a model E Spinco ultracentrifuge equipped with schlieren optics.

The product of the reaction was partially characterized and the results are presented in Table 9. The physical characteristics of the product of the reaction resembled those of authentic poly-L-phenylalanine, for, unlike many other polypeptides and proteins, both the product of the reaction and the polymer were resistant to hydrolysis by 6N HCl at 100° for 8 hr but were completely hydrolyzed by 12N HCl at 120–130° for 48 hr.

Poly-L-phenylalanine is insoluble in most solvents[25] but is soluble in 33 per cent

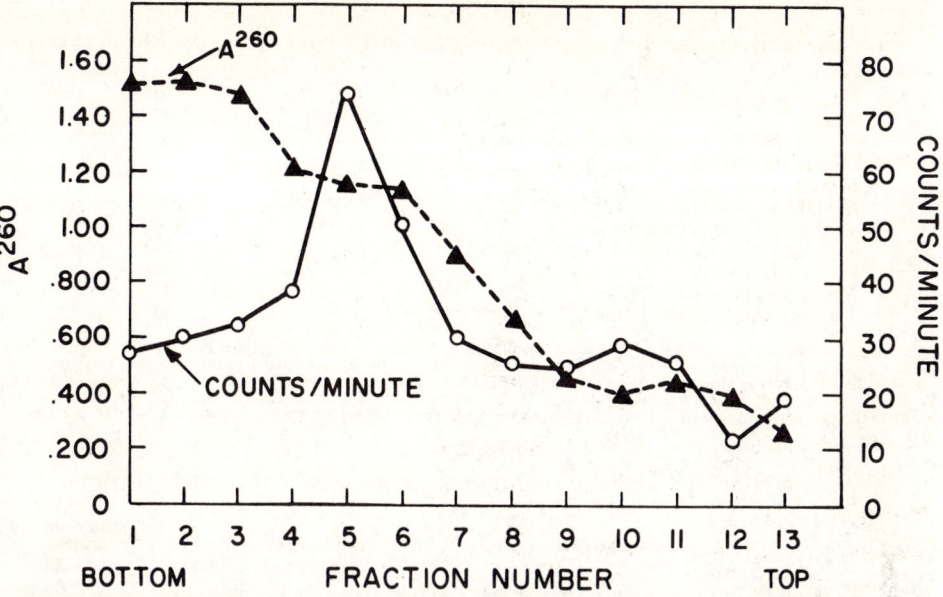

Fig. 5.—Sucrose density-gradient centrifugation of ribosomal RNA. A linear gradient of sucrose concentration ranging from 20 per cent at the bottom to 5 per cent at the top of the tube was prepared.[23] The sucrose solutions (4.4 ml total volume) contained 0.01 M Tris, pH 7.8, 0.01 M Mg acetate and 0.06 M KCl. 0.4 ml of ribosomal RNA (4.6 mg) was layered on top of each tube which was centrifuged at 38,000 × g for 4.5 hours at 3° in a swinging bucket rotor, Spinco type SW-39, using a Spinco Model L ultracentrifuge. 0.30 ml fractions were collected after piercing the bottom of the tube.[24]

0.025 ml aliquots diluted to 0.3 ml with H_2O were used for A^{260} measurements. 0.25 ml aliquots were used for amino acid incorporation assays. Reaction mixtures contained the components presented in Table 1. 0.7 mg of *E. coli* soluble RNA and 2.2 mg Incubated-S-30 protein were added. Control assays plus 0.25 ml 12.5 per cent sucrose in place of fractions gave 79 counts/min. This figure was subtracted from each value. Total volume was 0.7 ml. Samples were incubated at 35° for 20 min.

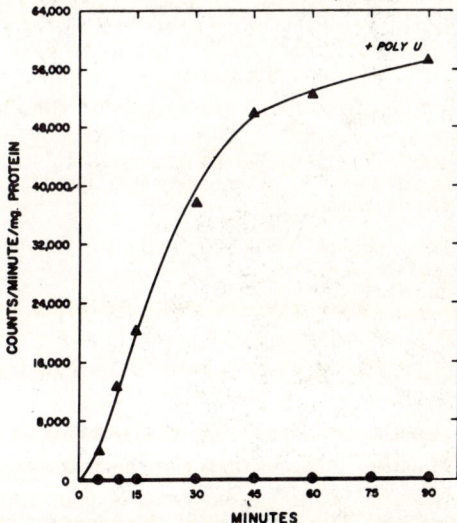

Fig. 6.—Stimulation of U-C^{14}-L-phenylalanine incorporation by polyuridylic acid. ● without polyuridylic acid; ▲ 10 μg polyuridylic acid added. The components of the reaction mixtures and the incubation conditions are given in Table 1. 0.024 μmole U-C^{14}-L-phenylalanine (∼500,000 counts/min) and 2.3 mg Incubated-S-30 protein were added/ml of reaction mixture.

TABLE 7

CHARACTERISTICS OF POLYURIDYLIC ACID–DEPENDENT PHENYLALANINE INCORPORATION

Additions	Counts/min/mg protein
Minus polyuridylic acid	70
None	29,500
Minus 100,000 × g supernatant solution	106
Minus ribosomes	52
Minus ATP, PEP, and PEP kinase	83
+ 0.02 µmoles puromycin	7,100
+ 0.31 µmoles chloramphenicol	12,550
+ 6 µg RNAase	120
+ 6 µg DNAase	27,600
Minus amino acid mixture	31,700
Deproteinized at zero time	30

The components of the reaction mixtures are presented in Table 1. 10 µg of polyuridylic acid were added to all samples except the specified one. 2.3 mg of Incubated-S-30 protein were added to each reaction mixture except those in which ribosomes alone and 100,000 × g supernatant solution alone were tested. 0.7 mg W-Rib protein and 1.3 mg S-100 protein were used respectively. 0.02 µmoles U-C^{14}-L-phenylalanine, Sp. Act. = 10.3 mC/mM (~125,000 counts/minute) were added to each reaction mixture. Samples were incubated at 35° for 60 min.

TABLE 8

SPECIFICITY OF AMINO ACID INCORPORATION STIMULATED BY POLYURIDYLIC ACID

Experiment no.	C^{14}-amino acids present	Additions	Counts/min/mg protein
1	Phenylalanine	Deproteinized at zero time	25
		None	68
		+ 10 µg polyuridylic acid	38,300
2	Glycine, alanine, serine, aspartic acid, glutamic acid	Deproteinized at zero time	17
		None	20
		+ 10 µg polyuridylic acid	33
3	Leucine, isoleucine, threonine, methionine, arginine, histidine, lysine, tyrosine, tryptophan, proline, valine	Deproteinized at zero time	73
		None	276
		+ 10 µg polyuridylic acid	899
4	S^{35}-cysteine	Deproteinized at zero time	6
		None	95
		+ 10 µg polyuridylic acid	113

Components of the reaction mixtures are presented in Table 1. The unlabeled amino acid mixture was omitted. 0.015 µM of each labeled amino acid was used. The specific activities of the labeled amino acids are present in the *Methods and Materials* section. 2.3 mg of protein of preincubated S-30 enzyme fraction were added to each reaction mixture. All samples were incubated at 35° for 30 min.

TABLE 9

COMPARISON OF CHARACTERISTICS OF PRODUCT OF REACTION AND POLY-L-PHENYLALANINE

Treatment	Product of reaction	Poly-L-phenylalanine
6 N HCl for 8 hours at 100°	Partially hydrolyzed	Partially hydrolyzed
12 N HCl for 48 hours at 120–130°	Completely hydrolyzed	Completely hydrolyzed
Extraction with 33% HBr in glacial acetic acid	Soluble	Soluble
Extraction* with the following solvents: H$_2$O, benzene, nitrobenzene, chloroform, N,N-dimethylformamide, ethanol, petroleum ether, concentrated phosphoric acid, glacial acetic acid, dioxane, phenol, acetone, ethyl acetate, pyridine, acetophenone, formic acid	Insoluble	Insoluble

* The product was said to be insoluble if <0.002 gm of product was soluble in 100 ml of solvent at 24°. Extractions were performed by adding 0.5 mg of authentic poly-L-phenylalanine and the C^{14}-product of a reaction mixture (1800 counts/min) to 5.0 ml of solvent. The suspensions were vigorously shaken for 30 min at 24° and were centrifuged. The precipitates were plated and their radioactivity was determined.

HBr in glacial acetic acid.§ The product of the reaction had the same apparent solubility as authentic poly-L-phenylalanine. The product of the reaction was purified by means of its unusual solubility behavior. Reaction mixtures were deproteinized after incubation, and precipitated proteins were washed in the usual manner according to the method of Siekevitz.[22] Dried protein pellets containing added carrier poly-L-phenylalanine were then extracted with 33 per cent HBr in glacial acetic acid, and the large amount of insoluble material was discarded. Polyphenylalanine was then precipitated from solution by the addition of H_2O and was washed several times with H_2O. Seventy per cent of the total amount of C^{14}-L-phenylalanine incorporated into protein due to the addition of polyuridylic acid could be recovered by this procedure. Complete hydrolysis of the purified reaction product with $12N$ HCl followed by paper electrophoresis** demonstrated that the reaction product contained C^{14}-phenylalanine. No other radioactive spots were found.

Discussion.—In this investigation, we have demonstrated that template RNA is a requirement for cell-free amino acid incorporation. Addition of soluble RNA could not replace template RNA in this system. In addition, the density-gradient centrifugation experiments showed that the active fractions in the ribosomal RNA preparations sedimented much faster than soluble RNA. It should be noted that ribosomal RNA is qualitatively different from soluble RNA, since bases such as pseudouracil, methylated guanines, etc., found in soluble RNA, are not present in ribosomal RNA.[5]

The bulk of the RNA in our ribosomal RNA fractions may be inactive as templates, for tobacco mosaic virus RNA was 20 times as active in stimulating amino acid incorporation as equivalent amounts of *E. coli* ribosomal RNA. In addition, preliminary fractionation of ribosomal RNA indicated that only a portion of the total RNA was active.

It should be emphasized that ribosomal RNA could not substitute for ribosomes, indicating that ribosomes were not assembled from the added RNA *in toto*. The function of ribosomal RNA remains an enigma, although at least part of the total RNA is thought to serve as templates for protein synthesis and has been termed "messenger" RNA.[12-14] Alternatively, a part of the RNA may be essential for the synthesis of active ribosomes from smaller ribosomal particles.[15-21]

Ribosomal RNA may be an aggregate of subunits which can dissociate after proper treatment.[6-8] Phenol extraction of *E. coli* ribosomes yields two types of RNA molecules with S_w^{20} of 23 and 16 (Fig. 4), equivalent to molecular weights of 1,000,000 and 560,000, respectively.[9,10] These RNA species can be degraded by boiling to products having sedimentation coefficients of 13.1, 8.8, and 4.4, corresponding to molecular weights of 288,000, 144,000, and 29,000. Although the sedimentation distributions of the latter preparations suggest a high degree of homogeneity among the molecules of each class, these observations do not eliminate the possibility that the subunits are linked to one another *via* covalent bonds.[8] Preliminary evidence indicates that the subunits may be active in our system, since the supernatant solution obtained after boiling *E. coli* ribosomal RNA for 10 min and centrifugation at $105,000 \times g$ for 60 min was active. Examination of boiled ribosomal RNA with the Spinco Model E ultracentrifuge showed a dispersed peak with a sedimentation coefficient of 4–8. This may be the same material found in the sucrose density-gradient experiment (using non-boiled RNA preparations), where a small peak of activity somewhat heavier than soluble RNA was usually noted (Fig. 5).

In our system, at low concentrations of ribosomal RNA, amino acid incorporation into protein was proportional to the amount of ribosomal RNA added, suggesting a stoichiometric rather than a catalytic action of ribosomal RNA. In contrast, soluble RNA has been shown to act in a catalytic fashion.[11]

The results indicate that polyuridylic acid contains the information for the synthesis of a protein having many of the characteristics of poly-L-phenylalanine. This synthesis was very similar to the cell-free protein synthesis obtained when naturally-occurring template RNA was added, i.e., both ribosomes and 100,000 × g supernatant solutions were required, and the incorporation was inhibited by puromycin or chloramphenicol. One or more uridylic acid residues therefore appear to be the code for phenylalanine. Whether the code is of the singlet, triplet, etc., type has not yet been determined. Polyuridylic acid seemingly functions as a synthetic template or messenger RNA, and this stable, cell-free *E. coli* system may well synthesize any protein corresponding to meaningful information contained in added RNA.

Summary.—A stable, cell-free system has been obtained from *E. coli* in which the amount of incorporation of amino acids into protein was dependent upon the addition of heat-stable template RNA preparations. Soluble RNA could not replace template RNA fractions. In addition, the amino acid incorporation required both ribosomes and 105,000 × g supernatant solution. The correlation between the amount of incorporation and the amount of added RNA suggested stoichiometric rather than catalytic activity of the template RNA. The template RNA–dependent amino acid incorporation also required ATP and an ATP-generating system, was stimulated by a complete mixture of L-amino acids, and was markedly inhibited by puromycin, chloramphenicol, and RNAase. Addition of a synthetic polynucleotide, polyuridylic acid, specifically resulted in the incorporation of L-phenylalanine into a protein resembling poly-L-phenylalanine. Polyuridylic acid appears to function as a synthetic template or messenger RNA. The implications of these findings are briefly discussed.

Note added in proof. The ratio between uridylic acid units of the polymer required and molecules of L-phenylalanine incorporated, in recent experiments, has approached the value of 1:1. Direct evidence for the number of uridylic acid residues forming the code for phenylalanine as well as for the eventual stoichiometric action of the template is not yet established. As polyuridylic acid codes the incorporation of L-phenylalanine, polycytidylic acid‡ specifically mediates the incorporation of L-proline into a TCA-precipitable product. Complete data on these findings will be included in a subsequent publication.

* Supported by a NATO Postdoctoral Research Fellowship.

† Dr. Frankel-Conrat, personal communication.

‡ We thank Drs. Leon A. Heppel and Maxine F. Singer for samples of these polyribonucleotides, and Dr. George Rushizky for TMV-RNA.

§ We thank Dr. Michael Sela for this information.

** We thank Drs. William Dreyer and Elwood Bynum for performing the high-voltage electrophoretic analyses.

[1] Matthaei, J. H., and M. W. Nirenberg, these PROCEEDINGS, **47**, 1580 (1961).
[2] Matthaei, J. H., and M. W. Nirenberg, *Biochem. & Biophys. Res. Comm.*, **4**, 404 (1961).
[3] Matthaei, J. H., and M. W. Nirenberg, *Fed. Proc.*, **20**, 391 (1961).
[4] Crestfield, A. M., K. C. Smith, and F. W. Allen, *J. Biol. Chem.*, **216**, 185 (1955).
[5] Davis, F. F., A. F. Carlucci, and I. F. Roubein. *ibid.*, **234**, 1525 (1959).
[6] Hall, B. D., and P. Doty, *J. Mol. Biol.*, **1**, 111 (1959).
[7] Osawa, S., *Biochim. Biophys. Acta*, **43**, 110 (1960).
[8] Aronson, A. I., and B. J. McCarthy, *Biophys. J.*, **1**, 215 (1961).

[9] Kurland, C. G., *J. Mol. Biol.*, 2, 83 (1960).
[10] Littauer, U. Z., H. Eisenberg, *Biochim. Biophys. Acta*, 32, 320 (1959).
[11] Hoagland, M. B., and L. T. Comly, these PROCEEDINGS, 46, 1554 (1960).
[12] Volkin, E., L. Astrachan, and J. L. Countryman, *Virology*, 6, 545 (1958).
[13] Nomura, M., B. D. Hall, and S. Spiegelman, *J. Mol. Biol.*, 2, 306 (1960).
[14] Hall, B. D., and S Spiegelman, these PROCEEDINGS, 47, 137 (1961).
[15] Bolton, E. T., B. H. Hoyen, and D. B. Ritter, in *Microsomal Particles and Protein Synthesis*, ed. R. B. Roberts (New York: Pergamon Press, 1958), p. 18.
[16] Tissières, A., J. D. Watson, D. Schlessinger, and B. R. Hollingworth, *J. Mol. Biol.*, 1, 221 (1959).
[17] Tissières, A., D. Schlessinger, and F. Gros, these PROCEEDINGS, 46, 1450 (1960).
[18] McCarthy, B. J., and A. I. Aronson, *Biophys. J.*, 1, 227 (1961).
[19] Hershey, A. D., *J. Gen. Physiol.*, 38, 145 (1954).
[20] Siminovitch, L., and A. F. Graham, *Canad. J. Microbiol.*, 2, 585 (1956).
[21] Davern, C. I., and M. Meselson, *J. Mol. Biol.*, 2, 153 (1960).
[22] Siekevitz, P., *J. Biol. Chem.*, 195, 549 (1952).
[23] Britten, R. J., and R. B. Roberts, *Science*, 131, 32 (1960).
[24] Martin, R., and B. Ames, *J. Biol. Chem.*, 236, 1372 (1961).
[25] Bamford, C. H., A. Elliott, and W. E. Hanby, *Synthetic Polypeptides* (New York: Academic Press, 1956), p. 322.

RNA Codewords and Protein Synthesis

The Effect of Trinucleotides upon the Binding of sRNA to Ribosomes

Marshall Nirenberg and Philip Leder

Although many properties of the RNA code and protein synthesis have been clarified with the use of synthetic polynucleotides containing randomly ordered bases, a more comprehensive understanding of certain aspects of the code clearly requires investigation with nucleic acid templates of demonstrated structure. Since oligonucleotides of known base sequence are readily prepared and characterized, we have tried, in many ways, to use defined oligonucleotide fractions for studies relating to the RNA code. In this article we describe a simple, direct method which should provide a general method for determining the genetic function of triplets of known sequence. The system is based upon interactions between ribosomes, aminoacyl sRNA (1), and mRNA which occur during the process of codeword recognition, prior to peptide-bond formation.

The binding of sRNA to ribosomes has been observed in many studies (2, 3); however, this interaction is not fully understood. An exchangeable binding of sRNA to ribosomes was reported by Cannon, Krug, and Gilbert (4). However, the addition of polyU induced, with specificity, Phe-sRNA binding to ribosomes, as demonstrated in the laboratories of Schweet (5–7) and Lipmann (8), by Kaji and Kaji (9, 10), and by Spyrides (11). Binding was reported to be dependent upon GTP (6, 7) and the first transfer enzyme (5–7), but not upon peptide-bond synthesis. However, the mechanism of binding and the possibility of a prior, nonenzymatic binding of amino-acyl sRNA induced by mRNA have not been clarified. The second transfer enzyme was shown to be required for peptide bond formation (6, 7).

To determine the minimum chain length of mRNA required for codeword recognition and to test the ability of chemically defined oligonucleotides to induce C^{14}-aminoacyl-sRNA binding to ribosomes, we have devised a rapid method of detecting this interaction and have found that trinucleotides are active as templates.

Methods

Preparation, purification, and characterization of oligonucleotides. To obtain oligonucleotides with different terminal groups, polyA, polyU, and polyC (12) were digested as follows: (i) *Oligonucleotides with 5′-terminal phosphate*; 100 mg of polynucleotide were incubated at 37°C for 18 hours in a 28-ml reaction mixture containing 29mM tris, pH 7.2; 0.18mM $MgCl_2$; 0.23mM 2-mercaptoethanol; 8.0 mg crystalline bovine albumin and 0.5 mg pork liver nuclease (13). (ii) *Oligonucleotides with 3′(2′)-terminal phosphate*; 100 mg of polynucleotide were incubated at 37°C for 24 hours in 20 ml of 7.0M NH_4OH. (iii) *Oligonucleotides without terminal phosphate*; Oligonucleotides with terminal phosphate were treated with *Escherichia coli* alkaline phosphatase (14) free of diesterase activity as described by Heppel *et al.* (15).

Oligonucleotide fractions were separated on Whatman 3 MM paper by chromatography with solvent A (H_2O:*n*-propanol:NH_3, 35:55:10, by volume) for 36 hours (fractions with terminal phosphate) or for 18 hours (fractions without terminal phosphate). This procedure fractionates oligonucleotides containing less than eight nucleotide residues according to chain length. Oligonucleotides were eluted with H_2O and further purified on Whatman 3 MM paper by electrophoresis at pH 2.7 (0.05M ammonium formate, 80 v/cm for 15 to 30 minutes).

After elution the purity of each fraction was estimated by subjecting 2.5 A^{290} units of each to paper chromatography (Whatman 54 paper) both with solvent A and with solvent B (40 g ammonium sulfate dissolved in 100 ml 0.1M sodium phosphate, pH 7.0). In addition 3.0 A^{290} units of each oligonucleotide were subjected to chromatography on Whatman DE 81 (DEAE) paper with solvent C (0.1M ammonium formate), and 3.0 A^{290} units with solvent D (0.3M ammonium formate). The four chromatographic systems described separate homologous series of oligonucleotides according to chain length. Contaminating oligonucleotides present in amounts greater than 2 percent could be detected. Several preparations of each oligonucleotide were used during the course of this study. In almost all preparations, no contaminants were detected. The following preparations, specified in legends of figures or tables when used, contained contaminants in the proportions indicated: No. 591, (Ap)$_4$ [(Ap)$_3$, 11 percent]; No. 599, (pA)$_6$ [(pA)$_5$, 37 percent]; No. 610, (pU)$_2$ [(pU)$_3$, 10 percent]; No. 613 (pU)$_3$ [(pU)$_6$, 14 percent]; No. 617, (Up)$_4$ [(Up)$_3$, 12 percent].

Base composition and position of terminal phosphate were determined by digesting 2.0 A^{290} units of each oligonucleotide preparation with 3.5 × 10^9 units of T$_2$ ribonuclease (16) in 0.1M NH_4HCO_3 at 37°C for 2.5 hours. The nucleotide and nucleoside products were separated by electrophoresis at pH 2.7 and identified by their mobilities and ultraviolet spectra at pH 2.0. Oligonucleotides with 5′-terminal phosphate yielded the appropriate 5′-3′(2′)-nucleoside diphosphate, 3′(2′)-nucleoside monophosphate, and nucleoside. From the ratio of these compounds, the average chain length of the parent oligonucleotide was calculated. Since oligonucleotides with 3′(2′)-terminal phosphate yielded only the appropriate 3′(2′)-nucleoside monophosphate (confirming its structure), terminal and total inorganic phosphate was determined (15, 17) in order to estimate the average chain length of each.

Oligo-d(pT) and oligo-d(pA) fractions were prepared and characterized by B. F. C. Clark as described previously (18). The UpUpUp with 3′-

---

The authors are affiliated with the Section of Biochemical Genetics of the National Heart Institute, National Institutes of Health, Bethesda, Maryland. These data were presented at the VIth International Congress of Biochemistry, 26 July–1 August 1964, New York City.

Table 1. Characteristics of the system. Complete reactions contained the components described in the text, 15 mμmole uridylic acid residues in polyU, and 20.6 $\mu\mu$mole C^{14}-Phe-sRNA (2050 count/min, 0.714 A^{260} units). Incubation was at 0°C for 60 minutes. Deacylated sRNA was added either at zero time or after 50 minutes of incubation, as indicated.

Modifications	C^{14}-Phe-sRNA bound to ribosomes ($\mu\mu$ mole)
Complete	5.99
— PolyU	0.12
— Ribosomes	0
— Mg^{++}	0.09
+sRNA (deacylated) at 50 min	
0.500 A^{260} units	5.69
2.500 A^{260} units	5.36
+sRNA (deacylated) at zero time	
0.500 A^{260} units	4.49
2.500 A^{260} units	2.08

terminal phosphate only was prepared and characterized by M. Bernfield (*19*).

Assay of ribosomal bound C^{14}-aminoacyl-sRNA. Each 50-μl reaction mixture contained: 0.1M tris-acetate, pH 7.2; 0.02M magnesium acetate; 0.05M KCl; 2.0 A^{260} units of ribosomes (washed three times) and, as indicated for each experiment, oligo- or polynucleotide, and C^{14}-aminoacyl sRNA. Tubes were kept at 0°C and C^{14}-aminoacyl sRNA was added last to initiate binding (less binding was obtained if polynucleotide was added last). Incubation for 20 minutes at 24°C was often convenient for studies requiring maximum binding, and incubation for 3 minutes at 24°C or 30 minutes at 0°C for rate studies.

After incubation, tubes were placed in ice and each reaction was immediately diluted with 3 ml of buffer containing 0.10M tris-acetate, pH 7.2; 0.02M magnesium acetate; and 0.05M KCl, at 0° to 3°C. A cellulose nitrate filter (HA Millipore filter, 25 mm diameter, 0.45μ pore size) in a stainless steel holder was washed under gentle suction with 5 ml of buffer at 0° to 3°C. The diluted reaction mixture was immediately poured on the filter under suction and washed to remove unbound C^{14}-aminoacyl sRNA with three, 3-ml portions of buffer at 0° to 3°C. Ribosomes and bound sRNA remained on the filter. Since reaction mixtures are not deproteinized, it is important to dilute and wash the ribosomes immediately after incubation, to use cold buffer, and to allow relatively little air to be pulled through the filter during the washing procedure. The filter was removed from the holder, glued with rubber cement to a disposable planchette, and dried. Radioactivity was determined in a thin-window, gas-flow counter (*20*) with a C^{14}-counting efficiency of 23 percent. In some experiments, radioactivity was determined in a liquid scintillation counter with a C^{14}-counting efficiency of 65 percent (*21*) and dried filters (not glued) were placed in vials containing 10 ml of

Table 2. Polynucleotide specificity. Reaction mixtures containing C^{14}-Phe- and C^{14}-Lys-sRNA were incubated for 60 minutes at 0°C; mixtures containing C^{14}-Pro-sRNA were incubated for 20 minutes at 24°C. In addition to the components described in the text, reaction mixtures contained, in a final volume of 50 μl, the specified polynucleotide and C^{14}-aminoacyl-sRNA (14.7 $\mu\mu$mole C^{14}-Phe-sRNA, 2015 count/min, 0.960 A^{260} units; 16.5 $\mu\mu$mole C^{14}-Lys-sRNA, 1845 count/min, 0.530 A^{260} units; 30 $\mu\mu$mole C^{14}-Pro-sRNA, 2750 count/min, 1.570 A^{260} units).

Polynucleotide (mμmole base residues)	C^{14}-Aminoacyl-sRNA bound to ribosomes ($\mu\mu$mole)		
	C^{14}-Phe-sRNA	C^{14}-Lys-sRNA	C^{14}-Pro-sRNA
None	0.19	0.99	0.25
PolyU, 25	6.00	.67	.15
PolyA, 16	0.22	4.35	.17
PolyC, 19	.21	0.72	.80

toluene-PPO-POPOP phosphor solution (*22*).

Preparation of sRNA. Except where noted *E. coli* B sRNA (*23*), was used. Uniformly labeled C^{14}-L-phenylalanine, C^{14}-L-lysine, and C^{14}-L-leucine with specific radioactivities of 250, 305, and 160 μc/μmole, respectively, were obtained commercially (*24*). The *E. coli* W 3100 sRNA was prepared as described by Zubay (*25*) from cells grown to late log phase in 0.9 percent Difco nutrient broth, containing 1 percent glucose. The C^{14}-amino acyl-sRNA was prepared by modifications of meth-

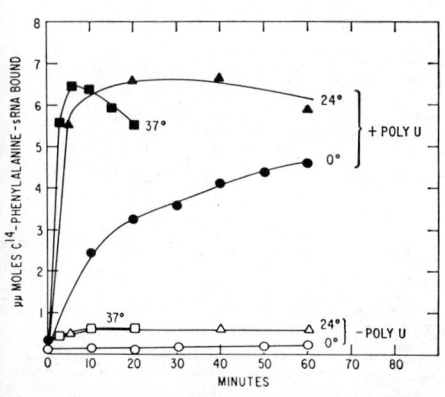 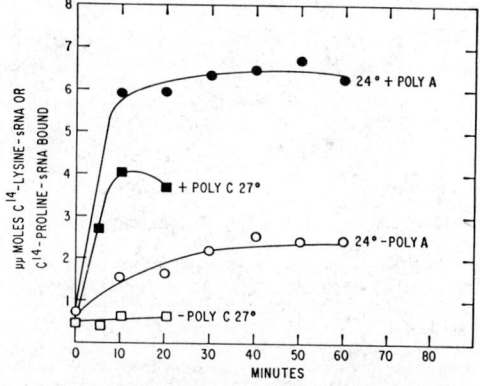

Fig. 1 (left). Effect of polyU upon the rate of C^{14}-Phe-sRNA binding to ribosomes at 0°, 24°, and 37°C. Each point represents a 50 μl reaction mixture incubated for the time and at the temperature indicated. Reaction mixtures contain the components described under *Methods*; 9.65 $\mu\mu$mole of C^{14}-Phe-sRNA (1180 count/min, 0.380 A^{260}); and polyU, 25 mμmole base residues, where specified. Fig. 2 (right). The effect of polyA and polyC upon the rates of C^{14}-Lys- and C^{14}-Pro-sRNA binding to ribosomes, respectively. Each point represents a 50-μl reaction mixture as described under *Methods*. The reactions specified contained 27.5 $\mu\mu$mole C^{14}-Lys-sRNA (3080 count/min, 0.880 A^{260} units) and polyA, 25 mμmole base residues, or 11.8 $\mu\mu$mole C^{14}-Pro-sRNA (2660 count/min, 0.905 A^{260} units) and polyC, 25 mμmole base residues. The temperature and the time of incubation are shown in the figure.

ods described previously (26). Unless otherwise specified, sRNA was acylated with one C^{14}-amino acid plus 19 C^{12}-amino acids. The formation of C^{14}-aminoacyl sRNA was catalyzed by the supernatant solution obtained by centrifugation of *E. coli* (W-3100) extracts at 100,000g.

Elution and characterization of C^{14}-phenylalanine product bound to ribosomes. Reaction mixtures (0.5 ml) incubated at 24°C for 10 minutes with C^{14}-Phe-sRNA and polyU were washed on cellulose nitrate filters in the usual manner. The ribosomal bound C^{14}-product was eluted from filters by washing with 0.01M tris-acetate, pH 7.2; $10^{-5}M$ magnesium acetate; and 0.05M KCl at 0°C.

The C^{14}-product eluted from ribosomes was precipitated in 10 percent TCA at 3°C in the presence of 200 μg of bovine serum albumin. Specified samples were heated in 10 percent TCA at 90° to 95°C for 20 minutes and then were chilled. Precipitates were washed on filters with 5 percent TCA at 3°C.

Aminoacyl sRNA was deacylated in 0.1M ammonium carbonate solution adjusted with NH_4OH to pH 10.2 to 10.5 by incubation at 37°C for 60 minutes.

Digestions with ribonuclease were performed by incubating 0.4-ml portions (each containing 1500 count/min precipitable by 10 percent TCA at 3°C) with and without 10 μg of purified pancreatic ribonuclease A (purified chromatographically) (27) at 37°C for 15 minutes.

Results and Discussion

Assay of ribosomal bound C^{14}-aminoacyl-sRNA. The assay is based upon the retention of ribosomes and C^{14}-aminoacyl sRNA bound to ribosomes by cellulose nitrate filters. After unbound C^{14}-aminoacyl sRNA is removed by washing with buffered salts solution, as already described, the radioactivity remaining on the filter is determined. Thirty reaction mixtures can be washed per hour easily. The sensitivity of the assay is limited primarily by the specific radioactivity of the aminoacyl-sRNA used. With sRNA which has accepted a C^{14}-amino acid of specific radioactivity 100 to 300 μc/μmole, the binding to ribosomes of 0.2 μμmoles of C^{14}-aminoacyl sRNA readily can be detected. A filter 25 mm in diameter with a pore size of 0.45 μ retains up to 1200 μg of *E. coli* ribosomes. The use of larger filters or columns packed with cellulose nitrate may be useful for preparative procedures.

The retention of ribosomes by cellulose nitrate filters may be the result of absorption rather than of filtration, for filters with pores 100 times larger than *E. coli* 70S ribosomes can be used. The rapidity of this assay, compared to others which depend upon the centrifugation of ribosomes, has greatly simplified this study.

The data of Table 1 show that little C^{14}-Phe-sRNA was retained on filters after incubation with ribosomes in the absence of polyU. Incubation in the presence of both polyU and ribosomes resulted in marked retention of C^{14}-Phe-sRNA by the filter. Ribosomes, polyU, and Mg^{++} were required for retention of C^{14}-Phe-sRNA. Spyrides (28) and Conway (29) have reported that polyU-directed binding of Phe-sRNA to ribosomes is dependent upon K^+ or NH_4^+.

The addition of deacylated sRNA to reactions shortly before incubation was terminated, after C^{14}-Phe-sRNA binding had ceased, had little effect upon ribosomal bound C^{14}-Phe-sRNA. The bound C^{14}-Phe-sRNA fraction apparently is not readily exchangeable.

In contrast, the addition of deacylated sRNA at the start of incubation inhibited C^{14}-Phe-sRNA binding. In other experiments, the extent of inhibition was found to be affected by the ratio of deacylated to acylated sRNA.

Deacylated sRNA added near the end of incubation often reduces background binding without polynucleotide and may afford a way of differentiating between exchangeable and nonexchangeable binding. It should be noted that the presence of a polynucleotide which is not recognized by a C^{14}-aminoacyl sRNA (for example, polyA and C^{14}-Phe-sRNA) also reduces background sRNA binding, perhaps by saturating ribosomal sites with specified nonexchangeable sRNA.

Characteristics of binding. The assay was validated further by demonstrating that the binding of sRNA to ribosomes was directed with specificity by different polynucleotides. As shown in Table 2, polyU, polyA and polyC specifically directed the binding to ribosomes of C^{14}-Phe-, C^{14}-Lys-, and C^{14}-Pro-sRNA, respectively. These data agree well with specificity data obtained with a sucrose-gradient centrifugation assay, reported by Nakamoto *et al.* (8) and Kaji and Kaji (9, 10) (also compare 5–7, 11) and with data on their specificity for

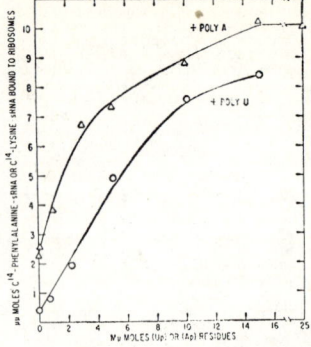

Fig. 3. Relation between polynucleotide concentration and C^{14}-aminoacyl-sRNA binding to ribosomes. Each point represents a 50-μl reaction mixture with the components described under *Methods*; △, polyA as specified and 27.5 μμmole C^{14}-Lys-sRNA (3080 count/min, 0.880 A^{260} units); ○, polyU as specified and 15 μμmole of C^{14}-Phe-sRNA (2050 count/min, 0.714 A^{260} units). Incubation was at 24°C for 20 minutes.

directing amino acid incorporation into protein (30, 31).

The rate of binding of Phe-sRNA to ribosomes at 0°, 24° and 37°C, in the presence and absence of polyU, is shown in Fig. 1. During incubation at each temperature polyU markedly stimulated C^{14}-Phe-sRNA binding; how-

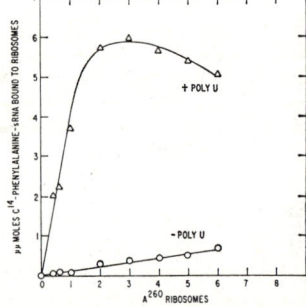

Fig. 4. Relation between ribosome concentration and C^{14}-Phe-sRNA binding. Each point represents a 50-μl reaction mixture containing the amount of ribosomes indicated; ○ No additions, (-poly U); △ + poly U, 25 mμmole base residues. In addition, reaction mixtures contain the components described under *Methods* and 9.65 μμmole of C^{14}-Phe-sRNA (1180 count/min, 0.380 A^{260}). Incubations were for 10 minutes at 24°C.

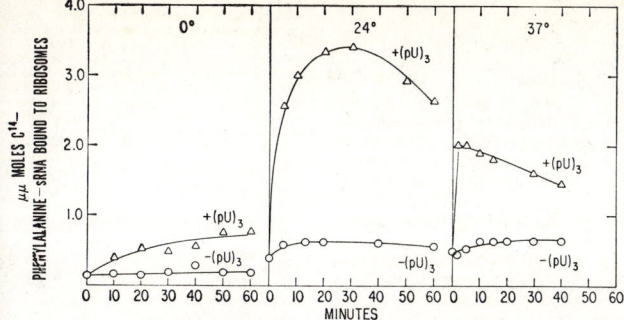

Fig. 5. Effect of pUpUpU upon the rate of C¹⁴-Phe-sRNA binding to ribosomes at 0°, 24°, and 37°C. ○, No addition; △, addition of 3.67 mμmole pUpUpU. Each point represents a 50-μl reaction mixture incubated for the time and at the temperature specified. Reaction mixtures contain the components described under *Methods*; 9.65 μμmole of C¹⁴-Phe-sRNA (1180 count/min, 0.380 A^{260}); and oligoU as specified.

ever, the rate of binding increased as the temperature of incubation was raised. Although polyU induced C¹⁴-Phe-sRNA binding at 0°C, maximum binding was not observed after 60 minutes of incubation. Maximum binding concentrations of C¹⁴-Phe-sRNA, 50 minutes of incubation at 24°C and after 6 minutes at 37°C. In this experiment equal amounts of C¹⁴-Phe-sRNA were bound at 24° and 37°C. In other experiments, with limiting concentrations of C¹⁴-Phe-sRNA, 50 to 98 percent of the C¹⁴-Phe-sRNA was induced to bind to ribosomes by polyU. Kaji and Kaji have suggested the possible utility of this system for the purification of sRNA species (9).

In the absence of polyU, relatively little C¹⁴-Phe-sRNA associated with ribosomes. Such binding may be due to endogenous mRNA on ribosomes or in sRNA preparations. Alternatively, this binding may be nonspecific, possibly similar to that described by Cannon, Krug, and Gilbert (4).

The effect of polyA and polyC upon the rates of C¹⁴-Lys- and C¹⁴-Pro-sRNA binding to ribosomes is shown in Fig. 2. Maximum stimulation of C¹⁴-Lys-sRNA binding by polyA, and of C¹⁴-Pro-sRNA binding by polyC, occurred after 10 minutes of incubation at 24° and 27°C, respectively. In this experiment, C¹⁴-Lys-sRNA binding observed in the absence of polyA was higher than that found in experiments with other sRNA preparations.

The relation between polyU or polyA concentration and the amount of C¹⁴-Phe- or C¹⁴-Lys-sRNA bound to ribosomes is shown in Fig. 3. Binding of sRNA was proportional to polynucleotide concentration in both cases.

In experiments not presented here, the effect of *p*H upon sRNA binding was studied. PolyU directed C¹⁴-Phe-

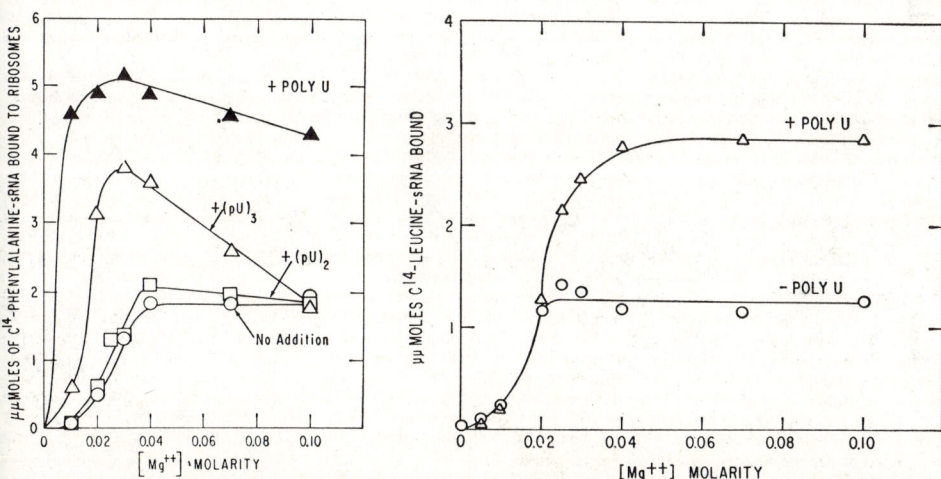

Fig. 6 (left). Effect of (Mg⁺⁺) concentration upon C¹⁴-Phe-sRNA binding to ribosomes. Each symbol represents a 50-μl reaction mixture containing: ○, polymer (no addition); □, pUpU, 10 mμmole base residues; △, pUpUpU, 10 mμmole base residues; ▲, polyU, 25 mμmole base residues; and the components described under *Methods*; the magnesium acetate concentrations of reaction mixtures and the washing buffer are shown in the figure; 3.5 A^{260} units of ribosomes; and 8.63 μμmole C¹⁴-Phe-sRNA (1025 count/min, 0.357 A^{260} units. Incubation was at 24°C for 20 minutes. Fig. 7 (right). The polyU-leucine ambiguity. The effect of polyU upon C¹⁴-Leu-sRNA binding to ribosomes at different Mg⁺⁺ concentrations. Each 50-μl reaction contained the components described under *Methods*; the magnesium acetate concentration specified in the figure 23.7 μμmole C¹⁴-Leu-sRNA (1710 count/min, 0.424 A^{260} units of *E. coli* W 3100 sRNA, charged only with C¹⁴-leucine); and polyU, 25 mμmole base residues where specified. Incubation was at 24°C for 20 minutes. Each reaction mixture was washed with a buffer containing Mg⁺⁺ at the concentration present during incub...

Table 3. Oligonucleotide specificity. Reaction mixtures containing either C^{14}-Phe-, C^{14}-Lys-, or C^{14}-Pro-sRNA, and oligonucleotide were incubated at 24°C for 20 minutes. Components of reaction mixtures are described in the legend accompanying Table 2. The numbers in parentheses are millimicromoles base residues.

Oligonucleotide (mμmoles base residues)	C^{14}-Aminoacyl-sRNA bound to ribosomes (μμmole)		
	C^{14}-Phe-sRNA	C^{14}-Lys-sRNA	C^{14}-Pro-sRNA
None	0.34	0.80	0.24
pUpUpU (10)	1.56	0.56	0.20
pApApA (7)	0.20	6.13	0.18
pCpCpC (8)	0.30	0.60	0.73

sRNA binding to ribosomes throughout the pH range tested, from 5.5 to 7.8. Binding was maximum in reactions buffered either with 0.1M tris, pH 7.2, or with 0.05M cacodylate, pH 6.5.

The addition of GTP, PEP, and PEP kinase to reactions, with or without polyU, did not stimulate C^{14}-Phe-sRNA binding; however, further study is necessary to determine whether binding in this system is dependent upon enzymatic catalysis or GTP or both. Polyphenylalanine synthesis (that is, radioactivity in the fraction precipitable with hot TCA) was not detected in reaction mixtures incubated under optimum conditions for C^{14}-Phe-sRNA binding in the presence of polyU. In additional experiments, C^{14}-Phe-sRNA induced by polyU to bind to ribosomes was eluted with $10^{-5}M$ Mg^{++} as already described. Although 50 to 98 percent of the ribosomal-bound C^{14}-product was eluted (in different experiments), few ribosomes were released from filters. The C^{14}-product was insoluble in TCA at 3°C, but was converted quantitatively to a soluble product by (i) incubation at pH 10.2 for 60 minutes at 37°C, (ii) heating in TCA, or (iii), incubation with pancreatic ribonuclease. Further characterization of the bound C^{14}-product is necessary; however, these preliminary results are in accord with the demonstration by Schweet and his co-workers that the binding of Phe-sRNA to ribosomes induced by polyU is not dependent upon peptide-bond formation (6, 7).

The relation between ribosome concentration and the amount of bound C^{14}-Phe-sRNA is illustrated in Fig. 4. C^{14}-Phe-sRNA binding was stimulated markedly by polyU and was proportional to the ribosome concentration, within the range 0 to 1.0 A^{260} units of ribosomes. The number of 70S E. coli ribosomes and ribosomal-bound C^{14}-Phe-sRNA molecules can be estimated from such data; however, various factors, such as the inhibitory effects of deacylated sRNA (see above) and mRNA terminal phosphate (described below) undoubtedly reduce the accuracy of such calculations. However, in the presence of polyU approximately 4.0 μμmole of C^{14}-Phe-sRNA became bound to 23.2 μμmoles of 70S ribosomes (4); therefore, the C^{14}-Phe-sRNA ribosome ratio was 1 : 5.8. Arlinghaus et al. (7) and Warner and Rich (32) recently reported two binding sites for each ribosome for sRNA. One site is thought to hold the nascent polypeptide chain to the ribosome; the other, to bind the next aminoacyl-sRNA molecule specified by mRNA.

Effect of oligonucleotides on the binding of C^{14}-aminoacyl sRNA to ribosomes. OligoU preparations of different chain length were prepared, and their effect upon C^{14}-Phe-sRNA binding to ribosomes was determined. The effect of the trinucleotide, pUpUpU, upon C^{14}-Phe-sRNA binding to ribosomes, at 0°, 24°, and 37°C, is shown in Fig. 5. The C^{14}-Phe-sRNA binding was stimulated by pUpUpU at each temperature; however, binding was maximum in reactions incubated at 24°C for 20 to 30 minutes. These results demonstrate that a trinucleotide can direct C^{14}-aminoacyl-sRNA binding to ribosomes and suggest a general method of great simplicity for determining the genetic function of other trinucleotide sequences.

The binding of C^{14}-Phe-, C^{14}-Lys-, and C^{14}-Pro-sRNA was induced with apparent specificity by pUpUpU, pApApA, and pCpCpC, respectively (Table 3). In additional experiments, each trinucleotide had no discernible effect upon the binding to ribosomes of 15 aminoacyl-sRNA preparations, each charged with a different C^{14}-amino acid (C^{14}-asparagine and C^{14}-glutamine-sRNA were not tested). Therefore, the specificity of each trinucleotide for inducing sRNA binding to ribosomes was high and clearly paralleled that of the corresponding polynucleotide.

The T_m of the interaction between pApApA and polyU is 17°C (33). Therefore, hydrogen-bonding between a triplet codeword in mRNA and a complementary "anticodeword" in sRNA would not by itself appear sufficient to account for the stability of the interaction between C^{14}-Phe-sRNA, polyU, and ribosomes. An interaction between

Table 4. Template activity of oligodeoxynucleotides. The components of each 50 μl reaction mixture are presented in the text. In addition, each reaction mixture in Expt. 1 contained 9.65 μμmoles of C^{14}-Phe-sRNA (1370 count/min, 0.380 A^{260} units); and in Expt. 2, 16.5 μμmoles C^{14}-Lys-sRNA (1845 count/min, 0.530 A^{260} units) and the oligonucleotides specified. Mixtures were incubated at 24°C for 10 minutes.

Oligonucleotide (mμmole)	C^{14}-Aminoacyl-sRNA bound to ribosomes (μμmole)
Experiment 1: C^{14}-Phe-sRNA	
None	0.29
1.00 pUpUpU	1.29
3.33 pUpUpU	2.40
6.67 pUpUpU	2.90
3.33 oligo d(pT)$_3$	0.35
6.67 oligo d(pT)$_3$	0.39
10.00 oligo d(pT)$_3$	0.40
1.67 oligo d(pT)$_{12}$	0.31
2.50 oligo d(pT)$_{12}$	0.44
Experiment 2: C^{14}-Lys-sRNA	
None	1.27
0.17 pApApA	3.48
1.37 pApApA	5.99
3.60 pApApA	7.40
0.65 oligo d(pA)$_8$	1.45
1.30 oligo d(pA)$_8$	1.68
2.00 oligo d(pA)$_8$	1.61

the -CpCpA end of sRNA and ribosomes is possible, for several laboratories (3, 4, 34) have reported that removal of the terminal adenosine of sRNA greatly reduces its ability to bind to ribosomes. The participation of an enzyme in the binding process also must be considered.

The data of Table 4 indicate that the 2′-hydroxyl of RNA codewords may be necessary for codeword recognition. Oligodeoxynucleotides such as oligo-d(pT)$_{3-12}$ and oligo-d(pA)$_8$ apparently were inactive as templates. In additional experiments not presented here, the effects of time and temperature of incubation (0°, 24°, 37°, and 43°C), template concentration, and chain length were studied. No template activity was found.

The template activities of pUpU, pUpUpU and polyU at different concentrations of Mg^{++} are shown in Fig. 6. Both tri- and polyU induced maximal binding at approximately 0.03M Mg^{++} (0.02 to 0.03M in other experiments). Although the dinucleotide, pUpU, stimulated binding slightly, it is not known whether the activity of pUpU indicates partial recognition of a triplet codeword as previously suggested (35). At 0.02M Mg^{++}, the concentration used throughout our study, little binding of C^{14}-Phe-sRNA was found in the absence of polyU. However, at higher Mg^{++} concentrations, Phe-sRNA binding in the absence of template RNA increased;

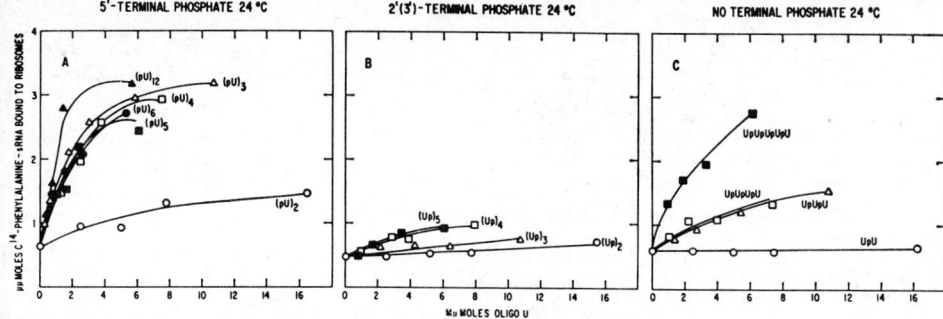

Fig. 8. *A*, *B*, and *C*, Relation between template activity and oligoU chain-length, concentration and end-group. The activities of oligoU with 5'-terminal phosphate are shown in Fig. 8*A*; with 2'(3')-terminal phosphate in Fig. 8*B*; and without terminal phosphate in Fig. 8*C*. Symbols represent oligoU chain-lengths as follows: ○ 2; △ 3; □ 4; ■ 5; ● 6; ▲ 12. Each reaction mixture contained, in a volume of 50 μl, the components specified in the *Methods* section; oligoU preparations and concentrations specified; and 10.8 μμmole C^{14}-Phe-sRNA (1880 count/min, 0.714 A^{260}). Oligonucleotide preparations (pU)$_2$, No. 610; (pU)$_5$, No. 613; (Up)$_4$, No. 617; with the contaminants specified under *Methods* were used in these experiments. Incubations were at 24°C for 30 minutes.

whereas binding induced by tri- and polyU, decreased. These data suggest that certain ribosomal binding sites become saturated with sRNA at Mg^{++} concentrations greater than 0.03M, but are not saturated at lower concentrations.

Leucine-polyU ambiguity. Since polyU is known to direct some leucine incorporation into protein in cell-free systems (36), especially at high Mg^{++} concentrations (37), the effect of polyU upon the binding of C^{14}-Leu-sRNA was determined (Fig. 7). In the absence of polyU, background binding saturated at 0.02M Mg^{++}. The addition of polyU clearly stimulated the binding of C^{14}-Leu-sRNA. It is possible that the Mg^{++}-dependent leucine-polyU ambiguity occurs before peptide-bond synthesis.

As shown in Table 5, C^{14}-Leu-sRNA binding was stimulated with specificity by polyU, but not by the trinucleotide, pUpUpU, polyA, or polyC at 0.07M Mg^{++}. In additional experiments, pUpUpU had no effect upon the binding of C^{14}-Leu-sRNA at other Mg^{++} concentrations. These data suggest that pUpUpU may be recognized by aminoacyl sRNA with greater specificity than polyU is recognized.

Effect of oligonucleotide chain length, concentration, and terminal phosphate. In Fig. 8 the template activity of oligo-U preparations differing in chain length and position of terminal phosphate are compared at different oligoU concentrations. The activity of oligoU with 5'-terminal phosphate is shown in Fig. 8*A*; preparations with 2'(3')-terminal phosphate in Fig 8*B*; and preparations without terminal phosphate, in Fig. 8*C*.

As shown in Fig. 8*A*, the dinucleotide with 5'-terminal phosphate, pUpU, had little template activity, whereas the trinucleotide, pUpUpU, markedly stimulated C^{14}-Phe-sRNA binding. This ob-

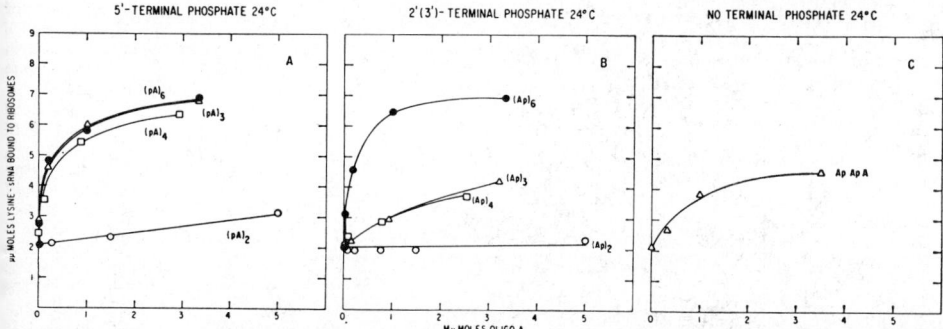

Fig. 9. *A*, *B*, and *C*, Relation between template activity and oligoA concentration, chain-length, and end group. The activity of oligoA with 5'-terminal phosphate is shown in Fig. 9*A*; preparation without terminal phosphate in Fig. 9*C*. The symbols indicate the chain-length of oligoA as follows: ○ 2; △ 3; □ 4; ● 6. Each 50 μl reaction mixture contained the components specified under methods; 15.9 μμmole of C^{14}-Lys-sRNA (1780 count/min, 0.562 A^{260} units); and the oligoA preparation specified. Incubation was at 24°C for 20 minutes. Oligonucleotide preparations (pA)$_6$, No. 599 and (Ap)$_4$, No. 591, containing the contaminants specified under *Methods*, were used in this experiment.

servation provides direct experimental support for a triplet code for phenylalanine and is in full accord with earlier genetic and biochemical studies (30, 38).

In addition, the data demonstrate that the template activity of pUpUpU equals that of the corresponding tetra-, penta-, and hexanucleotides.

In striking contrast to these results, the trinucleotide, UpUpUp, with 2'(3')-terminal phosphate, induced little or no binding of C¹⁴-Phe-sRNA to ribosomes (Fig. 8B). The template activity of the tetra- and pentauridylic acid fractions with 2'(3')-terminal phosphate also were markedly reduced when compared to similar fractions with 5'-terminal phosphate.

As shown in Fig. 8C, UpUpU and UpUpUpU, without terminal phosphate, induced C¹⁴-Phe-sRNA binding, but less actively than pUpUpU. The template activity of the pentamer, UpUpUpUpU, was almost equal to that of pUpUpU.

Since the sensitivity of an oligonucleotide to digestion by a nuclease often is influenced by terminal phosphate, the relative stability of pUpUpU, UpUpUp, and UpUpUU incubated with ribosomes at 37°C for 60 minutes was estimated by recovering mono- and oligonucleotides from reaction mixtures and separating them by paper chromatography. In each case, the expected trinucleotide was the only component found after incubation. Hydrolysis of oligoU was not observed.

The template activities of oligoA fractions with different end groups are shown in Fig. 9, A, B, and C. The results were similar to those obtained with oligoU; however, pApApA induced maximum C¹⁴-Lys-sRNA binding at one-fifth the oligonucleotide concentration required previously (compare with Fig. 8A). The hexamer with 3'-terminal phosphate induced as much C¹⁴-Lys-sRNA binding to ribosomes as the trimer with 5'-terminal phosphate, pApApA. When reaction mixtures were incubated at 0°C (Fig. 10) the difference between the template activities of ApApAp and pApApA was more marked than when incubations were at 24°C.

Since each oligonucleotide preparation with 2'(3')-terminal phosphate is a mixture of molecules, some chains terminating with 2'-phosphate and others with 3'-phosphate, a trinucleotide, UpUpUp, with 3'-terminal phosphate only, was prepared and found to be inactive as a template for C¹⁴-Phe-sRNA.

Attachment of ribosomes to mRNA. It is not known whether ribosomes attach to 5'-ends, 3'-ends, or internal positions of mRNA. The template activity of trinucleotides indicates that (i) ribosomes can attach to terminal codewords of mRNA, (ii) terminal codewords are capable of specifying the first and the last amino acids to be incorporated into protein, and (iii) attachment of a ribosome to *only* the terminal triplet of mRNA may provide the minimum stability necessary for codeword recognition, and possibly for the initiation of protein synthesis.

The demonstration that terminal and internal codeword phosphates strongly influence the codeword recognition process indicates that phosphate may take part in the binding of codewords to ribosomes. Watson has suggested interaction between phosphate of mRNA and amino groups of ribosomes, because 30S ribosomes treated with formaldehyde were found by Moore and Asano to bind less polyU than did ribosomes not so treated (39).

Although terminal codeword phosphate is not required for the recognition of a codeword on a ribosome, the observation that the template activity of trinucleotides with 5'-terminal phosphate equals that of tetra-, penta-, and hexanucleotides, even at limiting concentrations, suggests that 5'-terminal codewords may *attach to sites on ribosomes where codewords are recognized, in correct phase to be read.* A preferential, phased recognition of either terminal codeword by ribosomes would provide a simple mechanism for selecting the polarity of reading, the first word to be read, and the phase of reading. Since 5'-terminal codewords of mRNA most actively induce sRNA binding, such codewords would appear to serve these functions best. Although the polarity with which mRNA is read may be from the 5'- towards the 3'-terminal codeword, further work is necessary to clarify this point. The opposite polarity has been suggested (45).

We have reported (40) that trinucleotides can be used in this system to determine the base sequence of codewords and have shown that the sequence of the valine RNA codeword is GpUpU. Codewords are recognized with polarity in this system, for GpUpU induced C¹⁴-Val-sRNA binding to ribosomes, whereas UpUpG did not.

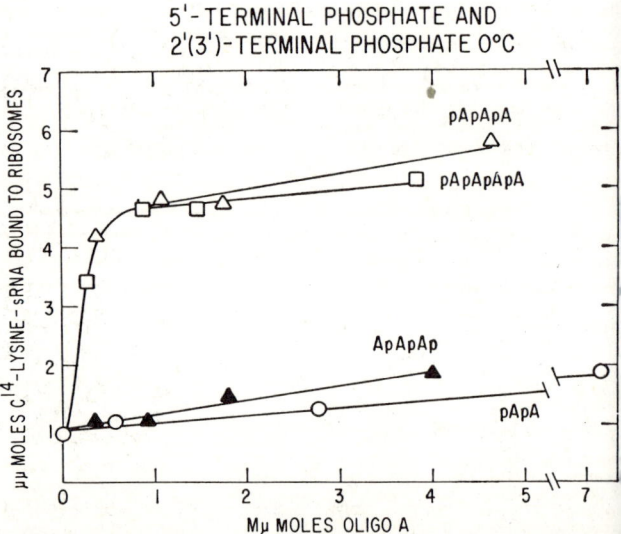

Fig. 10. Template activity of oligoA with 5'- or 2'(3')-terminal phosphate in reaction mixtures incubated at 0°C for 60 minutes. OligoA concentrations, chain-lengths, and end-groups are shown in the figure. Each 50-μl reaction mixture contained the components specified under *Methods*; 11.8 μμmole C¹⁴-Lys-sRNA (2840 count/min, 0.767 A²⁶⁰ units); and the oligoA preparation specified.

7

Regulatory Codewords

Terminal codewords. We suggest that 5'-terminal, 3'-terminal and internal codewords of RNA and DNA constitute separate classes of codewords, for each differs in chemical structure as shown diagrammatically in Fig. 11. The 5'-terminal codeword contains a 5'-terminal hydroxyl, the internal codeword is attached to adjacent codewords on both sides by way of (3'-5')-phosphodiester bonds, and the 3'-terminal codeword contains 2'- and 3'-terminal hydroxyl groups. It should be noted that phosphate, linked to a terminal hydroxyl is a monoester, whereas, an internal phosphate is a diester. Therefore, the 5'-terminal, 3'-terminal, and internal codewords differ in chemical structure. To avoid confusion between the three codeword classes, the codeword with free or substituted 5'-hydroxyl will be designated the 5'-terminal codeword, and the codeword with a free or substituted 3'-hydroxyl, will be designated the 3'-terminal codeword. Internal codewords will not be designated by position. These differences raise the possibility that codewords may occur in three chemically distinct forms and that RNA and DNA may contain either one, two, or three forms of any triplet codeword, depending upon the position of the codeword in the molecule.

Sense-missense-nonsense codewords.

Genetic evidence suggests that certain mutations result in the conversion of readable into nonreadable codewords, that is, sense-nonsense interchanges (*41*). The addition of terminal phosphate to a 3'-terminal codeword similarly changes a readable into a nonreadable codeword and resembles a sense-nonsense interchange. Two additional mechanisms for converting readable into nonreadable words have been found in cell-free systems; that is, an increase in secondary structure (*37*, *42*) and specific base methylations (*37*). It should be noted that each type of sense-nonsense interchange involves a modification of codewords rather than modification of a component required for codeword recognition.

It is possible that the synthesis of certain proteins may be regulated in vivo by sense-nonsense interchanges involving either modification of a codeword or, as proposed by Ames and Hartman (*43*), modification of sRNA, that is, codeword recognition. It seems probable that terminal codewords may have special functions in addition to directing amino acids into protein. For example, in mRNA they may specify (i) attachment and detachment of ribosomes, (ii) the first codeword to be read, (iii) the phase of reading, and (iv) the sensitivity of the message to degradation by exonucleases. Similarly, terminal DNA codewords may influence the rate with which DNA is copied

Table 5. Specificity of polyU-leucine-sRNA bound to ribosomes. Reaction mixtures contained 0.07M magnesium acetate; other components as described in the text; 3.1 $\mu\mu$moles of C^{14}-Leu-sRNA (2280 count/min, 0.565 A^{200} units of *E. coli* W 3100 sRNA acylated only with C^{14}-Leucine); and oligo- or polynucleotides as specified. Incubations were at 24°C for 20 minutes. The magnesium acetate concentration of the solution used to wash ribosomes on filters after incubation was 0.07M.

Addition (base residues) (mμmole)	C^{14}-Leu-sRNA bound to ribosomes ($\mu\mu$mole)
None	1.00
25 PolyU	2.02
15 pUpUpU	0.70
25 PolyA	.99
25 PolyC	.85

by DNA or RNA polymerase. Experimental observations support this possibility, for DNA without terminal phosphate has been found to serve as a template for DNA polymerase, whereas DNA with 3'-terminal phosphate has no template activity (*44*).

Terminal words with 3'-phosphate may be members of a larger class of DNA and mRNA nonsense words, with substituted 5'-, 3'-, or 2'-hydroxyl groups. Many enzymes have been described which catalyze the transfer of nucleotides, amino acids, methyl groups, carbohydrates, and other molecules, to or from mononucleotides or terminal ribose or deoxyribose of nucleic acids. Such enzymes may recognize terminal bases or conformations of nucleic acids and catalyze group transfer reactions with great specificity.

The data of Figs. 8 to 10, and the chemical and biological considerations described, suggest that *codeword modification* may serve a regulatory or operator function. Modification of codewords, at both terminal and internal positions, may regulate the reading of DNA or RNA by converting a readable word into one read incorrectly or not read. It should be noted that both 3'- and 5'-terminal codewords could serve, in different ways, as operator words.

The capacity of trinucleotides to direct the binding of sRNA to ribosomes and the ease with which the process can be assayed should provide a general method of great simplicity for studying the base sequence and genetic functions of each triplet codeword. In addition, this method should permit the detailed study of interactions between codewords, sRNA, and ribosomes during the codeword recognition process.

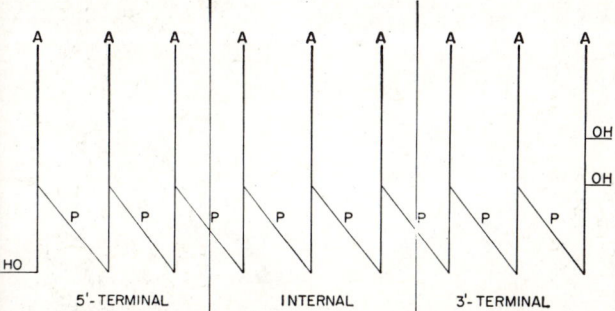

Fig. 11. The structures of three classes of codewords, 5'-terminal, internal, and 3'-terminal, are illustrated diagramatically. The figure represents oligoA of chain-length 9, ApApApApApApApApA. Although the three triplet codewords have identical base-sequences, AAA, each differs in structure. The 5'-terminal codeword contains a 5'-terminal hydroxyl; the internal codeword is linked to adjacent codewords on either side by way of 3', 5'-phosphodiester bonds; and the 3'-terminal codeword contains 2', 3'-terminal hydroxyl groups (3'-hydroxyl only if deoxyribose). Phosphate, when present at 5'-, 3'-, or 2'-terminal hydroxyl positions, may be a phosphomonoester, and therefore may differ from an internal phosphate which is a phosphodiester.

Summary

A rapid, sensitive method is described for measuring C^{14}-aminoacyl-sRNA interactions with ribosomes which are specifically induced by the appropriate RNA codewords prior to peptide-bond formation. Properties of the codeword recognition process and the minimum oligonucleotide chain length required to induce such interactions are presented. The trinucleotides, pUpUpU, pApApA, and pCpCpC, but not dinucleotides, specifically direct the binding to ribosomes of phenylalanine-, lysine-, and proline-sRNA, respectively.

Since 5'-terminal, 3'-terminal, and internal codewords differ in chemical structure, three corresponding classes of codewords are proposed. The recognition of each class in this system is described. The template efficiency of trinucleotide codewords is modified greatly by terminal phosphate. Triplets with 5'-terminal phosphate are more active as templates than triplets without terminal phosphate. Triplets with 3'- or 3' (2')-terminal phosphate are markedly less active as templates. These findings are discussed in relation to the probable functions of terminal codewords. The modification of RNA and DNA codewords, converting sense into missense or nonsense codewords, is suggested as a possible regulatory mechanism in protein synthesis.

References and Notes

1. The following abbreviations and symbols are used: Phe-, phenylalanine; Leu-, leucine; Lys-, lysine; Pro-, proline; Val-, valine; polyU, polyuridylic acid; polyC, polycytidylic acid; polyA, polyadenylic acid; TCA, trichloracetic acid; d(pT), deoxythymidylic acid; d(pA), deoxyadenylic acid; sRNA, transfer RNA; mRNA, messenger RNA; DEAE, diethylaminoethyl cellulose; PPO, 2,5-diphenyloxazole; POPOP, 1,4-bis-2'-(5'-phenyloxazolyl) benzene; PEP, phosphoenolpyruvate; T_m, melting temperature; A^{260}, at 260 mμ. For mono- and oligonucleotides of specific structure, the p to the left of a terminal nucleoside initial indicates a 5'-terminal phosphate; the p to the right, a 2' (3')-terminal phosphate. Internal phosphates of oligonucleotides are (3'-5')-linkages.
2. A Von der Decken and T. Hultin, *Exptl. Cell. Res.* **15**, 254, (1958); L. Bosch, H. Bloemendal, M. Sluyser, *Biochim. Biophys. Acta* **34**, 272 (1959); M. B. Hoagland and L. T. Comly, *Proc. Natl. Acad. Sci. U.S.* **46**, 1554 (1960).
3. M. Takanami, *Biochim. Biophys. Acta* **55**, 132 (1962); L. Bosch, F. Huizinga, H. Bloemendal, *ibid.* **61**, 220 (1962).
4. M. Cannon, R. Krug, W. Gilbert, *J. Mol. Biol.* **7**, 360 (1963).
5. R. Arlinghaus, G. Favelukes, R. Schweet, *Biochem. Biophys. Res. Commun.* **11**, 92 (1963).
6. B. Hardesty, R. Arlinghaus, J. Schaeffer, R. Schweet, in *Cold Spring Harbor Symp. Quant. Biol.* **28**, 215 (1963).
7. R. Arlinghaus, J. Shaeffer, R. Schweet, *Proc. Natl. Acad. Sci. U.S.* **51**, 1291 (1964).
8. T. Nakamoto, T. Conway, J. Allende, G. Spyrides, F. Lipmann, in *Cold Spring Harbor Symp. Quant. Biol.* **28**, 227 (1963).
9. A. Kaji and H. Kaji, *Biochem. Biophys. Res. Commun.* **13**, 186 (1963).
10. ———, *Federation Proc.* **23**, 478 (1964).
11. G. J. Spyrides. Ph.D. dissertation, Rockefeller Institute (1963); *Federation. Proc.* **23**, 318 (1964).
12. Miles Chemical Company.
13. L. A. Heppel, personal communication.
14. Worthington Biochemical Corporation.
15. L. A. Heppel, D. Harkness, R. Hilmoe, *J. Biol. Chem.* **237**, 841 (1962).
16. G. W. Rushizky and H. A. Sober, *ibid.* **238**, 371 (1963).
17. P. S. Chen, Jr., T. Y. Toribara, H. Warner, *Anal. Chem.* **28**, 1756 (1956); B. N. Ames and D. T. Dubin, *J. Biol. Chem.* **235**, 769 (1960).
18. P. Leder, B. F. C. Clark, W. S. Sly, S. Pestka, M. W. Nirenberg, *Proc. Natl. Acad. Sci. U.S.* **50**, 1135 (1963).
19. M. Bernfield and M. W. Nirenberg, *Abstracts, 148th Meeting, American Chemical Society*, Chicago, Ill., August 1964.
20. Nuclear Chicago Corporation.
21. Packard Instrument Company.
22. J. R. Sherman and J. Adler, *J. Biol. Chem.* **238**, 873 (1963).
23. General Biochemicals Corporation.
24. New England Nuclear Corporation.
25. G. Zubay, *J. Mol. Biol.* **4**, 347 (1962).
26. M. W. Nirenberg, J. H. Matthaei, O. W. Jones, *Proc. Natl. Acad. Sci. U.S.* **48**, 104 (1962).
27. Sigma Chemical Company.
28. G. J. Spyrides, *Proc. Natl. Acad. Sci. U.S.* **51**, 1220 (1964).
29. T. W. Conway, *ibid.* **51**, 1216 (1964).
30. M. W. Nirenberg, O. W. Jones, P. Leder, B. F. C. Clark, W. S. Sly, S. Pestka, in *Cold Spring Harbor Symp. Quant. Biol. Vol.* **28**, 549 (1963).
31. J. S. Speyer, P. Lengyel, C. Basilio, A. J. Wahba, R. S. Gardner, S. Ochoa, *ibid.* **28**, 559 (1963).
32. J. R. Warner and A. Rich, *Proc. Natl. Acad. Sci. U.S.* **51**, 1134 (1964).
33. M. N. Lipsett, L. A. Heppel, D. F. Bradley, *J. Biol. Chem.* **236**, 857 (1961).
34. H. Bloemendal, F. Huizinga, M. de Vries, L. Bosch, *Biochim. Biophys. Acta* **61**, 209 (1962).
35. M. W. Nirenberg and O. W. Jones, in *Symposium on Informational Macromolecules*, H. Vogel, V. Bryson, J. Lempen, Eds. (Academic Press, New York, 1963); O. W. Jones and M. W. Nirenberg, *Proc. Natl. Acad. Sci. U.S.* **48**, 2115 (1962); R. S. Gardner, A. J. Wahba, C. Basilio, R. S. Miller, P. Lengyel, J. F. Speyer, *ibid.*, p. 2087.
36. J. H. Matthaei, O. W. Jones, R. G. Martin, M. W. Nirenberg, *Proc. Natl. Acad. Sci. U.S.* **48**, 666 (1962); M. S. Bretscher and M. Grunberg-Manago, *Nature* **195**, 283 (1962).
37. M. Grunberg-Manago and A. M. Michelson, *Biochim. Biophys. Acta* **80**, 431 (1964); W. Szer and S. Ochoa, *J. Mol. Biol.* **8**, 823 (1964).
38. F. H. C. Crick, L. Barnett, S. Brenner, R. J. Watts-Tobin, *Nature* **192**, 1227 (1961).
39. J. D. Watson, personal communication.
40. P. Leder and M. W. Nirenberg, *Proc. Natl. Acad. Sci. U.S.* **51** (1964), in press.
41. S. Benzer and S. P. Champe, *ibid.* **48**, 1114 (1962); A. Garen and O. Siddiqi, *ibid.*, p. 1121.
42. M. W. Nirenberg and J. H. Matthaei, *ibid.* **47**, 1588 (1961); O. W. Jones, R. G. Martin. S. H. Barondes, *Federation Proc. Symp.* **22**, 55 (1963); M. F. Singer, O. W. Jones. M. W. Nirenberg, *Proc. Natl. Acad. Sci. U.S.* **49**, 392 (1962); R. Haselkorn and V. A. Fried, *ibid.* **51**, 308 (1964).
43. B. N. Ames and P. E. Hartman, in *Cold Spring Harbor Symp. Quant. Biol.* **28**, 349 (1963).
44. C. C. Richardson, C. L. Schildkraut, H. V. Aposhian, A. Kornberg, W. Bodner, J. Lederberg, in *Informational Macromolecules*, H. Vogel, V. Bryson, J. Lampen, Eds. (Academic Press, New York, 1963).
45. A. J. Wahla, C. Basilio, J. F. Speyer. P. Lengyel, R. S. Miller, S. Ochoa, *Proc. Natl. Acad. Sci. U.S.* **48**, 1683 (1962); presented by P. Doty, R. E. Thach, and T. A. Sundararajan at the VIth International Congress of Biochemistry (1964).
46. Thank Miss Norma Zabriskie and Mrs. Theresa Caryk for invaluable technical assistance; L. A. Heppel for his advice regarding preparation of oligonucleotides; and J. C. Keresztesy and D. Rogerson for providing *E. coli*.

The RNA Code and Protein Synthesis

M. Nirenberg, T. Caskey, R. Marshall, R. Brimacombe, D. Kellogg, B. Doctor[†],
D. Hatfield, J. Levin, F. Rottman, S. Pestka, M. Wilcox, and F. Anderson

Laboratory of Biochemical Genetics, National Heart Institute, National Institutes of Health, Bethesda, Maryland and
[†] *Division of Biochemistry, Walter Reed Army Institute of Research, Walter Reed Army Medical Center, Washington, D.C.*

Many properties of the RNA code which were discussed at the 1963 Cold Spring Harbor meeting were based on information obtained with randomly ordered synthetic polynucleotides. Most questions concerning the code which were raised at that time related to its fine structure, that is, the order of the bases within RNA codons. After the 1963 meetings a relatively simple means of determining nucleotide sequences of RNA codons was devised which depends upon the ability of trinucleotides of known sequence to stimulate AA-sRNA binding to ribosomes (Nirenberg and Leder, 1964). In this paper, information obtained since 1963 relating to the following topics will be discussed:

(1) The fine structure of the RNA code
(2) Factors affecting the formation of codon-ribosome-AA-sRNA complexes
(3) Patterns of synonym codons for amino acids and purified sRNA fractions
(4) Mechanism of codon recognition
(5) Universality
(6) Unusual aspects of codon recognition as potential indicators of special codon functions
(7) Modification of codon recognition due to phage infection.

FINE STRUCTURE OF THE RNA CODE

Formation of Codon-Ribosome-AA-sRNA Complexes

The assay for base sequences of RNA codons depends, first upon the ability of trinucleotides to serve as templates for AA-sRNA binding to

ABBREVIATIONS

The following abbreviations are used: Ala-, alanine-; Arg-, arginine-; Asn-, asparagine-; Asp-, aspartic acid-; Cys-, cysteine-; Glu-, glutamic acid-, Gln-, glutamine-, Gly-, glycine-, His-, histidine-, Ile-, isoleucine-, Leu-, leucine-, Lys-, lysine-, Met-, methionine-, Phe-, phenylalanine-, Pro-, proline-, Ser-, serine-, Thr-, threonine-, Trp-, tryptophan-, Tyr-, tyrosine-, and Val-, valine-; sRNA, transfer RNA; AA-sRNA, aminoacyl-sRNA; sRNA^{Phe}, deacylated phenylalanine-acceptor sRNA; Ala-sRNA^{Yeast}, acylated alanine-acceptor sRNA from yeast. U, uridine; C, cytidine; A, adenosine; G, guanosine; I, inosine; rT, ribothymidine; ψ, pseudouridine; DiHU, dihydro-uridine; MAK, methylated albumin kieselguhr; F-Met, N-formyl-methionine. For brevity, trinucleoside diphosphates are referred to as trinucleotides. Internal phosphates of trinucleotides are (3',5')-phosphodiester linkages.

Table 1. Characteristics of AA-sRNA Binding to Ribosomes

Modifications	C^{14}-Phe-sRNA bound to ribosomes ($\mu\mu$mole)
Complete	5.99
− Poly U	0.12
− Ribosomes	0.00
− Mg^{++}	0.09
+ deacylated sRNA at 50 min	
0.50 A^{260} units	5.69
2.50 A^{260} units	5.39
+ deacylated sRNA at zero time	
0.50 A^{260} units	4.49
2.50 A^{260} units	2.08

Complete reactions in a volume of 0.05 ml contained the following: 0.1 M Tris acetate (pH 7.2) (in other experiments described in this paper 0.05 M Tris acetate, pH 7.2 was used), 0.02 M magnesium acetate, 0.05 M potassium chloride (standard buffer); 2.0 A^{260} units of *E. coli* W3100 70 S ribosomes (washed by centrifugation 3 times); 15 mμmoles of uridylic acid residues of poly U; and 20.6 $\mu\mu$moles C^{14}-Phe-sRNA (0.71 A^{260} units). All components were added to tubes at 0°C. C^{14}-Phe-sRNA was added last to initiate binding reactions.

Incubation was at 0°C for 60 min (in all other experiments described in this paper, reactions were incubated at 24° for 15 min). Deacylated sRNA was added either at zero time or after 50 min of incubation, as indicated. After incubation, tubes were placed in ice and each reaction was immediately diluted with 3 ml of standard buffer at 0° to 3°C. A cellulose nitrate filter (HA type, Millipore Filter Corp., 25 mm diameter, 0.45 μ pore size) in a stainless steel holder was washed with gentle suction with 5 ml of the cold standard buffer. The diluted reaction mixture was immediately poured on the filter under suction and washed to remove unbound C^{14}-Phe-sRNA with three 3-ml and one 15-ml portions of standard buffer at 3°. Ribosomes and bound sRNA remained on the filter (Nirenberg and Leder, 1964). The filters were then dried, placed in vials containing 10 ml of a scintillation fluid (containing 4 gm 2,5-diphenyloxazole and 0.05 gm 1,4-bis-2-(5-phenyloxazolyl)-benzene per liter of toluene) and counted in a scintillation spectrometer.

ribosomes prior to peptide bond formation, and second, upon the observation that codon-ribosome-AA-sRNA complexes are retained by cellulose nitrate filters (Nirenberg and Leder, 1964). Results shown in Table 1 illustrate characteristics of codon-ribosome-sRNA complex formation. Ribosomes, Mg^{++}, and poly U are required for the binding of C^{14}-Phe-sRNA to ribosomes. The addition of deacylated sRNA to reactions at zero time greatly reduces the binding of C^{14}-Phe-sRNA (Table 1), since poly U specifically stimulates the binding of both deacylated sRNA^{Phe} and C^{14}-Phe-sRNA to ribosomes. Ribosomal bound

C^{14}-Phe-sRNA is not readily exchangeable with unbound Phe-sRNA or deacylated sRNA^{Phe} except at low Mg^{++} concentrations (Levin and Nirenberg, in prep.). Later in this volume Dr. Dolph Hatfield discusses the characteristics of exchange of ribosomal bound with unbound AA-sRNA when trinucleotides are present.

Two enzymatic methods were devised for oligonucleotide synthesis, since most trinucleotide sequences had not been isolated or synthesized earlier. One procedure employed polynucleotide phosphorylase to catalyze the synthesis of oligonucleotides from dinucleoside monophosphate primers and nucleoside diphosphates (Leder, Singer, and Brimacombe, 1965; Thach and Doty, 1965); the other approach (Bernfield, 1966) was based upon the demonstration (Heppel, Whitfeld, and Markham, 1955) that pancreatic RNase catalyzes the synthesis of oligonucleotides from uridine- or cytidine-2′,3′-cyclic phosphate and acceptor moieties. Elegant chemical procedures for oligonucleotide synthesis devised by Khorana and his associates (see Khorana et al., this volume) also are available.

Template Activity of Oligonucleotides with Terminal and Internal Substitutions

The trinucleotides, UpUpU and ApApA, but not the corresponding dinucleotides, stimulate markedly the binding of C^{14}-Phe- and C^{14}-Lys-sRNA, respectively. Such data directly demonstrate a triplet code and also show that codons contain three *sequential* bases. The template activity of triplets with 5′-terminal phosphate, pUpUpU, equals that of the corresponding tetra- and pentanucleotides; whereas, oligo U preparations with 2′,3′-terminal phosphate are much less active. Hexa-A preparations, with and without 3′-terminal phosphate, are considerably more active as templates than the corresponding pentamers; thus, one molecule of hexa-A may be recognized by two Lys-sRNA molecules bound to adjacent ribosomal sites (Rottman and Nirenberg, 1966).

An extensively purified doublet with 5′-terminal phosphate, pUpC, serves as a template for Ser-sRNA (but not for Leu- or Ile-sRNA), whereas a doublet without terminal phosphate, UpC, is inactive (see Figs. 1a and b). However, the template activity of pUpC is considerably lower than that of the triplet, UpCpU. The relation between Mg^{++} concentration and template activity is shown in Fig. 1b. pUpC and UpCpU stimulate Ser-sRNA binding in reactions containing 0.02–0.08 M Mg^{++}. These results demonstrate that a doublet with 5′-terminal phosphate can serve as a specific, although relatively weak, template for AA-sRNA. It is particularly intriguing to relate recognition of a doublet to the

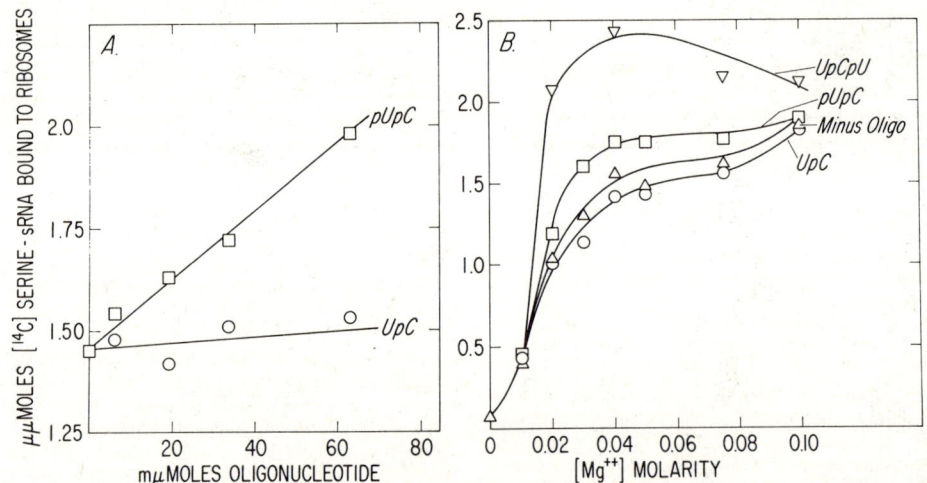

FIGURE 1a, b. The effects of UpC and pUpC on the binding of C^{14}-Ser-sRNA to ribosomes. The relation between oligonucleotide concentration and C^{14}-Ser-sRNA binding to ribosomes at 0.03 M Mg^{++} is shown in Fig. 1a. It should be noted that the ordinate begins at 1.25 μμmoles of C^{14}-Ser-sRNA. The relation between Mg^{++} concentration and C^{14}-Ser-sRNA binding to ribosomes is shown in Fig. 1b. As indicated, 50 mμmoles of UpC or pUpC, or 15 mμmoles of UpCpU, were added to each reaction. Each point in parts a and b represents a 50 μl reaction containing the components described in the legend to Table 1 except for the following: 14.3 μμmoles C^{14}-Ser-sRNA (0.42 A^{260} units); 1.1 A^{260} units of ribosomes. Incubations were for 15 min at 24°C. (Data from Rottman and Nirenberg, 1966.)

TABLE 2. RELATIVE TEMPLATE ACTIVITY OF SUBSTITUTED OLIGONUCLEOTIDES

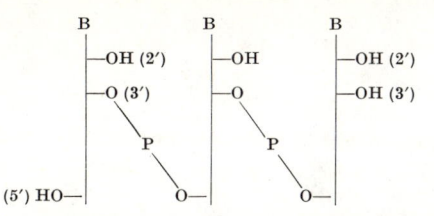

Oligonucleotide	Relative template activity
p-5'-UpUpU	510
UpUpU	100
CH_3O-pUpUpU	74
UpUpU-3'-p	48
UpUpUp-OCH_3	18
UpUpU-2',3'-cyclic p	17
(2'-5')-UpUpU	0
Oligodeoxy T	0
p-5'-ApApA	181
ApApA	100
ApApA-3'-p	57
ApApA-2'-p	15
(2'-5')-ApApA	0
Oligodeoxy A	0

Relative template activities are approximations obtained by comparing the amount of AA-sRNA bound to ribosomes in the presence of limiting concentrations of oligonucleotides (0.50 or 0.12 mμmoles of oligonucleotides containing U or A, respectively) compared to either UpUpU, for C^{14}-Phe-sRNA; or ApApA, for C^{14}-Lys-sRNA (each assumed to be 100%). Data are from Rottman and Nirenberg (1966) except results with oligodeoxynucleotides which are from Nirenberg and Leder (1964).

possibility that only two out of three bases in a triplet may be recognized occasionally during protein synthesis, and also to the possibility that a triplet code evolved from a more primitive doublet code.

Further studies on template activities of oligonucleotides with terminal and internal modifications are summarized in Table 2. At limiting oligonucleotide concentrations, the relative template activities of oligo U preparations are as follows: p-5'-UpUpU > UpUpU > CH_3O-p-5'-UpUpU > UpUpU-3'-p > UpUpU-3'-p-OCH_3 > UpUpU-2',3'-cyclic phosphate. Trimers with (2'-5') phosphodiester linkages, (2'-5')-UpUpU and (2'-5')-ApApA, do not serve as templates for Phe- or Lys-, sRNA respectively. The relative template efficiencies of oligo A preparations are as follows: p-5'-ApApA > ApApA > ApApA-3'-p > ApApA-2'-p.

These studies led to the proposal that RNA and DNA contain three classes of codons, differing in structure; 5'-terminal, 3'-terminal, and internal codons (Nirenberg and Leder, 1964). Certainly the first base of a 5'-terminal codon and the third base of a 3'-terminal codon may be recognized with less fidelity than an internal codon, for in the absence of a nucleotide neighbor a terminal base may have a greater freedom of movement on the ribosome. Substitution of 5'- or 3'-terminal hydroxyl groups may impose restrictions upon the orientation of terminal bases during codon recognition. 5'-Terminal and perhaps also 3'-terminal codons possibly serve, together with neighboring codons, as operator regions.

Since many enzymes have been described which catalyze the transfer of nucleotides, amino acids, phosphate, and other molecules to or from terminal ribose or deoxyribose of nucleic acids, modification of sugar hydroxyl groups was proposed as a possible mechanism for regulating the reading of RNA or DNA (Nirenberg and Leder, 1964).

NUCLEOTIDE SEQUENCES OF RNA CODONS

A summary of nucleotide sequences of RNA codons by *E. coli* AA-sRNA is shown in Table 3

TABLE 3. NUCLEOTIDE SEQUENCES OF RNA CODONS

1st Base	2nd Base				3rd Base
	U	C	A	G	
U	PHE*	SER*	TYR*	CYS*	U
	PHE*	SER*	TYR*	CYS	C
	leu*?	SER	TERM?	cys?	A
	leu*, f-met	SER*	TERM?	TRP*	G
C	leu*	pro*	HIS*	ARG*	U
	leu*	pro*	HIS*	ARG*	C
	leu	PRO*	GLN*	ARG*	A
	LEU	PRO	gln*	arg	G
A	ILE*	THR*	ASN*	SER	U
	ILE*	THR*	ASN*	SER*	C
	ile*	THR*	LYS*	arg*	A
	MET*, F-MET	THR	lys	arg	G
G	VAL*	ALA*	ASP*	GLY*	U
	VAL	ALA*	ASP*	GLY*	C
	VAL*	ALA*	GLU*	GLY*	A
	VAL	ALA	glu	GLY	G

Nucleotide sequences of RNA codons were determined by stimulating binding of *E. coli* AA-sRNA to *E. coli* ribosomes with trinucleotide templates. Amino acids shown in capitals represent trinucleotides with relatively high template activities compared to other trinucleotide codons corresponding to the same amino acid. Asterisks (*) represent base compositions of codons which were determined previously by directing protein synthesis in *E. coli* extracts with synthetic randomly-ordered polynucleotides (Speyer et al., 1963; Nirenberg et al., 1963). F-Met, represents N-formyl-Met-sRNA which may recognize initiator codons. TERM represents possible terminator codons. Question marks (?) indicate uncertain codon function. Data are from Nirenberg et al., 1965; Brimacombe et al., 1965; also see articles by Khorana et al., Söll et al., and Matthaei et al., in this volume.

TABLE 4. PATTERNS OF DEGENERATE CODONS FOR AMINO ACIDS

●● U C A G	●● U C A G	●● U C A G	●● U C (A)	●● U C	●● A G	●● A G	U C A (G) ●●
●● U C	●● G (A?)						
SER	ARG	GLY	CYS	ASP	GLU	MET	F-MET
LEU	LEU	ALA	ILE	ASN	GLN	TRP	
		VAL		HIS	LYS		
		THR		TYR	TERM?		
		PRO		PHE			

Solid circles represent the first and second bases of trinucleotides; U, C, A, and G indicate bases which may occupy the remaining position of degenerate codons. In the case of F-Met (N-formylmethionine), circles represent the second and third bases. Parentheses indicate codons with relatively low template activities.

and patterns of degeneracy in Table 4. Almost every trinucleotide was assayed for template specificity with 20 AA-sRNA preparations (unfractionated sRNA acylated with one labeled and 19 unlabeled amino acids). It is important to test trinucleotide template specificity with 20 AA-sRNA preparations, since relative responses of AA-sRNA are then quite apparent. In surveying trinucleotide specificity, unfractionated AA-sRNA should be used initially because altering ratios of sRNA species often influences the fidelity of codon recognition.

Almost all triplets correspond to amino acids; furthermore, patterns of codon degeneracy are logical. Six degenerate codons correspond to serine, five or six to arginine and also to leucine, and from one to four to each of the remaining amino acids. Alternate bases often occupy the third positions of triplets comprising degenerate codon sets. In all cases triplet pairs with 3'-terminal pyrimidines (XYU and XYC, where X and Y represent the first and second bases, respectively, in the triplet) correspond to the same amino acid; often XYA and XYG correspond to the same amino acid; sometimes XYG alone corresponds to an amino acid. For eight amino acids, U, C, A, or G may occupy the third position of synonym codons. Alternate bases also may occupy the first position of synonyms, as for N-formyl-methionine.

One consequence of logical degeneracy is that many single base replacements in DNA may be silent and thus not result in amino acid replacement in protein (cf. Sonneborn, 1965). Also, the code is arranged so that the effects of some errors may be minimized, since amino acids which are structurally or metabolically related often correspond to similiar RNA codons (for example, Asp-codons, GAU, and GAC, are similar to Glu-codons, GAA, and GAG). When various amino acids are grouped according to common biosynthetic precursors, close relationships among their synonym codons sometimes are observed. For example, codons for amino acids derived from aspartic acid begin with A: Asp, GAU, GAC; Asn, AAU, AAC; Lys, AAA, AAG; Thr, ACU, ACC, ACA, ACG; Ile, AUU, AUC, AUA; Met, AUG. Likewise, aromatic amino acids have codons beginning with U; Phe, UUU, UUC; Tyr, UAU, UAC; Trp, UGG. Such relationships may reflect either the evolution of the code or direct interactions between amino acids and bases in codons (see Woese et al., this volume).

At the time of the 1963 meeting at Cold Spring Harbor, 53 base compositions of RNA codons had been estimated (14 tentatively) in studies with randomly-ordered synthetic polynucleotides and a cell-free protein synthesizing system derived from *E. coli* (Speyer et al., 1963; Nirenberg et al., 1963). Forty-six base composition assignments now are confirmed by base sequence studies with trinucleotides (shown in Table 3). Thus, codon base compositions and base sequence assignments, obtained by assaying protein synthesis and AA-sRNA binding, respectively, agree well with one another. In addition, codon base sequences are confirmed by most amino acid replacement data obtained in vivo (see Yanofsky et al.; Wittman et al., this volume).

PATTERNS OF SYNONYM CODONS RECOGNIZED BY PURIFIED sRNA FRACTIONS

Table 5 contains a summary of synonym codons recognized by purified sRNA fractions obtained either by countercurrent distribution or by MAK column chromatography. The following patterns of codon recognition involving alternate bases in the third positions of synonym codons were found; C = U; A = G; G; U = C = A; A = G = (U) For example, Val-sRNA$_3$ recognizes GUU and GUC, whereas the major peak of Val-sRNA (fractions 1 and 2) recognizes GUA, GUG and, to a lesser extent, GUU. The possibility that the latter Val-sRNA fraction contains two or more Val-sRNA components has not been excluded. Met-sRNA,

THE RNA CODE AND PROTEIN SYNTHESIS

TABLE 5. CODON PATTERNS RECOGNIZED BY PURIFIED sRNA FRACTIONS

Alternate acceptable bases in 3rd or 1st positions of triplet					
C / U	A / G	G	U / C / A	A / G / (U)	Possibly only 2 bases recognized
$TYR_{1,2}$ UA_U^C	LYS AA_G^A	LEU_2 CUG	ALA^{yeast} GCC (U, A)	ALA_1 GCG (A, (U))	LEU_3 $CU_{(C)}^{(U)}$
VAL_3 GU_U^C		LEU_5 UUG	$SER_{2,3}^{yeast}$ UCC (U, A)	$VAL_{1,2}$ GUG (A, (U))	$LEU_{4a,b}$ $UU_{(C)}^{(U)}$
		MET_2 AUG	$F\text{-}MET_1$ UG (U, C, A)		LEU_1 $(U)UG$
			TRP_2 CGG (U, (A))		

Patterns of degenerate codons recognized by purified AA-sRNA fractions. sRNA fractions are from *E. coli* B, unless otherwise specified. At the top of the table are shown the alternate bases which may occupy the third or first positions of degenerate codon sets. Purified sRNA fractions and corresponding codons are shown below. Parentheses indicate codons with relatively low template activity. sRNA fractions were obtained by counter-current distribution (Kellogg et al., 1966), unless otherwise specified. Yeast Ser-sRNA fractions 2 and 3 (Connelly and Doctor, 1966) are thought to be equivalent to yeast Ser-sRNA fractions 1 and 2, respectively, discussed by Zachau et al. in this volume. Yeast Ala-sRNA was the gift of R. W. Holley; results are from Leder and Nirenberg (unpubl.). Results obtained with Val-, Met-, and Ala-sRNA$^{E.\ coli}$ fractions are from Kellogg et al. (1966). For additional results with Tyr-sRNA fractions, see Doctor, Loebel and Kellogg, this volume. Leu-sRNA fractions (see Fig. 6 and Sueoka et al., this volume) and Lys-sRNA (Kellogg, Doctor, and Nirenberg, unpubl.) were obtained by MAK column chromatography. Three Leu-sRNA fractions also were obtained by counter-current distribution (Nirenberg and Leder, 1964). Reactions contained the usual components (see legend to Table 1) and 0.01 or 0.02 M Mg^{++}. Incubation was at 24° for 15 min.

responds to UUG, CUG, AUG and, to a lesser extent, GUG, and can be converted enzymatically to N-formyl-Met-sRNA, whereas, Met-sRNA$_2$ responds primarily to AUG and does not accept formyl moieties (see later discussion). Unfractionated Trp-sRNA responds only to UGG; however one fraction of Trp-sRNA, after extensive purification, responds to UGG, CGG and AGG. Possibly the latter responses depend upon the removal of sRNA for other amino acids (e.g., Arg-sRNA) which also may recognize CGG or AGG. Yeast Ala- and Ser-sRNA$_{2,3}$ fractions recognize synonyms containing U, C, or A in the third position. Leu-sRNA$_{1,3,4}$ bind to ribosomes in response to polynucleotide templates but not to trinucleotides. Possibly, only two of the three bases are recognized by these Leu-sRNA fractions.

MECHANISM OF CODON RECOGNITION

Crick (1966; also this volume) has suggested that certain bases in anticodons may form alternate hydrogen bonds, via a wobble mechanism, with corresponding bases in mRNA codons. This hypothesis and further experimental findings are discussed below.

Yeast Ala-sRNA of known base sequence and of high purity (>95%) was the generous gift of Dr. Robert Holley. In Figs. 2 and 3 are shown the responses of purified yeast and unfractionated *E. coli* C^{14}-Ala-sRNA, respectively, to synonym Ala-codons as a function of Mg^{++} concentration. Purified yeast C^{14}-Ala-sRNA responds well to GCU, GCC, and GCA, but only slightly to GCG. Similar results were obtained with unfractionated Ala-sRNA^{Yeast}. In contrast, unfractionated *E. coli* C^{14}-Ala-sRNA responds best to GCG and GCA, less well to GCU, and only slightly to GCC.

In Fig. 4a and b, the relation between concentration of yeast or *E. coli* C^{14}-Ala-sRNA and response to synonym Ala-codons is shown. At *limiting* concentrations of purified yeast C^{14}-Ala-sRNA, at least 59, 45, 45, and 3% of the available C^{14}-Ala-sRNA molecules bind to ribosomes in response to GCU, GCC, GCA, and GCG, respectively. The response of unfractionated *E. coli* C^{14}-Ala-sRNA to each codon was 18, 2, 38, and 64%, respectively. Similar results have been obtained by Keller and Ferger (1966) and Söll et al. (this volume). Since the purity of the yeast Ala-sRNA was greater than 95%, the extent of binding at limiting Ala-sRNA concentrations indicates that one molecule of Ala-sRNA recognizes 3, possibly 4, synonym codons. In addition, the data demonstrate marked differences between the relative responses of yeast and *E. coli* Ala-sRNA to synonym codons.

Correlating the base sequences of yeast Ala-sRNA with corresponding mRNA codons also provides insight into the structure of the Ala-sRNA

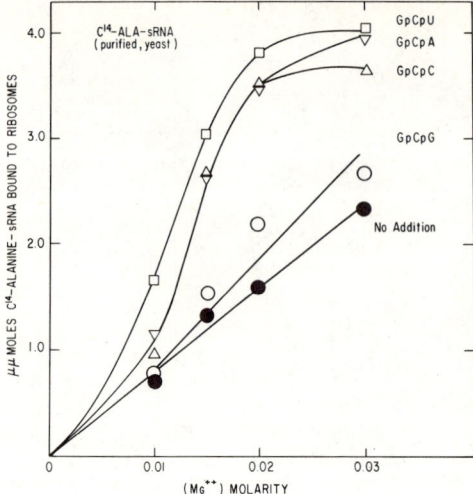

FIGURE 2. The relation between Mg^{++} concentration and binding to ribosomes of purified yeast C^{14}-Ala-sRNA of known base sequence (Holley et al., 1965) in response to trinucleotides. Each point represents a 50 μl reaction containing the components described in the legend to Table 1 except for the following: 1.5 A^{260} units of *E. coli* ribosomes, 11.2 μμmoles of purified yeast C^{14}-Ala-sRNA (0.038 A^{260} units); and 0.1 A^{260} units of trinucleotide as specified. Reactions were incubated at 24° for 15 min (Leder and Nirenberg, unpubl.).

anti-codon and the mechanism of codon recognition. Possible anticodon or enzyme recognition sequences in Ala-sRNA^{Yeast} are –IGC MeI– and DiHU–CGG–DiHU (Fig. 5; Holley et al., 1965). Each site potentially comprises a single-stranded loop region at the end of a hairpin-like double-stranded segment. If CGG were the anticodon, *parallel* hydrogen bonding with GCU, GCC, GCA codons would be expected. If IGC were the anticodon, *antiparallel* Watson-Crick hydrogen bonding between GC in the anticodon and GC in the first and second positions of codons, and alternate pairing of inosine in the anticodon with U, C, or A, but not G, in the third position of Ala-codons, would be expected. All of the available evidence is consistent with an IGC Ala-anticodon. Zachau has shown that Ser-sRNA$^{Yeast}_{1\ and\ 2}$ contain, in appropriate positions, IGA sequences (Zachau, Dütting, and Feldmann, 1966), and we find that Ser-sRNA^{Yeast} fractions 2 and 3 (believed to correspond to fractions 1 and 2 of Zachau) recognize UCU, UCC, and UCA, but not UCG (see Table 5). A purified Val-sRNA^{Yeast} fraction contains the sequence IAC which corresponds to three Val-codons, GUU, GUC, and GUA (Ingram and Sjöqvist, 1963). In addition, the sequence, GψA, is found at the postulated anticodon site of Tyr-sRNA^{Yeast} which corresponds to the Tyr-codons, UAU and UAC (Madison, Everett, and Kung, 1966).

Crick's wobble hypothesis and patterns of synonym codons found experimentally are in full agreement. In Table 6 are shown bases in anticodons which form alternate hydrogen bonds, via the wobble mechanism, with bases usually occupying the third positions of mRNA codons. U in the sRNA anticodon may pair alternately with A or G in mRNA codons; C may pair with G; A with U; G with C or U; and I with U, C, or A. In addition, we suggest that ribo T in the anticodon may hydrogen bond more strongly with A, and perhaps with G also, than with U; and ψ in the anticodon may hydrogen bond alternately with A, G or, less well, U.

Dihydro U in an anticodon may be unable to hydrogen bond with a base in mRNA but may be repelled less by pyrimidines than by purines.

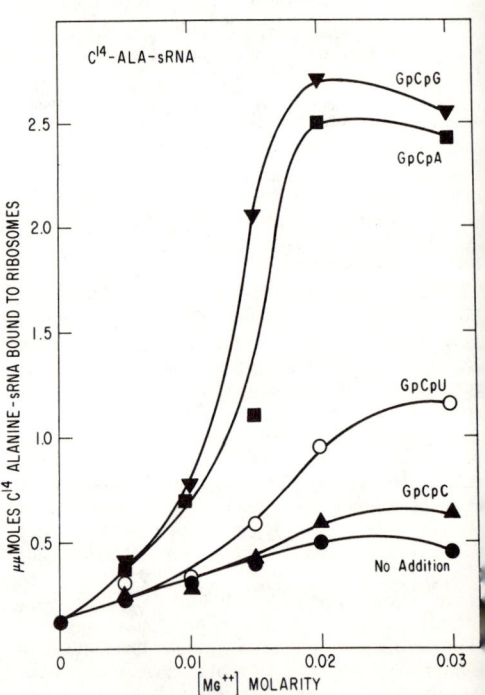

FIGURE 3. Relation between Mg^{++} concentration and binding of unfractionated *E. coli* C^{14}-Ala-sRNA to ribosomes in response to trinucleotides. Each point represents a 50 μl reaction containing the components described in the legend to Table 1, 2.0 A^{260} units of ribosomes; 18.8 μμmoles of unfractionated *E. coli* C^{14}-Ala-sRNA (0.54 A^{260} units); and 0.1 A^{260} unit of trinucleotide, as specified (Leder and Nirenberg, unpubl.).

THE RNA CODE AND PROTEIN SYNTHESIS

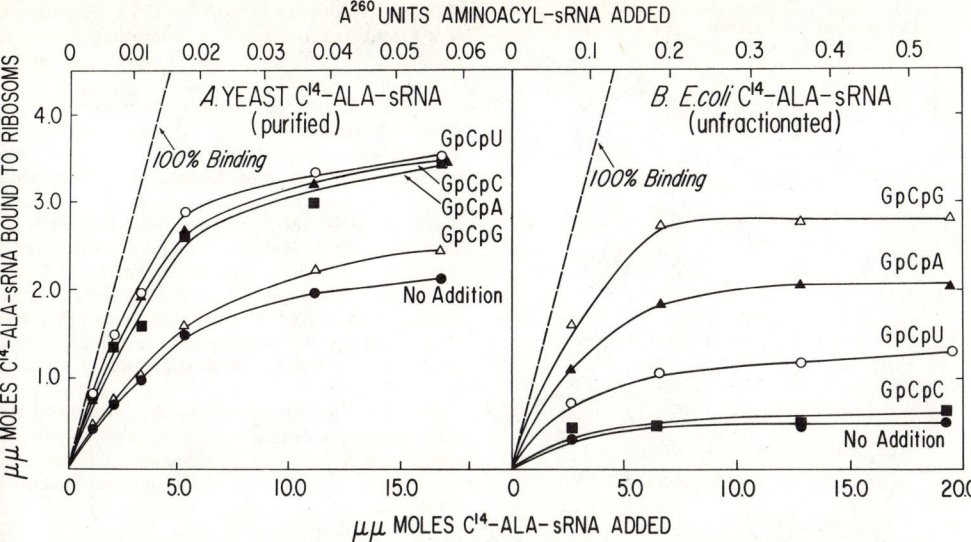

FIGURE 4a, b. Relation between the template activities of trinucleotides and the concentrations of purified yeast C^{14}-Ala-sRNA (part a) and unfractionated *E. coli* C^{14}-Ala-sRNA (part b). Each point represents a 50 μl reaction containing the components described in the legend of Table 1, and the following components: 0.02 M magnesium acetate; 0.1 A^{260} unit of trinucleotide as specified; 1.1 A^{260} units of *E. coli* ribosomes (part a) and 2.0 A^{260} units of *E. coli* ribosomes (part b); and C^{14}-Ala-sRNA as indicated on the abscissa (Leder and Nirenberg, unpubl.).

Possibly, hydrogen bonds then form between the two remaining bases of the codon (bases 1 and 2, or 2 and 3) and the corresponding bases in the anticodon. Only two out of three bases in a codon would then be recognized. This possibility is supported by the studies of Rottman and Cerutti (1966) and Cerutti, Miles, and Frazier, (1966). Possibly, some synonym codon patterns may be due to the formation of two rather than three base pairs per triplet, particularly if both are

RECOGNITION OF ALA-CODONS BY YEAST ALA-sRNA

	ME	Di Di
sRNA	CUUIGCIψGG	H H
		UAGUCGGUAGC
mRNA	GCU	GCU
	GCC	GCC
	GCA	GCA
	(GCG)	(GCG)

FIGURE 5. Base sequences from yeast Ala-sRNA shown in the upper portion of the figure represent possible anticodons. Base sequences of synonym RNA Ala-codons are shown in the lower portion of the figure. The first and second bases of Ala-codons on the left would form antiparallel Watson-Crick hydrogen bonds with the anticodon, while those on the right would form parallel hydrogen bonds. See text for further details.

TABLE 6. ALTERNATE BASE PAIRING

sRNA Anticodon	mRNA Codon
U	A G
C	G
A	U
G	C U
I	U C A
rT	A G
ψ	A G (U)
DiHU	No base pairing

The base in an sRNA anticodon shown in the left-hand column forms antiparallel hydrogen bonds with the base(s) shown in the right-hand column, which usually occupy the third position of degenerate mRNA codons. Relationships for U, C, A, G, and I of anticodons are "wobble" hydrogen bonds suggested by Crick (1966; also this volume). See text for further details.

TABLE 7. NUCLEOTIDE SEQUENCES OF RNA CODONS RECOGNIZED BY AA-sRNA FROM BACTERIA AND AMPHIBIAN AND MAMMALIAN LIVER

	U	C	A	G	
U	PHE	SER	TYR	cys	U
	PHE	SER	TYR	cys	C
	leu?	SER	TERM?	[cys]	A
	leu, F-MET	[SER]	TERM?	trp	G
C	leu	PRO	HIS	ARG	U
	leu	PRO	HIS	ARG	C
	leu	PRO	gln	ARG	A
	leu	PRO	gln	[ARG]	G
A	ILE	THR	asn	[SER]	U
	ILE	THR	asn	[SER]	C
	[ILE]	THR	LYS	[ARG†]	A
	MET, F-MET?	THR	[LYS]	[ARG]	G
G	VAL	ALA	ASP	GLY	U
	VAL	ALA	ASP	GLY	C
	VAL	ALA	GLU	gly	A
	VAL	[ALA]	GLU	gly	G

Universality of the RNA code. Nucleotide sequences and relative template activities of RNA codons determined with trinucleotides and AA-sRNA from *E. coli*, *Xenopus laevis* and guinea pig liver. Rectangles represent trinucleotides which are active templates for AA-sRNA from one organism, but not from another. Assignments in capitals indicate that the trinucleotide was assayed with AA-sRNAs from *E. coli*, *Xenopus laevis* liver, and guinea pig liver. Assignments in lower case indicate that the trinucleotide was assayed only with *E. coli* AA-sRNA (with the exception of cys-codons which were assayed with both *E. coli* and guinea pig liver Cys-sRNA).

†Söll et al. (1965) reported that both AGA and AGG stimulate yeast Arg-sRNA binding to ribosomes. The trinucleotide, AGA, however, has little or no effect upon the binding of *E. coli*, *Xenopus laevis* or guinea pig Arg-sRNA to ribosomes.

Reactions contained components described in the legend to Table 1, 0.01 or 0.02 M Mg^{++}, *E. coli* ribosomes, and 0.150 A^{260} units of trinucleotides (data from Marshall, Caskey, and Nirenberg, in prep.).

(C) · (G) pairs (also see earlier discussion concerning template activity of pUpC).

In summary, patterns for amino acids often represent the sum of two or more codon patterns recognized by different sRNA species. Specific sRNA patterns, in turn, often result from alternate pairing between bases in the codon and anticodon or, possibly, from the formation of only two base pairs if the remaining bases do not greatly repel one another.

UNIVERSALITY

The results of many studies indicate that the RNA code is largely universal. However, translation of the RNA code can be altered in vivo by extragenic suppressors and in vitro by altering components of reactions or conditions of incubation. Thus, cells sometimes differ in specificity of codon translation.

To investigate the fine structure of the code recognized by AA-sRNA from different organisms, nucleotide sequences and relative template activities of RNA codons recognized by bacterial, amphibian, and mammalian AA-sRNA (*E. coli*, *Xenopus laevis* and guinea pig liver, respectively) were determined (Marshall, Caskey, and Nirenberg, submitted for publication). Acylation of sRNA was catalyzed in all cases by aminoacyl-sRNA synthetases from corresponding organisms and tissues. *E. coli* ribosomes were used for binding studies. Therefore, the specificities of sRNA and AA-sRNA synthetases were investigated.

The results are shown in Table 7. Almost identical translations of nucleotide sequences to amino acids were found with bacterial, amphibian, and mammalian AA-sRNA. In addition, similar sets of synonym codons usually were recognized by AA-sRNA from each organism. However, *E. coli* AA-sRNA sometimes differed strikingly from *Xenopus* and guinea pig liver AA-sRNA in relative response to synonym codons. Differences in codon recognition are shown in Table 8. The following

TABLE 8. SPECIES DEPENDENT DIFFERENCES IN RESPONSE OF AA-sRNA TO TRINUCLEOTIDE CODONS

		sRNA		
	Codon	Bacterial (*E. coli*)	Amphibian (*Xenopus laevis*)	Mammalian (Guinea pig liver)
ARG	AGG	±	+ + + +	+ + +
	CGG	±	+ + + +	+ + + +
MET	UUG	+ +	±	±
ALA	GCG	+ + + +	±	+ +
ILE	AUA	±	+ +	+ +
LYS	AAG	±	+ + + +	+ + + +
SER	UCG	+ + + +	±	+ +
	AGU	±	+ + +	+ + +
	AGC	±	+ + +	+ + +
CYS	UGA	±		+ + +

Possible differences: ACG, THR; AUC, ILE; CAC, HIS; GUC, VAL; and GCC, ALA

No differences found: ASP, GLY, GLU, PHE, PRO, and TYR.

The following scale indicates the approximate response of AA-sRNA to a trinucleotide relative to the responses of the same AA-sRNA preparation to all other trinucleotides for that amino acid (except Gly-sRNA which was assayed only with GGU and GGC).

+ + + +	70–100%
+ + +	50–70%
+ +	20–50%
±	0–20%

trinucleotides had little or no detectable template activity for unfractionated *E. coli* AA-sRNA but served as active templates with *Xenopus* and guinea pig AA-sRNA: AGG, CGG, arginine; AUA, isoleucine; AAG, lysine; AGU, AGC, serine; and UGA, cysteine. Those trinucleotides with high template activity for *E. coli* AA-sRNA but low activity for *Xenopus* or guinea pig liver AA-sRNA were: UUG, N-formyl-methionine; GCG, alanine; and UCG, serine. Possible differences also were observed with ACG, threonine; AUC, isoleucine; CAC, histidine; GCC, alanine; and GUC, valine. No species dependent differences were found with Asp-, Gly-, Glu-, Phe-, Pro-, and Tyr-codons.

Thus, some degenerate trinucleotides were active templates with sRNA from each species studied, whereas others were active with sRNA from one species but not from another.

UAA and UAG do not appreciably stimulate binding of unfractionated *E. coli* AA-sRNA (AA-sRNA for each amino acid tested); *Xenopus* Arg-, Phe-, Ser-, or Tyr-sRNA; or guinea pig Ala-, Arg-, Asp-, His-, Ile-, Met-, Pro-, Ser-, or Thr-sRNA.

Nucleotide sequences recognized by *Xenopus* skeletal muscle Arg-, Lys-, Met-, and Ser-sRNA were determined and compared with sequences recognized by corresponding *Xenopus* liver AA-sRNA preparations. No differences between liver and muscle AA-sRNA were detected, either in nucleotide sequences recognized or in relative responses to synonym codons.

Fossil records of bacteria 3.1 billion years old have been reported (Barghoorn and Schopf, 1966). The first vertebrates appeared approximately 510 million years ago, and amphibians and mammals, 355 and 181 million years ago, respectively. The presence of bacteria 3 billion years ago may indicate the presence of a functional genetic code at that time. Almost surely the code has functioned for more than 500 million years. The remarkable similarity in codon base sequences recognized by bacterial, amphibian, and mammalian AA-sRNA suggest that most, if not all, forms of life on this planet use almost the same genetic language, and that the language has been used, possibly with few major changes, for at least 500 million years.

UNUSUAL ASPECTS OF CODON RECOGNITION AS POTENTIAL INDICATORS OF SPECIAL CODON FUNCTIONS

Most codons correspond to amino acids; however, some codons serve in other capacities, such as initiation, termination or regulation of protein synthesis. Although only a few codons have been assigned special functions thus far, we think it likely that many additional codons eventually may be found to serve special functions. Unusual properties of codon recognition sometimes may indicate special codon functions. For example, the properties of initiator and terminator codons, during codon recognition, are quite distinctive (see below). We find that approximately 20 codons have unusual properties related either to codon position, template activity, specificity, patterns of degeneracy, or stability of codon-ribosome-sRNA complexes. Until more information is available these observations will be considered as *possible* indicators of special codon functions.

Conclusions will be stated first to provide a frame of reference for discussion:

(1) A codon may have alternate meanings. (For example, UUG at or near the 5′-terminus of mRNA may correspond to N-formyl-methionine; whereas, an internal UUG codon may correspond to leucine.)

(2) A codon may serve multiple functions simultaneously. (For example, a codon may specify both initiation and an amino acid, perhaps via AA-sRNA with high affinity for peptidyl-sRNA sites on ribosomes.)

(3) Codon function sometimes is subject to modification.

(4) Degenerate codons for the same amino acid often differ markedly in template properties.

Codon Frequency and Distribution

Often, multiple species of sRNA corresponding to the same amino acid recognize different synonym codons. Degenerate codon usage in mRNA sometimes is nonrandom (Garen, pers. comm.; also von Ehrenstein; Weigert et al., this volume). The possibility that different sets of sRNA may be required for the synthesis of two proteins with the same amino acid composition suggests that protein synthesis sometimes may be regulated by codon frequency and distribution coupled with differential recognition of degenerate codons. Possibly, the rates of synthesis of certain proteins may be regulated simultaneously by alterations which affect the apparatus recognizing one degeneracy but not another (see reviews by Ames and Hartman, 1963; and Stent, 1964).

Codon Position

As discussed in an earlier section, the template properties of 5′-terminal-, 3′-terminal-, and internal- codons may differ. Regulatory mechanisms based on such differences have been suggested. Reading of mRNA probably is initiated at or near the 5′-terminal codon and then proceeds toward the 3′-terminus of the RNA chain (Salas, Smith,

Stanley, Jr., Wahba, and Ochoa, 1965). It is not known whether mechanisms of 5′-terminal and internal initiation in polycistronic messages are similar. Also, internal- and 3′-terminal mechanisms of termination remain to be defined.

N-formyl-Met-sRNA may serve as an initiator of protein synthesis in *E. coli* (Clark and Marcker, 1966; Adams and Capecchi, 1966; Webster, Englehardt, and Zinder, 1966; Thach, Dewey, Brown, and Doty, 1966). Met-sRNA$_1$ can be converted enzymatically to N-formyl-Met-sRNA$_1$ and responds to UUG, CUG, AUG and, to a lesser extent, GUG. Met-sRNA$_2$ does not accept formyl-moieties and responds primarily to AUG (Clark and Marcker, 1966; Marcker et al., this volume; also Kellogg, Doctor, Loebel, and Nirenberg, 1966). In *E. coli* extracts protein synthesis is initiated in at least two ways: by initiator codons specifying N-formyl-Met-sRNA or, at somewhat higher Mg^{++} concentrations, by another means, probably not dependent upon N-formyl-Met-sRNA since many synthetic polynucleotides without known initiator codons direct cell-free protein synthesis (Nakamoto and Kolakofsky, 1966). Poly U, for example, directs di- as well as polyphenylalanine synthesis (Arlinghaus, Schaeffer, and Schweet, 1964). Probably codons for N-formyl-Met-sRNA initiate protein synthesis with greater accuracy than codons which serve as initiators only at relatively high Mg^{++} concentrations.

UAA and UAG may function as terminator codons (Brenner, Stretton, and Kaplan, 1965; Weigert and Garen, 1965). The trinucleotides UAA and UAG do not stimulate binding appreciably of *unfractionated E. coli* AA-sRNA to ribosomes. However, sRNA fraction(s) corresponding to UAA and/or UAG are not ruled out.

Extragenic suppressors may affect the specificity of UAA and/or UAG recognition (see review by Beckwith and Gorini, 1966). The efficiencies of ochre suppressors (UAA) are relatively low compared to that of amber suppressors (UAG). Since amber suppressors do not markedly affect the rate of cell growth, and ochre suppressors with high efficiency have not been found, UAA may specify chain termination in vivo more frequently than UAG. In a study of great interest, Newton, Beckwith, Zipser and Brenner (1965) have shown that the synthesis of protein (probably mRNA also) is regulated by the relative position in the RNA message of codons sensitive to amber suppressors. Therefore, a codon may perform a regulatory function at one position but not at another.

Template Activity

Trinucleotides with little activity for AA-sRNA (in studies thus far) are: UAA, UAG, and UUA, (perhaps CUA also). In addition, the following trinucleotides are active templates with AA-sRNA from one organism, but not from another: AGG, AGA, CGG, arginine; UUG, (N-formyl-)-methionine; GCG, alanine; AUA, isoleucine; AAG, lysine, UCG, AGU, AGC, serine; and UGA, cysteine (see Universality Section and Table 9). However, some inactive trinucleotides possibly function as active codons at internal positions. For example, the following codon base compositions were estimated with synthetic polynucleotides and a cell-free protein synthesizing system from *E. coli*; AUA, isoleucine; AGA, arginine; and AGC, serine (Nirenberg et al., 1963; Speyer et al., 1963; also see Jones, Nishimura, and Khorana, 1966, for results with AGA). Among the many possible explanations for low template activities of trinucleotides in binding assays are: special codon function; codon position; appropriate species of sRNA absent or in low concentration; competition for codons or for ribosomal sites by additional species of sRNA; high ratio of deacylated to AA-sRNA; cryptic (non-acylatable) sRNA; reaction conditions, e.g., low concentration of Mg^{++} or other components, time or temperature of incubation.

Codon Specificity

Often synonym trinucleotides differ strikingly in template specificity. Such observations may indicate that template specificities of terminal- and internal-codons differ, or that special function codons or suppressors are present. At 0.010–0.015 M Mg^{++}, trinucleotide template specificity is high, in many cases higher than that of a polynucleotide; for example, poly U, but not UUU, stimulates binding of Ile-sRNA to ribosomes. However, at 0.03 M Mg^{++} ambiguous recognitions of tri- and polynucleotides are observed more frequently.

Relative template activities of synonym trinucleotides in reactions containing 0.01 or 0.03 M Mg^{++} are shown in Table 9. In some cases, only one or two trinucleotides in a synonym set are active templates at 0.01 M Mg^{++}; whereas all degeneracies are active at 0.03 M Mg^{++} (e.g., Glu, Lys, Ala, Thr). In other cases either all synonym trinucleotides are active at 0.01 M Mg^{++} as well as at 0.03 M Mg^{++} (e.g., Val), or none are active at the lower Mg^{++} concentration (e.g., Tyr, His, Asn). Such data suggest that codon-ribosome-AA-sRNA complexes formed with degenerate trinucleotides often differ in stability.

MODIFICATION OF CODON RECOGNITION DUE TO PHAGE INFECTION

N. and T. Sueoka (1964; also see Sueoka et al., this volume) have shown that infection of *E. coli*

TABLE 9. TEMPLATE ACTIVITY OF TRINUCLEOTIDES IN 0.01 OR 0.03 M Mg++

	U	C	A	G	
	PHE	[SER]	TYR	[CYS]	U
U	PHE	SER	TYR	[CYS]	C
		(SER)			A
	[F-MET]	[SER]		(TRP)	G
		PRO	HIS	[ARG]	U
C		PRO	HIS	ARG	C
		(PRO)	GLN	[ARG]	A
	LEU	(PRO)	[GLN]	ARG	G
	ILE	[THR]	ASN	SER, CYS	U
A	ILE	THR	ASN	SER, CYS	C
		THR	[LYS]		A
	[MET]	[THR]	LYS		G
	VAL	ALA	[ASP]	[GLY]	U
G	VAL	ALA	[ASP]	GLY	C
	[VAL]	[ALA]	[GLU]	(GLY)	A
	VAL	ALA	GLU	(GLY)	G

Legend:

	0.01 M Mg	0.03 M Mg
[ ] =	+	+
No Box =	−	+
( ) =	not tested	

Relative template activities of trinucleotides in reactions containing 0.01 or 0.03 M Mg++. A plus (+) sign in the legend means that the trinucleotide stimulates AA-sRNA binding to ribosomes at that magnesium concentration; a minus (−) sign means it is relatively inactive as a template. The results refer to AA-sRNA from *E. coli* strains B and/or W3100. The data are from Anderson, Nirenberg, Marshall, and Caskey (1966).

by T2 bacteriophage results, within one to three minutes, in the modification of one or more species of Leu-sRNA present in the *E. coli* host. Concomitantly, *E. coli*, but not viral protein synthesis is inhibited. Protein synthesis is required, however, for modification of Leu-sRNA.

In collaboration with N. and T. Sueoka, modification of Leu-sRNA has been correlated with codon recognition specificity. sRNA preparations were isolated from *E. coli* before phage infection and at one and eight min after infection. After acylation, Leu-sRNA preparations were purified by MAK column chromatography and the binding of each pooled Leu-sRNA fraction to ribosomes in response to templates was determined (Fig. 6). The profile of Leu-sRNA (eight min after infection) acylated with yeast, rather than *E. coli*, Leu-sRNA synthetase is shown also (Fig. 6D); thus, both anticodon and enzyme recognition sites were monitored. In Fig. 7 the approximate chromatographic mobility on MAK columns of each Leu-sRNA fraction is shown diagrammatically, together with the relative response of each fraction to tri- and polynucleotide templates and acylation specificity of *E. coli* and yeast Leu-sRNA synthetase preparations.

Within one minute after infection, a marked decrease was observed in Leu-sRNA$_2$, responding to CUG, and a corresponding increase was seen in Leu-sRNA$_1$, responding to poly UG, but not to the trinucleotides, UUU, UUG, UGU, GUU, UGG, GUG, GGU, CUU, CUC, CUG, UAA, UAG, UGA, or to poly U or poly UC. However, Leu-sRNA$_1$ was not detected 8 min. after infection.

A marked increase in the response of Leu-sRNA$_5$ to UUG was observed one minute after infection, and an even greater increase was seen eight minutes after infection.

Greater responses of Leu-sRNA$_3$ and Leu-sRNA$_{4a,b}$ to poly UC also were observed eight minutes after phage infection. Leu-sRNA fractions 3 and 4 differ in chromatographic mobility and in acylation specificity by yeast and *E. coli* Leu-sRNA synthetase preparations. Thus, Leu-sRNA$_3$ and a component in fraction 4 differ, although both fractions 3 and 4 respond to poly UC. The multiple responses of Leu-sRNA$_{4a,b}$ to poly U, poly UC, and the trinucleotides, CUU and CUC, suggest that fraction 4 may contain two or more Leu-sRNA species. Striking increases in response of fraction 4 to poly U were observed one and eight minutes after infection.

Leu-sRNA fractions 1, 2, and 3 are related, for each is recognized by yeast as well as by *E. coli* Leu-sRNA synthetase preparations. In contrast, Leu-sRNA$_{4a,b}$ and Leu-sRNA$_5$ are recognized by *E. coli*, but not yeast Leu-sRNA synthetase; thus, fraction 4 is related to fraction 5. Two *different* cistrons of Leu-sRNA are predicted: Leu-sRNA fractions 1, 2, and 3 may be products of one cistron; whereas, fractions 4 and 5 may be products of a different cistron. In this regard, Berg, Lagerkvist, and Dieckman (1962) have shown that *E. coli* Leu-sRNA contains two base sequences at the 4th, 5th, and 6th base positions from the 3′-terminus of the sRNA.

The data suggest the following sRNA precursor-product relationships. Leu-sRNA$_2$ is a product of "cistron A"; the decrease in Leu-sRNA$_2$ and the simultaneous increase in Leu-sRNA$_1$ (within one minute after infection) suggests that Leu-sRNA$_2$ is the precursor of Leu-sRNA$_1$. The data also suggest that Leu-sRNA$_2$ is a precursor of Leu-sRNA$_3$. The following anticodons and mRNA

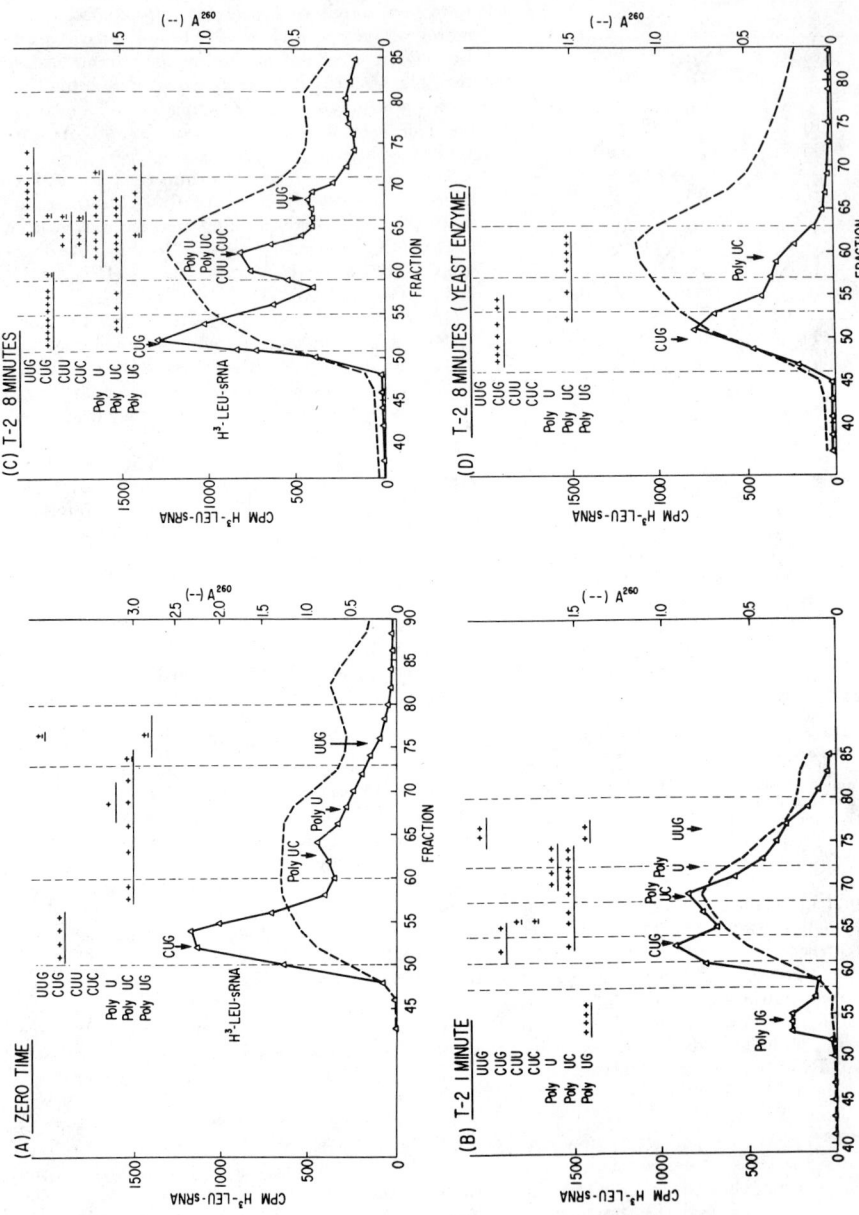

FIGURE 6. The graphs represent MAK column fractions of H³-Leu-sRNA from *E. coli* B before infection (a) and at 1 min (b) and 8 min (c and d) after infection with T2 phage. sRNA was acylated prior to chromatography with H³-leucine using *E. coli* (a, b, c) or yeast (d) synthetase preparations. Column eluates were pooled as indicated by the vertical broken lines; dialyzed against 5 × 10⁻⁴ M potassium cacodylate, pH 5.5, and lyophilized. Then binding of each fraction to ribosomes in response to tri- or polynucleotide templates was determined. At the top of each graph relative responses of Leu-sRNA fractions to templates are shown. Approximate relative responses are indicated as follows: No symbol, no detectable response of Leu-sRNA; ±, possible response; +, slight response; and ++ to +++++, moderate to strong responses. Profiles represented by broken lines indicate A²⁶⁰ units; △—△, represent H³-Leu-sRNA. Data are from Kano-Sueoka, Nirenberg, and Sueoka (unpubl.). Also see Sueoka et al., this volume.

THE RNA CODE AND PROTEIN SYNTHESIS

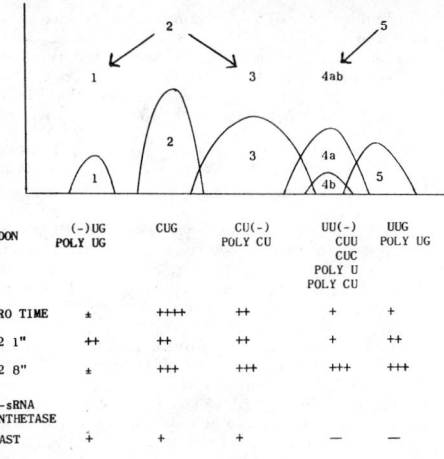

FIGURE 7. Diagrammatic representation of the data shown in Fig. 6. The relative mobilities of multiple species of Leu-sRNA, before and after phage infection, fractionated by MAK column chromatography, are shown at the top. Leu-sRNA peaks are numbered. Arrows represent predicted Leu-sRNA precursor-product relationships (Fractions 2 and 5 possibly are products of different cistrons).

Tri- and polynucleotide codons recognized by each Leu-sRNA peak are shown below. Approximate relative responses of Leu-sRNA$_{1-5}$ to codons are indicated as follows: $\pm$, possible response, $+$ to $+++$, slight to strong responses.

On the bottom are shown the specificities of E. coli (zero time, 1 and 8 min after infection) and yeast (8 min after infection only) Leu-sRNA synthetase preparations for sRNA^{Leu} fractions.

codons are suggested for Leu-sRNA fractions 2, 3, and 1, respectively (note: asterisks represent modifications of a nucleotide base; codon and anticodon sequences are written with 3',5'-phosphodiester linkages; antiparallel hydrogen bonding between codon and anticodons is assumed): Leu-sRNA$_2$-product of "cistron A", CAG anticodon, [CUG codon]; Leu-sRNA$_3$- derived from fraction 2, C*AG anticodon, [CU(−) codon]; Leu-sRNA$_1$- derived from fraction 2, CAG** anticodon, [(−)UG codon].

Leu-sRNA$_5$ is a product of "cistron B", and differs from Leu-sRNA$_2$ in anticodon and Leu-sRNA synthetase recognition sites. The sequence, CAA, is suggested for the Leu-sRNA$_5$ anticodon, corresponding to a UUG mRNA codon. Leu-sRNA$_{4a,b}$ are derived from fraction 5. Possible anticodons and codons are: C*AA anticodon, [UU(−) codon]; C*IA anticodon, [UU(−), UC(−), UA(−) codons]; C*AI anticodon, [UU(−), CU(−), AU(−) codons].

Since modification of Leu-sRNA after phage infection is dependent upon protein synthesis, enzyme(s) may be needed to modify bases in Leu-sRNA fractions.

The inhibition of host E. coli, but not viral protein synthesis following viral infection may result from modification of Leu-sRNA fractions. N-formyl-Met-sRNA$_1$ serves as an initiator of protein synthesis in E. coli and responds to two trinucleotides, UUG and CUG, which are also recognized by Leu-sRNA fractions (see previous discussion on special function codons). Possibly, initiation or termination of E. coli, but not viral protein synthesis is affected. Further studies are needed, however, to elucidate the mechanism of viral induced inhibition of host protein synthesis.

ACKNOWLEDGMENTS

It is a pleasure to thank Miss Norma Zabriskie, Mrs. Theresa Caryk, Mr. Taysir M. Jaouni, and Mr. Wayne Kemper for their invaluable assistance. D. Kellogg is a Postdoctoral fellow of the Helen Hay Whitney Foundation. J. Levin is supported by USPHS grant 1-F2-GM-6369-01. F. Rottman is supported by grant PF-244 from the American Cancer Society.

REFERENCES

ADAMS, J. M., and M. R. CAPECCHI. 1966. N-formyl-methionyl-sRNA as the initiator of protein synthesis. Proc. Natl. Acad. Sci. 55: 147–155.

AMES, B. N., and P. E. HARTMAN. 1963. The histidine operon. Cold Spring Harbor Symp. Quant. Biol. 28: 349–356.

ANDERSON, W. F., M. W. NIRENBERG, R. E. MARSHALL, and C. T. CASKEY. 1966. RNA codons and protein synthesis: Relative activity of synonym codons. Fed. Proc. 25: 404.

ARLINGHAUS, R., J. SHAEFFER, and R. SCHWEET. 1964. Mechanism of peptide bond formation in polypeptide synthesis. Proc. Natl. Acad. Sci. 51: 1291–1299.

BARGHOORN, E. S., and J. W. SCHOPF. 1966. Microorganisms three billion years old from the precambrian of South Africa. Science 152: 758–763.

BECKWITH, J. R., and L. GORINI. 1966. Suppression. Ann. Rev. Microbiol., in press.

BERG, P., U. LAGERKVIST, and M. DIECKMANN. 1962. The enzymic synthesis of amino acyl derivatives of ribonucleic acid. VI. Nucleotide sequences adjacent to the ...pCpCpA end groups of isoleucine- and leucine-specific chains. J. Mol. Biol. 5: 159–171.

BERNFIELD, M. 1966. Ribonuclease and oligoribonucleotide synthesis. II. Synthesis of oligonucleotides of specific sequence. J. Biol. Chem. 241: 2014–2023.

BRENNER, S., A. O. W. STRETTON, and S. KAPLAN. 1965. Genetic Code: the 'nonsense' triplets for chain termination and their suppression. Nature 206: 994–998.

BRIMACOMBE, R., J. TRUPIN, M. NIRENBERG, P. LEDER, M. BERNFIELD, and T. JAOUNI. 1965. RNA codewords and protein synthesis. VIII. Nucleotide sequences of synonym codons for arginine, valine, cysteine and alanine. Proc. Natl. Acad. Sci. 54: 954–960.

CERUTTI, P., H. T. MILES, and J. FRAZIER. 1966. Interaction of partially reduced polyuridylic acid with a polyadenylic acid. Biochem. Biophys. Res. Commun. 22: 466–472.

CLARK, B., and K. MARCKER. 1966. The role of N-formyl-methionyl-sRNA in protein biosynthesis. J. Mol. Biol., 17: 394–406.

CONNELLY, C. M., and B. P. DOCTOR. 1966. Purification of two yeast serine transfer ribonucleic acids by counter-current distribution. J. Biol. Chem. 241: 715–719.

CRICK, F. H. C. 1966. Codon-Anticodon Pairing: The wobble hypothesis. J. Mol. Biol., 19: 548–555.

HEPPEL, L. A., P. R. WHITFIELD, and R. MARKHAM. 1955. Nucleotide exchange reactions catalyzed by ribonuclease and spleen phosphodiesterase. 2. Synthesis of polynucleotides. Biochem. J. 60: 8–15.

HOLLEY, R. W., J. APGAR, G. A. EVERETT, J. T. MADISON, M. MARQUISEE, S. H. MERRILL, J. R. PENSWICK, and A. ZAMIR. 1965. Structure of a ribonucleic acid. Science 147: 1462–1465.

INGRAM, V. M., and J. A. SJÖQUIST. 1963. Studies on the structure of purified alanine and valine transfer RNA from yeast. Cold Spring Harbor Symp. Quant. Biol. 28: 133–138.

JONES, D. S., S. NISHIMURA, and H. G. KHORANA. 1966. Studies on polynucleotides LVI. Further syntheses, in vitro, of copolypeptides containing two amino acids in alternating sequence dependent upon DNA-like polymers containing two nucleotides in alternating sequence. J. Mol. Biol. 16: 454–472.

KELLER, E. B., and M. F. FERGER. 1966. Alanyl-sRNA in the aminoacyl polymerase system of protein synthesis. Fed. Proc. 25: 215.

KELLOGG, D. A., B. P. DOCTOR, J. E. LOEBEL, and M. W. NIRENBERG. 1966. RNA codons and protein synthesis, IX. Synonym codon recognition by multiple species of valine-, alanine-, and methionine-sRNA. Proc. Natl. Acad. Sci. 55: 912–919.

LEDER, P., M. F. SINGER, and R. L. C. BRIMACOMBE. 1965. Synthesis of trinucleoside diphosphates with polynucleotide phosphorylase. Biochem. 4: 1561–1567.

MADISON, J. T., G. A. EVERETT, and H. KUNG. 1966. Nucleotide sequence of a yeast tyrosine transfer RNA. Science 153: 531–534.

NAKAMOTO, T., and D. KOLAKOFSKY. 1966. A possible mechanism for initiation of protein synthesis. Proc. Natl. Acad. Sci. 55: 606–613.

NEWTON, W. A., J. R. BECKWITH, D. ZIPSER, and S. BRENNER. 1965. Nonsense mutants and polarity in the Lac operon of Escherichia coli. J. Mol. Biol. 14: 290–296.

NIRENBERG, M. W., O. W. JONES, P. LEDER, B. F. C. CLARK, W. S. SLY, and S. PESTKA. 1963. On the coding of genetic information. Cold Spring Harbor Symp. Quant. Biol. 28: 549–557.

NIRENBERG, M., and P. LEDER. 1964. RNA codewords and protein synthesis. I. The effect of trinucleotides upon the binding of sRNA to ribosomes. Science 145: 1399–1407.

NIRENBERG, M., P. LEDER, M. BERNFIELD, R. BRIMACOMBE, J. TRUPIN, F. ROTTMAN, and C. O'NEAL. 1965. RNA codewords and protein synthesis, VII. On the general nature of the RNA code. Proc. Natl. Acad. Sci. 53: 1161–1168.

ROTTMAN, F., and P. CERUTTI. 1966. Template activity of uridylic acid-dihydrouridylic acid copolymers. Proc. Natl. Acad. Sci. 55: 960–966.

ROTTMAN, F., and M. NIRENBERG. 1966. Regulatory mechanisms and protein synthesis XI. Template activity of modified RNA codons. J. Mol. Biol., in press.

SALAS, M., M. A. SMITH, W. M. STANLEY, JR., A. J. WAHBA, and S. OCHOA. 1965. Direction of reading of the genetic message. J. Biol. Chem. 240: 3988–3995.

SÖLL, D., E. OHTSUKA, D. S. JONES, R. LOHRMANN, H. HAYATSU, S. NISHIMURA, and H. G. KHORANA. 1965. Studies on polynucleotides, XLIX. Stimulation of the binding of aminoacyl-sRNA's to ribosomes by ribotrinucleotides and a survey of codon assignments for 20 amino acids. Proc. Natl. Acad. Sci. 54: 1378–1385.

SONNEBORN, T. M. 1965. Degeneracy of the genetic code: Extent, nature and genetic implications. pp. 377–397. In: V. Bryson and H. J. Vogel (ed.) Evolving Genes and Proteins. Academic Press, New York.

SPEYER, J., P. LENGYEL, C. BASILIO, A. WAHBA, R. GARDNER, and S. OCHOA. 1963. Synthetic polynucleotides and the amino acid code. Cold Spring Harbor Symp. Quant. Biol. 28: 559–567.

STENT, G. S. 1964. The operon: On its third anniversary. Science 144: 816–820.

SUEOKA, N., and T. KANO-SUEOKA. 1964. A specific modification of Leucyl-sRNA of Escherichia coli after phage T2 infection. Proc. Natl. Acad. Sci. 52: 1535–1540.

THACH, R. E., K. F. DEWEY, J. C. BROWN, and P. DOTY. 1966. Formylmethionine codon AUG as an initiator of polypeptide synthesis. Science 153: 416–418.

THACH, R. E., and P. DOTY. 1965. Enzymatic synthesis of tri- and tetranucleotides of defined sequence. Science 148: 632–634.

WEBSTER, R. E., D. L. ENGELHARDT, and N. D. ZINDER. 1966. In vitro protein synthesis: Chain initiation. Proc. Natl. Acad. Sci. 55: 155–161.

WEIGERT, M., and A. GAREN. 1965. Base composition of nonsense codons in E. coli; Evidence from amino-acid substitutions at a tryptophan site in alkaline phosphatase. Nature 206: 992–994.

ZACHAU, H., D. DÜTTING, and H. FELDMANN. 1966. Nucleotide sequences of two serine-specific transfer ribonucleic acids (1). Angew. Chem. 5: 422, English Edition.

Codon—Anticodon Pairing:

The Wobble Hypothesis

F. H. C. CRICK

Medical Research Council, Laboratory of Molecular Biology

Hills Road, Cambridge, England

(Received 14 February 1966)

It is suggested that while the standard base pairs may be used rather strictly in the first two positions of the triplet, there may be some wobble in the pairing of the third base. This hypothesis is explored systematically, and it is shown that such a wobble could explain the general nature of the degeneracy of the genetic code.

Now that most of the genetic code is known and the base-sequences of sRNA molecules are coming out, it seems a proper time to consider the possible base-pairing between codons on mRNA and the presumed anticodons on the sRNA.

The obvious assumption to adopt is that sRNA molecules will have certain common features, and that the ribosome will ensure that all sRNA molecules are presented to the mRNA in the same way. In short, that the pairing between one codon–anticodon matching pair will to a first approximation be "equivalent" to that between any other matching pair.

As far as I know, if this condition has to be obeyed, and if all four bases must be distinguished in any one position in the codon, then the pairing in this position is *highly likely* to be the standard one; that is:†

$$G ==== C$$
$$\text{and} \quad A ==== U$$

or some equivalent ones such as, for example,

$$I ==== C$$
$$\text{and} \quad A ==== T$$

since this is the only type of pairing which allows all four bases to be distinguished in a strictly equivalent way.

We now know enough of the genetic code to say that in the *first two* positions of the codon the four bases are clearly distinguished; certainly in many cases, and probably in all of them. I thus deduce that the pairings in the first two positions are likely to be the standard ones.

---

† Throughout this paper the sign $====$ is used to mean "pairs with". If two bases are equivalent in their coding properties, this is written $\begin{matrix}U\\C\end{matrix}$ or $\left.\begin{matrix}U\\C\end{matrix}\right\}$

However, what we know about the code has already suggested two generalizations about the third place of the codon. These are:

(1) $\left.\begin{array}{l}\text{U}\\\text{C}\end{array}\right\}$† this already appears true in about a dozen cases out of the possible 16, and there are no data to suggest any exceptions.

(2) $\left.\begin{array}{l}\text{A}\\\text{G}\end{array}\right\}$ probably true in about half of the possible 16 cases, but the evidence suggests it may perhaps be incorrect in several other cases.

The detailed experimental evidence is rather complicated and will not be discussed here. (For details of the code see, for example, Nirenberg *et al.*, 1965; and Söll *et al.*, 1965.) It suffices that these rules *may* be true, as suggested by Eck (1963) a little time ago. Alternatively, only the first one may be true.

This naturally raises the question: Does *one* sRNA molecule recognize more than one codon, e.g. both UUU *and* UUC. Some evidence for this was first presented by Bernfield & Nirenberg (1965). They showed that *all* the sRNA for phenylalanine can be bound by poly U, although this sRNA also recognizes the triplet UUC, at least in part. More recent evidence along these lines is presented in Söll *et al.* (1966) and Kellogg *et al.* (1966). Again I do not wish to discuss here the evidence in detail, but simply to ask: If one sRNA codes both XYU and XYC, how is this done?

Now if we do not know anything about the geometry of the situation, it might be thought that almost any base pairs might be used, since it is well known that the bases can be paired (i.e. form at least two hydrogen bonds) in many different ways. However, it occurred to me that if the first two bases in the codon paired in the standard way, the pairing in the third position might be *close* to the standard ones.

We therefore ask: How many base pairs are there in which the glycosidic bonds occur in a position close to the standard one? Possible pairs are:

$$G ==== A \tag{1}$$

In my opinion this will not occur, because the NH_2 group of guanine cannot make one of its hydrogen bonds, even to water (see Fig. 1).

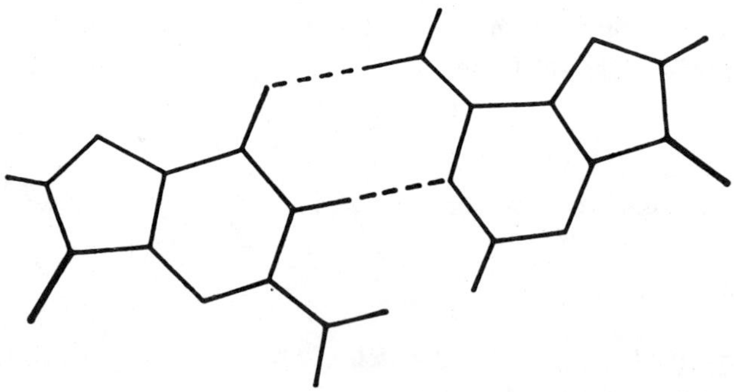

Fig. 1. The unlikely pair guanine–adenine.

$$U ==== C \tag{2}$$

This brings the two keto groups rather close together and also the two glycosidic bonds, but it may be possible (see Fig. 2).

† This symbol implies that both U and C code the same amino acid.

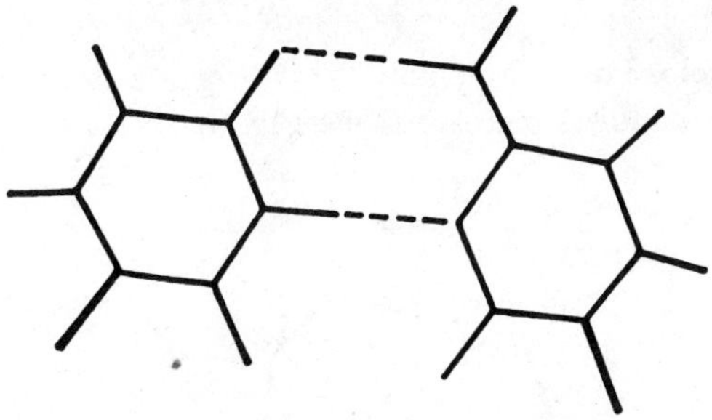

FIG. 2. The close pair uracil–cytosine.

$$U ==== U \qquad (3)$$

Again rather close together (see Fig. 3).

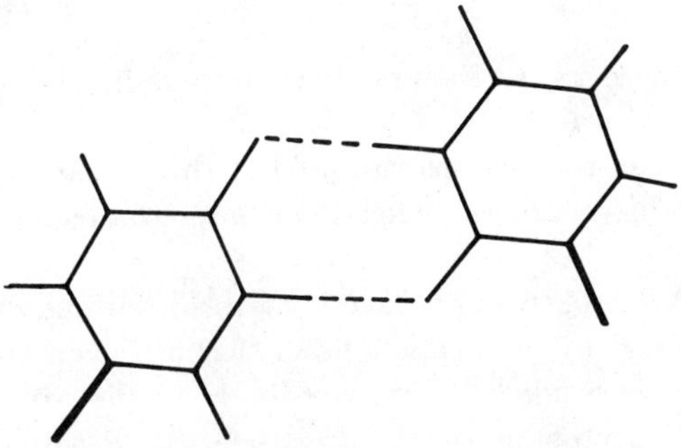

FIG. 3. The close pair uracil–uracil.

$$G ==== U \qquad (4)$$
$$\text{or } I ==== U$$

These only require the bond to move about 2·5 Å from the standard position (see Fig. 4).

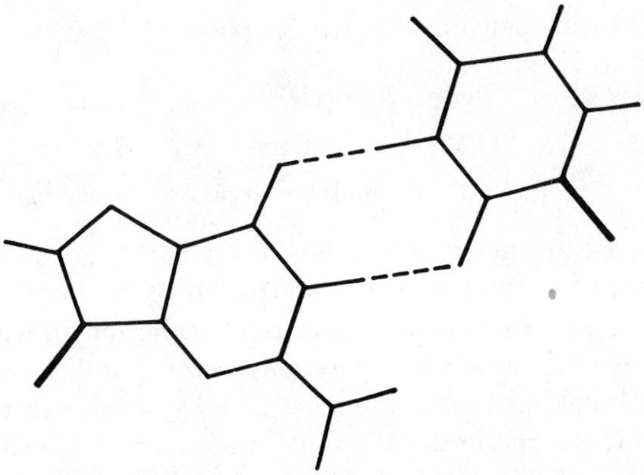

FIG. 4. The pair guanine–uracil (the pair inosine–uracil is similar).

"WOBBLE HYPOTHESIS"

$$I ==== A \qquad (5)$$

This is perfectly possible. Poly I and poly A will form a double helix. The distance between the glycosidic bonds is increased (see Fig. 5).

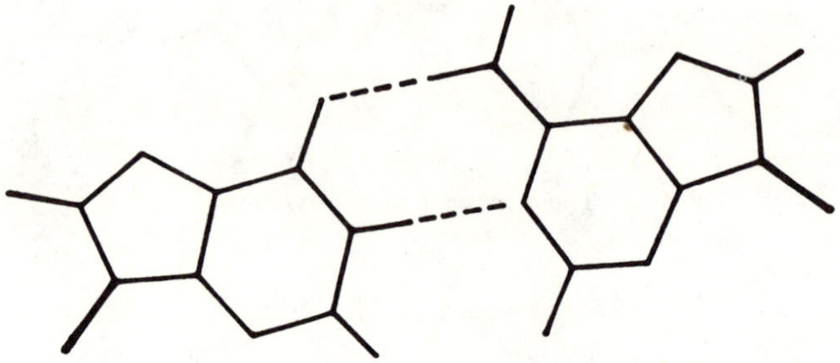

Fig. 5. The pair inosine–adenine.

As far as I know, these are all the possible solutions if it is assumed that the bases are in their usual tautomeric forms.

I now postulate that in the base-pairing of the third base of the codon there is a certain amount of play, or wobble, such that more than one position of pairing is possible.

As can be seen from Fig. 6, there are seven possible positions which might be reached by wobbling. However, it by no means follows that all seven are accessible, since the molecular structure is very likely to impose limits to the wobble. We should therefore strictly consider all possible *combinations of allowed positions*. There are 127 of these, but most of them are trivial. If we adopt the rule that *all four bases* on the codon (in the third position) must be recognized (that is, paired with) we are left with 51 different combinations. This is too many for easy consideration, but fortunately we can eliminate most of them by only accepting combinations which do not violate the broad features of the code. If we assume:

(a) that all four bases must be recognizable;

(b) that the code must *in some cases* distinguish between $\left.\begin{array}{l}U\\C\end{array}\right\}$ and $\left.\begin{array}{l}A\\G\end{array}\right\}$ as it appears to do for the pairs

| Phe | Tyr | His | Asn | Asp |
| Leu | C.T.† | Gln | Lys | Glu |

(not all of which are likely to be wrong)

then by strictly logical argument it can be shown both that the standard position must be used, and that the three positions on the left of Fig. 6 cannot be used.

This leaves us with only four possible sites to consider one of which—the standard one—must be included. There are therefore only seven possible combinations. I have examined all these, but I shall restrict myself here to the case in which all four positions are used, as this is structurally the most likely and also seems to give the code (called code 4 in the note privately circulated) which best fits the experimental data.

† C.T., Chain termination.

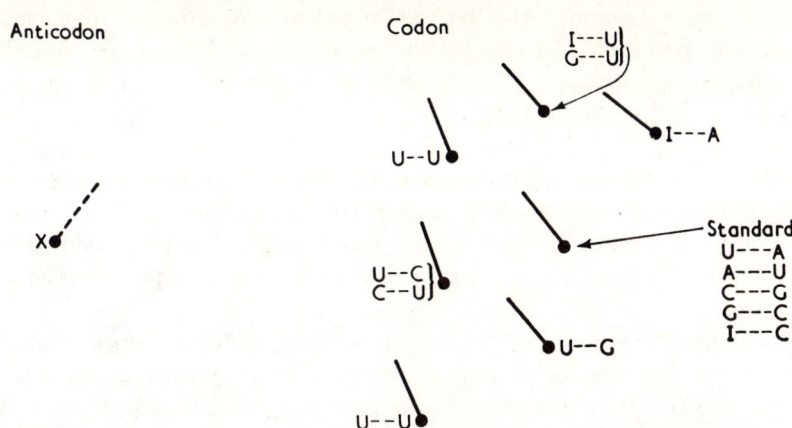

Fig. 6. The point X represents the position of the C_1' atom of the glycosidic bond (shown dotted) in the anticodon. The other points show where the C_1' atom and the glycosidic bond fall for the various base pairs. (Pairs with inosine in the codon have been omitted for simplicity.) The wobble code suggested uses the four positions to the right of the diagram, but not the three close positions.

The rules for pairing between the third base on the codon and the corresponding base on the anticodon are set out in Table 1. It can be seen that these rules make several strong predictions:

(1) it is not possible to code for either C alone, or for A alone.

For example, at the moment the codon UGA has not been decisively allocated. Wobble theory states that UGA might either:

(a) code for cysteine, which has UGU and UGC; or
(b) code for trypotophan, which has UGG; or
(c) not be recognized.

TABLE 1

Pairing at the third position of the codon

Base on the anticodon	Bases recognized on the codon
U	A, G
C	G
A†	U
G	U, C
I	U, C, A

† It seems likely that inosine will be formed enzymically from an adenine in the nascent sRNA. This may mean that A in this position will be rare or absent, depending upon the exact specificity of the enzyme(s) involved.

However it does *not* permit UGA to code for any amino acid other than cysteine or tryptophan. This rule could also explain why no suppressor has yet been found which suppresses only *ochre* mutants (UAA), although suppressors exist which suppress both *ochre* and *amber* mutants (UA$_G^A$).

(2) If an sRNA has inosine in the place at the relevant position on the anticodon (i.e. enabling it to pair with the third base of the codon), then it must recognize U, C and A in the third place of the codon. Conversely, those amino acids coded only by XY$_C^U$ (such as Phe, Tyr, His, etc.) cannot have inosine in that place on their sRNA.

(3) Wobble theory does not state exactly how many different types of sRNA will actually be found for any amino acid. However if an amino acid is coded for by all four bases in the third position (as are Pro, Thr, Val, etc.), then wobble theory predicts that there will be at least two sRNA's. These can have the recognition pattern:

$$\left.\begin{matrix}U\\C\end{matrix}\right\} \text{ plus } \left.\begin{matrix}A\\G\end{matrix}\right\}$$

or

$$\left.\begin{matrix}U\\C\\A\end{matrix}\right\} \text{ plus } G$$

Note that the sets actually used for any amino acid may well vary from species to species.

The Anticodons

At this point it is useful to examine the experimental evidence for the anticodon. In the sRNA for alanine from yeast, Holley *et al.* (1965) have the following sequences:

$$--- \text{pUpUpIp Gp CpMeIp}\Psi\text{p} ---$$

position --- 36 37 38 ---

Zachau and his colleagues (Dütting, Karan, Melchers & Zachau, 1965) have for one of the serine sRNA's from yeast:

$$--- \text{p}\Psi\text{pUpIpGpApA}^+\text{p}\Psi\text{p} ---$$

(A$^+$ stands for a modified A)

For the valine sRNA from yeast, Ingram & Sjöquist (1963) have shown that the only inosine occurs in the sequence:

$$--- \text{pIpApCp} ---$$

Holley *et al.* (1965) have already pointed out that IGC is a possible anticodon for alanine, and the additional evidence makes it almost certain to my mind that this is correct, and that the anticodons are as given in the Table below†:

† *Note added 26 April 1966.* Drs J. T. Madison, G. A. Everett and H. Kung (personal communication) have completed the sequence of the tyrosine sRNA from yeast. The sequence strongly suggests that the anticodon in this case is GΨA, corresponding to the known codons UA^U. Since Ψ can form the same base pairs as U, this is in excellent agreement with the previous data.

Yeast sRNA

	Anticodon	Codon
Ala	I G C	G C ?
Ser	I G A	U C ?
Val	I A C	G U ?

remembering that the pairing proposed between codon and anticodon is *anti*-parallel. Thus I confidently predict: the anticodon is a triplet at (or very near) positions 36–37–38 on every sRNA, and that the *first two bases* in the codon pair with this (in an anti-parallel manner) *using the standard base pairs*.

However, inosine does not occur in every sRNA. In particular Holley *et al.* (1963) (and personal communication) have reported that the tyrosine sRNA has two peaks, neither of which contains inosine. Moreover, Sanger (personal communication) tells me that there is rather little inosine in the total sRNA from *E. coli*.

Testing the Theory

Two obvious tests present themselves:

(1) To find which triplets are bound by any one type of sRNA. This is being done by Khorana and his colleagues (Söll *et al.*, 1966), and also by Nirenberg's group (Kellogg, Doctor, Loebel & Nirenberg, 1966). The difficulty here is to be sure that the sRNA used is pure, and not a mixture.

(2) To discover unambiguously the position of the anticodon on sRNA, and to find further anticodons. This will certainly happen as our knowledge of the base sequence of sRNA molecules develops. The absence of inosine from any anticodon is obviously of special interest.

In conclusion it seems to me that the preliminary evidence seems rather favourable to the theory. I shall not be surprised if it proves correct.

I thank my colleagues for many useful discussions and the following for sending me material in advance of publication: Dr M. W. Nirenberg, Dr H. G. Khorana, Dr G. Streisinger, Dr W. Holley, Dr J. Fresco, Dr H. G. Zachau, Dr C. Yanofsky, Dr H. G. Wittmann, Dr H. Lehmann and Dr J. D. Watson.

REFERENCES

Bernfield, M. R. & Nirenberg, M. W. (1965). *Science*, **147**, 479.
Dütting, D., Karan, W., Melchers, F. & Zachau, H. G. (1965). *Biochim. biophys. Acta*, **108**, 194.
Eck, R. V. (1963). *Science*, **140**, 477.
Holley, R. W., Apgar, J., Everett, G. A., Madison, J. T., Marquisee, M., Merrill, S. H., Penswick, J. R. & Zamir, A. (1965). *Science*, **147**, 1462.
Holley, R. W., Apgar, J. Everett, G. A., Madison, J. T., Merrill, S. H. & Zamir, A. (1963). *Cold Spr. Harb. Symp. Quant. Biol.* **28**, 117.
Ingram, V. M. & Sjöquist, J. A. (1963). *Cold. Spr. Harb. Symp. Quant. Biol.* **28**, 133.

Kellogg, D. A., Doctor, B. P., Loebel, J. E. & Nirenberg, M. W. (1966). *Proc. Nat. Acad. Sci., Wash.* **55**, 912.
Nirenberg, M., Leder, P., Bernfield, M., Brimacombe, R., Trupin, J., Rottman, F. & O'Neal, C. (1965). *Proc. Nat. Acad. Sci., Wash.* **53**, 1161.
Söll, D., Jones, D. S., Ohtsuka, E., Faulkner, R. D., Lohrmann, R., Hayatsu, H., Khorana, H. G., Cherayil, J. D., Hampel, A. & Bock, R. M. (1966). *J. Mol. Biol.* **19**, 556.
Söll, D., Ohtsuka, E., Jones, D. S., Lohrmann, R., Hayatsu, H., Nishimura, S. & Khorana, H. G. (1965). *Proc. Nat. Acad. Sci., Wash.* **54**, 1378.

Conformation of the Anticodon Loop in tRNA

by
W. FULLER
A. HODGSON
Biophysics Department and
MRC Biophysics Research Unit,
King's College, University of London

A molecular model for the anticodon arm is proposed which is compatible with chemical, X-ray and genetic evidence. It provides a stereochemical basis for Crick's "wobble" hypothesis.

NUCLEOTIDE sequences determined for a number of amino-acid specific tRNA molecules[1-4] have led to the suggestion that these molecules have a "clover leaf" structure (Fig. 1). This was because, despite their different nucleotide sequences, there are striking structural homologies when the tRNA molecules are folded so that the number of intramolecular Watson–Crick base-pairs is a maximum (Fig. 1). Diagrams like Fig. 1, however, indicate little of the three-dimensional appearance of such structures and their implications. Therefore we have constructed three-dimensional models, and here describe a molecular model-building study of the anticodon arm.

Using chemical information about nucleotide sequence and X-ray evidence on the conformation of base-paired regions in the tRNA, the maintenance of reasonable stereochemical constraints leads to a model for the anticodon arm. This model accounts for the observed degeneracy in the reading of the third position of the codon and also makes a prediction about the site of the distortion required to accommodate this degeneracy.

Model Building Technique

We used Corey, Pauling and Koltun[5] spacefilling models and also skeletal models with a scale of 4 cm to 1 Å (ref. 6). The former ensure that short van der Waals contacts are avoided during preliminary investigations. Because, however, the atomic centres in them are inaccessible, we used skeletal models when preliminary study suggested that a particular conformation merited detailed analysis. Lengths and angles of covalent bonds and short van der Waals contacts were calculated from atomic co-ordinates measured on skeletal models and the co-ordinates were adjusted until acceptable stereochemistry was obtained, that is lengths of covalent bonds within 0·05 Å of accepted values, covalent angles within 6° and no non-covalently bonded contacts more than 0·4 Å short of the sum of the atomic van der Waals radii. We do not necessarily believe that our models describe the actual molecular conformations to an accuracy of a few hundredths of an angstrom, but the analysis shows that a model with the general characteristics we propose can be built with acceptable stereochemistry. Only if model building is treated as a rigid discipline with strict attention paid to detailed stereochemistry can the results of a study such as this be considered reliable and meaningful.

Conformation of the Anticodon Arm

X-ray diffraction suggests that the molecules of tRNA (ref. 7), in common with all RNA molecules so far studied by this method, contain helical regions with a conformation similar to that determined for two-stranded reovirus RNA (ref. 8). We have assumed that the Watson–Crick base-paired regions in the clover leaf structure have a conformation like the eleven-fold double-helical structure of reovirus RNA (rather than the less favoured ten-fold possibility). In the anticodon arm there is a loop of seven nucleotides at the end of the helical region. From considerations of biological function, the structural homologies in the different tRNA species might be expected to extend to the conformation of this loop.

The characteristic features of polynucleotide secondary structure are provided by interbase hydrogen bonding and base-stacking. The tRNA nucleotide sequences so far determined do not suggest an intramolecular base-pairing scheme which would give a similar structure for all the anticodon loops (Fig. 2). Therefore we searched for conformations of this loop which maximized single-

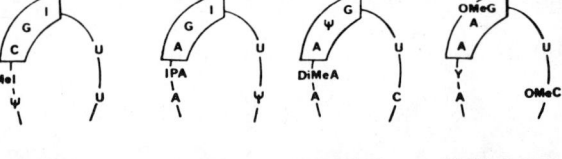

Fig. 2. The nucleotide sequences in yeast tRNA anticodon loops. The anticodon is shown boxed. The symbols ψ, MeI, C, G, I, U, A, IPA, DiMeA, Y, OMeG, OMeC stand for respectively: pseudouracil, 1-methyl-inosine, cytosine, guanine, inosine, uracil, adenine, isopentenyl-adenine, dimethyl-adenine, a so-far unidentified purine, O-methyl guanine, O-methyl cytosine. (The references from which the sequences were taken are in the caption to Fig. 1.)

stranded base-stacking. In doing this we also attempted to: (a) avoid negatively charged phosphate groups coming closer to each other than in accurately determined crystalline fibrous structures; (b) ensure that hydrogen bond donor groups on unpaired bases and ribose sugars were not buried in the structure, so they were unavailable for hydrogen bonding; (c) maintain single bond orientations in the polynucleotide chain (for example, the conformation at the glycosidic link) within the limits of values observed in model compounds and other polynucleotides.

When these stereochemical constraints are maintained, model-building studies suggest that the polynucleotide chain has surprisingly little conformational freedom. Furthermore, orientation of the single bonds in the only two polyribonucleotides whose structures have been determined in detail by X-ray analysis (two-stranded helical RNA (ref. 8), and two-stranded polyadenylic acid[9]) are rather similar. Therefore if stacking is to be maintained, the polynucleotide chain might be expected to have a conformation similar to that in one of the structures described for ribopolynucleotides. Stacking as much as possible of the anticodon loop on top of the double helical region of the anticodon arm might be expected to "nucleate" the structure of the single-stranded region so that its nucleotide conformation is similar to that in the double helical region, that is that of the eleven-fold model for reovirus RNA. (We refer to this conformation as standard.) There is some support from physical studies on solutions of polynucleotides and dinucleotides for postulating that the conformation of a single-stranded polynucleotide with base-stacking is similar to the conformation it would have as one of the strands in a two-stranded structure[10,11].

Figs. 3 and 4 illustrate the structure which stacks the greatest number of the nucleotides in the anticodon loop. Five nucleotides are stacked in the standard conformation so that they lie on the same helix as that chain in the anticodon arm double-helix nearer the tRNA 3' end. This structure represents a unique solution to the problem of maximizing base-stacking in the anticodon loop. Conformations with slightly different base tilt and rotation and translation of each nucleotide (for example if the standard nucleotide conformation was that of the ten-fold rather than eleven-fold RNA model) could of course give a similar degree of base-stacking. Stacking combinations of nucleotides other than those stacked in this structure, however, result in less than five nucleotides being stacked. In particular five nucleotides cannot be arranged so that they lie on the same helix as that chain of the anticodon arm double-helix nearer the tRNA 5' end. This is shown in Fig. 4 where A and B are closer together than they would be for a structure with bases perpendicular to the helix axis. If, however, the five bases rested on the chain of the two-stranded helix nearer the tRNA 5' end, the base tilt would make the distance to be spanned by the two non-standard nucleotides greater than for a structure with bases perpendicular to the helix axis. In addition to this increased distance, the two nucleotides would have to span the RNA groove containing the 2-keto

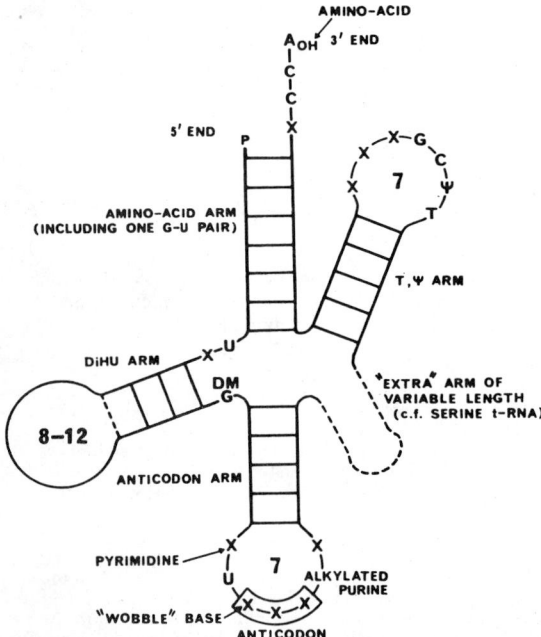

Fig. 1. Generalized clover-leaf structure for yeast tRNA based on sequences determined for tRNAs specific for alanine[1], serine[2], tyrosine[3] and phenylalanine[4]. A base-pair is indicated by a line linking two parts of the RNA chain (a dashed line indicates a base-pair occurring in some tRNAs but not in others); X indicates a nucleotide which varies with the tRNA species; nucleotides which occur at an equivalent position in all sequences are denoted as follows—uracil (U), dimethylguanine (DMG), adenine (A), cytosine (C), thymine (T), pseudouracil (ψ). The number at the centre of each loop indicates the number of nucleotides in the loop.

groups rather than that containing the 6-keto groups which is spanned in the model illustrated in Figs. 3 and 4. The RNA conformation is such that spanning the groove containing the 2-keto groups requires a much longer polynucleotide chain.

In the structure illustrated in Figs. 3 and 4 the two nucleotides of the anticodon loop not in the standard conformation have the planes of their bases approximately parallel, with some overlap of their hydrophobic surfaces. There is, however, some flexibility in this region of the structure and this conformation should be regarded as typical of a number of related possibilities. All the tRNA nucleotide sequences so far determined are compatible with the poorer stacking of these two nucleotides as compared with that in the standard helix because these bases are always pyrimidines (of which at least one is uracil). These bases are generally thought to stack least well. In all these sequences, the second of the five nucleotides in the single stranded helix (that is 7 in Fig. 4) has a chemically modified hydrogen bonding donor group on the base. In addition to inhibiting base-pairing which might favour alternative structures to that in Figs. 3 and 4, this group may increase hydrophobic stabilization of this stacked conformation.

The structure illustrated in Figs. 3 and 4 could describe the conformation of the loop of 7 unpaired bases in the TΨ loop (Fig. 1). The occurrence of uracil and cytosine, however, at what would be positions 6 and 7 in the single-stranded helix (Fig. 4) make it a rather less attractive solution than it is for the anticodon loop.

A loop at the end of an RNA double helix need be no longer than three nucleotides[12]. The base-stacking in such a structure, however, is probably much less perfect than that in the standard conformation and such a loop would only be expected to occur if it contained nucleotides with poor stacking interactions, for example the loop with UUU and UCU at the end of the extra arm in the two serine tRNAs (Fig. 1).

Codon–Anticodon Interactions

From consideration of the likely anticodon in a number of tRNAs and a knowledge of the different codons which will

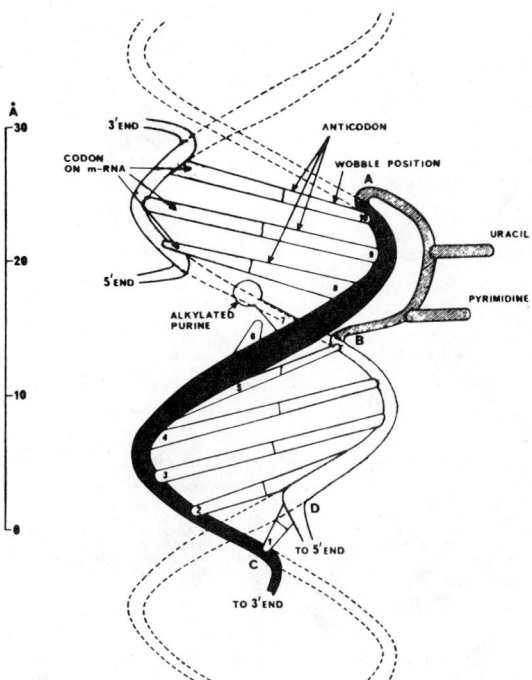

Fig. 4. Schematic diagram of the tRNA anticodon loop illustrating its relationship to the codon and the helical character of the structure. The letters A, B, C and D identify the same points on the structure as in Fig. 3. The bases in nucleotides 1 to 10 are stacked on one another and follow the regular helix which is shown black. The chain of the anticodon double helix between D and B is shaded like the codon to indicate that they follow the same helix. This helix is complementary to the black one. The two nucleotides not in the standard conformation are represented by dark line shading. The representation of their conformation is very schematic because they lie behind nucleotides 8, 9 and 10 in the black chain. The dotted lines indicate the generic helix from which the structure can be imagined to be derived.

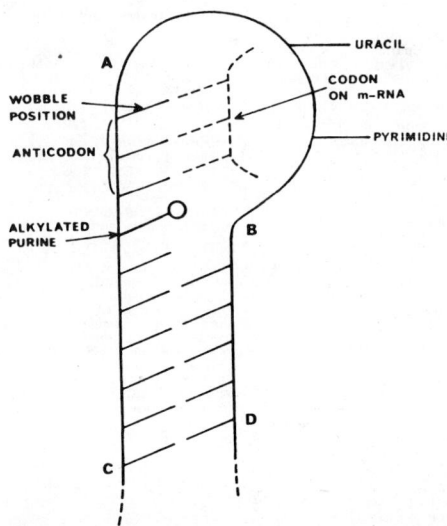

Fig. 3. Schematic diagram of the model for the tRNA anticodon arm illustrating its relationship to the codon. The helical regions are shown as straight in this diagram. CD is the first base-pair in the double helical region of the anticodon arm and all the bases between A and C are stacked on one another and follow a regular helix. The companion set between B and D and the set of three bases in the codon follow the complementary helix. In space (see Fig. 4) A and B are quite close together because five nucleotide pairs is about half a turn of the helix.

recognize a particular tRNA species, Crick[13] has proposed a hypothesis for codon–anticodon recognition. This involves standard Watson–Crick base-pairing between the bases in the codon and anticodon triplets while allowing the possibility of "wobble" or limited alternative pairing in the third position. When codon and anticodon are paired in this way, the atomic sequence in one chain is the reverse of that in the other and the two triplets can be arranged as the two strands in a regular RNA double helix. In our model the anticodon triplet occupies positions 8, 9 and 10 of the anticodon helix (Figs. 3 and 4) and a messenger RNA (mRNA) codon can be base-paired to it without steric hindrance between the rest of the anticodon arm and adjacent mRNA codons. In fact simultaneous recognition of two anticodon arms by adjacent codons is stereochemically possible (Fig. 5). There is not much flexibility in the relative position and orientation of two anticodon arms when they are interacting simultaneously with adjacent mRNA codons. They can interact, however, with each other in a way similar to neighbouring helices in crystalline fibres of reovirus RNA (ref. 8) (see caption to Fig. 5). The possible biological significance of intermolecular hydrogen bonds between the sugar hydroxyl of one helix and the phosphate oxygen of another has been noted[8]. One such hydrogen bond is formed between the two anticodon arms when arranged as in Fig. 5. Preliminary studies suggest that it is possible to arrange the clover leaf arms of the generalized tRNA molecule (Fig. 1) so that there is no steric hindrance between two tRNA molecules whose anticodon arms are interacting in this way.

It might be asked if mistakes in translation could occur by codon "recognition" of bases at positions 7, 8 and 9

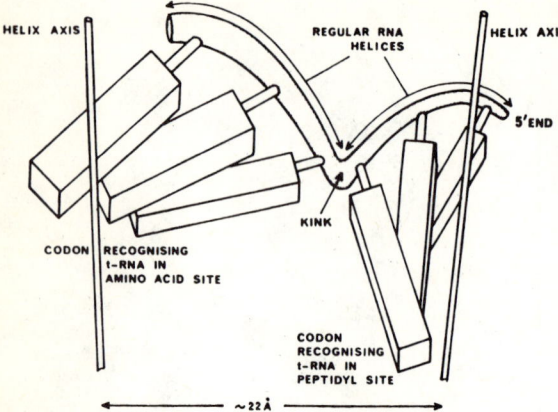

Fig. 5. Schematic diagram of successive codons in mRNA simultaneously recognizing anticodon arms. Each codon of the mRNA has the conformation illustrated in Figs. 3 and 4. The operation required to move the anticodon arm from the amino-acid site to the peptidyl site involves a rotation θ about and a translation t along the anticodon helix axis and a translation d perpendicular to the helix axis along a line joining the two helix axes illustrated in this figure. If the anticodon helices are linked by a hydrogen bond and have a stereochemical relationship like two reovirus RNA helices in the crystalline fibre then the symmetry operation can be defined precisely as follows (otherwise it is an approximate description). $\theta = 87.3°$ (that is $120-32.7$), $t = -2.73$ Å, $d = 22$ Å. Values of θ and t which would move nucleotide 9 into the position occupied by nucleotide 10 (Fig. 4) are taken as positive.

(or even 6, 7 and 8) rather than 8, 9 and 10 of the anticodon helix. Studies with spacefilling models suggest that recognition of both 7, 8 and 9 and 6, 7 and 8 could not be excluded even if it is insisted that adjacent codons recognize anticodon arms simultaneously. Recognition of 6, 7 and 8 is the less plausible stereochemically. If the geometry of the tRNA–mRNA–ribosome interaction is inadequate to prevent mis-reading of this kind, it may be that it is prevented by the chemical modification which occurs at nucleotide 7 (Fig. 4) in all sequences determined so far[1-4].

Stereochemical Aspects of the "Wobble" Hypothesis

We have considered possible distortions of our model for codon–anticodon interaction which would accommodate the alternative base-pairings described by Crick in his "wobble" hypothesis[13] (Table 1). The separation and relative orientation of the glycosidic links in these alternative base-pairs differ from that in the standard Watson–Crick pairs. The position of the wobble base-pair in our model is illustrated in Figs. 3 and 4. Accommodation of adenine-inosine in this position requires extension of the sugar–phosphate chain linking the second and third bases of the anticodon (at positions 9 and 10) or compression of the chain joining the second and third bases of the codon. The chain in the RNA helix is already rather compressed (about 5·6 Å between successive phosphates as compared to about 7 Å in a completely extended chain) and further compression results in steric hindrance between the 2′ hydroxyl (and the sugar carbon to which it is attached) and the base of the previous nucleotide. Therefore the principal distortion involved in accommodating adenine-inosine (and any other alternative pairs with an interglycosidic link separation larger than the standard pair) must occur at the anticodon.

In contrast accommodation of an alternative pair with an interglycosidic link separation smaller than the standard pair would require either extension of the sugar–phosphate chain between the second and third bases of the codon or compression of the chain between the second and third bases of the anticodon. Significant compression of the chain can be excluded, and so it appears that the principal distortion involved in accommodating pairs with an interglycosidic link separation significantly shorter than standard must occur at the codon.

Site of Distortion in "Wobble" Pairing

Using skeletal models we have found that all the alternative pairing required to account for the genetic evidence on degeneracy in the third position of the codon–anticodon interaction can be accommodated in our model by distortion of the anticodon alone (Table 1). Distortion of the codon conformation is not required. Further, our model building studies indicate that a uracil–uracil pairing can be accommodated if distortion is allowed at the codon. Therefore, because the genetic evidence excludes such pairing in this position, we can conclude that it does not occur because the codon conformation cannot be significantly distorted. It should be noted, however, that our criteria for an acceptable pairing relate to the geometry of the interbase hydrogen bonds and the stereochemistry of the sugar–phosphate chain. While these are clearly necessary requirements, other considerations may also be relevant to the occurrence of a particular alternative pair, for example interbase dipole–dipole interactions. (It may be that non-occurrence of the uracil–uracil pair is the result of such effects rather than of the codon being rigidly held.)

The assignment of "wobble" distortion to the anticodon rather than the codon seems reasonable from general considerations because one might expect each codon to be held on the 30S ribosome in a way which is independent of its position in the mRNA and therefore through bonds involving groups near to or part of the sugar–phosphate chain of the codon currently being read. Such bonds would be expected to limit the conformational flexibility of the codon as compared with the anticodon (which is a relatively small part of the tRNA molecule and not necessarily close to the ribosomal binding site on the tRNA) in a way which is compatible with the above assignment of the distortion

Table 1. THE ACCOMMODATION OF ALTERNATIVE BASE-PAIRS AT THE "WOBBLE" POSITION IN THE CODON–ANTICODON COMPLEX AS A FUNCTION OF WHETHER THE DISTORTION IS ALLOWED IN THE CODON OR ANTICODON CONFORMATION

Alternative codon–anticodon pairs (groups involved in interbase hydrogen bonding)	Genetic evidence for its occurrence (— denotes occurrence) (X denotes non-occurrence)	Stereochemistry of the polynucleotide chain according to the site of distortion (— denotes acceptable stereochemistry) (X denotes unacceptable stereochemistry)		
		Distortion at anticodon only	Distortion at codon only	Distortion at codon and anticodon
Adenine-inosine (6-amino to 6-keto and $N1$ to $N1$)	—	(The torsion angle of the inosine glycosidic link is about 5° outside the acceptable range)	X	—
Guanine-uracil (6-keto to $N1$ and $N1$ to 2-keto)	—	(There is a hydrogen-oxygen non-bonded contact of about 2·2 Å, i.e., about 0·3 Å less than the sum of the van der Waals radii of these atoms)	X	—
Uracil-guanine ($N1$-to 6-keto and 2-keto to $N1$)	—	(There is a hydrogen-oxygen non-bonded contact of about 2·2 Å, i.e., about 0·3 Å less than the sum of the van der Waals radii of these atoms)	X	—
Uracil-uracil (6-keto to $N1$ and $N1$ to 2-keto)	X	X	X	X
Uracil-uracil ($N1$ to 6-keto and 2-keto to $N1$)	X	X	—	—
Uracil-cytosine (6-keto to 6-amino and $N1$ to $N1$)	X	X	X	X

For none of the pairings denoted as "stereochemically acceptable" is the stereochemistry quite as satisfactory as that in the undistorted standard conformation. The departures from acceptable stereochemistry are noted and are small enough for it to be concluded that the codon–anticodon complex could be distorted to accommodate these pairs. In contrast the stereochemistry of the pairs denoted "stereochemically unacceptable" is quite unacceptable with non-covalently bonded contacts 1 or 2 Å less than normal values and with torsion angles 40 to 50° outside the range of observed values.

associated with "wobble" pairing to the anticodon. Further, the occurrence of the "wobble" base at the top of the single strand anticodon helix allows distortion in the part of the sugar-phosphate chain to which it is attached to be absorbed in the conformational flexibility of the two unstacked pyrimidines next to it.

The model we propose for the anticodon arm of tRNA allows codon–anticodon interaction through Watson–Crick base-pairing. The codon and anticodon nucleotide triplets have the conformation of the two strands in a regular RNA double helix. The alternative or "wobble" pairings suggested for the third base of the anticodon can be accommodated in this model by distortion of the anticodon conformation. It is not necessary to postulate distortion of the anticodon conformation. The observation that uracil–uracil is not a wobble pairing suggests that codon conformation distortion does not take place. This model of the anticodon arm allows adjacent mRNA codons to simultaneously recognize anticodon arms.

In a study such as this it is important to identify clearly the principal assumptions on which the model building is based. Essentially the only assumption we make is that the number of stacked bases in the anticodon loop should be a maximum; this leads to a unique solution for the conformation of the loop. The model receives support from the base sequences which have been determined for the anticodon loop in a number of tRNAs (Fig. 2): the pyrimidines (mainly uracil) are in the irregular part of the loop, the wobble base is at that position in the stacked part of the structure which has the most conformational flexibility, and the modified purine is at a position which could prevent a wrong set of three nucleotides in the anticodon being recognized by the codon. There is no structural or genetic evidence in conflict with this model and, while the model building study does not prove it to be correct, its stereochemical neatness and the manner in which it accounts for what is known about the anticodon region of tRNA suggest that it is essentially correct.

One of us (A. H.) is the holder of a Medical Research Council award. We thank Professor Sir John Randall, Professor M. H. F. Wilkins, and Dr F. H. C. Crick for their interest, Miss A. Kernaghan for preparing the figures and Mr Z. Gabor for carrying out photographic work.

Received June 16; revised August 3, 1967.

[1] Holley, R. W., Apgar, J., Everett, G. A., Madison, J. T., Marquisee, M., Merrill, S. H., Penswick, J. R., and Zamir, A., *Science*, **147**, 1462 (1965).
[2] Zachau, H., Dütting, D., and Feldman, M., *Angew Chemie*, **78**, 393 (1966).
[3] Madison, J. T., Everett, G. A., and Kung, H., *Science*, **153**, 531 (1966).
[4] RajBhandary, U. L., Chang, S. M., Stuart, A., Faulkner, R. D., Hoskinson, R. H., and Khorana, H. G., *Proc. US Nat. Acad. Sci.*, **57**, 751 (1967).
[5] Koltun, W. L., *Biopolymers*, **3**, 665 (1965).
[6] Langridge, R., Marvin, D. A., Seeds, W. E., Wilson, H. R., Hooper, C. W., Wilkins, M. H. F., and Hamilton, L. D., *J. Mol. Biol.*, **2**, 38 (1960).
[7] Dover, S. D., Spencer, M., Wilkins, M. H. F., and Fuller, W. (in preparation).
[8] Arnott, S., Hutchinson, F., Spencer, M., Wilkins, M. H. F., Fuller, W., and Langridge, R., *Nature*, **211**, 227 (1966).
[9] Rich, A., Davies, D. R., Crick, F. H. C., and Watson, J. D., *J. Mol. Biol.*, **3**, 71 (1961).
[10] McDonald, C. C., Phillips, W. D., and Lazar, J. (in the press).
[11] Buch, C. A., and Tinoco, jun., I., *J. Mol. Biol.*, **23**, 601 (1967).
[12] Spencer, M., Fuller, W., Wilkins, M. H. F., and Brown, G. L., *Nature*, **194**, 1014 (1962).
[13] Crick, F. H. C., *J. Mol. Biol.*, **19**, 548 (1966).

BIOSYNTHESIS OF THE COAT PROTEIN OF COLIPHAGE f2 BY E. COLI EXTRACTS*

By D. Nathans,† G. Notani, J. H. Schwartz, and N. D. Zinder

THE ROCKEFELLER INSTITUTE

Communicated by Fritz Lipmann, June 15, 1962

Nirenberg and Matthaei[1] have discovered an assay system in which RNA serves as an activator of protein synthesis in *E. coli* extracts. RNA fractions from cells,[1] synthetic polyribonucleotides,[2,3] and viral RNA[1,4] can all stimulate amino acid incorporation into acid-insoluble products in *E. coli* extracts. Although in each case the product formed is presumed to be a protein or polypeptide whose structure is uniquely determined by the RNA, the products have not as yet been completely analyzed. A recent paper by Tsugita et al. presents evidence suggesting that the coat protein of TMV is synthesized by E. coli extracts.[5] With the availability of a bacteriophage attacking E. coli and containing RNA (coliphage f2),[6] it became possible to use the RNA isolated from this phage to stimulate amino acid incorporation into protein and to identify at least part of the product as the coat protein of the phage.

Materials and Methods.—Phage RNA and protein: Large amounts of the phage f2 were prepared by scaling up the procedure of Loeb and Zinder.[6] RNA was prepared by deproteinization of the phage suspension by shaking with phenol and precipitation of the RNA from the aqueous phase by the addition of two volumes of ethanol. The RNA was kept at $-20°$ in distilled water and used as needed. Its concentration was determined by measuring the optical density at 260 mμ and assuming a specific absorption of 24 per mg. Phage protein was prepared by extraction of purified f2 in 67 per cent acetic acid followed by dialysis and lyophilization. The following properties of the protein are pertinent to this paper. On the basis of chromatography on DEAE-cellulose, starch gel electrophoresis, and ultracentrifugation, the protein appears to be homogeneous. It has a probable molecular weight of 20,000 and contains 10 leucine, 5 lysine, 3 arginine, and no histidine residues per molecule.[7] Digestion of the oxidized protein by trypsin gives nine ninhydrin-reacting components which are resolved by two-dimensional electrophoresis; one of these has the mobility of free lysine. The details of the digestion procedure used in these experiments will be described below.

E. coli extracts: Preparation of the E. coli extracts was based on the procedure described by Nirenberg and Matthaei.[1] E. coli B was harvested in the log phase of growth. The washed, frozen cell paste was ground with $2^1/_2$ parts of Alumina (Alcoa A-303) and was extracted with $1^1/_2$ volumes of 0.01 M pH 7.4 Tris buffer that was 0.01 M in magnesium acetate. After the addition of 5 μg/ml of DNase, the extract was centrifuged at 15,000 $\times$ g for 10 min. To the resulting supernatant, β-mercaptoethanol was added to a final concentration of 0.005 M, and then it was centrifuged at 30,000 $\times$ g for 30 min. The 30,000 $\times$ g supernatant fraction (S-30) was incubated at 35° for 30 min with the components required for amino acid incorporation into protein: 0.003 M ATP, 0.0002 M GTP, 0.01 M phosphoenolpyruvate, 30 μg/ml pyruvate kinase, 0.01 M glutathione, 0.011 M magnesium acetate, 0.03 M KCl, 0.05 M Tris HCl pH 7.8, 4 $\times$ 10^{-5} M of each amino acid, 5 μg/ml DNase, 1 mg/ml sRNA, and a volume of S-30 equal to half the total incubation volume. After overnight dialysis against 0.01 M Tris HCl pH 7.8, 0.01 M magnesium acetate, 0.03 M KCl, and 0.005 M β-mercaptoethanol, the preincubated mixture was stored at $-20°$.

For the incorporation experiments to be described, the components present during preincubation (minus additional sRNA and DNase) were incubated at 35° with the appropriate C^{14}-amino acids using an amount of preincubated S-30 (containing 4.6 mg of ribosomal RNA/ml) equal to a quarter of the total volume. Amino acid incorporation into proteins was determined by counting the hot acid-precipitate (5 per cent trichloroacetic acid, 90° for 15 min) after washing with TCA and ethanol-ether.

Electrophoresis: High-voltage electrophoresis[8] of tryptic digests of protein (fingerprinting) was performed on Whatman 3 MM paper in a Servonuclear Company tank under Varsol (Standard Oil Co., N.J.). The first dimension was run for 1 hr in pH 4.7 buffer containing by volume 25 ml of acetic acid and 25 ml of pyridine/liter of water. The field strength was 50 volts/cm. At the completion of the run, the paper was dried and the strip containing the peptides cut out and sewn onto another piece of paper. The second dimension was run for 3–4 hr in pH 1.9 buffer containing 87 ml of acetic acid and 25 ml of 88 per cent formic acid/liter of water. The field strength was 20 volts/cm.

Radioactive amino acids: C^{14}-leucine, -arginine, and -lysine with specific activities of 144 mC/mmole and H^3-leucine with specific activity of 5 C/mmole were obtained from New England Nuclear Corporation, Boston, Mass. C^{14} counting was done with a windowless gas flow counter. H^3 and C^{14} in the same sample were determined by counting with and without a thin window and then making the appropriate corrections.

Results.—Characteristics of the amino acid incorporating system: The following

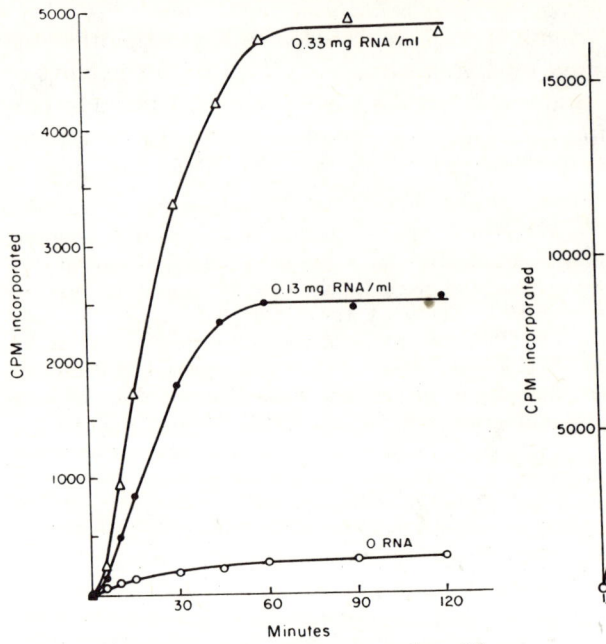

FIG. 1.—Kinetics of leucine incorporation with different concentrations of f2 RNA. The incubation mixtures were as described under *Methods* and contained C^{14}-leucine (1.5×10^7 cpm/μmole) and the concentration of RNA indicated. Each point represents the acid-insoluble counts in 0.20 ml.

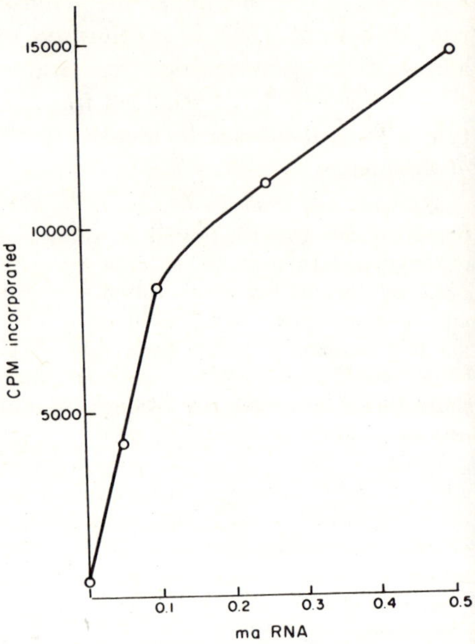

FIG. 2.—Effect of increasing concentration of f2 RNA on leucine incorporation. The components given in *Methods* were incubated with C^{14}-leucine (1.5×10^7 cpm/μmole) at 35° for 90 min. Total volume was 0.50 ml.

experiments show the extent of the stimulation of amino acid incorporation by *E. coli* extracts when exposed to f2 RNA. Figure 1 shows the kinetics of incorporation of leucine at different RNA concentrations. Both the rate of incorporation and the final amount of product vary directly with the RNA concentration. In Figure 2, the effect of increasing RNA concentration on the amount of leucine incorporated is shown. With the concentration of components used, there is a linear response up to 200 μg/ml and then a sharp break in the curve. Above this concentration, a smaller response per unit of RNA added is observed. Saturation is not achieved even at our highest RNA concentration, which stimulates some 40-fold. In the linear portion of the concentration curve, 0.53 mμ moles of leucine is incorporated per 100 μg of RNA. From this value, one can calculate that about 30 nucleotide residues are required for each molecule of amino acid incorporated. This estimation is based on the specific activity of the leucine, on the percentage composition of the leucine in the phage protein, and on the assumption that any other proteins synthesized have the same percentage of leucine as does the f2 protein.

To demonstrate that the reaction has the properties expected of a protein-synthesizing system, the antibiotics chloramphenicol and puromycin were added to the system. Both of these inhibitors of protein synthesis inhibited the incorporation of leucine (Table 1).

Identification of the product: Two kinds of experiments are to be described by which we characterize the product. The first involves chromatography of the

TABLE 1
Effect of Puromycin and Chloramphenicol on RNA-Dependent Protein Synthesis

Conditions	C^{14}-leucine (cpm incorporated)
Complete system	3650
+ puromycin (4×10^{-4} M)	44
+ chloramphenicol (3×10^{-4} M)	163
(Complete system minus f2 RNA	235)

The components given in *Methods* were incubated with C^{14}-leucine (1.5×10^7 cpm/μmole) at 35° for 60 min. Total volume was 0.25 ml, containing 50 μg of f2 RNA.

product, the other analysis of tryptic digests of the partially purified product. Of the counts incorporated into the acid-insoluble fraction, 45–55 per cent remains bound to the ribosomes when the ribosomes are removed by centrifugation, and the remainder is found in the 105,000 × g supernatant. It is this latter fraction only that we are studying. Analyses were done only when there was about a 40-fold stimulation over the background incorporation by the extracts so that the unstimulated product may essentially be neglected.

In the first experiment, the product was prepared by incubating C^{14}-leucine with the components described in *Methods*. The ribosomes were removed by centrifugation at 105,000 × g for 2 hr; 3 mg of f2 protein were added to an aliquot of the supernatant and this material was chromatographed on DEAE-cellulose in 3 M urea (in the absence of urea, f2 protein cannot be eluted) with a gradient of 0.1 to 1.0 M potassium phosphate, pH 7.4. Of the 3,000 acid-precipitable counts applied, 60 per cent was recovered of which 45 per cent was in the f2 protein peak.

The next two experiments take advantage of the reproducibility of the fingerprints of f2 protein. In the first of these experiments, the product was prepared by incubating the components described in *Methods* with 0.5 ml of preincubated S-30, 0.8 mg of f2 RNA, and C^{14}-arginine and lysine in a total volume of 2 ml at 35° for 90 min. After removal of the ribosomes by centrifugation at 105,000 × g for 2 hr, the supernatant contained 640,000 cpm in the hot acid-precipitable fraction. To an aliquot containing 140,000 cpm was added 3 mg of f2 protein and a large excess of C^{12}-arginine and lysine. The f2 protein was precipitated with ammonium sulfate at 20 per cent saturation, and the precipitate was washed with ammonium sulfate and suspended in water. By this procedure, 98 per cent of the acid-insoluble counts was precipitated with the carrier f2 protein. Although only 2 per cent of the product was in the supernatant, 3 mg of carrier f2 protein was added to this fraction, and it was carried through the procedure described below as a control for the adsorption of free C^{14}-amino acids by the carrier protein.

The two fractions were prepared for trypsin digestion by first precipitating and extracting with 5 per cent TCA at 90° for 15 min, followed by two TCA washes at room temperature, and then washing with ethanol-ether and ether. The dried product was subjected to performic acid oxidation by a modification of Hirs's procedure.[9] The oxidized protein was then digested for 3 hr at 25° in 1 ml of 0.05 M ammonium bicarbonate pH 7.9 containing 4 per cent trypsin by weight of substrate. After lyophilization to dryness, the digest was dissolved in 20 μl of water, an aliquot containing 105,000 cpm was applied to paper, and the peptides were separated by two-dimensional electrophoresis.

The electropherograms were radioautographed for five days and stained with ninhydrin to identify the peptides. Figure 3 shows the electropherogram of the

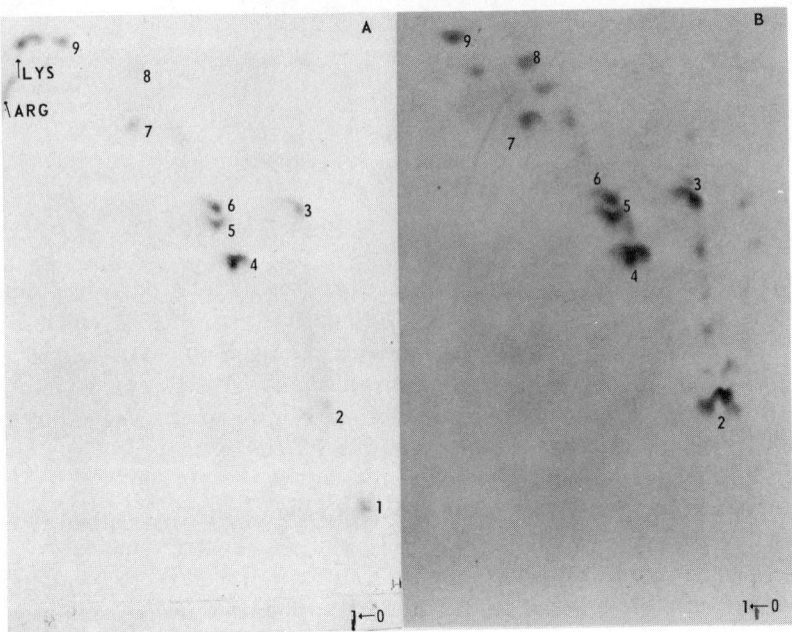

Fig. 3.—Electropherogram and radioautogram of the tryptic peptides from C^{14}-lysine- and arginine-labeled product and carrier f2 protein. A is the ninhydrin-stained paper, B is the radioautogram. "Arg" and "lys" represent markers; O indicates the origin.

ammonium sulfate precipitate. As the product was labeled with C^{14}-lysine and arginine, and since trypsin cleaves specifically at the carbonyl groups of lysine and arginine, we would expect that all the peptides would be labeled with the exception of the carboxy-terminal peptide. The nine peptides of f2 are marked 1 through 9. It may be noted that all peptides specifically identified with f2 except peptide number 1 are labeled. The congruence of the ninhydrin spots with the radioactive spots is immediately apparent even though the film is not superimposed here on the electropherogram. On the other hand, the electropherogram of the ammonium sulfate supernatant had the same ninhydrin peptides but the radioautogram was blank. This result was to be expected since, as already mentioned, 98 per cent of the TCA-precipitable counts was found in the original ammonium sulfate precipitate. This negative result does show, however, that the correspondence of ninhydrin and radioactive spots is not due to adsorption of C^{14}-lysine and arginine.

A similar control experiment was carried out using egg white lysozyme (Worthington) as carrier in place of f2 protein. In this instance, it was necessary to make the solution 60 per cent saturated with ammonium sulfate in order to precipitate the lysozyme. In other respects, the experiment proceeded as described above except that 50,000 cpm of the lysine and arginine-labeled product were applied to the paper. Although we obtained a whole new spectrum of ninhydrin spots in the electropherogram, the spots that appeared on the radioautogram (Fig. 4) corresponded with those of f2 protein and not with the lysozyme.

Note added in proof: Using C^{14}-lysine and -arginine, we have repeated the experiments described above with RNA isolated from *E. coli* K12 W6, starved of methionine (courtesy of L. Mandel and E. Borek), and with RNA isolated from

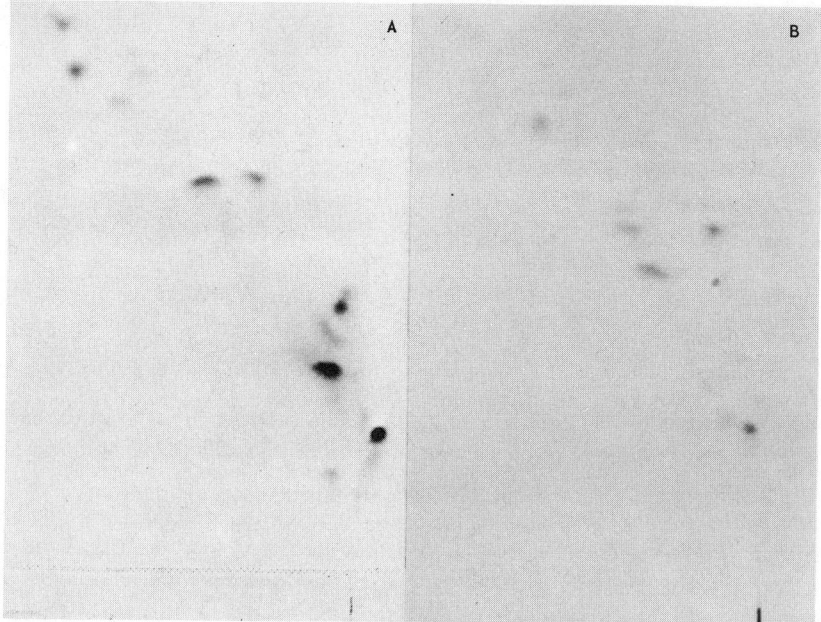

FIG. 4.—Electropherogram and radioautogram of the tryptic peptides from C^{14}-lysine- and arginine-labeled product and carrier lysozyme. *A* is the ninhydrin-stained paper, *B* is the radioautogram.

TMV (obtained from D. Caspar). Fingerprints of each product with f2 protein added as carrier showed no correspondence of radioactivity with the ninhydrin spots. In addition, the product prepared with f2 RNA was fingerprinted with TMV protein. Again there were no overlapping spots. We conclude that the product formed in the presence of f2 RNA is peculiar to the nucleic acid of the bacteriophage.

The next experiment used a second isotopic label to identify those f2 peptides containing leucine. Carrier protein labeled with H^3-leucine was to be mixed with the product labeled with C^{14}-leucine. If the product was identical to the phage protein, those peptides with H^3, i.e., containing leucine, would also have C^{14}, and those peptides without H^3, i.e., containing no leucine, would not have C^{14}.

A small amount of H^3-leucine-containing phage was made by growing f2 on a leucine-requiring strain of the host bacteria in the presence of high-specific-activity H^3-leucine. This phage was added to a large amount of pure, unlabeled phage, and protein was prepared. Fingerprinting of this material followed by elution and tritium counting showed that only peptides 2 and 4 contain leucine. C^{14}-leucine-labeled product was prepared as described above except for the change in C^{14}-amino acid. In this instance, the $105,000 \times g$ supernatant contained 810,000 acid-insoluble cpm. An aliquot with 100,000 cpm (C^{14}) was added to the mixed carrier protein containing 200,000 cpm (H^3), and this material was precipitated and digested as previously described. In this experiment, only the ammonium sulfate precipitate was analyzed since, again, nearly all of the TCA precipitable counts were in this fraction. The ninhydrin-stained peptides were eluted from the electropherogram with hot water and an aliquot counted to determine the C^{14} and H^3 content. In Table 2, we can see that only peptides 2 and 4 contain substantial

TABLE 2

RADIOACTIVITY OF f2 PEPTIDES

Peptide	C^{14} (cpm)	H^3 (cpm)	C^{14}/H^3
1	200	300	...
2	1,100	2,510	0.44
3	355	0	...
4	2,960	10,600	0.28
5	0	80	...
6	40	0	...
7	0	24	...
8	232	0	...
Origin	4,400	7,000	0.63
Total	9,290	20,500	...
cpm applied	47,000	104,000	0.45
% recovered	20	20	...

radioactivity and each has both C^{14} and H^3. The ratio of the two isotopes in each peptide is approximately that of the original material applied to the paper. Only a few faint spots were seen on the radioautogram to indicate other C^{14}-containing material; these had no corresponding ninhydrin spots nor did they contain H^3. The product thus resembles f2 protein. It may be noted (Table 2) that only about 20 per cent of the counts applied to the paper were recovered. The balance of the counts could not be located even when elutions were performed with 6 N HCl, nor were they found in the solvents used for the electrophoresis. Since both product and carrier are lost proportionately, these unexplained losses do not alter the conclusion that f2 protein is being made.

Discussion.—The experiments presented leave little room for doubt that the synthesis of f2 coat protein can be mediated by f2 RNA when it is added to *E. coli* extracts. Since the isotopic labels appear primarily in the f2 peptides, we conclude that the coat protein forms a major component of the product that is released from the ribosomes. Preliminary evidence based on the fact that f2 RNA also stimulates the incorporation of histidine, an amino acid which is lacking in the coat protein, indicates that other protein is also being made. In this instance, more than half of the radioactive protein remains bound to the ribosomes.

An important question raised by these results is whether viral RNA acts directly as a template or through some intermediate such as its complementary strand. Although the hypothesis that RNA acts directly is more appealing, the latter possibility cannot, as yet, be excluded.

With the demonstration that viral RNA directs the synthesis of the coat protein in cell extracts, problems relating to viral replication can now be studied in an experimental system more easily controlled than the infected cell.

* Supported by grants from the National Foundation, the National Science Foundation, and the National Cancer Institute, National Institutes of Health, U.S. Public Health Service.

† Post-doctoral fellow of the U.S. Public Health Service. Present address: Department of Microbiology, The Johns Hopkins University School of Medicine, Baltimore, Md.

[1] Nirenberg, M. W., and J. H. Matthaei, these PROCEEDINGS, **47**, 1588 (1961).

[2] Matthaei, J. H., O. W. Jones, R. G. Martin, and M. W. Nirenberg, these PROCEEDINGS, **48**, 666 (1962).

[3] Lengyel, P., J. F. Speyer, C. Basilio, and S. Ochoa, these PROCEEDINGS, **48**, 282 (1962).

[4] Ofengand, J., and R. Haselkorn, *Biochem. Biophys. Res. Comm.*, **6**, 469 (1962).

[5] Tsugita, A., H. Fraenkel-Conrat, M. W. Nirenberg, and J. H. Matthaei, these PROCEEDINGS, **48**, 846 (1962).

[6] Loeb, T., and N. D. Zinder, these PROCEEDINGS, **47,** 282 (1961).
[7] Notani, G., and N. D. Zinder, unpublished observations.
[8] Ingram, V. M., *Nature,* **178,** 792 (1956); Katz, A. M., W. J. Dreyer, and C. B. Anfinsen, *J. Biol. Chem.,* **234,** 2897 (1959).
[9] Hirs, C. H. W., S. Moore, and W. H. Stein, *J. Biol. Chem.,* **219,** 623 (1956). Two ml of formic acid were mixed with 0.1 ml of 30 per cent hydrogen peroxide (superoxal) and allowed to stand for 2 hr at 0°. One-tenth ml of this mixture was added to the dried protein and was incubated for 1 hr at 0°. Oxidation was stopped by the addition of 1 ml of water and the protein was lyophilized.

Gene Order in the Bacteriophage R17 RNA : 5'–A Protein–Coat Protein–Synthetase–3'

by
P. G. N. JEPPESEN
J. ARGETSINGER STEITZ
MRC Laboratory of Molecular Biology,
Hills Road, Cambridge

R. F. GESTELAND
Cold Spring Harbor Laboratory
of Quantitative Biology,
Cold Spring Harbor,
Long Island, New York 11724

P. F. SPAHR
Institut de Biologie Moléculaire
1211 Genève 4, Switzerland

The sequencing of the ribosome binding sites and parts of the coat protein cistron has enabled the order of the three R17 genes to be established by direct chemical means. Translation may be influenced by the conformation of the RNA.

THE related coliphages R17, MS2, fr and f2 appear to encode only three proteins in their single-stranded RNA genome[1-4]. These are: (1) the coat protein, which is the major protein component of the virus particle (129 amino-acids[5,6]); (2) the maturation or A protein, a minor component of the virus particle (approximately 350 amino-acids[7]); and (3) the RNA replicase or synthetase, which is not present in the mature phage particle (a minimum of about 450 amino-acids[8]). The phage genome contains 3,300–3,500[9-11] nucleotide residues, of which the three cistrons account for approximately 2,800.

Conditional lethal mutants of these bacteriophages fall into three complementation groups,[1,2] which have been correlated with the three known phage proteins, but, because of the inability to obtain genetic recombination between these mutants, no genetic map of the phage has been constructed.

It is possible to detect the *in vitro* synthesis of the three phage proteins directed by the phage RNA[12-15]. This has led several groups to attempt the isolation of different forms and fragments of the RNA molecule, in order to determine their messenger activities and to infer the order of the cistrons along the RNA genome[14,16-21,25]. Their conclusions, however, are conflicting unless one admits the unlikely possibility that the closely related RNA bacteriophages have different gene orders. Moreover, all previously determined gene orders are subject to the criticism that they are based on indirect measurements and rely on various assumptions about the messenger properties *in vitro* of the RNA and its fragments.

The elucidation of the nucleotide sequences of the three ribosomal binding sites[22] and of parts of the coat protein cistron[23,24] has now made possible the determination of gene order by direct chemical means. On the basis of the distribution of these sequences between two specific fragments of the R17 RNA, we conclude that the gene order is: 5'–A protein—coat protein—synthetase–3'.

Preparation of Specific Fragments of R17 RNA

A partially purified endonuclease from *Escherichia coli* MRE600, referred to as ribonuclease IV, cleaves R17 RNA at a position about 40 per cent of the way into the molecule from the 5'-end[16,25,26]. Further cuts are introduced only slowly, permitting isolation of two specific

fragments. *In vitro* translation studies in extracts from
E. coli have shown that the 60 per cent fragment from
the 3'-end directs efficient synthesis of the synthetase
protein, while the 40 per cent fragment is relatively inactive as messenger[16,25]. The coat protein, which is the
major product of *in vitro* protein synthesis with the
native R17 RNA, is not efficiently produced by either
fragment.

Investigation of the location of genes by RNA sequence
methods requires ^{32}P-labelled fragments of high specific
activity (about 10^6 c.p.m./μg). Because of radiation
damage, RNA of such specific activity extracted from
phage purified by conventional procedures is not intact,
and ribonuclease IV cannot cleave it into clean fragments.
By using chromatography on DEAE-cellulose for the
initial step of phage purification (see legend to Fig. 1),
however, damage can be minimized allowing isolation of
intact RNA.

Treatment with ribonuclease IV as described[16,25,26]
produces 15S and 22S fragments from the 26S native
R17 RNA (Fig. 1). The 15S preparation represents the
5' terminal 1,300–1,400 nucleotides (40 per cent) of the
molecule[26]; its yield of the terminal pppGp residue[32] is
nearly equivalent to that of the intact RNA (Table 1).
This fragment must therefore be relatively free of material
arising from other parts of the genome, as confirmed by
its apparent homogeneity on polyacrylamide gels. The 22S
RNA, containing 2,000–2,100 nucleotides (60 per cent)[26],
comes from the 3'-end of the R17 RNA molecule; no
more than 10 per cent of its chains terminate with pppGp
(Table 1). On gels the 60 per cent preparation appears
approximately 80 per cent homogeneous; its contaminants
are smaller RNA fragments, some exhibiting the same
mobility as the 40 per cent piece.

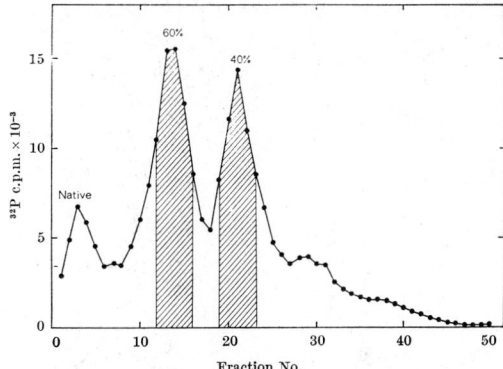

Fig. 1. Sucrose gradient preparation of fragments. ^{32}P-labelled RNA (10^6 c.p.m./μg) was digested with ribonuclease IV as described previously[25] and sedimented for 19 h in a 'Spinco SW 27' rotor at 4° C through a 5–20 per cent sucrose gradient in 0·1 M NaCl, 0·01 M sodium acetate, pH 5·0. Fractions of 0·75 ml. were collected and 1 μl. samples were counted directly in a liquid scintillation counter. Fractions 12–16 and 19–23 (cross-hatched) were pooled to give, respectively, the 60 per cent and 40 per cent preparations used in the experiments described in Figs. 3, 5 and 6 and Tables 1, 3 and 4. R17 phage was grown on ribonuclease I$_B$ in low phosphate medium as described previously[9] except that half the normal amount of casamino acids was used. 20 mCi of ^{32}P-orthophosphate was added to 400 ml. of culture 12 min before phage infection. The phage were purified using a new procedure that will be described in detail elsewhere (R. F. Gesteland and P. F. Spahr, in preparation). Briefly, the phages were concentrated by passage of the clarified lysate through a 20 ml. bed volume column of DEAE-cellulose. After washing the column with SSC (0·15 M NaCl, 0·015 M Na citrate, pH 7·0), the phages were eluted with 4 × SSC and banded to equilibrium in CsCl. The phages were freed of small RNA fragments and CsCl by filtration through 'Sephadex G-150'. The RNA extracted from R17 purified in this manner was intact as demonstrated by sedimentation analysis after formaldehyde treatment.

Table 1. ANALYSES OF 5' TERMINAL pppGp

	Native RNA	60 per cent	40 per cent
Total c.p.m. [32]P	612,800	1,520,000	1,173,000
pppGp c.p.m.	487	302	2,180
Percentage ^{32}P in pppGp	0·079	0·020	0·186
Percentage P in pppGp (theoretical)	0·114	0·19	0·286
Percentage chains with pppGp	69	(10)	65

The pppGp analyses were carried out after alkaline hydrolysis according to Roblin[32]. The calculations are made assuming 3,500 nucleotides in the intact R17 RNA molecule. The percentage of chains with pppGp reflects the fraction of chains of that size which could have 5' terminal pppGp.

Ribosome Binding to the Fragments

Ribosomes from *E. coli* recognize and shield from
nuclease digestion the initiation regions at the beginning
of the three cistrons in the R17 RNA[22]. The nucleotide
sequences of these sites have been elucidated and the
correspondence of each to one of the known R17 proteins
has been established.

The distribution of the initiation regions between the
40 per cent and 60 per cent R17 RNA fragments was
analysed using the procedure described previously for the
whole RNA[22]: MRE600 ribosomes were bound to the
radioactive fragments in the presence of formylated
charged transfer RNA, initiation factors, and GTP at
5 mM Mg^{2+} concentration (see legend to Fig. 2). After
incubation, the ends of the messenger not protected by the
initiation complex were trimmed with pancreatic ribonuclease and the digested complex was isolated by sucrose
gradient sedimentation. Ribosome binding sites were
detected by fingerprint analysis of the radioactive RNA
recovered from the 70S region of the gradient.

One set of T_1 ribonuclease fingerprints of initiation sites
from the 40 per cent and 60 per cent fragments is shown in
Fig. 2. Although the amount of radioactive RNA recovered
was insufficient to allow accurate measurement of the
relative amounts of the binding sites present in the two
fragments, the visual examination of the fingerprints
obtained from two sets of fragment preparations reveals
that: (1) Oligonucleotides derived from the synthetase
binding site (marked S in Fig. 2) are present in higher yield
in the 60 per cent than in the 40 per cent fragment (the
approximate ratio of 60 per cent to 40 per cent yields is
of 3 or 4 : 1). (2) Oligonucleotides specific for the A
protein binding site (marked A) are detected in the 40
per cent fragment only, and there at a very low level.
(3) Oligonucleotides arising from the coat protein initiation
region (marked C) appear primarily in the fingerprint of
the binding reaction performed with the 40 per cent
fragment (approximate ratio of 3 : 1 or 4 : 1, 40 : 60 per
cent).

Allocation of T_1 Oligonucleotides to the Fragments

If ^{32}P-labelled R17 RNA is exhaustively digested with T_1
ribonuclease, it is possible to fractionate many of the larger
oligonucleotides by a two dimensional fingerprinting
system using ionophoresis followed by homochromatography on thin layers of DEAE-cellulose[28]. The characterization of one such T_1 oligonucleotide coding for
amino-acids 57 to 61 of the R17 coat protein[5,6] has
been described[23].

To isolate further large T_1 oligonucleotides whose sequences could be related to known positions in the R17
RNA molecule, it was necessary to obtain greater resolution
in the homochromatography system. This was achieved
by partially degrading the RNA of the homomixture
and chromatographing on longer thin layer plates of
DEAE-cellulose, as described in the legend to Fig. 3.
An autoradiograph of the separation obtained for the
T_1 digestion products of the whole R17 RNA molecule is
shown in Fig. 3a. In addition to the large oligonucleotides
many smaller products are resolved if their mobility in
the first dimension is particularly high or low.

When the sequences of the well separated spots from
such a homochromatogram were investigated, certain of
them were found to come from recognizable regions of
the R17 genome. Eight of these T_1 oligonucleotides
are indicated in Fig. 3a and their sequences and origins

T₁ RNase fingerprint of initiation sites from 40 per cent fragment

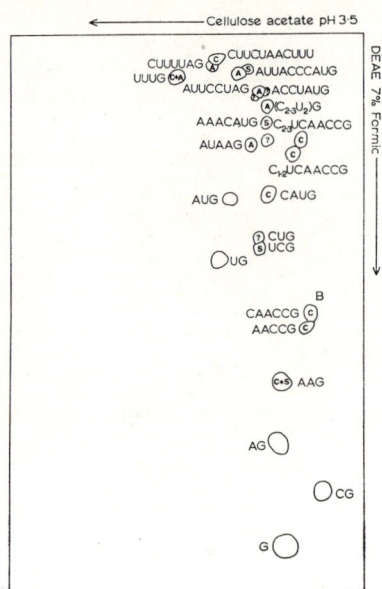

T₁ RNase fingerprint of initiation sites from 60 per cent fragment

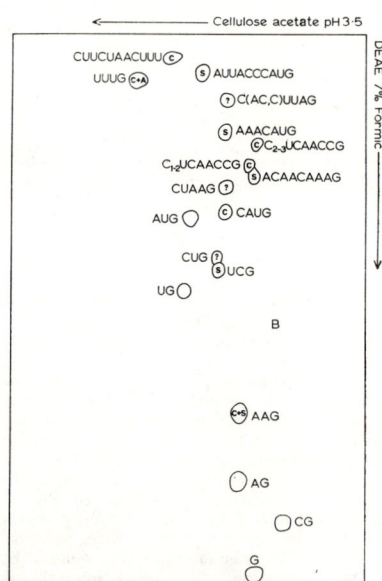

Fig. 2. T₁ ribonuclease fingerprints of ribosomal binding sites from the 40 per cent and 60 per cent fragments. Parallel binding reactions were performed at 37° C in 30 µl. of solution containing: 0·1 M Tris chloride (pH 7·4), 0·05 M NH₄Cl, 0·005 M magnesium acetate, 4·0 $A_{260\ nm}$ units of MRE600 ribosomes, 60 µg crude initiation factors, 0·0002 M GTP, 2·8 $A_{260\ nm}$ units of formylated charged mixed tRNA, and either 16 µg (1·1 × 10⁷ c.p.m.) of the 60 per cent fragment or 11 µg (7·7 × 10⁶ c.p.m.) of the 40 per cent fragment (incubated in 5 µl. of 0·001 M EDTA for 7 min at 37° C before addition). Ribosomes, initiation factors and fMet-tRNA were prepared as described previously[22]. After 12 min of incubation, the binding reaction mixtures were cooled; 0·15 µg of pancreatic ribonuclease was added and incubation at 22° C was continued for 15 min. Following sucrose gradient fractionation[22] the 70S regions were pooled and extracted with phenol in the presence of 0·1 per cent sodium dodecyl sulphate. The resulting 150 µg of RNA was digested with 7 µg of T₁ ribonuclease and fingerprinted as described elsewhere[22]. The identity of each oligonucleotide was checked by analysis of its pancreatic ribonuclease digestion products. With the exception of a few spots of unknown origin, each oligonucleotide in the fingerprints is seen to derive from one or another of the three established ribosome binding regions (Fig. 7) as indicated: A = A protein, C = coat protein, S = synthetase. Several oligonucleotides marked A do not appear in the established sequences (Fig. 7); they seem to originate from more extended ribosome-bound fragments and have been identified as linked to the A protein initiation site.

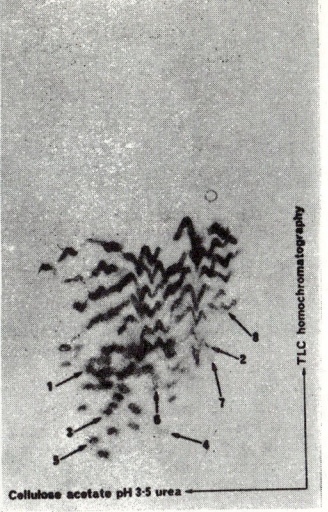 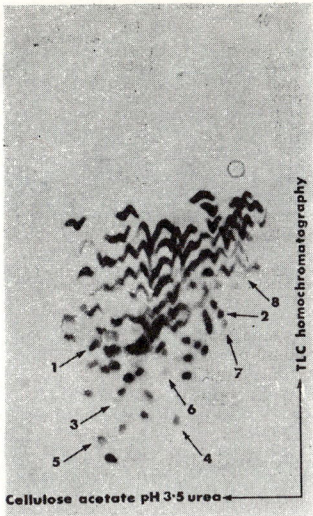 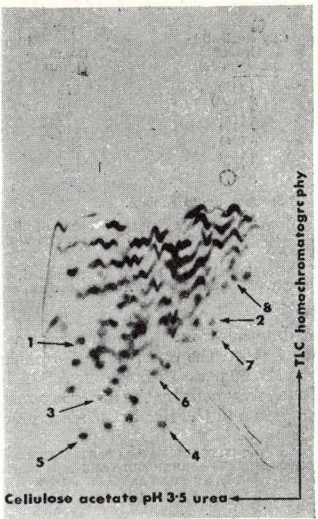

Fig. 3. Two-dimensional thin layer homochromatography fingerprints of T₁ digests of intact R17 RNA (a), the 40 per cent fragment (b) and the 60 per cent fragment (c). For the intact molecule, 20 μg of ³²P-RNA (about 10⁷ c.p.m.) was digested with 1 μg of ribonuclease T₁ in 3–4 μl. of 0·01 M Tris-HCl, pH 7·4, containing 0·001 M EDTA, at 37° C for 30 min. The products were fractionated by ionophoresis on cellulose acetate in 7 M urea, pH 3·5, as a first dimension, followed by homochromatography²⁸, using homomixture c with DEAE-cellulose and cellulose in the ratio 1 : 7·5 on a 20 cm × 40 cm plate. For the fragments about 10⁶ c.p.m. of each was made up to 20 μg with non-radioactive carrier tRNA; T₁ digestion and fractionation of products were performed as above. The numbered spots are identified in Table 2.

listed in Table 2. Sequence *1* is obtained by J. M. Adams and S. Cory (personal communication) in an R17 RNA fragment of approximately 75 nucleotides containing pppGp, the 5′ terminus of the RNA molecule. As shown in Fig. 4, sequences *3*, *4* and *5* are related to amino-acid sequences in the coat protein[5,6] *3* being an extension of the coat protein ribosome binding site[22] (Fig. 7). (Sequence *3* is the same as that proposed by Robinson, Frist and Kaesberg[29] from work with a strain of R17 having an amber mutation at position 6 of the coat protein.) Sequence *6* is homologous with the beginning of the synthetase binding site[22] (Fig. 7) and is found in an RNA fragment containing, and extending beyond, the 3′-terminus of the coat protein cistron[24]. Sequence *8* is the 3′-end of the R17 RNA molecule[30]. Owing to their size and composition, it was not possible to locate any of the A protein binding site T₁ oligonucleotides as pure spots on this homochromatographic system.

Figs. 3b and 3c are fingerprints of the 40 per cent and 60 per cent fragments, analogous to that of the whole molecule (Fig. 3a). With the exception of oligonucleotide *3* which appears in low yield in both fragments (the significance of which will be discussed later), each of the well

Table 2. SEQUENCES AND LOCATION OF SPECIFIC T₁ OLIGONUCLEOTIDES

Oligo-nucleotide (see Fig. 3)	Sequence	Location
1	CCAUUUUUAAUG	Approximately 50 residues from the 5′-end
2	(C)CCUCAACCG*	Coat protein ribosome binding site, proximal to initiating AUG
3	CUUCUAACUUUACUCAG	Coat protein cistron (amino-acids 1–6)
4	CAAAUACACCAUUAAAG	Coat protein cistron (amino-acids 57–61)
5	AAUUAACUAUUCCAAUUUUCG	Coat protein cistron (amino-acids 89–95)
6	CCAUUCAAACAUG	Synthetase ribosome binding site
7	ACAACAAAG	Synthetase ribosome binding site
8	UUACCACCCAoH	3′-end

* It is not clear whether this oligonucleotide has two or three C residues at the 5′-end, although the best data at present suggest there are three. The methods used to verify each of the above sequences will be published elsewhere (P. G. N. J., unpublished results).

3 (G)CU UCU AAC UUU ACU CAG
      f-Met Ala Ser Asn Phe Thr Gln ...
      1   5

4 (G)C AAA UAC ACC AUU AAA G
... Arg Lys Tyr Thr Ile Lys Val ...
         57   60

5 (G)AA UUA ACU AUU CCA AUU UUC G
... Glu Leu Thr Ile Pro Ile Phe Ala ...
         90   95

Fig. 4. Sequences of T₁ oligonucleotides *3*, *4* and *5*, showing their relationship to amino-acid sequences in the R17 coat protein[5,6].

resolved oligonucleotides occurs in major yield in one or other of the fragment preparations, but not in both. This finding supports the idea of a single primary site of cleavage by ribonuclease IV in the R17 RNA molecule, giving rise to the two fragments. The presence of sequence *1* as a major spot in the 40 per cent fingerprint, and similarly of sequence *8* in the 60 per cent fingerprint, confirms the fragment orientation in the original RNA molecule based on pppGp analyses[16,25,26].

A quantitative measure of the yields of oligonucleotides *1–8* in the 40 per cent and 60 per cent fragments was obtained by scraping the appropriate spots off the two fingerprints and counting them directly (Table 3, columns (a) and (c). The identity of each was then checked by analysis of its pancreatic ribonuclease digestion products. Sequences *1* and *8* can conveniently be used as standards with which to compare the yields of the other oligonucleotides as tabulated in columns (e) and (f) of Table 3. Sequence *2* is present in the 40 per cent fragment in 87 per cent yield whereas *3* to *8* appear in 25 per cent yield or less. Conversely, the 60 per cent fragment contains oligonucleotides *4* to *7* in better than 80 per cent yield and *1* to *3* in less than 23 per cent yield. The relative recovery of some oligonucleotides may be low, for the efficiency of transfer from the first dimension to the DEAE-cellulose thin layer[28] may vary both along the length of the cellulose acetate strip and with the size and/or composition of the oligonucleotide. Nevertheless,

Table 3. YIELDS OF SPECIFIC T_1 OLIGONUCLEOTIDES FROM THE 40 PER CENT AND 60 PER CENT FRAGMENTS

Oligonucleotide (No. of phosphates)	(a) 40 per cent C.p.m. (total)	(b) C.p.m. per phosphate	(c) 60 per cent C.p.m. (total)	(d) C.p.m. per phosphate	(e) Percentage yield (see text) 40 per cent relative to 1	(f) 60 per cent relative to 8
1 (12)	575	47·9	76	6·3	—	23
2 (10*)	416	41·6	57	5·7	87	21
3 (17)	86	5·1	59	3·5	11	13
4 (17)	146	8·6	376	22·1	18	80
5 (21)	232	11·0	538	25·6	23	92
6 (13)	130	10·0	286	22·0	21	80
7 (9)	107	11·9	249	27·7	25	100
8 (9)	78	8·7	249	27·7	18	—

* See footnote to Table 2.

Glass sinters were counted before and after taking up the DEAE-cellulose corresponding to the marker T_1 oligonucleotides in a 'Unilux I' (Nuclear Chicago) scintillation counter, without the addition of scintillator (setting C50). The differences between these two readings are tabulated in columns (a) and (c).

After its radioactivity had been counted, each oligonucleotide was eluted and its identity checked by analysis of its pancreatic ribonuclease digestion products. The radioactivity from spots observed in minor yield was shown to come from the oligonucleotides in question rather than from contaminants.

the data allow unequivocal assignment of oligonucleotides 2 to 7 (with the exception of 3) either to the 40 per cent or to the 60 per cent fragment. The level of contamination of each fragment by the other is consistent with the pppGp analyses and the results of ribosome binding.

Table 4. RELATIVE YIELDS OF A PROTEIN INITIATION SITE OLIGONUCLEOTIDES FROM COMPLEXES FORMED BETWEEN B. stearothermophilus RIBOSOMES AND THE 40 PER CENT AND 60 PER CENT FRAGMENTS

Oligonucleotide	Counts per min 40 per cent	60 per cent	Relative yield 40 per cent : 60 per cent
UUUG	6,450	1,200	5·4
CUUUUAG	7,350	1,500	4·9
ACCUAUG	10,400	1,850	5·6

Oligonucleotide spots were cut from the fingerprints shown in Fig. 6 and counted in a toluene scintillator. The actual counts per min obtained for the 60 per cent oligonucleotides have been multiplied by a factor of 1·75 to adjust for the fact that non-equimolar amounts of the two fragments were used in the initial binding reactions. The 40 per cent : 60 per cent ratio for CUUUUAG is probably least accurate because this oligonucleotide appears near the 3'-end of the A protein ribosomal binding site (see Fig. 7) and therefore often experiences internal cleavage upon pancreatic ribonuclease digestion of the initiation complex.

Location of A Protein Initiation Site

Because no unique T_1 oligonucleotide from the A protein initiation site could be identified in the homochromatogram of a complete T_1 ribonuclease digest of

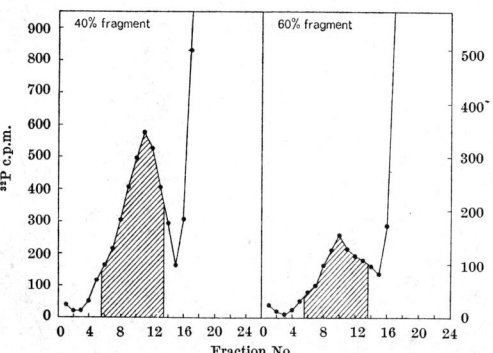

Fig. 5. Initiation complexes formed between ribosomes from B. stearothermophilus and the 40 per cent and 60 per cent fragments. Parallel binding reactions were performed at 66° C in 40 µl. of solution containing: 0·1 M NH_4 cacodylate (pH 7·1), 0·1 M NH_4Cl, 0·008 M magnesium acetate, 0·0002 M GTP, 4·5 A_{260} nm units of B. stearothermophilus ribosomes (prepared as described previously[22]), 4·0 A_{260} nm units of formylated $_1$charged mixed tRNA, and either 42 µg (1·4 × 10^7 c.p.m.) of the 40 per cent fragment or 36 µg (1·2 × 10^7 c.p.m.) of the 60 per cent fragment. After 8 min of incubation, the binding reaction mixtures were cooled; 0·2 µg of pancreatic ribonuclease was added and incubation continued for 15 min at 22° C. Sucrose gradient centrifugation was performed as described previously[22] and 1/40 volume aliquots of the fractions were counted. In the diagram the ^{32}P c.p.m. scales have been adjusted because non-equimolar amounts of the two fragments were used in the binding reactions. The fraction of A protein sites bound was calculated assuming that the average length of an initiator fragment is 30 nucleotides. Because 1 per cent of the radioactivity initially present in the 40 per cent fragment binding reaction and 0·3 per cent of that in the 60 per cent reaction appear in the 70S region of the gradients, a total of about 60 per cent of the theoretical number of A sites was recovered.

R17 RNA, an alternative method was used to estimate the amounts of A protein initiation site in the 40 per cent and 60 per cent fragments. When ribosomes from Bacillus stearothermophilus, rather than from E. coli, are used for the in vitro translation of f2 RNA, only the A protein is initiated and synthesized[31]. Similarly with R17 RNA[22], the A protein initiation site alone binds to B. stearothermophilus ribosomes; no recognition of the coat protein or synthetase sites can be detected. Moreover, at 65° C, B. stearothermophilus ribosomes attach to the beginning of the A protein cistron with extremely high efficiency. At least 50 per cent of the theoretical number of A sites can be recovered from B. stearothermophilus ribosomes as compared with approximately 3 per cent from E. coli ribosomes (J. A. S., unpublished observation).

Portions of the same 40 per cent and 60 per cent fragment preparations used for pppGp analyses and homochromatography (Fig. 3 and Tables 1 and 3) were incubated with B. stearothermophilus ribosomes, treated with nuclease and sedimented through sucrose gradients as shown in Fig. 5. It is clear that significantly more radioactive RNA is protected when the initiation complex is made with the 40 per cent fragment than with the 60 per cent fragment. Furthermore, fingerprint analyses (Fig. 6) of the material indicated as cross-hatched in Fig. 5 demonstrate that virtually all the ribosome bound radioactive RNA from the 40 per cent fragment derives from the A protein initiation site (see Fig. 7). The material from the 60 per cent fragment gives a less clean fingerprint, containing a higher background of non-specific oligonucleotides. The three large T_1 oligonucleotides arising from the interior of the A protein binding site were cut from the two fingerprints and counted directly (Table 4). In this experiment, in which approximately 60 per cent of the theoretical number of A sites were recovered, a five-fold higher yield of the A protein initiation region from the 40 per cent than from the 60 per cent fragment is observed.

Biochemical Map for R17 RNA

From the data given in Tables 3 and 4, it is possible to construct the biochemical map of the R17 genome illustrated in Fig. 7. The crucial assignment of the coat protein gene to the middle cistron is unambiguous, for T_1 oligonucleotide 2, which arises from the 5' end of the coat protein ribosomal binding site, is found in good yield in the 40 per cent fragment, whereas oligonucleotides 4 and 5, from the interior of the coat protein gene, are recovered primarily from the 60 per cent fragment. Thus ribonuclease IV most often cleaves the R17 RNA somewhere between the beginning of the coat gene and the nucleotides coding for amino-acids 57 to 61 of this protein. The low yield of oligonucleotide 3 from both the 40 per cent and the 60 per cent fragment preparations further pinpoints the principal site of ribonuclease IV attack to the region coding for the first six amino-acids of the coat protein and locates this oligonucleotide at a position about 40 per cent of the way along the R17 RNA molecule.

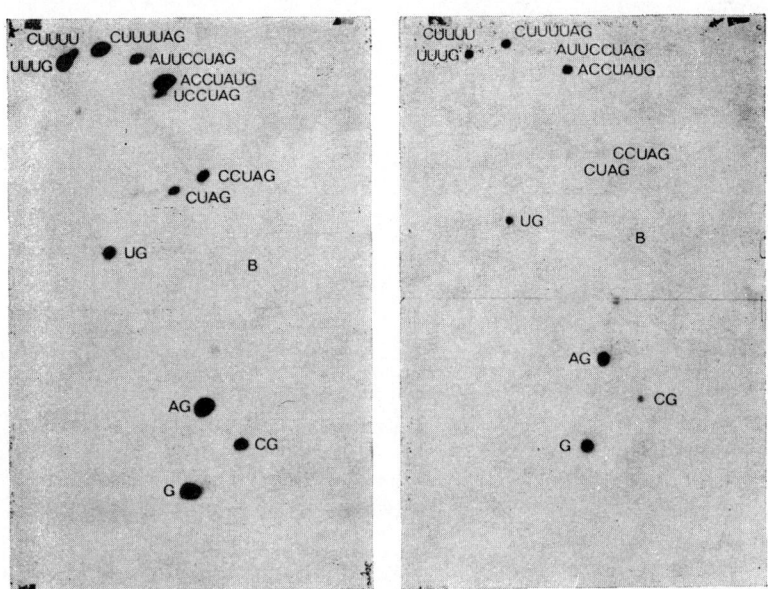

Fig. 6. T₁ ribonuclease fingerprints of initiation complexes formed between *B. stearothermophilus* ribosomes and the 40 per cent and 60 per cent fragments. Left: 40 per cent; right: 60 per cent. Fractions 6–13 from the sucrose gradients illustrated in Fig. 5 were pooled and extracted with phenol in the presence of 0.1 per cent sodium dodecyl sulphate. T₁ ribonuclease digestion and fingerprint analyses of the resulting RNA were performed as described previously[23]. Because non-equimolar amounts of the two fragments were used in the initial binding reactions, the 40 per cent fingerprint was exposed for 30 h, and the 60 per cent for 52 h to allow direct comparison.

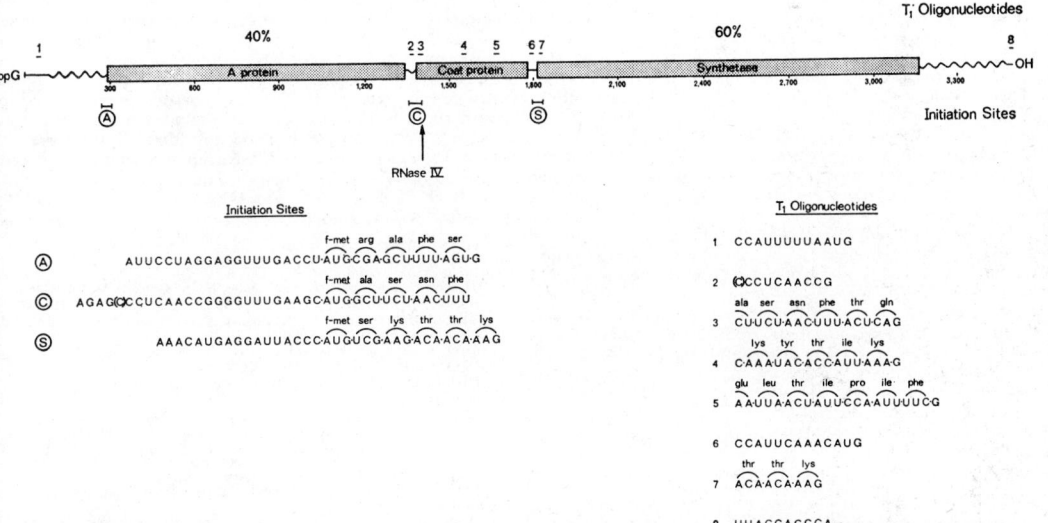

Fig. 7. Biochemical map of the R17 RNA (drawn to scale). An RNA chain length of 3,500 nucleotides has been assumed and the best molecular weight estimates for the A protein and synthetase were used. The eight specific T₁ oligonucleotides and the ribosomal binding sites have been located as accurately as possible along the genome. Wavy lines indicate that the actual lengths of the corresponding regions are unknown.

Once the position of the coat protein cistron has been assigned, it follows that each of the remaining genes must be wholly contained in one or the other of the two R17 RNA fragments. The high yields in the 60 per cent fragment of oligonucleotides 6 and 7, which arise from the initiation region of the synthetase cistron, place this gene at the 3′ end of the RNA; and the results of the binding experiments performed with *B. stearothermophilus* ribosomes strongly support the location of the A cistron nearest the 5′ terminus.

The recent elucidation of the nucleotide sequence of the entire intercistronic region between the end of the coat protein and the beginning of the synthetase genes[22,24] eliminates any doubt that the synthetase cistron directly follows the coat protein cistron in the R17 genome. Our knowledge of the sequence also allows us to draw this region to scale (36 nucleotides) in the biochemical map. The size of the intercistronic region between the A protein and the coat protein cistrons has been assumed to be the same.

The untranslated regions at the two ends of the R17 genome are the major uncertainties remaining in the map. The first 75 nucleotides (and probably the first 100 nucleotides) do not overlap the A protein initiation site (J. M. Adams and S. Cory, personal communication) and at least 11 nucleotides at the 3′ end of the RNA are not included in the synthetase cistron[30]. But because the RNA has been variously estimated to contain 3,300[9,10] to 3,500[11] nucleotides and the molecular weights of the A and synthetase proteins have not been precisely determined, both the total lengths and the manner of distribution of the remaining untranslated material between the termini of the R17 RNA molecule are unknown.

Ribonuclease Cleavage and RNA Conformation

Identification of the principal site of ribonuclease IV attack in the R17 RNA molecule clarifies many results previously obtained in studies of the 40 per cent and 60 per cent fragments[16,25]. Although the complete coat protein ribosomal binding site is usually retained by the 40 per cent fragment (preliminary results of J. M. Adams and S. Cory (personal communication) suggest that ribonuclease IV cleavage may occur specifically after the phenylalanine codon UUU), any polypeptides produced would be too short to be detected in the conventional assay systems. This, in addition to the low level of A protein synthesis in extracts from *E. coli*[14,15], explains the relative inactivity of the 40 per cent fragment as a messenger *in vitro*[16,25]. The 60 per cent fragment, on the other hand, efficiently directs synthetase production; the amount of coat protein it initiates (20 per cent to 25 per cent that of the intact RNA[16,25]) might suggest a second site of attack by ribonuclease N (ref. 25), but the degree of purity of the fragment preparations examined here is not sufficient to reveal more than the one major site of attack.

The fact that ribonuclease IV cleaves in the vicinity of the coat protein initiator strengthens previous suggestions that this area lies in an exposed region of the intact R17 RNA molecule[25]. Because several nucleases in addition to ribonuclease IV yield fragments of about 22S and 15S as the first cleavage products (refs. 33–35 and personal communication of H. Boedtker and H. D. Robertson), it seems that isolated bacteriophage RNAs may possess not only ordered regions of secondary structure[22–24,27] but also a well defined tertiary structure. The beginning of the coat gene may be most readily accessible not only to nuclease attack but also to ribosome attachment, possibly contributing to the high relative frequency of initiation of coat protein synthesis.

Translation and Gene Order

The genome of the RNA bacteriophages, either *in vivo* or *in vitro*, directs the synthesis of very different amounts of the three proteins. For example, the molar amount of coat protein made can exceed that of synthetase and A protein by three and twenty-fold respectively[14]. Translation is also subject to a polar effect whereby an amber mutation at position 6 of the coat protein drastically lowers the production of synthetase[36–39,42]. Because both these phenomena are demonstrable *in vitro*, they must reflect intrinsic properties of the phage RNA molecule, and any explanation of them must be consistent with the gene order.

Polarity and the high relative rate of coat protein synthesis have suggested that the RNA bacteriophage genome is read sequentially[38,40], starting with the coat cistron. The gene order established here, however, proves that translation of the R17 RNA cannot be wholly sequential. Moreover, because the A protein (which is made in smallest amount), is encoded before the coat protein, independent entry of ribosomes at the beginning of the coat gene is obligatory. The actual levels of synthesis of the three proteins very likely reflect different rates of initiation, which in turn may be influenced not only by the nucleotide sequences of the initiation regions but also by the secondary and tertiary structure of the RNA molecule.

Because ribosomes have independent access to initiation sites on fragmented RNA[14,22,41], polarity also cannot depend on sequential reading of the phage genome. In other words, the fact that the synthetase gene follows the coat cistron may be irrelevant; polarity could arise from defined secondary or tertiary structure in the intact phage RNA molecule[14,22,25,36,37,39]. The initially inaccessible synthetase binding site could be made accessible by changes in the folding of the RNA brought about either by the ribosome as it translates the coat cistron or by the removal of large segments of the molecule through nuclease action (suggested by the fact that position 6 amber mutants in the 60 per cent fragment seem to have lost their polarity[25]).

In conclusion we stress that the RNA bacteriophage genome may not be the typical operon that is so frequently assumed. The messenger activity appears to be subject to strict secondary and tertiary structure constraints, imposed possibly because it must fold efficiently into a spherical virus particle. Whether similar mechanisms operate in bacterial and other messengers is not known.

We thank Dr F. Sanger for his interest and encouragement, M. S. Bretscher for help in preparing the manuscript, R. Crouch and B. G. Barrell for helpful discussions, and Mrs J. Michelbank for technical assistance. This research was supported by a Medical Research Council scholarship (to P. G. N. J.), postdoctoral fellowships from the US National Science Foundation and the Jane Coffin Childs Memorial Fund for Medical Research (to J. A. S.), a grant from the US National Science Foundation and a Public Health Service career development award (to R. F. G.), and the Fonds National Suisse de la Recherche Scientifique (to P. F. S.).

Received February 12, 1970.

[1] Gussin, G. N., *J. Mol. Biol.*, **21**, 435 (1966).
[2] Horiuchi, K., Lodish, H. F., and Zinder, N. D., *Virology*, **28**, 438 (1966).
[3] Viñuela, E., Algranati, S., and Ochoa, S., *Europ. J. Biochem.*, **1**, 1 (1967).
[4] Nathans, D., Oeschger, M. P., Eggen, K., and Shimura, Y., *Proc. US Nat. Acad. Sci.*, **56**, 1844 (1966).
[5] Weber, K., and Konigsberg, W., *J. Biol. Chem.*, **242**, 3563 (1967).
[6] Weber, K., *Biochemistry*, **6**, 3144 (1967).
[7] Steitz, J. A., *J. Mol. Biol.*, **33**, 923 (1968).
[8] Capecchi, M. R., *J. Mol. Biol.*, **21**, 173 (1966).
[9] Gesteland, R. F., and Boedtker, H., *J. Mol. Biol.*, **8**, 496 (1964).
[10] Strauss, J. H., and Sinsheimer, R. L., *J. Mol. Biol.*, **7**, 43 (1963).
[11] Fiers, W., Van Montagu, M., DeWachter, R., Gillis, E., Min Jou, W., Messens, E., Ramaut, E., and Vandenberghe, A., *Cold Spring Harbor Symp. Quant. Biol.*, **34** (in the press, 1969).
[12] Eggen, K., Oeschger, M. P., and Nathans, D., *Biochem. Biophys. Res. Commun.*, **28**, 587 (1967).
[13] Viñuela, S., Salas, M., and Ochoa, S., *Proc. US Nat. Acad. Sci.*, **57**, 729 (1967).
[14] Lodish, H. F., *Nature*, **220**, 345 (1968).
[15] Lodish, H. F., and Robertson, H. D., *J. Mol. Biol.*, **45**, 9 (1969).
[16] Spahr, P. F., and Gesteland, R. F., *Proc. US Nat. Acad. Sci.*, **59**, 876 (1968).

[17] Shimura, Y., Kaizer, H., and Nathans, D., *J. Mol. Biol.*, **38**, 453 (1968).
[18] Engelhardt, D. L., Robertson, H. D., and Zinder, N. D., *Proc. US Nat. Acad. Sci.*, **59**, 972 (1968).
[19] Robertson, H. D., and Zinder, N. D., *Nature*, **220**, 69 (1968).
[20] Robertson, H. D., and Zinder, N. D., *J. Biol. Chem.*, **244**, 5790 (1969).
[21] Webster, R. E., Robertson, H. D., and Zinder, N. D., *Cold Spring Harbor Symp. Quant. Biol.*, **34** (in the press).
[22] Steitz, J. A., *Nature*, **224**, 957 (1969).
[23] Adams, J. M., Jeppesen, P. G. N., Sanger, F., and Barrell, B. G., *Nature*, **223**, 1009 (1969).
[24] Nichols, J. N., *Nature*, **225**, 147 (1970).
[25] Gesteland, R. F., and Spahr, P. F., *Cold Spring Harbor Symp. Quant. Biol.*, **34** (in the press).
[26] Spahr, P. F., Farber, M., and Gesteland, R. F., *Nature*, **222**, 455 (1969).
[27] Billeter, M. A., Dahlberg, J. E., Goodman, H. M., Hindley, J., and Weissmann, C., *Nature*, **224**, 1083 (1969).
[28] Brownlee, G. G., and Sanger, F., *Europ. J. Biochem.*, **11**, 395 (1969).
[29] Robinson, W. E., Frist, R. H., and Kaesberg, P., *Science*, **166**, 1291 (1969).
[30] Dahlberg, J. E., *Nature*, **220**, 548 (1968).
[31] Lodish, H. F., *Nature*, **224**, 867 (1969).
[32] Roblin, R., *J. Mol. Biol.*, **31**, 51 (1969).
[33] Shimura, Y., and Nathans, D., *Bact. Proc.*, **161** (1967).
[34] Bassel, B. A., and Spiegelman, S., *Proc. US Nat. Acad. Sci.*, **58**, 1155 (1967).
[35] Thach, S., and Boedtker, H., *J. Mol. Biol.*, **45**, 451 (1969).
[36] Gussin, G. N., Capecchi, M. R., Adams, J. M., Argetsinger, J. E., Tooze, J., Weber, K., and Watson, J. D., *Cold Spring Harbor Symp. Quant. Biol.*, **31**, 257 (1966).
[37] Zinder, N., Engelhardt, D. L., and Webster, R. E., *Cold Spring Harbor Symp. Quant. Biol.*, **31**, 251 (1966).
[38] Engelhardt, D. L., Webster, R. E., and Zinder, N. D., *J. Mol. Biol.*, **29**, 45 (1967).
[39] Gussin, G. N., *J. Mol. Biol.*, **21**, 435 (1966).
[40] Ohtaka, Y., and Spiegelman, S., *Science*, **142**, 493 (1963).
[41] Lodish, H. F., *J. Mol. Biol.*, **32**, 681 (1968).
[42] Capecchi, M. R., *J. Mol. Biol.*, **30**, 213 (1967).

Direction of Translation of the Lysozyme Gene of Bacteriophage T4 Relative to the Linkage Map

George Streisinger, Joyce Emrich, Yoshimi Okada†,
Akira Tsugita and Masayori Inouye

Institute of Molecular Biology, University of Oregon, Eugene, U.S.A.

and

*Laboratory of Molecular Genetics, University of Osaka Medical School
Osaka, Japan*

(*Received 7 July 1967*)

The lysozyme gene of bacteriophage T4 has been shown to be translated in the same direction as the *rII* gene and in a direction opposite to that of the head gene. The time of translation of genes of bacteriophage T4 seems not to be determined by the direction of translation.

1. Introduction

The direction of translation, relative to the circular linkage map, has been established for a number of genes of phage T4. In the case of gene 23 (Epstein *et al.*, 1963), the structural gene for head protein (Sarabhai, Stretton, Brenner & Bolle, 1964), the direction of translation was determined directly, by examining the peptides synthesized by strains of T4 carrying mutations that result in chain termination (amber mutations) (Sarabhai *et al.*, 1964). The directions of translation of a number of phage T4 genes that control various structural proteins have been deduced from the direction of polarity of amber mutations with respect to neighboring genes (Stahl, Murray, Nakata & Crasemann, 1966). In the *rII* A and B genes (Benzer, 1957), the direction was determined on the basis of the effect of frameshift mutations on the function of one of the two genes joined to the other by a deletion that removed the intergenic divide (Crick, Barnett, Brenner & Watts-Tobin, 1961).

The *rII* B gene controls a function that is expressed within two or three minutes after infection, while the head gene, and some others that control components of the phage structure, are expressed later (Levinthal, Hosoda & Shub, 1967). The direction of translation of the *rII* gene is opposite to that of the head gene and to other genes that seem to be expressed late.

Lysozyme begins to appear at about the same time after infection as head protein and other components of the structure of phage particles. Thus it is of interest to compare the direction of translation of the lysozyme gene to that of the other genes.

In this paper we establish that the direction of translation of the lysozyme gene is opposite to that of the head gene and some other genes whose products appear late.

† Present address: Institute for Plant Virus Research, Chiba, Japan.

2. Materials and Methods

(a) Bacterial strains

Escherichia coli B, Bphr^- and 011' were used. Bphr^-, obtained from W. Sauerbier, a strain in which photoreactivation does not occur, is used in all crosses, since in the course of some crosses u.v. light was used in order to enhance the frequency of recombinants; 011' is a permissive derivative of *E. coli* B obtained from F. W. Stahl and originally isolated by S. Brenner.

(b) Bacteriophage strains

T4D *amA494*, a mutant in gene 1, was received from R. S. Edgar; the lysozyme (*e*) mutant strains will be described elsewhere.

(c) Media

The media used have been described previously (Okada *et al.*, 1966).

(d) Spot tests for recombination

About 10^7 phage carrying any of several deletion mutations in the lysozyme gene were plated on bottom agar in the usual way with top agar containing B bacteria. About 20 min after plating (when the top agar had solidified) a small drop (containing about 0·005 ml. of phage at a concentration of about 10^8/ml.) of each of the lysozyme mutants to be spot-tested was deposited on the plate, and the plates were kept at room temperature, uncovered, until the spots had dried (about 30 min). Control plates (with spotted phage but with no deletion mutant phage) were also prepared. After overnight incubation at 30 or 43°C (depending on the mutants tested) the spots on the control plates were compared to those with the deletion mutant; the appearance of an excess of plaques in the spots against the deletion mutant was interpreted as resulting from recombination. The control spots in most cases gave no plaques or else very few plaques.

(e) Scoring of amA494 versus am$^+$ plaques

E. coli 011' and B were grown to a concentration of 2×10^8/ml. in aerated broth at 37°C. To 0·25 ml. of 011' at 37°C, 0·05 ml. of a suitable concentration of phage was added and, after incubation at 37°C for about 8 min, 0·25 ml. of B and 2·5 ml. of top agar were added and the contents of the tube were poured on bottom agar. Both top and bottom agar contain 0·05 M-Tris–HCl, pH 8·3. The plates were incubated at 43°C overnight; only e^+ phage form plaques and *am* plaques appear turbid while am^+ plaques appear clear.

3. Results

(a) The order of relevant mutant sites within the e gene

The order of the mutations discussed in this paper is given in Fig. 1. The qualitative results of simplified spot-test crosses of the mutants to a set of deletions, given in Fig. 1, define both the extent of the deletions and the relative positions of most of the other mutations in the lysozyme gene. The mutations *eJ42* and *eJ17* have previously been shown to be very closely linked. The relative positions of the mutations *eJ201* and *eJ42* are established by a three-factor cross described below. The isolation of the deletions, and more detailed mapping experiments, will be described elsewhere.

(b) The orientation of the e gene with respect to the genetic map of T4

In order to orient mutant sites within the *e* gene with respect to the genetic map of T4, the three-factor crosses 1, 2 and 3 described in Table 1 were undertaken. In these crosses the double mutant strain *eJ42 amA494* was crossed to various *e* mutants, whose positions relative to *eJ42* had been established by deletion mapping as described above. Plaques produced by e^+ (recombinant) phage among the progeny were scored as *am* or *am*+ (Table 1). Since *amA494* is known to be in gene 1 (outside of the *e* gene), the ratio *am* : *am*+ establishes the relative order of the three markers in any one cross. Cross 4 (Table 1) serves to establish the order of *eJ201* and *eJ42* relative to the other markers. Similar three factor crosses, using *rI* instead of *amA494*, confirm this orientation and will be described elsewhere.

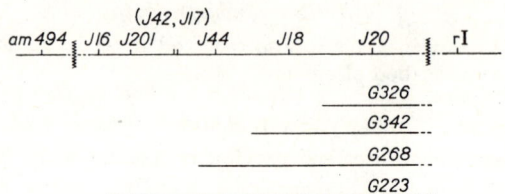

Mutant / Deletion	J16	J201	J42	J44	J18	J20
G326	+	+	+	+	+	0
G342	+	+	+	+	0	0
G268	+	+	+	0	0	0
G223	+	0	0	0	0	0

FIG. 1. The relative order of mutant sites in the *e* gene.

The order shown on the map is established by the results of the spot tests shown in the matrix and the results of the crosses presented in Table 1. A + in the matrix indicates plaques, due to recombination, in the spot test; a 0 indicates no plaques due to recombination.

TABLE 1

The order of e *mutants relative to* amA494

Cross	Among the progeny $ame^+/ame^+ + am^+e^+$
(1) $eJ16 \times eJ42\ amA494$	0·70
(2) $eJ18 \times eJ42\ amA494$	0·44
(3) $eJ20 \times eJ42\ amA494$	0·41
(4) $eJ201 \times eJ42\ amA494$	0·73

E. coli B*phr*⁻ were grown to a concentration of 10^8/ml. in aerated broth at 37°C and concentrated, by centrifugation, to $1·5 \times 10^9$/ml. To 0·5 ml. of the concentrated bacteria at 37°C was added 0·5 ml. of a mixture containing $7·5 \times 10^9$/ml. of each of the parental phages. 3 min after infection the bacteria were diluted in broth at 37°C and 19 min after infection were lysed by adding 1/10 vol. chloroform and 10 μg/ml. of egg white lysozyme. The lysate was further diluted and plated as described in Materials and Methods.

(c) *The orientation of the lysozyme molecule with respect to the map of the* e *gene*

We have previously described the lysozyme produced by two pseudo-wild strains carrying the pairs of frameshift mutations *eJ42eJ44* (Terzaghi *et al.*, 1966) and *eJ17eJ44* (Okada *et al.*, 1966). The mutations *eJ42* and *eJ17* were shown to be very closely linked (Okada *et al.*, 1966); the changed sequences of amino acids in the two double mutant strains were different at the end nearer the N-terminal end of the lysozyme molecule and were identical at the end nearer the C-terminal end. This result orients the *(J42 J17)–J44* order as being parallel to the N-terminal to C-terminal order in the lysozyme molecule.

The total sequence of amino acids in the lysozyme molecule has been established (Inouye & Tsugita, 1966), and the changed regions in a number of double mutant strains have been identified. The mutations *eJ16* (Okada, 1966), *eJ201* (Inouye, Akaboshi, Tsugita, Streisinger & Okada, 1967) and *eJ44* (Terzaghi *et al.*, 1966), affect base sequences that code for the amino acids in the neighborhoods (1 to 4), (25 to 26) and amino acid 41 respectively, counting amino acids from the N-terminal end. This result independently establishes the orientation *eJ16–eJ201–eJ44* as paralleling the N-terminal to C-terminal orientation of the lysozyme molecule.

4. Discussion

We have oriented a number of mutations within the lysozyme gene with respect to the N-terminal to C-terminal direction of the lysozyme molecule. Previously published results have established that the 5′ to 3′ direction of the messenger RNA molecule corresponds to the N-terminal to C-terminal direction of the lysozyme molecule (Terzaghi *et al.*, 1966). It has been established by others (Dintzis, 1961) that proteins are synthesized from the N-terminal towards the C-terminal end, and that messenger RNA is synthesized from the 5′ towards the 3′ end (Bremer, Konrad, Gaines & Stent, 1965). On this basis we can establish the direction of transcription (from DNA to RNA), and the direction of translation (from RNA to protein) with respect to the map of the lysozyme gene. These relationships are summarized in Fig. 2.

The relative positions of a large number of genes have been established for phage T4; Fig. 3 shows the orientation of the lysozyme gene with respect to certain of these genes. The solid arrows represent directions of translation that have been established by means of structural studies of the mutant protein as described here for the *e* gene and by Sarabhai *et al.* (1964), for gene 23. The dotted arrows represent directions

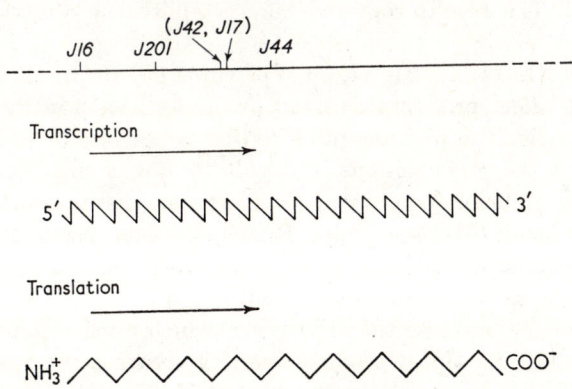

Fig. 2. The direction of transcription and translation relative to mutant sites in the *e* gene.

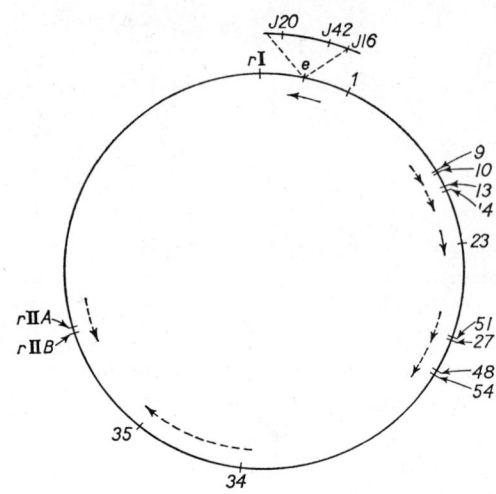

Fig. 3. A map of certain genes of phage T4.

The arrows indicate directions of translation established by analyses of the primary structure of mutant proteins (solid arrow) or by polarity of amber and other types of mutations (dotted arrow).

of translation established on the basis of studies of the polar behavior of mutants in the gene pairs 51-27, 34-35 (Stahl et al., 1966), 9-10, 13-14 and 48-54 (Stahl & Crasemann, personal communication) and by studies of Crick et al. (1961), of the effect of frameshift mutations in the *rII* A gene on the product of the *rII* B gene in a strain where these two genes are joined by a deletion. Previously (Streisinger, 1964) a direction of translation opposite to the one established here was suggested. That suggestion was based on the observation that some frameshift mutations located near one end of the *e* gene produced low levels of lysozyme activity. It seemed reasonable to assume that these mutations were located in the region of the *e* gene that specified the C-terminal region of the lysozyme molecule, the low enzyme activity being due to lysozyme defective in the C-terminal portion of the molecule. The mutations in question have now been shown to be located near that end of the lysozyme gene that specifies the N-terminal end of the lysozyme molecule and other hypotheses are being explored that would account for the low level of lysozyme activity produced. The results reported here establish the direction of translation directly.

The genes 9, 10, 13, 14, 23, 27, 34, 35 (Epstein et al., 1963), 48, 51 and 54 (Edgar & Wood, 1966) all affect products involved in the formation of the structure of the mature phage particle. The proteins produced by genes 10 and 23 have been shown to appear at about the same time as lysozyme by direct measurements (Levinthal et al., 1967), while indirect experiments suggest that many structural components begin to appear at about this time (Boice, Russel, & Edgar, personal communication). In contrast, the product of the *rII* gene appears within three minutes after infection (McClain & Champe, personal communication).

Our results thus demonstrate that lysozyme, which appears late, is translated in the same direction as the *rII* B product, which appears early, and in the opposite direction of other genes whose products appear late. The time of translation is thus not determined by the direction of translation.

Differences in the time-course of transcription of the *e* gene and the *rII* gene have been reported (Bautz, Kasai, Reilly & Bautz, 1966) and classes of messengers appearing early and late have been distinguished (Hall, Nygaard & Green, 1964). It is becoming clear that more than two classes of proteins exist with respect to the time of translation (Levinthal *et al.*, 1967) and transcription (Salser, personal communication) and little can be said at this time concerning the pertinent question of possible relationships between the time and direction of transcription.

The direction of translation of other genes of phage T4 is not yet known. It will be of interest to determine whether the genes between *e* and *rII* are translated in the same direction as these two, and whether there is a general correlation between direction of translation and map position.

This work was supported by a grant from the National Science Foundation (466 GB2261) to one of us (G. S.), by a Jane Coffin Childs Memorial Fund for Medical Research grant to another of us (Y. O.) and by a Jane Coffin Childs Memorial Fund for Medical Research grant and a grant from the National Institutes of Health (GM10982) to a third (A. T.).

REFERENCES

Bautz, E. K. F., Kasai, T., Reilly, E. & Bautz, F. A. (1966). *Proc. Nat. Acad. Sci., Wash.* **55**, 1081.
Benzer, S. (1957). In *The Chemical Basis of Heredity*, ed. by W. D. McElroy & B. Glass. Baltimore: The Johns Hopkins Press.
Bremer, H., Konrad, M. W., Gaines, K. & Stent, G. S. (1965). *J. Mol. Biol.* **13**, 540.
Crick, F. H. C., Barnett, L., Brenner, S. & Watts-Tobin, R. J. (1961). *Nature*, **192**, 1227.
Dintzis, H. M. (1961). *Proc. Nat. Acad. Sci., Wash.* **47**, 247.
Edgar, R. S. & Wood, W. B. (1966). *Proc. Nat. Acad. Sci., Wash.* **55**, 498.
Epstein, R. H., Bolle, A., Steinberg, C. M., Kellenberger, E., Boy de la Tour, E., Chevalley, R., Edgar, R. S., Sussman, M., Denhardt, G. H. & Lielausis, A. (1963). *Cold Spr. Harb. Symp. Quant. Biol.* **28**, 375.
Hall, B. D., Nygaard, A. P. & Green, M. H. (1964). *J. Mol. Biol.* **9**, 143.
Inouye, M., Akaboshi, E., Tsugita, A., Streisinger, G. & Okada, Y. (1967). *J. Mol. Biol.* **30**, 39.
Inouye, M. & Tsugita, A. (1966). *J. Mol. Biol.* **22**, 193.
Levinthal, C., Hosoda, J. & Shub, D. (1967). In *Molecular Biology of Viruses*, New York: Academic Press.
Okada, Y. (1966). Quoted in Streisinger, G., Okada, Y., Emrich, J., Newton, J., Tsugita, A., Terzaghi, E. & Inouye, M. (1966). *Cold Spr. Harb. Symp. Quant. Biol.* **31**, 77.
Okada, Y., Terzaghi, E., Streisinger, G., Emrich, J., Inouye, M. & Tsugita, A. (1966). *Proc. Nat. Acad. Sci., Wash.* **56**, 1692.
Sarabhai, A. S., Stretton, A. O. W., Brenner, S. & Bolle, A. (1964). *Nature*, **201**, 13.
Stahl, F. W., Murray, N. E., Nakata, A. & Crasemann, J. M. (1966). *Genetics*, **54**, 223.
Streisinger, G. (1964). In *Int. Symp. on Genes and Chromosomes, Structure and Function*, Nat. Cancer Inst. Monograph, no. 18, p. 1.
Terzaghi, E., Okada, Y., Streisinger, G., Emrich, J., Inouye, M. & Tsugita, A. (1966). *Proc. Nat. Acad. Sci., Wash.* **56**, 500.

Sense and Nonsense in the Genetic Code

Three exceptional triplets can serve as both chain-terminating signals and amino acid codons.

Alan Garen

The recent elucidation of the genetic code, shown in Table 1, marks a notable milestone in biology (1). This code designates the relations between the 64 possible codons (2) present in messenger RNA and the 20 amino acids present in proteins. The RNA codons are derived by transcription of complementary codons in DNA, which is the primary genetic material of most organisms (the only exceptions known are certain viruses in which messenger RNA is used directly as the genetic material).

Most of our present knowledge about the code has been obtained from studies with *Escherichia coli*, in which synthetic polyribonucleotides (rather than natural messenger RNA) are added to cell extracts containing the components required for protein biosynthesis in vitro (and presumably in vivo) (3). The polynucleotides in such experiments are either triplets, which can bind a specific transfer RNA species to ribosomes, or longer chain polymers (with random or defined base sequences), which can direct the incorporation of amino acids into polypeptides. A critical assumption for this approach to the deciphering of the code is that the coding properties of polynucleotide codons in vitro are the same as those of messenger RNA codons in vivo, allowing the extrapola-

The author is professor in the department of molecular biophysics at Yale University.

tion from in vitro to in vivo coding assignments. There is convincing support for this assumption from two lines of evidence, one showing that amino acid substitutions occurring in proteins as a result of mutations can be attributed to base changes which are consistent with the coding assignments for the amino acids (4–6), and another showing that the RNA component of an RNA phage acts in vitro as well as in vivo as a messenger for the coat protein of the phage (7). It should be noted that a coding assignment based on results in vitro does not necessarily prove that the codon is actually used in vivo; there is evidence that an organism can have the capacity to translate a codon but not incorporate it into its own code (see 6).

It is implicit in a triplet code, which provides a potential surplus of triplets for 20 amino acids, that the code contains degenerate codons (that is, different triplets coding for the same amino acid) or nonsense triplets (triplets which do not code for any amino acid), or both. It is immediately apparent from a glance at Table 1 that the code exhibits extensive degeneracy, as much as sixfold for some amino acids. There also are three triplets, UAG, UAA, and UGA, designated as nonsense, and it is this aspect of the code which is the main subject of this review.

Nonsense Mutants

Protein biosynthesis is a sequential process during which a peptide chain grows unidirectionally, by increments of one amino acid, from the amino-terminal toward the carboxy terminal residue (8). Accordingly, if a nonsense triplet is present at any of the positions in messenger RNA which code for the amino acid residues of a peptide chain, a gap will appear in the chain and cause premature termination of chain growth. Nonsense triplets do not normally occur within the coding regions of messenger RNA, but they can be generated from certain codons by mutation. The resulting mutants are called nonsense mutants, as distinguished from missense mutants which result from the transformation of a codon for one amino acid into a codon for another amino acid.

Since nonsense mutants cannot be produced selectively, a procedure is required for their identification in a population that may contain other classes such as missense, frame-shift, and deletion mutants. The problem can be simplified to some extent by restricting attention to mutants that can be reverted to the parental type by exposure to the base-analog mutagens 2-aminopurine or bromouracil (9); these mutants should comprise only the nonsense and missense classes.

To distinguish nonsense mutants from missense mutants, four procedures have been used with bacteriophage and bacteria. Evidence obtained by these procedures has firmly established the existence of nonsense mutants and has confirmed the hypothesis that the mutants produce chain-terminating nonsense triplets. The experimental details are as follows.

1) *Pleiotropic mutant phenotype.* Because of the polarity of messenger RNA, which is translated unidirectionally starting from the 5′-end of the molecule (5), it is possible for one nonsense triplet to block translation of an extended region of the RNA molecule. The extent of the block will depend on

Table 1. The genetic code. Each amino acid listed in the table is coded for by an RNA triplet (codon) which has a nucleotide sequence as follows. The 5′-terminal nucleotide appears in the column on the left, the middle nucleotide appears in the top row, and the 3′-terminal nucleotide appears in the column on the right. The symbols N1 (amber), N2 (ochre) and N3 designate the three nonsense triplets, UAG, UAA, and UGA. The symbols in the tables are explained in reference (2).

5′-terminal	Middle nucleotide				3′-terminal
	U	C	A	G	
U	Phe	Ser	Tyr	Cys	U
	Phe	Ser	Tyr	Cys	C
	Leu	Ser	N2 (Ochre)	N3	A
	Leu	Ser	N1 (Amber)	Trp	G
C	Leu	Pro	His	Arg	U
	Leu	Pro	His	Arg	C
	Leu	Pro	Gln	Arg	A
	Leu	Pro	Gln	Arg	G
A	Ile	Thr	Asn	Ser	U
	Ile	Thr	Asn	Ser	C
	Ile	Thr	Lys	Arg	A
	Met	Thr	Lys	Arg	G
G	Val	Ala	Asp	Gly	U
	Val	Ala	Asp	Gly	C
	Val	Ala	Glu	Gly	A
	Val	Ala	Glu	Gly	G

the distance between the nonsense triplet and the next site in the RNA at which translation can again be initiated. In the extreme case where there is no site for reinitiating translation, the block will extend over the entire region of the RNA molecule between the nonsense triplet and the 3′-terminal end.

The first evidence of this kind for the occurrence of nonsense mutants was obtained by Benzer and Champe working with mutants of bacteriophage T4, involving the two contiguous A and B genes of the rII region (10). In the standard phage strain, the two genes are probably transcribed as separate messenger RNA molecules, which are translated independently into separate A and B proteins. Consequently, any mutation (including the nonsense type) occurring in one gene will have no effect on the physiological function controlled by the other gene. There exists an exceptional rII mutant in which portions of both genes are deleted; the remaining portions of the two genes appear to act as a single gene, presumably transcribed as a single messenger RNA specifying one protein molecule (Fig. 1). This protein retains the function of the B protein, but has lost the function of the A protein.

The special significance of the deletion mutant is that it can be used to test for the effect on the B gene of mutations occurring in the remaining portion of the A gene. If, in the deletion mutant, a single messenger RNA for the A and B genes is translated in the sequence A to B [as suggested by results of Crick et al. (11)], a nonsense triplet appearing in the A region of the RNA could block the subsequent translation of the B region (Fig. 1). Benzer and Champe found that some of the base-analog revertible mutations in the A gene eliminated the function controlled by the B gene. They concluded that these were nonsense mutations acting as shown in Fig. 1. Although their evidence did not provide the proof for this conclusion, it was sufficiently persuasive to arouse widespread interest in the subject of genetic nonsense, and it introduced a valuable methodology for the identification of nonsense mutations.

The results obtained with rII mutants of bacteriophage T4, as summarized in Fig. 1, suggest that a nonsense triplet in one region of a messenger RNA can prevent the translation of another region of the RNA molecule. In the T4 rII case, both regions are transcribed from a segment of DNA which functions as a single gene (the combined A and B genes which are integrated by a deletion). It has been shown with bacteria that a cluster of several genes can operate as a polygenic unit of transcription, forming a messenger RNA molecule containing information for specifying more than one peptide chain (12, 13). If the entire RNA molecule is translated sequentially, a nonsense triplet causing premature chain termination of one of the peptide chains could exert a pleiotropic effect, blocking the formation of one or more of the other peptide chains specified by the molecule (Fig. 2). This effect will be polarized, preventing continuation of growth in one direction only, toward the regions that are subsequently translated. Thus the existence of polygenic species of messenger RNA, transcribed as a unit from several genes, provides a means of extending the pleiotropic test for nonsense mutations; instead of the test being limited to a single gene, as in the T4 rII experiments, the responses of different genes in a cluster to a mutation in one of the genes can be examined.

It has been found in studies of several gene clusters in bacteria—notably the lactose (14), tryptophan (13), histidine (15), and pyruvate dehydrogenase (16) clusters—that certain mutations, revertible by base-analog mutagens, produce a pleiotropic mutant phenotype involving more than one of the genes in the cluster. There is polarity associated with the pleiotropic effect which is generally in accordance with the map order of the genes; only the genes located along one direction from the mutation usually are affected (17). This is the expected result for nonsense mutations which act as shown in Fig. 2, but it is not a proof of the postulated mechanism. Pleiotropic mutations have also been described for an RNA phage, in which the genetic material acts directly as a polygenic messenger RNA (18).

A detailed quantitative analysis of pleiotropic mutations in the Z gene of the lactose cluster showed that the extent of the block of the function controlled by another gene in the cluster varied with the relative position of the mutation within the Z gene. The block became progressively weaker as the position of the mutation was shifted closer to that of the adjoining gene (14; for similar results with the tryptophan cluster, 13). A practical consequence of this position effect is that the pleiotropic action of a nonsense mutation located close to an adjoining gene might not be detected. An interpretation of the position effect, according to the mechanism illustrated in Fig. 2, is that premature chain termination occurring in the first gene to be translated in a polygenic messenger RNA has a certain probability of blocking reinitiation of translation of the RNA; this probability is directly proportional to the distance between the point of chain termination and the beginning of the next gene in the sequence.

Thus, on the basis of the polarized pleiotropic effect of some of the base-analog revertible mutations in phage and bacteria, it has been tentatively concluded that these are nonsense mu-

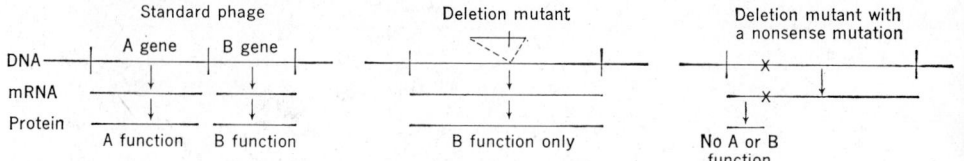

Fig. 1. Pleiotropic effect of a nonsense mutation in a T4 rII deletion mutant. In the standard bacteriophage strain, the A and B genes of the rII region are shown to be transcribed as separate messenger RNA molecules which are translated into separate proteins. In the deletion mutant, a single messenger RNA molecule is formed which is translated into a single protein capable of performing the B function but not the A function. In the deletion mutant containing a nonsense mutation in the A gene, a protein fragment is produced which cannot perform either the A or B functions (10).

tations. Other experiments confirming this conclusion are discussed later in relation to suppression; they show that the pleiotropic mutations respond to certain suppressor genes in *Escherichia coli*, which reverse the chain-terminating action of nonsense triplets.

2) *Extreme negative-mutant phenotype*. Another approach to the problem of distinguishing between a nonsense mutant and a missense mutant is to assay for the capacity of the mutant to perform certain functions controlled by the gene in which the mutation occurs. Two of the functions which are especially useful for this purpose are the formation of immunological cross-reacting material (CRM) and intragenic complementation. These functions are not likely to be retained by a nonsense mutant which can synthesize only a fragment of the protein required for the functions, while a missense mutant which synthesizes the protein intact might retain at least one of the functions. In a study done contemporaneously with the T4 rII experiments of Benzer and Champe, Garen and Siddiqi (19) showed that some of the alkaline phosphatase structural mutants of *Escherichia coli*, which are revertible by base-analog mutagens, have an extreme negative phenotype totally lacking in phosphatase enzymatic activity, CRM, and complementation activity, and that they are therefore possible nonsense mutants. These properties alone are not sufficient to identify a nonsense mutant because some missense mutants could also have an extreme negative phenotype. The significant aspect of the studies with alkaline phosphatase is that all of the mutants belonging to a certain suppressible class (which is a criterion for nonsense mutants discussed in the following section) have the extreme negative phenotype (19). This generalization applies not only to the suppressible phosphatase mutants but also to the same class of suppressible mutants involving other genes of bacteria (13, 15) and T4 phage (20). Furthermore, most of the extreme negative mutants which can be reverted by base-analog mutagens appear to be nonsense mutants by the same criterion of suppression.

3) *Suppression.* Two independent criteria for identifying the mutations in phage and bacteria which appear to be nonsense were described above. A connecting link between these criteria is the phenomenon of suppression of a mutation by the action of certain specialized bacterial suppressor genes, resulting in the restoration of a normal (or partially normal) phenotype despite the persistence of the original mutation. Both the T4 rII mutations and the alkaline phosphatase mutations, tentatively classified as nonsense (see sections 1 and 2, above), are suppressible in certain bacterial strains (10, 19). When the T4 rII mutations were tested in a suppressor strain active with alkaline phosphatase mutations, a positive response was obtained (21), an indication that mutations involving two genes with unrelated functions can be suppressed by the same strain (known to contain a single active suppressor gene). Evidently, suppression is specific for the mutation and not for the gene in which the mutation occurs. (This generalization is not strictly valid, since the mechanism of suppression, as discussed below, acts during translation and therefore cannot affect genes which are not translated, such as the structural genes for transfer RNA.)

Suppression is now the preferred method for identifying nonsense mutations in phage and bacterial genes. The principal reason for the preference is that the genetic and biochemical properties of the suppressor genes involved are well defined; and, as discussed in the following section, the genes act by reversing the chain-terminating effect of nonsense triplets in messenger RNA. Thus, suppression by these genes is a valid criterion for a nonsense mutation. Another reason is that the techniques of testing for suppression are simple and rapid. In describing the techniques, it is helpful to distinguish two ways of designing the suppression test: one which is applicable only to mutations in genes having an essential, indispensable role in the growth of an organism, and the other which is applicable only to mutations in genes that, under suitable culture conditions, are unessential

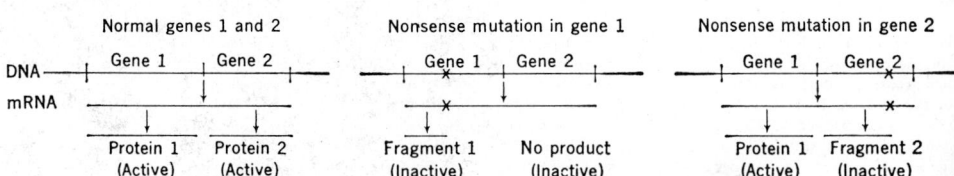

Fig. 2. Polarized pleiotropic effect of a nonsense mutation in a polygenic messenger RNA. A hypothetical example of a cluster of two genes which are transcribed as a single messenger RNA. The direction of translation is assumed to be from 1 to 2.

Table 2. Triplets related by a single base substitution to the *amber* and *ochre* triplets.

Nonsense triplet	Related triplet		
UAG (*amber*)	AAG (Lys)	UCG (Ser)	UAC (Tyr)
	CAG (Gln)	UUG (Leu)	UAU (Tyr)
	GAG (Glu)	UGG (Trp)	UAA (*ochre*)
UAA (*ochre*)	AAA (Lys)	UCA (Ser)	UAC (Tyr)
	CAA (Gln)	UUA (Leu)	UAU (Tyr)
	GAA (Glu)	UGA	UAG (*amber*)

for growth. For technical reasons, the first way of testing for suppression is feasible with bacteriophage but not with bacteria, while the second has been used both with phage and bacteria. The experimental details, in brief, are as follows.

For the first suppression test involving essential phage genes, mutants are produced and propagated in a host strain containing an active suppressor gene (Su^+ strain) and tested for growth in a host strain without an active suppressor gene (Su^- strain). The evidence for suppression is the inability of the mutant to grow in the Su^- host (20). A reverse procedure is required for the second test involving mutations in unessential genes of bacteria or phage. As

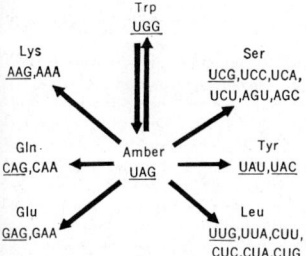

Fig. 3. Identification of an *amber* nonsense triplet as UAG. The diagram summarizes the pattern of amino acid substitutions in the alkaline phosphatases produced by revertants of a phosphatase *amber* mutant (32). Each substitution occurred at the same position, which is occupied by a tryptophan residue in the standard phosphatase molecule. All of the triplets designated from studies in vitro as codons for the substituted amino acids are listed; the underlined codons are those related to UAG by a single base change. It is evident that UAG is the only triplet which has this relationship to at least one of the codons for each of the seven substituted amino acids. See (2) for abbreviations.

applied to bacteria, the defective mutants (for example, phosphatase-negative) are isolated in an Su^- strain, and afterward the mutant genes are transferred, by means of mating or transduction, to an Su^+ strain. The evidence for suppression is the appearance of the mutant phenotype only in the Su^- strain, whereas in the Su^+ strain the normal phenotype (for example, phosphatase-positive) is at least partially restored (19).

An important feature of the first method of isolating nonsense mutants is that it is generally applicable to all genes which specify an essential protein. The nonsense mutants obtained in this way with the bacteriophage T4 (20), with a single-stranded DNA phage (22), and with an RNA phage (23), can be used to identify most, and possibly all, of the genes of these organisms.

4) *Fragment formation*. Although the three tests for nonsense mutations described above have given consistent results, they are based on indirect criteria. What is lacking is direct evidence for premature chain termination caused by a nonsense mutation, namely, the isolation of a protein fragment specified by a mutant gene. Fragments are difficult to detect because the assays developed for the intact protein generally do not work with a fragment. The first successful isolation of fragments was achieved with nonsense mutants of bacteriophage T4, involving the gene that specifies the major subunit of the head protein of the phage (24). The special characteristic of this protein, and the key to the success of the experiment, is that it represents the major component synthesized in cells infected with T4, and it can therefore be detected directly without prior purification. The experiment was performed with a group of the head protein mutants classified as nonsense by the criterion of suppression. The results were dramatic and convincing. The cells infected with a nonsense mutant produced a precisely defined fragment of the head protein, the size of the fragment correlating with the relative position of the nonsense mutation in the head protein gene. In subsequent work with nonsense mutations in the coat-protein gene of the RNA bacteriophage f2 (18), in the structural genes for β-galactosidase of *Escherichia coli* (25), and in the structural gene of alkaline phosphatase of *E. coli* (25a), fragments of the proteins specified by these genes have also been found.

The chain-terminating mutations that cause fragment formation in bacterio-

phages T4 and f2 are suppressed in a suppressor strain of bacteria, with the result that an intact protein is synthesized instead of a fragment (18, 24). Other experiments on the biochemistry of suppression (see below) show that the intact protein synthesized in a suppressor strain can have an amino acid substitution at a position in the protein which corresponds to the position of the chain-terminating mutation (26-28). These results indicate that suppression occurs at the stage of translation of a triplet which functions, in the absence of suppression, as a chain-terminating nonsense triplet. Thus, the first goal in the study of nonsense in the genetic code has been achieved with the demonstration of the occurrence of mutations which produce nonsense triplets in messenger RNA.

Nonsense Triplets

With the availability of well-characterized nonsense mutants, the problem shifts to the identification of the nonsense triplet (or triplets) causing premature chain termination. The first mutants selected for the analysis of nonsense triplets were classified together on the basis of their capacity to respond to three of the suppressor genes of *Escherichia coli*, *Su1*, *Su2*, and *Su3* (29, 30). The mutants in this class are called *amber* when they occur in T4 phage (20, 31) and *N1* when they occur in the alkaline phosphatase gene of *E. coli* (29); since the phage terminology has gained wide acceptance, it is adopted for this review.

Techniques adequate for nucleotide-sequence determination with cellular DNA or messenger RNA have not yet been developed. Therefore, indirect methods are required to identify a nonsense triplet. Two different methods have been employed for this purpose, the choice being determined by the potentialities and limitations of the materials involved. One method used with alkaline phosphatase nonsense mutants (32) is based on the triplet assignments in the genetic code (Table 1); in this method, the nonsense triplet is altered by mutations, and the new triplets derived in this way are identified by the amino acid substitutions they cause in the alkaline phosphatase molecule. If the triplets related to the nonsense triplet by a base substitution are known, it is possible to deduce the nonsense triplet. In the other method, chemical mutagens which appear to produce specific

4

base substitutions in T4 DNA are used (*31*); the nonsense triplet is deduced by an analysis of the pattern of induction and reversion of nonsense mutations by the mutagenic agents. At present, the genetic code probably provides a more reliable frame of reference than the specificity of chemical mutagenesis. As expressed in a review on mutagenesis (*9*): "At one time it was hoped that mutation studies would lead to a solution of the coding problem, but, with the development of in vitro protein-synthesizing systems, the situation is reversed." It was encouraging to find that results with the two methods were in agreement in identifying the *amber* nonsense triplet as UAG (*31, 32*). The evidence obtained for this assignment from experiments with an alkaline phosphatase *amber* mutant is summarized in Fig. 3. The assignment of UAG as the *amber* triplet has been made at five different mutant sites in the alkaline phosphatase gene (*6*) and at eight different sites in the head-protein gene of phage T4 (*33*). Thus, UAG appears to be the only *amber* triplet.

Further experiments on suppression of nonsense mutants, involving additional suppressor genes, revealed a new class of mutants which could be suppressed by some, but not by all, of the suppressor genes for *amber* mutants (*31, 34*). The new mutants are called *ochre* when they occur in T4, and *N2* when they occur in the alkaline phosphatase gene of *Escherichia coli;* as before, the phage nomenclature is used in this review. *Ochre* mutants can be transformed into *amber* mutants by mutation, indicating that different nonsense triplets are involved and that they are interrelated by a single base substitution (*31, 34*). Identification of the *ochre* triplet as UAA was accomplished by the same two methods used for the *amber* triplet, namely chemical mutagenesis of T4 (*31*) and amino acid substitution determinations with alkaline phosphatase (*35*). The experiments with alkaline phosphatase are discussed in detail because of certain interesting features they reveal about the code.

The nine triplets related by a single-base substitution to the *amber* triplet UAG are tabulated in the first part of Table 2. Eight of the triplets can be accounted for as codons for the amino acid substitutions shown in Fig. 3 (on the assumption that tyrosine substitutions are specified in some cases by UAC and in others by UAU). The remaining codon, UAA, appears not to have coding activity in vitro, and, since

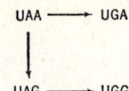

Fig. 4. A procedure for producing the triplets UAA, UAG, UGG, and UGA at the same position in alkaline phosphatase mRNA. The UAG triplet was derived from UAA by isolating a phosphatase *amber* mutant from a phosphatase *ochre* mutant. By further mutation of the *amber* mutant, the UGG triplet can be derived from UAG, and similarly, by mutation of the *ochre* mutant, the UGA triplet can be derived from UAA.

it is related to UAG, it is the prime candidate for the *ochre* triplet. In the second part of Table 2, the nine triplets related to UAA are listed. One is the *amber* triplet. Two others, the tyrosine codons UAC and UAU, are also related to UAG. The remaining six triplets are related only to UAA and not to UAG; however, five of these specify amino acids which are also specified by triplets related to UAG. The exception is UGA which does not appear to have coding activity in vitro (*3*). Thus, only one difference can be expected when *amber* and *ochre* nonsense mutants are analyzed by the amino acid substitution method, namely the appearance of tryptophan among the substitutions derived from an *amber* mutant but not from an *ochre* mutant. This expectation was realized in experiments with alkaline phosphatase *amber* and *ochre* mutants (*6, 35*). However, it is always a possibility in this kind of analysis that a particular substitution might not appear because it is incompatible with the activity of the protein molecule. To examine this possibility, the procedure outlined in Fig. 4 was followed. Both *amber* and *ochre* triplets were produced at the same positions in the alkaline phosphatase gene; each triplet was then altered by mutations, and the resulting alkaline phosphatase protein species were analyzed for amino acid substitutions occurring at the same position in the protein. Tryptophan substitutions did occur by alteration of the *amber* triplet but not of the *ochre* triplet; in contrast, the other six amino acids listed in Table 2 occurred as substitutions by alteration of both *amber* and *ochre* triplets. These results show that the alkaline phosphatase molecule can accept a tryptophan substitution at the positions analyzed. The fact that a tryptophan substitution did not occur with *ochre* mutants despite an extensive search is, for the reasons previously cited, consistent with the assignment of UAA as the *ochre* triplet.

From the position that the triplet UGA occupies in the genetic code (Table 1), it might be expected to code either for tryptophan or cysteine. That UGA is not a tryptophan codon is indicated by the results, discussed in the preceding section, of amino acid substitution analyses with alkaline phosphatase *ochre* mutants (*6, 35*) and also from experiments with T4 *ochre* mutants (*36*). Furthermore, there is no evidence that UGA can code in *Escherichia coli* for any other amino acid either in vitro (*3*) or in vivo (*6, 35*). The possibility that UGA might be a third nonsense triplet was examined by means of mutations in the T4 rII region which, on the basis of chemical mutagenesis criteria, appear to produce a UGA triplet (*37*). The principal finding was that one of these mutations in the *rIIA* gene exerted a pleiotropic effect on the *rIIB* gene when tested by the technique of Benzer and Champe (Fig. 1). Since this behavior conforms to one of the well-established criteria for nonsense mutations, UGA has been assigned as a third nonsense triplet in *E. coli*. However, in another organism, the guinea pig (liver cells), coding experiments in vitro suggest that UGA may be an effective cysteine codon (*3*).

Thus, the three triplets UAG, UAA, and UGA, produced by mutations in certain strains of bacteriophage and bacteria, can act as nonsense triplets. The coding results in vitro (Table 1) indicate that there are no additional nonsense triplets in *Escherichia coli*, since all remaining triplets have a coding assignment. However, nonsense is not an absolute property of a triplet but depends on a number of experimental factors that enter into an in vitro system, such as the organism used to test for translation of a triplet, the suppressor genotype of the selected strain of the organism, the ionic composition of the medium, and the sensitivities of the assay techniques for measuring coding activity. Therefore, the possibility remains that one or more triplets other than UAG, UAA, and UGA might be found to act as a nonsense triplet in some organisms.

Suppressor Genes for

Nonsense Mutations

The coding behavior of nonsense triplets is genetically controlled in bacteria by certain suppressor genes which

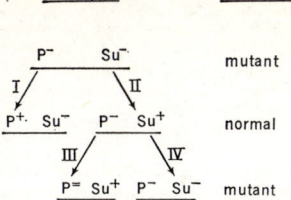

Fig. 5. A procedure (29) for the genetic analysis of suppressor genes for phosphatase nonsense mutations, showing pathways I, II, III, and IV which are discussed in the text.

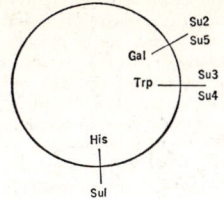

Fig. 6. Genetic map locations of five *Escherichia coli* suppressor genes for nonsense mutations. The *Su1*, *Su2*, and *Su3* suppressor genes (29) correspond to the suppressor genes designated by other investigators as suI, suII and suIII (33).

determine the capacity of the cell to translate the triplets. The properties of bacterial suppressor genes have been studied by the procedure outlined in Fig. 5 for obtaining suppressor mutants (29). For these studies, nonsense mutations in the structural gene for alkaline phosphatase were used as a marker to detect the suppressor mutants (nonsense mutations in other genes can also be used for this purpose, since the specificity of suppression pertains to the mutation and not to the gene in which the mutation occurs). The procedure starts with a suppressor-negative Su^- strain containing a phosphatase nonsense mutation. The first step is the isolation of revertant strains having a normal phosphatase phenotype. Genetic mapping shows that the revertants can occur in either of two ways. One is by a mutation which transforms the nonsense triplet into another triplet which is a codon in the Su^- strain (pathway I in Fig. 5). Revertants obtained in this way provided the alkaline phosphatase protein used for deducing the *amber* and *ochre* nonsense triplets by the amino acid substitution method (Fig. 3 and Table 2). The second way in which revertants occur is by a suppressor mutation that produces a suppressor-positive Su^+ strain capable of translating the nonsense triplet (pathway II). The revertants obtained in the second way comprise the strains needed to analyze the genetic and biochemical properties of suppressor genes for nonsense mutations. In addition to revertants that suppress phosphatase nonsense mutations, revertants have been isolated on the basis of their suppressor activity for nonsense mutations in other genes of *Escherichia coli* and T4 (30, 38).

The revertants which suppress *amber* or *ochre* mutations occur at several different map positions. In this review the three positions shown in Fig. 6, involving five suppressor genes, are discussed. Each suppressor gene designates a region of the map controlling a single suppressor function. The two genes *Su2* and *Su5*, which are located at one of the map positions, are closely linked, but for the biochemical reasons discussed below they appear to be functionally distinct. The evidence for separate *Su3* and *Su4* genes is that their map positions, although closely linked, are sufficiently separated to make it unlikely that only a single gene is involved (34).

The five suppressor genes in Fig. 6 differ in certain physiological and biochemical properties. All five of the genes suppress *amber* mutations, but only two, *Su4* and *Su5*, also suppress *ochre* mutations (Table 3). None of the genes suppress the UGA class of mutations. Another difference is in the efficiency of suppression of nonsense mutations, that is, the extent of restoration of the synthesis of the intact protein molecule. The efficiencies range from 5 to about 60 percent among the five genes (Table 3). These values are a measure of the relative rates of translation of a nonsense triplet in different suppressor strains. A third difference is in the amino acid residue incorporated into a protein as a result of suppression. The remarkable finding is that four of the five genes specify the incorporation of different amino acids (Table 3).

Thus, a suppressor gene controls not only the capacity of a cell to translate a nonsense triplet but also the coding specificity of the triplet. For example, the same *amber* triplet can either be nonsense if all the suppressor genes are Su^-, or it can specify any one of the four amino acids, serine, glutamine, tyrosine, or a basic residue, depending on which of the suppressor genes is Su^+ (Table 4) (39). There are also other *Escherichia coli* suppressor genes for *amber* and *ochre* nonsense mutations, in addition to the five described in Table 3; it is likely that biochemical studies with these suppressors will reveal new amino acids which can be specified by the nonsense triplets.

The efficiency of suppression of a nonsense mutation can depend on the relative position of the mutation within a gene as well as on the potency of a suppressor gene. Such a position effect has been demonstrated for one of the *amber* mutations in the structural gene for alkaline phosphatase; that mutation responds poorly to suppression in comparison to other *amber* mutations (6). The efficiency of translation of the *amber* triplet (and also other triplets) may be influenced by its nearest-neighbor nucleotide sequence.

A suppressor gene for mutations producing the UGA nonsense triplet has been found in *Escherichia coli* (40). The efficiency of suppression of the UGA class of mutations is high, about 60 percent, but there is no suppressor activity for either *amber* or *ochre* mutations. At present there is no information about either the map position of this suppressor gene or the amino acid which is specified by UGA as a result of its suppressor activity.

The mutational pathway II (Fig. 5), by which suppressor revertants are obtained, apparently generates a unique

Table 3. Response of *amber* and *ochre* nonsense mutations to suppression. The amino acid inserted by a suppressor gene was determined by analysis of the protein produced as a result of suppression. Experimental details are reported in the following references: $Su1^+$ (26), $Su2^+$ and $Su3^+$ (27), $Su4^+$ (28), $Su5^+$ (32). The tabulated efficiencies of suppression were estimated by the amount of alkaline phophatase protein (measured as CRM) produced as a result of suppression of a phosphatase-negative mutant (29, 34). These efficiencies have also been estimated by the extent of reversal of chain termination in an *amber* mutant of phage T4 (30, 33).

Suppressor gene	Nonsense mutation	Amino acid inserted	Efficiency of suppression (%)
$Su1^+$	Amber	Ser	28
	Ochre	0	0
$Su2^+$	Amber	Gln	14
	Ochre	0	0
$Su3^+$	Amber	Tyr	55
	Ochre	0	0
$Su4^+$	Amber	Tyr	16
	Ochre	Tyr	12
$Su5^+$	Amber	(Basic)	5
	Ochre	(Basic)	6

change in each suppressor gene, since independently isolated Su^+ revertants for the same gene fail to produce Su^- recombinants in pairwise genetic crosses. This genetic behavior can now be understood in terms of the biochemistry of the suppressor gene product (see below). The mutational pathway IV in Fig. 5, by which Su^- mutations are induced in an Su^+ strain, yields a different result; studies of the $Su1^+$ gene have shown that Su^- mutations occur at several separate sites within the suppressor gene (29). The Su^- mutants derived in this way provide material for fine-structure genetic analyses of suppressor genes and, as discussed below, also for biochemical analyses of the suppressor gene product.

Biochemical Basis of Suppression

Having established that suppressor genes control the specificity of translation of certain triplets, the desirability of studying suppression with the techniques of protein biosynthesis in vitro becomes evident. The experiments in vitro were initiated before the nonsense triplets were identified, and for that reason synthetic polynucleotides were not used initially as the message in vitro. Instead, the experiments were designed to use nonsense mutants of an RNA phage, since it had been shown that the RNA of the phage acts as a message in vitro for the coat protein of the phage (7). For this purpose *amber* mutants, suppressible in an $Su1^+$ host strain, were isolated (41). The availability of the phage *amber* mutants, combined with the knowledge that the amino acid specified by the *amber* triplet in the $Su1^+$ suppressor strain is serine (Table 3), set the stage for a test in vitro of suppression. Succesful results were obtained by two groups, by Cappecchi and Gussin working in J. D. Watson's laboratory (42) and by a collaborative effort between the laboratories of N. D. Zinder and A. Garen (43). Both groups were able to demonstrate that a transfer RNA (tRNA) for serine was the active suppressor component in an $Su1^+$ strain. The results of the latter group showed that in the absence of the suppressor tRNA a small fragment of the coat protein was formed, and that the addition of tRNA from the $Su1^+$ suppressor strain resulted in the formation of a complete protein molecule with a serine residue replacing a glutamine residue normally present in the

Table 4. Effects of five suppressor genes on the translation of a messenger RNA containing the *amber* triplet UAG. The original RNA is shown to contain the tryptophan codon UGG, from which the *amber* triplet is derived by mutation. The amino acid specfied by the *amber* triplet as a result of suppression by the $Su5^+$ gene is a basic residue which has not been identified but probably is lysine. See (2) for abbreviations.

Messenger RNA	Polypeptide	Suppressor genotype				
		$Su1$	$Su2$	$Su3$	$Su4$	$Su5$
AUGUACU*G*GGUCGUU	Met.Tyr.Trp.Gly.Val	−	−	−	−	−
AUGUACU*A*GGUCGUU	Met.Tyr	−	−	−	−	−
AUGUACU*A*GGUCGUU	Met.Tyr.Ser.Gly.Val	+	−	−	−	−
AUGUACU*A*GGUCGUU	Met.Tyr.Gln.Gly.Val	−	+	−	−	−
AUGUACU*A*GGUCGUU	Met.Tyr.Tyr.Gly.Val	−	−	+	−	−
AUGUACU*A*GGUCGUU	Met.Tyr.Tyr.Gly.Val	−	−	−	+	−
AUGUACU*A*GGUCGUU	Met.Tyr.(Lys).Gly.Val	−	−	−	−	+

molecule. Other suppressor tests in vitro on the tRNA derived from the remaining four suppressor strains listed in Table 2 suggest that the mechanism of suppression is the same in all of the strains tested, a specific tRNA from each suppressor strain apparently being the active suppressor component (44).

The results of the suppressor tests in vitro prove that certain species of tRNA in the Su^+ strains analyzed are the cellular components directly responsible for suppression. What remains unresolved by these analyses is the role of the suppressor gene in the formation of the tRNA. One possibility is that a suppressor gene acts directly as a structural gene for the tRNA. An alternative possibility is that a suppressor gene acts indirectly by specifying some product, perhaps an enzyme, which transforms an inactive tRNA into one with suppressor activity.

The function of a suppressor gene has been revealed by recent biochemical studies involving the $Su3$ gene. This gene occurs in the region of the *Escherichia coli* genome which is transducible by the phage $\phi 80$. Transduction by $\phi 80$ is of the "restricted" type in which certain bacterial genes from one region either replace some of the phage genes of the transducing particle, or are added to a complete phage genome. The $\phi 80$ particles with transducing activity for the $Su3^+$ gene provide a source of phage DNA, containing the $Su3^+$ gene, which can be used for hybridization tests between DNA and tRNA. The results of such hybridization tests reported by two laboratories (45) showed that $Su3^+$ DNA could hybridize specifically with tyrosyl tRNA. Since tyrosine is the amino acid specified by the *amber* triplet as a result of suppression by the $Su3^+$ gene, this is the expected result if the $Su3$ gene is a structural gene for a tyrosyl tRNA.

Purified fractions of two major species of tyrosyl tRNA, called species I and II, were equally effective in the hybridization tests. Thus, as a method for identifying the product of a suppressor gene, hybridization cannot distinguish between the different species of tRNA molecules with the same acceptor activity, presumably because of the close similarity of their base sequences. Furthermore, the tyrosyl tRNA obtained from both an $Su3^+$ and $Su3^-$ strain proved equally effective in hybridizing with the DNA of the transducing phage. This is not a surprising result in view of the fact that the two strains differ by a single suppressor mutation, and therefore the two tRNA molecules presumably differ by only a single base substitution.

Further biochemical studies of the tRNA specified by a suppressor gene require purified material. An attempt to purify the seryl tRNA, specified by the $Su1$ gene, was only partially successful and led to the discouraging conclusion that this tRNA was present as a minor species containing less than 10 percent of the amount of seryl tRNA in a major species (46). A similar conclusion was reached in the case of the tyrosyl tRNA specified by the $Su3$ gene (33). The purification of such minor components presents a formidable experimental obstacle. This obstacle was overcome by an ingenious use of the $\phi 80$ transducing phage carrying the $Su3$ gene (33). It was found that, in a cell infected with both the transducing phage and a nontransducing $\phi 80$ phage, conditions could be established in which the tyrosyl tRNA specified by the $Su3$ gene becomes the predominant species synthesized by the infected cell, thereby greatly simplifying its purification. With this procedure, enriched samples of tyrosyl tRNA specified by the $Su3^-$ and $Su3^+$ genes (the two genes differing by a single mutation) were prepared, and these were tested for their

Table 5. A model for the biochemical effect of certain suppressor mutations. The general assumption is that each suppressor gene is a structural gene for a tRNA species, and that an Su^+ mutation alters the anticodon of the tRNA. In the case of the tRNA specified by the $Su3^-$ gene, it has been shown that the anticodon is GUA, rather than the alternative possibility AUA (47). For further details see the text and reference (56).

Suppressor gene	Anticodon in tRNA	Matching codon in mRNA	Amino acid specified
$Su1^-$	CGA	UCG	Ser
$Su1^+$	CUA	UAG	
$Su2^-$	CUG	CAG	Gln
$Su2^+$	CUA	UAG	
$Su3^-$	GUA	UAU, UAC	Tyr
$Su3^+$	CUA	UAG	
$Su4^-$	AUA or GUA	UAU and/ or UAC	Tyr
$Su4^+$	UUA	UAA, UAG	
$Su5^-$	UUU	AAA, AAG	Lys
$Su5^+$	UUA	UAA, UAG	

capacity to bind to ribosomes in the presence of various nucleotide triplets (33). The triplets UAU and UAC, which are standard codons for tyrosine, stimulated binding of the tRNA specified by the $Su3^-$ gene, while the *amber* triplet UAG stimulated binding of the tRNA specified by the $Su3^+$ gene, and the *ochre* triplet UAA did not stimulate binding of either of these tRNA species. These results suggest that the two tyrosyl tRNA species have different anticodons, the species specified by the $Su3^-$ gene containing the anticodon GUA and the suppressor species specified by the $Su3^+$ gene containing the anticodon CUA. (The convention for designating an anticodon is the same as is used for a codon, placing the 5′-terminal nucleotide at the left and the 3′-terminal nucleotide at the right, although the codon-anticodon pairing occurs in an antiparallel orientation.)

This difference in the anticodons of the tyrosyl tRNA specified by the $Su3^-$ and $Su3^+$ genes was demonstrated by comparative base-sequence analyses of the purified tRNA. The tRNA specified by the $Su3^-$ gene contains the triplet GUA in the anticodon loop, which is replaced in the tRNA from the $Su3^+$ gene by the triplet CUA (47).

The impressive biochemical results obtained with the tyrosyl tRNA of the $Su3$ gene support the general mechanism of suppressor gene function shown in Table 5. The basic premise is that, for all of the suppressor genes that act on *amber* or *ochre* mutations, the phenotypic difference between the Su^- parental strain and the Su^+ revertant strain results from a change in the sequence of the suppressor tRNA. It is further assumed that the anticodon of the tRNA specified by the Su^- gene is complementary to one of the standard codons for the amino acid which the tRNA carries. This would account for the finding that the four amino acids known to be specified by *amber* and *ochre* triplets as a result of suppression (Table 3) are normally coded for by triplets which are related to the two nonsense triplets by a single base substitution. Accordingly, there probably exist other, still unidentified, suppressor genes for both *amber* and *ochre* mutations which specify leucyl and glutamyl tRNA species, and a suppressor gene for *amber* mutations which specifies a tryptophanyl tRNA.

The mechanism of suppression shown in Table 5, in conjunction with the "wobble" hypothesis of codon-anticodon specificity proposed by Crick (48), provides an adequate explanation for the fact that the $Su1^+$, $Su2^+$, and $Su3^+$ genes suppress only *amber* mutations, while the $Su4^+$ and $Su5^+$ genes suppress both *amber* and *ochre* mutations (Table 3). According to Crick's hypothesis, a CUA anticodon should be capable of interacting only with UAG because of the postulated stringency in the GC pairing when C is at the 5′ end of the anticodon, in contrast to a UUA anticodon which should be capable of interacting both with UAG and UAA because of the reduced stringency (that is, "wobble") permissible when uracil occupies the 5′-end position in the anticodon.

Suppressor mutations which act as shown in Table 5 are potentially dangerous to a cell, since these may produce alterations in anticodons required for the translation of some of the standard codons. Unless there is available to the cell an additional tRNA species containing an anticodon which is identical (or functionally equivalent) to the anticodon altered in the Su^+ strain, a suppressor mutation would be a lethal event. In the case of the tyrosyl tRNA specified by the $Su3^-$ gene, the GUA anticodon is also present in two other species of tyrosyl tRNA which are unaffected by the suppressor mutation and remain available to the $Su3^+$ strain for translating the standard tyrosine codons UAC and UAU (47).

There is no reason to believe that alteration of tRNA is the only cellular mechanism available for suppressing a nonsense mutation. The complex sequence of events involved in the translation of a triplet provides other possibilities. For example, alteration of ribosomes can affect the specificity and fidelity of translation (49). There has not yet been any conclusive demonstration that suppression of nonsense mutations does occur by alteration of components other than tRNA, although suggestive observations along these lines have been reported (50).

Fine-Structure Analysis of Suppressor tRNA

The genetic and biochemical methods that have been developed for studying the fine-structure details of suppression are also applicable to an analysis of the relation between the structure and function of the tRNA that is specified by a suppressor gene. Consider the mutational pathway IV in Fig. 5, by which Su^- mutations are induced in an Su^+ gene. The phenotypic effect of an Su^- mutation is the reduction or complete elimination of suppressor activity (29). When a suppressor gene is a structural gene for tRNA, this phenotypic effect must result from base substitutions occurring in the suppressor tRNA which affect its capacity to participate in the translation of a nonsense triplet. There are several ways that mutations might inhibit the suppressor activity of a tRNA. One is by alteration of the anticodon. The fact that Su^- mutations occur at more than three different genetic sites within a suppressor gene (29) indicates that the anticodon cannot be the only region of the suppressor tRNA affected by the mutations. Evidently, other essential functions of the tRNA molecule, such as its amino acid acceptor and transfer activities, are subject to inactivation by mutation. Thus, Su^- mutations can provide a family of differently altered forms of the same species of tRNA, involving base substitutions at various sites in the molecule, which is ideal material for biochemical studies of the structural requirements for the various functions of the molecule.

Minor Species of tRNA

A striking conclusion from fractionation experiments with suppressor tRNA is that these tRNA species are, at least in the two case analyzed (33, 46), minor components containing less than 10 percent of the tRNA in a major species having the same amino acid acceptor activity. The small amounts of suppressor tRNA may explain why the efficiency of suppression is less than 100 percent (Table 3). The rate of translation of a nonsense triplet is probably limited by the amount of suppressor tRNA available.

Identification of a minor species of tRNA is difficult to achieve without a specific assay such as suppression. For example, the seryl tRNA specified by the $Su1$ gene and the tyrosyl tRNA specified by the $Su3$ gene fractionate along with major species having the same amino acid acceptor activity (46, 47). These components might not have been detected without the availability of the suppressor assay. Probably other minor species of tRNA also exist.

The presence of both major and minor species of tRNA indicates that there is a mechanism in *Escherichia coli* for regulating the rates of transcription of the structural genes for tRNA. It remains to be shown whether the regulation of tRNA synthesis involves the same kind of control mechanism operating in protein sythesis (51). Analysis of tRNA regulation would be considerably advanced if regulatory mutants were available. Such mutants are difficult to identify because a change in the amount of tRNA synthesized is not readily detectable. The assay for suppression could provide a method of screening for mutants with altered rates of suppressor tRNA synthesis, since the efficiency of suppression, which is a measurable quantity, appears to be correlated with the amount of a suppressor tRNA.

Chain Termination
of Protein Biosynthesis

During the process of protein biosynthesis, the growing peptide chain is covalently linked to a tRNA which is attached (probably noncovalently) to the messenger-ribosome complex. The finished protein molecule is ultimately released unlinked to tRNA. When premature chain termination is induced by a nonsense triplet, either *amber* or *ochre*, the protein fragment apparently undergoes the complete reaction sequence involved in normal chain termination. The evidence now available, all from experiments in vitro, indicates that the fragment is not associated with any tRNA (18, 52).

Because virtually nothing is known about the mechanism of normal chain termination, which is one of the critical steps in protein biosynthesis, attention has been focused on premature chain termination in nonsense mutants as a potential model system for understanding the normal process. The central question is whether any of the nonsense triplets are signals for normal chain termination. There are two reasons why it appears necessary to assume that certain sequences are reserved for this purpose. First, in a polygenic messenger RNA which specifies more than one protein, a terminating signal is needed to prevent the linkage of one protein to another. The chain-initiating triplet AUG, which specifies formylmethionine (53), is, in principle, sufficient for this purpose since formylmethionine cannot form a peptide bond with the preceding amino acid specified by the message. However, when AUG occurs in an internal position in a message rather than at the 5′-terminal position, it acts as a codon for methionine and not for formylmethionine (3). Therefore, additional information must be associated with AUG to identify it as a chain-initiating formylmethionine codon for the internal genes of a polygenic messenger RNA.

A second reason for assuming that a chain-terminating signal exists is that a mechanism is needed for releasing newly synthesized polypetide chains from the messenger-ribosome complex. There is no information at present about how a release mechanism might operate; possibly additional components, such as special tRNA or protein species, are involved.

The sequence in messenger RNA which normally signals chain termination in *Escherichia coli* has not been identified. However, there are experiments bearing indirectly on the possible role of nonsense triplets in chain termination. If nonsense triplets do perform this essential function, suppression of the function could be harmful to the cell. For two of the nonsense triplets, UAG and UGA, suppression can be as efficient as 60 percent and still not affect cell growth (29, 33, 40), a finding which argues against a normal chain-terminating role for these triplets. By the same reasoning, UAA is a candidate for this role, since even low levels of suppression of the chain-terminating action of this triplet are inhibitory to cell growth (33, 34).

Another pertinent experiment involves a new class of mutants of the RNA phage. The mutants grow in an Su^- strain but are inhibited in Su^+ strains in which chain termination by UAG is strongly suppressed (54). One explanation for this mutant phenotype is that the mutation has transformed a UAA triplet, normally used for chain termination, into UAG. Accordingly, since UAG can cause chain termination as well as UAA in an Su^- strain, phage growth is normal in this strain; in the Su^+ strain there is an inhibition of phage growth because a critical chain-terminating site is exposed to the action of a strong suppressor.

Thus, there is suggestive evidence from experiments on suppression that UAG and UGA do not normally serve as chain-terminating triplets in *Escherichia coli*. Furthermore, it is a reasonable guess that these triplets are entirely absent from all messenger RNA. The suppression experiments also suggest that UAA is needed as a nonsense triplet, and therefore that it might be a normal signal for chain termination. The present evidence on these points, however, is inconclusive.

Dispensable and Indispensable
Suppressor Genes

It has been shown, as discussed in a preceding section, that the $Su3$ suppressor gene is a structural gene for a minor species of tyrosyl tRNA. This tRNA species, and also the minor species of seryl tRNA specified by the $Su1$ gene, appear to be dispensable components, since suppressor mutations which alter the anticodon or affect the activity of the molecules in other (unidentified) ways are not lethal to the cell. In addition to these dispensable tRNA species, there probably are also indispensable tRNA species which cannot be altered without causing cell death, notably the species required for translation of the standard codons of a cell. However, in a diploid cell containing two copies of the structural gene for an indispensable tRNA, a mutation in one of the genes need not be lethal as long as the cell retains a normal copy of the gene. *Escherichia coli* cells are usually haploid and therefore cannot tolerate such mutations, but some of the episomal strains of *E. coli*, which have multiple copies of certain regions of the bacterial chromosome, might contain structural genes for indispensable tRNA species in duplicate. In that case, it should be possible to obtain viable mutants of the episomal strains with altered forms of indispensable tRNA species. The problem remains of identifying the mutants.

Consider the mechanism of suppression outlined in Table 5. There are seven amino acids with codons which are related to the *amber* triplet UAG by a single base substitution (Table 2). The anticodons of the major tRNA species normally involved in translating these seven codons can be altered by mutation to yield suppressor tRNA which can translate the *amber* triplet. If the suppressor mutant is isolated in an appropriate episomal strain, it should be possible to obtain a heterozygote suppressor strain (that is, Su^+/Su^-) in which the suppressor gene is a structural gene for a major species of tRNA. Suppressor strains of this kind can provide genetic and biochemical information about the structural genes for some of the major, indispensable tRNA species. Experiments along these lines are now in progress.

Hindsight and Foresight

The study of nonsense triplets and their suppression has involved two related but experimentally distinct problems, one concerned with the nature of the genetic code and the mechanism of its translation, and the other with the chemistry of transfer RNA. With regard to the first problem, the results of genetic and biochemical experiments on nonsense triplets have contributed to an understanding of the genetic code in several ways. The identification of the three nonsense triplets in *Escherichia coli*, UAG, UAA, and UGA, filled in certain gaps which were left unresolved by coding experiments in vitro and thus helped to complete the code. The demonstration of a polarized chain-terminating effect of nonsense triplets in protein biosynthesis confirmed the sequential mechanism of translation of messenger RNA, and revealed an interdependence in the translation of different genes in a polygenic messenger RNA molecule. The elucidation of the mechanism of suppression of nonsense triplets showed that suppressor mutations, by altering the anticodons of certain minor species of tRNA, could introduce a limited degree of variability into the code. [In this review the important work on suppression of certain missense mutations in bacteria has not been discussed. The role of tRNA in the suppression of the missense mutations is discussed in reference (55).]

There are three outstanding questions related to nonsense triplets and suppression which remain unanswered. (i) Are any of the nonsense triplets involved in the normal process of chain termination of protein biosynthesis, and, if not, what is the signal for chain termination? (ii) What is the physiological role of the minor species of tRNA specified by a suppressor gene? These species appear to be redundant in the Su^- strain which also contains major tRNA species with the same anticodons and amino acid acceptor activities. (iii) Can suppression of nonsense triplets occur by genetic alteration of components other than tRNA, such as ribosomes and activating enzymes? Needless to say, these intriguing questions are currently under intensive study.

With regard to the second problem —the chemistry of tRNA—there has emerged from the study of suppression an elegant and powerful methodology for correlating the structure and function of this important macromolecular component of cells. By introducing mutations into a suppressor gene (Fig. 5), altered forms of a single species of tRNA can be obtained which differ structurally by a single base substitution in the polynucleotide chain, and functionally in the reaction specificities of the molecule. With the sophisticated techniques of tRNA chemistry now available to identify these base substitutions, it should be possible to elucidate in fine-structure detail the sites in a tRNA molecule which determine its capacity to accept a particular amino acid and to incorporate it into a growing protein chain.

References and Notes

1. For a comprehensive picture of the extensive work on this subject, see "The genetic code," *Cold Spring Harbor Symp. Quant. Biol.* **31**, entire volume (1966).
2. A glossary of some of the specialized terms used in this review follows. *Transcription*, the biosynthesis of an RNA containing a base sequence specified by a DNA template; *translation*, the biosynthesis of a protein containing an amino acid sequence specified by a messenger RNA template; *messenger RNA*, the cellular RNA species which act as templates for protein biosynthesis (as distinguished from ribosomal and transfer RNA); *complementary base pairs*, the specific pairing relationships between bases in double-stranded nucleic acids, which pair adenine with thymine or uracil, and guanine with cytosine; *codon*, the fundamental coding unit in DNA or RNA (a nucleotide triplet) which specifies the incorporation of an amino acid into protein; *deletion mutant*, a mutant resulting from the deletion of a segment of the genetic material; *frame-shift mutant*, a mutant resulting from the addition or deletion of one, or a few, nucleotides in the genetic material, which shifts the "reading frame" of the messenger RNA (it should be noted that a deletion mutant could also be classified as a frame-shift mutant if the deleted region is not extensive and does not contain three, or a multiple of three, nucleotides); *polygenic messenger RNA*, a species of messenger RNA which contains, in a single molecule, the information transcribed from more than one gene—a gene in this context is meant to designate a sequence of genetic information which is translated as a single polypeptide; *intragenic complementation*, the capacity of two functionally defective forms of the same gene (containing different mutations) to function cooperatively when present together in a heterozygous cell—complementation can occur when the gene product is a polymeric protein composed of identical subunits, allowing for the formation in a heterozygous cell of hybrid molecules containing subunits derived from different mutant forms of the gene; *mating*, the transfer of genetic material from a donor to a recipient bacterium by direct cell-cell contact; *transduction*, the indirect transfer of genetic material between bacteria by means of a phage vector. Abbreviations used are U, uracil; C, cytosine; A, adenine; G, guanine; tRNA, transfer RNA; mRNA, messenger RNA; the amino acids are referred to in the table and in sequences as follows: Ala, alanine; Arg, arginine; Asp, aspartic acid; Asn, asparagine; Cys, cysteine; Glu, glutamic acid; Gln, glutamine; Gly, glycine; His, histidine; Ile, isoleucine; Leu, leucine; Lys, lysine; Met, methionine; Phe, phenylalanine; Pro, proline; Ser, serine; Thr, threonine; Trp, tryptophan; Tyr, tyrosine; and Val, valine.
3. M. Nirenberg, R. Caskey, R. Marshall, D. Brimacombe, D. Kellogg, B. Doctor, D. Hatfield, J. Levin, F. Rottman, S. Pestka, M. Wilcox, F. Anderson, *Cold Spring Harbor Symp. Quant. Biol.* **36**, 11 (1966); H. G. Khorana, H. Buchi, N. Ghosh, T. M. Gupta, H. Jacob, H. Kossel, R. Morgan, S. A. Narang, E. Ohtsuka, R. D. Wells, *ibid.*, p. 39; J. H. Matthaei, H. P. Voigt, G. Heller, R. Neth, G. Schoch, H. Kubler, F. Amelunxen, G. Sander, A. Parmeggiani, *ibid.*, p. 25.
4. C. Yanofsky, J. Ito, V. Horn, *ibid.*, p. 151; H. G. Wittmann and B. Wittmann-Liebold, *ibid.*, p. 163.
5. G. Streisinger, U. Okada, J. Enrich, J. Newton, A. Tsugita, E. Terzaghi, M. Inouye, *ibid.*, p. 77.
6. M. G. Weigert, E. Gallucci, E. Lanka, A. Garen, *ibid.*, p. 145.
7. D. Nathans, G. Notani, J. H. Schwartz, N. D. Zinder, *Proc. Nat. Acad. Sci. U.S.* **48**, 1424 (1962).
8. H. Dintzis, *ibid.* **47**, 247 (1961); J. Bishop, J. Leahy, R. Schweet, *ibid.* **46**, 1030 (1960); A. Goldstein and B. J. Brown, *Biochim. Biophys. Acta* **53**, 438 (1961).
9. L. E. Orgel, *Advances Enzymol.* **27**, 289 (1965).
10. S. Benzer and S. P. Champe, *Proc. Nat. Acad. Sci. U.S.* **48**, 1114 (1962).
11. F. H. C. Crick, L. Barnett, S. Brenner, R. J. Watts-Robin, *Nature* **192**, 1227 (1961).
12. G. Attardi, S. Naono, J. Rouvière, F. Jacob, F. Gros, *Cold Spring Harbor Symp. Quant. Biol.* **28**, 363 (1963).
13. F. Imamoto, J. Ito, C. Yanofsky, *ibid.* **31**, 235 (1966).
14. W. A. Newton, J. R. Beckwith, D. Zipster, S. Brenner, *J. Mol. Biol.* **14**, 290 (1965); A. Newton, *Cold Spring Harbor Symp. Quant. Biol.* **31**, 181 (1966).
15. R. C. Martin, H. J. Whitfield, Jr., D. B. Berkowitz, M. J. Voll, *Cold Spring Harbor Symp. Quant. Biol.* **31**, 215 (1966).
16. U. Henning, G. Dennert, R. Hertel, W. S. Shipp, *ibid.*, p. 227.
17. An "antipolarity" effect of certain nonsense mutations has been found; it acts in a reverse direction, affecting the gene immediately preceding the gene containing the mutation (13). This puzzling phenomenon might be caused by interactions between different protein components subsequent to translation of the messenger RNA, and therefore it might be unrelated to the polarity effects discussed in the text, which block translation.
18. N. D. Zinder, D. L. Engelhardt, R. E. Webster, *Cold Spring Harbor Symp. Quant. Biol.* **36**, 251 (1966).
19. A. Garen and O. Siddiqi, *Proc. Nat. Acad. Sci. U.S.* **48**, 1121 (1962).
20. R. H. Epstein, A. Bolle, C. M. Steinberg, E. Kellenberger, E. Boy De la Tour, R. Chevalley, R. S. Edgar, M. Susman, G. H. Denhardt, A. Lielausis, *Cold Spring Harbor Symp. Quant. Biol.* **28**, 375 (1963).
21. S. Benzer, S. Champe, A. Garen, O. Siddiqi, unpublished results.
22. D. S. Ray in *Molecular Basis of Virology*, F. Fraenkel-Conrat, Ed. (Reinhold, New York, in press).

10

23. K. Horiuchi, H. Lodish, N. D. Zinder, *Virology* **27**, 139 (1966).
24. A. S. Sarabhai, A. O. W. Stretton, S. Brenner, A. Bolle, *Nature* **201**, 13 (1964).
25. A. V. Fowler and I. Zabin, *Science* **154**, 1027 (1966).
25a. T. Suzuki and A. Garen, in preparation.
26. M. G. Weigert and A. Garen, *J. Mol. Biol.* **12**, 448 (1965); G. W. Notani, D. L. Engelhardt, W. Konigsberg, N. D. Zinder, *ibid.*, p. 439; A. O. W. Stretton and S. Brenner, *ibid.*, p. 456.
27. M. G. Weigert, E. Lanka, A. Garen, *ibid.* **14**, 522 (1965); S. Kaplan, A. O. W. Stretton, S. Brenner, *ibid.*, p. 528.
28. M. G. Weigert, E. Lanka, A. Garen, *ibid.* **23**, 401 (1967).
29. A. Garen, S. Garen, R. C. Wilhelm, *ibid.* **14**, 167 (1965).
30. S. Kaplan, A. O. W. Stretton, S. Brenner, *ibid.*, p. 528.
31. ———, *Nature* **206**, 994 (1965).
32. M. G. Weigert and A. Garen, *ibid.*, p. 992.
33. J. D. Smith *et al.*, *Cold Spring Harbor Symp. Quant. Biol.* **31**, 479 (1966).
34. E. Gallucci and A. Garen, *J. Mol. Biol.* **15**, 193 (1966).
35. M. G. Weigert, E. Lanka, A. Garen, *ibid.* **23**, 391 (1967).
36. A. Sarabhai and S. Brenner, *ibid.* **26**, 141 (1967).
37. S. Brenner, L. Barnett, E. R. Katz, F. H. C. Crick, *Nature* **213**, 449 (1967).
38. N. M. Schwartz, *J. Bacteriol.* **89**, 712 (1965); E. Orias and T. K. Gartner, *Proc. Nat. Acad. Sci. U.S.* **52**, 859 (1964); G. Eggertsson and E. A. Adelberg. *Genetics* **52**, 319 (1965); J. R. Beckwith, *Biochim. Biophys. Acta* **76**, 162 (1963).
39. The first evidence that suppression of a nonsense mutation could alter the protein specified by the gene containing the mutation was reported by M. L. Dirksen, J. C. Hutson, J. M. Buchanan [*Proc. Nat. Acad. Sci. U.S.* **50**, 507 (1963)]. They showed that, as a result of suppression of a T4 nonsense mutant defective in the enzyme deoxycytidiylate hydroxymethylase, which is specified by a phage gene, an abnormal heat-labile form of the enzyme was produced.
40. J. F. Sambrook, D. P. Fan, S. Brenner, *Nature* **214**, 452 (1967).
41. N. D. Zinder and S. Cooper, *Virology* **23**, 152 (1964).
42. M. Capecchi and G. Gussin, *Science* **149**, 417 (1965).
43. D. L. Engelhardt, R. E. Webster, R. C. Wilhelm, N. D. Zinder, *Proc. Nat. Acad. Sci. U.S.* **54**, 1791 (1965).
44. R. C. Wilhelm, *Cold Spring Harbor Symp. Quant. Biol.* **31**, 496 (1966); R. F. Gesteland, W. Salser, A. Bolle, *Proc. Nat. Acad. Sci. U.S.* **58**, 2036 (1967). For an informative analysis of suppressor genes by mutagenic techniques, see M. Osborn, S. Person, S. Phillips, F. Funk, *J. Mol. Biol.* **26**, 437 (1967).
45. T. Andoh and H. Ozcki, *Proc. Nat. Acad. Sci. U.S.*, in press; A. Landy, J. N. Abelson, H. M. Goodman, J. D. Smith, *J. Mol. Biol.*, in press.
46. T. Andoh and A. Garen, *J. Mol. Biol.* **24**, 129 (1967).
47. J. D. Smith, J. N. Abelson, H. M. Goodman, A. Landy, S. Brenner, *Symp. Fed. European Biochem. Soc. 1967*, in press.
48. F. H. C. Crick, *J. Mol. Biol.* **19**, 548 (1966).
49. L. Gorini and J. R. Beckwith, *Ann. Rev. Microbiol.* **20**, 401 (1966).
50. P. J. Reid, E. Orias, R. K. Gartner, *Biochem. Biophys. Res. Commun.* **21**, 66 (1965); A. Bollen and A. Gerzog, *Arch. Int. Physiol. Biochim.* **73**, 139 (1965); A. Herzog and A. Bollen, *ibid.*, p. 526; A. Bollen, A. Herzog, R. Thomas, *ibid.*, p. 557.
51. W. Gilbert and B. Mueller-Hill, *Proc. Nat. Acad. Sci. U.S.* **56**, 1891 (1966); M. Ptashne, *Nature* **214**, 232 (1967).
52. M. C. Ganoza, *Cold Spring Harbor Symp. Quant. Biol.* **31**, 273 (1966).
53. K. Marcker and F. Sanger, *J. Mol. Biol.* **8**, 835 (1964); J. M. Adams and M. R. Capecchi, *Proc. Nat. Acad. Sci. U.S.* **55**, 147 (1966); R. E. Webster, D. L. Engelhardt, N. D. Zinder, *ibid.*, p. 155.
54. K. Horiuchi and N. D. Zinder, *Science* **156**, 1618 (1967).
55. J. P. Carbon, P. Berg, C. Yanofsky, *Cold Spring Harbor Symp. Quant. Biol.* **31**, 487 (1966).
56. P. Lengyel, *J. Gen. Physiol.* **49**, 305 (1966).
57. A. Garen and S. Garen, *J. Mol. Biol.* **7**, 13 (1963); M. Schlesinger and C. Levinthal, *ibid.*, p. 1.
58. I thank the National Science Foundation and National Institutes of Health for their generous support of my research on this subject. This review was completed in October 1967.

Structure and Function of Ribosomes and Their Molecular Components

M. NOMURA, S. MIZUSHIMA, M. OZAKI, P. TRAUB,* AND C. V. LOWRY

Department of Genetics, University of Wisconsin, Madison, Wisconsin

In this article, we summarize our recent work aimed at elucidating the functional role of ribosomal components. This work is based on the recently developed technique of reconstitution of the 30 S ribosomal subunit. In the article following this one we shall discuss in more detail the mechanism of this in vitro self-assembly reaction and in the third article, we shall describe our studies of ribosome assembly in vivo.

The gross structure of the ribosomal particles from *E. coli* was initially studied by Tissières, Watson, and their collaborators. We now know that the 30 S ribosomal subunit contains one 16 S RNA molecule and that the 50 S subunit contains one 23 S RNA molecule and one 5 S RNA molecule. In contrast to the RNA components, the protein composition of the ribosome is complex. Recent work from several laboratories has shown that there are about 20 different proteins in the 30 S subunit and about 35 in the 50 S subunit (Traut, Moore, Delius, Noller, and Tissières, 1967; Kaltschmidt, Dzionara, Donner, and Wittmann, 1967; Fogel and Sypherd, 1968a; Hardy, Kurland, Voynow, and Mora, 1969; and Traut et al., in this volume). Although these proteins have been purified and their general chemical properties are now known, the studies on their functional role have lagged far behind. Moreover, positive identification of which of the proteins are genuine ribosomal components, rather than tightly bound protein contaminants (as in the case of RNase I), has not been made. The problem of the functional role of ribosomal RNA has also been totally resistant to experimental attack. Thus, a method for the functional analysis of ribosomal components has been greatly needed.

Our recent success in the total reconstitution of the 30 S ribosomal particle from RNA and protein (Traub and Nomura, 1968a) has provided such a method. Table 1 describes the standard procedure for reconstitution. 16 S RNA and a mixture of proteins are derived from purified 30 S particles. They are mixed under appropriate conditions and incubated at 40°C. The reconstituted particles prove to be almost identical to the original 30 S ribosomes with respect to sedimentation properties, protein composition, and the several functional abilities which can be tested. Thus in this in vitro system, the 30 S ribosome, complex as it is, can be self-assembled, and it must be concluded that all the information needed for correct assembly is contained in the structure of the molecular components and not in extraribosomal factors.

PURIFICATION OF INDIVIDUAL 30 S RIBOSOMAL PROTEINS AND RECONSTITUTION FROM RNA AND THE SEPARATED PROTEIN COMPONENTS

In order to accomplish unambiguous identification and functional analysis of all the essential components of the 30 S subunit, we have separated and purified each of the proteins contained in the 30 S subunit. The methods used involve phosphocellulose column chromatography at several pH's and chromatography on DEAE cellulose and on

* Present address: Department of Biology, University of California, LaJolla, California.

TABLE 1. THE STANDARD PROCEDURE FOR RECONSTITUTION OF 30 S RIBOSOMAL SUBUNITS

30 S ─────────────→ 16 S RNA
    bentonite, duponol, phenol

30 S ─────────────→ 30 S ribosomal proteins
    4 M urea, 2 M LiCl

16 S RNA + 30 S ribosomal proteins ($P_1 + P_2 + \cdots + P_{20(?)}$)

─────────────→ 30 S ribosomal particles recovered by centrifugation

phosphate buffer (pH 7.8, 5×10^{-3} M)
$MgCl_2$ (2×10^{-2} M), KCl (0.3 M)
β-mercaptoethanol (6×10^{-3} M)
40°C, 20 min.

49

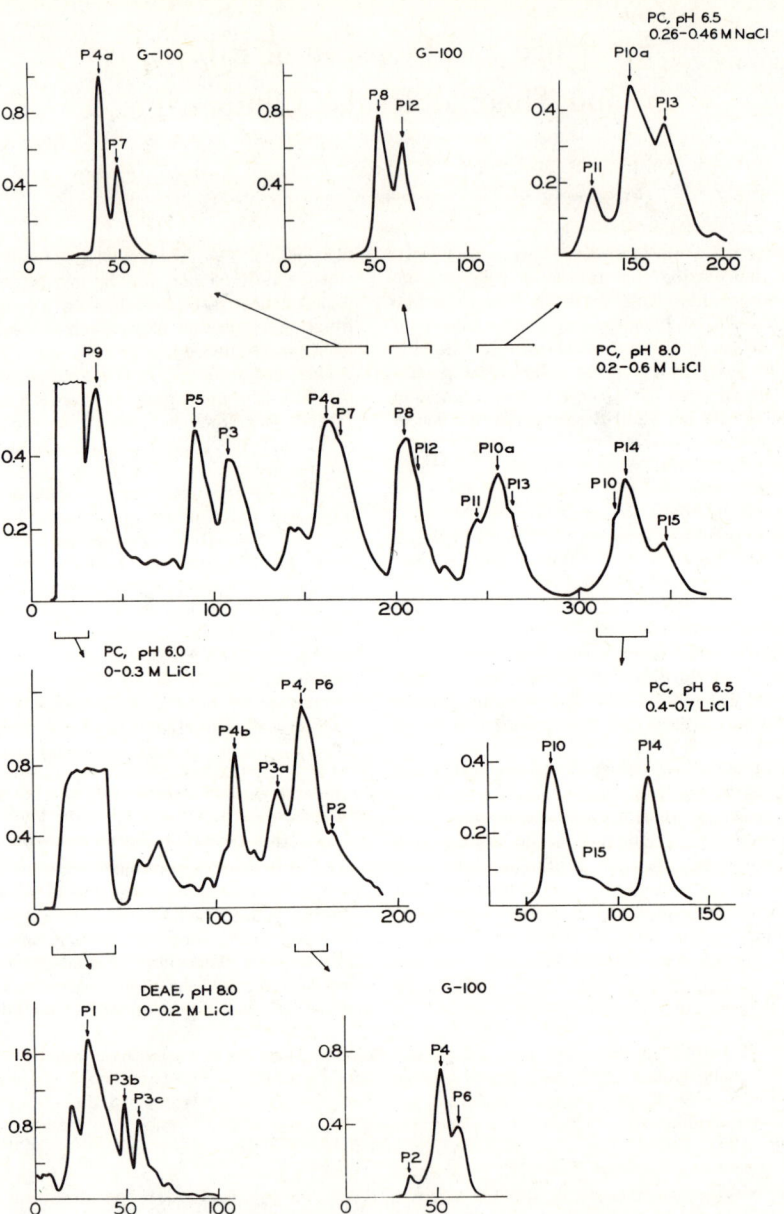

FIGURE 1. Separation of 30 S ribosomal proteins into each of the protein components. Fraction numbers are indicated on the abscissa. An aliquot from each fraction was analyzed for its protein content using the Folin reaction (Lowry et al., 1951) and the protein content in optical density units is plotted on the ordinate. A detailed description of the separation methods will be published elsewhere. PC stands for phosphocellulose; G-100 stands for Sephadex G-100; and DEAE for DEAE-cellulose.

Sephadex G100. Figure 1 summarizes the separation scheme. The purity of each fraction was tested by polyacrylamide gel electrophoresis (Fig. 2). Those proteins which migrate together under standard conditions (see Fig. 3) were also checked at lower gel concentrations, in which they are separable (Fogel and Sypherd, 1968b). Each of the 19 components purified showed a single major band in these analyses and were judged to be reasonably pure. Figure 3 shows the gel electrophoretic pattern of the original unfractionated 30 S ribosomal proteins and indicates the positions of the purified components. Although our methods of protein fractionation are not identical to those used by Kurland and his co-workers (Hardy et al., 1969), and hence the correspondence of our purified proteins to theirs is not unequivocal, we have attempted a comparison on the basis of behavior on phosphocellulose columns and in polyacrylamide gel electrophoresis. Table 7 shows our tentative conclusions. We have failed to find their protein 5, a minor component detected in their column chromatographic system. We also do not see their protein 7. However, because of the similarity in behavior of their proteins 6 and 7, (Hardy et al., 1969), it is possible that their proteins 6 and 7 together constitute our P9 fraction. We have also purified two proteins (P3b and P3c) present only in small amounts in the acidic protein fraction. We have also observed one peak (M) on the phosphocellulose column that contains several protein components apparently identical to ones found in other peaks. It is likely that the M peak represents an aggregate of these proteins, perhaps bound to some nonprotein compounds.

After obtaining the 21 separated components, we asked whether they comprised the full complement of 30 S ribosomal proteins. To do this, we compared the degree of reconstitution using a mixture of the 19 purified components (without including P3b and P3c) with that obtained using unfractionated proteins. As shown in Table 2, the efficiency of the reconstitution was only about half that obtained with unfractionated proteins. It is thus possible that we lost some essential components during the fractionation. However, other explanations are possible. Some protein components may have been inactivated during the purification procedure; or the relative proportions of the purified proteins in the reconstitution mixture may have been unbalanced, with some components under-represented and others present in excess. In any event, the fact that we could obtain as much as 50% reconstitution encouraged us to undertake studies of the functional role of these separated protein components.

In addition, the availability of the fractionated proteins and the reconstitution technique has made it possible to identify and study mutationally altered ribosomal components. We first investigated the protein responsible for sensitivity or resistance to the antibiotic streptomycin (Sm), and then, in collaboration with Bollen and Davies, the protein responsible for sensitivity or resistance to spectinomycin (Bollen, Davies, Ozaki, and Mizushima, 1969). The latter work is discussed by Bollen et al. elsewhere in this volume.

IDENTIFICATION OF THE PROTEIN CONTROLLED BY THE STREPTOMYCIN RESISTANT LOCUS IN *E. coli*

Two main effects of Sm have been observed in in vitro protein synthesizing systems: (a) Sm inhibits polypeptide synthesis directed by natural messenger RNA (mRNA), such as RNA from phage f2, and, to a lesser extent, that directed by certain synthetic polynucleotides (Speyer, Lengyel, and Basilio, 1962; Flaks, Cox, Witting, and White, 1962; Luzzato, Apirion and Schlessinger, 1968) (b) Sm causes misreading of synthetic mRNA (Davies, Gilbert, and Gorini, 1964). The latter phenomenon is presumably related to the observation that Sm suppresses a number of mutations in vivo (Gorini and Kataja, 1964). Mutations at the streptomycin locus (*str* locus) cause alterations in the 30 S ribosomal subunit. It has been observed that both of the above in vitro effects are abolished or greatly reduced in a cell-free system using 30 S ribosomal subunits from Sm-resistant strains (Cox, White, and Flaks, 1964; Davies, 1964; 1966). Thus the component controlled by the *str* locus appears to be important in the ribosomal functions of polypeptide synthesis *per se* and also in the control of translational fidelity.

We first showed that the component controlled by the *str* locus is a protein of the 30 S ribosome, and not the 16 S RNA moiety (Traub and Nomura, 1968b). We then fractionated the proteins from both streptomycin-sensitive and streptomycin-resistant strains. Reconstitutions were performed by combining fractions from the streptomycin-resistant strain with complementary fractions from the streptomycin-sensitive strain, and mixing them with 16 S RNA under the standard conditions for reconstitution. The streptomycin sensitivity of the resultant reconstituted particles was assayed. By this procedure we found that protein P10 is responsible for the streptomycin phenotype (Ozaki, Mizushima, and Nomura, 1969). Four different assay methods were used: inhibition of f2 RNA-directed ^{14}C-valine incorporation by Sm, inhibition of poly U-directed ^{14}C-phenylalanine incorporation by Sm, misreading of poly U induced by Sm, and binding of ^{3}H-dihydrostreptomycin (DHSm). Table 3 summarizes the results of the critical experiments

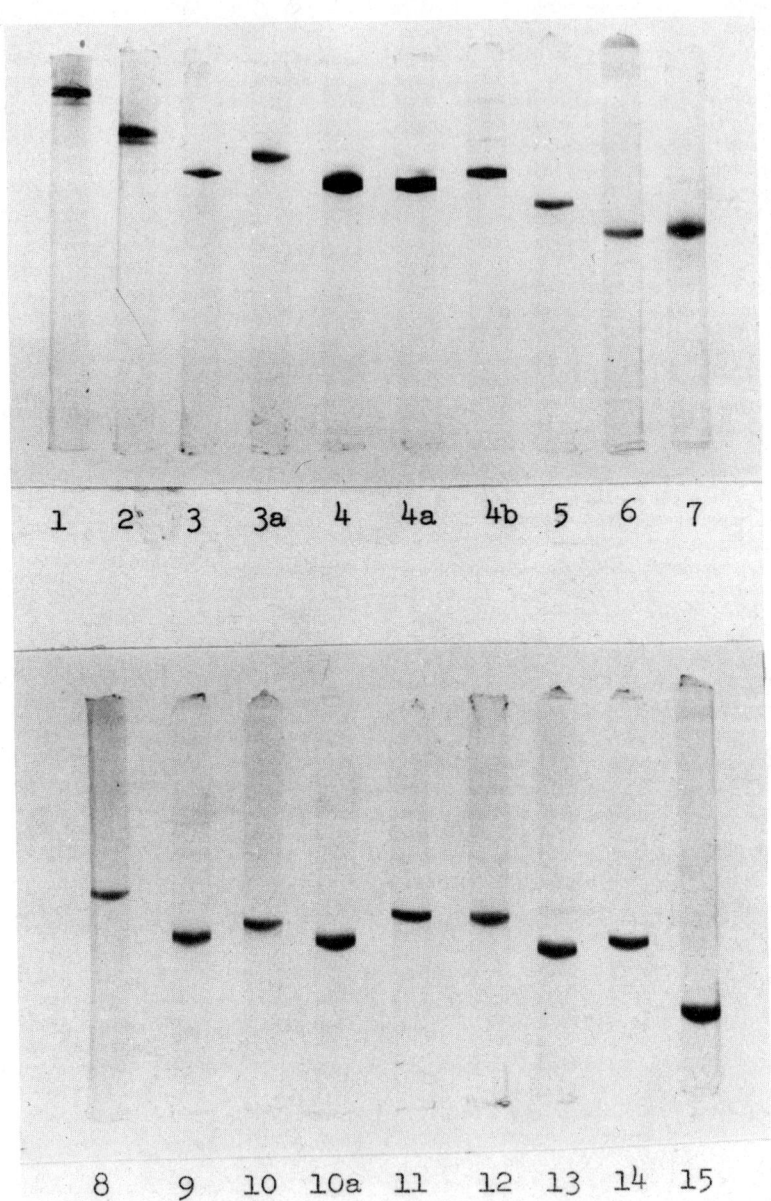

FIGURE 2. Polyacrylamide gel electrophoresis of purified 30 S ribosomal proteins at pH 4.5. Methods are described in a previous paper (Traub and Nomura, 1968c).

RIBOSOMAL COMPONENTS

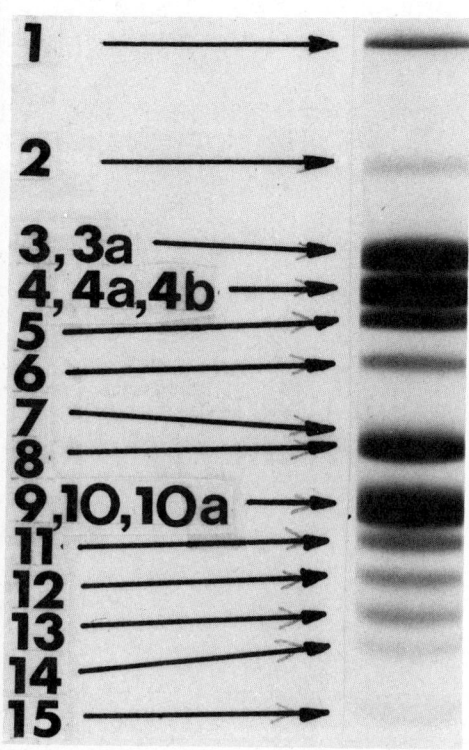

FIGURE 3. Polyacrylamide gel electrophoresis of total 30 S ribosomal proteins. Conditions were the same as in Fig. 2.

which identified P10 as the protein controlled by the *str* locus. More detailed experimental data are published elsewhere (Ozaki et al., 1969).

We then examined the functional role of P10. Reconstitutions were performed using 16 S RNA and a mixture of fractionated proteins from which P10 was omitted. The recovered particles were indistinguishable from control 30 S ribosomes in sucrose gradient sedimentation analysis; thus we could conclude that P10 is not essential for the assembly of 16 S RNA and other ribosomal proteins into a compact particle. The P10-deficient particles were found to be active in poly U-directed polyphenylalanine synthesis at 0.015 to 0.02 M Mg^{++} concentration, but only weakly active in polypeptide synthesis directed by natural mRNA. P10 was also found to be important in influencing the frequency of translational errors (Ozaki et al., 1969). These results will be discussed below in more detail together with the results of similar functional analyses of the other proteins.

FUNCTIONAL ANALYSIS OF 30 S RIBOSOMAL PROTEINS

As mentioned before, the reconstitution system has opened the way to the functional analysis of 16 S RNA and each of the individual 30 S ribosomal proteins. Several different approaches are possible. One which we have taken is to perform the reconstitution with one component omitted, and then to determine whether physically intact 30 S particles are formed and, if so, whether they are functionally active. In principle, we could perform this analysis using the appropriate mixture of highly purified proteins. However, in order to conserve our purified protein preparations and to improve the efficiency of the reconstitution, we have used four different protein fractions (mix A, B, C, and D; see Table 4) in combination with several purified proteins. The experimental design is described in Table 4. Reconstitutions were performed using 16 S RNA and a protein mixture with one component omitted at a time. The resulting particles were compared with those from the control reconstitution using the complete protein mixture. The reconstituted particles were recovered by high speed centrifugation sufficient to sediment the added 16 S RNA. The sedimented

TABLE 2. RECONSTITUTION OF 30 S RIBOSOMAL PARTICLES FROM 16 S RNA AND PURIFIED PROTEIN COMPONENTS

Proteins	Recovery (%)	Polypeptide synthesis		Poly U-Phe-tRNA binding
		Poly U Phe	f2 RNA Val	
Unfractionated proteins	75	100	100	100
19 purified proteins	63	42	61	51

Reconstitution was done according to the standard method. The recovery of RNA in the isolated particle preparation was calculated. The reconstituted particles were assayed for their activities in the following reactions: (a) poly U-directed incorporation of phenylalanine into protein, (b) valine incorporation into protein directed by RNA from phage f2, and (c) poly U-directed Phe-tRNA binding (for methods, see Traub and Nomura, 1968a). The activity of the particles reconstituted with unfractionated proteins is taken as 100% in each case.

TABLE 3. IDENTIFICATION OF P10 AS THE PROTEIN CONTROLLED BY THE *str* LOCUS

Origin of proteins used for reconstitution		Inhibition of amino acid incorporation by Sm (%)		Relative degree of misreading of poly U message (%)	Binding of ^{3}H-DHSm (Relative values)
Σ Pi − P10	P10	f2 RNA ^{14}C-valine	poly U- ^{14}C-phenylalanine		
s	s	74	35	(100)	(100)
s	r	10	7	8	5
r	r	3	4	3	5
r	s	78	35	(100)	(100)
Control 30 S (s)		69	45	(100)	(100)
Control 30 S (r)		10	2	7	4

s, derived from Sm-sensitive cells; r, derived from Sm-resistant cells.
The control 30 S particles are undissociated ribosomes.
This is a summary of experimental data published in the paper by Ozaki et al. (1969).

particles were then dissolved in the standard buffer TMA I (Traub and Nomura, 1968a). Any insoluble aggregates which formed were removed by a low speed centrifugation, and the recovery of RNA in the final preparation was determined. The recoveries are expressed as per cent of the value obtained with the control reconstitution, which itself varied from 60 to 100% in the various experiments (Fig. 4). The sedimentation behavior of the recovered particles was then examined by sucrose gradient sedimentation analysis together with a standard ribosome preparation containing radioactive 30 S and 50 S subunits as a reference. The sedimentation coefficient of the major peak of the reconstituted particles was then roughly estimated. The results of this analysis are summarized in Fig. 4. The sedimentation patterns observed are of three general types. The first type is essentially the same as that of the reference 30 S ribosomes. In the second type the major peak sediments close to 30 S, but a little behind it (27–29 S). In the third type the major peak sediments at 20–25 S. If the omission of a given protein results in a particle of the first type, we say that it is not required for assembly (even though it may be required for function). A protein whose omission leads to the second type of particle population is said to be partially required for assembly. And finally, a protein whose omission leads to the third type is termed essential for assembly. Assembly, according to the criterion of sedimentation, presumably indicates the formation of a compact, completely folded particle.

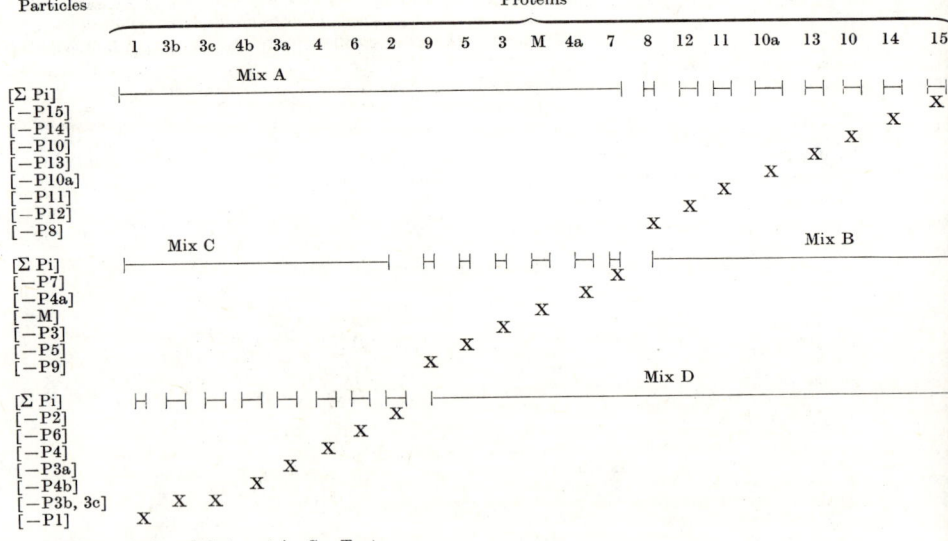

TABLE 4. PREPARATION OF VARIOUS SUBRIBOSOMAL PARTICLES

X signifies omission of that protein. See Text.

RIBOSOMAL COMPONENTS

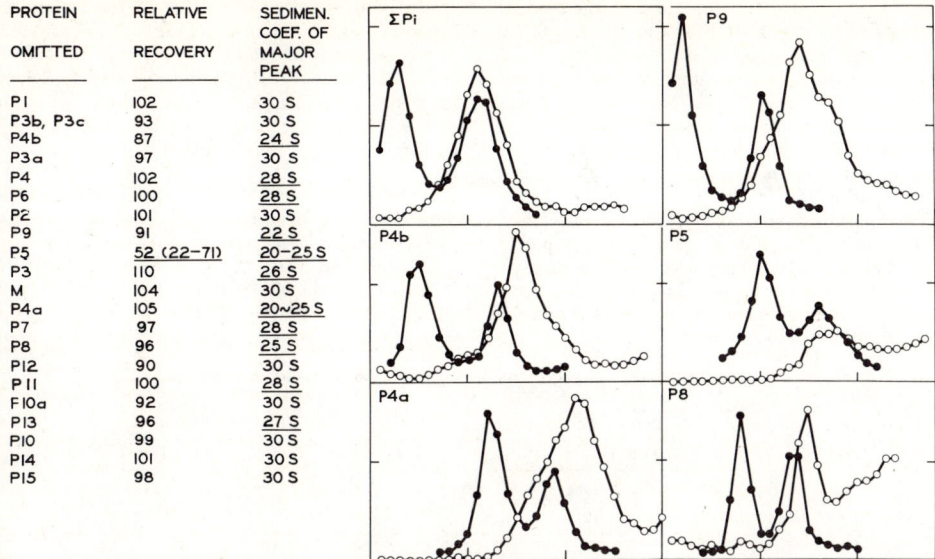

PROTEIN OMITTED	RELATIVE RECOVERY	SEDIMEN. COEF. OF MAJOR PEAK
P1	102	30 S
P3b, P3c	93	30 S
P4b	87	24 S
P3a	97	30 S
P4	102	28 S
P6	100	28 S
P2	101	30 S
P9	91	22 S
P5	52 (22–71)	20–25 S
P3	110	26 S
M	104	30 S
P4a	105	20~25 S
P7	97	28 S
P8	96	25 S
P12	90	30 S
P11	100	28 S
F10a	92	30 S
P13	96	27 S
P10	99	30 S
P14	101	30 S
P15	98	30 S

FIGURE 4. Protein requirements for the assembly of 30 S ribosomal particles. Reconstitution was done in the absence of one protein component. Relative recovery of the particles was calculated, and the average values from several experiments are shown. In the case of omission of P5, the numbers in parentheses indicate the range of recovery values obtained in 5 experiments. The recovered particles were analyzed by sucrose gradient sedimentation. Six patterns obtained with various protein-deficient particles are shown in the figure. ^{33}P-labeled marker 30 S and 50 S ribosomes were included (filled circles). The optical density of each fraction was measured to locate the position of the reconstituted particles (open circles). The ordinate shows radioactivity or optical density values and the abscissa, fraction number.

The proteins in the first group are P1, P2, P3a, P3b, P3c, P10, P10a, P12, P14 and P15. While only the [−P10] particle has been analyzed in detail (Ozaki et al., 1969), we expect that these particles contain all the proteins except the one omitted in each case. Polyacrylamide gel electrophoretic analysis of the proteins from [−P10] particles indicated that all the other proteins were present. As will be described below, [−P10] particles, while physically intact by all our available criteria, show a drastic decrease in some 30 S functions. The addition of P10 to the isolated particles under the conditions of reconstitution restored full activity (Ozaki et al., 1969). Thus the [−P10] particles contain all the functionally essential proteins except P10.

The second group is comprised of proteins P4, P6, P7, P11, P13, and, tentatively, P3, whose omission gave somewhat variable results. Further studies are needed to determine whether the absence of any of these proteins results in deficiency of additional proteins.

The third group includes P4a, P4b, P5, P8, and P9. A particularly drastic effect was observed with the omission of P5. The recovery of RNA was small and variable, and the recovered particles

sedimented at 20–25 S in a somewhat heterogeneous pattern. The peak observed in the sedimentation analysis of the other particles in the 20–25 S class was more well defined. We analyzed the protein composition of the recovered [−P4b] particle and found that at least three proteins beside P4b were missing. Hence the presence of P4b is essential for the binding of some other proteins. Although we have not yet analyzed the protein composition of the other particles in this group, it is reasonable to assume that they are also deficient in several proteins beside the one omitted in each case.

Next, we examined the activity of the protein-deficient reconstituted particles in several known 30 S functions. The following assays were performed: poly U-directed phenylalanine incorporation, poly U-directed incorporation of wrong amino acids (a mixture of isoleucine, tyrosine, and serine) in the presence of Sm, phage f2 RNA-directed incorporation of valine, poly U-directed Phe-tRNA binding, and AUG triplet-directed binding of F-met-tRNA. The last assay was done in two ways. In the first method, binding was measured at 0°C in the presence and absence of initiation factor F2 and the amount of stimulation by F2 was used as a measure of function. In the second, we measured binding at 30°C in the presence of F2 with and without F1, to determine stimulation by F1. By comparing these results with those obtained for ribosomes reconstituted with the complete protein complement, we hoped to identify any ribosomal proteins which specifically interact with these initiation factor proteins. Fractions F1 and F2 were purified according to the method of Thach (Hershey, Dewey, and Thach, 1969; Kolakofsky, Dewey, and Thach, 1969). Preparations with purity comparable to theirs were used in these assays. The results of these analyses obtained from several reconstitution experiments were averaged and are summarized in Table 5.

It should be noted that in these reconstitution experiments the proteins were added in an amount corresponding to about 1.5 equivalents of the RNA used. The use of batch-eluted protein fractions (Mix A, B, C, and D, see Fig. 4) and the use of an excess of proteins increases the possibility of contamination by the "omitted" protein. Hence, the observed activity of the protein-deficient particles described in Table 5 is probably a maximum value. In fact, the activities observed with [−P6] particles are generally somewhat higher than those observed with the corresponding particles prepared previously by partial reconstitution (Traub et al., 1967). P3, P4, P6, P8, and P11 correspond to B1, B2, B3, B4, and B5 in our

TABLE 5. FUNCTIONAL CAPACITY OF VARIOUS PROTEIN-DEFICIENT PARTICLES
(PER CENT RELATIVE ACTIVITY)

	Polypeptide synthesis			f2 RNA	tRNA binding		
	Poly U				Poly U-Phe-tRNA	AUG-F-met-tRNA	
Particles	Phe (1)	Ile, Tyr, Ser (+ Sm) (2)	Ratio (2)/(1)	Val		Δ with F2	Δ with F1
[Σ P_i]	100	100	1.0	100	100	100	100
[−P_1]	93	98	1.1	115	97	99	104
[−P_{3b}, P_{3c}]	90	108	1.2	83	81	50	85
[−P_{4b}]*	13	16	1.2	10	14	19	9
[−P_{3a}]	92	93	1.0	102	97	70	107
[−P_4]	50	61	1.2	50	67	37	60
[−P_6]	12	18	1.5	43	15	18	45
[−P_2]	47	90	1.9	131	78	66	90
[Σ P_i]	100	100	1.0	100	100	100	100
[−P_9]*	19	16	0.84	13	23	19	17
[−P_5]*	21	50	2.4	29	35	21	33
[−P_3]	48	48	1.0	47	33	31	57
[−M]	84	109	1.3	121	91	75	78
[−P_{4a}]*	23	38	1.7	38	37	17	23
[−P_7]	24	155	[6.5]	22	33	25	25
[Σ P_i]	100	100	1.0	100	100	100	100
[−P_8]*	29	32	1.1	22	45	30	30
[−P_{12}]	43	72	1.7	43	54	66	57
[−P_{11}]	7	15	2.1	3	12	12	19
[−P_{13}]	27	33	1.2	24	33	24	65
[−P_{10a}]	77	103	1.3	63	101	35	71
[−P_{10}]	43	19	[0.4]	24	96	8	3
[−P_{14}]	60	111	1.9	76	96	22	21
[−P_{15}]	48	83	1.7	66	70	17	83

Both poly U-directed phenylalanine incorporation and phage f2 RNA-directed valine incorporation were assayed at 0.01 M Mg^{++} in the presence of 50 S ribosomes. Poly U-directed incorporation of wrong amino acids (a mixture of isoleucine, tyrosine, and serine) as well as Poly U-directed Phe-tRNA binding was assayed at 0.02 M Mg^{++} in the presence of 50 S ribosomes. AUG-directed F-met-tRNA binding was assayed at 0.01 M Mg^{++} in the *absence* of 50 S ribosomes as described in the text.

The values are averages of results obtained from two to four reconstitution experiments. The values which are regarded as being significantly below the control values are underlined. The uniquely altered miscoding ratios "(2)/(1)" observed with [−P7] and [−P10] particles are boxed. The starred proteins appear to be essential for the assembly reaction.

original split protein nomenclature (see Table 7). While the present results are generally consistent with the previous results for partial reconstitution, one serious discrepancy from the previous data is obtained with [−P8] particles.

The conclusion obtained in this study that omission of P8 (called B4 in our previous paper) affects all the 30 S functions in a drastic way, does not agree with our previous results obtained with partial reconstitution using the 23 S core particles and other split proteins. Two possible explanations can be made: (1) P8 contaminated the 23 S core fraction and hence the effect of its omission from the split protein complement was not apparent. (2) In the previous experiments, the reconstitution mixture was dialyzed against a buffer with low salt concentration and the particles are then recovered by centrifugation. It is known that, under such conditions, there is nonspecific adsorption of ribosomal proteins to the particles. It is conceivable that P8 is essential only for the assembly and the [−P8] particles recovered directly from the buffer with high salt concentration in the present study are deficient in some other essential proteins, and hence are inactive, whereas the [−P8] particles obtained in the previous study contained all other essential proteins adsorbed and were able to achieve an active configuration under the conditions of the functional assays.

There are several conclusions one can draw from our results: (a) most of the proteins, except P1, P3a, and fraction M, showed some effect. That is, omission of these proteins, one at a time, caused at least a significant decrease in some functional activities. Omission of both P3b and P3c together caused a significant reduction only in the F-met-tRNA binding stimulated by the initiation factor F2. These proteins are acidic proteins obtained in small amounts and are not present in the 30 S ribosomal protein list obtained by Kurland's group (Hardy et al., 1969). It is not clear whether they are genuine ribosomal components. On the other hand, both P1 and P3a are definite ribosome-associated proteins and have been detected and isolated by other groups. Two possibilities can be considered: (1) these proteins are genuine ribosomal proteins but have functions which cannot be detected by the present activity assays used. (2) These proteins are not genuine ribosomal proteins but are proteins tightly bound to the 30 S ribosomal particles, as is RNase I. This latter possibility is more likely with the protein P1, since according to Kurland (Kurland et al., this volume), there is only 0.1 to 0.14 molecules of P1 per 30 S particle. In any event, out of 19 purified proteins (without counting P 3b and P 3c), only two proteins have failed to show significant functional requirements. In all likelihood, the remaining 17 represent genuine ribosomal protein components.

(b) The proteins (P4a, P4b, P5, P8, and P9) which are essential for physical assembly, were also shown to be essential in these functional tests. As described above, particles deficient in these proteins sediment mainly at about 20–25 S rather than at 30 S. It is not yet known if the residual activity observed in most of the assays (10 to 30 % of the complete system) is associated with a small amount of fully active 30 S particles formed due to contamination by the "omitted" protein in other fractions or whether it is associated with the deficient particles themselves. The fact that almost all the functions are drastically affected by omission of any one of these proteins (except the misreading of poly U by [−P5] particles, see below) is consistent with the conclusion that each is required for the assembly reaction itself.

(c) Some proteins (P2, P10, P10a, P12, P14, and P15) apparently not required for assembly, are still required for some or all 30 S functions. One of the primary objectives of our approach has been to identify proteins essential for function by correlating their omission with loss or alteration of function. In order to make a strong inference from such a correlation it must be shown that the effect of the omission is direct. Normal sedimentation behavior does not guarantee that all the proteins initially added are present in the final particle, and it is possible that the omission of a protein which is not essential *per se* could prevent the binding of a protein which is. In order to rule out this possibility it is necessary to show that all the proteins beside the one being omitted are present in the deficient particle. This can be done in two ways, either by direct analysis of the protein composition as revealed in polyacrylamide gel electrophoresis, or by showing that function can be restored by addition of the protein to the isolated deficient particle under reconstitution conditions. While neither of these tests has yet been applied to most of the particles, they have been done for the [−P4b] and [−P10] particles. In the former case it was already clear that P4b was essential for assembly, and both tests confirm that other proteins are missing; hence, no conclusion about whether it has a specific functional role can be drawn. In the case of P10, however, we can conclude that this protein is itself essential for certain 30 S functions. As mentioned above, the observed decrease in the activity of [−P10] particles can be restored by subsequent addition of P10, and this together with the electrophoretic analysis shows that P10 is the only protein missing from the [−P10] particle. Hence, the function of P10 can be defined more clearly.

TABLE 6. RELATIVE DEGREE OF MISREADING OF POLY U MESSAGE INDUCED BY VARIOUS AGENTS

Drugs Particles	(1) Phe (10^{-2} M Mg) (%)	(2) Phe (2×10^{-2} M Mg) (%)	(3) Ile, Tyr, Ser (2×10^{-2} M Mg) (%)				Ratio (3)/(2)			
	—	—	—	Sm	Nm	EtOH	—	Sm	Nm	EtOH
[Σ P_i]	100	100	100	100	100	100	1.0	1.0	1.0	1.0
[−P_{10}]	48	75	42	18	17	19	0.56	0.24	0.23	0.25
[Σ P_i]	100	100	100	100	100	100	1.0	1.0	1.0	1.0
[−P_7]	29	52	115	184	164	347	2.2	3.5	3.2	6.7

Incorporation of a mixture of isoleucine, tyrosine, and serine directed by poly U was measured in the presence of streptomycin (Sm), neomycin B (Nm), and ethyl alcohol (EtOH), respectively, or in their absence, and compared with the incorporation of phenylalanine directed by poly U. The data on [−P10] particles are taken from the paper by Ozaki et al. (1969). Experimental procedures are described in this paper.

The properties of P10-deficient particles have been studied in detail and the results published elsewhere (Ozaki et al., 1969). In the presence of 0.015 M or higher Mg^{++} concentration, [−P10] particles are 75 to 90% as active as the control [Σ Pi] particles in both amino acid incorporation and tRNA binding directed by synthetic mRNA. However, they are very weakly active in the initiation function, as judged by f2 RNA-directed valine incorporation, F-met-tRNA binding directed by triplet AUG or f2 RNA, and F-met-puromycin formation directed by AUG. Another unique feature of [−P10] particles is reduced misreading of poly U in the presence of Sm. The ratio of Sm-induced misreading of poly U at 0.02 M Mg^{++} to the normal reading at 0.01 M Mg^{++} decreased by a factor of 2.5 (Table 5), and the ratio of the misreading to normal reading at 0.02 M Mg^{++} decreased by a factor of 4 (Table 6). The data in Table 5 shows that the [−P10] particle is the only one which showed any significant reduction in the degree of misreading. On the other hand, [−P7] particles showed a drastic increase in misreading of the poly U message, even though normal reading of poly U

TABLE 7. NOMENCLATURES OF 30 S RIBOSOMAL PROTEINS AND TENTATIVE CONCLUSION OF FUNCTIONAL REQUIREMENTS

Our code	Kurland's code	Requirement for		Note
		assembly	function	
1 (A_1)	1	—	—	
2 (A_2)	4a	—	+	
3 (B_1)	9	±	++	
3a	2	—	—	
4 (B_2)	3	±	+	Spc sensitivity
4a	10	+	(++)	
4b	2a	+	(++)	
5	8	+	(++)	K character
6 (B_3)	4	±	++	
7	11	±	++	Fidelity of translation
8 (B_4)	12	+	(++)	
9	6 (and 7?)	+	(++)	
10	15	—	++	Sm sensitivity; Ambiguity of translation
10a	14	—	+	
11 (B_5)	12b	±	++	
12	12a	—	+	
13	13	±	++	
14	16	—	+	
15	15a	—	+	
3b		—	(±)	
3c		—	(±)	
	5			
	7			

Symbols for assembly requirement: +, omission of the protein produces particles sedimenting at 20–25 S; ±, omission of the protein produces a 27–28 S particle; −, the protein is not required for formation of the 30 S particle.
Symbols for functional requirement: (++), strong requirement presumably because of requirement in the assembly reaction; ++, strong requirement; +, partial requirement; −, no requirement demonstrated; (±), weak effect on the initiation reaction.
For details see the text.

was significantly reduced (Table 5 and 6). We conclude that two proteins, P10 and P7, are directly involved in the Sm-induced ambiguity and apparently act in opposite ways.

Table 7 summarizes the results discussed so far of the omission of the separated 30 S proteins.

P10 AND P7 PROTEINS AND THE FIDELITY OF TRANSLATION

Previous workers have demonstrated that ribosomes from Sm-resistant strains show a reduction of Sm-induced ambiguity. However, they still misread in the presence of other agents, such as other aminoglycoside antibiotics, neomycin C, ethyl alcohol, or Mg^{++} at high concentration (Davies et al., 1964). In view of the specificity exhibited by Sm-resistant ribosomes, it might also be expected that [−P10] particles would not misread in the presence of Sm but still show the normal ambiguity induced by other agents. So we tested misreading of poly U by [−P10] particles in the presence of several known ambiguity-inducing agents: kanamycin A, neomycin B, paromomycin, ethyl alcohol, and high concentration of Mg^{++}. Contrary to our initial expectation, the experimental results showed that [−P10] particles have lost both the Sm-induced misreading property and also that induced by the other agents.

A similar situation in reverse was observed with respect to protein P7. As described in the previous section, [−P7] particles showed an increase in the Sm-induced misreading of poly U. Misreading was also tested in the presence of neomycin C and ethyl alcohol at 0.02 M Mg^{++}. As shown in Table 6, [−P7] particles showed high misreading in every case. Even in the absence of the drugs, relative misreading at 0.02 M Mg^{++} increased by a factor of two over that by control [Σ Pi] particles. Although the altered properties of [−P10] particles are clearly due to the lack of P10 since "normal" properties can be restored by the addition of P10, similar experiments have not yet been performed with P7. The [−P7] particles sediment slightly more slowly (at about 28 S) than the normal 30 S particles, and might be deficient in other proteins, beside P7. However, the fact that the omission of no other single protein leads to particles of similar fidelity properties implies that P7 is directly involved in determining such properties. In any event, the results suggest that both P10 and P7 proteins are unique proteins, having a function of influencing error frequency in translation. Since both proteins are also required for some other ribosomal functions, they are normal and essential ribosomal components. The data support the concept that translational ambiguity is an inherent property of bacterial ribosomes. Moreover, two proteins act apparently in opposite ways; P10 increases ambiguity and P7 decreases it. Mutational alterations in these proteins may give various additional specificities in the function of these proteins in the translational fidelity. Thus it is conceivable that somewhat sophisticated and subtle control over translational ambiguity takes place genetically as well as physiologically through these two proteins. In this connection, it is interesting to note the recent work done by Rosset and Gorini (1969). They have isolated mutants of *E. coli* (*ram* mutants), which have simultaneously acquired both increased ability to suppress mutations in vivo and altered ribosomes with higher translational error in vitro. The component altered by the *ram* mutation is shown to be in the 30 S subunit. Moreover, they showed that effects caused by mutation at the *ram* locus are antagonized by additional mutations at the *str* locus. Thus the relationship between the *str* locus and the *ram* locus closely parallels the relationship between P10 and P7 observed in our work. Since P10 is the protein controlled by the *str* locus, it is tempting to predict that the protein controlled by the *ram* locus is P7.

FUNCTIONAL ANALYSIS OF 16 S RIBOSOMAL RNA

The functional role of the RNA in the ribosome has remained unclear. With the 30 S ribosome reconstitution system now available, we are in a position to investigate this problem. Although we have done only a few experiments so far, and much future study is needed, we shall summarize the available information here (Table 8).

We first studied RNA specificity in the reconstitution of functionally active 30 S ribosomal particles. We confirmed that RNA must be present for the assembly reaction to occur, since in the absence of 16 S RNA, no soluble particles resembling the ribosome are formed. Moreover, 16 S ribosomal RNA from yeast or 18 S ribosomal RNA from rat liver cannot replace *E. coli* 16 S RNA in the formation of particles sedimenting at 30 S. Only very heterogeneous particles or insoluble aggregates are formed under these conditions (Traub and Nomura, 1968a).

It has been found, however, that some 16 S RNAs from distantly related *bacterial* species, such as *Azotobacter vinelandii* or *Bacillus stearothermophilus*, can replace *E. coli* 16 S RNA and form functionally active hybrid 30 S particles with *E. coli* 30 S proteins (Nomura, Traub, and Bechmann, 1968). Reverse combinations, that is, 30 S ribosomal proteins from *A. vinelandii* or *B. stearothermophilus* and 16 S RNA from *E. coli* were also examined and formation of active hybrid 30 S particles was again demonstrated. Although 16 S RNAs from these three different bacterial species have some portions of

TABLE 8. RECONSTITUTION OF 30 S RIBOSOMAL PARTICLES FROM *E. coli* 30 S PROTEINS AND VARIOUS RNAs

RNA	Recovery of RNA-containing particles	Activity of recovered particles
E. coli 16 S RNA	++	++
E. coli 16 S RNA degraded (ca. 6 S)	+	−
E. coli 23 S RNA degraded (ca. 16 S)	+	−
Yeast 16 S RNA	+	−
Rat liver 16 S RNA	−	(−)
A. vinelandii 16 S RNA	++	++
B. stearothermophilus 16 S RNA	++	++
E. coli 16 S RNA treated with HNO_2 (ca. 10 deaminations)	++	−
RNA from HNO_2-treated 30 S ribosomes (1–4% surviving activity)	++	++

++, good (60 to 100%) recovery or good (80 to 100%) activity per unit amount of recovered particles; −, no recovery or no activity; +, partial recovery. For details, see the text.

their base sequences in common, large portions are different. Thus the requirement for a specific base sequence in ribosomal RNA is not absolute. We tentatively concluded that only certain small regions of 16 S RNA are directly involved in the specific interaction with the ribosomal proteins in the assembly of ribosomal particles and this region represents "conserved" regions, that is, regions having base sequences which are common or very similar among different bacterial species.

To examine further the functional significance of the base sequence in the ribosomal RNA, we have initiated studies on effects of chemical modification of 16 S ribosomal RNA on reconstitution. We have found that the activity of isolated *E. coli* 16 S RNA is very sensitive to nitrous acid treatment (Nomura et al., 1968). Chemical analysis of base alterations combined with an assay of the biological activity of the treated RNA, showed that only a few (6 to 8) base alterations are sufficient to destroy the reconstitution activity of one 16 S RNA molecule. The inactive particles reconstituted with nitrous acid-treated RNA sedimented more slowly than normal 30 S particles. These and other results show that nitrous acid destroys the ability of the RNA to be folded into the normal compact particle. Since the inactivation could be explained merely on the basis of defective physical assembly, the nitrous acid data provide no direct information about RNA function.

Thus the only positively identified function of 16 S RNA is its role in the assembly of the ribosomal particle. On the other hand, it is quite conceivable that some parts of 16 S RNA are involved in the binding of mRNA and/or transfer RNA. Moore (1966) examined the effects of treatment of the ribosome with several chemical reagents, including nitrous acid, on the ribosome function, specifically on its ability to bind mRNA, and suggested that amino groups of bases in the ribosomal RNA are involved in the mRNA binding function of the ribosome. However, these experiments are not decisive, and similar experiments should be repeated with the 30 S ribosome using the reconstitution technique to prove that the crucial functional group is really the amino group of RNA and not that of proteins. In preliminary experiments (Bowman and Nomura, unpubl., cf. Table 8) we have done nitrous acid treatment of the 30 S ribosomal particles, and found that the inactivation of the 30 S particles is *not* due to inactivation of RNA, but due to inactivation of proteins. We feel that direct involvement of ribosomal RNA in any of the known ribosomal functions has yet to be demonstrated experimentally.

CONCLUDING REMARKS

Although our studies on the functional role of each of 30 S ribosomal components are still in the preliminary stage, we have already obtained some useful information, especially with respect to the role of protein components. Most of the proteins isolated can be designated genuine ribosomal proteins, though a few others must wait until it can be clearly demonstrated whether or not they are required for function. We have found that some proteins are essential for the assembly of a compact functional particle. Some proteins are important for several or all of the functions assayed, but are apparently not required for assembly. These results are consistent with the concept that the ribosomes are a highly organized multicomponent functional structure and both their assembly process and their functional activity are highly cooperative in nature. We have also found two proteins, P10 and P7, which are important in influencing fidelity of the translation. Further systematic studies using this reconstitution system, coupled with the techniques of protein and nucleic acid chemistry, should soon lead to a comprehensive characterization of all the molecular components of the 30 S ribosomal subunit, and to much needed information on their relation to the three-dimensional ribosomal structure. Finally, it is our hope that knowledge and experience obtained in the reconstitution of 30 S ribosomal subunit will be helpful in achieving the next goal, the total reconstitution of the 50 S ribosomal subunit.

ACKNOWLEDGMENTS

We thank Dr. M. Susman for his critical reading of the manuscript; Miss D. Becker for her excellent technical assistance. This investigation was supported by United States Public Health Research grant (GM-15422) from the National Center for Urban and Industrial Health, and by National Science Foundation grant (GB-6594). Operation of the Biochemistry Department's Pilot Plant is supported by U.S. Public Health Research grant FR-00226. This is paper No. 1319 from the Laboratory of Genetics, University of Wisconsin, Madison, Wisconsin.

REFERENCES

BOLLEN, A., J. DAVIES, M. OZAKI, and S. MIZUSHIMA. 1969. Ribosomal protein conferring sensitivity to the antibiotic spectinomycin in *Escherichia coli*. Science 165: 85.

COX, E. C., J. R. WHITE, and J. G. FLAKS. 1964. Streptomycin action and the ribosome. Proc. Nat. Acad. Sci. 51: 703.

DAVIES, J. 1964. Studies on the ribosomes of streptomycin-sensitive and resistant strains of *Escherichia coli*. Proc. Nat. Acad. Sci. 51: 659.

———. 1966. Streptomycin and the genetic code. Cold Spring Harbor Symposium Quant. Biol. 31: 665.

DAVIES, J., W. GILBERT, and L. GORINI. 1964. Streptomycin, suppression, and the code. Proc. Nat. Acad. Sci. 51: 883.

FLAKS, J. G., E. C. COX, M. L. WITTING, and J. R. WHITE. 1962. Polypeptide synthesis with ribosomes from streptomycin-resistant and dependent *E. coli*. Biochem. Biophys. Res. Commun. 7: 390.

FOGEL, S., and P. S. SYPHERD. 1968a. Chemical basis for heterogeneity of ribosomal proteins. Proc. Nat. Acad. Sci. 59: 1329.

———, ———. 1968b. Extraction and isolation of individual ribosomal proteins from *Escherichia coli*. J. Bacteriol. 96: 358.

GORINI, L., and E. KATAJA. 1964. Phenotypic repair by streptomycin of defective genotypes in *E. coli*. Proc. Nat. Acad. Sci. 51: 487.

HARDY, S. J. S., C. G. KURLAND, P. VOYNOW, and G. MORA. 1969. The ribosomal proteins of *Escherichia coli*. I. Purification of the 30 S ribosomal proteins. Biochemistry 8: 2897.

HERSHEY, J. W. B., K. F. DEWEY, and R. E. THACH. 1969. Purification and properties of initiation factor f-1. Nature 222: 944.

KALTSCHMIDT, E., M. DZIONARA, D. DONNER, and H. G. WITTMANN. 1967. Ribosomal proteins. I. Isolation, amino acid composition, molecular weights and peptide mapping of proteins from *E. coli* ribosomes. Mol. Gen. Genet. 100: 364.

KOLAKOFSKY, D., K. F. DEWEY, and R. E. THACH. 1969. Purification and properties of initiation factor f-2. Nature 223: 694.

LOWRY, O. H., N. J. ROSEBROUGH, A. L. FARR, and R. J. RANDALL. 1951. Protein measurement with the Folin phenol reagent. J. Biol. Chem. 193: 265.

LUZZATTO, L., D. APIRION, and D. SCHLESSINGER. 1968. Mechanism of action of streptomycin in *E. coli*: Interruption of the ribosome cycle at the initiation of protein synthesis. Proc. Nat. Acad. Sci. 60: 873.

MOORE, P. B. 1966. Studies on the mechanism of messenger ribonucleic acid attachment to ribosomes. J. Mol. Biol. 22: 145.

NOMURA, M., P. TRAUB, and H. BECHMANN. 1968. Hybrid 30 S ribosomal particles reconstituted from components of different bacterial origins. Nature 219: 793.

OZAKI, M., S. MIZUSHIMA, and M. NOMURA. 1969. Identification and functional characterization of the protein controlled by the streptomycin-resistant locus in *Escherichia coli*. Nature 222: 333.

ROSSET, R., and L. GORINI. 1969. A ribosomal ambiguity mutation. J. Mol. Biol. 39: 95.

SPEYER, J. F., P. LENGYEL, and C. BASILIO. 1962. Ribosomal localization of streptomycin sensitivity. Proc. Nat. Acad. Sci. 48: 684.

TRAUB, P., and M. NOMURA. 1968a. Structure and function of *E. coli* ribosomes. V. Reconstitution of functionally active 30 S ribosomal particles from RNA and protein. Proc. Nat. Acad. Sci. 59: 777.

———, ———. 1968b. Streptomycin resistance mutation in *Escherichia coli*: Altered ribosomal proteins. Science 160: 198.

———, ———. 1968c. Structure and function of *Escherichia coli* ribosomes. I. Partial fractionation of the functionally active ribosomal proteins and reconstitution of artificial subribosomal particles. J. Mol. Biol. 34: 575.

TRAUB, P., K. HOSOKAWA, G. R. CRAVEN, and M. NOMURA. 1967. Structure and function of *E. coli* ribosomes. IV. Isolation and characterization of functionally active ribosomal proteins. Proc. Nat. Acad. Sci. 58: 2430.

TRAUT, R., P. B. MOORE, H. DELIUS, H. NOLLER, and A. TISSIÈRES. 1967. Ribosomal proteins of *Escherichia coli* I. Demonstration of different primary structures. Proc. Nat. Acad. Sci. 57: 1294.

SECTION IV

Regulation of Nucleic Acid and Protein Synthesis

Our understanding of the mechanism regulating nucleic acid and protein synthesis comes chiefly from studies on bacteria and bacteriophage. This is primarily due to the fact that with these microorganisms it has been possible to combine the techniques of biochemistry and genetics, demonstrating the importance of enzymatic steps especially when alternate pathways can be taken.

In this section regulation in bacteria will be discussed first.

Regulation of DNA Synthesis in Bacteria

The duplication of DNA in the chromosome is followed by segregation of new and old DNA as if the chromosome contained a single, long double helix which replicates semi-conservatively. Thus, each newly synthesized chromatid contains one parental DNA chain and one daughter DNA chain. The actual replication process is initiated at more than one point along the chromosomes of higher forms but in *E. coli*, which contains one circular chromosome, there is normally only one initiation point for replication. Maaløe and Hanawalt first proposed the idea that replication of the bacterial chromosome originates at a specific site and proceeds sequentially to the end of the chromosome (*J. Mol. Biol.* 1961, 3, 144). This concept was developed further in the *replicon* hypothesis of Jacob, Brenner and Cuzin (*Cold Spring Harb. Symp. Quant. Biol.* 1963, 28, 329). The replicon hypothesis states that initiation results from the interaction of an initiator substance with a unique region of the DNA. The initiator is assumed to be a diffusible gene product. The hypothesis also suggests that in bacteria the replicating unit is attached to the cell membrane in a way that assures segregation of the daughter chromosomes containing one old and one newly synthesized strand, into the two daughter cells. Worcel *et al.*, (*Cold Spring Harbor Symp. Quant. Biol.* 1973, 38 in the press) have recently shown that intact, folded chromosomes extracted from *E. coli* spheroplasts are attached to membrane fragments.

Helmstetter and Cooper (*J. Mol. Biol.* 1968, 31, 519) have helped to clarify the relationship between the initiation and termination of DNA synthesis as a function of rate of growth and cell division by varying the richness of the growth media. Exponentially growing cells were pulse-labeled with ^{14}C-thymidine, and the amount of label incorporated into cells of different ages was determined by sorting out the labeled cells according to age. This was accomplished by binding the cells to a membrane filter at the end of the labeling period and collecting the newborn cells which were eluted continuously from the membrane. These experiments have led to a general model for comparing DNA growth rates and cell division rates (*paper 56*). The cell age at which DNA replication begins is variable and depends upon the growth rate; under adequate growth conditions, the time for a complete round of replication is constant and independent of growth rate. In very rapidly growing *E. coli* a new round of replication begins before the previous round has ended. Reinitiation begins at the same point on the chromosome, and DNA synthesis is symmetrical, occuring at the origin of *both* branches of the replicating chromosome (Fritsch and Worcel, *J. Mol. Biol.* 1971, 59, 207; O'Sullivan and Sueoka, *J. Mol. Biol.* 1972, 69, 237). If DNA synthesis is first inhibited and then allowed to resume, new rounds of DNA synthesis are initiated; however, this abnormal type of initiation leads to asymmetric DNA synthesis (Schwartz and Worcel, *J. Mol. Biol.* 1971, 61, 329).

These results indicate that the rate of initiation (but not the rate of propagation) is particularly sensitive to growth conditions. This is true under normal conditions where the principal means of controlling the rate of DNA synthesis is by regulating the rate of chromosome initiation, a conclusion which is consistent with the replicon hypothesis. An *in vivo* and *in vitro* approach to the study of the regulation of DNA initiation has been undertaken using temperature sensitive bacterial mutants that cannot reinitiate DNA synthesis at the elevated, non-permissive temperature. Since the replication of the DNA of the single-stranded DNA phages both *in vivo* and *in vitro* requires some of the same protein(s) as are involved in the initiation of replication of *E. coli* DNA (from *ts* DNA mutant studies), the temperature sensitive initiation mutants and the replication of single stranded DNA phages have proven to be very useful in studies on the indentification of proteins involved in initiation. The assay consists of having the defective protein in the mutant extracts complemented by purified fractions from wild type cells so that single-stranded phage DNA will be synthesized successfully *in vitro* (Schekman *et al.*, Proc. Nat. Acad. Sci. U. S. 1972, **69**, 2691).

There is another kind of DNA synthesis involving preferential chromosome synthesis which has thus far only been observed in eukaryotes (e.g. Dawid, Brown and Reeder, *J. Mol. Biol.* 1970, **51**, 341; Malva *et al.*, *Nature New Biology*, 1972, **239**, 135). Parts of chromosome encoding ribosomal genes become replicated, amplifying the pool of DNA for rRNA transcription, but apparently not functioning as a genetic part of the DNA. This synthesis takes place in close association with the nucleolus which contains many copies of the ribosomal genes. It now appears (Hourcade, *Cold Spring Harbor Symp. Quant. Biol.* 1973, **38**, in the press) that the amplification of ribosomal RNA cistrons proceeds via a rolling circle intermediate.

Regulation of Transcription in Bacteria

As in DNA-directed DNA synthesis or in DNA-directed RNA synthesis it appears that the rate of initiation is the main control for the rate of synthesis. In the latter case there is much more evidence to support this notion. Moreover, in bacteria such as *E. coli*, the rate of protein synthesis may be controlled by the rate of RNA synthesis since the lifetime of most messenger RNAs is very short. Thus, if the synthesis of a particular messenger is halted, the corresponding protein(s) synthesis will stop within a few minutes.

With respect to the regulation of its synthesis, mRNAs can be divided into two categories according to whether they are synthesized constitutively or adaptively. Constitutive systems include those mRNAs whose synthesis does not vary more than a factor of 2 or so with growth conditions. Constitutive systems are believed to have no special control systems so that the rate of initiation is controlled mainly by the interaction of the σ-core RNA polymerase at or close to the promotor. It seems likely that constitutive genes are those which must be expressed all or most of the time regardless of the precise conditions; these would be expected to include genes for the synthesis of the nucleotide coenzymes NAD and FAD since these factors are required whether catabolism or anabolism is predominating. Not enough is known about any genes to classify them as constitutive as defined here. Implicit in the concept of constitutive genes is the notion that a considerable expenditure of energy and evolutionary development is necessary to create and maintain a gene regulatory mechanism. If the need for certain gene products does not vary appreciably, it would probably be more economical to synthesize a constant limited excess of the gene product rather than to create and maintain a sophisticated control system. Related to this notion is the fact that controlling the rate of protein synthesis only provides a *coarse* control for *enzyme activity*. *Fine control* of *enzyme activity* usually results from the modifying influence of small molecule modulators acting directly on the enzymes to increase or decrease their activity.

Enzymes whose activity varies greatly, from several-fold to a thousand-fold or more depending upon growth conditions, are called *adaptive* or *inducible*. It is obvious that special mechanisms for controlling the rates of initiation of RNA synthesis must exist for all adaptive genes. Most adaptive systems use at least one regulator protein specific for the gene system and one small molecule modulator, which either induces or represses gene activity. Inducibility is the rule for genes with a catabolic function, and repressibility is the rule for genes with an anabolic function. Inducible systems will be discussed first.

One of the most intensively studied and best understood inducible gene clusters is the lactose (or *lac*) operon (*papers 57–63*) which is involved in the breakdown of lactose to its component monosaccharides, galactose, and glucose (Reznifoff, *Ann. Rev. Genet.* 1972, 6, 133). The *lac* operon consists of three structural genes (which code for β-galactosidase, lactose permease, and transacetylase) as well as a promotor and an operator. The first protein hydrolyzes lactose to its constituent monosaccharides, the second concentrates lactose from outside the cell, and the third catalyzes the acetylation of β-galactosides, a reaction of unknown biologic value. The structural genes for these three proteins are adjacent to one another, and the controlling elements are located at one end of the gene cluster. This arrangement of controlling elements together with the structural genes is called an *operon*. Initiation of RNA synthesis occurs at or near the controlling element, referred to as the promoter locus (p), and continues to the end of the operon. In a mutant in which the p locus is deleted, the *lac* operon is not expressed unless the deletion is large enough for the *lac* operon to become attached to a neighboring promoter which is part of another operon (*paper 29*). Several factors are required for initiation of RNA synthesis at the *lac* promoter locus. In addition to the RNA polymerase and nucleotide triphosphates, 3′, 5′ cyclic AMP and a protein called the catabolite gene activator protein (CAP) are required. Mutants that are defective in cyclic AMP or CAP synthesis have been shown to be incapable of making not only *lac* operon enzymes but a wide variety of other inducible enzymes as well (*paper 61*). For this reason, the controlling mechanism for the *lac* promoter is believed to serve a large number of other genes as well. These genes are known as catabolite sensitive genes. Cyclic AMP and CAP, it is believed, form a complex with the p locus which is part of the initiation complex for RNA synthesis. Another condition for initiation of the *lac* promoter is that the other controlling element, referred to as the operator (o), is not occupied by repressor. Although not part of the operon, a regulatory protein (or repressor protein) coded for by the i gene (located adjacent to the *lac* operon) binds in a very specific manner (von Hippel and McGee, *Ann. Rev. Biochem.* 1972, 41, 231) with the o locus of the *lac* operon (*papers 57 and 58*). The dissociation constant for this complex is of the order of 10^{-12} moles liters^{-1} from which it can be calculated that little free operator is present at any given time in normal cells in the absence of induction. The binding of repressor to the o locus is reversed by certain small molecules. For instance, lactose, a substrate of β-galactosidase, when administered to growing cells leads to induction of *lac* mRNA and protein synthesis. Jacob and Monod (*J. Mol. Biol.* 1961, 3, 318) postulated that lactose (or, more probably, a chemically modified derivative of lactose, allolactose) exerts this effect by combining with the *lac* repressor protein, producing a structural alteration in the repressor favoring dissociation of the repressor-operator complex. Once the repressor is removed, RNA synthesis takes place if cAMP, CAP and RNA polymerase are available (*paper 62*). In saturating doses of inducer, *lac* mRNA and the corresponding proteins increase about a thousand-fold *in vivo*. Mutants with defective operators, called constitutive operator (or o^c) mutants, bind the repressor poorly and as a consequence produce elevated levels of *lac* mRNA even in the absence of inducer. Binding experiments with purified components have been made which verify the postulated mode of action of repressor and small molecule inducer. The *lac* repressor has been purified and shown to have a high affinity for DNA containing the *lac* operator as well as for the inducer of the *lac* operon. Sufficient concentration of inducer causes dissociation of the repressor-DNA complex. The base sequence of the operator region has now been determined by Gilbert and his associates (*Cold Spring Harbor Symp. Quant. Biol.* 1973, 38, in the press). The region is about 23 base pairs in length with symmetric base sequence regions at either end.

The arabinose (*ara*) operon, like the *lac* operon, is an inducible cluster of structural genes and regulatory elements concerned with a particular catabolic function, that is, the three step conversion of L-arabinose to D-xylulose 5-phosphate (*paper 64*). An unlinked permease gene (E) is involved in the active transport of L-arabinose from the external medium. The *ara* operon and the E gene are both induced when L-arabinose is present in the growth medium. Because of this, it seems likely that some, if not all, of the control factors from the two genetic sites are the same. The model for control of the *ara* operon, postulated by Englesberg and his coworkers, contains both a *positive* and a *negative* control site, as illustrated in *paper 64*. According to this proposal, the C gene encodes a specific regulator protein. In the absence of L-arabinose, this protein acts as a repressor, P_1, binding to the o locus. Thus, the o locus is a point of negative control like the *o* locus in the *lac* operon. P_1 is displaced from *o* by L-arabinose. L-arabinose stimulates the conversion of P_1 to an alternate conformation P_2. P_2 has a high affinity for the I locus on the DNA. The binding of P_2 to I leads to a high level of gene expression for the *ara* operon; the I locus is called a site for positive gene control. The *ara* operon, like the *lac* operon, requires cyclic AMP and the CAP protein (providing *positive* control) for gene expression. The promoter element is presumed to reside somewhere in the I region as does the site for binding of CAP.

Less is known about most other catabolite sensitive operons, although most of them require cyclic AMP and CAP for expression. Catabolite sensitive operons are either of the *positive control type* like the *ara* operon or of the *negative control type* like the *lac* operon. All catabolite sensitive operons show two types of control at the gene level. One of these control sites, the operator, is switched on by specific substrate induction; the other, in some way related to the promoter, is switched on by cyclic AMP. The cyclic AMP level is controlled by the concentration of catabolites; if the level is adequate, as when bacteria are fed on glucose or glucose related derivatives, the cyclic AMP concentration drops. Since the products of catabolic operon enzymes are themselves catabolites, the cyclic AMP control mechanism can be thought of as a type of general feedback inhibition.

We turn now to a consideration of repressible gene systems. The best examples of repressible genes which can be discussed are those that serve for the biosynthesis of amino acids (Umbarger, *Ann. Rev. Biochem.* 1967, 38, 323). The simplest metabolic pattern is that in which the amino acid causes repression of the genes for some or all of the enzymes involved in the particular amino acid synthesis. This is characteristic of *unifunctional biosynthetic pathways*. For example, the operon for three specific leucine forming enzymes in the bacterium *S. typhimurium* is repressed by leucine. Exactly the same result is achieved in the control of that pathway in the fungus *Neurospora crassa*, but by a different kind of pattern. In *N. crassa*, leucine represses the formation of the first enzyme in the pathway, whereas the product of the first enzyme, γ-isopropylmalate somehow induces the second and third enzymes. One might anticipate that a number of variations of such schemes will be found.

The operons for the unifunctional pathways leading to histidine or tryptophan synthesis contain a control site at one end of a cluster of structural genes (see *Metabolic Regulation*, Vol. 5, H. J. Vogel (ed.), New York: Academic Press, 1971). A promoter site to the left of the tryptophan *o* locus has been identified; no comparable promoter has yet been found for the histidine operon. In the presence of sufficient specific amino acid, added extracellularly, mRNA synthesis for these operons is inhibited. As a consequence, the biosynthesis of all the operon associated enzymes is coordinately inhibited. The order of genes in the operon and the order of the enzymes in the corresponding biochemical pathway is usually not the same. The way in which the amino acid leads to repression is clear only for the *trp* operon (Rose *et al.*, *Nature New Biology*, 1973, in the press); it has been suggested that in some cases an amino acyl-tRNA, rather than the amino acid itself, is the active co-repressor. The co-repressor, it is believed interacts with a specific protein repressor molecule encoded by an unlinked gene site and forms a strong complex at the operator site. The level of transcription can be increased in any of three ways: (1) by lowering the concentration of amino acid in the growth medium; (2) by mutation of the repressor gene to a defective state; (3) by mutation of the operator *o* locus to the o^c state. If any of these three conditions exists the operon is said to be in a derepressed state.

While the simple scheme shown above seems to adequately to explain the regulation of many amino acid related operons, the situation for the histidine operon is more complicated. Mutations in any of six loci can produce derepression of the histidine operon. Three of these loci have been assigned definite functions: *his* O is an operator, *his* S is the gene specifying the histidyl-tRNA synthetase, and *his* R is probably a gene for histidyl-tRNA. The other three genes, *his* U, *his* W, and *his* T, appear to be involved in the maturation of histidine tRNA. None of the regulatory loci unlinked to the operon are thought to function solely as repressor genes. This is because none of these mutants give as high a level of derepression as some of the *his* O mutants, and because all are capable of being further derepressed. There is recent evidence that the first enzyme in the histidine biosynthetic pathway may play a role in repression (Kovach *et al.*, *J. Bact.* 1969, 97, 1283).

When a single enzyme catalyzes a step in a common pathway (that is common to two or more biosyntheses), two means exist for effective control of its formation. One is multivalent repression resulting from the concerted action of the end products of the branching pathways. For example, isoleucine, valine, and leucine are all required for repression of the genes for the isoleucine-valine biosynthetic enzymes and both isoleucine and threonine are required for repression of genes for several threonine-forming enzymes. The other pattern of repression in branched pathways is one in which partial inhibition by the end products of the individual branch pathways occurs.

Regulation of Ribosomal and Transfer RNA Synthesis

Although the mechanism for regulating the transcription of the ribosomal RNA (rRNA) and transfer RNA (tRNA) genes is far from being understood, some interesting correlations have been observed. Under normal growth conditions, these RNA species are synthesized much more rapidly than mRNAs. This is probably because of a greater rate of reinitiation due to strong promotors rather than to a greater rate of propagation. Whether RNA polymerase associated factors specifically stimulate the synthesis of rRNA or tRNA is not clear.

Under conditions of amino acid starvation, the synthesis of rRNA and tRNA is severely inhibited. The inhibition is controlled either directly or indirectly by a protein coded for by the rel locus. The rel locus can be mutated from the normal (stringent) to the relaxed state ($rel^+ \longrightarrow rel^-$). In the rel^- state, RNA and tRNA synthesis continues, for an appreciable period during amino acid starvation. Many differences exist between cells differing only at the rel locus. One of the most intriguing changes observed first by Cashel and Gallant (Nature, 1969, 221, 838) is that amino acid starvation causes a rapid accumulation of two unusual guanine nucleotides called MSI and MSII (paper 65). Cashel and others have made attempts to attribute the inhibition of rRNA and tRNA synthesis to a direct effect of MSI, a guanosine tetraphosphate. However, the DNA directed cell-free synthesis of tRNA shows no specific inhibitory effect by MSI (paper 34). Quite unexpectedly, it has also been observed that MSI does stimulate the DNA-directed cell-free synthesis of protein from the lac and ara operons discussed above (Zubay, Gielow and Englesberg, Nature New Biology, 1971, 233, 464). It now appears likely that MSI and MSII will ultimately be assigned key regulatory roles in the regulation of RNA synthesis. For this reason a paper has been included which discusses their cell-free synthesis (paper 65; also see Pedersen, Lund and Kjeldgaard, Nature New Biology, 1973, 243, 13).

Although rRNA and tRNA genes show parallel behavior as far as the rel locus is concerned, there are indications of other differences. Thus, the rate of tRNA synthesis is roughly proportional to DNA synthesis under most conditions of growth, whereas the rate of rRNA synthesis can be much faster or much slower (see Maaløe and Kjeldgaard in Control of Macromolecular Synthesis, W. A. Benjamin, 1966). The synthesis of the latter seems to be more nearly proportional to the gross rate of protein synthesis.

Regulation of Amino Acid Biosynthesis

Regulation of amino acid biosynthesis is accomplished first, by controlling the rate of enzyme synthesis, as described above for certain amino acid pathways, and, second, by controlling the enzyme activities of preexisting enzymes as described below. By comparison with gene level regulation of enzyme synthesis, this might be called fine level control of enzyme activity. The former provides the gross enzyme level while the latter determines precisely the percent of active enzyme. Between the two types of control, a broad range of enzyme activity may be obtained with great precision. End product inhibition is found for most amino acid pathways; thus the amino acid itself inhibits the first reaction in the pathway for its biosynthesis (Imamoto, Prog. Nuc. Acid Res. Mol. Biol. 1973, 13, 339).

Frequently, biosynthetic pathways for amino acids diverge so that some of the relevant enzymatic reactions are required for more than one function. The control of such multifunctional pathways has been achieved in several ways, and, interestingly, even for the same pathway in more than one way in different organisms. A situation in which more than one end product inhibits the common enzyme is called multivalent end product inhibition. A good example of multivalent inhibition is seen in some of the common enzymes used for the biosynthesis of isoleucine, valine and leucine. The enzyme α-aceto-α-hydroxyacid synthetase is required directly by two pathways and indirectly by the third pathway. This enzyme is severely inhibited only in the presence of an excess of all three amino acid end products of the divergent pathways. This *cooperative multivalent end product inhibition* assures a supply of precursors as long as there is an insufficient supply of any one of the amino acids.

A novel form of control exists for the aspartokinases of *E. coli*. Aspartokinase is required for the biosynthesis of four amino acids—threonine, isoleucine, methionine, and lysine (see G.N. Cohen, *The Regulation of Cell Metabolism*, New York: Holt, Rinehart and Winston, 1967, p. 112). Instead of having one enzyme sensitive to inhibition by all these amino acids, three aspartokinases exist in the same *E. coli* cells, each subject to different gene repressors and allosteric inhibitors. A well-regulated flow of the common intermediates is insured by the existence of three "isofunctional" enzymes, each of which is subject to separate control by repression, and two of which are in addition allosterically regulated.

In addition to multivalent end product inhibition of the first enzyme in the common pathway, one often finds monovalent inhibition of the first enzyme after the branch point. The pathway for aromatic amino acid synthesis in *E. coli* has several branches starting with a common pathway leading in 6 steps to chorismate. The first step in this common pathway is inhibited by either of the end products L-phenylalanine or L-tyrosine. Chorismate is converted into either anthranilate or prephenate. The formation of anthranilate is inhibited by L-tryptophan, the end product of this branch. The formation of prephenate is inhibited by either phenylalanine or tyrosine, the end products of this branch. Prephenate conversion to phenylpyruvate is inhibited by phenylalanine while prephenate conversion to p-hydroxyphenylpyruvate is inhibited by tyrosine.

Regulation of Nucleotide Biosynthesis*

We turn now from amino acid biosynthesis to an examination of some aspects of nucleotide biosynthesis. Regulation of a variety of metabolic processes is required to ensure optimal amounts and a balanced distribution of the many different forms of purine and pyrimidine derivatives used by the cell. The first complete nucleotide to be formed in the purine biosynthetic pathway is inosine-5'-phosphate (IMP), while that in the pyrimidine pathway is uridine-5'-phosphate (UMP). The two pathways are operationally separate and distinct; they are metabolically related only in their sharing of some common participants, such as glutamine, CO_2, aspartate, and phosphoribosyl pyrophosphate (PRPP). The major control of these biosynthetic pathways operates through end product regulation, either by repression of enzyme synthesis at the level of gene expression, or by feedback inhibition of the activity of early enzymes. Although the end product of each of the pathways exerts negative control on itself, it may also act as a positive effector of the other pathway. Regulatory complications are introduced in both pathways by branch points from which the synthesis of other metabolites diverges. The pyrimidine pathway shares common intermediates with the arginine pathway, and an intermediate in the purine pathway is also used for thiamine synthesis. In addition, folic acid, riboflavin and histidine are derived directly from purine nucleotides.

After the synthesis of the first nucleotides, metabolic regulation is directed toward the control of interconversions to yield the proper varieties and balance of the required nucleotide classes. Ring substitutions are made through the processes of oxidation, reduction, amination, and deamination. Feedback loops and recycling mechanisms are marshalled for the control of these interconversions. Both negative and positive feedback loops also regulate the reductive conversions of ribonucleotides to their deoxyribose counterparts. Finally, the so-called salvage and scavenging pathways, whereby nucleosides are rescued and returned to the nucleotide pools, play an important role in the overall regulation.

The control of the activity of the enzymes involved in the synthesis of the pyrimidine precursors of RNA and DNA occurs for at least six different sites: (1) the carbamylation of aspartate transcarbamylase; (2) the reduction of CDP to dCDP by the deoxycytidine reductase system; (3) the deamination of deoxycytidylate by deoxycytidylate aminohydrolase; (4) and (5) the direct phosphorylation of uridine and deoxythymidine by the corresponding kinases; (6) amination of UTP to CTP by CTP synthetase.

In *E. coli*, three major control mechanisms cooperate in regulating the overall rate of purine nucleotide synthesis and the relative rates of synthesis of the two end products AMP and GMP. The first control is exerted on the

---

*Gots, in *Metabolic Regulation*, vol. 5, *ed.* H. J. Vogel, New York, Academic Press, 1971, p. 225.

early reaction step leading to the transfer of an amino group to 5-phosphoribosyl-l-pyrophosphate. The amidotransferase catalyzing this reaction is a multivalent regulatory enzyme; it is inhibited by ATP, or AMP, or by GTP, GDP, or GMP, each series of nucleotides apparently acting at a separate control site on the enzyme. Thus, whenever the purine mononucleotides accumulate, the first step in their synthesis undergoes feedback inhibition.

The second control mechanism is unusual. Whereas the reaction pathway leading from inosinic to guanylic acid requires ATP as a cofactor, the pathway leading from inosinic acid to AMP requires GTP. Thus, whenever there is an excess of ATP, the pathway to GMP is accelerated; and whenever there is an excess of GTP, the synthesis of AMP is accelerated.

There third control mechanism is provided by the fact that an excess of GMP brings about allosteric inhibition of the conversion of inosinic acid to adenylosuccinic acid and thus inhibits synthesis of AMP.

Regulation of Nucleic Acid and Protein Synthesis in Bacteriophage*

Phages of varying complexity thrive on *E. coli* and these have been studied in the greatest detail. They all possess a single nucleic acid molecule, either RNA or DNA, but never both. The smallest are the RNA bacteriophages which contain only three genes in a single-stranded RNA genome. The single-stranded DNA phages (ϕX174 and fd) contain about 8 genes. At a more complex level, there are double-stranded DNA phages, such as T7 and λ, each containing sufficient DNA for about 30 and 50 or more genes, respectively. There are also more complex viruses, such as phage T4 which codes for over 150 genes, and some viruses of animal origin whose nucleic acids are comparable in size to those of the largest bacteriophages.

The Simple Single-stranded RNA Viruses of the f2 Family (paper 66)**

The life cycle of a single-stranded RNA phage has been briefly considered in Section II (for a

---

*Bautz, *Prog. Nuc. Acid Res. Mol. Biol.* 1972, **12**, 129.
**The order of genes in the RNA phage is incorrect in this paper. See *paper 52* for correct gene order.

recent review of this subject see Kozak and Nathans, *Bacteriol. Rev* 1972, **36**, 109). The infectious RNA strand ("+" strand), stripped of its protein coat, enters the host *E. coli* cell. The plus strand attaches to the ribosome and translation starts with the synthesis of the A (maturation) protein and of the coat protein and eventually with the synthesis of the RNA synthetase. After a sufficient pool has been made of the polypeptide chain required for the phage specified RNA-dependent RNA synthetase, formation of the virus minus ("–") strand begins. There is then a preferential synthesis of plus strands on minus strand templates by the phage specified RNA synthetase. This enzyme is unique in its very high specificity, being incapable of replicating cellular RNA or the RNA of similar, but unrelated *E. coli* RNA phages. Some newly synthesized plus strands attach to ribosomes with the resulting synthesis of many copies of the coat protein. Coat proteins attach to the plus strands and eventually form a complete shell about the plus strands. The cell eventually lyses, releasing a crop of from 1,000 to 10,000 new phage. Biochemical studies show that synthesis of the A protein and coat protein begin immediately upon entry of the plus strand into the host, but the translation of the third cistron coding for the RNA-dependent RNA synthetase is delayed. Sequence studies on the plus strand suggest that secondary structure accounts for this difference in time of translation.

The precise nucleotide sequences for several regions of the RNA of phage R17, MS2 and Qβ, including the entire sequence of the coat protein gene (Mills, Kramer and Spiegelman, *Science*, 1973, **180**, 916; Min Jou *et al*, *Nature*, 1972, **237**, 82), have recently been determined. From these studies a probable secondary structure configuration can be inferred. The secondary structure results from folds or hairpin turns in the polynucleotide strand, with Watson-Crick base pairs forming between most, but not all, opposing bases (Ball, J. Theor. *Biol.* 1972, **36**, 313). Most likely, these regions twist into a helical conformation similar to double helix DNA. The AUG initiation codon for coat protein appears to be conveniently located directly on top of a hairpin turn. This probably accounts for the fact that the coat protein initiation site has the highest affinity for the ribosome. However, the AUG initiation codon for the RNA polymerase cistron is buried by base-pairing. The A protein initiation region binds poorly to the ribosome and does not bind at all to synthetase initiation region. Biochemical studies show that the coat protein is produced in about a 40 times greater concentration than the A protein, and that the translation of the synthetase gene requires prior translation of the adjacent coat protein gene. Thus, regulation of the time and rates of production of the three virus proteins is believed to result from the affinities of the initiation sites for the ribosome. These affinities in turn are controlled by the primary and secondary structures of the RNA at the initiation sites. The intercistronic distance in phages R17 and MS2 varies from 26 to 36 nucleotides (Contreros *et al.*, *Nature New Biology*, 1973, **241**, 99).

Little is known about the regulation of virus RNA replication, but it seems likely that the relative affinities of the synthetase (complexed with several host proteins) for the ends of the plus and minus strands is a crucial factor in determining the synthesis of excess plus strands.

Other Possibilities for Translation Control

The RNA viruses provides us the the only direct evidence of translational control in which a principal factor is the affinity of the initiation site for ribosome attachment. It seems appropriate at this point to speculate on the existence of translational control mechanisms in other systems. Due to the short-lived nature of bacterial messengers it seems unlikely that translation control will be a major regulatory device in bacteria. However, in higher forms which often contain long-lived messengers translation control may be very important. Indeed, Tomkins and coworkers (*Nature New Biology*, 1972, **239**, 9) have obtained evidence for a translational repressor which regulates the synthesis of tyrosine aminotransferase in cultured rat hepatoma cells.

There are at least three ways in which protein synthesis might be controlled at the level of translation once the RNA has been transferred from the nucleus to the cytoplasm: (1) by the primary or secondary structure of the messenger RNA, (2) by the structure of initiation factors (or the presence of inhibitors) that affect the binding of messenger to ribosome; (3) by the presence or absence of certain tRNAs.

(1) Messenger breakdown and messenger adsorption to ribosomes could both be influenced by the primary or secondary structure at the 5' end. (2) The f_3 protein is required for the initial binding of the bacterial messenger to the ribosome. One can imagine a variety of such factors or inhibitors which might influence complex formation between messenger and ribosome and thereby the extent of messenger translation. For example, the structure of an f_3 factor could be altered by certain small molecule cofactors or by covalent bond alteration. Such possibilities have been briefly discussed in the introduction to section III. (3) There is more than one tRNA for encoding most amino acids. This allows for a form of control of translation which could act at the level of propagation. Some messengers may have unusual codons which require unusual tRNAs for translation. The presence and charging of these tRNAs would be a necessary condition for translation of such messengers.

Single-stranded DNA Viruses

The single-stranded DNA phages of *E. coli* are classified into two groups, depending on their shape (Pratt, *Ann. Rev. Genet.* 1969 **3**, 343; Marvin and Hohn, *Bact. Revs.* 1969, **33**, 172). The filamentous viruses include phages such as f1 and fd and the spherical viruses are ϕX174 and S13 all of which infect male cells only. These DNA phages possess a circular genome, contain about 6,000 nucleotides and probably eight genes or cistrons. In the case of ϕX174, four of these genes code for structural proteins with molecular weights of 60,000, 36,000, 19,000 and 5,000. One of the remaining four genes is involved in the replication of the double helical, intracellular DNA intermediate; another functions in the production of the progeny single-stranded circles and still another brings about cell lysis.

Synthesis of viral specific mRNA begins after the double helical (RF) intermediates are made. All the phage specific mRNA is complementary to the viral minus strand; among the transcripts synthesized there appears to be no temporal control in the transcription of RF. The mechanism and control of synthesis of the single-stranded DNA phages *in vitro* and *in vivo* is now being studied very intensively since they could provide models of how more complex chromosomes replicate (Marvin and coworkers, *J. Virol.* 1972, **10**, 362, 371, 384, 392; Alberts *et al.*, *J. Mol. Biol.* 1972, **68**, 139).

T4 and T7 bacteriophages

The seven T bacteriophages are composed solely of double helix DNA and protein in approximately equal proportions. The morphology and genetics of the T-even bacteriophages, in particular T4, have been the most extensively investigated in the past (for a recent reference see Eiserling and Dickson, *Ann Rev. Biochem.* 1972, **41**, 467); however, bacteriophage T7 as well as the related phage T3 are currently receiving a good deal of attention (Summers, *Ann. Rev. Genet.* 1972, **6**, 191) due in part to their less complex structure and relatively small, double-stranded genome (about 25×10^6 daltons).

Immediately after the T4 chromosome enters a host cell, a group of about 25 "pre-early" genes are transcribed (principally from the *l* strand) (Guha *et al*, *J. Mol. Biol.* 1971 **59**, 329) by host specific RNA polymerase, specifying enzymes involved in phage DNA synthesis and in the shutdown of host specific macromolecular synthesis. Other T4 genes have been classified according to the times at which they are transcribed early and/or late. After T4 infection, an important early function is the synthesis of new small polypeptides, encoded by the phage genome, that become associated with the host DNA-dependent RNA polymerase. Their molecular weights range from 10,000 to 22,000. Some of the RNA polymerase associated polypeptides are absent when the enzyme is isolated from cells infected with phage mutants carrying maturation defective genes (Stevens, *Proc. Nat. Acad. Sci. U.S.* 1972, **69**, 603). Also modifications in the subunits of the host RNA polymerase occur which include changes in size, adenylation, and phosphorylation. Most likely, these modifications in the RNA polymerase play a major role in determining which phage genes will be transcribed (*paper 67*). Using rifamycin resistant or temperature sensitive host RNA polymerase mutants, it was found that following phage infection the synthesis of phage specific messenger RNA was rifamycin resistant or temperature sensitive, just like the host mutant. These findings showed that at least part of the host RNA polymerase (the β polypeptide subunit in particular) is necessary for phage DNA transcription. If the host RNA polymerase is not modified, it will only transcribe early mRNA *in vitro* from phage DNA, mimicking the *in vivo* selectivity. In addition to the modification of the host RNA polymerase, Geiduschek and his coworkers have shown that continued phage DNA synthesis and certain T4 gene products (in particular the gene 55 product and to a lesser extent the product of gene 33) are required for turning on and maintaining transcription of late genes (Bolle *et al.*, *J. Mol. Biol.* 1968, **33**, 339). Late transcription takes place principally from the *r* strand of the phage DNA (see reference 28 at end of this Section).

Other T-bacteriophage, such as T3 and T7, synthesze their own RNA polymerase molecules soon after infection (*paper 68*). The host σ-core complex transcribes some early phage genes (about 20 percent of the total) immediately after infection. One of these early genes is for a phage specified RNA polymerase which then takes over the transcription of non-early genes. In the case of *in vitro* transcription of the T3 and of T7 phage genomes, all the messenger RNA is transcribed from only one of the DNA strands (Summers and Szybalski, *Virology* 1968, **34**, 9). In the case of T-even phage DNA transcription early genes are transcribed from one strand, while late genes are transcribed from the complementary strand. No repressor proteins have yet been described which exert negative control, in the same way that the *lac* or the phage λ repressor does, either early or late in the transcription of the T-even phage or T3 or T7 genomes.

Another case where the modification of cellular RNA polymerase may prove to be important in regulating development is during sporulation in *Bacillus* species (*paper 69*). The current hypothesis is that early during sporulation the RNA polymerase undergoes a change in template specificity (Linn *et al.*, *Proc. Nat. Acad. Sci. U.S.* 1973, **70**, 1865). The change in template specificity observed *in vivo* and *in vitro* might be related to the loss of sigma activity of RNA polymerase early in sporulation. Leighton (*Proc. Nat. Acad. Sci. U.S.* 1973, **70**, 1179) has recently characterized a single site, temperature sensitive, *B. subtilis* RNA polymerase mutant that gives rise to temperature sensitive sporulation at non-permissive temperatures. These leads are very promising, however, the temporal correlation between the changes in the host RNA polymerase and the extent of the change of the enzyme and the transcriptional changes occuring after commitment to sporulation have not yet been worked out.

A number of bacteriophages (T-even and T5, for example) carry genes for tRNA molecules. In addition, phage infection sometimes leads to modification of one or more host tRNA molecules (such as one of the leucine tRNA species) as well as some of their synthetases. When T4 infects *E. coli* it modifies the valine tRNA synthetase (Chrispielo et al., *J. Mol. Biol.* 1968, 31, 463). However, the significance of these modifications in the translational machinery is still unclear. Recent experiments have shown that the phage mutants that do not cause these alterations to take place are viable.

The λ bacteriophage (paper 70)

The λ bacteriophage is a temperate phage since it can replicate either lytically or as a prophage, the latter integrated into the host genome (Echols, *Ann. Rev. Biochem.* 1971, 40, 827; Echols, *Ann. Rev. Genet.* 1972, 6, 157). In the lysogenic state it can be passed from one cell generation to the next without the appearance of any mature viral particles. In the vegatative state, it replicates and produces cell lysis and death with the appearance of 50 or more virus particles per cell. When phage λ infects *E. coli* there is an equal chance that the phage genome will replicate lytically or be converted to the prophage state.

There are at least four regulatory genes located on the λ genome, Q, N, *c1*, and *tof*, which probably function as follows. In the lysogenic state the repressor, coded by the *c1* genes, binds to two sites on either side of the *c1* region of the viral DNA, referred to as the left and right operators (o_L and o_R respectively). The *c1* gene is usually active when the phage genome is in the prophage state, and therefore, it is probably transcribed by the host polymerase alone. The resulting low level of *c1* repressor maintains the phage in a dormant, lysogenic state. When the *c1* repressor becomes inactivated (usually accomplished either indirectly by brief ultraviolet light exposure, treatment with the drug mitomycin, or directly by exposure of the lysogenized cells to an elevated temperature if the *c1* phage repressor protein is temperature sensitive), transcription starts toward the left from the p_L (the lefthand promoter), transcribing the N gene, and toward the right from p_R (the righthand promoter), transcribing the *tof* gene. In the absence of N gene product, reading of the early left gene stops after the N gene, and reading of the early right genes stop after the *tof* gene, as discussed in *papers 28* and *70*. This transcriptional arrest probably results from the presence of a base sequence on the DNA which in the presence of the bacterial factor causes the polymerase to stop and release its nascent RNA chain. When sufficient N gene product is present it acts as a positive (or rather anti-negative) regulator, turning on transcription of other phage genes. The N product acts on several sites: the transcription of the *exo-int* region from the *l* strand and the *O-P-Q* region from the *r* strand. The O and P gene products are required for λ DNA replication which starts at the *ori* site, just to the left of gene O. The N gene product acts as an anti-terminator permitting transcription to take place beyond the initial termination points for early right and early left transcription. This view is held since it appears as though the *exo-int* genes are part of the same transcriptional unit as the N gene: transcription beyond the N gene requires the N gene product. The Q gene product is also a positive regulator which acts at a single site between Q and S to turn on the late λ phage genes. The product of the *tof* gene has two known functions, one, preventing synthesis of *c1* repressor during phage growth, and the other, turning off genes in the N operon. The sites of action of the *tof* gene product appear to be the o_L o_R operators which are also the sites of action of the *c1* product. A more detailed discussion of the controls of λ transcription can be found in an article by Szybalski, "Transcription and replication in *E. coli* bacteriophage lambda" in *Uptake of Informative Molecules by Living Cells*, L.G.H. Ledoux (ed.), Amsterdam: North-Holland Publ. Co., 1972, pp. 59—82.

This concludes the discussion of the regulatory mechanisms in bacteria and bacteriophage. It remains to be seen if these simple mechanisms are used in varying combinations by plant and animal systems. The probability seems great since one is dealing with the same basic commodities, namely, polynucleotide and polypeptide chains.

Relationship of Flac Replication and Chromosome Replication

(DNA/*Escherichia coli*/plasmid/division cycle)

STEPHEN COOPER

Department of Microbiology, University of Michigan, Ann Arbor, Mich. 48104

Communicated by James V. Neel, June 30, 1972

ABSTRACT The time of replication of a bacterial plasmid, F*lac*, during the division cycle of *Escherichia coli* has been estimated in exponentially growing cultures and at various times after a shift from minimal medium to a richer medium (a shift-up). There is a variation in the cell age at which the capacity to synthesize β-galactosidase (β-D-galactoside galactohydrolase, EC 3.2.1.23) doubles (assumed to be a measure of the time at which the F*lac* plasmid replicates) when this capacity is measured at various times during the shift-up, and with increasing steady-state exponential growth rate. Cells growing at slow and moderate growth rates exhibit F*lac* replication in the middle of the division cycle. With increasing time after a shift-up or with increasing growth rate the plasmid replicates at earlier times, eventually at cell division, and finally in the older cells. This variation in the cell age at which the plasmid replicates is similar to the variation in cell age at which chromosome initiation occurs during a shift-up, although plasmid replication occurs slightly before initiation of chromosome replication.

Bacterial plasmids are relatively small extra-chromosomal pieces of DNA that are maintained in various strains of bacteria. The very fact that they are maintained in a growing culture implies that they must, on the average, double in number every generation. The mechanism by which this replication is controlled is not understood at the moment.

Bazarel and Helinski (1) have presented evidence suggesting that colE₁ replicates by a random replication event with a pool of about 20–40 colE₁ plasmids. Similar results have been reported for RTF plasmid DNA (2). Zeuthen and Pato (3) studied F*lac* replication and found that in contrast to the random replication during the division cycle of the colE₁ and RTF plasmids, the F*lac* plasmid replicates at a specific time during the division cycle. This difference may result, in part, from the fact that there is only a small number of F*lac* plasmids per cell, while there is a demonstrably larger pool of colE₁ and RTF plasmids in each cell.

A second conclusion of Zeuthen and Pato (3) is that at all growth rates studied, replication appeared to occur in the middle of the division cycle. This proposal is contradicted by the data presented below that demonstrate that F*lac* replication can occur at cell division as well as the middle of the division cycle. While the model of Zeuthen and Pato precludes any observable relationship between F*lac* replication and any chromosomal replication event (initiation, termination, etc.) the data presented below suggest that control of F*lac* replication may be similar to control of chromosome initiation. It will be shown that the cell age at which F*lac* replication occurs varies in a manner similar to the variation in cell age at which initiation of chromosome replication occurs. This result suggests that the mechanisms initiating replication of the chromosome and the plasmid may be similar. This implies that biochemical analysis of the events leading to initiation of F*lac* replication may be important in understanding chromosome replication.

EXPERIMENTAL RATIONALE

The pattern of chromosome replication of *Escherichia coli* B/r growing at 37° is determined by two constants: C, the time for a round of replication to proceed from the origin to the terminus of the genome (about 42 min) and D, the time between termination of a round of replication and cell division (about 21 min) (4). As growth rates vary depending on changes in medium composition, the pattern of replication during the division cycle changes, with a gap present in cells growing slower than C min per doubling, and multiple forks present for a portion of the division cycle when cells grow faster than C min per doubling. More important for the consideration of this paper, however, is that the combination of the two parameters, C and D, leads to a variation in the cell-cycle age at which initiation of replication occurs.

As illustrated in Fig. 1, in cells doubling every 40 min initiation occurs 1.5 [(C+D)/doubling time] generations before a cell division, or in the middle of a division cycle. With increasing growth rate (decreasing doubling time) initiation occurs in younger cells, at 1.7, 1.9, and finally 2.0 generations before a cell division (that is, at a cell division). This variation is observed as a continuous decrease in the cell age at which initiation occurs. With further increases in the growth rate initiation occurs 2.1 or more generations before a cell division, or in the oldest cells in a division cycle. This sudden change in the cell age at initiation of replication from the youngest to the oldest cells with only a slight increase in growth rate is only an apparent discontinuity. It is part of a continuous variation when related to the fact that initiation occurs at a contant time [(C+D) min] before a cell division (4, 5), and as the generation time decreases, the number of generations intervening between initiation and cell division increases. The question to be answered is "Does the cell age at which F*lac* replicates vary in the way that the cell age at initiation of chromosome replication varies?"

Zeuthen and Pato (3) varied the cell age at which the chromosome is initiated by using exponentially growing cells with different steady-state growth rates. Although it is possible to cover a range of growth rates, one problem with this approach is that it is difficult to prepare cells growing with any particular growth rate (or replication pattern) merely by changing the medium composition.

2706

Another approach to this problem is to study cells undergoing a shift-up. A shift-up is the transfer of bacteria to a richer medium that supports a faster rate of growth. It is accomplished by addition of the rich medium to the minimal medium. The pattern of chromosome replication at each point in time in a culture undergoing a shift-up from one steady-state growth rate to a faster steady-state growth rate is identical to the pattern of some steady-state growth rate intermediate between the original and final steady-state growth rates (6). The shifted culture, during the transition period, recapitulates each replication pattern of the infinite number of steady-state growth rates between two growth rates. With reference to Fig. 1, each line could also be considered the pattern at some instant during a shift-up. Using these shift-up cultures, as well as cultures with various steady-state growth rates, I have been able to demonstrate that plasmid replication can occur at cell division as well as in the middle of the cell cycle, and it does vary in a manner similar to the variation in cell age at which chromosome initiation occurs.

MATERIALS AND METHODS

Bacteria. The bacteria used were *E. coli* B/r (4, 5) and B/r *lac⁻/Flac⁺* (from D. Davis). The plasmid nature of the *lac⁺* gene in the second strain was monitored by heat curing (7). No integration of the plasmid *lac⁺* gene was observed.

Growth of Bacteria. C medium was used with 0.2% glucose or succinate (5). Amino acids were present at concentrations of 20 μg/ml each or as 0.2% casamino acids. Conditioned medium for elution was prepared by mixing three parts of fresh medium with one part of medium from a stationary phase culture that was filtered to remove the cells. When a shift-up was performed, 18 amino acids (leucine and cysteine omitted) were added. Cultures were analyzed directly or "shifted-up" after at least 15 hr (or with faster cultures after at least 6 hr) of unlimited steady-state growth.

Determination of DNA and Plasmid Replication Patterns. The method used is similar to that described (8, 9). Cells grown in a shaking water bath were induced with 1 mM isopropylthiogalactoside and labeled with radioactive thymidine (^{14}C or ^{3}H, 0.1 μg/ml). After 2–6 min about 10^{10} cells were filtered onto a membrane filter (10), washed with conditioned medium, and elution was started by pumping conditioned medium at a rate of 15 ml/min for 2 min and then lowering the rate to about 3 ml/min. At 2-min intervals, fractions were collected, and

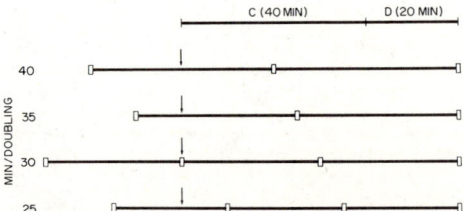

FIG. 1. Variation in the cell age at which chromosome replication is initiated due to variation in the growth rate. The *arrows* indicate initiation of a round of replication. For simplicity the figure is drawn with C = 40 and D = 20 min. The *open boxes* indicate cell division.

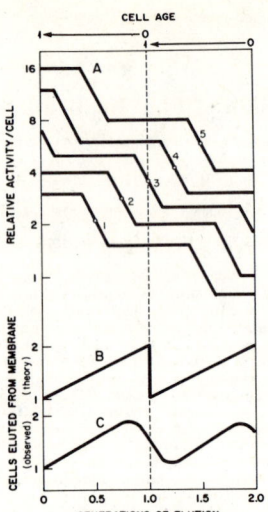

FIG. 2. Analysis of elution patterns. The curves in (*A*) are the expected patterns for events (doubling in the capacity to synthesize an enzyme, increases in the rate of isotope incorporation, etc.) occurring (*1*) in the middle of the division cycle, (*2*) in young cells, (*3*) at cell division, (*4*) in old cells, and (*5*) in the middle of the division cycle. If the time required for one generation of elution is known (*vertical dashed line*), then one can get the pattern of activity during a division cycle. This is indicated by the "cell age" at the top of the figure. Note that the cell age is read from right to left. This is because the first cells to be eluted are the progeny of the oldest cells, and with time, the progeny of younger and younger cells are eluted. The time for one generation to be eluted is determined by the cell elution pattern that is ideally represented in (*B*). This pattern is due to the fact that there are twice as many old cells as young cells in an exponentially growing culture. An observed elution pattern is given in (*C*). This pattern is due to the dispersion of interdivision times about the mean interdivision time. The midpoint of the decrease in a cell-elution curve is a good approximation of the mean generation time.

from these 0.1–0.4 ml were added to 10 ml of saline–formalin for cell count, 2.0 ml were added to 5% Cl$_3$CCOOH for determination of radioactivity, and triplicate 1.0-ml samples were incubated for an additional 20–30 min for complete expression of induced enzyme activity. Automatic pipetting devices were used to remove the samples. The reproducible sample size appears to be an improvement over the use of pipettes. After enzyme expression the cells were chilled and 100 μg/ml chloramphenicol was added. When the experiment was completed toluene was added to all samples, which were then incubated 1 hr at 37°. Enzyme activity was assayed with orthonitrophenylgalactoside.

Analysis of Elution Patterns. The analysis of elution patterns has been described (4–6) and the main points will be summarized here. Pulse-induced and labeled cells are grown attached to a membrane. In the first minutes after binding, the newborn cells eluted from the membrane are the progeny of the

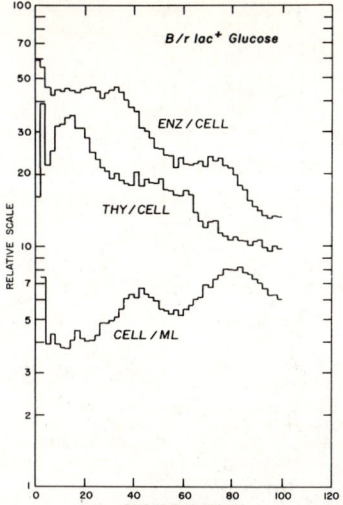

FIG. 3. Elution pattern of *E. coli* B/r *lac*+ induced and labeled during steady-state growth in C medium with glucose. Cells were induced with isopropylthiogalactoside and labeled with [³H]thymidine for 2 min and analyzed as described in *Methods*.

oldest labeled cells. With time, younger and younger bound cells divide and contribute progeny to the eluate. By measuring the enzyme and radioactivity per cell in different fractions, one gets a measure of the enzyme activity and the amount of thymidine incorporation per cell for different cell ages at the time of induction and labeling. This is illustrated in Fig. 2. Peaks and decreases in the cell-elution curve are due to the age distribution of exponentially grown cells, which have about twice as many new-born cells as cells about to divide. The middle of a decrease in the cell-elution curve indicates one generation of elution (5). Therefore, any decrease in the enzyme per cell, or radioactivity per cell, that occurs at a decrease in the cell elution curve is indicative of an event (gene doubling or initiation of a round of replication, respectively) occurring at cell division in the batch culture. In the shift-up expriments, the analysis is similar since the age distribution does not change.

It should be mentioned that many experiments exhibit a small cell-elution peak soon after elution begins. This peak is of unknown origin (it may be due to physical factors of elution, variations in temperature, etc.) but it does not affect the analysis of the graphs.

RESULTS

β-Galactosidase (EC 3.2.1.23) synthetic capacity and chromosome replication during the division cycle of *E. coli* B/r *lac*+

Fig. 3 shows the results obtained when *E. coli* B/r *lac*+ was induced with isopropylthiogalactoside, labeled with [³H] thymidine, and analyzed by the standard membrane technique (10). The decreases in the curve for enzyme per cell occur at the same time as the peaks in the cell elution curve, while the decreases in the thymidine curve occur in the middle of the increases in cell number. This has been analyzed and reported (8, 9) and is interpreted (see Fig. 2) as replication of the chromosomal gene for β-galactosidase at time of cell division, and initiation of chromosome replication in the middle of the division cycle.

Analysis of this graph, while quite clear, should also suggest that any precise positioning of the various events during the division cycle is tenuous because of their dispersion in time. Therefore, rather than refer to a particular time during the division cycle at which an event occurs, the data will be discussed in terms of the relative positioning of an event with respect to the fall in thymidine incorporation (initiation of chromosome replication) and the peaks and decreases in the cell-elution curve (indicating completion of one generation of elution).

F*lac* replication during the division cycle during a shift-up

If the time during the division cycle when the F*lac* plasmid replicated varied with the time of chromosome initiation, then there should be a time during the shift-up when F*lac* replication occurred at cell division, a situation not predicted by the model of Zeuthen and Pato (3). This conjecture was easily confirmed as illustrated in Fig. 4A and B, where cells are analyzed 30 and 35 min after a shift-up for the cell age when chromosome replication is initiated and plasmid replication occurs. The decreases in the curves for enzyme and thymidine per cell occur at the peaks or at the decreases in the cell-elution curve. As noted above, this means that these events, chromosome initiation and plasmid replication, occur around cell division in the shifted culture. This conclusion is strengthened by the observation in Fig. 4A and B of the initial portion of the elution curve that shows decreases in enzyme activity in the

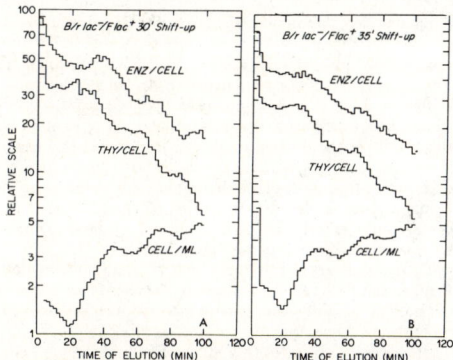

FIG. 4. Elution patterns of B/r *lac*−/F*lac*+ induced and labeled during a shift-up. Cells growing in C medium with glucose were enriched with 18 amino acids. (*A*) After 29 min the cells were labeled with [³H]thymidine and induced with isopropyl-thiogalactoside for 2 min, filtered onto a membrane, and eluted as described in *Methods*. (*B*) Same as (*A*) except that the labeling was between 34 and 36 min after a shift-up. Note that decreases in the enzyme per cell and the cells per ml occur at about 48 min of elution in both of these figures. This indicates that the increased capacity to synthesize the enzyme appeared at cell division.

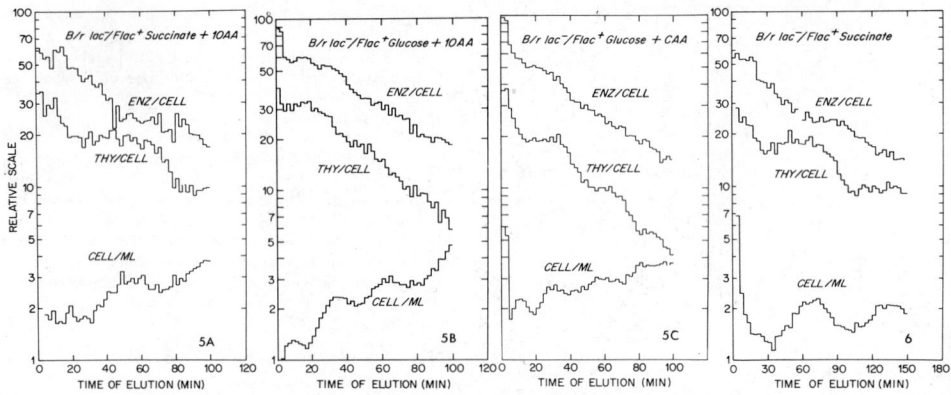

FIG. 5. Elution pattern of B/r $lac^-/Flac^+$ induced and labeled during steady-state growth. Cells grown in (A) C medium with succinate and 10 amino acids (47 min/doubling), (B) C medium with glucose plus 10 amino acids (35 min/doubling), or (C) C medium with glucose plus casamino acids (27 min/doubling), were induced, labeled, and analyzed as described in *Methods*. In (B) the decreases in enzyme per cell and cell per ml occur at about 40 min. This result indicates that the increased capacity to synthesize the enzyme appeared at cell division.

FIG. 6. Elution pattern of slow growing B/r $lac^-/Flac^+$. Cells growing in C medium with succinate at a doubling time of 108 min were analyzed as described in *Methods*.

first cells eluted from the membrane. This indicates that the events occurred in the oldest and youngest cells of the culture, that is, at the time of cell division. Examination of these experiments also indicates that Flac replication precedes initiation of chromosome replication.

Flac replication during the division cycle of cells growing at different steady-state growth rates

Supported by the results of the shift-up experiments, I searched for a steady-state growth rate that would illustrate the predicted plasmid replication at the time of cell division. This is illustrated in Fig. 5B, flanked by two other experiments analyzing slower (Fig. 5A) and faster (Fig. 5C) growing cells. These three experiments demonstrate plasmid replication in the middle of the division cycle, at cell division, and again in the middle of the division cycle, respectively. Again, in these experiments there is evidence that replication of the plasmid occurs slightly earlier than initiation of chromosome replication. Other experiments at different growth rates are consistent with the general pattern presented here.

These results are in agreement with the data presented by Zeuthen and Pato (3). My conclusion is different because they did not find a steady-state growth rate at which the plasmid replicated at cell division.

At slower growth rates (Fig. 6) plasmid replication occurs in the middle of the division cycle, although initiation of chromosome replication occurs at cell division. This is similar to the observation of Zeuthen and Pato (3). Whatever the explanation of this dissociation of chromosome initiation from plasmid replication, it should be noted that at slow growth rates the values for C and D are increased (10, 11).

DISCUSSION

The results in this paper are based upon three assumptions: (a) the time at which the capacity to synthesize β-galactosidase doubles indicates the time at which the plasmid replicates, (b) the time for a plasmid to replicate is fairly short, and (c) the membrane elution method used to analyze events during the cell cycle is valid.

The first assumption is supported by the finding that the activity of β-galactosidase is proportional to the number of copies of the gene (12, 13) and by the finding that the time at which the capacity to synthesize β-galactosidase from a chromosomal gene doubles (Fig. 3) coincides with the time at which that gene replicates (8, 9). The second assumption is supported by the observation that plasmids that undergo a density shift do not exhibit intermediate densities (1, 2). The last assumption has been discussed and analyzed (4–6) and has been supported by other data (14).

The main point of this paper is the presentation of evidence indicating that the time at which Flac replicates during the division cycle can vary. Flac replication can occur in the middle of the cycle, as noted by Zeuthen and Pato (3), as well as at cell division. The general conclusion is that plasmid replication varies in a manner similar to that of initiation of chromosome replication and that any models pertaining to variation in cell size and chromosome initiation (15–17) may apply equally to plasmid replication. In agreement with this proposal, D. Davis (personal communication) has evidence that the cell size when plasmid replication occurs is fairly constant as growth rate varies. One explanation for the fact that Flac replicates slightly before chromosome initiation is that there may be some competition between plasmid and chromosome for some hypothetical initiator substance, and in a sense, initiation of chromosome replication may be delayed due to the presence of the plasmid.

In Fig. 7 the relationship between the cell age at initiation and completion of a round of chromosome replication is schematically compared with the cell age at Flac replication. The general variation is similar although the plasmid replica-

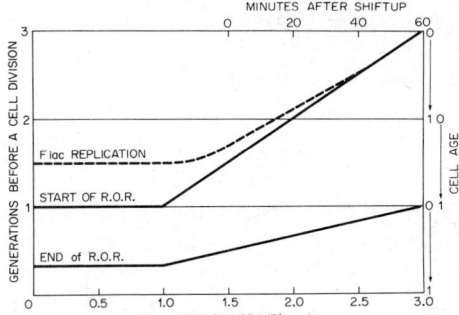

FIG. 7. Summary of results illustrating proposed variation of cell age at the time of Flac replication as a function of increasing time after a shift-up and as a function of increasing growth rate. Calculations of expected shift-up patterns of chromosome replication have been described (6). R.O.R., round of chromosome replication.

tion curve is higher than the chromosome initiation curve. This summarizes the experiments presented here that indicate that, at all but the fastest growth rates, Flac replication precedes chromosome initiation.

The model presented in Fig. 7 is independent of the absolute number of plasmids per cell, although the model predicts that the number of plasmids per cell at moderate and slow growth rates [μ(doublings/hr) of 1.5 or less] is 50% the number of plasmids per cell for cells growing more rapidly (μ of 2.5) with plasmid replication occurring in the middle of the division cycle in both cases. There is some disagreement regarding the number of plasmids per cell, although it is agreed that the number is low. Frame and Bishop (18) determined that there were two episomes per chromosome for cells growing with a 50-min doubling time. This may be calculated to be about 3–4 plasmids per cell when accounting for the number of chromosomes in these cells (4). Hohn and Korn (19) estimated the number of Fgal plasmids as about 1.35 per cell. If it is assumed that the slow-growing cells have the minimal number of plasmids per cell, then new-born cells contain one plasmid that replicates in the middle of the division cycle. As the growth rate increased, plasmid replication would occur in younger cells, and eventually a growth rate would exist where Flac replication occurred at cell division. These cells would contain two Flac molecules throughout the division cycle. More rapidly growing cells, therefore, would contain two plasmids in the new-born cells that would replicate in the middle of the division cycle to give four plasmids in the older cells. Any higher multiple plasmid content would also be consistent with the data presented in this paper. The important point is that the number of plasmids per cell would increase with increasing growth rate. In contrast, the model of Zeuthen and Pato (3) would predict that the number of plasmids per cell is independent of growth rate.

Aside from the results reported above I would like to emphasize one technical innovation. This is the use of the shift-up to study events during the division cycle. Rather than vary the medium composition to achieve a desired replication pattern, one merely has to choose an instant during a shift-up to have a particular replication pattern.

This work was supported by Grant AI10059-02 from the National Institutes of Health. Therese Reuttinger provided excellent technical assistance.

1. Bazaral, M. & Helinski, D. R. (1970) "Replication of a bacterial plasmid and an episome in *Escherichia coli*," *Biochemistry* 9, 399–406.
2. Terawaki, Y. & Rownd, R. (1972) "Replication of the R factor Rts1 in *Proteus mirabilis*," *J. Bacteriol.* 109, 492–498.
3. Zeuthen, J. & Pato, M. L. (1971) "Replication of the Flac sex factor in the cell cycle of *Escherichia coli*" *Mol. Gen. Genet.* 111, 242–255.
4. Cooper, S. & Helmstetter, C. E. (1968) "Chromosome replication and the division cycle of *Escherichia coli* B/r," *J. Mol. Biol.* 31, 519–540.
5. Helmstetter, C. E. & Cooper, S. (1968) "DNA synthesis during the division cycle of rapidly growing *Escherichia coli* B/r," *J. Mol. Biol.* 31, 507–518.
6. Cooper, S. (1969) "Cell division and DNA replication following a shift to a richer medium," *J. Mol. Biol.* 43, 1–11.
7. Stadler, J. & Adelberg, E. A. (1972) "Temperature dependence of sex-factor maintenance in *Escherichia coli* K12," *J. Bacteriol.* 109, 447–449.
8. Helmstetter, C. E. (1968) "Origin and sequence of chromosome replication in *Escherichia coli* B/r," *J. Bacteriol.* 95, 1634–1641.
9. Pato, M. L. & Glaser, D. A. (1968) "The origin and direction of replication of the chromosome of *E. coli* B/r," *Proc. Nat. Acad. Sci. USA* 60, 1268–1274.
10. Helmstetter, C. E. (1967) "Rate of DNA synthesis during the division cycle of *Escherichia coli* B/r," *J. Mol. Biol.* 24, 417–427.
11. Pierucci, O. (1972) "Chromosome replication and cell division in *Escherichia coli* at various temperatures of growth," *J. Bacteriol.* 109, 848–854.
12. Nishi, A. & Horiuchi, T. (1966) "β-Galactosidase formation controlled by an episomal gene during the cell cycle of *Escherichia coli*," *J. Biochem.* 60, 338–340.
13. Pittard, J. & Ramakrishran, T. (1964) "Gene transfer by F' strains of *Escherichia coli* IV. Effect of chromosomal deletion on chromosomal transfer," *J. Bacteriol.* 88, 367–373.
14. Helmstetter, C. E. & Pierucci, O. (1968) "Cell division during inhibition of deoxyribonucleic acid synthesis in *Escherichia coli*," *J. Bacteriol.* 95, 1627–1633.
15. Pritchard, R. H., Barth, P. T. & Collins, J. (1969) "Control of DNA synthesis in bacteria," *Symp. Soc. Gen. Microbiol.* 19, 263–297.
16. Donachie, W. D. (1968) "Relationship between cell size and time of initiation of DNA replication," *Nature* 219, 1077–1079.
17. Helmstetter, C. E. (1969) in *The Cell Cycle: Gene-Enzyme Interactions*, eds. Padilla, G. M., Whitson, G. L. & Cameron, I. L. (Adademic Press, New York), pp. 15–35.
18. Frame, R. & Bishop, J. O. (1971) "The number of sex factors per chromosome in *Escherichia coli*," *J. Biochem.* 121, 93–102.
19. Hohn, B. & Korn, O. (1969) "Cosegregation of a sex factor with *Escherichia coli* chromosome during curing by acridine orange," *J. Mol. Biol.* 24, 417–427.

ISOLATION OF THE LAC REPRESSOR

By Walter Gilbert and Benno Müller-Hill

DEPARTMENTS OF PHYSICS AND BIOLOGY, HARVARD UNIVERSITY

Communicated by J. D. Watson, October 24, 1966

The realization that the synthesis of proteins is often under the control of repressors[1,2] has posed a central question in molecular biology: What is the nature of the controlling substances? The scheme of negative control proposed by Jacob and Monod envisages that certain genes, regulatory genes, make products that can act through the cytoplasm to prevent the functioning of other genes. These other genes are organized into operons with cis-dominant operators, such operators behaving as acceptors for the repressor. Appropriate small molecules act either as inducers, by preventing the repression, or as corepressors, leading to the presence of active repressor. The simplest explicit hypothesis for inducible systems is that the direct product of the control gene is itself the repressor and that this repressor binds to the operator site on a DNA molecule to prevent the transcription of the operon. The inducer would combine with the repressor to produce a molecule which can no longer bind to the operator, and the synthesis of the enzymes made by the operon would begin. However, other models will also fit the data. Repressors could have almost any target that would serve as a block to any of the initiation processes required to make a protein. A molecular understanding of the control process has waited on the isolation of one or more repressors.

We have developed an assay for the lactose repressor, the product of the control gene (i gene) of the lactose operon. The assay detects and quantitates this repressor by measuring its binding to an inducer, as seen in this case by equilibrium dialysis against radioactive IPTG (isopropyl-thio-galactoside).

In order to seek the lactose repressor, we desired some means that would not depend on the specific models that might be imagined for the actual mechanism of repression. The minimal assumption on which the assay is based is that there should be some interaction between the repressor and the inducer. However, inducing substances added to the cell are often modified before they can trigger induction. Such is the case with lactose which must be split by β-galactosidase in order to induce,[3] but the thiogalactosides appear to be true gratuitous inducers; no chemical modification has yet been detected associated with their action as inducers. The ability of IPTG to stabilize certain temperature-sensitive i-gene mutants[4] argues strongly that the i-gene product interacts directly with this inducer. Furthermore, the existence of a competitive inhibitor of induction, ONPF (o-nitrophenyl-fucoside), which can also stabilize certain leaky i mutants and which behaves as though it drives the repressor into the form that shuts off the operator, also supports this thesis.[5]

Design of the Experiment.—In order to detect the binding of IPTG to the repressor by equilibrium dialysis, one must achieve concentrations of repressor that are comparable to the dissociation constant of the complex. What affinity does the repressor have for the inducer? Half-maximal induction occurs at $2 \cdot 10^{-4}$ M IPTG in a permeaseless strain. This fact alone does not lead directly to an estimate of the equilibrium constant because the enzyme level changes by a factor of 1,000 on induction. Since the rate of enzyme synthesis varies inversely with the first power of the repressor concentration,[4] the concentration of free repressor must have dropped by a factor of 1,000. Because the interaction between the repressor and inducer is quadratic in the inducer concentration, as is shown by the shape of the induction curve,[6] the inducer concentration must be a factor of the square root of 1,000 above the equilibrium constant at half-maximal induction. Thus, the wild-type repressor should have an affinity of the order of $6 \cdot 10^{-6}$ M for IPTG, and if any binding is to be detectable, the repressor concentration must exceed a few times 10^{-6} M. Since the cell pellet is about 10^{-9} M in cells, and since one does not expect many copies of the repressor per operator site, it seems likely that one would have to fractionate and concentrate the repressor by at least a factor of 100, working blindly, before detecting any effect. To improve the chances for success, we decided to isolate a mutant bacterium in which the i gene produces a product that binds more tightly to the inducer. Fortunately, such a mutant, an i^t (tight-binding) mutant, was found.

The i^t Mutant.—We enriched for an i^t mutant by using a technique which will select preinduced cells out of an uninduced population. A challenge with a very low level of inducer will trigger only those cells that are unusually sensitive. Such cells will be selected for along with constitutive mutants, but the constitutives can then be selected against by growth in the presence of TONPG, a compound that inhibits cells with an expressed permease (see *Experimental Details*). A super-inducible mutant was found. The induction curve of a permeaseless derivative of this strain is shown in Figure 1. The basal level is raised, in comparison to the wild type, but the midpoint of the induction curve is pulled lower than would be expected on the basis of the changed basal level alone. Furthermore, in contrast to the wild type, the induction kinetics are *linear* at the lowest levels of inducer (shown in the insert), the basal level being doubled at $7 \cdot 10^{-7}$ M IPTG, and the behavior only becoming quadratic as the level of inducer rises toward that necessary for full induction. The linear behavior suggests that this range shows single-site binding of IPTG to the repressor and that the point at which the basal level is doubled corresponds to the condition that half the repressor is bound to IPTG and half is free to repress. This yields a naïve estimate for the K_m of $7 \cdot 10^{-7}$ M. The rest of the curve can be interpreted as showing that there are two (or more) sites for the binding of IPTG on the molecule and that in order to scavenge the last repressor

TABLE 1

THE REVERSIBLE BINDING OF C^{14} IPTG

Sample	Cpm/0.100 ml	Corrected value	Excess bound as % of outside concentration
Outside concentration of C^{14} IPTG	475	—	—
1—Inside after 30 min dialysis	841	1025	116
2—Inside after 1 hr dialysis	950	1125	137
3—Inside a sac after 1 hr further of dialysis against buffer without IPTG	78	—	—
4—Inside after another 30 min dialysis against C^{14} IPTG	795	1035	118

Four identical 0.1-ml samples of protein were dialyzed against C^{14} IPTG in TMS buffer at 4°. At the end of 30 min, the IPTG concentration inside sac #1 was measured. At 1 hr the concentration inside sac #2 was measured, and the others were transferred to unlabeled buffer. At the end of the next hour, sac #3 was read, and sac #4 was transferred back to the original flask for another 30 min. The raw data, cpm/0.100 ml, are given in the first column. The volume of each sample was measured, to correct for the water uptake during the dialysis and that carried on the walls of the dialysis sac, and the corrected numbers, expressed as cpm/0.100 ml, are given in the second column.

off the operator, both IPTG sites must be fully loaded. (If the two-site nature of the curve is taken into account, the estimate for the binding constant would rise to about $1.2 \cdot 10^{-6}$ M.)

Detection of an Effect.—Since, even for the mutant strain, the repressor must be more concentrated than it is in the cell in order to display any binding of IPTG, we made a diploid derivative of the i^t mutant strain and proceeded to fractionate cell extracts with spermine and ammonium sulfate and to examine the concentrated protein fractions. The first detection of any effect was marginal; the spermine precipitate from the mutant strain yielded a 4 per cent binding (1000 cpm/0.1 ml outside a dialysis sac, 1040 cpm within). Further purification, however, immediately produced greater effects. The procedure described in the *Experimental Details* yields material that will draw the inducer into a dialysis sac to a concentration 1.5–2 times that outside the sac (50–100% excess binding) at a protein concentration of 10 mg/ml.

The labeled IPTG is bound reversibly; it can be freely dialyzed into, out of, and even into the sac again. This is illustrated in Table 1. At 4°C, the dialysis goes essentially to completion in about 30 minutes; assays were run for periods ranging up to eight hours. That the observed binding is not due to a contaminant in the radioactive IPTG was confirmed by using two different preparations of labeled IPTG (a C^{14}-labeled commercial preparation and a H^3-labeled Wilzbach preparation) and by examining the competition with unlabeled IPTG.

Negative Controls.—Is the material that binds IPTG really the i-gene product? The most critical controls are to examine a diploid amber-suppressor-sensitive i^- strain, which should have only a fragment of the i-gene product, and to examine a diploid i^s strain, which should have an i-gene product that is unable to recognize the inducer.[7] The i^s strain that was used, like many such strains, is slightly inducible at 0.1 M IPTG. This affinity of the i^s repressor for IPTG should be completely undetectable by the assay. The necessary diploid strains, isogenic to the i^t mutant strain, were constructed by F-duction. An identical parallel purification was run on 50-gm lots of control cells and i^t mutant cells. At the end of the purification, samples of each protein solution were dialyzed against several different IPTG concentrations. Figure 2 shows the result for the i^s control; no binding was observed. This finding implies that the substance to which the IPTG binds is either the i-gene product or else some other material of the *lac* operon whose synthesis would be blocked in this strain. This second possibility is ruled out by the control shown in Figure 3. No binding was found with the fraction isolated from the i^- strain. Since this strain is wild type with respect to the lactose enzymes, no binding to any of them was being observed.

Other less specific negative controls have been done. No binding was found in an isogenic diploid deletion strain carrying the Beckwith deletion M116, which has cut out the i gene as well as the beginning of the β-galactosidase gene. Mixing experiments were done with the extracts from this delection and the *sus* i^- strains mixed with the i^t extract to show that no inhibitors were present in the negative controls. Furthermore, we examined, in haploid strains, a deletion of the entire *lac* region carried in the Lederberg strain W-4032 and the i^- strains ML 308 and 2.340; no binding was found. The haploid amount of wild-type i-gene product is detectable. Our yields, however, are not sufficiently consistent to guarantee a factor of two

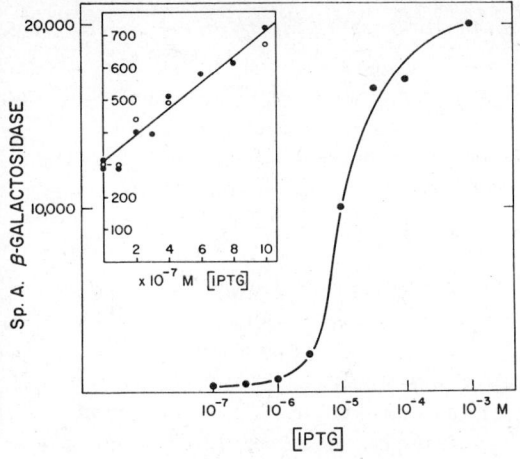

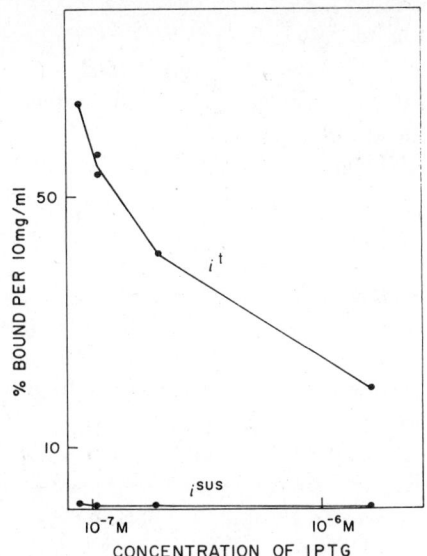

FIG. 1.—The induction of β-galactosidase in the i^t mutant. A permease-negative derivative of the i^t mutant, carrying an $F'\ gal^+$ factor, was grown at 35°C in M56-glycerol medium for ten generations in the presence of various concentrations of IPTG. The β-galactosidase activity was measured on toluenized samples. The insert shows that the induction kinetics are linear at the lowest concentrations of inducer. The open and closed circles represent independent experiments done with different permeaseless derivatives.

FIG. 3.—The binding ability of a suppressible i^- strain. The repressor fraction was isolated in parallel from isogenic $F'i^{sus}/i^{sus}$ and $F'i^t/i^t$ cells and assayed as described for Fig. 2.

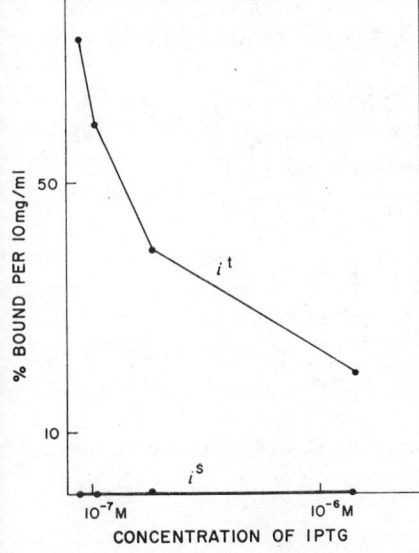

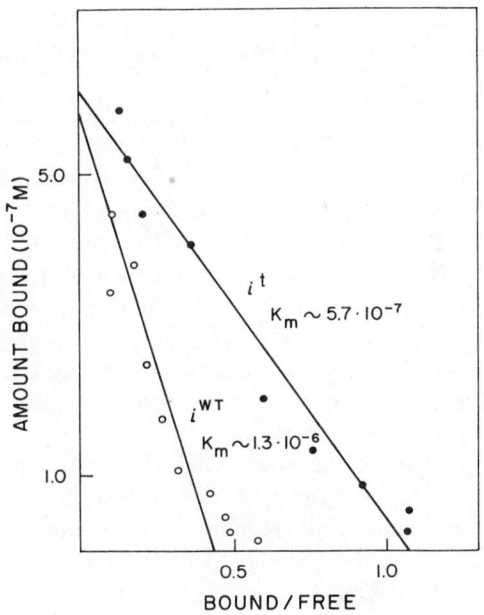

FIG. 2.—The binding ability of an i^s strain. The repressor fraction was isolated in parallel from 50-gm lots of isogenic $F'\ i^s/i^s$ and $F'\ i^t/i^t$ cells. Samples of each extract were dialyzed against different concentrations of C^{14} IPTG for 8 hr at 4°C in TMS buffer as described in the *Experimental Details*. The excess label bound inside each dialysis sac, normalized to a protein concentration of 10 mg/ml, is plotted against the IPTG concentration outside the sac. The corrections for protein concentration are of the order of 20%.

FIG. 4.—The binding constants of the wild-type and i^t mutant strains. Samples isolated from isogenic wild-type and i^t diploid strains were dialyzed against many different concentrations of C^{14} IPTG as described under Fig. 2. The amount of IPTG bound at each external IPTG concentration, normalized to a protein concentration of 10 mg/ml, is plotted against the ratio of bound to free IPTG. Single-side binding should obey the relation: (Amount bound) = (number of sites) − K_m (bound/free).

difference that might be due to gene dosage, so we cannot usefully compare haploid and diploid strains.

The amount of repressor is not increased if the cells are fully induced. Neither is the purification affected if done in the presence of a high concentration of IPTG.

Positive Controls.—Since the binding of IPTG to the wild-type repressor is detectable, the binding constants of the i^t mutant and wild-type strains can be compared *in vitro*. By measuring the amount of IPTG bound at various IPTG concentrations and plotting the amount bound against the ratio of bound to free IPTG, one would get a straight line if there is a single type of noninteracting binding site. The intercept on the y axis will be the molarity of binding sites and the slope will be the negative of the binding constant. Such plots are shown in Figure 4 for the mutant and wild strains. The binding is linear, and the measured binding constants (K_m) *in vitro* at 4°C are $6 \cdot 10^{-7} M$ for the i^t mutant and $1.3 \cdot 10^{-6} M$ for the wild type. The *in vivo* estimates led us to expect that these binding constants would differ by a factor of 4–10. We find only a factor of two difference and can only suggest that temperature effects, buffer effects, or the indirect nature of the *in vivo* estimates might account for this discrepancy. The important finding is that the mutation in the i gene that produces the superinducible phenotype, and presumably a repressor more sensitive to inducer, alters similarly the substance observed *in vitro*.

What affinities does the binding site have for other compounds? One can measure K_i's by dialyzing the repressor against a mixture of radioactive IPTG and varying amounts of unlabeled competitors. Such estimates are compiled in Table 2. Competition with cold IPTG yields essentially the same binding constant as was obtained with the labeled IPTG. This confirms that it is truly the IPTG that is binding. TMG (thio-methyl-galactoside), which is a weaker inducer than IPTG, has a weaker binding constant. ONPF, which is a competitive inhibitor of induction (and behaves as though it drives an allosteric equilibrium toward the form that binds to the operator), shows a 40-fold weaker binding than IPTG, again about the value that would be anticipated from its behavior *in vivo*.[8]

The interaction with galactose is of interest because the i^t represssor *in vivo* shows a sensitivity to galactose. The enhanced basal level of β-galactosidase in the i^t strain is doubled if the i^t gene is in a strain that lacks galactokinase. Such strains have internal galactose pools of the order of $2.5 \cdot 10^{-4} M$ produced from UDPGal.[9] Thus one expects an interaction between the mutant repressor and galactose in this concentration range.

Glucose shows very little affinity for this site. There is a 100-fold discrimination between glucose and galactose. The magnitudes of all these binding constants further support the identification of this material as the repressor.

Properties of the Repressor.—The ability to bind IPTG is not attacked by RNase or DNase. It is destroyed by pronase. The binding site can be inactivated by temperatures above 50°C. These are properties of the part of the molecule that interacts with the inducer. As yet there is no way of seeing the rest of the molecule.

TABLE 2

COMPETITION FOR THE INDUCER BINDING SITE

Competitor	K_i for the i^t gene product
IPTG	$5 \cdot 10^{-7} M$
TMG	$2 \cdot 10^{-6} M$
ONPF	$2 \cdot 10^{-5} M$
Galactose	$5 \cdot 10^{-4} M$
Glucose	$\geq 3 \cdot 10^{-2} M$

Binding constants inferred by competition of the unlabeled material with labeled IPTG. Protein samples were dialyzed against $1.2 \cdot 10^{-7} M$ C^{14} IPTG and three different concentrations of each of the competitors for 2 hr at 4°C. The amount of IPTG bound was measured, corrected for any changes in the protein concentration, and plotted to permit an estimate of the K_i.

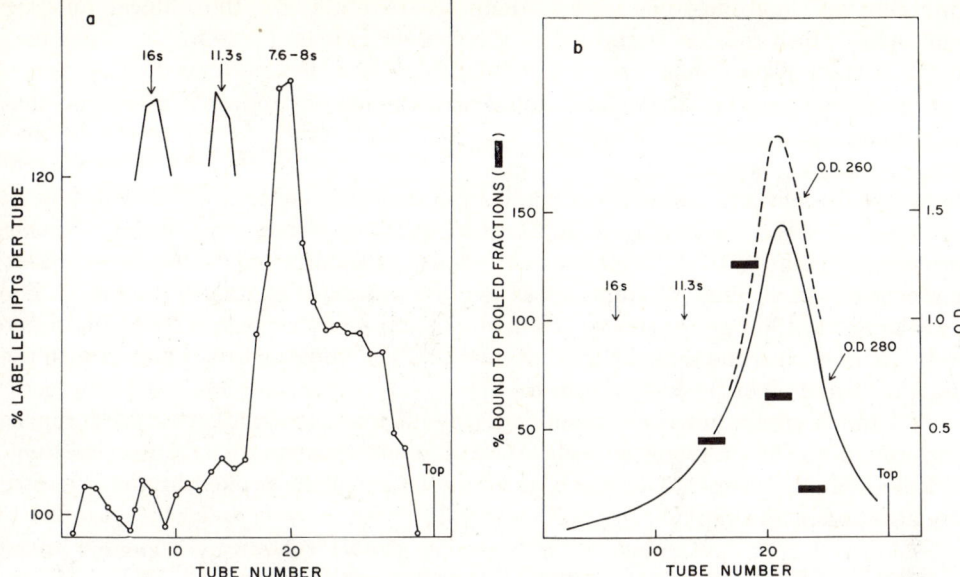

FIG. 5.—Sedimentation of the repressor. Samples of 0.2 ml of a concentrated repressor fraction were layered on 5-ml gradients, from 5 to 30% glycerol, in TMS buffer. The gradients were spun for 8.5 hr at 50,000 rpm at 4°C. In one tube, catalase and β-galactosidase were run as sedimentation markers. For the experiment shown in (a) the repressor fraction was brought to $2.5 \cdot 10^{-7}$ M C^{14} IPTG and layered on a gradient containing $1.2 \cdot 10^{-7}$ M C^{14} IPTG. After the centrifugation, two-drop samples were collected directly into scintillation vials and counted with Bray's solution. The background level in the gradient is about 1000 cpm; the counts are plotted as a per cent of the background level (open circles). For the experiment shown in (b), a parallel gradient was run without IPTG. The optical density was measured on each two-drop sample at 260 mμ (- - - - -) and 280 mμ (———). Three-tube samples were pooled and placed in dialysis sacs. Their volumes were reduced by burying the sacs under dry G 200 Sephadex for 1 hr. Then the samples were dialyzed against C^{14} IPTG. The excess bound, corrected to 0.100 ml final volume inside the dialysis sac, is plotted as a bar. Superimposed on both experimental curves is the position of the markers taken from a parallel gradient.

The sedimentation of the repressor was followed on glycerol gradients. By sedimenting the material through a gradient in the presence of a uniform level of labeled IPTG, one can observe a peak of label bound in equilibrium. This is shown in Figure 5a. In a parallel tube, catalase and β-galactosidase were used as markers. The repressor sediments at 7–8S; we estimate its molecular weight to be 150,000–200,000. Figure 5b shows the repressor seen by assaying after a gradient without IPTG. The tubes were pooled; the samples were concentrated and dialyzed against radioactive IPTG. The 280- and 260-mμ absorption is indicated, to underline the impurity of our preparation. The material that binds IPTG sediments again at 7–8S.

We estimate the amount of repressor in the cell by the yield that is obtained after the first steps of purification. The most recent procedure, breaking the cells with glass beads in a Waring Blendor, and fractionating directly with ammonium sulfate, yields all of the binding ability in the 0–35 per cent cut. The highest recoveries are about 100 sites per cell for the binding of IPTG. These are diploid cells in a tryptone-yeast extract medium harvested late in the log phase. These cells may have four to six gene copies, so we interpret the figure as 20 sites per gene copy and probably ten molecules per gene if there are two sites for the inducer on each molecule. With this last assumption, the repressor corresponds to about one part in 10^4 of the cell's proteins.

Conclusions and Outlook.—Our findings that the i-gene product is a protein, that it is uninducible, and that it occurs in a small number of copies serve to confirm many of the expectations that have grown up over the years. The discovery of temperature-sensitive mutants in the i gene implied that the i gene coded for a protein.[10, 11] The isolation of amber-suppressor-sensitive i^- mutants further proved the point.[12, 13] The estimate of a small number of copies of the repressor has been the traditional explanation of the phenomenon of escape synthesis.[14] The positioning of the i gene outside the operon[15] and an *in vivo* experiment on i-gene induction[16] both argue that the level of the i product would not rise and fall with the state of induction of the lactose enzymes.

An explicit assay, however, unambiguously demonstrates these points and opens the way to a full physical and chemical characterization of the i-gene product. Furthermore, experiments designed to ask which steps are blocked by the repressor are now possible *in vitro*.

Summary.—The *lac* repressor binds radioactive IPTG strongly enough to be visible by equilibrium dialysis. This property serves as an assay to detect the repressor, to quantitate it, and to guide a purification. It is a protein molecule, about 150,000–200,000 in molecular weight, occurring in about ten copies per gene. That the assay detects the product of the regulatory i gene is confirmed by the unusually high affinity shown for IPTG, by the difference in affinity of the substances isolated from the wild-type and a superinducible i-gene mutant, and by the absence of binding in fractions from i^-, i-deletion, and i^s strains.

Experimental Details.—Buffers: Two buffers were used. TMEM buffer contained 0.01 M tris HCl, pH 7.4, 0.01 M Mg^{++}, 0.006 M β-mercaptoethanol, and 10^{-4} M EDTA; TMS buffer was 0.2 M KCl in TMEM buffer. Double-distilled water was used throughout.

The assay: A 0.1-ml protein sample was dialyzed at 4°C with shaking against 10 ml of TMS buffer containing $1.2 \cdot 10^{-7}$ M C^{14} IPTG (about 500 cpm/0.1 ml). Number 20 visking tubing was prepared by boiling three times in 10^{-3} M EDTA and stored in 10^{-4} M EDTA. A sac about 3–5 cm long was used for the dialysis. After the dialysis had gone to completion, the contents of the sac were squeezed out into a tube. A 0.100-ml portion was put directly into Bray's[17] solution and counted in a scintillation counter. A standard was made by taking a 0.100-ml sample of the external fluid. An additional 0.020 ml of the dialyzed protein sample was used for Biuret determination of the protein concentration, to correct for variations in liquid uptake from sample to sample. C^{14} IPTG, 25 mC/mM, was obtained from Calbiochem.

The Purification: Cells were grown in 8 gm Tryptone, 1 gm yeast extract, and 5 gm NaCl per liter of medium. They were harvested in late log phase at $1.5 \cdot 10^9$ cells/ml. The cultures were poured on ice; the cells were washed in TMEM and stored frozen in 50–60-gm lots. For most of the experiments described in this paper, the following purification was used. A 50-gm lot of cells was broken in the Hughes Press, and taken up in 200 ml of TMEM. After a DNase treatment (Worthington, electrophoretically pure) at 2 μg/ml and a low-speed spin, the extract was brought to 2 mg/ml with spermine. The spermine precipitate was dissolved in 50 ml of TMS buffer and the ribosomes were removed by a 90-min, 40,000-rpm spin. The high-speed supernatant was brought to 35% saturation with solid ammonium sulfate, at pH 7.0, and the precipitate collected, dissolved in, and dialyzed against TMS buffer. All these operations were carried out at 4°C.

Our latest procedure consists in breaking the cells by blending with glass beads in a Waring Blendor. Cells (150 gm) are blended with 150 ml of TMS buffer and 450 gm of glass beads for 15 min. The extract is made up to 5 vol with TMS buffer. After a DNase treatment and a low-speed spin, two ammonium sulfate precipitations are done: the 0–23% cut is discarded; the

23–33% cut is saved. The repressor is then applied to a DEAE Sephadex column in 0.075 M KCl in TMEM and eluted during a gradient at 0.15 M KCl.

Selection of the i^t mutant: A Sm^R derivative of W3102 (F^- gal k^-), obtained from M. Meselson, was used. After mutagenesis in N-methyl-N'-nitro-N-nitrosoguanidine (100 µg/ml in mineral medium M56[18] for 1 hr at 37°) and segregation, the cells were given a maintenance challenge. They were grown in M56 with $10^{-2} M$ glycerol, $8 \cdot 10^{-4} M$ leucine, and 10 µg/ml B_1 at 42° for six generations, challenged with $10^{-6} M$ IPTG during the last two generations of growth, then chilled at $5 \cdot 10^8$ cells/ml, washed twice with preconditioned medium (made by growing W3102 to glycerol exhaustion in M56), and inoculated at $5 \cdot 10^6$ cells/ml into 100 ml M56, preconditioned, containing $10^{-3} M$ melibiose and $1.5 \cdot 10^{-3} M$ ONPF.[8, 13] After 36 hr the bacteria came up, about 50% constitutives. For the back selection, the bacteria were deadapted in M56-glycerol, then inoculated at 10^6/ml into M56 containing $10^{-2} M$ glycerol and $3-6 \cdot 10^{-3} M$ TONPG which, as we found, inhibits the growth of bacteria whose *lac* permease is expressed. After three cycles of forward and backward selection, the bacterial colonies were screened with ONPG on plates containing $10^{-6} M$ IPTG.

Enzyme assays: The lactose enzymes were assayed as described previously.[8]

We wish to thank Christina Weiss and Susan Michener for their technical assistance; Drs. Jonathan Beckwith, Salvador Luria, and Matthew Meselson for providing bacterial strains; and the National Institutes of Health (GM 09541-05) and the NSF (GB 4369) for their support of this work.

Abbreviations used: IPTG, isopropyl-1-thio-β-D-galactopyranoside; ONPF, o-nitrophenyl-β-D-fucopyranoside; ONPG, o-nitrophenyl-β-D-galactopyranoside; TMG, methyl-1-thio-β-D-galactopyranoside; TONPG, o-nitrophenyl-1-thio-β-D-galactopyranoside.

[1] Pardee, A. B., F. Jacob, and J. Monod, *J. Mol. Biol.*, **1**, 165 (1959).
[2] Jacob, F., and J. Monod, *J. Mol. Biol.*, **3**, 318 (1961).
[3] Burstein, C., M. Cohn, A. Kepes, and J. Monod, *Biochim. Biophys. Acta*, **95**, 634 (1965).
[4] Sadler, J. R., and A. Novick, *J. Mol. Biol.*, **12**, 305 (1965).
[5] Kunthala Jayaraman, B. Müller-Hill, and H. V. Rickenberg, *J. Mol. Biol.*, **18**, 339 (1966).
[6] Boezi, J. A., and D. B. Cowie, *Biophys. J.*, **1**, 639 (1961).
[7] Willson, C., D. Perrin, M. Cohn, F. Jacob, and J. Monod, *J. Mol. Biol.*, **8**, 582 (1964).
[8] Müller-Hill, B., H. V. Rickenberg, and K. Wallenfels, *J. Mol. Biol.*, **10**, 303 (1964).
[9] Wu, H. C. P., and H. M. Kalckar, manuscript in preparation.
[10] Horiuchi, T., and A. Novick, in *Cold Spring Harbor Symposia on Quantitative Biology*, vol. 26 (1961), p. 247.
[11] Novick, A., E. S. Lennox, and F. Jacob, in *Cold Spring Harbor Symposia on Quantitative Biology*, vol. 28 (1963), p. 397.
[12] Bourgeois, S., M. Cohn, and L. E. Orgel, *J. Mol. Biol.*, **14**, 300 (1965).
[13] Müller-Hill, B., *J. Mol. Biol.*, **15**, 374 (1966).
[14] Revel, H. R., and S. E. Luria, in *Cold Spring Harbor Symposia on Quantitative Biology*, vol. 28 (1963), p. 403.
[15] Jacob, F., and J. Monod, *Biochem. Biophys. Res. Commun.*, **18**, 693 (1965).
[16] Novick, A., J. M. McCoy, and J. R. Sadler, *J. Mol. Biol.* **12**, 328 (1965).
[17] Bray, G. A., *Anal. Biochem.*, **1**, 279 (1960).
[18] Monod, J., G. Cohen-Bazire, and M. Cohn, *Biochim. Biophys. Acta*, **7**, 585 (1951).

THE LAC OPERATOR IS DNA

By Walter Gilbert and Benno Müller-Hill

DEPARTMENTS OF PHYSICS AND BIOLOGY, HARVARD UNIVERSITY

Communicated by J. D. Watson, October 25, 1967

How repressors act at the molecular level to turn off genes is only now beginning to be worked out. Most vital to this understanding is whether the operator, defined genetically as the site for the action of a repressor, would turn out to be part of a DNA molecule, a region of a messenger RNA molecule, or even a protein. Now that two specific repressors (lactose and λ) are available,[1, 2] it is possible to attack this problem directly. This was first done by Ptashne,[3] who showed that the λ phage repressor, a 30,000-mol-wt protein, binds specifically only to that region of a λ-DNA molecule where the genetic receptors (operators) lie. Here we report experiments, with the lactose repressor, that further show that the operator is DNA. This repressor binds specifically to DNA molecules that carry the lactose operon, attaching only to that unique region of the DNA molecule where the mutations that characterize the operator lie. Furthermore, this repressor is *released* from the operator by inducers, such as IPTG (isopropyl-1-thio-β-D-galactoside).

The Principle of the Experiment.—The assay for the *lac* repressor used the fact that this repressor could bind radioactive IPTG tightly enough to be detected by equilibrium dialysis. Since the relevant affinity is on the order of $10^{-6}\ M$, only repressor concentrations in this range are detectable. This assay cannot be used immediately to study the interaction of the repressor with the operator because attainable gene concentrations are so small. Even if one uses *lac* genes carried on the DNA isolated from a defective phage, one set of genes for each 3×10^7 mol wt, a 3 mg/ml solution of DNA is only $10^{-7}\ M$. The binding of repressor to such DNA would only be barely visible by the IPTG binding assay. An alternative approach is to prepare radioactive repressor, to follow the molecule directly. The IPTG binding assay has been used to guide a several thousandfold purification of unlabeled repressor. With this knowledge one could try to mimic this purification on a small scale with very highly labeled proteins—a blind purification, since the specific labeling is so high and the physical scale of the radioactive preparation so small that one cannot follow the purification by the IPTG binding assay. A complete purification is unnecessary; all that is required is a sufficient enrichment of the *lac* repressor so that it represents a reasonable fraction of the labeled material, while other proteins that bind to DNA are removed so that specific effects can be observed. By including a sizing step, isolating only 7–8S material that includes the *lac* repressor, one can easily distinguish later a small fraction of the label binding to and sedimenting with 35S *dlac* phage DNA.

The details of the purification are given in the experimental methods. Sulfur-labeled proteins from a triploid strain, carrying three copies of the *lac* genes, are fractionated with ammonium sulfate and then run on a DEAE Sephadex column using a step elution. The material is then concentrated and run upon a glycerol gradient. Since the repressor, as determined by its binding to IPTG, sediments near 7.6S, samples are taken from this region of the gradient, determined by an aldolase marker. When this radioactive material is mixed with phage DNA carry-

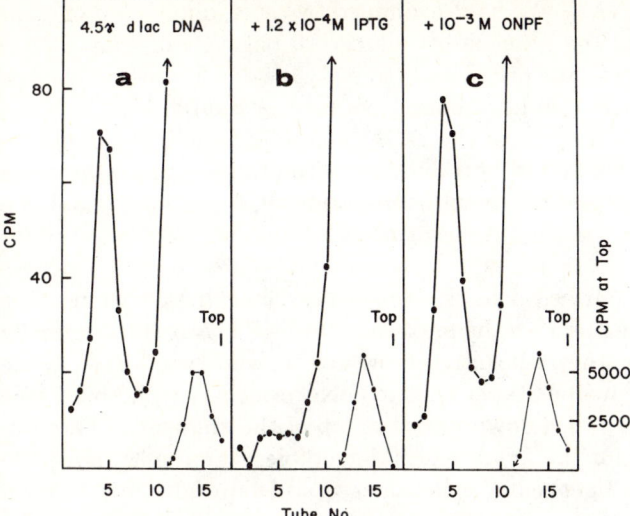
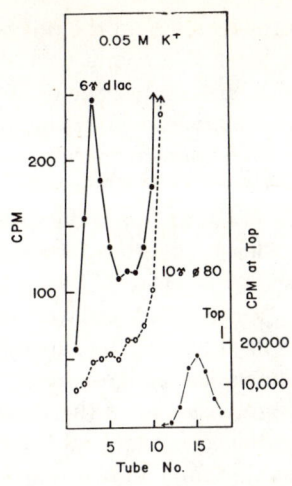

FIG. 1.—The binding of the *lac* repressor to *dlac* phage DNA and its release by inducer. Three identical mixtures of 4.5 γ of *dlac* phage DNA and 8S radioactive protein were sedimented on glycerol gradients containing TMEM buffer and run for 2 hr and 20 min at 65,000 rpm as described in the *Experimental Details*. The DNA by itself would form a sharp peak at tubes 4 and 5. The left panel (*a*) shows that a distinct peak of label sticks to the DNA and sediments down the gradient. The DNA is in at least tenfold excess; all the radioactivity that can bind to DNA at this concentration has bound to it. The center panel (*b*) shows that if the gradient solution contains $1.2 \times 10^{-4}\ M$ IPTG, this binding is abolished. The right-hand panel (*c*) shows that if the gradient contains $10^{-3}\ M$ ONPF, there is no effect on the binding.

FIG. 2.—The specificity of the binding. In parallel gradients the same repressor preparation was run with two different DNA's. One reaction mixture contained 6 γ of pure *dlac* phage DNA; the other contained 10 γ of the parental ϕ80–λ hybrid. The gradients contained 0.05 M KCl in TMEM.

ing the *lac* region, the mixture incubated and then sedimented on a glycerol gradient, a small peak of radioactivity moves out of the 8S region and sediments with the DNA around 35 to 40S. Figure 1*a* shows such a gradient pattern in 0.01 M Mg^{++}. Only 1 per cent of the radioactivity moves with the DNA, even though the DNA is in excess. That the label binding to DNA represents the *lac* repressor, and not just sticky proteins, is shown by the material being released by IPTG. Figure 1*b* shows that if $10^{-4}\ M$ IPTG is put throughout the gradient, no binding is observed. This effect of IPTG is specific: ONPF (*o*-nitrophenyl-β-D-fucoside), a substance which binds to the repressor but does not induce, has no effect. Figure 1*c* shows that even $10^{-3}\ M$ ONPF does not interfere with the binding.

RNase has no effect on this binding. When 75 γ/ml of RNase was added to the binding mixture, for a 20-minute incubation at 30°C, no effect on the binding to DNA was observed. Unlabeled, purified *lac* repressor competes for this binding.

The Specificity of the Binding to DNA.—If the repressor interacts with the operator region, the repressor should bind only to DNA carrying the lactose operon itself and, specifically, only to that region at the beginning of the lactose operon which is characterized through mutations as the genetic operator. In fact, no binding is found with phage DNA not carrying the *lac* genes. Figure 2 shows such an experiment in 0.05 M KCl. Furthermore, one can ask the more specific question, Does the repressor bind to the operator region by using operator-constitutive (o^c) mutants carried on the phage DNA? We have examined two such mutants: one has a level of enzyme activity in the absence of inducer 20 per cent of that attainable in the presence of inducer; the other, an extremely low-level constitutive, has an enzyme level in the absence of inducer only 1 per cent of the full level. If the active operator region itself is a region of the DNA molecule, the affinity of the repressor for DNA would be changed in both of these o^c mutants. Since the basal level of enzyme is only 0.1 per cent of the fully induced level, the affinity should be at least a factor of 10 weaker for the 1 per cent o^c and a factor of 200 weaker for the 20 per cent o^c.

When DNA isolated from purified defective phages carrying these o^c mutations is used in the experiment, one observes the patterns shown in Figure 3. Figure 3a shows the control binding of the radioactive repressor to wild-type DNA, while Figure 3b shows the binding to the 20 per cent o^c. No peak is visible, but some radioactive material has been pulled down from the top of the gradient. The residual affinity of the repressor for the DNA is still detectable. Figure 3c shows the affinity of the repressor for the 1 per cent o^c. In this case, still more label moves down the gradient, but the affinity of the repressor for this mutant DNA is less than the affinity for the wild-type DNA.

These experiments demonstrate that the repressor binds to a unique sequence on this DNA molecule, the operator region. Furthermore, all attempts to demonstrate binding to denatured DNA have failed. One infers that the binding is to double-stranded DNA.

The Magnitude of the Binding Constants.—One can obtain rough estimates for the affinity of the repressor for the operator region by observing the shape of the peaks riding on the DNA. These experiments have all been done with an excess of DNA, but the DNA concentration falls as the band moves down the gradient, dropping by a factor of 6 from its initial value in the reaction mixture to its final value when the gradient is collected. If the DNA is run separately in a parallel gradient, the recovery is about 70 per cent and the peak concentration, with 4.5 µg of DNA as an input, is only 2.5 µg/ml (only $8 \times 10^{-11} M$). From the sharpness of the peaks shown in Figure 1, one would infer that the DNA concentration must be at least a factor of 10 higher than the binding constant. Since a peak of similar sharpness is obtained when only 0.9 µg of DNA is used, one would estimate that the affinity is on the order of $2 \times 10^{-12} M$ in $0.01 M$ Mg^{++}.

The tightness of the binding is influenced by the salt concentration. As Ptashne has observed for the λ repressor, when the salt concentration rises, the affinity for the DNA weakens. Since the DNA concentrations that are used are close to the affinity constants, dissociation plays a role, and a slight change in the salt alters the experimental picture. The profile shown in Figure 2, taken in $0.05 M$ KCl, can be interpreted as showing that 20 per cent of the bound material trails immediately behind the DNA peak due to a weakened affinity. At $0.15 M$ KCl, the binding to our standard amount of DNA has been essentially abolished. Figure 4 shows, however, that the binding can be easily observed again by raising the DNA concentration a factor of 4. Table 1 collects estimates for the binding constants of the wild type and of operator-constitutive mutants.

What affinities does one expect *in vivo*? The affinity of the repressor for the operator can be estimated from the magnitude of the basal level of enzyme synthe-

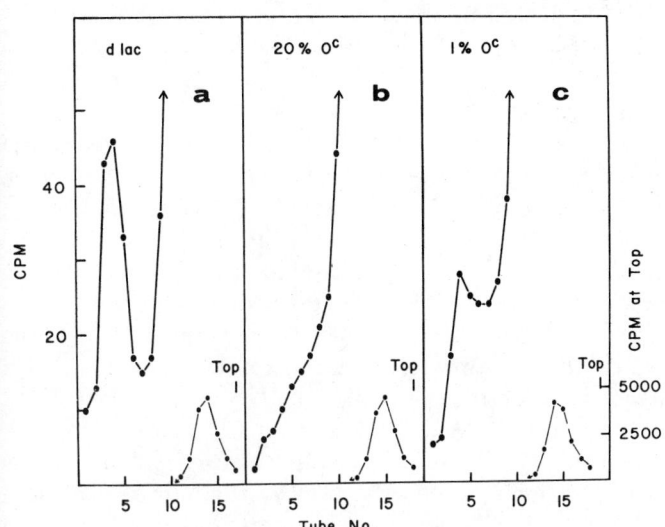
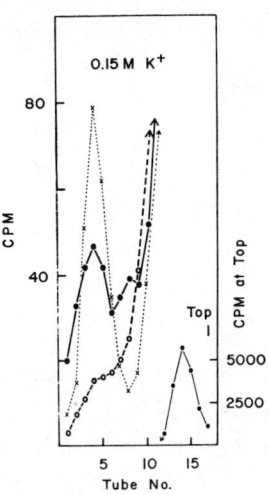

FIG. 3.—The *lac* repressor binds specifically to the *lac* operator. In parallel gradients the same repressor preparation was run with three different purified *dlac* phage DNA's in TMEM. The left-hand panel (*a*) shows the profile obtained with 4.5 γ of wild-type *dlac* DNA. The center panel (*b*) shows that the same amount of DNA carrying an o^c mutation that produces 20% of the full level of enzyme does not bind repressor. The third panel (*c*) shows the binding to an o^c that produces 1% of the full amount (10 times the basal level).

FIG. 4.—The effect of salt. The three superimposed profiles correspond to three parallel gradients: (1) repressor bound to 4.5 γ of *dlac* DNA run in 0.15 *M* KCl and TMEM, O---O---O; (2) repressor bound to 18 γ of *dlac* DNA run in 0.15 *M* KCl and TMEM, ●—●—●; and (3) a control of 4.5 γ of *dlac* DNA run in TMEM, x···x···x.

sis. To the extent that the rate of enzyme synthesis is simply proportional to the amount of DNA free of repressor and obeys mass action kinetics,[4] the basal rate's being only one thousandth of the full rate means that the repressor concentration is about one thousand times the dissociation constant for the operator. On the basis of the isolation procedures, our current estimates are of the order of 10–20 repressor molecules per haploid cell. Thus the repressor concentration in the cell is of the order of $1-2 \times 10^{-8} M$, and the affinity of the repressor for the operator should be $1-2 \times 13^{-11} M$. One does not know how to duplicate the ionic conditions within the cell, nor does one know that the interpretation of the basal level as simply determined by the repressor concentration is true. Furthermore, the *in vitro* estimates are made at high pressure on a gradient. Nonetheless, the estimates *in vitro* for the affinity, ranging from $2 \times 10^{-12} M$ in low salt to several times $10^{-10} M$ in higher salt, are in reasonable agreement with the *in vivo* estimate. The weakening of the binding with the o^c's is in the right direction and consistent, roughly, with the difference of the basal rates of synthesis of these two mutations.

TABLE 1

ESTIMATED REPRESSOR-OPERATOR DISSOCIATION CONSTANTS *in vitro*

DNA	$0.01 M Mg^{++}$	$0.05 M K^+ +$ $0.01 M Mg^{++}$	$0.15 M K^+ +$ $0.01 M Mg^{++}$
dlac o^+	$2-4 \times 10^{-12} M$	$2 \times 10^{-11} M$	$3 \times 10^{-10} M$
1% o^c	$10^{-10} M$	$>10^{-9} M$	
20% o^c	$4 \times 10^{-10} M$	$>10^{-9} M$	

Is the magnitude of this binding physically reasonable? A binding of the order of 10^{-11} M requires some 15 or 16 kcal of binding energy. This energy could arise through the formation of many weak or four to five moderately strong bonds. The repressor must be able to recognize a stretch of at least 11 to 12 bases to select a unique site on the *E. coli* chromosome. (A 12-base sequence selects one out of 1.6×10^7 possible locations, while there are about 3×10^6 base pairs in the chromosome.) To recognize this number of bases individually would require at least 11 or 12 bonds, and thus a free energy change easily in the 15-kcal range. The recognition region will span a considerable distance along the DNA molecule, at least one turn of the helix, a 35-Å stretch; however, the *lac* repressor, 150,000 mol wt, is large enough. If the recognition were not as efficient as possible, the region would be larger.

These dissociation constants imply that the repressor takes hours to fall off the operator. Because the forward rate of formation of the operator-repressor complex will be limited by diffusion and steric factors to be only 10^8 M^{-1} sec^{-1} at the most, a dissociation constant in the 10^{-11} to 10^{-12} M range requires that the rate of decay of the complex be 10^{-3} to 10^{-4} sec^{-1}. How then is it possible that enzyme synthesis begins only minutes after the addition of inducer? Clearly the inducer will bind to the repressor-DNA complex in times that are short compared to 10^4 seconds and trigger the release of the repressor from the operator. Since the rate of release will depend on the amount of inducer, one expects a lag in enzyme induction at low levels of inducer. This lag can be estimated on very general principles and shown to depend only on the slow decay rate of the repressor-operator complex (k_s) and on the ratio of the final rate of enzyme synthesis to the basal rate (r_E/r_B). The argument given in the appendix shows that the time delay (T_c) in the induction curve can be written

$$T_c = (1/k_s)(r_B/r_E) + \text{a constant}.$$

Such time delays have been observed for the induction of the lactose enzymes by Boezi and Cowie.[5] Their data fit this formula with a k_s of 2.2×10^{-4} sec^{-1}, an entirely independent estimate for the decay time *in vivo*.

Summary.—The experiments reported here demonstrate that the *lac* repressor binds specifically to the operator region, that its binding to the operator is weakened by mutations in that region which produce o^c's, and that it is released from the operator by the inducer. These experiments completely support the model of repression which proposes that the repressor, on binding to the operator, hinders the transcription of the adjacent genes into RNA and thus prevents their functioning.

Experimental Details.—Buffers and general methods: TMEM buffer is 0.01 M tris, pH 7.4, 0.01 M magnesium acetate, 10^{-4} M ethylenediaminetetraacetate (EDTA), and 0.007 M β-mercaptoethanol. TMS buffer is TMEM with 0.2 M KCl. All tubes and centrifuge tubes were boiled in EDTA. Other methods were described previously.[1]

Labeled repressor: The bacterial strain carried three sets of *lac* genes, one on ϕ80 *dlac* carried as a single defective lysogen at the ϕ80 attachment site, one at the normal *lac* site, and one on an F *lac* episome. For the sulfur labeling the cells were grown in minimal medium: 0.1 M potassium phosphate buffer, pH 7.4, 2 gm/liter NH$_4$Cl, 3 gm/liter NaCl, 2×10^{-4} M Mg^{++}, and 10^{-4} M sulfate, to glycerol starvation at a few times 10^8/ml. They were diluted back and grown to sulfur starvation in 6×10^{-5} M radioactive sulfate. Ten ml of cells were labeled with 20 mc of S^{35}, harvested, and diluted with 1 gm of unlabeled cells. The cells were ground with alumina, the extract suspended in TMS buffer to a final volume of 5 ml, and the debris spun out. The extract was brought to 35% saturation with solid ammonium sulfate, and the precipitate collected and back-extracted with 2 × 1-ml portions of TMS buffer at 28% and 23% of saturation with ammonium sulfate. The two 23% extractions were pooled, and the ammonium sulfate dialyzed out against TMEM containing 0.1 M KCl. The sample was applied to a 2-ml DEAE Sephadex column in the same buffer, and a cut was taken between 0.12 M and 0.17 M KCl in TMEM. Column fractions (0.5 ml) were collected into tubes containing 0.1 mg of aldolase to provide protective protein and a marker during the centrifugation. The material was concentrated by drying down a dialysis sac with G200 Sephadex, layered on a 5–30% glycerol gradient in TMS containing 0.1 mg/ml BSA, and centrifuged for 16 hr at 45,000 rpm, 4°C. The aldolase marker was located by optical density, and samples from the tubes at and immediately following the aldolase peak were used for the binding experiment.

Phage DNA: The *lac* DNA that was used was isolated from a defective *lac* phage made by V. Rybtchine and Ethan Signer. The phage is derived from a $\phi 80$–λ hybrid,[6] $h_{80}i^{\lambda}$, contains the $c_1 857$ temperature-inducible mutation, and carries the *lac* genes as a replacement of late phage functions. The o^c mutants used were also made in this phage by Signer. They all have a functioning i gene. The defective phages are 0.005 gm/cc denser than the parental hybrid. A double lysogen was grown at 34°C in a glucose casein–amino acid medium buffered with 0.1 M tris, pH 7.5, and containing more magnesium ion than phosphate. At 5–8 × 10^8 cells/ml the culture was heat-shocked to 42°C for 15 min, chilled to 37°C, and shaken at 37°C until lysis. The pH was maintained at 7.5. Titers were 1–2 × 10^{11}/ml, after chloroforming. The phage were harvested in the Spinco 30 head, purified on a block CsCl gradient, and then banded in an equilibrium CsCl gradient overnight in the 40 head. DNA was prepared from the purified defective phage by rolling the phage stocks with phenol. The DNA was dialyzed overnight against 0.01 M tris, pH 7.4, 0.05 M KCl, and 10^{-4} M EDTA, and stored at about 100 γ/ml at 4°C.

DNA binding assay: The DNA, handled in 0.1-ml pipettes, was heated to 70°C for 5 min and chilled quickly, to break aggregates. The reaction mixture contained, in a final volume of 0.25–0.3 ml, TMEM buffer with 0.05 M KCl, 150 γ of BSA, generally about 5 γ of DNA, and radioactive repressor. After a 20-min incubation at 30°C, the mixture was layered on a 5-ml, 5–30% glycerol gradient containing 500 γ/ml BSA and the specified buffer, either TMEM or TMEM with KCl. After a 2-hr 20-min spin at 65,000 rpm, 8°C, 4-drop samples were collected into scintillation vials and counted with Bray's solution.

Mathematical Appendix.—One can calculate how rapidly the repressor will be driven off the operator by the inducer in a variety of specific models for the induction process. The results are identical so we shall give a general, rough argument. The rate of change of the amount of the repressor-operator complex ($[D\text{-}R]$) after the addition of inducer will be given by

$$\frac{d}{dt}[D\text{-}R] = k_s [D\text{-}R] + k_f [D\text{-}R\text{-}I],$$

where $[D\text{-}R\text{-}I]$ is the amount of the repressor-inducer-operator complex and k_s and k_f are the slow and fast decay rates of the two complexes, respectively. One expects k_f to be at least 10^4 times larger than k_s. Furthermore, the equilibration of the inducer, a small molecule, with the various complexes should be very rapid. As we shall see, $[D\text{-}R\text{-}I]$ itself is in general negligible compared to $[D\text{-}R]$. One does not know directly what the affinity of the inducer for the DNA-repressor complex would be, but each of these complexes has a dissociation constant

$$\frac{[D][R]}{[D\text{-}R]} = K_0 \quad \text{and} \quad \frac{[D][R\text{-}I]}{[D\text{-}R\text{-}I]} = \tilde{K}_0.$$

Each of these dissociation constants is the ratio of a decay rate to an association rate. To the extent that the shape of the repressor is not greatly changed by complexing with the inducer, the diffusion constant and steric factors will not be changed and the two association rates, k_1 and k_2, will be equal. Thus

$$\frac{[D\text{-}R\text{-}I]}{[D\text{-}R]} = \frac{K_0 [R\text{-}I]}{\tilde{K}_0 [R]} = \frac{k_s/k_1}{k_f/k_2} \frac{[R\text{-}I]}{[R]} \approx \frac{k_s}{k_f} \frac{[R\text{-}I]}{[R]}.$$

This last is the crucial statement: the amount of DNA-repressor-inducer complex is small just in proportion to its instability. The decay of the repressor-operator complex can be rewritten as

$$\frac{d}{dt}[D\text{-}R] = [D\text{-}R]\, k_s \left(1 + \frac{[R\text{-}I]}{[R]}\right).$$

Thus the complex decays exponentially with a time constant

$$T = \frac{1}{k_s}\left(\frac{[R]}{[R] + [R\text{-}I]}\right).$$

Since there is a reasonable excess of repressor over operators, $[R] + [R\text{-}I]$ is approximately the total amount of repressor in the cell, and the quantity in parentheses is just $(r_B/r_E) - r_B$ because the rate of enzyme synthesis is given by

$$r_E = 1/(1 + [R]/K_0),$$

if one assumes that this rate is proportional to the amount of DNA free of repressor.

We wish to thank Christine Weiss for her excellent technical assistance, Drs. Ethan Signer and Jonathan Beckwith for providing bacteria and phage strains, and the National Institutes of Health (GM 09541-06) for their support of this work.

[1] Gilbert, W., and B. Müller-Hill, these PROCEEDINGS, **56**, 1891 (1966).
[2] Ptashne, M., these PROCEEDINGS, **57**, 306 (1967).
[3] Ptashne, M., *Nature*, **214**, 232 (1967).
[4] Sadler, J. R., and A. Novick, *J. Mol. Biol.*, **12**, 305 (1965).
[5] Boezi, J. A., and D. B. Cowie, *Biophys. J.*, **1**, 639 (1961).
[6] Signer, E., *Virology*, **22**, 650 (1964).

New Controlling Element in the *Lac* Operon of *E. coli*

by
KARIN IPPEN*
JEFFREY H. MILLER
JOHN SCAIFE†
JONATHAN BECKWITH
Department of Bacteriology and Immunology,
Harvard Medical School,
Boston, Massachusetts

The *lac* promoter maps between the repressor *i* gene and the operator *o* gene, so that the operator gene is probably transcribed.

THE genetic elements (*lac*) determining lactose metabolism in *E. coli* map very close together in a small region of the chromosome (Fig. 1)[1]. Three structural genes, *z* (for β-galactosidase), *y* (for galactoside permease) and *a* (for thiogalactoside transacetylase), are regulated by the repressor product of the *i* gene. In the absence of β-galactoside inducers of the system, the repressor interacts with the *lac* operator (*o*) to prevent synthesis of the three proteins. Jacob and Monod have suggested that repression involves an inhibition of transcription of *lac* operon DNA into a messenger-RNA (*m*RNA) copy[1]. This model for regulation has recently received strong support from the experiments of Gilbert and Muller-Hill, who have isolated the repressor protein and have found that it binds specifically to *lac* operator double-stranded DNA[2]. Thus it seems likely that this repressor–operator complex prevents transcription of the *lac* operon.

There is genetic evidence indicating that it is not the operator itself which serves as a transcription initiation site. Whereas one would expect mutations in an initiation site to alter the potential for operon expression, operator-constitutive (*o*ᶜ) mutations, including some well characterized deletions (unpublished results of Davies and Jacob), do not affect the rate at which the operon can be expressed[3]. Such an initiation site must therefore lie outside the operator, either between *i* and *o* or between *o* and *z*. Jacob, Ullman and Monod[3] have proposed that such a site, the promoter (*p*), maps between *o* and *z*.

We have reported the isolation and characterization of mutants which have all the properties of mutants in an initiation site of the *lac* operon[4]. These mutants, which reduce the maximum rate of *lac* operon expression, do not alter either the operator or the *i* gene. On the basis of genetic mapping, we had concluded that these mutants altered a site between *o* and *z*, possibly identical to the promoter. This conclusion was based on the use of certain *i⁻oᶜ* mutants thought to be deletions extending from the operator into or beyond the *i* gene[3]. Davies and Jacob have now shown, however, that these mutants are not overlapping deletions (unpublished results). In fact, it now seems likely that most or all of them behave as either negatively complementing *i⁻* mutations (personal communication from J. Davies and W. Gilbert) or as double mutants (unpublished results). The former class would be mutations in the *i* gene, which produce a defective subunit of the repressor, impairing the normal function of wild-type repressor sub-units. The findings of Davies and Jacob have led us to re-examine the mapping of our mutants.

* Harold C. Ernst Research Fellow.
† On leave from MRC Microbial Genetics Research Unit, London.

We present here conclusive evidence that the genetic site altered in our mutants maps between *i* and *o*. This evidence thus identifies a new controlling element determining the expression of the *lac* operon. We propose that it is in this region that the initiation of transcription takes place. Because, in the original formulation of the promoter concept, it was considered most likely that the site defined as promoter was an initiation point for *m*RNA transcription[3], we prefer to retain the term promoter to describe this new region (Jacob and Monod, personal communication, agree with this use of the term promoter). It now seems probable that, in fact, the region between *o* and *z* corresponds only to an initiation site for translation.

We have characterized four ultraviolet-induced mutants which co-ordinately reduce the rate of expression of the *z*, *y* and *a* genes. Some of their properties are described in Table 1. Three of these, although independently isolated, are almost identical in their characteristics, and at least two, *L*8 and *L*37, do not recombine with each other (unpublished results of J. Beckwith). *L*8 and *L*37 do, however, seem to be point mutants. The lesion in the fourth mutant, *L*1, confers an additional property on the strain, and, although the operator in *L*1 is normal, the activity of its *i* gene is reduced.

Table 1

	Induced levels of *lac* enzymes as percentage of wild-type	Regulation	*cis*-Dominant
*L*1	2	Partly constitutive	Yes
*L*8, *L*37	6	Normally inducible	Yes
*L*29	4	Normally inducible	Yes

Three of these mutants, *L*1, *L*8 and *L*37, have been described before[4]. The fourth, *L*29, was isolated as one of two *lac*⁻ mutants found after ultraviolet mutagenesis of a *lac*⁺ strain. The *o* and *i* characters of *L*1 were determined in diploid studies.

These mutations are *cis*-dominant in that they only reduce the rate of expression of the operon on the same chromosome and these effects are not relieved by the introduction of a second *lac* region into the cell[4]. An explanation for the properties of these mutants is that they have altered a site essential for the initiation of either transcription or translation of the operon. The result is a reduced efficiency of reading. In order to determine conclusively the position of the mutants relative to the operator and the *i* and *z* genes, we have employed a deletion analysis of this region. To avoid confusion we shall call these promoter mutants.

We have described a genetic system in which it is possible to isolate deletions removing the distal end (*a*

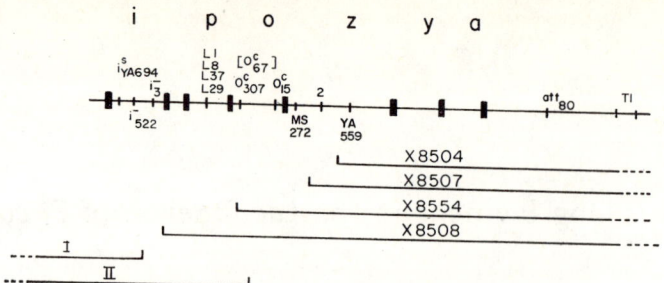

Fig. 1. The *lac* operon transposed (in φ80*dlac*) to the φ80 site on the bacterial chromosome. The $T1^R$ mutations were selected as before[7]. All *i* gene and o^c mutations were isolated by Jacob, Monod and co-workers. Lac^-_{MS272} is an extremely polar mutant isolated by Malamy[8]. Lac^-_2 and lac^-_{YA559} have been described before[9]. The orders of *i* and *o* mutations which are not established by deletion mapping have been determined by Davies and Jacob (results in preparation).

end) of the operon and extending varying distances toward and past the operator (Fig. 1)[5]. These deletions are isolated in a strain in which a φ80*dlac* transducing phage is integrated in the chromosome near the locus (*T1*) determining sensitivity to the bacteriophages *T1* and φ80. In the strain described in Fig. 1, the *lac* region is inverted from its normal orientation on the chromosome so that the *i* gene is furthest from the *T1* locus. By selecting $T1^R$ mutants of such a strain, it is possible to isolate deletions extending from the *T1* locus or beyond into the defective prophage[5,6]. These deletions are presumed to include all the genetic material between the *T1* locus and the determined deletion end, for they remove all point mutant sites in the *y* gene and in the *z* gene as indicated.

The deletions isolated in this way, which we have used in our mapping studies, are described in Fig. 1. Table 2 shows the frequency of recombination of these deletions with mutants affecting various components of the *lac* operon and the *i* gene. These mutants include an i^s mutation, two i^- mutants, four promoter mutants, three o^c mutants and three z^- mutants. One of the mutants, o^c_{307}, was shown by Davies and Jacob (unpublished results) to map the furthest to the left of any known o^c mutant. In most cases, the deletion strains were infected with an F′ factor (F-*lac-pro*) carrying the mutation to be mapped. Cultures of these partial diploids were then screened for recombinants. It should be noted therefore that the frequencies reported in Table 2 do not correspond to recombination frequencies measured in the standard way.

The crucial deletion for determining the position of the promoter mutants relative to the operator is *X8554*. Two promoter mutants, *L8* and *L37*, give frequencies of 0·065–0·086 per cent recombinants with this deletion. In contrast, no recombinants have been detected between this deletion and any of the o^c mutants as indicated in Table 2. Either this deletion covers the entire operator region, or else its terminus lies very close to the end of the operator nearest the *i* gene. In general, the other frequencies presented in Table 2 are entirely consistent with the order of mutations as determined by their ability to recombine with a series of other point mutations and deletions. The map order is also in complete agreement with respect to the order of *i* and *o* mutations determined by Davies and Jacob (unpublished results). Furthermore, the results unambiguously show that the position of all four promoter mutations is between *i* and *o*.

We have presented genetic evidence showing that mutants affecting the maximum level of *lac* operon expression map in the region between *i* and *o*. The properties of these mutants can be accounted for by two different general explanations. (1) These mutants define a site necessary for expression of the *lac* operon, which can be altered by mutation to reduce operon activity. (2) The mutants map outside all controlling sites for the operon, but in some undefined way affect operon expression negatively. Although it is very difficult to conceive of any simple mechanisms for the second explanation, it is, of course, imperative to show directly that these mutants define an essential site.

There is some evidence indicating that the region between *i* and *o* is essential to operon expression. It is possible to isolate *lac*-constitutive mutations which are caused by deletions (Table 2) cutting into either the *i* gene (class I) or the *o* region (class II) from a point outside the *i* end of the *lac* region[5]. In such a system we have isolated a large number of i^- *lac*+ deletions. In addition,

Table 2. PERCENTAGE OF RECOMBINANTS AMONG TOTAL COLONIES

Mutant	Deletion X8504	X8507	X8554	X8508
z^-_{YA559}	0	0	0	0
z^-_2	+	0	0	0
z^-_{MS272}	+	+	0	0
o^c_{15}	0·26	0·037	< 0·001	n.t.
o^c_{307}	n.t.	n.t.	< 0·001	n.t.
o^c_{67}	n.t.	n.t.	< 0·002	n.t.
L1, L29	+	+	+	0
L8	+	+	0·065	0
L37	0·79	0·34	0·086	0
i^-_3	n.t.	n.t.	0·19	0·069
i^-_{522}	n.t.	n.t.	0·16	0·094
i^s_{YA694}	+	+	0·24	+

The deletions are isolated in a strain carrying the φ80*lac* as described in the text and also carrying a *lac-proA,B* deletion. To estimate the percentage of recombinants we constructed strains in which the mutant to be mapped against the deletion was carried on an F-*lac-proA,B*+ episome. With the dominant i^- mutation (i_{522}) and the o^c mutants, aliquots of a culture of a strain partially diploid, in this way, for the *lac* region, were spread on to minimal medium (M63) containing glucose as carbon source, 5-bromo-4-chloro-indoxyl-β-D-galactoside (Xg; 0·004 per cent) and 0·004 M sodium citrate, necessary for the growth of $T1^R$ strains on this medium. Because Xg is not an inducer[4], the constitutive diploid strains were blue and any i^-o^- recombinants were white on these plates. White colonies appearing on such medium were purified and assayed for β-galactosidase to ensure that they carried a wild-type *lac* region. The same method was used to map the i^- deletion X8508 against i_3. The frequency of the *i* gene mutations with X8554 is actually the frequency of homogenotization of the i^+ or i^- allele. To map the promoter mutants, L8 and L37, cultures of partially diploid strains were spread on *lac*-EMB plates and pure *lac*+ (wild-type) colonies were picked and verified by assay. The percentage of recombinants is based on samples from many independent diploid cultures; the frequencies are not seriously biased by "jackpots" caused by recombination events occurring early in the growth of the culture. Some mutations (in F′ donors) were mapped by crossing against the F- *lac* Sr strains and selecting *lac*+Sr recombinants (spot tests). Positive results are indicated as + and negative as 0. These spot tests detect recombinants at a much higher efficiency than the scoring techniques described here. Some combinations were not tested (n.t.).

we have isolated one deletion of class II. This deletion removes the i gene and the promoter region, but leaves part of the operator intact. As expected, the expression of the operon is impaired by the deletion of the promoter region. The properties of this strain will be described elsewhere (unpublished results of J. H. Miller, J. R. Beckwith and E. R. Signer). In addition to confirming the mapping results described here, the properties of this deletion also point to the importance of the promoter region in the expression of the lac operon. Thus it seems most likely that the mutants, $L1$, $L8$, $L37$ and $L29$, do lie in an essential region (the promoter) determining the initiation of operon expression.

The evidence pointing to a new site essential to lac operon expression which maps between i and o provides strong indication that the transcription process is initiated before the operator region. This conclusion does not exclude the possibility that the initiation of operon translation also occurs in this region. If protein synthesis does begin before o, however, then the operator should be translated; there is no evidence to support this prediction. We therefore favour the view that there are two sites essential to the expression of the lac operon. One would be the promoter, lying between i and o, which is the site of initiation of mRNA synthesis. Initiation of translation may then take place at a second site in the o–z boundary region, defined by Jacob, Ullman and Monod[3].

With this picture of the organization of the controlling sites of the lac operon, it is possible to visualize a fairly straightforward mechanism for the expression and regulation of operon activity. According to this scheme, the promoter region serves as the initiation point for transcription, possibly by acting as a binding site for RNA polymerase. Promoter mutants would reduce the site's affinity for the polymerase. While it is possible that the repressor hinders binding of RNA polymerase, the results presented here suggest a very simple alternative. In binding to the operator, the repressor could directly block the progress of the RNA polymerase into the structural genes of the lac operon.

This work was supported by a grant from the US Public Health Service and a grant from the US National Science Foundation. J. Miller was supported by a Public Health Service training grant to the Department of Biochemistry and Molecular Biology, Harvard University.

Received February 12, 1968.

[1] Jacob, F., and Monod, J., *J. Mol. Biol.*, **3**, 318 (1961).
[2] Gilbert, W., and Muller-Hill, B., *Proc. US Nat. Acad. Sci.*, **56**, 1891 (1966); Gilbert, W., and Muller-Hill, B., *Proc. US Nat. Acad. Sci.*, **58**, 2415 (1967).
[3] Jacob, F., Ullman, A., and Monod, J., *Compt. Rend.*, **258**, 3125 (1964).
[4] Scaife, J., and Beckwith, J. R., *Cold Spring Harbor Symp. Quant. Biol.*, **31**, 403 (1967).
[5] Beckwith, J. R., Signer, E. R., and Epstein, W., *Cold Spring Harbor Symp. Quant. Biol.*, **31**, 393 (1967).
[6] Franklin, N. C., Dove, W. F., and Yanofsky, C., *Biochem. Biophys. Res. Commun.*, **18**, 910 (1965).
[7] Beckwith, J. R., and Signer, E. R., *J. Mol. Biol.*, **19**, 254 (1966).
[8] Malamy, M., *Cold Spring Harbor Symp. Quant. Biol.*, **31**, 89 (1967).
[9] Newton, W. A., Beckwith, J. R., Zipser, D., and Brenner, S., *J. Mol. Biol.*, **14**, 290 (1965).

Evidence for Two Sites in the *lac* Promoter Region

An analysis of promoter mutants of the *lac*† operon indicates that there are two distinct sites in the promoter. One of these is a site which normally promotes a low level (2%) of *lac* transcription, possibly by RNA polymerase holoenzyme alone. The second site, defined by the promoter mutations analyzed so far, is a site through which CAP protein and 3′5′-cyclic AMP stimulate *lac* transcription. We propose that the function of the CAP/cyclic AMP complex is to bind to a site in the promoter, thus stimulating the initiation by RNA polymerase at the normally weak initiation site.

The promoter of the lactose operon has been defined by several different mutations (Fig. 1). Point mutations in the promoter, L8, L29 and L37, reduce the level of *lac* operon expression by 15-fold. A deletion of a part of the promoter which also extends into the *i* gene, L1, results in a 50-fold reduction in *lac* operon expression (Scaife & Beckwith, 1966). None of the point mutants recombines with the deletion L1 and, therefore, must map in the region covered by it (Miller, Ippen, Scaife & Beckwith, 1968). It was proposed that these mutations affected the process of initiation‡ of transcription of the *lac* operon. In support of this proposal, the promoter mutation L1 has been shown to reduce the rate of transcription of the *lac* operon *in vitro* (Eron & Block, 1971).

A complexity to transcription initiation in the *lac* promoter region has been suggested by the discovery of a protein factor which is needed in addition to RNA polymerase for proper initiation of transcription. It has been shown *in vivo* and in *in vitro* transcription systems that CAP protein and 3′5′-cyclic AMP are necessary in addition to RNA polymerase for high levels of expression of the *lac* operon (Perlman & Pastan, 1968; Schwartz & Beckwith, 1970; Chen *et al.*, 1971; Eron & Block, 1971). The precise mechanism whereby CAP protein and cyclic AMP promote proper transcription initiation is not known.

In this paper we present data which are consistent with the proposal that the *lac* promoter region can be subdivided into a site through which CAP action is mediated

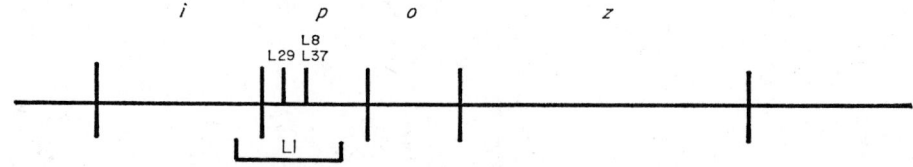

Fig. 1. Mutants of the *lac* promoter region. See Miller *et al.* (1968).

† Abbreviations used: *lac*, genes determining metabolism of lactose; *galE*, gene determining the structure of uridine-diphosphogalactose-4-epimerase; *trp*, *pro*, genes determining tryptophan and proline biosynthesis; *strA*, gene determining sensitivity to streptomycin; *p*, *o*, *i*, *z*, *y*, the *lac* promoter and operator and the structural genes for *lac* repressor, β-galactosidase and galactoside-permease, respectively; *cap*, structural gene for CAP protein; and *cya*, gene determining the synthesis of adenyl cyclase.

‡ Initiation as used in this paper "includes all of the processes involved in transcription up until the point at which the first RNA nucleotide is copied from the DNA" (Miller, 1970).

and an RNA polymerase holoenzyme initiation site (Fig. 2). Evidence is presented that there is a weak non-CAP-dependent transcription initiation site in the *lac* promoter.

Suggestion for a subdivision of the promoter came from the properties of various *lac* regulatory mutants. First, the *lac* promoter mutant L1, although it is a deletion of the initial section of the *lac* promoter, still makes approximately 2% of the wild-type levels of *lac* enzymes (Scaife & Beckwith, 1966). This residual promotion is determined by the remaining segment of the promoter rather than by the operator or *i* gene (Reznikoff, Miller, Scaife & Beckwith, 1969). Furthermore, this 2% appears to be insensitive to catabolite repression, a condition which is thought to lower the cyclic AMP concentrations in the cell (Silverstone, Magasanik, Reznikoff, Miller & Beckwith, 1969). Second, two mutants which eliminate CAP protein activity and one which eliminates the enzyme responsible for cyclic AMP synthesis (adenyl cyclase) still have levels of *lac* expression of around 2% (Schwartz & Beckwith, 1970). These results indicate that there might be a site in the *lac* promoter region where transcription could begin with a low frequency in the absence of CAP and cyclic AMP. This site, then, would still remain in the p^- deletion L1 and would be revealed in p^+ strains when CAP and/or cyclic AMP were absent. Alternative explanations would be equally satisfactory. For example, the 2% level seen in cap^- and cya^- mutants might be due to leakiness of the mutants used.

To distinguish between some of the alternative explanations and to obtain further information on the low-level promotion, we have assayed the levels of *lac* operon expression of a series of *lac* promoter mutants in a bacterial strain which carries both a cap^- and a cya^- mutation. Analysis in this strain should eliminate the possibility of leakiness of mutants.

Strains were constructed which were doubly mutant (cap 7900, cya 7902) (Schwartz & Beckwith, 1970), carried a deletion of the *lac* and *proA,B* genes, and harbored F-*lac-pro* episomes with different *lac* regions. The four strains differed only in the promoter region of the *lac* operon. One carried a wild-type *lac* operon (p^+) and the other three, *lac* p^- mutants L1 and L8 and a p^r mutation L8.UV5. The mutation L8.UV5 is a revertant of L8 which makes 60 to 70% of wild-type levels of *lac* enzymes and is insensitive to catabolite repression (Silverstone, Arditti & Magasanik, 1970). *In vitro* evidence indicates that the L8.UV5 promoter is altered so that it no longer requires CAP and cyclic AMP but only RNA polymerase holoenzyme for *lac* transcription (Eron & Block, 1971).

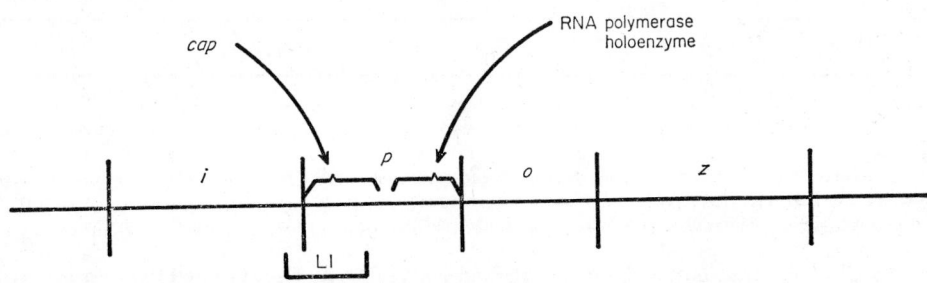

FIG. 2. Two sites in the *lac* promoter region. See text.

TABLE 1

Levels of β-galactosidase in wild-type and cap^- cya^- *genetic background*

lac promoter	cap, cya genotype	
	cap^+cya^+	cap^-cya^-
p^+	100	2
p^-L1	2	2
p^-L8	6	2
p^R.L8.UV5	67	70

The cap^- cya^- double mutant was constructed by transducing strain CA-7902 (F$^-cya^-$7902 Sm^s) to Sm^R with a P1 lysate grown on X-7901 (F$^-$ cap^-7900 Sm^R). Since cap is linked to $strA$, a fraction of the transductants were cap^-. A lac-pro deletion (X 111) was introduced into the double mutant strain by an Hfr cross, looking for white colonies on plates containing glucose synthetic media, streptomycin and 5-bromo-4-chloro-3-indolyl-β-D-galactoside (Miller et al., 1968). The various lac regions were crossed into the Δ lac-pro cap^-cya^- strain from donors carrying F-lac-pro^+ episomes. Two isolates of each strain were grown up in glucose/M63 minimal medium and assayed according to Pardee, Jacob & Monod (1959). The error on duplicate assays was no greater than 5%. Activities are presented as percentage of a wild-type control.

The results of β-galactosidase assays done on these strains are presented in Table 1. The results show that there is a 2% level of lac operon expression which functions independently of CAP protein and cyclic AMP in strains carrying p$^+$, L8 and L1. These results indicate that a portion of the promoter can serve as a CAP-independent initiation site. This site is intact in wild-type strains, in a partial deletion of the promoter, L1, and in the point mutant L8.

In addition, these and other data suggest that the UV5 mutation may have altered the base sequence of the CAP-independent initiation site, so that it alone can serve as a high-efficiency promoter†. We would predict, then, that the UV5 mutation, which maps extremely close to L8 (Silverstone et al., 1970), maps outside the region of the promoter covered by the deletion L1. The data in Tables 2 and 3 and Fig. 3 show that this is the case. Further, we are able to recover by recombination the double mutant L1.UV5, which makes high levels of the lac enzymes (Table 3). The mutation UV5

TABLE 2

Recovery of L8 from a P1 transduction between L8.UV5 and L1

Donor strain	Recipient	Frequency of L8 among $proC^+$ transductants	Corrected frequency of recombination
L8.UV5	L1 $proC^-galE^-$	2/6100	0·13%

The transductions, selections and scoring for recombinants were carried out as described previously (Arditti, Scaife & Beckwith, 1968). In the $galE^-$ background, only transductants carrying the L8 mutation survive on glycerol/minimal agar containing melibiose. The two recombinants were assayed for induced and uninduced levels of β-galactosidase and found to be identical to L8. The relatively high recombination frequency may be explained by the high frequencies always seen in crosses with L1 (Miller et al., 1968).

† The UV5 mutation induced by ultraviolet light is unlikely to have resulted from an insertion of genetic material, since ultraviolet light does not raise the frequency of insertion mutations (M. Malamy, personal communication).

TABLE 3

Recovery of L1.UV5 in an Hfr-F⁻ cross between L8.UV5 and L1

Donor	Recipient	Frequency of PG⁺ among $proC^+$ (%)
Hfr-L1	F⁻-L1	0·002
Hfr-L8.UV5	F⁻-L8.UV5	0·004
Hfr-L1	F⁻-L8.UV5	0·04

In these crosses, donors were $proC^+$ and recipients $proC^-$. To select recombinants with the expected properties of L1.UV5, we selected on minimal agar containing phenyl-β-D-galactoside (PG) as the sole carbon source. Since phenyl-β-D-galactoside is not an inducer of the *lac* operon and requires high levels of β-galactosidase for its metabolism, neither parent will grow on this medium. The PG⁺ colonies found in the control crosses are presumably revertants to higher levels of L1 or constitutive mutants of L8.UV5. Two of the PG⁺ recombinants from the experimental cross presumed to be genetically L1.UV5 were purified and assayed for induced and uninduced levels of β-galactosidase. The induced levels were identical to that of an induced isogenic L8.UV5 strain and the uninduced levels were about 60% of the induced. This induction ratio is identical to that seen with the low level of L1 expression in strains carrying L1 alone. It is due to a partially active repressor (Miller, Platt & Weber, 1970).

probably maps in the region between the L1 deletion and the operator. An alternative, that the UV5 mutation maps in the *i* gene or the operator, is made unlikely by the normal inducibility of strains carrying the mutation L8.UV5 and by other studies on the regulation of this system (Reznikoff *et al.*, 1969).

These results are consistent with the following model for initiation of *lac* operon transcription. There exists a weak transcription-initiation site between the region defined by the deletion L1 and the *lac* operator. In the absence of CAP and cyclic AMP, RNA polymerase holoenzyme interacts with this site to give 2% of the normal rate of transcription initiation. The role of CAP and cyclic AMP is to increase greatly the interaction of RNA polymerase with this site, by binding to the adjacent site defined by L1. There are other explanations for these data. For example, this low activity site may be just a weak promoter which has no role in the normal expression of the *lac* operon. This would then be analogous to the weak internal promoter found

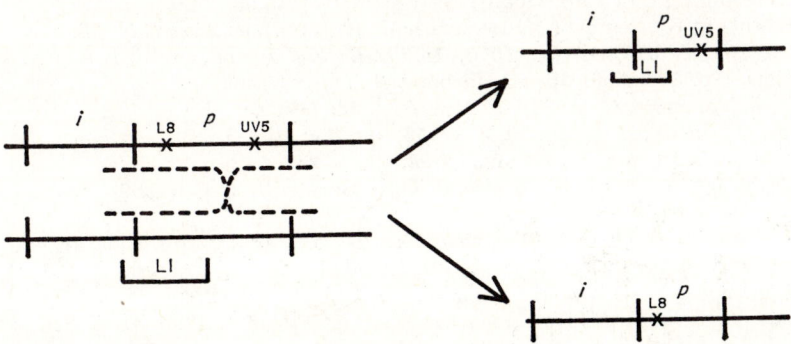

FIG. 3. The products of the recombination between promoter mutants L1 and L8.UV5. The two reciprocal recombination products are shown. If mutation UV5 mapped under the L1 deletion, neither of these products would be obtained.

normally with the *trp* operon of *Salmonella typhimurium* and *Escherichia coli* (Bauerle & Margolin, 1967; Morse & Yanofsky, 1968).

If the first explanation above is correct, then, so far, we have only been examining one class (class I) of promoter mutants, namely those which alter the CAP site. We would then predict a second class (class II) of promoter mutants which have no effect on CAP interaction, but reduce the initiation by RNA polymerase. These are now being sought and should be easily distinguishable from class I promoter mutants first by their behavior in a cap^-, cya^- genetic background.

We have previously suggested on the basis of properties of the known promoter mutants that there was probably little overlap between the *lac* operator and promoter (Miller *et al.*, 1968). Now, however, if our interpretation of these results is correct, we have been examining the wrong class of promoter mutants for operator effects. All we have shown is that the site through which CAP action is mediated does not overlap with the operator. Smith & Sadler (1971) have seen some effects of presumed operator mutants on maximal levels of *lac* operon expression. On the other hand, transcription studies indicate that, at least *in vitro*, RNA polymerase and repressor do not compete for binding to the *lac* operon (Chen *et al.*, 1971). We believe this still remains an open question.

Department of Microbiology and Molecular Genetics
Harvard Medical School
Boston, Mass. 02115, U.S.A.

JON BECKWITH
TERRI GRODZICKER
RITA ARDITTI

Received 10 January 1972, and in revised form 15 March 1972.

This work was supported by grants from the National Science Foundation and from the National Institutes of Health (GM13017) and a Career Development Award to one of us (J.B.) from the National Institutes of Health.

REFERENCES

Arditti, R., Scaife, J. & Beckwith, J. (1968). *J. Mol. Biol.* **38**, 421.
Bauerle, R. H. & Margolin, P. (1967). *J. Mol. Biol.* **26**, 389.
Chen, B., deCrombrugghe, B., Anderson, W. B., Gottesman, M. E., Pastan, I. & Perlman, R. L. (1971). *Nature*, **233**, 67.
Eron, L. & Block, R. (1971). *Proc. Nat. Acad. Sci., Wash.* **68**, 1828.
Miller, J. (1970). In *The Lactose Operon*, ed. by J. R. Beckwith and D. Zipser, p. 173. Cold Spring Harbor Lab. Quant. Biol.
Miller, J., Ippen, K., Scaife, J. & Beckwith, J. R. (1968). *J. Mol. Biol.* **38**, 413.
Miller, J., Platt, T. & Weber, K. (1970). In *The Lactose Operon*, ed. by J. R. Beckwith and D. Zipser, p. 343. Cold Spring Harbor Lab. Quant. Biol.
Morse, D. & Yanofsky, C. (1968). *J. Mol. Biol.* **38**, 447.
Pardee, A. B., Jacob, F. & Monod, J. (1959). *J. Mol. Biol.* **1**, 165.
Perlman, R. L. & Pastan, I. (1968). *J. Biol. Chem.* **243**, 5420.
Reznikoff, W. S., Miller, J. H., Scaife, J. G. & Beckwith, J. R. (1969). *J. Mol. Biol.* **43**, 201.
Scaife, J. & Beckwith, J. R. (1966). *Cold. Spr. Harb. Symp. Quant. Biol.* **31**, 403.
Schwartz, D. O. & Beckwith, J. R. (1970). In *The Lactose Operon*, ed. by J. R. Beckwith and D. Zipser, p. 417. Cold Spring Harbor Lab. Quant. Biol.
Silverstone, A. E., Arditti, R. & Magasanik, B. (1970). *Proc. Nat. Acad. Sci., Wash.* **66**, 773.
Silverstone, A. E., Magasanik, B., Reznikoff, W. S., Miller, J. H. & Beckwith, J. R. (1969). *Nature*, **221**, 1012.
Smith, T. F. & Sadler, J. R. (1971). *J. Mol. Biol.* **59**, 273.

Note added in proof: We regret not having noted the paper of Yudkin (1971) in which it is shown that the two promoter mutants, L8 and L29, both appear to be insensitive to transient repression. In fact, the mutants *are* probably sensitive to transient repression, but the rate of synthesis of β-galactosidase under transient repression is identical to that seen under catabolite repression, i.e. 2% of wild type. Therefore, transient repression exists, but would not be observed.

REFERENCE

Yudkin, M. D. (1971). *Biochem. J.* **123**, 579.

Mechanism of Activation of Catabolite-Sensitive Genes: A Positive Control System*

Geoffrey Zubay,† Daniele Schwartz,‡ and Jon Beckwith‡

COLUMBIA UNIVERSITY, NEW YORK, NEW YORK, AND HARVARD MEDICAL SCHOOL, BOSTON, MASSACHUSETTS

Communicated by Sol Spiegelman, February 26, 1970

Abstract. Catabolite repression is defined as the inhibition of enzyme induction by glucose or related substances. In the bacterium *E. coli*, the effect of glucose appears to be due to a lowering of the cyclic AMP level. A DNA-directed cell-free system for β-galactosidase synthesis has served as a model system for studying the mechanism of action of cyclic AMP. Previously, it was reported that in this system cyclic AMP is required for normal initiation of mRNA synthesis. A protein factor which acts in conjunction with the cyclic AMP has been partially purified. This protein factor has a high affinity for cyclic AMP. These and other results presented herein lead us to the conclusion that cyclic AMP and a protein factor called the catabolite gene activator protein are part of a positive control system for activating catabolite-sensitive genes.

Introduction. Catabolite repression involves the inhibition of enzyme induction by glucose or biochemically related compounds such as glucose 6-phosphate, fructose, or glycerol.[1] The extent of inhibition, which varies from a few per cent to more than 90%, is dependent on the bacterial strain, the source of the catabolite, and the growth conditions. Enzymes which show the repression effect include enzymes of glycerol regulation, the *gal* operon, the arabinose operon, the *lac* operon, tryptophanase, D-serine deaminase, and histidase. In fact, most enzymes classified as inducible show some effect. Although observations on catabolite repression date back to the turn of the century,[2] it is only recently that significant understanding of the mechanism of the reaction has been achieved.

For a long time the most serious impediment to progress was in determining, out of the multiplicity of effects of glucose, the ones which were directly linked to catabolite repression. A turning point in our understanding of this phenomenon was provided when Makman and Sutherland[3] showed that a rapid decrease in the intracellular level of cyclic 3′:5′-adenosine monophosphate (cyclic AMP) occurs in the presence of glucose. From that time on it has become increasingly clear that the glucose effect is due to a lowering of the cyclic AMP level. Thus Perlman and Pastan[4] and Ullman and Monod[5] showed that the repressing effect of glucose on the synthesis of enzymes could be reversed by the addition of cyclic AMP. Having determined that cyclic AMP plays a key role in catabolite repression, furrher understanding of the phenomenon requires that two questions be answered: (1) How does glucose control the cyclic AMP level and (2) How does cyclic AMP control the enzyme level? Our efforts are directed toward answering the second question.

Methods. (*a*) **Isolation of *E. coli* strain X7901:** The starting strain was CA-8000, an Hfr Hayes prototroph. After mutagenic treatment with nitrosoguanidine, approximately 500 bacteria are spread on tetrazolium agar containing arabinose and maltose. Mutants incapable of metabolizing the two carbon sources in the media give rise to red colonies with a frequency of about 0.5%. Most of the ara^- mal^- are also lac^-. Some of these show phenotypic reversion to lac^+ in the presence of cyclic AMP, others do not. One of the latter (CA-7900) was crossed with an F^- strain which was arg^-, met B^-, pyr F^-, trp^-, and arg^+ met^+ recombinants selected. One recombinant (X-7900) which was arg^+ met^+ trp^+ pyr F^- was selected for further crosses. An Hfr (CA-7033) carrying the lac-pro A, B deletion, X-111, was crossed with X-7900; pyr F^+ trp^- recombinants were selected and scored for the pro^- character. A pro^- trp^- recombinant, (X7901), carrying the lac-pro deletion was isolated. This strain is ara^-, mal^-, lac^- and is unaffected by cyclic AMP in the growth medium. Work to be reported elsewhere has shown that these deficiencies are due to a single gene alteration at a distant point from the operons.

(*b*) **Procedures for β-galactosidase synthesis and assay:** Except for slight modifications described herein, all procedures used for synthesis, enzyme assay, and preparation of bacterial extracts and DNA have been described in detail elsewhere. The procedures for synthesis and assay will be reviewed here. The incubation mixture contains per ml: 44 μmoles Tris-acetate, pH 8.2; 1.37 μmoles dithiothreitol; 55 μmoles KAc; 27 μmoles NH$_4$Ac; 14.7 μmoles MgAc$_2$; 7.4 μmoles CaCl$_2$; 0.22 μmole amino acids; 2.2 μmoles ATP; 0.55 μmole each GTP, CTP, UTP; 21 μmoles phosphoenolpyruvic acid; 100 μg tRNA; 27 μg pyridoxine HCl; 27 μg triphosphopyridine nucleotide; 27 μg flavine adenine dinucleotide; 11 μg p-amino-benzoic acid. The above ingredients are preincubated for 3 min at 37°C with 50 γ/ml DNA with shaking before 6.5 mg S-30 protein extract is added. When catabolite gene activator protein (CAP) (see (*c*) below) is present, addition is made by mixing the protein extract with S-30. When cyclic AMP is present, the concentration is 5×10^{-4} M; the cyclic AMP is added to the mixture of salts and cofactors described above. Incubations with shaking are allowed to continue for 60 min at 37°C. The enzyme assays are performed at 28°C in 0.1 M sodium phosphate buffer; pH 7.3, 0.14 M β-mercaptoethanol, and 0.35 mg O-nitrophenyl β-D-galactoside (ONPG)/ml. Samples (0.2 ml) of the incubation mixture are mixed with 1.5 ml of the ONPG solution. At the end of the incubation with the substrate ONPG, 1 drop of glacial acetic acid is added to each tube to precipitate the protein, thus decreasing the background absorption and preventing errors due to turbidity. The tubes are quickly stirred and chilled in ice, then centrifuged in the cold for 15 min at $2000 \times g$. The supernatant is transferred to a clean tube and an equal volume of 1 M Na$_2$CO$_3$ is added. The optical density is determined at 420 mμ.

(*c*) **Partial purification of CAP protein:** About 200 gm of frozen *E. coli* strain 514 is homogenized in 700 ml of buffer I (0.01 M Tris-acetate, pH 8.2, 0.01 M Mg(Ac)$_2$, 0.06 M KAc, 1.4 mM dithiothreitol) and centrifuged for 30 min at $16,000 \times g$. The sediment containing the bacteria is homogenized and recentrifuged. The final sediment is resuspended in 260 ml of buffer I. The suspension of cells is lysed in an Aminco pressure cell at pressures between 4000 and 8000 psi. The lysate is centrifuged for 30 min at $30,000 \times g$ in a small Sorvall rotor. The resulting supernatant is dialyzed for 16 hr against buffer II (0.01 M KH$_2$PO$_4$—KOH, pH 7.7 + 0.01 M mercaptoethanol). The dialyzed extract is passed over a 5×20 cm DEAE-cellulose column previously equilibrated with buffer II. A linear gradient containing buffer II and increasing NaCl concentration is run to 0.25 M NaCl. Tubes around 0.1 M NaCl containing appreciable CAP activity are pooled and dialyzed for 16 hr against buffer III (0.01 M K$_2$HPO$_4$—HAc, pH 7.0, + 0.01 M mercaptoethanol). This material is passed over a 2.5×15 cm phosphocellulose column   A linear gradient is executed starting with buffer III + 0.15 M KCl and finishing with buffer III + 0.45 M KCl. Tubes around 0.3 M KCl containing the main peak of activity are dialyzed for 16 hr against buffer I, quick frozen and stored at $-90°C$ until ready for use.

(*d*) **Dialysis-binding studies:** The partially purified CAP described above is concentrated to about 0.5% protein by burying a dialysis sac in G-200 Sephadex for 18 hr at 5°C. Concentrated extract (0.7 ml) is placed in another dialysis sac and dialyzed for 18 hr against buffer I and the appropriate concentration of ^{3}H-cyclic AMP. A 0.3-ml aliquot from inside and outside the dialysis sac are dissolved in 3 ml of formic acid; 1-ml aliquots are plated and counted in a windowless gas flow counter.

Results and Discussion. The direct involvement of cyclic AMP in triggering catabolite genes was made clearest by the finding that this compound is required in a DNA-directed cell-free system for normal initiation[6-8] leading to the synthesis of the enzymes of the *lac* operon. Since normal initiation of mRNA synthesis occurs at the promoter locus of the operon, it seems likely that cyclic AMP or a closely related derivative interacts at this gene site.[9] This stimulatory action of cyclic AMP is most probably mediated by a protein since a single mononucleotide would not be expected to interact strongly and specifically with a DNA molecule. With this in mind it was suggested[6] that cyclic AMP triggers the synthesis of catabolite enzymes such as β-galactosidase by interacting with the RNA polymerase. A more elaborate working hypothesis, currently favored by us, is that cyclic AMP binds to a protein subunit, which we shall call the catabolite gene activator protein (CAP), and thereby stimulates the binding of this protein subunit to RNA polymerase, producing a complex active in initiation at the promoter locus.

Detection of the CAP protein has been greatly aided by the finding of a mutant bacterial strain which appears to produce defective CAP. This mutant was obtained by Schwartz and Beckwith[10] by isolation and analysis of a number of variant strains of *E. coli* that could not grow on either lactose or arabinose. These variants all produce low levels of most catabolite repressible enzymes tested. Some are defective in cyclic AMP production as evidenced by their phenotypic reversion under the influence of this compound. Others are not, and one of the latter mutants has been found to be lacking a protein factor required for "turning on" the *lac* operon. The assignment of this defect to a protein factor has been made possible through use of the cell-free system for β-galactosidase synthesis described below.

A DNA-directed cell-free system for synthesis of *lac* operon proteins has been developed which contains DNA with the *lac* operon, a cell-free extract of *E. coli*, and all the cofactors and substrates essential for RNA and protein synthesis (see *Methods* for details). In the cell-free synthesis studies presented here two strains, 514 and X7901, have been used to produce the cell-free bacterial extracts. Both strains contain a deletion of the *lac* region including the repressor so that the results would not be complicated by the presence of enzymes of the *lac* operon or *lac* operon repressor at the beginning of synthesis. Strain 514 is normal in other respects and strain X7901 is the defective strain isolated by Schwartz and Beckwith (described above). In all cases the DNA used to stimulate synthesis was derived from φ80d*lac* virus containing a normal *lac* operon region—a region which includes promoter, operator, structural gene for β-galcatosidase, and other structural genes of the *lac* operon, in that order.

In all our studies we have found that the amount of β-galactosidase produced by the cell-free system is directly related to the amount of active *lac* operon present.[7,8] When the strain believed to contain the defective CAP factor is used as the source of the cell-free extract, only about 5% of the usual level of β-galactosidase is made (Table 1, compare lines 2 and 6). This level of activity is

TABLE 1. *Effect of cyclic AMP and CAP protein on DNA-directed cell-free synthesis of β-galactosidase.*

	Source of cell-free bacterial extract	Cyclic AMP	CAP protein	β-galactosidase activity (relative values*)
1	X7901	−	−	1
2	"	+	−	1
3	"	−	+	1
4	"	+	+	5
5	514	−	−	1
6	"	+	−	20
7	"	−	+	1
8	"	+	+	24

* A unit of 1 on this scale is equivalent to 5×10^{-4} International Units. One unit of β-galactosidase is defined as that amount of enzyme producing 1 μmole of O-nitrophenol/min at 28°C and pH 7.3. All measurements are the average of duplicate determinations. Conditions for cell-free synthesis are described in *Methods*.

When present during synthesis, the concentration of cyclic AMP is 5×10^{-4} M and the concentration of CAP-containing extract is 10 γ/ml. See text for explanation of results.

interpreted as resulting from abnormal initiations in transcription, since the presence of cyclic AMP, which usually stimulates synthesis (Table 1, compare lines 5 and 6), has no effect (Table 1, compare lines 1 and 2), and DNA with a defective *lac* promoter yields about the same level of activity. The addition of unfractionated extract from a normal strain stimulates the amount of β-galactosidase synthesized but only in the presence of cyclic AMP and normal DNA. An extract can be tested for the presence of CAP by this stimulation effect. With this assay, CAP protein has been purified about 200-fold relative to the total protein in a crude extract. The purification procedure (described in *Methods*) consists of subjecting the crude bacterial lysate to a high speed centrifugation, fractionation of the resulting supernatant by gradient elution on a diethylaminoethyl cellulose column (DEAE), and further fractionation of the eluate containing most of the activity by gradient elution on phosphocellulose. Most of the DNA and ribosomes are removed in the high-speed centrifugation and the remaining nucleic acid is removed in the DEAE step.

The stimulation effect of the partially purified CAP protein on β-galactosidase synthesis is most pronounced when the defective strain X7901 is used as the source of the cell-free extract; this stimulation effect requires the presence of cyclic AMP (Table 1, compare lines 1 and 2 with 3 and 4). Little stimulation by added CAP is seen when the normal strain is used to make the cell-free extract (Table 1, compare lines 6 and 8). This is undoubtedly because the normal extract contains a large supply of CAP protein. Thus far, the CAP protein has been used to augment cell-free synthesis of β-galactosidase tenfold, with the upper limit yet to be determined. Over the range of concentration studied, the amount of stimulation is proportional to the amount of CAP protein added (Fig. 1). This linear response to CAP suggests that gene activation requires a single molecule of the CAP protein.

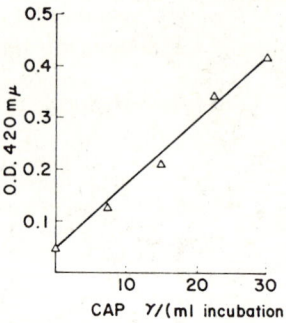

FIG. 1.—By using the cell-free system described in *Methods* in conjunction with cell-free extract prepared from strain X7901 and otherwise standard conditions, the β-galactosidase synthesized as a function of varying levels of partially purified CAP protein was determined. The β-galactosidase activity is expressed in arbitrary units.

Some of the physicochemical properties of the partially purified factor have been studied. The total extract has an absorption maximum in the ultraviolet at 278 mµ. The activity is completely destroyed by heating for 5 min at 60°C even though longer exposures to 50°C have no effect. The size of the CAP protein has been estimated by its flow rate on G-100 Sephadex (Fig. 2). It elutes in a volume about 60% greater than the column exclusion volume, and bovine serum albumin of molecular weight 6.8×10^4 elutes in a volume about 43% greater than the exclusion volume. A molecular weight of 4.5×10^4 is estimated for CAP on the basis of its elution behavior;[11] we do not know whether the molecule is composed of one or more polypeptide chains.

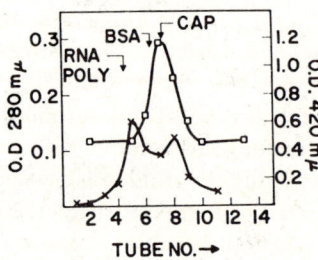

FIG. 2.—Elution diagram of partially purified CAP protein on G-100 Sephadex. Column dimensions, 1.5×30 cm; flow rate, 18 ml/hr; 4 ml/fraction. Column and samples were equilibrated in a buffer containing 0.01 M Tris-acetate, pH 8.2, 0.04 M MgAc$_2$, 0.06 M KAc, and 1.4 mM dithiothreitol. The ultraviolet absorption profile (*lower curve*) of the eluted fractions is indicated on the left and the corresponding stimulation effect on β-galactosidase synthesis (in arbitrary units) is on the right. The elution pattern of RNA polymerase and bovine serum albumin was determined in separate runs. Arrows indicate the position of elution of *E. coli* RNA polymerase and bovine serum albumin.

The binding of cyclic AMP to the enriched CAP-containing extracts has been examined extensively. Whereas crude extracts of *E. coli* show no measurable affinity for cyclic AMP, the partially purified and concentrated preparations of CAP show a substantial affinity. With ^{3}H-labeled cyclic AMP, the binding has been measured by equilibrium dialysis over a broad range of cyclic AMP concentrations (from about 10^{-9} to $10^{-5} M$). The standard equation relating bound to free ligand, when it is assumed that there exists only one type of binding site, is $I_b/I_f = K_f n(\text{CAP}) - K_f I_b$, where I_b is the concentration of bound ligand, I_f the concentration of free ligand, CAP the concentration of CAP protein, n the number of ligand binding sites per CAP protein, and K_f the formation constant for the CAP-cyclic AMP complex defined by the equation:

$$K_f = \frac{(CAP^* + \text{cyclic AMP})}{(CAP^*)(\text{cyclic AMP})}.$$

In this equation (CAP*) is the concentration of ligand binding sites. If $n = 1$, then (CAP*) = (CAP); if $n = 2$, then (CAP*) = 2(CAP); and so on. All concentrations are expressed as molar amounts. When I_b/I_f is plotted against I_b (as in Fig. 3) the slope is equal to $-K_f$. The numerical value obtained for K_f is 0.6×10^5 liters moles^{-1}.

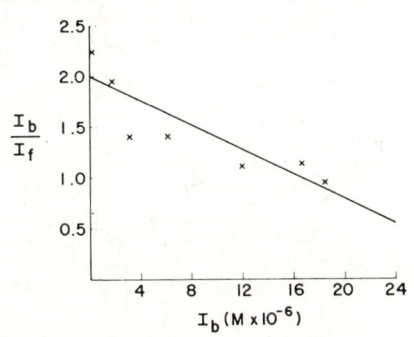

FIG. 3.—Binding curve for ^{3}H-labeled cyclic AMP to CAP-containing extract. The preparation of partially purified CAP and technique for dialysis binding are described in *Methods*. Calculations made in the text assume the concentration of free cyclic AMP, I_f, is proportional to the radioactivity found outside the dialysis sac and the concentration of bound cyclic AMP, I_b, is proportional to the difference in radioactivity found inside and outside the sac. Protein concentration is determined on an aliquot of material inside the sac by the Lowry method with a serum albumin standard. All plotted measurements have been normalized to a protein concentration of 1%. The experimental point plotted closest to the ordinate represents an average of 12 independent measurements between I_f values of 10^{-9} and 10^{-7} M. Other points each represent single experiments.

Since the CAP-containing extracts are impure, we have been concerned with the possibility that species other than the CAP are binding the cyclic AMP. Previously it had been found that cyclic GMP inhibited the cyclic AMP stimulatory effect on β-galactosidase synthesis at comparable concentrations.[7] In parallel binding studies, the level of ^{3}H-cyclic AMP was fixed at 5×10^{-6} M and the concentration of competing nucleotide was fixed at 10^{-4} M. Cyclic GMP inhibits 70% of the binding whereas either 3'-GMP or 5'-GMP have less than an 8% inhibitory effect. The parallelism between cyclic GMP in inhibiting the stimulatory action of cyclic AMP in cell-free synthesis and in inhibiting the binding of cyclic AMP to CAP-containing extracts supports the view that most of the cyclic AMP binding in the CAP-containing extracts is due to the CAP protein itself, but it does not eliminate other possibilities. The following discussion assumes that most of the cyclic AMP binding is to the CAP.

The formation constant for the CAP-cyclic AMP complex can also be used to estimate the concentration of CAP present from the amount of bound cyclic AMP. Only approximate calculations are possible since we do not know n, the number of cyclic AMP binding sites per CAP molecule. Assuming $n = 1$ we calculate that the molarity of CAP at the point of maximum stimulation in Figure 1 is 6×10^{-8} M. This is about 35 times the molarity of the *lac* operon containing DNA and about twice the molarity of the RNA polymerase.[12] It seems unlikely that there are many other catabolite-sensitive gene promoter sites on the $\phi 80 dlac$ DNA. Therefore, the large excess of CAP over *lac* operon makes it most unlikely that CAP functions by forming a strong stoichiometric complex with the *lac* operon promoter. The formation of a strong stoichiometric complex between CAP and RNA polymerase is also unlikely since such an event would cause an appreciable decrease in the pool of free polymerase, thereby causing departure from linearity over the range of CAP concentrations studied in Figure 1. We conclude that CAP does not form a strong complex with either DNA or RNA polymerase. In quantitative terms, the K_f's for the complexes must be less than 10^8. Since the hypothesized function of CAP is to trigger initiation, a complex with more than transient existence may not be needed. We are in the process of studying the interaction between CAP, RNA polymerase, and DNA by direct means; it remains to be demonstrated that CAP functions by forming such a complex.

We are indebted to Tetteh Blankson for technical assistance.

* This work was supported by grants from the National Institute of Health, 5-R01-GM-16648-02, and the American Cancer Society, E-545.

† Department of Biological Sciences, Columbia University, New York, N.Y.

‡ Department of Bacteriology and Immunology, Harvard Medical School, Boston, Mass.

[1] For reviews of the subject of catabolite repression see Magasanik, B., in *Cold Spring Harbor Symposia on Quantitative Biology*, vol 26, (1961), p 249; Magasanik, B., in *The Lac Operon*, ed. D. Zipser and J. R. Beckwith, *Cold Spring Harbor Laboratory on Quantitative Biology* (1970).

[2] Dennert, F., *Ann. Inst. Pasteur*, **14**, 139 (1900).

[3] Makman, R. S., and E. W. Sutherland, *J. Biol. Chem.*, **240**, 1309 (1965).

[4] Perlman, R. L., and I. Pastan, *J. Biol. Chem.*, **243**, 5420 (1968).

[5] Ullman, A., and J. Monod, *FEBS Letters*, **2**, 57 (1968).

[6] Chambers, D. A., and G. Zubay, these PROCEEDINGS, **63**, 118 (1969).

[7] Zubay, G., D. Chambers, and L. Cheong, in *The Lac Operon*, ed. D. Zipser and J. Beckwith, *Cold Spring Harbor Laboratory on Quantitative Biology* (1970).

[8] Zubay, G., and D. Chambers, in *Cold Spring Harbor Symposia on Quantitative Biology* (1969), vol. 64, in press.

[9] Silverstone, A. E., B. Magasanik, W. S. Reznikoff, J. H. Miller, and J. Beckwith, *Nature*, **221**, 1012 (1969).

[10] Schwartz, D., and J. Beckwith, in *The Lac Operon*, ed. D. Zipser and J. Beckwith, *Cold Spring Harbor Laboratory on Quantitative Biology* (1970).

[11] For the method of molecular weight determination, see Murphy, W. H., G. Barrie Kitto, J. Everse, and N. O. Kaplan, *Biochemistry*, **6**, 603 (1967).

[12] The molarity of the operon is calculated from the molarity of $\phi 80 dlac$ DNA added with a molecular weight of 30×10^6. The molarity of RNA polymerase is estimated from the total protein present in the cell free extract, 6500 γ/ml, and assuming RNA polymerase comprises 1 part per 1000 of the protein with a molecular weight of 4.95×10^5 as suggested by Burgess, A., *J. Biol. Chem.*, **244**, 6168 (1969).

Proc. Nat. Acad. Sci. USA
Vol. 68, No. 8, pp. 1828–1832, August 1971

Mechanism of Initiation and Repression of *In Vitro* Transcription of the *Lac* Operon of *Escherichia coli*

(cyclic AMP/RNA polymerase/sigma factor/rho factor/cAMP-binding protein)

LARRY ERON* AND RICARDO BLOCK

* Department of Microbiology and Molecular Genetics, Harvard Medical School, Boston, Massachusetts 02115; and The Biological Laboratories, Harvard University, Cambridge, Mass. 02138

Communicated by Boris Magasanik, May 21, 1971

ABSTRACT A cyclic AMP-binding protein (CAP protein), cyclic AMP, and RNA polymerase holoenzyme are shown to initiate *lac* transcription at the *lac* promoter. *Lac* repressor appears to control transcription by preventing RNA polymerase and/or CAP protein from binding to the *lac* promoter. Results support the idea that the *lac* promoter is composed of two sites that interact with CAP protein and RNA polymerase holoenzyme. The promoter can be altered by mutation so that holoenzyme alone can initiate *lac* transcription correctly.

The expression of the lactose (*lac*) and other operons of *Escherichia coli* subject to catabolite repression requires adenosine 3':5'-cyclic monophosphate (cyclic AMP) (1, 2) and a cyclic AMP-binding protein (CAP protein) (3, 4). In a purified *in vitro* transcription system (5, 6), cyclic AMP and CAP protein have been shown to act at the level of transcription, and the effect is dependent on the presence of σ, a subunit of RNA polymerase necessary for initiation of transcription at phage promoters (7); this result indicates that CAP protein acts in a manner different from σ.

Surprisingly, however, in this system, which employs as template DNA extracted from a transducing phage (φ80p*lac*) carrying *lac* in place of the early genes of the phage in the orientation shown in Fig. 1, effects of *lac* repressor and *lac* promoter mutations on *lac* transcription could not be demonstrated (6). However, control of *lac* transcription by *lac* repressor has been obtained (8, 27) by use of a two-step hybridization assay of RNA synthesized from a phage template, similar to φ80d*lac* in Fig. 1, carrying *lac* in place of the late genes of the phage in an orientation opposite to that in φ80 p*lac*. Therefore, we decided to reinvestigate the transcription of *lac* from the φ80d*lac* template by a simple one-step hybridization assay (5, 6).

Using a simple one-step hybridization procedure (5, 6), we confirm that in a purified *in vitro* transcription system, *lac* operon transcription is stimulated asymmetrically from the correct DNA strand by cyclic AMP and a cyclic AMP-binding protein (CAP protein), and is controlled by repressor (8, 27). This effect is dependent on the presence of σ factor, indicating that CAP protein acts in a manner different from σ. In addition, we show that the transcription initiates at the *lac* promoter, since it is affected by *lac* promoter mutants on the transcription template. One of these mutants, p^r uv-5, allows initiation of *lac* transcription in the absence of CAP

Abbreviation: CAP protein, a cyclic AMP-binding protein, referred to elsewhere as CRP (8, 27).; *lac*, the lactose operon of *E. coli*.

protein, although it retains its requirement for σ. Apparently, the *lac* promoter has been altered so that it can initiate in the absence of transcription factors other than σ. Our results suggest that the *lac* repressor acts by preventing RNA polymerase and/or CAP protein from binding to the *lac* promoter.

For template in the purified *in vitro* transcription system, we extracted DNA from three types of φ80 phages transducing *lac*: φ80d*lac*$_I$, φ80d*lac*$_{III}$, and φ80p*lac* (see Fig. 1). The first two carry *lac* in an orientation opposite to that of φ80p*lac*, transcribed *in vivo* from the L-strand, while on φ80-p*lac*, *lac* is transcribed from the H-strand (9). To assay *lac* transcription, RNA synthesized from these templates is directly hybridized to separated strands of λp*lac* DNA. Only *lac* RNA should anneal to λp*lac* DNA, since there is less than 0.5% homology between φ80 RNA and λDNA (5, 6). Although the three phages used as template DNA carry bacterial genes adjacent to *lac* (5), this should not interfere with the assay of *lac* sequences, because all non-*lac* bacterial genes have been deleted in the pλ*lac* phage phage (9). Furthermore, correctly initiated *lac* RNA should anneal to the λp*lac*$_L$

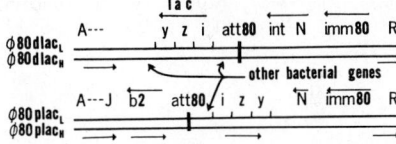

FIG. 1. Direction of transcription of genes on *lac* transducing phages. The *arrows* indicate the direction of transcription and are placed closest to that strand from which the RNA is transcribed (11, 12). The subscripts *H* and *L* refer to the heavy and light DNA strands after CsCl equilibrium density gradient ultracentrifugation after annealing with poly(rU, G) (13). The origin and marker notation of the phages are described elsewhere (6, 9, 10). φ80d*lac*$_{III}$ was derived from the λ–φ80 hybrid phage (20), λh80d*lac* (5), by recombination with φ80. Lysogens of λ h80d*lac* were plated overnight at 30°C with 10^6 φ80/plate. The lysates were harvested and transduced into M182, a *lac* deletion strain, on lactose minimal medium at 42°C. 50% of the *lac*⁺ transductants were lysogens of φ80/φ80d*lac*$_{III}$. RNA synthesized *in vitro* from φ80d*lac*$_{III}$ has less than 0.5% homology with separated strands of λDNA (L.E., unpublished results). φ80d*lac*$_I$ is an independently isolated defective transducing phage carrying *lac* in the same location and orientation as φ80d*lac*$_{III}$ (21). φ80p*lac* is an infectious transducing phage carrying *lac* in a different location and orientation (see text and ref. 6).

1828

Proc. Nat. Acad. Sci. USA 68 (1971) Regulation of Lac Transcription 1829

TABLE 1. *β-galactosidase synthesis in vitro directed by phage templates carrying lac*

Template	A_{420} with CAP protein$^+$ S-30		A_{420} with CAP protein$^-$ S-30	
	−cyclic AMP	+cyclic AMP	CAP protein −	CAP protein +
$\phi 80 dlac_{III}$	0.04	0.35	—	—
$\phi 80 dlac_{III}\ p^s$	0.06	10.00	0.44	7.15
$\phi 80 dlac_{III} p^r uv$-5	18.10	34.60	20.20	30.90
$\phi 80 dlac_I$	0.07	0.73	—	—
$\phi 80 dlac_I L1$	0.07	0.06	—	—
$\phi 80 plac$	0.01	0.06	—	—

The experimental procedure, S-30 preparation, reaction mixture, and β-galactosidase assay are described in detail elsewhere (18). Reaction mixtures (75 μl) contained, where indicated, cyclic AMP (0.5 mM) and CAP protein (25 μl), purified 200-fold, the gift of G. Zubay (3). CAP protein (2 μg/ml), purified to homogeneity (gift of W. Anderson, R. Perlman, and I. Pastan), behaved identically with the partially purified CAP protein. Although the pure CAP protein is referred to elsewhere as CRP (4, 8), we have adopted a unified nomenclature for clarity. Except where indicated in Fig. 2 and Tables 3 and 4, we have used partially purified CAP protein. β-galactosidase synthesis is normalized to absorbance units at 420 nm of product formed per 200-μl reaction mixture incubated for 20 hr. Strains used to prepare S-30 fractions are RV, a *lac* deletion (X74) strain that is CAP protein$^+$, and X7901, a *lac-pro A-pro B* deletion strain that is CAP protein$^-$. The p^s (the gift of M. Gottesman), $p^r uv$-5, and L1 mutations are described in the text.

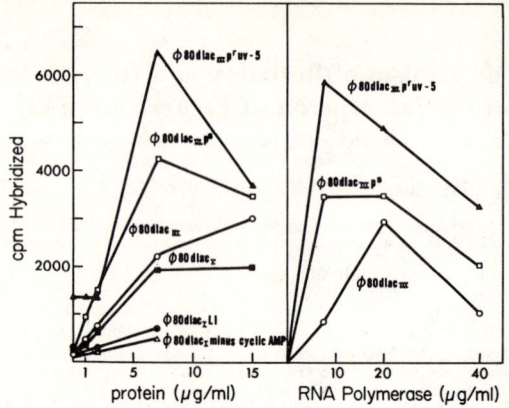

FIG. 2. Effect of CAP protein and RNA polymerase concentration on *lac* transcription. (*a*) Conditions were as in Table 2, except that pure CAP protein was used at the concentrations indicated with cyclic AMP (1 mM) and the indicated templates. [^{3}H]RNA (45,000 cpm/tube) was annealed to λ $plac_L$ DNA. (*b*) Conditions were as in Table 2 except that RNA polymerase concentration was varied with the indicated templates. Pure CAP protein (7 μg/ml) and cyclic AMP (1 mM) were present. Total RNA synthesis was linear with RNA polymerase concentration, and the input of [^{3}H]RNA was normalized to 45,000 cpm/tube and annealed to λ$plac_L$ DNA.

Asymmetric stimulation of *lac* transcription by CAP protein and cyclic AMP

The first two transducing phages, $\phi 80 dlac_I$ and $\phi 80 dlac_{III}$, are efficient DNA templates for β-galactosidase synthesis in a crude cell-free system (Table 1). In the purified transcription system, they produce essentially no *lac* RNA in the absence of CAP protein and cyclic AMP (Table 2; Fig. 2) even in the absence of transcription termination factor ρ (11), indicating the absence of read-through in our system. When

strand only, since this is the strand transcribed *in vivo* (9). Using this assay system, we have examined the fidelity of *lac* transcription *in vitro* according to four criteria: (*a*) *Lac* RNA synthesis should be dependent on the presence of cyclic AMP and CAP protein (1–4), (*b*) should occur asymmetrically, that is, anneal to λ$plac_L$ strand only (9, 14), (*c*) should initiate at the *lac* promoter (15, 16, 23), and (*d*) should be controlled by *lac* repressor (17).

TABLE 2. *Effect of CAP protein, cyclic AMP, and σ factor on lac transcription*

Template	CAP protein	cyclic AMP	σ	cpm Hybridized to		% *lac* RNA
				λ$plac_L$–λ$_L$	λ$plac_H$–λ$_H$	
$\phi 80 dlac_{III}$	−	−	+	1	22	<0.1
$\phi 80 dlac_{III}$	+	−	+	189	58	0.4
$\phi 80 dlac_{III}$	+	+	+	2,289	124	5
$\phi 80 dlac_{III}$	+	+	−	85	57	(2)*
$\phi 80 dlac_{III} p^s$	+	+	+	3,210	101	7
$\phi 80 dlac_{III} p^s$	+	+	−	150	132	(4)*
$\phi 80 dlac_{III} p^s$	+	−	−	147	125	(4)*
$\phi 80 dlac_{III} p^r uv$-5	+	−	+	1,708	73	4
$\phi 80 dlac_{III} p^r uv$-5	+	+	+	3,945	54	9
$\phi 80 dlac_{III} p^r uv$-5	+	+	−	108	65	(3)*

[^{3}H]RNA is synthesized in a reaction mixture described in detail elsewhere (5, 6) with the following modifications: DNA (50 μg/ml), KCl (120 mM), IPTG (1 mM), [^{3}H]ATP (0.1 mM), and RNA polymerase (20 μg/ml), purified to homogeneity either as core or holoenzyme from a rifampicin-resistant strain (gift of K. Weber). CAP protein (25 μl) and cyclic AMP (1 mM) were incubated for 1 min at 37°C with the reaction mixture before the addition of RNA polymerase. The reaction was incubated 10 min at 37°C, terminated, extracted, and hybridized (5, 6, 22). [^{3}H]RNA (45,000 cpm/tube with holoenzyme polymerase; 3,500 cpm/tube with core polymerase) was annealed to separated strands of λ$plac$ and λDNA; the difference is expressed above; cpm annealing to λ separated strands was always less than 0.5% of the [^{3}H]RNA input.

* % *lac* RNA is enclosed in brackets for experiments with core polymerase because it does not reflect correct initiation (see text).

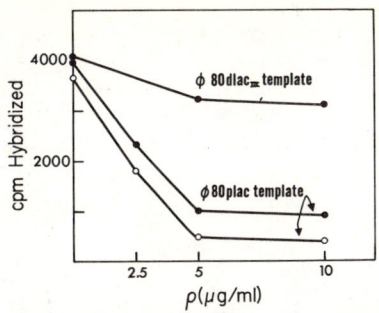

FIG. 3. Effect of ρ on *lac* transcription from φ80d*lac*_III and φ80p*lac* templates. Conditions were as in Table 2, except that ρ (purified to homogeneity, free of RNAase III, the gift of N. Minkley) was added as indicated to reaction mixtures containing CAP protein (25 μl) and cyclic AMP (1 mM). The addition of ρ produces a two- to three-fold depression of total RNA synthesis. [³H]RNA (in the absence of ρ, 60,000 cpm/tube for φ80d*lac*_III RNA and 50,000 cpm/tube for φ80p*lac* RNA; 25,000 and 20,000 cpm/tube, respectively, in the presence of excess ρ) was annealed to (●) λp*lac*_L DNA and (○) λp*lac*_H DNA.

CAP protein and cyclic AMP are added, *lac* transcription is stimulated in an asymmetric fashion from the correct DNA strand to maximal concentrations of 5% of the total RNA synthesized from the template.

The above situation contrasts sharply with *lac* transcription from the φ80p*lac* template. As we have reported (5, 6), *lac* RNA synthesis from this template is stimulated asymmetrically by CAP protein and cyclic AMP, but is not subject to control by the *lac* promoter and repressor. The fact that transcription termination factor ρ decreases CAP protein-dependent *lac* transcription (see Fig. 3) suggests that *lac* transcription from this phage template is the result of transcription initiation in a nearby, similarly oriented gene, with subsequent read-through into the *lac* genes. *In vivo* studies in-

TABLE 3. *Repression of lac transcription*

| | | cpm Hybridized to | | |
Template	Repressor	IPTG	λp*lac*_L–λ_L	% *lac* RNA
φ80d*lac*_III	−	+	3,106	7
φ80d*lac*_III	+	−	742	1.5
φ80d*lac*_III	+	+	2,739	6
φ80d*lac*_IIIp^ruv-5	−	+	3,784	8.5
φ80d*lac*_IIIp^ruv-5	+	−	1,501	3
φ80d*lac*_IIIp^ruv-5	+	+	3,450	8
φ80d*lac*_IIIp^s	−	+	3,468	8
φ80d*lac*_IIIp^s	+	−	987	2
φ80d*lac*_IIIp^s	+	+	3,560	8

Conditions were as in Table 2, except that *lac* repressor (5 μg/ml, purified to homogeneity, a gift of T. Platt) was incubated 2 min at 37°C with the template DNA in 50 mM KCl–10 mM MgCl₂ before addition of CAP protein. Pure CAP protein (7 μg/ml) and cyclic AMP (1 mM) were incubated 1 min at 37°C with the repressor–DNA complex before RNA polymerase (20 μg/ml) was added. IPTG was added where indicated with CAP protein. [³H]RNA (45,000 cpm/tube) was annealed to λp*lac*_L-DNA and λ_LDNA; the difference is expressed above.

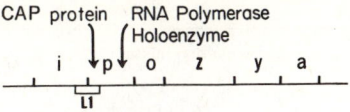

FIG. 4. Model for the interaction of CAP protein and RNA polymerase holoenzyme with the *lac* promoter. Genes *p* and *o* refer to the *lac* promoter and operator regions. *i* codes for *lac* repressor, and *z*, *y*, and *a* for the *lac* enzymes.

dicate that such read-through should not be subject to *lac* promoter control, and that repression might be less efficient (23). Furthermore, this phage is an extremely poor template for β-galactosidase synthesis in a crude, cell-free system (Table 1), although it produces normal levels of β-galactosidase in lysogens *in vivo* (5). In addition, using DNA extracted from λp*lac*, a phage carrying *lac* in place of the phage late genes in an orientation opposite to that of φ80p*lac*, we could not demonstrate effects of cyclic AMP and CAP protein (6). These discrepancies may be due to the orientation and location of the *lac* genes with respect to other genes on the phage template.

Lac transcription is initiated at the lac promoter

Initiation of CAP protein- and cyclic AMP-dependent *lac* transcription occurs *in vivo* at the *lac* promoter (15, 16, 24). If the *in vitro* transcription system is behaving properly, *lac* transcription should initiate at the *lac* promoter also. To ascertain this, we have assayed *lac* transcription with various promoter mutations on the φ80d*lac* template DNA. First, we have used L1, a small deletion of part of the promoter and the adjacent *i* gene that results in a 100-fold reduction in β-galactosidase synthesis *in vivo* (24), and a 10-fold reduction in the crude, cell-free system (Table 1). In the purified transcription system, it produces from 3- to 10-fold less *lac* RNA, depending on the conditions (Fig. 2). Moreover, *in vivo* studies indicate that L1 is insensitive to CAP protein and cyclic AMP (25). In the cell-free system, L1 shows no dependence on cyclic AMP (Table 1). In the purified transcription system, only a slight stimulation of L1 *lac* transcription occurs when pure CAP protein and cyclic AMP are added (Fig. 2).

We have examined another category of *lac* promoter mutants termed "super-promoters" (8), because they produce 10- to 30-fold more β-galactosidase in the cell-free system (Table 1). One super-promoter, designated p^ruv-5, is a revertant of the *lac* promoter mutant L8, and is a second-site mutation within the *lac* promoter (26). While L8 produces 15-fold less β-galactosidase *in vivo* (24), p^ruv-5 produces normal levels of β-galactosidase. In addition, p^ruv-5 has lost its requirement for CAP protein and cyclic AMP both *in vivo* (26) and in the cell-free system (Zubay, unpublished results; Table 1). The other super-promoter, p^s, isolated as a revertant of a CAP protein⁻ cell, retains a partial requirement for CAP protein and cyclic AMP (8). In the purified transcription system, *lac* RNA levels, synthesized from template DNA carrying super-promoter mutations, are elevated from 2- to 10-fold over wild-type depending on the conditions (Fig. 2). The striking feature about *lac* transcription from the φ80-d*lac*_IIIp^ruv-5 DNA template is that a substantial amount occurs in the absence of CAP protein (although CAP protein does stimulate it).

TABLE 4. *Effect of repressor on lac RNA synthesis by prebound RNA polymerase*

Tube	Preincubation mixture				cpm of [^{3}H]RNA/tube	cpm Hybridized to $\lambda plac_L - \lambda_L$	% lac RNA
	XTP-UTP	CAP protein + cyclic AMP	Repressor	IPTG			
1	+	−	−	+	30,000	191	0.6
2	+	+	+	+	20,000	1,090	5
3	+	+	+	−	20,000	1,188	6
4	−	+	+	+	5,000	545	11
5	−	+	+	−	6,000	741	12
6	−	+	+ (before polymerase)	−	5,000	105	2

The reaction mixture in tubes 1–3 was as in Table 2 with $\phi 80 dlac_{III} p^s$ DNA (50 µg/ml) and rifampicin-sensitive RNA polymerase (20 µg/ml), except that UTP was omitted from the preincubation mixture. This mixture was incubated 10 min at 20°C. Repressor (5 µg/ml), IPTG (1 mM), rifampicin (5 µg/ml), and finally UTP (0.15 mM) were then added and the reaction was further incubated 10 min at 37°C. In tube 1, CAP protein and cyclic AMP were omitted from the preincubation mixture. The reaction mixture in tubes 4–6 was as above, except that all four nucleoside triphosphates were omitted from the preincubation, and were added after rifampicin. In tube 6, repressor was added before RNA polymerase. [^{3}H]RNA at the indicated input/tube was annealed to $\lambda plac_L$ and λ_L DNA; the difference is expressed above.

Since L1, p^ruv-5, and p^s control transcription in the purified system, we conclude that CAP protein and cyclic AMP stimulate *lac* transcription predominantly at the *lac* promoter with $\phi 80 dlac$·DNA as template in the purified system. If this is true, then *lac* transcription from this template should not be depressed by the addition of transcription termination factor ρ, contrary to the case of the $\phi 80 plac$ template (5), where *lac* transcription results from read-through as described in the previous section. Indeed, only a 10% decrease in *lac* transcription is observed in the presence of ρ with the $\phi 80 dlac$ template, as compared to an 80% decrease with the $\phi 80 plac$ template (Fig. 3). That ρ has little or no effect on *lac* transcription from the $\phi 80 dlac$ template confirms that *lac* transcription from the $\phi 80 dlac$ template initiates predominantly at the *lac* promoter.

As we have shown previously (5, 6), σ acts in a manner different from CAP protein. Core polymerase (without σ) transcribes *lac* symmetrically and shows no requirement for cyclic AMP, indicating that it is not initiating correctly at the *lac* promoter (Table 2).

Repression of *lac* transcription

As others have shown (8, 27), when *lac* repressor is added to the transcription system, *lac* transcription is repressible from 50 to 95%, the mode being 75% (Table 3). This repression is completely reversible by the inducer, IPTG.

Bacteriophage λ repressor has been proposed to act by competing with RNA polymerase for binding to the λ regulatory elements (12). Our results indicate that *lac* repressor acts in the same way (Table 4). RNA polymerase, CAP protein, cyclic AMP, and *lac* p^s DNA are incubated in the absence of either one nucleoside triphosphate (to allow polymerase binding and initiation in tubes 2 and 3) or in the absence of all four nucleoside triphosphates (to allow polymerase binding, but not initiation in tubes 4 and 5). After this incubation, rifampicin is added to inhibit further initiation. Repressor is then added, either with or without IPTG. The results in Table 4 indicate that repressor cannot block *lac* transcription once CAP protein and RNA polymerase binding or initiation has occurred, while, if added before binding, it blocks the formation of an initiation complex. This suggests that repressor acts by preventing the binding of CAP protein and/or RNA polymerase to the *lac* promoter.

DISCUSSION

The *lac* promoter mutation L1 supports the idea that the *lac* promoter is composed of at least two sites (27). Since strains with L1 produce β-galactosidase at an equally low rate in the presence and absence of CAP protein, it appears that the site of the CAP protein effect has been deleted (Fig. 4). However, the remaining low level of *lac* expression, which cannot be due to read-through from another gene (23), indicates that the part of the promoter that is intact can still serve as a weak transcription initiation site. This part of the promoter may be the site where RNA polymerase holoenzyme (core plus σ) normally binds (Fig. 4), and CAP protein would greatly enhance this binding by interacting with the adjacent region. On this model, super-promoter p^ruv-5 would alter the holoenzyme binding site so that the promoter can bind holoenzyme in the absence of CAP protein.

We thank Dr. Jonathan Beckwith for his advice and helpful suggestions and Dr. Walter Gilbert for critical examination of this manuscript. We also thank Penelope Wood for her excellent technical assistance, G. Zubay and R. Perlman, W. Anderson, and I. Pastan for CAP protein, M. Gottesman for the p^s mutant, N. Minkley for ρ, K. Weber for RNA polymerase, and T. Platt for repressor. This work was supported in part by grants from the National Science Foundation, The American Cancer Society, and the Jane Coffin Childs Memorial Fund for Medical Research to J. Beckwith, and in part from the NIH Grant GM09541 to J. D. Watson. R. B. received a scholarship from the Instituto Nacional de la Investigacion Cientifica, Mexico.

1. Perlman, R. L., and I. Pastan, *J. Biol. Chem.*, **243**, 5420 (1969).
2. Ullman, A., and J. Monod, *FEBS Lett.*, **2**, 57 (1968).
3. Zubay, G., D. Schwartz, and J. Beckwith, *Proc. Nat. Acad. Sci. USA*, **66**, 104 (1970).
4. Emmer, M., B. deCrombrugghe, I. Pastan, and R. Perlman, *Proc. Nat. Acad. Sci. USA*, **66**, 480 (1970).
5. Arditti, R. R., L. Eron, G. Zubay, G. Tocchini-Valentini, S. Connaway, and J. R. Beckwith, *Cold Spring Harbor Symp. Quant. Biol.*, **35**, 437 (1970).
6. Eron, L., R. Arditti, G. Zubay, S. Connaway, and J. R. Beckwith, *Proc. Nat. Acad. Sci. USA*, **68**, 215 (1971).
7. Travers, A., *Nature*, **229**, 69 (1971).

8. deCrombrugghe, B., B. Chen, M. Gottesman, I. Pastan, H. E. Varmus, M. Emmer, and R. L. Perlman, *Nature*, **230**, 37 (1971).
9. Shapiro, J., L. MacHattie, L. Eron, G. Ihler, K. Ippen, and J. Beckwith, *Nature*, **224**, 768 (1969).
10. Signer, E., and J. Beckwith, *J. Mol. Biol.*, **22**, 33 (1966).
11. Roberts, Jeffrey W., *Nature*, **224**, 1168 (1969).
12. Steinberg, R., and M. Ptashne, *Nature*, **230**, 76 (1971).
13. Hradecna, A., and W. Szybalski, *Virology*, **32**, 633 (1967).
14. Kumar, S., and W. Szybalski, *J. Mol. Biol*, **40**, 145 (1969).
15. Eron, L., J. Beckwith, and F. Jacob, in *The Lac Operon*, ed D. Zipser and J. Beckwith (Cold Spring Harbor Laboratory, N.Y., 1970) p. 353.
16. Miller, J., K. Ippen, J. Scaife, and J. R. Beckwith, *J. Mol. Biol.*, **38**, 413 (1968).
17. Gilbert, W., and B. Muller-Hill, *Proc. Nat. Acad. Sci., USA*, **58**, 2415 (1967).
18. Zubay, G., D. A. Chambers, and L. C. Cheong, in *The Lac Operon*, ed D. Zipser and J. Beckwith (Cold Spring Harbor Laboratory, N.Y., 1970). p. 375.
19. Gilbert, W., and B. Muller-Hill, *Proc. Nat. Acad. Sci. USA*, **56**, 1891 (1966).
20. Szpirer, J., R. Thomas, and C. M. Radding, *Virology*, **37**, 585 (1969).
21. Beckwith, J. R., and E. R. Signer, *J. Mol. Biol.*, **19**, 254 (1966).
22. Gillespie, D., and S. Spiegelman, *J. Mol. Biol.*, **12**, 829 (1965).
23. Reznikoff, W., J. Miller, J. Scaife, and J. Beckwith, *J. Mol. Biol.*, **43**, 201 (1969).
24. Ippen, K., J. Miller, J. Scaife, and J. Beckwith, *Nature*, **217**, 825 (1968).
25. Silverstone, A., B. Magasanik, W. Reznikoff, J. Miller, and J. Beckwith, *Nature*, **221**, 1012 (1970).
26. Silverstone, A., R. R. Arditti, and B. Magasanik, *Proc. Nat. Acad. Sci. USA*, **66**, 773 (1970).
27. deCrombrugghe, B., B. Chen, W. Anderson, P. Nissley, M. Gottesman, R. Perlman, and I. Pastan, *Nature* **231**, 139, (1971).

Purification and DNA-Binding Properties of the Catabolite Gene Activator Protein

(phosphocellulose/DEAE-cellulose/*lac* operon DNA/cyclic AMP)

A. D. RIGGS, G. REINESS, AND G. ZUBAY

Department of Biology, City of Hope National Medical Center, 1500 East Duarte Road, Duarte, California 91010; and Department of Biological Sciences, Columbia University, New York, N.Y. 10036

Communicated by Norman Davidson, March 29, 1971

ABSTRACT A protein required for the activation of the *lac* operon has been extensively purified and partly characterized. This protein, called CGA protein (catabolite gene activator protein, sometimes named CAP), is a dimer with subunits of 22,000 daltons. Purified CGA protein has a substantial affinity for DNA; this affinity is greatly strengthened by cAMP and strongly inhibited by cGMP. Other studies have shown that these cyclic nucleotides compete for a binding site on CGA protein. The opposing effects of the two cyclic compounds in DNA–CGA protein binding show a parallel behavior to their effects on the expression of the *lac* operon. Thus cAMP, in addition to CGA protein, is required for expression of the *lac* operon, whereas cGMP inhibits the expression. The obvious inference is that CGA protein activates the *lac* operon by binding to the DNA under the influence of cAMP. Thus, CGA protein seems to be a new type of regulatory protein: a DNA-binding activator.

The phenomenon of catabolite repression in *Escherichia coli* (1) has led to the delineation of a positive control system sensitive to the intracellular and extracellular level of adenosine cyclic 3':5'-monophosphate (cAMP) (2). Recently, a protein called the catabolite gene activator protein (CGA protein) has been shown (3) to stimulate the DNA-directed synthesis of the enzymes of the *lac* operon, a catabolite-sensitive operon. The CGA protein is effective only in the presence of cAMP. Mutants defective in CGA protein can express *in vivo* neither *lac* nor other catabolite-sensitive genes (4), which suggests a common mechanism of activation by CGA protein. Studies with altered promoters of the *lac* operon show that certain properties of an intact promoter are needed for effective action of CGA protein (5). Since the action of CGA protein is intimately involved with the promoter part of the operon, it seems likely that CGA protein is involved in the initiation of messenger synthesis from the *lac* operon, perhaps by forming a part of the initiation complex. Previous attempts to detect interaction between CGA protein and RNA polymerase have been unsuccessful. Here, we report the complete purification of CGA protein and demonstrate it to be a DNA-binding protein. This result suggests that binding of CGA protein to promoter is required for expression of the *lac* operon.

METHODS

Purification of CGA protein

The procedure reported here is a significant modification of an already reported procedure (3). About 200 g of frozen *E. coli* strain 514 is homogenized in 700 ml of buffer I (0.01 M Tris-acetate (pH 8.2)–0.01 M Mg(OAc)$_2$–0.06 M KCl–6 mM mercaptoethanol) and centrifuged for 30 min at 16,000 × *g*. The sediment containing the bacteria is homogenized and recentrifuged. The final sediment is resuspended in 260 ml of buffer I. The suspension of cells is lysed in an Aminco pressure cell at pressures between 4000 and 8000 psi. The lysate is centrifuged for 30 min at 30,000 × *g* in a small Sorvall rotor. The resulting supernatant is centrifuged for 4 hr at 30,000 rpm in a Spinco no. 30 rotor. The resulting supernatant is dialyzed for 16 hr against buffer II [0. M K$_2$HPO$_4$–CH$_3$COOH (pH 7.0)–6 mM mercaptoethanol]. This solution is passed over a 2.5 × 15 cm phosphocellulose column (Whatman P11, medium fibrous powder, 7.4 meq/g), previously equilibrated with buffer II. After the column is rinsed progressively with 200 ml of buffer II and 100 ml of buffer II + 0.4 M KCl, the active fraction, which constitutes about 1% of the protein put on the column, is eluted in buffer II + 0.50 M KCl. The active fraction, detected by ultraviolet absorption, is pooled and dialyzed against buffer III [0.01 M KH$_2$PO$_4$–KOH (pH 7.7)–6 mM mercaptoethanol] overnight. This solution is passed over a 1.4 × 13 cm DEAE-cellulose column previously equilibrated with buffer III. The active fraction, which constitutes about 10% of the protein put on this column, passes through the column with no holdback. The remainder of the protein is retained by the column. Total yield of protein is 300–1000 μg in about 20 ml.

Acrylamide gel electrophoresis

The CGA protein solution was concentrated about 5-fold prior to electrophoresis by placing a dialysis sac containing the CGA protein solution in a beaker of dry G-200 Sephadex for 16 hr. 10% polyacrylamide gels containing 0.1% sodium dodecyl sulfate (SDS) were prepared by the procedure of Weber and Osborn (6) and prerun without samples for 2 hr at 3 mA per gel in Weber and Osborn's running buffer. CGA protein preparations, obtained as above, were layered on the gels in 100 μl of sample buffer [containing 0.1% SDS–0.01

Abbreviation: SDS, sodium dodecyl sulfate; CGA protein, catabolite gene activator protein. This protein has frequently been referred to as CAP, a less logical choice since it implies activation of the *catabolite*. The word "protein" is necessary in the abbreviation to eliminate possible confusion with a trinucleotide (cytidine-guanosine-adenosine). In general, this journal prefers 3-letter to Greek-letter designations of this kind of compound.

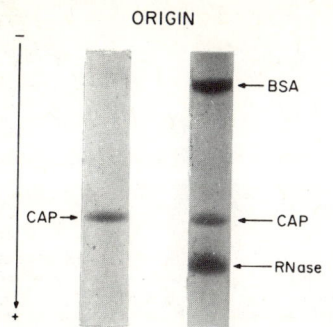

FIG. 1. Acrylamide gel electrophoresis of CGA protein. 10% polyacrylamide gels containing 0.1% SDS were prepared according to Weber and Osborn (6). The gel at the *left* contains only a CGA protein preparation. To the gel at the *right* were added, in addition to CGA protein, the marker proteins bovine serum albumin (BSA; 5 μg) and RNase (2.5 μg).

M sodium phosphate (pH 7.2)–10% (v/v) glycerol–0.002% bromophenol blue–0.14 M β-mercaptoethanol] that had been incubated at 65°C for 30 min. The gels were run for 2.5–3 hr at 8 mA per gel, after which time the tracking dye had migrated about 50 mm. Gels were removed from the tubes and stained for 4 hr in 0.25% Coomassie Brilliant Blue in methanol–water–acetic acid 5:5:1. Gels were then soaked overnight in 7.5% acetic acid–5% methanol and destained electrophoretically in the same solution. Mobilities of protein bands were determined according to Weber and Osborn and the molecular weight of CGA protein was estimated.

DNA-binding methods

The membrane filter technique developed for *lac* repressor (7) was used with only minor changes. A typical experiment will be described here. Any variations in this basic procedure are indicated in the figure legends. The appropriate volume of CGA protein solution was mixed with 0.05 μg of λh80*dlac* [^{32}P]DNA in a total volume of 1.3 ml that contained buffer IV [10 mM KCl–3 mM Mg(OAc)$_2$–0.1 mM EDTA–0.1 mM dithiothreitol–50 μg/ml of bovine serum albumin–5% dimethyl sulfoxide–10 mM Tris·HCl (pH 7.4 at 24°C)]. After incubation at room temperature (about 24°C) for 30 min, a time more than adequate to reach equilibrium, 0.4-ml samples were filtered in duplicate through 25-mm nitrocellulose membrane filters (Schleicher and Schuell, B-6). The filtering rate was such that the sample passed through in about 15 sec. The filters were washed two times with 0.4 ml of buffer IV without either bovine serum albumin or dithiothreitol. The data points shown here represent the average of two filters. The filters had been treated with 0.5 M KOH for 30 min at room temperature to help reduce DNA binding in the absence of CGA protein (11). For most experiments, this background was less than 5% of the total counts filtered (see Fig. 2).

[^{32}P]DNA from λh80*dlac* was prepared as was described (7). DNA concentrations were measured spectrophotometrically at 260 nm; an extinction coefficient of 0.02 cm^2/μg was assumed. CGA protein was prepared as described above;

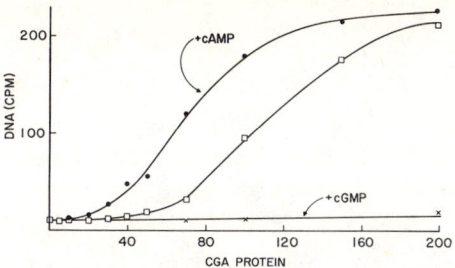

FIG. 2. Bound DNA as a function of the amount of CGA protein. A fixed amount of λh80*dlac* [^{32}P]DNA (0.05 μg) was mixed with the indicated volume of CGA protein solution in a final volume of 1.3 ml, containing buffer IV. The abscissa units are μl of a CGA protein solution that contained about 0.4 μg/ml of CGA protein. Each filter received 0.4 ml, containing a total of 260 cpm. (□) Buffer alone; (●) 3.7×10^{-4} M cAMP; (×) 3.7×10^{-4} M cGMP.

the concentrations of CGA protein given here are based on total protein estimated from the UV absorption at 280 nm; an extinction coefficient of 0.5 cm^2/mg was assumed. Low protein concentrations and the presence of dithiothreitol make other methods difficult.

RESULTS AND DISCUSSION

CGA protein purification and properties

Only two major purification steps, phosphocellulose and DEAE-cellulose chromatography, are needed to give apparently pure CGA protein. This fortunate result was the first hint that CGA protein interacts with DNA, since many other DNA-binding proteins, among them the *lac* repressor (8), bind to phosphocellulose, whereas more than 90% of *E. coli* proteins do not. As shown in Fig. 1, electrophoresis of the purified CGA protein in SDS–acrylamide gels yields only a single band, corresponding to a polypeptide of 22,000 daltons. In the absence of SDS, the protein has a molecular weight of about 45,000 (3); thus, it seems to be a dimer. CGA protein binds cAMP with a bimolecular formation constant, K_f, of about 0.6×10^5 liters/mol, as measured by equilibrium dialysis. cGMP inhibits the binding of cAMP to CGA protein and also antagonizes its stimulation of an *in vitro* system that synthesizes β-galactosidase (3).

During purification, about 90% of the activity that stimulates the *in vitro* synthesis of β-galactosidase is lost. It is not possible to measure cAMP binding throughout the preparation, because binding cannot be detected in crude extracts. However, the cAMP binding of partially purified or highly purified extracts indicates that most of the CGA protein molecules bind cAMP. The "activation" activity of CGA protein thus seems to be more labile than its cAMP-binding activity, an important fact in light of results to be presented below.

CGA protein is a DNA-binding protein

A membrane-filter technique was used to study DNA–CGA protein complex formation. This method consists of mixing unlabeled protein with radioactively labeled DNA and passing the solution through a nitrocellulose filter. Little of

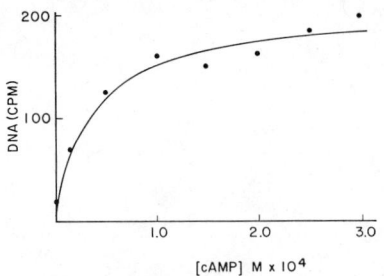

FIG. 3. Bound DNA as a function of cAMP concentration. CGA protein and λh80dlac [^{32}P]DNA were present at 0.01 and 0.05 μg, respectively. Only the concentration of cAMP in buffer IV was varied. Each filter received 0.4 ml, containing a total of 706 cpm.

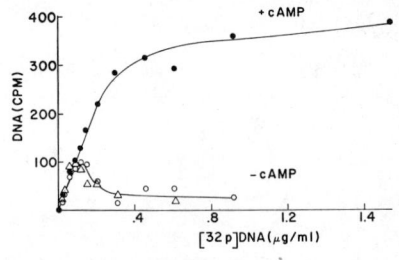

FIG. 4. Binding curves with CGA protein fixed and DNA varied. In these experiments, the CGA protein concentration was fixed at 0.04 μg/1.3 ml and the λh80dlac [^{32}P]DNA concentration was varied as indicated. In one set of experiments, (●), cAMP was present at a concentration of 1.6×10^{-4} M; in the others, (○) and (△), no cAMP was present. Each filter received 0.4 ml, containing [^{32}P]DNA at the concentration indicated. For experiments (●) and (○), the DNA used had a specific activity of 3.4×10^3 cpm/μg. For experiment (△), the DNA had a specific activity of 1.0×10^5 cpm/μg, but to facilitate comparison, the data have been normalized to the same specific activity as for the other curves. For each DNA concentration, the DNA retained in the absence of CGA protein was determined and subtracted. Only the binding due to CGA protein is shown.

the free, native DNA binds to the filter, but DNA bound to protein adheres strongly to the filter. The amount of DNA bound to protein is measured by the retention of radioactivity on the filter. This simple and sensitive technique has proven to be very effective in studies of interaction between DNA and the lac repressor (7, 9).

The binding of λh80dlac DNA as a function of CGA protein concentration is shown in Fig. 2. Binding occurs in buffer alone, but it is clearly stimulated by cAMP. In the early portion of the curve, when DNA is in excess of CGA protein (see below), five to ten times more DNA is bound in the presence of 3.7×10^{-4} M cAMP than in its absence. cGMP, on the other hand, eliminates DNA binding. The 5'-mononucleotides, AMP and GMP, have no significant effect on the binding curves, even at higher concentrations (6×10^{-4} M).

The dependence of DNA binding on cAMP concentrations was studied at a low weight ratio of CGA protein to DNA, where a marked enhancement of binding is seen. These results are shown in Fig. 3. Half-maximum binding of DNA to protein occurs at a cAMP concentration of about 2×10^{-5} M. The dependence of binding on cAMP concentration is quantitatively very similar to the dependence shown by the DNA-directed cell-free system for the synthesis of β-galactosidase of the lac operon (10), and is also in agreement with equilibrium dialysis data (3).

cAMP not only increases the amount of DNA–CGA protein complex, but it produces a complex that can be qualitatively distinguished from that formed in its absence. The relevant data are presented in Fig. 4, in which the binding is studied as a function of DNA concentration at a fixed concentration of CGA protein. The curves with and without cAMP are indistinguishable at very low DNA concentrations. As the DNA concentration is increased, however, a marked difference begins to appear between the two curves. Thus, when cAMP is present, the amount of filter-bound DNA increases to a plateau. In the absence of cAMP, a much lower maximum amount of filter-bound DNA is obtained, and there is a sharp decrease in DNA bound as the DNA concentration is further increased. The curve with cAMP present is consistent with the notion that one molecule of CGA protein binding to DNA (or several molecules binding cooperatively) is suf-

ficient to make it adhere to the filter. In contrast, when no cAMP is present, the binding curve can be explained most simply by assuming that the binding of more than one molecule of CGA protein is required to make a molecule of DNA adhere to the filter. As the amount of DNA is increased, so that there is only one or no molecule of CGA protein bound to each DNA molecule, the DNA no longer adheres to the filter. Certainly the complex formed in the absence of cAMP is different from that formed when cAMP is present.

Since the curves in Fig. 2 are distinctly sigmoidal, it is worth pointing out that at least two interpretations are consistent with our present data: (a) there is adjacent cooperative binding, or (b) CGA protein reversibly dissociates into subunits, but only oligomeric CGA protein binds to DNA. Additional work will be required to distinguish between these two possibilities.

Only a small fraction of CGA protein is involved in the formation of the DNA complex

The data shown in Fig. 4 can be used to estimate the concentration of active CGA protein. That a plateau is reached means that essentially all molecules capable of binding DNA have done so. Yet, in the presence of cAMP, 0.4 μg/ml of DNA is sufficient (extrapolating from the early linear portion of the curve) to saturate 0.031 μg/ml of CGA protein. If we assume that one CGA protein per DNA molecule is sufficient to cause retention, 0.4 μg/ml of DNA of molecular weight 30×10^6 corresponds to only 6×10^{-4} μg/ml of a protein of molecular weight 45,000. Thus, only about 2% of the CGA protein is active in binding DNA. If there is cooperative binding of several molecules, then a higher percentage is active. Nevertheless, it seems reasonably certain that most molecules cannot bind to DNA. This result is consistent with our other observations that a considerable inactivation of CGA protein occurs during purification.

Lack of expected specificity

Mutants defective in CGA protein are viable; it is only certain catabolite-sensitive genes that can not be expressed (4). Thus, our working hypothesis was, and remains, that *in vivo* CGA protein binds specifically with catabolite-sensitive promoters. However, under the conditions of our binding experiments, we have not been able to demonstrate such specificity. It was hoped that the only catabolite-sensitive promoter in λh80*dlac* would be that of the *lac* operator. Thus, CGA protein might not have bound to DNA from λh80, but in fact we have observed no difference between the binding of [^{32}P]DNA from λh80*dlac* and [^{33}P]DNA from λh80. Unlabeled λh80 DNA also competes well with [^{32}P]DNA from λh80*dlac* for CGA protein. In such competition experiments, no significant differences were seen between the DNAs from λh80*dlac*, λh80, salmon sperm, *Clostridium perfringens*, *Micrococcus luteus*, and poly(dA-dT). *E. coli* rRNA and tRNA, however, compete much less than DNA, so there is specificity for DNA. Under the conditions of low ionic strength of the assay used here, the *lac* repressor also binds to DNA other than the *lac* operator, though some specificity for λh80*dlac* is still observed (11). The *lac* repressor shows greater specificity, though weaker binding, at higher ionic strengths, so a most pressing problem for the future will be to study DNA–CGA protein complex formation at higher ionic strengths.

However, since we have been unable to demonstrate specificity, one should keep in mind that specificity of action need not be based solely on specificity of binding. Perhaps only catabolite-sensitive promoters *require* CGA protein binding.

Conclusion: a new type of regulatory protein

The opposing effects of cAMP and cGMP on DNA–CGA protein complex formation parallels the effects of these cyclic compounds on the activation of *in vitro* synthesis of β-galactosidase from λh80*dlac* DNA. For the latter system, cAMP is a required activator and cGMP inhibits the activation by cAMP (10). Other studies have shown that cAMP and cGMP compete for binding sites on CGA protein. The activation and DNA-binding properties of CGA protein are also similar, in that both seem to be preferentially labile during purification when compared to cAMP-binding activity. Another regulatory protein, the *lac* repressor, has often shown a similar preferential loss of DNA-binding activity during purification (7). It thus seems almost certain that the biochemically significant action of cAMP is to stimulate CGA protein complex formation with DNA, and that this leads to both increased mRNA (2) and enzyme induction (3). It now appears that there are three distinct types of regulatory proteins: Type I includes the DNA-binding repressors such as *lac*, phage λ, and phage 434 (7, 12, 14). Type II is the σ factor, which interacts with RNA polymerase and activates transcription (13). Type III is CGA protein, an activator that interacts with DNA.

NOTE ADDED IN PROOF

Recent preparations of CGA protein, made with care and and assayed promptly, give linear rather than sigmoidal DNA binding curves and show an almost total dependence on cAMP. However, we are still unable to demonstrate specificity for the *lac* promoter.

We are indebted to Tetteh Blankson and to Joan Roberts for excellent technical assistance. This work was supported by grants from the National Institutes of Health (5-R01-GM-16648-03), the American Cancer Society (E-545), and the National Science Foundation (B01 8733-000) to G. Zubay, and by a grant from the National Institutes of Health (1-R01-HD-04420-01) to A. D. Riggs. G. R. thanks the National Institutes of Health for a predoctoral fellowship.

1. For reviews of the subject see Magasanik, B., in *Cold Spring Harbor Symp. Quant. Biol.*, **26**, 249 (1961); Magasanik, B., in *The Lac Operon*, ed. D. Zipser and J. R. Beckwith, Cold Spring Harbor, N.Y., Laboratory of Quantitative Biology (1970).
2. Pastan, I., and R. Perlman, *Science*, **24**, 339 (1970).
3. Zubay, G., D. Schwartz, and J. Beckwith, *Proc. Nat. Acad. Sci. USA*, **66**, 104 (1970).
4. Schwartz, D., and J. Beckwith, in *The Lac Operon*, ed. D. Zipser and J. R. Beckwith, Cold Spring Harbor, N.Y., Laboratory of Quantitative Biology (1970).
5. Silverstone, A. E., R. R. Arditti, and B. Magasanik, *Proc. Nat. Acad. Sci. USA*, **66**, 773 (1970).
6. Weber, K., and M. Osborn, *J. Biol. Chem.*, **244**, 4406 (1969).
7. Riggs, A. D., H. Suzuki, and S. Bourgeois, *J. Mol. Biol.*, **48**, 67 (1970).
8. Riggs, A. D., and S. Bourgeois, *J. Mol. Biol.*, **34**, 361 (1968).
9. Riggs, A. D., R. F. Newby, and S. Bourgeois, *J. Mol. Biol.*, **51**, 303 (1970).
10. Zubay, G., D. A. Chambers, and L. C. Cheong, in *The Lac Operon*, ed. D. Zipser and J. R. Beckwith, Cold Spring Harbor, N.Y., Laboratory of Quantitative Biology (1970).
11. Lin, S., and A. D. Riggs, *Nature*, in press.
12. Pirrotta, V., and M. Ptashne, *Nature*, **222**, 541 (1969).
13. Burgess, R. R., A. A. Travers, J. J. Dun, and E. K. F. Bautz, *Nature*, **221**, 43 (1969).
14. Gilbert, W., and B. Müller-Hill, *Proc. Nat. Acad. Sci. USA*, **58**, 2415 (1967).

*THE L-ARABINOSE OPERON IN ESCHERICHIA COLI B/r:
A GENETIC DEMONSTRATION OF TWO FUNCTIONAL
STATES OF THE PRODUCT OF A REGULATOR GENE**

BY ELLIS ENGLESBERG, CRAIG SQUIRES, AND FRANK MERONK, JR.

DEPARTMENT OF BIOLOGICAL SCIENCES, UNIVERSITY OF CALIFORNIA (SANTA BARBARA)

Communicated by Charles Yanofsky, January 2, 1969

Abstract.—The product of the regulator gene *araC* in the L-arabinose gene complex exists in two functional states: P1, the repressor, and P2, the activator, presumably in equilibrium with each other, and with P1 and P2 attached to their respective controlling sites, *araO*, the operator, and *araI*, the initiator. The controlling sites are linked in the following order with respect to genes *araB* and *araC*: *BIOC*. Two *C* gene deletions (Δ719 and Δ766) serve to define the newly described *araO* site, and to place it adjacent to the left end of the *C* gene. We have suggested that deletion 719 deletes, in addition to the *C* gene, part or all of the *araO* site. Deletion 766 leaves the *araO* site intact. Complementation analysis employing stable merodiploids indicates that the repressor-operator site function is epistatic over the activator-initiator site function. It is necessary both for activator (P2) to be present and for repressor (P1) to be absent at their respective controlling sites (*araI* and *araO*) for full expression of the L-arabinose operon.

The L-arabinose gene–enzyme complex in *Escherichia coli* B/r has been shown to consist of structural genes *araD*, *araA*, *araB*, controlling site *araI* (the initiator), and regulatory gene *araC*, linked in that order between the markers *thr* and *leu* and the unlinked gene, *araE*, concerned with the active transport of L-arabinose, with its corresponding controlling sites.[1–3] Several lines of evidence have been presented indicating that gene *araC* controls in a positive fashion the expression of the structural genes in this system. A detailed analysis of this evidence is presented in another publication.[4] Briefly, it has been shown that deletions[1–3] and nonsense mutations[5] of the *C* gene lead to the production of a pleiotropic-negative phenotype (C^-) which is recessive (*cis* and *trans*) to both the wild-type (C^+) and constitutive (C^c) alleles of that gene. The findings that unlinked suppressors restore activity of C^- mutants and that some revertants of these C^- mutants form a thermolabile *C* gene product indicate that the activator is at least partially a protein molecule.[5] No support has been found for the existence of another L-arabinose regulatory gene producing a repressor specific for the L-arabinose system.[4,5] Several lines of evidence indicate that the controlling sites are located in the region between genes *B* and *C*: (1) polarity is in the direction *BAD*;[6,7] (2) gene *araC* is not part of the *BAD* operon, since nonsense and deletion mutations in the *araC* gene[1–3,5] and deletion mutations that end within the *araC* gene and the *leu* operon do not affect the L-arabinose-gene *araC* control of the L-arabinose operon, while deletion mutations that end within the *araB* gene and the *leu* operon remove the remaining structural genes of the operon from the control of L-arabinose and gene *araC*

1100

(as demonstrated in merodiploids) and place them under the control of the *leu* regulator gene;[8] and (3) initiator constitutive mutants (I^c) map within this region and produce low-constitutive, *cis*-dominant phenotypes (I^c is dominant to I^+ with no *trans* effect) in a genetic background lacking an active *C* gene.[4]

The I^c mutants were isolated as revertants of the Ara⁻ deletion mutant (Δ)719, a deletion encompassing all known *ara*⁻ mutant sites in the regulator gene, *araC*.[4] The revertants contain the original deletion plus the secondary I^c mutation. In the complementation analysis with merodiploids of the type F′A2I⁺C⁺/A⁺I^{c}Δ719 and F′A2I⁺C⁺/A⁺I⁺Δ719, it was observed that the constitutive L-arabinose isomerase levels (see legend, Fig. 2) of most of the former and the basal level of the latter were significantly higher than those found for the corresponding F⁻ haploid strains. When a C^- allele was substituted for the C^+ in the exogenote (F′A2C3), the L-arabinose isomerase levels were reduced to those characteristic of the respective F⁻ haploid strains. Thus, it appears that the product of the C^+ allele, in the absence of inducer, does affect an increase in expression of the *araA* gene, the structural gene for the L-arabinose isomerase, *cis* to Δ719.

In the model for positive control,[1] it was proposed that the initial product of the regulator gene, *araC*, is an allosteric protein (P1) that is converted by inducer into P2, the activator, which reacts at the initiator site and "turns on" gene expression. The existence of P1 was first indicated by the demonstrated dominance of C^+ to C^c.[1-3] It was suggested that P1 might be a true repressor and that P1 competes with P2 for the same functional site, *araI*, or has a separate site of attachment and thus prevents the expression of the operon in the presence of the constitutive activator.[1-3]

In this paper, we compare the results of complementation experiments performed with strains containing Δ719 as compared with those performed with Δ766, a deletion whose left end terminates within the *C* gene. We present evidence that (1) the *C* gene product, P1, produced in the absence of inducer, does exist as a true repressor with a specific site of attachment, the operator (*araO*), and is presumed to be in equilibrium with P2, the activator, and P1 and P2 attached to their respective controlling sites *araO* and *araI*; (2) the order of controlling sites in reference to genes *araB* and *araC* is *araB*, *araI*, *araO*, *araC*; and (3) that the repressor-operator site function is epistatic over the activator-initiator site function. It is necessary both for P2 to be present and for P1 to be absent at their respective controlling sites for full expression of the L-arabinose operon. The stimulation of the expression of gene *araA cis* to Δ719 in merodiploids containing a *trans* C^+ allele is explained on the basis that Δ719 extends into the region between *araC* and *araB* and excises all or a portion of *araO*. Thus, this strand is sensitized to the small amounts of P2 produced by the C^+ allele in the *trans* position that would otherwise have remained cryptic.

Materials and Methods.—With the exceptions given below, the materials and methods employed in the experiments described in this paper are essentially the same as previously described.[4] The strains used are described in Table 1.

The isolation and characterization of deletion 719 has already been presented.[4] Deletion 766 was originally isolated in strain SB5088 F⁻ *thr*⁻ *D*139 *str*^{r} as a result of a spontaneous mutation making the strain resistant to the L-arabinose inhibition.[1,3,9] To isolate the deletion free of the $D139$ marker, phage P1bt grown on the wild type was crossed to strain SB1064 F$^-$ thr^- $D139\Delta766$ str^r and selection was for Thr$^+$ on mineral-glucose agar plates. Thr$^+$ transductants were scored by matings with F' Ara$^-$ homogenotes for the $D139$ marker and for the deletion. Transductants of the genotype thr^+ $araD^+\Delta766$ str^r were isolated in pure culture and a nonlysogenic strain (SB1114) was employed in this study. The extent of deletion 766 was determined by matings with pertinent F' Ara$^-$ homogenotes. Strain SB1114 failed to yield any Ara$^+$ recombinants with F' $araC19/araC19$; F'$araC3/araC3$; F'$araC5/araC5$; but did yield Ara$^+$ recombinants with F'$araC12/araC12$; F'$araC101/C101$; and F'$araB27/araB27$. This places the left end of 766 within the C gene between mutant sites $araC5$ and $araC12$.

TABLE 1. *List of strains.*

Strain	Mating type	Genotype L-arabinose	thr1	leuB1	str	Origin, source or reference
SB5088	F$^-$	$D139$	−	+	r	1, 3, 9
SB1064	F$^-$	$D139\Delta766$	−	+	r	Spontaneous mutation
SB1114	F$^-$	$\Delta766$	+	+	r	From SB1064 by transduction (this paper)
SB1094	F$^-$	$\Delta719$	−	+	r	4
SB2012	F$^-$	$I^c13\Delta719$	−	+	r	4
SB2018	F$^-$	$I^c19\Delta719$	−	+	r	4
SB5316	F$^-$	$I^c13\Delta766$	+	+	r	By transduction (this paper)
SB5317	F$^-$	$I^c19\Delta766$	+	+	r	By transduction (this paper)
SB1509	F$^-$	$\Delta1109$	+	−	r	8
SB3101	F'	F'$A2/A2$	+/+	+/+	s	3
SB3107	F'	F'$B24/B24$	+/+	+/+	s	3
SB3114	F'	F'$C3/C3$	+/+	+/+	s	3
SB3115	F'	F'$C5/C5$	+/+	+/+	s	3
SB3116	F'	F'$C12/C12$	+/+	+/+	s	3
SB3129	F'	F'$B27/B27$	+/+	+/+	s	3
SB3139	F'	F'$C19/C19$	+/+	+/+	s	3
SB3141	F'	F'$C101/C101$	+/+	+/+	s	3
SB3147	F'	F'$A2C3/A2C3$	+/+	+/+	s	4
UP1010	F$^-$	$C3$	+	+	s	10
UP1009	F$^-$	$A2$	+	+	s	10
UP1029	F$^-$	$B26$	+	+	s	10
SB5161	F$^-$	$C101$	+	+	s	4

Symbols and abbreviations: A, structural gene for L-arabinose isomerase; B, structural gene for L-ribulokinase; C, regulator gene in the L-arabinose system; D, structural gene for L-ribulose 5-phosphate 4-epimerase; Δ, deletion; leu, leucine; thr, threonine; str, streptomycin; r, resistant; s, sensitive; plus, ability to synthesize or utilize; minus, inability to synthesize or utilize.

I^c13 and I^c19 mutant sites[4] were crossed into F$^-\Delta766$ (SB1114) in two steps. First, phage P1bt grown on strains SB2012 (I^c13) and SB2018 (I^c19) was used to infect F$^-\Delta1109$ (SB1509) (a deletion of $araC$ extending into the leucine operon[8]). We selected Ara$^+$ (slow-growing) transductants and scored for those that were Leu$^-$, indicating transfer of the I^c mutation to $\Delta1109$.[4] Second, phage P1bt was grown on the two $I^c\Delta1109$ derivatives and used to infect F$^-\Delta766$ (SB1114). We selected for transfer of the I^c's on mineral-arabinose agar without leucine. The only progeny that could grow received the I^c-mutation from the phage while retaining the leu^+ (and thus the linked $\Delta766$) of the recipient strain. The two $I^c\Delta766$ strains isolated SB5316 (I^c13) and SB5317 (I^c19) grew slowly on mineral-arabinose agar and showed the recombination pattern of $\Delta766$ when cross-streaked against F'C^- homogenotes.

Construction of merodiploids: Matings were carried out as previously described.[3] In crosses between strain SB3101 and strains SB1094, SB2012, SB2018, SB1114, SB5316, and SB5317, the mating mixture was streaked out on eosin methylene blue (EMB)

L-arabinose agar plates. In a normal mating, an appreciable fraction of the population is Ara$^+$. Ara$^+$ clones are isolated and their genotype determined by noting the segregation of negative clones and picking several positive colonies (usually ten) from each positive clone to L-broth and demonstrating the segregation of Ara$^-$ clones (note: $I^c\Delta719$ and $I^c\Delta766$ appear Ara$^-$ on EMB plates) containing either araA2 and $\Delta719$ or $\Delta766$. The latter was determined by cross-streaking against the following homogenotes on mineral L-arabinose threonine agar plates: (strain SB3101) F'A2, (strain SB3107) F'B24, (strain SB3141) F'C101, and (strain SB3139)F'C19. The I^c character was scored by the typical slow growth on mineral-arabinose agar plates. This character is always associated with the deletion.

In crosses between F'A2C3 homogenote (SB3147) and SB1094, SB2012, SB2018, SB1114, SB5316, and SB5317, contraselection of the F' donor occurred on mineral-glucose-streptomycin plates. Individual clones were isolated and tested for transfer of the episome by cross-streaking against appropriate F$^-$ haploid strains F$^-$A2, F$^-$B26, F$^-$C3, and F$^-$101. Merodiploids of the type F'A2C3/A$^+$I$^+\Delta766$ and F'A2C3/A$^+$I$^+\Delta719$ do not grow on mineral arabinose and when cross-streaked against F$^-$A2, yield a small number of Ara$^+$ recombinants as a result of episomal mobilization of the A^+ allele on the chromosome. When cross-streaked against F$^-$B26, a heavy complementation reaction is evident, whereas a completely negative response is elicited with F$^-$C3. When cross-streaked against F$^-$C101, a small number of recombinants is apparent as a result of recombination between araC3 and araC101.

Merodiploids of the type F'A2C3/A$^+$I$^c\Delta766$ and F'A2C3/A$^+$I$^c\Delta719$ were tested as described above and behaved in a similar manner except that slow growth on mineral arabinose due to the I^c allele was apparent in each case, verifying the presence of the I^c marker.

In the preparation of cell-free extracts in the case of the F'A2C$^+$/$\Delta766$ (I^c or I^+) or F'A2C$^+$/$\Delta719$ (I^c or I^+) merodiploids, a sample of the cells employed in the preparation of the extracts was streaked on EMB L-arabinose and scored for Ara$^-$ segregants. In addition, usually ten Ara$^+$ clones were analyzed as described above for the identity of Ara$^-$ segregants. With merodiploid cultures of the type $A^-C^-/\Delta766$ and $A^-C^-/\Delta719$, besides cross-streaking the cultures against the F$^-$ strains listed above, ten clones isolated on nutrient agar were tested in a similar manner. The enzymatic data reported here are the result of experiments in which segregation was 20% or less and in which the genotype of the exogenote and endogenote was verified as present in at least eight out of ten of the tested clones.

Results.—*A comparison of the trans effect of C^+ and C^- with mutants containing deletions 719 and 766:* As shown in Table 2, and as previously demonstrated,[4] as a result of the expression of the araA gene *cis* to deletion 719, there is a large constitutive increase (35×) in L-arabinose isomerase activity in a merodiploid containing A^-C^+ in the exogenote, over the isomerase activity of the original haploid F$^-$ deletion strain or a merodiploid containing A^-C^- in the exogenote. These experiments demonstrate that it is the product of the C^+ allele (in the absence of the inducer) that causes the increased expression of gene araA *cis*

TABLE 2. *A comparison of the trans effect of C^+ and C^- on mutants containing deletions 719 and 766.*

	Endogenote	
	$A^+I^+\Delta719$	$A^+I^+\Delta766$
Exogenote	L-Arabinose Isomerase*	
None (haploid)	0.10	0.09
F' A2C$^+$	3.53	0.20
F' A2C3	0.12	0.13

* L-Arabinose isomerase activity is expressed as μmoles of ribulose formed per hour per milligram of protein.

to deletion 719. When similar experiments were conducted with a strain containing deletion 766, the left end of which terminates within the C gene, there was a twofold increase in isomerase activity in the merodiploid containing the C^+ allele over that of the F^- haploid deletion strain, or a merodiploid containing a C^- allele in the exogenote. Thus the deletions modify the effect of the C gene product.

A comparison of the trans effect of C^+ and C^- on the constitutive expression of I^c alleles in strains containing deletion 719 and deletion 766: There is a striking difference in the effect that the C^+ allele has on the constitutive expression of I^c alleles in strains containing deletions 766 and 719. In merodiploids $F'A2I^+C^+/A^+I^c13\Delta719$ and $F'A2I^+C^+/A^+I^c19\Delta719$, there is an increase in constitutive isomerase activity over that found in the F^- haploid strains. This increase in isomerase activity disappears if a C^- allele is substituted for the C^+ allele in the exogenote (Tables 3 and 4). These results are similar to those previously reported.[4] This is a further demonstration of the *cis* dominance of I^c to I^+ and the role of the product of the C^+ allele in stimulating the expression of the structural genes *cis* to deletion 719, as observed in the merodiploids described above.

In contrast to the results with deletion 719, with similar merodiploids constructed with deletion 766 ($F'A2I^+C^+/A^+I^c13\Delta766$, $F'A2I^+C^+/A^+I^c19\Delta766$), there is a severe repression of the expression of the I^c alleles. This repression of constitutive synthesis of isomerase by the *araA* gene *cis* to the deletion is removed when a C^- allele is substituted for the C^+ allele in the exogenote. Thus, in merodiploids containing deletion 766, there is a severe epistatic effect of the C^+ allele on the expression of the isomerase gene *cis* to the deletion. As with deletion 719, with merodiploids containing the C^- allele in the exogenote, the *cis* dominance of I^c to I^+ is demonstrated.

Discussion.—The phenotypic expression of gene *araA*, the structural gene for L-arabinose isomerase, is affected differently by a C^+ allele in a *trans* position, depending upon whether the *araA* gene is linked to deletion 719 or 766. The

TABLE 3. *A comparison of the trans effect of C^+ on I^c in mutants containing deletions 719 and 766.*

Exogenote	Endogenote	
	$A^+I^c13\Delta719$	$A^+I^c13\Delta766$
	L-Arabinose Isomerase*	
None (haploid)	4.0	6.9
$F'A2I^+C^+$	6.2	0.93
$F'A2I^+C3$	4.9	4.4

* See Table 2 for L-arabinose isomerase activity of $F^-A^+I^+\Delta719$ and $F^-A^+I^+\Delta766$.

TABLE 4. *A comparison of the trans effect of C^+ on I^c in mutants containing deletions 719 and 766.*

Exogenote	Endogenote	
	$A^+I^c19\Delta719$	$A^+I^c19\Delta766$
	L-Arabinose Isomerase*	
None (haploid)	4.5	5.0
$F'A2I^+C^+$	9.1	0.75
$F'A2I^+C3$	3.4	3.3

* See Table 2 for L-arabinose isomerase activity of $F^-A^+I^+\Delta719$ and $F^-A^+I^+\Delta766$.

mapping of these two deletions indicates that the left end of deletion 766 terminates within the C gene between mutant sites $araC5$ and $araC12$. On the other hand, deletion 719 covers all known point mutant sites in gene $araC$. This leaves the termination of the left end of $\Delta 719$ undetermined, somewhere between $araB27$ and $araC101$. Mutants containing either deletion fail to recombine with $araC19$, the ara^- mutant site farthest to the right in the C gene, and are both Leu$^+$. This places the right end of both deletions somewhere between $araC19$ and the leucine operon. It has been shown that it is possible to delete the entire region from $araC$ to the leucine operon and, except for the leucine requirement, the phenotype characteristic of such deletion mutants with respect to the L-arabinose operon BAD is identical to that of $araC^-$-point mutants.[8] Therefore, since there is no evidence to indicate that a difference in the termination point of the right arm of these deletions, if there were any, would make any difference in our analysis of the effect of these deletions on the function of this ara operon, at this point we choose to ignore the possibility.

A modified model for positive control designed to explain the experimental data presented in this paper is shown in Figure 1. In essence, this model is similar to one proposed earlier in which the dominance of C^+ to C^c was explained

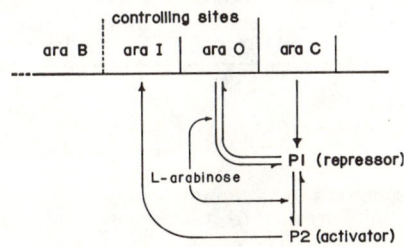

Fig. 1.—Positive control model.

on the basis that P1, the product of the C gene produced in the absence of inducer, acts as a true repressor with a separate site of action, the operator. It differs in that we can now more precisely define and locate the operator site. To review the model briefly, it is proposed that $araC$ is a regulator gene which produces, in the absence of inducer, a repressor, P1, which is at least partially a protein molecule. P1 exists in equilibrium with P1 attached to an operator site (O) located between $araC$ and $araI$ (the initiator site), P2 (the activator), and P2 attached to $araI$. The inducer removes P1 from the operator[11, 12] and shifts the equilibrium to P2. P2 reacts at the initiator site and turns on the expression of the structural genes cis to the initiator. The basal level of expression of the L-arabinose operon $OIBAD$ probably is determined by the relative effective concentrations of P1 and P2 and their respective affinities for $araO$ and $araI$. In this positive control system, it is not sufficient to remove P1 from the operator in order to achieve the expression of the structural genes concerned. P2 must be produced and react at I to stimulate the expression of these genes.[1, 3] This model now provides a logical explanation for the dominance of C^+ to C^c and the different effects of C^+ on the expression of structural genes cis to $I^c\Delta 719$

and $I^c\Delta 766$ if we assume in the latter case that deletion 719 terminates outside of the C gene and encompasses $araO$ but leaves intact $araI$. Experiments have shown that deletion 766 does terminate within the C gene and therefore must leave intact the postulated $araO$ site (Fig. 2).

The repressive epistatic effect of the C^+ allele on the constitutive expression of the I^c alleles in strains containing deletion 766 and the absence of such a repressive effect on the I^c alleles *cis* to deletion 719 can now be understood on the basis that P1 produced by the C^+ allele (in the absence of inducer) reacts with the operator site *cis* to deletion 766 and prevents the complete expression of the I^c alleles. Such a repressive effect does not occur in strains containing deletion 719 because this deletion excises the operator site. The large increase in isomerase activity in the merodiploid of $F^-I^+\Delta 719$ containing a *trans* C^+ allele as compared to only a minor increase with a similar merodiploid of $F^-I^+\Delta 766$ is also consistent with the model. Strains containing $I^+\Delta 719$, lacking an operator

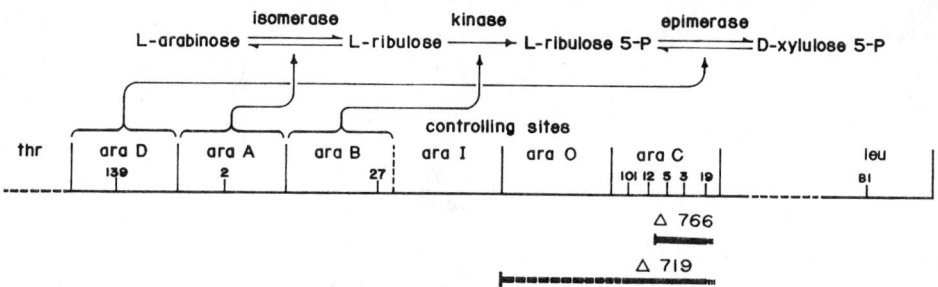

Fig. 2.—The L-arabinose gene–enzyme complex.
Structural genes: $araB$, L-ribulokinase; $araA$, L-arabinose isomerase; $araD$, L-ribulose 5-phosphate 4-epimerase.
Controlling sites: $araI$, initiator site (this is the position of I^c mutations and the site for activator (P2) function); $araO$, operator site (this is the site for repressor (P1) function).
The numbers indicate the mutants employed in this study.
Deletions: Solid lines indicate the portion of the genome excised by the deletion as determined by genetic mapping with F' Ara⁻ homogenotes. The dashed portions of the lines extending the deletion are based upon complementation and enzymatic analysis.

region, are able to detect the small quantities of P2 (activator) that exist in equilibrium with P1. This small quantity of P2 is able to partially "turn on" the expression of the genes *cis* to this deletion. No large increase in isomerase activity would be expected with the strain containing $I^+\Delta 766$ with the operator and initiator sites both present.

The simplest explanation of the dominance of C^+ to C^{c} [1, 3] in the light of the evidence that we have presented in this paper is that the C^+ allele produces a repressor which attaches to the operator site on both the exogenote and the endogenote and prevents the constitutive activator from functioning. The presence of P1 at the operator site takes precedence over the action of the constitutive activator at the initiator site.

Our evidence therefore indicates that the C gene product has a dual function. In the presence of an operator site and in the absence of the inducer, the C gene product is mainly present in the configurational state of a functional repressor but with detectable amounts of activator present that probably represent another configurational state of the molecule. In the presence of inducer and an initiator site, repressor function is no longer detectable and full activation of the operon occurs probably as a result of a shift in equilibrium of the C product to the activator configuration.

* This work was supported in part by National Science Foundation grant 5392 and a contract between the University of California, Santa Barbara, and the Office of Naval Research.

[1] Englesberg, E., J. Irr, J. Power, and N. Lee, *J. Bacteriol.*, **90**, 946 (1965).
[2] Sheppard, D., and E. Englesberg, in *Cold Spring Harbor Symposia on Quantitative Biology*, vol. 31 (1966), p. 345.
[3] Sheppard, D., and E. Englesberg, *J. Mol. Biol.*, **25**, 443 (1967).
[4] Englesberg, E., D. Sheppard, C. Squires, and F. Meronk, *J. Mol. Biol.*, in press.
[5] Irr, J., and E. Englesberg, *Bacteriol. Proc.* (1967), p. 54.
[6] Katz, L., and E. Englesberg, *Bacteriol. Proc.* (1967), p. 50.
[7] Hogg, R., and E. Englesberg, unpublished data.
[8] Kessler, D., and E. Englesberg, *J. Bacteriol.*, in press.
[9] Englesberg, E., R. L. Anderson, R. Weinberg, N. Lee, P. Hoffee, G. Huttenhauer, and H. Boyer, *J. Bacteriol.*, **84**, 137 (1962).
[10] Gross, J., and E. Englesberg, *Virology*, **9**, 314 (1959).
[11] Biggs, A. D., S. Bourgeois, R. F. Newby, and M. Cohn, *J. Mol. Biol.*, **34**, 365 (1968).
[12] Bourgeois, S., personal communication. As a result of binding experiments with purified *lac* repressor and *lac* DNA, it is apparent that the inducer must react with the repressor-DNA complex and, in this manner, remove the repressor from the operator site of attachment. We feel that such a possibility should be left open in the L-arabinose system.

MSI and MSII made on Ribosome in Idling Step of Protein Synthesis

WILLIAM A. HASELTINE & RICARDO BLOCK
Committee on Biophysics of Harvard University

WALTER GILBERT & KLAUS WEBER
Department of Biochemistry and Molecular Biology, Harvard University Biological Laboratories, 16 Divinity Avenue, Cambridge, Massachusetts 02138

The unusual guanosine nucleotides, MSI and MSII, accumulated *in vivo* during amino-acid starvation of stringent strains of *E. coli* are shown to be synthesized *in vitro* as a product of an idling step in protein synthesis occurring on the ribosomes.

THE synthesis of stable RNA species in many *Escherichia coli* strains is sharply curtailed by deprivation of a required amino-acid[1-4]. This control mechanism, called the stringent response, depends on the function of the *rel* gene. Mutations in this gene give rise to relaxed mutants[1], which continue to accumulate RNA for a considerable time after amino-acid deprivation. Cashel and Gallant[5,6] found that amino-acid starvation causes a rapid accumulation of two unusual guanosine nucleotides, called "magic spots" I and II (MSI and MSII), in stringent (rel^+) but not in relaxed (rel^-) strains. They postulated[5] that high intracellular concentrations of the MS compounds led to the cessation of RNA accumulation and to the other characteristics of the stringent response: inability to incorporate uracil from the medium[7], shrinkage of the nucleoside triphosphate pools[7-9], and inhibition of the synthesis of gylcolytic esters[10]. Cashel and Kalbacher[11] have identified MSI as a guanosine tetraphosphate (ppGpp, 5′ diphosphate 3′ or 2′ diphosphate guanosine) and MSII as a guanosine pentaphosphate.

We report here that MSI and MSII can be synthesized *in vitro* on the ribosome using GTP (or GDP) and ATP as substrates, and that the difference between relaxed and stringent strains with respect to MS accumulation during amino-acid starvation is due to an alteration in a protein factor present in the 0.5 M NH_4Cl ribosomal wash.

MSI and MSII synthesized *in vitro*

MSI and MSII are synthesized *in vitro* in a reaction containing the components listed in Table 1. This simple reaction requires, in addition to buffer and salts, only the substrates GTP (or GDP) and ATP, high salt washed ribo-

Table 1 Composition of *in vitro* MSI and MSII Synthesizing System

Component	Amount/50 µl. reaction
Tris-acetate, pH 7.8	42 mm
Dithiothreitol	1.4 mm
Magnesium-acetate	11.4 mm
Ammonium-acetate	27.0 mm
GTP	0.55 mm
ATP	2.2 mm
Potassium-acetate	10.0 mm
High salt ribosomes	100 µg
α-^{32}P-GTP	1 µCi
0.5 M NH_4Cl ribosomal wash	7 µg

The ribosomes and ribosomal washes were prepared as follows. CP 78 rel^+ (obtained from the strain collection at Yale through Peter Lengyel) was grown to a cell density of 5×10^8 cells/ml. at 34° C in a medium containing (per litre) 5.6 g KH_2PO_4 (anhydrous); 28.9 g K_2HPO_4 (anhydrous); 10 g yeast extract; 10 mg thiamine; and 1% glucose (the medium described by Zubay *et al.*[25]). The cells were chilled and collected at 7,000*g* for 20 min and washed twice by resuspending the pellet in four times the pellet volume of buffer A (0.01 M Tris-acetate, pH 7.8, 0.014 M Mg-acetate, 0.06 M K-acetate and 10^{-4} M dithiothreitol) and recentrifuging. Finally, the cells were resuspended in buffer B (buffer A, with 10^{-3} M dithiothreitol); 4 µg/ml. of DNAase (Worthington) added, and the cells broken in an 'Aminco' pressure cell using pressures between 5,000 and 10,000 p.s.i. The lysate was stirred gently for 30 min at 4° C and centrifuged at 30,000*g* for 20 min to remove debris. These "S-30" crude extracts were centrifuged in a 'Beckman angle 40' rotor for 3 h at 40,000 r.p.m. at 4° C. The supernatant was decanted and the ribosomal pellet rinsed with buffer B and resuspended in the same buffer. The resuspended ribosomes were layered above an equal volume of buffer B made 40% in sucrose and centrifuged at 40,000 r.p.m. for 12 h in an 'International A-170' rotor at 4° C. The supernatant was decanted and the pellet washed an additional time with buffer B by resuspending it and centrifuging at 40,000 r.p.m. in an angle 40 rotor for 2 h. This pellet was resuspended in one-fourth the original "S-30" volume to give a final concentration of ribosomes between 30 and 60 mg/ml.
The 0.5 M NH_4Cl wash was made by centrifuging ribosomes (60 mg/ml.) at 10,000*g* to remove aggregates and then pelleted for 2 h. They were then resuspended at 60 mg/ml. in buffer B made 0.5 M in NH_4Cl. The resuspended ribosomes were gently shaken for 2 h at 4° C and pelleted again. The supernatant was decanted and centrifuged once again for 2 h at 40,000 r.p.m. in the angle 40 rotor. This supernatant was the source of the 0.5 M NH_4Cl ribosomal wash. The ribosomal pellet was washed twice more in a large volume of the 0.5 M NH_4Cl buffer, and finally resuspended at 120 mg/ml. in buffer B and used as the 0.5 M NH_4Cl washed ribosomes.

somes which retain some G factor activity, and a factor present in the 0.5 M NH₄Cl wash of ribosomes extracted from stringent cells; we shall call this the stringent (str) factor. The preparation of the high salt washed ribosomes and the 0.5 M NH₄Cl wash is described in Table 1. The conversion of the substrates to MSI and MSII is followed by chromatography of the reaction products on thin layer polyethyleneimine (PEI) plates, which separate guanosine nucleotides according to the number of phosphates. In most experiments the GTP was labelled in the α position with ^{32}P.

Fig. 1 demonstrates that both the high-salt washed ribosomes and the 0.5 M NH₄Cl wash are needed for MSI and MSII synthesis. Neither of these two components by themselves can mediate the synthesis of these guanosine compounds. Only the 0.5 M NH₄Cl ribosomal wash from the rel^+ cells contains the stringent factor. Furthermore, the addition of uncharged tRNA, Qβ RNA and poly U does not affect the amount of MS compounds synthesized.

1 2 3 4 5 6 7 8

Fig. 1 A factor present in the 0.5 M NH₄Cl ribosomal wash from stringent strains is necessary for MS synthesis.

Reaction	0.5 M NH₄Cl washed ribosomes		0.5 M NH₄Cl M ribosomal wash	
	413 (rel^+)	415 (rel^-)	413 (rel^+)	415 (rel^-)
1	−	−	+	−
2	−	−	−	+
3	+	−	−	−
4	−	+	−	−
5	+	−	+	−
6	−	+	+	−
7	+	−	−	+
8	−	+	−	+

High salt washed ribosomes and the 0.5 M NH₄Cl ribosomal wash were prepared as in Table 1. The reactions contained the components listed in Table 1 in 50 µl., and were incubated at 37° C for 30 min. The reactions were terminated by chilling in ice and adding 1 µl. of 88% formic acid. The reaction tubes were mixed vigorously on a vortex mixer and centrifuged in a clinical centrifuge at 3,000g for 5 min. The clear supernatant was transferred to clean tubes and 1.0 µl. samples spotted on 'Brinkmann' polyethyleneimine (PEI) thin layer plates. The samples were dried, the plate rinsed with distilled water, dried again and developed with 1.5 M KH₂PO₄, pH 3.4, as described by Cashel et al.³⁰. The plates were dried and autoradiographed by exposure to 'Kodak No-Screen' medical X-ray film. The spots in ascending order were MSII, MSI, GTP, GDP, GMP, and free phosphate. The identity of MSI and MSII has been confirmed by co-chromatography with authentic MSI and MSII prepared in bulk from cells as described by Cashel and Kalbacher¹¹ in two dimensions, first with 3.3 M ammonium formate and 4.2% boric acid (adjusted to pH 7.0 with NH₄OH) and then in 1.5 M KH₂PO₄, pH 3.4. The rate of synthesis of the MS compounds is, as calculated from an experiment not pictured here, about 0.8 molecules of MS compounds/s/ribosome, assuming all ribosomes are active. The high salt washed ribosomes and 0.5 M NH₄Cl ribosomal wash were obtained from the stringent and relaxed isogenic pair of strains 413 (rel^+) and 415 (rel^-) (from the collection of D. Morse).

The elongation factor G¹²,¹³ is also necessary for the synthesis of the MS compounds. When the synthesis is attempted with ribosomes washed seven times in high salt, added G factor (provided by H. Weissbach) stimulated the synthesis of MSI and MSII up to eight times. Stringent factor and the seven times washed ribosomes in our standard reaction mixture synthesized 0.9 nmol of MS compounds; when 6 µg of G factor was added, 7 nmol of MS compounds was synthesized. A small amount of G factor probably remains bound to the five times washed ribosomes, as has been reported by Nishizuka and Lipmann¹². The washed ribosomes and G factor alone do not make MSI and MSII without the stringent factor. Separate experiments have shown that the elongation factors Tu and Ts cannot substitute for the stringent factor.

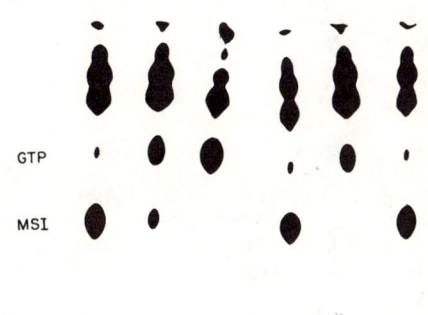

Fig. 2 Inhibition of MS synthesis by the antibiotics fusidic acid and thiostrepton.

Reaction	Fusidic acid	Thiostrepton
1	−	−
2	4×10^{-4} M	−
3	1×10^{-3} M	−
4	−	−
5	−	1×10^{-6} M
6	−	4×10^{-7} M

The conditions were as those described for Table 1. However, the ribosomes were not washed with high salt and retained the ability to synthesize the MS compounds. Fusidic acid was a gift of D. Goldéz and R. T. Garvin and thiostrepton a gift of B. Wallace. Both drugs were dissolved in 50% dimethylsulphoxide and the final concentration of this solvent in all reaction mixtures was 1%.

The dependence of the reaction on active G factor was further confirmed by inhibition experiments using fusidic acid or thiostrepton. Fig. 2 shows that concentrations of fusidic acid and thiostrepton which inhibit GTP hydrolysis and in vitro protein synthesis prevent the conversion of GTP to MS compounds. Tocchini-Valentini et al.¹⁵ and Kinoshita et al.¹⁶ have shown that fusidic acid interacts directly with the G factor, by finding that mutants resistant to fusidic acid have an altered G factor. Fusidic acid inhibits protein synthesis by stabilizing the G-GDP complex on the ribosome¹⁷ and thus preventing the binding of the Tu-aminoacyl tRNA-GTP complex to the acceptor (A) site¹⁸. Thiostrepton binds to the 50S subunit¹⁹ and blocks enzymatic and non-enzymatic translocation²⁰. This drug also inhibits the binding of both the Tu-aminoacyl tRNA-GTP and the G factor-GTP (or G-GDP) complexes to the ribosome²¹⁻²⁴. Thus, inhibition by both fusidic acid and thiostrepton and the requirement for G factor suggest that the synthesis of MSI and MSII is associated with the translocation machinery.

How MSI and MSII are Made

MSI and MSII are synthesized in the in vitro system by transfer of phosphates from ATP to GDP and GTP. ATP is absolutely required for the synthesis of the MS compounds. We propose that MSII is the product of the reaction 5' pppG + 5' pppA 5' → 5' pppGpp (2' or 3') + 5' pA, and the MSI is synthesized in the reaction 5' ppG + 5' pppA → 5' ppGpp (2' or 3') + 5' pA.

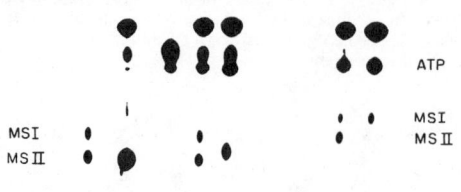

Fig. 3 MSI and MSII are synthesized from GDP and GTP using ATP.

Reaction	Unlabelled substrates	Labelled substrates
1	ATP	α-^{32}P GTP
2	ATP	γ-^{32}P GTP
3	GTP	α-^{32}P ATP
4	GTP	γ-^{32}P ATP
5	pcppG	γ-^{32}P ATP
6	GTP	γ-^{32}P ATP
7	GDP	γ-^{32}P ATP

Reaction conditions as in Table 1. The stringent factor was further purified by precipitation with 33% $(NH_4)_2SO_4$ saturation. The γ-^{32}P labelled 5'GTP and 5'ATP were prepared by E. Minkley and the α-^{32}P labelled 5'ATP and 5'GTP were purchased from ICN and New England Nuclear. The β-γ methylene guanosine 5' triphosphate was purchased from Sigma Biochemicals and used at a concentration of 0.55 mM. The concentration of 5' GDP is 0.55 mM in the experiment in which it replaces GTP. The specific activity of the ^{32}P labelled substrates was approximately 0.2 Ci/mmol.

A direct demonstration that phosphates are transferred from ATP to GDP and GTP is shown in Fig. 3. When γ-^{32}P-ATP is added to a reaction mixture containing high salt washed ribosomes, partially purified stringent factor and cold GTP, the product of the reaction is predominantly MSII. When GDP is added instead of GTP, only MSI is formed. The α phosphate of ^{32}P-ATP is not transferred to either MSI or MSII. The absence of labelled MSI in a reaction involving γ-^{32}P-GTP and nonradioactive ATP demonstrates that GTP does not donate the additional phosphates to form the MS compounds. The MSI produced in the reaction involving GTP arises in part because GTP is hydrolysed to GDP, but could stem from the degradation of MSII. When the β-γ bond is modified in the GTP analogue, pcpopoG (β-γ methylene guanosine 5' triphosphate), the product is a single compound with a mobility on PEI plates (developed with 1.5 M KH_2PO_4, pH 3.4) between those of MSI and MSII. This compound probably corresponds to the analogue of MSII (because in these chromatography conditions the methylene analogue, pcpopoG, has a higher mobility than GTP. In our most highly purified system, over 90% of the product of reaction involving ATP and GTP is MSII.

The Stringent Factor

We discovered the stringent factor because the coupled in vitro protein synthesizing system developed by Zubay et al.[25] mimics (with respect to synthesis of MS compounds) the cellular response to amino-acid starvation. Table 2 illustrates an experiment in which extracts are prepared from a pair of strains which contain a deletion of the lactose operon, and are isogenic except for the rel locus.. This makes it possible to assay for protein synthesis by measuring the amount of β-galactosidase synthesized when the DNA of a lac transducing phage is added as template to the reaction mixture. Extracts prepared from both the stringent and relaxed strain were active in protein synthesis. When a full complement of amino-acids is present in the reaction mixtures, both extracts make approximately the same amounts of β-galactosidase and no MS compounds accumulate. When the amino-acids were left out of the reaction mixtures, MSI and MSII accumulated when the extracts are prepared from stringent strains, not from relaxed strains; in both cases no detectable amounts of β-galactosidase are synthesized.

Table 2 Synthesis of MS Compounds in a Coupled in vitro Protein Synthesizing System

Reaction	Amino-acids	λplac$_5$	Strain	OD_{420}/20 h	nmol MSI and MSII
1	+	−	rel$^+$	0.03	0.05
2	+	+	rel$^+$	6.12	0.05
3	+	+	rel$^-$	5.40	0.05
4	−	−	rel$^+$	0.03	2.0
5	−	+	rel$^+$	0.03	3.0
6	−	+	rel$^-$	0.03	0.05
7	+	−	rel$^-$	0.02	0.05
8	−	−	rel$^-$	0.03	0.05

100 μl. in vitro protein synthesizing reactions were as described by Zubay et al.[25]. S-30 crude extracts were made from the isogenic pair CP 78 lac$^-$ rel$^+$ and CP 79 lac$^-$ rel$^-$ containing a deletion of the lactose operon by grinding with alumina. A full complement of amino-acids was 0.22 mM; otherwise all 20 are left out. λplac$_5$ was used as an exogenous template. β-Galactosidase assays were made with 50 μl. of the reaction mixtures and normalized to OD_{420}/20 h, as in Chambers and Zubay[31] for a 200 μl. reaction mixture. After PEI chromatography and autoradiography the spots were cut out and counted and the amount of MSI and MSII calculated by determining the percentage of the total radioactive phosphate present in these spots.

Table 3 Rel$^+$ is Dominant to rel$^-$ in vitro

Reaction	413 (rel$^+$) (mg/ml.)	415 (rel$^-$) (mg/ml.)	mmol MS
1	6.5	—	20
2	3.2	—	2
3	3.2	3.2	12
4	1.5	—	0.05
5	1.5	5.0	4
6	0.4	—	0.05
7	0.4	6.1	0.05
8	—	6.5	0.05

100 μl. in vitro protein synthesizing reaction mixtures were done as described by Zubay et al.[25] except that no amino-acids were present in the final reactions. The crude extracts were made from the isogenic pair 413 (rel$^+$) and 415 (rel$^-$) as described in Table 2 and mixed as specified. The reactions were incubated for 60 min at 37° C, and the amount of MSI and MSII was quantitated as for Table 2.

In an in vitro mixing experiment (Table 3), the rel$^+$ effect is dominant. MS compounds are synthesized when extracts from stringent and relaxed strains are mixed and incubated without amino-acids. This agrees with the observation that the rel$^+$ allele is dominant in heterozygotes[26].

This difference was observed in three different stringent and relaxed pairs of strains, isogenic except for the rel locus

(413 (rel^+) 415 (rel^-) D. Morse; CP 78 (rel^+) CP 79 (rel^-) P. Lengyel; NP 29 (rel^+) NP 290291 (rel^-) F. C. Neidhardt). Fig. 4 shows that the difference between relaxed and stringent strains rests only on the ribosomes. The addition of a high speed supernatant from stringent cells to "relaxed" ribosomes does not stimulate the synthesis of MSI and MSII. Furthermore, the high speed supernatant from relaxed cells does not inhibit synthesis of MS compounds by "stringent" ribosomes.

Fig. 4 The high speed supernant fraction is not needed for the synthesis of MS.

Reaction	Ribosomes before high salt wash		S-100	
	413 (rel^+)	415 (rel^-)	413 (rel^+)	415 (rel^-)
1	+	−	−	−
2	+	−	+	−
3	+	−	−	+
4	−	+	−	−
5	−	+	+	−
6	−	+	−	+
7	−	−	+	−
8	−	−	−	+

Reaction conditions as in Table 1. The ribosomes were made from S-30 extracts prepared as described by Zubay *et al.*[25] from the isogenic relaxed stringent pair 413 (rel^+), 415 (rel^-) by centrifuging the S-30s in a 'Beckman angle 40' rotor for 2 h at 40,000 r.p.m at 4° C. The ribosomal pellet was suspended and washed three times with buffer B. The high speed supernant from the first centrifugation was further centrifuged. After the final wash the ribosomes were suspended in one-eighth the original volume in buffer B and used at a final concentration of 100 μg/ml. The reactions were incubated for 60 min at 37° C.

Discussion

Cashel and Gallant[5,8] have proposed that during the stringent response a reaction normally involved in protein biosynthesis idles and produces the MS compounds. Our finding that MSI and MSII can be synthesized on the ribosome, and that a factor present in the 0.5 M NH_4Cl ribosomal wash extracted from the stringent but not relaxed strains is necessary for the reaction to occur, supports their hypothesis. The direct involvement of the translocation factor G in the production of the MS compounds suggests that the idling reaction substitutes for the translocation reaction during the stringent response. Because neither supernatant factors nor tRNA are required for the reaction, the signal which triggers the idling reaction may be an unoccupied A site on the ribosome. The A site should remain empty after a translocation event if the appropriate aminoacyl tRNA species is not available. This should be the case following the stimuli which induce the stringent response such as amino-acid starvation or inactivation of an aminoacyl synthetase. If the empty A site is the signal that induces the idling reaction, then the stringent factor may be involved in the recognition of the empty site or may be required as an essential enzyme, or it may serve both functions. The stringent factor is either the product of the *rel* gene or a protein modified by the action of the *rel* gene.

The possibility that the MS compounds arise as a product of an idling translocation reaction raises the question of whether MSI and MSII are normal intermediates of protein biosynthesis, or arise as the product of a side reaction which occurs only when the appropriate aminoacyl tRNA species are not available. If the MS compounds are an integral part of protein biosynthesis, then the absolute requirement for ATP suggests that ATP may play a heretofore unaccounted role in protein biosynthesis.

The reaction described here may not be the only one in the cell which can synthesize MSI and MSII. Even though relaxed strains do not accumulate MS compounds in response to amino-acid deprivation, they do have normal basal levels of MSI and MSII, and they do accumulate these compounds in response to other changes in cell conditions such as carbon source downshift or deprivation[27,28] and high salt shock[29]. This implies that either there is more than one mechanism for MSI and MSII synthesis, or that the one we have described is the only one and relaxed cells are mutated such that, while they do not respond to the signal present during amino-acid starvation, they do respond to other signals and synthesize the MS compounds on the ribosome.

We thank Douglas Bowen for technical assistance. This work was supported by a grant from the US National Institutes of Health. R. B. received a scholarship from the Instituto Nacional de la Investigacíon Cientifíca, México.

Received May 15; revised June 2, 1972.

[1] Stent, G. S., and Brenner, S., *Proc. US Nat. Acad. Sci.*, **47**, 2005 (1961).
[2] Lazzarini, R. A., and Winslow, R. M., *Cold Spring Harbor Symp. Quant. Biol.*, **35**, 383 (1970).
[3] Primakoff, P., and Berg, P., *Cold Spring Harbor Symp. Quant. Biol.*, **35**, 391 (1970).
[4] Ikemura, T., and Dahlberg, J., *Fed. Proc. Abs.*, **31**, 868 (1972).
[5] Cashel, M., and Gallant, J., *Nature*, **221**, 838 (1969).
[6] Cashel, M., *J. Biol. Chem.*, **245**, 2309 (1969).
[7] Edlin, G., and Neuhard, J., *J. Mol. Biol.*, **24**, 225 (1967).
[8] Cashel, M, and Gallant, J., *J. Mol. Biol.*, **34**, 317 (1968).
[9] Gallant, J., and Harada, B., *J. Biol. Chem.*, **244**, 3125 (1969).
[10] Irr, J., and Gallant, J., *J. Biol. Chem.*, **244**, 2233 (1969).
[11] Cashel, M., and Kalbacher, B., *J. Biol. Chem.*, **245**, 2309 (1970).
[12] Nishizuka, Y., and Lipmann, F., *Proc. US Nat. Acad. Sci.*, **55**, 212 (1966).
[13] Leder, P., Skogerson, L. E., and Nau, M. N., *Proc. US Nat. Acad. Sci.*, **62**, 454 (1969).
[14] Nishizuka, Y., and Lipmann, F., *Arch. Biochem. Biophys.*, **116**, 344 (1966).
[15] Tocchini-Valentini, G. P., Felicetti, L., and Rinaldi, G. M., *Cold Spring Harbor Symp. Quant. Biol.*, **34**, 463 (1969).
[16] Kinoshita, T., Kawano, G., and Tanaka, N., *Biochem. Biophys. Res. Comm.*, **33**, 769 (1969).
[17] Bodley, J. W., and Lin, L., *Nature*, **227**, 60 (1970).
[18] Cundliff, E., *Biochem. Biophys. Res. Comm.*, **46**, 421 (1972).
[19] Pestka, S., *Biochem. Biophys. Res. Comm.*, **40**, 667 (1970).
[20] Pestka, S., and Brot, M. J., *J. Biol. Chem.*, **246**, 7715 (1971).
[21] Bodley, J. W., Lin, L., and Highland, T. H., *Biochem. Biophys. Res. Comm.*, **41**, 1406 (1970).
[22] Highland, J. H., Lin, L., and Bodley, J. W., *Biochemistry*, **10**, 4404 (1970).
[23] Modolell, J., Carber, B., Parmeggiani, A., and Vazquez, D., *Proc. US Nat. Acad. Sci.*, **68**, 1796 (1971).
[24] Modolell, J., Vazquez, D., and Monro, P. E., *Nature New Biology*, **230**, 109 (1971).
[25] Zubay, G., Chambers, D. A., and Cheong, L. C., *The Lactose Operon*, 375 (Cold Spring Harbor Laboratory, New York, 1970).
[26] Fiil, N., *J. Mol. Biol.*, **45**, 195 (1969).
[27] Lazzarini, R. A., Cashel, M., and Gallant, J., *J. Biol. Chem.*, **246**, 4381 (1971).
[28] Harshman, R. B., and Yamazaki, H., *Biochemistry*, **10**, 3981 (1971).
[29] Harshman, R. B., and Yamazaki, H., *Biochemistry*, **11**, 615 (1972).
[30] Cashel, M., Lazzarini, R. A., and Kalbacher, B., *J. Chromatog.*, **40**, 103 (1969).
[31] Chambers, D. A., and Zubay, G., *Proc. US Nat. Acad. Sci.*, **63**, 118 (1969).

Bacteriophage f2 RNA: Control of Translation and Gene Order

by
HARVEY F. LODISH*
MRC Laboratory of Molecular Biology,
Cambridge

The use of a novel protein fingerprinting technique has shown that, in a cell-free system, the three genes of bacteriophage f2 can be translated independently. A crude genetic map of f2 RNA was constructed by using specific fragments of f2 RNA as messenger.

ONE way in which an organism can regulate the synthesis of specific proteins is by translation of the different genes on a polygenic messenger RNA at different rates and at different times. An ideal system for studying this type of regulation *in vitro* is the single stranded RNA from small bacteriophages such as f2, for it is the only purified polycistronic messenger that makes defined proteins in a cell-free system. With f2 RNA the predominant reaction product is the phage coat protein[1-5]. f2 RNA (molecular weight 1.1×10^6) also codes for two other proteins[6,7]: maturation or "A" protein (a minor component of the phage particle[8]) and an RNA polymerase. Two limitations, however, have hampered interpretations of experiments on *in vitro* synthesis of these proteins. First, there is the inability to identify all of the six to nine polypeptides detected when the reaction product is fractionated on acrylamide gels[3-5]. It is, moreover, usually not possible to detect synthesis of maturation protein. Second, in the absence of recombination between RNA phage mutants we do not know the relative positions of the genes on the RNA chromosome.

Here I report a new fingerprinting technique which detects the amino terminal sequences of proteins synthesized in the cell-free system. f2 RNA initiates the synthesis of only three proteins, which can be tentatively assigned to the three known phage genes. This system allows investigation of the factors which regulate initiation of translation of the different genes. The chief conclusions are first that, in certain conditions, synthesis of all of the f2 proteins can be initiated independently of each other. This result suggests that the secondary structure of f2 RNA usually regulates initiation of protein synthesis.

Second, a fragment of f2 RNA derived predominantly from the normal 3' end directs the synthesis of only the two non-coat proteins. This result suggests that the coat protein gene is closest to the 5' end of normal f2 RNA. Spahr and Gesteland[9] reached a similar conclusion using fragments of R17 RNA obtained in a different manner.

* Present address: Department of Biology, Massachusetts Institute of Technology, Cambridge, Massachusetts.

Amino-terminal Peptides of f2 Specific Proteins

All phage specific proteins synthesized *in vitro* are believed to begin with the amino-acid N-formyl methionine (FMet)[4,10-12]. I have therefore used N-formyl-^{35}S-methionyl-*t*RNA[13-15] to label the amino termini of f2 proteins produced in a standard *E. coli* cell-free system. That this technique will label all phage proteins depends on the assumptions (1) that indeed all proteins begin with FMet and (2) that at least the amino-terminal ^{35}S-methionine is not removed by enzymes in the cell extract.

The proteins were digested with a mixture of trypsin and chymotrypsin and the radioactive peptides separated by a two-dimensional fingerprinting technique. Figs. 1*a* and *b* show an autoradiogram of product synthesized by wild type f2 RNA. The evidence discussed below and in Table 1 indicates that more than 99 per cent of the radioactivity in peptides is derived from the amino-terminal peptides of three phage proteins: I, II and III.

The six major peptides, A–F, in Fig. 1*A* accounted for 96.5 per cent of the radioactivity (Table 1). Because there is an enzyme in extracts of *E. coli* which removes the amino-terminal formyl residue from nascent proteins[16], it is reasonable to expect that from each protein two ^{35}S-methionine peptides can be isolated, one having lost the formyl group. Indeed, digestion with 0.25 N HCl : dioxane (1 : 1) at 37° C for 24 h[12], conditions which remove only an N-formyl residue from peptides, established that peptides B, D and F are derivatives, respectively, of peptides A, C and E which have lost the formyl group, and are thus derived from the same protein (A and B from I; C and D, II; E and F, III; Table 1, fourth column).

The formylated peptides A, C and E were digested with pronase, and the identity of the N-formyl-^{35}S-methionine peptides formed was used to establish the amino-acid sequence at the beginning of the three phage proteins (Table 1, fifth column). Protein I (peptide A) begins with N-formyl methionylalanyl serine (FMet Ala Ser) and protein II (peptide C) with N-formyl methionyl serine (FMet Ser). Pronase digestion of peptide E (protein III) yielded only

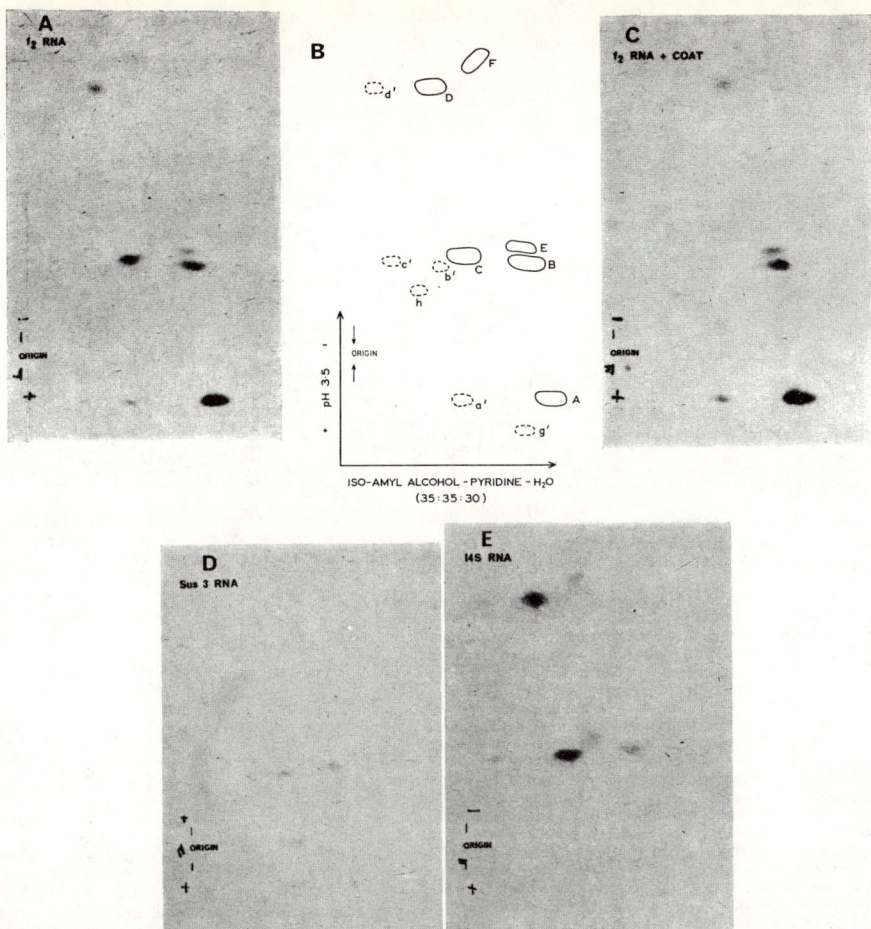

Fig. 1. Autoradiogram of digest of protein formed *in vitro* and labelled with N-formyl-^{35}S-methionyl-tRNA. (*A*) 27*S* f2 RNA; (*B*) diagram of (*A*); (*C*) 27*S* f2 RNA plus f2 coat protein; (*D*) 27*S* sus 3 RNA; (*E*) 14*S* RNA. Reaction mixtures (300 μl.) were the same as described previously[24], except that they contained all twenty non-radioactive amino-acids (100 mμmoles/ml.) and 350 μg/ml. N-formyl-^{35}S-methionyl tRNA (5 × 10^4 c.p.m./mg). The latter material was prepared (personal communication from M. Bretscher) by charging unfractionated *E. coli* tRNA with ^{35}S-methionine (specific activity about 6,000 mCi/mmole[27]) using S100 as source of activating enzyme and formyl tetrahydrofolic acid as formyl donor[14]. The isolated tRNA was subjected to enzyme pyrophosphorolysis using S100, a procedure which removes from tRNA methionine but not N-formyl methionine[14]. Digestion with pancreatic and T1 ribonucleases followed by paper electrophoresis[13,14] showed that over 98 per cent of the radioactivity was in N-formyl-^{35}S-methionyl tRNA. Phage RNAs were used at a concentration of 145 μg/ml.; all "27*S*" RNA used was prefractionated on a sucrose gradient, and more than 95 per cent of the material sedimented as a single peak at 27*S*. (*C*) Contained also 48 μg/ml. f2 coat protein, equivalent to 20 molecules per molecule of f2 RNA. Incubation was at 35° C for 11 min, and the reaction was terminated by addition of ribonuclease and EDTA as described previously[24]. An aliquot of the reaction mixture was precipitated with 5 per cent trichloroacetic acid and filtered through a 'Millipore' filter, and was used to calculate the total incorporation into protein (measured in a scintillation counter at 62 per cent efficiency): (*A*) 72,000 c.p.m.; (*C*) 49,500 c.p.m.; (*D*) 5,600 c.p.m.; (*E*) 30,200 c.p.m.; without added phage RNA, 900. The reaction mixture was precipitated with trichloroacetic acid and the protein precipitate isolated as described earlier[24]. During this step between 10 per cent and 30 per cent of the radioactive protein was lost. The precipitate was suspended in 0·6 ml. of 1 per cent ammonium bicarbonate and 50 μg of trypsin was added and incubated at 37° C for 20 min. Chymotrypsin (50 μg) was added and incubation was continued for 3 h. The digest was lyophilized and washed twice with water. The entire digest was subjected to descending chromatography on Whatman No. 3*MM* paper for 14 h. The strip was sewn to a new sheet of 3*MM* paper and subjected to electrophoresis at *p*H 3·5, using pyridine acetic acid buffer, for 1 h at 50 V/cm. The paper was exposed to Kodak autoprocess X-ray film for 16 days. All spots were cut out and counted in a scintillation counter, and the results used to calculate the amounts of the three phage proteins made (compare text and Table 1). From a reaction mixture without added phage RNA there was only one very faint spot near the position of peptide c'.

N-formyl methionine, even when the digestion was performed in the presence of a large excess of N-formyl methionyl alanine (FMet Ala), FMet Ser and FMet Ala Ser. The ionophoretic mobility of F, the deformylated derivative of E, indicated that peptide E is probably the dipeptide N-formyl methionyl lysine (FMet Lys) or N-formyl methionyl arginine (FMet Arg).

Of the minor peptides in Fig. 1*A*, four (a', b', c' and d') are derivatives of major peptides in which the methionine residue is oxidized, probably to the sulphoxide (Table 1). Peptide g' is FMet Ala Ser and was produced from peptide A by impurities, presumably carboxypeptidase A, in the chymotrypsin. The only unidentified peptide is h' (Table 1).

Table 1. QUANTITATION AND CHARACTERIZATION OF ^{35}S-FMET PEPTIDES

Protein	Peptide	Fraction of total radioactivity	HCl : dioxane	Pronase	Air oxidation	Sequence
I (Coat)	A	0.444	B	60 FMet Ala	a'	FMet Ala Ser . . .
				35 FMet Ala Ser		
				5 FMet		
	B	0.263	unch	Met	b'	Met Ala Ser . . .
	a'	0.013	b'	60 FMet(SO) Ala	unch	FMet(SO) Ala Ser . . .
				40 FMet(SO) Ala Ser		
	b	0.006	unch	Met(SO)	unch	Met(SO) Ala Ser . . .
	gg'	0.004	Met Ala Ser	unch (FMet Ala Ser)	ND	FMet Ala Ser
	SUM	0.730				
II (Polymerase)	C	0.151	D	90 FMet Ser	c'	FMet Ser . . .
				10 FMet		
	D	0.063	unch	Met	d'	Met Ser . . .
	c'	0.005	d'	ND	unch	FMet(SO) Ser . . .
	d'	0.002	unch	ND	unch	Met(SO) Ser . . .
	SUM	0.221				
III (Maturation)	E	0.026	F	FMet	ND	FMet Lys or FMet Arg
	F	0.015	unch	Met	ND	Met Lys or Met Arg
	SUM	0.041				
?	hh'	0.008	Gains + charge	Unidentified	ND	?

FMet(SO), N-formyl methionine sulphoxide. ND, not done; unch, electrophoretic and chromatographic mobility unchanged.

Peptides from the fingerprint of Fig. 1A were cut out and counted in a scintillation counter. A total of 53,000 c.p.m. was recovered from the paper, equivalent to 97 per cent of the radioactivity spotted on the chromatogram, and to 74 per cent of the acid-precipitable radioactivity in the original reaction mixture. For characterization of the peptides a fingerprint from a 4.0 ml. of reaction mixture was used; the peptides were eluted from paper with 0.02 M aqueous triethylamine, lyophilized twice, and resuspended in water. Pronase digestion required 6 µg of enzyme in 25 µl. 0.5 per cent ammonium bicarbonate solution, left for 2 h at 37° C. Peptides were oxidized by drying them on filter paper and treating them with a gentle stream of air at room temperature for 24 h. Other reaction conditions are described in the text. Radioactive peptides in the digest were compared with standard peptides by paper ionophoresis at pH 3.5 and 6.5 and by paper chromatography in the system described in the legend to Fig. 1. These conditions were sufficient to differentiate all the peptides in Fig. 1A and also all forty-eight formyl methionine and amino-terminal methionine dipeptides and tripeptides available. In particular, ionophoresis at pH 3.5 for 2 h at 50 V/cm is sufficient to separate completely FMet, FMet Ser, FMet Ala Ser, FMet Ala and peptides A–F. Most of these marker peptides were purchased from Mann and then chemically formylated. Oxidized derivatives of FMet, FMet Ser and FMet Ala Ser were prepared by drying the peptides on filter paper and treated with a gentle stream of air. Standard peptides were detected by the platinic iodide stain or, when oxidized, by a chlorine starch stain.

Correlation with Known Phage Genes

Protein I (peptide A) is the coat protein. Its amino-terminal sequence, FMet Ala Ser . . ., is the same as that of the phage coat protein synthesized in vitro[10,11] (in the phage particle the coat begins Ala Ser . . ., having lost the amino-terminal methionine[17]). As expected, protein I is the predominant product of the cell-free system.

Protein III (peptide E) is probably the maturation protein. Digestion with trypsin of purified maturation protein (obtained from phage particles[8]) labelled with ^{35}S-methionine yielded five radioactive peptides. One of these was inseparable from peptide F (the deformylated derivative of E) by paper chromatography and ionophoresis and is believed to be the amino-terminal peptide of the maturation protein (my unpublished work).

An indirect experiment suggests that protein II (peptide C) is the RNA polymerase. Several groups have shown that when f2 RNA accompanied by 6–20 molar equivalents of coat protein is used as messenger in vitro a normal amount of coat protein is produced, but only a very small amount of most of the non-coat polypeptides, including the one believed to be completed RNA polymerase[5,18]. This inhibition of polymerase synthesis is considered important in regulation of polymerase production during phage infection[19]. Fig. 1C is a fingerprint of N-formyl-^{35}S-methionine protein produced by a complex of twenty coat protein molecules per f2 RNA. Inhibition by coat protein was extremely specific; there was no effect on synthesis of coat protein (peptides A, B, a', b', g') or of maturation protein (peptides E, F), but production of protein II (peptides C, D, c', d') was completely prevented. This suggests that II is the RNA polymerase.

Polarity and the Control of RNA Translation

With wild type f2 RNA the three proteins were synthesized in the ratio 100 coat protein (I) : 30 polymerase (II) : 5.5 maturation protein (III) (Table 1).

Because our fingerprinting system measures only the amino-terminal peptides of f2 proteins, it can be used to study the factors which regulate initiation of translation of the three phage genes. Some insights into this problem come from using sus 3 RNA as messenger, which contains an amber mutation at the site corresponding to the sixth amino-acid of the coat protein cistron[20]. Fig. 1D shows that sus 3 RNA directs the synthesis of only 25 per cent as much polymerase as does wild type f2 RNA (peptides C and D; compare Fig. 1A). By contrast, sus 3 RNA directs the synthesis of 20 per cent more maturation protein than does f2 RNA (peptides E and F; compare Fig. 1D and A). (The major product of sus 3 RNA is a hexa-peptide containing ^{35}S-FMet and the next five amino-acids of the coat protein which is soluble in trichloroacetic acid and is not seen in Fig. 1A. Thus all of the coat peptides, A, B, a', b' and g', are missing.) These results confirm the polar effect of the sus 3 mutation on polymerase synthesis[21,22] and mean that translation beyond the sixth amino-acid of the coat protein cistron is necessary for efficient initiation of polymerase synthesis. By contrast, coat translation is not needed for normal initiation of translation of maturation protein, a result suggesting that initiation of maturation protein synthesis is independent of that of the coat protein.

Kinetics of Protein Synthesis

These conclusions on control of initiation are supported by experiments on the kinetics of protein synthesis directed by wild type f2 RNA (Fig. 2a). The time course of synthesis of the amino-terminal peptides of the three phage proteins was calculated from a series of fingerprints similar to those of Fig. 1A prepared from protein made after different times of incubation. The initiation of translation of RNA polymerase (protein II) lags 1 to 2 min after that of the coat protein (I) (see legend to Fig. 2). This extends previous observations that synthesis of polymerase follows that of coat protein[1-3], and is consistent with the above conclusions that initiation of polymerase translation requires translation of part of the coat gene. In sharp contrast, the kinetics of synthesis of coat and maturation protein are the same, both being made after a 1.5 min lag, except that much more coat protein is made. This also suggests that translation of maturation protein is independent of that of the coat protein.

A Fragment of f2 RNA

Because initiation of polymerase translation is normally dependent on synthesis of at least part of the coat protein, one might predict that a fragment of f2 RNA which contains the gene for RNA polymerase but not that for coat protein should be inactive in directing synthesis of RNA polymerase. In fact, such a fragment is extremely efficient in directing synthesis of RNA polymerase, a result which demonstrates that initiation of RNA polymerase, as well as of maturation protein, can be independent of that of the coat protein.

14S RNA, a fragment of f2 RNA that has lost the coat gene, has been obtained[24] from the defective phage particles produced by infection of Su- hosts by sus 4, an f2 amber mutant in the maturation protein[25-27]. RNA from the particles with lightest density in CsCl ($\rho = 1.42$ g/cm³ compared with 1.45 g/cm³ for f2) was fractionated by sedimentation through a sucrose gradient (Fig. 3a), and

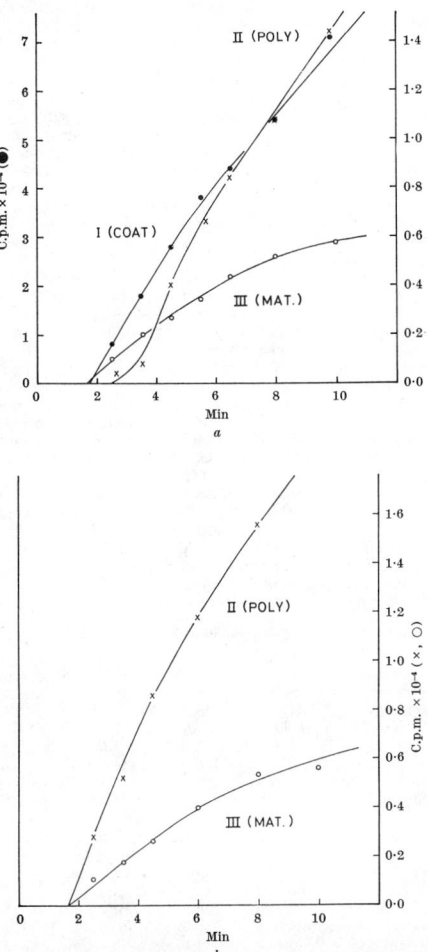

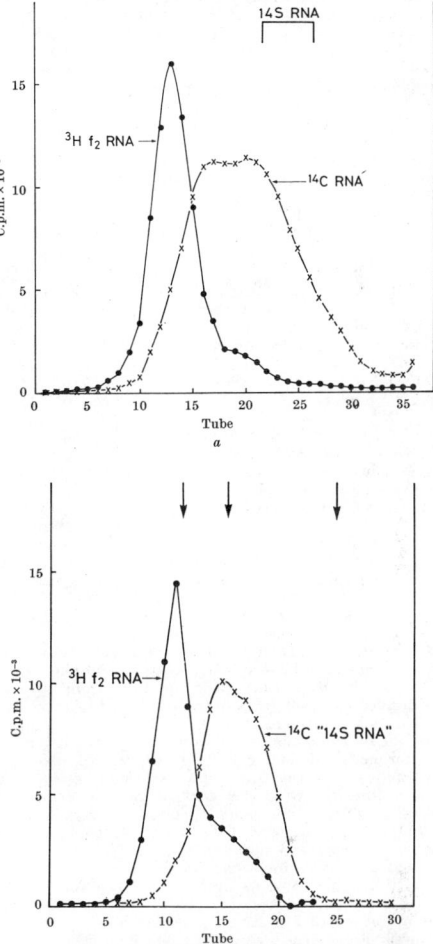

Fig. 2. Kinetics of synthesis of phage proteins (a) 27S f2 RNA; (b) 14S RNA. Two 5·0 ml. reaction mixtures of identical composition to that of Fig. 1A and E were used; the temperature was 25° C. Samples of different sizes were taken so that at each time about 30,000 c.p.m. of acid-precipitable radioactivity were used for fingerprinting. The amount of radioactivity in the peptides from each protein was normalized to 1·0 ml. of reaction mixture. Note that the scales for coat protein and for maturation and polymerase protein are different. In all fingerprints only the peptides found in Fig. 1A were observed. The difference in temperature between this experiment and those of Fig. 1 and Table 1 may explain the different relative amounts of the proteins produced. The lag in synthesis of polymerase relative to coat and maturation protein (part a) is significant, for the results for all three proteins were derived from the same fingerprints. For example, in the fingerprint from the 2·5 min reaction the ratio of counts in peptides from coat : polymerase : maturation proteins was 100 : 5·6 : 12, while from the 8·0 min reaction the ratio was 100 : 20 : 9·6. Furthermore, the lag cannot be caused by the necessity to synthesize a longer part of protein II before it becomes precipitable in acid, for the lag is abolished when 14S RNA is used (part b).

Fig. 3. (a) Sucrose gradient of RNA from defective phage. RNA labelled with ¹⁴C, taken from sus 4 defective phage, was mixed with a small amount of tritiated f2 27S RNA and layered on a 5–20 per cent linear sucrose gradient (28 cm³, 0·10 M NaCl, 0·05 M tris, pH 7·5). Centrifugation was for 20 h at 25,000 r.p.m. at 4° C in the SW 25 rotor of the Spinco L4 ultracentrifuge. Samples from each fraction were precipitated with trichloroacetic acid and counted in a scintillation counter; fractions pooled as 14S RNA are indicated. Centrifugation is from right to left. (b) Formaldehyde sedimentation[33] of 14S RNA. A small amount of tritiated f2 27S RNA ¹⁴C-14S RNA, and 4 A_{260} units of E. coli RNA in 160 μl. of solution containing 0·09 M Na₂HPO₄, 0·01 M NaH₂PO₄, and 1 M formaldehyde was heated in a closed tube at 63° C for 10 min. After cooling rapidly in ice the solution was layered on a 4·7 cm³ linear 5–20 per cent sucrose gradient containing the same concentration of phosphate buffer and formaldehyde and also 0·1 per cent sodium dodecyl sulphate. Centrifugation was at 18° C for 3 h at 65,000 r.p.m. Optical density (260 mμ) and acid-precipitable radioactivity were determined on each fraction. The arrows represent, from left to right, the peaks of E. coli "23S" and "16S" ribosomal RNA and "4S" soluble RNA as determined from the optical density profile. Note that "27S" f2 RNA and "23S" ribosomal RNA, which have the same molecular weight, have the same sedimentation velocity in these conditions[32]. Some of the slower sedimenting ³H and ¹⁴C radioactivity may be produced by bond scission and also exposure of pre-existing bond cleavages during heating.

a fraction with a mean sedimentation coefficient of 14S (compared with 27S for f2 RNA) was isolated.

A fingerprint of the protein products directed by 14S RNA is shown in Fig. 1E. The important result is that per microgram 14S RNA directs the synthesis of the same amount of both polymerase (peptides C, D, c' and d') and of maturation protein (E, F) as does 27S RNA, but less than 3 per cent of the amount of coat protein (peptides A and B; compare with Fig. 1A). This agrees with previous work on 14S RNA[24], and suggests that this RNA has in fact lost the gene for coat protein.

With 14S RNA initiation of translation of the polymerase gene occurs in the complete absence of synthesis of the coat protein. This is supported by the kinetics of protein synthesis (Fig. 2b). Initiation of polymerase translation directed by 14S RNA proceeds immediately and without an additional lag period, in contrast to the result obtained with 27S f2 RNA (Fig. 2a).

Gene Order

A further understanding of the factors which regulate f2 RNA translation requires knowledge of the order of the genes on the RNA chromosome, and some information on this point can be obtained from the results with 14S RNA.

14S RNA consists predominantly of material derived from the 3' end of f2 RNA. Alkaline digests of 27S and 14S RNAs labelled with ^{14}C-adenine and ^{14}C-guanine were fractionated by paper ionophoresis at pH 3·5. From 27S RNA I obtained, in addition to the 3' monophosphates of adenosine and guanosine, between 0·95 and 1·07 moles of ^{14}C-guanosine tetraphosphate and of ^{14}C-adenosine per mole ($1·1 \times 10^6$ molecular weight units (MWU)) of RNA. These are derived, respectively, from the 5' and 3' ends[28-31]. From 14S RNA only one mole of ^{14}C-guanosine tetraphosphate was obtained per 6,000–9,000 moles of guanosine 3' monophosphate, equivalent to 0·10–0·15 moles of guanosine tetraphosphate per $1·1 \times 10^6$ MWU of RNA. However, one mole of radioactive adenosine per 490 of adenosine monophosphate was recovered, while from f2 RNA, as expected, the recovery was one per 780 residues. Because no radioactive guanosine was ever detected, it seems very likely that all of the adenosine from 14S RNA was derived from the 3' end of normal f2 RNA.

An estimate of the molecular weight of 14S RNA comes from the sedimentation velocity of the RNA after reaction with formaldehyde at 63° C for 15 min[32]. Fig. 3b shows that after such treatment the major component of 14S RNA sediments 78–82 per cent as rapidly as does f2 RNA, indicating[32] that it has a molecular weight between 55 and 70 per cent that of f2 RNA. But 14S RNA is very heterogeneous, containing a significant fraction of material of smaller molecular weight.

14S RNA thus consists predominantly of single polynucleotide strands containing the 3' end and 55–70 per cent of normal f2 27S RNA. Because 14S RNA directs synthesis of only the two non-coat proteins (Fig. 1E) we can conclude that with normal f2 27S RNA the coat gene is located closest to the 5' end. This conclusion must be considered tentative until a well defined fragment from the 5' end of f2 RNA has been examined. Furthermore, because the maturation protein is made in very small amounts it is extremely difficult to establish the location of its gene with certainty; 14S RNA is heterogeneous and contains some material derived from the 5' end of f2 RNA.

Regulation of RNA Translation

The results obtained with 14S RNA demonstrate that at least the guanosine triphosphate residue at the 5' end of f2 RNA[28,29] is unnecessary for efficient initiation of protein synthesis, as Spahr and Gesteland[9] showed. They also show that the dependence of polymerase initiation on synthesis of the coat protein observed with 27S RNA (Fig. 2a) cannot be due to some intrinsic property of the nucleotide sequence at the beginning of the polymerase gene.

One plausible interpretation of my results is that changes in secondary structure of messenger RNA regulate initiation of translation. Such models predict that ribosomes can attach independently at several sites on a polycistronic messenger. On intact (27S) f2 RNA ribosomes would be able to attach at the sites for initiation of translation of both coat and maturation proteins. Possibly because of differences in local secondary structure the attachment to the site at the beginning of the coat would be more efficient. But the site for initiation of polymerase synthesis would usually not be exposed. During polymerization of the amino-acids of the coat protein the secondary structure of the RNA would change in such a manner as to expose the polymerase initiation signal to ribosome attachment. Alternatively, physical removal of the 5' end and the coat gene (14S RNA) would also expose this initiation signal. A second model postulates that initiation of RNA translation requires a free 5' end, and that the messenger is translated sequentially beginning with the gene (in f2 RNA, the coat) closest to the 5' end. According to such a model, in 14S RNA the initiation signal for polymerase translation would be near the new 5' end and would be translated immediately. It is, however, difficult to explain why an amber mutation near the middle of the coat gene (sus 11), which results in complete inhibition of synthesis of the C-terminal region of the coat protein[20], does not affect the initiation of polymerase synthesis[21]. Furthermore, this model does not explain how translation of the maturation protein proceeds independently of that of the coat and polymerase proteins with both 27S and 14S RNA, nor does it explain how addition of coat protein to f2 RNA blocks completely initiation of polymerase translation but does not affect synthesis of maturation protein. This model would explain the results on synthesis of maturation protein if some specific fragmentation product of f2 RNA formed in the cell-free system directed translation of this protein. This, however, appears unlikely, because the kinetics of synthesis of maturation protein are linear (Fig. 2a).

Effects of Coat Protein on RNA Translation

14S RNA retains some control over the initiation of polymerase synthesis. Addition of 20 molar equivalents of coat protein to 14S RNA results in the complete loss of synthesis of RNA polymerase, but does not affect the normal synthesis of maturation protein. Taken together with the results of Fig. 1C, this means that coat protein binds to f2 RNA in such a manner as to inhibit initiation of polymerase translation. The site of f2 RNA which binds coat protein is located on the two thirds nearest the 3' end, perhaps at the site of initiation of polymerase synthesis. It is of considerable interest that polymerase initiation is specifically blocked whether (14S RNA) or not (27S RNA) the initiation signal for polymerase synthesis is directly available for ribosome attachment.

I thank my colleagues, especially Dr S. Brenner, for advice and encouragement. I also thank Dr H. Boedtker for a copy of her manuscript before publication, and Mr A. Smith for ^{35}S-methionine. Drs F. H. C. Crick, B. Davis, R. Russell and J. Argetsinger-Steitz criticized the manuscript. I was supported by a postdoctoral fellowship from the American Cancer Society.

Received July 17; revised September 13, 1968.

[1] Nathans, D., Notani, G., Schwartz, J., and Zinder, N., *Proc. US Nat. Acad. Sci.*, **48**, 1424 (1962).
[2] Ohtaka, Y., and Spiegelman, S., *Science*, **142**, 493 (1963).
[3] Eggen, K., Oeschger, M., and Nathans, D., *Biochem. Biophys. Res. Commun.*, **28**, 587 (1967).
[4] Vinuela, E., Salas, M., and Ochoa, S., *Proc. US Nat. Acad. Sci.*, **57**, 729 (1967).
[5] Sugiyama, T., and Nakada, D., *J. Mol. Biol.*, **31**, 431 (1968).
[6] Horiuchi, K., Lodish, H., and Zinder, N., *Virology*, **28**, 438 (1966).
[7] Gussin, G., *J. Mol. Biol.*, **21**, 435 (1966).

[8] Steitz, J., *J. Mol. Biol.*, **33**, 923 (1968).
[9] Spahr, P., and Gesteland, R., *Proc. US Nat. Acad. Sci.*, **59**, 876 (1968).
[10] Adams, J., and Capecchi, M., *Proc. US Nat. Acad. Sci.*, **55**, 147 (1966).
[11] Webster, R., Engelhardt, D., and Zinder, N., *Proc. US Nat. Acad. Sci.*, **55**, 155 (1966).
[12] Clark, B., and Marcker, K., *Nature*, **211**, 378 (1966).
[13] Marcker, K., and Sanger, F., *J. Mol. Biol.*, **8**, 835 (1964).
[14] Marcker, K., *J. Mol. Biol.*, **14**, 63 (1965).
[15] Clark, B., and Marcker, K., *J. Mol. Biol.*, **17**, 394 (1966).
[16] Adams, J., *J. Mol. Biol.*, **33**, 571 (1968).
[17] Weber, K., Notani, G., Wikler, M., and Konigsberg, W., *J. Mol. Biol.*, **20**, 423 (1966).
[18] Eggen, K., and Nathans, D., *Fed. Proc.*, **26**, 449 (1967).
[19] Lodish, H., and Zinder, N., *J. Mol. Biol.*, **19**, 333 (1966).
[20] Webster, R., Engelhardt, D., Zinder, N., and Konigsberg, W., *J. Mol. Biol.*, **29**, 27 (1967).
[21] Engelhardt, D., Webster, R., and Zinder, N., *J. Mol. Biol.*, **29**, 45 (1967).
[22] Capecchi, M., *J. Mol. Biol.*, **30**, 213 (1967).
[23] Tooze, J., and Weber, K., *J. Mol. Biol.*, **28**, 311 (1967).
[24] Lodish, H., *J. Mol. Biol.*, **32**, 681 (1968).
[25] Lodish, H., Horiuchi, K., and Zinder, N., *Virology*, **27**, 139 (1965).
[26] Argetsinger, J., and Gussin, G., *J. Mol. Biol.*, **21**, 421 (1966).
[27] Heisenberg, M., *J. Mol. Biol.*, **17**, 136 (1966).
[28] Roblin, R., *J. Mol. Biol.*, **31**, 51 (1968).
[29] Watanabe, M., and August, J., *Proc. US Nat. Acad. Sci.*, **59**, 513 (1968).
[30] De Wachter, R., and Fiers, W., *J. Mol. Biol.*, **30**, 507 (1967).
[31] Weith, H., and Gilham, P., *J. Amer. Chem. Soc.*, **89**, 5473 (1967).
[32] Boedtker, H., *J. Mol. Biol.*, **35**, 61 (1968).
[33] Vinuela, E., Algranati, I., Feix, G., Garwes, D., Weissmann, C., and Ochoa, S., *Biochim. Biophys. Acta*, **155**, 558 (1968).
[34] Steitz, J., *J. Mol. Biol.*, **33**, 937 (1968).
[35] Robertson, H., Webster, R., and Zinder, N., *Nature*, **218**, 533 (1968).
[36] Lodish, H., Cooper, S., and Zinder, N., *Virology*, **24**, 60 (1964).
[37] Sanger, F., Bretscher, M., and Hocquard, E., *J. Mol. Biol.*, **8**, 38 (1964).

A Correlation of Changes in Host and T_4 Bacteriophage Specific RNA Synthesis with Changes of DNA-Dependent RNA Polymerase in *Escherichia coli* Infected with Bacteriophage T_4

Melitta SCHACHNER, Wilfried SEIFERT, and Wolfram ZILLIG
with the technical assistance of I. HOLZ

Max-Planck-Institut für Biochemie, München

(Received March 9/June 28, 1971)

1. After infection with bacteriophage T_4 the DNA-dependent RNA polymerase of the *Escherichia coli* host is changed in two consecutive steps: alteration and modification.
The alteration takes place immediately after phage adsorption and is independent of protein synthesis. It leads to the shut off of *E. coli* specific RNA synthesis and permits the transcription of the immediate early phage genes.
Under conditions allowing phage specific protein synthesis the *E. coli* core polymerase is modified. The first step of modification is completed within 4 min after phage infection (25 °C). The purified modified enzyme does not contain σ-factor and exhibits structural changes in all the subunits of the core.

2. Under conditions which do not allow DNA synthesis (infection with T_4 am N82) the net rate *in vivo* of immediate early, followed by delayed early gene transcription shows one early burst.
In normal phage development immediate early and delayed early transcription display a second rise preceding replication. Late RNA is made after the onset of DNA synthesis.

3. In crude extracts prepared from *E. coli* cells infected under conditions which allow as well as under conditions which do not allow protein synthesis, one burst of transcriptional capacity follows infection. The length of this burst is sufficient for the transcription of approximately one reading unit. Although transcription *in vivo* continues under conditions allowing protein synthesis, polymerase activity in crude extracts *in vitro* is very low after the burst even when exogenous template DNA is added.

The sequential transcriptional events during bacteriophage T_4 development occur in an inherent timing [1—3]. They consist in the shut off of bacterial messenger RNA synthesis [4,5] within a few minutes after phage infection and in the differential transcription of immediate early, delayed early, and late genes [6]. It was observed in our laboratory that after infection with bacteriophage T_4 the host's transcriptional system undergoes a number of consecutive changes [7,8].

The first of these which does not require protein synthesis and is also induced by phage ghosts has been called alteration. It was first detected as a reduction of the ratio of activities of the purified RNA polymerase on T_4 over calf thymus DNA as templates. This is indicative of a reduced σ activity which either could be due to an alteration of σ itself or to a reduced σ affinity of an altered core polymerase. Concomitant structural changes have been observed.

The second change called modification requires protein synthesis. It leads to a structural change of all subunits [9] and to a complete loss of σ from the purified enzyme. A concomitant further reduction of the template activity ratio is observed. It is not known yet whether the modification of α which has been detected first, follows a different kinetics than that of β and β', but it is evident from labelling experiments [10] and fingerprints [9] that all these changes conserve the basic structure of the host enzyme subunits.

In order to correlate changes in host and phage-specific RNA synthesis with the described changes of the host's DNA-dependent RNA polymerase we studied the dependence of the pattern of RNA synthesis *in vivo* and of the activity of RNA polymerase *in vitro* at various times after infection. Infection was carried out at 25 °C under three differ-

---

The results presented in this paper have been reported at the 8th International Congress of Biochemistry in Montreux, Switzerland (3rd —9th September 1970).
Enzyme. DNA-dependent RNA polymerase (EC 2.7.7.6).

ent conditions: (a) inhibition of protein synthesis by chloramphenicol (100 µg/ml) allowing alteration only; (b) defective synthesis of phage DNA (T_4 am N82 mutant was used) and (c) normal phage development.

Under each of these three conditions the change of the net rate of total RNA synthesis *in vivo* and of the rates of synthesis *in vivo* of different classes of T_4-RNAs (immediate early, delayed early and late RNA) with time after T_4-infection was compared with the concomitant change of the activity of RNA polymerase *in vitro* in crude extracts and in extracts from which endogenous template was removed by chromatography on DEAE-cellulose.

MATERIAL AND METHODS

Bacteriophage T_4D^+ and T_4 am N82 were multiplied on *Escherichia coli* B and *E. coli* CR63, respectively. Cultures of *E. coli* B (5×10^8 cells/ml in the medium of Fraser and Jerrel [11]) were infected with a multiplicity of infection of 10 at 25 °C. In all experiments the amount of surviving bacteria was less than 1%. If infection was carried out under inhibition of protein synthesis, chloramphenicol (100 µg/ml) was added 3 min before phage infection.

For the DNA-dependent RNA polymerase assay aliquots of the bacterial culture were removed before and at various times after infection and killed on ice with 10 mM NaN_3. Bacterial extracts were prepared by vibration homogenization with glass beads in Tris-Mg buffer (0.01 M Tris, 0.01 M Mg acetate, 0.022 M NH_4Cl, 0.25 mM EDTA, 1 mM mercaptoethanol, pH 7.3) and tested for RNA polymerase activity without and with the addition of exogenous template DNAs (T_4 and thymus DNA). The RNA polymerase assay has been described [12]. Polymerase activity was estimated by the actinomycin-inhibitable incorporation of [^{14}C]UMP into trichloroacetic acid-precipitable material. 1 m-unit of enzyme activity corresponds to the incorporation of 1 nmole [^{14}C]UMP into RNA in 1 min at 37 °C under the described conditions. DNA from bacteriophage T_4 was prepared by the phenol method. Calf thymus DNA was purchased from Worthington Biochemicals.

RNA was labelled *in vivo* before and at various times after infection at 25 °C by removing aliquots of the same bacterial culture from which samples for the preparation of crude extracts were taken and by incubating them for 30 sec with [^{3}H]uridine (1 µCi/ml, 31 Ci/mmole). The labelling pulse was stopped by pouring the cells onto ice with 10 mM NaN_3. Unlabelled RNA was prepared from bacteria harvested (a) 3 min after infection under chloramphenicol (100 µg/ml) at 37 °C, (b) 8 min after infection with T_4 am N82 at 37 °C or (c) 20 min after infection with T_4D^+ at 37 °C.

RNA was extracted from the labelled and unlabelled cultures according to the method of Bolle *et al.* [3]. In some cases extraction of labelled and unlabelled RNA species was carried out in the presence of bentonite (2 mg/ml). Hybridisation and hybridisation competition experiments in DNA excess were carried out as described by Bolle *et al.* [3]. The rates of synthesis of the T_4 specific RNA classes were calculated from the plateaux of the competition curves (giving their fractions of the total T_4 specific RNA) and the rates of total phage specific RNA synthesis, as measured by hybridisation to excess T_4 DNA in the absence of competitor.

If the nucleotide precursor pools remain constant throughout infection, the specific activity of RNA labelled during a short pulse (30 sec) may be considered proportional to the net rate of messenger RNA synthesis, since ribosomal and transfer RNA amounting to more than 90% of the cell's total RNA are fairly constant [13] and since decay of messenger RNA can be neglected [14].

Phage DNA synthesis was measured by the incorporation of [^{3}H]thymidine into alkali-stable, trichloroacetic acid-precipitable material.

RESULTS AND DISCUSSION

Infection under Conditions not Allowing Protein Synthesis

As can be seen from Fig. 1 the net rate of RNA synthesis *in vivo* in the presence of chloramphenicol (100 µg/ml) drops shortly after infection and then increases again reaching a maximum at 3 min after infection at 25 °C. This maximum is immediately followed by a sharp drop to a low background rate. Hybridisation of the newly synthesized RNA with *E. coli* and T_4 DNA (Fig. 2) shows that the net synthesis of *E. coli* specific RNA is reduced within 1 min to 50% of its rate before infection and reaches a constant residual level of 5% at 3 min after infection. The sharp maximum of RNA synthesis at 3 min after infection is caused by the transcription of phage-genes. A small, but significant rate of transcription of the T_4 genome is still detectable even 25 min after the maximum.

In order to characterize which of the phage-specific gene classes were transcribed at various times after infection, hybridisation-competition experiments were performed using two different species of competitor RNA. One species was obtained from *E. coli* cells harvested 3 min after T_4 infection in the presence of chloramphenicol at 37 °C, the other consisted of the total prereplicative species. Table 1 indicates that towards later times after infection some early RNA species are transcribed which are not contained in the "chloramphenicol" competitor RNA.

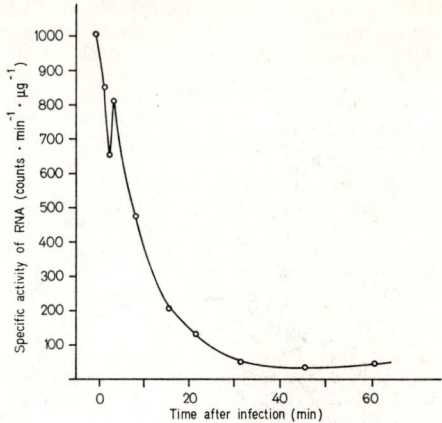

Fig.1. *Specific activity of total RNA extracted from bacteria labelled before and at various times after infection (30 sec pulse, 25 °C) under conditions which do not allow protein synthesis (100 µg/ml chloramphenicol)*

Table 1. *Hybridisation of RNA (50 µg) labelled at various times after infection (30 sec pulse, 25 °C) under conditions not allowing protein synthesis with excess of T_4 DNA (10 µg) in the presence of competitor RNA*

Time after infection	Hybridised RNA in the presence of			
	total prereplicative competitor RNA		"chloramphenicol" competitor RNA [a]	
	50 µg	750 µg	50 µg	750 µg
min	%	%	%	%
0.5	63	9.4	40	7.3
1.5	65	8.3	39	7.5
2.5	61	7.9	43	8.2
3.5	63	7.6	41	9.6
4.5	67	6.2	37	8.9
8.5	64	9.4	39	7.6
15.5	50	7.3	48	9.7
21.5	42	6.7	52	10.1
31.5	38	5.0	49	11.2
45.5	40	4.3	53	10.7
60.5	41	4.9	53	12.3

[a] "Chloramphenicol" competitor RNA was extracted from bacteria harvested 3 min after infection under chloramphenicol (100 µg/ml) at 37 °C.

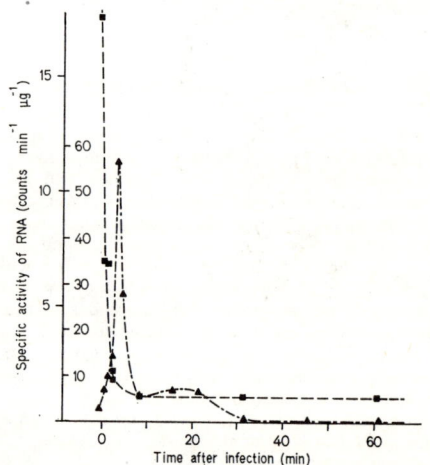

Fig. 2. *Hybridisation of RNA labelled before and at various times after infection (30 sec pulse, 25 °C) under conditions which do not allow protein synthesis (100 µg/ml chloramphenicol) with E. coli (■, left hand scale) and T_4 (▲, right hand scale) DNA*

Salser *et al.* defined the immediate early class of messenger RNA as the only one synthesized in the presence of chloramphenicol [6]. Our experiments indicate, however, that even in the presence of chloramphenicol newly synthesized RNA made later in infection contains a small amount of delayed early RNA species but no late RNA.

RNA polymerase activity *in vitro* in the crude extracts shows a maximum 3—4 min after infection. This maximum coincides with the peak of immediate early RNA synthesis *in vivo* and is followed by an abrupt decrease which is seen also when exogenous template DNA is added (Fig. 3). This reduction of polymerase activity is not due to a loss of σ-activity during the observed round of immediate early transcription but to a block of polymerase itself. Two observations support this conclusion:

a) The background activity observed in extracts from this phase is only slightly increased by the addition of exogenous templates, *e.g.* thymus DNA, the transcription of which proceeds with high efficiency even in complete absence of σ. This shows clearly that little free polymerase is available for the transcription of exogenous template. The ratio of template activities of T_4 over calf thymus DNA, which is a measure of σ activity increases towards the end of the burst of immediate early gene transcription indicating an even higher availability of σ during this phase.

b) The template ratio of polymerase activity on T_4 over calf thymus DNA after chromatography of crude extracts on DEAE-cellulose, which drops drastically within 1 min after phage infection, remains then at the constant value of 0.5 throughout later times after infection. This ratio is characteristic of the completely purified altered *E. coli* polymerase and is distinguished from that of normal *E. coli* polymerase which is 1.5—2.0 and that of T_4-modified enzyme which is 0.2. Thus a further

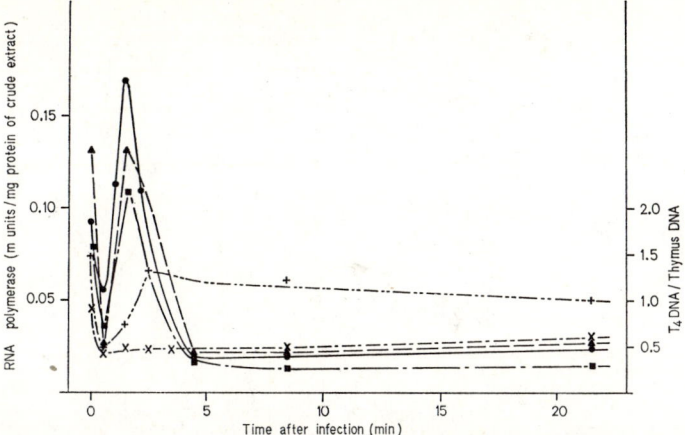

Fig. 3. *Activity of DNA-dependent RNA polymerase in crude extracts and in extracts partially purified by chromatography on DEAE-cellulose.* The extracts were prepared from cells harvested before and at various times after infection (25 °C) under conditions which do not allow protein synthesis (100 μg/ml chloramphenicol). ■, endogenous DNA; ●, thymus DNA; ▲, T_4 DNA; +, ratio T_4 DNA to thymus DNA in crude extract; ×, ratio T_4 DNA to thymus DNA after chromatography on DEAE-cellulose.

reduction of σ activity during the burst of immediate early transcription is ruled out. The alteration of *E. coli* RNA polymerase, which appears as a reduction of σ activity [15], may be caused either by a change of σ itself or by a change of the affinity of an altered core to an unchanged σ or to both these reasons. Recent observations support the latter possibility.

The low enzymic activity in crude extracts prepared at different times after the burst of early transcription could be caused by the arrest of polymerase, for example by an impaired termination. As polymerase molecules initiate transcription one after another, all of the cell's molecules (approximately 5000) could be packed onto the immediate early genes. At later times after infection a minimal amount of enzyme might leak past the termination block so that it can reinitiate immediate early transcription or read the delayed early genes, as was shown above to take place *in vivo*. The small but significant transcription of delayed early genes can be explained by read-through, which has been observed *in vitro* before [16].

The temporal coincidence of the completion of alteration with the block of host RNA synthesis indicates that the alteration is responsible for the shut off of host RNA synthesis. Thus the rapid cessation of bacterial RNA and protein synthesis after T_4 infection, which even takes place under conditions not allowing protein synthesis and which is also effected by phage ghosts [5,8,17] can be attributed to a rapid change in the host's RNA polymerase.

Infection under Conditions not Allowing Synthesis of Phage DNA (T_4 am N82)

In the absence of DNA synthesis (infection with T_4 am N82) RNA is synthesized *in vivo* over a longer time period than under conditions not allowing protein synthesis (Fig. 4). Fig. 5 indicates that this broader burst of RNA synthesis is caused by the consecutive transcription of immediate early and delayed early genes.

As under conditions which do not allow protein synthesis, a striking reduction of RNA synthesis is observed which in this case, however, follows delayed early transcription. The synthesis of delayed early RNA starts with a lag after the transcription of immediate early genes. The peaks of both gene classes are clearly separated but level out together.

This observation favours the hypothesis that immediate and delayed early genes are located adjacent to each other and that both are transcribed from the same promoters. In contrast to the situation in the absence of protein synthesis, a protein factor may be synthesized under these conditions which releases the block of polymerase molecules at the termination sites of the immediate early genes and thus allows read-through into the delayed early genes or reinitiation at the immediate early genes. Finally all polymerase molecules are arrested again

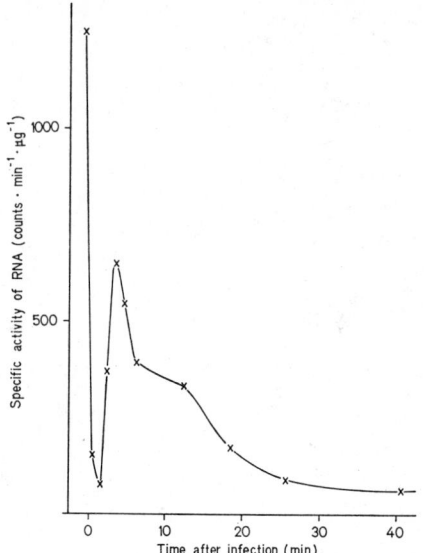

Fig.4. *Specific activity of total RNA extracted from bacteria labelled before and at various times after infection (30 sec pulse, 25 °C) under conditions which do not allow phage DNA synthesis (T_4 am N82 was used)*

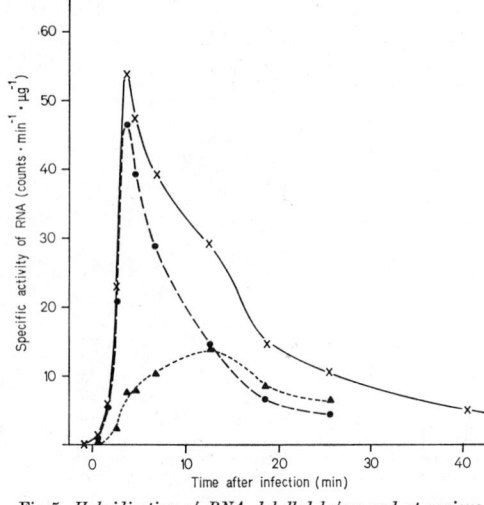

Fig.5. *Hybridisation of RNA, labelled before and at various times after infection (30 sec pulse, 25 °C) under conditions which do not allow phage DNA synthesis (T_4 am N82 was used), with T_4 DNA. The amount of immediate (●) and delayed (▲) early RNA in each sample was calculated as described in Material and Methods. ×, total T_4 specific RNA synthesis. Unlabelled competitor RNA was extracted from infected bacteria killed at various times after infection as specified under Material and Methods*

by an impaired termination at the delayed early termination sites either as a consequence of the inhibition of DNA synthesis or of a direct effect of the amber N82 mutation.

As under conditions not allowing protein synthesis polymerase activity in crude extracts (Fig. 6) again coincides with the peak of the synthesis of immediate early RNA *in vivo*. The lack of coincidence with delayed early transcription *in vivo* may be interpreted as a deficiency of our crude extracts to synthesize delayed early RNA, for example as a consequence of a lability of the antitermination system.

Infection with T_4 Wild Type Phage

The absolute rate of RNA synthesis in T_4 wild type infected cells shows two waves (Fig. 7). The first is caused mainly by immediate early RNA synthesis, the second by a sharp rise in synthesis of immediate and delayed early RNA (Fig. 8). Since synthesis of phage DNA under our conditions (25 °C) becomes detectable only 15 min after infection, this second wave of RNA synthesis cannot be caused by a gene dosage effect. Some prereplicative event in normal phage development seems to trigger the second wave. It is difficult to understand why it is not observed in T_4 am N82 cells. We have not yet investigated whether this discrepancy between normal RNA synthesis and that with the phage defective in DNA synthesis is a general phenomenon or due to the special defect of the T_4 am N82 mutant, whose impaired function unfortunately is not known.

The increased capacity of crude extracts to synthesise RNA coincides with the first peak of immediate early transcription *in vivo*. The second wave of early transcription is, however, not reflected *in vitro*. In analogy to the interpretation of the immediate delayed early switch under conditions not allowing DNA synthesis, the second wave of early transcription may be triggered by a specific anti-termination system, which acts at the delayed early termination sites. The deficiency of the crude extracts to show a corresponding second burst of activity *in vitro* may be due to a lability of the antitermination system or to greatly changed requirements (pH, ionic strength *etc.*) for optimal transcription. Alternatively the second wave could be caused by a newly synthesized polymerase with different requirements, as has been shown for bacteriophage T_7 [18].

In normal phage development (Fig.9), as in the case of infection under conditions not allowing DNA

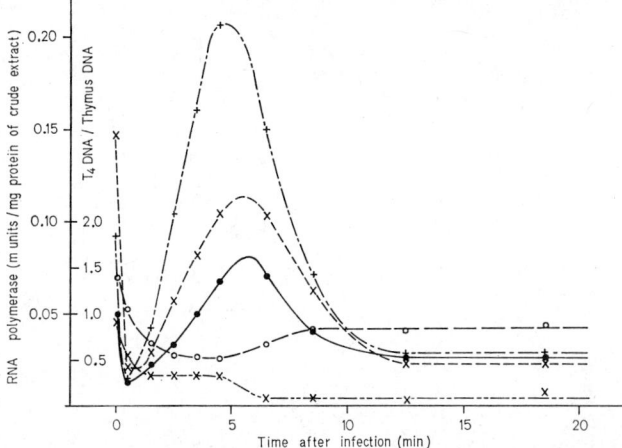

Fig.6. *Activity of DNA-dependent RNA polymerase in crude extracts and in extracts partially purified by chromatography on DEAE-cellulose.* The extracts were prepared from cells harvested before and at various times after infection (25 °C) under conditions which do not allow phage DNA synthesis (T_4 am N82 was used). ●, Endogenous DNA; +, thymus DNA; ×---×, T_4 DNA; ○, ratio T_4 DNA to thymus DNA in crude extract; ×-----×, ratio T_4 DNA to thymus DNA after chromatography on DEAE-cellulose

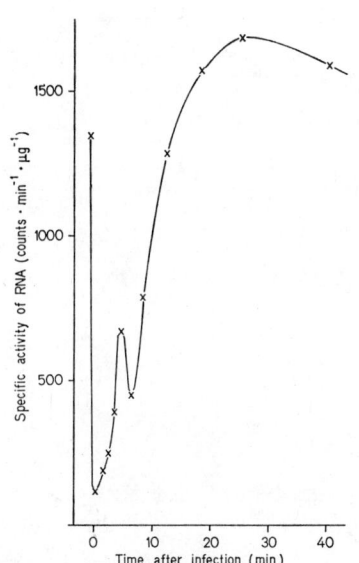

Fig.7. *Specific activity of total RNA extracted from bacteria labelled before and at various times after infection (30 sec pulse, 25 °C) under conditions which allowed normal phage development*

synthesis (Fig.6), the enzymatic activity of the polymerase purified by chromatography on DEAE-cellulose, drops about 5 min after infection from the T_4 DNA to thymus DNA template ratio of the altered enzyme to the ratio characteristic of the T_4-modified polymerase. The temporal coincidence of modification with the second wave of early RNA synthesis suggests that the modification may be one of the causes for the second wave.

The kinetics of enzyme activity (Fig.9) show that the modification lags behind the onset of delayed early gene transcription and precedes the second wave of early transcription. These observations do not allow conclusions as to the role of the modification process. One might speculate that modification is required for the reduction of immediate early transcription which has been postulated previously [6]. Our results, however, show clearly that the reduction of the relative amounts of immediate early messenger at later times is not due to a decrease of the rate of immediate early RNA synthesis but to an increase of the amount of other RNA species. The suggestion that a reduction of the rate of immediate early transcription is the consequence of a reduced affinity of the modified core for the *E. coli* σ-factor [19] is therefore not valid. We have indeed shown that the transcription of excess T_4 DNA by modified core is stimulated by *E. coli* σ-factor not only to the same final extent but even following the same saturation curve as that by unmodified core enzyme (Fig.10).

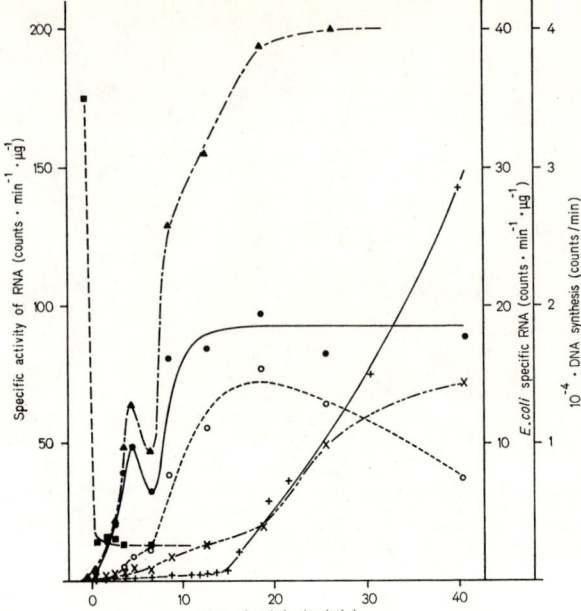

Fig. 8. *Hybridisation of RNA labelled before and at various times after infection (30 sec pulse, 25 °C) under conditions which allowed normal phage development with* E. coli *and T_4 DNA.* The amount of immediate early (●); delayed early (○); and late (×) RNA in each sample was calculated as described in Materials and Methods. ■, Total *E. coli* specific RNA synthesis; ▲, total T_4 specific RNA synthesis; +, total DNA synthesis. Unlabelled competitor RNA was extracted from infected bacteria killed at various times after infection as specified under Material and Methods

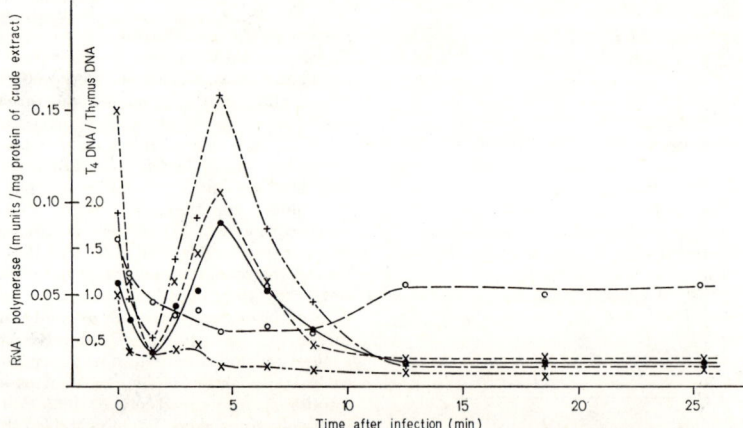

Fig. 9. *Activity of DNA-dependent RNA polymerase in crude extracts and in extracts partially purified by chromatography on DEAE-cellulose.* The extracts were prepared from cells harvested before and at various times after infection (25 °C) under conditions which allowed normal phage development. ●, Endogenous DNA; +, thymus DNA; ×——×, T_4 DNA; ○, ratio T_4 DNA to thymus DNA in crude extract; ×----×, ratio T_4 DNA to thymus DNA after chromatography on DEAE-cellulose

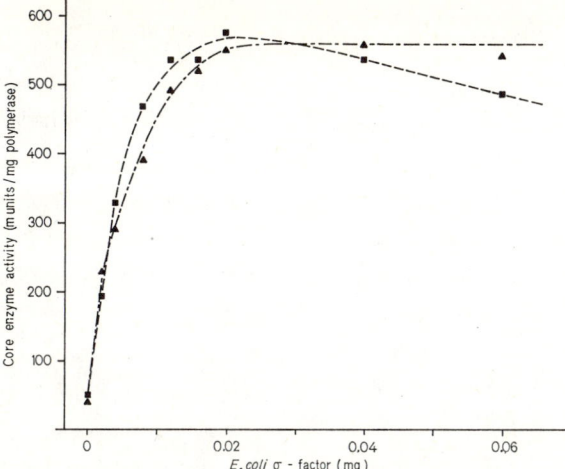

Fig. 10. *Stimulation of* E. coli *(■) and* T_4 *(▲) modified core enzyme activity with increasing amounts of* E. coli *specific σ factor on* T_4-*DNA. The weight ratio of DNA to polymerase was 10*

On the other hand highly purified modified enzyme does not contain any σ-factor. This could either be due to a loss of *E. coli* σ early after infection or to a reduction of the affinity of *E. coli* σ to modified core which does not influence its activity but leads to its complete removal during purification. The σ content of altered enzyme is not only greatly reduced but there is evidence that an altered σ appears shortly after infection under conditions not allowing protein synthesis. These findings favour the first possibility. The fate of *E. coli* σ, however, remains obscure until more direct evidence is obtained by analysis not of purified enzyme but of the crude extract, for example with immunochemical techniques. The so called σT_4 which has been described by Travers might either be a product of an alteration of *E. coli* σ, or as postulated, a new phage gene dependent σ-factor with different specificity. It could also be one of the antitermination factors required for the release of the block either between immediate and delayed early transcription or before the second wave of early transcription. It is indeed not impossible that the same factor might either act as σ on free polymerase effecting its attachment to promotors or as an antitermination factor on polymerase arrested at termination sites enabling it to enter the next reading unit without intermediate release. Whereas *in vitro* normal *E. coli* σ stimulates immediate early RNA synthesis, exclusively when termination factor ϱ is present, the so called σT_4 stimulates both delayed and immediate early transcription. Instead of assuming that σT_4 is a σ-factor specific for two types of promotors we favour the interpretation that "σT_4" contains normal or altered *E. coli* σ for the initiation from immediate early promotors and in addition an anti-termination factor for the immediate delayed early switch.

The comparative analysis of RNA synthesis in the infected cell on the one hand and the processes of polymerase alteration and modification on the other has thus not clearly answered the question as to the role of these processes for the regulation of transcription after infection. It has however furnished strong evidence for the idea that alteration is responsible for the shut off of host transcription preceding immediate early RNA synthesis. Although *in vitro* immediate early genes can be transcribed by normal *E. coli* enzyme containing normal *E. coli* σ factor, *in vivo* immediate early transcription is carried out by the altered enzyme.

Another striking result of our experiments is that the continuation and induction of phage controlled transcription processes are triggered by factors either newly synthesized or activated by prereplicative phage gene dependent processes.

More generally we are now forced to ask whether ϱ termination normally leads to an arrest of polymerase at ϱ termination sites, the release of which requires specific antitermination factors. Our failure to demonstrate their action in the crude extracts points to their lability or to quite specific structural or other requirements.

Further studies on the nature of the alteration of α and σ, of the modification of all the subunits of the core enzyme, on the mechanisms by which they are produced and on their influence on the properties of the enzyme are under way.

Thanks are due to Professor Dr A. Butenandt, to the *Deutsche Forschungsgemeinschaft* and to the *Sonderforschungsbereich 51* for generous support of this work.

REFERENCES

1. Hall, B. D., Nygaard, A. P., and Green, M. H., *J. Mol. Biol.* 9 (1964) 143.
2. Khesin, R. B., and Shemyakin, M. F., *Biokhimiya*, 27 (1962) 761.
3. Bolle, A., Epstein, R. H., Salser, W., and Geiduschek, E. P., *J. Mol. Biol.* 31 (1968) 325.
4. Hayward, W. S., and Green, M. H., *Proc. Nat. Acad. Sci. U. S. A.* 54 (1965) 1675.
5. Nomura, M., Witten, C., Mantei, N., and Echols, H., *J. Mol. Biol.* 17 (1966) 273.
6. Salser, W., Bolle, A., and Epstein, R., *J. Mol. Biol.* 49 (1970) 271.
7. Walter, G., Seifert, W., and Zillig, W., *Biochem. Biophys. Res. Commun.* 30 (1968) 240.
8. Seifert, W., Qasba, P., Walter, G., Palm, P., Schachner, M., and Zillig, W., *Eur. J. Biochem.* 9 (1969) 319.
9. Schachner, M., and Zillig, W., *Eur. J. Biochem.* 22 (1971) 513.
10. Palm, P., unpublished results.
11. Fraser, D., and Jerrel, E. A., *J. Biol. Chem.* 205 (1953) 291.
12. Zillig, W., Fuchs, E., and Millette, R. L., in *Procedures in Nucleic Acid Research* (edited by G. L. Cantoni and D. R. Davies), Harper and Row, New York 1966, p. 323.
13. Landy, A., and Spiegelman, S., *Biochemistry*, 7 (1968) 585.
14. Adesnik, M., and Levinthal, C., *J. Mol. Biol.* 48 (1970) 187.
15. Schachner, M., and Seifert, W., *Hoppe-Seyler's Z. Physiol. Chem.*, in press.
16. Milanesi, G., Brody, E. N., and Geiduschek, E. P., *Nature (London)*, 223 (1969) 1014.
17. Rouvière, J., Wyngaarden, J., Cantoni, J., Gros, F., and Kepes, A., *Biochim. Biophys. Acta*, 166 (1968) 94.
18. Chamberlin, M., McGrath, J., and Waskell, L., *Nature (London)*, 228 (1970) 227.
19. Travers, A., *Nature (London)*, 223 (1969) 1107.

M. Schachner's present address:
Neuropathology Department, Harvard Medical School,
25 Shattuck Street, Boston, Massachusetts 02115, U.S.A.

W. Seifert's present address:
The Salk Institute,
P.O. Box 1809, San Diego, California 92112, U.S.A.

W. Zillig
Max-Planck-Institut für Biochemie
BRD-8000 München 15, Goethestraße 31
German Federal Republic

Bacteriophage T7

Genetic and biochemical analysis of this simple phage gives information about basic genetic processes.

F. William Studier

Bacteriophages are among the simplest biological entities known, and yet they carry out biological processes that are basic to even the most complex organisms. In trying to understand such processes, phage-host systems have one big advantage over more complex systems: they are very easy to manipulate genetically. Mutations can be obtained in virtually any phage function (and in many host functions as well), and biochemical analysis of mutant strains can often reveal the molecular interactions that make up a biological process. Nucleic acid replication, genetic recombination, regulation of gene expression, and assembly of complex structures can all be studied in bacteriophage systems.

Among phages that contain double-stranded DNA, T4 and lambda and their relatives have been favorite objects of study. Phage T7 appears to be even simpler, and is currently receiving considerable attention. A strictly virulent phage, T7 (along with its relative, T3) is the smallest of the seven T phages originally described by Demerec and Fano (1) and Delbruck (2). It has a polyhedral head, a small, simple tail (3), and contains a single piece of double-stranded DNA of molecular weight 25×10^6, about one-fourth the size of the DNA from T even phages (4). The DNA from T7 contains the four usual DNA bases (5), and its base sequence is not circularly permuted across the population, as it is in the T even phage DNA's (6). Mature T7 DNA has a terminal repetition, but, unlike lambda DNA (7), the ends are double stranded and do not associate unless first treated with an exonuclease (6). After infection, T7 specifies approximately 30 proteins, which account for virtually all of the coding capacity of T7 DNA (8). Thus, T7 seems to be a manageable size, and it may well be possible to find a mutation in each of its genes, to determine the function of each of its proteins, and to define the molecular details of the processes directed by T7 after infection.

Genetic Analysis of T7

Conventional genetic techniques include the isolation of mutant strains, the analysis of their patterns of recombination and complementation, and the construction of a genetic map. In T7 it is also possible to analyze the end products of gene expression, the protein chains. The combining of physical and genetic techniques has permitted an unusually thorough genetic analysis of T7.

T7 proteins. The proteins of T7 are analyzed by electrophoresis on polyacrylamide gels in the presence of the detergent sodium dodecyl sulfate (SDS), a simple yet powerful technique (8, 9). A culture of host cells growing in minimal medium is infected with T7, and ^{14}C-labeled amino acids are added to label the proteins that are synthesized. At the end of the labeling period the infected cells are harvested by centrifugation, suspended in a buffer solution containing SDS, and heated briefly to 100°C. This treatment solubilizes virtually all of the proteins in the cell and dissociates them to individual protein chains, each complexed with large amounts of SDS. Since SDS is negatively charged, the protein-SDS complexes are also negatively charged, and all migrate in the same direction in an electric field. Electrophoresis through a polyacrylamide gel resolves the protein-SDS complexes according to their size (10). After electrophoresis, the gels can be dried and autoradiographed to determine the positions of the labeled proteins. Thus, all proteins that incorporate ^{14}C-labeled amino acids during a labeling period are displayed, and their sizes can be estimated from their positions on the gel. Autoradiograms showing the time course of protein synthesis during a normal T7 infection are given in Fig. 1a.

In a normal infection, any host proteins that are being synthesized at the same time as the T7 proteins are also labeled and appear on the autoradiogram. However, if uninfected host cells are irradiated sufficiently with ultraviolet light, they become unable to synthesize their own proteins but retain the ability to synthesize T7 proteins after infection (8). In this system, only T7 proteins incorporate added ^{14}C-labeled amino acids, and it is possible to visualize T7 proteins without any interference from host proteins. The time of appearance of T7 proteins seems to be unaffected by the prior irradiation of the host (Fig. 1b).

Approximately 30 T7 proteins can be identified on SDS-polyacrylamide gels. Their molecular weights range between approximately 7,000 and 150,000, and together they account for virtually all of the coding capacity of T7 DNA (8, 11).

T7 mutations. Amber mutants and deletion mutants are particularly useful in the genetic analysis of T7. Not only are they easily isolated but both types of mutation alter the size of any protein they affect, thus permitting the protein to be identified electrophoretically.

For most of the genetic analysis of T7, amber mutants have been utilized (11–14). Amber mutants make a protein of normal size when grown on a permissive host, but make a shorter fragment, the amber peptide, when grown on a restrictive host (15). If the affected protein is essential to the growth of the phage, the mutant will be unable to grow on the restrictive host but will usually be able to grow in the permissive host. Such mutants are referred to as conditional-lethal mutants (16). These mutants can be isolated by plating a phage stock that had been treated with a mutagen on the permissive host and then testing individual plaques to find strains that do not grow on the restrictive host. Conditional-lethal amber mutations are easy to map genetically. The mutants are crossed in the permissive host, and the progeny are plated on both the permissive and restrictive hosts. All of the progeny

---

The author is biophysicist in the biology department, Brookhaven National Laboratory, Upton, New York 11973.

grow on the permissive host, but only wild-type recombinants can grow on the restrictive host; thus, the ratio of plaques found on the two hosts is readily converted to percent recombination.

Deletion mutants, or strains of T7 in which a portion of the DNA has been deleted, are also easy to isolate. Parkinson and Huskey (17) found that phage particles containing a piece of DNA shorter than normal length are more resistant to disruption at high temperatures (perhaps because the smaller amount of DNA exerts less pressure on the head structure). This makes it possible to select deletion mutants at random by heating a solution of T7 particles to the point where most are inactivated, and then growing the survivors. Many of the heat-stable strains isolated from T7 in this way are deletion mutants (18, 19). Conventional genetic mapping of deletions is difficult unless the deletion causes a change in plaque morphology or some easily measured character. However, the location and extent of a deletion can be determined in the DNA molecule itself. This physical mapping is done by hybridizing DNA molecules and looking at them in the electron microscope (20). When a wild-type strand of DNA is hybridized to a complementary strand from a molecule containing a deletion, the unmatched region of the wild-type strand will loop out, marking the position and length of the deletion.

Deletion mutants are viable only if they delete nonessential regions of the DNA, that is, nonessential genes, nonessential portions of a gene, or nonessential extragenic DNA. (In this article the term gene refers to that portion of a DNA molecule which specifies a single protein chain.) Thus, deletion mutations are complementary to conditional-lethal amber mutations in the sense that deletions are found only in nonessential genes and conditional lethals only in essential genes. The combination should, theoretically, yield a mutation in every T7 gene.

The protein affected by an amber mutation or a deletion should disappear from its normal position in the patterns obtained upon electrophoresis of extracts of infected cells on SDS-polyacrylamide gels. This is because the protein chains are resolved on the basis of size, and both types of mutation affect the size of the protein synthesized. The size of the amber peptide will depend on the location of the amber mutation, and this property has been used to determine the direction of translation (and thus

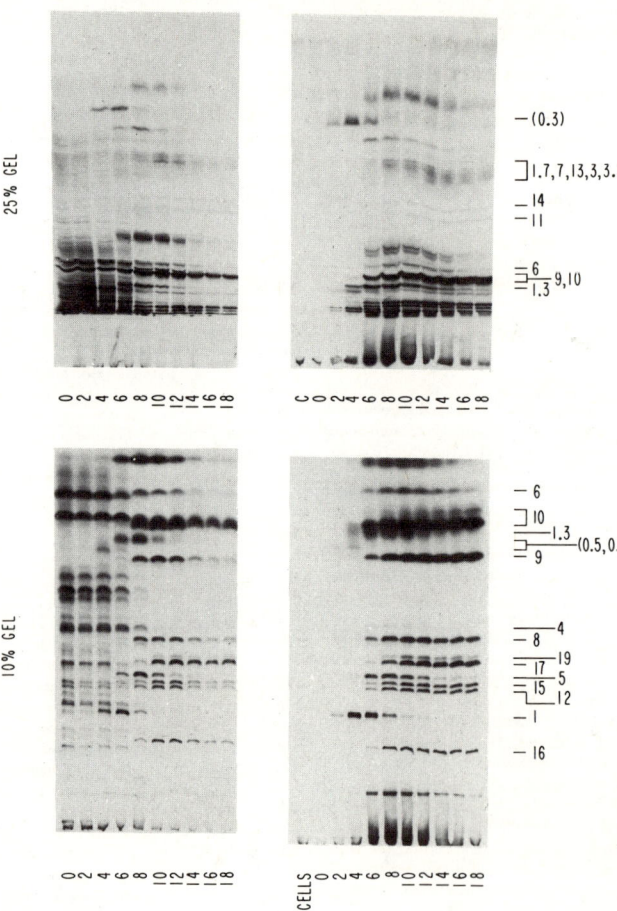

Fig. 1. Time course of protein synthesis in normal (a) and ultraviolet-irradiated (b) *E. coli* B. Cells were grown and infected in M9 minimal medium at 30°C. At 2-minute intervals, samples of infected culture were added to a mixture of ^{14}C-labeled amino acids. Cultures that had been irradiated with ultraviolet light received ten times as much ^{14}C as those that had not been irradiated. After 2 minutes in the presence of ^{14}C, a great excess of unlabeled amino acids was added in order to greatly reduce further incorporation of ^{14}C into proteins. The samples were incubated another 4 minutes to allow any ^{14}C-containing protein chains to be completed, chilled to 0°C, and centrifuged. The cells were suspended in 0.05M tris-Cl, pH 6.8, containing 1 percent SDS, 1 percent mercaptoethanol, and 10 percent glycerol. After being heated for 2 minutes in a boiling water bath, the samples were subjected to electrophoresis in a discontinuous, SDS-containing buffer system (47) on slabs of 10 and 25 percent polyacrylamide gel [a modification of the system of Reid and Bieleski (52)]. The origin of electrophoresis is at the bottom, and the farther a protein moves the smaller its molecular weight. The time after infection at which the pulse of ^{14}C began is given beneath each pattern. The number of the gene which specifies each band (where known) is given at the right of the patterns. The numbering of genes 0.3, 0.5, and 0.7 is tentative, since it has not been definitely established whether these protein bands are the product of two genes or three genes nor what the gene order is (see text).

transcription) relative to the genetic map (8). The size of any new protein chain that might appear in a deletion mutant will depend on whether the deletion eliminates all or a part of a gene or fuses two genes. Presumably, all of the proteins coded for by T7 are displayed in the protein patterns; therefore, it should be possible to know when a mutation has been found in every T7 gene.

T7 genes. Amber mutants have been isolated in 19 essential genes of T7 (12–14). These mutations fall on a linear genetic map, and the 19 genes are numbered in order from left to right. More than 800 conditional-lethal amber mutants have been characterized, but no new gene has been found since mutant number 107 (11); either all of the essential genes have been identified, or amber mutations in the remaining essential genes are relatively rare. The proteins specified by 17 of these 19 genes have been identified in the protein patterns on SDS-polyacrylamide gels (see Fig. 2) (8, 11).

Amber mutations can also be found in nonessential genes if a test for gene function is available, for example, an assay for enzyme activity. In this case, strains are isolated randomly from a heavily mutagenized stock and are tested for the enzyme activity. Since T7 has only 30 or so genes, if a stock were mutagenized to an average of several mutations per phage particle, only a moderate number of strains should have to be tested before the desired mutant is obtained. A strain containing an amber mutation in the gene for T7 ligase was isolated in this way by Masamune, Frenkel, and Richardson (21). It was found among the first 20 strains tested.

T7 ligase mutants grow in normal *Escherichia coli* hosts, but not in a ligase-deficient host (21, 22). Thus it is possible to map ligase mutants relative to conditional-lethal amber mutations by using a restrictive host which is also ligase deficient. The ligase gene falls between genes 1 and 2 (11). An advantage of the genetic map of T7 is that it makes it possible to recognize the position of a gene by its number. To preserve this feature, new genes are given a decimal number which indicates their relative position. Thus, ligase is designated gene 1.3, since it falls between genes 1 and 2, and, as discussed below, at least one other gene is located between it and gene 2. An amber mutation in lysozyme has also been found by enzymatic test (11), and it maps between genes 3 and 4; thus, the lysozyme gene is designated 3.5. Both the ligase and lysozyme proteins have been identi-

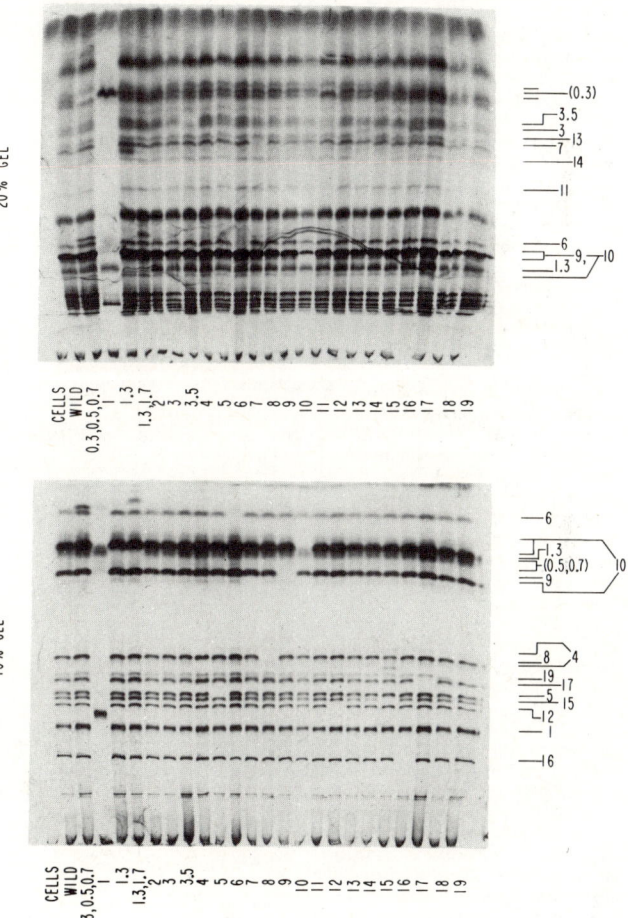

Fig. 2. Protein patterns of different T7 amber and deletion mutants. Patterns were obtained essentially as described in the legend to Fig. 1, with the following exceptions: the upper patterns were obtained on a 20 percent gel instead of a 25 percent gel; twice as much sample was analyzed on the 20 percent gel as on the 10 percent gel, in order to increase the relative intensity of the fainter bands; and ^{14}C-labeled amino acids were present from 2 to 20 minutes after infection so that all T7 proteins would be labeled. The host cells were irradiated with ultraviolet light before infection; therefore, only T7 proteins appear. The patterns are oriented as in Fig. 1. The number or numbers beneath each pattern identifies the gene or genes in which each strain has mutated. Those patterns labeled with a single gene number are amber mutants; those with more than one are deletions which cover the numbered genes. The numbers to the right of the patterns identify the genes which specify individual proteins. (A slight distortion has shifted the outside patterns down and away from the center patterns; the lines on the right point to the bands in the center patterns.) Except for the pattern given by the mutant defective in gene 1, each pattern lacks only the protein band specified by the gene itself. Amber peptides can be seen on the 10 percent gel in the patterns from gene 1 and gene 15, and deletion peptides can be seen in the patterns from genes 0.3, 0.5, 0.7 and from genes 1.3, 1.7. Possible amber peptides are more difficult to distinguish on the 20 percent gel, but may be present in the patterns for genes 1.3, 11, 13, and 16.

fied in the protein patterns on SDS-polyacrylamide gels (*11*).

Deletions have been found in two different regions of the T7 DNA molecule (*19*). Since the T7 ligase is not essential in the normal hosts, it has been possible to isolate deletions which include at least part of the ligase gene. Ligase has been mapped genetically, and thus ligase deletions provide a link between the genetic and physical maps, identifying the left end of the T7 DNA molecule. Genetic mapping shows that some ligase deletions come very close to the right end of gene 1, and these seem to lack only the ligase protein; others come very close to gene 2, and these lack at least one additional protein, which indicates that at least one gene lies between ligase and gene 2. This gene is designated 1.7.

Physical position on the T7 DNA molecule is given in terms of percent of the total length, measured from the left end. Deletions in the region between genes 1 and 2 (including ligase and gene 1.7) extend from approximately 15 to 24 percent (*19*). A second series of deletions has been found between approximately 3 and 8 percent (*19, 23*). These deletions must fall to the left of gene 1, as indicated by the following combination of facts: (i) the protein specified by gene 1 is essential to the growth of T7, which means that little, if any, of gene 1 could be deleted; (ii) genetic mapping places gene 1 to the left of ligase; (iii) the gene 1 protein has a molecular weight of approximately 100,000 (*8, 24*), which requires a coding capacity of approximately 8 percent the length of T7 DNA; and (iv) there is barely enough room for gene 1 to fit between the two sets of deletions, and not enough room for it to be to the left of the 3 percent mark. Thus, gene 1 must be located between approximately 8 and 15 percent, and nonessential regions lie to either side of it.

The deletions to the left of gene 1 affect three different protein bands in the polyacrylamide gel, an intense one at approximately 9,000 daltons and a weak double band near 40,000 daltons (*19*). Of the mutants with deletions in this region that have been tested, all eliminate both of the bands at 40,000 daltons, and some eliminate the band at 9,000 daltons as well. Preliminary results place the gene which specifies the protein at 9,000 daltons to the left of the gene or genes specifying the 40,000-dalton protein. Since no deletion mutant which eliminates only one of the bands at 40,000 daltons has been found, it is not clear whether these bands are the product of one gene or two. Provisionally, the gene for the band at 9,000 daltons is numbered 0.3 and those for the two bands at 40,000 daltons 0.5 and 0.7, but this numbering may have to be revised after more is known about this region.

Protein patterns of mutants in the 25 genes of T7 for which mutants are available are given in Fig. 2. These patterns were obtained in cells that had been irradiated with ultraviolet light before infection, so that only T7 proteins would be displayed, and ^{14}C-labeled amino acids were present throughout the infective cycle so that all T7 proteins would be labeled. All mutants are either ambers or deletions, so the proteins directly affected by the mutation should be missing from the pattern. Two patterns are given for each mutant, one on a 10 percent gel under conditions which resolve proteins of molecular weight 25,000 and higher, and a second on a 20 percent gel under conditions which resolve proteins of molecular weight less than 25,000.

All patterns except that from the gene 1 mutant are substantially the same as wild type, except that they lack the protein(s) directly affected by the mutation. A mutation in gene 1 affects most of the proteins specified by T7; this behavior will be discussed in the next section. In some patterns, amber peptides or deletion peptides can be found, but most of the amber mutations used are near the left end of the gene and the amber peptides would be very small. Mutations have been identified for most of the bigger proteins of T7, but several bands appear on the 20 percent gel for which no mutation is yet available. It may be possible to obtain amber or deletion mutants affecting these proteins by analyzing protein patterns of strains isolated randomly from a heavily mutagenized stock, in much the same way strains defective in ligase were found. Gene numbers and approximate molecular weights of the proteins affected by T7 mutants currently available are given in the first two columns of Table 1. The proteins that have been identified with specific genes account for approximately 90 percent of the coding capacity of T7 DNA.

Regulation of Gene Expression by T7

In a normal infection, T7 proteins are synthesized according to a characteristic time course (Fig. 1), and at least three classes of proteins can be distinguished (*8, 11*): class I, those synthesized between approximately 4 and 8 minutes after infection; class II, those synthesized between approximately 6 and 15 minutes after infection; and class III, those synthesized from approximately 6 to 8 minutes after infection until lysis. The class I proteins are the same ones that are synthesized at normal rates in gene 1 mutants (Fig. 2). As far as can be determined, these three classes are strictly correlated with position on the genetic map (*11*): class I includes those proteins specified by genes 0.3 to 1.3, the left-most genes of T7; class II includes the next group of proteins, specified by genes 1.7 to 6; and class III includes the right arm of the genetic map, the proteins specified by genes 7 to 19. (Gene 1.3, ligase, is assigned to class I because it is synthesized at normal rates in gene 1 mutants; however, the ligase protein usually continues to be synthesized until class II

Table 1. Approximate molecular weights (M.W.) and functions of the proteins specified by T7 genes.

Gene	~ M.W.*	Function †
0.3 ‡	8,700	Nonessential
0.5	40,000	Nonessential
0.7	42,000	Nonessential
1	100,000	RNA polymerase
1.3	40,000	Ligase
1.7	17,000	Nonessential
2		Reduced DNA synthesis
3	13,500	Endonuclease
3.5	13,000	Lysozyme
4	67,000	Reduced DNA synthesis
5	81,000	DNA polymerase
6	31,000	Exonuclease
7	14,700	Found in phage particle
8	62,000	Head protein
9	40,000	Head assembly protein
10	38,000	Major head protein
11	21,000	Tail protein
12	86,000	Tail protein
13	14,000	Found in phage particle
14	18,000	Head protein
15	83,000	Head protein
16	150,000	Head protein
17	76,000	Tail protein
18		DNA maturation
19	73,000	DNA maturation

* Molecular weights are estimated from relative positions on SDS-polyacrylamide gels (*11*). Some proteins change relative position with change in electrophoresis buffer or in concentration of acrylamide, but most seem to retain approximately the same relative position. The values given, particularly at low molecular weights, are rather crude approximations, which will be refined as better marker proteins become available and as molecular weights of specific T7 proteins are determined by absolute methods. † See text for discussion and references. ‡ The numbering of genes 0.3, 0.5, and 0.7 is tentative, since it has not been definitely established whether these protein bands are the product of two genes or three, nor what the gene order is.

protein synthesis ends.) Thus, the T7 genes seem to be expressed according to their map position, left being early and right being late. The orderly appearance of T7 proteins implies some type of control over gene expression, which could be mediated at the level of transcription or translation or both.

All transcription from T7 DNA after infection proceeds from left to right (8, 25), the same direction as the early to late polarity in T7. Summers (26) has identified 12 or 13 species of T7 messenger RNA in infected cells, which together can account for the total capacity of T7 DNA. If these represent nonoverlapping natural messengers, then a number of them must contain information for more than one protein chain. However, the polar effects on protein synthesis that might be expected from amber mutations in a polygenic messenger RNA (27) are not very pronounced, if present at all (Fig. 2). The time course of T7 RNA synthesis has not yet been studied in sufficient detail to determine whether three classes of messenger RNA molecules, corresponding to the three classes of proteins, can be distinguished. However, as discussed in the next section, Siegel and Summers (28) and Summers (29) have identified the class I messenger RNA's. Summers (30) has also found that T7 messenger RNA synthesized between 6 and 8 minutes after infection seems to be stable, unlike the messenger RNA of uninfected cells which has a half-life of 2 to 3 minutes. This suggests that the messenger RNA's for class III proteins, and perhaps other T7 messenger RNA's as well, are not degraded.

Gene 1 action. As was mentioned in the previous section, gene 1 mutants are the only ones isolated so far that seem to affect the regulation of gene expression after T7 infection (8, 11). When gene 1 is inactive, all class I proteins are synthesized at normal rates, but synthesis of class II and class III proteins is greatly depressed. By analyzing the messenger RNA produced after infection with normal T7, using gene 1 mutants or the presence of chloramphenicol (which prevents the synthesis of any T7 proteins, including the gene 1 protein), Siegel and Summers (28) have shown that gene 1 action in vivo is at the level of transcription. They identified four or five messenger RNA molecules, ranging in size from 200,000 to 1,000,000 daltons, which must be the messenger RNA's for the class I proteins.

Chamberlin et al. (24) have shown that the gene 1 product is a new RNA polymerase. They have purified it to near homogeneity and shown that it differs from host RNA polymerase in several respects, including size, template preference, conditions for optimal activity, and sensitivity to inhibitors. Thus, it seems likely that the host RNA polymerase transcribes only a limited region of the T7 DNA in vivo (the class I genes), and that the gene 1 RNA polymerase is necessary for efficient transcription of the class II and class III genes. Summers and Siegel (31) have shown that partially purified T7 RNA polymerase does transcribe the late regions of T7 DNA in vitro.

Transcription by host RNA polymerase. When gene 1 does not function in vivo, all transcription is presumably by the host RNA polymerase. Apparently the host RNA polymerase can transcribe only the left-most genes of T7 and cannot proceed past a point in the T7 DNA just to the right of gene 1.3 (ligase).

Experiments in vitro by Davis and Hyman (32) indicate that T7 DNA has only a single starting point for transcription by host RNA polymerase. They used the electron microscope to look at complexes of T7 DNA and purified E. coli RNA polymerase which had been allowed to synthesize RNA for varying lengths of time, and concluded that RNA synthesis starts near one end of the T7 DNA molecule. Examining hybrid molecules made by annealing newly synthesized RNA to T7 DNA, they found a single initiation point approximately 1.3 percent from the left end of the T7 DNA, which would be to the left of the class I genes. Hyman (33) has examined hybrids between isolated T7 DNA and class I RNA made in vivo. He found that the hybrid region began approximately 1 percent from the left end, in good agreement with the results in vitro.

A specific signal in the T7 DNA itself stops transcription by host RNA polymerase at a point just to the right of ligase, and deletions which extend through the right end of the ligase gene seem to remove the stop signal (11, 19). This can be seen in double mutants containing a mutation in gene 1 plus a ligase deletion; the time course of protein synthesis after infection by two such double mutants is shown in Fig. 3. In the first double mutant (1, LG3), class II and III proteins are made in the small amounts characteristic of any gene 1 mutant, but in the second (1, LG37) they appear in much greater amounts, which suggests that the stop signal has been deleted. The amounts of class II and III proteins are not as great as the amounts found during infection with wild-type T7, but one might expect the T7 RNA polymerase to be more active than host RNA polymerase in transcribing these genes.

The physical position of the stop signal in vivo in the T7 DNA can be determined from the positions of the right-hand ends of ligase deletions which do and do not delete the stop signal. Only a few such deletions have been analyzed so far (19) and results indicate that the stop signal is located between approximately 20 and 20.8 percent from the left end of T7 DNA. Hyman's mapping of the class I RNA made in vivo (33) places the stop signal approximately 20 percent from the left end, in excellent agreement with the data from the deletion mutants. The four or five class I proteins could account for as much as 18.5 percent of the coding capacity of T7 DNA, which would leave little room for anything else to the left of the stop signal.

Experiments in vitro have given conflicting results on what is required for termination of transcription at the end of the class I region. Using isolated T7 DNA and purified host RNA polymerase, Davis and Hyman (32) did not find specific termination unless rho factor (34) was added to the reaction. Without rho (a termination factor) transcription proceeded all the way to the right end of the T7 DNA; in the presence of rho, most transcription did not proceed past a point approximately 18 percent from the left end, and the RNA synthesized was not uniform in size. These authors suggested that termination at the right end of the class I region requires rho and that the host RNA polymerase is prevented from moving past this point. They further suggested that there are termination sites within the first 18 percent of the T7 DNA molecule at which rho causes the release of RNA chains but does not prevent the RNA polymerase from proceeding.

On the other hand, the results of Millette et al. (35) suggest that rho is not needed for termination at the right end of the class I region. Again using purified host RNA polymerase, these workers found that, in the absence of rho, RNA molecules of approximately 2.2×10^6 daltons are produced. Almost all of this RNA has uracil as the 3' terminal base, which suggests that all of these molecules may have terminated at a unique point. An RNA molecule this size would represent almost 18 percent of the length of T7 DNA, approximately the length of the class I region.

These two sets of experiments do not conflict on the possibility that rho-mediated termination events may occur in the first 18 percent of the T7 DNA, but they disagree on what is required to prevent transcription past the stop signal at the end of the class I region. It is possible that unrecognized factors play a role. It should also be pointed out that termination at the end of the class I region is not completely effective in vivo (see below).

After infection with mutants which lack gene 1 activity and in which the stop signal is deleted, the order of appearance of the T7 proteins is the same as their order on the genetic map (Fig. 3). Once the host RNA polymerase passes the stop signal, it apparently transcribes the genes in order all the way to the right end, which indicates that the rest of the T7 DNA is free of stop signals for host RNA polymerase. Thus, the time of appearance of a protein in such mutants provides a crude measure of its position on the DNA molecule. The last protein appears approximately 13 minutes after infection; therefore, the minimum rate of transcription of T7 DNA (25×10^6 daltons) by $E.$ $coli$ RNA polymerase in vivo at 30°C is approximately 10^6 daltons of RNA per minute, or 50 nucleotides per second. This is almost twice as fast as previous estimates for the rate of transcription in uninfected $E.$ $coli$ (36).

The stop signal to the right of ligase is apparently not 100 percent effective in vivo. Small amounts of class II and III proteins are made after infection with gene 1 mutants even when the stop signal is present (Fig. 3). This indicates that there is some transcription beyond the stop signal. Such transcription could be caused by a small amount of active T7 RNA polymerase or by some host RNA polymerase molecules which transcribe past the stop signal. The latter possibility is probably correct for two reasons: (i) double mutants in gene 1 produce the same small amounts of class II and III proteins (11), and (ii) the time course of appearance of the class II and III proteins in gene 1 mutants is the same whether or not the stop signal is present (only the amounts are different), and it is very different from the time course observed in a wild-type infection (see Figs. 1 and 3).

On the whole, experiments in vivo and in vitro agree quite well, and their combined results give the following picture of transcription after T7 infection. The DNA of T7 is first transcribed by the host RNA polymerase, which starts at a single point close to the left end of T7 DNA and stops at a point approximately 20 percent from the left end. The messenger RNA is found in four or five pieces, which may arise from specific termination events, perhaps mediated by rho or some other termination factor. The class I proteins are made, and they account for virtually all of the coding capacity of this region. One of these proteins is the gene 1 RNA polymerase, which in turn transcribes the rest of the T7 DNA. The stop signal at approximately 20 percent from the left end is not completely effective, a certain fraction of the host RNA polymerase molecules reading through all the way to the right end.

Other regulation. The interpretation just described explains the dramatic effect of gene 1 action, but leaves unanswered a number of questions concerning other aspects of gene expression after T7 infection. Why does the synthesis of host proteins and some class I proteins stop approximately 8 minutes after infection? It is tempting to speculate that inactivation of host RNA polymerase could be responsible for both. [If this were the case, the ligase gene (1.3) should be transcribed by the gene 1 RNA polymerase as well as by the host RNA polymerase, since ligase continues to be synthesized until 15 minutes after infection, shutting off with the class II proteins.] Gene 1 mutants are the only ones found so far that are defective in shutting off the synthesis of host proteins and class I proteins (11). The gene 1 RNA polymerase could actively participate in this process, but it seems more likely that a class II or III gene is responsible and the effect of the gene 1 mutation is indirect. However, none of the class II or III genes identified so far seems to be responsible (11).

A similar unanswered question is: why does the synthesis of class II proteins stop midway through the infective cycle? In some experiments, including that shown in Fig. 1, the synthesis of class III proteins slows at the same time. No mutant has been found in which this process does not occur, but the shutoff seems to be much less pronounced in cells irradiated with ultraviolet light (Fig. 1). It is not known whether this process is regulated at the level of transcription or translation.

Another element in the control of gene expression concerns the mechanism for determining the relative rates of synthesis of different proteins. Much of the difference in rates among class II and III proteins disappears when transcription is directed by host RNA polymerase (compare Figs. 1 and 3). In fact, when transcription is by the host RNA polymerase there seems to be a gradient in the rate of synthesis of class II and III proteins, the rate generally decreasing from left to right (Fig. 3). This suggests that a major determinant of the rate of protein synthesis may be the number of copies of messenger RNA present, and Summers (26) did observe widely varying amounts of different T7 messenger RNA's in a normal infection. Specific messenger RNA's for individual genes have not been identified, so it remains to be determined whether a correlation exists between the number of copies of a messenger RNA and the rate of protein synthesis. If the relative rate of protein synthesis does depend on the relative quantities of different messenger RNA's, then the question becomes: how is the rate of synthesis of different messenger RNA's regulated? This presumably depends on the number and location of specific initiation and termination points for T7 RNA polymerase, whether the enzyme is more active at some sites than others, and whether other factors which interact with the enzyme or the DNA might be involved. Again, this information is not yet available. There may also be factors which regulate rates of protein synthesis at the level of translation, and Morrison and Malamy (37) have suggested that the bacterial sex factor interferes with T7 gene expression at the translational level.

Functions of T7 Genes

Genes with related functions tend to cluster along the genetic map in T7 (13), as in lambda (38) and T4 (16). The functions of genes 0.3 to 0.7 are not known, but they are not essential for growth in normal hosts. Gene 1 specifies the T7 RNA polymerase, as was discussed before. Gene 1.3 specifies the T7 ligase, which presumably has a role in DNA metabolism after infection, although the host ligase can substitute for it. The function of gene 1.7 is unknown, but it is not essential for growth in normal hosts. All of genes 2 to 6 affect the kinetics of DNA synthesis (*12–14*). Genes 7 to 17 specify proteins that are found in mature T7 particles or in partly assembled particles (*8, 11, 13*). Genes 18 and 19 seem to be involved in the maturation of T7 DNA from the replicating intermediate (*14, 39, 40*). Thus, early functions are to the left and late functions to the right: genes located between 1.3 and 6 seem to be mostly if not entirely involved in the early steps of DNA metabolism; genes 7 to 19 seem to participate in phage assembly, including the late stages of DNA metabolism. Known functions of T7 genes are summarized in Table 1.

DNA metabolism. The breakdown of host DNA and the synthesis of T7 DNA are among the early steps of DNA metabolism after infection by T7. The kinetics of DNA synthesis after infection (at 30°C) suggest the following (*12–14*): synthesis of host DNA con-

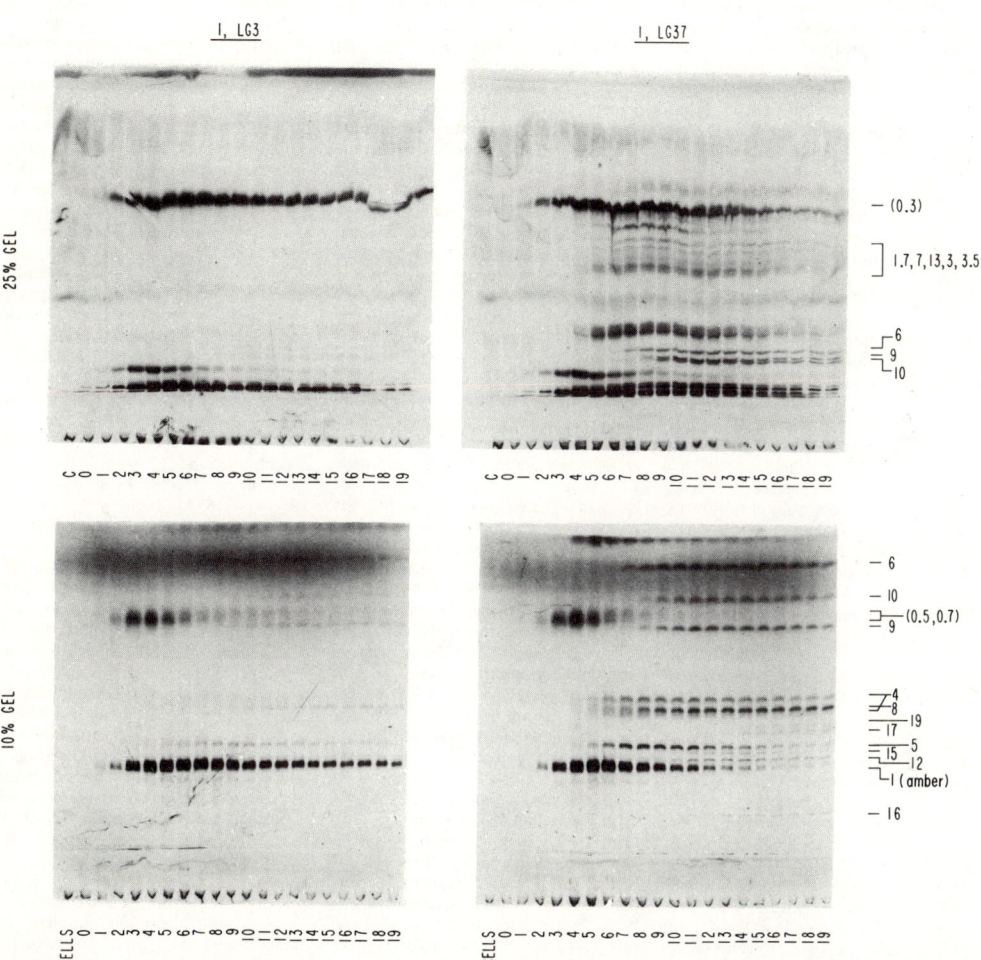

Fig. 3. Time course of protein synthesis in two double mutants of T7 (1, LG3 and 1, LG37), each containing an amber mutation near the right end of gene 1 plus a deletion in the ligase region. Patterns were obtained essentially as described in the legend to Fig. 1, except that the cultures were exposed to ^{14}C-labeled amino acids for 1 minute instead of 2 minutes and the autoradiograms were exposed almost three times as long. The host cells were irradiated with ultraviolet light before infection; therefore, only T7 proteins appear. The patterns are oriented as in Fig. 1, and the numbers beneath the patterns give the time after infection at which the pulse of ^{14}C began. The numbers to the right identify the genes which specify individual proteins; the amber peptide from gene 1 is also identified. The ligase protein (gene 1.3) is missing from the patterns because of the deletions. The right end of the deletion in the double mutant 1, LG3 is approximately 20 percent from the left end of T7 DNA; the right end of the deletion in 1, LG37 is approximately 22 percent from the left end of T7 DNA (*19*). The protein patterns indicate that the LG3 deletion apparently does not delete the stop signal for host RNA polymerase whereas the LG37 deletion apparently does (see text).

tinues at the preinfection rate for approximately 5 minutes after infection and then stops between 5 and 10 minutes after infection; synthesis of T7 DNA begins between 5 and 10 minutes after infection, reaches its maximum rate (perhaps 5 to 10 times the preinfection rate) between 15 and 20 minutes after infection, and shuts off just before or along with lysis, which begins between 25 and 30 minutes after infection. Replication of T7 DNA proceeds through intermediate forms that are longer than the piece of DNA found in phage particles (39, 40). It is not known how these replicative forms arise, but they could presumably be produced by replication, recombination, or both (40). It is thought that mature T7 DNA is cut from the intermediate forms during phage assembly in a process which also generates the terminal repetition found in mature T7 DNA (40).

Phage T7 is very efficient at breaking down host DNA to utilize nucleotides for its own DNA. Eighty-five to 90 percent of the nucleotides found in mature T7 phage particles were present in host DNA at the time of infection (41). A typical burst of 250 phage particles per host cell produces 6.3×10^9 daltons of T7 DNA, which would require that most of the host DNA be used to make T7 DNA. The nucleotides of host DNA are presumably incorporated into T7 DNA about as fast as they are released, since almost none of the host DNA can be found as acid-soluble material during the normal infective cycle (12, 42, 43).

Genes 3 and 6 participate in the breakdown of host DNA after T7 infection (42, 43). Gene 3 specifies an endonuclease (42–44) and gene 6 an exonuclease (45), and both are required for any significant degradation of host DNA. Synthesis of T7 DNA stops prematurely in mutants of genes 3 or 6 (13), presumably because little T7 DNA can be made without the normal supply of nucleotides from host DNA.

A third function of T7 may also be connected with the degradation of host DNA. Sadowski and Kerr (43) have shown that release of host DNA from its "membrane complex" occurs at an early stage of degradation. Such release occurs in mutants defective in genes 3 or 6 but not in gene 1 mutants (which are deficient in all class II and III proteins). This suggests that there may be a class II protein which causes release, perhaps to facilitate breakdown by the gene 3 and 6 nucleases. The release function cannot be ascribed to the products of genes 2, 3, 4, 5, or 6 (43). It is not yet known whether the products of genes 1.7 or 3.5, the other class II genes for which mutants are available, could be responsible for release, but T7 lysozyme, the product of gene 3.5, seems a possible candidate.

There are several indications that T7 lysozyme may have a role in DNA metabolism rather than lysis (11). The lysozyme gene lies in the middle of the class II region, among genes involved in the early steps of DNA metabolism. In common with other class II proteins, lysozyme is synthesized between 5 and 15 minutes after infection, an unusual time course for a function concerned with lysis, which begins approximately 25 minutes after infection. Furthermore, the T7 lysozyme does not seem to be required for lysis, since cultures infected with an amber mutant in lysozyme lyse completely, even though no detectable lysozyme is produced (less than 2 percent of the amount produced by the wild type). Lysis is slightly delayed, but this is a property of most mutants in class II genes (13). Gene 1 mutants, on the other hand, do not lyse the host even though they can ultimately produce 10 to 20 percent of the amount of lysozyme produced by the wild-type T7 (because a fraction of the host RNA polymerase molecules transcribes past the normal stop signal). The only lysozyme amber mutant available so far grows poorly in the restrictive host; it has a reduced rate of DNA synthesis and produces few progeny. This mutant has not yet been tested for its ability to cause release of host DNA from its "membrane complex."

Normal replication of T7 DNA requires the action of genes 2, 4, and 5 in addition to those discussed above. Mutants with defects in genes 4 or 5 make little if any T7 DNA (12–14), and a defect in gene 2 leads to premature cessation of DNA synthesis (13). Gene 5 has been shown to specify a DNA polymerase (46). The lack of replication in gene 5 mutants suggests that this enzyme may participate directly in replication. So far, no specific functions have been assigned to genes 2 and 4.

Assembly. Functions of class III proteins are primarily, if not exclusively, connected with assembly of the T7 phage particle (8, 11). At least 11 different protein chains can be resolved when purified T7 particles are subjected to electrophoresis on SDS-polyacrylamide gels (11). These proteins have molecular weights of 13,000 to 150,000 and account for almost half of the coding capacity of T7 DNA. Genes 7, 8, and 10 to 17 specify these 11 proteins, and gene 9 specifies a protein that is found in empty head structures but not in any DNA-containing structures analyzed so far. Thus, genes 7 to 17 all specify proteins found in phage structures.

Analysis of protein compositions, electron microscopy, and complementation of partly assembled particles have been used to determine which proteins make up the head and which the tail of the T7 particle (8, 11, 13). Particles containing DNA are produced when mutants in genes 7, 11 to 13, and 17 are grown under restrictive conditions; therefore, none of the proteins specified by these genes is necessary for the production of a filled head. Lysates of wild-type T7 contain unequal amounts of two types of empty head. The minor component contains the proteins specified by genes 8, 10, and 14 to 16, the same proteins which are necessary to make filled heads; the major component contains an additional protein, the one specified by gene 9. Thus, it would appear that the head structure of the T7 particle contains the proteins specified by genes 8, 10, and 14 to 16.

Are the empty heads precursors to filled heads or a by-product? It seems likely that the major species, which contains the protein specified by gene 9, is a precursor to filled heads. In T7 [but not in T4 (16) or lambda (38)], late proteins are synthesized in normal amounts whether or not phage DNA is synthesized (11). A mutant with a defect in gene 5, for example, makes no T7 DNA but makes normal amounts of class II and III proteins; and it also makes a large amount of empty heads. The empty heads made in the absence of T7 DNA synthesis appear identical to the major species of empty head found in wild-type lysates; that is, they contain the proteins specified by genes 8 to 10 and 14 to 16. Virtually none of them lacks the gene 9 protein. It seems likely that gene 9 protein is somehow lost during the filling of the heads, and the empty heads which lack the gene 9 protein may be the by-products of abortive attempts to fill the heads.

Gene 10 specifies the major subunit of the head, a protein having a molecular weight of approximately 38,000 which accounts for more than 60 percent of the mass of the phage particle (8). However, mutants with defects in gene 10 also fail to synthesize at least one additional protein, which has a molecular weight of approximately 45,000 (Fig. 2). This protein is also found in the head, but in an amount equal to approximately 10 percent that of the major subunit (11). The two gene 10 proteins may be related by a cleavage step, as is found for certain proteins in T4 head assembly (47, 48), but such a relationship has not been shown directly.

The tail of T7 contains at least three proteins, those specified by genes 11, 12, and 17 (8, 11). Noninfectious, DNA-containing particles from mutants in genes 11 or 12 lack all three proteins, and also lack any visible tail structure when examined in the electron microscope. Particles from mutants defective in gene 17, on the other hand, contain the proteins specified by genes 11 and 12 but not the protein specified by gene 17; and they possess a tail structure which looks similar, but not identical to that of normal T7. These particles can all be complemented in vitro to produce infectious T7 particles (13), by procedures similar to those introduced by Edgar and Wood (49) in the study of T4 assembly.

It is not clear where the proteins specified by genes 7 and 13 fit into the structure of the T7 phage particle. The DNA-containing particles produced by mutants with defects in genes 11, 12, and 17 have not been carefully tested for the presence of the proteins specified by genes 7 and 13, so it is not known whether one or both may form a part of the tail structure. However, the DNA-containing particles produced by mutants defective in genes 7 or 13 seem to lack only the protein specified by gene 7 or 13, and they do have a tail structure in electron micrographs (8, 11). Attempts to complement the gene 7 or 13 particles in vitro have been unsuccessful (13). Apparently, the T7 particle can be assembled without the proteins specified by genes 7 and 13, but these proteins cannot be added after assembly is complete.

As was mentioned previously, replication of T7 DNA proceeds through intermediate forms that are subsequently processed to give mature T7 DNA molecules (39, 40). Mutants in any of genes 8, 9, 10, 18, or 19 do not produce mature T7 DNA from the intermediate forms (14, 39, 50). Thus, maturation of T7 DNA seems to require a minimal head structure (containing the proteins specified by genes 8, 9, and 10) and the functions of genes 18 and 19. This is similar to the situation in lambda and T4, where some kind of head structure seems to be required for maturation of the DNA (38, 51). As yet, no specific function has been assigned to either gene 18 or 19.

Summary

Amber mutations or deletion mutations have been found in 25 genes of T7, and the proteins specified by 23 of these genes have been identified by electrophoresis on polyacrylamide gels in the presence of SDS. These genes account for more than 90 percent of the coding capacity of T7 DNA, but there are a few genes in which mutations have yet to be found.

The genetic map of T7 is linear, and all transcription and translation proceed from left to right. Three classes of T7 genes can be distinguished by their times of expression, and the different classes lie in distinct regions of the T7 DNA: class I takes up the left-most 20 percent, class II the next 20 to 25 percent, and class III the right-hand 55 to 60 percent. Class I proteins are the first to appear after infection, followed by class II and III proteins. Class II proteins may appear slightly earlier than those of class III, but the main distinction is that the synthesis of class II proteins stops midway through the infection whereas class III proteins continue to be synthesized until lysis.

After infection, the class I genes are transcribed by the RNA polymerase of the host. A stop signal located approximately 20 percent from the left end of the T7 DNA normally prevents the host RNA polymerase from continuing down the molecule to transcribe the class II and III genes. One of the class I genes, gene 1, specifies a new RNA polymerase which transcribes the class II and III genes.

Something is known about the function of most of the T7 genes. Class I and II genes seem to be concerned mainly with the control of gene expression and with DNA metabolism, that is, breakdown of host DNA and synthesis of T7 DNA. The proteins of the mature phage particle are all specified by class III genes, as are some proteins involved in the assembly of the phage particle.

The availability of mutants defective in most T7 functions and the prospect of soon having a mutation in every T7 gene makes this phage ideal for the study of molecular mechanisms underlying control of gene expression, synthesis of DNA, genetic recombination, and the assembly of virus particles.

References and Notes

1. M. Demerec and U. Fano, *Genetics* **30**, 119 (1945).
2. M. Delbrück, *Biol. Rev. Cambridge Phil. Soc.* **21**, 30 (1946).
3. D. Fraser and R. C. Williams, *J. Bacteriol.* **65**, 458 (1953).
4. S. B. Dubin, G. B. Benedek, F. C. Bancroft, D. Freifelder, *J. Mol. Biol.* **54**, 547 (1970).
5. K. D. Lunan and R. L. Sinsheimer, *Virology* **2**, 455 (1956).
6. D. A. Ritchie, C. A. Thomas, Jr., L. A. MacHattie, P. C. Wensink, *J. Mol. Biol.* **23**, 365 (1967).
7. A. D. Hershey, E. Burgi, L. Ingraham, *Proc. Nat. Acad. Sci. U.S.* **49**, 748 (1963); H. Ris and B. L. Chandler, *Cold Spring Harbor Symp. Quant. Biol.* **28**, 1 (1963); L. A. MacHattie and C. A. Thomas, Jr., *Science* **144**, 1142 (1964).
8. F. W. Studier and J. V. Maizel, Jr., *Virology* **39**, 575 (1969).
9. J. V. Maizel, Jr., in *Fundamental Techniques in Virology*, K. Habel and N. P. Salzman, Eds. (Academic Press, New York, 1969), p. 334.
10. A. L. Shapiro, E. Vinuela, J. V. Maizel, Jr., *Biochem. Biophys. Res. Commun.* **28**, 815 (1967).
11. F. W. Studier, unpublished observations.
12. R. Hausmann and B. Gomez, *J. Virol.* **1**, 779 (1967).
13. F. W. Studier, *Virology* **39**, 562 (1969).
14. ——— and R. Hausmann, *ibid.*, p. 587.
15. A. S. Sarabhai, A. O. W. Stretton, S. Brenner, S. Bolle, *Nature* **201**, 13 (1964); A. O. W. Stretton and S. Brenner, *J. Mol. Biol.* **12**, 456 (1965).
16. R. H. Epstein, A. Bolle, C. M. Steinberg, E. Kellenberger, E. Boy de la Tour, R. Chevalley, R. S. Edgar, M. Susman, G. H. Denhardt, A. Lielausis, *Cold Spring Harbor Symp. Quant. Biol.* **28**, 375 (1963).
17. J. S. Parkinson and R. J. Huskey, *J. Mol. Biol.* **56**, 369 (1971).
18. D. A. Ritchie and F. E. Malcolm, *J. Gen. Virol.* **9**, 35 (1970).
19. F. W. Studier and M. N. Simon, unpublished observations.
20. R. W. Davis and N. Davidson, *Proc. Nat. Acad. Sci. U.S.* **60**, 243 (1968); B. C. Westmoreland, W. Szybalski, H. Ris, *Science* **163**, 1343 (1969); R. W. Davis, M. N. Simon, N. Davidson, in *Methods in Enzymology*, L. Grossman and K. Moldave, Eds. (Academic Press, New York, 1971), vol. 21, part D, p. 413.
21. Y. Masamune, G. D. Frenkel, C. C. Richardson, *J. Biol. Chem.* **246**, 7874 (1971).
22. M. Gellert and M. L. Bullock, *Proc. Nat. Acad. Sci. U.S.* **67**, 1580 (1970).
23. R. W. Davis and R. W. Hyman, personal communication.
24. M. Chamberlin, J. McGrath, L. Waskell, *Nature* **228**, 227 (1970).
25. W. C. Summers and W. Szybalski, *Virology* **34**, 9 (1968).
26. W. C. Summers, *ibid.* **39**, 175 (1969).
27. W. A. Newton, J. R. Beckwith, D. Zipser, S. Brenner, *J. Mol. Biol.* **14**, 290 (1965); C. Yanofsky and J. Ito, *ibid.* **21**, 313 (1966); R. G. Martin, D. F. Silbert, D. W. E. Smith, H. J. Whitfield, Jr., *ibid.*, p. 357.
28. R. B. Siegel and W. C. Summers, *ibid.* **49**, 115 (1970).
29. W. C. Summers, personal communication.
30. ———, *J. Mol. Biol.* **51**, 671 (1970).
31. ——— and R. B. Siegel, *Nature* **228**, 1160 (1970).
32. R. W. Davis and R. W. Hyman, *Cold Spring Harbor Symp. Quant. Biol.* **35**, 269 (1970).
33. R. W. Hyman, *J. Mol. Biol.* **61**, 369 (1971).
34. J. W. Roberts, *Nature* **224**, 1168 (1969).
35. R. L. Millette, C. D. Trotter, P. Herrlich, M. Schweiger, *Cold Spring Harbor Symp. Quant. Biol.* **35**, 135 (1970).

36. H. Bremer and D. Yuan, *J. Mol. Biol.* **38**, 163 (1968); H. Manor, D. Goodman, G. S. Stent, *ibid.* **39**, 1 (1969); J. K. Rose, R. D. Mosteller, C. Yanofsky, *ibid.* **51**, 541 (1970).
37. T. G. Morrison and M. H. Malamy, *Nature New Biol.* **231**, 37 (1971).
38. W. F. Dove, *J. Mol. Biol.* **19**, 187 (1966).
39. R. Hausmann and K. LaRue, *J. Virol.* **3**, 278 (1969).
40. T. J. Kelly, Jr., and C. A. Thomas, Jr., *J. Mol. Biol.* **44**, 459 (1969).
41. L. W. Labaw, *J. Bacteriol.* **62**, 169 (1951); *ibid.* **66**, 429 (1953); F. W. Putnam, D. Miller, L. Palm, E. A. Evans, Jr., *J. Biol. Chem.* **199**, 177 (1952).
42. M. S. Center, F. W. Studier, C. C. Richardson, *Proc. Nat. Acad. Sci. U.S.* **65**, 242 (1970).
43. P. D. Sadowski and C. Kerr, *J. Virol.* **6**, 149 (1970).
44. M. S. Center and C. C. Richardson, *J. Biol. Chem.* **245**, 6285 (1970); *ibid.*, p. 6292; P. D. Sadowski, *ibid.* **246**, 209 (1970).
45. G. D. Frenkel and C. C. Richardson, personal communication; C. Kerr and P. D. Sadowski, *J. Biol. Chem.* **247**, 305 (1972); *ibid.*, p. 311.
46. P. Grippo and C. C. Richardson, *J. Biol. Chem.* **246**, 6867 (1971).
47. U. K. Laemmli, *Nature* **227**, 680 (1970).
48. J. Hosada and R. Cone, *Proc. Nat. Acad. Sci. U.S.* **66**, 1275 (1970); R. C. Dickson, S. L. Barnes, F. A. Eiserling, *J. Mol. Biol.* **53**, 461 (1970).
49. R. S. Edgar and W. B. Wood, *Proc. Nat. Acad. Sci. U.S.* **55**, 498 (1966).
50. R. Schlegel, personal communication.
51. F. R. Frankel, *Proc. Nat. Acad. Sci. U.S.* **59**, 131 (1968); A. G. Mackinlay and A. D. Kaiser, *J. Mol. Biol.* **39**, 679 (1969).
52. M. S. Reid and R. L. Bieleski, *Anal. Biochem.* **22**, 374 (1968).
53. Research was carried out at Brookhaven National Laboratory under the auspices of the U.S. Atomic Energy Commission.

Role of RNA Polymerase in Sporulation

RICHARD LOSICK, ABRAHAM L. SONENSHEIN,[*] ROSALIND G. SHORENSTEIN, AND CAROLINE HUSSEY

The Biological Laboratories, Harvard University, Cambridge, Massachusetts

[*] *Department of Biology, Massachusetts Institute of Technology, Cambridge, Massachusetts*

Bacterial sporulation represents a simple example of cellular differentiation. As cells pass from the vegetative state to the sporulating state, they undergo dramatic changes in morphology and physiology. Sporulating cells synthesize enzymes and messenger RNA (mRNA) molecules not found in vegetative cells (Deutscher and Kornberg, 1969; Bach and Gilvarg, 1966; Doi and Igarashi, 1964; Yamagishi and Takahashi, 1968). In addition, many vegetative enzymes and mRNA molecules disappear early during spore formation. It is likely that these changes are the result of the expression of new classes of genes during sporulation, and the turn-off of certain vegetative genes, though little is known about the molecular mechanism of this changeover.

Recently, Travers (1969, 1970) and Summers and Siegel (1969) have suggested that changes in the template specificity of RNA polymerase after phage infection are responsible for the expression of new phage genes and possibly the turnoff of host genes. After infection of *E. coli* by phage T4, a new RNA polymerase σ factor is made which replaces the σ factor of uninfected cells and directs the transcription of "delayed-early" phage genes.

We have suggested that a similar mechanism controls gene expression during sporulation (Losick and Sonenshein, 1969). We postulated that a vegetative RNA polymerase σ factor directs the transcription of vegetative genes. During sporulation this factor does not function and therefore vegetative genes are turned off. On the other hand, the synthesis of sporulation σ factors would direct the transcription of genes responsible for spore formation. To test this hypothesis we have taken advantage of the properties of the virulent *B. subtilis* phage ϕe (Sonenshein and Roscoe, 1969). When this phage infects vegetative cells, it gives rise to a large burst of progeny. During sporulation the host bacteria cease to support multiplication of ϕe or the expression of at least three early phage genes. The failure of ϕe to grow on sporulating cells is not due to the inability of the phage to inject its DNA or to the destruction of the genome in sporulating cells since the phage genome is incorporated into the mature spore and expressed after germination. Furthermore, this failure is specifically associated with the sporulation process since certain non-sporulating mutants of *B. subtilis* continue to support the growth of phage ϕe during late stationary phase.

One interpretation of the behavior of this phage is that transcription of the ϕe genome, like transcription of some vegetative genes of *B. subtilis*, is dependent on a vegetative RNA polymerase σ factor which does not function in sporulating cells. It has been possible to use ϕe DNA to demonstrate that RNA polymerase changes in template specificity early in the process leading to sporulation (Losick and Sonenshein, 1969). While RNA polymerase purified from vegetative cells actively transcribes ϕe DNA, sporulation polymerase is inactive with the same template. We have been able to provide evidence that this change is critical for sporulation since mutations that alter the RNA polymerase in such a way that it retains the ability to transcribe ϕe DNA cause failure to sporulate. These mutants also support the growth of ϕe during late stationary phase.

The vegetative RNA polymerase depends on a σ factor of 57,000 mol wt for the transcription of ϕe DNA. We have reversibly fractionated vegetative polymerase into the σ factor and a core enzyme by phosphocellulose chromatography. The core enzyme transcribes the synthetic template poly dAT (poly[deoxyadenylate-thymidylate]) but not ϕe DNA; transcription of the phage DNA is restored by addition of the σ factor to the vegetative core. Phosphocellulose enzyme from sporulating cells contains an altered β polypeptide and does not transcribe ϕe DNA even after addition of σ factor from vegetative RNA polymerase. The failure of sporulation phosphocellulose enzyme to complement vegetative σ factor may be responsible for the loss of vegetative template specificity.

CHANGE IN THE TEMPLATE SPECIFICITY OF RNA POLYMERASE DURING SPORULATION

Time course of the change in template specificity. DNA-dependent RNA polymerase was purified from cells of *B. subtilis* 3610 harvested at various times during growth and sporulation. The cells were disrupted by sonication, centrifuged at high speed, and RNA polymerase partially purified by ammonium sulfate fractionation. The enzyme was assayed for its ability to transcribe φe DNA and the synthetic template, poly dAT. Other samples of the culture were infected with phage φe and the average burst size was determined. Cells grew logarithmically for 2.5 hr, and then they entered stationary phase and began the process of sporulation. Prespores could be detected 6 hr after the end of logarithmic growth. The burst size of phage φe decreased rapidly during the first hour of stationary phase and, after an additional 2 hr, was less than two phage per cell.

The specific activity of RNA polymerase with φe DNA as template also decreased to nearly zero early in stationary phase, and the loss of ability of RNA polymerase to transcribe φe DNA in vitro closely parallels the decrease in phage burst size (Fig. 1). In contrast, RNA polymerase retained the ability to transcribe poly dAT at all times. The specific activity with the synthetic polymer as template was never less than 30% of the value for the vegetative enzyme and, at a late time, it rose to a value higher than in the vegetative phase. We conclude that the inability of φe to grow on sporulating cells might arise from the failure of sporulation RNA polymerase to transcribe φe DNA and that the change in template specificity occurs early in sporulation.

A rifampicin-sensitive component of RNA polymerase is conserved during sporulation. We have suggested that RNA polymerase loses vegetative template specificity during sporulation because the vegetative σ factor does not function or is destroyed in sporulating cells while the core enzyme, the machinery for polymerizing RNA, is conserved. Alternatively, an entirely new RNA polymerase might be synthesized during spore formation which replaces the vegetative enzyme. To rule out this possibility, rifampicin-resistant mutants of *B. subtilis* were isolated. Rifampicin is a specific antagonist of RNA polymerase core from both vegetative and sporulating cells of *B. subtilis*, and rifampicin-resistant mutants of *B. subtilis* contain an altered RNA polymerase which is resistant to the drug in vitro (Losick and Sonenshein, unpubl. results; Geiduschek and Sklar, 1969). Cells of one such mutant grown in the absence of rifampicin were harvested during sporulation (Losick and Sonenshein, 1969). Sporulation polymerase purified from these mutant cells was completely resistant to rifampicin. Thus a mutation that renders vegetative RNA polymerase resistant to rifampicin also results in a resistant sporulation polymerase. This strongly suggests that, while the template specificity of RNA polymerase changes during sporulation, the rifampicin-sensitive component of the vegetative enzyme is conserved.

Difference in template specificity is inherent in the RNA polymerases. Vegetative and sporulation RNA polymerases purified by ammonium sulfate fractionation and DEAE-cellulose column chromatography were analyzed on glycerol density

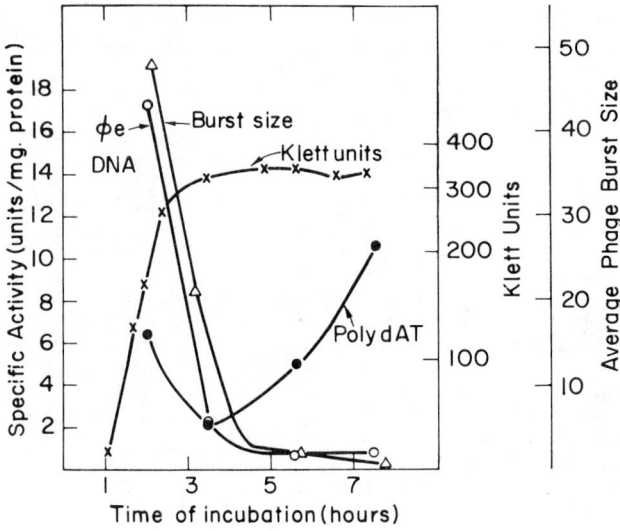

FIGURE 1. Time course of change in template specificity during sporulation. Cells were grown in 2 liters of medium 121 B (Sonenshein and Roscoe, 1969) after synchronization by dilution. Growth was followed by measuring turbidity at 540 mμ with a Klett-Summerson photoelectric colorimeter. The appearance of prespores was measured by phase contrast microscopy. Samples of 450 ml were harvested at 2, 3.5, 5.6 and 7.5 hr, disrupted by sonication, purified by ammonium sulfate fractionation, and assayed with poly dAT or φe DNA as previously described (Losick and Sonenshein, 1969). Specific activity is defined in the legend to Table 1. Samples of 0.9 ml were taken at 2.2, 3.2, 5.7 and 7.8 hr, infected with φe at low multiplicity, and the burst size was determined according to the method of Sonenshein and Roscoe (1969).

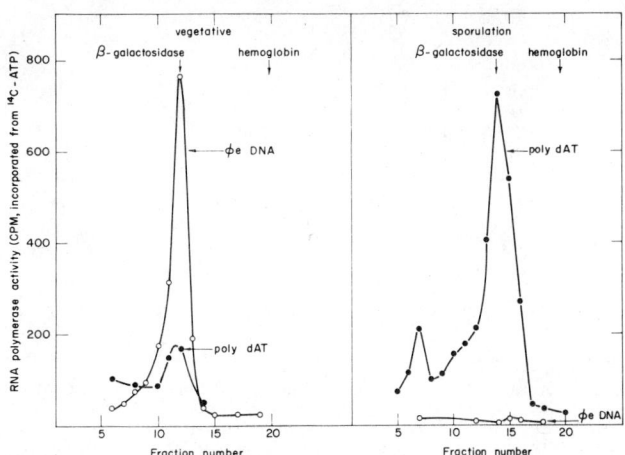

FIGURE 2. Zone centrifugation of RNA polymerase. Samples containing, in 0.5 ml, 40 units (1 unit incorporates 1 mμmole ^{14}C-AMP in 10 min) of vegetative RNA polymerase or 80 units of sporulation enzyme, both purified through the DEAE-cellulose column chromatography step (see legend to Table 1), were layered on 12 ml linear gradients of 10 to 30% glycerol in buffer containing 0.01 M Tris-HCl pH 7.9, 0.01 M $MgCl_2$, 0.0001 M EDTA and 0.0001 M dithiothreitol. Following centrifugation for 13 hr at 40,000 rpm in an International SB 283 rotor at 4°C, fractions were collected and 0.025 ml of each fraction was assayed as described previously (Losick and Sonenshein, 1969). Two molecular weight markers, hemoglobin (4.6 S) and β-galactosidase (16 S), were centrifuged on a parallel gradient.

gradients (Fig. 2). A sedimentation coefficient of 15.4 S was estimated for the vegetative enzyme. Most of the sporulation polymerase sediments at 15 S, but there is another peak at 24 S which may be an aggregate of the 15 S enzyme. In the case of the sporulation polymerase, those fractions of the glycerol gradients which transcribe poly dAT are without activity with φe DNA. The vegetative enzyme remains active with poly dAT and φe DNA after zone centrifugation. In addition, if a sample of the sporulation glycerol gradient enzyme is mixed with an equal amount of vegetative enzyme, there is no inhibition of the ability of vegetative enzyme to transcribe phage φe DNA. These results suggest that the difference in template specificity between vegetative and sporulation polymerases is due to an inherent difference in the enzymes themselves rather than to the presence in sporulating cells of a hypothetical inhibitor of φe DNA transcription.

The vegetative enzyme has been highly purified by the use of calf-thymus DNA-cellulose column chromatography after ammonium sulfate fractionation and DEAE-cellulose column chromatography. After centrifugation of the polymerase through a glycerol density gradient, enzyme activity corresponds to a protein peak. Table 1 shows that the highly purified vegetative enzyme actively transcribes both φe DNA and poly dAT. The sporulation enzyme does not bind well to calf-thymus DNA-cellulose but has been substantially purified by ammonium sulfate fractionation, elution from a DEAE-cellulose column with a linear KCl gradient, and zone centrifugation. After sedimentation of the polymerase through a glycerol gradient, enzyme activity corresponds to a protein shoulder. Table 1 shows that while the purified sporulation enzyme is active with poly dAT as a template, its specific

TABLE 1. SPECIFIC ACTIVITY OF VEGETATIVE AND SPORULATION RNA POLYMERASE

Enzyme	Specific activity (units/mg protein)	
	φe DNA	poly dAT
Vegetative	198	77
Vegetative core	18	320
Sporulation	5	87
Sporulation core	6	250

Vegetative cells were obtained by growing B. subtilis 3610 in medium 121 B (Sonenshein and Roscoe, 1969) and harvesting during mid-log phase. Cells were disrupted by sonication and vegetative RNA polymerase purified by high speed centrifugation, ammonium sulfate fractionation, and DEAE-cellulose column chromatography as previously described (Losick and Sonenshein, 1969). The vegetative enzyme was further purified by calf-thymus DNA-cellulose column chromatography using stepwise elution with 0.15–0.6 M KCl according to the method of Alberts et al. (1968). Finally, the enzyme was sedimented through a 10 to 30% glycerol gradient containing 0.15 M KCl. Vegetative core enzyme was obtained by applying glycerol gradient enzyme to a phosphocellulose column and eluting stepwise between 0.25 and 0.4 M KCl (Burgess et al., 1969). Sporulating cells were also grown in 121 B (Sonenshein and Roscoe, 1969) medium and harvested 6 hr after the end of logarithmic growth. Cells were disrupted by sonication and sporulation enzyme purified by high speed centrifugation and ammonium sulfate fractionation (Losick and Sonenshein, 1969). The enzyme was next applied to a DEAE-cellulose column and eluted with a 0.1 to 0.3 M KCl gradient. The enzyme was further purified by glycerol density gradient centrifugation. Sporulation core enzyme was obtained by phosphocellulose chromatography of the DEAE-cellulose enzyme with a linear KCl gradient. Enzymes were assayed with either φe DNA or poly dAT as previously described (Losick and Sonenshein, 1969). One unit of specific activity is that amount of enzyme which incorporated 1 mμmole of ^{14}C-AMP in 10 min.

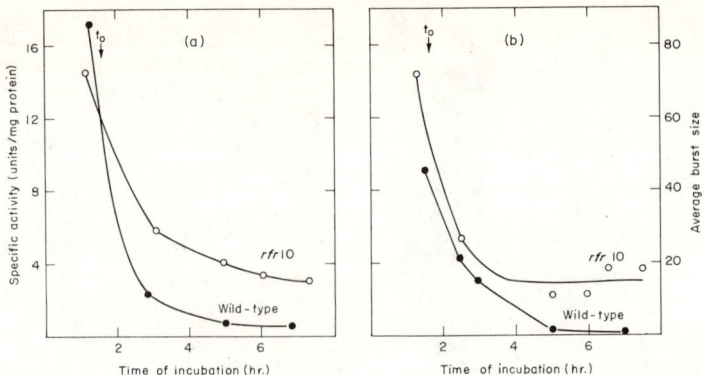

FIGURE 3. Time course for *rfr* 10. Methods are the same as in the legend to Figure 1. RNA polymerase was assayed with ϕe DNA as a template. t_0 indicates the end of logarithmic growth.

activity with ϕe DNA is much lower than in the case of the purified vegetative enzyme.

Thus, early in passing from the vegetative to the sporulating state, the template specificity of the RNA polymerase is altered. This change is not to be attributed to an entirely different enzyme because at least a rifampicin-sensitive component of the enzyme in both states is coded by the same gene. Nor is this change due to a specific antagonist of ϕe DNA transcription in sporulating cells. Instead, early during sporulation, there is an alteration of the RNA polymerase enzyme itself.

RNA Polymerase Mutant Blocking Sporulation

In the course of isolating mutants resistant to rifampicin, we discovered that a frequent class of such mutants also failed to sporulate. This was in contrast to the rifampicin-resistant mutant described above which sporulates normally. The finding that these non-sporulating strains were common among mutants spontaneously resistant to rifampicin suggested that a single mutation caused both drug resistance and inability to sporulate. To show that these were indeed point mutants, revertants that could sporulate were selected by centrifugation through a Renographin gradient (Sonenshein and Roscoe, 1969; Tamir and Gilvarg, 1966). Revertants arose at a frequency consistent with a single mutation (between 10^{-5} and 10^{-6}) and were once again sensitive to rifampicin. One of the non-sporulating rifampicin-resistant mutants known as *rfr* 10 is completely resistant to rifampicin in vitro. Figure 3 shows that RNA polymerase from *rfr* 10 cells harvested at late stationary phase retains substantial ability to transcribe ϕe DNA. *Rfr* 10 cells also continue to support the growth of phage ϕe at times when ϕe fails to grow on wild-type cells (Fig. 3b). Figure 4 shows the activity with ϕe DNA of RNA polymerase from *rfr* 10 and wild-type cells harvested 6 hr into stationary phase. It is clear that RNA polymerase from the mutant cells is much more active with ϕe DNA than polymerase from wild-type cells. RNA polymerase from a sporulating revertant of *rfr* 10 known as *rfr* 10 *rev* 8 is once again sensitive to rifampicin. In addition, RNA polymerase from late stationary phase cells of *rfr* 10 *rev* 8 has much lower activity with ϕe DNA as a template

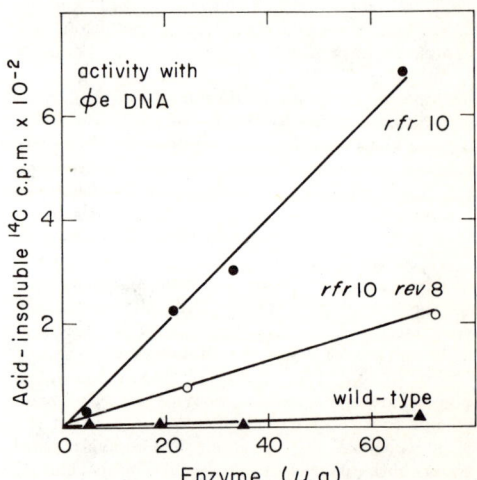

FIGURE 4. Activity of RNA polymerase from stationary phase cells of *rfr* 10 with ϕe DNA. RNA polymerase was purified through the ammonium sulfate step (see legend to Table 1) from cells of *rfr* 10, *rfr* 10 *rev* 8, and wild-type harvested 6 hr into stationary phase. RNA polymerase was assayed with ϕe DNA as previously described (Losick and Sonenshein, 1969).

than RNA polymerase from the parent cells (Fig. 4). Thus, an RNA polymerase mutant of *B. subtilis* retains vegetative template specificity, continues to support the growth of phage φe during late stationary phase, and fails to sporulate as the result of a single mutation. This finding provides direct evidence that the loss of vegetative template specificity by RNA polymerase is responsible for the failure of phage φe to grow on sporulating cells and is essential for the process of sporulation itself.

STRUCTURAL ALTERATION OF RNA POLYMERASE DURING SPORULATION

Transcription of φe DNA requires an RNA polymerase σ factor. Vegetative polymerase was applied to a phosphocellulose column and eluted between 0.25 and 0.4 M KCl. This procedure is known to separate the σ factor of *E. coli* from the core enzyme (Burgess et al., 1969). Table 1 shows that after phosphocellulose chromatography, vegetative polymerase fails to transcribe φe DNA. The vegetative phosphocellulose enzyme has a sedimentation coefficient of 13.5 S compared to 15.4 S for vegetative DNA-cellulose enzyme. This finding also suggests that a component of RNA polymerase has been removed during phosphocellulose chromatography. Figure 5 shows that addition of flow-through of phosphocellulose chromatography restores transcription of φe DNA by vegetative core enzyme. Flow-through was obtained from vegetative enzyme which had been purified either through only the DEAE-cellulose step or the DNA-cellulose step. *E. coli* RNA polymerase transcribes φe DNA and it seemed possible that the *E. coli* σ factor would complement vegetative *B. subtilis* core enzyme. Figure 5 shows that, indeed, *E. coli* σ factor will restore the transcription of φe DNA by vegetative enzyme. Thus vegetative RNA polymerase contains a σ factor similar to the *E. coli* σ factor and necessary for the transcription of φe DNA. Kerjan and Szulmajster (1969) have similarly shown that phosphocellulose chromatography releases a factor from *B. subtilis* polymerase.

Subunit composition of vegetative RNA polymerase. Purified vegetative polymerase was analyzed by SDS-polyacrylamide gel electrophoresis, a procedure which separates polypeptides on the basis of molecular weight. Figure 6 shows the gel pattern for vegetative enzyme and vegetative core enzyme. Vegetative enzyme preparations contain polypeptides of 155,000, 120,000, 57,000 and 45,000 daltons. In gels of phosphocellulose enzyme, the 57,000 dalton subunit is the only one missing. It is found, instead, in gels of the flow-through of phosphocellulose chromatography. The 57,000 dalton subunit, therefore, appears to be the σ factor of vegetative RNA polymerase. Avila et al. (1970) have reported a similar molecular weight for the *B. subtilis* σ factor. In view of our finding that both *B. subtilis* and *E. coli* factors complement *B. subtilis* core enzyme, it is interesting to note that the *B. subtilis* factor is considerably smaller than the *E. coli* factor (95,000 daltons; Burgess et al., 1969).

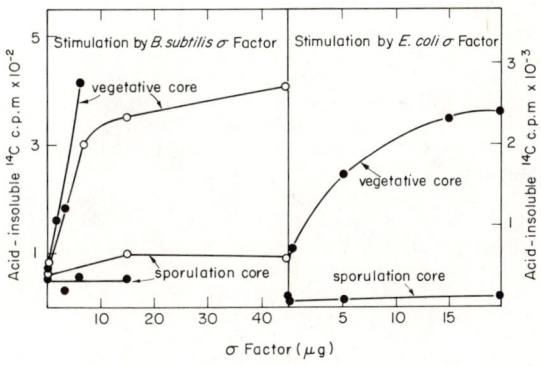

FIGURE 5. Stimulation of core enzyme by σ factor. *B. subtilis* σ factor was obtained by phosphocellulose chromatography of vegetative enzyme purified either only through the DEAE-cellulose step (—○—) or through the DNA-cellulose step (—●—). Varying amounts of flow-through from the phosphocellulose chromatography were added to 3 µg of vegetative core enzyme or 4 µg of sporulation core enzyme and assayed as previously described with φe DNA as a template (Losick and Sonenshein, 1969). In the case of σ factor derived from DEAE-cellulose enzyme, a background value of 7 count/min/µg of flow-through has been subtracted due to transcription of φe DNA in the absence of added core enzyme. A value of 39 count/min/µg of flow-through has been subtracted in the case of σ factor derived from DNA-cellulose enzyme. This background incorporation is due to small amounts of core enzyme present in the phosphocellulose flow-through. *E. coli* σ factor was the gift of R. R. Burgess and was assayed in the presence of 3 µg of vegetative core or 4 µg of sporulation core. A background value of 11 count/min/µg of σ factor was subtracted.

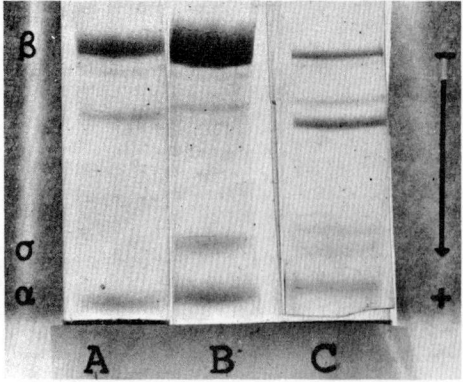

FIGURE 6. SDS-polyacrylamide gel electrophoresis patterns of vegetative and sporulation RNA polymerase. From left to right: (A) vegetative core; (B) vegetative glycerol gradient enzyme; (C) sporulation core. Samples were carboxymethylated according to the method of Pringle (1970) before electrophoresis. Carboxymethylation destroys proteases which may be present in trace amounts and are resistant to SDS and attack other proteins during denaturation in SDS. Gels containing 0.1 % SDS and 5% acrylamide were run for 3 hr at 8 ma per gel according to the procedure of Weber and Osborn (1969). The gels were then stained with Coomassie brilliant blue and destained electrophoretically according to Weber and Osborn (1969).

Electrophoresis for extended times resolves the 155,000 dalton band into two polypeptides (Fig. 7) which probably correspond to the β and β' polypeptides of *E. coli* RNA polymerase (about 160,000 daltons; Burgess, 1969). The 45,000 dalton band presumably represents the α polypeptide (40,000 daltons in *E. coli*; Burgess, 1969). The molar ratio of the 45,000 and 155,000 dalton bands calculated from densitometer tracings is 1:1.1, which is in good agreement with the $\alpha_2\beta\beta'$ structure of *E. coli* RNA polymerase (Burgess, 1969). The 120,000 dalton polypeptide is probably a contaminating protein: it is present in variable amounts in different preparations and is absent in at least some preparations that are active with both poly dAT and ϕe DNA. It appears, therefore, that vegetative RNA polymerase from *B. subtilis* has a subunit composition similar to that of the *E. coli* enzyme, consisting of α, β, β' and σ polypeptides.

Vegetative σ factor does not complement sporulation "core" enzyme. Sporulation enzyme purified through the DEAE-cellulose step was applied to a phosphocellulose column. The flow-through of the column did not stimulate transcription of ϕe DNA by vegetative core enzyme. Thus, if sporulation enzyme contained vegetative σ factor, this factor is not released by phosphocellulose chromatography. Sporulation enzyme was eluted from the column with a linear KCl gradient. Enzyme activity coincided with a protein peak; the enzyme is, therefore, highly purified. We shall provisionally call the activity purified by phosphocellulose sporulation "core" enzyme. Figure 5 shows that sporulation core enzyme does not transcribe phage ϕe DNA even after addition of σ factor derived from either vegetative *B. subtilis* polymerase or from *E. coli* polymerase.

Sporulation RNA polymerase contains an altered β polypeptide. Sporulation core enzyme contains polypeptides of 155,000 and 45,000 daltons and small amounts on a molar basis of a 120,000 mol wt species (Fig. 6). These polypeptides are similar in molecular weight to corresponding polypeptides of vegetative core enzyme. In addition, there is in the sporulation core enzyme a polypeptide of 110,000 mol wt. The following evidence indicates that this new polypeptide is present in place of one of the β polypeptides: (a) the 155,000 dalton band of sporulation enzyme does not separate into two polypeptides after extended electrophoresis as does the vegetative β band (Fig. 7); (b) the molar ratio of the 45,000 dalton band to the 110,000 and the 155,000 dalton polypeptides is 1:0.54:0.47. Thus, there is approximately only half as much of the 155,000 subunit species in the sporulation enzyme as in the vegetative enzyme. These findings suggest that one of the 155,000 dalton polypeptides is missing and is replaced in sporulation polymerase by a new subunit of 110,000 mol wt. Sporulation core also contains small amounts of material of about 60,000 daltons. We therefore cannot conclude that sporulation core enzyme is entirely free of either vegetative or sporulation σ factors.

To test the possibility that the structural alteration of RNA polymerase takes place in vitro

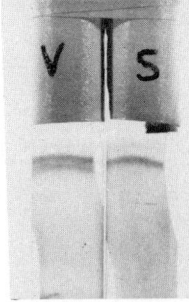

FIGURE 7. Separation of β and β' by polyacrylamide gel electrophoresis for extended time. (V) Vegetative enzyme; (S) sporulation enzyme. SDS gels were run as described in the legend to Fig. 6, except that electrophoresis was for 8 hr. Only the 155,000 dalton bands are shown.

φe DNA, however, rRNA synthesis continues during stationary phase. One possibility is that expression of the rRNA genes requires the vegetative template specificity of RNA polymerase and that rfr 10 does not turn off the rRNA genes during stationary phase because it retains the vegetative template specificity. Assuming that the vegetative RNA polymerase σ factor which we have described directs the transcription of these vegetative genes, the molecular event responsible for the turnoff of vegetative genes might be a structural alteration of RNA polymerase. On the other hand, many new genes are almost certainly transcribed during sporulation. This transcription may be directed by specific sporulation factors. We are currently searching for such factors.

ACKNOWLEDGMENTS

We thank K. Weber, S. E. Luria, R. R. Burgess, A. A. Travers, J. L. Strominger, and J. Pero for helpful discussions. R. L. is a Junior Fellow of the Harvard Society of Fellows, R. G. S. is a predoctoral fellow of the National Science Foundation, and A. L. S. is a postdoctoral fellow of the American Cancer Society. This work was supported by a U.S. National Science Foundation grant to R. L. and J. D. Watson and grants from the U.S. National Institutes of Health and the U.S. NSF to S. E. Luria.

REFERENCES

ALBERTS, B. M., F. J. AMODIO, M. JENKINS, E. D. GUTMANN, and R. L. FERRIS. 1968. Studies with DNA-cellulose chromatography, 1. DNA-binding proteins from *Escherichia coli*. Cold Spring Harbor Symp. Quant. Biol. *33:* 289.

AVILA, J., J. J. HERMOSO, E. VINUELA, and M. SALAS. 1970. Subunit composition of *B. subtilis* RNA polymerase. Nature *226:* 1244.

BACH, M. and C. GILVARG. 1966. Biosynthesis of dipicolinic acid in sporulating *Bacillus megaterium*. J. Biol. Chem. *241:* 4563.

BURGESS, R. R. 1969. Separation and characterization of the subunits of ribonucleic acid polymerase. J. Biol. Chem. *244:* 6168.

BURGESS, R. R., A. A. TRAVERS, J. J. DUNN, and E. K. F. BAUTZ. 1969. Factor stimulating transcription by RNA polymerase. Nature *221:* 43.

DEUTSCHER, M. P. and A. KORNBERG. 1969. Biochemical studies of bacterial sporulation and germination. VIII. Patterns of enzyme development during growth and sporulation of *Bacillus subtilis*. J. Biol. Chem. *243:* 4653.

DOI, R. H. and R. T. IGARASHI. 1964. Genetic transcription during morphogenesis. Proc. Nat. Acad. Sci. *52:* 755.

GEIDUSCHEK, E. P. and J. SKLAR. 1969. Role of host RNA polymerase in phage development. Nature *221:* 833.

KERJAN, P. and J. SZULMAJSTER. 1969. DNA-dependent RNA polymerase from vegetative cells and from spores of *B. subtilis*. III. Isolation of a stimulating factor. FEBS Letters *5:* 288.

LOSICK, R. and A. L. SONENSHEIN. 1969. Change in the template specificity of RNA polymerase during sporulation. Nature *224:* 35.

PRINGLE, J. R. 1970. The molecular weight of the undegraded polypeptide chain of yeast hexokinase. Biochem. Biophys. Res. Commun. *39:* 46.

RABUSSAY, D. and W. ZILLIG. 1969. A rifampicin-resistant RNA polymerase from *E. coli* altered in the β-subunit. FEBS Letters *5:* 104.

SADOFF, H. L. and E. CELIKKOL. 1968. Modification of the fructose 1,6-diphosphate aldolase of *Bacillus cereus* by limited proteolysis, p. 25. Bacteriol. Proc.

SONENSHEIN, A. L. and D. H. ROSCOE. 1969. The course of phage φe infection in sporulating cells of *Bacillus subtilis* 3610. Virology *39:* 265.

SUMMERS, W. and R. B. SIEGEL. 1969. Control of template specificity of *E. coli* RNA polymerase by a phage-coded protein. Nature *223:* 1111.

TAMIR, H. and C. GILVARG. 1966. Density gradient centrifugation for the separation of sporulating forms of bacteria. J. Biol. Chem. *241:* 1085.

TRAVERS, A. A. 1969. Bacteriophage sigma factor for RNA polymerase. Nature *223:* 1108.

———. 1970. Positive control of transcription by bacteriophage sigma factor. Nature *225:* 1009.

WALTER, G., W. SEIFERT, and W. ZILLIG. 1968. Modified DNA-dependent RNA polymerase from *E. coli* infected with bacteriophage T4. Biochem. Biophys. Res. Commun. *30:* 240.

WEBER, K. and M. OSBORN. 1969. The reliability of molecular weight determinations by dodecyl sulfate polyacrylamide gel electrophoresis. J. Biol. Chem. *244:* 4401.

YAMAGISHI, H. and I. TAKAHASHI. 1968. Genetic transcription during morphogenesis. Biochim. Biophys. Acta *155:* 150.

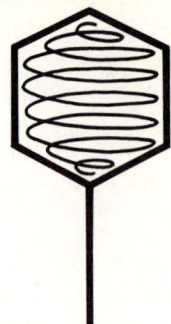

Chapter 11

Repressor and Its Action

MARK PTASHNE

*The Biological Laboratories, Harvard University,
Cambridge, Massachusetts*

IN 1961 Jacob and Monod elaborated their operon concept, and described the phenomenon of lysogeny as a typical manifestation of gene control by repressors. They proposed that the product of the cI gene of phage λ directly blocks early functions required for DNA synthesis. Furthermore, the same repressor required for maintenance of the lysogenic state would also confer immunity against superinfecting phages of the prophage type by turning off their genes. Induction of the prophage would follow upon inactivation of the repressor, the presumed consequence of treatment with inducing agents such as ultraviolet light.

The argument for the existence of the λ repressor was analogous to that advanced for the existence of the *lac* repressor (the product of the i gene of the *lac* operon). Genetic experiments with both *lac* and λ supported these assertions: (1) there is in each case a single regulatory gene, (2) the active allele of the regulatory gene turns off the genes of the corresponding operon ("negative" control), and (3) *cis*-dominant mutations are found that modify the molecular targets (operators) of the repressors so that the genes ordinarily under repressor control can function even in the presence of repressor. Jacob and Monod inferred that the product of the regulatory gene acts directly on an operator to repress the genes of the repressor-controlled operon. They speculated that the operators might be DNA, and that repressors might block transcription. They at first suggested, on the basis of reports that repressor synthesis proceeds in the presence of inhibitors of protein synthesis, that repressor might be RNA.

The genetic data of Jacob and Monod were subject to alternative explanations that did not require the cI or i products to interact with operators. Brenner (1965) pointed out, for example, that the cI and i products could be enzymes that synthesize corepressors, the "true" repressor genes being as yet undetected. Given the repressor hypothesis in its general form, one was still faced with the problem of whether the repressor acts directly on the DNA to block transcription or whether it prevents translation of messenger RNA

221

(Jacob and Monod, 1961; Stent, 1964). Proof of the repressor hypothesis therefore required a biochemical demonstration that the product of a regulatory gene specifically recognizes the operator. Identification of the operator would in turn show at what molecular level the repressor works.

Three repressors have now been isolated, the *lac* repressor (Gilbert and Müller-Hill, 1966), the λ repressor (Ptashne, 1967a), and the 434 repressor (Pirrotta and Ptashne, 1969). Each is a protein, and each binds specifically and with high affinity to regions of DNA known to contain the appropriate operators (Ptashne, 1967b; Gilbert and Müller-Hill, 1967; Pirrotta and Ptashne, 1969). The λ repressor binds independently to the two operators o_L and o_R (Fig. 1) (Ptashne and Hopkins, 1968). In vitro this binding blocks transcription of two operons read in opposite directions beginning at the promoters p_L and p_R (Chadwick et al., 1970; Steinberg and Ptashne, 1971). One class of virulent mutants of λ has been shown to bear mutations in both o_L and o_R that decrease the affinities of these operators for repressor (Ptashne and Hopkins, 1968; Steinberg and Ptashne, 1971).

An Overview of Gene Control in λ

Our current picture of gene control in λ explains how repressor acting at only two sites turns off all the other phage genes. Figure 1 is a simplified map of the λ chromosome. The arrows show the direction of transcription for each region of DNA. Note the following three groups of genes: the late genes, which determine phage structural components; the recombination genes, which determine enzymes that catalyze recombination between λ DNA molecules and between the λ and host chromosomes; and genes O and P, the products of which are required for λ DNA replication. (Gene R, another late gene, encodes a lysin.) In addition, there are four regulatory genes shown on the map, Q, N, cI, and *tof*, which probably function as follows. N and Q products are positive regulators, that is, they turn on transcription of other phage genes. In particular, Q acts at a single site between Q and R to turn on the late genes (Herskowitz and Signer, 1970). (λ DNA becomes circular after infection, and hence the other late genes of Fig. 1 are found

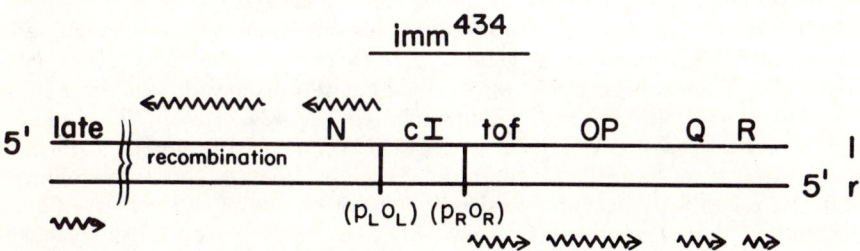

FIGURE 1. A simplified genetic and physical map of λ. The arrows show the directions of transcription of some of the genes from the DNA strands labeled l and r. The N and *tof* messengers are synthesized beginning at the promoters p_L and p_R, respectively. Repressor blocks this transcription by binding to the operators o_L and o_R. imm^{434} delineates the segment of 434 inserted into λimm^{434}.

11. Repressor

adjacent to R during transcription.) N product acts at three places: to the left of Q to turn on Q; to the left of N to turn on the recombination genes; and to the right of *tof* to turn on genes O and P (Hopkins, 1970; Butler and Echols, 1970). In the absence of N product, therefore, the only messages synthesized are the transcripts of genes N and *tof*. These genes are transcribed in opposite directions from different DNA strands beginning at the promoters p_L (leftward promoter) and p_R (rightward promoter) (Roberts, 1969). It is these two operons that are repressed by the cI product. See Chapter 12 for a discussion of the role of *tof*, and Chapters 10 and 13 for a more complete discussion of gene control in λ.

Before describing the experiments with isolated phage repressors, I will review some of the genetic experiments that made the isolation possible.

Lysogeny and Immunity

In 1957 Kaiser found that mutations in any of the three cistrons cI, cII, or $cIII$ specifically inhibit lysogenization; cII and $cIII$ mutants lysogenize rarely, and cI mutants not at all. These mutants form clear or faintly turbid plaques on sensitive cells but do not grow in λ lysogens, which shows that although they themselves cannot lysogenize they remain sensitive to immunity. Mixed infection of a sensitive cell with wild-type λ and any of these clear mutants yields cells lysogenic for both phages; thus the wild-type alleles are dominant over the mutant alleles. Furthermore, from cells infected with pairs of mutants one may isolate lysogens carrying cII or $cIII$ mutants only, but not cI mutants only. These results suggested to Kaiser that the products of all three genes are needed to establish lysogeny, but that only cI product is required for its maintenance. This interpretation, now generally accepted, was later supported by the finding that genes cII and $cIII$ do not function in immune lysogens (Bode and Kaiser, 1965).

In 1957 Kaiser and Jacob showed that cI plays the determining role in the production of immunity. They analyzed the genetic differences between λ and the closely related phage 434. These two phages are said to be heteroimmune because cells carrying 434 prophage are immune to superinfection with 434, but remain sensitive to λ, and vice versa for λ lysogens. Kaiser and Jacob found that the 434 immunity determinant could be genetically recombined with any known gene of λ with one exception: cI. They constructed a phage whose DNA was derived entirely from λ except for a short segment of the chromosome which included the 434 cI gene. This phage, λimm^{434}, has the immunity of phage 434. The DNA segment from 434 substituted into λimm^{434}, the immunity region, includes the repressor gene and the operators that are sensitive to that repressor. Kaiser and Jacob concluded: "the cI region controls not only a reaction involved in the lysogenization process, but also the immunity pattern.... The question arises, therefore, whether these properties are but different expressions of the same function."

One source of confusion remained to be resolved: among phages related to λ, the specificity of immunity seemed to be correlated with the location of the prophage in the bacterial chromosome, raising the possibility that immunity might be the direct effect of occupation of the site of prophage attachment (Kaiser and Jacob, 1957). However, the fact that cells diploid for the λ attachment site and carrying only one prophage were immune argued against this alternative (Jacob et al., 1960), and the discovery of a phage mutant ($\lambda b2$) that lysogenizes bacteria without attaching to the host chromosome argued strongly that chromosomal attachment of the prophage is not required for repression of the phage genes or for the establishment of immunity (Zichichi and Kellenberger, 1963). Bertani (1956, 1958) had reached this conclusion earlier on the basis of his experiments with phage P2. It is now clear, of course, that prophage insertion and immunity depend on different molecular processes (Chapter 6).

The proposition that prophage genes are controlled by repressor explained the earlier observation that introduction of a prophage into a nonlysogenic cell during bacterial mating results in induction of phage growth (zygotic induction) whereas no induction occurs if the recipient cell is lysogenic (Jacob and Wollman, 1956).

Several workers (Attardi et al., 1963; Isaacs et al., 1965; Skalka et al., 1967; Sly et al., 1965; Naono and Gros, 1966) showed that control of phage functions could be described in terms of specific messenger RNA synthesis. Thus, induction of a lysogen causes increased synthesis of specific messenger RNA, superinfection of λ lysogens with λ does not, but superinfection of λimm^{434} lysogens with λ does, and so on. In 1967 Taylor et al. analyzed λ messenger by hybridizing RNA with the separated strands of λ DNA. They discovered that, immediately after induction of a lysogen, messenger RNAs are synthesized in opposite directions, probably beginning in or near the immunity region. Other regions of the DNA were transcribed only if protein synthesis was allowed to proceed, whereas these messengers were synthesized in the presence of agents that block protein synthesis. This observation, confirmed by Kourilsky et al. (1968), was consistent with the idea that the repressor directly controls two operons, a possibility raised earlier by Hogness et al. (1966) and supported by the results of Pereira da Silva and Jacob (1967).

The cI Gene

The cI gene and its product were the subject of extensive study before repressor isolation in 1967. The existence of suppressor-sensitive cI mutations (Jacob et al., 1962; Thomas and Lambert, 1962) showed that cI product is at least partly protein, a conclusion supported by the existence of cI temperature-sensitive mutants (Sussman and Jacob, 1962) and by reports that chloramphenicol prevents synthesis of λ repressor (Horiuchi and Inokuchi, 1966; Green, 1966). In 1959, Jacob and Campbell described the important mutant $cIind^-$, which renders the repressor insensitive to the inducing effects of ultraviolet light. More recently Horiuchi and Inokuchi (1967) have isolated $cIind^s$ mutants, which are unusually sensitive to ultraviolet induction. Neither the ind^- nor the ind^s mutation affects the intracellular concentration of repressor (Ogawa and Tomizawa, 1967), and the ind^- mutation has been shown to modify repressor structure (Ptashne, 1967b). These properties of the ind^- and ind^s repressors suggest that cI product interacts directly with an inducer produced by treatments such as ultraviolet irradiation of lysogens.

Lieb (1966a) and Horiuchi and Inokuchi (1967) found that cI temperature-sensitivity mutations fall into two classes. Prophages bearing mutations of type B (class II) are induced only if heated under conditions allowing protein synthesis, whereas prophages bearing mutations of type A (class I) are induced when heated in buffer. Lieb mapped the A mutations in the left part of cI, and the B in the right. Naono and Gros (1966) and Ogawa and Tomizawa (1967) explained the results of these heat-induction experiments by showing that cIB mutants produce repressors that renature if cooled after heat induction, whereas cIA products renature more slowly, if at all. Hence, for thermal induction of type B mutants to occur, phage development must proceed to some irreversible stage before the temperature is lowered. Guha and Szybalski found that all the B repressors they tested renatured completely, albeit at different rates (cited in Szybalski, 1969). In particular they studied the mutant ts1, with a mutation near the center of cI, and found that its repressor renatures unusually slowly. This may explain the reports of Naono and Gros (1966) and Green (1966) that λ mRNA was synthesized in heat-induced ts1 lysogens following cooling.

It is not clear why there is a correlation between renaturability and map position among the temperature-sensitive mutants. There are strong reasons for believing that mutations of types A and B are located in a single cistron. No complementation has been observed between amber mutants located in the two regions, nor between amber mutants and temperature-sensitive mutants located anywhere in cI (Lieb, personal communication). Moreover, as described below, missense mutations in the right and left parts of cI modify the same polypeptide. Therefore, the complementation between cI temperature-sensitive mutants reported by Green (1967) and Lieb (1966a) is probably intracistronic. Horiuchi and Inokuchi (1967) failed to detect complementation among cI temperature-sensitive mutants.

The original ind^- mutation, as well as a mutation producing both ts and ind^- characters, are located in the A region of cI (Lieb, 1966a). In addition, Lieb (1966a) and Horiuchi and Inokuchi (1967) have described temperature mutations in cI that heighten the sensitivity of lysogens to the inducing effects of ultraviolet light. These mutations are also of type A according to Lieb, but Horiuchi and Inokuchi found no such correlation among their mutants.

A simple experiment showed that repressor is not unstable during growth. Cells lysogenic for a temperature-sensitive cI prophage were superinfected with homoimmune wild-type phage and grown first at low temperature and then at the inducing temperature. Ten to twenty times as many cells survived as carried the superinfecting chromosome, evidence that one cI gene makes a large excess of repressor, which persists for several generations (Lieb, 1966b; Ptashne, unpublished; see also Zichichi and Kellenberger,

1963). Ogawa and Tomizawa (1967) reached similar conclusions by studying immune nonlysogenic segregants arising from infection of sensitive cells with $\lambda b2$. Lieb's (1966b) experiments also showed that superinfecting homoimmune phage begins to make repressor at once.

The genetic experiments argued strongly that cI encodes a single polypeptide responsible for both maintenance of lysogeny and immunity of lysogens.

Virulent Mutants

In 1954, Jacob and Wollman described a virulent mutant of λ called λvir that was insensitive to repressor; it could grow in λ lysogens. They concluded on the basis of genetic analysis that this phage had acquired at least four mutations, one of them in cI, two flanking cI, and a fourth somewhere farther to the left. In 1961 Jacob and Monod explained these as operator mutations that decrease affinity of operator for repressor. The complexity of the virulent mutant, however, especially the possibility of the existence of a mutation outside the immunity region where, according to the developing ideas, the repressor did not act, discouraged further analysis for some time.

In 1968 Nancy Hopkins and I repeated Jacob and Wollman's experiments, a task greatly simplified by the availability of many precisely mapped λ markers. We found that three mutations, located in the immunity region, are required for virulence: $v2$, which lies near N, and $v1$ and $v3$, which are closely linked and lie to the right of cI (Hopkins and Ptashne, this volume).

Horiuchi et al. (1969) and Koga et al. (1970) isolated a "weak virulent" mutant, which grows on cells carrying a λ prophage that makes a temperature-sensitive repressor. This mutant, which does not grow in cI^+ lysogens, has acquired two mutations, $virL$ and $virR$, located within the immunity region to the left and to the right of cI. The mutant gains full virulence by acquiring a third mutation, $virC$. Several different $virR$ mutations were mapped just to the right of a single $virC$ mutation. The fact that $\lambda v2virC$ is fully virulent (Koga et al., 1970) shows that it is possible to form a virulent mutant by introducing just one mutation on each side of cI.

Ordal (this volume) has selected virulent mutants by growing $\lambda v2v3$ on λ lysogens. Some of these grow on λdv carriers (Matsubara and Kaiser, 1968) and are called supervirulent. Ordal's results suggest that the supervirulent mutants contain, in addition to $v2$ and $v3$, a third mutation located at either of two sites (both called vs), closely linked and just to the right of $v3$. Ordal has mapped two presumed promoter mutations between $v3$ and vs, a result suggesting that o_R and p_R overlap.

Ptashne and Hopkins (1968) tested the genetic components of λvir for ability to synthesize the products of genes N and O in the presence of repressor. A λ lysogen was infected with the phage to be tested, and the culture was split into two parts, which were infected separately with either $\lambda imm^{434}susN$ or $\lambda imm^{434}susO$. Phage $\lambda v2$ preferentially increased the yield of $\lambda imm^{434}susN$, whereas $\lambda v3$ had the opposite effect. Phage $\lambda v1v3$ gave no differential effect, probably because it replicates extensively and produces N gene product when the repressor supply is exhausted (Sly et al., this volume). On the basis of similar experiments Pereira da Silva and Jacob (1968 and personal communication) concluded that $\lambda susNv1v3$ and $\lambda v2v3$ synthesize O and N products, respectively, in the presence of repressor. Koga et al. (1970) mention similar experiments showing that virL mutants synthesize N product, and virC and virR mutants O product, in the presence of repressor. Ordal (this volume) has performed a similar experiment with λvs.

Virulence of the type described here depends on mutations in sites lying on each side of cI; the left- and right-hand mutations separately decrease the effect of repressor on N and O gene expression, respectively. This decreased sensitivity to repressor might be conferred by operator mutations or by mutations that create new promoters that are insensitive to repressor. Experiments with isolated repressor show that $v1$, $v2$, $v3$, and vs are operator mutations.

Radiochemical Purification of Repressor

We set about isolating the λ repressor, assuming nothing about its mechanism of action, and using, therefore, no functional assay for its detection (Ptashne, 1967a). Rather, we sought conditions under which the relative rate of synthesis of the cI gene product would be greatly increased over its estimated value of about 1 part in 10^4 of the cell's protein synthesis. We thought we might then be able to detect the cI product by differential labeling with radioactive isotopes. This was accomplished by irradiating cells lysogenic for the noninducible prophage λind^- with massive doses of ultraviolet light, and superinfecting with many wild-type phage particles per bacterium. We used lysogenic cells because we expected the repressor would prevent synthesis of phage proteins other than the cI product itself (Horiuchi and Inokuchi, 1966; Lieb, 1966b; Ptashne, unpublished). The irradiation decreased 5000-fold the synthesis of bacterial proteins, but left relatively intact the capacity to synthesize phage proteins after infection. The cI product was first detected in a mixture of two extracts, one prepared from irradiated su^- cells fed [^{3}H]leucine after superinfection with wild-type λ, and the other from irradiated cells fed [^{14}C]leucine after superinfection with suppressor-sensitive cI mutants. Preliminary fractionation revealed a component preferentially labeled with ^{3}H and comprising 5–10% of the incorporated tritium. It was separated on a DEAE-cellulose column as a single protein labeled with ^{3}H but not with ^{14}C (Fig. 2). The protein was identified as the cI product by these criteria: it was missing from cells infected with phage bearing any of four suppressible mutations whose map positions span the known length of the cI gene; and it was produced in a modified form by the missense mutants ind^-, cI50, and various temperature-sensitive cI mutants (including cI857, contrary to our original, mistaken, report). The mutations cIts2 and cI50, which are located at the opposite extremities of cI, each modified the structure of the isolated polypeptide, an indication that we had isolated the

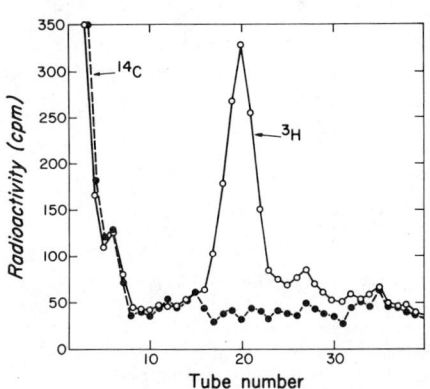

FIGURE 2. Radioactive labeling of the λ repressor. $E.$ $coli$ su^- cells lysogenic for λind^- were heavily irradiated with ultraviolet light and infected, in separate portions, with λ bearing in one case a wild-type cI gene and in the other a suppressor-sensitive cI mutation ($cIsus10$). (Both phages were also mutant in gene N.) The λcI^+-infected cells were fed [^{3}H]leucine, and those infected with $\lambda cIsus10$, [^{14}C]leucine. Shown here is a DEAE elution profile of an extract of a mixture of these cells. Further experiments (see text) confirmed that the single protein labeled with ^{3}H, but not ^{14}C, is the product of the cI gene. [From Ptashne, 1967a, by permission of Proceedings of the National Academy of Sciences.]

product of the entire cI gene. This isolation procedure provided about 10^{-12} moles of repressor of high isotopic purity but of low chemical purity. The irradiation method has also been used to isolate other λ proteins (Hendrix, this volume).

The 434 phage repressor was isolated in a fashion similar to that used for isolation of the λ repressor (Pirrotta and Ptashne, 1969). For this purpose λimm^{434} lysogens were irradiated and infected with λimm^{434} phage particles, and the repressor was isolated by means of a phosphocellulose column. During the course of these experiments an unexpected difference was detected between strains of phage λimm^{434} in common use. One obtained from Dr. J. Tomizawa (called $\lambda imm^{434}T$) was found to produce about 10 times more labeled repressor in superinfected cells than the original one of Kaiser and Jacob obtained from Dr. M. Meselson. The amount of repressor produced by $\lambda imm^{434}T$ is similar to that produced by λ.

Phage $\lambda imm^{434}T$ apparently produces more active repressor than does λimm^{434} in the lysogenic state as well. Stocks of λimm^{434} and $\lambda imm^{434}T$ contain frequent mutants able to grow in λimm^{434} lysogens, but these mutants require additional mutations before they can form plaques on $\lambda imm^{434}T$ lysogens. The restricted mutants contain a mutation in o_R (Pirrotta, unpublished) which permits phage growth in the presence of the low repressor levels characteristic of λimm^{434} lysogens. Possibly $\lambda imm^{434}T$ has acquired a mutation in its cI promoter which increases transcription of this gene.

A related phenomenon has been observed with λ. Lysogens carrying a mutant prophage called rim are not immune to $\lambda v1v3$. Strack et al. (1970) explained this phenomenon by assuming that the mutant produces less repressor than does the wild type. Following a similar line of reasoning, Horiuchi and co-workers have isolated weakly virulent mutants of λ that do not form plaques on ordinary λ lysogens but do on lysogens whose prophage contains a temperature-sensitive cI mutation (Horiuchi et al., 1969; Koga et al., 1970). Apparently the temperature mutation decreases the amount or

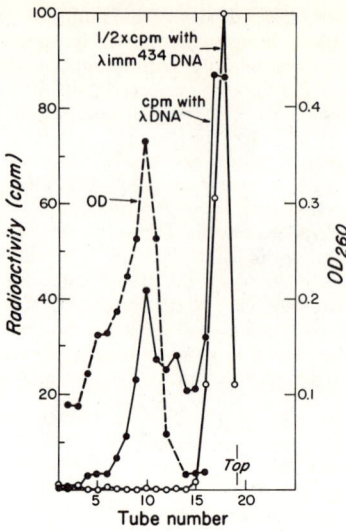

FIGURE 3. Specific binding of λind^- repressor to λ DNA. ^{14}C-labeled λind^- repressor was sedimented in sucrose gradients with λ and with λimm^{434} DNA. The UV absorption peak shows the position of the DNA after sedimentation. Repressor sedimented with the λ DNA, but remained near the top of the gradient when spun with λimm^{434} DNA. [From Ptashne, 1967b, by permission of Macmillan (Journals) Limited.]

activity of repressor and permits growth of operator mutants that do not grow in ordinary lysogens.

Properties of the Isolated Repressors

The λ and 434 repressors isolated as described have molecular weights of about 27,000 and 26,000, respectively, as judged by sedimentation rates in sucrose and by mobility in gel electrophoresis. Both are monomers. As described below, however, the monomers exist in equilibrium with oligomers that are the DNA-binding forms. The 434 repressor is much more basic than the λ repressor.

That repressor binds to DNA was first shown by sedimentation analysis of mixtures of repressor and DNA (Ptashne 1967b). In the presence of λ DNA, labeled λ repressor sediments at the rate of the DNA (30 s), whereas alone or in the presence of λimm^{434} DNA it sediments at about 28 s (Fig. 3). Similarly, the 434 repressor binds to λimm^{434} DNA but not to λ DNA (Pirrotta and Ptashne, 1969). In both cases the affinity of the repressor for DNA decreases as the ionic strength is increased. From the amount of DNA used in the binding experiments with λ repressor, I estimated a dissociation constant of 10^{-10} mole liter^{-1}—an overestimate based on the erroneous assumption that the monomer binds to DNA. Neither repressor binds to denatured DNA.

DNA isolated from phage λvir was found to have a lower affinity than wild-type DNA for λ repressor. Furthermore, binding experiments with DNAs bearing the $v2$, $v3$, and $v1$ mutations showed that each decreases the affinity of DNA for repressor (Ptashne and Hopkins, 1968). These results, combined with those described previously indicating that repressor blocks expression of genes on either side of cI (Taylor et al., 1967; Ptashne and

Hopkins, 1968), showed that the repressor binds independently to two operators (o_L and o_R) and controls two separate operons read in opposite directions from origins in the immunity region. The effect of repressor on transcription of these two operons in vitro is described below.

Purification of the Repressors

Riggs and Bourgeois (1968) showed that DNA bound to the *lac* repressor is retained on nitrocellulose filters. They used this effect to study the binding and to detect *lac* repressor in crude extracts. A similar assay was developed for the λ and 434 repressors, and was used to guide purifications (Pirrotta et al., 1970). We currently isolate repressor from strains that produce 7-10 times more repressor than do ordinary lysogens (Chadwick et al., 1970). The isolation steps include NH_4SO_4 precipitation, DEAE-cellulose and phosphocellulose chromatography, and glycerol-gradient sedimentation (Pirrotta et al., 1971). We have isolated about 50 mg of λ repressor, contaminated with less than 5 mg of other protein, from 600 liters of culture. We have also isolated smaller amounts of 434 repressor. In this volume Wu et al. describe a method of purification similar to ours.

The Active Form of Repressor

The equilibrium curve showing the amount of ^{32}P-labeled λ DNA bound to repressor in the presence of various amounts of repressor, as measured by the filter assay, is sigmoid (Pirrotta et al., 1970; Chadwick et al., 1970). A curve of similar shape is obtained if, instead of ordinary λ DNA, a mutant λ DNA carrying only one operator is used (Fig. 4). A similar curve also describes the binding of 434 repressor to its operators. The shape of the curve reflects

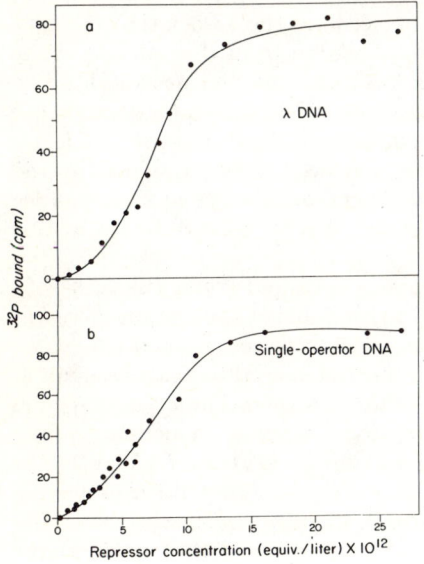

FIGURE 4. Binding of λind^- repressor to λ DNA as a function of repressor concentration. Increasing amounts of highly purified λind^- repressor were incubated with 10^{-12} M ^{32}P-labeled DNA. Aliquots were then filtered, washed, and counted on membrane filters. One equivalent of repressor binds one mole of λ DNA. a. DNA was λ wild type. b. DNA was $\lambda dbio$ M30-7pf5 (Court and Sato, 1969); one of the two operators, o_L, has been deleted from this phage. [From Pirrotta et al., 1970, by permission of Macmillan (Journals) Limited.]

a concentration-dependent interaction of repressor monomers to form the active, DNA-binding form of the repressor, according to the reactions

$$R_n \rightleftharpoons nR \text{ and } R_nO \rightleftharpoons R_n + O,$$

where R and O denote repressor monomer and operator, respectively. These reactions can be characterized by the equilibrium constants $K_1 = (R)^n/(R_n)$, $K_2 = (R_n)(O)/(R_nO)$, $K = K_1K_2 = (R)^n(O)/(R_nO)$.

As implied by this formulation, the repressor oligomer forms independently of the DNA and subsequently binds to it (Pirrotta et al., 1970). The sedimentation constant of both repressors increases from about 2.4 to about 6.2 s as the concentration is increased from about 10^{-10} M to 10^{-5} M. The ratio of sedimentation rates suggests formation of a tetramer at high concentration. Careful analysis of the binding curve, however, indicates that the active form is a dimer. The λ and 434 monomers do not form mixed aggregates, whether tested by a DNA-binding experiment or by oligomer formation in sucrose gradients (Pirrotta et al., 1970). The following constants describing the binding of the λind^- repressor to o_R have been measured (Chadwick et al., 1970): $K_1 = 7 \times 10^{-9}$ mole liter^{-1} and $K = 2 \times 10^{-22}$ mole^2 liter^{-2}, from which K_2 is 3×10^{-14} mole liter^{-1}.

The small size of K_2 indicates enormously tight binding of the dimer to DNA, similar to that of the *lac* repressor for its operator (about 10^{-13} mole liter^{-1}) (Riggs et al., 1970a). The half-life of the *lac* repressor-operator complex is 5–20 min at 0.05 M KCl, and remains relatively constant over a range of temperatures from 0°C to 30°C (Riggs et al., 1970b). In contrast, the half-life of the λ repressor-operator complex increases from 7 min at 20°C to about 175 min at 0°C, or a factor of about 5 for every 10°C decrease in temperature. The forward rate constant describing the association of the λ repressor dimer with its operator may be calculated from the half-life of the operator-repressor complex and the value of K_2, or it may be measured, given K_1, from the forward rate of the reaction at low DNA concentration. Either of these methods gives a forward rate constant of about 3×10^{10} mole/sec (Chadwick, unpublished), implying a rate too fast to be explained by the assumption that the repressor finds the operator by free diffusion. (See Gilbert and Müller-Hill, 1970, and Riggs et al., 1970b, for discussion of this phenomenon.)

The value of K for the λ repressor increases 10-fold when the KCl concentration is increased from 0.05 M to 0.15 M. Similarly, the constant increases about 10-fold for each increase of 0.5 unit in pH between 6.5 and 8.5.

The amount of λind^- repressor protein detected in crude extracts is equivalent to about 25 dimers per cell; this requires an intracellular equilibrium characterized by roughly ten times as many dimers as monomers in the cytoplasm. The measured half-life of the ind^- dimer is about 7 sec at 20°C.

Reports of intracistronic complementation among temperature-sensitive cI mutants suggested that the λ repressor is an oligomer (Lieb, 1966a and personal communication; Green, 1967). Lieb also found that subinducing doses of ultraviolet light caused $cIts$B lysogens to behave as $cIts$A lysogens. From this she argued that the λ repressor is an oligomer (Lieb, 1969).

Repression in Vitro

Echols et al. (1968) reported that a protein fraction isolated from λ lysogens specifically reduced to one-third the rate of transcription of λ DNA. This effect was seen only if ribosomes were present in the reaction mixture, possibly because they contained active M factor or some other component able to improve initiation specificity (Davison et al., 1969). In this volume Wu et al. report a larger effect of more highly purified repressor on synthesis of early messenger RNA. Efficient repression is observed when the reaction is carried out under conditions believed to favor specific initiation by the polymerase, namely, with addition of rifampicin to preformed polymerase-DNA complexes (Bautz and Bautz, 1970). The repressor has no effect on transcription from λimm^{434} DNA, and ribosomes are not required for repression.

Experiments with highly purified λind^- repressor have provided a clear demonstration that the repressor blocks transcription of the two operons, read in opposite directions from the immunity region, by binding to the two operators o_L and o_R (Chadwick et al., 1970; Steinberg and Ptashne, 1971). These experiments were made possible by Roberts' (1969) description of an assay for the transcripts initiated at the promoters p_L and p_R. In the presence of the termination factor ρ, and with $\lambda b2$ as template, these two messengers form a significant fraction of the total RNA synthesized in vitro, and they are easily displayed as two separate peaks sedimenting at 12 and 7 s in sucrose gradients. The 12 s and 7 s RNAs have several of the predicted properties of the N and tof gene messengers, respectively. Thus, the 12 s RNA hybridizes to the l DNA strand isolated from phages λ and λimm^{434} (which have identical N genes); it is a leftward transcript. Synthesis of this RNA is greatly reduced if the template DNA carries a mutation called *sex*, located near the left boundary of the immunity region, which decreases the affinity of p_L for RNA polymerase (Chadwick et al., 1970). On the other hand, the 7 s RNA hybridizes with the r strand of λ DNA but not with that of λimm^{434}, showing that this RNA is transcribed rightward from within the immunity region as expected for the *tof* messenger. Synthesis of the 7 s RNA is greatly reduced if the template DNA carries a mutation believed to be located in p_R.

Addition of highly purified repressor in a 10- to 15-fold molar excess over the operator concentration abolishes synthesis of the 12 s and 7 s messengers (Fig. 5). The λ repressor has no effect on synthesis of the corresponding messengers directed by λimm^{434} DNA. Somewhat higher repressor concentrations are required for complete repression of synthesis of the 7 s RNA than of 12 s RNA (Steinberg and Ptashne, 1971). This suggests that o_R and o_L may not be identical, but no conclusive evidence on this point has been presented.

The transcription of DNA bearing the virulence mutations confirms that $v2$ decreases the affinity of o_L for repressor, whereas $v1$ and $v3$ have a similar effect on o_R. The mutations do not abolish the effect of repressor, but increase the concentration required to produce an effect (Steinberg and Ptashne, 1971). The operator o_L bearing $v2$ retains considerably more affinity for repressor than does o_R bearing $v1$ and $v3$, a result consistent with other observations (Sly et al., 1965; Kumar and Szybalski, 1970).

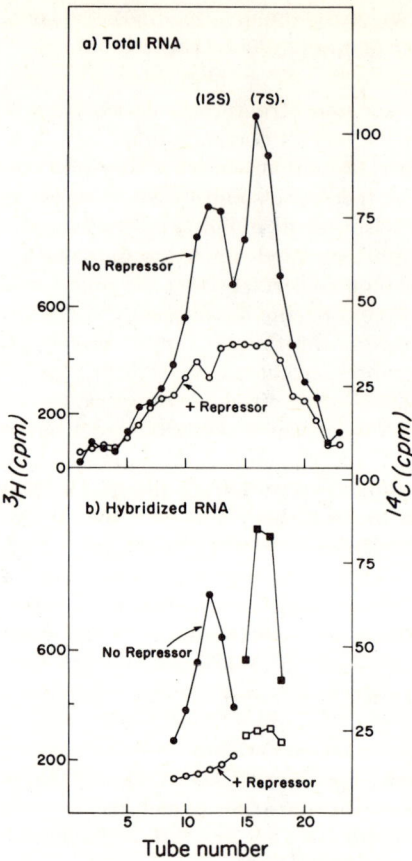

FIGURE 5. Lambda $b2$ DNA was transcribed by *E. coli* RNA polymerase in the presence of the termination factor rho, with and without repressor. The products were labeled with [^{3}H]UTP and with [^{14}C]UTP, respectively. The pooled RNAs were sedimented through sucrose gradients. One gradient (a) was assayed for radioactivity directly. Fractions from the 12 s and 7 s regions of the other gradient (b) were collected and hybridized with separated l and r strands of λ DNA, respectively; only counts in the hybridized RNA are shown here. Hybridization selected those RNA molecules that were transcribed from the correct strand in each case, and revealed that the repression was highly efficient. The small peak of 7 s RNA synthesized in the presence of repressor is further reduced if the repressor concentration in the reactions is increased. The residual 12 s and 7 s RNA synthesized in the presence of repressor is then mainly nonspecific background material that hybridizes with λimm^{21} DNA, which bears different N and *tof* genes. [From Steinberg and Ptashne, 1971, by permission of Macmillan (Journals) Limited.]

It is likely that the repressor blocks some very early step in the transcription, possibly by preventing binding of the RNA polymerase to its promoter (Hayward and Green, 1969; Steinberg and Ptashne, 1971). The latter idea is supported by the demonstration that binding of RNA polymerase to a promoter blocks binding of the repressor to the adjacent operator (Chadwick et al., 1970). The repressor has no effect on nascent or completed RNA molecules, and the template properties of DNA are unaltered if it is transcribed in the presence of repressor and then repurified (Steinberg and Ptashne, 1971).

In addition to confirming that N and *tof* are under direct repressor control, the experiments with purified cI protein make it likely that the cI product is the functional repressor, no other macromolecular factors being required for efficient repression.

Regulation of Repressor Function

Treatment of lysogenic cells with various agents (ultraviolet light, for example) activates the previously repressed prophage genes. A large body of evidence shows that known inducing agents do not act directly on the repressor, but elicit the formation of a secondary inducer (Tomizawa and Ogawa, 1967a; Rosner et al., 1968, and references therein).

A priori, the actual inducer might interact with either operator or repressor. Chadwick (unpublished) has found that DNA-binding repressor activity, measured in cell extracts, is destroyed if the cultures are induced by ultraviolet light before the extracts are prepared. The mechanism of inactivation is unknown. The fact that both λ and 434 repressors bind to DNA in the form of oligomers that are in rapid, concentration-dependent equilibrium with the monomers raises the possibility that the inducer might promote dissociation of the oligomer. The inducer isopropyl-thiogalactoside does not have any such effect on the *lac* repressor (Gilbert and Müller-Hill, 1966).

What Else Does the Repressor Do?

In Chapter 12, Eisen and I note the possibility that the repressor may function as an inducer for transcription of the cI gene. Here are some other possibilities.

The repressor may participate in excluding rII mutants of phage T4 from growth in λ lysogens. T4 rII mutants grow on λimm^{434} lysogens but not on λ lysogens. Howard (1967) isolated λ prophage mutants (called rex^-) that permit growth of rII, and located one of these mutations in the immunity region and to the left of cI. He proposed that rex is a gene, separate from cI, that is expressed in lysogens. R. Freedman (unpublished) and G. Kayajanian (unpublished) showed that double lysogens of the type $suscIrex^+/cI^+rex^-$ exclude rII. This result strongly supports the conclusion that rex is a gene separate from cI.

Tomizawa and Ogawa (1967b) have shown, however, that cells containing temperature-sensitive repressors of mutant type A lose the ability to exclude rII when heated. Repressors of mutant type B do not (see also Kasatiya and Lieb, 1964), possibly because repressors of this type are not fully denaturable (Szybalski, 1969). For unknown reasons, neither induction by ultraviolet light nor induction by mitomycin destroys the ability of lysogens to exclude rII (Tomizawa and Ogawa, 1967b).

These data taken together suggest that the repressor collaborates with the rex product to exclude rII mutants.

The repressor may be involved in transfer RNA methylation. Wainfan (1968, and Wainfan and Visser, 1969) claims that the transfer RNA methylating activity in crude extracts of $cI857$ lysogens is temperature sensitive. The activity is stable, however, in extracts of ordinary lysogens. To the best of my knowledge, no confirmation or refutation of this remarkable report has appeared.

The repressor probably does not directly block DNA replication. This question was first raised by Thomas and Bertani (1964), who showed that the repressor prevents replication of a homoimmune phage even in a cell in which a heteroimmune phage is multiplying. It is now quite clear, however, that this may be explained as a secondary effect of repression of rightward transcription beginning at p_R (Dove et al., this volume).

REFERENCES

Attardi, G., S. Naono, J. Rouviere, F. Jacob, and F. Gros. 1963. Production of messenger RNA and regulation of protein synthesis. Cold Spring Harbor Symp. Quant. Biol. *28:* 363.

Bautz, E. K. F., and F. A. Bautz. 1970. Initiation of RNA synthesis: the function of σ in the binding of RNA polymerase to promoter sites. Nature *226:* 1219.

Bertani, G. 1956. The role of phage in bacterial genetics. Brookhaven Symp. Biol. *8:* 50.

Bertani, G. 1958. Lysogeny. Advan. Virus Res. *5:* 151.

Bode, V. C., and A. D. Kaiser. 1965. Repression of the cII and cIII cistrons of phage λ in a lysogenic bacterium. Virology *25:* 111.

Brenner, S. 1965. Theories of gene regulation. Brit. Med. Bull. *21:* 244.

Butler, B., and H. Echols. 1970. Regulation of bacteriophage λ development by gene N: properties of a mutation that bypasses N control of late protein synthesis. Virology *14:* 22.

Chadwick, P., V. Pirrotta, R. Steinberg, N. Hopkins, and M. Ptashne. 1970. The λ and 434 phage repressors. Cold Spring Harbor Symp. Quant. Biol. *35:* 283.

Court, D., and K. Sato. 1969. Studies of novel transducing variants of lambda: dispensability of genes N and Q. Virology *39:* 348.

Davison, J., L. M. Polarski, and H. Echols. 1969. A factor that stimulates RNA synthesis by purified RNA polymerase. Proc. Nat. Acad. Sci. *63:* 168.

Echols, H., L. Polarski, and P. Y. Cheng. 1968. In vitro repression of phage λ DNA transcription by a partially purified repressor from lysogenic cells. Proc. Nat. Acad. Sci. *59:* 1016.

Gilbert, W., and B. Müller-Hill. 1966. Isolation of the *lac* repressor. Proc. Nat. Acad. Sci. *56:* 1891.

Gilbert, W., and B. Müller-Hill. 1967. The *lac* operator is DNA. Proc. Nat. Acad. Sci. *58:* 2415.

Gilbert, W., and B. Müller-Hill. 1970. The lactose repressor, p. 93–109. In J. R. Beckwith and D. Zipser [ed.] The lactose operon. Cold Spring Harbor Laboratory, Cold Spring Harbor, N.Y.

Green, M. H. 1966. Inactivation of the prophage λ repressor without induction. J. Mol. Biol. *16:* 134.

Green, M. H. 1967. Regulation of the development of the temperate phage lambda, p. 139–158. In J. S. Colter and W. Paranchych [ed.] The molecular biology of viruses. Academic Press, New York.

Hayward, W. S., and M. H. Green. 1969. Effect of the lambda repressor on the binding of RNA polymerase to DNA. Proc. Nat. Acad. Sci. *64:* 962.

Herskowitz, I., and E. R. Signer. 1970. A site essential for expression of all late genes in bacteriophage λ. J. Mol. Biol. *47:* 545.

Hogness, D. S., W. Doerfler, J. B. Egan, and L. W. Black. 1966. The position and orientation of genes in λ and λdg DNA. Cold Spring Harbor Symp. Quant. Biol. *31:* 129.

Hopkins, N. 1970. Bypassing a positive regulator: isolation of a λ mutant that does not require N to grow. Virology *40:* 223.

Horiuchi, T., and H. Inokuchi. 1966. Inhibition of the formation of superinfecting bacteriophage λ repressor with chloramphenicol. J. Mol. Biol. *15:* 674.

Horiuchi, T., and H. Inokuchi. 1967. Temperature-sensitive regulation system of prophage λ induction. J. Mol. Biol. *23:* 217.

Horiuchi, T., H. Koga, H. Inokuchi, and J. Tomizawa. 1969. Lambda phage mutants insensitive to temperature-sensitive repressor. I. Isolation and genetic analysis of weak-virulent mutants. Molec. Gen. Genet. *104:* 51.

Howard, B. 1967. Phage λ mutants deficient in r_{II} exclusion. Science *158:* 1588.

Isaacs, L. N., H. Echols, and W. S. Sly. 1965. Control of λ messenger RNA by the cI immunity region. J. Mol. Biol. *13:* 963.

JACOB, F., and A. CAMPBELL. 1959. Sur le système de répression assurant l'immunité chez les bactéries lysogènes. Compt. Rend. Acad. Sci. *248:* 3219.
JACOB, F., and J. MONOD. 1961. Genetic regulatory mechanisms in the synthesis of proteins. J. Mol. Biol. *3:* 318.
JACOB, F., P. SCHAEFFER and E. L. WOLLMAN. 1960. Episomic elements in bacteria, p. 67–91. *In* W. Hayes and R. C. Clowes [ed.] Microbial genetics, 10th Symp. Soc. Gen. Microbiol. Cambridge University Press, London and New York.
JACOB, F., R. SUSSMAN, and J. MONOD. 1962. Sur la nature du répresseur assurant l'immunité des bactéries lysogènes. Compt. Rend. Acad. Sci. *254:* 4214.
JACOB, F., and E. L. WOLLMAN. 1954. Étude génétique d'un bactériophage tempéré d'*Escherichia coli*. I. Le système génétique du bactériophage λ. Ann. Inst. Pasteur *87:* 653.
JACOB, F., and E. L. WOLLMAN. 1956. Sur les processus de conjugaison et de recombinaison chez *Escherichia coli*. I. L'induction par conjugaison ou induction zygotique. Ann. Inst. Pasteur *91:* 486.
KAISER, A. D. 1957. Mutations in a temperate bacteriophage affecting its ability to lysogenize *E. coli*. Virology *3:* 42.
KAISER, A. D., and F. JACOB. 1957. Recombination between related temperate bacteriophages and the genetic control of immunity and prophage localization. Virology *4:* 509.
KASATIYA, S., and M. LIEB. 1964. Effect of heat shock on T4rII multiplication in *Escherichia coli*. J. Bacteriol. *88:* 1585.
KOGA, H., T. MIYAUCHI, and T. HORIUCHI. 1970. λ phage mutants insensitive to temperature-sensitive repressor. II. Genetic character of $\lambda virC$ mutant. Molec. Gen. Genet. *106:* 114.
KOURILSKY, P., L. MARCAUD, P. SHELDRICK, D. LUSSATI, and F. GROS. 1968. Studies on the messenger RNA of bacteriophage λ. I. Various species synthesized early after induction of the prophage. Proc. Nat. Acad. Sci. *61:* 1013.
KUMAR, S., and W. SZYBALSKI. 1970. Transcription of the λdv plasmid and inhibition of λ phages in λdv carrier cells of *Escherichia coli*. Virology *41:* 665.
LIEB, M. 1966a. Studies of heat-inducible lambda phage. I. Order of genetic sites and properties of mutant prophage. J. Mol. Biol. *16:* 149.
LIEB, M. 1966b. Studies of heat-inducible λ phage. II. Production of cI product by superinfecting λ^+ in heat inducible lysogens. Virology *29:* 367.
LIEB, M. 1969. Allosteric properties of the λ repressor. J. Mol. Biol. *39:* 379.
MATSUBARA, K., and A. D. KAISER. 1968. λ dv: An autonomously replicating DNA fragment. Cold Spring Harbor Symp. Quant. Biol. *33:* 769.
NAONO, S., and F. GROS. 1966. Control and selectivity of λ DNA transcription in lysogenic bacteria. Cold Spring Harbor Symp. Quant. Biol. *31:* 363.
OGAWA, T., and J. TOMIZAWA. 1967. Abortive lysogenization of bacteriophage $\lambda b2$ and residual immunity of non-lysogenic segregants. J. Mol. Biol. *23:* 225.
PEREIRA DA SILVA, L. H., and F. JACOB. 1967. Induction of C_{II} and O functions in early defective lambda prophages. Virology *33:* 618.
PEREIRA DA SILVA, L. H., and F. JACOB. 1968. Étude génétique d'une mutation modifiant la sensibilité a l'immunité chez le bactériophage lambda. Ann. Inst. Pasteur *115:* 145.
PIRROTTA, V., P. CHADWICK, and M. PTASHNE. 1970. The active form of two coliphage repressors. Nature *227:* 41.
PIRROTTA, V., and M. PTASHNE. 1969. Isolation of the 434 phage repressor. Nature *222:* 541.
PIRROTTA, V., M. PTASHNE, P. CHADWICK, and R. STEINBERG. 1971. Isolation of repressors. *In* G. L. Cantoni and D. R. Davies [ed.] Procedures in nucleic acid research. Harper and Row, New York and London.
PTASHNE, M. 1967a. Isolation of the λ phage repressor. Proc. Nat. Acad. Sci. *57:* 306.
PTASHNE, M. 1967b. Specific binding of the λ phage repressor to λ DNA. Nature *214:* 232.

PTASHNE, M., and N. HOPKINS. 1968. The operators controlled by the λ phage repressor. Proc. Nat. Acad. Sci. *60:* 1282.

RIGGS, A. D., and S. BOURGEOIS. 1968. On the assay, isolation and characterization of the *lac* repressor. J. Mol. Biol. *34:* 361.

RIGGS, A. D., H. SUZUKI, and S. BOURGEOIS. 1970a. The *lac* repressor-operator interaction. I. Equilibrium studies. J. Mol. Biol. *48:* 67.

RIGGS, A. D., S. BOURGEOIS, and M. COHN. 1970b. The *lac* repressor-operator interaction. III. Kinetic studies. J. Mol. Biol. *53:* 401.

ROBERTS, J. 1969. Termination factor for RNA synthesis. Nature *224:* 1168.

ROSNER, J. L., L. R. KASS, and M. B. YARMOLINSKY. 1968. Parallel behavior of F and P1 in causing indirect induction of lysogenic bacteria. Cold Spring Harbor Symp. Quant. Biol. *33:* 785.

SKALKA, A., B. BUTLER, and H. ECHOLS. 1967. Genetic control of transcription during development of phage λ. Proc. Nat. Acad. Sci. *58:* 576.

SLY, W. S., H. ECHOLS, and J. ADLER. 1965. Control of viral messenger RNA after λ phage infection and induction. Proc. Nat. Acad. Sci. *53:* 378.

STEINBERG, R. A., and M. PTASHNE. 1971. *In vitro* repression of RNA synthesis by purified λ phage repressor. Nature, New Biol. 230: 76.

STENT, G. S. 1964. The operon on its third anniversary. Science *144:* 816.

STRACK, H. B., M. KAYSER, and S. HOLDER. 1970. Reduced immunity in lysogens of bacteriophage λ due to a mutation in the prophage. Virology *42:* 707.

SUSSMAN, R., and F. JACOB. 1962. Sur un système de répression thermosensible chez le bacteriophage λ d'*Escherichia coli*. Compt. Rend. Acad. Sci. *254:* 1517.

SZYBALSKI, W. 1969. Initiation and patterns of transcription during phage development. Proc. Can. Cancer Res. Conf. *8:* 183.

TAYLOR, K., Z. HRADECNA, and W. SZYBALSKI. 1967. Asymmetric distribution of the transcribing regions on the complementary strands of the coliphage λ DNA. Proc. Nat. Acad. Sci. *57:* 1618.

THOMAS, R., and L. E. BERTANI. 1964. On the control of the replication of temperate bacteriophages superinfecting immune hosts. Virology *24:* 241.

THOMAS, R., and L. LAMBERT. 1962. On the occurrence of bacterial mutations permitting lysogenization by clear variants of temperate bacteriophages. J. Mol. Biol. *5:* 373.

TOMIZAWA, J., and T. OGAWA. 1967a. Effect of ultraviolet irradiation on bacteriophage lambda immunity. J. Mol. Biol. *23:* 247.

TOMIZAWA, J., and T. OGAWA. 1967b. Inhibition of growth of *r*II mutants of bacteriophage T4 by immunity substance of bacteriophage λ. J. Mol. Biol. *23:* 277.

WAINFAN, E. 1968. Development of transfer RNA methylating enzymes with altered properties during heat induction of *Escherichia coli* K12 ($\lambda C_1 857$). Virology *35:* 282.

WAINFAN, E., and D. W. VISSER. 1969. Expression of a lambda gene in uninduced cells of *Escherichia coli*. Virology *37:* 148.

ZICHICHI, M. L., and G. KELLENBERGER. 1963. Two distinct functions in the lysogenization process: the repression of phage multiplication and the incorporation of the prophage in the bacterial genome. Virology *19:* 450.

LIBRARY OF D